STUDENT'S SOLUTIONS MANUAL

BEVERLY FUSFIELD

CALCULUS FOR THE LIFE SCIENCES

SECOND EDITION

Raymond N. Greenwell

Hofstra University

Nathan P. Ritchey

Edinboro University

Margaret L. Lial

American River College

PEARSON

Boston Columbus Indianapolis New York San Francisco Upper Saddle River
Amsterdam Cape Town Dubai London Madrid Milan Munich Paris Montréal Toronto
Delhi Mexico City São Paulo Sydney Hong Kong Seoul Singapore Taipei Tokyo

Reproduced by Pearson from electronic files supplied by the author.

Publishing as Pearson, 75 Arlington Street, Boston, MA 02116.

ISBN-13: 978-0-321-96383-3
ISBN-10: 0-321-96383-0

1 2 3 4 5 6 OPM 17 16 15 14

www.pearsonhighered.com

CONTENTS

CHAPTER 5 GRAPHS AND THE DERIVATIVE

CHAPTER 6 APPLICATIONS OF THE DERIVATIVE

CHAPTER 7 INTEGRATION

CHAPTER 8 FURTHER TECHNIQUES AND APPLICATIONS OF INTEGRATION

CHAPTER 9 MULTIVARIABLE CALCULUS

CHAPTER 10 MATRICES

CHAPTER 11 DIFFERENTIAL EQUATIONS

CHAPTER 12 PROBABILITY

CHAPTER 13 PROBABILITY AND CALCULUS

CHAPTER 14 Discrete Dynamical Systems

Chapter R

ALGEBRA REFERENCE

R.1 Polynomials

1. $(2x^2 - 6x + 11) + (-3x^2 + 7x - 2)$
$= 2x^2 - 6x + 11 - 3x^2 + 7x - 2$
$= (2-3)x^2 + (7-6)x + (11-2)$
$= -x^2 + x + 9$

2. $(-4y^2 - 3y + 8) - (2y^2 - 6y - 2)$
$= (-4y^2 - 3y + 8) + (-2y^2 + 6y + 2)$
$= -4y^2 - 3y + 8 - 2y^2 + 6y + 2$
$= (-4y^2 - 2y^2) + (-3y + 6y) + (8 + 2)$
$= -6y^2 + 3y + 10$

3. $-6(2q^2 + 4q - 3) + 4(-q^2 + 7q - 3)$
$= (-12q^2 - 24q + 18) + (-4q^2 + 28q - 12)$
$= (-12q^2 - 4q^2) + (-24q + 28q) + (18 - 12)$
$= -16q^2 + 4q + 6$

4. $2(3r^2 + 4r + 2) - 3(-r^2 + 4r - 5)$
$= (6r^2 + 8r + 4) + (3r^2 - 12r + 15)$
$= (6r^2 + 3r^2) + (8r - 12r) + (4 + 15)$
$= 9r^2 - 4r + 19$

5. $\left(0.613x^2 - 4.215x + 0.892\right) - 0.47\left(2x^2 - 3x + 5\right)$
$= 0.613x^2 - 4.215x + 0.892 - 0.94x^2 + 1.4x - 0.235$
$= \left(0.613x^2 - 0.94x^2\right) + \left(-4.215x + 1.4x\right) + \left(0.892 - 02.35\right)$
$= -0.327x^2 - 2.805x - 1.458$

6. $0.5(5r^2 + 3.2r - 6) - (1.7r^2 - 2r - 1.5)$
$= (2.5r^2 + 1.6r - 3) + (-1.7r^2 + 2r + 1.5)$
$= (2.5r^2 - 1.7r^2) + (1.6r + 2r) + (-3 + 1.5)$
$= 0.8r^2 + 3.6r - 1.5$

7. $-9m(2m^2 + 3m - 1)$
$= -9m(2m^2) - 9m(3m) - 9m(-1)$
$= -18m^3 - 27m^2 + 9m$

8. $6x(-2x^3 + 5x + 6) = -12x^4 + 30x^2 + 36x$

9. $(3t - 2y)(3t + 5y)$
$= (3t)(3t) + (3t)(5y) + (-2y)(3t) + (-2y)(5y)$
$= 9t^2 + 15ty - 6ty - 10y^2$
$= 9t^2 + 9ty - 10y^2$

10. $(9k + q)(2k - q)$
$= (9k)(2k) + (9k)(-q) + (q)(2k) + (q)(-q)$
$= 18k^2 - 9kq + 2kq - q^2$
$= 18k^2 - 7kq - q^2$

11. $(2 - 3x)(2 + 3x)$
$= (2)(2) + (2)(3x) + (-3x)(2) + (-3x)(3x)$
$= 4 + 6x - 6x - 9x^2$
$= 4 - 9x^2$

12. $(6m + 5)(6m - 5)$
$= (6m)6m) + (6m)(-5) + (5)(6m) + (5)(-5)$
$= 36m^2 - 30m + 30m - 25$
$= 36m^2 - 25$

13. $\left(\frac{2}{5}y + \frac{1}{8}z\right)\left(\frac{3}{5}y + \frac{1}{2}z\right)$
$= \left(\frac{2}{5}y\right)\left(\frac{3}{5}y\right) + \left(\frac{2}{5}y\right)\left(\frac{1}{2}z\right) + \left(\frac{1}{8}z\right)\left(\frac{3}{5}y\right) + \left(\frac{1}{8}z\right)\left(\frac{1}{2}z\right)$
$= \frac{6}{25}y^2 + \frac{1}{5}yz + \frac{3}{40}yz + \frac{1}{16}z^2$
$= \frac{6}{25}y^2 + \left(\frac{8}{40} + \frac{3}{40}\right)yz + \frac{1}{16}z^2$
$= \frac{6}{25}y^2 + \frac{11}{40}yz + \frac{1}{16}z^2$

14. $\left(\frac{3}{4}r-\frac{2}{3}s\right)\left(\frac{5}{4}r+\frac{1}{3}s\right)$
$=\left(\frac{3}{4}r\right)\left(\frac{5}{4}r\right)+\left(\frac{3}{4}r\right)\left(\frac{1}{3}s\right)+\left(-\frac{2}{3}s\right)\left(\frac{5}{4}r\right)+\left(-\frac{2}{3}s\right)\left(\frac{1}{3}s\right)$
$=\frac{15}{16}r^2+\frac{1}{4}rs-\frac{5}{6}rs-\frac{2}{9}s^2$
$=\frac{15}{16}r^2-\frac{7}{12}rs-\frac{2}{9}s^2$

15. $(3p-1)(9p^2+3p+1)$
$=(3p-1)(9p^2)+(3p-1)(3p)+(3p-1)(1)$
$=3p(9p^2)-1(9p^2)+3p(3p)-1(3p)+3p(1)-1(1)$
$=27p^3-9p^2+9p^2-3p+3p-1$
$=27p^3-1$

16. $(3p+2)(5p^2+p-4)$
$=(3p)(5p^2)+(3p)(p)+(3p)(-4)+(2)(5p^2)+(2)(p)+(2)(-4)$
$=15p^3+3p^2-12p+10p^2+2p-8$
$=15p^3+13p^2-10p-8$

17. $(2m+1)(4m^2-2m+1)$
$=2m(4m^2-2m+1)+1(4m^2-2m+1)$
$=8m^3-4m^2+2m+4m^2-2m+1$
$=8m^3+1$

18. $(k+2)(12k^3-3k^2+k+1)$
$=k(12k^3)+k(-3k^2)+k(k)+k(1)+2(12k^3)+2(-3k^2)+2(k)+2(1)$
$=12k^4-3k^3+k^2+k+24k^3-6k^2+2k+2$
$=12k^4+21k^3-5k^2+3k+2$

19. $(x+y+z)(3x-2y-z)$
$=x(3x)+x(-2y)+x(-z)+y(3x)+y(-2y)+y(-z)+z(3x)+z(-2y)+z(-z)$
$=3x^2-2xy-xz+3xy-2y^2-yz+3xz-2yz-z^2$
$=3x^2+xy+2xz-2y^2-3yz-z^2$

20. $(r+2s-3t)(2r-2s+t)$
$=r(2r)+r(-2s)+r(t)+2s(2r)+2s(-2s)+2s(t)-3t(2r)-3t(-2s)-3t(t)$
$=2r^2-2rs+rt+4rs+2st-6rt+6st-3t^2+2rt-st+t^2$
$=2r^2+2rs-5rt-4s^2+8st-3t^2$

21. $(x+1)(x+2)(x+3)$
$=[x(x+2)+1(x+2)](x+3)$
$=[x^2+2x+x+2](x+3)$
$=[x^2+3x+2](x+3)$
$=x^2(x+3)+3x(x+3)+2(x+3)$
$=x^3+3x^2+3x^2+9x+2x+6$
$=x^3+6x^2+11x+6$

22. $(x-1)(x+2)(x-3)$
$=[x(x+2)+(-1)(x+2)](x-3)$
$=(x^2+2x-x-2)(x-3)$
$=(x^2+x-2)(x-3)$
$=x^2(x-3)+x(x-3)+(-2)(x-3)$
$=x^3-3x^2+x^2-3x-2x+6$
$=x^3-2x^2-5x+6$

23. $(x+2)^2=(x+2)(x+2)$
$=x(x+2)+2(x+2)$
$=x^2+2x+2x+4$
$=x^2+4x+4$

24. $(2a-4b)^2=(2a-4b)(2a-4b)$
$=2a(2a-4b)-4b(2a-4b)$
$=4a^2-8ab-8ab+16b^2$
$=4a^2-16ab+16b^2$

25. $(x-2y)^3$
$=[(x-2y)(x-2y)](x-2y)$
$=(x^2-2xy-2xy+4y^2)(x-2y)$
$=(x^2-4xy+4y^2)(x-2y)$
$=(x^2-4xy+4y^2)x+(x^2-4xy+4y^2)(-2y)$
$=x^3-4x^2y+4xy^2-2x^2y+8xy^2-8y^3$
$=x^3-6x^2y+12xy^2-8y^3$

26. $(3x+y)^3=(3x+y)(3x+y)^2$
$=(3x+y)(9x^2+6xy+y^2)$
$=27x^3+18x^2y+3xy^2+9x^2y+6xy^2+y^3$
$=27x^3+27x^2y+9xy^2+y^3$

R.2 Factoring

1. $7a^3 + 14a^2 = 7a^2 \cdot a + 7a^2 \cdot 2$
$= 7a^2(a+2)$

2. $3y^3 + 24y^2 + 9y = 3y \cdot y^2 + 3y \cdot 8y + 3y \cdot 3$
$= 3y(y^2 + 8y + 3)$

3. $13p^4q^2 - 39p^3q + 26p^2q^2$
$= 13p^2q \cdot p^2q - 13p^2q \cdot 3p + 13p^2q \cdot 2q$
$= 13p^2q(p^2q - 3p + 2q)$

4. $60m^4 - 120m^3n + 50m^2n^2$
$= 10m^2 \cdot 6m^2 - 10m^2 \cdot 12mn + 10m^2 \cdot 5n^2$
$= 10m^2(6m^2 - 12mn + 5n^2)$

5. $m^2 - 5m - 14 = (m-7)(m+2)$
because $(-7)(2) = -14$ and $-7 + 2 = -5$.

6. $x^2 + 4x - 5 = (x+5)(x-1)$
because $5(-1) = -5$ and $-1 + 5 = 4$.

7. $z^2 + 9z + 20 = (z+4)(z+5)$
because $4 \cdot 5 = 20$ and $4 + 5 = 9$.

8. $b^2 - 8b + 7 = (b-7)(b-1)$
because $(-7)(-1) = 7$ and $-7 + (-1) = -8$.

9. $a^2 - 6ab + 5b^2 = (a-b)(a-5b)$
because $(-b)(-5b) = 5b^2$ and
$-b + (-5b) = -6b$.

10. $s^2 + 2st - 35t^2 = (s-5t)(s+7t)$
because $(-5t)(7t) = -35t^2$ and $7t + (-5t) = 2t$.

11. $y^2 - 4yz - 21z^2 = (y+3z)(y-7z)$
because $(3z)(-7z) = -21z^2$ and
$3z + (-7z) = -4z$.

12. $3x^2 + 4x - 7$
The possible factors of $3x^2$ are $3x$ and x and the possible factors of -7 are -7 and 1, or 7 and -1. Try various combinations until one works.
$3x^2 + 4x - 7 = (3x+7)(x-1)$

13. $3a^2 + 10a + 7$
The possible factors of $3a^2$ are $3a$ and a and the possible factors of 7 are 7 and 1. Try various combinations until one works.
$3a^2 + 10a + 7 = (a+1)(3a+7)$

14. $15y^2 + y - 2 = (5y+2)(3y-1)$

15. $21m^2 + 13mn + 2n^2 = (7m+2n)(3m+n)$

16. $6a^2 - 48a - 120 = 6(a^2 - 8a - 20)$
$= 6(a-10)(a+2)$

17. $3m^3 + 12m^2 + 9m = 3m(m^2 + 4m + 3)$
$= 3m(m+1)(m+3)$

18. $4a^2 + 10a + 6 = 2(2a^2 + 5a + 3)$
$= 2(2a+3)(a+1)$

19. $24a^4 + 10a^3b - 4a^2b^2$
$= 2a^2(12a^2 + 5ab - 2b^2)$
$= 2a^2(4a-b)(3a+2b)$

20. $24x^4 + 36x^3y - 60x^2y^2$
$= 12x^2(2x^2 + 3xy - 5y^2)$
$= 12x^2(x-y)(2x+5y)$

21. $x^2 - 64 = x^2 - 8^2 = (x+8)(x-8)$

22. $9m^2 - 25 = (3m)^2 - (5)^2$
$= (3m+5)(3m-5)$

23. $10x^2 - 160 = 10(x^2 - 16) = 10(x^2 - 4^2)$
$= 10(x+4)(x-4)$

24. $9x^2 + 64$ is the *sum* of two perfect squares. It cannot be factored. It is prime.

25. $z^2 + 14zy + 49y^2 = z^2 + 2 \cdot 7zy + 7^2y^2$
$= (z+7y)^2$

26. $s^2 - 10st + 25t^2 = s^2 - 2 \cdot 5st + (5t)^2$
$= (s-5t)^2$

27. $9p^2 - 24p + 16 = (3p)^2 - 2 \cdot 3p \cdot 4 + 4^2$
$= (3p-4)^2$

28. $a^3 - 216 = a^3 - 6^3$
$= (a-6)[(a)^2 + (a)(6) + (6)^2]$
$= (a-6)(a^2 + 6a + 36)$

29. $27r^3 - 64s^3 = (3r)^3 - (4s)^3$
$= (3r - 4s)(9r^2 + 12rs + 16s^2)$

30. $3m^3 + 375 = 3(m^3 + 125)$
$= 3(m^3 + 5^3)$
$= 3(m + 5)(m^2 - 5m + 25)$

31. $x^4 - y^4 = (x^2)^2 - (y^2)^2$
$= (x^2 + y^2)(x^2 - y^2)$
$= (x^2 + y^2)(x + y)(x - y)$

32. $16a^4 - 81b^4 = (4a^2)^2 - (9b^2)^2$
$= (4a^2 + 9b^2)(4a^2 - 9b^2)$
$= (4a^2 + 9b^2)[(2a)^2 - (3b)^2]$
$= (4a^2 + 9b^2)(2a + 3b)(2a - 3b)$

R.3 Rational Expressions

1. $\frac{5v^2}{35v} = \frac{5 \cdot v \cdot v}{5 \cdot 7 \cdot v} = \frac{v}{7}$

2. $\frac{25p^3}{10p^2} = \frac{5 \cdot 5 \cdot p \cdot p \cdot p}{2 \cdot 5 \cdot p \cdot p} = \frac{5p}{2}$

3. $\frac{8k + 16}{9k + 18} = \frac{8(k + 2)}{9(k + 2)} = \frac{8}{9}$

4. $\frac{2(t - 15)}{(t - 15)(t + 2)} = \frac{2}{t + 2}$

5. $\frac{4x^3 - 8x^2}{4x^2} = \frac{4x^2(x - 2)}{4x^2} = x - 2$

6. $\frac{36y^2 + 72y}{9y} = \frac{36y(y + 2)}{9y}$
$= \frac{9 \cdot 4 \cdot y(y + 2)}{9 \cdot y}$
$= 4(y + 2)$

7. $\frac{m^2 - 4m + 4}{m^2 + m - 6} = \frac{(m - 2)(m - 2)}{(m - 2)(m + 3)}$
$= \frac{m - 2}{m + 3}$

8. $\frac{r^2 - r - 6}{r^2 + r - 12} = \frac{(r - 3)(r + 2)}{(r + 4)(r - 3)}$
$= \frac{r + 2}{r + 4}$

9. $\frac{3x^2 + 3x - 6}{x^2 - 4} = \frac{3(x + 2)(x - 1)}{(x + 2)(x - 2)} = \frac{3(x - 1)}{x - 2}$

10. $\frac{z^2 - 5z + 6}{z^2 - 4} = \frac{(z - 3)(z - 2)}{(z + 2)(z - 2)} = \frac{z - 3}{z + 2}$

11. $\frac{m^4 - 16}{4m^2 - 16} = \frac{(m^2 + 4)(m + 2)(m - 2)}{4(m + 2)(m - 2)}$
$= \frac{m^2 + 4}{4}$

12. $\frac{6y^2 + 11y + 4}{3y^2 + 7y + 4} = \frac{(3y + 4)(2y + 1)}{(3y + 4)(y + 1)} = \frac{2y + 1}{y + 1}$

13. $\frac{9k^2}{25} \cdot \frac{5}{3k} = \frac{3 \cdot 3 \cdot 5k^2}{5 \cdot 5 \cdot 3k} = \frac{3k^2}{5k} = \frac{3k}{5}$

14. $\frac{15p^3}{9p^2} \div \frac{6p}{10p^2} = \frac{15p^3}{9p^2} \cdot \frac{10p^2}{6p}$
$= \frac{150p^5}{54p^3}$
$= \frac{25 \cdot 6p^5}{9 \cdot 6p^3} = \frac{25p^2}{9}$

15. $\frac{3a + 3b}{4c} \cdot \frac{12}{5(a + b)} = \frac{3(a + b)}{4c} \cdot \frac{3 \cdot 4}{5(a + b)}$
$= \frac{3 \cdot 3}{c \cdot 5} = \frac{9}{5c}$

16. $\frac{a - 3}{16} \div \frac{a - 3}{32} = \frac{a - 3}{16} \cdot \frac{32}{a - 3}$
$= \frac{a - 3}{16} \cdot \frac{16 \cdot 2}{a - 3}$
$= \frac{2}{1} = 2$

17. $\frac{2k - 16}{6} \div \frac{4k - 32}{3} = \frac{2k - 16}{6} \cdot \frac{3}{4k - 32}$
$= \frac{2(k - 8)}{6} \cdot \frac{3}{4(k - 8)}$
$= \frac{1}{4}$

18. $\frac{9y - 18}{6y + 12} \cdot \frac{3y + 6}{15y - 30} = \frac{9(y - 2)}{6(y + 2)} \cdot \frac{3(y + 2)}{15(y - 2)}$
$= \frac{27}{90} = \frac{3 \cdot 3}{10 \cdot 3} = \frac{3}{10}$

19. $\frac{4a + 12}{2a - 10} \div \frac{a^2 - 9}{a^2 - a - 20}$
$= \frac{4(a + 3)}{2(a - 5)} \cdot \frac{(a - 5)(a + 4)}{(a - 3)(a + 3)}$
$= \frac{2(a + 4)}{a - 3}$

20. $\dfrac{6r-18}{9r^2+6r-24}\cdot\dfrac{12r-16}{4r-12}$
$=\dfrac{6(r-3)}{3(3r^2+2r-8)}\cdot\dfrac{4(3r-4)}{4(r-3)}$
$=\dfrac{6(r-3)}{3(3r-4)(r+2)}\cdot\dfrac{4(3r-4)}{4(r-3)}$
$=\dfrac{6}{3(r+2)}=\dfrac{2}{r+2}$

21. $\dfrac{k^2+4k-12}{k^2+10k+24}\cdot\dfrac{k^2+k-12}{k^2-9}$
$=\dfrac{(k+6)(k-2)}{(k+6)(k+4)}\cdot\dfrac{(k+4)(k-3)}{(k+3)(k-3)}$
$=\dfrac{k-2}{k+3}$

22. $\dfrac{m^2+3m+2}{m^2+5m+4}\div\dfrac{m^2+5m+6}{m^2+10m+24}$
$=\dfrac{m^2+3m+2}{m^2+5m+4}\cdot\dfrac{m^2+10m+24}{m^2+5m+6}$
$=\dfrac{(m+1)(m+2)}{(m+4)(m+1)}\cdot\dfrac{(m+6)(m+4)}{(m+3)(m+2)}$
$=\dfrac{m+6}{m+3}$

23. $\dfrac{2m^2-5m-12}{m^2-10m+24}\div\dfrac{4m^2-9}{m^2-9m+18}$
$=\dfrac{2m^2-5m-12}{m^2-10m+24}\cdot\dfrac{m^2-9m+18}{4m^2-9}$
$=\dfrac{(2m+3)(m-4)(m-6)(m-3)}{(m-6)(m-4)(2m-3)(2m+3)}$
$=\dfrac{m-3}{2m-3}$

24. $\dfrac{4n^2+4n-3}{6n^2-n-15}\cdot\dfrac{8n^2+32n+30}{4n^2+16n+15}$
$=\dfrac{(2n+3)(2n-1)}{(2n+3)(3n-5)}\cdot\dfrac{2(2n+3)(2n+5)}{(2n+3)(2n+5)}$
$=\dfrac{2(2n-1)}{3n-5}$

25. $\dfrac{a+1}{2}-\dfrac{a-1}{2}=\dfrac{(a+1)-(a-1)}{2}$
$=\dfrac{a+1-a+1}{2}$
$=\dfrac{2}{2}=1$

26. $\dfrac{3}{p}+\dfrac{1}{2}$

Multiply the first term by $\frac{2}{2}$ and the second by $\frac{p}{p}$.

$\dfrac{2\cdot 3}{2\cdot p}+\dfrac{p\cdot 1}{p\cdot 2}=\dfrac{6}{2p}+\dfrac{p}{2p}=\dfrac{6+p}{2p}$

27. $\dfrac{6}{5y}-\dfrac{3}{2}=\dfrac{6\cdot 2}{5y\cdot 2}-\dfrac{3\cdot 5y}{2\cdot 5y}=\dfrac{12-15y}{10y}$

28. $\dfrac{1}{6m}+\dfrac{2}{5m}+\dfrac{4}{m}=\dfrac{5\cdot 1}{5\cdot 6m}+\dfrac{6\cdot 2}{6\cdot 5m}+\dfrac{30\cdot 4}{30\cdot m}$
$=\dfrac{5}{30m}+\dfrac{12}{30m}+\dfrac{120}{30m}$
$=\dfrac{5+12+120}{30m}$
$=\dfrac{137}{30m}$

29. $\dfrac{1}{m-1}+\dfrac{2}{m}=\dfrac{m}{m}\left(\dfrac{1}{m-1}\right)+\dfrac{m-1}{m-1}\left(\dfrac{2}{m}\right)$
$=\dfrac{m+2m-2}{m(m-1)}$
$=\dfrac{3m-2}{m(m-1)}$

30. $\dfrac{5}{2r+3}-\dfrac{2}{r}=\dfrac{5r}{r(2r+3)}-\dfrac{2(2r+3)}{r(2r+3)}$
$=\dfrac{5r-2(2r+3)}{r(2r+3)}=\dfrac{5r-4r-6}{r(2r+3)}$
$=\dfrac{r-6}{r(2r+3)}$

31. $\dfrac{8}{3(a-1)}+\dfrac{2}{a-1}=\dfrac{8}{3(a-1)}+\dfrac{3}{3}\left(\dfrac{2}{a-1}\right)$
$=\dfrac{8+6}{3(a-1)}$
$=\dfrac{14}{3(a-1)}$

32. $\dfrac{2}{5(k-2)}+\dfrac{3}{4(k-2)}=\dfrac{4\cdot 2}{4\cdot 5(k-2)}+\dfrac{5\cdot 3}{5\cdot 4(k-2)}$
$=\dfrac{8}{20(k-2)}+\dfrac{15}{20(k-2)}$
$=\dfrac{8+15}{20(k-2)}$
$=\dfrac{23}{20(k-2)}$

33. $$\frac{4}{x^2+4x+3}+\frac{3}{x^2-x-2}$$
$$=\frac{4}{(x+3)(x+1)}+\frac{3}{(x-2)(x+1)}$$
$$=\frac{4(x-2)}{(x-2)(x+3)(x+1)}+\frac{3(x+3)}{(x-2)(x+3)(x+1)}$$
$$=\frac{4(x-2)+3(x+3)}{(x-2)(x+3)(x+1)}$$
$$=\frac{4x-8+3x+9}{(x-2)(x+3)(x+1)}$$
$$=\frac{7x+1}{(x-2)(x+3)(x+1)}$$

34. $$\frac{y}{y^2+2y-3}-\frac{1}{y^2+4y+3}$$
$$=\frac{y}{(y+3)(y-1)}-\frac{1}{(y+3)(y+1)}$$
$$=\frac{y(y+1)}{(y+3)(y+1)(y-1)}-\frac{1(y-1)}{(y+3)(y+1)(y-1)}$$
$$=\frac{y(y+1)-(y-1)}{(y+3)(y+1)(y-1)}$$
$$=\frac{y^2+y-y+1}{(y+3)(y+1)(y-1)}$$
$$=\frac{y^2+1}{(y+3)(y+1)(y-1)}$$

35. $$\frac{3k}{2k^2+3k-2}-\frac{2k}{2k^2-7k+3}$$
$$=\frac{3k}{(2k-1)(k+2)}-\frac{2k}{(2k-1)(k-3)}$$
$$=\left(\frac{k-3}{k-3}\right)\frac{3k}{(2k-1)(k+2)}-\left(\frac{k+2}{k+2}\right)\frac{2k}{(2k-1)(k-3)}$$
$$=\frac{(3k^2-9k)-(2k^2+4k)}{(2k-1)(k+2)(k-3)}$$
$$=\frac{k^2-13k}{(2k-1)(k+2)(k-3)}$$
$$=\frac{k(k-13)}{(2k-1)(k+2)(k-3)}$$

36. $$\frac{4m}{3m^2+7m-6}-\frac{m}{3m^2-14m+8}$$
$$=\frac{4m}{(3m-2)(m+3)}-\frac{m}{(3m-2)(m-4)}$$
$$=\frac{4m(m-4)}{(3m-2)(m+3)(m-4)}-\frac{m(m+3)}{(3m-2)(m-4)(m+3)}$$
$$=\frac{4m(m-4)-m(m+3)}{(3m-2)(m-4)(m+3)}$$
$$=\frac{4m^2-16m-m^2-3m}{(3m-2)(m+3)(m-4)}$$
$$=\frac{3m^2-19m}{(3m-2)(m+3)(m-4)}$$
$$=\frac{m(3m-19)}{(3m-2)(m+3)(m-4)}$$

37. $$\frac{2}{a+2}+\frac{1}{a}+\frac{a-1}{a^2+2a}$$
$$=\frac{2}{a+2}+\frac{1}{a}+\frac{a-1}{a(a+2)}$$
$$=\left(\frac{a}{a}\right)\frac{2}{a+2}+\left(\frac{a+2}{a+2}\right)\frac{1}{a}+\frac{a-1}{a(a+2)}$$
$$=\frac{2a+a+2+a-1}{a(a+2)}$$
$$=\frac{4a+1}{a(a+2)}$$

38. $$\frac{5x+2}{x^2-1}+\frac{3}{x^2+x}-\frac{1}{x^2-x}$$
$$=\frac{5x+2}{(x+1)(x-1)}+\frac{3}{x(x+1)}-\frac{1}{x(x-1)}$$
$$=\left(\frac{x}{x}\right)\left(\frac{5x+2}{(x+1)(x-1)}\right)+\left(\frac{x-1}{x-1}\right)\left(\frac{3}{x(x+1)}\right)-\left(\frac{x+1}{x+1}\right)\left(\frac{1}{x(x-1)}\right)$$
$$=\frac{x(5x+2)+(x-1)(3)-(x+1)(1)}{x(x+1)(x-1)}$$
$$=\frac{5x^2+2x+3x-3-x-1}{x(x+1)(x-1)}$$
$$=\frac{5x^2+4x-4}{x(x+1)(x-1)}$$

R.4 Equations

1. $2x+8=x-4$
$x+8=-4$
$x=-12$
The solution is $x=-12$.

2. $5x+2=8-3x$
$8x+2=8$
$8x=6$
$x=\frac{3}{4}$
The solution is $x=\frac{3}{4}$.

3. $0.2m-0.5=0.1m+0.7$
$10(0.2m-0.5)=10(0.1m+0.7)$
$2m-5=m+7$
$m-5=7$
$m=12$
The solution is $m=12$.

4. $\frac{2}{3}k-k+\frac{3}{8}=\frac{1}{2}$
Multiply both sides of the equation by 24.
$24\left(\frac{2}{3}k\right)-24(k)+24\left(\frac{3}{8}\right)=24\left(\frac{1}{2}\right)$
$16k-24k+9=12$
$-8k+9=12$
$-8k=3$
$k=-\frac{3}{8}$
The solution is $k=-\frac{3}{8}$.

5. $3r+2-5(r+1)=6r+4$
$3r+2-5r-5=6r+4$
$-3-2r=6r+4$
$-3=8r+4$
$-7=8r$
$-\frac{7}{8}=r$
The solution is $r=-\frac{7}{8}$.

6. $5(a+3)+4a-5=-(2a-4)$
$5a+15+4a-5=-2a+4$
$9a+10=-2a+4$
$11a+10=4$
$11a=-6$
$a=-\frac{6}{11}$
The solution is $a=-\frac{6}{11}$.

7. $2[3m-2(3-m)-4]=6m-4$
$2[3m-6+2m-4]=6m-4$
$2[5m-10]=6m-4$
$10m-20=6m-4$
$4m-20=-4$
$4m=16$
$m=4$
The solution is $m=4$.

8. $4[2p-(3-p)+5]=-7p-2$
$4[2p-3+p+5]=-7p-2$
$4[3p+2]=-7p-2$
$12p+8=-7p-2$
$19p+8=-2$
$19p=-10$
$p=-\frac{10}{19}$
The solution is $p=-\frac{10}{19}$.

9. $x^2+5x+6=0$
$(x+3)(x+2)=0$
$x+3=0$ or $x+2=0$
$x=-3$ or $x=-2$
The solutions are $x=-3$ and $x=-2$.

10. $x^2=3+2x$
$x^2-2x-3=0$
$(x-3)(x+1)=0$
$x-3=0$ or $x+1=0$
$x=3$ or $x=-1$
The solutions are $x=3$ and $x=-1$.

11. $m^2=14m-49$
$m^2-14m+49=0$
$(m)^2-2(7m)+(7)^2=0$
$(m-7)^2=0$
$m-7=0$
$m=7$
The solution is $m=7$.

12. $2k^2-k=10$
$2k^2-k-10=0$
$(2k-5)(k+2)=0$
$2k-5=0$ or $k+2=0$
$k=\frac{5}{2}$ or $k=-2$
The solutions are $k=\frac{5}{2}$ and $k=-2$.

13. $12x^2 - 5x = 2$

$$12x^2 - 5x - 2 = 0$$
$$(4x+1)(3x-2) = 0$$
$$4x+1 = 0 \quad \text{or} \quad 3x-2 = 0$$
$$4x = -1 \quad \text{or} \quad 3x = 2$$
$$x = -\frac{1}{4} \quad \text{or} \quad x = \frac{2}{3}$$

The solutions are $x = -\frac{1}{4}$ and $x = \frac{2}{3}$.

14. $m(m-7) = -10$

$$m^2 - 7m + 10 = 0$$
$$(m-5)(m-2) = 0$$
$$m-5 = 0 \quad \text{or} \quad m-2 = 0$$
$$m = 5 \quad \text{or} \quad m = 2$$

The solutions are $m = 5$ and $m = 2$.

15. $4x^2 - 36 = 0$

Divide both sides of the equation by 4.

$$x^2 - 9 = 0$$
$$(x+3)(x-3) = 0$$
$$x+3 = 0 \quad \text{or} \quad x-3 = 0$$
$$x = -3 \quad \text{or} \quad x = 3$$

The solutions are $x = -3$ and $x = 3$.

16. $z(2z+7) = 4$

$$2z^2 + 7z - 4 = 0$$
$$(2z-1)(z+4) = 0$$
$$2z-1 = 0 \quad \text{or} \quad z+4 = 0$$
$$z = \frac{1}{2} \quad \text{or} \quad z = -4$$

The solutions are $z = \frac{1}{2}$ and $z = -4$.

17. $p^2 + p - 1 = 0$

$$12y(y) - 12y(4) = 0$$
$$12y(y-4) = 0$$
$$12y = 0 \quad \text{or} \quad y-4 = 0$$
$$y = 0 \quad \text{or} \quad y = 4$$

The solutions are $y = 0$ and $y = 4$.

18. $3x^2 - 5x + 1 = 0$

Use the quadratic formula.

$$x = \frac{-(-5) \pm \sqrt{(-5)^2 - 4(3)(1)}}{2(3)}$$
$$= \frac{5 \pm \sqrt{25-12}}{6}$$
$$x = \frac{5+\sqrt{13}}{6} \quad \text{or} \quad x = \frac{5-\sqrt{13}}{6}$$
$$\approx 1.4343 \qquad \approx 0.2324$$

The solutions are $x = \frac{5+\sqrt{13}}{6} \approx 1.4343$ and $x = \frac{5-\sqrt{13}}{6} \approx 0.2324$.

19. $2m^2 - 4m = 3$

$$2m^2 - 4m - 3 = 0$$
$$m = \frac{-(-4) \pm \sqrt{(-4)^2 - 4(2)(-3)}}{2(2)}$$
$$= \frac{4 \pm \sqrt{40}}{4} = \frac{4 \pm \sqrt{4 \cdot 10}}{4}$$
$$= \frac{4 \pm \sqrt{4}\sqrt{10}}{4}$$
$$= \frac{4 \pm 2\sqrt{10}}{4} = \frac{2 \pm \sqrt{10}}{2}$$

The solutions are $m = \frac{2+\sqrt{10}}{2} \approx 2.5811$ and $m = \frac{2-\sqrt{10}}{2} \approx -0.5811$.

20. $p^2 + p - 1 = 0$

$$p = \frac{-1 \pm \sqrt{1^2 - 4(1)(-1)}}{2(1)}$$
$$= \frac{-1 \pm \sqrt{5}}{2}$$

The solutions are $p = \frac{-1+\sqrt{5}}{2} \approx 0.6180$ and $p = \frac{-1-\sqrt{5}}{2} \approx -1.6180$.

21. $k^2 - 10k = -20$

$$k^2 - 10k + 20 = 0$$
$$k = \frac{-(-10) \pm \sqrt{(-10)^2 - 4(1)(20)}}{2(1)}$$
$$k = \frac{10 \pm \sqrt{100-80}}{2}$$
$$k = \frac{10 \pm \sqrt{20}}{2}$$
$$k = \frac{10 \pm \sqrt{4}\sqrt{5}}{2}$$
$$k = \frac{10 \pm 2\sqrt{5}}{2}$$
$$k = \frac{2(5 \pm \sqrt{5})}{2}$$
$$k = 5 \pm \sqrt{5}$$

The solutions are $k = 5 + \sqrt{5} \approx 7.2361$ and $k = 5 - \sqrt{5} \approx 2.7639$.

22. $5x^2 - 8x + 2 = 0$

$$x = \frac{-(-8) \pm \sqrt{(-8)^2 - 4(5)(2)}}{2(5)} = \frac{8 \pm \sqrt{24}}{10} = \frac{8 \pm \sqrt{4 \cdot 6}}{10} = \frac{8 \pm \sqrt{4}\sqrt{6}}{10} = \frac{8 \pm 2\sqrt{6}}{10} = \frac{4 \pm \sqrt{6}}{5}$$

The solutions are $x = \frac{4+\sqrt{6}}{5} \approx 1.2899$ and $x = \frac{4-\sqrt{6}}{5} \approx 0.3101.$

23.
$$2r^2 - 7r + 5 = 0$$
$$(2r - 5)(r - 1) = 0$$
$$2r - 5 = 0 \quad \text{or} \quad r - 1 = 0$$
$$2r = 5$$
$$r = \frac{5}{2} \quad \text{or} \quad r = 1$$

The solutions are $r = \frac{5}{2}$ and $r = 1$.

24. $2x^2 - 7x + 30 = 0$

$$x = \frac{-(-7) \pm \sqrt{(-7)^2 - 4(2)(30)}}{2(2)}$$
$$x = \frac{7 \pm \sqrt{49 - 240}}{4}$$
$$x = \frac{7 \pm \sqrt{-191}}{4}$$

Since there is a negative number under the radical sign, $\sqrt{-191}$ is not a real number. Thus, there are no real number solutions.

25. $3k^2 + k = 6$

$$3k^2 + k - 6 = 0$$
$$k = \frac{-1 \pm \sqrt{1 - 4(3)(-6)}}{2(3)} = \frac{-1 \pm \sqrt{73}}{6}$$

The solutions are $k = \frac{-1+\sqrt{73}}{6} \approx 1.2573$ and $k = \frac{-1-\sqrt{73}}{6} \approx -1.5907.$

26. $5m^2 + 5m = 0$

$$5m(m + 1) = 0$$
$$5m = 0 \quad \text{or} \quad m + 1 = 0$$
$$m = 0 \quad \text{or} \quad m = -1$$

The solutions are $m = 0$ and $m = -1$.

27. $\frac{3x - 2}{7} = \frac{x + 2}{5}$

$$35\left(\frac{3x-2}{7}\right) = 35\left(\frac{x+2}{2}\right)$$
$$5(3x - 2) = 7(x + 2)$$
$$15x - 10 = 7x + 14$$
$$8x = 24 \Rightarrow x = 3$$

The solution is $x = 3$.

28. $\frac{x}{3} - 7 = 6 - \frac{3x}{4}$

Multiply both sides by 12, the least common denominator of 3 and 4.

$$12\left(\frac{x}{3} - 7\right) = 12\left(6 - \frac{3x}{4}\right)$$
$$12\left(\frac{x}{3}\right) - (12)(7) = (12)(6) - 12\left(\frac{3x}{4}\right)$$
$$4x - 84 = 72 - 9x$$
$$13x - 84 = 72$$
$$13x = 156$$
$$x = 12$$

The solution is $x = 12$.

29. $\frac{4}{x-3} - \frac{8}{2x+5} + \frac{3}{x-3} = 0$

$$\frac{4}{x-3} + \frac{3}{x-3} - \frac{8}{2x+5} = 0$$
$$\frac{7}{x-3} - \frac{8}{2x+5} = 0$$

Multiply both sides by $(x-3)(2x+5)$. Note that $x \neq 3$ and $x \neq \frac{5}{2}$.

$$(x-3)(2x+5)\left(\frac{7}{x-3} - \frac{8}{2x+5}\right) = (x-3)(2x+5)(0)$$
$$7(2x+5) - 8(x-3) = 0$$
$$14x + 35 - 8x + 24 = 0$$
$$6x + 59 = 0$$
$$6x = -59$$
$$x = -\frac{59}{6}$$

Note: It is especially important to check solutions of equations that involve rational expressions. Here, a check shows that $x = -\frac{59}{6}$ is a solution.

30. $\frac{5}{p-2} - \frac{7}{p+2} = \frac{12}{p^2-4}$

$\frac{5}{p-2} - \frac{7}{p+2} = \frac{12}{(p-2)(p+2)}$

Multiply both sides by $(p-2)(p+2)$. Note that $p \neq 2$ and $p \neq -2$.

$$(p-2)(p+2)\left(\frac{5}{p-2} - \frac{7}{p+2}\right) = (p-2)(p+2)\left(\frac{12}{(p-2)(p+2)}\right)$$

$$(p-2)(p+2)\left(\frac{5}{p-2}\right) - (p-2)(p+2)\left(\frac{7}{p+2}\right) = (p-2)(p+2)\left(\frac{12}{(p-2)(p+2)}\right)$$

$$(p+2)(5) - (p-2)(7) = 12$$
$$5p + 10 - 7p + 14 = 12$$
$$-2p + 24 = 12$$
$$-2p = -12$$
$$p = 6$$

The solution is $p = 6$.

31. $\frac{2m}{m-2} - \frac{6}{m} = \frac{12}{m^2-2m}$

$\frac{2m}{m-2} - \frac{6}{m} = \frac{12}{m(m-2)}$

Multiply both sides by $m(m-2)$. Note that $m \neq 0$ and $m \neq 2$.

$$m(m-2)\left(\frac{2m}{m-2} - \frac{6}{m}\right) = m(m-2)\left(\frac{12}{m(m-2)}\right)$$

$$m(2m) - 6(m-2) = 12$$
$$2m^2 - 6m + 12 = 12$$
$$2m^2 - 6m = 0$$
$$2m(m-3) = 0$$
$$2m = 0 \quad \text{or} \quad m - 3 = 0$$
$$m = 0 \quad \text{or} \quad m = 3$$

Since $m \neq 0$, 0 is not a solution. The solution is $m = 3$.

32. $\frac{2y}{y-1} = \frac{5}{y} + \frac{10-8y}{y^2-y}$

$\frac{2y}{y-1} = \frac{5}{y} + \frac{10-8y}{y(y-1)}$

Multiply both sides by $y(y-1)$. Note that $y \neq 0$ and $y \neq 1$.

$$y(y-1)\left(\frac{2y}{y-1}\right) = y(y-1)\left[\frac{5}{y} + \frac{10-8y}{y(y-1)}\right]$$

$$y(y-1)\left(\frac{2y}{y-1}\right) = y(y-1)\left(\frac{5}{y}\right) + y(y-1)\left[\frac{10-8y}{y(y-1)}\right]$$

$$y(2y) = (y-1)(5) + (10-8y)$$
$$2y^2 = 5y - 5 + 10 - 8y$$
$$2y^2 = 5 - 3y$$
$$2y^2 + 3y - 5 = 0$$
$$(2y+5)(y-1) = 0$$
$$2y + 5 = 0 \quad \text{or} \quad y - 1 = 0$$
$$y = -\frac{5}{2} \quad \text{or} \quad y = 1$$

Since $y \neq 1$, 1 is not a solution. The solution is $y = -\frac{5}{2}$.

33. $\frac{1}{x-2} - \frac{3x}{x-1} = \frac{2x+1}{x^2-3x+2}$

$\frac{1}{x-2} - \frac{3x}{x-1} = \frac{2x+1}{(x-2)(x-1)}$

Multiply both sides by $(x-2)(x-1)$. Note that $x \neq 2$ and $x \neq 1$.

$$(x-2)(x-1)\left(\frac{1}{x-2} - \frac{3x}{x-1}\right) = (x-2)(x-1) \cdot \left[\frac{2x+1}{(x-2)(x-1)}\right]$$

$$(x-2)(x-1)\left(\frac{1}{x-2}\right) - (x-2)(x-1) \cdot \left(\frac{3x}{x-1}\right) = \frac{(x-2)(x-1)(2x+1)}{(x-2)(x-1)}$$

$$(x-1) - (x-2)(3x) = 2x + 1$$
$$x - 1 - 3x^2 + 6x = 2x + 1$$
$$-3x^2 + 7x - 1 = 2x + 1$$
$$-3x^2 + 5x - 2 = 0$$
$$3x^2 - 5x + 2 = 0$$
$$(3x-2)(x-1) = 0$$
$$3x - 2 = 0 \quad \text{or} \quad x - 1 = 0$$
$$x = \frac{2}{3} \quad \text{or} \quad x = 1$$

1 is not a solution since $x \neq 1$. The solution is $x = \frac{2}{3}$.

34. $\frac{5}{a}+\frac{-7}{a+1}=\frac{a^2-2a+4}{a^2+a}$

$$a(a+1)\left(\frac{5}{a}+\frac{-7}{a+1}\right) = a(a+1)\left(\frac{a^2-2a+4}{a^2+a}\right)$$

Note that $a \neq 0$ and $a \neq -1$.

$$5(a+1)+(-7)(a)=a^2-2a+4$$
$$5a+5-7a=a^2-2a+4$$
$$5-2a=a^2-2a+4$$
$$5=a^2+4$$
$$0=a^2-1$$
$$0=(a+1)(a-1)$$
$$a+1=0 \quad \text{or} \quad a-1=0$$
$$a=-1 \quad \text{or} \quad a=1$$

Since -1 would make two denominators zero, $a = 1$ is the only solution.

35. $\frac{5}{b+5}-\frac{4}{b^2+2b}=\frac{6}{b^2+7b+10}$

$$\frac{5}{b+5}-\frac{4}{b(b+2)}=\frac{6}{(b+5)(b+2)}$$

Multiply both sides by $b(b+5)(b+2)$. Note that $b \neq 0$, $b \neq -5$, and $b \neq -2$.

$$b(b+5)(b+2)\left(\frac{5}{b+5}-\frac{4}{b(b+2)}\right) = b(b+5)(b+2)\left(\frac{6}{(b+5)(b+2)}\right)$$

$$5b(b+2)-4(b+5)=6b$$
$$5b^2+10b-4b-20=6b$$
$$5b^2-20=0$$
$$b^2-4=0$$
$$(b+2)(b-2)=0$$
$$b+2=0 \quad \text{or} \quad b-2=0$$
$$b=-2 \quad \text{or} \quad b=2$$

Since $b \neq -2$, -2 is not a solution. The solution is $b = 2$.

36. $\frac{2}{x^2-2x-3}+\frac{5}{x^2-x-6}=\frac{1}{x^2+3x+2}$

$$\frac{2}{(x-3)(x+1)}+\frac{5}{(x-3)(x+2)} = \frac{1}{(x+2)(x+1)}$$

Multiply both sides by $(x-3)(x+1)(x+2)$. Note that $x \neq 3$, $x \neq -1$, and $x \neq -2$.

$$(x-3)(x+1)(x+2)\left(\frac{2}{(x-3)(x+1)}\right) + (x-3)(x+1)(x+2)\left(\frac{5}{(x-3)(x+2)}\right) = (x-3)(x+1)(x+2)\left(\frac{1}{(x+2)(x+1)}\right)$$

$$2(x+2)+5(x+1)=x-3$$
$$2x+4+5x+5=x-3$$
$$7x+9=x-3$$
$$6x+9=-3$$
$$6x=-12$$
$$x=-2$$

However, $x \neq -2$. Therefore there is no solution.

37. $\frac{4}{2x^2+3x-9}+\frac{2}{2x^2-x-3}=\frac{3}{x^2+4x+3}$

$$\frac{4}{(2x-3)(x+3)}+\frac{2}{(2x-3)(x+1)} = \frac{3}{(x+3)(x+1)}$$

Multiply both sides by $(2x-3)(x+3)(x+1)$. Note that $x \neq \frac{3}{2}$, $x \neq -3$, and $x \neq -1$.

$$(2x-3)(x+3)(x+1) \cdot \left(\frac{4}{(2x-3)(x+3)}+\frac{2}{(2x-3)(x+1)}\right) = (2x-3)(x+3)(x+1)\left(\frac{3}{(x+3)(x+1)}\right)$$

$$4(x+1)+2(x+3)=3(2x-3)$$
$$4x+4+2x+6=6x-9$$
$$6x+10=6x-9$$
$$10=-9$$

This is a false statement. Therefore, there is no solution.

R.5 Inequalities

1. $x < 4$

Because the inequality symbol means "less than," the endpoint at 4 is not included. This inequality is written in interval notation as $(-\infty, 4)$. To graph this interval on a number line, place an open circle at 4 and draw a heavy arrow pointing to the left.

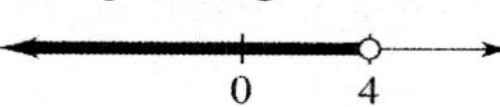

2. $x \geq -3$
Because the inequality sign means "greater than or equal to," the endpoint at -3 is included. This inequality is written in interval notation as $[-3, \infty)$. To graph this interval on a number line, place a closed circle at -3 and draw a heavy arrow pointing to the right.

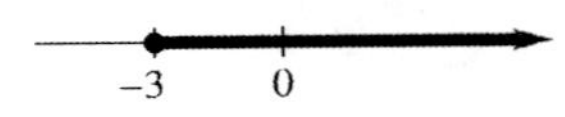

3. $1 \leq x < 2$
The endpoint at 1 is included, but the endpoint at 2 is not. This inequality is written in interval notation as $[1, 2)$. To graph this interval, place a closed circle at 1 and an open circle at 2; then draw a heavy line segment between them.

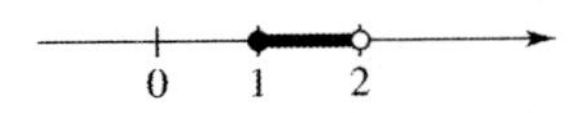

4. $-2 \leq x \leq 3$
The endpoints at -2 and 3 are both included. This inequality is written in interval notation as $[-2, 3]$. To graph this interval, place an open circle at -2 and another at 3 and draw a heavy line segment between them.

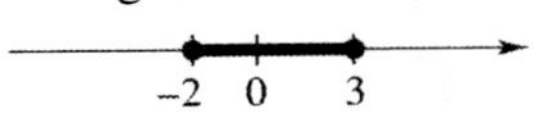

5. $-9 > x$
This inequality may be rewritten as $x < -9$, and is written in interval notation as $(-\infty, -9)$. Note that the endpoint at -9 is not included. To graph this interval, place an open circle at -9 and draw a heavy arrow pointing to the left.

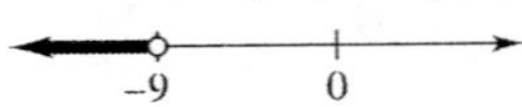

6. $6 \leq x$
This inequality may be written as $x \geq 6$, and is written in interval notation as $[6, \infty)$. Note that the endpoint at 6 is included. To graph this interval, place a closed circle at 6 and draw a heavy arrow pointing to the right.

7. $[-7, -3]$
This represents all the numbers between -7 and -3, including both endpoints. This interval can be written as the inequality $-7 \leq x \leq -3$.

8. $[4, 10)$
This represents all the numbers between 4 and 10, including 4 but not including 10. This interval can be written as the inequality $4 \leq x < 10$.

9. $(-\infty, -1]$
This represents all the numbers to the left of -1 on the number line and includes the endpoint. This interval can be written as the inequality $x \leq -1$.

10. $(3, \infty)$
This represents all the numbers to the right of 3, and does not include the endpoint. This interval can be written as the inequality $x > 3$.

11. Notice that the endpoint -2 is included, but 6 is not. The interval shown in the graph can be written as the inequality $-2 \leq x < 6$.

12. Notice that neither endpoint is included. The interval shown in the graph can be written as $0 < x < 8$.

13. Notice that both endpoints are included. The interval shown in the graph can be written as $x \leq -4$ or $x \geq 4$.

14. Notice that the endpoint 0 is not included, but 3 is included. The interval shown in the graph can be written as $x < 0$ or $x \geq 3$.

15.
$$\begin{aligned} 6p + 7 &\leq 19 \\ 6p &\leq 12 \\ \left(\frac{1}{6}\right)(6p) &\leq \left(\frac{1}{6}\right)(12) \\ p &\leq 2 \end{aligned}$$
The solution in interval notation is $(-\infty, 2]$.

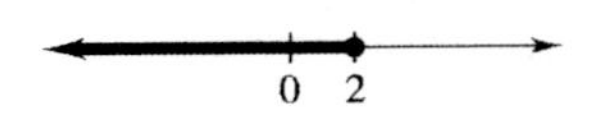

16.
$$\begin{aligned} 6k - 4 &< 3k - 1 \\ 6k &< 3k + 3 \\ 3k &< 3 \\ k &< 1 \end{aligned}$$
The solution in interval notation is $(-\infty, 1)$.

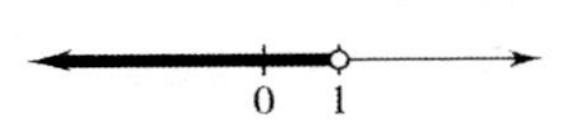

17.
$$\begin{aligned} m - (3m - 2) + 6 &< 7m - 19 \\ m - 3m + 2 + 6 &< 7m - 19 \\ -2m + 8 &< 7m - 19 \\ -9m + 8 &< -19 \\ -9m &< -27 \\ -\frac{1}{9}(-9m) &> -\frac{1}{9}(-27) \\ m &> 3 \end{aligned}$$
The solution is $(3, \infty)$.

0 3

18. $-2(3y-8) \geq 5(4y-2)$
$$-6y+16 \geq 20y-10$$
$$-6y+16+(-16) \geq 20y-10+(-16)$$
$$-6y \geq 20y-26$$
$$-6y+(-20y) \geq 20y+(-20y)-26$$
$$-26y \geq -26$$
$$-\frac{1}{26}(-26)y \leq -\frac{1}{26}(-26)$$
$$y \leq 1$$
The solution is $(-\infty, 1]$.

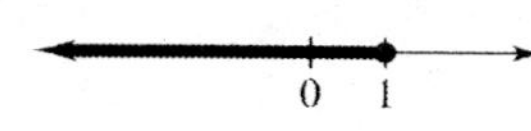

19. $3p-1<6p+2(p-1)$
$$3p-1<6p+2p-2$$
$$3p-1<8p-2$$
$$-5p-1<-2$$
$$-5p<-1$$
$$-\frac{1}{5}(-5p)>-\frac{1}{5}(-1)$$
$$p>\frac{1}{5}$$
The solution is $\left(\frac{1}{5}, \infty\right)$.

20. $x+5(x+1)>4(2-x)+x$
$$x+5x+5>8-4x+x$$
$$6x+5>8-3x$$
$$6x>3-3x$$
$$9x>3$$
$$x>\frac{1}{3}$$
The solution is $\left(\frac{1}{3}, \infty\right)$.

21. $-11<y-7<-1$
$$-11+7<y-7+7<-1+7$$
$$-4<y<6$$
The solution is $(-4, 6)$.

22. $8 \leq 3r+1 \leq 13$
$$8+(-1) \leq 3r+1+(-1) \leq 13+(-1)$$
$$7 \leq 3r \leq 12$$
$$\frac{1}{3}(7) \leq \frac{1}{3}(3r) \leq \frac{1}{3}(12)$$
$$\frac{7}{3} \leq r \leq 4$$
The solution is $\left[\frac{7}{3}, 4\right]$.

23. $-2<\frac{1-3k}{4} \leq 4$
$$4(-2)<4\left(\frac{1-3k}{4}\right) \leq 4(4)$$
$$-8<1-3k \leq 16$$
$$-9<-3k \leq 15$$
$$-\frac{1}{3}(-9)>-\frac{1}{3}(-3k) \geq -\frac{1}{3}(15)$$
Rewrite the inequalities in the proper order.
$-5 \leq k<3$
The solution is $[-5, 3)$.

24. $-1 \leq \frac{5y+2}{3} \leq 4$
$$3(-1) \leq 3\left(\frac{5y+2}{3}\right) \leq 3(4)$$
$$-3 \leq 5y+2 \leq 12$$
$$-5 \leq 5y \leq 10$$
$$-1 \leq y \leq 2$$
The solution is $[-1, 2]$.

25. $\frac{3}{5}(2p+3) \geq \frac{1}{10}(5p+1)$
$$10\left(\frac{3}{5}\right)(2p+3) \geq 10\left(\frac{1}{10}\right)(5p+1)$$
$$6(2p+3) \geq 5p+1$$
$$12p+18 \geq 5p+1$$
$$7p \geq -17 \Rightarrow p \geq -\frac{17}{7}$$
The solution is $\left[-\frac{17}{7}, \infty\right)$.

26. $\frac{8}{3}(z-4) \le \frac{2}{9}(3z+2)$

$(9)\frac{8}{3}(z-4) \le (9)\frac{2}{9}(3z+2)$

$24(z-4) \le 2(3z+2)$

$24z - 96 \le 6z + 4$

$24z \le 6z + 100$

$18z \le 100$

$z \le \frac{100}{8} \Rightarrow z \le \frac{50}{9}$

The solution is $\left(-\infty, \frac{50}{9}\right]$.

27. $(m-3)(m+5) < 0$

Solve $(m-3)(m+5) = 0$.

$(m-3)(m+5) = 0$

$m = 3$ or $m = -5$

Intervals: $(-\infty, -5), (-5, 3), (3, \infty)$

For $(-\infty, -5)$, choose -6 to test for m.

$(-6-3)(-6+5) = -9(-1) = 9 \not< 0$

For $(-5, 3)$, choose 0.

$(0-3)(0+5) = -3(5) = -15 < 0$

For $(3, \infty)$, choose 4.

$(4-3)(4+5) = 1(9) = 9 \not< 0$

The solution is $(-5, 3)$.

28. $(t+6)(t-1) \ge 0$

Solve $(t+6)(t-1) = 0$.

$(t+6)(t-1) = 0$

$t = -6$ or $t = 1$

Intervals: $(-\infty, -6), (-6, 1), (1, \infty)$

For $(-\infty, -6)$, choose -7 to test for t.

$(-7+6)(-7-1) = (-1)(-8) = 8 \ge 0$

For $(-6, 1)$, choose 0.

$(0+6)(0-1) = (6)(-1) = -6 \not\ge 0$

For $(1, \infty)$, choose 2.

$(2+6)(2-1) = (8)(1) = 8 \ge 0$

Because the symbol $\ge$ is used, the endpoints -6 and 1 are included in the solution. The solution is $(-\infty, -6] \cup [1, \infty)$.

29. $y^2 - 3y + 2 < 0$

$(y-2)(y-1) < 0$

Solve $(y-2)(y-1) = 0$.

$y = 2$ or $y = 1$

Intervals: $(-\infty, 1), (1, 2), (2, \infty)$

For $(-\infty, 1)$, choose $y = 0$.

$0^2 - 3(0) + 2 = 2 \not< 0$

For $(1, 2)$, choose $y = \frac{3}{2}$.

$\left(\frac{3}{2}\right)^2 - 3\left(\frac{3}{2}\right) + 2 = \frac{9}{4} - \frac{9}{2} + 2$

$= \frac{9 - 18 + 8}{4} = -\frac{1}{4} < 0$

For $(2, \infty)$, choose 3.

$3^2 - 3(3) + 2 = 2 \not< 0$

The solution is $(1, 2)$.

30. $2k^2 + 7k - 4 > 0$

Solve $2k^2 + 7k - 4 = 0$.

$2k^2 + 7k - 4 = 0$

$(2k-1)(k+4) = 0 \Rightarrow k = \frac{1}{2}$ or $k = -4$

Intervals: $(-\infty, -4), \left(-4, \frac{1}{2}\right), \left(\frac{1}{2}, \infty\right)$

For $(-\infty, -4)$, choose -5.

$2(-5)^2 + 7(-5) - 4 = 11 > 0$

For $\left(-4, \frac{1}{2}\right)$, choose 0.

$2(0)^2 + 7(0) - 4 = -4 \not> 0$

For $\left(\frac{1}{2}, \infty\right)$, choose 1.

$2(1)^2 + 7(1) - 4 = 5 > 0$

The solution is $(-\infty, -4) \cup \left(\frac{1}{2}, \infty\right)$.

31. $x^2 - 16 > 0$

Solve $x^2 - 16 = 0$.

$x^2 - 16 = 0$

$(x+4)(x-4) = 0 \Rightarrow x = -4$ or $x = 4$

Intervals: $(-\infty, -4), (-4, 4), (4, \infty)$

(continued on next page)

(continued)

For $(-\infty, -4)$, choose -5.

$(-5)^2 - 16 = 9 > 0$

For $(-4, 4)$, choose 0.

$0^2 - 16 = -16 \not> 0$

For $(4, \infty)$, choose 5.

$5^2 - 16 = 9 > 0$

The solution is $(-\infty, -4) \cup (4, \infty)$.

32. $2k^2 - 7k - 15 \le 0$

Solve $2k^2 - 7k - 15 = 0$.

$$2k^2 - 7k - 15 = 0$$
$$(2k+3)(k-5) = 0$$
$$k = -\frac{3}{2} \quad \text{or} \quad k = 5$$

Intervals: $\left(-\infty, -\frac{3}{2}\right), \left(-\frac{3}{2}, 5\right), (5, \infty)$

For $\left(-\infty, -\frac{3}{2}\right)$, choose -2.

$2(-2)^2 - 7(-2) - 15 = 7 \not\le 0$

For $\left(-\frac{3}{2}, 5\right)$, choose 0.

$2(0)^2 - 7(0) - 15 = -15 \le 0$

For $(5, \infty)$, choose 6.

$2(6)^2 - 7(6) - 15 \not\le 0$

The solution is $\left[-\frac{3}{2}, 5\right]$.

33. $x^2 - 4x \ge 5$

Solve $x^2 - 4x = 5$.

$$x^2 - 4x = 5$$
$$x^2 - 4x - 5 = 0$$
$$(x+1)(x-5) = 0$$
$$x + 1 = 0 \quad \text{or} \quad x - 5 = 0$$
$$x = -1 \quad \text{or} \quad x = 5$$

Intervals: $(-\infty, -1), (-1, 5), (5, \infty)$

For $(-\infty, -1)$, choose -2.

$(-2)^2 - 4(-2) = 12 \ge 5$

For $(-1, 5)$, choose 0.

$0^2 - 4(0) = 0 \not\ge 5$

For $(5, \infty)$, choose 6.

$(6)^2 - 4(6) = 12 \ge 5$

The solution is $(-\infty, -1] \cup [5, \infty)$.

34. $10r^2 + r \le 2$

Solve $10r^2 + r = 2$.

$$10r^2 + r = 2$$
$$10r^2 + r - 2 = 0$$
$$(5r-2)(2r+1) = 0 \Rightarrow r = \frac{2}{5} \quad \text{or} \quad r = -\frac{1}{2}$$

Intervals: $\left(-\infty, -\frac{1}{2}\right), \left(-\frac{1}{2}, \frac{2}{5}\right), \left(\frac{2}{5}, \infty\right)$

For $\left(-\infty, -\frac{1}{2}\right)$, choose -1.

$10(-1)^2 + (-1) = 9 \not\le 2$

For $\left(-\frac{1}{2}, \frac{2}{5}\right)$, choose 0.

$10(0)^2 + 0 = 0 \le 2$

For $\left(\frac{2}{5}, \infty\right)$, choose 1.

$10(1)^2 + 1 = 11 \not\le 2$

The solution is $\left[-\frac{1}{2}, \frac{2}{5}\right]$.

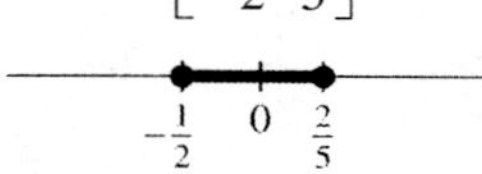

35. $3x^2 + 2x > 1$

Solve $3x^2 + 2x = 1$.

$$3x^2 + 2x = 1$$
$$3x^2 + 2x - 1 = 0$$
$$(3x-1)(x+1) = 0 \Rightarrow x = \frac{1}{3} \quad \text{or} \quad x = -1$$

Intervals: $(-\infty, -1), \left(-1, \frac{1}{3}\right), \left(\frac{1}{3}, \infty\right)$

For $(-\infty, -1)$, choose -2.

$3(-2)^2 + 2(-2) = 8 > 1$

For $\left(-1, \frac{1}{3}\right)$, choose 0.

$3(0)^2 + 2(0) = 0 \not> 1$

For $\left(\frac{1}{3}, \infty\right)$, choose 1.

$3(1)^2 + 2(1) = 5 > 1$

(continued on next page)

(continued)

The solution is $(-\infty, -1) \cup \left(\frac{1}{3}, \infty\right)$.

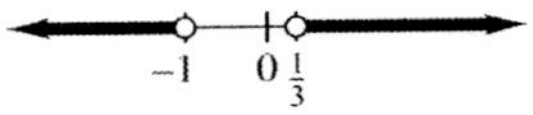

36. $3a^2 + a > 10$

Solve $3a^2 + a = 10$.

$$3a^2 + a = 10$$
$$3a^2 + a - 10 = 0$$
$$(3a-5)(a+2) = 0$$
$$a = \frac{5}{3} \quad \text{or} \quad a = -2$$

Intervals: $(-\infty, -2), \left(-2, \frac{5}{3}\right), \left(\frac{5}{3}, \infty\right)$

For $(-\infty, -2)$, choose -3.

$3(-3)^2 + (-3) = 24 > 10$

For $\left(-2, \frac{5}{3}\right)$, choose 0.

$3(0)^2 + 0 = 0 \not> 10$

For $\left(\frac{5}{3}, \infty\right)$, choose 2.

$3(2)^2 + 2 = 14 > 10$

The solution is $(-\infty, -2) \cup \left(\frac{5}{3}, \infty\right)$.

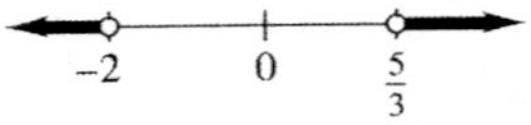

37. $9 - x^2 \le 0$

Solve $9 - x^2 = 0$.

$$9 - x^2 = 0$$
$$(3+x)(3-x) = 0$$
$$x = -3 \quad \text{or} \quad x = 3$$

Intervals: $(-\infty, -3), (-3, 3), (3, \infty)$

For $(-\infty, -3)$, choose -4.

$9 - (-4)^2 = -7 \le 0$

For $(-3, 3)$, choose 0.

$9 - (0)^2 = 9 \not\le 0$

For $(3, \infty)$, choose 4.

$9 - (4)^2 = -7 \le 0$

The solution is $(-\infty, -3] \cup [3, \infty)$.

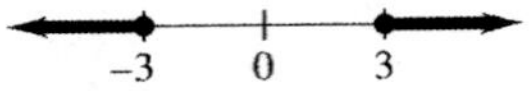

38. $p^2 - 16p > 0$

Solve $p^2 - 16p = 0$.

$$p^2 - 16p = 0$$
$$p(p-16) = 0$$
$$p = 0 \quad \text{or} \quad p = 16$$

Intervals: $(-\infty, 0), (0, 16), (16, \infty)$

For $(-\infty, 0)$, choose -1.

$(-1)^2 - 16(-1) = 17 > 0$

For $(0, 16)$, choose 1.

$(1)^2 - 16(1) = -15 \not> 0$

For $(16, \infty)$, choose 17.

$(17)^2 - 16(17) = 17 > 0$

The solution is $(-\infty, 0) \cup (16, \infty)$.

39. $x^3 - 4x \ge 0$

Solve $x^3 - 4x = 0$.

$$x^3 - 4x = 0$$
$$x(x^2 - 4) = 0$$
$$x(x+2)(x-2) = 0$$
$$x = 0, \quad \text{or} \quad x = -2, \quad \text{or} \quad x = 2$$

Intervals: $(-\infty, -2), (-2, 0), (0, 2), (2, \infty)$

For $(-\infty, -2)$, choose -3.

$(-3)^3 - 4(-3) = -15 \not\ge 0$

For $(-2, 0)$, choose -1.

$(-1)^3 - 4(-1) = 3 \ge 0$

For $(0, 2)$, choose 1.

$(1)^3 - 4(1) = -3 \not\ge 0$

For $(2, \infty)$, choose 3.

$(3)^3 - 4(3) = 15 \ge 0$

The solution is $[-2, 0] \cup [2, \infty)$.

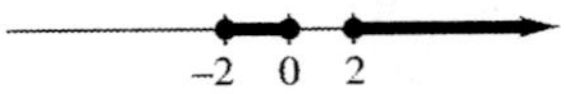

40. $x^3 + 7x^2 + 12x \le 0$

Solve $x^3 + 7x^2 + 12x = 0$.

$$x^3 + 7x^2 + 12x = 0$$
$$x(x^2 + 7x + 12) = 0$$
$$x(x+3)(x+4) = 0$$
$$x = 0, \quad \text{or} \quad x = -3, \quad \text{or} \quad x = -4$$

Intervals: $(-\infty, -4), (-4, -3), (-3, 0), (0, \infty)$

(continued on next page)

(continued)

For $(-\infty, -4)$, choose -5.

$$(-5)^3 + 7(-5)^2 + 12(-5) = -10 \le 0$$

For $(-4, -3)$, choose $-\frac{7}{2}$.

$$\left(-\frac{7}{2}\right)^3 + 7\left(-\frac{7}{2}\right)^2 + 12\left(-\frac{7}{2}\right) = \frac{7}{8} \not\le 0$$

For $(-3, 0)$, choose -1.

$$(-1)^3 + 7(-1)^2 + 12(-1) = -6 \le 0$$

For $(0, \infty)$, choose 1.

$$(1)^3 + 7(1)^2 + 12(1) = 20 \not\le 0$$

The solution is $(-\infty, -4] \cup [-3, 0]$.

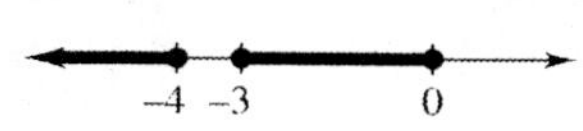

41. $2x^3 - 14x^2 + 12x < 0$

Solve $2x^3 - 14x^2 + 12x = 0$.

$$2x^3 - 14x^2 + 12x = 0$$
$$2x(x^2 - 7x + 6) = 0$$
$$2x(x-1)(x-6) = 0$$

$x = 0$, or $x = 1$, or $x = 6$

Intervals: $(-\infty, 0), (0, 1), (1, 6), (6, \infty)$

For $(-\infty, 0)$, choose -1.

$$2(-1)^3 - 14(-1)^2 + 12(-1) = -28 < 0$$

For $(0, 1)$, choose $\frac{1}{2}$.

$$2\left(\frac{1}{2}\right)^3 - 14\left(\frac{1}{2}\right)^2 + 12\left(\frac{1}{2}\right) = \frac{11}{4} \not< 0$$

For $(1, 6)$, choose 2.

$$2(2)^3 - 14(2)^2 + 12(2) = -16 < 0$$

For $(6, \infty)$, choose 7.

$$2(7)^3 - 14(7)^2 + 12(7) = 84 \not< 0$$

The solution is $(-\infty, 0) \cup (1, 6)$.

42. $3x^3 - 9x^2 - 12x > 0$

Solve $3x^3 - 9x^2 - 12x = 0$.

$$3x^3 - 9x^2 - 12x = 0$$
$$3x(x^2 - 3x - 4) = 0$$
$$3x(x-4)(x+1) = 0$$

$x = 0$, or $x = 4$, or $x = -1$

Intervals: $(-\infty, -1), (-1, 0), (0, 4), (4, \infty)$

For $(-\infty, -1)$, choose -2.

$$3(-2)^3 - 9(-2)^2 - 12(-2) = -36 \not> 0$$

For $(-1, 0)$, choose $-\frac{1}{2}$.

$$3\left(-\frac{1}{2}\right)^3 - 9\left(-\frac{1}{2}\right)^2 - 12\left(-\frac{1}{2}\right) = \frac{27}{8} > 0$$

For $(0, 4)$, choose 1.

$$3(1)^3 - 9(1)^2 - 12(1) = -18 \not> 0$$

For $(4, \infty)$, choose 5.

$$3(5)^3 - 9(5)^2 - 12(5) = 90 > 0$$

The solution is $(-1, 0) \cup (4, \infty)$.

43. $\frac{m-3}{m+5} \le 0$

Solve $\frac{m-3}{m+5} = 0$.

$$(m+5)\frac{m-3}{m+5} = (m+5)(0)$$
$$m - 3 = 0$$
$$m = 3$$

Set the denominator equal to 0 and solve.

$$m + 5 = 0$$
$$m = -5$$

Intervals: $(-\infty, -5), (-5, 3), (3, \infty)$

For $(-\infty, -5)$, choose -6.

$$\frac{-6-3}{-6+5} = 9 \not\le 0$$

For $(-5, 3)$, choose 0.

$$\frac{0-3}{0+5} = -\frac{3}{5} \le 0$$

For $(3, \infty)$, choose 4.

$$\frac{4-3}{4+5} = \frac{1}{9} \not\le 0$$

Although the $\le$ symbol is used, including -5 in the solution would cause the denominator to be zero. The solution is $(-5, 3]$.

44. $\frac{r+1}{r-1} > 0$

Solve the equation $\frac{r+1}{r-1} = 0$.

$$\frac{r+1}{r-1} = 0$$
$$(r-1)\frac{r+1}{r-1} = (r-1)(0)$$
$$r+1 = 0 \Rightarrow r = -1$$

Find the value for which the denominator equals zero.

$r-1 = 0 \Rightarrow r = 1$

Intervals: $(-\infty, -1), (-1, 1), (1, \infty)$

For $(-\infty, -1)$, choose -2.

$$\frac{-2+1}{-2-1} = \frac{-1}{-3} = \frac{1}{3} > 0$$

For $(-1, 1)$, choose 0.

$$\frac{0+1}{0-1} = \frac{1}{-1} = -1 \not> 0$$

For $(1, \infty)$, choose 2.

$$\frac{2+1}{2-1} = \frac{3}{1} = 3 > 0$$

The solution is $(-\infty, -1) \cup (1, \infty)$.

45. $\frac{k-1}{k+2} > 1$

Solve $\frac{k-1}{k+2} = 1$.

$$k-1 = k+2$$
$$-1 \neq 2$$

The equation has no solution. Solve

$k+2 = 0 \Rightarrow k = -2$

Intervals: $(-\infty, -2), (-2, \infty)$

For $(-\infty, -2)$, choose -3.

$$\frac{-3-1}{-3+2} = 4 > 1$$

For $(-2, \infty)$, choose 0.

$$\frac{0-1}{0+2} = -\frac{1}{2} \not> 1$$

The solution is $(-\infty, -2)$.

46. $\frac{a-5}{a+2} < -1$

Solve the equation $\frac{a-5}{a+2} = -1$.

$$\frac{a-5}{a+2} = -1$$
$$a-5 = -1(a+2)$$
$$a-5 = -a-2$$
$$2a = 3 \Rightarrow a = \frac{3}{2}$$

Set the denominator equal to zero and solve for a.

$a+2 = 0 \Rightarrow a = -2$

Intervals: $(-\infty, -2), \left(-2, \frac{3}{2}\right), \left(\frac{3}{2}, \infty\right)$

For $(-\infty, -2)$, choose -3.

$$\frac{-3-5}{-3+2} = \frac{-8}{-1} = 8 \not< -1$$

For $\left(-2, \frac{3}{2}\right)$, choose 0.

$$\frac{0-5}{0+2} = \frac{-5}{2} = -\frac{5}{2} < -1$$

For $\left(\frac{3}{2}, \infty\right)$, choose 2.

$$\frac{2-5}{2+2} = \frac{-3}{4} = -\frac{3}{4} \not< -1$$

The solution is $\left(-2, \frac{3}{2}\right)$.

47. $\frac{2y+3}{y-5} \le 1$

Solve $\frac{2y+3}{y-5} = 1$.

$$2y+3 = y-5$$
$$y = -8$$

Solve $y-5 = 0 \Rightarrow y = 5$.

Intervals: $(-\infty, -8), (-8, 5), (5, \infty)$

For $(-\infty, -8)$, choose $y = -10$.

$$\frac{2(-10)+3}{-10-5} = \frac{17}{15} \not\le 1$$

For $(-8, 5)$, choose $y = 0$.

$$\frac{2(0)+3}{0-5} = -\frac{3}{5} \le 1$$

For $(5, \infty)$, choose $y = 6$.

$$\frac{2(6)+3}{6-5} = \frac{15}{1} \not\le 1$$

The solution is $[-8, 5)$.

48. $\frac{a+2}{3+2a} \le 5$

Solve the equation $\frac{a+2}{3+2a} = 5$.

$$\frac{a+2}{3+2a} = 5$$
$$a+2 = 5(3+2a)$$
$$a+2 = 15+10a$$
$$-9a = 13 \Rightarrow a = -\frac{13}{9}$$

(continued on next page)

(*continued*)

Set the denominator equal to zero and solve for a.

$3+2a=0 \Rightarrow 2a=-3 \Rightarrow a=-\frac{3}{2}$

Intervals: $\left(-\infty,-\frac{3}{2}\right),\left(-\frac{3}{2},-\frac{13}{9}\right),\left(-\frac{13}{9},\infty\right)$

For $\left(-\infty,-\frac{3}{2}\right)$, choose -2.

$\frac{-2+2}{3+2(-2)}=\frac{0}{-1}=0 \le 5$

For $\left(-\frac{3}{2},-\frac{13}{9}\right)$, choose -1.46.

$\frac{-1.46+2}{3+2(-1.46)}=\frac{0.54}{0.08}=6.75 \not\le 5$

For $\left(-\frac{13}{9},\infty\right)$, choose 0.

$\frac{0+2}{3+2(0)}=\frac{2}{3} \le 5$

The value $-\frac{3}{2}$ cannot be included in the solution since it would make the denominator zero. The solution is $\left(-\infty,-\frac{3}{2}\right) \cup \left[-\frac{13}{9},\infty\right)$.

49. $\frac{2k}{k-3} \le \frac{4}{k-3}$

Solve $\frac{2k}{k-3}=\frac{4}{k-3}$.

$$\frac{2k}{k-3}=\frac{4}{k-3}$$
$$\frac{2k}{k-3}-\frac{4}{k-3}=0$$
$$\frac{2k-4}{k-3}=0 \Rightarrow 2k-4=0 \Rightarrow k=2$$

Set the denominator equal to 0 and solve for k.

$k-3=0 \Rightarrow k=3$

Intervals: $(-\infty, 2), (2, 3), (3, \infty)$

For $(-\infty, 2)$, choose 0.

$\frac{2(0)}{0-3}=0$ and $\frac{4}{0-3}=-\frac{4}{3}$, so $\frac{2(0)}{0-3} \not\le \frac{4}{0-3}$.

For $(2, 3)$, choose $\frac{5}{2}$.

$\frac{2\left(\frac{5}{2}\right)}{\frac{5}{2}-3}=\frac{5}{-\frac{1}{2}}=-10$ and $\frac{4}{\frac{5}{2}-3}=\frac{4}{-\frac{1}{2}}=-8$, so $\frac{2\left(\frac{5}{2}\right)}{\frac{5}{2}-3} \le \frac{4}{\frac{5}{2}-3}$.

For $(3, \infty)$, choose 4.

$\frac{2(4)}{4-3}=8$ and $\frac{4}{4-3}=4$, so $\frac{2(4)}{4-3} \not\le \frac{4}{4-3}$.

The solution is $[2, 3)$.

50. $\frac{5}{p+1} > \frac{12}{p+1}$

Solve the equation $\frac{5}{p+1}=\frac{12}{p+1}$.

$\frac{5}{p+1}=\frac{12}{p+1} \Rightarrow 5=12$

The equation has no solution.

Set the denominator equal to zero and solve for p. $p+1=0 \Rightarrow p=-1$

Intervals: $(-\infty,-1), (-1,\infty)$

For $(-\infty,-1)$, choose -2.

$\frac{5}{-2+1}=-5$ and $\frac{12}{-2+1}=-12$, so $\frac{5}{-2+1} > \frac{12}{-2+1}$.

For $(-1,\infty)$, choose 0.

$\frac{5}{0+1}=5$ and $\frac{12}{0+1}=12$, so $\frac{5}{0+1} \not> \frac{12}{0+1}$.

The solution is $(-\infty,-1)$.

51. $\frac{2x}{x^2-x-6} \ge 0$

Solve $\frac{2x}{x^2-x-6}=0$.

$\frac{2x}{x^2-x-6}=0 \Rightarrow 2x=0 \Rightarrow x=0$

Set the denominator equal to 0 and solve for x.

$$x^2-x-6=0$$
$$(x+2)(x-3)=0 \Rightarrow x=-2 \text{ or } x=3$$

Intervals: $(-\infty,-2), (-2, 0), (0, 3), (3, \infty)$

For $(-\infty,-2)$, choose -3.

$\frac{2(-3)}{(-3)^2-(-3)-6}=-1 \not\ge 0$

For $(-2, 0)$, choose -1.

$\frac{2(-1)}{(-1)^2-(-1)-6}=\frac{1}{2} \ge 0$

For $(0, 3)$, choose 2.

$\frac{2(2)}{2^2-2-6}=-1 \not\ge 0$

(*continued on next page*)

(continued)

For $(3, \infty)$, choose 4.

$$\frac{2(4)}{4^2-4-6} = \frac{4}{3} \geq 0$$

The solution is $(-2, 0] \cup (3, \infty)$.

52. $\dfrac{8}{p^2+2p} > 1$

Solve the equation $\dfrac{8}{p^2+2p} = 1$.

$$\frac{8}{p^2+2p} = 1$$
$$8 = p^2 + 2p$$
$$0 = p^2 + 2p - 8$$
$$0 = (p+4)(p-2)$$
$$p+4=0 \quad \text{or} \quad p-2=0$$
$$p=-4 \quad \text{or} \quad p=2$$

Set the denominator equal to zero and solve for p.

$$p^2+2p=0$$
$$p(p+2)=0$$
$$p=0 \quad \text{or} \quad p+2=0 \Rightarrow p=-2$$

Intervals:
$(-\infty, -4), (-4, -2), (-2, 0), (0, 2), (2, \infty)$

For $(-\infty, -4)$, choose -5.

$$\frac{8}{(-5)^2+2(-5)} = \frac{8}{15} \ngtr 1$$

For $(-4, -2)$, choose -3.

$$\frac{8}{(-3)^2+2(-3)} = \frac{8}{9-6} = \frac{8}{3} > 1$$

For $(-2, 0)$, choose -1.

$$\frac{8}{(-1)^2+2(-1)} = \frac{8}{-1} = -8 \ngtr 1$$

For $(0, 2)$, choose 1.

$$\frac{8}{(1)^2+2(1)} = \frac{8}{3} > 1$$

For $(2, \infty)$, choose 3.

$$\frac{8}{(3)^2+(2)(3)} = \frac{8}{15} \ngtr 1$$

The solution is $(-4, -2) \cup (0, 2)$.

53. $\dfrac{z^2+z}{z^2-1} \geq 3$

Solve

$$\frac{z^2+z}{z^2-1} = 3.$$
$$z^2+z = 3z^2-3$$
$$-2z^2+z+3=0$$
$$-1(2z^2-z-3)=0$$
$$-1(z+1)(2z-3)=0$$
$$z=-1 \quad \text{or} \quad z=\frac{3}{2}$$

Set $z^2-1=0$.

$z^2=1 \Rightarrow z=-1$ or $z=1$

Intervals: $(-\infty, -1), (-1, 1), \left(1, \frac{3}{2}\right), \left(\frac{3}{2}, \infty\right)$

For $(-\infty, -1)$, choose $x=-2$.

$$\frac{(-2)^2+3}{(-2)^2-1} = \frac{7}{3} \ngeq 3$$

For $(-1, 1)$, choose $x=0$.

$$\frac{0^2+3}{0^2-1} = -3 \ngeq 3$$

For $\left(1, \frac{3}{2}\right)$, choose $x=\frac{3}{2}$.

$$\frac{\left(\frac{3}{2}\right)^2+3}{\left(\frac{3}{2}\right)^2-1} = \frac{21}{5} \geq 3$$

For $\left(\frac{3}{2}, \infty\right)$, choose $x=2$.

$$\frac{2^2+3}{2^2-1} = \frac{7}{3} \ngeq 3$$

The solution is $\left(1, \frac{3}{2}\right]$.

54. $\dfrac{a^2+2a}{a^2-4} \leq 2$

Solve the equation $\dfrac{a^2+2a}{a^2-4} = 2$.

$$\frac{a^2+2a}{a^2-4} = 2$$
$$a^2+2a = 2(a^2-4)$$
$$a^2+2a = 2a^2-8$$
$$0 = a^2-2a-8$$
$$0 = (a-4)(a+2)$$
$$a-4=0 \quad \text{or} \quad a+2=0$$
$$a=4 \quad \text{or} \quad a=-2$$

(continued on next page)

(continued)

But −2 is not a possible solution. Set the denominator equal to zero and solve for a.

$$a^2-4=0 \quad a+2=0 \quad \text{or} \quad a-2=0$$
$$(a+2)(a-2)=0 \quad a=-2 \quad \text{or} \quad a=2$$

Intervals: $(-\infty,-2)$, $(-2, 2)$, $(2, 4)$, $(4,\infty)$

For $(-\infty,-2)$, choose −3.

$$\frac{(-3)^2+2(-3)}{(-3)^2-4}=\frac{9-6}{9-4}=\frac{3}{5}\le 2$$

For $(-2, 2)$, choose 0.

$$\frac{(0)^2+2(0)}{0-4}=\frac{0}{-4}=0\le 2$$

For $(2, 4)$, choose 3.

$$\frac{(3)^2+2(3)}{(3)^2-4}=\frac{9+6}{9-5}=\frac{15}{4}\not\le 2$$

For $(4,\infty)$, choose 5.

$$\frac{(5)^2+2(5)}{(5)^2-4}=\frac{25+10}{25-4}=\frac{35}{21}\le 2$$

The value 4 will satisfy the original inequality, but the values −2 and 2 will not since they make the denominator zero. The solution is $(-\infty,-2)\cup(-2, 2)\cup[4,\infty)$.

R.6 Exponents

1. $8^{-2}=\frac{1}{8^2}=\frac{1}{64}$

2. $3^{-4}=\frac{1}{3^4}=\frac{1}{81}$

3. $5^0=1$, by definition.

4. $\left(-\frac{3}{4}\right)^0=1$, by definition.

5. $-(-3)^{-2}=-\frac{1}{(-3)^2}=-\frac{1}{9}$

6. $-(-3^{-2})=-\left(-\frac{1}{3^2}\right)=-\left(-\frac{1}{9}\right)=\frac{1}{9}$

7. $\left(\frac{1}{6}\right)^{-2}=\frac{1}{\left(\frac{1}{6}\right)^2}=\frac{1}{\frac{1}{36}}=36$

8. $\left(\frac{4}{3}\right)^{-3}=\frac{1}{\left(\frac{4}{3}\right)^3}=\frac{1}{\frac{64}{27}}=\frac{27}{64}$

9. $\frac{4^{-2}}{4}=4^{-2-1}=4^{-3}=\frac{1}{4^3}=\frac{1}{64}$

10. $\frac{8^9\cdot 8^{-7}}{8^{-3}}=8^{9+(-7)-(-3)}=8^{9-7+3}=8^5$

11.
$$\frac{10^8\cdot 10^{-10}}{10^4\cdot 10^2}=\frac{10^{8+(-10)}}{10^{4+2}}=\frac{10^{-2}}{10^6}=10^{-2-6}=10^{-8}=\frac{1}{10^8}$$

12.
$$\left(\frac{7^{-12}\cdot 7^3}{7^{-8}}\right)^{-1}=(7^{-12+3-(-8)})^{-1}=(7^{-12+3+8})^{-1}=(7^{-1})^{-1}=7^{(-1)(-1)}=7^1=7$$

13. $\frac{x^4\cdot x^3}{x^5}=\frac{x^{4+3}}{x^5}=\frac{x^7}{x^5}=x^{7-5}=x^2$

14. $\frac{y^{10}\cdot y^{-4}}{y^6}=y^{10-4-6}=y^0=1$

15.
$$\frac{(4k^{-1})^2}{2k^{-5}}=\frac{4^2k^{-2}}{2k^{-5}}=\frac{16k^{-2-(-5)}}{2}=8k^{-2+5}=8k^3=2^3k^3$$

16.
$$\frac{(3z^2)^{-1}}{z^5}=\frac{3^{-1}(z^2)^{-1}}{z^5}=\frac{3^{-1}z^{2(-1)}}{z^5}=\frac{3^{-1}z^{-2}}{z^5}=3^{-1}z^{-2-5}=3^{-1}z^{-7}=\frac{1}{3}\cdot\frac{1}{z^7}=\frac{1}{3z^7}$$

17.
$$\frac{3^{-1}\cdot x\cdot y^2}{x^{-4}\cdot y^5}=3^{-1}\cdot x^{1-(-4)}\cdot y^{2-5}=3^{-1}\cdot x^{1+4}\cdot y^{-3}=\frac{1}{3}\cdot x^5\cdot\frac{1}{y^3}=\frac{x^5}{3y^3}$$

18. $\dfrac{5^{-2}m^2y^{-2}}{5^2m^{-1}y^{-2}} = \dfrac{5^{-2}}{5^2}\cdot\dfrac{m^2}{m^{-1}}\cdot\dfrac{y^{-2}}{y^{-2}}$
$= 5^{-2-2}m^{2-(-1)}y^{-2-(-2)}$
$= 5^{-2-2}m^{2+1}y^{-2+2}$
$= 5^{-4}m^3y^0 = \dfrac{1}{5^4}\cdot m^3\cdot 1$
$= \dfrac{m^3}{5^4}$

19. $\left(\dfrac{a^{-1}}{b^2}\right)^{-3} = \dfrac{(a^{-1})^{-3}}{(b^2)^{-3}} = \dfrac{a^{(-1)(-3)}}{b^{2(-3)}}$
$= \dfrac{a^3}{b^{-6}} = a^3b^6$

20. $\left(\dfrac{c^3}{7d^{-2}}\right)^{-2} = \dfrac{(c^3)^{-2}}{7^{-2}(d^{-2})^{-2}}$
$= \dfrac{c^{(3)(-2)}}{7^{-2}d^{(-2)(-2)}} = \dfrac{c^{-6}}{7^{-2}d^4}$
$= \dfrac{7^2}{c^6d^4} = \dfrac{49}{c^6d^4}$

21. $a^{-1}+b^{-1} = \dfrac{1}{a}+\dfrac{1}{b}$
$= \left(\dfrac{b}{b}\right)\left(\dfrac{1}{a}\right)+\left(\dfrac{a}{a}\right)\left(\dfrac{1}{b}\right)$
$= \dfrac{b}{ab}+\dfrac{a}{ab}$
$= \dfrac{b+a}{ab} = \dfrac{a+b}{ab}$

22. $b^{-2}-a = \dfrac{1}{b^2}-a$
$= \dfrac{1}{b^2}-a\left(\dfrac{b^2}{b^2}\right)$
$= \dfrac{1}{b^2}-\dfrac{ab^2}{b^2}$
$= \dfrac{1-ab^2}{b^2}$

23. $\dfrac{2n^{-1}-2m^{-1}}{m+n^2} = \dfrac{\frac{2}{n}-\frac{2}{m}}{m+n^2}$
$= \dfrac{\frac{2}{n}\cdot\frac{m}{m}-\frac{2}{m}\cdot\frac{n}{n}}{(m+n^2)}$
$= \dfrac{2m-2n}{mn(m+n^2)}$
or $\dfrac{2(m-n)}{mn(m+n^2)}$

24. $\left(\dfrac{m}{3}\right)^{-1}+\left(\dfrac{n}{2}\right)^{-2} = \left(\dfrac{3}{m}\right)^1+\left(\dfrac{2}{n}\right)^2$
$= \dfrac{3}{m}+\dfrac{4}{n^2}$
$= \left(\dfrac{3}{m}\right)\left(\dfrac{n^2}{n^2}\right)+\left(\dfrac{4}{n^2}\right)\left(\dfrac{m}{m}\right)$
$= \dfrac{3n^2}{mn^2}+\dfrac{4m}{mn^2}$
$= \dfrac{3n^2+4m}{mn^2}$

25. $(x^{-1}-y^{-1})^{-1} = \dfrac{1}{\frac{1}{x}-\frac{1}{y}}$
$= \dfrac{1}{\frac{1}{x}\cdot\frac{y}{y}-\frac{1}{y}\cdot\frac{x}{x}}$
$= \dfrac{1}{\frac{y}{xy}-\frac{x}{xy}}$
$= \dfrac{1}{\frac{y-x}{xy}} = \dfrac{xy}{y-x}$

26. $(x\cdot y^{-1}-y^{-2})^{-2} = \left(\dfrac{x}{y}-\dfrac{1}{y^2}\right)^{-2}$
$= \left[\left(\dfrac{x}{y}\right)\left(\dfrac{y}{y}\right)-\dfrac{1}{y^2}\right]^{-2}$
$= \left(\dfrac{xy}{y^2}-\dfrac{1}{y^2}\right)^{-2} = \left(\dfrac{xy-1}{y^2}\right)^{-2}$
$= \left(\dfrac{y^2}{xy-1}\right)^2 = \dfrac{(y^2)^2}{(xy-1)^2}$
$= \dfrac{y^4}{(xy-1)^2}$

27. $121^{1/2} = (11^2)^{1/2} = 11^{2(1/2)} = 11^1 = 11$

28. $27^{1/3} = \sqrt[3]{27} = 3$

29. $32^{2/5} = (32^{1/5})^2 = 2^2 = 4$

30. $-125^{2/3} = -(125^{1/3})^2 = -5^2 = -25$

31. $\left(\dfrac{36}{144}\right)^{1/2} = \dfrac{36^{1/2}}{144^{1/2}} = \dfrac{6}{12} = \dfrac{1}{2}$

This can also be solved by reducing the fraction first.

$\left(\dfrac{36}{144}\right)^{1/2} = \left(\dfrac{1}{4}\right)^{1/2} = \dfrac{1^{1/2}}{4^{1/2}} = \dfrac{1}{2}$

32. $\left(\frac{64}{27}\right)^{1/3} = \frac{64^{1/3}}{27^{1/3}} = \frac{4}{3}$

33. $8^{-4/3} = (8^{1/3})^{-4} = 2^{-4} = \frac{1}{2^4} = \frac{1}{16}$

34. $625^{-1/4} = \frac{1}{625^{1/4}} = \frac{1}{5}$

35. $\left(\frac{27}{64}\right)^{-1/3} = \frac{27^{-1/3}}{64^{-1/3}} = \frac{64^{1/3}}{27^{1/3}} = \frac{4}{3}$

36. $\left(\frac{121}{100}\right)^{-3/2} = \frac{1}{\left(\frac{121}{100}\right)^{3/2}} = \frac{1}{\left[\left(\frac{121}{100}\right)^{1/2}\right]^3}$

$= \frac{1}{\left(\frac{11}{10}\right)^3} = \frac{1}{\frac{1331}{1000}} = \frac{1000}{1331}$

37. $3^{2/3} \cdot 3^{4/3} = 3^{(2/3)+(4/3)} = 3^{6/3} = 3^2 = 9$

38. $27^{2/3} \cdot 27^{-1/3} = 27^{(2/3)+(-1/3)}$

$= 27^{2/3-1/3}$

$= 27^{1/3} = 3$

39. $\frac{4^{9/4} \cdot 4^{-7/4}}{4^{-10/4}} = 4^{9/4-7/4-(-10/4)}$

$= 4^{12/4} = 4^3 = 64$

40. $\frac{3^{-5/2} \cdot 3^{3/2}}{3^{7/2} \cdot 3^{-9/2}} = 3^{(-5/2)+(3/2)-(7/2)-(-9/2)}$

$= 3^{-5/2+3/2-7/2+9/2}$

$= 3^0 = 1$

41. $\left(\frac{x^6y^{-3}}{x^{-2}y^5}\right)^{1/2} = (x^{6-(-2)}y^{-3-5})^{1/2}$

$= (x^8y^{-8})^{1/2}$

$= (x^8)^{1/2}(y^{-8})^{1/2}$

$= x^4y^{-4}$

$= \frac{x^4}{y^4}$

42. $\left(\frac{a^{-7}b^{-1}}{b^{-4}a^2}\right)^{1/3} = (a^{-7-2}b^{-1-(-4)})^{1/3}$

$= (a^{-9}b^3)^{1/3}$

$= (a^{-9})^{1/3}(b^3)^{1/3}$

$= a^{-3}b^1 = \frac{b}{a^3}$

43. $\frac{7^{-1/3} \cdot 7r^{-3}}{7^{2/3} \cdot (r^{-2})^2} = \frac{7^{-1/3+1}r^{-3}}{7^{2/3} \cdot r^{-4}}$

$= 7^{-1/3+3/3-2/3}r^{-3-(-4)}$

$= 7^0r^{-3+4} = 1 \cdot r^1 = r$

44. $\frac{12^{3/4} \cdot 12^{5/4} \cdot y^{-2}}{12^{-1} \cdot (y^{-3})^{-2}} = \frac{12^{3/4+5/4} \cdot y^{-2}}{12^{-1} \cdot y^{(-3)(-2)}}$

$= \frac{12^{8/4} \cdot y^{-2}}{12^{-1} \cdot y^6} = \frac{12^2 \cdot y^{-2}}{12^{-1}y^6}$

$= 12^{2-(-1)} \cdot y^{-2-6} = 12^3y^{-8}$

$= \frac{12^3}{y^8}$

45. $\frac{3k^2 \cdot (4k^{-3})^{-1}}{4^{1/2} \cdot k^{7/2}} = \frac{3k^2 \cdot 4^{-1}k^3}{2 \cdot k^{7/2}}$

$= 3 \cdot 2^{-1} \cdot 4^{-1}k^{2+3-(7/2)}$

$= \frac{3}{8} \cdot k^{3/2} = \frac{3k^{3/2}}{8}$

46. $\frac{8p^{-3}(4p^2)^{-2}}{p^{-5}} = \frac{8p^{-3} \cdot 4^{-2}p^{(2)(-2)}}{p^{-5}}$

$= \frac{8p^{-3}4^{-2}p^{-4}}{p^{-5}}$

$= 8 \cdot 4^{-2}p^{(-3)+(-4)-(-5)}$

$= 8 \cdot 4^{-2}p^{-3-4+5}$

$= 8 \cdot 4^{-2}p^{-2}$

$= 8 \cdot \frac{1}{4^2} \cdot \frac{1}{p^2} = 8 \cdot \frac{1}{16} \cdot \frac{1}{p^2}$

$= \frac{8}{16p^2} = \frac{1}{2p^2}$

47. $\frac{a^{4/3}}{a^{2/3}} \cdot \frac{b^{1/2}}{b^{-3/2}} = a^{4/3-2/3}b^{1/2-(-3/2)}$

$= a^{2/3}b^2$

48. $\frac{x^{3/2} \cdot y^{4/5} \cdot z^{-3/4}}{x^{5/3} \cdot y^{-6/5} \cdot z^{1/2}}$

$= x^{3/2-(5/3)} \cdot y^{4/5-(-6/5)} \cdot z^{-3/4-(1/2)}$

$= x^{-1/6} \cdot y^2 \cdot z^{-5/4}$

$= \frac{y^2}{x^{1/6}z^{5/4}}$

49. $\dfrac{k^{-3/5}\cdot h^{-1/3}\cdot t^{2/5}}{k^{-1/5}\cdot h^{-2/3}\cdot t^{1/5}}$
$= k^{-3/5-(-1/5)}h^{-1/3-(-2/3)}t^{2/5-1/5}$
$= k^{-3/5+1/5}h^{-1/3+2/3}t^{2/5-1/5}$
$= k^{-2/5}h^{1/3}t^{1/5} = \dfrac{h^{1/3}t^{1/5}}{k^{2/5}}$

50. $\dfrac{m^{7/3}\cdot n^{-2/5}\cdot p^{3/8}}{m^{-2/3}\cdot n^{3/5}\cdot p^{-5/8}}$
$= m^{7/3-(-2/3)}n^{-2/5-(3/5)}p^{3/8-(-5/8)}$
$= m^{7/3+2/3}n^{-2/5-3/5}p^{3/8+5/8}$
$= m^{9/3}n^{-5/5}p^{8/8}$
$= m^3n^{-1}p^1$
$= \dfrac{m^3p}{n}$

51. $3x^3(x^2+3x)^2 - 15x(x^2+3x)^2$
$= 3x\cdot x^2(x^2+3x)^2 - 3x\cdot 5(x^2+3x)^2$
$= 3x(x^2+3x)^2(x^2-5)$

52. $6x(x^3+7)^2 - 6x^2(3x^2+5)(x^3+7)$
$= 6x(x^3+7)(x^3+7) - 6x(x)(3x^2+5)(x^3+7)$
$= 6x(x^3+7)[(x^3+7) - x(3x^2+5)]$
$= 6x(x^3+7)(x^3+7-3x^3-5x)$
$= 6x(x^3+7)(-2x^3-5x+7)$

53. $10x^3(x^2-1)^{-1/2} - 5x(x^2-1)^{1/2}$
$= 5x\cdot 2x^2(x^2-1)^{-1/2} - 5x(x^2-1)^{-1/2}(x^2-1)^1$
$= 5x(x^2-1)^{-1/2}[2x^2-(x^2-1)]$
$= 5x(x^2-1)^{-1/2}(x^2+1)$

54. $9(6x+2)^{1/2} + 3(9x-1)(6x+2)^{-1/2}$
$= 3\cdot 3(6x+2)^{-1/2}(6x+2)^1$
$\quad + 3(9x-1)(6x+2)^{-1/2}$
$= 3(6x+2)^{-1/2}[3(6x+2)+(9x-1)]$
$= 3(6x+2)^{-1/2}(18x+6+9x-1)$
$= 3(6x+2)^{-1/2}(27x+5)$

55. $x(2x+5)^2(x^2-4)^{-1/2} + 2(x^2-4)^{1/2}(2x+5)$
$= (2x+5)^2(x^2-4)^{-1/2}(x)$
$\quad + (x^2-4)^1(x^2-4)^{-1/2}(2)(2x+5)$
$= (2x+5)(x^2-4)^{-1/2}$
$\quad \cdot [(2x+5)(x)+(x^2-4)(2)]$
$= (2x+5)(x^2-4)^{-1/2}\cdot(2x^2+5x+2x^2-8)$
$= (2x+5)(x^2-4)^{-1/2}(4x^2+5x-8)$

56. $(4x^2+1)^2(2x-1)^{-1/2}$
$\quad + 16x(4x^2+1)(2x-1)^{1/2}$
$= (4x^2+1)(4x^2+1)(2x-1)^{-1/2}$
$\quad + 16x(4x^2+1)(2x-1)^{-1/2}(2x-1)$
$= (4x^2+1)(2x-1)^{-1/2}$
$\quad \cdot [(4x^2+1)+16x(2x-1)]$
$= (4x^2+1)(2x-1)^{-1/2}$
$\quad \cdot (4x^2+1+32x^2-16x)$
$= (4x^2+1)(2x-1)^{-1/2}(36x^2-16x+1)$

R.7 Radicals

1. $\sqrt[3]{125} = 5$ because $5^3 = 125$.

2. $\sqrt[4]{1296} = \sqrt[4]{6^4} = 6$

3. $\sqrt[5]{-3125} = -5$ because $(-5)^5 = -3125$.

4. $\sqrt{50} = \sqrt{25\cdot 2} = \sqrt{25}\sqrt{2} = 5\sqrt{2}$

5. $\sqrt{2000} = \sqrt{4\cdot 100\cdot 5}$
$= 2\cdot 10\sqrt{5}$
$= 20\sqrt{5}$

6. $\sqrt{32y^5} = \sqrt{(16y^4)(2y)}$
$= \sqrt{16y^4}\sqrt{2y}$
$= 4y^2\sqrt{2y}$

7. $\sqrt{27}\cdot\sqrt{3} = \sqrt{27\cdot 3} = \sqrt{81} = 9$

8. $\sqrt{2}\cdot\sqrt{32} = \sqrt{2\cdot 32} = \sqrt{64} = 8$

9. $7\sqrt{2} - 8\sqrt{18} + 4\sqrt{72}$
$= 7\sqrt{2} - 8\sqrt{9\cdot 2} + 4\sqrt{36\cdot 2}$
$= 7\sqrt{2} - 8(3)\sqrt{2} + 4(6)\sqrt{2}$
$= 7\sqrt{2} - 24\sqrt{2} + 24\sqrt{2}$
$= 7\sqrt{2}$

10. $4\sqrt{3} - 5\sqrt{12} + 3\sqrt{75}$
$= 4\sqrt{3} - 5(\sqrt{4}\sqrt{3}) + 3(\sqrt{25}\sqrt{3})$
$= 4\sqrt{3} - 5(2\sqrt{3}) + 3(5\sqrt{3})$
$= 4\sqrt{3} - 10\sqrt{3} + 15\sqrt{3}$
$= (4-10+15)\sqrt{3} = 9\sqrt{3}$

11. $4\sqrt{7} - \sqrt{28} + \sqrt{343}$
$= 4\sqrt{7} - \sqrt{4}\sqrt{7} + \sqrt{49}\sqrt{7}$
$= 4\sqrt{7} - 2\sqrt{7} + 7\sqrt{7}$
$= (4-2+7)\sqrt{7}$
$= 9\sqrt{7}$

12. $3\sqrt{28}-4\sqrt{63}+\sqrt{112}$
$=3(\sqrt{4}\sqrt{7})-4(\sqrt{9}\sqrt{7})+(\sqrt{16}\sqrt{7})$
$=3(2\sqrt{7})-4(3\sqrt{7})+(4\sqrt{7})$
$=6\sqrt{7}-12\sqrt{7}+4\sqrt{7}$
$=(6-12+4)\sqrt{7}$
$=-2\sqrt{7}$

13. $\sqrt[3]{2}-\sqrt[3]{16}+2\sqrt[3]{54}$
$=\sqrt[3]{2}-(\sqrt[3]{8\cdot 2})+2(\sqrt[3]{27\cdot 2})$
$=\sqrt[3]{2}-\sqrt[3]{8}\sqrt[3]{2}+2(\sqrt[3]{27}\sqrt[3]{2})$
$=\sqrt[3]{2}-2\sqrt[3]{2}+2(3\sqrt[3]{2})$
$=\sqrt[3]{2}-2\sqrt[3]{2}+6\sqrt[3]{2}$
$=5\sqrt[3]{2}$

14. $2\sqrt[3]{5}-4\sqrt[3]{40}+3\sqrt[3]{135}$
$=2\sqrt[3]{5}-4\sqrt[3]{8\cdot 5}+3\sqrt[3]{27\cdot 5}$
$=2\sqrt[3]{5}-4(2)\sqrt[3]{5}+3(3)\sqrt[3]{5}$
$=2\sqrt[3]{5}-8\sqrt[3]{5}+9\sqrt[3]{5}$
$=3\sqrt[3]{5}$

15. $\sqrt{2x^3y^2z^4}=\sqrt{x^2y^2z^4\cdot 2x}=xyz^2\sqrt{2x}$

16. $\sqrt{160r^7s^9t^{12}}$
$=\sqrt{(16\cdot 10)(r^6\cdot r)(s^8\cdot s)(t^{12})}$
$=\sqrt{(16r^6s^8t^{12})(10rs)}$
$=\sqrt{16r^6s^8t^{12}}\sqrt{10rs}$
$=4r^3s^4t^6\sqrt{10rs}$

17. $\sqrt[3]{128x^3y^8z^9}=\sqrt[3]{64x^3y^6z^9\cdot 2y^2}$
$=\sqrt[3]{64x^3y^6z^9}\sqrt[3]{2y^2}$
$=4xy^2z^3\sqrt[3]{2y^2}$

18. $\sqrt[4]{x^8y^7z^{11}}=\sqrt[4]{(x^8)(y^4\cdot y^3)(z^8z^3)}$
$=\sqrt[4]{(x^8y^4z^8)(y^3z^3)}$
$=\sqrt[4]{x^8y^4z^8}\sqrt[4]{y^3z^3}$
$=x^2yz^2\sqrt[4]{y^3z^3}$

19. $\sqrt{a^3b^5}-2\sqrt{a^7b^3}+\sqrt{a^3b^9}$
$=\sqrt{a^2b^4ab}-2\sqrt{a^6b^2ab}+\sqrt{a^2b^8ab}$
$=ab^2\sqrt{ab}-2a^3b\sqrt{ab}+ab^4\sqrt{ab}$
$=(ab^2-2a^3b+ab^4)\sqrt{ab}$
$=ab\sqrt{ab}(b-2a^2+b^3)$

20. $\sqrt{p^7q^3}-\sqrt{p^5q^9}+\sqrt{p^9q}$
$=\sqrt{(p^6p)(q^2q)}-\sqrt{(p^4p)(q^8q)}+\sqrt{(p^8p)q}$
$=\sqrt{(p^6q^2)(pq)}-\sqrt{(p^4q^8)}\sqrt{(pq)}+\sqrt{(p^8)}\sqrt{pq}$
$=\sqrt{p^6q^2}\sqrt{pq}-\sqrt{p^4q^8}\sqrt{pq}+\sqrt{p^8}\sqrt{pq}$
$=p^3q\sqrt{pq}-p^2q^4\sqrt{pq}+p^4\sqrt{pq}$
$=p^2pq\sqrt{pq}-p^2q^4\sqrt{pq}+p^2p^2\sqrt{pq}$
$=p^2\sqrt{pq}(pq-q^4+p^2)$

21. $\sqrt{a}\cdot\sqrt[3]{a}=a^{1/2}\cdot a^{1/3}$
$=a^{1/2+(1/3)}$
$=a^{5/6}=\sqrt[6]{a^5}$

22. $\sqrt{b^3}\cdot\sqrt[4]{b^3}=b^{3/2}\cdot b^{3/4}$
$=b^{3/2+(3/4)}=b^{9/4}$
$=\sqrt[4]{b^9}=\sqrt[4]{b^8\cdot b}$
$=\sqrt[4]{b^8}\sqrt[4]{b}=\left(b^2\right)\sqrt[4]{b}$

23. $\sqrt{16-8x+x^2}$
$=\sqrt{(4-x)^2}$
$=|4-x|$

24. $\sqrt{9y^2+30y+25}=\sqrt{(3y+5)^2}=|3y+5|$

25. $\sqrt{4-25z^2}=\sqrt{(2+5z)(2-5z)}$
This factorization does not produce a perfect square, so the expression $\sqrt{4-25z^2}$ cannot be simplified.

26. $\sqrt{9k^2+h^2}$
The expression $9k^2+h^2$ is the sum of two squares and cannot be factored. Therefore, $\sqrt{9k^2+h^2}$ cannot be simplified.

27. $\dfrac{5}{\sqrt{7}}=\dfrac{5}{\sqrt{7}}\cdot\dfrac{\sqrt{7}}{\sqrt{7}}=\dfrac{5\sqrt{7}}{7}$

28. $\dfrac{5}{\sqrt{10}}=\dfrac{5}{\sqrt{10}}\cdot\dfrac{\sqrt{10}}{\sqrt{10}}=\dfrac{5\sqrt{10}}{\sqrt{100}}$
$=\dfrac{5\sqrt{10}}{\sqrt{100}}=\dfrac{5\sqrt{10}}{10}$
$=\dfrac{\sqrt{10}}{2}$

29. $\frac{-3}{\sqrt{12}} = \frac{-3}{\sqrt{4 \cdot 3}}$
$= \frac{-3}{2\sqrt{3}} \cdot \frac{\sqrt{3}}{\sqrt{3}} = \frac{-3\sqrt{3}}{6} = -\frac{\sqrt{3}}{2}$

30. $\frac{4}{\sqrt{8}} = \frac{4}{\sqrt{8}} \cdot \frac{\sqrt{2}}{\sqrt{2}} = \frac{4\sqrt{2}}{\sqrt{16}} = \frac{4\sqrt{2}}{4} = \sqrt{2}$

31. $\frac{3}{1-\sqrt{2}} = \frac{3}{1-\sqrt{2}} \cdot \frac{1+\sqrt{2}}{1+\sqrt{2}}$
$= \frac{3(1+\sqrt{2})}{1-2}$
$= -3(1+\sqrt{2})$

32. $\frac{5}{2-\sqrt{6}} = \frac{5}{2-\sqrt{6}} \cdot \frac{2+\sqrt{6}}{2+\sqrt{6}}$
$= \frac{5(2+\sqrt{6})}{4+2\sqrt{6}-2\sqrt{6}-\sqrt{36}}$
$= \frac{5(2+\sqrt{6})}{4-\sqrt{36}} = \frac{5(2+\sqrt{6})}{4-6}$
$= \frac{5(2+\sqrt{6})}{-2} = -\frac{5(2+\sqrt{6})}{2}$

33. $\frac{6}{2+\sqrt{2}} = \frac{6}{2+\sqrt{2}} \cdot \frac{2-\sqrt{2}}{2-\sqrt{2}}$
$= \frac{6(2-\sqrt{2})}{4-2\sqrt{2}+2\sqrt{2}-\sqrt{4}}$
$= \frac{6(2-\sqrt{2})}{4-2} = \frac{6(2-\sqrt{2})}{2}$
$= 3(2-\sqrt{2})$

34. $\frac{\sqrt{5}}{\sqrt{5}+\sqrt{2}} = \frac{\sqrt{5}}{\sqrt{5}+\sqrt{2}} \cdot \frac{\sqrt{5}-\sqrt{2}}{\sqrt{5}-\sqrt{2}}$
$= \frac{\sqrt{5}(\sqrt{5}-\sqrt{2})}{\sqrt{25}-\sqrt{10}+\sqrt{10}-\sqrt{4}}$
$= \frac{5-\sqrt{10}}{5-2}$
$= \frac{5-\sqrt{10}}{3}$

35. $\frac{1}{\sqrt{r}-\sqrt{3}} = \frac{1}{\sqrt{r}-\sqrt{3}} \cdot \frac{\sqrt{r}+\sqrt{3}}{\sqrt{r}+\sqrt{3}}$
$= \frac{\sqrt{r}+\sqrt{3}}{r-3}$

36. $\frac{5}{\sqrt{m}-\sqrt{5}} = \frac{5}{\sqrt{m}-\sqrt{5}} \cdot \frac{\sqrt{m}+\sqrt{5}}{\sqrt{m}+\sqrt{5}}$
$= \frac{5(\sqrt{m}+\sqrt{5})}{\left(\sqrt{m}\right)^2+\sqrt{5m}-\sqrt{5m}-\sqrt{25}}$
$= \frac{5(\sqrt{m}+\sqrt{5})}{\left(\sqrt{m}\right)^2-\sqrt{25}}$
$= \frac{5(\sqrt{m}+\sqrt{5})}{m-5}$

37. $\frac{y-5}{\sqrt{y}-\sqrt{5}} = \frac{y-5}{\sqrt{y}-\sqrt{5}} \cdot \frac{\sqrt{y}+\sqrt{5}}{\sqrt{y}+\sqrt{5}}$
$= \frac{(y-5)(\sqrt{y}+\sqrt{5})}{y-5}$
$= \sqrt{y}+\sqrt{5}$

38. $\frac{\sqrt{z}-1}{\sqrt{z}-\sqrt{5}} = \frac{\sqrt{z}-1}{\sqrt{z}-\sqrt{5}} \cdot \frac{\sqrt{z}+\sqrt{5}}{\sqrt{z}+\sqrt{5}}$
$= \frac{\left(\sqrt{z}\right)^2+\sqrt{5z}-\sqrt{z}-\sqrt{5}}{\left(\sqrt{z}\right)^2+\sqrt{5z}-\sqrt{5z}-\sqrt{25}}$
$= \frac{z+\sqrt{5z}-\sqrt{z}-\sqrt{5}}{z-5}$

39. $\frac{\sqrt{x}+\sqrt{x+1}}{\sqrt{x}-\sqrt{x+1}} = \frac{\sqrt{x}+\sqrt{x+1}}{\sqrt{x}-\sqrt{x+1}} \cdot \frac{\sqrt{x}+\sqrt{x+1}}{\sqrt{x}+\sqrt{x+1}}$
$= \frac{x+2\sqrt{x(x+1)}+(x+1)}{x-(x+1)}$
$= \frac{2x+2\sqrt{x(x+1)}+1}{-1}$
$= -2x-2\sqrt{x(x+1)}-1$

40. $\frac{\sqrt{p}+\sqrt{p^2-1}}{\sqrt{p}-\sqrt{p^2-1}}$
$= \frac{\sqrt{p}+\sqrt{p^2-1}}{\sqrt{p}-\sqrt{p^2-1}} \cdot \frac{\sqrt{p}+\sqrt{p^2-1}}{\sqrt{p}+\sqrt{p^2-1}}$
$= \frac{\left(\sqrt{p}\right)^2+2\sqrt{p}\sqrt{p^2-1}+\left(\sqrt{p^2-1}\right)^2}{\left(\sqrt{p}\right)^2+\sqrt{p}\sqrt{p^2-1}-\sqrt{p}\sqrt{p^2-1}-\left(\sqrt{p^2-1}\right)^2}$
$= \frac{p+2\sqrt{p}\sqrt{p^2-1}+(p^2-1)}{p-(p^2-1)}$
$= \frac{p^2+p+2\sqrt{p(p^2-1)}-1}{-p^2+p+1}$

41. $\frac{1+\sqrt{2}}{2}=\frac{\left(1+\sqrt{2}\right)\left(1-\sqrt{2}\right)}{2\left(1-\sqrt{2}\right)}=\frac{1-2}{2\left(1-\sqrt{2}\right)}$
$=-\frac{1}{2\left(1-\sqrt{2}\right)}$

42. $\frac{3-\sqrt{3}}{6}=\frac{3-\sqrt{3}}{6}\cdot\frac{3+\sqrt{3}}{3+\sqrt{3}}$
$=\frac{9+3\sqrt{3}-3\sqrt{3}-\sqrt{9}}{6\left(3+\sqrt{3}\right)}$
$=\frac{9-3}{6\left(3+\sqrt{3}\right)}=\frac{6}{6\left(3+\sqrt{3}\right)}$
$=\frac{1}{3+\sqrt{3}}$

43. $\frac{\sqrt{x}+\sqrt{x+1}}{\sqrt{x}-\sqrt{x+1}}$
$=\frac{\sqrt{x}+\sqrt{x+1}}{\sqrt{x}-\sqrt{x+1}}\cdot\frac{\sqrt{x}-\sqrt{x+1}}{\sqrt{x}-\sqrt{x+1}}$
$=\frac{x-(x+1)}{x-2\sqrt{x}\cdot\sqrt{x+1}+(x+1)}$
$=\frac{-1}{2x-2\sqrt{x(x+1)}+1}$

44. $\frac{\sqrt{p}-\sqrt{p-2}}{\sqrt{p}}$
$=\frac{\sqrt{p}-\sqrt{p-2}}{\sqrt{p}}\cdot\frac{\sqrt{p}+\sqrt{p-2}}{\sqrt{p}+\sqrt{p-2}}$
$=\frac{\left(\sqrt{p}\right)^2+\sqrt{p}\sqrt{p-2}-\sqrt{p}\sqrt{p-2}-\left(\sqrt{p-2}\right)^2}{\left(\sqrt{p}\right)^2+\sqrt{p}\sqrt{p-2}}$
$=\frac{p-(p-2)}{p+\sqrt{p(p-2)}}$
$=\frac{2}{p+\sqrt{p(p-2)}}$

Chapter 1

FUNCTIONS

1.1 Lines and Linear Functions

1. Find the slope of the line through (4, 5) and (−1, 2).

$$m = \frac{5-2}{4-(-1)} = \frac{3}{5}$$

3. Find the slope of the line through (8, 4) and (8, −7).

$$m = \frac{4-(-7)}{8-8} = \frac{11}{0}$$

The slope is undefined; the line is vertical.

5. $y = x$

Using the slope-intercept form, $y = mx + b$, we see that the slope is 1.

7. $5x - 9y = 11$

Rewrite the equation in slope-intercept form.

$$9y = 5x - 11$$
$$y = \frac{5}{9}x - \frac{11}{9}$$

The slope is $\frac{5}{9}$.

9. $x = 5$

This is a vertical line. The slope is undefined.

11. $y = 8$

This is a horizontal line, which has a slope of 0.

13. Find the slope of a line parallel to $6x - 3y = 12$.

Rewrite the equation in slope-intercept form.

$$-3y = -6x + 12$$
$$y = 2x - 4$$

The slope is 2, so a parallel line will also have slope 2.

15. The line goes through (1, 3), with slope $m = -2$. Use point-slope form.

$$y - 3 = -2(x - 1)$$
$$y = -2x + 2 + 3$$
$$y = -2x + 5$$

17. The line goes through $(-5, -7)$ with slope $m = 0$. Use point-slope form.

$$y - (-7) = 0[x - (-5)]$$
$$y + 7 = 0$$
$$y = -7$$

19. The line goes through (4, 2) and (1, 3). Find the slope, then use point-slope form with either of the two given points.

$$m = \frac{3-2}{1-4} = -\frac{1}{3}$$
$$y - 3 = -\frac{1}{3}(x - 1)$$
$$y = -\frac{1}{3}x + \frac{1}{3} + 3$$
$$y = -\frac{1}{3}x + \frac{10}{3}$$

21. The line goes through $\left(\frac{2}{3}, \frac{1}{2}\right)$ and $\left(\frac{1}{4}, -2\right)$.

$$m = \frac{-2 - \frac{1}{2}}{\frac{1}{4} - \frac{2}{3}} = \frac{-\frac{4}{2} - \frac{1}{2}}{\frac{3}{12} - \frac{8}{12}}$$
$$m = \frac{\frac{-5}{2}}{-\frac{5}{12}} = \frac{60}{10} = 6$$
$$y - (-2) = 6\left(x - \frac{1}{4}\right)$$
$$y + 2 = 6x - \frac{3}{2}$$
$$y = 6x - \frac{3}{2} - 2$$
$$y = 6x - \frac{7}{2}$$

23. The line goes through (−8, 4) and (−8, 6).

$m = \frac{4-6}{-8-(-8)} = \frac{-2}{0}$, which is undefined. This is a vertical line; the value of x is always −8. The equation of this line is $x = -8$.

25. The line has x-intercept -6 and y-intercept -3. Two points on the line are $(-6, 0)$ and $(0, -3)$. Find the slope; then use slope-intercept form.

$$m = \frac{-3-0}{0-(-6)} = \frac{-3}{6} = -\frac{1}{2}$$
$$b = -3$$
$$y = -\frac{1}{2}x - 3$$
$$2y = -x - 6$$
$$x + 2y = -6$$

27. The line is vertical, through $(-6, 5)$. The line has an equation of the form $x = k$, where k is the x-coordinate of the point. In this case, $k = -6$, so the equation is $x = -6$.

29. Write an equation of the line through $(-4, 6)$, parallel to $3x + 2y = 13$.

Rewrite the equation of the given line in slope-intercept form.

$$3x + 2y = 13$$
$$2y = -3x + 13$$
$$y = -\frac{3}{2}x + \frac{13}{2}$$

The slope is $-\frac{3}{2}$.

Use $m = -\frac{3}{2}$ and the point $(-4, 6)$ in the point-slope form.

$$y - 6 = -\frac{3}{2}[x - (-4)]$$
$$y = -\frac{3}{2}(x + 4) + 6$$
$$y = -\frac{3}{2}x - 6 + 6$$
$$y = -\frac{3}{2}x$$
$$2y = -3x$$
$$3x + 2y = 0$$

31. Write an equation of the line through $(3, -4)$, perpendicular to $x + y = 4$.

Rewrite the equation of the given line as $y = -x + 4$.

The slope of this line is -1. To find the slope of a perpendicular line, solve $-1m = -1 \Rightarrow m = 1$.

Use $m = 1$ and $(3, -4)$ in the point-slope form.

$$y - (-4) = 1(x - 3)$$
$$y = x - 3 - 4$$
$$y = x - 7$$
$$x - y = 7$$

33. Write an equation of the line with y-intercept 4, perpendicular to $x + 5y = 7$.

Find the slope of the given line.

$$x + 5y = 7$$
$$5y = -x + 7$$
$$y = -\frac{1}{5}x + \frac{7}{5}$$

The slope is $-\frac{1}{5}$, so the slope of the perpendicular line will be 5. If the y-intercept is 4, then using the slope-intercept form we have

$$y = mx + b$$
$$y = 5x + 4, \quad \text{or} \quad 5x - y = -4$$

35. Do the points $(4, 3)$, $(2, 0)$, and $(-18, -12)$ lie on the same line?

Find the slope between $(4, 3)$ and $(2, 0)$.

$$m = \frac{0-3}{2-4} = \frac{-3}{-2} = \frac{3}{2}$$

Find the slope between $(4, 3)$ and $(-18, -12)$.

$$m = \frac{-12-3}{-18-4} = \frac{-15}{-22} = \frac{15}{22}$$

Since these slopes are not the same, the points do not lie on the same line.

37. A parallelogram has 4 sides, with opposite sides parallel. The slope of the line through $(1, 3)$ and $(2, 1)$ is $m = \frac{3-1}{1-2} = \frac{2}{-1} = -2$.

The slope of the line through $\left(-\frac{5}{2}, 2\right)$ and $\left(-\frac{7}{2}, 4\right)$ is $m = \dfrac{2-4}{-\frac{5}{2}-\left(-\frac{7}{2}\right)} = \frac{-2}{1} = -2$.

Since these slopes are equal, these two sides are parallel.

The slope of the line through $\left(-\frac{7}{2}, 4\right)$ and $(1, 3)$ is $m = \dfrac{4-3}{-\frac{7}{2}-1} = \dfrac{1}{-\frac{9}{2}} = -\frac{2}{9}$.

Slope of the line through $\left(-\frac{5}{2}, 2\right)$ and $(2, 1)$ is

$$m = \frac{2-1}{-\frac{5}{2}-2} = \frac{1}{-\frac{9}{2}} = -\frac{2}{9}.$$

Since these slopes are equal, these two sides are parallel.

Since both pairs of opposite sides are parallel, the quadrilateral is a parallelogram.

39. The line goes through (0, 2) and (−2, 0)

$$m = \frac{2-0}{0-(-2)} = \frac{2}{2} = 1$$

The correct choice is (a).

41. The line appears to go through (0, 0) and (−1, 4).

$$m = \frac{4-0}{-1-0} = \frac{4}{-1} = -4$$

43. **(a)** See the figure in the textbook. Segment MN is drawn perpendicular to segment PQ. Recall that MQ is the length of segment MQ.

$$m_1 = \frac{\Delta y}{\Delta x} = \frac{MQ}{PQ}$$

From the diagram, we know that $PQ = 1$.

Thus, $m_1 = \frac{MQ}{1}$, so MQ has length m_1.

(b) $m_2 = \frac{\Delta y}{\Delta x} = \frac{-QN}{PQ} = \frac{-QN}{1} \Rightarrow$

$QN = -m_2$

(c) Triangles MPQ, PNQ, and MNP are right triangles by construction. In triangles MPQ and MNP, angle M = angle M, and in the right triangles PNQ and MNP, angle N = angle N.

Since all right angles are equal, and since triangles with two equal angles are similar, triangle MPQ is similar to triangle MNP and triangle PNQ is similar to triangle MNP. Therefore, triangles MNQ and PNQ are similar to each other.

(d) Since corresponding sides in similar triangles are proportional,

$MQ = k \cdot PQ$ and $PQ = k \cdot QN$.

$$\frac{MQ}{PQ} = \frac{k \cdot PQ}{k \cdot QN} \Rightarrow \frac{MQ}{PQ} = \frac{PQ}{QN}$$

From the diagram, we know that $PQ = 1$.

$$MQ = \frac{1}{QN}$$

From (a) and (b), $m_1 = MQ$ and $-m_2 = QN$.

Substituting, we get $m_1 = \frac{1}{-m_2}$.

Multiplying both sides by m_2, we have $m_1 m_2 = -1$.

45. $y = x - 1$

Three ordered pairs that satisfy this equation are (0, −1), (1, 0), and (3, 2). Plot these points and draw a line through them.

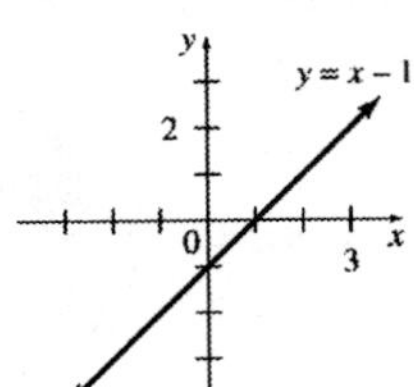

47. $y = -4x + 9$

Three ordered pairs that satisfy this equation are (0, 9), (1, 5), and (2, 1). Plot these points and draw a line through them.

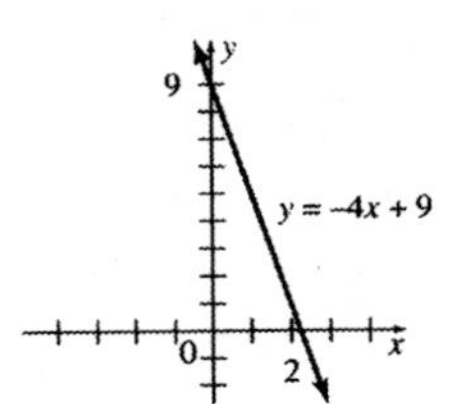

49. $2x - 3y = 12$

Find the intercepts.

If $y = 0$, then

$2x - 3(0) = 12 \Rightarrow 2x = 12 \Rightarrow x = 6$

so the x-intercept is 6.

If $x = 0$, then

$2(0) - 3y = 12 \Rightarrow -3y = 12 \Rightarrow y = -4$

so the y-intercept is −4.

Plot the ordered pairs (6, 0) and (0, −4) and draw a line through these points. (A third point may be used as a check.)

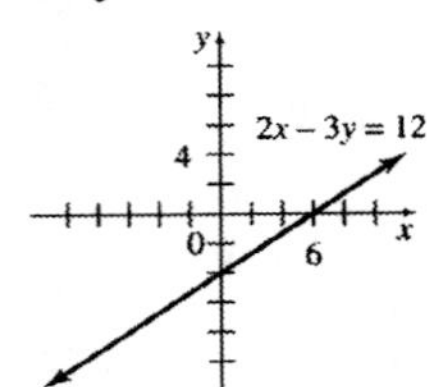

51. $3y - 7x = -21$

Find the intercepts.

If $y = 0$, then

$3(0) + 7x = -21 \Rightarrow -7x = -21 \Rightarrow x = 3$

so the x-intercept is 3.

If $x = 0$, then

$3y - 7(0) = -21 \Rightarrow 3y = -21 \Rightarrow y = -7$

So the y-intercept is −7.

(continued on next page)

(*continued*)

Plot the ordered pairs (3, 0) and (0, −7) and draw a line through these points. (A third point may be used as a check.)

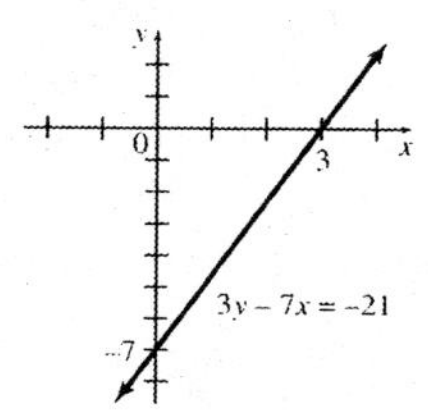

53. $y = -2$

The equation $y = -2$, or, equivalently, $y = 0x - 2$, always gives the same y-value, −2, for any value of x. The graph of this equation is the horizontal line with y-intercept −2.

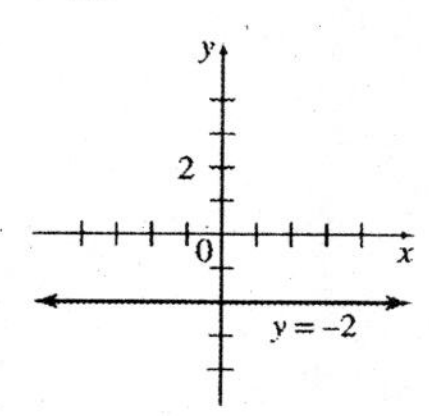

55. $x + 5 = 0$

This equation may be rewritten as $x = -5$. For any value of y, the x-value is −5. Because all ordered pairs that satisfy this equation have the same first number, this equation does not represent a function. The graph is the vertical line with x-intercept −5.

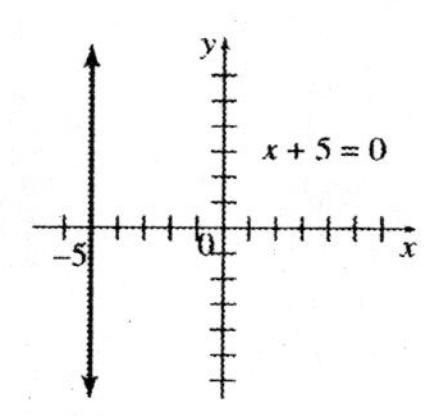

57. $y = 2x$

Three ordered pairs that satisfy this equation are (0, 0), (−2, −4), and (2, 4). Use these points to draw the graph.

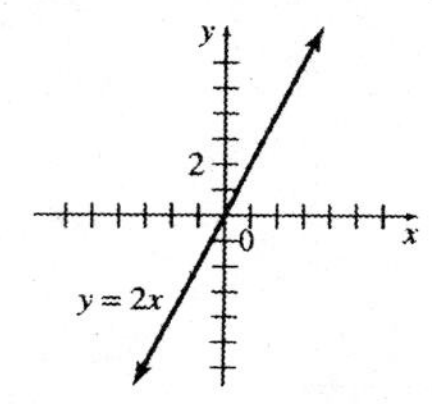

59. $x + 4y = 0$

If $y = 0$, then $x = 0$, so the x-intercept is 0. If $x = 0$, then $y = 0$, so the y-intercept is 0. Both intercepts give the same ordered pair, (0, 0). To get a second point, choose some other value of x (or y). For example if $x = 4$, then

$$x + 4y = 0$$
$$4 + 4y = 0$$
$$4y = -4$$
$$y = -1,$$

giving the ordered pair (4, −1). Graph the line through (0, 0) and (4, −1).

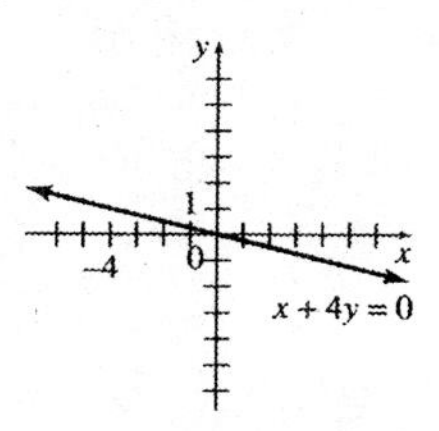

61. $f(2) = 7 - 5(2) = 7 - 10 = -3$

63. $f(-3) = 7 - 5(-3) = 7 + 15 = 22$

65. $g(1.5) = 2(1.5) - 3 = 3 - 3 = 0$

67. $g\left(-\frac{1}{2}\right) = 2\left(-\frac{1}{2}\right) - 3 = -1 - 3 = -4$

69. $f(t) = 7 - 5(t) = 7 - 5t$

71. **(a)** Let x = age.

$$u = 0.85(200 - x) = 187 - 0.85x$$
$$l = 0.7(200 - x) = 154 - 0.7x$$

(b) $u = 187 - 0.85(20) = 170$
$l = 154 - 0.7(20) = 140$

The target heart rate zone is 140 to 170 beats per minute.

(c) $u = 187 - 0.85(40) = 153$
$l = 154 - 0.7(40) = 126$

The target heart rate zone is 126 to 153 beats per minute.

(d)
$$154 - 0.7x = 187 - 0.85(x + 36)$$
$$154 - 0.7x = 187 - 0.85x - 30.6$$
$$154 - 0.7x = 156.4 - 0.85x$$
$$0.15x = 2.4$$
$$x = 16$$

The younger woman is 16; the older woman is $16 + 36 = 52$. $l = 0.7(220 - 16) \approx 143$ beats per minute.

73. Let $x = 0$ correspond to 1900. Then the "life expectancy from birth" line contains the points (0, 46) and (108, 78.1).

$m = \frac{78.1 - 46}{108 - 0} = \frac{32.1}{108} \approx 0.297$

Since (0, 46) is one of the points, the line is given by the equation $y = 0.297x + 46$.

The "life expectancy from age 65" line contains the points (0, 76) and (108, 83.8).

$m = \frac{83.8 - 76}{108 - 0} = \frac{7.8}{108} \approx 0.072$

Since (0, 76) is one of the points, the line is given by the equation $y = 0.072x + 76$.

Set the two expressions for y equal to determine where the lines intersect. At this point, life expectancy should increase no further.

$$0.297x + 46 = 0.072x + 76$$
$$0.225x = 30$$
$$x \approx 133.3$$

Determine the y-value when $x \approx 133.3$. Use the first equation.

$y = 0.297(133.3) + 46 \approx 85.6$

Thus, the maximum life expectancy for humans is about 86 years.

75. (a) The number of years of healthy life is increasing linearly at a rate of about 28 million years every 10 years, so the slope of the line is $m = \frac{28}{10} = 2.8$. Because 35 million years of healthy life was lost in 1900, it follows that $y = 35$ when $t = 0$. So, the y-intercept is $b = 35$. Therefore, $y = mt + b = 2.8t + 35$ is the number of years (in millions) of healthy life lost globally to tobacco t years after 1900.

(b) The number of years lost to diarrhea is declining linearly at a rate of 22 million years every 10 years, so the slope of the line is $m = \frac{-22}{10} = -2.2$. Because 100 million years of healthy life was lost in 1990, it follws that $y = 100$ when $t = 0$. So, the y-intercept is $b = 100$. Therefore, $y = -2.2t + 100$ is the number of years lost (in millions) to diarrhea t years after 1990.

(c)
$$2.8t + 35 = -2.2t + 100$$
$$5.0t = 65 \Rightarrow t = 13$$

The amount of healthy life lost to tobacco will exceed the amount of healthy life lost to diarrhea 13 years after 1990, or in 2003.

77. (a) The function is of the form $y = f(t) = mt + b$ and contains the point (5, 8.2) and (17, 33.34). The slope is

$m = \frac{33.34 - 8.2}{17 - 5} = \frac{25.14}{12} = 2.095 \approx 2.1$

Using the point-slope form, we have

$$y - 8.2 = 2.1(t - 5)$$
$$y - 8.2 = 2.1t - 10.5$$
$$y = 2.1t - 2.3$$

So, the number of male alates in an ant colony that is t years old is given by $y = 2.1t - 2.3$.

(b) $f(1) = 2.1(1) - 2.3 = -0.2 \approx 0$, so the function predicts 0 alates in a one-year-old colony.

(c) Set $f(t) = 40$ and solve for x.

$$2.1t - 2.3 = 40$$
$$2.1t = 42.3$$
$$t = \frac{42.3}{2.1} \approx 20.1$$

Assuming the linear function continues to be accurate, we would expect a colony to be about 20 years old before it has approximately 40 male aletes.

79. (a) If the temperature rises 0.3C° per decade, it rises 0.03C° per year. Therefore, $m = 0.03$.

$b = 15$, since a point is (0, 15). If T is the average global temperature in degrees Celsius, we have $T = 0.03t + 15$.

(b) Let $T = 19$. Find t.

$$19 = 0.03t + 15$$
$$4 = 0.03t$$
$$t = 133.3 \approx 133$$

So, $1970 + 133 = 2103$.

The temperature will rise to 19°C in about the year 2103.

81. The cost to use the first thermometer on x patients is $y = 2x + 10$ dollars. The cost to use the second thermometer on x patients is $y = 0.75x + 120$ dollars. If these two costs are equal, then

$$2x + 10 = 0.75x + 120$$
$$1.25x = 110$$
$$x = \frac{110}{1.25} = 88$$

Therefore, the costs of the two thermometers are equal when each is used for 88 patients.

83. **(a)** The line (for the data for men) goes through (0, 24.7) and (30, 28.2).

$$m = \frac{28.2 - 24.7}{30 - 0} \approx 0.117$$

Use the point (0, 24.7) and the point-slope form.

$$y - 24.7 = 0.117(t - 0)$$
$$y = 0.117t + 24.7$$

(b) The line (for the data for women) goes through (0, 22.0) and (30, 26.1).

$$m = \frac{26.1 - 22.0}{30 - 0} \approx 0.137$$

Use the point (0, 22.0) and the point-slope form.

$$y - 22.0 = 0.137(t - 0)$$
$$y = 0.137t + 22.0$$

(c) Since $0.137 > 0.117$, women seem to have the faster increase in median age at first marriage.

(d) Let $y = 30$.

$$30 = 0.117t + 24.7$$
$$5.3 = 0.117t$$
$$45.3 \approx t$$

The median age at first marriage for men will reach 30 in the year $1980 + 45 = 2025$ or $1980 + 46 = 2026$, depending on how the computations were rounded.

(e) Let $t = 45$.

$$y = 0.137(45) + 22.0$$
$$y \approx 28.2$$

The median age at first marriage for women will be 28.2 when the median age for men is 30. (The answer will be 28.3 if the year $t = 46$ is used as the answer for part (d).)

85. **(a)** The line goes through (90, 88) and (110, 57).

$$m = \frac{57 - 88}{110 - 90} = -1.55$$

Use the point (90, 88) and the point-slope form.

$$y - 88 = -1.55(t - 90)$$
$$y - 88 = -1.55t + 139.5$$
$$y = -1.55t + 227.5$$

(b) Let $y = 40$.

$$40 = -1.55t + 227.5$$
$$-187.5 = -1.55t$$
$$121 \approx t$$

The mortality rate will drop to 40 or below in the year $1900 + 121 = 2021$.

87. If the temperatures are numerically equal, then $F = C$.

$$F = \frac{9}{5}C + 32$$
$$C = \frac{9}{5}C + 32$$
$$-\frac{4}{5}C = 32$$
$$C = -40$$

The Celsius and Fahrenheit temperatures are numerically equal at $-40°$.

1.2 The Least Squares Line

1. The correlation coefficient measures the degree to which two variables are linearly related. A positive correlation coefficient does not necessarily mean that an increase in one of the quantities causes the other to increase also. There are two reasons why this is the case:

- We don't know the direction of the cause. Does X cause Y or does Y cause X?
- Another variable, or variables, may be involved that is responsible for the change.

A positive correlation coefficient means that as one quantity increases, the other quantity also increases.

3. **(a)**

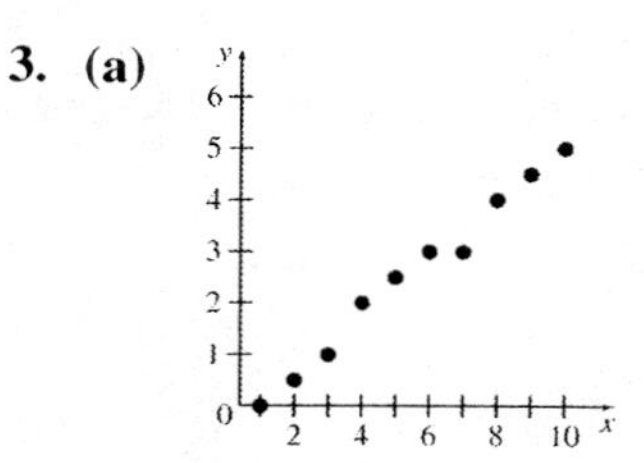

(b)

x	y	xy	x^2	y^2
1	0	0	1	0
2	0.5	1	4	0.25
3	1	3	9	1
4	2	8	16	4
5	2.5	12.5	25	6.25
6	3	18	36	9
7	3	21	49	9
8	4	32	64	16
9	4.5	40.5	81	20.25
10	5	50	100	25
55	25.5	186	385	90.75

(continued on next page)

(continued)

$$r = \frac{n(\Sigma xy) - (\Sigma x)(\Sigma y)}{\sqrt{n(\Sigma x^2) - (\Sigma x)^2} \cdot \sqrt{n(\Sigma y^2) - (\Sigma y)^2}}$$
$$= \frac{10(186) - (55)(25.5)}{\sqrt{10(385) - (55)^2} \cdot \sqrt{10(90.75) - (25.5)^2}}$$
$$\approx 0.993$$

(c) The least squares line is of the form $Y = mx + b$. First solve for m.

$$m = \frac{n(\Sigma xy) - (\Sigma x)(\Sigma y)}{n(\Sigma x^2) - (\Sigma x)^2}$$
$$= \frac{10(186) - (55)(25.5)}{10(385) - (55)^2}$$
$$= 0.5545454545 \approx 0.555$$

Now find b.

$$b = \frac{\Sigma y - m(\Sigma x)}{n}$$
$$= \frac{25.5 - 0.5545454545(55)}{10} = -0.5$$

Thus, $Y = 0.555x - 0.5$.

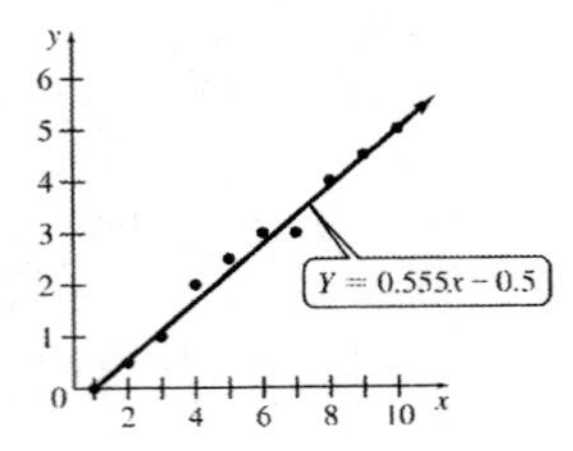

(d) Let $x = 11$. Find Y.
$Y = 0.55(11) - 0.5 \approx 5.6$

5.

x	y	xy	x^2	y^2
1	1	1	1	1
1	2	2	1	4
2	1	2	4	1
2	2	4	4	4
9	9	81	81	81
15	15	90	91	91

(a) $n = 5$

$$m = \frac{n(\Sigma xy) - (\Sigma x)(\Sigma y)}{n(\Sigma x^2) - (\Sigma x)^2}$$
$$= \frac{5(90) - (15)(15)}{5(91) - (15)^2}$$
$$= 0.9782608 \approx 0.9783$$

$$b = \frac{\Sigma y - m(\Sigma x)}{n}$$
$$= \frac{15 - (0.9782608)(15)}{5} \approx 0.0652$$

Thus, $Y = 0.9783x + 0.0652$.

$$r = \frac{n(\Sigma xy) - (\Sigma x)(\Sigma y)}{\sqrt{n(\Sigma x^2) - (\Sigma x)^2} \cdot \sqrt{n(\Sigma y^2) - (\Sigma y)^2}}$$
$$= \frac{5(90) - (15)(15)}{\sqrt{5(91) - (15)^2} \cdot \sqrt{5(91) - (15)^2}}$$
$$\approx 0.9783$$

(b)

x	y	xy	x^2	y^2
1	1	1	1	1
1	2	2	1	4
2	1	2	4	1
2	2	4	4	4
6	6	9	10	10

$n = 4$

$$m = \frac{n(\Sigma xy) - (\Sigma x)(\Sigma y)}{n(\Sigma x^2) - (\Sigma x)^2}$$
$$= \frac{4(9) - (6)(6)}{4(10) - (6)^2} = 0$$
$$b = \frac{\Sigma y - m(\Sigma x)}{n} = \frac{6 - (0)(6)}{4} = 1.5$$

Thus, $Y = 0x + 1.5$, or $Y = 1.5$.

$$r = \frac{n(\Sigma xy) - (\Sigma x)(\Sigma y)}{\sqrt{n(\Sigma x^2) - (\Sigma x)^2} \cdot \sqrt{n(\Sigma y^2) - (\Sigma y)^2}}$$
$$= \frac{4(9) - (6)(6)}{\sqrt{4(10) - (6)^2} \cdot \sqrt{4(10) - (6)^2}}$$
$$= 0$$

(c)

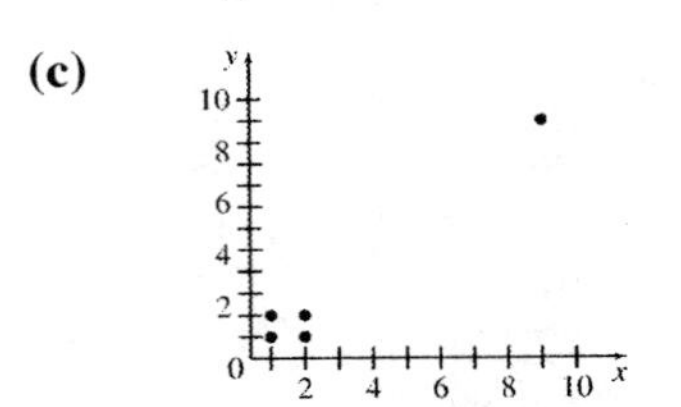

The point $(9, 9)$ is an outlier that has a strong effect on the least squares line and the correlation coefficient.

7.

x	y	xy	x^2	y^2
1	1	1	1	1
2	1	2	4	1
3	1	3	9	1
4	1.1	4.4	16	1.21
10	4.1	10.4	30	4.21

(a) $n = 4$

$$r = \frac{n(\sum xy)-(\sum x)(\sum y)}{\sqrt{n\left(\sum x^2\right)-\left(\sum x\right)^2}\cdot\sqrt{n\left(\sum y^2\right)-\left(\sum y\right)^2}}$$

$$= \frac{4(10.4)-(10)(4.1)}{\sqrt{4(30)-(10)^2}\cdot\sqrt{4(4.21)-(4.1)^2}}$$

$$= 0.7745966 \approx 0.7746$$

(b)

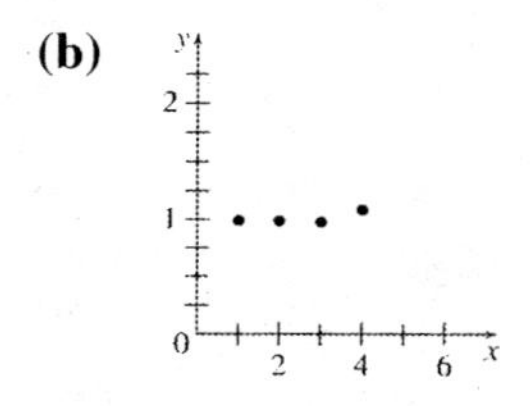

(c) Yes; because the data points are either on or very close to the horizontal line $y = 1$, it seems that the data should have a strong linear relationship. The correlation coefficient does not describe well a linear relationship if the data points fit a horizontal line.

9. $nb + (\sum x)m = \sum y$

$$(\sum x)b + (\sum x^2)m = \sum xy$$

$$nb + (\sum x)m = \sum y$$

$$nb = (\sum y) - (\sum x)m$$

$$b = \frac{\sum y - m(\sum x)}{n}$$

$$(\sum x)\left(\frac{\sum y - m(\sum x)}{n}\right) + (\sum x^2)m = \sum xy$$

$$(\sum x)[(\sum y) - m(\sum x)] + nm(\sum x^2) = n(\sum xy)$$

$$(\sum x)(\sum y) - m(\sum x)^2 + nm(\sum x^2) = n(\sum xy)$$

$$nm(\sum x^2) - m(\sum x)^2 = n(\sum xy) - (\sum x)(\sum y)$$

$$m[n(\sum x^2) - (\sum x)^2] = n(\sum xy) - (\sum x)(\sum y)$$

$$m = \frac{n(\sum xy) - (\sum x)(\sum y)}{n(\sum x^2) - (\sum x)^2}$$

11. (a)

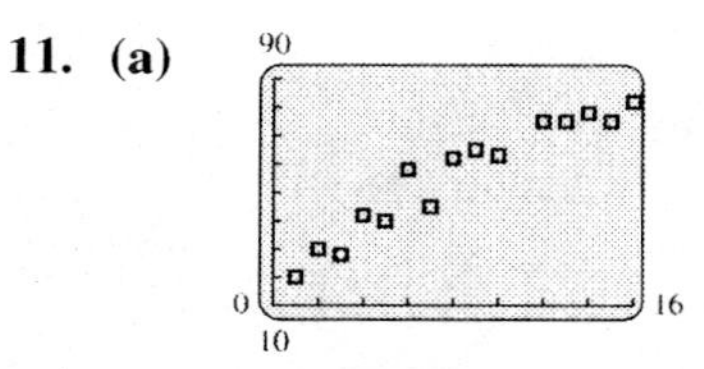

Yes, the points lie in a linear pattern.

(b) Using a calculator's STAT feature, the correlation coefficient is found to be $r \approx 0.959$. This indicates that the percentage of successful hunts does trend to increase with the size of the hunting party.

(c) $Y = 3.98x + 22.7$

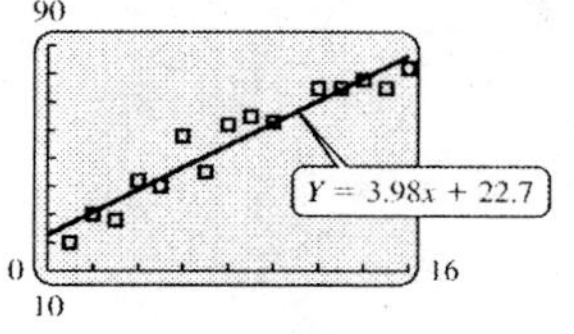

13. (a) Using a TI-84 Plus graphing calculator, the linear regression function gives the coefficient of correlation as $r \approx 0.9940$.

(b) Using a TI-84 Plus graphing calculator, the linear regression function gives the least squares line as $Y = 1.3525x - 2.51$.

(c)

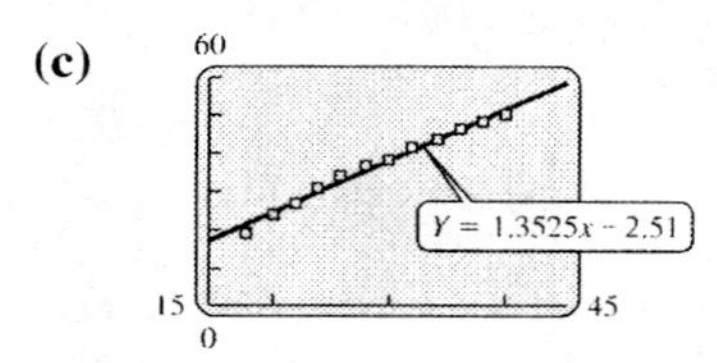

Yes, the line accurately fits the data.

(d) The slope of the least squares line is $m = 1.3525$, so the fetal stature is increasing by about 1.3525 cm each week.

(e) When $x = 45$, the linear model gives $Y = 1.3525(45) - 2.51 \approx 58.35$ cm.

15. (a) Using a TI-84 Plus graphing calculator, the linear regression function gives the least squares line as $Y = 0.212x - 0.309$.

(b) Let $x = 73$; find Y.

$Y = 0.212(73) - 0.309 \approx 15.2$

If the temperature were 73° F, you would expect to hear 15.2 chirps per second.

(c) Let $Y = 18$. Find x.

$$18 = 0.212x - 0.309$$
$$18.309 = 0.212x$$
$$86.4 \approx x$$

When the crickets are chirping 18 times per second, the temperature is about 86.4°F.

(d) Using a TI-84 Plus graphing calculator, the linear regression function gives the coefficient of correlation as $r \approx 0.835$.

17. (a) Skaggs's average speed was

$$\frac{100.5}{23.3833} \approx 4.298 \text{ miles per hour.}$$

(b)

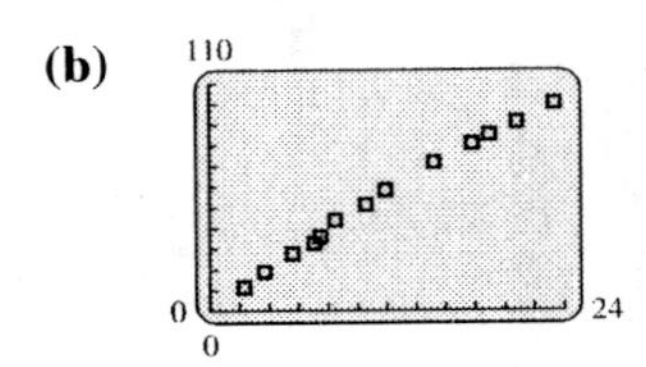

The data appear to lie approximately on a straight line.

(c) Using a graphing calculator, $Y = 4.317x + 3.419$.

(d) Using a graphing calculator, $r \approx 0.9971$
Yes, the least squares line is a very good fit to the data.

(e) A good value for Skaggs' average speed would be the slope of the least squares line, or $m = 4.317$ miles per hour. This value is faster than the average speed found in part (a). The value 4.317 miles per hour is most likely the better value.

19. (a) Using a graphing calculator, we have $Y = -0.08915x + 74.28$ and $r \approx -0.1035$. The slope suggests that the taller the student, the shorter the ideal partner's height is.

(b) Females: Using a graphing calculator, we have $Y = 0.6674x + 27.89$ and $r \approx 0.9459$.
Males: Using a graphing calculator, we have $Y = 0.4348x + 34.04$ and $r \approx 0.7049$.

(c)

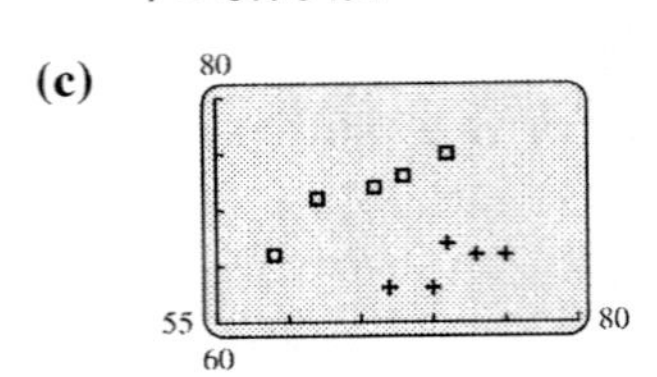

There is no linear relationship among all 10 data pairs. However, there is a linear relationship among the first five data pairs (female students) and a separate linear relationship among the second five data pairs (male students).

21. (a)

(b)

L	T	LT	L^2	T^2
1.0	1.11	1.11	1	1.2321
1.5	1.36	2.04	2.25	1.8496
2.0	1.57	3.14	4	2.4649
2.5	1.76	4.4	6.25	3.0976
3.0	1.92	5.76	9	3.6864
3.5	2.08	7.28	12.25	4.3264
4.0	2.22	8.88	16	4.9284
17.5	12.02	32.61	50.75	21.5854

$$m = \frac{n(\sum xy) - (\sum x)(\sum y)}{n(\sum x^2) - (\sum x)^2}$$
$$m = \frac{7(32.61) - (17.5)(12.02)}{7(50.75) - 17.5^2}$$
$$m = 0.3657142857$$
$$\approx 0.366$$
$$b = \frac{\sum T - m(\sum L)}{n}$$
$$b = \frac{12.02 - 0.3657142857(17.5)}{7}$$
$$\approx 0.803$$
$$Y = 0.366x + 0.803$$

The line seems to fit the data.

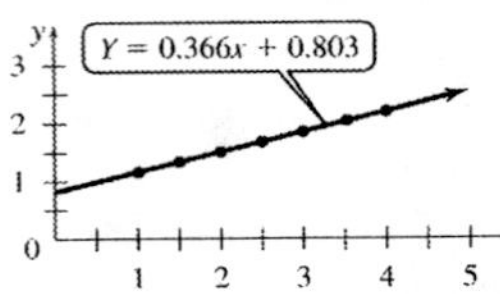

(c) $$r = \frac{7(32.61) - (17.5)(12.02)}{\sqrt{7(50.75) - 17.5^2} \cdot \sqrt{7(21.5854) - 12.02^2}}$$
$$= 0.995$$

This is a good fit and confirms the conclusion in part (b).

1.3 Properties of Functions

1.3 Exercises

1. The x-value of 82 corresponds to two y-values, 93 and 14. In a function, each value of x must correspond to exactly one value of y.
   The rule is not a function.

3. Each x-value corresponds to exactly one y-value.The rule is a function.

5. $y = x^3 + 2$
   Each x-value corresponds to exactly one y-value. The rule is a function.

7. $x = |y|$
   Each value of x (except 0) corresponds to two y-values. The rule is not a function.

9. $y = 2x + 3$

x	−2	−1	0	1	2	3
y	−1	1	3	5	7	9

Pairs: $(-2,-1)$, $(-1, 1)$, $(0, 3)$, $(1, 5)$, $(2, 7)$, $(3, 9)$
Range: $\{-1, 1, 3, 5, 7, 9\}$

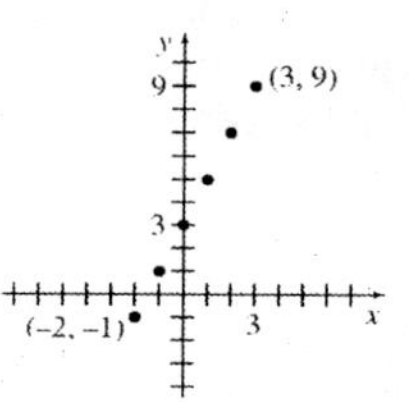

11. $2y - x = 5$

$$2y = 5 + x$$
$$y = \frac{1}{2}x + \frac{5}{2}$$

x	−2	−1	0	1	2	3
y	$\frac{3}{2}$	2	$\frac{5}{2}$	3	$\frac{7}{2}$	4

Pairs: $\left(-2, \frac{3}{2}\right)$, $(-1, 2)$, $\left(0, \frac{5}{2}\right)$, $(1, 3)$, $\left(2, \frac{7}{2}\right)$, $(3, 4)$.

Range: $\left\{\frac{3}{2}, 2, \frac{5}{2}, 3, \frac{7}{2}, 4\right\}$

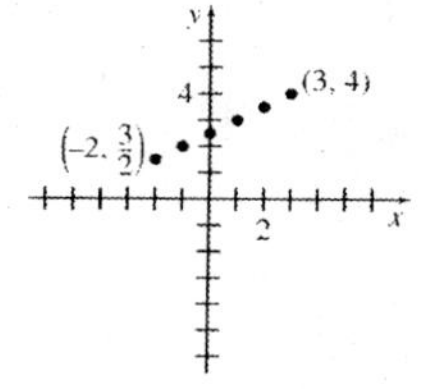

13. $y = x(x + 2)$

x	−2	−1	0	1	2	3
y	0	−1	0	3	8	15

Pairs: $(-2, 0)$, $(-1, -1)$, $(0, 0)$, $(1, 3)$, $(2, 8)$, $(3, 15)$
Range: $\{-1, 0, 3, 8, 15\}$

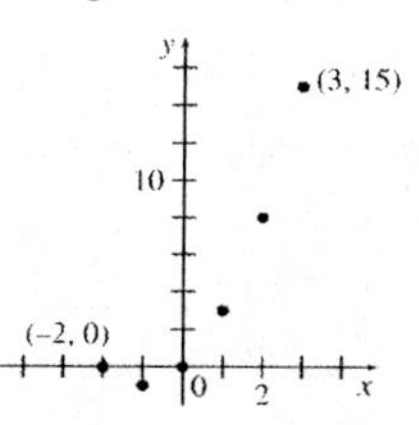

15. $y = x^2$

x	−2	−1	0	1	2	3
y	4	1	0	1	4	9

Pairs: $(-2, 4)$, $(-1, 1)$, $(0, 0)$, $(1, 1)$, $(2, 4)$, $(3, 9)$
Range: $\{0, 1, 4, 9\}$

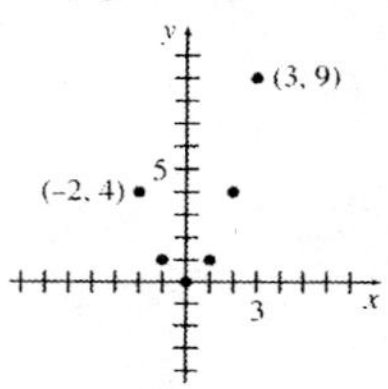

17. $f(x) = 2x$
    x can take on any value, so the domain is the set of real numbers, $(-\infty, \infty)$.

19. $f(x) = x^4$
    x can take on any value, so the domain is the set of real numbers, $(-\infty, \infty)$.

21. $f(x) = \sqrt{4 - x^2}$
    For $f(x)$ to be a real number, $4 - x^2 \geq 0$.
    Solve $4 - x^2 = 0$.
    $(2 - x)(2 + x) = 0 \Rightarrow x = 2$ or $x = -2$
    The numbers form the intervals $(-\infty, -2)$, $(-2, 2)$, and $(2, \infty)$.
    Values in the interval $(-2, 2)$ satisfy the inequality; $x = 2$ and $x = -2$ also satisfy the inequality. The domain is $[-2, 2]$.

23. $f(x) = (x - 3)^{1/2} = \sqrt{x - 3}$
    For $f(x)$ to be a real number,
    $x - 3 \geq 0 \Rightarrow x \geq 3$.
    The domain is $[3, \infty)$.

25. $f(x) = \frac{2}{1-x^2} = \frac{2}{(1-x)(1+x)}$
Since division by zero is not defined, $(1-x)\cdot(1+x) \neq 0$.
When $(1-x)(1+x) = 0$,
$1-x=0$ or $1+x=0$
$x=1$ or $x=-1$
Thus, x can be any real number except ± 1. The domain is $(-\infty,-1)\cup(-1,1)\cup(1,\infty)$.

27. $f(x) = -\sqrt{\frac{2}{x^2-16}} = -\sqrt{\frac{2}{(x-4)(x+4)}}$.
$(x-4)(x+4) > 0$, since $(x-4)(x+4) < 0$ would produce a negative radicand and $(x-4)\cdot(x+4) = 0$ would lead to division by zero.
$(x-4)(x+4) = 0 \Rightarrow x = 4$ or $x = -4$
Use the values -4 and 4 to divide the number line into 3 intervals, $(-\infty, -4)$, $(-4, 4)$ and $(4, \infty)$. Only the values in the intervals $(-\infty, -4)$ and $(4, \infty)$ satisfy the inequality.
The domain is $(-\infty, -4)\cup(4,\infty)$.

29. $f(x) = \sqrt{x^2-4x-5} = \sqrt{(x-5)(x+1)}$
See the method used in Exercise 21.
$(x-5)(x+1) \geq 0$ when $x \geq 5$ and when $x \leq -1$. The domain is $(-\infty,-1]\cup[5,\infty)$.

31. $f(x) = \frac{1}{\sqrt{3x^2+2x-1}} = \frac{1}{\sqrt{(3x-1)(x+1)}}$
$(3x-1)(x+1) > 0$, since the radicand cannot be negative and the denominator of the function cannot be zero.
Solve $(3x-1)(x+1) = 0$.
$3x-1=0$ or $x+1=0$
$x=\frac{1}{3}$ or $x=-1$

Use the values -1 and $\frac{1}{3}$ to divide the number line into 3 intervals, $(-\infty,-1)$, $\left(-1, \frac{1}{3}\right)$ and $\left(\frac{1}{3}, \infty\right)$. Only the values in the intervals $(-\infty, -1)$ and $\left(\frac{1}{3}, \infty\right)$ satisfy the inequality.

The domain is $(-\infty,-1)\cup\left(\frac{1}{3},\infty\right)$.

33. By reading the graph, the domain is all numbers greater than or equal to -5 and less than 4. The range is all numbers greater than or equal to -2 and less than or equal to 6.
Domain: $[-5, 4)$; range: $[-2, 6]$

35. By reading the graph, x can take on any value, but y is less than or equal to 12.
Domain: $(-\infty, \infty)$; range: $(-\infty, 12]$

37. The domain is all real numbers between the end points of the curve, or $[-2, 4]$.
The range is all real numbers between the minimum and maximum values of the function or $[0, 4]$.

(a) $f(-2) = 0$

(b) $f(0) = 4$

(c) $f\left(\frac{1}{2}\right) = 3$

(d) From the graph, $f(x) = 1$ when $x = -1.5, 1.5,$ or 2.5.

39. The domain is all real numbers between the endpoints of the curve, or $[-2, 4]$.
The range is all real numbers between the minimum and maximum values of the function or $[-3, 2]$.

(a) $f(-2) = -3$

(b) $f(0) = -2$

(c) $f\left(\frac{1}{2}\right) = -1$

(d) From the graph, $f(x) = 1$ when $x = 2.5$.

41. $f(x) = 3x^2 - 4x + 1$

(a) $f(4) = 3(4)^2 - 4(4) + 1$
$= 48 - 16 + 1 = 33$

(b) $f\left(-\frac{1}{2}\right) = 3\left(-\frac{1}{2}\right)^2 - 4\left(-\frac{1}{2}\right) + 1$
$= \frac{3}{4} + 2 + 1 = \frac{15}{4}$

(c) $f(a) = 3(a)^2 - 4(a) + 1$
$= 3a^2 - 4a + 1$

(d) $f\left(\frac{2}{m}\right)=3\left(\frac{2}{m}\right)^2-4\left(\frac{2}{m}\right)+1$
$=\frac{12}{m^2}-\frac{8}{m}+1$
or $\frac{12-8m+m^2}{m^2}$

(e)
$f(x)=1$
$3x^2-4x+1=1$
$3x^2-4x=0$
$x(3x-4)=0$
$x=0$ or $x=\frac{4}{3}$

43. $f(x)=\begin{cases}\frac{2x+1}{x-4} & \text{if } x\neq 4\\ 7 & \text{if } x=4\end{cases}$

(a) $f(4)=7$

(b) $f\left(-\frac{1}{2}\right)=\frac{2\left(-\frac{1}{2}\right)+1}{\left(-\frac{1}{2}\right)-4}=\frac{0}{-\frac{9}{2}}=0$

(c) $f(a)=\frac{2a+1}{a-4}$ if $a\neq 4$
$f(a)=7$ if $a=4$

(d) $f\left(\frac{2}{m}\right)=\frac{2\left(\frac{2}{m}\right)+1}{\frac{2}{m}-4}=\frac{\frac{4}{m}+1}{\frac{2}{m}-4}$
$=\frac{\frac{4+m}{m}}{\frac{2-4m}{m}}=\frac{4+m}{2-4m}$ if $m\neq\frac{1}{2}$
$f\left(\frac{2}{m}\right)=7$ if $m=\frac{1}{2}$

(e)
$\frac{2x+1}{x-4}=1$
$2x+1=x-4$
$x=-5$

45. $f(x)=6x^2-2$
$f(t+1)=6(t+1)^2-2$
$=6(t^2+2t+1)-2$
$=6t^2+12t+6-2$
$=6t^2+12t+4$

47. $g(r+h)$
$=(r+h)^2-2(r+h)+5$
$=r^2+2hr+h^2-2r-2h+5$

49. $g\left(\frac{3}{q}\right)=\left(\frac{3}{q}\right)^2-2\left(\frac{3}{q}\right)+5$
$=\frac{9}{q^2}-\frac{6}{q}+5=\frac{9-6q+5q^2}{q^2}$

51. $f(x)=2x+1$

(a) $f(x+h)=2(x+h)+1$
$=2x+2h+1$

(b) $f(x+h)-f(x)$
$=2x+2h+1-2x-1$
$=2h$

(c) $\frac{f(x+h)-f(x)}{h}$
$=\frac{2h}{h}=2$

53. $f(x)=2x^2-4x-5$

(a) $f(x+h)$
$=2(x+h)^2-4(x+h)-5$
$=2(x^2+2hx+h^2)-4x-4h-5$
$=2x^2+4hx+2h^2-4x-4h-5$

(b) $f(x+h)-f(x)$
$=2x^2+4hx+2h^2-4x-4h-5$
$-(2x^2-4x-5)$
$=2x^2+4hx+2h^2-4x-4h-5$
$-2x^2+4x+5$
$=4hx+2h^2-4h$

(c) $\frac{f(x+h)-f(x)}{h}$
$=\frac{4hx+2h^2-4h}{h}$
$=\frac{h(4x+2h-4)}{h}$
$=4x+2h-4$

55. $f(x)=\frac{1}{x}$

(a) $f(x+h)=\frac{1}{x+h}$

(b) $f(x+h)-f(x)=\frac{1}{x+h}-\frac{1}{x}$
$=\left(\frac{x}{x}\right)\frac{1}{x+h}-\frac{1}{x}\left(\frac{x+h}{x+h}\right)$
$=\frac{x-(x+h)}{x(x+h)}=\frac{-h}{x(x+h)}$

(c) $\frac{f(x+h)-f(x)}{h}=\frac{1}{h}\left[\frac{-h}{x(x+h)}\right]$
$=\frac{-1}{x(x+h)}$

57. A vertical line drawn anywhere through the graph will intersect the graph in only one place. The graph represents a function.

59. A vertical line drawn through the graph may intersect the graph in two places. The graph does not represent a function.

61. A vertical line drawn anywhere through the graph will intersect the graph in only one place. The graph represents a function.

63. $f(x)=3x-7,\quad g(x)=4-x$

(a) $g(1)=4-1=3$
$f(g(1))=f(3)=3(3)-7=2$

(b) $f(1)=3(1)-7=-4$
$g(f(1))=g(-4)=4-(-4)=8$

(c) $(f\circ g)(x)=f(g(x))=f(4-x)$
$=3(4-x)-7$
$=12-3x-7=5-3x$

(d) $(g\circ f)(x)=g(f(x))=g(3x-7)$
$=4-(3x-7)=11-3x$

65. $f(x)=2x^2+5x+1,\quad g(x)=3x-1$

(a) $g(1)=3(1)-1=2$
$f(g(1))=f(2)$
$=2(2)^2+5(2)+1$
$=8+10+1=19$

(b) $f(1)=2(1)^2+5(1)+1=8$
$g(f(1))=g(8)=3(8)-1=23$

(c) $(f\circ g)(x)=f(g(x))=f(3x-1)$
$=2(3x-1)^2+5(3x-1)+1$
$=2(9x^2-6x+1)+15x-5+1$
$=18x^2-12x+2+15x-5+1$
$=18x^2+3x-3$

(d) $(g\circ f)(x)=g(f(x))$
$=g(2x^2+5x+1)$
$=3(2x^2+5x+1)-1$
$=6x^2+15x+3-1$
$=6x^2+15x+2$

67. (a) The curve in the graph crosses the point with x-coordinate 17:37 and y-coordinate of approximately 140. So, at time 17 hours, 37 minutes the whale reaches a depth of about 140 m.

(b) The curve in the graph crosses the point with x-coordinate 17:39 and y-coordinate of approximately 240. So, at time 17 hours, 39 minutes the whale reaches a depth of about 250 m.

69. (a) (i) By the given function f, a muskrat weighing 800 g expends

$f(800)=0.01(800)^{0.88}$
≈ 3.6, or approximately

3.6 kcal/km when swimming at the surface of the water.

(ii) A sea otter weighing 20,000 g expends

$f(20{,}000)=0.01(20{,}000)^{0.88}$
≈ 61, or approximately

61 kcal/km when swimming at the surface of the water.

(b) If z is the number of kilograms of an animal's weight, then $x=g(z)=1000z$ is the number of grams since 1 kilogram equals 1000 grams.

(c) $f(g(z))=f(1000z)$
$=0.01(1000z)^{0.88}$
$=0.01(1000^{0.88})z^{0.88}$
$\approx 4.4z^{0.88}$

71. (a) Let w = the width of the field and let l = the length.
The perimeter of the field is 6000 ft, so
$2l+2w=6000$
$l+w=3000$
$l=3000-w.$
Thus, the area of the field is given by
$A=lw$
$A=(3000-w)w.$

(b) Since $l = 3000 - w$ and w must be positive, $0 < w \leq 3000$.
The domain of A is $0 < w \leq 3000$.

(c)

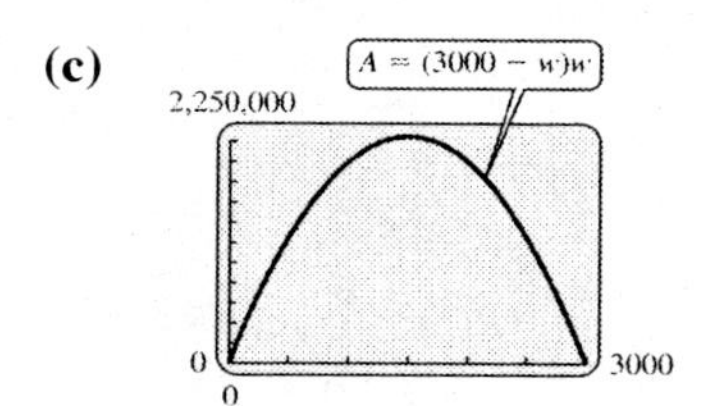

(d) Answers will vary.

73. (a) The independent variable is the year.

(b) The dependent variable is the number of Internet users.

(c) $f(2010) = 1971$ million users

(d) Domain: [2008, 2012]
Range: [1574, 2405]

1.4 Quadratic Functions; Translation and Reflection

1. If $a \geq 1$, then the graph of $y = ax^2$ becomes narrower as the value of a increases. If $0 < a < 1$, then the graph of $y = ax^2$ becomes wider as the value of a decreases.

3. The graph of $y = x^2 - 3$ is the graph of $y = x^2$ translated 3 units downward.
This is graph **D**.

5. The graph of $y = (x-3)^2 + 2$ is the graph of $y = x^2$ translated 3 units to the right and 2 units upward.
This is graph **A**.

7. The graph of $y = -(3-x)^2 + 2$ is the same as the graph of $y = -(x-3)^2 + 2$. This is the graph of $y = x^2$ reflected in the x-axis, translated 3 units to the right and 2 units upward.
This is graph **C**.

9.
$$\begin{aligned} y &= 3x^2 + 9x + 5 \\ &= 3(x^2 + 3x) + 5 \\ &= 3\left(x^2 + 3x + \frac{9}{4}\right) + 5 - 3\left(\frac{9}{4}\right) \\ &= 3\left(x + \frac{3}{2}\right)^2 - \frac{7}{4} \end{aligned}$$

The vertex is $\left(-\frac{3}{2}, -\frac{7}{4}\right)$.

11.
$$\begin{aligned} y &= -2x^2 + 8x - 9 \\ &= -2(x^2 - 4x) - 9 \\ &= -2(x^2 - 4x + 4) - 9 - (-2)(4) \\ &= -2(x-2)^2 - 1 \end{aligned}$$

The vertex is $(2, -1)$.

13. $y = x^2 + 5x + 6$
$y = (x+3)(x+2)$

Set $y = 0$ to find the x-intercepts.

$0 = (x+3)(x+2)$
$x = -3, x = -2$

The x-intercepts are -3 and -2. Set $x = 0$ to find the y-intercept.

$y = 0^2 + 5(0) + 6$
$y = 6$

The y-intercept is 6.
The x-coordinate of the vertex is

$$x = \frac{-b}{2a} = \frac{-5}{2} = -\frac{5}{2}.$$

Substitute to find the y-coordinate.

$$y = \left(-\frac{5}{2}\right)^2 + 5\left(-\frac{5}{2}\right) + 6 = \frac{25}{4} - \frac{25}{2} + 6 = -\frac{1}{4}$$

The vertex is $\left(-\frac{5}{2}, -\frac{1}{4}\right)$

The axis is $x = -\frac{5}{2}$, the vertical line through the vertex.

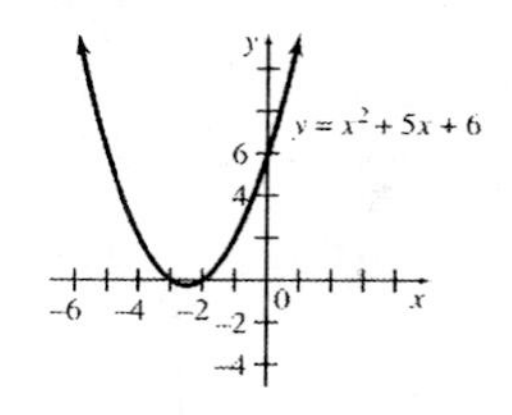

15. $y=-2x^2-12x-16$
$=-2(x^2+6x+8)$
$=-2(x+4)(x+2)$

Let $y=0$.

$0=-2(x+4)(x+2)$
$x=-4,\ x=-2$

-4 and -2 are the x-intercepts. Let $x=0$.

$y=-2(0)^2+12(0)-16$

-16 is the y-intercept.

Vertex:

$x=\frac{-b}{2a}=\frac{12}{-4}=-3$

$y=-2(-3)^2-12(-3)-16$
$=-18+36-16=2$

The vertex is $(-3,\ 2)$.

The axis is $x=-3$, the vertical line through the vertex.

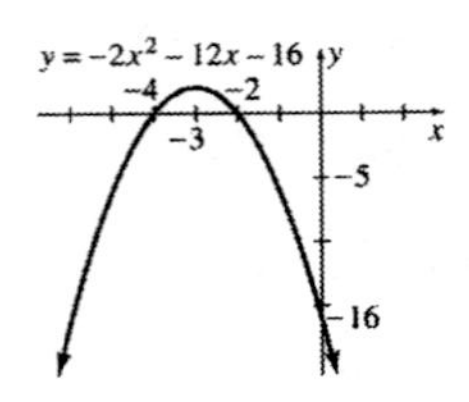

17. $y=2x^2+8x-8$

Let $y=0$.

$2x^2+8x-8=0$
$2\left(x^2+4x-4\right)=0$
$x^2+4x-4=0$

$x=\frac{-4\pm\sqrt{4^2-4(1)(-4)}}{2(1)}$

$=\frac{-4\pm\sqrt{32}}{2}=\frac{-4\pm4\sqrt{2}}{2}=-2\pm2\sqrt{2}$

The x-intercepts are $-2\pm2\sqrt{2}\approx0.83$ or -4.83.

Let $x=0$.

$y=2(0)^2+8(0)-8=-8$

The y-intercept is -8.

The x-coordinate of the vertex is

$x=\frac{-b}{2a}=-\frac{8}{4}=-2.$

If $x=-2$,

$y=2(-2)^2+8(-2)-8$
$=8-16-8=-16.$

The vertex is $(-2,\ -16)$.

The axis is $x=-2$.

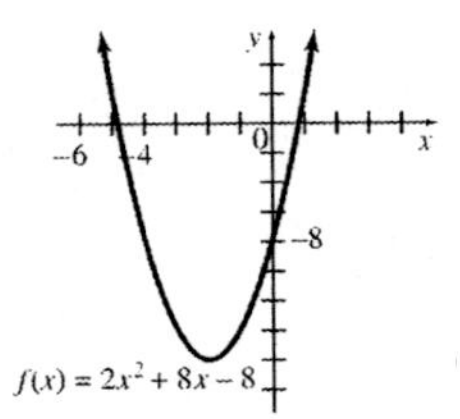

19. $f(x)=2x^2-4x+5$

Let $f(x)=0$.

$0=2x^2-4x+5$

$x=\frac{-(-4)\pm\sqrt{(-4)^2-4(2)(5)}}{2(2)}$

$=\frac{4\pm\sqrt{16-40}}{4}$

$=\frac{4\pm\sqrt{-24}}{4}$

Since the radicand is negative, there are no x-intercepts.

Let $x=0$.

$y=2(0)^2-4(0)+5$
$y=5$

The y-intercept is 5.

Vertex: $x=\frac{-b}{2a}=\frac{-(-4)}{2(2)}=\frac{4}{4}=1$

$y=2(1)^2-4(1)+5=2-4+5=3$

The vertex is $(1,\ 3)$.

The axis is $x=1$.

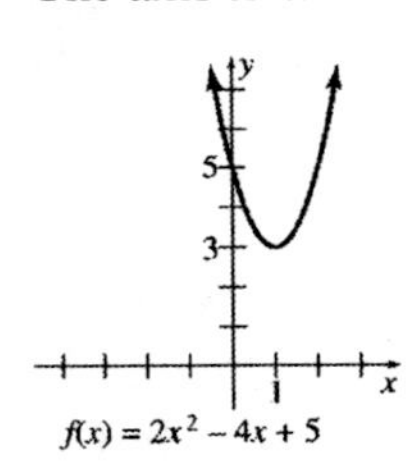

21. $f(x)=-2x^2+16x-21$

Let $f(x)=0$

Use the quadratic formula.

$x=\frac{-16\pm\sqrt{16^2-4(-2)(-21)}}{2(-2)}$

$=\frac{-16\pm\sqrt{88}}{-4}=\frac{-16\pm2\sqrt{22}}{-4}=4\pm\frac{\sqrt{22}}{2}$

The x-intercepts are $4+\frac{\sqrt{22}}{2}\approx6.35$

and $4-\frac{\sqrt{22}}{2}\approx1.65.$

(*continued on next page*)

(*continued*)

Let $x = 0$.

$y = -2(0)^2 + 16(0) - 21 = -21$

The y-intercept is -21.

Vertex:

$$x = \frac{-b}{2a} = \frac{-16}{2(-2)} = \frac{-16}{-4} = 4$$

$$y = -2(4)^2 + 16(4) - 21$$
$$= -32 + 64 - 21 = 11$$

The vertex is (4, 11).

The axis is $x = 4$.

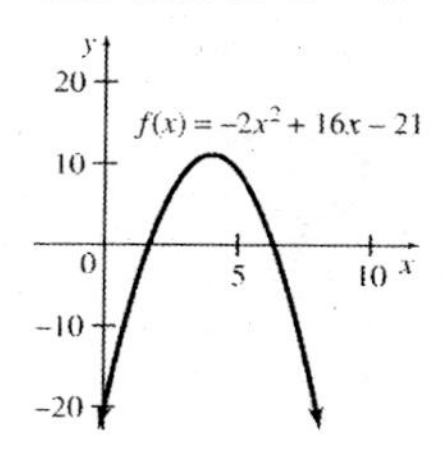

23. $y = \frac{1}{3}x^2 - \frac{8}{3}x + \frac{1}{3}$

Let $y = 0$.

$$0 = \frac{1}{3}x^2 - \frac{8}{3}x + \frac{1}{3}$$

Multiply by 3 to clear the fractions. Then use the quadratic formula.

$$0 = x^2 - 8x + 1$$

$$x = \frac{-(-8) \pm \sqrt{(-8)^2 - 4(1)(1)}}{2(1)}$$

$$= \frac{8 \pm \sqrt{64 - 4}}{2} = \frac{8 \pm \sqrt{60}}{2} = \frac{8 \pm 2\sqrt{15}}{2}$$

$$= 4 \pm \sqrt{15}$$

The x-intercepts are $4 + \sqrt{15} \approx 7.87$ and $4 - \sqrt{15} \approx 0.13$.

Let $x = 0$.

$$y = \frac{1}{3}(0)^2 - \frac{8}{3}(0) + \frac{1}{3}$$

$\frac{1}{3}$ is the y-intercept.

Vertex:

$$x = \frac{-b}{2a} = \frac{-\left(-\frac{8}{3}\right)}{2\left(\frac{1}{3}\right)} = \frac{\frac{8}{3}}{\frac{2}{3}} = 4$$

$$y = \frac{1}{3}(4)^2 - \frac{8}{3}(4) + \frac{1}{3}$$
$$= \frac{16}{3} - \frac{32}{3} + \frac{1}{3} = -\frac{15}{3} = -5$$

The vertex is (4, −5).

The axis is $x = 4$.

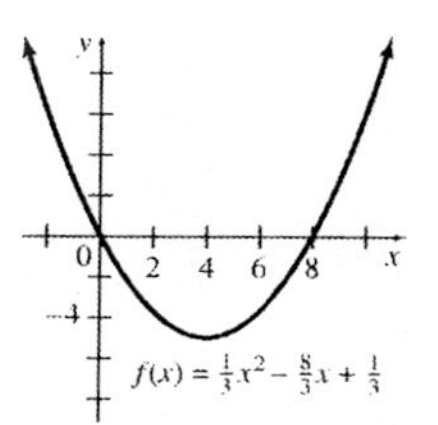

25. The graph of $y = \sqrt{x+2} - 4$ is the graph of $y = \sqrt{x}$ translated 2 units to the left and 4 units downward. This is graph **D**.

27. The graph of $y = \sqrt{-x+2} - 4$ is the graph of $y = \sqrt{-(x-2)} - 4$, which is the graph of $y = \sqrt{x}$ reflected in the y-axis, translated 2 units to the right, and translated 4 units downward. This is graph **C**.

29. The graph of $y = -\sqrt{x+2} - 4$ is the graph of $y = \sqrt{x}$ reflected in the x-axis, translated 2 units to the left, and translated 4 units downward. This is graph **E**.

31. The graph of $y = -f(x)$ is the graph of $y = f(x)$ reflected in the x-axis.

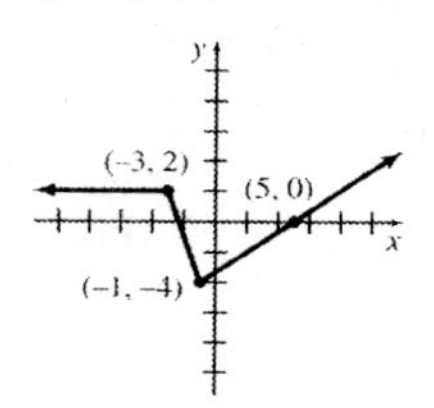

33. The graph of $y = f(-x)$ is the graph of $y = f(x)$ reflected across the y-axis.

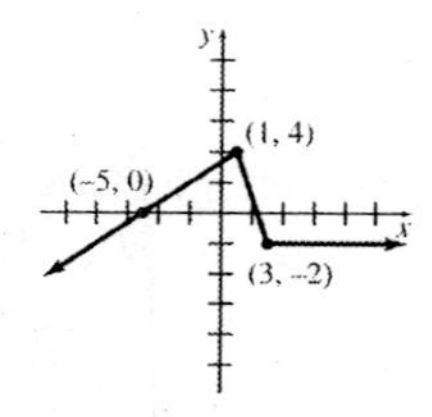

35. $f(x) = \sqrt{x-2} + 2$

Translate the graph of $f(x) = \sqrt{x}$ 2 units right and 2 units up.

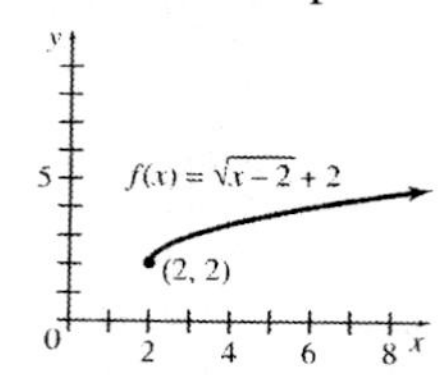

37. $f(x) = -\sqrt{2-x} - 2$
$= -\sqrt{-(x-2)} - 2$

Reflect the graph of $f(x)$ vertically and horizontally. Translate the graph 2 units right and 2 units down.

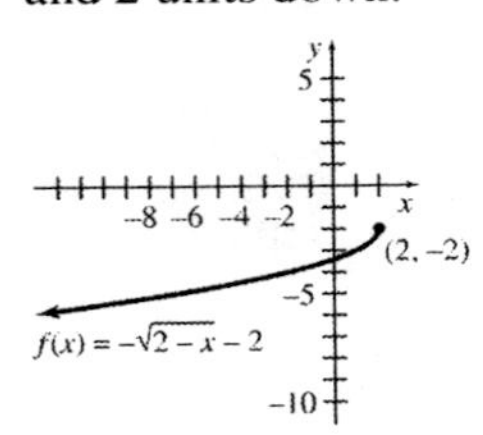

39. If $0 < a < 1$, the graph of $f(ax)$ will be the graph of $f(x)$ stretched horizontally.

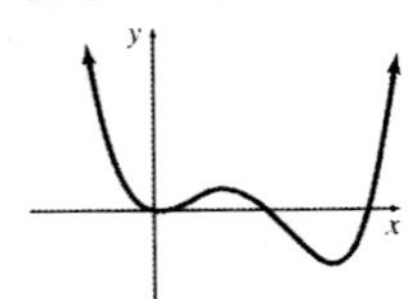

41. If $-1 < a < 0$, the graph of $f(ax)$ will be reflected horizontally, since a is negative. It will be stretched horizontally.

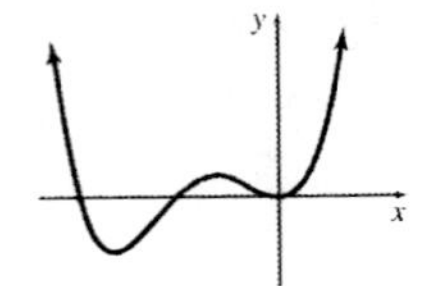

43. If $0 < a < 1$, the graph of $a\,f(x)$ will be flatter than the graph of $f(x)$. Each y-value is only a fraction of the height of the original y-values.

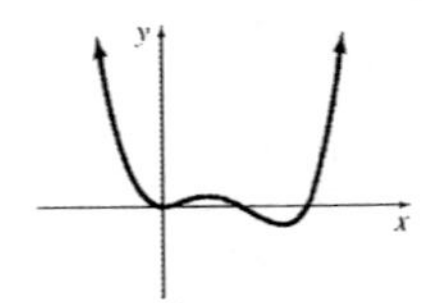

45. If $-1 < a < 0$, the graph will be reflected vertically, since a will be negative. Also, because a is a fraction, the graph will be flatter because each y-value will only be a fraction of its original height.

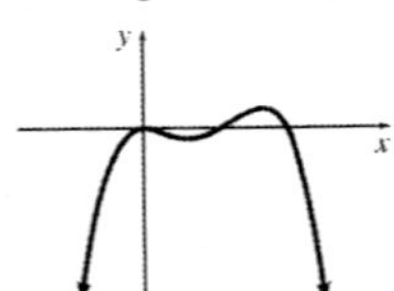

47. **(a)** Since the graph of $y = f(x)$ is reflected vertically to obtain the graph of $y = -f(x)$, the x-intercept is unchanged. The x-intercept of the graph of $y = f(x)$ is r.

(b) Since the graph of $y = f(x)$ is reflected horizontally to obtain the graph of $y = f(-x)$, the x-intercept of the graph of $y = f(-x)$ is $-r$.

(c) Since the graph of $y = f(x)$ is reflected both horizontally and vertically to obtain the graph of $y = -f(-x)$, the x-intercept of the graph of $y = -f(-x)$ is $-r$.

49. $S(x) = 1 - 0.058x - 0.076x^2$

(a) $0.50 = 1 - 0.058x - 0.076x^2$
$0.076x^2 + 0.058x - 0.50 = 0$
$76x^2 + 58x - 500 = 0$
$38x^2 + 29x - 250 = 0$

$$x = \frac{-29 \pm \sqrt{(29)^2 - 4(38)(-250)}}{2(38)}$$
$$= \frac{-29 \pm \sqrt{38{,}841}}{76} \approx -2.97 \text{ or } \approx 2.21$$

We ignore the negative value.
The value $x = 2.2$ represents 2.2 decades or 22 years, and 22 years after 65 is 87. The median length of life is 87 years.

(b) If nobody lives, $S(x) = 0$.
$1 - 0.058x - 0.076x^2 = 0$
$76x^2 + 58x - 1000 = 0$
$38x^2 + 29x - 500 = 0$

$$x = \frac{-29 \pm \sqrt{(29)^2 - 4(38)(-500)}}{2(38)}$$
$$= \frac{-29 \pm \sqrt{76{,}841}}{76} \approx -4.03 \text{ or } \approx 3.27$$

We ignore the negative value. The value $x = 3.3$ represents 3.3 decades or 33 years, and 33 years after 65 is 98.Virtually nobody lives beyond 98 years.

51. **(a)** The vertex of the quadratic function $y = 0.057x - 0.001x^2$ occurs when

$$x = -\frac{b}{2a} = -\frac{0.057}{2(-0.001)} = 28.5.$$

Since the coefficient of the leading term, −0.001, is negative, then the graph of the function opens downward, so a maximum is reached at 28.5 weeks of gestation.

(b) The maximum splenic artery resistance reached at the vertex is

$$y = 0.057(28.5) - 0.001(28.5)^2 \approx 0.81.$$

(c) The splenic artery resistance equals 0, when $y = 0$.

$$0.057x - 0.001x^2 = 0$$
$$x(0.057 - 0.001x) = 0$$
$$x = 0 \text{ or } 0.057 - 0.001x = 0$$
$$x = \frac{0.057}{0.001} = 57$$

So, the splenic artery resistance equals 0 at 0 weeks or 57 weeks of gestation. This is not reasonable because at $x = 0$ or 57 weeks, the fetus does not exist.

53. (a)

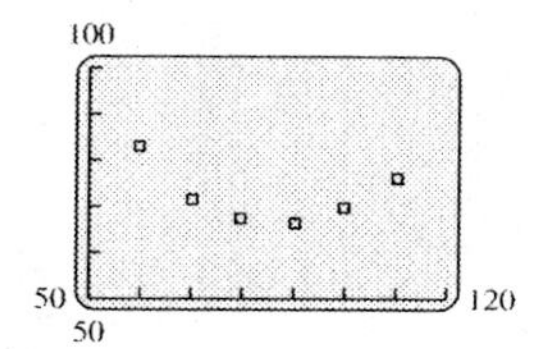

(b) By looking at the graph, it is clear that a quadratic function would be model these data.

(c) $f(x) = 0.021018x^2 - 3.6750x + 227.2$

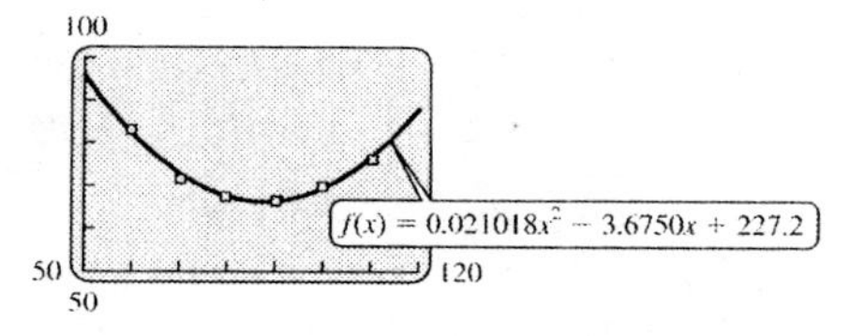

(d) Given that $(h, k) = (90, 67.2)$, the equation has the form $y = a(x - 90)^2 + 67.2$. The point (110, 77.0) is also on the curve.

$$77.0 = a(110 - 90)^2 + 67.2$$
$$9.8 = 400a$$
$$a = 0.0245$$

A quadratic function that models the data is $f(x) = 0.0245(x - 90)^2 + 67.2$.

(e)

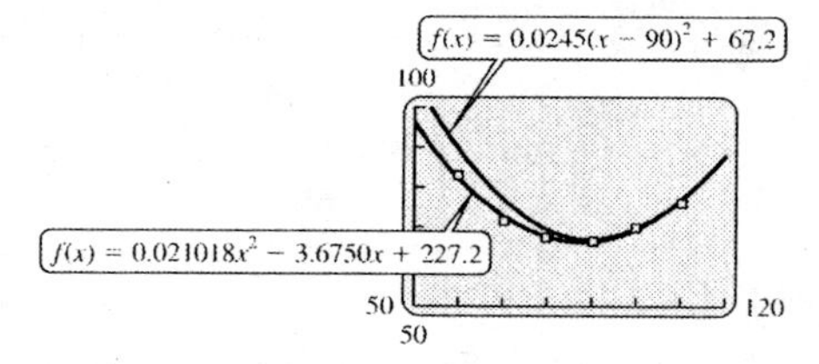

The two graphs are very close.

(f) $f(104)$

$$= 0.021018(104)^2 - 3.6750(104) + 227.2$$
$$\approx 72.3$$

$$f(104) = 0.0245(104 - 90)^2 + 67.2 \approx 72.0$$

The quadratic regression function that best fits the data is closer to the actual value of 71.7. That is the function from part (d).

55. (a)

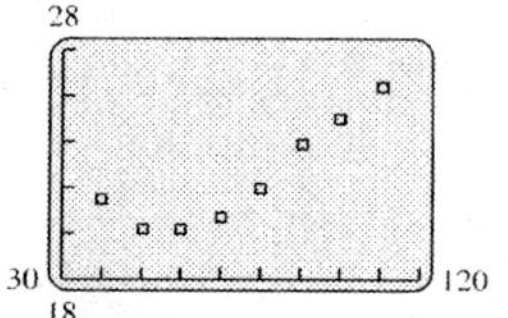

(b) By looking at the graph, it is clear that a quadratic function would be model these data.

(c) $f(x) = 0.002071x^2 - 0.2262x + 26.78$

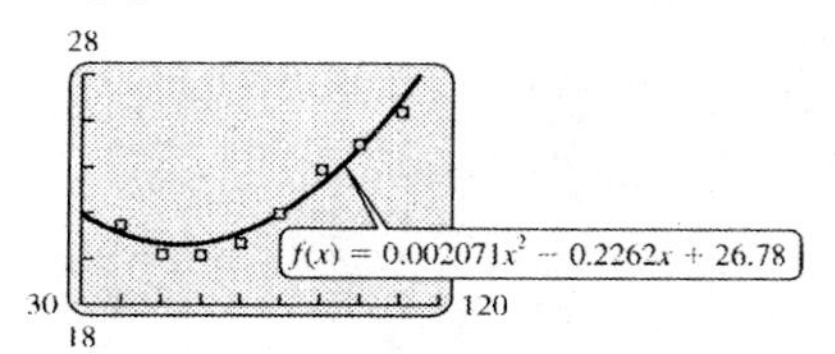

(d) Given that $(h, k) = (60, 20.3)$, the equation has the form $y = a(x - 60)^2 + 20.3$. The point (110, 26.5) is also on the curve.

$$26.5 = a(110 - 60)^2 + 20.3$$
$$6.2 = 2500a$$
$$a = 0.0248$$

A quadratic function that models the data is $y = 0.0248(x - 60)^2 + 20.3$.

(e)

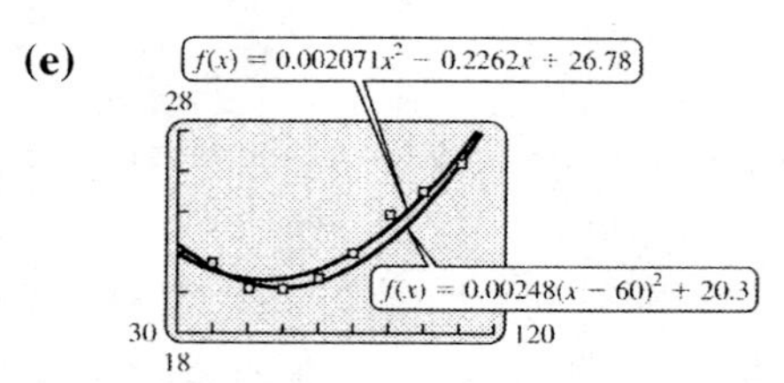

The two graphs are very close.

57. $h = 32t - 16t^2$
$= -16t^2 + 32t$

(a) Find the vertex.

$$x = \frac{-b}{2a} = \frac{-32}{-32} = 1$$
$$y = -16(1)^2 + 32(1) = 16$$

The vertex is (1, 16), so the maximum height is 16 ft.

(b) When the object hits the ground, $h = 0$, so

$$32t - 16t^2 = 0$$
$$16t(2 - t) = 0$$
$$t = 0 \quad \text{or } t = 2.$$

When $t = 0$, the object is thrown upward. The object hits the ground after 2 sec.

59. Let $x =$ the width.
Then $380 - 2x =$ the length.

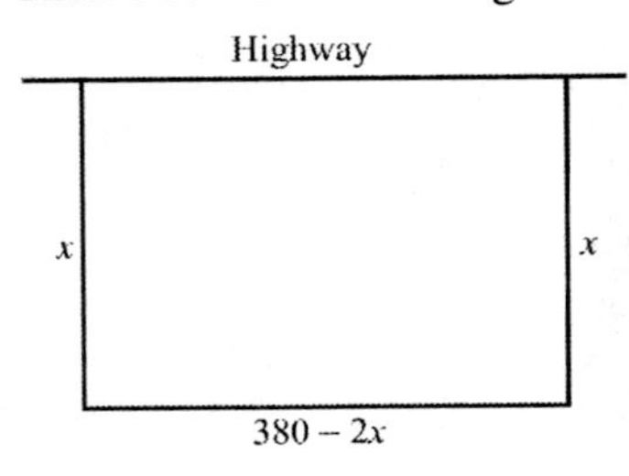

Area $= x(380 - 2x) = -2x^2 + 380x$

Find the vertex:

$$x = \frac{-b}{2a} = \frac{-380}{-4} = 95$$
$$y = -2(95)^2 + 380(95) = 18{,}050$$

The graph of the area function is a parabola with vertex (95,18,050).
The maximum area of 18,050 sq ft occurs when the width is 95 ft and the length is $380 - 2x = 380 - 2(95) = 190$ ft.

61. Draw a sketch of the arch with the vertex at the origin.

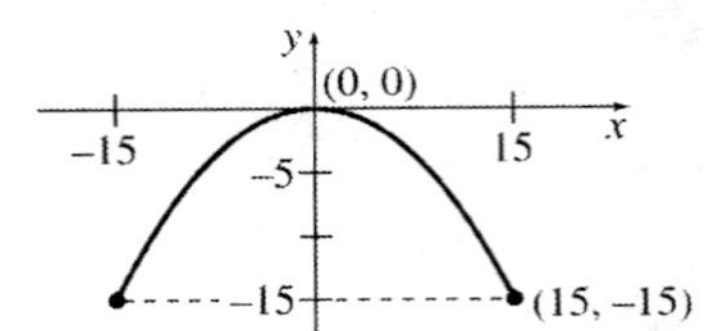

Since the arch is a parabola that opens downward, the equation of the parabola is the form $y = a(x - h)^2 + k$, where the vertex $(h, k) = (0, 0)$ and $a < 0$. That is, the equation is of the form $y = ax^2$. Since the arch is 30 meters wide at the base and 15 meters high, the points $(15, -15)$ and $(-15, -15)$ are on the parabola. Use $(15, -15)$ as one point on the parabola.

$$-15 = a(15)^2 \Rightarrow a = \frac{-15}{15^2} = -\frac{1}{15}$$

So, the equation is $y = -\frac{1}{15}x^2$.

Ten feet from the ground (the base) is at $y = -5$. Substitute -5 for y and solve for x.

$$-5 = -\frac{1}{15}x^2$$
$$x^2 = -5(-15) = 75$$
$$x = \pm\sqrt{75} = \pm 5\sqrt{3}$$

The width of the arch ten feet from the ground is then $5\sqrt{3} - (-5\sqrt{3}) = 10\sqrt{3} \approx 17.32$ meters.

1.5 Polynomial and Rational Functions

1. The graph of $y = -(x - 1)^4 + 2$ is the graph of $y = x^4$ shifted one unit right, reflected across the x-axis, and then shifted two units up.

3. The graph of $f(x) = (x - 2)^3 + 3$ is the graph of $y = x^3$ translated 2 units to the right and 3 units upward.

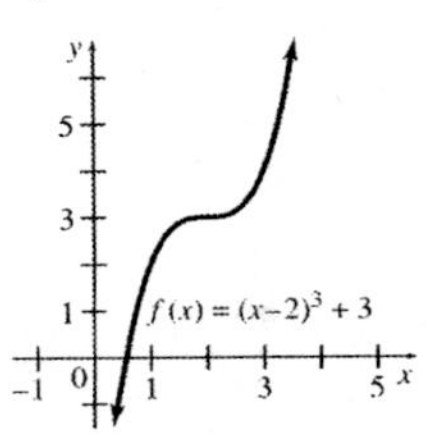

5. The graph of $f(x) = -(x + 3)^4 + 1$ is the graph of $y = x^4$ reflected horizontally, translated 3 units to the left, and translated 1 unit upward.

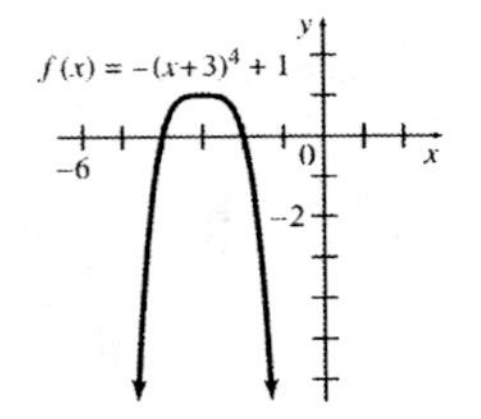

7. The graph of $y = x^3 - 7x - 9$ has the right end up, the left end down, at most two turning points, and a y-intercept of -9.
This is graph **D**.

9. The graph of $y = -x^3 - 4x^2 + x + 6$ has the right end down, the left end up, at most two turning points, and a y-intercept of 6.
This is graph **E**.

11. The graph of $y = x^4 - 5x^2 + 7$ has both ends up, at most three turning points, and a y-intercept of 7. This is graph **I**.

13. The graph of $y = -x^4 + 2x^3 + 10x + 15$ has both ends down, at most three turning points, and a y-intercept of 15. This is graph **G**.

15. The graph of $y=-x^5+4x^4+x^3-16x^2+12x+5$ has the right end down, the left end up, at most four turning points, and a y-intercept of 5. This is graph **A**.

17. The graph of $y=\dfrac{2x^2+3}{x^2+1}$ has no vertical asymptote, the line with equation $y = 2$ as a horizontal asymptote, and a y-intercept of 3. This is graph **D**.

19. The graph $y=\dfrac{-2x^2-3}{x^2+1}$ has no vertical asymptote, the line with equation $y=-2$ as a horizontal asymptote, and a y-intercept of –3. This is graph **E**.

21. The right end is up and the left end is up. There are three turning points. The degree is an even integer equal to 4 or more. The x^n term has a + sign.

23. The right end is up and the left end is down. There are four turning points. The degree is an odd integer equal to 5 or more. The x^n term has a + sign.

25. The right end is down and the left end is up. There are six turning points. The degree is an odd integer equal to 7 or more. The x^n term has a – sign.

27. $y=\dfrac{-4}{x+2}$

The function is undefined for $x = -2$, so the line $x = -2$ is a vertical asymptote.

x	−102	−12	−7	−5	−3	−1	8	98
$x+2$	−100	−10	−5	−3	−1	1	10	100
y	0.04	0.4	0.8	1.3	4	−4	−0.4	−0.04

The graph approaches $y = 0$, so the line $y = 0$ (the x-axis) is a horizontal asymptote.
Asymptotes: $y = 0$, $x = -2$
x-intercept: none
y-intercept: –2, the value when $x = 0$

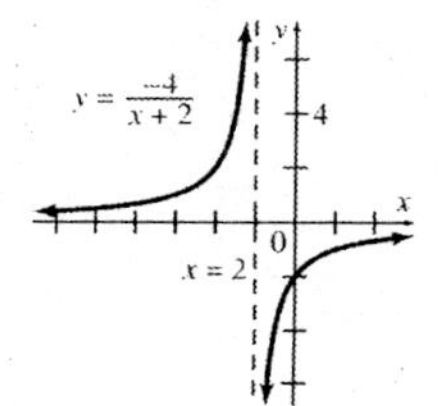

29. $y=\dfrac{2}{3+2x}$

$3+2x=0$ when $2x=-3$ or $x=-\dfrac{3}{2}$,

so the line $x=-\dfrac{3}{2}$ is a vertical asymptote.

x	−51.5	−6.5	−2	−1	3.5	48.5
$3+2x$	−100	−10	−1	1	10	100
y	−0.02	−0.2	−2	2	0.2	0.02

The graph approaches $y = 0$, so the line $y = 0$ (the x-axis) is a horizontal asymptote.

Asymptote: $y=0$, $x=-\dfrac{3}{2}$

x-intercept: none

y-intercept: $\dfrac{2}{3}$, the value when $x = 0$

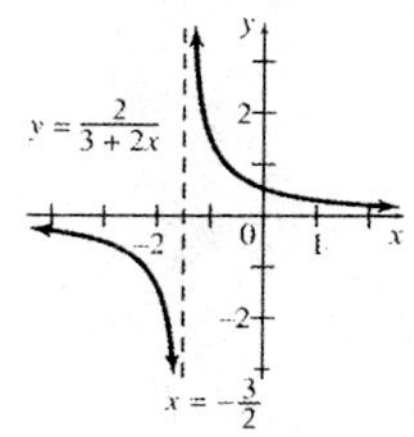

31. $y=\dfrac{2x}{x-3}$

$x-3=0$ when $x=3$, so the line $x=3$ is a vertical asymptote.

x	−97	−7	−1	1	2	2.5
$2x$	−194	−14	−2	2	4	5
$x-3$	−100	−10	−4	−2	−1	−0.5
y	1.94	1.4	0.5	−1	−4	−10

x	3.5	4	5	7	11	103
$2x$	7	8	10	14	22	206
$x-3$	0.5	1	2	4	8	100
y	14	8	5	3.5	2.75	2.06

As x increases, $\dfrac{2x}{x-3}\approx\dfrac{2x}{x}=2.$

Thus, $y = 2$ is a horizontal asymptote.
Asymptotes: $y = 2$, $x = 3$
x-intercept: 0, the value when $y=0$
y-intercept: 0, the value when $x=0$

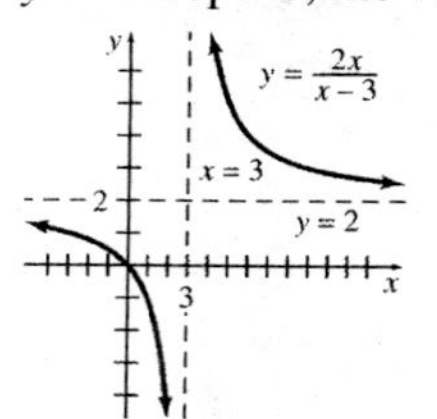

33. $y=\dfrac{x+1}{x-4}$

$x-4=0$ when $x=4$, so $x=4$ is a vertical asymptote.

x	−96	−6	−1	0	3
$x+1$	−95	−5	0	1	4
$x-4$	−100	−10	−5	−4	−1
y	0.95	0.5	0	−0.25	−4

x	3.5	4.5	5	14	104
$x+1$	4.5	5.5	6	15	105
$x-4$	−0.5	0.5	1	10	100
y	−9	11	6	1.5	1.05

As x gets larger, $\dfrac{x+1}{x-4}\approx\dfrac{x}{x}=1.$

Thus, $y=1$ is a horizontal asymptote.

Asymptotes: $y=1$, $x=4$

x-intercept: −1, the value when $y=0$

y-intercept: $-\dfrac{1}{4}$, the value when $x=0$

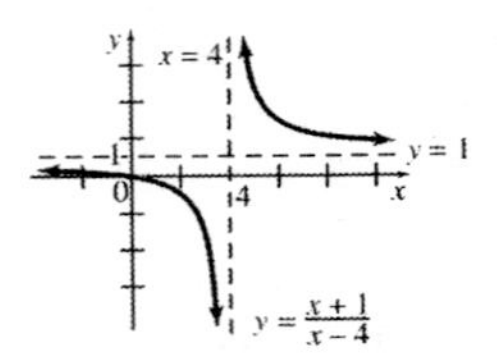

35. $y=\dfrac{3-2x}{4x+20}$

$4x+20=0$ when $4x=-20$, or $x=-5$. so the line $x=-5$ is a vertical asymptote.

x	−8	−7	−6	−4	−3	−2
$3-2x$	−26	−23	−20	−14	−11	−8
$4x+20$	−12	−8	−4	4	8	12
y	2.17	2.88	5	−3.5	−1.38	−0.67

As x increases, $\dfrac{3-2x}{4x+20}\approx\dfrac{-2x}{4x}=-\dfrac{1}{2}$. Thus, the line $y=-\dfrac{1}{2}$ is a horizontal asymptote.

Asymptotes: $x=-5$, $y=-\dfrac{1}{2}$

x-intercept: $\dfrac{3}{2}$, the value when $y=0$

y-intercept: $\dfrac{3}{20}$, the value when $x=0$

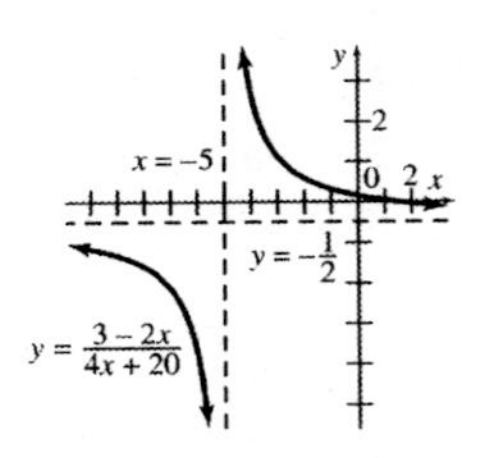

37. $y=\dfrac{-x-4}{3x+6}$

$3x+6=0$ when $3x=-6$ or $x=-2$, so the line $x=-2$ is a vertical asymptote.

x	−5	−4	−3	−1	0	1
$-x-4$	1	0	−1	−3	−4	−5
$3x+6$	−9	−6	−3	3	6	9
y	−0.11	0	0.33	−1	−0.67	−0.56

As x increases, $\dfrac{-x-4}{3x+6}\approx\dfrac{-x}{3x}=-\dfrac{1}{3}.$

The line $y=-\dfrac{1}{3}$ is a horizontal asymptote.

Asymptotes: $y=-\dfrac{1}{3}, x=-2$

x-intercept: −4, the value when $y=0$

y-intercept: $-\dfrac{2}{3}$, the value when $x=0$

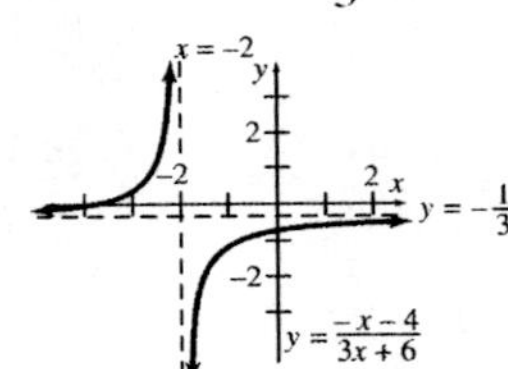

39. $y=\dfrac{x^2+7x+12}{x+4}=\dfrac{(x+3)(x+4)}{x+4}$

$=x+3,\ x\neq-4$

There are no asymptotes, but there is a hole at $x=-4$.

x-intercept: −3, the value when $y=0$.

y-intercept: 3, the value when $x=0$.

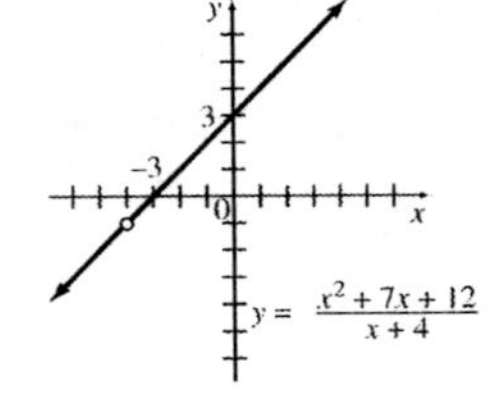

41. For a vertical asymptote at $x = 1$, put $x - 1$ in the denominator. For a horizontal asymptote at $y = 2$, the degree of the numerator must equal the degree of the denominator and the quotient of their leading terms must equal 2. So, $2x$ in the numerator would cause y to approach 2 as x becomes larger. So, one possible answer is $y = \frac{2x}{x-1}$.

43. $f(x) = (x-1)(x-2)(x+3)$,
$g(x) = x^3 + 2x^2 - x - 2$,
$h(x) = 3x^3 + 6x^2 - 3x - 6$

(a) $f(1) = (0)(-1)(4) = 0$

(b) $f(x)$ is zero when $x = 2$ and when $x = -3$.

(c)
$$\begin{aligned} g(-1) &= (-1)^3 + 2(-1)^2 - (-1) - 2 \\ &= -1 + 2 + 1 - 2 = 0 \\ g(1) &= (1)^3 + 2(1)^2 - (1) - 2 \\ &= 1 + 2 - 1 - 2 = 0 \\ g(-2) &= (-2)^3 + 2(-2)^2 - (-2) - 2 \\ &= -8 + 8 + 2 - 2 = 0 \end{aligned}$$

(d) $g(x) = [x - (-1)](x-1)[x - (-2)]$
$g(x) = (x+1)(x-1)(x+2)$

(e) $h(x) = 3g(x)$
$= 3(x+1)(x-1)(x+2)$

(f) If f is a polynomial and $f(a) = 0$ for some number a, then one factor of the polynomial is $x - a$.

45. $f(x) = \frac{1}{x^5 - 2x^3 - 3x^2 + 6}$

(a) In the graphing window [–3.4, 3.4] by [–3, 3], there appear to be two vertical asymptotes, one at $x = -1.4$ and one at $x = 1.4$. (Note the sign changes in the y-values.)

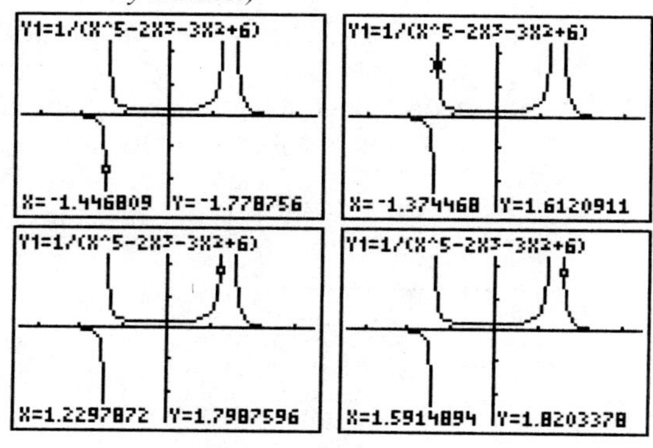

(b) In the graphing windows [–1.5, 1.4] by [–10, 10] and [1.4, 1.5] by [–1000, 1000], there appear to be three vertical asymptotes, one at $x = -1.414$, one at $x = 1.414$, and one at $x = 1.442$. (A vertical asymptote can be forced to appear on a TI-84 Plus graphing calculator by setting Xres in the WINDOW menu to 2. Note the sign changes in the y-values.)

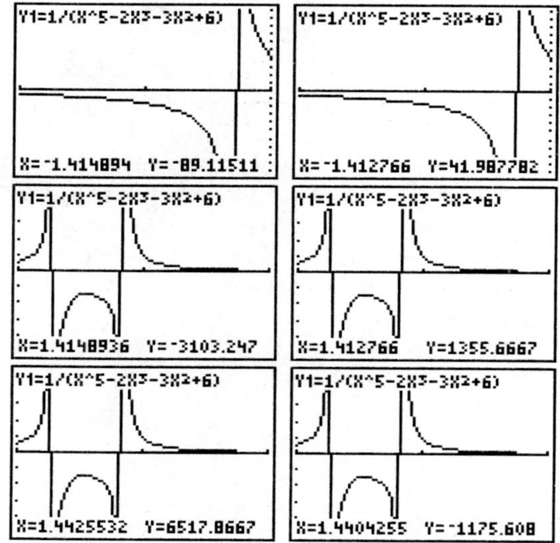

(c) Answers will vary.

47. $A(x) = 0.003631x^3 - 0.03746x^2 + 0.1012x + 0.009$

(a)

x	0	1	2	3	4	5
$A(x)$	0.009	0.076	0.091	0.073	0.047	0.032

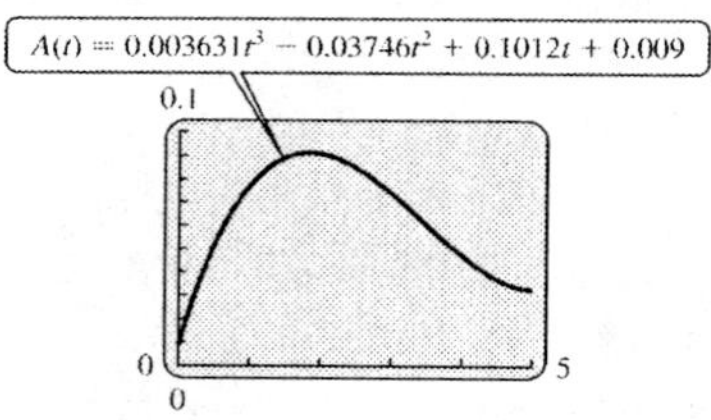

(b) The peak of the curve comes at about $x = 2$ hours.

(c) The curve rises to a y-value of 0.08 at about $x = 1.1$ hours and stays at or above that level until about $x = 2.7$ hours.

49. (a) When $c = 30, w = \frac{30^3}{100} - \frac{1500}{30} = 220,$ so the brain weights 220 g when its circumference measures 30 cm. When $c = 40, w = \frac{40^3}{100} - \frac{1500}{40} = 602.5,$ so the brain weighs 602.5 g when its circumference is 40 cm. When $c = 50,$

$w = \frac{50^3}{100} - \frac{1500}{50} = 1220,$ so the brain weighs 1220 g when its circumference is 50 cm.

(b) Set the window of a graphing calculator so you can trace to the positive x-intercept of the function. Using a "root" or "zero" program, this x-intercept is found to be approximately 19.68. Notice in the graph that positive c values less than 19.68 correspond to negative w values. Therefore, the answer is $c < 19.68$.

(c)

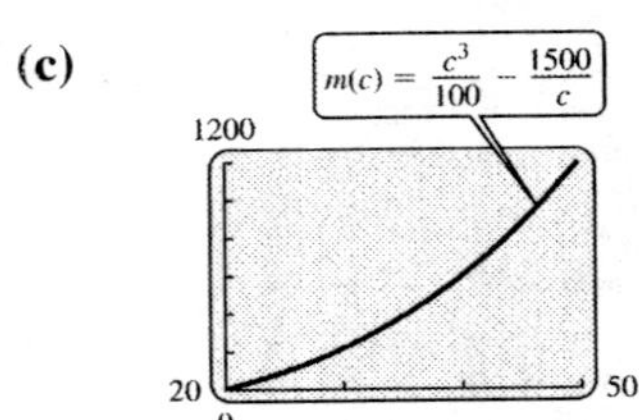

(d) One method is to graph the line $y = 700$ on the graph found in part (c) and use an "intersect" program to find the point of intersection of the two graphs. This point has the approximate coordinates (41.9, 700). Therefore, an infant has a brain weighing 700 g when the circumference measures 41.9 cm.

51. $f(x) = \frac{\lambda x}{1 + (ax)^b}$

(a) A reasonable domain for the function is $[0, \infty)$. Populations are not measured using negative numbers and they may get extremely large.

(b) If $\lambda = a = b = 1,$ the function becomes

$$f(x) = \frac{x}{1 + x}.$$

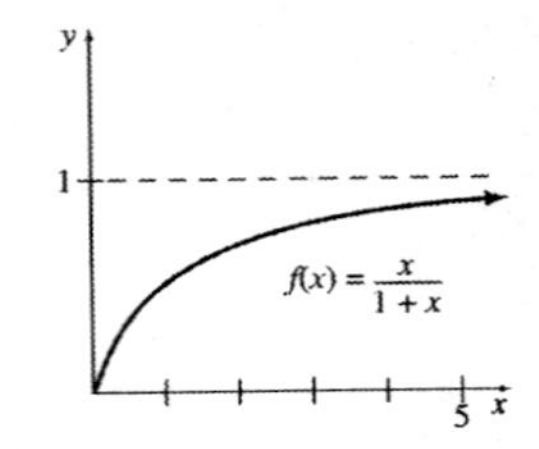

(c) If $\lambda = a = 1$ and $b = 2,$ the function becomes $f(x) = \frac{x}{1 + x^2}.$

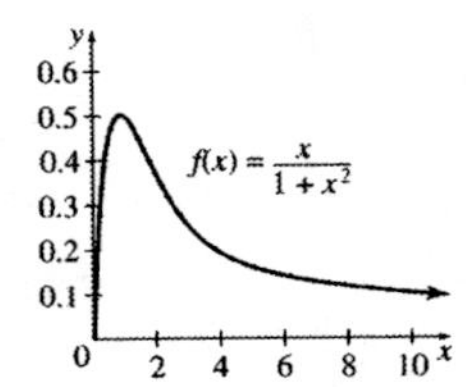

(d) As seen from the graphs, when b increases, the population of the next generation, $f(x)$, gets smaller when the current generation, x, is larger.

53. (a) The numerator represents total amount of acid-insoluble ash in the food and the soil.

(b) The numerator represents the fraction of non-digestible food in the soil.

(c)
$$y = \frac{b(1 - x) + cx}{1 - a(1 - x)}$$
$$y = \frac{0.025(1 - x) + 0.92x}{1 - 0.76(1 - x)}$$
$$= \frac{0.025 - 0.025x + 0.92x}{1 - 0.76 + 0.76x}$$
$$= \frac{0.025 + 0.895x}{0.24 + 0.76x}$$

(d)

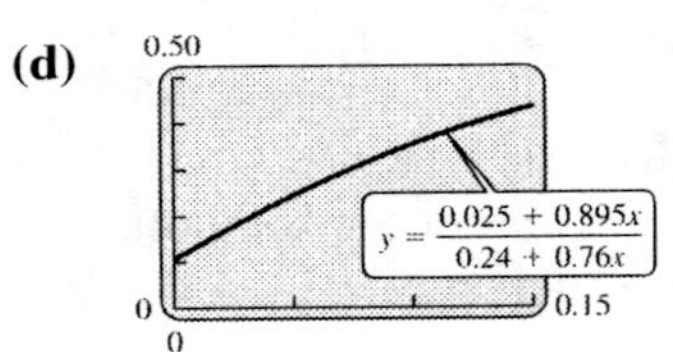

(e)
$$0.20 = \frac{0.025 + 0.895x}{0.24 + 0.76x}$$
$$0.20(0.24 + 0.76x) = 0.025 + 0.895x$$
$$0.048 + 0.152x = 0.025 + 0.895x$$
$$0.743x = 0.023$$
$$x = \frac{0.023}{0.743} \approx 0.031$$

One method for checking this answer with a graphing calculator is to find the intersection point of the function graphed in part (d) with the line $y = 0.20$.

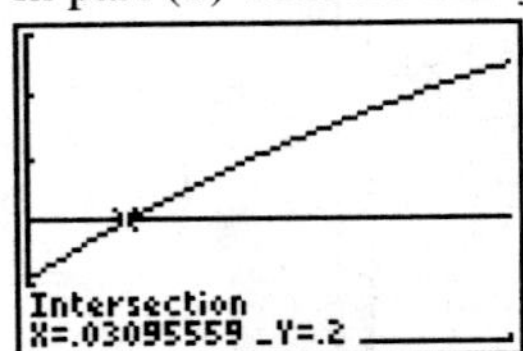

55. $y = \dfrac{6.7x}{100-x}$,
Let x = percent of pollutant and y = cost in thousands.

(a) $x = 50$: $y = \dfrac{6.7(50)}{100-50} = 6.7$
The cost is \$6700.
$x = 70$: $y = \dfrac{6.7(70)}{100-70} \approx 15.6$
The cost is \$15,600.
$x = 80$: $y = \dfrac{6.7(80)}{100-80} = 26.8$
The cost is \$26,800.
$x = 90$: $y = \dfrac{6.7(90)}{100-90} = 60.3$
The cost is \$60,300.
$x = 95$: $y = \dfrac{6.7(95)}{100-95}$
The cost is \$127,300.
$x = 98$: $y = \dfrac{6.7(98)}{100-98} = 328.3$
The cost is \$328,300.
$x = 99$: $y = \dfrac{6.7(99)}{100-99} = 668.3$
The cost is \$663,300.

(b) No, because $x = 100$ makes the denominator zero, so $x = 100$ is a vertical asymptote.

(c)

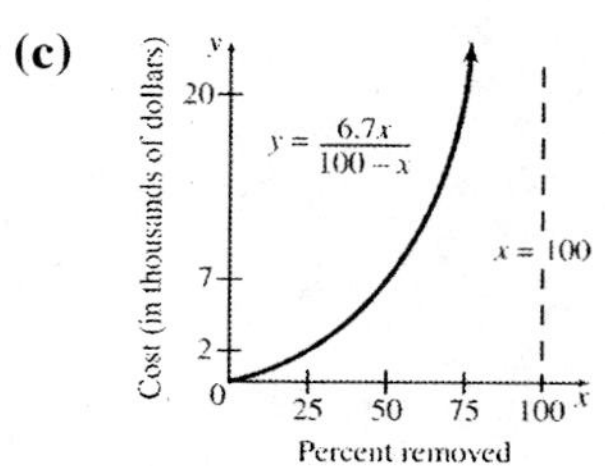

57. (a)

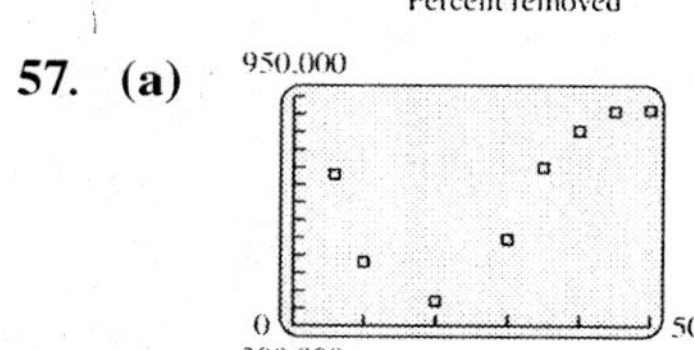

(b) $y = 531.27x^2 - 20,425x + 712,448$

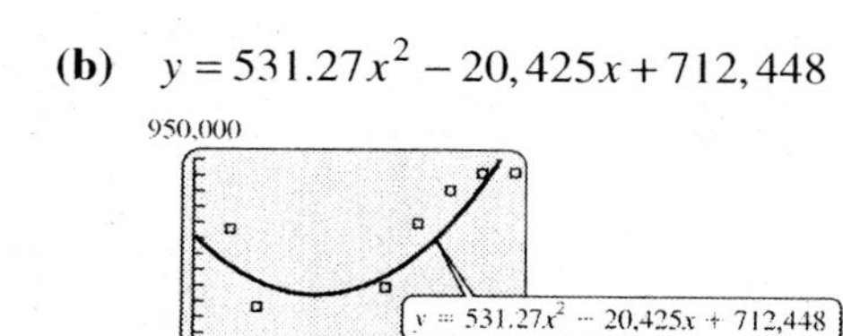

(c) $y = -49.713x^3 + 4785.53x^2 - 123,025x + 1,299,118$

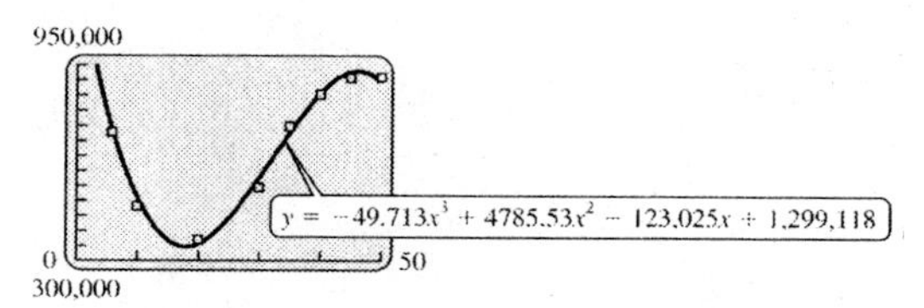

(d) From part (c), it is clear that the cubic function is a better fit for the data.

59. (a)

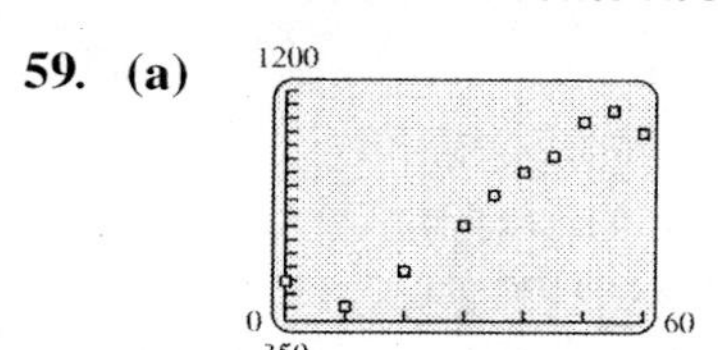

(b)

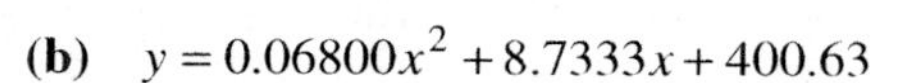

$y = 0.06800x^2 + 8.7333x + 400.63$

(c)

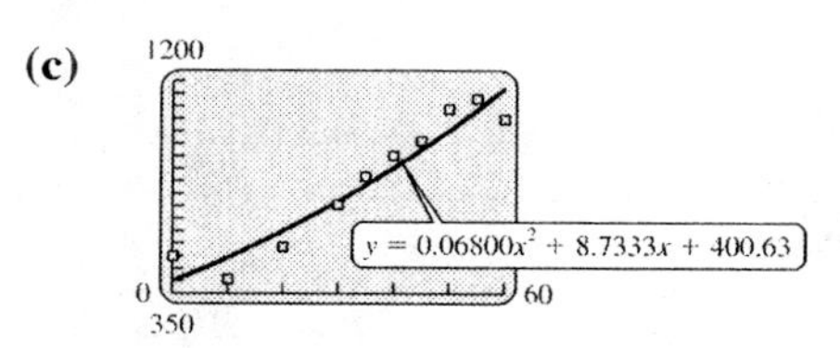

(d)

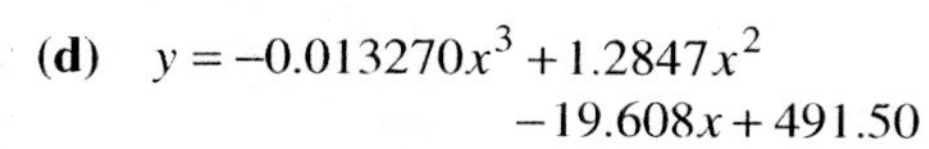

$y = -0.013270x^3 + 1.2847x^2 - 19.608x + 491.50$

(e)

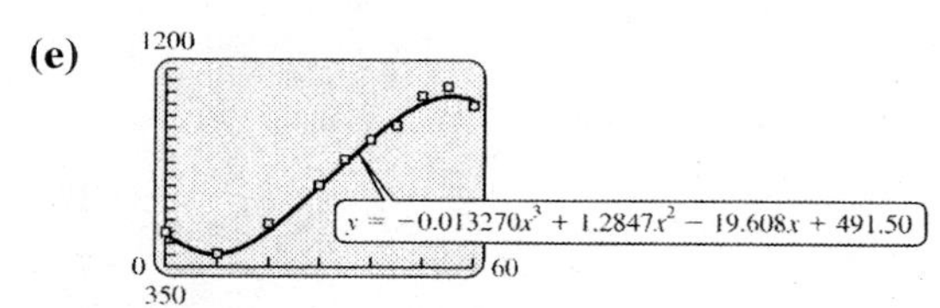

(f) The function in part (d) appears to be a better fit to the data.

Chapter 1 Review Exercises

1. False; a line can have only one slant, so its slope is unique.
2. False; the equation $y = 3x + 4$ has slope 3.
3. True; the point $(3, -1)$ is on the line because $-1 = -2(3) + 5$ is a true statement.
4. False; the points $(2, 3)$ and $(2, 5)$ do not have the same y-coordinate.
5. True; the points $(4, 6)$ and $(5, 6)$ do have the same y-coordinate, so the line is horizontal.
6. False; the x-intercept of the line $y = 8x + 9$ is $-\frac{9}{8}$.

7. True; $f(x) = \pi x + 4$ is a linear function because it is in the form $y = mx + b$, where m and b are real numbers.

8. False; $f(x) = 2x^2 + 3$ is not linear function because it isn't in the form $y = mx + b$, and it is a second-degree equation.

9. False; the line $y = 3x + 17$ has slope 3, and the line $y = -3x + 8$ has slope -3. Since $3 \cdot -3 \neq -1$, the lines cannot be perpendicular.

10. False; the line $4x + 3y = 8$ has slope $-\frac{4}{3}$, and the line $4x + y = 5$ has slope -4. Since the slopes are not equal, the lines cannot be parallel.

11. False; a correlation coefficient of zero indicates that there is no linear relationship among the data.

12. True; a correlation coefficient always will be a value between -1 and 1.

13. True

14. False; for example $f(x) = \frac{x}{x+1}$ is a rational function but not a polynomial function.

15. True

16. True

17. False; the vertical asymptote is at $x = 6$.

18. False; the domain includes all numbers except $x = 2$ and $x = -2$.

19. Marginal cost is the rate of change of the cost function; the fixed cost is the initial expenses before production begins.

20. To compute the coefficient of correlation, you need the following quantities:
$\sum x, \sum y, \sum xy, \sum x^2, \sum y^2$, and n.

21. A function is a rule that assigns to each element from one set exactly one element from another set. A linear function is a function that can be defined by $y = f(x) = mx + b$ for real numbers m and b. A quadratic function is a function that is defined by $f(x) = ax^2 + bx + c$, where a, b, and c are real numbers and $a \neq 0$.

A rational function is a function that can be defined by $f(x) = \frac{p(x)}{q(x)}$, where $p(x)$ and $q(x)$ are polynomial functions and $q(x) \neq 0$.

22. To find a vertical asymptote of a rational function, find the zeros of the denominator. If a number k makes the denominaotr 0, but does not make the numerator 0, then the line $x = k$ is a vertical asymptote.

23. If the degree of the numerator equals the degree of the denominator, then the horizontal asymptote is
$$y = \frac{\text{numerator's leading coefficient}}{\text{denominator's leading coefficient}}.$$
If the degree of the denominator is greater than the degree of the numerator, then the horizontal asymptote is the x-axis, or $y = 0$.
If the degree of the numerator is greater than the degree of the denominator, then there is no horizontal asymptote. Instead, there is a slant asymptote.

24. The end behavior can be determined.

25. Through $(-3, 7)$ and $(2, 12)$
$$m = \frac{12 - 7}{2 - (-3)} = \frac{5}{5} = 1$$

26. Through $(4, -1)$ and $(3, -3)$.
$$m = \frac{-3 - (-1)}{3 - 4} = \frac{-3 + 1}{-1} = \frac{-2}{-1} = 2$$

27. Through the origin and $(11, -2)$.
$$m = \frac{-2 - 0}{11 - 0} = -\frac{2}{11}$$

28. Through the origin and $(0, 7)$
$$m = \frac{7 - 0}{0 - 0} = \frac{7}{0}$$
The slope of the line is undefined.

29.
$$4x + 3y = 6$$
$$3y = -4x + 6$$
$$y = -\frac{4}{3}x + 2$$
Therefore, the slope is $m = -\frac{4}{3}$.

30.
$$4x - y = 7$$
$$y = 4x - 7$$
$$m = 4$$

31.
$$y + 4 = 9 \Rightarrow y = 5$$
$$y = 0x + 5 \Rightarrow m = 0$$

32. $3y - 1 = 14 \Rightarrow 3y = 14 + 1 \Rightarrow$
$3y = 15 \Rightarrow y = 5$

This is a horizontal line. The slope of a horizontal line is 0.

33. $y = 5x + 4$
$m = 5$

34. $x = 5y \Rightarrow \frac{1}{5}x = y$

$m = \frac{1}{5}$

35. Through $(5, -1)$; slope $\frac{2}{3}$

Use point-slope form.

$$y - (-1) = \frac{2}{3}(x - 5)$$
$$y + 1 = \frac{2}{3}(x - 5)$$
$$3(y + 1) = 2(x - 5)$$
$$3y + 3 = 2x - 10$$
$$3y = 2x - 13 \Rightarrow y = \frac{2}{3}x - \frac{13}{3}$$

36. Through (8, 0), with slope $-\frac{1}{4}$

Use point-slope form.

$$y - 0 = -\frac{1}{4}(x - 8) \Rightarrow y = -\frac{1}{4}x + 2$$

37. Through $(-6, 3)$ and $(2, -5)$

$$m = \frac{-5 - 3}{2 - (-6)} = \frac{-8}{8} = -1$$

Use point-slope form.

$$y - 3 = -1[x - (-6)]$$
$$y - 3 = -x - 6 \Rightarrow y = -x - 3$$

38. Through $(2, -3)$ and $(-3, 4)$

$$m = \frac{4 - (-3)}{-3 - 2} = -\frac{7}{5}$$

Use point-slope form.

$$y - (-3) = -\frac{7}{5}(x - 2)$$
$$y + 3 = -\frac{7}{5}x + \frac{14}{5}$$
$$y = -\frac{7}{5}x + \frac{14}{5} - 3$$
$$y = -\frac{7}{5}x - \frac{1}{5}$$

39. Through (2, –10), perpendicular to a line with undefined slope
A line with undefined slope is a vertical line. A line perpendicular to a vertical line is a horizontal line with equation of the form $y = k$. The desired line passes through (2, –10), so $k = -10$. Thus, an equation of the desired line is $y = -10$.

40. Through (–2, 5), with slope 0
Horizontal lines have 0 slope and an equation of the form $y = k$. The line passes through (–2, 5), so $k = 5$. An equation of the line is $y = 5$.

41. Through $(3, -4)$ parallel to $4x - 2y = 9$
Solve $4x - 2y = 9$ for y.

$$-2y = -4x + 9$$
$$y = 2x - \frac{9}{2}$$
$$m = 2$$

The desired line has the same slope. Use the point-slope form.

$$y - (-4) = 2(x - 3)$$
$$y + 4 = 2x - 6$$
$$y = 2x - 10$$
$$2x - y = 10$$

42. Through (0, 5), perpendicular to $8x + 5y = 3$
Find the slope of the given line first.

$$8x + 5y = 3$$
$$5y = -8x + 3 \Rightarrow y = \frac{-8}{5}x + \frac{3}{5}$$
$$m = -\frac{8}{5}$$

The perpendicular line has $m = \frac{5}{8}$.

Use point-slope form.

$$y - 5 = \frac{5}{8}(x - 0)$$
$$y = \frac{5}{8}x + 5$$
$$8y = 5x + 40$$
$$5x - 8y = -40$$

43. Through (–1, 4); undefined slope
Undefined slope means the line is vertical. The equation of the vertical line through (–1, 4) is $x = -1$.

44. Through $(7, -6)$, parallel to a line with undefined slope.
A line with undefined slope has the form $x = a$ (a vertical line). The vertical line that goes through (7, –6) is the line $x = 7$.

45. Through $(3, -5)$, parallel to $y = 4$
$y = 4$ is a horizontal line, the required line will also be a horizaontal line. In this case, the equation is $y = -5$.

46. Through $(-3, 5)$, perpendicular to $y = -2$
The given line, $y = -2$, is a horizontal line. A line perpendicular to a horizontal line is a vertical line with equation of the form $x = k$. The desired line passes through $(-3, 5)$, so $k = -3$. Thus, an equation of the desired line is $x = -3$.

47. $y = 4x + 3$
Let $x = 0$: $y = 4(0) + 3 \Rightarrow y = 3$
Let $y = 0$: $0 = 4x + 3 \Rightarrow -3 = 4x \Rightarrow -\frac{3}{4} = x$
Draw the line through (0, 3) and $\left(-\frac{3}{4}, 0\right)$.

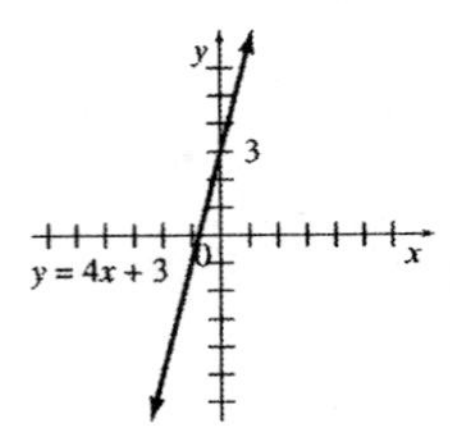

48. $y = 6 - 2x$
Let $x = 0$. Then, $y = 6 - 2(0) = 6$
Let $y = 0$. Then, $0 = 6 - 2x \Rightarrow 2x = 6 \Rightarrow x = 3$
Draw the line through (0, 6) and (3, 0).

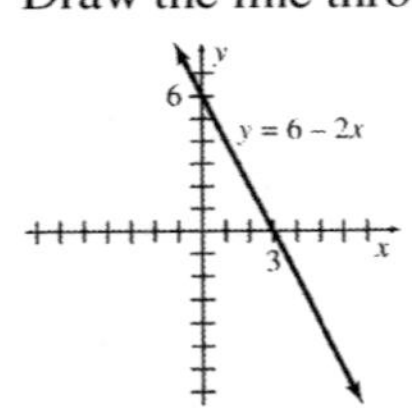

49. $3x - 5y = 15$
$$-5y = -3x + 15$$
$$y = \frac{3}{5}x - 3$$
When $x = 0$, $y = -3$.
When $y = 0$, $x = 5$.
Draw the line through $(0, -3)$ and $(5, 0)$.

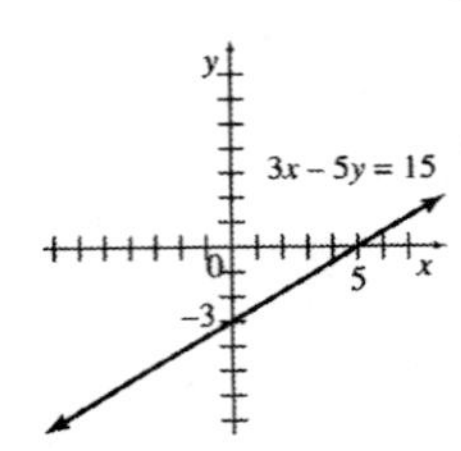

50. $4x + 6y = 12$
When $x = 0$, $y = 2$, so the y-intercept is 2.
When $y = 0, x = 3$, so the x-intercept is 3.
Draw the line through (0, 2) and (3, 0).

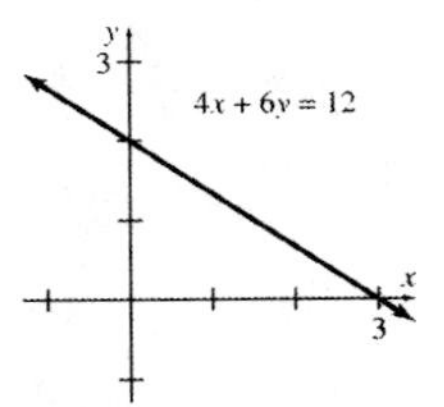

51. $x - 3 = 0 \Rightarrow x = 3$
This is the vertical line through (3, 0).

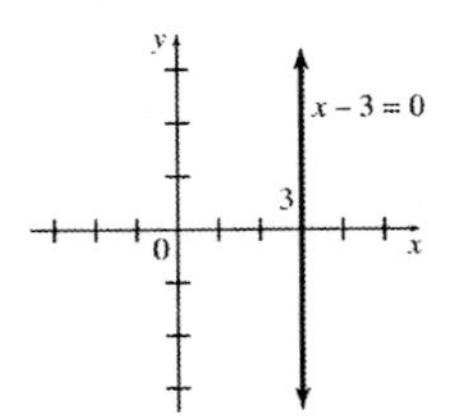

52. $y = 1$
This is the horizontal line passing through (0, 1).

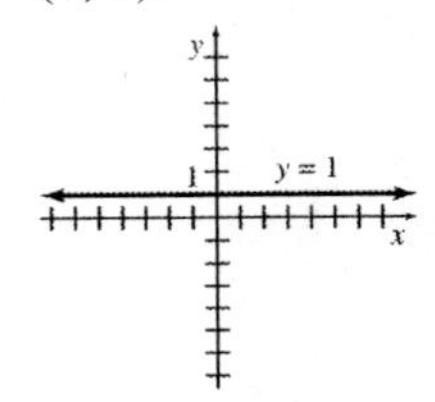

53. $y = 2x$
When $x = 0$, $y = 0$.
When $x = 1$, $y = 2$.
Draw the line through (0, 0) and (1, 2).

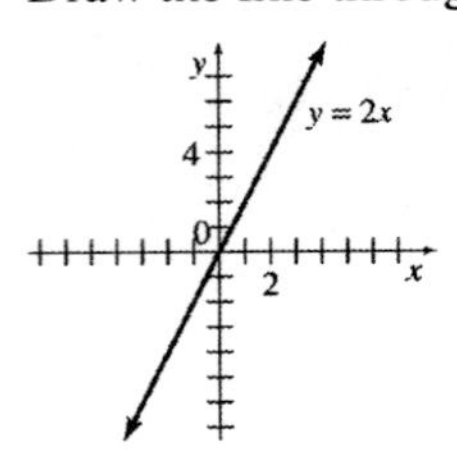

54. $x + 3y = 0$
When $x = 0$, $y = 0$.
When $x = 3$, $y = -1$.
Draw the line through (0, 0) and $(3, -1)$.

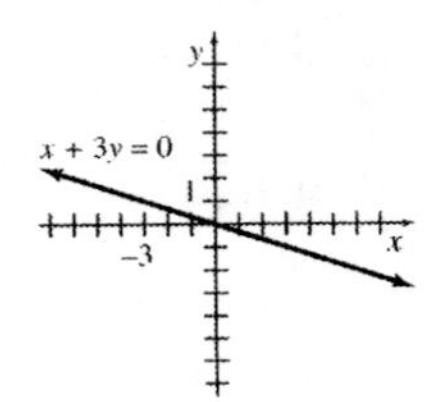

55. $y = (2x-1)(x+1) = 2x^2 + x - 1$

x	-3	-2	-1	0	1	2	3
y	14	5	0	-1	2	9	20

Pairs: $(-3, 14), (-2, 5), (-1, 0), (0, -1), (1, 2), (2, 9), (3, 20)$

Range: $\{-1, 0, 2, 5, 9, 14, 20\}$

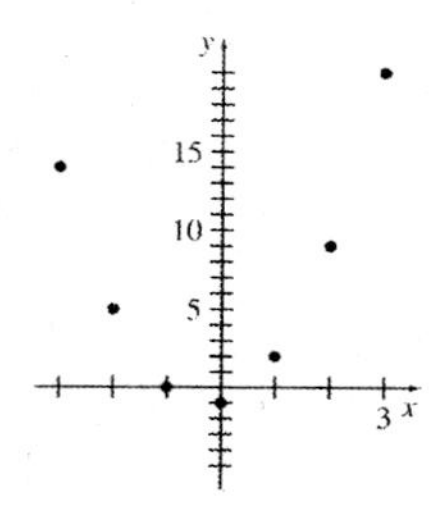

56. $y = \dfrac{x}{x^2+1}$

x	-3	-2	-1	0	1	2	3
y	$-\frac{3}{10}$	$-\frac{2}{5}$	$-\frac{1}{2}$	0	$\frac{1}{2}$	$\frac{2}{5}$	$\frac{3}{10}$

Pairs: $\left(-3, -\dfrac{3}{10}\right), \left(-2, -\dfrac{2}{5}\right), \left(-1, -\dfrac{1}{2}\right),$
$(0,0), \left(1, \dfrac{1}{2}\right), \left(2, \dfrac{2}{5}\right), \left(3, \dfrac{3}{10}\right)$

Range: $\left\{-\frac{1}{2}, -\frac{2}{5}, -\frac{3}{10}, 0, \frac{3}{10}, \frac{2}{5}, \frac{1}{2}\right\}$

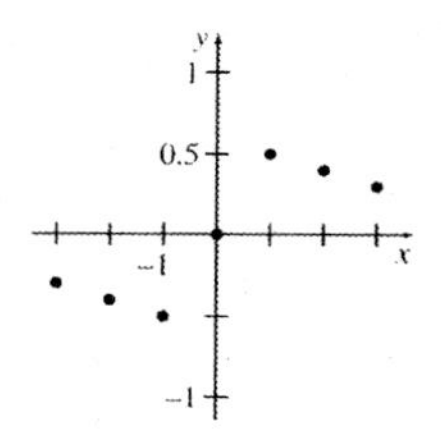

57. $f(x) = 5x^2 - 3$ and $g(x) = -x^2 + 4x + 1$

(a) $f(-2) = 5(-2)^2 - 3 = 17$

(b) $g(3) = -(3)^2 + 4(3) + 1 = 4$

(c) $f(-k) = 5(-k)^2 - 3 = 5k^2 - 3$

(d) $g(3m) = -(3m)^2 + 4(3m) + 1$
$= -9m^2 + 12m + 1$

(e) $f(x+h) = 5(x+h)^2 - 3$
$= 5(x^2 + 2xh + h^2) - 3$
$= 5x^2 + 10xh + 5h^2 - 3$

(f) $g(x+h) = -(x+h)^2 + 4(x+h) + 1$
$= -(x^2 + 2xh + h^2) + 4x + 4h + 1$
$= -x^2 - 2xh - h^2 + 4x + 4h + 1$

(g) $\dfrac{f(x+h) - f(x)}{h}$

$= \dfrac{5(x+h)^2 - 3 - (5x^2 - 3)}{h}$

$= \dfrac{5(x^2 + 2hx + h^2) - 3 - 5x^2 + 3}{h}$

$= \dfrac{5x^2 + 10hx + 5h^2 - 5x^2}{h}$

$= \dfrac{10hx + 5h^2}{h} = 10x + 5h$

(h) $\dfrac{g(x+h) - g(x)}{h}$

$= \dfrac{-(x+h)^2 + 4(x+h) + 1 - (-x^2 + 4x + 1)}{h}$

$= \dfrac{-(x^2 + 2xh + h^2) + 4x + 4h + 1 + x^2 - 4x - 1}{h}$

$= \dfrac{-x^2 - 2xh - h^2 + 4h + x^2}{h}$

$= \dfrac{-2xh - h^2 + 4h}{h}$

$= -2x - h + 4$

58. $f(x) = 2x^2 + 5$ and $g(x) = 3x^2 + 4x - 1$

(a) $f(-3) = 2(-3)^2 + 5 = 23$

(b) $g(2) = 3(2)^2 + 4(2) - 1 = 19$

(c) $f(3m) = 2(3m)^2 + 5 = 18m^2 + 5$

(d) $g(-k) = 3(-k)^2 + 4(-k) - 1$
$= 3k^2 - 4k - 1$

(e) $f(x+h) = 2(x+h)^2 + 5$
$= 2(x^2 + 2xh + h^2) + 5$
$= 2x^2 + 4xh + 2h^2 + 5$

(f) $g(x+h) = 3(x+h)^2 + 4(x+h) - 1$
$= 3(x^2 + 2xh + h^2) + 4x + 4h - 1$
$= 3x^2 + 6xh + 3h^2 + 4x + 4h - 1$

(g) $\frac{f(x+h)-f(x)}{h}$

$$= \frac{2(x+h)^2+5-(2x^2+5)}{h}$$
$$= \frac{2(x^2+2xh+h^2)+5-2x^2-5}{h}$$
$$= \frac{2x^2+4xh+2h^2+5-2x^2+5}{h}$$
$$= \frac{4xh+2h^2}{h}$$
$$= 4x+2h$$

(h) $\frac{g(x+h)-g(x)}{h}$

$$= \frac{3(x+h)^2+4(x+h)-1-(3x^2+4x-1)}{h}$$
$$= \frac{3(x^2+2xh+h^2)+4x+4h-1-3x^2-4x+1}{h}$$
$$= \frac{3x^2+6xh+3h^2+4x+4h-1-3x^2-4x+1}{h}$$
$$= \frac{6xh+3h^2+4h}{h}$$
$$= 6x+3h+4$$

59. $y = \frac{3x-4}{x} \Rightarrow x \neq 0$

Domain: $(-\infty, 0) \cup (0, \infty)$

60. $y = \frac{\sqrt{x-2}}{2x+3}$

$x-2 \geq 0$ and $2x+3 \neq 0$

$x \geq 2$ $\quad 2x \neq -3$

$x \neq -\frac{3}{2}$

Domain: $[2, \infty)$

61. $y = 2x^2+3x-1$

The graph is a parabola.

Let $y = 0$.

$$0 = 2x^2+3x-1$$
$$x = \frac{-3 \pm \sqrt{3^2-4(2)(-1)}}{2(2)}$$
$$= \frac{-3 \pm \sqrt{9+8}}{4} = \frac{-3 \pm \sqrt{17}}{4}$$

The x-intercepts are $\frac{-3+\sqrt{17}}{4} \approx 0.28$ and $\frac{-3-\sqrt{17}}{4} \approx -1.48$.

Let $x = 0$. Then, $y = 2(0)^2+3(0)-1$

The y-intercept is -1.

Vertex:

$$x = \frac{-b}{2a} = \frac{-3}{2(2)} = -\frac{3}{4}$$
$$y = 2\left(-\frac{3}{4}\right)^2 + 3\left(-\frac{3}{4}\right) - 1$$
$$= \frac{9}{8} - \frac{9}{4} - 1 = -\frac{17}{8}$$

The vertex is $\left(-\frac{3}{4}, -\frac{17}{8}\right)$.

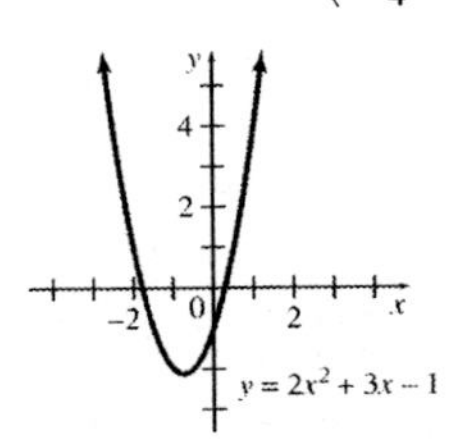

62. $y = -\frac{1}{4}x^2+x+2$

The graph is a parabola.

Let $y = 0$. Then, $0 = -\frac{1}{4}x^2+x+2$

Multiply by 4 to eliminate the fraction, then solve for x using the quadratic formula.

$$0 = -x^2+4x+8$$
$$x = \frac{-4 \pm \sqrt{4^2-4(-1)(8)}}{2(-1)}$$
$$= \frac{-4 \pm \sqrt{48}}{-2} = 2 \pm 2\sqrt{3}$$

The x-intercepts are $2+2\sqrt{3} \approx 5.46$ and $2-2\sqrt{3} \approx -1.46$.

Let $x = 0$. Then $y = -\frac{1}{4}(0)^2+0+2$

The y-intercept is $y = 2$.

Vertex:

$$x = \frac{-b}{2a} = \frac{-1}{2\left(-\frac{1}{4}\right)} = 2$$
$$y = -\frac{1}{4}(2)^2+2+2 = -1+4 = 3$$

The vertex is $(2, 3)$.

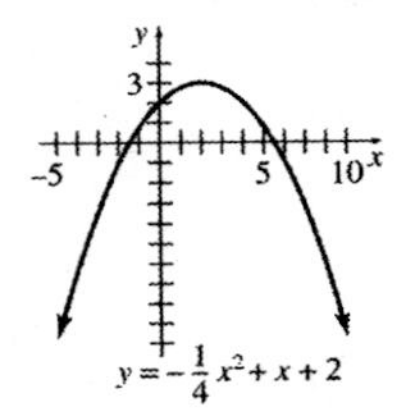

63. $y = -x^2 + 4x + 2$

Let $y = 0$.

$0 = -x^2 + 4x + 2$

$$x = \frac{-4 \pm \sqrt{4^2 - 4(-1)(2)}}{2(-1)} = \frac{-4 \pm \sqrt{24}}{-2}$$

$$= 2 \pm \sqrt{6}$$

The x-intercepts are $2 + \sqrt{6} \approx 4.45$ and $2 - \sqrt{6} \approx -0.45$.

Let $x = 0$. Then, $y = -0^2 + 4(0) + 2$

The y-intercept is 2.

Vertex:

$$x = \frac{-b}{2a} = \frac{-4}{2(-1)} = \frac{-4}{-2} = 2$$

$$y = -2^2 + 4(2) + 2 = 6$$

The vertex is (2, 6).

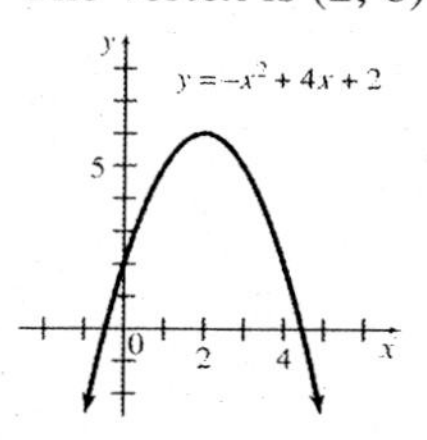

64. $y = 3x^2 - 9x + 2$

Let $y = 0$.

$0 = 3x^2 - 9x + 2$

$$x = \frac{-(-9) \pm \sqrt{(-9)^2 - 4(3)(2)}}{2(3)} = \frac{9 \pm \sqrt{57}}{6}$$

The x-intercepts are $\frac{9 + \sqrt{57}}{6} \approx 2.76$ and $\frac{9 - \sqrt{57}}{6} \approx 0.24$.

Let $x = 0$. Then $y = 3(0)^2 - 9(0) + 2 = 2$.

The y-intercept is 2.

Vertex: $x = \frac{-b}{2a} = \frac{-(-9)}{2(3)} = \frac{9}{6} = \frac{3}{2}$

$$y = 3\left(\frac{3}{2}\right)^2 - 9\left(\frac{3}{2}\right) + 2 = -\frac{19}{4}$$

The vertex is $\left(\frac{3}{2}, -\frac{19}{4}\right)$.

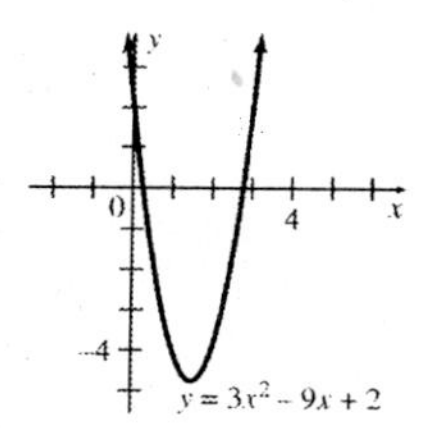

65. $f(x) = x^3 - 3$

Translate the graph of $f(x) = x^3$ 3 units down.

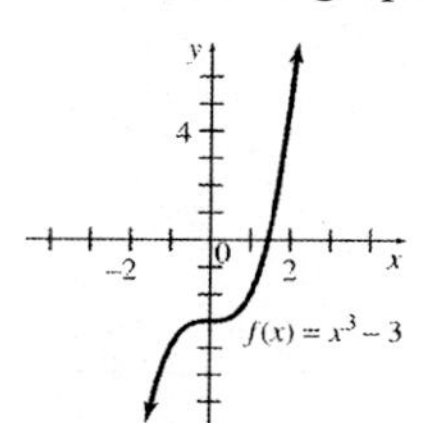

66. $f(x) = 1 - x^4 = -x^4 + 1$

Reflect the graph of $y = x^4$ vertically then translate 1 unit upward.

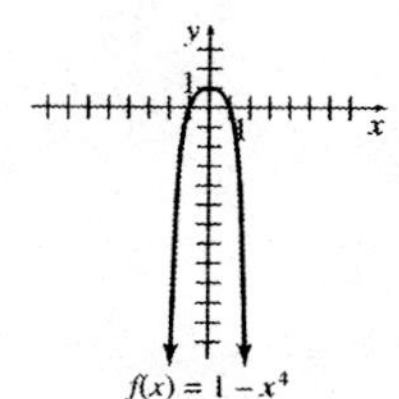

67. $y = -(x-1)^4 + 4$

Translate the graph of $y = x^4$ 1 unit to the right and reflect vertically. Translate 4 units upward.

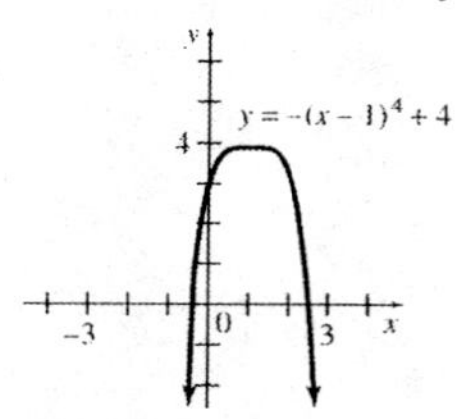

68. $y = -(x+2)^3 - 2$

Translate the graph of $y = x^3$ 2 units to the left and reflect vertically. Translate 2 units downward.

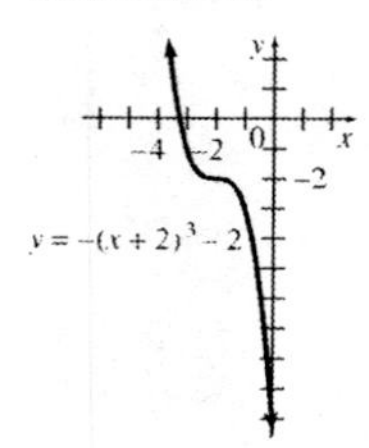

69. $f(x)=\frac{8}{x}$

Vertical asymptote: $x=0$

Horizontal asymptote:

$\frac{8}{x}$ approaches zero as x increases.

$y=0$ is an asymptote.

x	−4	−3	−2	−1	1	2	3	4
y	−2	−2.7	−4	−8	8	4	2.7	2

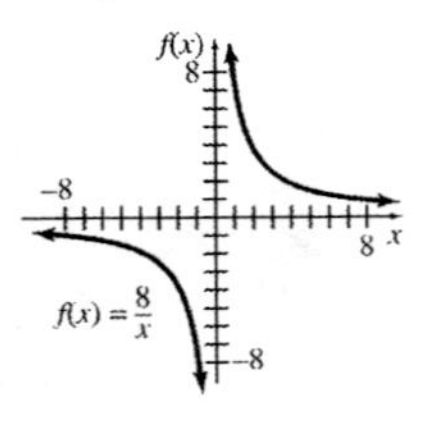

70. $f(x)=\frac{2}{3x-6}$

Vertical asymptote: $3x-6=0$ or $x=2$

Horizontal asymptote: $y=0$, since $\frac{2}{3x-6}$ approaches zero as x increases.

x	0	1	3	4	5
y	$-\frac{1}{3}$	$-\frac{2}{3}$	$\frac{2}{3}$	$\frac{1}{3}$	$\frac{2}{9}$

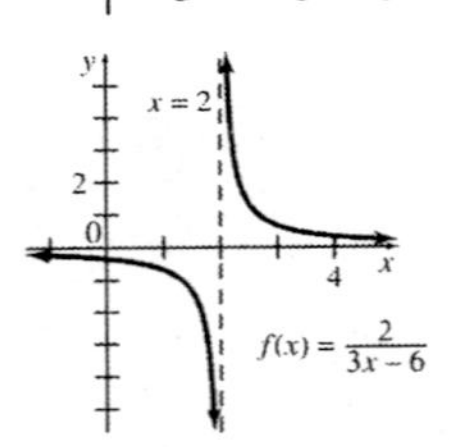

71. $f(x)=\frac{4x-2}{3x+1}$

Vertical asymptote: $3x+1=0\Rightarrow x=-\frac{1}{3}$

Horizontal asymptote: As x increases, $\frac{4x-2}{3x-1}\approx\frac{4x}{3x}=\frac{4}{3}$.

$y=\frac{4}{3}$ is an asymptote.

x	−3	−2	−1	0	1	2	3
y	1.75	2	3	−2	0.5	0.86	1

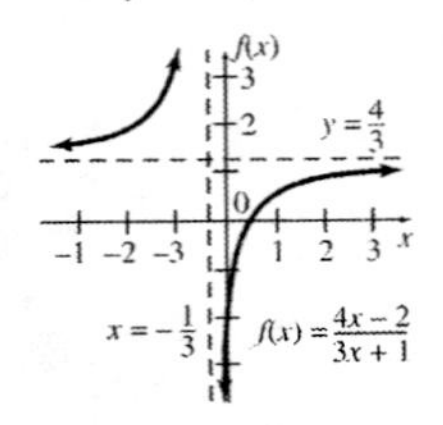

72. $f(x)=\frac{6x}{x+2}$

Vertical asymptote: $x=-2$

Horizontal asymptote: $y=6$

x	−5	−4	−3	−1	0	1	2
y	10	12	18	−6	0	2	3

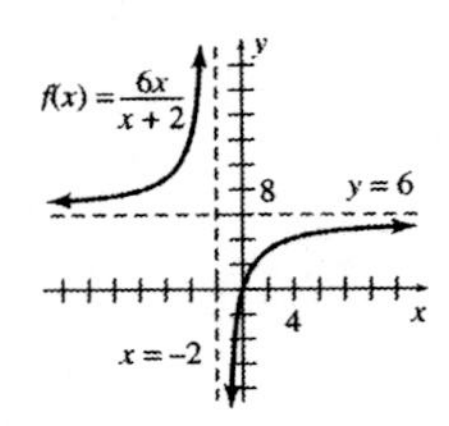

73. $f(x)=4x^2+3x,\quad g(x)=\frac{x}{x+1}$

$$\begin{aligned}(f\circ g)(x)&=f(g(x))=f\left(\frac{x}{x+1}\right)\\&=4\left(\frac{x}{x+1}\right)^2+3\left(\frac{x}{x+1}\right)\\&=\frac{4x^2}{(x+1)^2}+\frac{3x(x+1)}{(x+1)^2}\\&=\frac{4x^2+3x^2+3x}{(x+1)^2}=\frac{7x^2+3x}{(x+1)^2}\end{aligned}$$

$$\begin{aligned}(g\circ f)(x)&=g(f(x))=g\left(4x^2+3x\right)\\&=\frac{4x^2+3x}{\left(4x^2+3x\right)+1}=\frac{4x^2+3x}{4x^2+3x+1}\end{aligned}$$

74. $f(x)=3x+4,\quad g(x)=x^2-6x-7$

$$\begin{aligned}(f\circ g)(x)&=f(g(x))=f\left(x^2-6x-7\right)\\&=3\left(x^2-6x-7\right)+4\\&=3x^2-18x-21+4\\&=3x^2-18x-17\end{aligned}$$

$$\begin{aligned}(g\circ f)(x)&=g(f(x))=g(3x+4)\\&=(3x+4)^2-6(3x+4)-7\\&=9x^2+24x+16-18x-24-7\\&=9x^2+6x-15\end{aligned}$$

75. (a) The line includes the points (10, 2.74) and (20, 3.45). First find the slope.

$$m = \frac{3.45 - 2.74}{20 - 10} = 0.071$$

Now find the equation of the line using the point-slope form and the point (10, 2.74).

$$y = m(x - x_1) + y_1$$
$$y = 0.071(x - 10) + 2.74$$
$$y = 0.071x - 0.71 + 2.74$$
$$y = 0.071x + 2.03$$

(b) $f(14) = 0.071(14) + 2.03 = 3.024$

Based of the answer in part (a), there are 3.024 million employed nurses in 2014.

(c) The slope tells us that the number of employed nurses is increasing by 0.071 million, or 71,000, per year.

76. (a) Beef: The line includes the points (0, 64.5) and (20,56.7).

$$m = \frac{56.7 - 64.5}{20 - 0} = -0.39$$

We are given the y-intercept, so the equation is $b(t) = -0.39t + 64.5$.

Pork: The line includes the points (0, 47.8) and (20, 44.3).

$$m = \frac{44.3 - 47.8}{20 - 0} = -0.175$$

We are given the y-intercept, so the equation is $p(t) = -0.175t + 47.8$.

Chicken: The line includes the points (0, 42.4) and (20, 58.0).

$$m = \frac{58.0 - 42.4}{20 - 0} = 0.78$$

We are given the y-intercept, so the equation is $c(t) = 0.78t + 42.4$.

(b) The amount of beef consumed is decreasing at the rate of 0.39 pound per year.
The amount of pork consumed is decreasing at the rate of 0.175 pound per year.
The amount of chicken consumed is increasing at the rate of 0.78 pound per year.

(c) We are seeking the solution to the inequality $c(t) > p(t)$.

$$0.78t + 42.4 > -0.175t + 47.8$$
$$0.955t > 5.4$$
$$t > 5.7 \approx 6$$

The consumption of chicken surpassed the consumption of pork during 1996.

(d) The year 2015 is represented by $t = 25$.

$$b(25) = -0.39(25) + 64.5 = 54.75$$
$$p(t) = -0.175(25) + 47.8 = 43.425$$
$$c(t) = 0.78(25) + 42.4 = 61.9$$

If this trend were to continue, in 2015, the consumption of beef will be 54.75 pounds, pork, 43.425 pounds, and chicken, 61.9 pounds.

77. (a) $y = \frac{x - 1.1}{0.124} = \frac{1}{0.124}x - \frac{1.1}{0.124}$

$$\approx 8.06x - \frac{1.1}{0.124}$$

The slope is about 8.06. This means each additional local species per 0.5 square meter leads to an additional 8.06 regional species per 0.5 hectare.

(b) If $x = 6$, then $y = \frac{6 - 1.1}{0.124} \approx 39.5$. So, if the average local diversity is 6 species per 0.5 square meter, then the regional diversity must be 39.5 species per 0.5 hectare.

(c) If $y = 70$, then

$$70 = \frac{x - 1.1}{0.124} \Rightarrow 8.68 = x - 1.1 \Rightarrow 9.78 = x$$

A regional diversity of 70 species per 0.5 hectare means that there is a local diversity of 9.78 species per 0.5 meter.

(d) A coefficient of correlation of 0.82 means that as the average local diversity increases, so does the average regional diversity. The data points are close to the line.

78. (a) Using a graphing calculator, we find that the correlation coefficient is 0.8664. Yes, the data seem to fit a straight line.

(b)

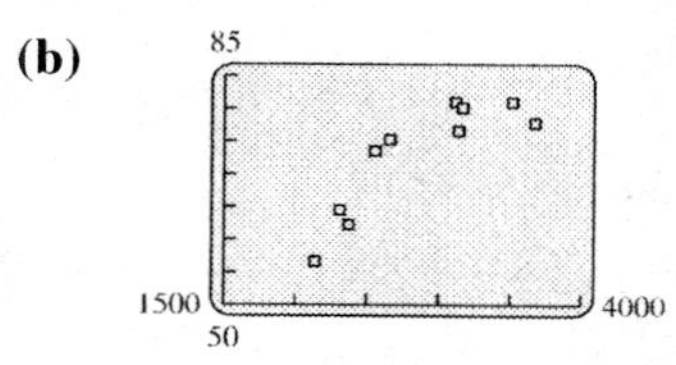

The data somewhat fit a straight line, but a curve would fit the data better.

(c) Using a graphing calculator, $Y = 0.01423x + 32.19$.

(d) Let $x = 3399$.

$Y = 0.01423(3399) + 32.19 \approx 80.56$

The predicted life expectancy in Canada, with a daily calorie supply of 3399, is about 80.56 years. This agrees with the actual value of 80.5 years.

(e) Answers will vary. The higher daily calorie supply most likely contains more healthy nutrients, which might result in a longer life expectancy. However, an American who eats 5000 calories a day, unless extremely active, would probably become obese.

(f) Answers will vary.

79. (a) $\sum x = 1394, \sum y = 1607, \sum xy = 291,990,$

$\sum x^2 = 255,214, \sum y^2 = 336,155$

$$m = \frac{n(\sum xy) - (\sum x)(\sum y)}{n(\sum x^2) - (\sum x)^2}$$

$$m = \frac{8(291,990) - (1394)(1607)}{8(225,214) - 1394^2}$$

$m = 0.9724399854 \approx 0.9724$

$$b = \frac{\sum y - m(\sum x)}{n}$$

$$b = \frac{1607 - 0.9724(1394)}{8} \approx 31.43$$

$Y = 0.9724x + 31.43$

(b) Let $x = 190$; find Y.

$Y = 0.9724(190) + 31.43 \approx 216.19 \approx 216$

The cholesterol level for a person whose blood sugar level is 190 would be about 216.

(c)
$$r = \frac{8(291,990) - (1394)(1607)}{\sqrt{8(255,214) - 1394^2} \cdot \sqrt{8(336,155) - 1607^2}}$$

$= 0.933814 \approx 0.93$

80. (a) Since $a = -1.5 < 0$, the vertex is the maximum. The vertex occurs at

$$t = -\frac{b}{2a} = -\frac{72}{2(-1.5)} = 24.$$

Therefore, the intake reaches a maximum at $t = 24$ minutes.

(b) The maximum intake is

$I = 27 + 72(24) - 1.5(24)^2 = 891$ grams.

(c) Since t represents time, it follows that t is nonnegative. Therefore, $t \geq 0$. Since the function gives cumulative intake, the values of I must get larger at t gets larger. The graph is a parabola that opens downward, so I increases as t increases from 0 to 24 and begins to decrease as t increases past 24. Therefore, the domain is [0, 24].

81. Since $a = -\frac{2}{3} < 0$, the vertex is the maximum.

The vertex occurs at

$$t = -\frac{b}{2a} = -\frac{\frac{14}{3}}{2\left(-\frac{2}{3}\right)} = \frac{7}{2} = 3.5.$$

Therefore, the fever reaches a maximum at day 3.5, which is on day 3.
The maximum fever is

$$F\left(\frac{7}{2}\right) = -\frac{2}{3}\left(\frac{7}{2}\right)^2 + \frac{14}{3}\left(\frac{7}{2}\right) + 96 \approx 104.2°.$$

82. (a) (i) Set the two formulas equal and solve for S.

$1486S^2 - 4106S + 4514 = 1486S - 825$

$1486S^2 - 5592S + 5339 = 0$

The discriminant of the function is

$b^2 - 4ac = (-5592)^2 - 4(1486)(5339)$
$= -464,552$

The discriminant is negative, so this quadratic equation has no real solution. Therefore, there is no value for S that will give the same value for RCV in the two formulas.

(ii) This result can also be shown by graphing the two functions and noting that the graphs do not intersect.

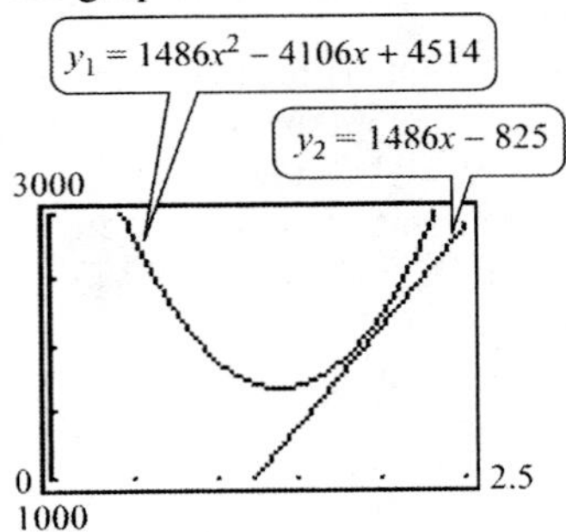

(b) Set the two formulas equal and solve for S.

$$1278S^{1.289} = 1395S$$
$$1278S^{1.289} - 1395S = 0$$
$$9S\left(142S^{0.289} - 155\right) = 0$$

$$9S = 0 \qquad S = 0$$

$$142S^{0.289} - 155 = 0$$
$$142S^{0.289} = 155$$
$$S^{0.289} = \frac{155}{142}$$
$$S = \left(\frac{155}{142}\right)^{1/0.289} \approx 1.35$$

The two formulas give the same answer when $S = 0$ square meters or $S \approx 1.35$ square meters.

$$PV|_{S=0} = 1395(0) = 0 = 1278(0)^{1.289}$$

$$PV|_{S=1.35} = 1395\left(\left(\frac{155}{142}\right)^{1/0.289}\right) \approx 1889$$
$$= 1278\left(\left(\frac{155}{142}\right)^{1/0.289}\right)^{1.289}$$

The predicted plasma volumes are $PV = 0$ ml when $S = 0$ and $PV \approx 1889$ ml when $S \approx 1.35$.

83. (a) $100\% - 87.5\% = 12.5\%$

$12.5\% = 0.125 = \frac{125}{1000} = \frac{1}{8}$

The amount of radiation let in is 1 over the SPF rating.

(b)

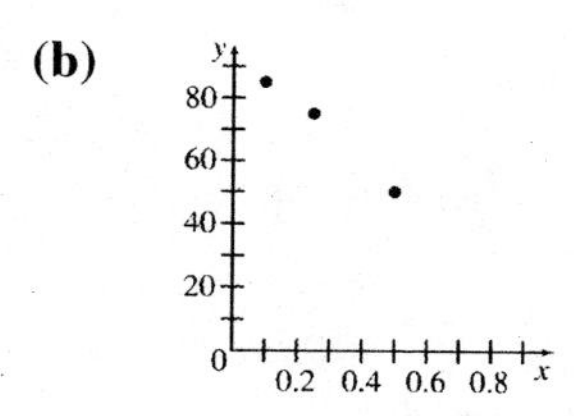

(c) $\text{UVB} = 1 - \frac{1}{\text{SPF}}$

(d) $1 - \frac{1}{8} = 87.5\%$

$1 - \frac{1}{4} = 75.0\%$

The increase is 12.5%.

(e) $1 - \frac{1}{30} = 96.\overline{6}\%$

$1 - \frac{1}{15} = 93.\overline{3}\%$

The increase is $3.\overline{3}\%$ or about 3.3%.

(f) The increase in percent protection decreases to zero.

84. $p(t) = \frac{1.79 \cdot 10^{11}}{(2026.87 - t)^{0.99}}$

(a) $p(2010) \approx 10.915$ billion

This is about 4.006 billion more than the estimate of 6.909 billion.

(b) $p(2020) \approx 26.56$ billion

$p(2025) \approx 96.32$ billion

(c) Answers will vary.

85. (a) $y = \frac{7(80)}{100 - 80} = 28$

The cost to remove 80% of a pollutant is \$28,000.

(b) $y = \frac{7(50)}{100 - 50} = 7$

The cost to remove 50% of a pollutant is \$7000.

(c) $y = \frac{7(90)}{100 - 90} = 63$

The cost to remove 90% of a pollutant is \$63,000.

(d)

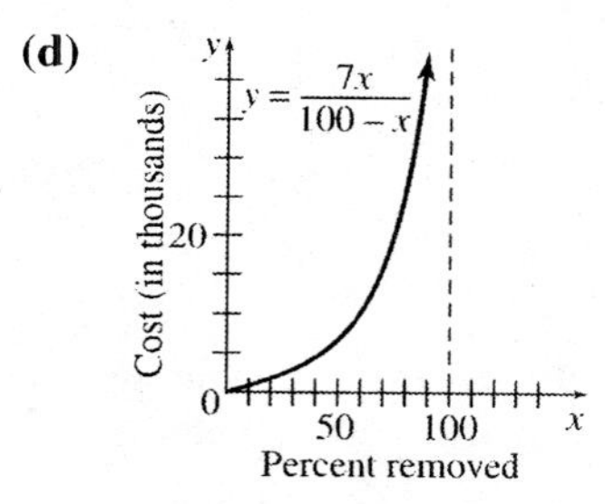

(e) No, all of the pollutant cannot be removed. As the function approaches 100, the denominator approaches 0.

86. (a)

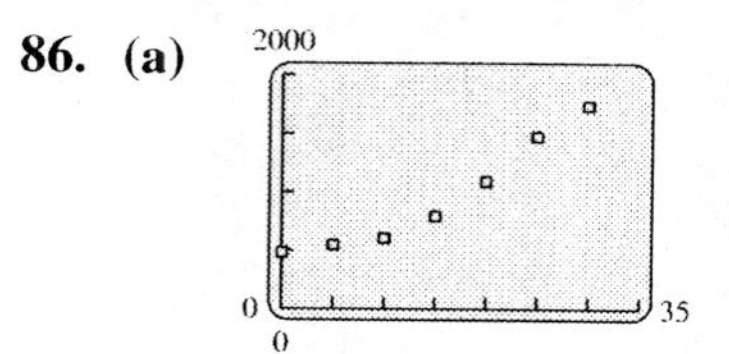

(b) Using a graphing calculator, the linear function is $y = 43.11t + 319.5$. The quadratic function is

$y = 1.35t^2 + 2.607t + 488.3$. The cubic function is

$y = -0.0362t^3 + 2.98t^2 - 15.5t + 515.5$.

(continued on next page)

(continued)

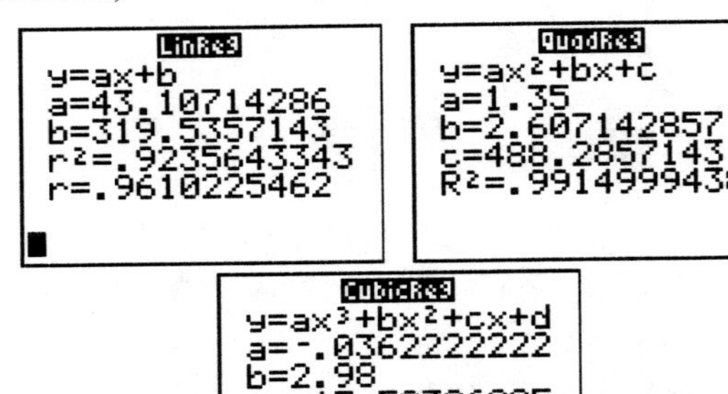

CubicReg
y=ax³+bx²+cx+d
a=-.0362222222
b=2.98
c=-15.50396825
d=515.452381
R²=.9946440311

(c) $y = 1.35t^2 - 2.607t + 488.3$; $y = -0.0362t^3 + 2.98t^2 - 15.5t + 515.5$; $y = 43.11t + 319.5$ (window: 0 to 35, 0 to 2000)

Both the quadratic function and the cubic function fit the data well.

(d) The year 2015 is represented by $t = 35$. Using the results stored in a graphing calculator, we have:

Linear: $f(35) \approx 1{,}828{,}000$ cases

Quadratic: $f(35) \approx 2{,}233{,}000$ cases

Cubic: $f(35) \approx 2{,}070{,}000$ cases

87. **(a)** $f(0) = 7.95(0)^2 - 8.85(0) + 447.9 = 447.9$

$f(8) = 7.95(8)^2 - 8.85(8) + 447.9 = 885.9$

From 2000 to 2008, the amount of organic farmland for grains has increased from 447,900 acres to 885,000 acres.

(b) $g(t) = 7.95(t - 2000)^2 - 8.85(t - 2000) + 447.9$

88. Using the points (5, 55) and (22, 77.3),

$$m = \frac{77.3 - 55}{22 - 5} \approx 1.31$$

$$y - 55 = 1.31(t - 5)$$
$$y - 55 = 1.31t - 6.55$$
$$y = 1.31t + 48.45$$

89. **(a)** Use the points (0, 6400) and (10, 9400) to find the slope.

$$m = \frac{9400 - 6400}{10 - 0} = 300$$

$$b = 6400$$

The linear equation for the average number of families below poverty level since 2000 is $y = 300t + 6400$.

(b) Use the points (5, 7657) and (10, 9400) to find the slope.

$$m = \frac{9400 - 7657}{10 - 5} = 348.6$$

$$y - 9400 = 348.6(t - 10)$$
$$y - 9400 = 348.6t - 3486$$
$$y = 348.6t + 5914$$

The linear equation for the average number of families below poverty level since 2000 is $y = 348.6t + 5914$.

(c) Using a graphing calculator, the least squares line is $Y = 237.9t + 6553.2$.

(d) Using a graphing calculator, the correlation coefficient is 0.9434. This indicates that the least squares line describes the data well.

(e)

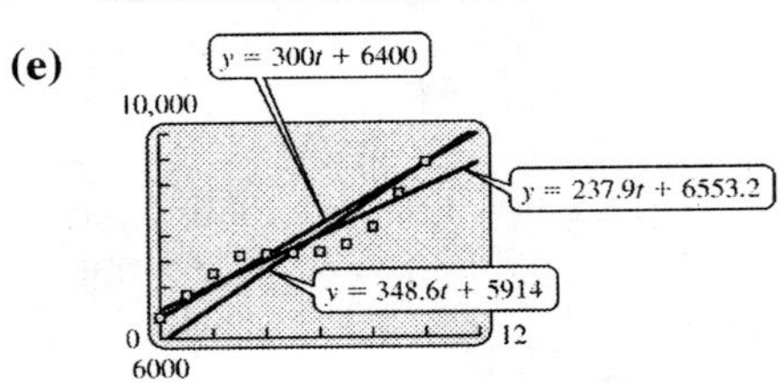

The least squares line seems to fit the data the best of the three linear functions.

(f) $y = 10.11t^3 - 141.87t^2 + 718.26t + 6336.28$

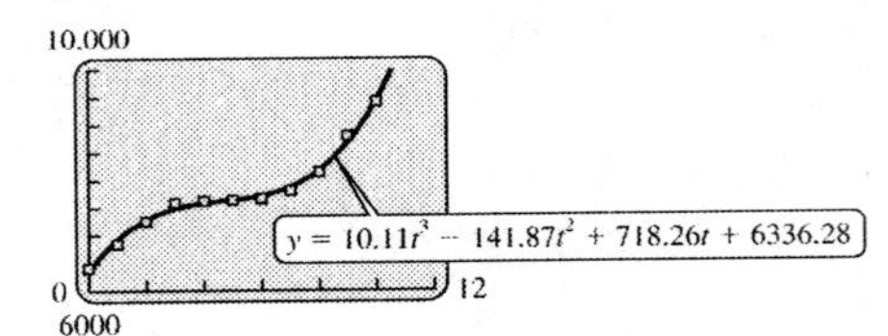

A cubic model appears to be more accurate than a linear model.

90. **(a)** Using a graphing calculator, $r \approx 0.7485$. The data seem to fit a line but the fit is not very good.

(b)

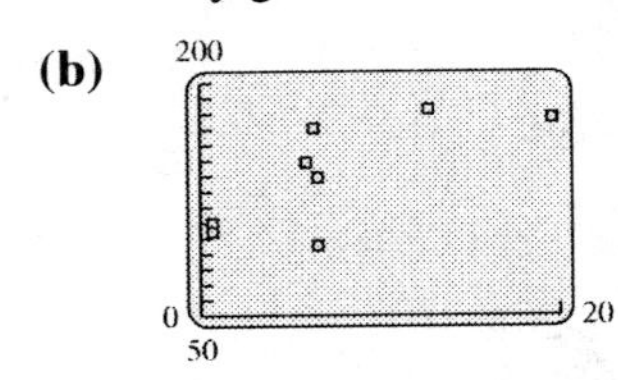

(c) Using a graphing calculator, $Y = 4.179x + 110.6$

(d) The slope is 4.179 thousand (or 4179). On average, the governor's salary increases \$4179 for each additional million in population.

(e) Answers will vary.

(f) Answers will vary.

91. (a) The speed in one direction is $v + w$ and the speed in the other direction is $v - w$, so the time in one direction is $\frac{d}{v+w}$ and the time in the other direction is $\frac{d}{v-w}$. Thus, the total time to make the round trip is $\frac{d}{v+w}+\frac{d}{v-w}$.

(b) The average speed is the total distance divided by the total time. So,

$$v_{aver} = \frac{2d}{\frac{d}{v+w}+\frac{d}{v-w}}.$$

(c)
$$\frac{2d}{\frac{d}{v+w}+\frac{d}{v-w}} = \frac{2d}{\frac{d}{v+w}+\frac{d}{v-w}}\cdot\frac{(v+w)(v-w)}{(v+w)(v-w)} = \frac{2d(v^2-w^2)}{d(v-w)+d(v+w)}$$
$$= \frac{2d(v^2-w^2)}{dv-dw+dv+dw} = \frac{2d(v^2-w^2)}{2dv} = \frac{v^2-w^2}{v} = v-\frac{w^2}{v}$$

(d) From parts (b) and (c), we have

$$v_{aver} = v-\frac{w^2}{v}.$$

The greatest average speed occurs when the wind speed is 0.

92. (a) $x = 0.9$ means that the speed is 10% slower on the return trip.
$x = 1.1$ means that the speed is 10% faster on the return trip.

(b) The speed in one direction is v and the speed in the other direction is xv, so the time in one direction is $\frac{d}{v}$ and the time in the other direction is $\frac{d}{xv}$. Thus, the total time to make the round trip is $\frac{d}{v}+\frac{d}{xv}$.
The average speed is the total distance divided by the total time. So,

$$v_{aver} = \frac{2d}{\frac{d}{v}+\frac{d}{xv}}.$$

$$v_{aver} = \frac{2d}{\frac{d}{v}+\frac{d}{xv}} = \frac{2d}{\frac{d}{v}+\frac{d}{xv}}\cdot\frac{xv}{xv} = \frac{2dxv}{dx+d}$$
$$= \frac{2dxv}{d(x+1)} = \frac{2xv}{x+1} = \left(\frac{2x}{x+1}\right)v$$

(c) The formula for v_{aver} is a rational function with a horizontal asymptote at $v_{aver} = 2v$. This means that as the return velocity becomes greater and greater, the average velocity approaches twice the velocity on the first part of the trip, and can never exceed twice that velocity.

93. (a)
$$P = kD^1$$
$$164.8 = k(30.1) \Rightarrow k = \frac{164.8}{30.1} \approx 5.48$$
For $n = 1$, $P = 5.48D$.
$$P = kD^{1.5}$$
$$164.8 = k(30.1)^{1.5} \Rightarrow k = \frac{164.8}{(30.1)^{1.5}} \approx 1.00$$
For $n = 1.5$, $P = 1.00D^{1.5}$.
$$P = kD^2$$
$$164.8 = k(30.1)^2$$
$$k = \frac{164.8}{(30.1)^2} \approx 0.182$$
For $n = 2$, $P = 0.182D^2$.

(b)

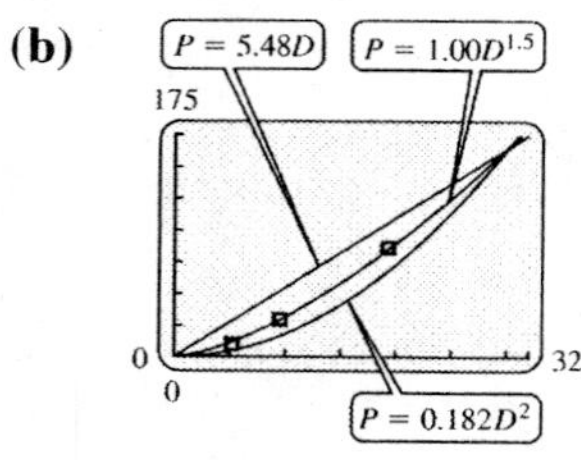

$P = 1.00D^{1.5}$ appears to be the best fit.

(c) $P = 1.00(39.5)^{1.5} \approx 248.3$ years

(d) We obtain $P = 1.00D^{1.5}$. This is the same as the function found in part (b).

Chapter 2

EXPONENTIAL, LOGARITHMIC, AND TRIGONOMETRIC FUNCTIONS

2.1 Exponential Functions

1.

number of folds	1	2	3	4	5 ...	10 ...	50
layers of paper	2	4	8	16	32 ...	1024 ...	2^{50}

$2^{50} = 1.125899907 \times 10^{15}$

3. The graph of $y = 3^x$ is the graph of an exponential function $y = a^x$ with $a > 1$. This is graph **E**.

5. The graph of $y = \left(\frac{1}{3}\right)^{1-x}$ is the graph of $y = \left(3^{-1}\right)^{1-x}$ or $y = 3^{x-1}$. This is the graph of $y = 3^x$ translated 1 unit to the right. This is graph **C**.

7. The graph of $y = 3(3)^x$ is the same as the graph of $y = 3^{x+1}$. This is the graph of $y = 3^x$ translated 1 unit to the left. This is graph **F**.

9. The graph of $y = 2 - 3^{-x}$ is the same as the graph of $y = -3^{-x} + 2$. This is the graph of $y = 3^x$ reflected in the x-axis, reflected across the y-axis, and translated up 2 units. This is graph **A**.

11. The graph of $y = 3^{x-1}$ is the graph of $y = 3^x$ translated 1 unit to the right. This is graph **C**.

13.
$$2^x = 32$$
$$2^x = 2^5$$
$$x = 5$$

15.
$$3^x = \frac{1}{81}$$
$$3^x = \frac{1}{3^4}$$
$$3^x = 3^{-4}$$
$$x = -4$$

17.
$$4^x = 8^{x+1}$$
$$(2^2)^x = (2^3)^{x+1}$$
$$2^{2x} = 2^{3x+3}$$
$$2x = 3x + 3$$
$$-x = 3$$
$$x = -3$$

19.
$$16^{x+3} = 64^{2x-5}$$
$$(2^4)^{x+3} = (2^6)^{2x-5}$$
$$2^{4x+12} = 2^{12x-30}$$
$$4x + 12 = 12x - 30$$
$$42 = 8x$$
$$\frac{21}{4} = x$$

21.
$$e^{-x} = (e^4)^{x+3}$$
$$e^{-x} = e^{4x+12}$$
$$-x = 4x + 12$$
$$-5x = 12$$
$$x = -\frac{12}{5}$$

23.
$$5^{-|x|} = \frac{1}{25}$$
$$5^{-|x|} = 5^{-2}$$
$$|x| = 2$$
$$x = 2 \text{ or } x = -2$$

25.
$$5^{x^2+x} = 1$$
$$5^{x^2+x} = 5^0$$
$$x^2 + x = 0$$
$$x(x+1) = 0$$
$$x = 0 \quad \text{or} \quad x + 1 = 0$$
$$x = 0 \quad \text{or} \quad x = -1$$

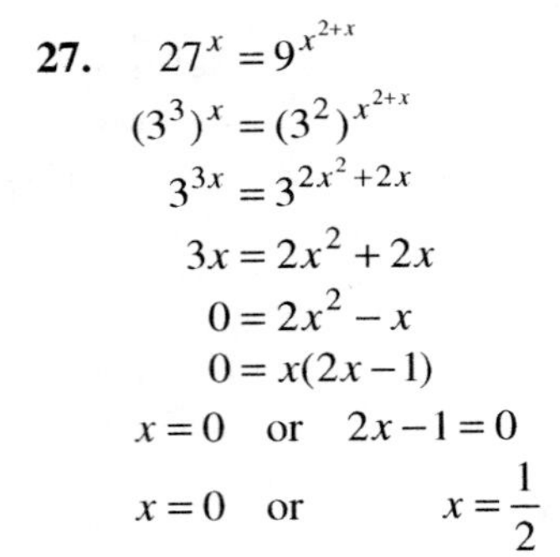

27.
$$27^x = 9^{x^{2+x}}$$
$$(3^3)^x = (3^2)^{x^{2+x}}$$
$$3^{3x} = 3^{2x^2+2x}$$
$$3x = 2x^2 + 2x$$
$$0 = 2x^2 - x$$
$$0 = x(2x-1)$$
$$x = 0 \quad \text{or} \quad 2x - 1 = 0$$
$$x = 0 \quad \text{or} \quad x = \frac{1}{2}$$

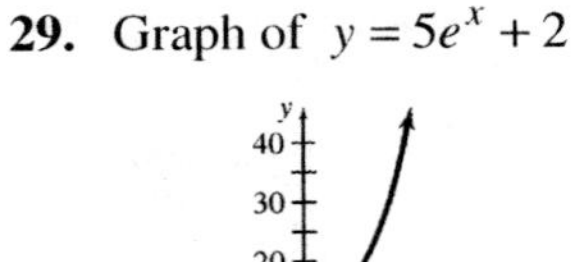

29. Graph of $y = 5e^x + 2$

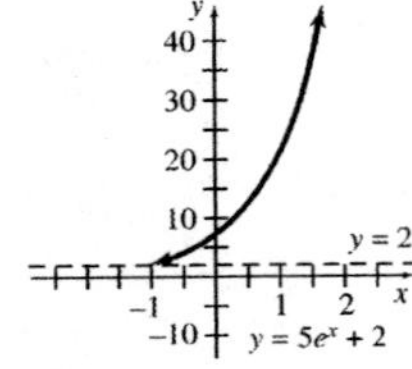

31. Graph of $y = -3e^{-2x} + 2$

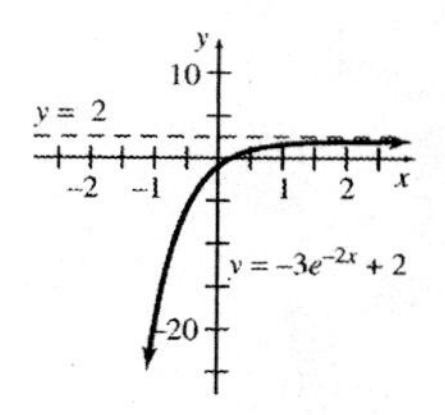

33. Answers will vary.

35. When $x > 0$, $3^x > e^x > 2^x$ because the functions are increasing. When $x < 0$,

$$3^x < e^x < 2^x \Rightarrow \left(\frac{1}{3}\right)^{-x} < \left(\frac{1}{e}\right)^{-x} < \left(\frac{1}{2}\right)^{-x}, \text{ and}$$

the functions are decreasing. The graphs below illustrate this.

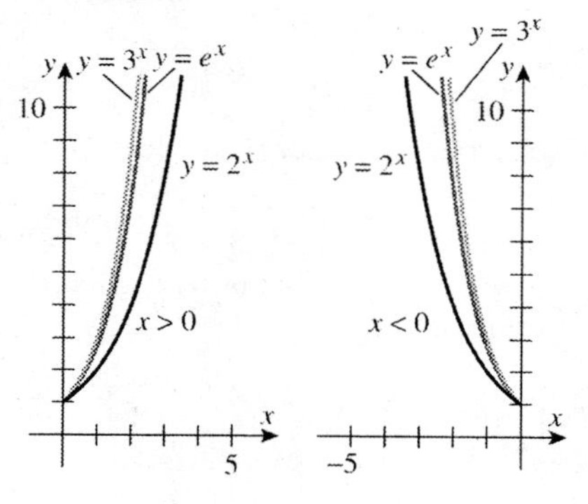

37. $A(t) = 3100e^{0.0166t}$

(a) 1970: $t = 10$

$$A(10) = 3100e^{(0.0166)(10)}$$
$$= 3100e^{0.166}$$
$$\approx 3659.78$$

The function gives a population of about 3660 million in 1970. This is very close to the actual population of about 3686 million.

(b) 2000: $t = 40$

$$A(40) = 3100e^{0.0166(40)} = 3100e^{0.664}$$
$$\approx 6021.90$$

The function gives a population of about 6022 million in 2000.

(c) 2015: $t = 55$

$$A(55) = 3100e^{0.0166(55)} = 3100e^{0.913}$$
$$= 7724.54$$

From the function, we estimate that the world population in 2015 will be 7725 million.

39. (a) Hispanic population:

$$h(t) = 37.79(1.021)^t$$
$$h(5) = 37.79(1.021)^5 \approx 41.93$$

The projected Hispanic population in 2005 is 41.93 million, which is slightly less than the actual value of 42.69 million.

(b) Asian population:

$$h(t) = 11.14(1.023)^t$$
$$h(5) = 11.14(1.023)^t \approx 12.48$$

The projected Asian population in 2005 is 12.48 million, which is very close to the actual value of 12.69 million.

(c) Annual Hispanic percent increase:
$1.021 - 1 = 0.021 = 2.1\%$
Annual Asian percent increase:
$1.023 - 1 = 0.023 = 2.3\%$
The Asian population is growing at a slightly faster rate.

(d) Black population:

$$b(t) = 0.5116t + 35.43$$
$$b(5) = 0.5116(5) + 35.43$$
$$\approx 37.99$$

The projected Black population in 2005 is 37.99 million, which is extremely close to the actual value of 37.91 million.

(e) Hispanic population:
Double the actual 2005 value is $2(42.69) = 85.38$ million.

$h(t) = 37.79(1.021)^t$
125
0
35
50
$y = 85.38$

The doubling point is reached when $t \approx 39$, or in the year 2039.
Asian population:
Double the actual 2005 value is $2(12.69) = 25.38$ million.

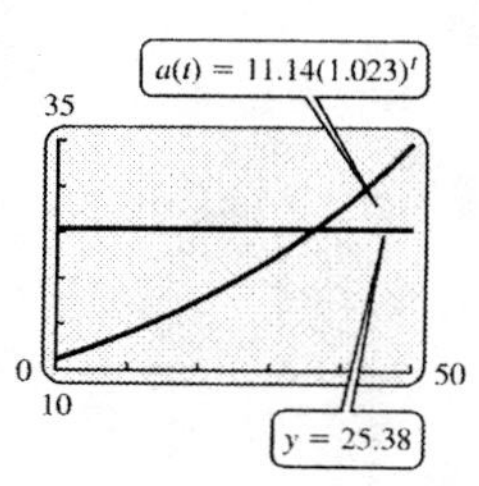

(continued on next page)

(continued)

The doubling point is reached when $t \approx 36$, or in the year 2036.

Black population:

Double the actual 2005 value is $2(37.91) = 75.82$ million.

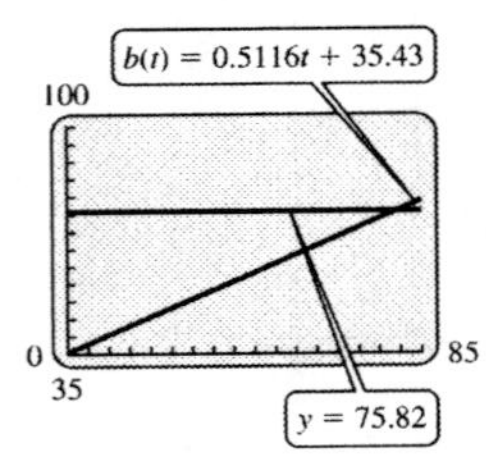

The doubling point is reached when $t \approx 79$, or in the year 2079.

41. (a)

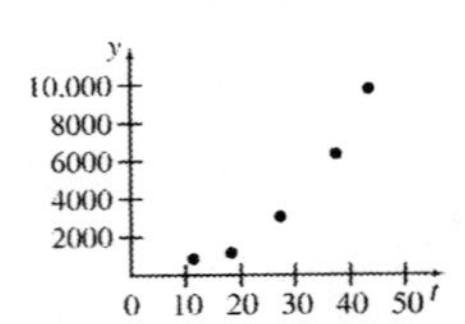

The data appear to fit an exponential curve.

(b) The year 1963 corresponds to $t = 0$ and the year 2006 corresponds to $t = 43$.

$$f(t) = f_0 a^t$$

$$f(0) = f_0 a^0 \qquad f(43) = f_0 a^{43}$$

$$487 = f_0 \qquad 9789 = f_0 a^{43}$$

$$f_0 = \frac{9789}{a^{43}}$$

The function passes through the points (0, 487) and $(43, 9789)$. Now solve for a.

$$487 = \frac{9789}{a^{43}}$$

$$a^{43} = \frac{9789}{487}$$

$$a = \left(\frac{9789}{487}\right)^{1/43} \Rightarrow a \approx 1.0723$$

Thus, $f(t) = 487(1.0723)^t$.

(c) From part (b), we have $a \approx 1.0723$, so the number of breeding pairs is about 1.0723 times the number of breeding pairs in the previous year. Therefore, the average annual percentage increase is about 7.2%.

(d) Using a TI-84 Plus, the exponential regression is $f(t) = 398.8(1.0762)^t$.

43. (a)

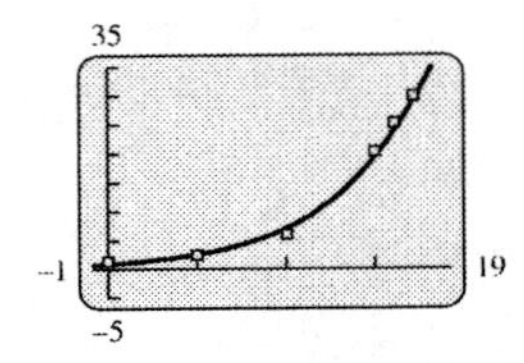

The function fits the data closely.

(b) $f(3) = 0.8454e^{0.2081(3)} \approx 1.578$

The risk for a man with a score of 3 is about 1.6%.

(c) $f(11) = 0.8454e^{0.2081(11)} \approx 8.341$

The risk increases about 6.8%

(d) $f(6) = 0.8454e^{0.2081(6)} \approx 2.947$

The risk for a man with a score of 6 is about 2.9%.

$f(14) = 0.8454e^{0.2081(14)} \approx 15.572$

The risk increases by about 12.6%.

(e) $f(3) = 0.1210e^{0.2249(3)} \approx 0.2376$

The risk for a woman with a score of 3 is about 0.24%.

(f) $f(12) = 0.1210e^{0.2249(12)} \approx 1.7983$

The risk increases by about 1.6%.

(g) $f(3) = 0.1210e^{0.2249(6)} \approx 0.4665$

The risk for a woman with a score of 6 is about 0.47%.

$f(15) = 0.1210e^{0.2249(15)} \approx 3.5308$

The risk increases by about 3.1%

45. $C = \dfrac{D \times a}{V(a-b)}\left(e^{-bt} - e^{-at}\right)$

(a) At time $t = 0$,

$$C = \frac{D \times a}{V(a-b)}\left(e^{-b(0)} - e^{-a(0)}\right)$$

$$= \frac{D \times a}{V(a-b)}\left(e^0 - e^0\right)$$

$$= \frac{D \times a}{V(a-b)}(0) = 0$$

This makes sense because no cortisone has been administered yet.

(b) As a large amount of time passes, the concentration is going to approach zero. The longer the drug is in the body, the lower the concentration.

(c) $D = 500, a = 8.5, b = 0.09, V = 3700$

$$C = \frac{500 \times 8.5}{3700(8.5 - 0.09)}\left(e^{-0.09t} - e^{-8.5t}\right) = \frac{4250}{31{,}117}\left(e^{-0.09t} - e^{-8.5t}\right)$$

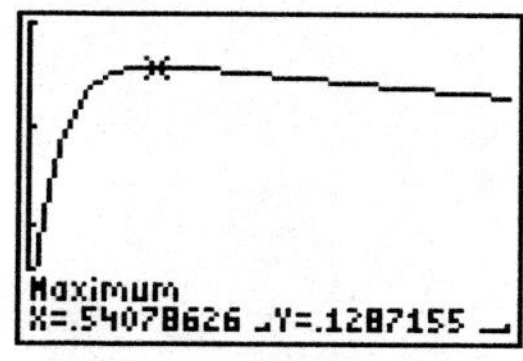

[0, 2] by [0, 0.15]

The maximum concentration occurs at about $t = 0.54$ hour.

47. $Q(t) = 1000(5^{-0.3t})$

(a) $Q(6) = 1000\left[5^{-0.3(6)}\right] \approx 55.189$

In 6 months, there will be about 55 grams.

(b)

$$8 = 1000(5^{-0.3t})$$
$$\frac{1}{125} = 5^{-0.3t}$$
$$5^{-3} = 5^{-0.3t}$$
$$-3 = -0.3t$$
$$10 = t$$

It will take 10 months to reduce the substance to 8 grams.

49. (a) $y = mt + b$

$b = 0.275$

Use the points $(0, 0.275)$ and $(24, 1900)$ to find m.

$$m = \frac{1900 - 0.275}{24 - 0} = \frac{1899.725}{24} \approx 79.155$$

$y = 79.155t + 0.275$

$y = at^2 + b$

$b = 0.275$

Use the point $(24, 1900)$ to find a.

$$1900 = a(24)^2 + 0.275$$
$$1899.925 = 576a$$
$$a = \frac{1899.725}{576} \approx 3.298$$

$y = 3.298t^2 + 0.275$

$y = ab^t$

$a = 0.275$

Use the point $(24, 1900)$ to find b.

$$1900 = 0.275b^{24}$$
$$b^{24} = \frac{1900}{0.275}$$
$$b = \sqrt[24]{\frac{1900}{275}} \approx 1.445$$

$y = 0.275(1.445)^t$

(b)

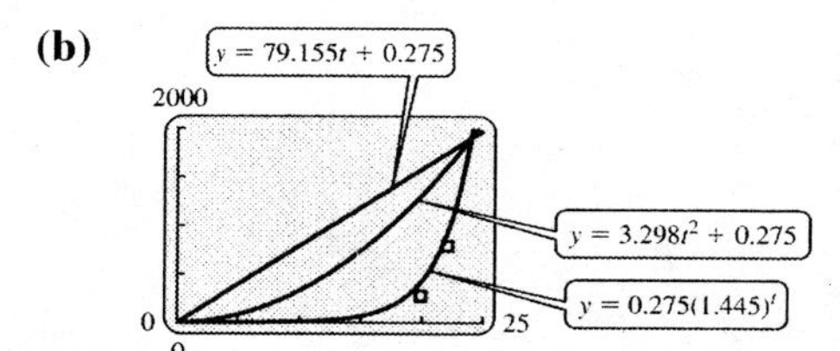

The function $y = 0.275(1.445)^t$ is the best fit.

(c) $x = 30$ corresponds to 2015.

$$y = 0.275(1.445)^{30} = 17193.65098 \approx 17{,}200$$

(d) The regression function is

$y = 0.1787(1.441)^t$.

This is close to the function in part (b).

51. $A = P\left(1 + \frac{r}{m}\right)^{tm}$, $P = 10{,}000$, $r = 0.04$, $t = 5$

(a) annually, $m = 1$

$$A = 10{,}000\left(1 + \frac{0.04}{1}\right)^{5(1)} = 10{,}000(1.04)^5 = \$12{,}166.53$$

$$\text{Interest} = \$12{,}166.53 - \$10{,}000 = \$2166.53$$

(b) semiannually, $m = 2$

$$A = 10{,}000\left(1 + \frac{0.04}{2}\right)^{5(2)} = 10{,}000(1.02)^{10} = \$12{,}189.94$$

$$\text{Interest} = \$12{,}189.94 - \$10{,}000 = \$2189.94$$

(c) quarterly, $m = 4$

$$A = 10{,}000\left(1 + \frac{0.04}{4}\right)^{5(4)} = 10{,}000(1.01)^{20} = \$12{,}201.90$$

$$\text{Interest} = \$12{,}201.90 - \$10{,}000 = \$2201.90$$

(d) monthly, $m = 12$

$$A = 10,000\left(1 + \frac{0.04}{12}\right)^{5(12)}$$
$$= 10,000(1.00\overline{3})^{60}$$
$$= \$12,209.97$$
$$\text{Interest} = \$12,209.97 - \$10,000$$
$$= \$2209.97$$

(e) continuously, $t = 5$

$$A = 10,000e^{(0.04)(5)} = \$12,214.03$$
$$\text{Interest} = \$12,214.03 - \$10,000$$
$$= \$2214.03$$

53. For 6% compounded annually for 2 years,

$$A = 18,000(1 + 0.06)^2 = 18,000(1.06)^2$$
$$= 20,224.80$$

For 5.9% compounded monthly for 2 years,

$$A = 18,000\left(1 + \frac{0.059}{12}\right)^{12(2)}$$
$$= 18,000\left(\frac{12.059}{12}\right)^{24} = 20,248.54$$

The 5.9% investment is better. The additional interest is
$\$20,248.54 - \$20,224.80 = \$23.74.$

55. $A = Pe^{rt}$

(a) $r = 3\%$
$A = 10e^{0.03(3)} \approx \$10.94$

(b) $r = 4\%$
$A = 10e^{0.04(3)} \approx \$11.27$

(c) $r = 5\%$
$A = 10e^{0.05(3)} \approx \$11.62$

57. $1200 = 500\left(1 + \frac{r}{4}\right)^{(14)(4)}$

$$\frac{1200}{500} = \left(1 + \frac{r}{4}\right)^{56}$$
$$2.4 = \left(1 + \frac{r}{4}\right)^{56}$$
$$1 + \frac{r}{4} = (2.4)^{1/56}$$
$$4 + r = 4(2.4)^{1/56}$$
$$r = 4(2.4)^{1/56} - 4$$
$$r \approx 0.0630$$

The required interest rate is 6.30%.

59. $y = (0.92)^t$

(a)

t	y
0	$(0.92)^0 = 1$
1	$(0.92)^1 = 0.92$
2	$(0.92)^2 \approx 0.85$
3	$(0.92)^3 \approx 0.78$
4	$(0.92)^4 \approx 0.72$
5	$(0.92)^5 \approx 0.66$
6	$(0.92)^6 \approx 0.61$
7	$(0.92)^7 \approx 0.56$
8	$(0.92)^8 \approx 0.51$
9	$(0.92)^9 \approx 0.47$
10	$(0.92)^{10} \approx 0.43$

(b)

y
1
0.5
0
5
10
t
$y = (0.92)^t$

(c) Let x = the cost of the house in 10 years. Then, $0.43x = 165,000 \Rightarrow x \approx 383,721$.
In 10 years, the house will cost about \$384,000.

(d) Let x = the cost of the book in 8 years. Then, $0.51x = 50 \Rightarrow x \approx 98$.
In 8 years, the textbook will cost about \$98.

2.2 Logarithmic Functions

1. $5^3 = 125$

Since $a^y = x$ means $y = \log_a x$, the equation in logarithmic form is $\log_5 125 = 3$.

3. $3^4 = 81$

Since $a^y = x$ means $y = \log_a x$, the equation in logarithmic form is $\log_3 81 = 4$.

5. $3^{-2} = \frac{1}{9}$

Since $a^y = x$ means $y = \log_a x$, the equation in logarithmic form is $\log_3 \frac{1}{9} = -2$.

7. $\log_2 32 = 5$

Since $y = \log_a x$ means $a^y = x$, the equation in exponential form is $2^5 = 32$.

9. $\ln \frac{1}{e} = -1$

Since $y = \ln x = \log_e x$ means $e^y = x$, the equation in exponential form is $e^{-1} = \frac{1}{e}$.

11.
$$\log 100{,}000 = 5$$
$$\log_{10} 100{,}000 = 5$$
$$10^5 = 100{,}000$$

When no base is written, $\log_{10} x$ is understood.

13. Let $\log_8 64 = x$. Then,

$8^x = 64 \Rightarrow 8^x = 8^2 \Rightarrow x = 2$

Thus, $\log_8 64 = 2$.

15. $\log_4 64 = x \Rightarrow 4^x = 64 \Rightarrow 4^x = 4^3 \Rightarrow x = 3$

17. $\log_2 \frac{1}{16} = x \Rightarrow 2^x = \frac{1}{16} \Rightarrow 2^x = 2^{-4} \Rightarrow x = -4$

19. $\log_2 \sqrt[3]{\frac{1}{4}} = x \Rightarrow 2^x = \left(\frac{1}{4}\right)^{1/3} \Rightarrow$

$2^x = \left(\frac{1}{2^2}\right)^{1/3} \Rightarrow 2^x = 2^{-2/3} \Rightarrow x = -\frac{2}{3}$

21. $\ln e = x$

Recall that $\ln y$ means $\log_e y$.

$e^x = e \Rightarrow x = 1$

23. $\ln e^{5/3} = x \Rightarrow e^x = e^{5/3} \Rightarrow x = \frac{5}{3}$

25. The logarithm to the base 3 of 4 is written $\log_3 4$. The subscript denotes the base.

27. $\log_5 (3k) = \log_5 3 + \log_5 k$

29.
$$\begin{aligned}\log_3 \frac{3p}{5k} &= \log_3 3p - \log_3 5k \\ &= (\log_3 3 + \log_3 p) - (\log_3 5 + \log_3 k) \\ &= 1 + \log_3 p - \log_3 5 - \log_3 k\end{aligned}$$

31.
$$\begin{aligned}\ln \frac{3\sqrt{5}}{\sqrt[3]{6}} &= \ln 3\sqrt{5} - \ln \sqrt[3]{6} \\ &= \ln 3 \cdot 5^{1/2} - \ln 6^{1/3} \\ &= \ln 3 + \ln 5^{1/2} - \ln 6^{1/3} \\ &= \ln 3 + \frac{1}{2}\ln 5 - \frac{1}{3}\ln 6\end{aligned}$$

33. $\log_b 32 = \log_b 2^5 = 5\log_b 2 = 5a$

35.
$$\begin{aligned}\log_b 72b &= \log_b 72 + \log_b b = \log_b 72 + 1 \\ &= \log_b \left(2^3 \cdot 3^3\right) + 1 \\ &= \log_b 2^3 + \log_b 3^2 + 1 \\ &= 3\log_b 2 + 2\log_b 3 + 1 \\ &= 3a + 2c + 1\end{aligned}$$

37. $\log_5 30 = \frac{\ln 30}{\ln 5} \approx \frac{3.4012}{1.6094} \approx 2.113$

39. $\log_{1.2} 0.95 = \frac{\ln 0.95}{\ln 1.2} \approx -0.281$

41.
$$\begin{aligned}\log_x 36 &= -2 \\ x^{-2} &= 36 \\ (x^{-2})^{-1/2} &= 36^{-1/2} \\ x &= \frac{1}{6}\end{aligned}$$

43.
$$\begin{aligned}\log_8 16 &= z \\ 8^z &= 16 \\ (2^3)^z &= 2^4 \\ 2^{3z} &= 2^4 \\ 3z &= 4 \\ z &= \frac{4}{3}\end{aligned}$$

45.
$$\begin{aligned}\log_r 5 &= \frac{1}{2} \\ r^{1/2} &= 5 \\ (r^{1/2})^2 &= 5^2 \\ r &= 25\end{aligned}$$

47.
$$\begin{aligned}\log_5 (9x-4) &= 1 \\ 5^1 &= 9x - 4 \\ 9 &= 9x \\ 1 &= x\end{aligned}$$

49.
$$\begin{aligned}\log_9 m - \log_9 (m-4) &= -2 \\ \log_9 \frac{m}{m-4} &= -2 \\ 9^{-2} &= \frac{m}{m-4} \\ \frac{1}{81} &= \frac{m}{m-4} \\ m - 4 &= 81m \\ -4 &= 80m \Rightarrow -0.05 = m\end{aligned}$$

This value is not possible since $\log_9(-0.05)$ does not exist. Thus, there is no solution to the original equation.

51. $\log_3 (x-2)+\log_3 (x+6)=2$
$$\log_3 [(x-2)(x+6)]=2$$
$$(x-2)(x+6)=3^2$$
$$x^2+4x-12=9$$
$$x^2+4x-21=0$$
$$(x+7)(x-3)=0$$
$$x=-7 \quad \text{or} \quad x=3$$

$x=-7$ is not a solution of the original equation because if $x=-7$, $x+6$ would be negative, and log (-1)does not exist. Therefore, $x=3$.

53. $\log_2 (x^2-1)-\log_2 (x+1)=2$
$$\log_2 \frac{x^2-1}{x+1}=2$$
$$2^2=\frac{x^2-1}{x+1}$$
$$4=\frac{x^2-1}{x+1}$$
$$4x+4=x^2-1$$
$$x^2-4x-5=0$$
$$(x-5)(x+1)=0$$
$$x=5 \quad \text{or} \quad x=-1$$

$x=-1$ is not a solution of the original equation because if $x=-1$, $x+1$ would not exist. Therefore, $x=5$.

55. $\ln x+\ln 3x=-1$
$$\ln 3x^2=-1$$
$$3x^2=e^{-1}$$
$$x^2=\frac{e^{-1}}{3}$$
$$x=\sqrt{\frac{e^{-1}}{3}}=\frac{1}{\sqrt{3e}}\approx 0.3502$$

57.
$$2^x=6$$
$$\ln 2^x=\ln 6$$
$$x\ln 2=\ln 6$$
$$x=\frac{\ln 6}{\ln 2}\approx 2.5850$$

59.
$$e^{k-1}=6$$
$$\ln e^{k-1}=\ln 6$$
$$(k-1)\ln e=\ln 6$$
$$k-1=\frac{\ln 6}{\ln e}$$
$$k-1=\frac{\ln 6}{1}$$
$$k=1+\ln 6\approx 2.7918$$

61.
$$3^{x+1}=5^x$$
$$\ln 3^{x+1}=\ln 5^x$$
$$(x+1)\ln 3=x\ln 5$$
$$x\ln 3+\ln 3=x\ln 5$$
$$x\ln 5-x\ln 3=\ln 3$$
$$x(\ln 5-\ln 3)=\ln 3$$
$$x=\frac{\ln 3}{\ln(5/3)}\approx 2.1507$$

63.
$$5(0.10)^x=4(0.12)^x$$
$$\ln[5(0.10)^x]=\ln[4(0.12)^x]$$
$$\ln 5+x\ln 0.10=\ln 4+x\ln 0.12$$
$$x(\ln 0.12-\ln 0.10)=\ln 5-\ln 4$$
$$x=\frac{\ln 5-\ln 4}{\ln 0.12-\ln 0.10}=\frac{\ln 1.25}{\ln 1.2}$$
$$\approx 1.2239$$

For exercises 65–68, use the formula $a^x=e^{(\ln a)x}$.

65. $10^{x+1}=e^{(\ln 10)(x+1)}$

67. $e^{3x}=(e^3)^x\approx 20.09^x$

69. $f(x)=\log(5-x)$
$5-x>0\Rightarrow -x>-5\Rightarrow x<5$
The domain of f is $x<5$ or $(-\infty, 5)$.

71. $\log A-\log B=0$
$$\log\frac{A}{B}=0$$
$$\frac{A}{B}=10^0=1$$
$$A=B$$
$$A-B=0$$

Thus, solving $\log A-\log B=0$ is equivalent to solving $A-B=0$

73. Let $m=\log_a \frac{x}{y}$, $n=\log_a x$, and $p=\log_a y$.

Then $a^m=\frac{x}{y}$, $a^n=x$, and $a^p=y$.

Substituting gives $a^m=\frac{x}{y}=\frac{a^n}{a^p}=a^{n-p}$.

So $m=n-p$. Therefore,
$$\log_a\frac{x}{y}=\log_a x-\log_a y.$$

75. **(a)** $t=\dfrac{\ln 2}{\ln(1+0.03)}\approx 23.4$ years

(b) $t=\dfrac{\ln 2}{\ln(1+0.06)}\approx 11.9$ years

(c) $t = \dfrac{\ln 2}{\ln(1+0.08)} \approx 9.0$ years

77. (a) $h(t) = 37.79(1.021)^t$

Double the 2005 population is $2(42.69) = 85.38$ million.

$$85.38 = 37.79(1.021)^t$$
$$\frac{85.38}{37.79} = (1.021)^t$$
$$\log_{1.021}\left(\frac{85.38}{37.79}\right) = t$$
$$t = \frac{\ln\left(\frac{85.38}{37.79}\right)}{\ln 1.021} \approx 39.22$$

The Hispanic population is estimated to double their 2005 population in 2039.

(b) $h(t) = 11.14(1.023)^t$

Double the 2005 population is $2(12.69) = 25.38$ million.

$$25.38 = 11.14(1.023)^t$$
$$\frac{25.38}{11.14} = (1.023)^t$$
$$\log_{1.023}\left(\frac{25.38}{11.14}\right) = t$$
$$t = \frac{\ln\left(\frac{25.38}{11.14}\right)}{\ln 1.023} \approx 36.21$$

The Asian population is estimated to double their 2005 population in 2036.

79. Set the exponential growth functions $y = 4500(1.04)^t$ and $y = 3000(1.06)^t$ equal to each other and solve for t.

$$4500(1.04)^t = 3000(1.06)^t$$
$$1.5(1.04)^t = 1.06^t$$
$$\ln\left(1.5(1.04)^t\right) = \ln(1.06)^t$$
$$\ln 1.5 + t\ln 1.04 = t\ln 1.06$$
$$\ln 1.5 = t(\ln 1.06 - \ln 1.04)$$
$$t = \frac{\ln 1.5}{\ln 1.06 - \ln 1.04} \approx 21.29$$

After about 21.3 years, the black squirrels will outnumber the gray squirrels. Verify this with a graphing calculator.

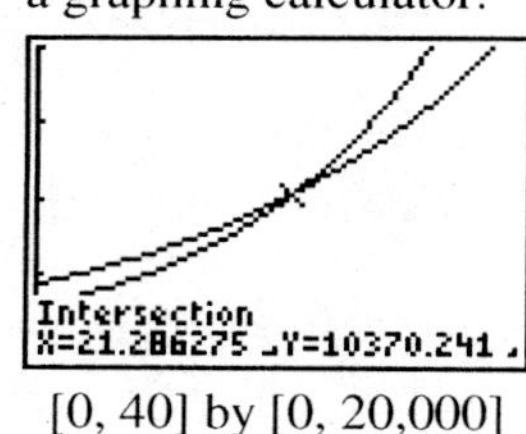

[0, 40] by [0, 20,000]

81. (a) The total number of individuals in the community is 50 + 50, or 100.

Let $P_1 = \dfrac{50}{100} = 0.5,\ P_2 = 0.5.$

$$H = -1[P_1 \ln P_1 + P_2 \ln P_2]$$
$$= -1[0.5\ln 0.5 + 0.5\ln 0.5]$$
$$\approx 0.693$$

(b) For 2 species, the maximum diversity is ln 2.

(c) Yes, $\ln 2 \approx 0.693$.

83. (a) 3 species, $\frac{1}{3}$ each:

$$P_1 = P_2 = P_3 = \frac{1}{3}$$
$$H = -(P_1 \ln P_1 + P_2 \ln P_2 + P_3 \ln P_3)$$
$$= -3\left(\frac{1}{3}\ln\frac{1}{3}\right) = -\ln\frac{1}{3}$$
$$\approx 1.099$$

(b) 4 species, $\frac{1}{4}$ each:

$$P_1 = P_2 = P_3 = P_4 = \frac{1}{4}$$
$$H = -(P_1 \ln P_1 + P_2 \ln P_2 + P_3 \ln P_3 + P_4 \ln P_4)$$
$$= -4\left(\frac{1}{4}\ln\frac{1}{4}\right) = -\ln\frac{1}{4}$$
$$\approx 1.386$$

(c) Notice that

$$-\ln\frac{1}{3} = \ln(3^{-1})^{-1} = \ln 3 \approx 1.099 \text{ and}$$
$$-\ln\frac{1}{4} = \ln(4^{-1})^{-1} = \ln 4 \approx 1.386$$

by Property **c** of logarithms, so the populations are at a maximum index of diversity.

85. $A = 4.688w^{0.8168-0.0154\log w}$

(a) $A = 4.688(4000)^{0.8168-0.0154\log 4000}$
$\approx 2590\text{ cm}^2$

(b) $A = 4.688(8000)^{0.8168-0.0154\log 8000}$
$\approx 4211\text{ cm}^2$

(c) Graph the equations

$Y_1 = 4.688x^{0.8168-0.0154\log x}$ and

$Y_2 = 4000$, then find the intersection to

solve $4000 = 4.688x^{0.8168-0.0154\log w}$.

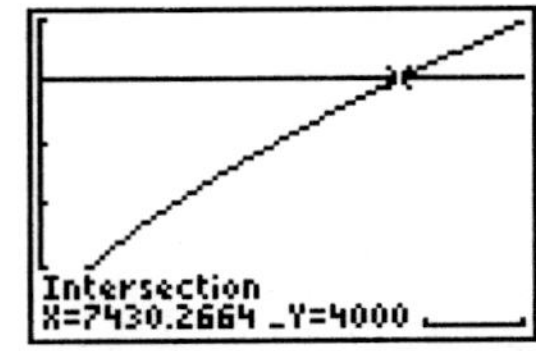

[0, 10,000] by [0, 5000]

An infant with a surface area of 4000 square cm weighs about 7430 g.

87. $C(t) = C_0 e^{-kt}$

When $t = 0$, $C(t) = 2$, and when $t = 3$, $C(t) = 1$.

$$2 = C_0 e^{-k(0)}$$
$$C_0 = 2$$
$$1 = 2e^{-3k}$$
$$\frac{1}{2} = e^{-3k}$$
$$-3k = \ln\frac{1}{2} = \ln 2^{-1} = -\ln 2$$
$$k = \frac{\ln 2}{3}$$
$$T = \frac{1}{k}\ln\frac{C_2}{C_1}$$
$$T = \frac{1}{\frac{\ln 2}{3}}\ln\frac{5C_1}{C_1} = \frac{3\ln 5}{\ln 2} \approx 7.0$$

The drug should be given about every 7 hours.

89. (a)

$$y(t) = y_0 e^{kt}$$
$$\frac{y(t)}{y_0} = e^{kt}$$
$$\ln\left(\frac{y(t)}{y_0}\right) = kt$$
$$\frac{1}{t}\ln\left(\frac{y(t)}{y_0}\right) = k, \text{ which is a constant.}$$

(b) The data in Section 2.1, exercise 40 is as follows:

Year	Demand for Physicians (in thousands)
2006	680.5
2015	758.6
2020	805.8
2025	859.3

Evaluate the expression for the years 2015 ($t = 9$), 2020 ($t = 14$), and 2025 $t = 19$)

$$2015: \frac{1}{9}\ln\left(\frac{758.6}{680.5}\right) \approx 0.01207$$
$$2020: \frac{1}{14}\ln\left(\frac{805.8}{680.5}\right) \approx 0.01207$$
$$2025: \frac{1}{19}\ln\left(\frac{859.3}{680.5}\right) \approx 0.01228$$

91. (a)

No, the data do not appear to lie along a straight line.

(b)

Yes, the graph appears to be more linear, especially if the first point is eliminated.

(c) Using a TI-84 Plus, the least squares line for the data in part (b) is $Y = 0.02940x + 9.348$.

(d) Using a TI-84 Plus, the exponential regression function for the data in part (a) is $Y = 11{,}471(1.0298)^x$.

(e)
$$\ln Y = \ln\left(11{,}471(1.0298)^x\right)$$
$$= \ln 11{,}471 + \ln\left(1.0298^x\right)$$
$$= \ln 11{,}471 + x\ln 1.0298$$
$$\approx 9.348 + 0.2940x$$

93. $N(r) = -5000\ln r$

(a) $N(0.9) = -5000\ln(0.9) \approx 530$

(b) $N(0.5) = -5000\ln(0.5) \approx 3500$

(c) $N(0.3) = -5000\ln(0.3) \approx 6000$

(d) $N(0.7) = -5000\ln(0.7) \approx 1800$

About 1800 years have elapsed since the split if 70% of the words of the ancestral language are common to both languages today.

(e) $-5000 \ln r = 1000$

$$\ln r = \frac{1000}{-5000}$$
$$\ln r = -\frac{1}{5}$$
$$r = e^{-1/5} \approx 0.8$$

95. Let I_1 be the intensity of the sound whose decibel rating is 85.

$$10 \log \frac{I_1}{I_0} = 85$$
$$\log \frac{I_1}{I_0} = 8.5$$
$$\log I_1 - \log I_0 = 8.5$$
$$\log I_1 = 8.5 + \log I_0$$

Let I_2 be the intensity of the sound whose decibel rating is 75.

$$10 \log \frac{I_2}{I_0} = 75$$
$$\log \frac{I_2}{I_0} = 7.5$$
$$\log I_2 - \log I_0 = 7.5$$
$$\log I_0 = \log I_2 - 7.5$$

Substitute for I_0 in the equation for $\log I_1$.

$$\log I_1 = 8.5 + \log I_0$$
$$= 8.5 + \log I_2 - 7.5$$
$$= 1 + \log I_2$$
$$\log I_1 - \log I_2 = 1$$
$$\log \frac{I_1}{I_2} = 1$$

Then $\frac{I_1}{I_2} = 10$, so $I_2 = \frac{1}{10} I_1$. This means the intensity of the sound that had a rating of 75 decibels is $\frac{1}{10}$ as intense as the sound that had a rating of 85 decibels.

97. $\text{pH} = -\log[\text{H}^+]$

(a) For pure water:

$7 = -\log[\text{H}^+] \Rightarrow -7 = \log[\text{H}^+] \Rightarrow$
$10^{-7} = [\text{H}^+]$

For acid rain:

$$4 = -\log[\text{H}^+]$$
$$-4 = \log[\text{H}^+]$$
$$10^{-4} = [\text{H}^+]$$
$$\frac{10^{-4}}{10^{-7}} = 10^3 = 1000$$

The acid rain has a hydrogen ion concentration 1000 times greater than pure water.

(b) For laundry solution:

$11 = -\log[\text{H}^+] \Rightarrow 10^{-11} = [\text{H}^+]$

For black coffee:

$$5 = -\log[\text{H}^+]$$
$$10^{-5} = [\text{H}^+]$$
$$\frac{10^{-5}}{10^{-11}} = 10^6 = 1{,}000{,}000$$

The coffee has a hydrogen ion concentration 1,000,000 times greater than the laundry mixture.

2.3 Applications: Growth and Decay

1. y_0 represents the initial quantity; k represents the rate of growth or decay.

3. The half-life of a quantity is the time period for the quantity to decay to one-half of the initial amount.

5. Assume that $y = y_0 e^{kt}$ is the amount left of a radioactive substance decaying with a half-life of T. From Exercise 4, we know $k = -\frac{\ln 2}{T}$, so

$$y = y_0 e^{(-\ln 2/T)t} = y_0 e^{-(t/T)\ln 2} = y_0 e^{(-\ln 2^{-t/T})}$$
$$= y_0 2^{-t/T} = y_0 \left[\left(\frac{1}{2}\right)^{-1}\right]^{-t/T} = y_0 \left(\frac{1}{2}\right)^{t/T}$$

7. (a) $y = 2y_0$ after 12 hours.

$$y = y_0 e^{kt}$$
$$2y_0 = y_0 e^{12k}$$
$$2 = e^{12k}$$
$$\ln 2 = \ln e^{12k}$$
$$12k = \ln 2$$
$$k = \frac{\ln 2}{12} \approx 0.05776$$
$$y = y_0 e^{0.05776t}$$

(b) $y = y_0 e^{(\ln 2/12)t}$

$$= y_0 e^{(\ln 2)(t/12)}$$
$$= y_0 [e^{\ln 2}]^{t/12}$$
$$= y_0 2^{t/12} \text{ since } e^{\ln 2} = 2$$

(c) For 10 days, $t = 10 \cdot 24$ or 240.

$$y = (1)2^{240/12} = 2^{20} = 1{,}048{,}576$$

For 15 days, $t = 15 \cdot 24$ or 360.

$$y = (1)2^{360/12} = 2^{30} = 1{,}073{,}741{,}824$$

9. $y = y_0 e^{kt}$

(a) $y = 20{,}000,\ y_0 = 50{,}000,\ t = 9$

$$20{,}000 = 50{,}000e^{9k}$$
$$0.40 = e^{9k}$$
$$\ln 0.4 = 9k$$
$$-0.102 \approx k$$

The equation is $y = 50{,}000e^{-0.102t}$.

(b) $\frac{1}{2}(50{,}000) = 25{,}000$

$$25{,}000 = 50{,}000e^{-0.102t}$$
$$0.5 = e^{-0.102t}$$
$$\ln 0.5 = -0.102t$$
$$6.8 \approx t$$

Half the bacteria remain after about 6.8 hours.

11. Use $y = y_0 e^{-kt}$.
When $t = 5,\ y = 0.37y_0$.

$$0.37y_0 = y_0 e^{-5k}$$
$$0.37 = e^{-5k}$$
$$-5k = \ln(0.37) \Rightarrow k = \frac{\ln(0.37)}{-5} \approx 0.1989$$

13.

$$A(t) = A_0\left(\frac{1}{2}\right)^{t/5600}$$
$$A(43{,}000) = A_0\left(\frac{1}{2}\right)^{43{,}000/5600} \approx 0.005A_0$$

About 0.5% of the original carbon 14 was present.

15. $A(t) = A_0 e^{kt}$

First, find k. Let $A(t) = \frac{1}{2}A_0$ and $t = 1.25$.

$$\frac{1}{2}A_0 = A_0 e^{1.25k}$$
$$\frac{1}{2} = e^{1.25k}$$
$$\ln\frac{1}{2} = 1.25k \Rightarrow k = \frac{\ln\frac{1}{2}}{1.25}$$

Now, let $A_0 = 1$ and let $t = 0.25$. (Note that the half-life is given in billions of years, so 250 million years is 0.25 billion year.)

$$A(0.25) = 1 \cdot e^{0.25\ln\frac{1}{2}/1.25} \approx 0.87$$

Therefore, about 87% of the potassium-40 remains from a creature that died 250 million years ago.

17.

$$A(t) = A_0 e^{kt}$$
$$0.60A_0 = A_0 e^{(-\ln 2/5600)t}$$
$$0.60 = e^{(-\ln 2/5600)t}$$
$$\ln 0.60 = -\frac{\ln 2}{5600}t$$
$$\frac{5600(\ln 0.60)}{-\ln 2} = t$$
$$4127 \approx t$$

The sample was about 4100 years old.

19. $\frac{1}{2}A_0 = A_0 e^{-0.00043t}$

$$\frac{1}{2} = e^{-0.00043t}$$
$$\ln\frac{1}{2} = -0.00043t$$
$$-\ln 2 = -0.00043t$$
$$t = \frac{\ln 2}{0.00043} \approx 1612$$

The half-life of radium 226 is about 1600 years.

21. (a)

$$A(t) = A_0\left(\frac{1}{2}\right)^{t/1620}$$
$$A(100) = 4.0\left(\frac{1}{2}\right)^{100/1620} \approx 3.83$$

After 100 years, about 3.8 grams will remain.

(b)

$$0.1 = 4.0\left(\frac{1}{2}\right)^{t/1620}$$
$$\frac{0.1}{4} = \left(\frac{1}{2}\right)^{t/1620}$$
$$\ln 0.025 = \frac{t}{1620}\ln\frac{1}{2}$$
$$t = \frac{1600\ln 0.025}{\ln\left(\frac{1}{2}\right)} \approx 8515$$

The half-life is about 8600 years.

23. (a) $y = y_0 e^{kt}$

When $t = 0$, $y = 25.0$, so $y_0 = 25.0$

When $t = 50$, $y = 19.5$

$$19.5 = 25.0e^{50k}$$

$$\frac{19.5}{25.0} = e^{50k}$$

$$50k = \ln\left(\frac{19.5}{25.0}\right)$$

$$k = \frac{\ln\left(\frac{19.5}{25.0}\right)}{50} \approx -0.00497$$

$$y = 25.0e^{-0.00497t}$$

(b) From part (a), we have $k = \dfrac{\ln\left(\frac{19.5}{25.0}\right)}{50}$.

$$\begin{aligned} y &= 25.0e^{kt} \\ &= 25.0e^{[(\ln 19.5/25.0)/50]t} \\ &= 25.0e^{\ln(19.5/25.0)\cdot(t/50)} \\ &= 25.0\,[e^{\ln(19.5/25.0)}]^{t/50} \\ &= 25.0(19.5/25.0)^{t/50} \\ &= 25(0.78)^{t/50} \end{aligned}$$

(c)

$$\frac{1}{2}y_0 = y_0 e^{-0.00497t}$$

$$\frac{1}{2} = e^{-0.00497t}$$

$$-0.00497t = \ln\left(\frac{1}{2}\right)$$

$$t = \frac{\ln\left(\frac{1}{2}\right)}{-0.00497} \approx 139.47$$

The half-life is about 139 days.

25. (a) Let t = the number of degrees Celsius.

$y = y_0 \cdot e^{kt}$

$y_0 = 10$ when $t = 0°$.

To find k, let $y = 11$ when $t = 10°$.

$$11 = 10e^{10k}$$

$$e^{10k} = \frac{11}{10}$$

$$10k = \ln 1.1$$

$$k = \frac{\ln 1.1}{10} \approx 0.0095$$

The equation is $y = 10e^{0.0095t}$.

(b) Let $y = 1.5$; solve for t.

$$15 = 10e^{0.0095t}$$

$$\ln 1.5 = 0.0095t$$

$$t = \frac{\ln 1.5}{0.0095} \approx 42.7$$

15 grams will dissolve at 42.7°C.

27. $f(t) = T_0 + Ce^{-kt}$

$$25 = 20 + 100e^{-0.1t}$$

$$5 = 100e^{-0.1t}$$

$$e^{-0.1t} = 0.05$$

$$-0.1t = \ln 0.05$$

$$t = \frac{\ln 0.05}{-0.1} \approx 30$$

It will take about 30 min.

2.4 Trigonometric Functions

1. $60° = 60\left(\dfrac{\pi}{180}\right) = \dfrac{\pi}{3}$

3. $150° = 150\left(\dfrac{\pi}{180}\right) = \dfrac{5\pi}{6}$

5. $270° = 270\left(\dfrac{\pi}{180}\right) = \dfrac{3\pi}{2}$

7. $495° = 495\left(\dfrac{\pi}{180}\right) = \dfrac{11\pi}{4}$

9. $\dfrac{5\pi}{4} = \dfrac{5\pi}{4}\left(\dfrac{180°}{\pi}\right) = 225°$

11. $-\dfrac{13\pi}{6} = -\dfrac{13\pi}{6}\left(\dfrac{180°}{\pi}\right) = -390°$

13. $\dfrac{8\pi}{5} = \dfrac{8\pi}{5}\left(\dfrac{180°}{\pi}\right) = 288°$

15. $\dfrac{7\pi}{12} = \dfrac{7\pi}{12}\left(\dfrac{180°}{\pi}\right) = 105°$

17. Let α = the angle with terminal side through $(-3, 4)$. Then $x = -3$, $y = 4$, and

$$r = \sqrt{x^2 + y^2} = \sqrt{(-3)^2 + (4)^2} = \sqrt{25} = 5.$$

$\sin\alpha = \dfrac{y}{r} = \dfrac{4}{5}$ $\quad\cot\alpha = \dfrac{x}{y} = -\dfrac{3}{4}$

$\cos\alpha = \dfrac{x}{r} = -\dfrac{3}{5}$ $\quad\sec\alpha = \dfrac{r}{x} = -\dfrac{5}{3}$

$\tan\alpha = \dfrac{y}{x} = -\dfrac{4}{3}$ $\quad\csc\alpha = \dfrac{r}{y} = \dfrac{5}{4}$

19. Let $\alpha =$ the angle with terminal side through $(7, -24)$. Then $x = 7$, $y = -24$, and

$r = \sqrt{x^2 + y^2} = \sqrt{49 + 576} = \sqrt{625} = 25.$

$\sin \alpha = \frac{y}{r} = -\frac{24}{25}$ $\quad\cot \alpha = \frac{x}{y} = -\frac{7}{24}$

$\cos \alpha = \frac{x}{r} = \frac{7}{25}$ $\quad\sec \alpha = \frac{r}{x} = \frac{25}{7}$

$\tan \alpha = \frac{y}{x} = -\frac{24}{7}$ $\quad\csc \alpha = \frac{r}{y} = -\frac{25}{24}$

21. In quadrant I, all six trigonometric functions are positive, so their sign is +.

23. In quadrant III, $x < 0$ and $y < 0$. Furthermore, $r > 0$.

$\sin \theta = \frac{y}{r} < 0$, so the sign is $-$.

$\cos \theta = \frac{x}{r} < 0$, so the sign is $-$.

$\tan \theta = \frac{y}{x} > 0$, so the sign is +.

$\cot \theta = \frac{x}{y} > 0$, so the sign is +.

$\sec \theta = \frac{r}{x} < 0$, so the sign is $-$.

$\csc \theta = \frac{r}{y} < 0$, so the sign is $-$.

25. When an angle θ of 30° is drawn in standard position, one choice of a point on its terminal side is $(x, y) = (\sqrt{3}, 1)$. Then

$r = \sqrt{x^2 + y^2} = \sqrt{3 + 1} = 2.$

$\tan \theta = \frac{y}{x} = \frac{1}{\sqrt{3}} = \frac{\sqrt{3}}{3}$

$\cot \theta = \frac{x}{y} = \sqrt{3}$

$\csc \theta = \frac{r}{y} = 2$

27. When an angle θ of 60° is drawn in standard position, one choice of a point on its terminal side is $(x, y) = (1, \sqrt{3})$. Then

$r = \sqrt{x^2 + y^2} = \sqrt{1 + 3} = 2.$

$\sin \theta = \frac{y}{r} = \frac{\sqrt{3}}{2}$

$\cot \theta = \frac{x}{y} = \frac{1}{\sqrt{3}} = \frac{\sqrt{3}}{3}$

$\csc \theta = \frac{r}{y} = \frac{2}{\sqrt{3}} = \frac{2\sqrt{3}}{3}$

29. When an angle θ of 135° is drawn in standard position, one choice of a point on its terminal side is $(x, y) = (-1, 1)$. Then

$r = \sqrt{x^2 + y^2} = \sqrt{1 + 1} = \sqrt{2}.$

$\tan \theta = \frac{y}{x} = -1$

$\cot \theta = \frac{x}{y} = -1$

31. When an angle θ of 210° is drawn in standard position, one choice of a point on its terminal side is $(x, y) = (-\sqrt{3}, -1)$. Then

$r = \sqrt{x^2 + y^2} = \sqrt{3 + 1} = 2.$

$\cos \theta = \frac{x}{r} = -\frac{\sqrt{3}}{2}$

$\sec \theta = \frac{r}{x} = \frac{2}{-\sqrt{3}} = -\frac{2\sqrt{3}}{3}$

33. When an angle of $\frac{\pi}{3}$ is drawn in standard position, one choice of a point on its terminal side is $(x, y) = (1, \sqrt{3})$. Then

$r = \sqrt{x^2 + y^2} = \sqrt{1 + 3} = 2.$

$\sin \frac{\pi}{3} = \frac{y}{r} = \frac{\sqrt{3}}{2}$

35. When an angle of $\frac{\pi}{4}$ is drawn in standard position, one choice of a point on its terminal side is $(x, y) = (1, 1)$.

$\tan \frac{\pi}{4} = \frac{y}{x} = 1$

37. When an angle of $\frac{\pi}{6}$ is drawn in standard position, one choice of a point on its terminal side is $(x, y) = (\sqrt{3}, 1)$. Then

$r = \sqrt{x^2 + y^2} = \sqrt{3 + 1} = 2.$

$\csc \frac{\pi}{6} = \frac{r}{y} = \frac{2}{1} = 2$

39. When an angle of 3π is drawn in standard position, one choice of a point on its terminal side is $(x, y) = (-1, 0)$. Then

$r = \sqrt{x^2 + y^2} = \sqrt{1} = 1.$

$\cos 3\pi = \frac{x}{r} = -1$

41. When an angle of $\frac{7\pi}{4}$ is drawn in standard position, one choice of a point on its terminal side is $(x, y) = (1, -1)$. Then

$$r = \sqrt{x^2 + y^2} = \sqrt{1+1} = \sqrt{2}.$$

$$\sin\frac{7\pi}{4} = \frac{y}{r} = \frac{-1}{\sqrt{2}} = -\frac{\sqrt{2}}{2}$$

43. When an angle of $\frac{5\pi}{4}$ is drawn in standard position, one choice of a point on its terminal side is $(x, y) = (-1, -1)$. Then

$$r = \sqrt{x^2 + y^2} = \sqrt{1+1} = \sqrt{2}.$$

$$\sec\frac{5\pi}{4} = \frac{r}{x} = \frac{\sqrt{2}}{-1} = -\sqrt{2}$$

45. When an angle of $-\frac{3\pi}{4}$ is drawn in standard position, one choice of a point on its terminal side is $(x, y) = (-1, -1)$. Then

$$\cot\left(-\frac{3\pi}{4}\right) = \frac{x}{y} = \frac{-1}{-1} = 1$$

47. When an angle of $-\frac{7\pi}{6}$ is drawn in standard position, one choice of a point on its terminal side is $(x, y) = (-\sqrt{3}, 1)$. Then

$$r = \sqrt{x^2 + y^2} = \sqrt{3+1} = 2.$$

$$\sin\left(-\frac{7\pi}{6}\right) = \frac{y}{r} = \frac{1}{2}$$

49. The cosine function is positive in quadrants I and IV. We know that $\cos(\pi/3) = 1/2$, so the solution in quadrant I is $\pi/3$. The solution in quadrant IV is $2\pi - (\pi/3) = 5\pi/3$. The two solutions of $\cos x = 1/2$ between 0 and 2π are $\pi/3$ and $5\pi/3$.

51. The tangent function is negative in quadrants II and IV. We know that $\tan(\pi/4) = 1$. The solution in quadrant II is $\pi - (\pi/4) = 3\pi/4$. The solution in quadrant IV is $2\pi - (\pi/4) = 7\pi/4$. The two solutions of $\tan x = -1$ between 0 and 2π are $3\pi/4$ and $7\pi/4$.

53. The secant function is negative in quadrants II and III. We know that $\sec(\pi/6) = 2/\sqrt{3}$, so the solution in quadrant II is $\pi - (\pi/6) = 5\pi/6$. The solution in quadrant III is $\pi + (\pi/6) = 7\pi/6$. The two solutions of $\sec x = -2/\sqrt{3}$ between 0 and 2π are $5\pi/6$ and $7\pi/6$.

55. $\sin 39° \approx 0.6293$

57. $\tan 123° \approx -1.5399$

59. $\sin 0.3638 \approx 0.3558$

61. $\cos 1.2353 \approx 0.3292$

63. $f(x) = \cos(3x)$ is of the form $f(x) = a\cos(bx)$ where $a = 1$ and $b = 3$. Thus, $a = 1$ and $T = \frac{2\pi}{b} = \frac{2\pi}{3}$.

65. $g(t) = -2\sin\left(\frac{\pi}{4}t + 2\right)$ is of the form $g(t) = a\sin(bt + c)$ where $a = -2$, $b = \frac{\pi}{4}$, and $c = 2$. Thus, $a = |-2| = 2$ and

$$T = \frac{2\pi}{b} = \frac{2\pi}{\frac{\pi}{4}} = 8.$$

67. The graph of $y = 2\sin x$ is similar to the graph of $y = \sin x$ except that it has twice the amplitude.

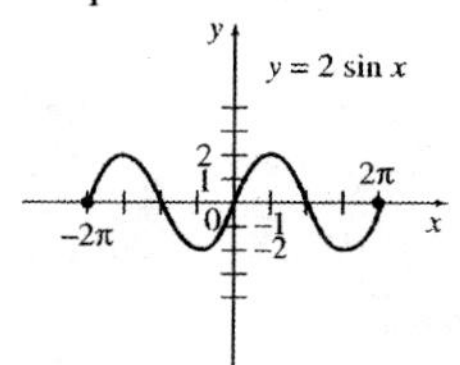

69. The graph of $y = -\sin x$ is similar to the graph of $y = \sin x$ except that it is reflected about the x-axis.

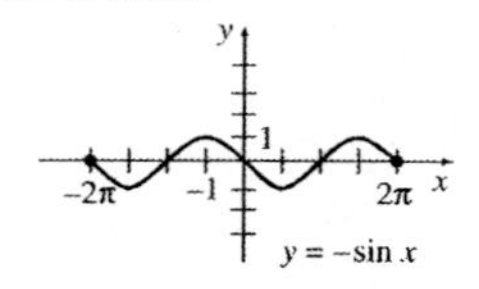

71. $y = 2\cos\left(3x - \frac{\pi}{4}\right) + 1$ has amplitude $a = 2$, period $T = \frac{2\pi}{b} = \frac{2\pi}{3}$, phase shift $\frac{c}{b} = \frac{-\pi/4}{3} = -\frac{\pi}{12}$, and vertical shift $d = 1$. Thus, the graph of $y = 2\cos\left(3x - \frac{\pi}{4}\right) + 1$ is similar to the graph of $f(x) = \cos x$ except that it has 2 times the amplitude, a third of the period, and is shifted 1 unit vertically.

(*continued on next page*)

(continued)

Also, $y = 2\cos\left(3x - \frac{\pi}{4}\right) + 1$ is shifted $\frac{\pi}{12}$ units to the right relative to the graph of graph of $g(x) = \cos 3x$.

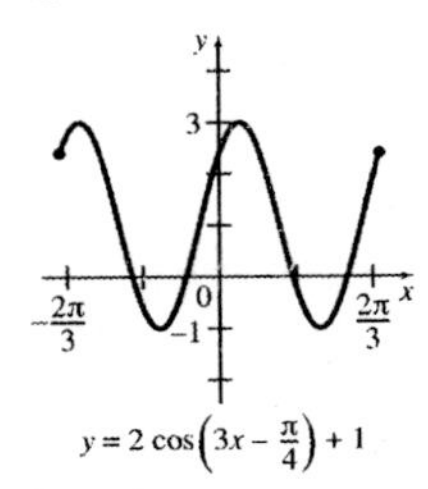

73. The graph of $y = \frac{1}{2}\tan x$ is similar to the graph of $y = \tan x$ except that the y-values of points on the graph are one-half the y-values of points on the graph of $y = \tan x$.

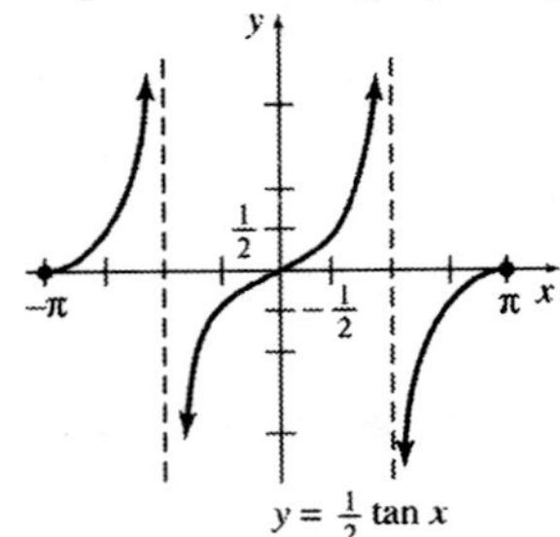

75. **(a)** Since the three angles θ are equal and their sum is 180°, each angle θ is 60°.

(b) The base angle on the left is still 60°. The bisector is perpendicular to the base, so the other base angle is 90°. The angle formed by bisecting the original vertex angle θ is 30°.

(c) Two sides of the triangle on the left are given in the diagram: the hypotenuse is 2, and the base is half of the original base of 2, or 1. The Pythagorean Theorem gives the length of the remaining side (the vertical bisector) as $\sqrt{2^2 - 1^2} = \sqrt{3}$.

77. **(a)** Since the amplitude is 2 and the period is 0.350, $a = 2$ and

$$0.350 = \frac{2\pi}{b} \Rightarrow \frac{1}{0.350} = \frac{b}{2\pi} \Rightarrow \frac{2\pi}{0.350} = b.$$

Therefore, the equation is $y = 2\sin(2\pi t/0.350)$, where t is the time in seconds.

(b) $2 = 2\sin\left(\frac{2\pi t}{0.350}\right)$

$$1 = \sin\left(\frac{2\pi t}{0.350}\right)$$

$$\frac{\pi}{2} = \frac{2\pi t}{0.350} \Rightarrow t = \frac{0.350\pi}{4\pi} = 0.0875$$

The image reaches its maximum amplitude after 0.0875 seconds.

(c) $t = 2$

$$y = 2\sin\left(\frac{2\pi(2)}{0.350}\right) \approx -1.95$$

The position of the object after 2 seconds is −1.95°.

79. $P(t) = 7(1 - \cos 2\pi t)(t + 10) + 100e^{0.2t}$

(a) Since January 1 of the base year corresponds to $t = 0$, the pollution level is

$$P(0) = 7(1 - \cos 0)(0 + 10) + 100e^0 = 7(0)(10) + 100 = 100.$$

(b) Since July 1 of the base year corresponds to $t = 0.5$, the pollution level is

$$P(0.5) = 7(1 - \cos \pi)(0.5 + 10) + 100e^{0.1} = 7(2)(10.5) + 100e^{0.1} \approx 258.$$

(c) Since January 1 of the following year corresponds to $t = 1$, the pollution level is

$$P(1) = 7(1 - \cos 2\pi)(1 + 10) + 100e^{0.2} = 7(0)(11) + 100e^{0.2} \approx 122$$

(d) Since July 1 of the following year corresponds to $t = 1.5$, the pollution level is

$$P(1.5) = 7(1 - \cos 3\pi)(1.5 + 10) + 100e^{0.3} = 7(2)(11.5) + 100e^{0.3} \approx 296.$$

81. **(a)**

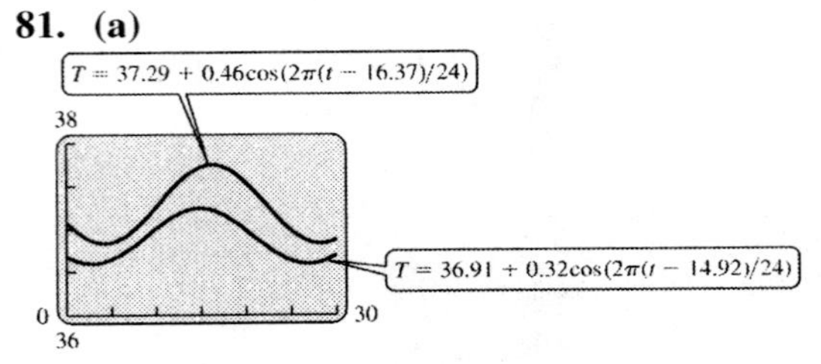

No, the functions never cross.

(b) For a patient without Alzheimer's, the heights temperature occurs when $t \approx 14.92$. $14.92 - 12 = 2.92$ and $0.92 \text{ hr} = 0.92 \cdot 60 \text{ min} \approx 55 \text{ min}$. So, $t \approx 14.92$ corresponds to about 2:55 P.M.

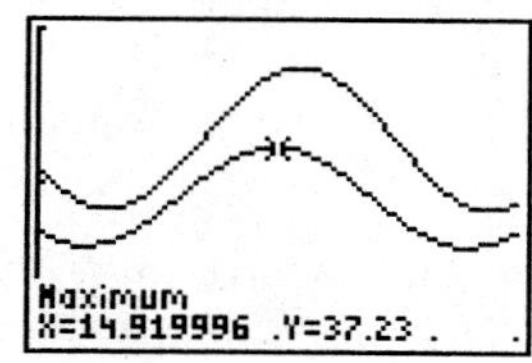

(c) For a patient without Alzheimer's, the heights temperature occurs when $t \approx 16.37$. $16.37 - 12 = 4.37$ and $0.37 \text{ hr} = 0.37 \cdot 60 \text{ min} \approx 22 \text{ min}$. So, $t \approx 16.37$ corresponds to about 4:22 P.M.

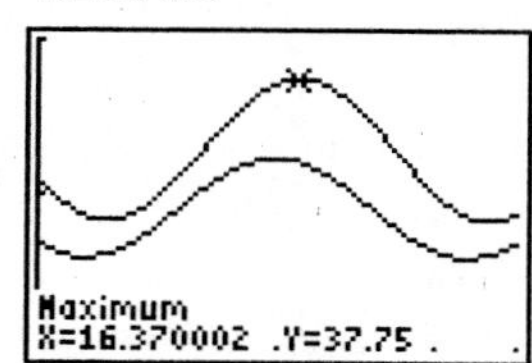

83. Solving $\frac{c_1}{c_2} = \frac{\sin \theta_1}{\sin \theta_2}$ for c_2 gives

$$c_2 = \frac{c_1 \sin \theta_2}{\sin \theta_1}.$$

When $\theta_1 = 39°, \theta_2 = 28°$, and $c_1 = 3 \times 10^8$,

$$c_2 = \frac{(3 \times 10^8)(\sin 28°)}{\sin 39°} \approx 2.2 \times 10^8 \text{ m/sec.}$$

85. Since the horizontal side of each square represents 30° and the sine wave repeats itself every 8 squares, the period of the sine wave is $8 \cdot 30° = 240°$.

87. (a)

P(t) = 0.002 sin (880πt)
0.004
0
0.003
−0.004

(b) Since the sine function is zero for multiples of π, we can determine the values (s) of t where $P = 0$ by setting $880\pi t = n\pi$, where n is an integer, and solving for t. After some algebraic manipulations, $t = \frac{n}{880}$ and $P = 0$ when $n = \ldots, -2, -1, 0, 1, 2, \ldots$. However, only values of $n = 0$, $n = 1$, or $n = 2$ produce values of t that lie in the interval $[0, 0.003]$. Thus, $P = 0$ when $t = 0$, $t = \frac{1}{880} \approx 0.0011$, and $t = \frac{1}{440} \approx 0.0023$. These values check with the graph in part (a).

(c) The period is $T = \frac{2\pi}{880\pi} = \frac{1}{440}$. Therefore, the frequency is 440 cycles per second.

89. $T(x) = 37 \sin\left[\frac{2\pi}{365}(t - 101)\right] + 25$

(a) $T(74) = 37 \sin\left[\frac{2\pi}{365}(-27)\right] + 25 \approx 8°\text{F}$

(b) $T(121) = 37 \sin\left[\frac{2\pi}{365}(20)\right] + 25 \approx 37°\text{F}$

(c) $T(250) = 37 \sin\left[\frac{2\pi}{365}(149)\right] + 25 \approx 45°\text{F}$

(d) $T(325) = 37 \sin\left[\frac{2\pi}{365}(224)\right] + 25 \approx 1°\text{F}$

(e) The maximum and minimum values of the sine function are 1 and −1, respectively. Thus, the maximum value of T is $37(1) + 25 = 62°\text{F}$ and the minimum value of T is $37(-1) + 25 = -12°\text{F}$.

(f) The period is $\frac{2\pi}{\frac{2\pi}{365}} = 365$.

91. Let h = the height of the building.

$$\tan 42.8° = \frac{h}{65} \Rightarrow h = 65 \tan 42.8° \approx 60.2$$

The height of the building is approximately 60.2 meters.

Chapter 2 Review Exercises

1. False; an exponential function has the form $f(x) = a^x$.

2. True

3. False; the logarithmic function $f(x) = \log_a x$ is not defined for $a = 1$.

4. False; $\ln(5+7) = \ln 12 \neq \ln 5 + \ln 7$

5. False; $(\ln 3)^4 \neq 4 \ln 3$ since $(\ln 3)^4$ means $(\ln 3)(\ln 3)(\ln 3)(\ln 3)$.

6. False; $\log_{10} 0$ is undefined since $10^x = 0$ has no solution.

7. True

8. False; $\ln(-2)$ is undefined.

9. False; $\dfrac{\ln 4}{\ln 8} = 0.6667$ and $\ln 4 - \ln 8 = \ln(1/2) \approx -0.6931$.

10. True

11. True

12. False; The period of cosine is 2π.

13. True

14. False. There's no reason to suppose that the Dow is periodic.

15. False; $\cos(a+b) = \cos a \cos b - \sin a \sin b$

16. True

17. A logarithm is the power to which a base must be raised in order to obtain a given number. It is the inverse of an exponential. We can write the definition mathematically as

    $y = \log_a x \Leftrightarrow a^y = x$, for $a > 0$, $a \neq 1$, and $x > 0$.

18. Exponential growth functions grow without bound while limited growth functions reach a maximum size that is limited by some external constraint.

19. One degree is $\frac{1}{360}$ of a complete rotation, while one radian is the measure of the central angle in the unit circle that intercepts an arc with length 1. In a circle with radius r, an angle measuring 1 radian intercepts an arc with length r. To convert from degree measure to radian measure, multiply the number of degrees by $\frac{\pi}{180°}$. To convert from radian measure to degree measure, multiply the number of radians by $\frac{180°}{\pi}$.

20. There's a nice answer to the question of when to use radians vs degrees at http://mathwithbaddrawings.com/2013/05/02/degrees-vs-radians/

21. Let (x, y) is a point on the terminal side of an angle θ in standard position, and let r be the distance from the origin to (x, y). Then

    $\sin\theta = \dfrac{y}{r}$ $\qquad \csc\theta = \dfrac{r}{y},\ y \neq 0$

    $\cos\theta = \dfrac{x}{r}$ $\qquad \sec\theta = \dfrac{r}{x},\ y \neq 0$

    $\tan\theta = \dfrac{y}{x},\ x \neq 0$ $\qquad \cot\theta = \dfrac{x}{y},\ y \neq 0$

22. The exact value for the trigonometric functions can be determined for any integer multiple of $\frac{\pi}{6}$ or $\frac{\pi}{4}$.

23. $y = \ln\left(x^2 - 9\right)$

    In order for the logarithm to be defined, $x^2 - 9 > 0$. So, $x^2 > 9 \Rightarrow x < -3$ or $x > 3$. The domain is $(-\infty, -3) \cup (3, \infty)$.

24. $y = \dfrac{1}{e^x - 1}$

    In order to the fraction to be defined, $e^x - 1 \neq 0$. So, $e^x - 1 \neq 0 \Rightarrow e^x \neq 1 \Rightarrow x \neq 0$. The domain is $(-\infty, 0) \cup (0, \infty)$.

25. $y = \dfrac{1}{\sin x - 1}$

    In order to the fraction to be defined, $\sin x - 1 \neq 0$. So, $\sin x - 1 \neq 0 \Rightarrow \sin x \neq 1$. The domain is

    $\left\{x \mid x \neq \dfrac{\pi}{2}, -\dfrac{3\pi}{2}, \dfrac{5\pi}{2}, -\dfrac{7\pi}{2}, \ldots\right\}$.

26. $y = \tan x = \dfrac{\sin x}{\cos x}$

    In order to the fraction to be defined, $\cos x \neq 0$. The domain is

    $\left\{x \mid x \neq \pm\dfrac{\pi}{2}, \pm\dfrac{3\pi}{2}, \pm\dfrac{5\pi}{2}, \ldots\right\}$.

27. $y = 4^x$

x	−2	−1	0	1	2
y	$\frac{1}{16}$	$\frac{1}{4}$	1	4	16

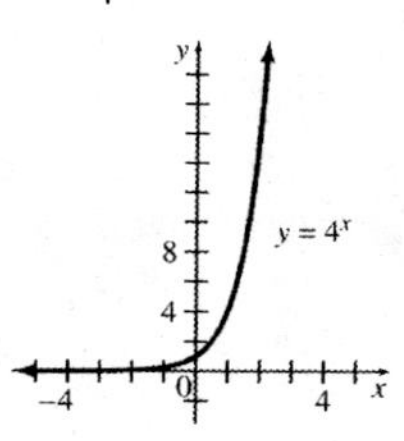

28. $y = 4^{-x} + 3$

x	−2	−1	0	1	2
y	19	7	4	$\frac{13}{4}$	$\frac{49}{16}$

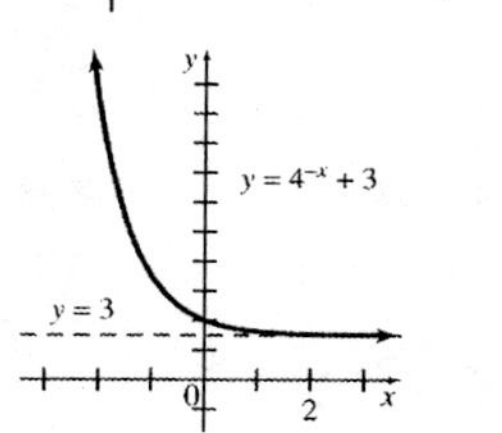

29. $y = \left(\frac{1}{5}\right)^{2x-3}$

x	0	1	2
y	125	5	$\frac{1}{5}$

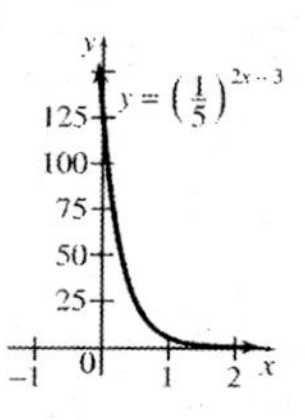

30. $y = \left(\frac{1}{2}\right)^{x-1}$

x	−2	−1	0	1	2
y	8	4	2	1	$\frac{1}{2}$

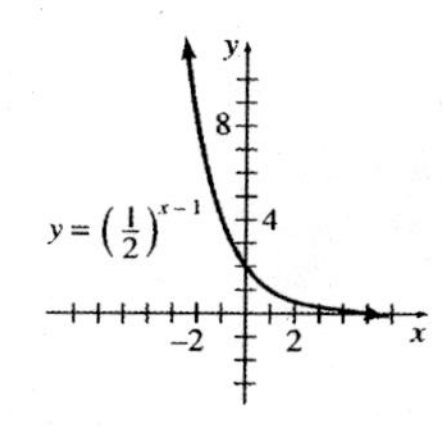

31. $y = \log_2(x-1)$

$2^y = x - 1$

$x = 1 + 2^y$

x	2	3	5	9
y	0	1	2	3

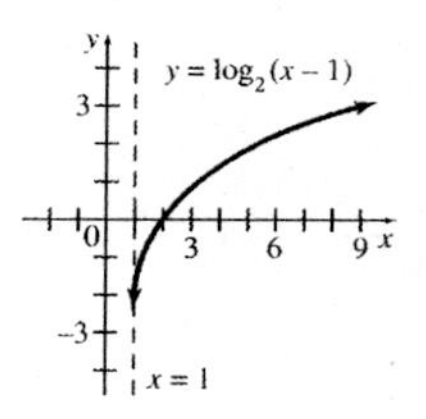

32. $y = 1 + \log_3 x$

$y - 1 = \log_3 x$

$3^{y-1} = x$

x	$\frac{1}{9}$	$\frac{1}{3}$	1	3	9
y	−1	0	1	2	3

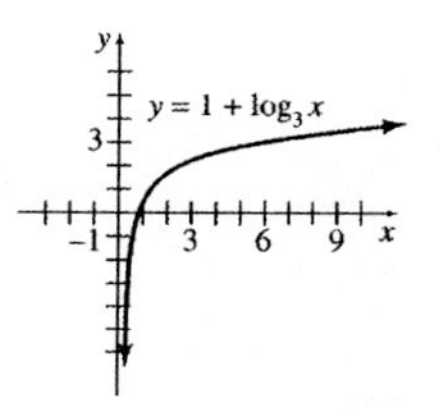

33. $y = -\ln(x+3)$

x	−2	−1	0	1	2
y	0	−0.69	−1.1	−1.4	−1.6

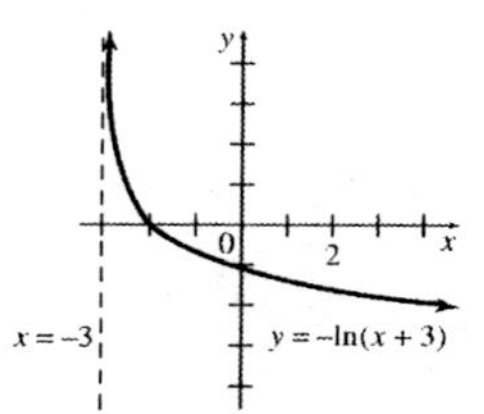

34. $y = 2 - \ln x^2$

x	−4	−3	−2	−1
y	−0.8	−0.2	0.6	2

x	1	2	3	4
y	2	0.6	−0.2	−0.8

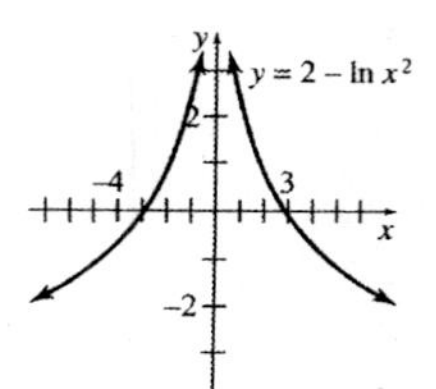

35. $2^{x+2}=\frac{1}{8}$

$2^{x+2}=\frac{1}{2^3}$

$2^{x+2}=2^{-3}$

$x+2=-3$

$x=-5$

36. $\left(\frac{9}{16}\right)^x=\frac{3}{4}$

$\left(\frac{3}{4}\right)^{2x}=\left(\frac{3}{4}\right)^1$

$2x=1$

$x=\frac{1}{2}$

37. $9^{2y+3}=27^y$

$(3^2)^{2y+3}=(3^3)y$

$3^{4y+6}=3^{3y}$

$4y+6=3y$

$y=-6$

38. $\frac{1}{2}=\left(\frac{b}{4}\right)^{1/4}$

$\left(\frac{1}{2}\right)^4=\frac{b}{4}$

$4\left(\frac{1}{2}\right)^4=b$

$4\left(\frac{1}{16}\right)=\frac{1}{4}=b$

39. $3^5=243$
The equation in logarithmic form is
$\log_3 243=5$.

40. $5^{1/2}=\sqrt{5}$
The equation in logarithmic form is
$\log_5\sqrt{5}=\frac{1}{2}$.

41. $e^{0.8}=2.22554$
The equation in logarithmic form is
$\ln 2.22554=0.8$.

42. $10^{1.07918}=12$
The equation in logarithmic form is
$\log 12=1.07918$.

43. $\log_2 32=5$

The equation in exponential form is $2^5=32$.

44. $\log_9 3=\frac{1}{2}$

The equation in exponential form is $9^{1/2}=3$.

45. $\ln 82.9=4.41763$
The equation in exponential form is
$e^{4.41763}=82.9$.

46. $\log 3.21=0.50651$
The equation in exponential form is
$10^{0.50651}=3.21$.
Recall that log x means $\log_{10} x$.

47. $\log_3 81=x$

$3^x=81$

$3^x=3^4$

$x=4$

48. $\log_{32} 16=x$

$32^x=16$

$2^{5x}=2^4$

$5x=4$

$x=\frac{4}{5}$

49. $\log_4 8=x$

$4^x=8$

$(2^2)^x=2^3$

$2x=3$

$x=\frac{3}{2}$

50. $\log_{100} 1000=x$

$100^x=1000$

$(10^2)^x=10^3$

$2x=3$

$x=\frac{3}{2}$

51. $\log_5 3k+\log_5 7k^3=\log_5 3k(7k^3)$
$=\log_5(21k^4)$

52. $\log_3 2y^3-\log_3 8y^2=\log_3\frac{2y^3}{8y^2}=\log_3\frac{y}{4}$

53. $4\log_3 y-2\log_3 x=\log_3 y^4-\log_3 x^2$
$=\log_3\left(\frac{y^4}{x^2}\right)$

54. $3\log_4 r^2-2\log_4 r=\log_4(r^2)^3-\log_4 r^2$
$=\log_4\left(\frac{r^6}{r^2}\right)=\log_4(r^4)$

55. $6^p=17$

$\ln 6^p=\ln 17$

$p\ln 6=\ln 17$

$p=\frac{\ln 17}{\ln 6}\approx 1.581$

56. $3^{z-2}=11$

$\ln 3^{z-2}=\ln 11$

$(z-2)\ln 3=\ln 11$

$z-2=\frac{\ln 11}{\ln 3}$

$z=\frac{\ln 11}{\ln 3}+2\approx 4.183$

57.
$$\begin{aligned} 2^{1-m} &= 7 \\ \ln 2^{1-m} &= \ln 7 \\ (1-m)\ln 2 &= \ln 7 \\ 1-m &= \frac{\ln 7}{\ln 2} \\ -m &= \frac{\ln 7}{\ln 2} - 1 \\ m &= 1 - \frac{\ln 7}{\ln 2} \approx -1.807 \end{aligned}$$

58.
$$\begin{aligned} 12^{-k} &= 9 \\ \ln 12^{-k} &= \ln 9 \\ -k \ln 12 &= \ln 9 \\ k &= -\frac{\ln 9}{\ln 12} \approx -0.884 \end{aligned}$$

59.
$$\begin{aligned} e^{-5-2x} &= 5 \\ \ln e^{-5-2x} &= \ln 5 \\ -5-2x &= \ln 5 \\ -2x &= \ln 5 + 5 \\ x &= \frac{\ln 5 + 5}{-2} \approx -3.305 \end{aligned}$$

60.
$$\begin{aligned} e^{3x-1} &= 14 \\ \ln(e^{3x-1}) &= \ln 14 \\ 3x-1 &= \ln 14 \\ 3x &= 1 + \ln 14 \\ x &= \frac{1+\ln 14}{3} \approx 1.213 \end{aligned}$$

61.
$$\begin{aligned} \left(1+\frac{m}{3}\right)^5 &= 15 \\ \left[\left(1+\frac{m}{3}\right)^5\right]^{1/5} &= 15^{1/5} \\ 1+\frac{m}{3} &= 15^{1/5} \\ \frac{m}{3} &= 15^{1/5} - 1 \\ m &= 3(15^{1/5} - 1) \approx 2.156 \end{aligned}$$

62.
$$\begin{aligned} \left(1+\frac{2p}{5}\right)^2 &= 3 \\ 1+\frac{2p}{5} &= \pm\sqrt{3} \\ 5+2p &= \pm 5\sqrt{3} \\ 2p &= -5 \pm \sqrt{3} \\ p &= \frac{-5 \pm 5\sqrt{3}}{2} \end{aligned}$$

$$p = \frac{-5+5\sqrt{3}}{2} \approx 1.830$$
$$\text{or } p = \frac{-5-5\sqrt{3}}{2} \approx -6.830$$

63.
$$\begin{aligned} \log_k 64 &= 6 \\ k^6 &= 64 \\ k^6 &= 2^6 \\ k &= 2 \end{aligned}$$

64.
$$\begin{aligned} \log_3(2x+5) &= 5 \\ 3^5 &= 2x+5 \\ 243 &= 2x+5 \\ 238 &= 2x \\ x &= 119 \end{aligned}$$

65.
$$\begin{aligned} \log(4p+1) + \log p &= \log 3 \\ \log[p(4p+1)] &= \log 3 \\ \log(4p^2 + p) &= \log 3 \\ 4p^2 + p &= 3 \\ 4p^2 + p - 3 &= 0 \\ (4p-3)(p+1) &= 0 \end{aligned}$$
$$4p-3=0 \quad \text{or} \quad p+1=0$$
$$p = \frac{3}{4} \qquad\qquad p=-1$$

p cannot be negative, so $p = \frac{3}{4}$.

66.
$$\begin{aligned} \log_2(5m-2) - \log_2(m+3) &= 2 \\ \log_2 \frac{5m-2}{m+3} &= 2 \\ \frac{5m-2}{m+3} &= 2^2 \\ 5m-2 &= 4(m+3) \\ 5m-2 &= 4m+12 \\ m &= 14 \end{aligned}$$

67. $f(x) = a^x;\ a > 0,\ a \neq 1$

(a) The domain is $(-\infty, \infty)$.

(b) The range is $(0, \infty)$.

(c) The y-intercept is 1.

(d) The x-axis, $y = 0$, is a horizontal asymptote.

(e) The function is increasing if $a > 1$.

(f) The function is decreasing if $0 < a < 1$.

68. $f(x) = \log_a x;\ a > 0,\ a \neq 1$

(a) The domain is $(0, \infty)$.

(b) The range is $(-\infty, \infty)$.

(c) The x-intercept is 1.

(d) The y-axis, $x = 0$, is a vertical asymptote.

(e) f is increasing if $a > 1$.

(f) f is decreasing if $0 < a < 1$.

69. The domain of $f(x) = a^x$ is the same as the range of $f(x) = \log_a x$ and the domain of $f(x) = \log_a x$ is the same as the range of $f(x) = a^x$. Both functions are increasing if $a > 1$. Both functions are decreasing if $0 < a < 1$. The functions are asymptotic to different axes.

70. $90° = 90\left(\frac{\pi}{180}\right) = \frac{90\pi}{180} = \frac{\pi}{2}$

71. $160° = 160\left(\frac{\pi}{180}\right) = \frac{8\pi}{9}$

72. $225° = 225\left(\frac{\pi}{180}\right) = \frac{5\pi}{4}$

73. $270° = 270\left(\frac{\pi}{180}\right) = \frac{3\pi}{2}$

74. $360° = 2\pi$

75. $405° = 405\left(\frac{\pi}{180}\right) = \frac{9\pi}{4}$

76. $5\pi = 5\pi\left(\frac{180°}{\pi}\right) = 900°$

77. $\frac{3\pi}{4} = \frac{3\pi}{4}\left(\frac{180°}{\pi}\right) = 135°$

78. $\frac{9\pi}{20} = \frac{9\pi}{20}\left(\frac{180°}{\pi}\right) = 81°$

79. $\frac{3\pi}{10} = \frac{3\pi}{10}\left(\frac{180°}{\pi}\right) = 54°$

80. $\frac{13\pi}{20} = \frac{13\pi}{20}\left(\frac{180°}{\pi}\right) = 117°$

81. $\frac{13\pi}{15} = \frac{13\pi}{15}\left(\frac{180°}{\pi}\right) = 156°$

82. When an angle $60°$ is drawn in standard position, one choice of a point on its terminal side is $(x, y) = (1, \sqrt{3})$. Then

$r = \sqrt{x^2 + y^2} = \sqrt{1+3} = 2$, so

$\sin 60° = \frac{y}{r} = \frac{\sqrt{3}}{2}$.

83. When an angle of $120°$ is drawn in standard position, $(x, y) = (-1, \sqrt{3})$ is one point on its terminal side, so $\tan 120° = \frac{y}{x} = -\sqrt{3}$.

84. When an angle of $-45°$ is drawn in standard position, one choice of a point on its terminal side is $(x, y) = (1, -1)$. Then

$r = \sqrt{x^2 + y^2} = \sqrt{1+1} = \sqrt{2}$, so

$\cos(-45°) = \frac{x}{r} = \frac{1}{\sqrt{2}} = \frac{\sqrt{2}}{2}$.

85. When an angle of $150°$ is drawn in standard position, one choice of a point on its terminal side is $(x, y) = (-\sqrt{3}, 1)$. Then

$r = \sqrt{x^2 + y^2} = \sqrt{3+1} = 2$, so

$\sec 150° = \frac{r}{x} = \frac{2}{-\sqrt{3}} = -\frac{2\sqrt{3}}{3}$.

86. When an angle of $120°$ is drawn in standard position, one choice of a point on its terminal side is $(x, y) = (-1, \sqrt{3})$. Then

$r = \sqrt{x^2 + y^2} = \sqrt{1+3} = 2$, so

$\csc 120° = \frac{r}{y} = \frac{2}{\sqrt{3}} = \frac{2\sqrt{3}}{3}$.

87. When an angle of $300°$ is drawn in standard position, $(x, y) = (1, -\sqrt{3})$ is one point on its terminal side, so $\cot 300° = \frac{x}{y} = -\frac{1}{\sqrt{3}} = -\frac{\sqrt{3}}{3}$.

88. When an angle of $\frac{\pi}{6}$ is drawn in standard position, one choice of a point on its terminal side is $(x, y) = (\sqrt{3}, 1)$. Then

$r = \sqrt{x^2 + y^2} = \sqrt{3+1} = 2$, so $\sin\frac{\pi}{6} = \frac{y}{r} = \frac{1}{2}$.

89. When an angle of $\frac{7\pi}{3}$ is drawn in standard position, one choice of a point on its terminal side is $(x, y) = (1, \sqrt{3})$. Then

$r = \sqrt{x^2 + y^2} = \sqrt{1+3} = 2$, so

$\cos \frac{7\pi}{3} = \frac{x}{r} = \frac{1}{2}$.

90. When an angle of $\frac{5\pi}{3}$ is drawn in standard position, one choice of a point on its terminal side is $(x, y) = (1, -\sqrt{3})$. Then

$r = \sqrt{x^2 + y^2} = \sqrt{1+3} = 2$, so

$\sec \frac{5\pi}{3} = \frac{r}{x} = \frac{2}{1} = 2$.

91. When an angle of $\frac{7\pi}{3}$ is drawn in standard position, $(x, y) = (1, \sqrt{3})$ is one point on its terminal side. Then $r = \sqrt{1+3} = 2$, so

$\csc \frac{7\pi}{3} = \frac{r}{y} = \frac{2}{\sqrt{3}} = \frac{2\sqrt{3}}{3}$.

92. $\sin 47° \approx 0.7314$

93. $\cos 72° \approx 0.3090$

94. $\tan 115° \approx -2.1445$

95. $\sin(-123°) \approx -0.8387$

96. $\sin 2.3581 \approx 0.7058$

97. $\cos 0.8215 \approx 0.6811$

98. $\cos 0.5934 \approx 0.8290$

99. $\tan 1.2915 \approx 3.4868$

100. The graph of $y = \cos x$ appears in Figure 27 in Section 4 of this chapter. To get $y = 4\cos x$, each value of y in $y = \cos x$ must be multiplied by 4. This gives a graph going through $(0, 4)$, $(\pi, -4)$ and $(2\pi, 4)$.

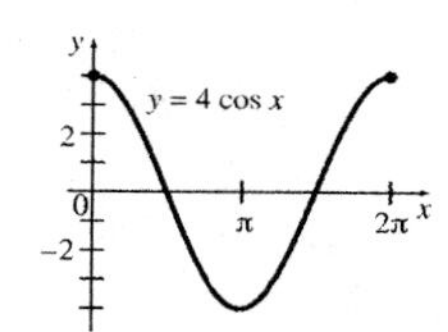

101. The graph of $y = \frac{1}{2}\tan x$ is similar to the graph of $y = \tan x$ except that each ordinate value is multiplied by a factor of $\frac{1}{2}$. Note that the points $\left(\frac{\pi}{4}, \frac{1}{2}\right)$, $(0, 0)$, and $\left(-\frac{\pi}{4}, -\frac{1}{2}\right)$ lie on the graph.

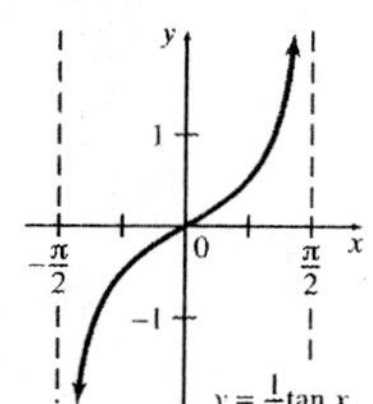

102. The graph of $y = \tan x$ appears in Figure 28 in Section 4 in this chapter. The difference between the graph of $y = \tan x$ and $y = -\tan x$ is that the y-values of points on the graph of $y = -\tan x$ are the opposites of the y-values of the corresponding points on the graph of $y = \tan x$.
A sample calculation:
When $x = \frac{\pi}{4}$, $y = -\tan\frac{\pi}{4} = -1$.

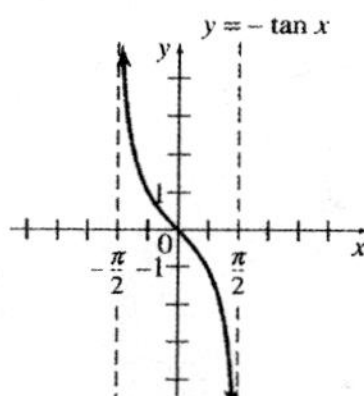

103. The graph of $y = -\frac{2}{3}\sin x$ is similar to the graph of $y = \sin x$ except that it has two-thirds the amplitude and is reflected about the x-axis.

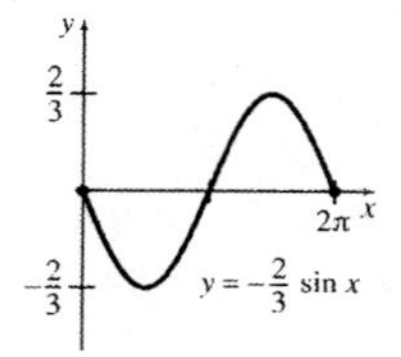

104. $y = 17{,}000$, $y_0 = 15{,}000$, $t = 4$

(a) $y = y_0 e^{kt}$

$17{,}000 = 15{,}000e^{4k} \Rightarrow \frac{17}{15} = e^{4k} \Rightarrow$

$\ln\left(\frac{17}{15}\right) = 4k \Rightarrow k = \frac{\ln\left(\frac{17}{15}\right)}{4} \approx 0.313$

So, $y = 15{,}000e^{0.0313t}$.

(b) $45,000 = 15,000e^{0.0313t}$

$3 = e^{0.0313t}$

$\ln 3 = 0.0313t$

$\dfrac{\ln 3}{0.0313} \approx 35.1 = k$

It would take about 35 years.

105. $I(x) \ge 1$

$I(x) = 10e^{-0.3x}$

$10e^{-0.3x} \ge 1$

$e^{-0.3x} \ge 0.1$

$-0.3x \ge \ln 0.1$

$x \le \dfrac{\ln 0.1}{-0.3} \approx 7.7$

The greatest depth is about 7.7 m.

106. Graph $y = c(t) = e^{-t} - e^{-2t}$ on a graphing calculator and locate the maximum point. A calculator shows that the x-coordinate of the maximum point is about 0.69, and the y-coordinate is exactly 0.25. Thus, the maximum concentration of 0.25 occurs at about 0.69 minutes.

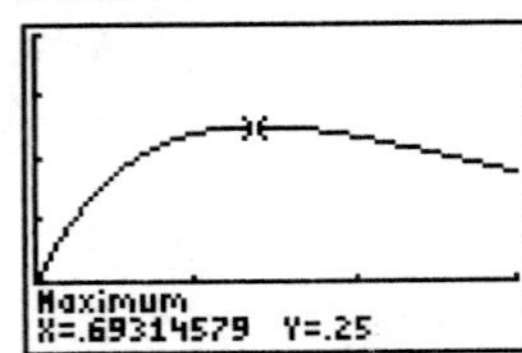

[0, 1.5] by [–0.1, 0.4]

107. $g(t) = \dfrac{c}{a} + \left(g_0 - \dfrac{c}{a}\right)e^{-at}$

(a) If $g_0 = 0.08$, $c = 0.1$, and $a = 1.3$, the function becomes

$g(t) = \dfrac{0.1}{1.3} + \left(0.08 - \dfrac{0.1}{1.3}\right)e^{-1.3t}.$

Graph this function on a graphing calculator.

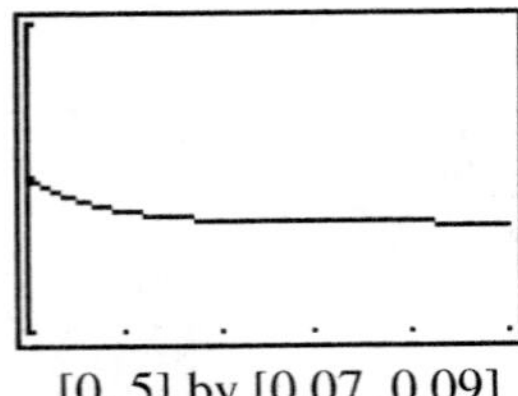

[0, 5] by [0.07, 0.09]

From the graph, we see that the maximum value of g for $t \ge 0$ occurs at $t = 0$, the time when the drug is first injected. The maximum amount of glucose in the bloodstream, given by $G(0)$ is 0.08 gram.

(b) From the graph, we see that the amount of glucose in the bloodstream decreases from the initial value of 0.08 gram, so it will never increase to 0.1 gram. We can also reach this conclusion by graphing $y_1 = G(t)$ and $y_2 = 0.1$ on the same screen with the window given in (a) and observing that the graphs of y_1 and y_2 do not intersect.

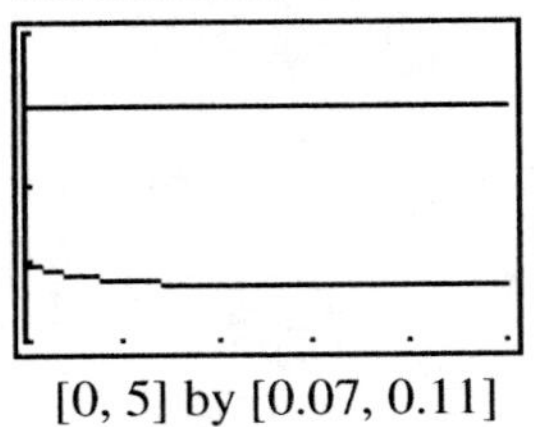

[0, 5] by [0.07, 0.11]

(c) Using the TRACE function and the graph in part (a), we see that as t increases, the graph of $y_1 = G(t)$ becomes almost horizontal, and $G(t)$ approaches approximately 0.0769. Note that $\dfrac{c}{a} = \dfrac{0.1}{1.3} \approx 0.0769$. The amount of glucose in the bloodstream after a long time approaches 0.0769 grams.

108. **(a)**

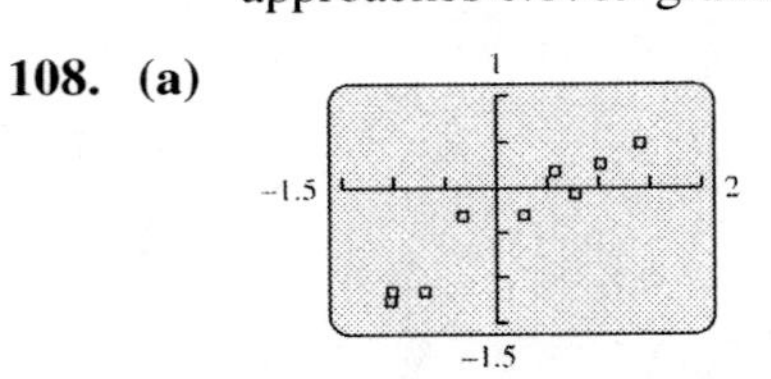

Yes, the data follows a linear trend.

(b)

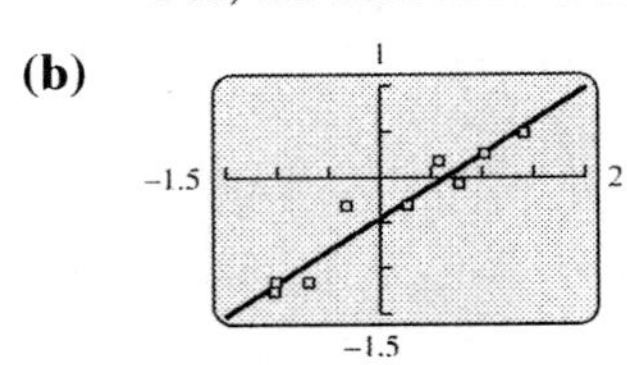

The least squares line is $\log y = 0.7097 \log x - 0.4480$.

(c) Solve the equation from part (b) for y.

$\log y = 0.7097 \log x - 0.4480$

$\log y = \log x^{0.7097} - 0.4480$

$10^{\log y} = 10^{\log x^{0.7097} - 0.4480}$

$y = 10^{\log x^{0.7097}} \cdot 10^{-0.4480}$

$y \approx 0.3565x^{0.7097}$

(d) Using a graphing calculator, the coefficient of correlation $r \approx 0.9625$.

109. (a) The volume of a sphere is given by $V = \frac{4}{3}\pi r^3$. The radius of a cancer cell is $\frac{2(10)^{-5}}{2} = 10^{-5}$ m. So,

$$V = \frac{4}{3}\pi r^3 = \frac{4}{3}\pi\left(10^{-5}\right)^3$$
$$\approx 4.19\times 10^{-15} \text{ cubic meters}$$

(b) The formula for the total volume of the cancer cells after t days is $V \approx \left(4.19\times 10^{-15}\right)2^t$ cubic meters.

(c) The radius of a tumor with a diameter of 1 cm = 0.01 m is 0.005 m, so $V = \frac{4}{3}\pi r^3 = \frac{4}{3}\pi(0.005)^3$.

Using the equation found in part (b), we have

$$\frac{4}{3}\pi(0.005)^3 = \left(4.19\times 10^{-15}\right)2^t$$
$$\frac{\frac{4}{3}\pi(0.005)^3}{4.19\times 10^{-15}} = 2^t$$
$$\ln\left(\frac{\frac{4}{3}\pi(0.005)^3}{4.19\times 10^{-15}}\right) = t\ln 2$$
$$t = \frac{\ln\left(\frac{\frac{4}{3}\pi(0.005)^3}{4.19\times 10^{-15}}\right)}{\ln 2} \approx 26.8969$$

It will take about 27 days for the cancer cell to grow to a tumor with a diameter of 1 cm.

110. $m(g) = e^{0.02+0.062g-0.000165g^2}$

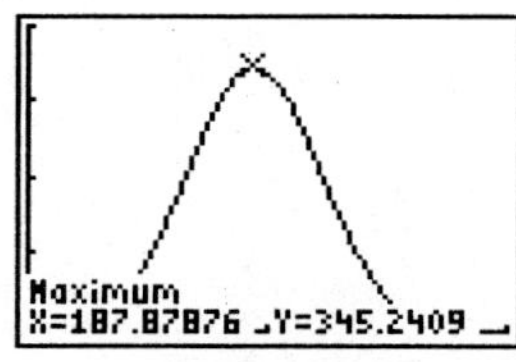

[0, 400] by [0, 400]

This function has a maximum value of $y \approx 345$ at $x \approx 187.9$. This is the largest value for which the formula gives a reasonable answer. The predicted mass of a polar bear with this girth is about 345 kg.

111. (a) The first three years of infancy corresponds to 0 months to 36 months, so the domain is [0, 36].

(b) In both cases, the graph of the quadratic function in the exponent opens upward and the x coordinate of the vertex is greater than 36 ($x \approx 47$ for the awake infants and $x \approx 40$ for the sleeping infants). So the quadratic functions are both decreasing over this time. Therefore, both respiratory rates are decreasing.

(c)

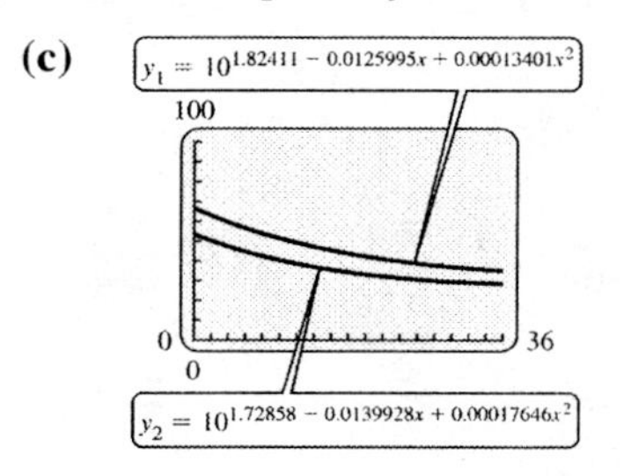

(d) When $x = 12$, the waking respiratory rate is $y_1 \approx 49.23$ breaths per minute, and the sleeping respiratory rate is $y_2 \approx 38.55$. Therefore, for a 1-year-old infant in the 95th percentile, the waking respiratory rate is approximately 49.23 – 38.55 ≈ 10.7 breaths per minute higher.

112. (a) $S = 21.35 + 104.6\ln A$

(b) $S = 85.49A^{0.3040}$

(c)

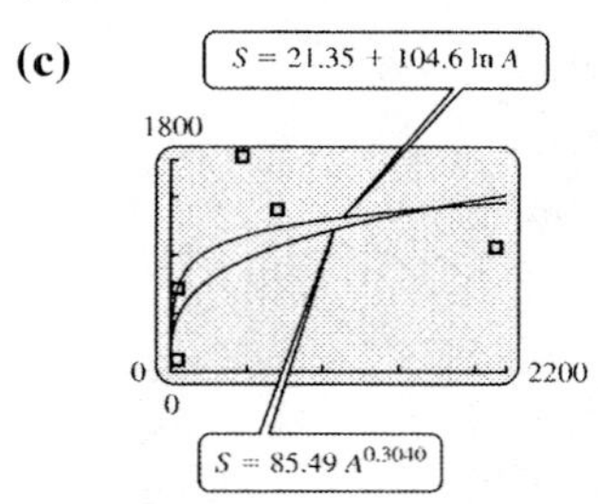

(d) $S = 21.35 + 104.6\ln(984.2) \approx 742.2$

$S = 85.49(984.2)^{0.3040} \approx 694.7$

Neither number is close to the actual number of 421.

(e) Answers will vary.

113. $P(t) = 90 + 15\sin 144\pi t$

The maximum possible value of $\sin\alpha$ is 1, while the minimum possible value is –1. Replacing α with $144\pi t$ gives

$$-1 \le \sin 144\pi t \le 1,$$
$$90 + 15(-1) \le P(t) \le 90 + 15(1)$$
$$75 \le P(t) \le 105.$$

(continued on next page)

(continued)

Therefore, the minimum value of $P(t)$ is 75 and the maximum value of $P(t)$ is 105.

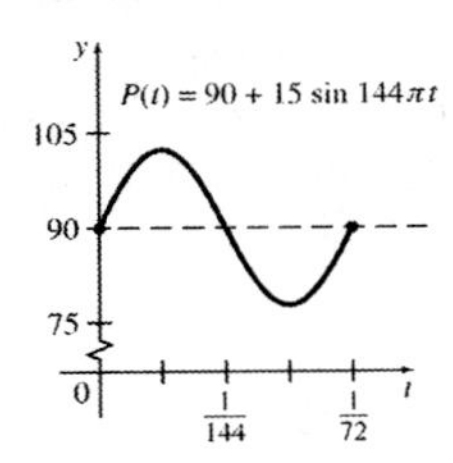

114. (a) The period is given by $\frac{2\pi}{b}$, so

$$\text{period} = \frac{2\pi}{\frac{2\pi}{24}} = 24 \text{ hr}$$

(b) Let $a = 546$, $C_0 = 511$, $C_1 = 634$, $t_0 = 20.27$, $t_1 = 6.05$, and $b = 24$. Then, the function is

$$C(t) = 546 + (634 - 546)e^{3[\cos((2\pi/24)(t-6.05))-1]} + (511 - 546)e^{3[\cos((2\pi/24)(t-20.27))-1]}$$
$$= 546 + 88e^{3[\cos((\pi/12)(t-6.05))-1]} - 35e^{3[\cos((\pi/12)(t-20.27))-1]}$$

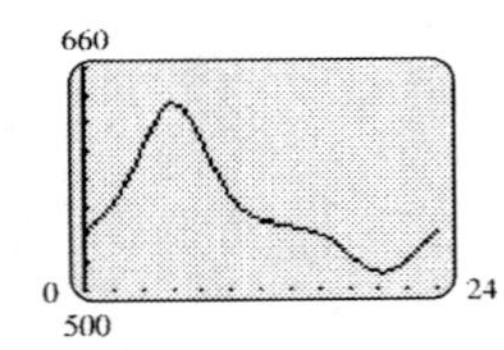

(c)

$$C(20.27) = 546 + 88e^{3[\cos((\pi/12)(20.27-6.05))-1]} - 35e^{3[\cos((\pi/12)(20.27-20.27))-1]}$$
$$\approx 511 \text{ ng/dl}$$

This is approximately the value of C_0.

(d)

$$C(6.05) = 546 + 88e^{3[\cos((\pi/12)(6.05-6.05))-1]} - 35e^{3[\cos((\pi/12)(6.05-20.27))-1]}$$
$$\approx 634 \text{ ng/dl}$$

This is approximately the value of C_1.

(e) No, $C(t_0)$ is very close to C_0 because, when $t = t_0$, the last term in C_0 and the first term combine to yield C_0. The middle term in $C(t_0)$ is small but not 0. It has the value of about 0.3570. Similarly, the first two terms in $C(t_1)$ combine to give C_1, while the last term is small, but not 0.

115. (a) The line passes through the points (60, 0.8) and (110, 2.2).

$$m = \frac{2.2 - 0.8}{110 - 60} = \frac{1.4}{50} = 0.028$$

Using the point (60, 0.8), we have

$$0.8 = 0.028(60) + b$$
$$0.8 = 1.68 + b$$
$$-0.88 = b$$

Thus, the equation is $g(t) = 0.028t - 0.88$.

(b)

Per capita grain harvests have been increasing, but at a slower rate and are leveling off at about 0.32 ton per person.

116. (a)

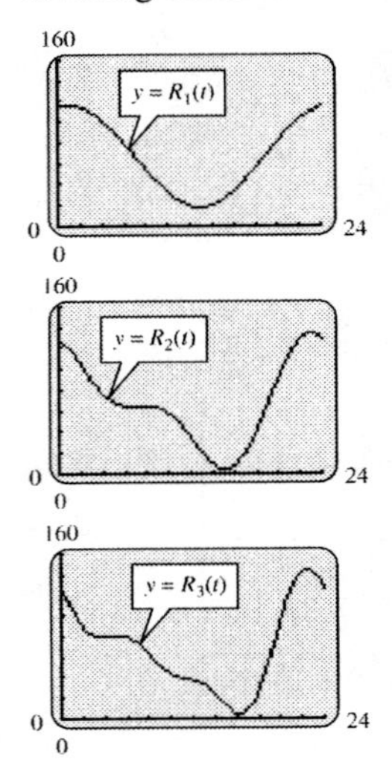

As more terms are added, the function has greater complexity.

(b) $R_1(12) = 67.75 + 47.72\cos\left(\frac{12\pi}{12}\right) + 11.79\sin\left(\frac{12\pi}{12}\right)$
$= 67.75 + 47.72\cos\pi + 11.79\sin\pi$
$= 67.75 + 47.72(-1) + 11.79(0)$
≈ 20.0

$R_2(12) = R_1(12) + 14.29\cos\left(\frac{12\pi}{6}\right) - 21.09\sin\left(\frac{12\pi}{6}\right)$
$= 20.0 + 14.29\cos 2\pi - 21.09\sin 2\pi$
$= 20.0 + 14.29(1) - 21.09(0)$
$= 34.3$

$R_3(12) = R_2(12) - 2.86\cos\left(\frac{12\pi}{4}\right) - 14.31\sin\left(\frac{12\pi}{4}\right)$
$= 34.3 - 2.86\cos 3\pi - 14.31\sin 3\pi$
$= 34.3 - 2.86(-1) - 14.31(0)$
≈ 37.2

R_3 gives the most accurate value.

117. $p = \$6902,\ r = 6\%,\ t = 8,\ m = 2$

$A = P\left(1+\frac{r}{m}\right)^{tm}$

$A = 6902\left(1+\frac{0.06}{2}\right)^{8(2)}$
$= 6902(1.03)^{16}$
$= \$11,075.68$

Interest $= A - P$
$= \$11,075.68 - \$6902 = \$4173.68$

118. $P = \$2781.36,\ r = 4.8\%,\ t = 6,\ m = 4$

$A = P\left(1+\frac{r}{m}\right)^{tm}$

$A = 2781.36\left(1+\frac{0.048}{4}\right)^{(6)(4)}$
$= 2781.36(1.012)^{24}$
$= \$3703.31$

Interest $= \$3703.31 - \$2781.36 = \$921.95$

119. $P = \$12,104,\ r = 6.2\%,\ t = 2$

$A = Pe^{rt}$
$= 12,104e^{0.062(2)} = \$13,701.92$

120. $P = \$12,104,\ r = 6.2\%,\ t = 4$

$A = Pe^{rt}$
$A = 12,104e^{0.062(4)} = 12,104e^{0.248}$
$\approx \$15,510.79$

121. $A = \$1500,\ r = 0.06,\ t = 9$

$A = Pe^{rt}$
$= 1500e^{0.06(9)} = 1500e^{0.54} \approx \$2574.01$

122. $P = \$12,000,\ r = 0.05,\ t = 8$

$A = 12,000e^{0.05(8)} = 12,000e^{0.40}$
$\approx \$17,901.90$

123. $1000 deposited at 6% compounded semiannually.

$A = P\left(1+\frac{r}{m}\right)^{tm}$

To double:

$2(1000) = 1000\left(1+\frac{0.06}{2}\right)^{t\cdot 2}$
$2 = 1.03^{2t}$
$\ln 2 = 2t \ln 1.03$
$t = \frac{\ln 2}{2\ln 1.03} \approx 12$ years

To triple:

$3(1000) = 1000\left(1+\frac{0.06}{2}\right)^{t\cdot 2}$
$3 = 1.03^{2t}$
$\ln 3 = 2t \ln 1.03$
$t = \frac{\ln 3}{2\ln 1.03} \approx 19$ years

124. $2100 deposited at 4% compounded quarterly.

$A = P\left(1+\frac{r}{m}\right)^{tm}$

To double:

$2(2100) = 2100\left(1+\frac{0.04}{4}\right)^{t\cdot 4}$
$2 = 1.01^{4t}$
$\ln 2 = 4t \ln 1.01$
$t = \frac{\ln 2}{4\ln 1.01} \approx 17.4$

Because interest is compounded quarterly, round the result up to the nearest quarter, which is 17.5 years or 70 quarters.

(*continued on next page*)

(*continued*)

To triple:

$$3(2100) = 2100\left(1+\frac{0.04}{4}\right)^{t\cdot 4}$$
$$3 = 1.01^{4t}$$
$$\ln 3 = 4t \ln 1.01$$
$$t = \frac{\ln 3}{4 \ln 1.01}$$
$$\approx 27.6$$

Because interest is compounded quarterly, round the result up to the nearest quarter, 27.75 years or 111 quarters.

125. $y = y_o e^{-kt}$

(a) $100,000 = 128,000e^{-k(5)}$

$$128,000 = 100,000e^{5k}$$
$$\frac{128}{100} = e^{5k}$$
$$\ln\left(\frac{128}{100}\right) = 5k$$
$$0.05 \approx k$$
$$y = 100,000e^{-0.05t}$$

(b) $70,000 = 100,000e^{-0.05t}$

$$\frac{7}{10} = e^{-0.05t}$$
$$\ln\frac{7}{10} = -0.05t$$
$$7.1 \approx t$$

It will take about 7.1 years.

126. $t = (1.26\times10^9)\dfrac{\ln\left[1+8.33\left(\frac{A}{K}\right)\right]}{\ln 2}$

(a) $A = 0, K > 0$

$$t = (1.26\times10^9)\frac{\ln[1+8.33(0)]}{\ln 2}$$
$$= (1.26\times10^9)(0) = 0 \text{ years}$$

(b) $t = (1.26\times10^9)\dfrac{\ln[1+8.33(0.212)]}{\ln 2}$

$$= (1.26\times10^9)\frac{\ln 2.76596}{\ln 2}$$
$$= 1,849,403,169$$

or about 1.85×10^9 years

(c) As r increases, t increases, but at a slower and slower rate. As r decreases, t decreases at a faster and faster rate

127. (a) Enter the data into a graphing calculator and plot.

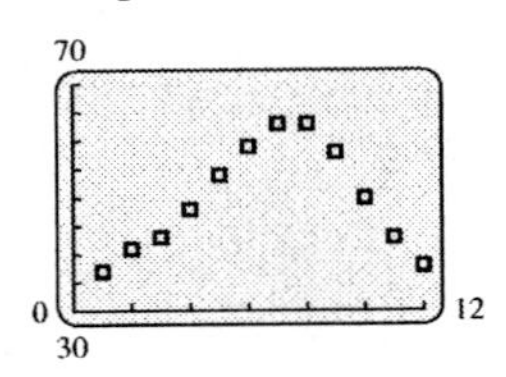

(b) The sine regression is
$y = 12.5680 \sin(0.54655t - 2.34990) + 50.2671$

(c) Graph the function with the data.

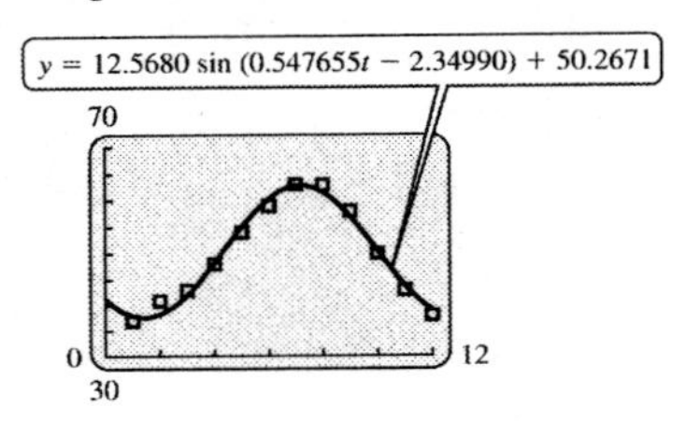

(d) $T = \dfrac{2\pi}{b} \approx 11.4729$

The period is about 11.5 months.

Chapter 3

THE DERIVATIVE

3.1 Limits

1. Since $\lim_{x\to 2^-} f(x)$ $f(x)$ does not equal $\lim_{x\to 2^+} f(x)$, $\lim_{x\to 2} f(x)$ does not exist. The answer is **c**.

3. Since $\lim_{x\to 4^-} f(x) = \lim_{x\to 4^+} f(x) = 6$, $\lim_{x\to 4} f(x) = 6$. The answer is **b**.

5. (a) By reading the graph, as x gets closer to 3 from the left or right, $f(x)$ gets closer to 3. $\lim_{x\to 3} f(x) = 3$

(b) By reading the graph, as x gets closer to 0 from the left or right, $f(x)$ gets closer to 1. $\lim_{x\to 0} f(x) = 1$.

7. (a) By reading the graph, as x gets closer to 0 from the left or right, $f(x)$ gets closer to 0. $\lim_{x\to 0} f(x) = 0$

(b) By reading the graph, as x gets closer to 2 from the left, $f(x)$ gets closer to -2, but as x gets closer to 2 from the right, $f(x)$ gets closer to 1. $\lim_{x\to 2} f(x)$ does not exist.

9. (a) (i) By reading the graph, as x gets closer to –2 from the left, $f(x)$ gets closer to –1.

$\lim_{x\to -2^-} f(x) = -1$

(ii) By reading the graph, as x gets closer to –2 from the right, $f(x)$ gets closer to $-\frac{1}{2}$.

$$\lim_{x\to -2^+} f(x) = -\frac{1}{2}$$

(iii) Since $\lim_{x\to -2^-} f(x) = -1$ and $\lim_{x\to -2^+} f(x) = -\frac{1}{2}$, $\lim_{x\to -2} f(x)$ does not exist.

(iv) $f(-2)$ does not exist since there is no point on the graph with an x-coordinate of –2.

(b) (i) By reading the graph, as x gets closer to -1 from the left, $f(x)$ gets closer to $-\frac{1}{2}$.

$$\lim_{x\to -1^-} f(x) = -\frac{1}{2}$$

(ii) By reading the graph, as x gets closer to -1 from the right, $f(x)$ gets closer to $-\frac{1}{2}$.

$$\lim_{x\to -1^+} f(x) = -\frac{1}{2}$$

(iii) Since $\lim_{x\to -1^-} f(x) = -\frac{1}{2}$ and $\lim_{x\to -1^+} f(x) = -\frac{1}{2}$, $\lim_{x\to -1} f(x) = -\frac{1}{2}$.

(iv) $f(-1) = -\frac{1}{2}$ since $\left(-1, -\frac{1}{2}\right)$ is a point of the graph.

11. By reading the graph, as x moves further to the right, $f(x)$ gets closer to 3. Therefore, $\lim_{x\to\infty} f(x) = 3$.

13. $\lim_{x\to 2^-} F(x)$ in Exercise 6 exists because $\lim_{x\to 2^-} F(x) = 4$ and $\lim_{x\to 2^+} F(x) = 4$

$\lim_{x\to -2} f(x)$ in Exercise 9 does not exist since $\lim_{x\to -2^-} f(x) = -1$, but $\lim_{x\to -2^+} f(x) = -\frac{1}{2}$.

15. From the table, as x approaches 1 from the left or the right, $f(x)$ approaches 4. $\lim_{x\to 1} f(x) = 4$

17. $k(x)=\dfrac{x^3-2x-4}{x-2}$; find $\lim_{x\to 2} k(x)$.

x	1.9	1.99	1.999
$k(x)$	9.41	9.9401	9.9940

x	2.001	2.01	2.1
$k(x)$	10.0060	10.0601	10.61

As x approaches 2 from the left or the right, $k(x)$ approaches 10.

$\lim_{x\to 2} k(x)=10$

19. $h(x)=\dfrac{\sqrt{x}-2}{x-1}$; find $\lim_{x\to 1} h(x)$.

x	0.9	0.99	0.999
$h(x)$	10.51317	100.50126	1000.50013

x	1.001	1.01	1.1
$h(x)$	−999.50012	−99.50124	−9.51191

$\lim_{x\to 1^-} =\infty,\ \lim_{x\to 1^+} =-\infty$

Thus, $\lim_{x\to 1} h(x)$ does not exist.

21. $f(x)=\dfrac{\sin x}{x}$; find $\lim_{x\to 0} f(x)$.

x	−0.1	−0.01	−0.001
$f(x)$	0.99833	0.99998	1.00000

x	0.001	0.01	0.1
$f(x)$	1.00000	0.99998	0.99833

$\lim_{x\to 0^-} f(x)=1,\ \lim_{x\to 0^+} f(x)=1$

Thus, $\lim_{x\to 0} f(x)=1$.

23. $\lim_{x\to 4}[f(x)-g(x)]=\lim_{x\to 4} f(x)-\lim_{x\to 4} g(x)$
$=9-27=-18$

25. $\lim_{x\to 4}\dfrac{f(x)}{g(x)}=\dfrac{\lim_{x\to 4} f(x)}{\lim_{x\to 4} g(x)}=\dfrac{9}{27}=\dfrac{1}{3}$

27. $\lim_{x\to 4}\sqrt{f(x)}=\lim_{x\to 4}[f(x)^{1/2}]=\left[\lim_{x\to 4} f(x)\right]^{1/2}$
$=9^{1/2}=3$

29. $\lim_{x\to 4} 2^{f(x)}=2^{\lim_{x\to 4} f(x)}=2^9=512$

31. $\lim_{x\to 4}\dfrac{f(x)+g(x)}{2g(x)}=\dfrac{\lim_{x\to 4}[f(x)+g(x)]}{\lim_{x\to 4} 2g(x)}$
$=\dfrac{\lim_{x\to 4} f(x)+\lim_{x\to 4} g(x)}{2\lim_{x\to 4} g(x)}$
$=\dfrac{9+27}{2(27)}=\dfrac{36}{54}=\dfrac{2}{3}$

33. $\lim_{x\to 4}\left[\sin\left(\dfrac{\pi}{18}\cdot f(x)\right)\right]=\sin\left(\dfrac{\pi}{18}\cdot\lim_{x\to 4} f(x)\right)$
$=\sin\left(\dfrac{\pi}{18}\cdot 9\right)$
$=\sin\dfrac{\pi}{2}=1$

35. $\lim_{x\to 3}\dfrac{x^2-9}{x-3}=\lim_{x\to 3}\dfrac{(x-3)(x+3)}{x-3}$
$=\lim_{x\to 3}(x+3)$
$=\lim_{x\to 3} x+\lim_{x\to 3} 3$
$=3+3=6$

37. $\lim_{x\to 1}\dfrac{5x^2-7x+2}{x^2-1}=\lim_{x\to 1}\dfrac{(5x-2)(x-1)}{(x+1)(x-1)}$
$=\lim_{x\to 1}\dfrac{5x-2}{x+1}$
$=\dfrac{5-2}{2}=\dfrac{3}{2}$

39. $\lim_{x\to -2}\dfrac{x^2-x-6}{x+2}=\lim_{x\to -2}\dfrac{(x-3)(x+2)}{x+2}$
$=\lim_{x\to -2}(x-3)$
$=\lim_{x\to -2} x+\lim_{x\to -2}(-3)$
$=-2-3=-5$

41. $\lim_{x\to 0}\dfrac{\frac{1}{x+3}-\frac{1}{3}}{x}=\lim_{x\to 0}\left(\dfrac{1}{x+3}-\dfrac{1}{3}\right)\left(\dfrac{1}{x}\right)$
$=\lim_{x\to 0}\left[\dfrac{3}{3(x+3)}-\dfrac{x+3}{3(x+3)}\right]\left(\dfrac{1}{x}\right)$
$=\lim_{x\to 0}\dfrac{3-x-3}{3(x+3)(x)}$
$=\lim_{x\to 0}\dfrac{-x}{3(x+3)x}=\lim_{x\to 0}\dfrac{-1}{3(x+3)}$
$=\dfrac{-1}{3(0+3)}=-\dfrac{1}{9}$

43. $$\lim_{x\to 25}\frac{\sqrt{x}-5}{x-25}=\lim_{x\to 25}\frac{\sqrt{x}-5}{x-25}\cdot\frac{\sqrt{x}+5}{\sqrt{x}+5}$$
$$=\lim_{x\to 25}\frac{x-25}{(x-25)(\sqrt{x}+5)}$$
$$=\lim_{x\to 25}\frac{1}{\sqrt{x}+5}=\frac{1}{\sqrt{25}+5}=\frac{1}{10}$$

45. $$\lim_{h\to 0}\frac{(x+h)^2-x^2}{h}$$
$$=\lim_{h\to 0}\frac{x^2+2hx+h^2-x^2}{h}$$
$$=\lim_{h\to 0}\frac{2hx+h^2}{h}=\lim_{h\to 0}\frac{h(2x+h)}{h}$$
$$=\lim_{h\to 0}(2x+h)=2x+0=2x$$

47. $$\lim_{x\to 0}\frac{1-\cos^2 x}{\sin^2 x}=\lim_{x\to 0}\frac{\sin^2 x}{\sin^2 x}=\lim_{x\to 0}1=1$$

49. $$\lim_{x\to\infty}\frac{3x}{7x-1}=\lim_{x\to\infty}\frac{\frac{3x}{x}}{\frac{7x}{x}-\frac{1}{x}}=\lim_{x\to\infty}\frac{3}{7-\frac{1}{x}}$$
$$=\frac{3}{7-0}=\frac{3}{7}$$

51. $$\lim_{x\to-\infty}\frac{3x^2+2x}{2x^2-2x+1}=\lim_{x\to-\infty}\frac{\frac{3x^2}{x^2}+\frac{2x}{x^2}}{\frac{2x^2}{x^2}-\frac{2x}{x^2}+\frac{1}{x^2}}$$
$$=\lim_{x\to-\infty}\frac{3+\frac{2}{x}}{2-\frac{2}{x}+\frac{1}{x^2}}$$
$$=\frac{3-0}{2-0+0}=\frac{3}{2}$$

53. $$\lim_{x\to\infty}\frac{3x^3+2x-1}{2x^4-3x^3-2}=\lim_{x\to\infty}\frac{\frac{3x^3}{x^4}+\frac{2x}{x^4}-\frac{1}{x^4}}{\frac{2x^4}{x^4}-\frac{3x^3}{x^4}-\frac{2}{x^4}}$$
$$=\lim_{x\to\infty}\frac{\frac{3}{x}+\frac{2}{x^3}-\frac{1}{x^4}}{2-\frac{3}{x}-\frac{2}{x^4}}$$
$$=\frac{0+0-0}{2-0-0}=0$$

55. $$\lim_{x\to\infty}\frac{2x^3-x-3}{6x^2-x-1}=\lim_{x\to\infty}\frac{\frac{2x^3}{x^2}-\frac{x}{x^2}-\frac{3}{x^2}}{\frac{6x^2}{x^2}-\frac{x}{x^2}-\frac{1}{x^2}}$$
$$=\lim_{x\to\infty}\frac{2x-\frac{1}{x}-\frac{3}{x^2}}{6-\frac{1}{x}-\frac{1}{x^2}}=\infty$$

The denominator approaches 6, while the numerator becomes a positive number that is greater and greater in magnitude, so the limit does not exist.

57. $$\lim_{x\to\infty}\frac{2x^2-7x^4}{9x^2+5x-6}=\lim_{x\to\infty}\frac{\frac{2x^2}{x^2}-\frac{7x^4}{x^2}}{\frac{9x^2}{x^2}+\frac{5x}{x^2}-\frac{6}{x^2}}$$
$$=\lim_{x\to\infty}\frac{2-7x^2}{9+\frac{5}{x}-\frac{6}{x^2}}$$

The denominator approaches 9, while the numerator becomes a negative number that is larger and larger in magnitude, so

$$\lim_{x\to\infty}\frac{2x^2-7x^4}{9x^2+5x-6}=-\infty \text{ (does not exist).}$$

59. $\lim_{x\to-1^-}f(x)=1$ and $\lim_{x\to-1^+}f(x)=1$.

Therefore $\lim_{x\to-1}f(x)=1$.

61. **(a)** $\lim_{x\to 3}f(x)=2$.

(b) $\lim_{x\to 5}f(x)$ does not exist since

$\lim_{x\to 5^-}f(x)=2$ and $\lim_{x\to 5^+}f(x)=8$.

63. Find $\lim_{x\to 3}f(x)$, where $f(x)=\frac{x^2-9}{x-3}$.

x	2.9	2.99	2.999	3.001	3.01	3.1
$f(x)$	5.9	5.99	5.999	6.001	6.01	6.1

$$\lim_{x\to 3}f(x)=\lim_{x\to 3}\frac{x^2-9}{x-3}=6.$$

65. Find $\lim_{x\to 1}f(x)$, where $f(x)=\frac{5x^2-7x+2}{x^2-1}$.

x	0.9	0.99	0.999	1.001	1.01	1.1
$f(x)$	1.316	1.482	1.498	1.502	1.517	1.667

$$\lim_{x\to 1}f(x)=\lim_{x\to 1}\frac{5x^2-7x+2}{x^2-1}=\frac{3}{2}=1.5$$

67. **(a)** $\lim_{x\to-2}\frac{3x}{(x+2)^3}$ does not exist since

$$\lim_{x\to-2^+}\frac{3x}{(x+2)^3}=-\infty \text{ and}$$
$$\lim_{x\to-2^-}\frac{3x}{(x+2)^3}=\infty.$$

(b) Since $(x+2)^3=0$ when $x=-2$, the vertical asymptote of the graph of $F(x)$ is $x=-2$.

(c) The two answers are related. Since $x = -2$ is a vertical asymptote, we know that $\lim_{x \to -2} F(x)$ does not exist.

69. Answers will vary.

71. (a) $\lim_{x \to -\infty} e^x = 0$ since, as the graph goes further to the left, e^x gets closer to 0.

(b) The graph of e^x has a horizontal asymptote at $y = 0$ since $\lim_{x \to -\infty} e^x = 0$.

73. (a) $\lim_{x \to 0^+} \ln x = -\infty$ (does not exist) since, as the graph gets closer to $x = 0$, the value of $\ln x$ decreases.

(b) The graph of $y = \ln x$ has a vertical asymptote at $x = 0$ since $\lim_{x \to 0^+} \ln x = -\infty$.

75. Answers will vary.

77. $\lim_{x \to 1} \dfrac{x^4 + 4x^3 - 9x^2 + 7x - 3}{x - 1}$

(a)

x	1.01	1.001	1.0001
$f(x)$	5.0908	5.009	5.0009
x	0.99	0.999	0.9999
$f(x)$	4.9108	4.991	4.9991

As $x \to 1^-$ and as $x \to 1^+$, we see that $f(x) \to 5$.

(b) Graph $y = \dfrac{x^4 + 4x^3 - 9x^2 + 7x - 3}{x - 1}$ on a graphing calculator. One suitable choice for the viewing window is [–6, 6] by [–10, 40] with Xscl = 1, Yscl = 10.

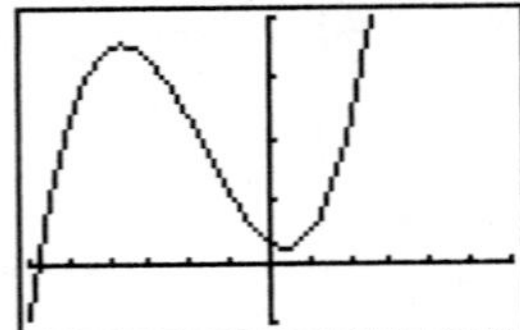

Because $x - 1 = 0$ when $x = 1$, we know that the function is undefined at this x-value. The graph does not show an asymptote at $x = 1$. This indicates that the rational expression that defines this function is not written in lowest terms, and that the graph should have an open circle to show a "hole" in the graph at $x = 1$. The graphing calculator doesn't show the hole, but if we try to find the value of the function at $x = 1$, we see that it is undefined. (Using the TABLE feature on a TI-84 Plus, we see that for $x = 1$, the y-value is listed as "ERROR.") By viewing the function near $x = 1$, and using the ZOOM feature, we see that as x gets close to 1 from the left or the right, y gets close to 5, suggesting that

$$\lim_{x \to 1} \frac{x^4 + 4x^3 - 9x^2 + 7x - 3}{x - 1} = 5.$$

79. $\lim_{x \to -1} \dfrac{x^{1/3} + 1}{x + 1}$

(a)

x	−1.01	−1.001	−1.0001
$f(x)$	0.33223	0.33322	0.33332
x	−0.99	−0.999	−0.9999
$f(x)$	0.33445	0.33344	0.33334

We see that as $x \to -1^-$ and as $x \to -1^+$, $f(x) \to 0.3333$ or $\frac{1}{3}$.

(b) Graph $y = \dfrac{x^{1/3} + 1}{x + 1}$.

One suitable choice for the viewing window is [–5. 5] by [0, 2].

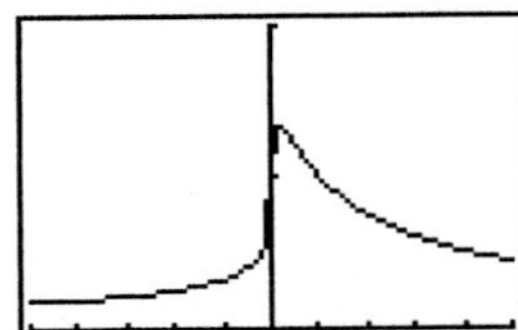

Because $x + 1 = 0$ when $x = -1$, we know that the function is undefined at this x-value. The graph does not show an asymptote at $x = -1$. This indicates that the rational expression that defined this function is not written lowest terms, and that the graph should have an open circle to show a "hole" in the graph at $x = -1$. The graphing calculator doesn't show the hole, but if we try to find the value of the function at $x = -1$, we see that it is undefined. (Using the TABLE feature on a TI-83, we see that for $x = -1$, the y-value is listed as "ERROR.") By viewing the function near $x = -1$ and using the ZOOM feature, we see that as x gets close to –1 from the left or right, y gets close to 0.3333, suggesting that

$$\lim_{x \to -1} \frac{x^{1/3} + 1}{x + 1} = 0.3333 \text{ or } \frac{1}{3}.$$

81. $\lim\limits_{x\to\infty} \dfrac{\sqrt{9x^2+5}}{2x}$

Graph the functions on a graphing calculator. A good choice for the viewing window is [–10, 10] by [–5, 5].

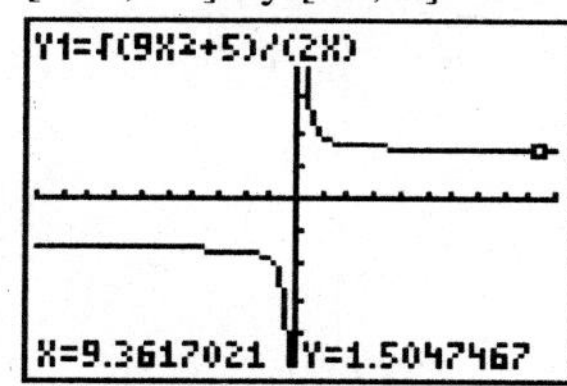

(a) The graph appears to have horizontal asymptotes at $y=\pm 1.5$. We see that as $x\to\infty$, $y\to 1.5$, so we determine that

$$\lim_{x\to\infty} \frac{\sqrt{9x^2+5}}{2x} = 1.5.$$

(b) As $x\to\infty$, $\dfrac{\sqrt{9x^2+5}}{2x} \to \dfrac{3|x|}{2x}$.

Since $x>0$, $|x|=x$, so $\dfrac{3|x|}{2x} = \dfrac{3x}{2x} = \dfrac{3}{2}$.

Thus, $\lim\limits_{x\to\infty} \dfrac{\sqrt{9x^2+5}}{2x} = \dfrac{3}{2} = 1.5$

83. $\lim\limits_{x\to-\infty} \dfrac{\sqrt{36x^2+2x+7}}{3x}$

Graph this function on a graphing calculator. A good choice for the viewing window is –10, 10] by [–5. 5].

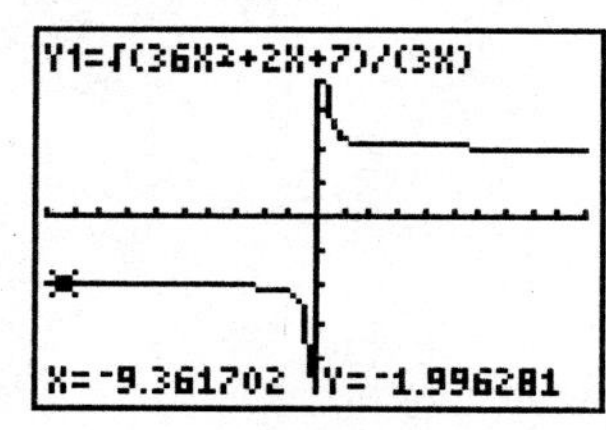

(a) The graph appears to have horizontal asymptotes at $y=\pm 2$. We see that as $x\to-\infty$, $y\to -2$, so we determine that

$$\lim_{x\to-\infty} \frac{\sqrt{36x^2+2x+7}}{3x} = -2.$$

(b) As $x\to-\infty$, $\dfrac{\sqrt{36x^2+2x+7}}{3x} \to \dfrac{6\,|\,x\,|}{3x}$.

Since $x<0$, $|\,x\,| = -x$, so

$\dfrac{6\,|\,x\,|}{3x} = \dfrac{6(-x)}{3x} = -2$. Thus,

$$\lim_{x\to-\infty} \frac{\sqrt{36x^2+2x+7}}{3x} = -2.$$

85. $\lim\limits_{x\to\infty} \dfrac{(1+5x^{1/3}+2x^{5/3})^3}{x^5}$

Graph this function on a graphing calculator. A good choice for the viewing window is [–20, 20] by [0, 20].

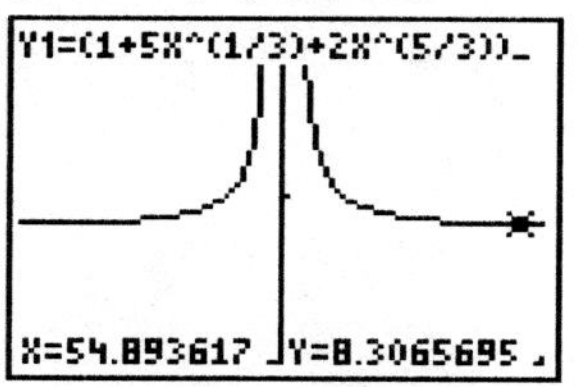

(a) The graph appears to have a horizontal asymptote at $y=8$. We see that as $x\to\infty$, $y\to 8$, so we determine that

$$\lim_{x\to\infty} \frac{(1+5x^{1/3}+2x^{5/3})^3}{x^5} = 8.$$

(b) As $x\to\infty$,

$$\frac{(1+5x^{1/3}+2x^{5/3})^3}{x^5} \to \frac{8x^5}{x^5} = 8.$$

Thus, $\lim\limits_{x\to\infty} \dfrac{(1+5x^{1/3}+2x^{5/3})^3}{x^5} = 8.$

87. Let $p(x) = a_1x^n + a_2x^{n-1} + \cdots + a_{n-1}x + a_n$ and $q(x) = b_1x^m + b_2x^{m-1} + \cdots + b_{m-1}x + b_m$.

(a) If $n<m$, then

$$\begin{aligned}\lim_{x\to\infty} \frac{p(x)}{q(x)} &= \lim_{x\to\infty} \frac{a_1x^n + a_2x^{n-1} + \cdots + a_{n-1}x + a_n}{b_1x^m + b_2x^{m-1} + \cdots + b_{m-1}x + b_m} \\ &= \lim_{x\to\infty} \frac{a_1}{b_1x^{m-n}} = \frac{a_1}{b_1} \lim_{x\to\infty} \frac{1}{x^{m-n}} \\ &= \frac{a_1}{b_1}\cdot\frac{1}{\infty} = 0\end{aligned}$$

(b) If $n=m$, then

$$\begin{aligned}\lim_{x\to\infty} \frac{p(x)}{q(x)} &= \lim_{x\to\infty} \frac{a_1x^n + a_2x^{n-1} + \cdots + a_{n-1}x + a_n}{b_1x^m + b_2x^{m-1} + \cdots + b_{m-1}x + b_m} \\ &= \lim_{x\to\infty} \frac{a_1}{b_1x^{m-n}} = \frac{a_1}{b_1}\end{aligned}$$

(c) If $n > m$, then

$$\lim_{x\to\infty} \frac{p(x)}{q(x)}$$
$$= \lim_{x\to\infty} \frac{a_1x^n + a_2x^{n-1} + \cdots + a_{n-1}x + a_n}{b_1x^m + b_2x^{m-1} + \cdots + b_{m-1}x + b_m}$$
$$= \lim_{x\to\infty} \frac{a_1x^{n-m}}{b_1} = \frac{a_1}{b_1} \lim_{x\to\infty} x^{n-m}$$
$= \infty$ or $-\infty$, depending on the sign of $\frac{a_1}{b_1}$.

89. **(a)** $\lim_{x\to 12^-} C(t) = 2 = \lim_{x\to 12^+} C(t)$, so $\lim_{x\to 12} C(t) = 2$ million units

(b) $\lim_{x\to 16^-} C(t) = 2.5$, but $\lim_{x\to 16^+} C(t) = 3.5$. So, $\lim_{x\to 16} C(t)$ does not exist.

(c) There is an open circle at (16, 3.5) and a closed circle at (16, 3.5), so $C(16) = 3.5$ million units.

(d) There is a sudden substantial increase in consumption of the antibiotic at $t = 16$ months. Therefore, the tipping point occurs at $t = 16$ months.

91. Note that

$$\bar{C}(n) = \frac{C(n)}{n} = \frac{15{,}000 + 60n}{n}$$
$$= \frac{15{,}000}{n} + 60$$

Therefore,

$$\lim_{n\to\infty} \bar{C}(n) = \lim_{n\to\infty} \left(\frac{15{,}000}{n} + 60\right)$$
$$= \lim_{n\to\infty} \left(\frac{15{,}000}{n}\right) + \lim_{n\to\infty} 60$$
$$= 0 + 60 = 60$$

The average cost per test approaches \$60.

93. **(a)** $N(65) = 71.8e^{-8.96e^{(-0.0685(65))}}$
≈ 64.68

To the nearest whole number, this species of alligator has approximately 65 teeth after 65 days of incubation by this formula.

(b) Since $\lim_{t\to\infty} (-8.96e^{-0.0685t}) = -8.96 \cdot 0 = 0$, it follows that

$$\lim_{t\to\infty} 71.8e^{-8.96e^{(-0.0685t)}} = 71.8e^0 = 71.8$$

So, to the nearest whole number, $\lim_{t\to\infty} N(t) \approx 72$. Therefore, by this model a newborn alligator of this species will have about 72 teeth.

95. $C = F(S) = \dfrac{1-S}{S(1+kS)^{f-1}}$

(a) $$\lim_{S\to 1} F(S) = \lim_{S\to 1} \frac{1-S}{S(1+kS)^{f-1}}$$
$$= \lim_{S\to 1} \frac{0}{(1+S)^{f-1}} = 0$$

(b) $$\lim_{S\to 0} F(S)$$
$$= \lim_{S\to 0} \frac{1-S}{S(1+kS)^{f-1}}$$
$$= \lim_{S\to 0} \left(\frac{1}{S(1+kS)^{f-1}} - \frac{S}{S(1+kS)^{f-1}}\right)$$
$$= \lim_{S\to 0} \frac{1}{S(1+kS)^{f-1}} - \lim_{S\to 0} \frac{S}{S(1+kS)^{f-1}}$$
$$= \lim_{S\to 0} \frac{1}{0} - \lim_{S\to 0} \frac{1}{(1+kS)^{f-1}} = \frac{1}{0} - 1 = \infty,$$

so the limit does not exist, and $F(S)$ increases without bound.

97. **(a)** $p_2 = \frac{1}{2} + \left(0.7 - \frac{1}{2}\right)[1 - 2(0.2)]^2 = 0.572$

(b) $p_4 = \frac{1}{2} + \left(0.7 - \frac{1}{2}\right)[1 - 2(0.2)]^4 = 0.526$

(c) $p_8 = \frac{1}{2} + \left(0.7 - \frac{1}{2}\right)[1 - 2(0.2)]^8 = 0.503$

(d) $$\lim_{n\to\infty} P_n = \lim_{n\to\infty} \left[\frac{1}{2} + \left(p_0 - \frac{1}{2}\right)(1-2p)^n\right]$$
$$= \frac{1}{2} + \lim_{n\to\infty} \left(p_0 - \frac{1}{2}\right)(1-2p)^n$$
$$= \frac{1}{2} + \left(p_0 - \frac{1}{2}\right) \lim_{n\to\infty} (1-2p)^n$$
$$= \frac{1}{2} + \left(p_0 - \frac{1}{2}\right) \cdot 0 = \frac{1}{2}$$

The number in parts (a), (b), and (c) represent the probability that the legislator will vote yes on the second, fourth, and eighth votes. In (d), as the number of roll calls increases, the probability gets close to 0.5, but is never less than 0.5.

3.2 Continuity

1. Discontinuous at $x = -1$

(a) $f(-1)$ does not exist.

(b) $\lim_{x\to -1^-} f(x) = \frac{1}{2}$

(c) $\lim_{x\to -1^+} f(x) = \frac{1}{2}$

(d) $\lim_{x\to -1} f(x) = \frac{1}{2}$ (since (a) and (b) have the same answers)

(e) $f(-1)$ does not exist.

3. Discontinuous at $x = 1$

(a) $f(1) = 2$

(b) $\lim_{x\to 1^-} f(x) = -2$

(c) $\lim_{x\to 1^+} f(x) = -2$

(d) $\lim_{x\to 1} f(x) = -2$ since parts (b) and (c) have the same answer.

(e) $\lim_{x\to 1} f(x) \neq f(1)$

5. Discontinuous at $x = -5$ and $x = 0$

(a) $f(-5)$ does not exist. $f(0)$ does not exist.

(b) $\lim_{x\to -5^-} f(x) = \infty$ (limit does not exist)
$\lim_{x\to 0^-} f(x) = 0$

(c) $\lim_{x\to -5^+} f(x) = -\infty$ (limit does not exist)
$\lim_{x\to 0^+} f(x) = 0$

(d) $\lim_{x\to -5} f(x)$ does not exist, since the answers to parts (b) and (c) are different. $\lim_{x\to 0} f(x) = 0$, since the answers to parts (b) and (c) are the same.

(e) $f(-5)$ does not exist and $\lim_{x\to -5} f(x)$ does not exist. $f(0)$ does not exist.

7. $f(x) = \frac{5+x}{x(x-2)}$

$f(x)$ is discontinuous at $x = 0$ and $x = 2$ since the denominator equals 0 at these two values.
$\lim_{x\to 0} f(x)$ does not exist since
$\lim_{x\to 0^-} f(x) = \infty$ and $\lim_{x\to 0^+} f(x) = -\infty$.
$\lim_{x\to 2} f(x)$ does not exist since
$\lim_{x\to 2^-} f(x) = -\infty$ and $\lim_{x\to 2^+} f(x) = \infty$.

9. $f(x) = \frac{x^2-4}{x-2}$

$f(x)$ is discontinuous at $x = 2$ since the denominator equals zero at that value.

For $x \neq 2$, $\frac{x^2-4}{x-2} = \frac{(x+2)(x-2)}{x-2} = x+2$, so $\lim_{x\to 2} f(x) = 2+2 = 4$.

11. $p(x) = x^2 - 4x + 11$

Since $p(x)$ is a polynomial function, it is continuous everywhere and thus discontinuous nowhere.

13. $p(x) = \frac{|x+2|}{x+2}$

$p(x)$ is discontinuous at $x = -2$ since the denominator is zero at that value.
Because $\lim_{x\to -2^-} p(x) = -1$ and $\lim_{x\to -2^+} p(x) = 1$, $\lim_{x\to -2} p(x)$ does not exist.

15. $k(x) = e^{\sqrt{x-1}}$

The function is undefined for $x < 1$, so the function is discontinuous for $a < 1$. The limit as x approaches any $a < 1$ does not exist because the function is undefined for $x < 1$.

17. $r(x) = \ln\left|\frac{x}{x-1}\right|$

As x approaches 0 from the left or the right, $\left|\frac{x}{x-1}\right|$ approaches 0 and $r(x) = \ln\left|\frac{x}{x-1}\right|$ goes to $-\infty$. So $\lim_{x\to 0} r(x)$ does not exist. As x approaches 1 from the left or the right, $\left|\frac{x}{x-1}\right|$ goes to ∞ and so does $r(x) = \ln\left|\frac{x}{x-1}\right|$. So $\lim_{x\to 1} r(x)$ does not exist.

19. $f(x) = \sin\left(\frac{x}{x+2}\right)$

$f(x)$ is discontinuous at $x = -2$ because the fraction is undefined at that value.

$\lim\limits_{x\to -2} \sin\left(\frac{x}{x+2}\right)$ does not exist because

$\lim\limits_{x\to -2^-} \sin\left(\frac{x}{x+2}\right)$ does not exist and

$\lim\limits_{x\to -2^+} \sin\left(\frac{x}{x+2}\right)$ does not exist.

21. $f(x) = \begin{cases} 1 & \text{if } x < 2 \\ x+3 & \text{if } 2 \le x \le 4 \\ 7 & \text{if } x > 4 \end{cases}$

(a)

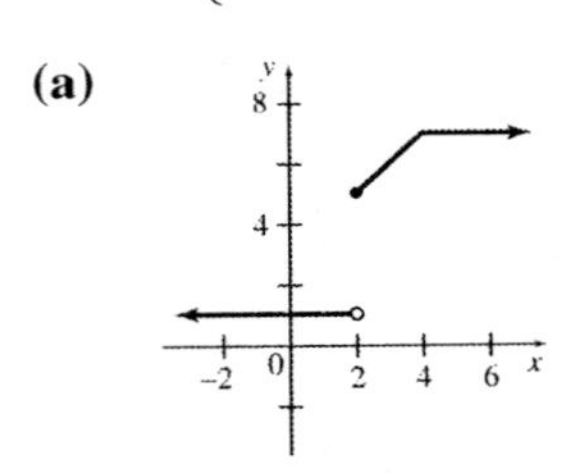

(b) $f(x)$ is discontinuous at $x = 2$.

(c) $\lim\limits_{x\to 2^-} f(x) = 1$

$\lim\limits_{x\to 2^+} f(x) = 5$

23. $g(x) = \begin{cases} 11 & \text{if } x < -1 \\ x^2 + 2 & \text{if } -1 \le x \le 3 \\ 11 & \text{if } x > 3 \end{cases}$

(a)

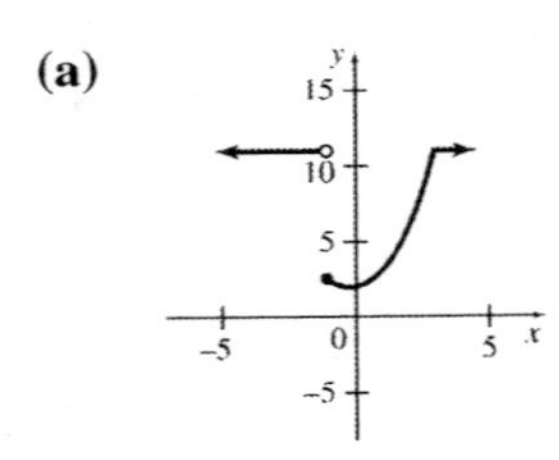

(b) $g(x)$ is discontinuous at $x = -1$.

(c) $\lim\limits_{x\to -1^-} g(x) = 11$

$\lim\limits_{x\to -1^+} g(x) = (-1)^2 + 2 = 3$

25. $h(x) = \begin{cases} 4x+4 & \text{if } x \le 0 \\ x^2 - 4x + 4 & \text{if } x > 0 \end{cases}$

(a)

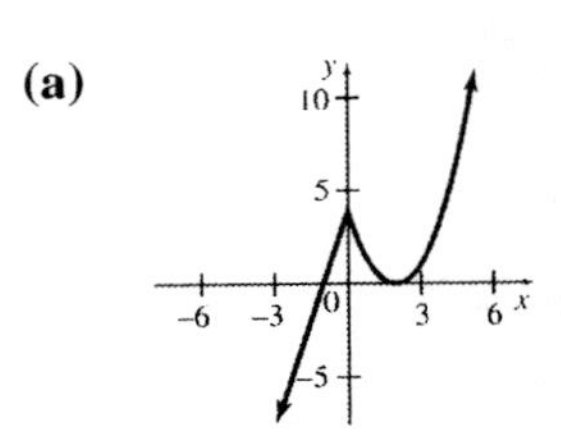

(b) There are no points of discontinuity.

27. Find k so that $kx^2 = x + k$ for $x = 2$.

$$k(2)^2 = 2 + k$$
$$4k = 2 + k$$
$$3k = 2$$
$$k = \frac{2}{3}$$

29. $\frac{2x^2 - x - 15}{x-3} = \frac{(2x+5)(x-3)}{x-3} = 2x+5$

Find k so that $2x + 5 = kx - 1$ for $x = 3$.

$$2(3) + 5 = k(3) - 1$$
$$6 + 5 = 3k - 1$$
$$11 = 3k - 1$$
$$12 = 3k$$
$$4 = k$$

31. Answers will vary.

33. $f(x) = \frac{x^2 + x + 2}{x^3 - 0.9x^2 + 4.14x - 5.4} = \frac{P(x)}{Q(x)}$

(a) Graph

$Y_1 = \frac{P(x)}{Q(x)} = \frac{x^2 + x + 2}{x^3 - 0.9x^2 + 4.14x - 5.4}$ on a graphing calculator. A good choice for the viewing window is [–3, 3] by [–10, 10].

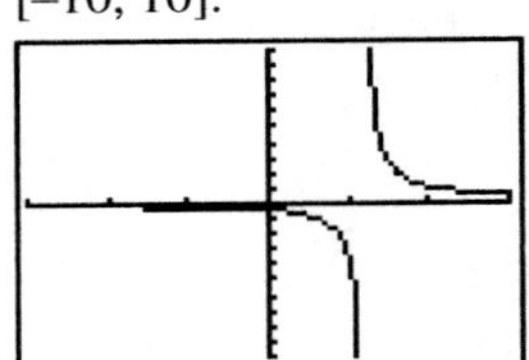

The graph has a vertical asymptote at $x = 1.2$, which indicates that f is discontinuous at $x = 1.2$.

(b) Graph

$Y_2 = Q(x) = x^3 - .09x^2 + 4.14x - 5.4$

using the same viewing window.

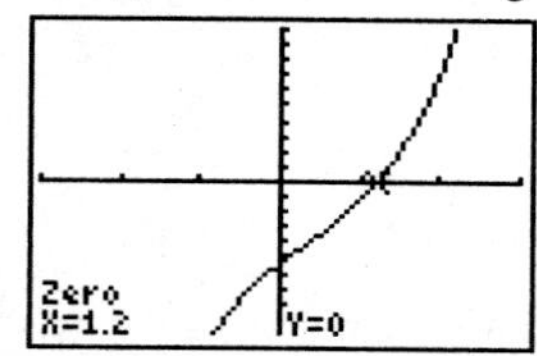

We see that this graph has one x-intercept, 1.2. This indicates that 1.2 is the only real solution of the equation $Q(x) = 0$. This result verifies our answer from part (a) because a rational function of the form $f(x) = \frac{P(x)}{Q(x)}$ will be discontinuous wherever $Q(x) = 0$.

The graph has a vertical asymptote at $x \approx -0.9$. It is difficult to read this value accurately from the graph.

35. $g(x) = \frac{x+4}{x^2+2x-8} = \frac{x+4}{(x-2)(x+4)}$

$= \frac{1}{x-2}, \ x \neq -4$

If $g(x)$ is defined so that $g(-4) = \frac{1}{-4-2} = -\frac{1}{6}$, then the function becomes continuous at -4. It cannot be made continuous at 2. The correct answer is (a).

37. $N(t) = 2^t$

(a) At diagnosis, $t = 40$.

$N(40) = 2^{40} \approx 1.0995 \times 10^{12}$

At diagnosis, there are about 1.0995×10^{12} tumor cells.

(b) If 99.9% of the cells are killed instantaneously, then 0.1% remain.

$N(t) = 0.001 \cdot 2^{40} \approx 1.0995 \times 10^9$

$\log_2 \left(1.0995 \times 10^9\right) \approx 30$

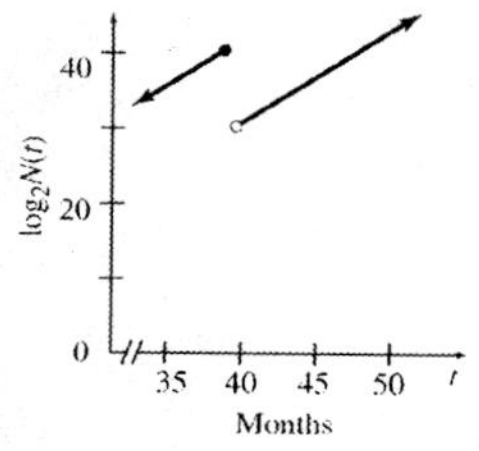

(c) The function described in part (b) is discontinuous at $t = 40$ because $\lim_{t \to 40} N(t)$ does not exist.

$\lim_{t \to 40^-} N(t) = \lim_{t \to 40^-} 2^t = 2^{40}$, while

$\lim_{t \to 40^+} N(t) \approx 1.0995 \times 10^9$

Thus, $\lim_{t \to 40^-} N(t) \neq \lim_{t \to 40^+} N(t)$, and $\lim_{t \to 40} N(t)$ does not exist.

39. Answers will vary, but the function is discontinuous.

41. $W(t) = \begin{cases} 48 + 3.64t + 0.6363t^2 + 0.00963t^3 & \text{if } 1 \le t \le 28 \\ -1004 + 65.8t & \text{if } 28 < t \le 56 \end{cases}$

(a) $W(25) = 48 + 3.64(25) + 0.6363(25)^2 + 0.00963(25)^3$

≈ 687.156

A male broiler at 25 days weighs about 687 grams.

(b) $W(t)$ is not a continuous function. At $t = 28$

$\lim_{t \to 28^-} W(t)$

$= \lim_{t \to 28^-} \left(48 + 3.64t + 0.6363t^2 + 0.00963t^3\right)$

$= 48 + 3.64(28) + 0.6363(28)^2 + 0.00963(28)^3$

≈ 860.18

and

$\lim_{t \to 28^-} W(t) \neq \lim_{t \to 28^+} (-1004 + 65.8t)$

$= -1004 + 65.8(28)$

$= 838.4$

so $\lim_{t \to 28^-} W(t) \neq \lim_{t \to 28^+} W(t)$. Thus $W(t)$ is discontinuous.

(c)

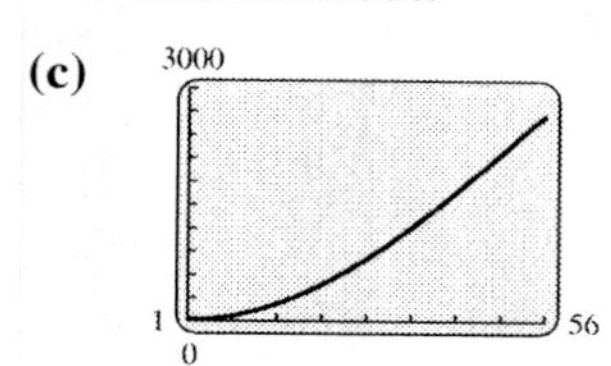

(d) Answers will vary.

43. **(a)** As x approaches 6 from the left or the right, the value of $P(x)$ for the corresponding point on the graph approaches 500. Thus, $\lim_{x\to 6} P(x) = \$500.$

(b) As x approaches 10 from the left, $P(x)$ approaches \$1500. Thus,
$\lim_{x\to 10^-} P(x) = \$1500.$

(c) As x approaches 10 from the right, $P(x)$ approaches \$1000. Thus,
$\lim_{x\to 10^+} P(x) = \$1000.$

(d) $\lim_{x\to 10} P(x)$ does not exist because
$\lim_{x\to 10^-} P(x) \neq \lim_{x\to 10^+} P(x).$

(e) The graph is discontinuous at $x = 10$. This may be a result of the change in shift.

(f) From the graph, it appears that the second shift will be as profitable as the first shift when 15 units are produced.

45. $F(x) = \begin{cases} 1.25x & \text{if } 0 < x \le 100 \\ 1.00x & \text{if } x > 100. \end{cases}$

(a) $F(80) = 1.25(80) = \$100$

(b) $F(150) = 1.00(150) = \$150$

(c) $F(100) = 1.25(100) = \$125$

(d) F is discontinuous at $x = 100$.

3.3 Rates of Change

1. $y = x^2 + 2x = f(x)$ between $x = 1$ and $x = 3$.

$$\text{Average rate of change} = \frac{f(3) - f(1)}{3 - 1} = \frac{15 - 3}{2} = 6$$

3. $y = -3x^3 + 2x^2 - 4x + 1 = f(x)$ between $x = -2$ and $x = 1$.

$$\text{Average rate of change} = \frac{f(1) - f(-2)}{1 - (-2)} = \frac{(-4) - (-41)}{1 - (-2)} = -15$$

5. $y = \sqrt{x} = f(x)$ between $x = 1$ and $x = 4$.

$$\text{Average rate of change} = \frac{f(4) - f(1)}{4 - 1} = \frac{2 - 1}{3} = \frac{1}{3}$$

7. $y = \frac{1}{x - 1}$ between $x = -2$ and $x = 0$

Average rate of change

$$= \frac{f(0) - f(-2)}{0 - (-2)} = \frac{-1 - \left(-\frac{1}{3}\right)}{2} = \frac{-\frac{2}{3}}{2} = -\frac{1}{3}$$

9. $y = e^x = f(x)$ between $x = -2$ and $x = 0$

$$\text{Average rate of change} = \frac{f(0) - f(-2)}{0 - (-2)} = \frac{1 - e^{-2}}{2} \approx 0.4323$$

11. $y = \sin x = f(x)$ between $x = 0$ and $x = \frac{\pi}{4}$

$$\text{Average rate of change} = \frac{f\left(\frac{\pi}{4}\right) - f(0)}{\frac{\pi}{4} - 0} = \frac{\sin\frac{\pi}{4} - \sin 0}{\frac{\pi}{4}} = \frac{\frac{\sqrt{2}}{2}}{\frac{\pi}{4}} = \frac{2\sqrt{2}}{\pi}$$

13. $s(t) = t^2 + 5t + 2$

$$\lim_{h\to 0} \frac{s(6 + h) - s(6)}{h}$$
$$= \lim_{h\to 0} \frac{(6 + h)^2 + 5(6 + h) + 2 - [6^2 + 5(6) + 2]}{h}$$
$$= \lim_{h\to 0} \frac{(36 + 12h + h^2) + (30 + 5h) + 2 - 68}{h}$$
$$= \lim_{h\to 0} \frac{h^2 + 17h + 68 - 68}{h} = \lim_{h\to 0} \frac{h^2 + 17h}{h}$$
$$= \lim_{h\to 0} \frac{h(h + 17)}{h} = \lim_{h\to 0} (h + 17) = 17$$

The instantaneous velocity at $t = 6$ is 17.

15. $s(t) = 5t^2 - 2t - 7$

$$\lim_{h\to 0} \frac{s(2+h) - s(2)}{h}$$
$$= \lim_{h\to 0} \frac{[5(2+h)^2 - 2(2+h) - 7] - [5(2)^2 - 2(2) - 7]}{h}$$
$$= \lim_{h\to 0} \frac{[20 + 20h + 5h^2 - 4 - 2h - 7] - [20 - 4 - 7]}{h}$$
$$= \lim_{h\to 0} \frac{9 + 18h + 5h^2 - 9}{h} = \lim_{h\to 0} \frac{18h + 5h^2}{h}$$
$$= \lim_{h\to 0} \frac{h(18+5h)}{h} = \lim_{h\to 0} (18 + 5h) = 18$$

The instantaneous velocity at $t = 2$ is 18.

17. $s(t) = t^3 + 2t + 9$

$$\lim_{x\to 0} \frac{s(1+h) - s(1)}{h}$$
$$= \lim_{h\to 0} \frac{[(1+h)^3 + 2(1+h) + 9] - [(1)^3 + 2(1) + 9]}{h}$$
$$= \lim_{h\to 0} \frac{[1 + 3h + 3h^2 + h^3 + 2 + 2h + 9] - [1 + 2 + 9]}{h}$$
$$= \lim_{h\to 0} \frac{h^3 + 3h^2 + 5h + 12 - 12}{h}$$
$$= \lim_{h\to 0} \frac{h^3 + 3h^2 + 5h}{h} = \lim_{h\to 0} \frac{h(h^2 + 3h + 5)}{h}$$
$$= \lim_{h\to 0} (h^2 + 3h + 5) = 5$$

The instantaneous velocity at $t = 1$ is 5.

19. $f(x) = x^2 + 2x$ at $x = 0$

$$\lim_{h\to 0} \frac{f(0+h) - f(0)}{h}$$
$$= \lim_{h\to 0} \frac{(0+h)^2 + 2(0+h) - [0^2 + 2(0)]}{h}$$
$$= \lim_{h\to 0} \frac{h^2 + 2h}{h} = \lim_{h\to 0} \frac{h(h+2)}{h}$$
$$= \lim_{h\to 0} h + 2 = 2$$

The instantaneous rate of change at $x = 0$ is 2.

21. $g(t) = 1 - t^2$ at $t = -1$

$$\lim_{h\to 0} \frac{g(-1+h) - g(-1)}{h}$$
$$= \lim_{h\to 0} \frac{1 - (-1+h)^2 - [1 - (-1)^2]}{h}$$
$$= \lim_{h\to 0} \frac{1 - (1 - 2h + h^2) - 1 + 1}{h}$$
$$= \lim_{h\to 0} \frac{2h - h^2}{h} = \lim_{h\to 0} \frac{h(2-h)}{h}$$
$$= \lim_{h\to 0} (2 - h) = 2$$

The instantaneous rate of change at $t = -1$ is 2.

23. $f(x) = \sin x$ at $x = 0$

h	
0.1	$\frac{f(0+0.1) - f(0)}{0.1} = \frac{\sin 0.1 - \sin 0}{0.1} = 0.99833$
0.01	$\frac{f(0+0.01) - f(0)}{0.01} = \frac{\sin 0.01 - \sin 0}{0.01} = 0.99998$
0.001	$\frac{f(0+0.001) - f(0)}{0.001} = \frac{\sin 0.001 - \sin 0}{0.001} = 1$
0.0001	$\frac{f(0+0.0001) - f(0)}{0.0001} = \frac{\sin 0.00001 - \sin 0}{0.0001} = 1$
0.00001	$\frac{f(0+0.00001) - f(0)}{0.00001} = \frac{\sin 0.000001 - \sin 0}{0.00001} = 1$
0.000001	$\frac{f(0+0.000001) - f(0)}{0.000001} = \frac{\sin 0.0000001 - \sin 0}{0.000001} = 1$

The instantaneous rate of change at $x = 0$ is 1.

25. $f(x) = x^x$ at $x = 2$

h	
0.01	$\frac{f(2+0.01)-f(2)}{0.01} = \frac{2.01^{2.01}-2^2}{0.01} = 6.84$
0.001	$\frac{f(2+0.001)-f(2)}{0.001} = \frac{2.001^{2.001}-2^2}{0.001} = 6.779$
0.0001	$\frac{f(2+0.0001)-f(2)}{0.00001} = \frac{2.0001^{2.0001}-2^2}{0.0001} = 6.773$
0.00001	$\frac{f(2+0.00001)-f(2)}{0.00001} = \frac{2.00001^{2.00001}-2^2}{0.00001} = 6.7727$
0.000001	$\frac{f(2+0.000001)-f(2)}{0.000001} = \frac{2.000001^{2.000001}-2^2}{0.000001} = 6.7726$

The instantaneous rate of change at $x = 2$ is 6.773.

27. $f(x) = x^{\ln x}$ at $x = 2$

h	
0.01	$\frac{f(2+0.01)-f(2)}{0.01} = \frac{2.01^{\ln 2.01}-2^{\ln 2}}{0.01} = 1.1258$
0.001	$\frac{f(2+0.001)-f(2)}{0.001} = \frac{2.001^{\ln 2.001}-2^{\ln 2}}{0.001} = 1.1212$
0.0001	$\frac{f(2+0.0001)-f(2)}{0.0001} = \frac{2.0001^{\ln 2.0001}-2^{\ln 2}}{0.0001} = 1.1207$
0.00001	$\frac{f(2+0.00001)-f(2)}{0.00001} = \frac{2.00001^{\ln 2.00001}-2^{\ln 2}}{0.00001} = 1.1207$

The instantaneous rate of change at $x = 2$ is 1.121.

29. Answers will vary.

31. $p(t) = t^2 + t$

(a) $p(1) = 1^2 + 1 = 2$
$p(4) = 4^2 + 4 = 20$
Average rate of change

$$= \frac{p(4)-p(1)}{4-1} = \frac{20-2}{3} = 6$$

The average rate of change is 6% per day.

(b)

$$\lim_{h\to 0} \frac{p(3+h)-p(3)}{h}$$
$$= \lim_{h\to 0} \frac{(3+h)^2 + (3+h) - (3^2+3)}{h}$$
$$= \lim_{h\to 0} \frac{9+6h+h^2+3+h-12}{h}$$
$$= \lim_{h\to 0} \frac{h^2+7h}{h} = \lim_{h\to 0} \frac{h(h+7)}{h}$$
$$= \lim_{h\to 0} (h+7) = 7$$

The instantaneous rate of change is 7% per day. The number of people newly infected on day 3 is about 7% of the total population.

33. (a) $P(1) = 3,\ P(2) = 5$

$$\text{Average rate of change} = \frac{P(2)-P(1)}{2-1} = \frac{5-3}{2-1} = \frac{2}{1} = 2$$

From 1 min to 2 min, the population of bacteria increases, on the average, 2 million per min.

(b) $P(2) = 5,\ P(3) = 4.2$

$$\text{Average rate of change} = \frac{P(3)-P(2)}{3-2} = \frac{4.2-5}{3-2} = -0.8$$

From 2 min to 3 min, the population of bacteria decreases, on the average, 0.8 million or 800,000 per min.

(c) $P(3) = 4.2,\ P(4) = 2$

$$\text{Average rate of change} = \frac{P(4)-P(3)}{4-3} = \frac{2-4.2}{4-3} = -2.2$$

From 3 min to 4 min, the population of bacteria decreases, on the average, 2.2 million per min.

(d) $P(4) = 2,\ P(5) = 1$

Average rate of change $= \dfrac{P(5) - P(4)}{5 - 4} = \dfrac{1 - 2}{5 - 4} = -1$

From 4 min to 5 min, the population decreases, on the average, –1 million per min.

(e) The population increased up to 2 min after the bactericide was introduced, but decreased after 2 min.

(f) The rate of decrease of the population slows down at about 3 min. The graph becomes less and less steep after that point.

35. $L(t) = -0.01t^2 + 0.788t - 7.048$

(a) $\dfrac{L(28) - L(22)}{28 - 22} = \dfrac{7.176 - 5.448}{6} = 0.288$

The average rate of growth during weeks 22 through 28 is 0.288 mm per week.

(b)
$$\begin{aligned}\lim_{h\to 0} \frac{L(t+h) - L(t)}{h} &= \lim_{h\to 0} \frac{L(22+h) - L(22)}{h} \\ &= \lim_{h\to 0} \frac{[-0.01(22+h)^2 + 0.788(22+h) - 7.048] - 5.448}{h} \\ &= \lim_{h\to 0} \frac{-0.01(h^2 + 44h + 484) + 17.336 + 0.788h - 12.496}{h} \\ &= \lim_{h\to 0} \frac{-0.01h^2 + 0.348h}{h} = \lim_{h\to 0} (-0.01h + 0.348) = 0.348\end{aligned}$$

The instantaneous rate of growth at exactly 22 weeks is 0.348 mm per week.

(c)

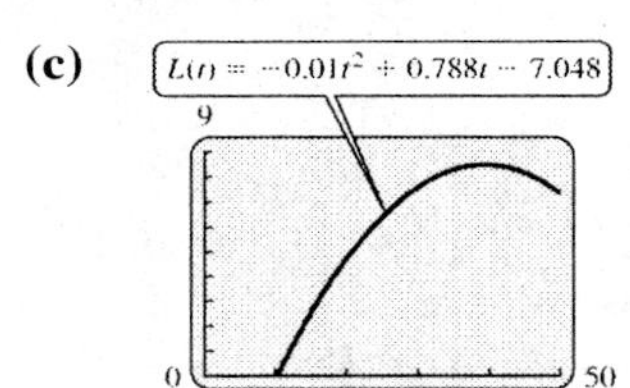

Answers will vary. Yes, a function that increases and then begins to decrease makes sense for this particular application. It appears that the molars do not begin to develop until week 11.

37. $H(t) = 37.791(1.021)^t,\ 0 \le t \le 50$

(a) $\dfrac{H(10) - H(0)}{10 - 0} = \dfrac{37.791(1.021)^{10} - 37.791(1.021)^0}{10 - 0} \approx \dfrac{46.521 - 37.791}{10} = 0.873$ million

Based on this model, the Hispanic population in the U.S. increased by 873,000 people per year, on average, between 2000 and 2010.

(b)

h	
0.1	$\frac{H(10+0.1)-H(10)}{0.1}=\frac{37.791(1.021)^{10+0.1}-37.791(1.021)^{10}}{0.1}=0.967823$
0.01	$\frac{H(10+0.01)-H(10)}{0.01}=\frac{37.791(1.021)^{10+0.01}-37.791(1.021)^{10}}{0.01}=0.966827$
0.001	$\frac{H(10+0.001)-H(10)}{0.001}=\frac{37.791(1.021)^{10+0.0001}-37.791(1.021)^{10}}{0.001}=0.966818$
0.0001	$\frac{H(10+0.0001)-H(10)}{0.0001}=\frac{37.791(1.021)^{10+0.0001}-37.791(1.021)^{10}}{0.0001}=0.966817$
0.00001	$\frac{H(10+0.00001)-H(10)}{0.00001}=\frac{37.791(1.021)^{10+0.00001}-37.791(1.021)^{10}}{0.00001}=0.96682$
0.000001	$\frac{H(10+0.000001)-H(10)}{0.000001}=\frac{37.791(1.021)^{10+0.000001}-37.791(1.021)^{10}}{0.000001}=0.96682$ million

The instantaneous rate of change in 2010 is about 967,000 people per year.

39. Let $D(t)$ represent the percent of students (8th, 10th, or 12th graders) who have used marijuana by the year t.

(a) 8th graders:

$$\frac{D(2008)-D(2004)}{2008-2004}=\frac{14.6-16.3}{4}=-0.425\% \text{ per year}$$

$$\frac{D(2012)-D(2008)}{2012-2008}=\frac{15.2-14.6}{4}=0.15\% \text{ per year}$$

$$\frac{D(2012)-D(2004)}{2012-2004}=\frac{15.2-16.3}{8}=-0.1375\% \text{ per year}$$

(b) 10th graders:

$$\frac{D(2008)-D(2004)}{2008-2004}=\frac{29.9-35.1}{4}=-1.3\% \text{ per year}$$

$$\frac{D(2012)-D(2008)}{2012-2008}=\frac{33.8-29.9}{4}=0.975\% \text{ per year}$$

$$\frac{D(2012)-D(2004)}{2012-2004}=\frac{33.8-35.1}{8}=-0.1625\% \text{ per year}$$

(c) 12th graders:

$$\frac{D(2008)-D(2004)}{2008-2004}=\frac{42.6-45.7}{4}=-0.775\% \text{ per year}$$

$$\frac{D(2012)-D(2008)}{2012-2008}=\frac{45.2-42.6}{4}=0.65\% \text{ per year}$$

$$\frac{D(2012)-D(2004)}{2012-2004}=\frac{45.2-45.7}{8}=-0.0625\% \text{ per year}$$

(d) Answers will vary.

41. (a) $\frac{287.6-381.6}{2012-2008}=-\frac{94}{4}=-23.5$

The approximate rate of change from 2008 to 2012 was a decrease of \$23.5 billion per year.

(b) $\frac{493.3-287.6}{2020-2012}=\frac{205.7}{8}\approx 25.7$

Based on this model, the Medicare Trust Fund will have an average rate of change of \$25.7 billion per year from 2012 to 2020.

43. (a) $\frac{T(3000)-T(1000)}{3000-1000}=\frac{23-15}{2000}=\frac{8}{2000}=\frac{4}{1000}$

From 1000 to 3000 ft, the temperature changes about 4° per 1000 ft; the temperature rises (on the average).

(b) $\frac{T(5000)-T(1000)}{5000-1000}=\frac{22-15}{4000}=\frac{7}{4000}=\frac{1.75}{1000}$

From 1000 to 5000 ft, the temperature changes about 1.75° per 1000 ft; the temperature rises (on the average).

(c) $\frac{T(9000)-T(3000)}{9000-3000} = \frac{17-23}{6000} = \frac{-6}{6000}$
$= \frac{-1}{1000}$
From 3000 to 9000 ft, the temperature changes about 1° per 1000 ft; the temperature falls (on the average).

(d) $\frac{T(9000)-T(1000)}{9000-1000} = \frac{17-15}{8000} = \frac{\frac{1}{4}}{1000}$
From 1000 to 9000 ft, the temperature changes about $\frac{1}{4}$° per 1000 ft; the temperature rises slightly (on the average).

(e) The temperature is highest at 3000 ft and lowest at 1000 ft. If 7000 ft is changed to 10,000 ft, the lowest temperature would be at 10,000 ft.

(f) The temperature at 9000 ft is the same as 1000 ft.

45. (a) Average rate of change from 0.5 to 1:
$\frac{f(1)-f(0.5)}{1-0.5} = \frac{55-30}{0.5} = 50$ mph
Average rate of change from 1 to 1.5:
$\frac{f(1.5)-f(1)}{1.5-1} = \frac{80-55}{0.5} = 50$ mph
Estimate of instantaneous velocity is
$\frac{50+50}{2} = 50$ mph.

(b) Average rate of change from 1.5 to 2:
$\frac{f(2)-f(1.5)}{2-1.5} = \frac{104-80}{0.5} = 48$ mph
Average rate of change from 2 to 2.5
$\frac{f(2.5)-f(2)}{2.5-2} = \frac{124-104}{0.5} = 40$ mph
Estimate of instantaneous velocity is
$\frac{48+40}{2} = 44$ mph.

47. $A(t) = 1000(1.05)^t$

(a) Average rate of change in the total amount from $t = 0$ to $t = 5$

$$\frac{A(5)-A(0)}{5-0} = \frac{1000(1.05)^5 - 1000(1.05)^0}{5} \approx 55.2563,$$

which is $55.26 per year.

(b) Average rate of change in the total amount from $t = 5$ to $t = 10$

$$\frac{A(10)-A(5)}{10-5} = \frac{1000(1.05)^{10} - 1000(1.05)^5}{5} \approx 70.5226,$$

which is $70.52 per year.

(c) Instantaneous rate of change for $t = 5$:

$$\lim_{h\to 0} \frac{1000(1.05)^{5+h} - 1000(1.05)^5}{h}$$

Use the TABLE feature on a TI-84 Plus calculator to estimate the limit.

h	$\frac{1000(1.05)^{5+h} - 1000(1.05)^5}{h}$
1	63.814
0.1	62.422
0.01	62.285
0.001	62.272
0.0001	62.270
0.00001	62.27

The limit seems to be approaching 62.27. So, the instantaneous rate of change for $t = 5$ is about $62.27 per year.

3.4 Definition of the Derivative

1. (a) $f(x) = 5$ is a horizontal line and has slope 0; the derivative is 0.

(b) $f(x) = x$ has slope 1; the derivative is 1.

(c) $f(x) = -x$ has slope of -1, the derivative is -1.

(d) $x = 3$ is vertical and has undefined slope; the derivative does not exist.

(e) $y = mx + b$ has slope m; the derivative is m.

3. $f(x) = \frac{x^2-1}{x+2}$ is not differentiable when $x + 2 = 0$ or $x = -2$ because the function is undefined and a vertical asymptote occurs there.

5. Using the points (5, 3) and (6, 5), we have $m = \frac{5-3}{6-5} = 2.$

7. Using the points (–2, 2) and (2, 3), we have $m = \frac{3-2}{2-(-2)} = \frac{1}{4}.$

9. Using the points (–3, –3) and (0, –3), we have

$$m = \frac{-3-(-3)}{0-3} = \frac{0}{-3} = 0.$$

11. $f(x) = 3x - 7$

Step 1 $f(x+h)$
$= 3(x+h) - 7$
$= 3x + 3h - 7$

Step 2 $f(x+h) - f(x)$
$= 3x + 3h - 7 - (3x - 7)$
$= 3x + 3h - 7 - 3x + 7$
$= 3h$

Step 3 $\dfrac{f(x+h)-f(x)}{h} = \dfrac{3h}{h} = 3$

Step 4 $f'(x) = \lim\limits_{h\to 0} \dfrac{f(x+h)-f(x)}{h}$
$= \lim\limits_{h\to 0} 3 = 3$

$f'(-2) = 3,\ f'(0) = 3\ f'(3) = 3$

13. $f(x) = -4x^2 + 9x + 2$

Step 1 $f(x+h)$
$= -4(x+h)^2 + 9(x+h) + 2$
$= -4(x^2 + 2xh + h^2) + 9x + 9h + 2$
$= -4x^2 - 8xh - 4h^2 + 9x + 9h + 2$

Step 2 $f(x+h) - f(x)$
$= -4x^2 - 8xh - 4h^2 + 9x + 9h + 2$
$\quad -(-4x^2 + 9x + 2)$
$= -8xh - 4h^2 + 9h$
$= h(-8x - 4h + 9)$

Step 3 $\dfrac{f(x+h)-f(x)}{h}$
$= \dfrac{h(-8x-4h+9)}{h}$
$= -8x - 4h + 9$

Step 4 $f'(x) = \lim\limits_{h\to 0} \dfrac{f(x+h)-f(x)}{h}$
$= \lim\limits_{h\to 0} (-8x - 4h + 9)$
$= -8x + 9$

$f'(-2) = -8(-2) + 9 = 25$
$f'(0) = -8(0) + 9 = 9$
$f'(3) = -8(3) + 9 = -15$

15. $f(x) = \dfrac{12}{x}$

Step 1 $f(x+h) = \dfrac{12}{x+h}$

Step 2 $f(x+h) - f(x) = \dfrac{12}{x+h} - \dfrac{12}{x}$
$= \dfrac{12x - 12(x+h)}{x(x+h)}$
$= \dfrac{12x - 12x - 12h}{x(x+h)}$
$= \dfrac{-12h}{x(x+h)}$

Step 3 $\dfrac{f(x+h)-f(x)}{h} = \dfrac{-12h}{hx(x+h)}$
$= \dfrac{-12}{x(x+h)}$
$= \dfrac{-12}{x^2 + xh}$

Step 4 $f'(x) = \lim\limits_{h\to 0} \dfrac{f(x+h)-f(x)}{h}$
$= \lim\limits_{h\to 0} \dfrac{-12}{x^2 + xh}$
$= \dfrac{-12}{x^2}$

$f'(-2) = \dfrac{-12}{(-2)^2} = \dfrac{-12}{4} = -3$

$f'(0) = \dfrac{-12}{0^2}$ which is undefined so $f'(0)$ does not exist.

$f'(3) = \dfrac{-12}{3^2} = \dfrac{-12}{9} = -\dfrac{4}{3}$

17. $f(x) = \sqrt{x}$

Steps 1–3 are combined.

$\dfrac{f(x+h)-f(x)}{h}$
$= \dfrac{\sqrt{x+h} - \sqrt{x}}{h}$
$= \dfrac{\sqrt{x+h} - \sqrt{x}}{h} \cdot \dfrac{\sqrt{x+h} + \sqrt{x}}{\sqrt{x+h} + \sqrt{x}}$
$= \dfrac{x + h - x}{h(\sqrt{x+h} + \sqrt{x})}$
$= \dfrac{1}{\sqrt{x+h} + \sqrt{x}}$

$f'(x) = \lim\limits_{h\to 0} \dfrac{f(x+h)-f(x)}{h}$
$= \lim\limits_{h\to 0} \dfrac{1}{\sqrt{x+h} + \sqrt{x}} = \dfrac{1}{2\sqrt{x}}$

$f'(-2) = \dfrac{1}{2\sqrt{-2}}$ which is undefined so $f'(-2)$ does not exist.

(continued on next page)

(*continued*)

$f'(0)=\frac{1}{2\sqrt{0}}=\frac{1}{0}$ which is undefined so $f'(0)$ does not exist.

$f'(3)=\frac{1}{2\sqrt{3}}$

19. $f(x)=2x^3+5$

Steps 1–3 are combined.

$$\frac{f(x+h)-f(x)}{h}$$
$$=\frac{2(x+h)^3+5-(2x^3+5)}{h}$$
$$=\frac{2(x^3+3x^2h+3xh^2+h^3)+5-2x^3-5}{h}$$
$$=\frac{2x^3+6x^2h+6xh^2+2h^3+5-2x^3-5}{h}$$
$$=\frac{6x^2h+6xh^2+2h^3}{h}=\frac{h(6x^2+6xh+2h^2)}{h}$$
$$=6x^2+6xh+2h^2$$

$$f'(x)=\lim_{h\to 0}(6x^2+6xh+2h^2)=6x^2$$

$$f'(-2)=6(-2)^2=24$$
$$f'(0)=6(0)^2=0$$
$$f'(3)=6(3)^2=54$$

21. (a) $f(x)=x^2+2x;\ x=3,\ x=5$

Slope of secant line

$$=\frac{f(5)-f(3)}{5-3}$$
$$=\frac{(5)^2+2(5)-[(3)^2+2(3)]}{2}$$
$$=\frac{35-15}{2}=10$$

Now use $m=10$ and $(3, f(3))=(3, 15)$ in the point-slope form.

$$y-15=10(x-3)$$
$$y-15=10x-30$$
$$y=10x-15$$

(b) $f(x)=x^2+2x;\ x=3$

$$\frac{f(x+h)-f(x)}{h}$$
$$=\frac{[(x+h)^2+2(x+h)]-(x^2+2x)}{h}$$
$$=\frac{(x^2+2hx+h^2+2x+2h)-(x^2+2x)}{h}$$
$$=\frac{2hx+h^2+2h}{h}=2x+h+2$$

$$f'(x)=\lim_{h\to 0}(2x+h+2)=2x+2$$

$f'(3)=2(3)+2=8$ is the slope of the tangent line at $x=3$.

Use $m=8$ and (3, 15) in the point-slope form.

$$y-15=8(x-3)$$
$$y=8x-9$$

23. (a) $f(x)=\frac{5}{x};\ x=2,\ x=5$

Slope of secant line $=\frac{f(5)-f(2)}{5-2}$

$$=\frac{\frac{5}{5}-\frac{5}{2}}{3}=\frac{1-\frac{5}{2}}{3}=-\frac{1}{2}$$

Now use $m=-\frac{1}{2}$ and $(5, f(5))=(5, 1)$ in the point-slope form.

$$y-1=-\frac{1}{2}[x-5]$$
$$y-1=-\frac{1}{2}x+\frac{5}{2}\Rightarrow y=-\frac{1}{2}x+\frac{7}{2}$$

(b) $f(x)=\frac{5}{x};\ x=2$

$$\frac{f(x+h)-f(x)}{h}=\frac{\frac{5}{x+h}-\frac{5}{x}}{h}$$
$$=\frac{\frac{5x-5(x+h)}{(x+h)x}}{h}$$
$$=\frac{5x-5x-5h}{h(x+h)(x)}$$
$$=\frac{-5h}{h(x+h)x}$$
$$=\frac{-5}{(x+h)x}$$

$$f'(x)=\lim_{h\to 0}\frac{-5}{(x+h)(x)}=-\frac{5}{x^2}$$

$f'(2)=\frac{-5}{2^2}=-\frac{5}{4}$ is the slope of the tangent line at $x=2$.

Now use $m=-\frac{5}{4}$ and $\left(2, \frac{5}{2}\right)$ in the point-slope form.

$$y-\frac{5}{2}=-\frac{5}{4}(x-2)$$
$$y-\frac{5}{2}=-\frac{5}{4}x+\frac{10}{4}$$
$$y=-\frac{5}{4}x+5$$

25. **(a)** $f(x)=4\sqrt{x};\ x=9,\ x=16$

Slope of secant line $=\dfrac{f(16)-f(9)}{16-9}$

$=\dfrac{4\sqrt{16}-4\sqrt{9}}{7}$

$=\dfrac{16-12}{7}=\dfrac{4}{7}$

Now use $m=\dfrac{4}{7}$ and $(9, f(9))=(9,12)$ in the point-slope form.

$$y-12=\frac{4}{7}(x-9)$$
$$y-12=\frac{4}{7}x-\frac{36}{7}$$
$$y=\frac{4}{7}x+\frac{48}{7}$$

(b) $f(x)=4\sqrt{x};\ x=9$

$$\frac{f(x+h)-f(x)}{h}$$
$$=\frac{4\sqrt{x+h}-4\sqrt{x}}{h}\cdot\frac{4\sqrt{x+h}+4\sqrt{x}}{4\sqrt{x+h}+4\sqrt{x}}$$
$$=\frac{16(x+h)-16x}{h(4\sqrt{x+h}+4\sqrt{x})}$$
$$f'(x)=\lim_{h\to 0}\frac{16(x+h)-16x}{h(4\sqrt{x+h}+4\sqrt{x})}$$
$$=\lim_{h\to 0}\frac{16h}{h(4\sqrt{x+h}+4\sqrt{x})}$$
$$=\lim_{h\to 0}\frac{4}{(\sqrt{x+h}+\sqrt{x})}=\frac{4}{2\sqrt{x}}=\frac{2}{\sqrt{x}}$$

$f'(9)=\dfrac{2}{\sqrt{9}}=\dfrac{2}{3}$ is the slope of the tangent line at $x=9$.

Use $m=\frac{2}{3}$ and $(9,12)$ in the point-slope form.

$$y-12=\frac{2}{3}(x-9)$$
$$y=\frac{2}{3}x+6$$

27. $f(x)=-4x^2+11x$

$f'(2)=-5$

$f'(16)=-117$

$f'(-3)=35$

29. $f(x)=e^x$

$f'(2)=7.3891$

$f'(16)=8{,}886{,}111$

$f'(-3)=0.0498$

31. $f(x)=-\dfrac{2}{x}$

$f'(2)=\dfrac{2}{2^2}=\dfrac{1}{2}$

$f'(16)=\dfrac{2}{16^2}=\dfrac{2}{256}=\dfrac{1}{128}.$

$f'(-3)=\dfrac{2}{(-3)^2}=\dfrac{2}{9}$

33. $f(x)=\sqrt{x}$

$f'(2)=\dfrac{1}{2\sqrt{2}}$

$f'(16)=\dfrac{1}{2\sqrt{16}}=\dfrac{1}{8}$

$f'(-3)=\dfrac{1}{2\sqrt{-3}}$ is not a real number, so $f'(-3)$ does not exist.

35. At $x=0$, the graph of $f(x)$ has a sharp point. Therefore, there is no derivative for $x=0$.

37. For $x=-3$ and $x=0$, the tangent to the graph of $f(x)$ is vertical. For $x=-1$, there is a gap in the graph of $f(x)$. For $x=2$, the function $f(x)$ does not exist. For $x=3$ and $x=5$, the graph of $f(x)$ has sharp points. Therefore, no derivative exists for $x=-3$, $x=-1$, $x=0$, $x=2$, $x=3$, and $x=5$.

39. **(a)** The rate of change of $f(x)$ is positive when $f(x)$ is increasing, that is, on $(a, 0)$ and (b, c).

(b) The rate of change of $f(x)$ is negative when $f(x)$ is decreasing, that is, on $(0, b,)$.

(c) The rate of change is zero when the tangent to the graph is horizontal, that is, at $x=0$ and $x=b$.

41. The zeros of graph (b) correspond to the turning points of graph (a), the points where the derivative is zero. Graph (a) gives the distance, while graph (b) gives the velocity.

43. $f(x) = x^x, a = 3$

(a)

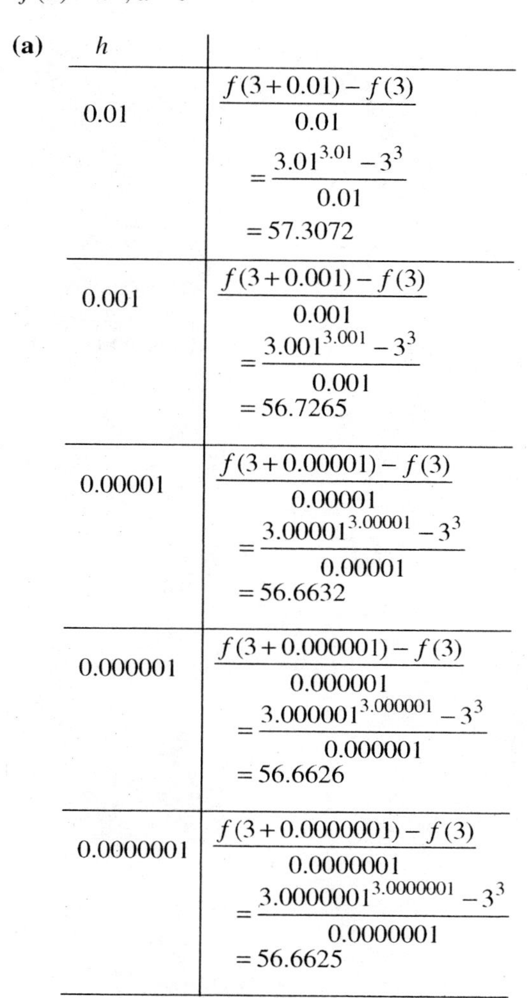

h	
0.01	$\frac{f(3+0.01)-f(3)}{0.01} = \frac{3.01^{3.01}-3^3}{0.01} = 57.3072$
0.001	$\frac{f(3+0.001)-f(3)}{0.001} = \frac{3.001^{3.001}-3^3}{0.001} = 56.7265$
0.00001	$\frac{f(3+0.00001)-f(3)}{0.00001} = \frac{3.00001^{3.00001}-3^3}{0.00001} = 56.6632$
0.000001	$\frac{f(3+0.000001)-f(3)}{0.000001} = \frac{3.000001^{3.000001}-3^3}{0.000001} = 56.6626$
0.0000001	$\frac{f(3+0.0000001)-f(3)}{0.0000001} = \frac{3.0000001^{3.0000001}-3^3}{0.0000001} = 56.6625$

It appears that $f'(3) \approx 56.66$.

(b) Graph the function on a graphing calculator and move the cursor to an x-value near $x = 3$. A good choice for the initial viewing window is [0, 4] by [0, 60]. Now zoom in on the function several times. Each time you zoom in, the graph will look less like a curve and more like a straight line. Use the TRACE feature to select two points on the graph, and record their coordinates. Use these two points to compute the slope. The result will be close to the most accurate value found in part (a), which is 56.66.

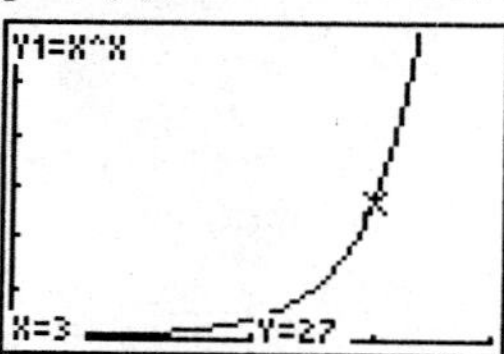

After zooming in twice:

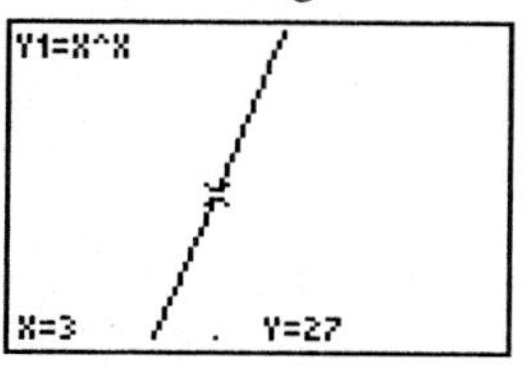

45. $f(x) = x^{1/x}, \ a = 3$

(a)

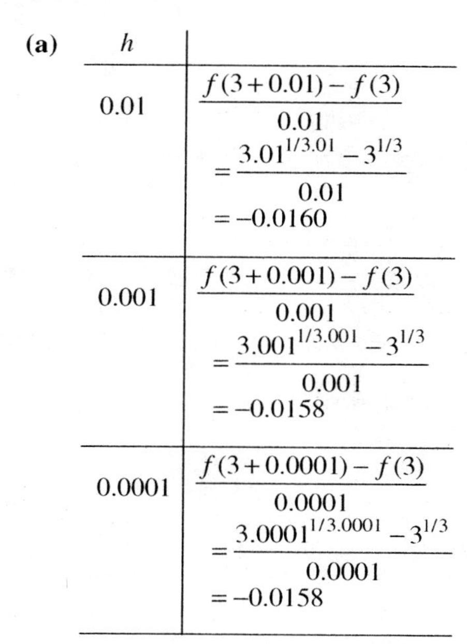

h	
0.01	$\frac{f(3+0.01)-f(3)}{0.01} = \frac{3.01^{1/3.01}-3^{1/3}}{0.01} = -0.0160$
0.001	$\frac{f(3+0.001)-f(3)}{0.001} = \frac{3.001^{1/3.001}-3^{1/3}}{0.001} = -0.0158$
0.0001	$\frac{f(3+0.0001)-f(3)}{0.0001} = \frac{3.0001^{1/3.0001}-3^{1/3}}{0.0001} = -0.0158$

It appears that $f'(3) = -0.0158$.

(b) Graph the function on a graphing calculator and move the cursor to an x-value near $x = 3$ A good choice for the initial viewing window is [0, 5] by [0, 3]. Follow the procedure outlined in the solution for Exercise 43, part (b). Note that near $x = 3$, the graph is very close to a horizontal line, so we expect that it slope will be close to 0. The final result will be close to the value found in part (a) of this exercise, which is –0.0158.

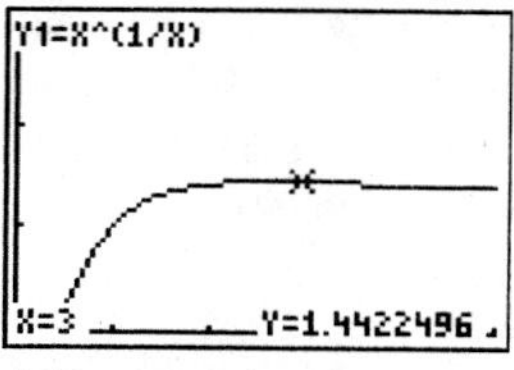

After zooming in once:

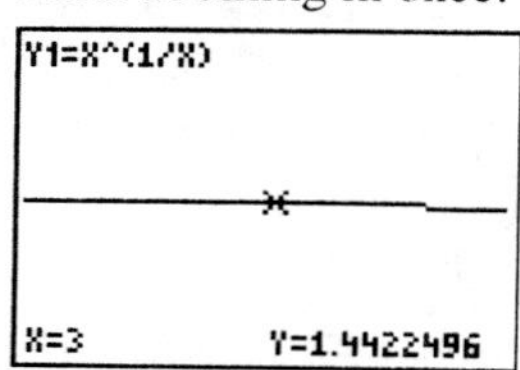

47. Answers will vary.

49. $l(x) = -3.6 + 0.17x$

(a) $l(70) = -3.6 + 0.17(70) = 8.3$

A snake of length 70 cm will prey on a catfish of length 8.3 cm.

(b) $$l'(x) = \lim_{h\to 0} \frac{l(x+h) - l(x)}{h} = \lim_{h\to 0} \frac{-3.6 + 0.17(x+h) - (-3.6 + 0.17x)}{h}$$
$$= \lim_{h\to 0} \frac{-3.6 + 0.17x + 0.17h + 3.6 - 0.17x}{h} = \lim_{h\to 0} \frac{0.17h}{h} = 0.17$$
$$l'(x) = 0.17$$

This means that longer snakes eat bigger catfish at the constant rate of 0.17 cm increase in length of prey to each 1 cm increase in the size of the snake.

(c)
$$l(x) > 0$$
$$-3.6 + 0.17x > 0$$
$$x > \frac{3.6}{0.17} \approx 21.176$$

The smallest snake for which the function makes sense has a length of about 22 cm.

51. $I(t) = 27 + 72t - 1.5t^2$

(a) $$I'(t) = \lim_{h\to 0} \frac{I(t+h) - I(t)}{h}$$
$$= \lim_{h\to 0} \frac{\left[27 + 72(t+5) - 1.5(t+5)^2\right] - \left[27 + 72t - 1.5t^2\right]}{h}$$
$$= \lim_{h\to 0} \frac{27 + 72t + 72h - 1.5t^2 - 3th - 1.5h^2 - 27 - 72t + 1.5t^2}{h}$$
$$= \lim_{h\to 0} \frac{72h - 3th - 1.5h^2}{h} = \lim_{h\to 0} 72 - 3t - 1.5h = 72 - 3t$$
$$I'(5) = 72 - 3(5) = 57$$

The rate of change of the intake of food 5 minutes into a meal is 57 grams per minute.

(b)
$$I'(24) \stackrel{?}{=} 0$$
$$72 - 3(24) \stackrel{?}{=} 0$$
$$0 = 0$$

24 minutes after the meal starts the rate of food consumption is 0.

(c) After 24 minutes the rate of food consumption is negative according to the function where a rate of zero is more accurate. A logical range for this function is $0 \le t \le 24$.

53. The derivative at (2, 4000) can be approximated by the slope of the line through (0, 2000) and (2, 4000).
The derivative is approximately
$$\frac{4000 - 2000}{2 - 0} = \frac{2000}{2} = 1000.$$
Thus, the shellfish population is increasing at a rate of 1000 shellfish per unit time.
The derivative at about (10, 10,300) can be approximated by the slope of the line through (10, 10,300) and (13, 12,000). The derivative is approximately
$$\frac{12,000 - 10,300}{13 - 10} = \frac{1700}{3} \approx 570.$$
The shellfish population is increasing at a rate of about 570 shellfish per unit time.
The derivative at about (13, 11,250) can be approximated by the slope of the line through (13, 11,250) and (16, 12,000). The derivative is approximately
$$\frac{12,000 - 11,250}{16 - 13} = \frac{750}{3} \approx 250.$$
The shellfish population is increasing at a rate of 250 shellfish per unit time.

55. The slope of the tangent line to the graph at the first point is found by finding two points on the tangent line.

$(x_1, y_1) = (1000, 13.5)$
$(x_2, y_2) = (0, 18.5)$

$$m = \frac{18.5 - 13.5}{0 - 1000} = \frac{5}{-1000} = -0.005$$

At the second point, we have

$(x_1, y_1) = (1000, 13.5)$
$(x_2, y_2) = (2000, 21.5)$

$$m = \frac{21.5 - 13.5}{2000 - 1000} = \frac{8}{1000} = 0.008$$

At the third point, we have

$(x_1, y_1) = (5000, 22.5)$
$(x_2, y_2) = (7000, 20).$

$$m = \frac{22.5 - 20}{5000 - 7000} = \frac{2.5}{-2000} = -0.00125$$

At 500 ft, the temperature decreases 0.005° per foot. At about 1500 ft, the temperature increases 0.008° per foot. At 5000 ft, the temperature decreases 0.00125° per foot.

57. (a) The slope of the graph at $x = 16$ looks horizontal. Thus, the derivative at $x = 16$ is 0. This means that for a 16 ounce bat, the velocity changes at about 0 mph per oz.

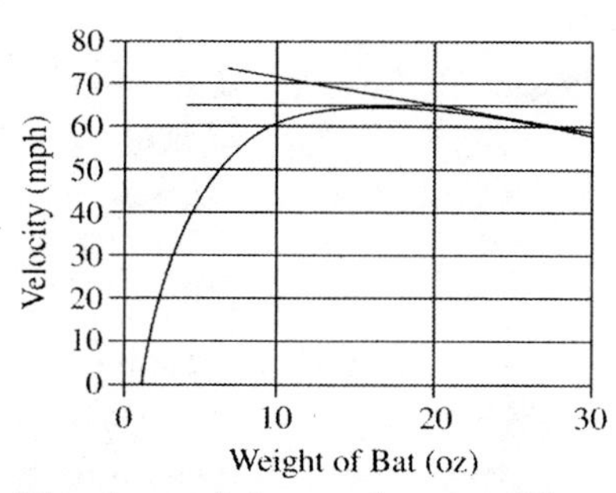

The slope of the graph at $x = 25$ can be estimated using the points (25, 63.4) and (26, 62.8).

$$\text{slope} = \frac{62.8 - 63.4}{26 - 25} = -0.6$$

Thus, the derivative at $x = 25$ is –0.6. This means that for a 25-ounce bat, the velocity changes at about –0.6 mph per oz.

(b) The optimal bat weight for this player is about 16 oz.

59. (a) $f(t) = 0.0000329t^3 - 0.00405t^2 + 0.0613t + 2.34$

$f(10) = 2.54$
$f(20) = 2.03$
$f(30) = 1.02$

(b) $Y_1 = 0.0000329x^3 - 0.00405x^2 + 0.0613x + 2.34$

$\text{nDeriv}(Y_1, x, 0) \approx 0.061$
$\text{nDeriv}(Y_1, x, 10) \approx -0.019$
$\text{nDeriv}(Y_1, x, 20) \approx -0.079$
$\text{nDeriv}(Y_1, x, 30) \approx -0.120$
$\text{nDeriv}(Y_1, x, 35) \approx -0.133$

(c) From year 0 through year 10, the graph of f increases. From year 10 through year 50, the graph decreases at an increasing rate every ten years.

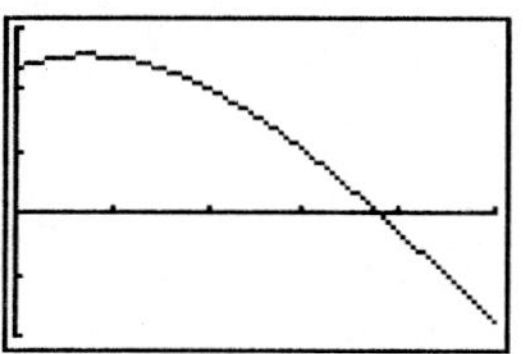

3.5 Graphical Differentiation

1. Graph the derivative of a function given the graph of the function by estimating the slope of the tangent line at various points, then connect the points with a smooth curve.

3. Since the x-intercepts of the graph of f' occur whenever the graph of f has a horizontal tangent line, Y_1 is the derivative of Y_2. Notice that Y_1 has two x-intercepts; each occurs at an x-value where the tangent line to Y_2 is horizontal. Note also that Y_1 is positive whenever Y_2 is increasing, and that Y_1 is negative whenever Y_2 is decreasing.

5. Since the x-intercepts of the graph of f' occur whenever the graph of f has a horizontal tangent line, Y_2 is the derivative of Y_1. Notice that Y_2 has one x-intercept which occurs at the x-value where the tangent line to Y_1 is horizontal. Also notice that the range on which Y_1 is increasing, Y_2 is positive and the range on which it is decreasing, Y_2 is negative.

7. To graph f', observe the intervals where the slopes of tangent lines are positive and where they are negative to determine where the derivative is positive and where it is negative. Also, whenever f has a horizontal tangent, f' will be 0, so the graph of f' will have an x-intercept. The x-values of the three turning point on the graph of f become the three x-intercepts of the graph of f. Estimate the magnitude of the slope at several points by drawing tangents to the graph of f.

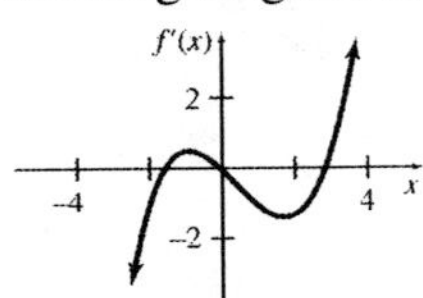

9. On the interval $(-\infty, -2)$, the graph of f is a horizontal line, so its slope is 0. Thus, on this interval, the graph of f' is $y = 0$ on $(-\infty, -2)$.
On the interval (–2, 0), the graph of f is a straight line, so its slope is constant. To find this slope, use the points (–2, 2) and (0, 0).

$$m = \frac{2-0}{-2-0} = \frac{2}{-2} = -1$$

On the interval (0, 1), the slope is also constant. To find this slope, use the points (0, 0) and (1, 1).

$$m = \frac{1-0}{1-0} = 1$$

On the interval $(1, \infty)$, the graph is again a horizontal line, so $m = 0$. The graph of f' will be made up of portions of the y-axis and the lines $y = -1$ and $y = 1$. Because the graph of f has "sharp points" or "corners" at $x = -2$, $x = 0$, and $x = 1$, we know that $f'(-2)$, $f'(0)$, and $f'(1)$ do not exist. We show this on the graph of f' by using open circles at the endpoints of the portions of the graph.

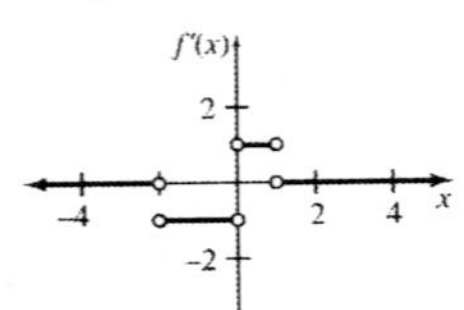

11. On the interval $(-\infty, -2)$, the graph of f is a straight line, so its slope is constant. To find this slope, use the points (–4, 2) and (–2, 0).

$$m = \frac{0-2}{-2-(-4)} = \frac{-2}{2} = -1$$

On the interval $(2, \infty)$, the slope of f is also constant. To find this slope, use the points (2, 0) and (3, 2).

$$m = \frac{2-0}{3-2} = \frac{2}{1} = 2$$

Thus, we have $f'(x) = -1$ on $(-\infty, -2)$ and $f'(x) = 2$ on $(2, \infty)$.
Because f is discontinuous at $x = -2$ and $x = 2$, we know that $f'(-2)$ and $f'(2)$ do not exist, which we indicate with open circles at (–2, –1) and (2, 2) on the graph of f'.
On the interval (–2, 2), all tangent lines have positive slopes, so the graph of f' will be above the y-axis. Notice that the slope of f (and thus the y-value of f') decreases on (–2, 0) and increases on (0, 2) with a minimum value on this interval of about 1 at $x = 0$.

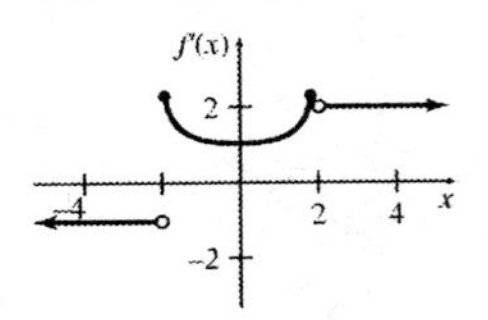

13. We observe that the slopes of tangent lines are positive on the interval $(-\infty, 0)$ and negative on the interval $(0, \infty)$, so the value of f' will be positive on $(-\infty, 0)$ and negative on $(0, \infty)$. Since f is undefined at $x = 0$, $f'(0)$ does not exist.
Notice that the graph of f becomes very flat when $|x| \to \infty$. The *value* of f approaches 0 and also the *slope* approaches 0. Thus, $y = 0$ (the x-axis) is a horizontal asymptote for both the graph of f and the graph of f'.

As $x \to 0^-$ and $x \to 0^+$, the graph of f gets very steep, so $|f'(x)| \to \infty$. Thus, $x = 0$ (the y-axis) is a vertical asymptote for both the graph of f and the graph of f'.

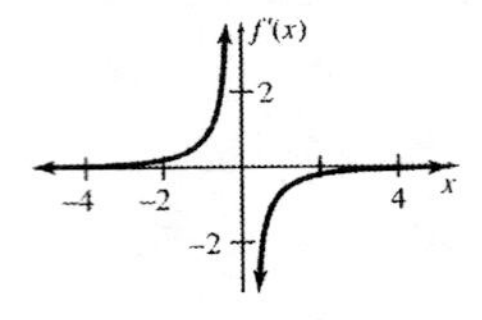

15. The slope of $f(x)$ is undefined at $x = -2$, $x = -1$, $x = 0$, $x = 1$, and $x = 2$, and the graph approaches vertical (unbounded slope) as x approaches those values. Accordingly, the graph of $f'(x)$ has vertical asymptotes at $x = -2$, $x = -1$, $x = 0$, $x = 1$, and $x = 2$. $f(x)$ has turning points (zero slope) at $x = -1.5$, $x = -0.5$, $x = 0.5$, and $x = 1.5$, so the graph of $f'(x)$ crosses the x-axis at those values.

Elsewhere, the graph of $f'(x)$ is negative where $f(x)$ is decreasing and positive where $f(x)$ is increasing.

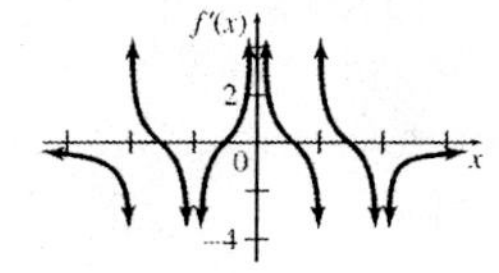

17. The growth rate of the function $y = f(x)$ is given by the derivative of this function $y' = f(x)$. We use the graph of f to sketch the graph of f'. First, notice as x increases, y increases throughout the domain of f, but at a slower and slower rate. The slope of f is positive but always decreasing, and approaches 0 as t gets large. Thus, y' will always be positive and decreasing. It will approach but never reach 0.
To plot points on the graph of f', we need to estimate the slope of f at several points. From the graph of f, we obtain the values given in the table.

t	y'
2	1000
10	700
13	250

Use these points to sketch the graph.

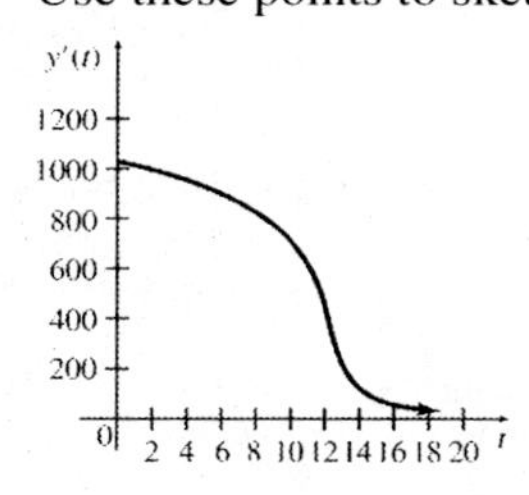

19. The remaining growth is about 9 cm. The rate of change of remaining growth is about 2.6 cm less per year.

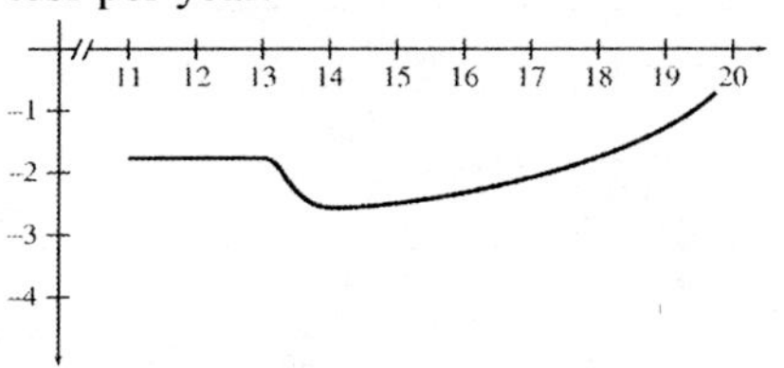

21. The graph rises steadily, with varying degrees of steepness. The graph is steepest around 1976 and nearly flat around 1950 and 1980. Accordingly, the rate of change is always positive, with a maximum value around 1976 and values near zero around 1950 and 1980.

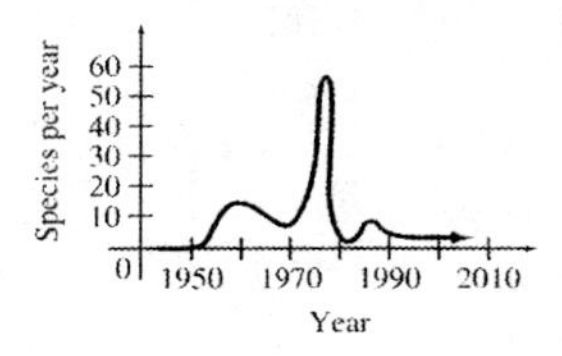

23. Both curves appear to be level (slope = 0), then they increase sharply (slope increase), then become level (slope = 0).

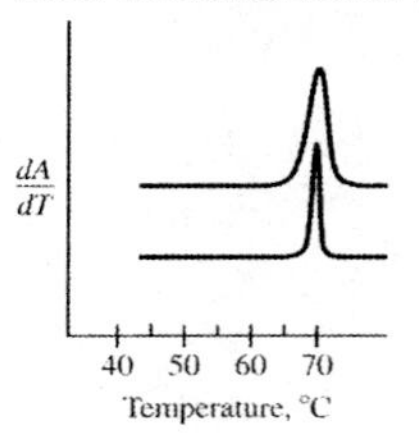

25. Let S' be the derivative function showing the rate of change of sales. On the interval [0, 1], the slope of the line segment between points (0, 0) and (1, 1) is 1. Thus, on [0, 1] the graph of S' is $y = 1$.
On the interval [1, 2], the slope of the line segment between points (1, 1) and (2, 3) is 2. Thus on [1, 2], the graph of S' is $y = 2$.
On [2, 3], the slope of the line segment between points (2, 3) and (3, 7) is 4. Thus on [2, 3], the graph of S' is $y = 4$.
Continuing in this fashion, the table gives the slopes of the line segments (which make up the graph of S') on the remaining intervals.

(*continued on next page*)

(continued)

Interval	Graph of S'
[0, 1]	$y = 1$
[1, 2]	$y = 2$
[2, 3]	$y = 4$
[3, 4]	$y = 3$
[4, 7]	$y = 0$
[7, 9]	$y = -2$
[9, 10]	$y = -3$
[10, 12]	$y = -1$

Because the graph of the annual sales function has "sharp points" or "corners" where the slopes of the line segments change, we know that S' does not exist at these points. We show this on the graph of S' by using open circles.

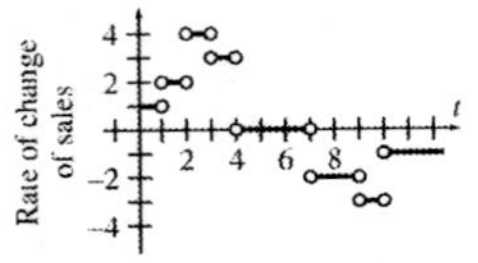

Rate of change of sales

Chapter 3 Review Exercises

1. True
2. True
3. True
4. False; for example, if $f(x) = \frac{x^2 - 4}{x+2}$, $\lim_{x \to -2} f(x) = -4$, but the graph of $f(x) = \frac{x^2 - 4}{x+2}$ has a hole at the point $(-2, -4)$.
5. False; for example, $f(x) = \tan x$ has discontinuities at multiples of $\frac{\pi}{2}$.
6. False; for example, the rational function $f(x) = \frac{5}{x+1}$ is discontinuous at $x = -1$.
7. False; the derivative gives the instantaneous rate of change of a function.
8. True
9. True
10. True
11. False; the slope of the tangent line gives the instantaneous rate of change.
12. False; for example, the function $f(x) = |x|$ is continuous at $x = 0$, but $f'(0)$ does not exist. The graph of $f(x) = |x|$ has a "corner" at $x = 0$.
13. Yes, a derivative is always a limit because the definition of the derivative is a limit. No, a limit is not always a derivative.
14. No, every continuous function is not differentiable. For example, $f(x) = |x|$ is continuous at $x = 0$, but $f'(0)$ does not exist. The graph of $f(x) = |x|$ has a "corner" at $x = 0$. No, every differentiable function is not continuous. For example, $f(x) = \tan x$ is differentiable, although not for all values of x.
15. Answers will vary.
16. The derivative can be used to find the instantaneous rate of change at a point on a function, and the slope of a tangent line at a point on a function.
17. **(a)** $\lim_{x \to -3^-} = 4$

    (b) $\lim_{x \to -3^+} = 4$

    (c) $\lim_{x \to -3} = 4$, since parts (a) and (b) have the same answer.

    (d) $f(-3) = 4$, since $(-3, 4)$ is a point of the graph.
18. **(a)** $\lim_{x \to -1^-} g(x) = -2$

    (b) $\lim_{x \to -1^+} g(x) = 2$

    (c) $\lim_{x \to -1} g(x)$ does not exist since parts (a) and (b) have different answers.

    (d) $g(-1) = -2$, since $(-1, -2)$ is a point on the graph.
19. **(a)** $\lim_{x \to 4^-} f(x) = \infty$

    (b) $\lim_{x \to 4^+} f(x) = -\infty$

(c) $\lim_{x\to 4} f(x)$ does not exist since limits in parts (a) and (b) do not exist.

(d) $f(4)$ does not exist since the graph has no point with an x-value of 4.

20. (a) $\lim_{x\to 2^-} h(x) = 1$

(b) $\lim_{x\to 2^+} h(x) = 1$

(c) $\lim_{x\to 2} h(x) = 1$

(d) $h(2)$ does not exist since the graph has no point with an x-value of 2.

21. $\lim_{x\to -\infty} g(x) = \infty$ since the y-value gets very large as the x-value gets very small.

22. $\lim_{x\to \infty} f(x) = -3$ since the line $y = -3$ is a horizontal asymptote for the graph.

23. $\lim_{x\to 6} \frac{2x+7}{x+3} = \frac{2(6)+7}{6+3} = \frac{19}{9}$

24. Let $f(x) = \frac{2x+5}{x+3}$.

x	−3.1	−3.01	−3.001
$f(x)$	12	102	1002
x	−2.9	−2.99	−2.999
$f(x)$	−8	−98	−998

As x approaches −3 from the left, $f(x)$ gets infinitely larger. As x approaches −3 from the right, $f(x)$ gets infinitely smaller.

Therefore, $\lim_{x\to -3} \frac{2x+5}{x+3}$ does not exist.

25. $$\lim_{x\to 4} \frac{x^2-16}{x-4} = \lim_{x\to 4} \frac{(x-4)(x+4)}{x-4} = \lim_{x\to 4}(x+4) = 4+4 = 8$$

26. $$\lim_{x\to 2} \frac{x^2+3x-10}{x-2} = \lim_{x\to 2} \frac{(x+5)(x-2)}{x-2} = \lim_{x\to 2}(x+5) = 2+5 = 7$$

27. $$\lim_{x\to -4} \frac{2x^2+3x-20}{x+4} = \lim_{x\to -4} \frac{(2x-5)(x+4)}{x+4} = \lim_{x\to -4}(2x-5) = 2(-4)-5 = -13$$

28. $$\lim_{x\to 3} \frac{3x^2-2x-21}{x-3} = \lim_{x\to 3} \frac{(3x+7)(x-3)}{x-3} = \lim_{x\to 3}(3x+7) = 9+7 = 16$$

29. $$\lim_{x\to 9} \frac{\sqrt{x}-3}{x-9} = \lim_{x\to 9} \frac{\sqrt{x}-3}{x-9} \cdot \frac{\sqrt{x}+3}{\sqrt{x}+3} = \lim_{x\to 9} \frac{x-9}{(x-9)(\sqrt{x}+3)} = \lim_{x\to 9} \frac{1}{\sqrt{x}+3} = \frac{1}{\sqrt{9}+3} = \frac{1}{6}$$

30. $$\lim_{x\to 16} \frac{\sqrt{x}-4}{x-16} = \lim_{x\to 16} \frac{\sqrt{x}-4}{x-16} \cdot \frac{\sqrt{x}+4}{\sqrt{x}+4} = \lim_{x\to 16} \frac{x-16}{(x-16)(\sqrt{x}+4)} = \lim_{x\to 16} \frac{1}{\sqrt{x}+4} = \frac{1}{\sqrt{16}+4} = \frac{1}{4+4} = \frac{1}{8}$$

31. $$\lim_{x\to \infty} \frac{2x^2+5}{5x^2-1} = \lim_{x\to \infty} \frac{\frac{2x^2}{x^2}+\frac{5}{x^2}}{\frac{5x^2}{x^2}-\frac{1}{x^2}} = \lim_{x\to \infty} \frac{2+\frac{5}{x^2}}{5-\frac{1}{x^2}} = \frac{2+0}{5-0} = \frac{2}{5}$$

32. $$\lim_{x\to \infty} \frac{x^2+6x+8}{x^3+2x+1} = \lim_{x\to \infty} \frac{\frac{x^2}{x^3}+\frac{6x}{x^3}+\frac{8}{x^3}}{\frac{x^3}{x^3}+\frac{2x}{x^3}+\frac{1}{x^3}} = \lim_{x\to \infty} \frac{\frac{1}{x}+\frac{6}{x^2}+\frac{8}{x^3}}{1+\frac{2}{x^2}+\frac{1}{x^3}} = \frac{0+0+0}{1+0+0} = 0$$

33. $$\lim_{x\to -\infty} \left(\frac{3}{8}+\frac{3}{x}-\frac{6}{x^2}\right) = \lim_{x\to -\infty} \frac{3}{8} + \lim_{x\to -\infty} \frac{3}{x} - \lim_{x\to -\infty} \frac{6}{x^2} = \frac{3}{8}+0-0 = \frac{3}{8}$$

34. $$\lim_{x\to -\infty} \left(\frac{9}{x^4}+\frac{10}{x^2}-6\right) = \lim_{x\to -\infty} \frac{9}{x^4} + \lim_{x\to -\infty} \frac{10}{x^2} - \lim_{x\to -\infty} 6 = 0+0-6 = -6$$

35. $\lim\limits_{x\to\frac{\pi}{2}} \dfrac{1-\sin^2 x}{\cos^2 x} = \lim\limits_{x\to\frac{\pi}{2}} \dfrac{\cos^2 x}{\cos^2 x} = 1$

36. $\lim\limits_{x\to\pi} \dfrac{1+\cos x}{1-\cos^2 x} = \lim\limits_{x\to\pi} \dfrac{1+\cos x}{(1+\cos x)(1-\cos x)}$

$= \lim\limits_{x\to\pi} \dfrac{1}{1-\cos x}$

$= \dfrac{1}{1-\cos\pi} = \dfrac{1}{1-(-1)} = \dfrac{1}{2}$

37. As shown on the graph, $f(x)$ is discontinuous at x_2 and x_4.

38. As shown on the graph, $f(x)$ is discontinuous at x_1 and x_4.

39. $f(x)$ is discontinuous at $x = 0$ and $x = -\frac{1}{3}$ since that is where the denominator of $f(x)$ equals 0. $f(0)$ and $f\left(-\frac{1}{3}\right)$ do not exist.

$\lim\limits_{x\to 0} f(x)$ does not exist because

$\lim\limits_{x\to 0^+} f(x) = -\infty$, but $\lim\limits_{x\to 0^-} f(x) = \infty$.

$\lim\limits_{x\to -\frac{1}{3}} f(x)$ does not exist because

$\lim\limits_{x\to -\frac{1}{3}^-} = -\infty$, but $\lim\limits_{x\to -\frac{1}{3}^+} f(x) = \infty$.

40. $f(x) = \dfrac{7-3x}{(1-x)(3+x)}$

The function is discontinuous at $x = -3$ and $x = 1$ because those values make the denominator of the fraction equal to zero.

$\lim\limits_{x\to -3} f(x)$ does not exist because

$\lim\limits_{x\to -3^-} f(x) = -\infty$ and $\lim\limits_{x\to -3^+} f(x) = \infty$.

$\lim\limits_{x\to 1} f(x)$ does not exist because

$\lim\limits_{x\to 1^-} f(x) = \infty$ and $\lim\limits_{x\to 1^+} f(x) = -\infty$.

$f(-3)$ and $f(1)$ do not exist since there is no point of the graph that has an x-value of -3 or 1.

41. $f(x)$ is discontinuous at $x = -5$ since that is where the denominator of $f(x)$ equals 0.

$f(-5)$ does not exist.

$\lim\limits_{x\to -5} f(x)$ does not exist because

$\lim\limits_{x\to -5^-} f(x) = \infty$, but $\lim\limits_{x\to -5^+} f(x) = -\infty$.

42. $f(x) = \dfrac{x^2-9}{x+3}$

The function is discontinuous at $x = -3$ because this value makes the denominator of the fraction equal to zero.

$\lim\limits_{x\to -3} \dfrac{x^2-9}{x+3} = \lim\limits_{x\to -3} \dfrac{(x+3)(x-3)}{x+3}$

$= \lim\limits_{x\to -3} (x-3)$

$= -3-3 = -6$

$f(-3)$ does not exist since there is no point on the graph with an x-value of -3.

43. $f(x) = x^2 + 3x - 4$ is continuous everywhere since f is a polynomial function.

44. $f(x) = 2x^2 - 5x - 3$ has no points of discontinuity since it is a polynomial function, which is continuous everywhere.

45. $f(x) = \cos\left(\dfrac{x}{x-1}\right)$ is discontinuous at $x = 1$.

$f(1)$ does not exist and $\lim\limits_{x\to 1} f(x)$ does not exist.

46. $f(x) = e^{\sin x}$ is continuous everywhere.

47. $f(x) = \begin{cases} 1-x & \text{if } x < 1 \\ 2 & \text{if } 1 \le x \le 2 \\ 4-x & \text{if } x > 2 \end{cases}$

(a)

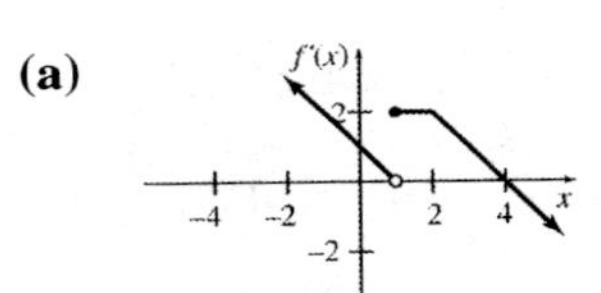

(b) The graph is discontinuous at $x = 1$.

(c) $\lim\limits_{x\to 1^-} f(x) = 0$; $\lim\limits_{x\to 1^+} f(x) = 2$

48. $f(x) = \begin{cases} 2 & \text{if } x < 0 \\ -x^2 + x + 2 & \text{if } 0 \le x \le 2 \\ 1 & \text{if } x > 2 \end{cases}$

(a)

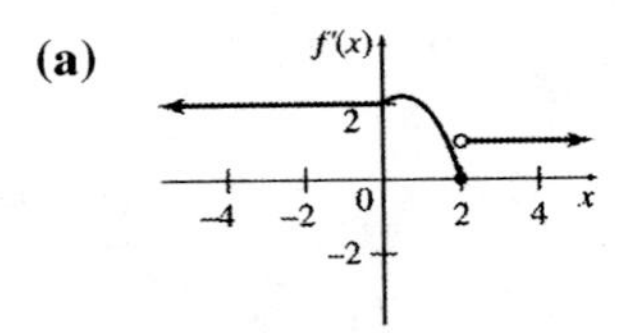

(b) The graph is discontinuous at $x = 2$.

(c) $\lim_{x\to 2^-} f(x) = -4+2+2 = 0$
$\lim_{x\to 2^+} f(x) = 1$

49. $f(x) = \dfrac{x^4+2x^3+2x^2-10x+5}{x^2-1}$

(a) Find the values of $f(x)$ when x is close to 1.

x	y	x	y
1.1	2.6005	0.99	1.94
1.01	2.06	0.999	1.994
1.001	2.006	0.9999	1.9994
1.0001	2.0006		

It appears that $\lim_{x\to 1} f(x) = 2.$

(b) Graph $y = \dfrac{x^4+2x^3+2x^2-10x+5}{x^2-1}$ on a graphing calculator. One suitable choice for the viewing window is [–3, 3] by [–40, 40]. Because $x^2-1=0$ when $x=-1$ or $x=1$, this function is discontinuous at these two x-values. There is a vertical asymptote at $x=-1$ but not at $x=1$. The graph should have an open circle to show a "hole" in the graph at $x=1$.The graphing calculator doesn't show the hole, but trying to find the value of the function of $x=1$ will show that this value is undefined. By viewing the function near $x=1$ and using the ZOOM feature, we see that as x gets close to 1 from the left or the right, y gets close to 2, suggesting that

$$\lim_{x\to 1}\frac{x^4+2x^3+2x^2-10x+5}{x^2-1} = 2.$$

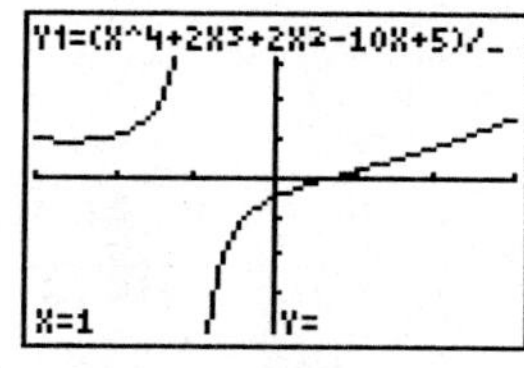

After zooming in once,

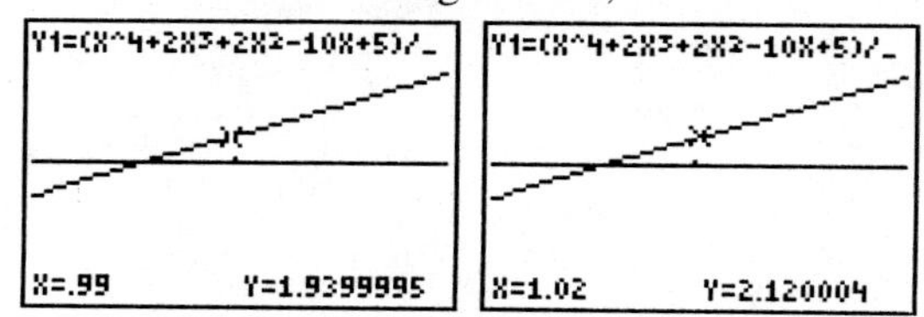

50. $f(x) = \dfrac{x^4+3x^3+7x^2+11x+2}{x^3+2x^2-3x-6}$

(a) Find values of $f(x)$ when x is close to –2.

x	$f(x)$
–2.01	–12.62
–2.001	–12.96
–2.0001	–13
–1.99	–13.41
–1.999	–13.04
–1.9999	–13

It appears that $\lim_{x\to -2} f(x) = -13.$

(b) Graph $y = \dfrac{x^4+3x^3+7x^2+11x+2}{x^3+2x^2-3x-6}$ on a graphing calculator. One suitable choice for the viewing window is [–5, 5] by [–15, 15]. The graph should have an open circle to show a "hole" in the graph at $x=-2$. The graphing calculator doesn't show the hole, but trying to find the value of the function of $x=-2$ will show that this value is undefined. By viewing the function near $x=-2$, we see that as x gets close to –2 from the left or the right, y gets close to –13, suggesting that

$$\lim_{x\to -2}\frac{x^4+3x^3+7x^2+11x+2}{x^3+2x^2-3x-6} = -13.$$

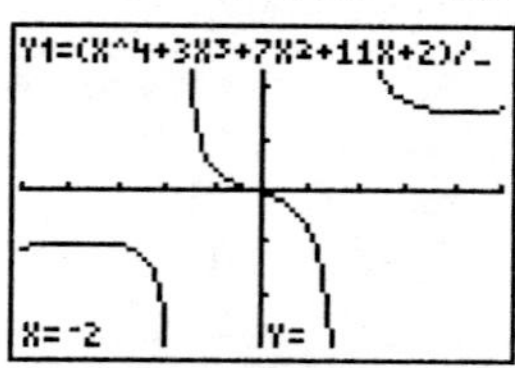

After zooming in once,

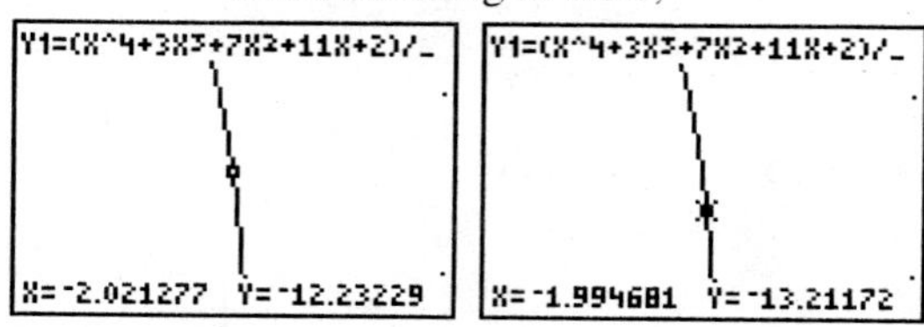

51. $y = 6x^3 + 2 = f(x)$ from $x = 1$ to $x = 4$

$f(4) = 6(4)^3 + 2 = 386$
$f(1) = 6(1)^3 + 2 = 8$

Average rate of change: $\dfrac{f(4) - f(1)}{4 - 1} = \dfrac{386 - 8}{4 - 1} = \dfrac{378}{3} = 126$

$$\frac{f(x+h) - f(x)}{h} = \frac{6(x+h)^3 + 2 - (6x^3 + 2)}{h} = \frac{6x^3 + 18x^2h + 18xh^2 + h^3 + 2 - 6x^3 - 2}{h}$$
$$= \frac{18x^2h + 18xh^2 + h^3}{h} = 18x^2 + 18xh + h^2$$

$f'(x) = \lim_{h \to 0}\left(18x^2 + 18xh + h^2\right) = 18x^2$

Instantaneous rate of change at $x = 1$: $f'(1) = 18(1) = 18$

52. $y = -2x^3 - 3x^2 + 8 = f(x)$ from $x = -2$ to $x = 6$

$f(6) = -2(6)^3 - 3(6)^2 + 8 = -532$
$f(-2) = -2(-2)^3 - 3(-2)^2 + 8 = 12$

Average rate of change: $= \dfrac{f(6) - f(-2)}{6 - (-2)} = \dfrac{-532 - 12}{6 + 2} = \dfrac{-544}{8} = -68$

$$\frac{f(x+h) - f(x)}{h} = \frac{-2(x+h)^3 - 3(x+h)^2 + 8 - (-2x^3 - 3x^2 + 8)}{h}$$
$$= \frac{-2x^3 - 6x^2h - 6xh^2 - 2h^3 - 3x^2 - 6xh - 3h^2 + 8 + 2x^3 + 3x^2 - 8}{h}$$
$$= \frac{-6x^2h - 6xh^2 - 6xh - 2h^3 - 3h^2}{h} = -6x^2 - 6xh - 6x - 2h^2 - 3h$$

$f'(x) = \lim_{h \to 0}\left(-6x^2 - 6xh - 6x - 2h^2 - 3h\right) = -6x^2 - 6x$

Instantaneous rate of change at $x = -2$: $f'(-2) = -6(-2)^2 - 6(-2) = -6(4) + 12 = -12$

53. $y = \dfrac{-6}{3x - 5} = f(x)$ from $x = 4$ to $x = 9$

$f(9) = \dfrac{-6}{3(9) - 5} = \dfrac{-6}{22} = -\dfrac{3}{11}$, $\quad f(4) = \dfrac{-6}{3(4) - 5} = -\dfrac{6}{7}$

Average rate of change: $\dfrac{f(9) - f(4)}{9 - 4} = \dfrac{\frac{-3}{11} - \left(-\frac{6}{7}\right)}{9 - 4} = \dfrac{\frac{-21 + 66}{77}}{5} = \dfrac{45}{5(77)} = \dfrac{9}{77}$

$$\frac{f(x+h) - f(x)}{h} = \frac{\frac{-6}{3(x+h) - 5} - \left(\frac{-6}{3x - 5}\right)}{h} = \frac{\frac{-6}{3x + 3h - 5} + \frac{6}{3x - 5}}{h} = \frac{-6(3x - 5) + 6(3x + 3h - 5)}{h(3x + 3h - 5)(3x - 5)}$$
$$= \frac{-18x + 30 + 18x + 18h - 30}{h(3x + 3h - 5)(3x - 5)} = \frac{18h}{h(3x + 3h - 5)(3x - 5)} = \frac{18}{(3x + 3h - 5)(3x - 5)}$$

$f'(x) = \lim_{h \to 0} \dfrac{18}{(3x + 3h - 5)(3x - 5)} = \dfrac{18}{(3x - 5)^2}$

Instantaneous rate of change at $x = 4$: $f'(4) = \dfrac{18}{(3 \cdot 4 - 5)^2} = \dfrac{18}{7^2} = \dfrac{18}{49}$

54. $y = \dfrac{x+4}{x-1} = f(x)$ from $x = 2$ to $x = 5$

$f(5) = \dfrac{5+4}{5-1} = \dfrac{9}{4}, \quad f(2) = \dfrac{2+4}{2-1} = 6$

Average rate of change: $= \dfrac{\frac{9}{4}-6}{5-2} = \dfrac{\frac{-15}{4}}{3} = -\dfrac{5}{4}$

$$\frac{f(x+h)-f(x)}{h} = \frac{\frac{(x+h)+4}{(x+h)-1} - \left(\frac{x+4}{x-1}\right)}{h} = \frac{\frac{(x+h+4)(x-1)-(x+4)(x+h-1)}{(x+h-1)(x-1)}}{h}$$

$$= \frac{x^2+xh+3x-h-4-\left(x^2+xh+3x+4h-4\right)}{h(x+h-1)(x-1)} = \frac{-5h}{h(x+h-1)(x-1)}$$

$$= \frac{-5}{(x+h-1)(x-1)}$$

$$f'(x) = \lim_{h\to 0} \frac{-5}{(x+h-1)(x-1)} = \frac{-5}{(x-1)^2}$$

Instantaneous rate of change at $x = 2$: $f'(2) = \dfrac{-5}{(2-1)^2} = \dfrac{-5}{1} = -5$

55. (a) $f(x) = 3x^2 - 5x + 7;\ x = 2,\ x = 4$

Slope of secant line

$= \dfrac{f(4)-f(2)}{4-2}$

$= \dfrac{[3(4)^2 - 5(4)+7]-[3(2)^2-5(2)+7]}{2}$

$= \dfrac{35-9}{2} = 13$

Now use $m = 13$ and $2, f(2) = (2,9)$ in the point-slope form.

$y - 9 = 13(x-2)$
$y - 9 = 13x - 26$
$y = 13x - 17$

(b) $f(x) = 3x^2 - 5x + 7;\ x = 2$

$\dfrac{f(x+h)-f(x)}{h}$

$= \dfrac{[3(x+h)^2 - 5(x+h)+7]-[3x^2-5x+7]}{h}$

$= \dfrac{3x^2+6xh+3h^2-5x-5h+7-3x^2+5x-7}{h}$

$= \dfrac{6xh+3h^2-5h}{h} = 6x+3h-5$

$f'(x) = \lim\limits_{h\to 0}(6x+3h-5) = 6x-5$

$f'(2) = 6(2)-5 = 7$

Now use $m = 7$ and $(2, f(2)) = (2,9)$ in the point-slope form.

$y - 9 = 7(x-2)$
$y - 9 = 7x - 14$
$y = 7x - 5$

56. (a) $f(x) = \dfrac{1}{x};\ x = \dfrac{1}{2},\ x = 3$

Slope of secant line $= \dfrac{f(3)-f\left(\frac{1}{2}\right)}{3-\frac{1}{2}}$

$= \dfrac{\frac{1}{3}-2}{3-\frac{1}{2}} = \dfrac{2-12}{18-3}$

$= \dfrac{-10}{15} = -\dfrac{2}{3}$

Now use $m = -\dfrac{2}{3}$ and $\left(\dfrac{1}{2}, f\left(\dfrac{1}{2}\right)\right) = \left(\dfrac{1}{2}, 2\right)$ in the point-slope form.

$y - 2 = -\dfrac{2}{3}\left(x - \dfrac{1}{2}\right)$

$y - 2 = -\dfrac{2}{3}x + \dfrac{1}{3}$

$y = -\dfrac{2}{3}x + \dfrac{7}{3}$

(b) $f(x)=\frac{1}{x};\ x=\frac{1}{2}$

$$\frac{f(x+h)-f(x)}{h}=\frac{\frac{1}{x+h}-\frac{1}{x}}{h}=\frac{x-(x+h)}{xh(x+h)}=-\frac{h}{xh(x+h)}=-\frac{1}{x(x+h)}$$

$$f'(x)=\lim_{h\to 0}-\frac{1}{x(x+h)}=-\frac{1}{x^2}$$

$$f'\left(\frac{1}{2}\right)=-\frac{1}{\left(\frac{1}{2}\right)^2}=-4$$

Now use $m=-4$ and $\left(\frac{1}{2}, f\left(\frac{1}{2}\right)\right)=\left(\frac{1}{2}, 2\right)$ in the point-slope form.

$$y-2=-4\left(x-\frac{1}{2}\right)$$
$$y-2=-4x+2$$
$$y=-4x+4$$

57. (a) $f(x)=\frac{12}{x-1};\ x=3,\ x=7$

Slope of secant line $=\frac{f(7)-f(3)}{7-3}=\frac{\frac{12}{7-1}-\frac{12}{3-1}}{4}=\frac{2-6}{4}=-1$

Now use $m=-1$ and $(3, f(x))=(3,6)$ in the point-slope form.

$$y-6=-1(x-3)$$
$$y-6=-x+3$$
$$y=-x+9$$

(b) $f(x)=\frac{12}{x-1};\ x=3$

$$\frac{f(x+h)-f(x)}{h}=\frac{\frac{12}{x+h-1}-\frac{12}{x-1}}{h}=\frac{12(x-1)-12(x+h-1)}{h(x-1)(x+h-1)}$$
$$=\frac{-12h}{h(x-1)(x+h-1)}=-\frac{12}{(x-1)(x+h-1)}$$

$$f'(x)=\lim_{h\to 0}-\frac{12}{(x-1)(x+h-1)}=-\frac{12}{(x-1)^2}$$

$$f'(3)=-\frac{12}{(3-1)^2}=-3$$

Now use $m=-3$ and $(3, f(x))=(3,6)$ in the point-slope form.

$$y-6=-3(x-3)$$
$$y-6=-3x+9$$
$$y=-3x+15$$

58. (a) $f(x)=2\sqrt{x-1};\ x=5,\ x=10$

Slope of secant line $=\frac{f(10)-f(5)}{10-5}$
$=\frac{2\sqrt{10-1}-2\sqrt{5-1}}{5}=\frac{2(3)-2(2)}{5}=\frac{2}{5}$

(*continued on next page*)

(*continued*)

Now use $m = \frac{2}{5}$ and $(5, f(x)) = (5, 4)$ in the point-slope form.

$$y - 4 = \frac{2}{5}(x - 5)$$
$$y - 4 = \frac{2}{5}x - 2$$
$$y = \frac{2}{5}x + 2$$

(b) $f(x) = 2\sqrt{x - 1};\ x = 5$

$$\frac{f(x+h) - f(x)}{h} = \frac{2\sqrt{x+h-1} - 2\sqrt{x-1}}{h} = \frac{2(\sqrt{x+h-1} - \sqrt{x-1})(\sqrt{x+h-1} + \sqrt{x-1})}{h(\sqrt{x+h-1} + \sqrt{x-1})}$$
$$= \frac{2(x+h-1-x+1)}{h(\sqrt{x+h-1} + \sqrt{x-1})} = \frac{2h}{h(\sqrt{x+h-1} + \sqrt{x-1})} = \frac{2}{\sqrt{x+h-1} + \sqrt{x-1}}$$

$$f'(x) = \lim_{h \to 0} \frac{2}{\sqrt{x+h-1} + \sqrt{x-1}} = \frac{2}{2\sqrt{x-1}} = \frac{1}{\sqrt{x-1}}$$

$$f'(5) = \frac{1}{\sqrt{5-1}} = \frac{1}{2}$$

Now use $m = \frac{1}{2}$ and $(5, f(x)) = (5, 4)$ in the point-slope form.

$$y - 4 = \frac{1}{2}(x - 5)$$
$$y - 4 = \frac{1}{2}x - \frac{5}{2}$$
$$y = \frac{1}{2}x + \frac{3}{2}$$

59. $y = 4x^2 + 3x - 2 = f(x)$

$$y' = \lim_{h \to 0} \frac{f(x+h) - f(x)}{h}$$
$$= \lim_{h \to 0} \frac{[4(x+h)^2 + 3(x+h) - 2] - [4x^2 + 3x - 2]}{h}$$
$$= \lim_{h \to 0} \frac{4(x^2 + 2xh + h^2) + 3x + 3h - 2 - 4x^2 - 3x + 2}{h}$$
$$= \lim_{h \to 0} \frac{4x^2 + 8xh + 4h^2 + 3x + 3h - 2 - 4x^2 - 3x + 2}{h}$$
$$= \lim_{h \to 0} \frac{8xh + 4h^2 + 3h}{h} = \lim_{h \to 0} \frac{h(8x + 4h + 3)}{h} = \lim_{h \to 0} (8x + 4h + 3) = 8x + 3$$

60. $y = 5x^2 - 6x + 7 = f(x)$

$$y' = \lim_{h\to 0} \frac{f(x+h) - f(x)}{h}$$
$$= \lim_{h\to 0} \frac{[5(x+h)^2 - 6(x+h) + 7] - [5x^2 - 6x + 7]}{h}$$
$$= \lim_{h\to 0} \frac{5(x^2 + 2xh + h^2) - 6x - 6h + 7 - 5x^2 + 6x - 7}{h}$$
$$= \lim_{h\to 0} \frac{5x^2 + 10xh + 5h^2 - 6x - 6h + 7 - 5x^2 + 6x - 7}{h}$$
$$= \lim_{h\to 0} \frac{10xh + 5h^2 - 6h}{h} = \lim_{h\to 0} \frac{h(10x + 5h - 6)}{h} = \lim_{h\to 0} (10x + 5h - 6) = 10x - 6$$

61. $f(x) = (\ln x)^x,\ x_0 = 3$

(a)

h	
0.01	$\frac{f(3+0.01) - f(3)}{0.01} = \frac{(\ln 3.01)^{3.01} - (\ln 3)^3}{0.01} = 1.3385$
0.001	$\frac{f(3+0.001) - f(3)}{0.001} = \frac{(\ln 3.001)^{3.001} - (\ln 3)^3}{0.001} = 1.3323$
0.0001	$\frac{f(3+0.0001) - f(3)}{0.0001} = \frac{(\ln 3.0001)^{3.0001} - (\ln 3)^3}{0.0001} = 1.3317$
0.00001	$\frac{f(3+0.00001) - f(3)}{0.00001} = \frac{(\ln 3.00001)^{3.00001} - (\ln 3)^3}{0.00001} = 1.3317$

It appears that $f'(3) \approx 1.332$.

(b) Graph the function on a graphing calculator and move the cursor to an x-value near $x = 3$. A good choice for the initial viewing window is [0, 4] by [0, 60]. Now zoom in on the function several times. Each time you zoom in, the graph will look less like a curve and more like a straight line. Use the TRACE feature to select two points on the graph, and record their coordinates. Use these two points to compute the slope. The result will be close to the most accurate value found in part (a), which is 1.33.

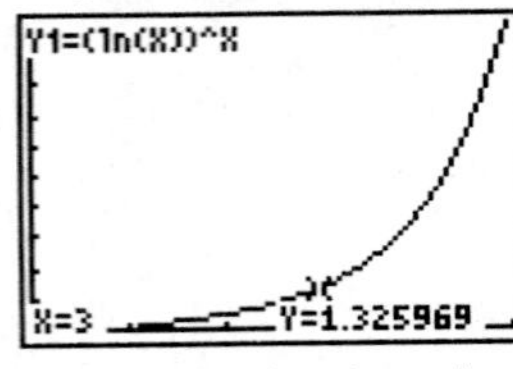

After zooming in twice:

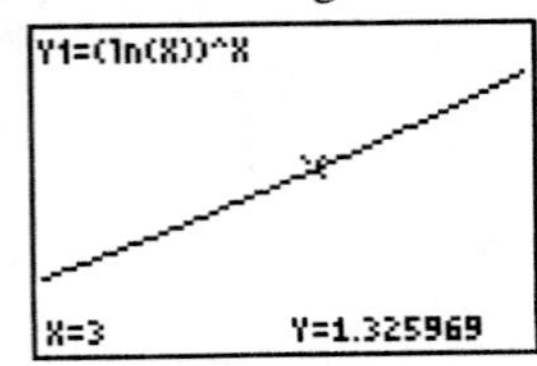

62. $f(x) = x^{\ln x}, x_0 = 2$

(a)

h	
0.01	$\frac{f(2+0.01)-f(2)}{0.01} = \frac{2.01^{\ln 2.01} - 2^{\ln 2}}{0.01} = 1.1258$
0.001	$\frac{f(2+0.001)-f(2)}{0.001} = \frac{2.001^{\ln 2.001} - 2^{\ln 2}}{0.001} = 1.1212$
0.0001	$\frac{f(2+0.0001)-f(2)}{0.0001} = \frac{2.0001^{\ln 2.0001} - 2^{\ln 2}}{0.0001} = 1.1207$
0.00001	$\frac{f(2+0.00001)-f(2)}{0.00001} = \frac{2.00001^{\ln 2.00001} - 2^{\ln 2}}{0.00001} = 1.1207$

It appears that $f'(2) = 1.121$.

(b) Graph the function on a graphing calculator and move the cursor to an x-value near $x = 2$. A good choice for the viewing window is [0, 4] by [0, 4]. Zoom in on the function until the graph looks like a straight line. Use the TRACE feature to select two points on the graph, and use these points to compute the slope. The result will be close to the most accurate value found in part (a), which is 1.121.

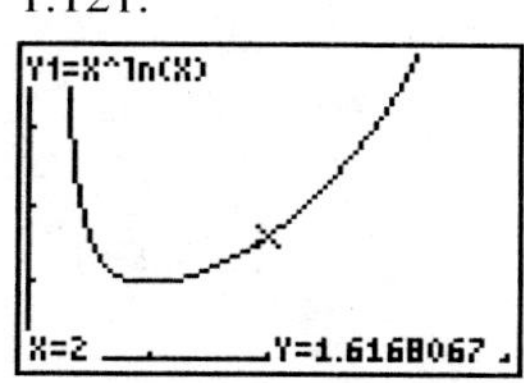

After zooming in twice:

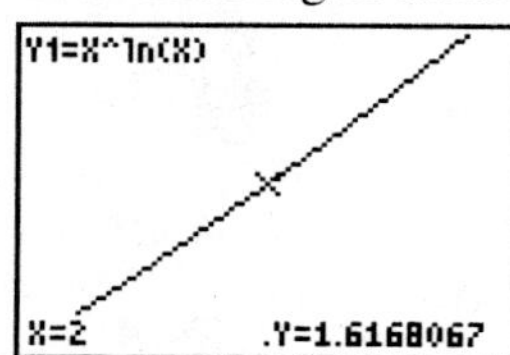

63. On the interval $(-\infty, 0)$, the graph of f is a straight line, so its slope is constant. To find this slope, use the points $(-2, 2)$ and $(0, 0)$.

$$m = \frac{0-2}{0-(-2)} = \frac{-2}{2} = -1$$

Thus, the value of f' will be -1 on this interval.

The graph of f has a sharp point at 0, so $f'(0)$ does not exist. To show this, we use an open circle on the graph of f' at $(0, -1)$.

We also observe that the slope of f is positive but decreasing from $x = 0$ to about $x = 1$, and then negative from there on. As $x \to \infty$, $f(x) \to 0$ and also $f'(x) = 0$.

Use this information to complete the graph of f'.

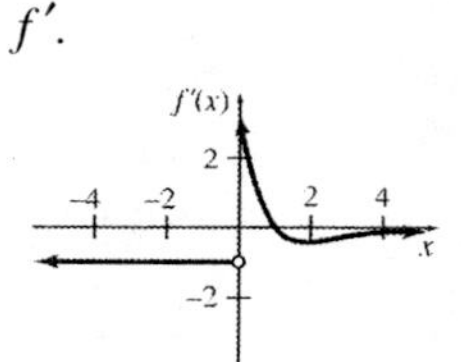

64. On the intervals $(-\infty, 0)$ and $(0, \infty)$, the slope of any tangent line will be positive, so the derivative will be positive. Thus, the graph of f' will lie above the y-axis. The slope of f and thus the value of f' approaches 0 when $x \to -\infty$ and $x \to \infty$ and approaches some particular but unknown positive value > 1 when $x \to 0^-$ and $x \to 0^+$.

Because f is discontinuous at $x = 0$, we know that $f'(0)$ does not exist, which we indicate with an open circle at $x = 0$ on the graph of f'.

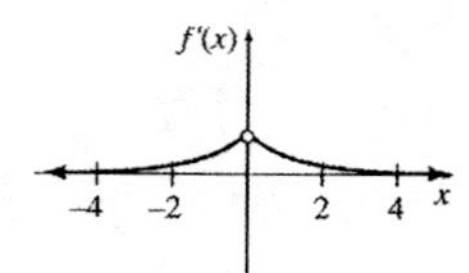

65. $$\lim_{x\to\infty}\frac{cf(x)-dg(x)}{f(x)-g(x)}$$
$$=\frac{\lim_{x\to\infty}[cf(x)-dg(x)]}{\lim_{x\to\infty}[f(x)-g(x)]}$$
$$=\frac{\lim_{x\to\infty}[cf(x)]-\lim_{x\to\infty}[dg(x)]}{\lim_{x\to\infty}[f(x)]-\lim_{x\to\infty}[g(x)]}$$
$$=\frac{c\lim_{x\to\infty}[f(x)]-d\lim_{x\to\infty}[g(x)]}{\lim_{x\to\infty}[f(x)]-\lim_{x\to\infty}[g(x)]}$$
$$=\frac{c\cdot c-d\cdot d}{c-d}=\frac{c^2-d^2}{c-d}=\frac{(c+d)(c-d)}{c-d}$$
$$=c+d$$

The answer is (e).

66. **(a)**

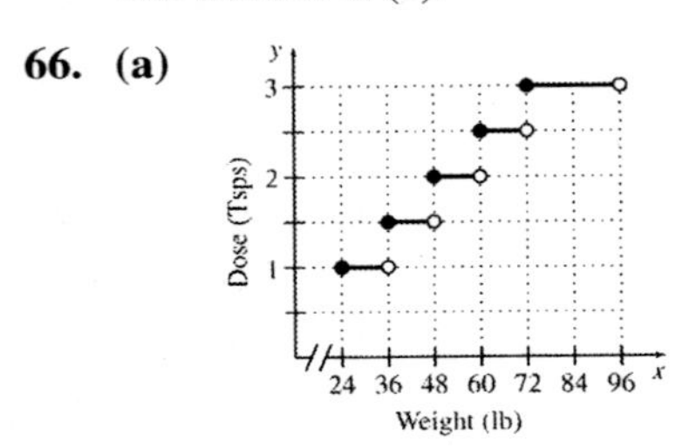

(b) $\lim_{x\to 48^-} D(x)=1.5$ tsp

(c) $\lim_{x\to 48^+} D(x)=2$ tsp

(d) $\lim_{x\to 48} D(x)$ does not exist

(e) $D(48)=2$ tsp

(f) No, the function is not continuous. There are discontinuities at $x=36$, 48, 60, and 72 pounds.

67. **(a)** The slope of the tangent line at $x=2000$ is about 0.13; the number of people aged 65 and over with Alzheimer's Disease is going up at a rate of about 0.13 million per year.

(b) The slope of the tangent line at $x=2040$ is about 0.34; the number of people aged 65 and over with Alzheimer's Disease is going up at a rate of about 0.34 million per year.

(c) $$\frac{A(2040)-A(2000)}{2040-2000}=\frac{11.1-4.6}{40}\approx 0.16$$

The average rate of change in the number of people 65 and over with Alzheimer's Disease over this interval is about 0.16 million people per year.

68. $V(t)=-t^2+6t-4$

(a)

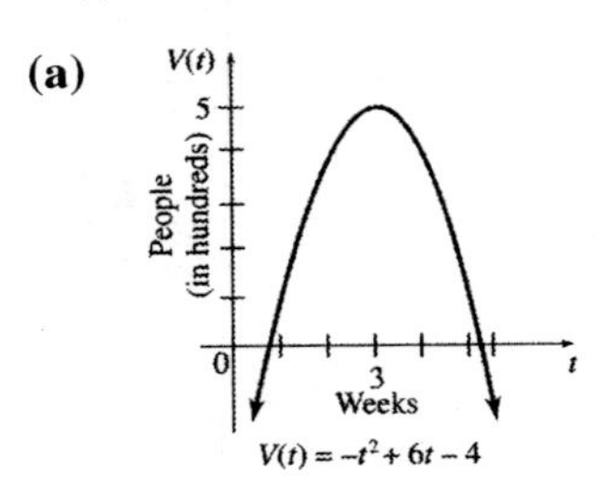

(b) The x-intercepts of the parabola are about 0.8 and 5.2, so a reasonable domain would be [0.8, 5.3], which represents the time period from 0.8 to 5.3 weeks.

(c) The number of cases reaches a maximum at the vertex.

$$x=\frac{-b}{2a}=\frac{-6}{-2}=3$$
$$V(3)=-3^2+6(3)-4=5$$

The vertex of the parabola is (3, 5). This represents a maximum at 3 weeks of 500 cases.

(d) The rate of change function is the derivative of $V(t)=-t^2+6t-4$.

$$\frac{V(t+h)-V(t)}{h}$$
$$=\frac{-(t+h)^2+6(t+h)-4-\left(-t^2+6t-4\right)}{h}$$
$$=\frac{-t^2-2th-h^2+6h-4+t^2-6t+4}{h}$$
$$=\frac{-2th-h^2+6h}{h}=-2t-h+6$$
$$V'(t)=\lim_{h\to 0}(-2t-h+6)=-2t+6$$

(e) The rate of change in the number of cases at the maximum is $V'(3)=-2(3)+6=0$.

(f) The sign of the rate of change up to the maximum is + because the function is increasing. The sign of the rate of change after the maximum is – because the function is decreasing.

69. **(a)** The rate can be estimated by estimating the slope of the tangent line to the curve at the given time. Answers will vary depending on the points used to determine the slope of the tangent line.

(i) The tangent line to the curve at 17:37 can be found by using the endpoints at (17:35.5, 0) and (17:40, 400). The rate the whale is descending at 17:37 is about

$$\frac{400-0}{17{:}40-17{:}35.5}=\frac{400}{4.5}\approx 90 \text{ m/min.}$$

(ii) The tangent line to the curve at 17:39 can be found by using the endpoints at (17:36.2, 0) and (17:40.8, 400). The rate the whale is descending at 17:39 is about

$$\frac{400-0}{17{:}40.8-17{:}36.2}=\frac{400}{4.6}\approx 85 \text{ m/min.}$$

(b) The whale appears to have 5 distinct rates at which it is descending.

Interval	Rate (meters per minute)
17:35–17:35.3	$\frac{f(17{:}35.3)-f(17{:}35)}{17{:}35.3-17{:}35}=\frac{10-0}{0.3}\approx 33$
17:36–17:37	$\frac{f(17{:}37)-f(17{:}36)}{17{:}37-17{:}36}=\frac{130-40}{1}\approx 90$
17:37.3–17:37.7	$\frac{f(17{:}37.7)-f(17{:}37.3)}{17{:}37.7-17{:}37.3}=\frac{150-150}{0.4}=0$
17:38.3–17:39.5	$\frac{f(17{:}39.5)-f(17{:}38.3)}{17{:}39.5-17{:}38.3}=\frac{300-200}{1.2}\approx 83$
17:40–17:41	$\frac{f(14{:}41)-f(14{:}40)}{17{:}41-17{:}40}=\frac{400-333}{1}=67$

Making smooth transitions between each interval, we get

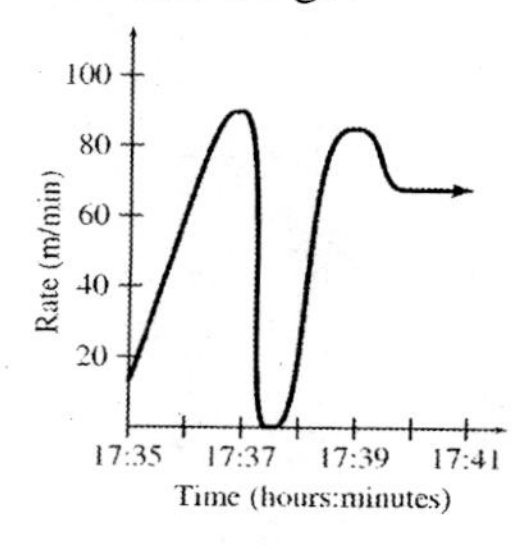

70. (a)

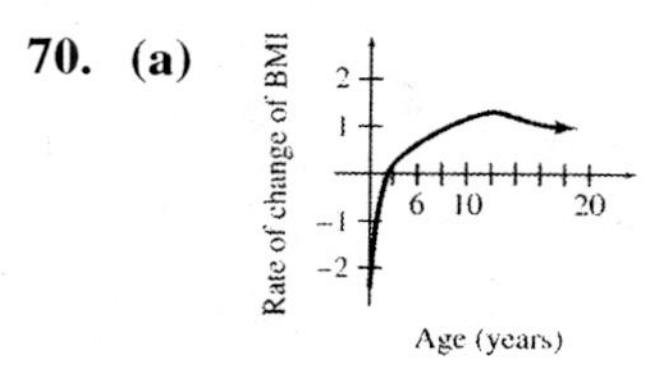

(b)

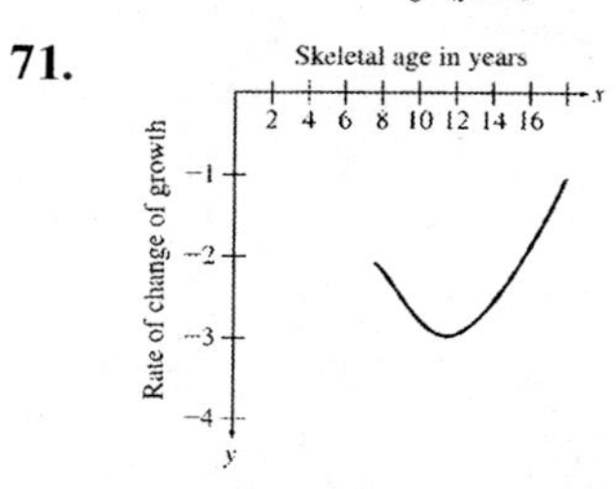

71.

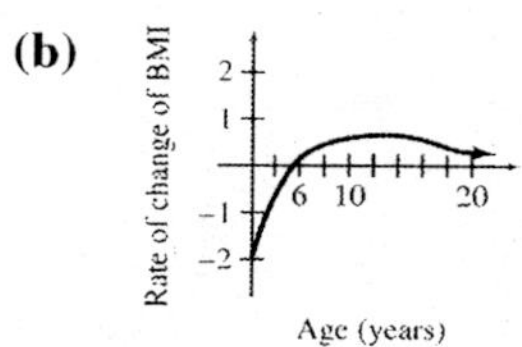

For a 10-year old girl, the remaining growth is about 14 cm and the rate of change is about −2.75 per year.

72. $p(t)=\left(\frac{a\left[e^{(b-a)t}-1\right]}{be^{(b-a)t}-a}\right)^N$

(a) (i) Because $b>a$, it follows that $a-b<0$, so $\lim_{t\to\infty} e^{(a-b)t}=0.$

Note that

$$p(t)=\left(\frac{a\left[e^{(b-a)t}-1\right]}{be^{(b-a)t}-a}\right)^N$$

$$=\frac{a^N\left(e^{(b-a)t}-1\right)^N}{b^N\left(e^{(b-a)t}-\frac{a}{b}\right)^N}$$

$$=\frac{a^N\left(e^{(b-a)t}-1\right)^N e^{(a-b)tN}}{b^N\left(e^{(b-a)t}-\frac{a}{b}\right)^N e^{(a-b)tN}}$$

$$=\frac{a^N\left(1-e^{(a-b)t}\right)^N}{b^N\left(1-\frac{a}{b}e^{(a-b)t}\right)^N}$$

(continued on next page)

(continued)

Therefore,

$$\lim_{t\to\infty} p(t) = \frac{a^N\left(1-\lim_{t\to\infty} e^{(a-b)t}\right)^N}{b^N\left(1-\frac{a}{b}\lim_{t\to\infty} e^{(a-b)t}\right)^N}$$

$$= \frac{a^N(1-0)^N}{b^N\left(1-\frac{a}{b}\cdot 0\right)^N}$$

$$= \frac{a^N}{b^N} = \left(\frac{a}{b}\right)^N$$

(ii) If $b < a$, then $b - a < 0$. Therefore, $\lim_{t\to\infty} e^{(b-a)t} = 0$. So,

$$\lim_{t\to\infty} p(t) = \left(\frac{a\left[e^{(b-a)t}-1\right]}{be^{(b-a)t}-a}\right)^N$$

$$= \left(\frac{a\left[\lim_{t\to\infty} e^{(b-a)t}-1\right]}{b\lim_{t\to\infty} e^{(b-a)t}-a}\right)^N$$

$$= \left(\frac{a(0-1)}{b\cdot 0-a}\right)^N = \left(\frac{-a}{-a}\right)^N$$

$$= 1^N = 1$$

(b) $$\lim_{t\to\infty} p(t) = \lim_{t\to\infty}\left[\left(\frac{at}{at+1}\right)^N \frac{\left(\frac{1}{at}\right)^N}{\left(\frac{1}{at}\right)^N}\right]$$

$$= \lim_{t\to\infty}\left(\frac{1}{1+\frac{1}{at}}\right)^N$$

$$= \frac{1}{\left(1+\lim_{t\to\infty}\left(\frac{1}{at}\right)\right)^N}$$

$$= \frac{1}{(1+0)^N} = 1$$

73. (a) The graph is discontinuous nowhere.

(b) The graph is not differentiable where the graph makes a sudden change, namely at $x = 50$, $x = 130$, $x = 230$, and $x = 770$.

(c)

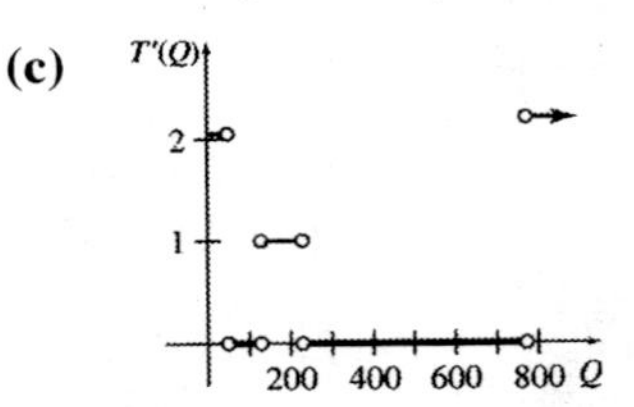

74. (a) The slope of the line is $\frac{C(5)-0}{5-0} = \frac{C(5)}{5}$.

(b) The value of x for which the average cost is smallest is $x = 7.5$. This can be found by drawing a line from the origin to any point of $C(x)$. At $x = 7.5$, you will get a line with the smallest slope.

(c) The marginal cost equals the average cost at the point where the average cost is smallest.

Chapter 4

CALCULATING THE DERIVATIVE

4.1 Techniques for Finding Derivatives

1. $y = 12x^3 - 8x^2 + 7x + 5$

$$\frac{dy}{dx} = 12(3x^{3-1}) - 8(2x^{2-1}) + 7x^{1-1} + 0$$
$$= 36x^2 - 16x + 7$$

3. $y = 3x^4 - 6x^3 + \frac{x^2}{8} + 5$

$$\frac{dy}{dx} = 3(4x^{4-1}) - 6(3x^{3-1}) + \frac{1}{8}(2x^{2-1}) + 0$$
$$= 12x^3 - 18x^2 + \frac{1}{4}x$$

5. $y = 6x^{3.5} - 10x^{0.5}$

$$\frac{dy}{dx} = 6(3.5x^{3.5-1}) - 10(0.5x^{0.5-1})$$
$$= 21x^{2.5} - 5x^{-0.5} \text{ or } 21x^{2.5} - \frac{5}{x^{0.5}}$$

7. $y = 8\sqrt{x} + 6x^{3/4} = 8x^{1/2} + 6x^{3/4}$

$$\frac{dy}{dx} = 8\left(\frac{1}{2}x^{1/2-1}\right) + 6\left(\frac{3}{4}x^{3/4-1}\right)$$
$$= 4x^{-1/2} + \frac{9}{2}x^{-1/4} \text{ or } \frac{4}{x^{1/2}} + \frac{9}{2x^{1/4}}$$

9. $y = 10x^{-3} + 5x^{-4} - 8x$

$$\frac{dy}{dx} = 10(-3x^{-3-1}) + 5(-4x^{-4-1}) - 8x^{1-1}$$
$$= -30x^{-4} - 20x^{-5} - 8 \text{ or } \frac{-30}{x^4} - \frac{20}{x^5} - 8$$

11. $f(t) = \frac{7}{t} - \frac{5}{t^3}$

$$= 7t^{-1} - 5t^{-3}$$
$$f'(t) = 7(-1t^{-1-1}) - 5(-3t^{-3-1})$$
$$= -7t^{-2} + 15t^{-4} \text{ or } \frac{-7}{t^2} + \frac{15}{t^4}$$

13. $y = \frac{6}{x^4} - \frac{7}{x^3} + \frac{3}{x} + \sqrt{5}$

$$= 6x^{-4} - 7x^{-3} + 3x^{-1} + \sqrt{5}$$
$$\frac{dy}{dx} = 6(-4x^{-4-1}) - 7(-3x^{-3-1})$$
$$+ 3(-1x^{-1-1}) + 0$$
$$= -24x^{-5} + 21x^{-4} - 3x^{-2}$$
$$\text{or } \frac{-24}{x^5} + \frac{21}{x^4} - \frac{3}{x^2}$$

15. $p(x) = -10x^{-1/2} + 8x^{-3/2}$

$$p'(x) = -10\left(-\frac{1}{2}x^{-3/2}\right) + 8\left(-\frac{3}{2}x^{-5/2}\right)$$
$$= 5x^{-3/2} - 12x^{-5/2} \text{ or } \frac{5}{x^{3/2}} - \frac{12}{x^{5/2}}$$

17. $y = \frac{6}{\sqrt[4]{x}} = 6x^{-1/4}$

$$\frac{dy}{dx} = 6\left(-\frac{1}{4}\right)x^{-5/4}$$
$$= -\frac{3}{2}x^{-5/4} \text{ or } -\frac{3}{2x^{5/4}}$$

19. $f(x) = \frac{x^3 + 5}{x} = x^2 + 5x^{-1}$

$$f'(x) = 2x^1 + 5\left(-1x^{-2}\right)$$
$$= 2x - 5x^{-2} \text{ or } 2x - \frac{5}{x^2}$$

21. $g(x) = (8x^2 - 4x)^2$

$$= 64x^4 - 64x^3 + 16x^2$$
$$g'(x) = 64(4x^3) - 64(3x^2) + 16(2x^1)$$
$$= 256x^3 - 192x^2 + 32x$$

23. A quadratic function has degree 2. When the derivative is taken, the power will decrease by 1 and the derivative function will be linear, so the correct choice is (b).

25. The slope of the tangent to a curve at a given point *a* equals the value of the derivative of the function at that point.

27. $D_x\left[9x^{-1/2}+\frac{2}{x^{3/2}}\right]$
$= D_x[9x^{-1/2}+2x^{-3/2}]$
$= 9\left(-\frac{1}{2}x^{-3/2}\right)+2\left(-\frac{3}{2}x^{-5/2}\right)$
$= -\frac{9}{2}x^{-3/2}-3x^{-5/2}$ or $-\frac{9}{2x^{3/2}}-\frac{3}{x^{5/2}}$

29. $f(x)=\frac{x^4}{6}-3x=\frac{1}{6}x^4-3x$
$f'(x)=\frac{1}{6}(4x^3)-3=\frac{2}{3}x^3-3$
$f'(-2)=\frac{2}{3}(-2)^3-3=-\frac{16}{3}-3=-\frac{25}{3}$

31. $y=x^4-5x^3+2;\ x=2$
$y'=4x^3-15x^2$
$y'(2)=4(2)^3-15(2)^2=-28$
The slope of tangent line at $x = 2$ is –28.
Use $m=-28$ and $(x_1, y_1)=(2,-22)$ to obtain the equation.
$y-(-22)=-28(x-2)$
$y=-28x+34$

33. $y=-2x^{1/2}+x^{3/2}$
$y'=-2\left(\frac{1}{2}x^{-1/2}\right)+\frac{3}{2}x^{1/2}$
$=-x^{-1/2}+\frac{3}{2}x^{1/2}=-\frac{1}{x^{1/2}}+\frac{3x^{1/2}}{2}$
$y'(9)=-\frac{1}{(9)^{1/2}}+\frac{3(9)^{1/2}}{2}=-\frac{1}{3}+\frac{9}{2}=\frac{25}{6}$
The slope of the tangent line at $x=9$ is $\frac{25}{6}$.

35. $f(x)=9x^2-8x+4$
$f'(x)=18x-8$
Let $f'(x)=0$ to find the point where the slope of the tangent line is zero.
$18x-8=0$
$18x=8$
$x=\frac{8}{18}=\frac{4}{9}$
Find the y-coordinate.
$f(x)=9x^2-8x+4$
$f\left(\frac{4}{9}\right)=9\left(\frac{4}{9}\right)^2-8\left(\frac{4}{9}\right)+4$
$=9\left(\frac{16}{81}\right)-\frac{32}{9}+4$
$=\frac{16}{9}-\frac{32}{9}+\frac{36}{9}=\frac{20}{9}$
The slope of the tangent line is zero at one point, $\left(\frac{4}{9},\frac{20}{9}\right)$.

37. $f(x)=2x^3+9x^2-60x+4$
$f'(x)=6x^2+18x-60$
If the tangent line is horizontal, then its slope is zero and $f'(x)=0$.
$6x^2+18x-60=0$
$6(x^2+3x-10)=0$
$6(x+5)(x-2)=0$
$x=-5$ or $x=2$
Thus, the tangent line is horizontal at $x = -5$ and $x = 2$.

39. $f(x)=x^3-4x^2-7x+8$
$f'(x)=3x^2-8x-7$
If the tangent line is horizontal, then its slope is zero and $f'(x)=0$.
$3x^2-8x-7=0$
$x=\frac{8\pm\sqrt{64+84}}{6}=\frac{8\pm\sqrt{148}}{6}$
$=\frac{8\pm2\sqrt{37}}{6}=\frac{4\pm\sqrt{37}}{3}$
Thus, the tangent line is horizontal at $x=\frac{4\pm\sqrt{37}}{3}$.

41. $f(x)=6x^2+4x-9$
$f'(x)=12x+4$
If the slope of the tangent line is –2, $f'(x)=-2$.
$12x+4=-2$
$12x=-6$
$x=-\frac{1}{2}$
$f\left(-\frac{1}{2}\right)=-\frac{19}{2}$
The slope of the tangent line is –2 at $\left(-\frac{1}{2},-\frac{19}{2}\right)$.

43. $f(x) = 2x^3 - 9x^2 - 12x + 5$

$f'(x) = 6x^2 - 18x - 12$

If the slope of the tangent line is 12, $f'(x) = 12$.

$6x^2 - 18x - 12 = 12$

$6x^2 - 18x - 24 = 0$

$6(x^2 - 3x - 4) = 0$

$6(x-4)(x+1) = 0$

$x = 4$ or $x = -1$

$f(4) = -59$ and $f(-1) = 6$

The slope of the tangent line is –12 at (4, –59) and (–1, 6).

45. $f(x) = 3g(x) - 2h(x) + 3$

$f'(x) = 3g'(x) - 2h'(x)$

$f'(5) = 3g'(5) - 2h'(5)$

$= 3(12) - 2(-3) = 42$

47. (a) From the graph, $f(1) = 2$, because the curve goes through (1, 2).

(b) $f'(1)$ gives the slope of the tangent line to *f* at 1. The line goes through (–1, 1) and (1, 2).

$$m = \frac{2-1}{1-(-1)} = \frac{1}{2}, \text{ so } f'(1) = \frac{1}{2}.$$

(c) The domain of *f* is $[-1, \infty]$ because the *x*-coordinates of the points of *f* start at $x = -1$ and continue infinitely through the positive real numbers.

(d) The range of *f* is $[0, \infty)$ because the *y*-coordinates of the points on *f* start at $y = 0$ and continue infinitely through the positive real numbers.

49. $\frac{f(x)}{k} = \frac{1}{k} \cdot f(x)$

Use the rule for the derivative of a constant times a function.

$$\frac{d}{dx}\left[\frac{f(x)}{k}\right] = \frac{d}{dx}\left[\frac{1}{k} \cdot f(x)\right] = \frac{1}{k} f'(x) = \frac{f'(x)}{k}$$

51. $N(t) = 0.00437t^{3.2}$

$N'(t) = 0.013984t^{2.2}$

(a) $N'(5) = 0.013984(5)^{2.2} \approx 0.4824$

(b) $N'(10) = 0.013984(10)^{2.2} \approx 2.216$

53. $M(t) = 4t^{3/2} + 2t^{1/2}$

(a) $M(16) = 4(16)^{3/2} + 2(16)^{1/2} = 264$

(b) $M(25) = 4(25)^{3/2} + 2(25)^{1/2} = 510$

(c) $M'(t) = 4\left(\frac{3}{2}\right)t^{1/2} + 2\left(\frac{1}{2}\right)t^{-1/2}$

$= 6t^{1/2} + t^{-1/2}$

$M'(16) = 6(16)^{1/2} + (16)^{-1/2} = 24 + \frac{1}{4}$

$= \frac{97}{4}$ or 24.25

The rate of change is about 24.25 matings per degree.

55. $V(t) = -2159 + 1313t - 60.82t^2$

(a) $V(3) = -2159 + 1313(3) - 60.82(3)^2$

$= 1232.62 \text{ cm}^3$

(b) $V'(t) = 1313 - 121.64t$

$V'(3) = 1313 - 121.64(3) = 948.08 \text{ cm}^3/\text{yr}$

57. $V = C(R_0 - R)R^2 = CR_0R^2 - CR^3$

$\frac{dV}{dR} = 2CR_0R - 3CR^2$

$2CR_0R - 3CR^2 = 0$

$CR(2R_0 - 3R) = 0$

$CR = 0$ or $2R_0 - 3R = 0$

$R = 0$ $\quad R = \frac{2}{3}R_0$

Discard $R = 0$, because a closed windpipe produces no airflow. Velocity is maximized when $R = \frac{2}{3}R_0$.

59. $l(t) = -2.318 + 0.2356t - 0.002674t^2$

(a) The problem states that a fetus this formula concerns is at least 18 weeks old. So, the minimum *x* value should be 18. Considering the gestation time of a human fetus in general, a meaningful range for this function is $18 \le t \le 44$.

(b) $l'(x) = 0.2356 - (2)0.002674t$

$= 0.2356 - 0.005348t$

(c) $l'(25) = 0.2356 - 0.005348(25)$

$= 0.1019$ cm/week

61. $f(m) = 40m^{-0.6}$

$f'(m) = 40(-0.6)m^{-0.6-1} = -24m^{-1.6}$

$f'(0.01) = -24(0.01)^{-1.6} \approx -38,037$

The number of species is decreasing by about 38,087 per gram of body mass, or about 38 for an increase in body mass of 1 microgram.

63. $v(x) = -35.98 + 12.09x - 0.4450x^2$

(a)

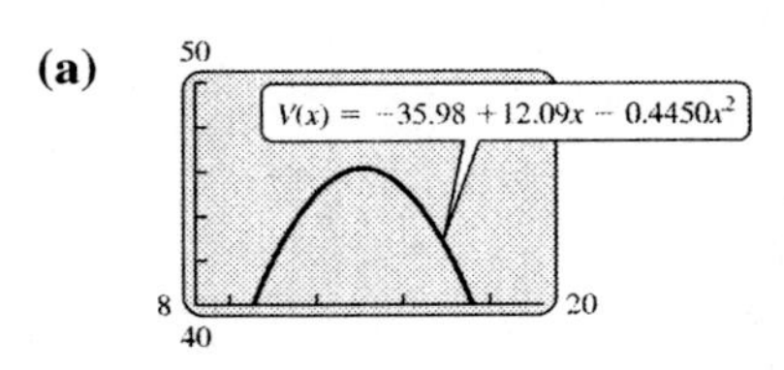

(b) The maximum value occurs at about $x \approx 13.58$.

The moisture content is about 13.58% when the volume is at its maximum.

(c) $v'(x) = 12.09 - 0.890x^2$

$v'(13.58) = 12.09 - 0.890(13.58) \approx 0$

65. $H = 137 + aD - bD^2$

$$\frac{dH}{dD} = a - 2bD$$

$$\frac{dH}{dD} = 0 \Rightarrow a - 2bD = 0 \Rightarrow$$

$$a = 2bD \text{ or } b = \frac{a}{2D}$$

Substitute $b = \frac{a}{2D}$ for b in the original equation with $H = H_{\max}$ and $D = D_{\max}$, and solve for a.

$$H_{\max} = 137 + aD_{\max} - \left(\frac{a}{2D_{\max}}\right)D_{\max}^2$$

$$H_{\max} = 137 + aD_{\max} - \frac{aD_{\max}}{2}$$

$$= 137 + \frac{aD_{\max}}{2}$$

$$H_{\max} - 137 = \frac{aD_{\max}}{2}$$

$$\frac{2(H_{\max} - 137)}{D_{\max}} = a$$

Substitute $a = 2bD$ for a in the original equation with $H = H_{\max}$ and $D = D_{\max}$, and solve for b.

$$H_{\max} = 137 + (2bD_{\max})D_{\max} - bD_{\max}^2$$

$$H_{\max} = 137 + 2bD_{\max}^2 - bD_{\max}^2$$

$$= 137 + bD_{\max}^2$$

$$H_{\max} - 137 = bD_{\max}^2$$

$$\frac{H_{\max} - 137}{D_{\max}^2} = b$$

67. $s(t) = 18t^2 - 13t + 8$

(a) $v(t) = s'(t) = 18(2t) - 13 + 0$

$= 36t - 13$

(b) $v(0) = 36(0) - 13 = -13$

$v(5) = 36(5) - 13 = 167$

$v(10) = 36(10) - 13 = 347$

69. $s(t) = -3t^3 + 4t^2 - 10t + 5$

(a) $v(t) = s'(t) = -3(3t^2) + 4(2t) - 10 + 0$

$= -9t^2 + 8t - 10$

(b) $v(0) = -9(0)^2 + 8(0) - 10 = -10$

$v(5) = -9(5)^2 + 8(5) - 10$

$= -225 + 40 - 10 = -195$

$v(10) = -9(10)^2 + 8(10) - 10$

$= -900 + 80 - 10 = -830$

71. $s(t) = -16t^2 + 64t$

(a) $v(t) = s'(t) = -16(2t) + 64 = -32t + 64$

$v(2) = -32(2) + 64 = -64 + 64 = 0$

$v(3) = -32(3) + 64 = -96 + 64 = -32$

The ball's velocity is 0 ft/sec after 2 seconds and –32 ft/sec after 3 seconds.

(b) As the ball travels upward, its speed decreases because of the force of gravity until, at maximum height, its speed is 0 ft/sec. In part (a), we found that $v(2) = 0$. It takes 2 seconds for the ball to reach its maximum height.

(c) $s(2) = -16(2)^2 + 64(2) = 64$

It will go 64 ft high.

73. (a) 1982 when $t = 50$:

$$C(50) = 0.007756(50)^2 - 0.04258(50) + 0.7585$$
$$= 18.0195 \approx 18¢ \text{ vs. } 20¢ \text{ actual cost}$$

2002 when $t = 70$:

$$C(70) = 0.007756(70)^2 - 0.04258(70) + 0.7585$$
$$= 35.78 \approx 36¢ \text{ vs. } 37¢ \text{ actual cost}$$

(b)
$$C'(t) = 0.015512t - 0.04258$$
$$C'(50) = 0.015512(50) - 0.04258$$
$$\approx 0.733¢ \text{ per year}$$
$$C'(70) = 0.015512(70) - 0.04258$$
$$\approx 1.04¢ \text{ per year}$$

These results mean that the cost of postage was increasing at a rate of about 0.0733¢ per year in 1982 and at a rate of about 1.04¢ per year in 2002.

(c) Using a graphing calculator, a cubic function that models the postage cost data is $C(t) = -0.0001429t^3 + 0.02637t^2 - 0.6322t + 3.186$

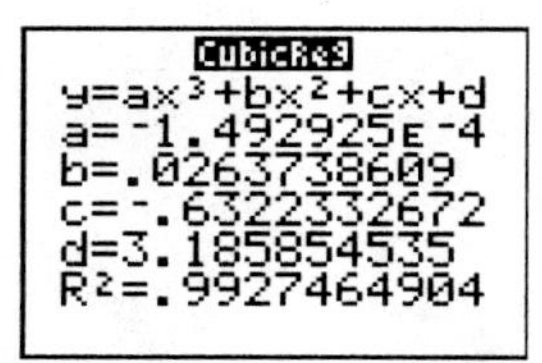

Store the function as Y_1 and then use nDeriv to find $C'(50)$ and $C'(70)$.

```
d/dx(Y1)|X=50
          .885458792
d/dx(Y1)|X=70
          .865506957
```

The rate of change for 1982 was about 0.885¢ per year and the rate of change for 2002 was about 0.866¢ per year.

(d)

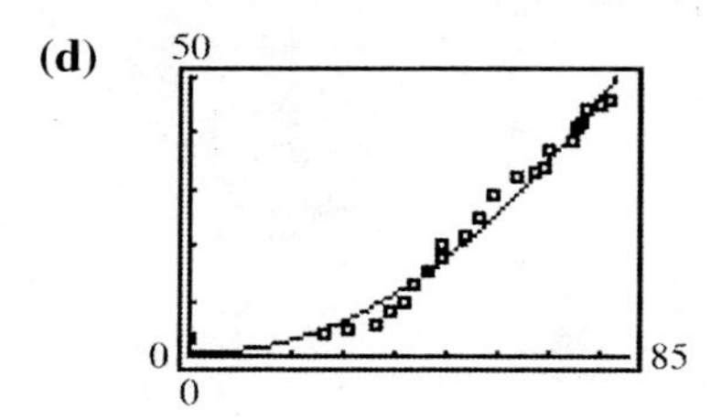

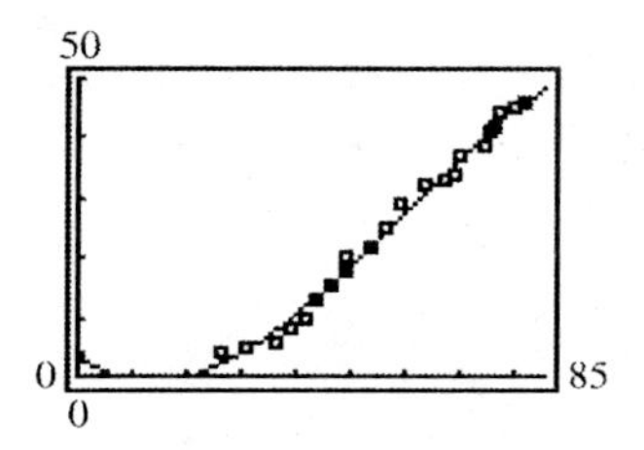

It appears that the cubic function best describes the data.

(e) Answers will vary.

75.
$$P(x) = -0.00209x^3 + 0.3387x^2 - 15.15x + 208.6$$
$$P'(x) = -0.00627x^2 + 0.6774x - 15.15$$

(a) $P(45) \approx 22.3$

$P'(45) \approx 2.64$

The probability of scoring 3 or more on the Calculus AB Advanced Placement Exam is about 22.3% if a student received a score of 45 on the PSAT/NMSQT. The rate that the probability is increasing is about 2.64.

(b) $P(75) \approx 95.8$

$P'(75) \approx 0.386$

The probability of scoring 3 or more on the Calculus AB Advanced Placement Exam is about 95.8% if a student received a score of 75 on the PSAT/NMSQT. The rate that the probability is increasing is about 0.386. **(c)** The probabilities in parts (a) and (b) seem to indicate that more students with higher PSAT/NMSQT scores will score 3 or better on the Calculus AB Advanced Placement Exam than will those with lower scores.

4.2 Derivatives of Products and Quotients

1.
$$y = (3x^2 + 2)(2x - 1)$$
$$\frac{dy}{dx} = (3x^2 + 2)(2) + (2x - 1)(6x)$$
$$= 6x^2 + 4 + 12x^2 - 6x$$
$$= 18x^2 - 6x + 4$$

3.
$$y = (2x - 5)^2$$
$$= (2x - 5)(2x - 5)$$
$$\frac{dy}{dx} = (2x - 5)(2) + (2x - 5)(2)$$
$$= 4x - 10 + 4x - 10$$
$$= 8x - 20$$

5. $k(t) = (t^2-1)^2 = (t^2-1)(t^2-1)$
$$k'(t) = (t^2-1)(2t) + (t^2-1)(2t) = 2t^3 - 2t + 2t^3 - 2t = 4t^3 - 4t$$

7. $y = (x+1)(\sqrt{x}+2) = (x+1)(x^{1/2}+2)$
$$\frac{dy}{dx} = (x+1)\left(\frac{1}{2}x^{-1/2}\right) + (x^{1/2}+2)(1) = \frac{1}{2}x^{1/2} + \frac{1}{2}x^{-1/2} + x^{1/2} + 2 = \frac{3}{2}x^{1/2} + \frac{1}{2}x^{-1/2} + 2$$
or $\dfrac{3x^{1/2}}{2} + \dfrac{1}{2x^{1/2}} + 2$

9. $p(y) = (y^{-1}+y^{-2})(2y^{-3}-5y^{-4})$
$$\begin{aligned} p'(y) &= (y^{-1}+y^{-2})(-6y^{-4}+20y^{-5}) + (-y^{2}-2y^{-3})(2y^{-3}-5y^{-4}) \\ &= -6y^{-5}+20y^{-6}-6y^{-6}+20y^{-7} - 2y^{-5}+5y^{-6}-4y^{-6}+10y^{-7} \\ &= -8y^{-5}+15y^{-6}+30y^{-7} \end{aligned}$$

11. $f(x) = \dfrac{6x+1}{3x+10}$
$$f'(x) = \frac{(3x+10)(6)-(6x+1)(3)}{(3x+10)^2} = \frac{18x+60-18x-3}{(3x+10)^2} = \frac{57}{(3x+10)^2}$$

13. $y = \dfrac{5-3t}{4+t}$
$$\frac{dy}{dx} = \frac{(4+t)(-3)-(5-3t)(1)}{(4+t)^2} = \frac{-12-3t-5+3t}{(4+t)^2} = \frac{-17}{(4+t)^2}$$

15. $y = \dfrac{x^2+x}{x-1}$
$$\frac{dy}{dx} = \frac{(x-1)(2x+1)-(x^2+x)(1)}{(x-1)^2} = \frac{2x^2+x-2x-1-x^2-x}{(x-1)^2} = \frac{x^2-2x-1}{(x-1)^2}$$

17. $f(t) = \dfrac{4t^2+11}{t^2+3}$
$$f'(t) = \frac{(t^2+3)(8t)-(4t^2+11)(2t)}{(t^2+3)^2} = \frac{8t^3+24t-8t^3-22t}{(t^2+3)^2} = \frac{2t}{(t^2+3)^2}$$

19. $g(x) = \dfrac{x^2-4x+2}{x^2+3}$
$$g'(x) = \frac{(x^2+3)(2x-4)-(x^2-4x+2)(2x)}{(x^2+3)^2} = \frac{2x^3-4x^2+6x-12-2x^3+8x^2-4x}{(x^2+3)^2} = \frac{4x^2+2x-12}{(x^2+3)^2}$$

21. $p(t) = \dfrac{\sqrt{t}}{t-1} = \dfrac{t^{1/2}}{t-1}$
$$p'(t) = \frac{(t-1)\left(\frac{1}{2}t^{-1/2}\right)-t^{1/2}(1)}{(t-1)^2} = \frac{\frac{1}{2}t^{1/2}-\frac{1}{2}t^{-1/2}-t^{1/2}}{(t-1)^2} = \frac{-\frac{1}{2}t^{1/2}-\frac{1}{2}t^{-1/2}}{(t-1)^2} = \frac{-\frac{\sqrt{t}}{2}-\frac{1}{2\sqrt{t}}}{(t-1)^2} \text{ or } \frac{-t-1}{2\sqrt{t}(t-1)^2}$$

23. $y = \dfrac{5x+6}{\sqrt{x}} = \dfrac{5x+6}{x^{1/2}} = 5x^{1/2}+6x^{-1/2}$
$$\frac{dy}{dx} = \frac{5}{2}x^{-1/2} - 3x^{-3/2} \text{ or } \frac{5x-6}{2x\sqrt{x}}$$

25. $h(z) = \dfrac{z^{2.2}}{z^{3.2}+5}$
$$h'(z) = \frac{(z^{3.2}+5)(2.2z^{1.2})-z^{2.2}(3.2z^{2.2})}{(z^{3.2}+5)^2} = \frac{2.2z^{4.4}+11z^{1.2}-3.2z^{4.4}}{(z^{3.2}+5)^2} = \frac{-z^{4.4}+11z^{1.2}}{(z^{3.2}+5)^2}$$

27. $f(x)=\dfrac{(3x^2+1)(2x-1)}{5x+4}$

$$f'(x)=\frac{(5x+4)[(3x^2+1)(2)+(6x)(2x-1)]-(3x^2+1)(2x-1)(5)}{(5x+4)^2}$$
$$=\frac{(5x+4)(18x^2-6x+2)-(3x^2+1)(10x-5)}{(5x+4)^2}$$
$$=\frac{90x^3-30x^2+10x+72x^2-24x+8-30x^3+15x^2-10x+5}{(5x+4)^2}=\frac{60x^3+57x^2-24x+13}{(5x+4)^2}$$

29. $h(x)=f(x)g(x)$
$h'(x)=f(x)g'(x)+g(x)f'(x)$
$h'(3)=f(3)g'(3)+g(3)f'(3)=9(5)+4(8)=77$

31. In the first step, the two terms in the numerator are reversed. The correct work follows.

$$D_x\left(\frac{2x+5}{x^2-1}\right)=\frac{(x^2-1)(2)-(2x+5)(2x)}{(x^2-1)^2}=\frac{2x^2-2-4x^2-10x}{(x^2-1)^2}=\frac{-2x^2-10x-2}{(x^2-1)^2}$$

33. $f(x)=\dfrac{x}{x-2}$, at (3, 3)

$$m=f'(x)=\frac{(x-2)(1)-x(1)}{(x-2)^2}=-\frac{2}{(x-2)^2}$$

At (3, 3), $m=-\dfrac{2}{(3-2)^2}=-2.$

Use the point-slope form.
$y-3=-2(x-3)$
$y=-2x+9$

35. **(a)** $f(x)=\dfrac{3x^3+6}{x^{2/3}}$

$$f'(x)=\frac{(x^{2/3})(9x^2)-(3x^3+6)(\frac{2}{3}x^{-1/3})}{(x^{2/3})^2}=\frac{9x^{8/3}-2x^{8/3}-4x^{-1/3}}{x^{4/3}}=\frac{7x^{8/3}-\frac{4}{x^{1/3}}}{x^{4/3}}=\frac{7x^3-4}{x^{5/3}}$$

(b) $f(x)=3x^{7/3}+6x^{-2/3}$

$$f'(x)=3\left(\frac{7}{3}x^{4/3}\right)+6\left(-\frac{2}{3}x^{-5/3}\right)=7x^{4/3}-4x^{-5/3}$$

(c) The derivatives are equivalent.

37. $f(x)=\dfrac{u(x)}{v(x)}$

$$f'(x)=\lim_{h\to0}\frac{f(x+h)-f(x)}{h}=\lim_{h\to0}\frac{\frac{u(x+h)}{v(x+h)}-\frac{u(x)}{v(x)}}{h}=\lim_{h\to0}\frac{u(x+h)v(x)-u(x)v(x+h)}{hv(x+h)v(x)}$$
$$=\lim_{h\to0}\frac{u(x+h)v(x)-u(x)v(x)+u(x)v(x)-u(x)v(x+h)}{hv(x+h)v(x)}$$
$$=\lim_{h\to0}\frac{v(x)[u(x+h)-u(x)]-u(x)[v(x+h)-v(x)]}{hv(x+h)v(x)}$$
$$=\lim_{h\to0}\frac{v(x)\frac{u(x+h)-u(x)}{h}-u(x)\frac{v(x+h)-v(x)}{h}}{v(x+h)v(x)}=\frac{v(x)\cdot u'(x)-u(x)v'(x)}{[v(x)]^2}$$

39. On a TI-84, graph the numerical derivative of $f(x) = Y_1 = (x^2 - 2)(x^2 - \sqrt{2})$ using the equation $Y_2 = \text{nDeriv}(Y_1, x, x)$ and the window [–2, 2] by [–12, 12]. The derivative crosses the x-axis at 0 and at approximately –1.307 and 1.307.

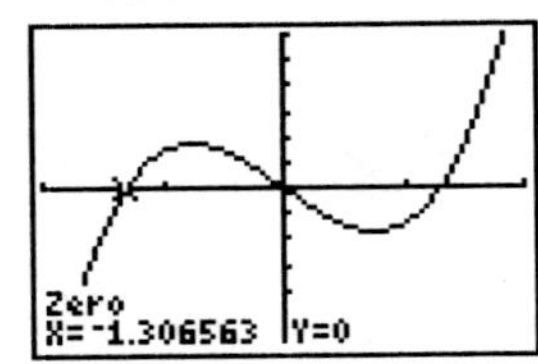

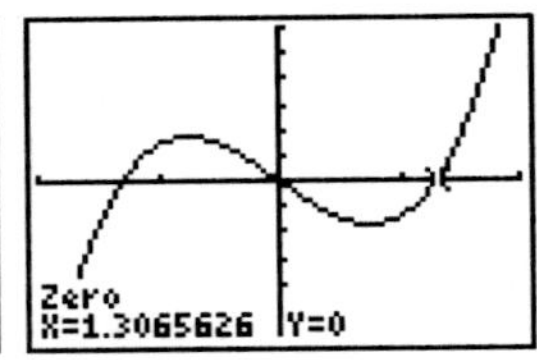

41. $s(x) = \dfrac{x}{m + nx}$; m and n constants

(a) $$s'(x) = \frac{(m+nx)1 - x(n)}{(m+nx)^2} = \frac{m + nx - nx}{(m+nx)^2} = \frac{m}{(m+nx)^2}$$

(b) $x = 50,\ m = 10,\ n = 3$

$$s'(50) = \frac{m}{(m+50n)^2} = \frac{10}{[10 + 50(3)]^2} = \frac{1}{2560} \approx 0.000391 \text{ mm per ml}$$

43. $N(t) = 3t(t - 10)^2 + 40$

(a) $$N(t) = 3t(t^2 - 20t + 100) + 40 = 3t^3 - 60t^2 + 300t + 40$$

$$N'(t) = 9t^2 - 120t + 300$$

(b) $N'(8) = 9(8)^2 - 120(8) + 300 = -84$

The rate of change is –84 million per hour.

(c) $N'(11) = 9(11)^2 - 120(11) + 300 = 69$

The rate of change is 69 million per hour.

(d) The population first declines, and then increases.

45. $W = \left(1 + \dfrac{20}{H - 0.93}\right)H = H + \dfrac{20H}{H - 0.93}$

(a) $$\frac{dW}{dH} = 1 + \frac{(H - 0.93)(20) - 20H(1)}{(H - 0.93)^2}$$

$$= \frac{(H - 0.93)^2}{(H - 0.93)^2} + \frac{20H - 18.6 - 20H}{(H - 0.93)^2}$$

$$= \frac{H^2 - 1.86H + 0.8649}{(H - 0.93)^2} + \frac{-18.6}{(H - 0.93)^2}$$

$$= \frac{H^2 - 1.86H - 17.7351}{(H - 0.93)^2}$$

(b) $\dfrac{dW}{dH} = 0$ when $H^2 - 1.86H - 17.7351 = 0$

By the quadratic formula,

$$H = \frac{1.86 \pm \sqrt{1.86^2 + 4(17.7351)}}{2} \approx -3.38 \text{ or } 5.24$$

Discarding the negative solution, W is minimized at $H = 5.24$ m.

(c) Crows apply optimal foraging techniques.

47. $R(c) = \dfrac{ac^2}{1 + bc^2} - kc$

$$R'(c) = \frac{(1 + bc^2)(2ac) - ac^2(2bc)}{(1 + bc^2)^2} - k$$

$$= \frac{2ac + 2abc^3 - 2abc^3}{(1 + bc^2)^2} - k$$

$$= \frac{2ac}{(1 + bc^2)^2} - k$$

$R'(c)$ is the rate of change of the release of calcium per change in the amount of free calcium.

49. $p = \dfrac{n_A}{n_A + n_a}$

We have $u(t) = n_A$ and $v(t) = n_A + n_a$.
Using the quotient rule gives

$$\frac{dp}{dt} = \frac{(n_A + n_a)\frac{d}{dt}n_A - n_A\frac{d}{dt}(n_A + n_a)}{(n_A + n_a)^2}$$

$$= \frac{n_A\frac{dn_A}{dt} + n_a\frac{dn_a}{dt} - n_A\frac{dn_A}{dt} - n_A\frac{dn_a}{dt}}{(n_A + n_a)^2}$$

$$= \frac{n_a\frac{dn_A}{dt} - n_A\frac{dn_a}{dt}}{(n_A + n_a)^2}$$

51. $f(x) = \frac{x^2}{2(1-x)}$

$$f'(x) = \frac{2(1-x)(2x) - x^2(-2)}{[2(1-x)]^2} = \frac{4x - 4x^2 + 2x^2}{4(1-x)^2} = \frac{4x - 2x^2}{4(1-x)^2} = \frac{2x(2-x)}{4(1-x)^2} = \frac{x(2-x)}{2(1-x)^2}$$

(a) $f'(0.1) = \frac{0.1(2-0.1)}{2(1-0.1)^2} \approx 0.1173$

(b) $f'(0.6) = \frac{0.6(2-0.6)}{2(1-0.6)^2} = 2.625$

4.3 The Chain Rule

In Exercises 1 through 6, $f(x) = 5x^2 - 2x$ and $g(x) = 8x + 3$.

1. $g(2) = 8(2) + 3 = 19$

$$f[g(2)] = f[19] = 5(19)^2 - 2(19) = 1805 - 38 = 1767$$

3. $f(2) = 5(2)^2 - 2(2) = 20 - 4 = 16$

$g[f(2)] = g[16] = 8(16) + 3 = 128 + 3 = 131$

5. $g(k) = 8k + 3$

$$f[g(k)] = f[8k+3] = 5(8k+3)^2 - 2(8k+3) = 5(64k^2 + 48k + 9) - 16k - 6 = 320k^2 + 224k + 39$$

7. $f(x) = \frac{x}{8} + 7;\ g(x) = 6x - 1$

$$f[g(x)] = \frac{6x-1}{8} + 7 = \frac{6x-1}{8} + \frac{56}{8} = \frac{6x+55}{8}$$

$$g[f(x)] = 6\left[\frac{x}{8} + 7\right] - 1 = \frac{6x}{8} + 42 - 1 = \frac{3x}{4} + 41 = \frac{3x}{4} + \frac{164}{4} = \frac{3x+164}{4}$$

9. $f(x) = \frac{1}{x};\ g(x) = x^2$

$$f[g(x)] = \frac{1}{x^2}$$

$$g[f(x)] = \left(\frac{1}{x}\right)^2 = \frac{1}{x^2}$$

11. $f(x) = \sqrt{x+2};\ g(x) = 8x^2 - 6$

$$f[g(x)] = \sqrt{(8x^2-6)+2} = \sqrt{8x^2 - 4}$$

$$g[f(x)] = 8(\sqrt{x+2})^2 - 6 = 8x + 16 - 6 = 8x + 10$$

13. $f(x) = \sqrt{x+1};\ g(x) = \frac{-1}{x}$

$$f[g(x)] = \sqrt{\frac{-1}{x} + 1} = \sqrt{\frac{x-1}{x}}$$

$$g[f(x)] = \frac{-1}{\sqrt{x+1}}$$

15. $y = (5 - x^2)^{3/5}$

If $f(x) = x^{3/5}$ and $g(x) = 5 - x^2$, then $y = f[g(x)] = (5 - x^2)^{3/5}$.

17. $y = -\sqrt{13 + 7x}$

If $f(x) = -\sqrt{x}$ and $g(x) = 13 + 7x$, then $y = f[g(x)] = -\sqrt{13 + 7x}$.

19. $y = (x^2 + 5x)^{1/3} - 2(x^2 + 5x)^{2/3} + 7$

If $f(x) = x^{1/3} - 2x^{2/3} + 7$ and $g(x) = x^2 + 5x$, then

$y = f[g(x)] = (x^2 + 5x)^{1/3} - 2(x^2 + 5x)^{2/3} + 7.$

21. $y = (8x^4 - 5x^2 + 1)^4$

Let $f(x) = x^4$ and $g(x) = 8x^4 - 5x^2 + 1$.

Then $(8x^4 - 5x^2 + 1)^4 = f[g(x)]$.

Use the alternate form of the chain rule.

$$\frac{dy}{dx} = f'[g(x)] \cdot g'(x)$$

$$f'(x) = 4x^3$$

$$f'[g(x)] = 4[g(x)]^3 = 4(8x^4 - 5x^2 + 1)^3$$

$$g'(x) = 32x^3 - 10x$$

$$\frac{dy}{dx} = 4(8x^4 - 5x^2 + 1)^3(32x^3 - 10x)$$

23. $k(x) = -2(12x^2 + 5)^{-6}$

Use the alternate form of the chain rule.

$u = 12x^2 + 5$, $n = -6$, and $u' = 24x$.

$$k'(x) = -2[-6(12x^2+5)^{-6-1} \cdot 24x] = -2[-144x(12x^2+5)^{-7}] = 288x(12x^2+5)^{-7}$$

25. $s(t) = 45(3t^3 - 8)^{3/2}$

Use the alternate form of the chain rule.

$u = 3t^3 - 8$, $n = \frac{3}{2}$, and $u' = 9t^2$.

$$s'(t) = 45\left[\frac{3}{2}(3t^3 - 8)^{1/2} \cdot 9t^2\right]$$
$$= 45\left[\frac{27}{2}t^2(3t^3 - 8)^{1/2}\right]$$
$$= \frac{1215}{2}t^2(3t^3 - 8)^{1/2}$$

27. $g(t) = -3\sqrt{7t^3 - 1}$
$$= -3\sqrt{(7t^3 - 1)^{1/2}}$$

Use the alternate form of the chain rule.

$u = 7t^3 - 1$, $n = \frac{1}{2}$, and $u' = 21t^2$.

$$g'(t) = -3\left[\frac{1}{2}(7t^3 - 1)^{-1/2} \cdot 21t^2\right]$$
$$= -3\left[\frac{21}{2}t^2(7t^3 - 1)^{-1/2}\right]$$
$$= \frac{-63}{2}t^2 \cdot \frac{1}{(7t^3 - 1)^{1/2}}$$
$$= \frac{-63t^2}{2\sqrt{7t^3 - 1}}$$

29. $m(t) = -6t(5t^4 - 1)^4$

Use the product rule and the power rule.

$$m'(t) = -6t[4(5t^4 - 1)^3 \cdot 20t^3] + (5t^4 - 1)^4(-6)$$
$$= -480t^4(5t^4 - 1)^3 - 6(5t^4 - 1)^4$$
$$= -6(5t^4 - 1)^3[80t^4 + (5t^4 - 1)]$$
$$= -6(5t^4 - 1)^3(85t^4 - 1)$$

31. $y = (3x^4 + 1)^4(x^3 + 4)$

Use the product rule and the power rule.

$$\frac{dy}{dx} = (3x^4 + 1)^4(3x^2)$$
$$+ (x^3 + 4)[4(3x^4 + 1)^3 \cdot 12x^3]$$
$$= 3x^2(3x^4 + 1)^4 + 48x^3(x^3 + 4)(3x^4 + 1)^3$$
$$= 3x^2(3x^4 + 1)^3[3x^4 + 1 + 16x(x^3 + 4)]$$
$$= 3x^2(3x^4 + 1)^3(3x^4 + 1 + 16x^4 + 64)$$
$$= 3x^2(3x^4 + 1)^3(19x^4 + 64x + 1)$$

33. $q(y) = 4y^2(y^2 + 1)^{5/4}$

Use the product rule and the power rule.

$$q'(y) = 4y^2 \cdot \frac{5}{4}(y^2 + 1)^{1/4}(2y) + 8y(y^2 + 1)^{5/4}$$
$$= 10y^3(y^2 + 1)^{1/4} + 8y(y^2 + 1)^{5/4}$$
$$= 2y(y^2 + 1)^{1/4}[5y^2 + 4(y^2 + 1)^{4/4}]$$
$$= 2y(y^2 + 1)^{1/4}(9y^2 + 4)$$

35. $y = \frac{-5}{(2x^3 + 1)^2} = -5(2x^3 + 1)^{-2}$

$$\frac{dy}{dx} = -5[-2(2x^3 + 1)^{-3} \cdot 6x^2]$$
$$= -5[-12x^2(2x^3 + 1)^{-3}]$$
$$= 60x^2(2x^3 + 1)^{-3} = \frac{60x^2}{(2x^3 + 1)^3}$$

37. $r(t) = \frac{(5t - 6)^4}{3t^2 + 4}$

$$r'(t) = \frac{(3t^2 + 4)[4(5t - 6)^3 \cdot 5] - (5t - 6)^4(6t)}{(3t^2 + 4)^2}$$
$$= \frac{20(3t^2 + 4)(5t - 6)^3 - 6t(5t - 6)^4}{(3t^2 + 4)^2}$$
$$= \frac{2(5t - 6)^3[10(3t^2 + 4) - 3t(5t - 6)]}{(3t^2 + 4)^2}$$
$$= \frac{2(5t - 6)^3(30t^2 + 40 - 15t^2 + 18t)}{(3t^2 + 4)^2}$$
$$= \frac{2(5t - 6)^3(15t^2 + 18t + 40)}{(3t^2 + 4)^2}$$

39. $y = \frac{3x^2 - x}{(2x - 1)^5}$

$$\frac{dy}{dx} = \frac{(2x - 1)^5(6x - 1) - (3x^2 - x)[5(2x - 1)^4 \cdot 2]}{[(2x - 1)^5]^2}$$
$$= \frac{(2x - 1)^5(6x - 1) - 10(3x^2 - x)(2x - 1)^4}{(2x - 1)^{10}}$$
$$= \frac{(2x - 1)^4[(2x - 1)(6x - 1) - 10(3x^2 - x)]}{(2x - 1)^{10}}$$
$$= \frac{12x^2 - 2x - 6x + 1 - 30x^2 + 10x}{(2x - 1)^6}$$
$$= \frac{-18x^2 + 2x + 1}{(2x - 1)^6}$$

41. Answers will vary.

43. **(a)** $D_x(f[g(x)])$ at $x=1$

$$= f'[g(1)]\cdot g'(1)$$
$$= f'(2)\cdot\left(\frac{2}{7}\right) = -7\left(\frac{2}{7}\right) = -2$$

(b) $D_x(f[g(x)])$ at $x=2$

$$= f'[g(2)]\cdot g'(2)$$
$$= f'(3)\cdot\left(\frac{3}{7}\right) = -8\left(\frac{3}{7}\right) = -\frac{24}{7}$$

45. $f(x)=\sqrt{x^2+16};\ x=3$

$$f(x)=(x^2+16)^{1/2}$$
$$f'(x)=\frac{1}{2}(x^2+16)^{-1/2}(2x)$$
$$f'(x)=\frac{x}{\sqrt{x^2+16}}$$
$$f'(3)=\frac{3}{\sqrt{3^2+16}}=\frac{3}{5}$$
$$f(3)=\sqrt{3^2+16}=5$$

We use $m=\frac{3}{5}$ and the point $P(3, 5)$in the point-slope form

$$y-5=\frac{3}{5}(x-3)$$
$$y-5=\frac{3}{5}x-\frac{9}{5}$$
$$y=\frac{3}{5}x+\frac{16}{5}$$

47. $f(x)=x(x^2-4x+5)^4;\ x=2$

$$f'(x)=x\cdot 4(x^2-4x+5)^3\cdot(2x-4)$$
$$+1\cdot(x^2-4x+5)^4$$
$$=(x^2-4x+5)^3$$
$$\cdot[4x(2x-4)+(x^2-4x+5)]$$
$$=(x^2-4x+5)^3(9x^2-20x+5)$$
$$f'(2)=\left(2^2-4(2)+5\right)^3\left(9\left(2^2\right)-20(2)+5\right)$$
$$=(1)^3(1)=1$$
$$f(2)=2(1)^4=2$$

We use $m = 1$ and the point $P(2, 2)$.

$$y-2=1(x-2)$$
$$y-2=x-2$$
$$y=x$$

49. $f(x)=\sqrt{x^3-6x^2+9x+1}$

$$f(x)=(x^3-6x^2+9x+1)^{1/2}$$
$$f'(x)=\frac{1}{2}(x^3-6x^2+9x+1)^{-1/2}$$
$$\cdot(3x^2-12x+9)$$
$$f'(x)=\frac{3(x^2-4x+3)}{2\sqrt{x^3-6x^2+9x+1}}$$

If the tangent line is horizontal, its slope is zero and $f'(x)=0$.

$$\frac{3(x^2-4x+3)}{2\sqrt{x^3-6x^2+9x+1}}=0$$
$$3(x^2-4x+3)=0$$
$$3(x-1)(x-3)=0$$
$$x=1 \text{ or } x=3$$

The tangent line is horizontal at $x = 1$ and $x = 3$.

51. Answers will vary. Note that the two answers are equivalent.

53. $P(x)=2x^2+1;\ x=f(a)=3a+2$

$$P[f(a)]=2(3a+2)^2+1$$
$$=2(9a^2+12a+4)+1$$
$$=18a^2+24a+9$$

55. $L(w)=2.472w^{2.571}$

$$w(t)=0.265+0.21t$$
$$L'(t)=L'(w)\cdot w'(t)$$
$$=(2.472)(2.571)w^{(2.571-1)}\cdot 0.21$$
$$\approx 1.335w^{1.571}$$
$$\approx 1.335(0.265+0.21t)^{1.571}$$
$$L'(25)\approx 1.335(0.265+0.21(25))^{1.571}$$
$$\approx 19.5 \text{ mm/wk}$$

57. $N(t)=2t(5t+9)^{1/2}+12$

$$N'(t)=(2t)\left[\frac{1}{2}(5t+9)^{-1/2}(5)\right]$$
$$+2(5t+9)^{1/2}+0$$
$$=5t(5t+9)^{-1/2}+2(5t+9)^{1/2}$$
$$=(5t+9)^{-1/2}[5t+2(5t+9)]$$
$$=(5t+9)^{-1/2}(15t+18)$$
$$=\frac{15t+18}{(5t+9)^{1/2}}$$

(a) $N'(0) = \dfrac{15(0)+18}{[5(0)+9]^{1/2}} = \dfrac{18}{9^{1/2}} = 6$

At time $t = 0$, the population is increasing at a rate of 6 million per hour.

(b) $N'\left(\dfrac{7}{5}\right) = \dfrac{15\left(\frac{7}{5}\right)+18}{\left[5\left(\frac{7}{5}\right)+9\right]^{1/2}} = \dfrac{21+18}{(7+9)^{1/2}} = \dfrac{39}{(16)^{1/2}} = \dfrac{39}{4} = 9.75$

At time $t = 7/5$, the population is increasing at a rate of 9.75 million per hour.

(c) $N'(8) = \dfrac{15(8)+18}{[5(8)+9]^{1/2}} = \dfrac{120+18}{(49)^{1/2}} = \dfrac{138}{7} \approx 19.71$

At time $t = 8$, the population is increasing at a rate of about 19.71 million per hour.

59. **(a)** $R(Q) = Q\left(C - \dfrac{Q}{3}\right)^{1/2}$

$$R'(Q) = Q\left[\frac{1}{2}\left(C - \frac{Q}{3}\right)^{-1/2}\left(-\frac{1}{3}\right)\right] + \left(C - \frac{Q}{3}\right)^{1/2}(1) = -\frac{1}{6}Q\left(C - \frac{Q}{3}\right)^{-1/2} + \left(C - \frac{Q}{3}\right)^{1/2}$$

$$= -\frac{Q}{6\left(C - \frac{Q}{3}\right)^{1/2}} + \left(C - \frac{Q}{3}\right)^{1/2}$$

(b) $R'(Q) = -\dfrac{Q}{6\left(C - \frac{Q}{3}\right)^{1/2}} + \left(C - \dfrac{Q}{3}\right)^{1/2}$

If $Q = 87$ and $C = 59$, then

$$R'(Q) = -\frac{87}{6\left(59 - \frac{87}{3}\right)^{1/2}} + \left(59 - \frac{87}{3}\right)^{1/2} = -\frac{87}{6(30)^{1/2}} + (30)^{1/2} = -\frac{87}{32.86} + 5.48 = 2.83$$

(c) Because $R'(Q)$ is positive, the patient's sensitivity to the drug is increasing.

61. $p(t) = \left(\dfrac{at}{at+1}\right)^N$

$$p'(t) = N\left(\frac{at}{at+1}\right)^{N-1}\left(\frac{a(at+1) - at(a)}{(at+1)^2}\right)$$

$$= N\left(\frac{at}{at+1}\right)^{N-1}\left(\frac{a}{(at+1)^2}\right)$$

$$= \frac{Na^N t^{N-1}}{(at+1)^{N+1}}$$

$p'(t)$ represents the rate of change of the probability of a population going extinct with respect to time.

63. $A = 1500\left(1 + \dfrac{r}{36{,}500}\right)^{1825}$

$\dfrac{dA}{dr}$ is the rate of change of A with respect to r.

$$\frac{dA}{dr} = 1500(1825)\left(1 + \frac{r}{36{,}500}\right)^{1824}\left(\frac{1}{36{,}500}\right)$$

$$= 75\left(1 + \frac{r}{36{,}500}\right)^{1824}$$

(a) For $r = 6\%$,

$$\frac{dA}{dr} = 75\left(1 + \frac{6}{36{,}500}\right)^{1824}$$

$\approx \$101.22$ per percent

(b) For $r = 8\%$,

$$\frac{dA}{dr} = 75\left(1 + \frac{8}{36{,}500}\right)^{1824}$$

$\approx \$111.86$ per percent

(c) For $r = 9\%$,

$$\frac{dA}{dr} = 75\left(1+\frac{9}{36,500}\right)^{1824}$$
$$\approx \$117.59 \text{ per percent}$$

65. (a) $y = ((x^2)^2)^2$

$$\frac{dy}{dx} = 2((x^2)^2)\cdot\frac{d}{dx}((x^2)^2)$$
$$= 2x^4\cdot 2(x^2)\cdot\frac{d}{dx}(x^2)$$
$$= 2x^4\cdot 2x^2\cdot 2x = 8x^7$$

(b) $y = ((x^2)^2)^2 = x^8$

$$\frac{dy}{dx} = 8x^{8-1} = 8x^7$$

4.4 Derivatives of Exponential Functions

1. $y = e^{4x}$

Let $g(x) = 4x$, so $g'(x) = 4$.

Thus, $\frac{dy}{dx} = 4e^{4x}$.

3. $y = -8e^{3x}$

Let $g(x) = 3x$, so $g'(x) = 3$.

Thus, $\frac{dy}{dx} = -8(3e^{3x}) = -24e^{3x}$.

5. $y = -16e^{2x+1}$

$g(x) = 2x+1$, so $g'(x) = 2$

Thus, $\frac{dy}{dx} = -16(2e^{2x+1}) = -32e^{2x+1}$.

7. $y = e^{x^2}$

$g(x) = x^2$, so $g'(x) = 2x$

$$\frac{dy}{dx} = 2xe^{x^2}$$

9. $y = 3e^{2x^2}$

$g(x) = 2x^2$, so $g'(x) = 4x$

$$\frac{dy}{dx} = 3\left(4xe^{2x^2}\right) = 12xe^{2x^2}$$

11. $y = 4e^{2x^2-4}$

$g(x) = 2x^2 - 4$, so $g'(x) = 4x$

$$\frac{dy}{dx} = 4\left[(4x)e^{2x^2-4}\right] = 16xe^{2x^2-4}$$

13. $y = xe^x$

Use the product rule.

$$\frac{dy}{dx} = xe^x + e^x\cdot 1 = e^x(x+1)$$

15. $y = (x+3)^2e^{4x}$

Use the product rule.

$$\frac{dy}{dx} = (x+3)^2(4)e^{4x} + e^{4x}\cdot 2(x+3)$$
$$= 4(x+3)^2e^{4x} + 2(x+3)e^{4x}$$
$$= 2(x+3)e^{4x}\left[2(x+3)+1\right]$$
$$= 2(x+3)(2x+7)e^{4x}$$

17. $y = \frac{x^2}{e^x}$

Use the quotient rule.

$$\frac{dy}{dx} = \frac{e^x(2x) - x^2e^x}{(e^x)^2} = \frac{xe^x(2-x)}{e^{2x}} = \frac{x(2-x)}{e^x}$$

19. $y = \frac{e^x + e^{-x}}{x}$

$$\frac{dy}{dx} = \frac{x(e^x - (-1)e^{-x}) - (e^x + e^{-x})(1)}{x^2}$$
$$= \frac{x(e^x - e^{-x}) - (e^x + e^{-x})}{x^2}$$

21. $p = \frac{10,000}{9+4e^{-0.2t}}$

$$\frac{dp}{dt} = \frac{(9+4e^{-0.2t})\cdot 0 - 10,000[0 + 4(-0.2)e^{-0.2t}]}{(9+4e^{-0.2t})^2}$$
$$= \frac{8000e^{-0.2t}}{(9+4e^{-0.2t})^2}$$

23. $f(z) = \left(2z + e^{-z^2}\right)^2$

$$f'(z) = 2\left(2z + e^{-z^2}\right)^1\left(2 - 2ze^{-z^2}\right)$$
$$= 4\left(2z + e^{-z^2}\right)\left(1 - ze^{-z^2}\right)$$

25. $y = 7^{3x+1}$

Let g $(x) = 3x+1$, with $g'(x) = 3$. Then

$$\frac{dy}{dx} = (\ln 7)(7^{3x+1})\cdot 3 = 3(\ln 7)7^{3x+1}$$

27. $y = 3\cdot 4^{x^2+2}$

Let $g(x) = x^2 + 2$, with $g'(x) = 2x$. Then

$$\frac{dy}{dx} = 3(\ln 4)4^{x^2+2}\cdot 2x = 6x(\ln 4)4^{x^2+2}$$

29. $s = 2 \cdot 3^{\sqrt{t}}$

Let $g(t) = \sqrt{t}$, with $g'(t) = \frac{1}{2\sqrt{t}}$. Then

$$\frac{ds}{dt} = 2(\ln 3)3^{\sqrt{t}} \cdot \frac{1}{2\sqrt{t}} = \frac{(\ln 3)3^{\sqrt{t}}}{\sqrt{t}}$$

31. $y = \frac{te^t + 2}{e^{2t} + 1}$

Use the quotient rule and product rule.

$$\frac{dy}{dt} = \frac{(e^{2t}+1)(te^t + e^t \cdot 1) - (te^t + 2)(2e^{2t})}{(e^{2t}+1)^2}$$
$$= \frac{(e^{2t}+1)(te^t + e^t) - (te^t + 2)(2e^{2t})}{(e^{2t}+1)^2}$$
$$= \frac{te^{3t} + e^{3t} + te^t + e^t - 2te^{3t} - 4e^{2t}}{(e^{2t}+1)^2}$$
$$= \frac{-te^{3t} + e^{3t} + te^t + e^t - 4e^{2t}}{(e^{2t}+1)^2}$$
$$= \frac{(1-t)e^{3t} - 4e^{2t} + (1+t)e^t}{(e^{2t}+1)^2}$$

33. $f(x) = e^{x\sqrt{3x+2}}$

Let $g(x) = x\sqrt{3x+2}$.

$$g'(x) = 1 \cdot \sqrt{3x+2} + x\left(\frac{3}{2\sqrt{3x+2}}\right)$$
$$= \sqrt{3x+2} + \frac{3x}{2\sqrt{3x+2}}$$
$$= \frac{2(3x+2)}{2\sqrt{3x+2}} + \frac{3x}{2\sqrt{3x+2}} = \frac{9x+4}{2\sqrt{3x+2}}$$
$$f'(x) = e^{x\sqrt{3x+2}} \cdot \left(\frac{9x+4}{2\sqrt{3x+2}}\right)$$

35. $y = y_o e^{kt}$

$$\frac{dy}{dx} = \frac{d}{dt}\left[y_o e^{kt}\right] = y_o k e^{kt} = k(y_o e^{kt}) = ky$$

37. Graph the function $y = e^x$, then sketch the lines tangent to the graph at $x = 1, 0, 1, 2$.

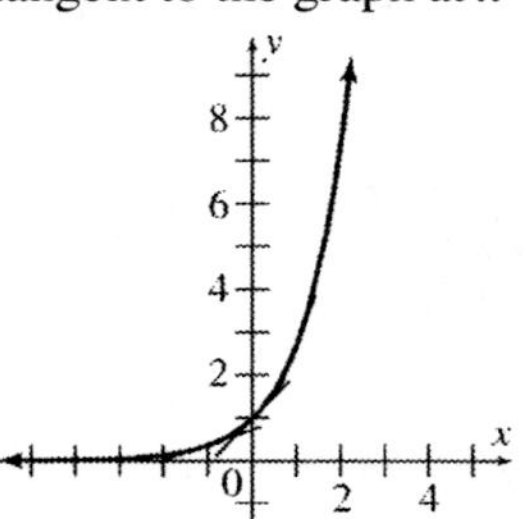

Estimate the slopes of the tangent lines at these points.

At $x = -1$, the slope is a little steeper than $\frac{1}{3}$ or approximately $0.\overline{3}$.

At $x = 0$, the slope is 1.

At $x = 1$ the slope is a little steeper than $\frac{5}{2}$ or 2.5.

At $x = 2$, the slope is a little steeper than $7\frac{1}{3}$ or $7.\overline{3}$.

Note that $e^{-1} \approx 0.36787944$, $e^0 = 1$, $e^1 = e \approx 2.7182812$, and $e^2 \approx 7.3890561$. The values are close enough to the slopes of the tangent lines to convince us that $\frac{de^x}{dx} = e^x$.

39. First substitute the expression for $G(t)$ in the expression for $\frac{dG}{dt}$.

$$G(t)=\frac{m}{1+\left(\frac{m}{G_0}+1\right)e^{-kmt}}$$

$$\frac{dG}{dt}=r\left(1-\frac{G}{m}\right)G=r\left(\frac{m-G}{m}\right)G$$

$$=r\left(\frac{m-\frac{m}{1+\left(\frac{m}{G_0}+1\right)e^{-kmt}}}{m}\right)\left(\frac{m}{1+\left(\frac{m}{G_0}+1\right)e^{-kmt}}\right)=r\left(\frac{m-\frac{m}{1+\left(\frac{m}{G_0}+1\right)e^{-kmt}}}{1+\left(\frac{m}{G_0}+1\right)e^{-kmt}}\right)$$

$$=r\left(\frac{\frac{m\left(1+\left(\frac{m}{G_0}+1\right)e^{-kmt}\right)-m}{1+\left(\frac{m}{G_0}+1\right)e^{-kmt}}}{1+\left(\frac{m}{G_0}+1\right)e^{-kmt}}\right)=r\left(\frac{m\left(1+\left(\frac{m}{G_0}+1\right)e^{-kmt}\right)-m}{\left(1+\left(\frac{m}{G_0}+1\right)e^{-kmt}\right)^2}\right)=\frac{r\left(\frac{m^2}{G_0}+1\right)e^{-kmt}}{\left(1+\left(\frac{m}{G_0}+1\right)e^{-kmt}\right)^2}$$

Now use $G(t)$ to find $\frac{dG}{dt}$ and show that the two expressions are equal.

$$\frac{dG}{dt}=\frac{\left(1+\left(\frac{m}{G_0}+1\right)e^{-kmt}\right)(0)-m\left(0-km\left(\frac{m}{G_0}+1\right)e^{-kmt}\right)}{\left(1+\left(\frac{m}{G_0}+1\right)e^{-kmt}\right)^2}=\frac{-km\left(\frac{m^2}{G_0}+1\right)e^{-kmt}}{\left(1+\left(\frac{m}{G_0}+1\right)e^{-kmt}\right)^2}$$

Thus, with $r=-km$, the two expressions are equal.

41. $h(t)=37.79(1.021)^t$

$h'(t)=37.79(\ln 1.021)(1.021)^t \quad =0.785(1.021)^t$

(a) For 2015, $t=15$: $h(15)=37.79(1.021)^{15}\approx 51.6$
The Hispanic population in the United States will be about 51,600,000 in 2015.

(b) $h'(15)=0.785(\ln 1.021)(1.021)^{15}\approx 1.07$ The Hispanic population in the United States will be increasing at the rate of 1,070,000 people per year at the end of the year 2015.

43. (a) First, find k. We know that $G_0=400$, $m=5200$, and $G(2)\approx 1000$.

$$G(t)=\frac{mG_0}{G_0+(m-G_0)e^{-kmt}}$$

$$1000=\frac{5200(400)}{400+(5200-400)e^{-(5200)(2)k}}=\frac{5200(400)}{400+4800e^{-10400k}}=\frac{5200}{1+12e^{-10400k}}$$

$$1+12e^{-10400k}=\frac{5200}{1000}=5.2$$

$$12e^{-10400k}=4.2\Rightarrow e^{-10400k}=0.35$$

$$-10400k=\ln 0.35\Rightarrow k=-\frac{\ln 0.35}{10,400}\approx -0.0001$$

Thus, the logistic equation is $G(t)=\frac{5200}{1+12e^{-0.0001(5200)t}}=\frac{5200}{1+12e^{-0.52t}}$

(b) $G'(t) = -5200(1+12e^{-0.52t})^{-2}(-6.24e^{-0.52t})$
$$= \frac{32,448e^{-0.52t}}{(1+12e^{-0.52t})^2}$$
$$G(1) = \frac{5200}{1+12e^{-0.52}} \approx 639$$
$$G'(1) = \frac{32,448e^{-0.52}}{(1+12e^{-0.52})^2} \approx 292$$
After 1 year, there are about 639 clams. The population is growing at about 292 clams per year.

(c)
$$G(4) = \frac{5200}{1+12e^{-2.08}} \approx 2081$$
$$G'(4) = \frac{32,448e^{-2.08}}{(1+12e^{-2.08})^2} \approx 649$$
After 4 years, there are about 2081 clams. The population is growing at about 649 clams per year.

(d)
$$G(10) = \frac{5200}{1+12e^{-5.2}} \approx 4877$$
$$G'(10) = \frac{32,448e^{-5.2}}{(1+12e^{-5.2})^2} \approx 157$$
After 10 years, there are about 4877 clams. The population is growing at about 157 clams per year.

(e) It increases for a while and then gradually decreases to 0. Verify by graphing $G'(t)$ in the window $[0, 15]$ by $[0, 700]$.

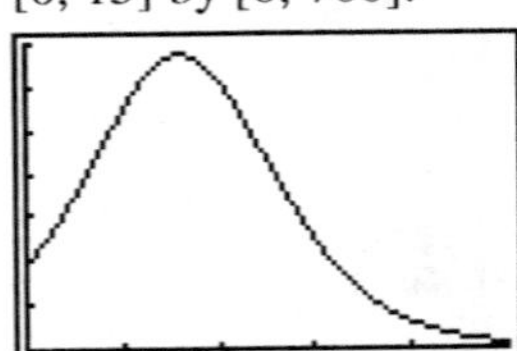

45. We want to know how long it will take for the first population to reach the size of the other. Let $G(0) = 2.1\times10^8$. We are given $m = 3.7\times10^8$ and $km = 0.55$, so
$$G(t) = \frac{m}{1+\left(\frac{m}{G_0}-1\right)e^{-kmt}}$$
$$= \frac{3.7\times10^8}{1+\left(\frac{3.7\times10^8}{2.1\times10^8}-1\right)e^{-0.55t}}$$
$$= \frac{3.7\times10^8}{1+\left(\frac{3.7}{2.1}-1\right)e^{-0.55t}}$$

Now let $G(t) = 2.6\times10^8$ solve for t.
$$2.6\times10^8 = \frac{3.7\times10^8}{1+\left(\frac{3.7}{2.1}-1\right)e^{-0.55t}}$$
$$1+\left(\frac{3.7}{2.1}-1\right)e^{-0.55t} = \frac{3.7\times10^8}{2.6\times10^8}$$
$$\left(\frac{3.7}{2.1}-1\right)e^{-0.55t} = \frac{3.7}{2.6}-1$$
$$e^{-0.55t} = \frac{\frac{3.7}{2.6}-1}{\frac{3.7}{2.1}-1}$$
$$-0.55t = \ln\left(\frac{\frac{3.7}{2.6}-1}{\frac{3.7}{2.1}-1}\right)$$
$$t = -\frac{\ln\left(\frac{\frac{3.7}{2.6}-1}{\frac{3.7}{2.1}-1}\right)}{0.55} \approx 1.070$$
It will take about 1.070 hours for the size of the first group to reach the size of the second group. To find how fast it was growing at that time, find $G'(1.070)$.
$$G'(t) = \frac{-3.7\times10^8 \frac{d}{dt}\left(1+\left(\frac{3.7}{2.1}-1\right)e^{-0.55t}\right)}{\left(1+\left(\frac{3.7}{2.1}-1\right)e^{-0.55t}\right)^2}$$
$$= \frac{-3.7\times10^8\left(0-0.55(0.7619)e^{-0.55t}\right)}{\left(1+0.7619e^{-0.55t}\right)^2}$$
$$\approx \frac{1.550\times10^8 e^{-0.55t}}{\left(1+0.7619e^{-0.55t}\right)^2}$$
$$G'(1.070) \approx 4.251\times10^7$$
The population was growing at the rate of about 4.251×10^7 cells per hour.

47. $P(t) = 0.00239e^{0.0957t}$

(a)
$$P(25) = 0.00239e^{0.0957(25)} \approx 0.026\%$$
$$P(50) = 0.00239e^{0.0957(50)} \approx 0.286\%$$
$$P(75) = 0.00239e^{0.0957(75)} \approx 3.130\%$$

(b)
$$\begin{aligned} P'(t) &= 0.00239e^{0.0957t}(0.0957) \\ &= 0.000228723e^{0.0957t} \\ P'(25) &= 0.000228723e^{0.0957(25)} \\ &\approx 0.0025\% \text{ per year} \\ P'(50) &= 0.000228723e^{0.0957(50)} \\ &\approx 0.0274\% \text{ per year} \\ P'(75) &= 0.000228723e^{0.0957(75)} \\ &\approx 0.300\% \text{ per year} \end{aligned}$$

(c) The percentage of people in each of the age groups that die in a given year is increasing as indicated by the answers in parts (a) and (b). A person who is 75 years old has a 3% chance of dying during year and the rate is increasing by almost 0.3%. The formula implies that everyone will be dead by age 112.

49. $M(t) = 3102e^{-e^{-0.022(t-56)}}$

(a)
$$\begin{aligned} M(200) &= 3102e^{-e^{-0.022(200-56)}} \\ &\approx 2974.15 \text{ grams,} \end{aligned}$$
or about 3 kilograms.

(b) As t gets very large, $-e^{-0.022(t-56)}$ goes to zero, $e^{-e^{-0.022(t-56)}}$ goes to 1, and $M(t)$ approaches 3102 grams or about 3.1 kilograms.

(c) 80% of 3102 is 2481.6.

$$\begin{aligned} 2481.6 &= 3102e^{-e^{-0.022(t-56)}} \\ -\ln\frac{2481.6}{3102} &= e^{-0.022(t-56)} \\ \ln\left(\ln\frac{3102}{2481.6}\right) &= -0.022(t-56) \\ t &= -\frac{1}{0.022}\ln\left(\ln\frac{3102}{2481.6}\right) + 56 \\ &\approx 124 \text{ days} \end{aligned}$$

(d) $D_t M(t)$
$$\begin{aligned} &= 3102e^{-e^{-0.022(t-56)}} D_t\left(-e^{-0.022(t-56)}\right) \\ &= 3102e^{-e^{-0.022(t-56)}}\left(-e^{-0.022(t-56)}\right)(-0.022) \\ &= 68.244e^{-e^{-0.022(t-56)}}e^{-0.022(t-56)} \end{aligned}$$
When $t = 200$, $D_t M(t) \approx 2.75$ g/day.

(e)

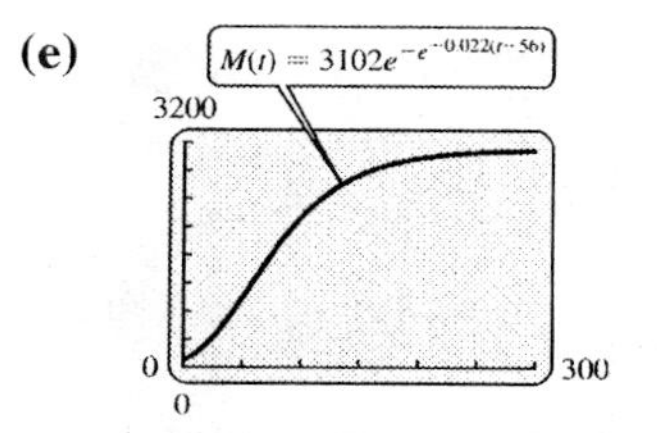

Growth is initially rapid, then tapers off.

(f)

Day	Weight	Rate
50	991	24.88
100	2122	17.73
150	2734	7.60
200	2974	2.75
250	3059	0.94
300	3088	0.32

51. $W(t) = 532\left(1 - 0.911e^{-0.0021t}\right)^{1.2466}$

(a) $W(t)$ is a strictly increasing function.
$$\begin{aligned} &\lim_{t\to\infty} W(t) \\ &= \lim_{t\to\infty} 532\left(1 - 0.911e^{-0.0021t}\right)^{1.2466} \\ &= 532(1-0) = 532 \end{aligned}$$
The maximum weight of the average Ayrshire cow is 532 kg.

(b)
$$\begin{aligned} 0.9(532) &= 532\left(1 - 0.911e^{-0.0021t}\right)^{1.2466} \\ 0.9 &= \left(1 - 0.911e^{-0.0021t}\right)^{1.2466} \\ 0.9^{1/1.2466} &= 1 - 0.911e^{-0.0021t} \\ e^{-0.0021t} &= \frac{1 - 0.9^{1/1.2466}}{0.911} \\ -0.0021t &= \ln\left(\frac{1 - 0.9^{1/1.2466}}{0.911}\right) \\ t &= \frac{\ln\left(\frac{1 - 0.9^{1/1.2466}}{0.911}\right)}{-0.0021} \approx 1152 \end{aligned}$$
A cow reaches 90% of the maximum weight at about 1152 days old.

(c) $W'(t) = 532(1.2466)\left(1-0.911e^{-0.0021t}\right)^{0.2466}(0.00019131)e^{-0.0021t}$

$\approx 1.26875\left(1-0.911e^{-0.0021t}\right)^{0.2466}e^{-0.0021t}$

$W'(1000) = 1.26875\left(1-0.911e^{-0.0021(1000)}\right)^{0.2466}e^{-0.0021(1000)} \approx 0.15$

When the cow is 1000 days old, it is gaining about 0.15 kilograms per day.

(d)

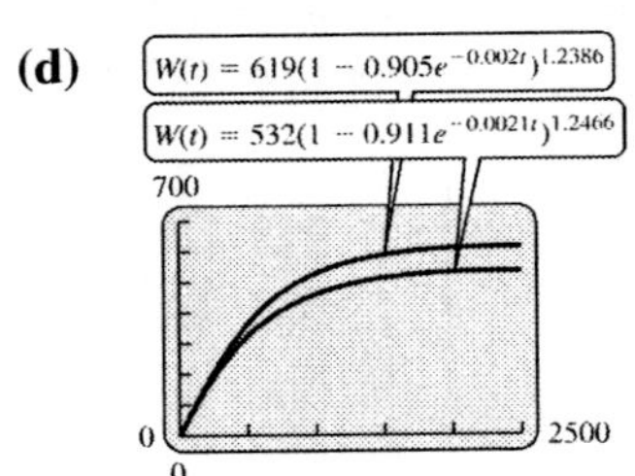

Both approach their maximums asymptotically.

53. $W_1(t) = 509.7(1-0.941e^{-0.00181t})$

$W_2(t) = 498.4(1-0.889e^{-0.00219t})^{1.25}$

(a) Both W_1 and W_2 are strictly increasing functions, so they approach their maximum values as t approaches ∞.

$\lim_{t\to\infty} W_1(t) = \lim_{t\to\infty} 509.7(1-0.941e^{-0.00181t}) = 509.7(1-0) = 509.7$

$\lim_{t\to\infty} W_2(t) = \lim_{t\to\infty} 498.4(1-0.889e^{-0.00219t})^{1.25} = 498.4(1-0)^{1.25} = 498.4$

So, the maximum values of W_1 and W_2 are 509.7 kg and 498.4 kg respectively.

(b)

Using $W_1(t)$

$0.9(509.7) = 509.7\left(1-0.941e^{-0.00181t}\right)$

$0.9 = 1-0.941e^{-0.00181t}$

$\frac{0.1}{0.941} \approx e^{-0.00181t}$

$t = \frac{\ln\left(\frac{0.1}{0.941}\right)}{-0.00181} \approx 1239$

Using $W_2(t)$

$0.9(498.4) = 498.4\left(1-0.889e^{-0.00219t}\right)^{1.25}$

$0.9 = \left(1-0.889e^{-0.00219t}\right)^{1.25}$

$0.9^{1/1.25} = 0.9^{0.8} = 1-0.889e^{-0.00219t}$

$\frac{1-0.9^{0.8}}{0.889} = e^{-0.00219t}$

$t = \frac{\ln\left(\frac{1-0.9^{0.8}}{0.889}\right)}{-0.00219} \approx 1095$

Respectively, it will take the average beef cow about 1239 days or 1095 days to reach 90% of its maximum.

(c) $W_1'(t) = (509.7)(-0.941)(-0.00181)e^{-0.00181t} \approx 0.868126e^{-0.00181t}$

$W_1'(750) \approx 0.868126e^{-0.00181(750)} \approx 0.22$ kg/day

$W_2'(t) = (498.4)(1.25)(1-0.889e^{-0.00219t})^{0.25}\cdot(-0.889)(-0.00219)e^{-0.00219t}$

$\approx 1.21292e^{-0.00219t}(1-0.889e^{-0.00219t})^{0.25}$

$W_2'(750) \approx 1.12192e^{-0.00219(750)}\cdot\left(1-0.889e^{-0.00219(750)}\right)^{0.25} \approx 0.22$ kg/day

Both functions yield a rate of change of about 0.22 kg per day.

(d) Looking at the graph, the growth patterns of the two functions are very similar.

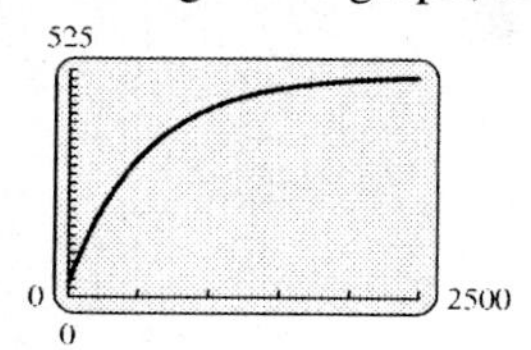

(e) The graphs of the rate of change of the two functions are also very similar.

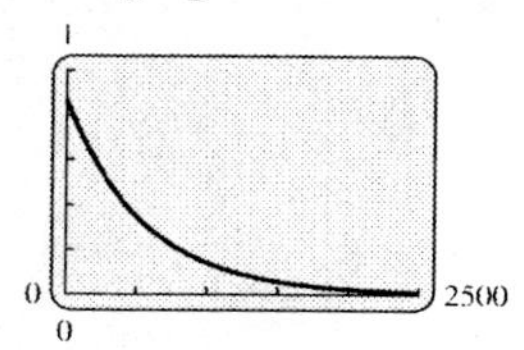

55. $p(x) = 0.001131e^{0.1268x}$

(a) $p(40) = 0.001131e^{0.1268(40)} \approx 0.180$

(b) When $p(x) = 1$,

$$0.001131e^{0.1268x} = 1 \Rightarrow e^{0.1268x} = \frac{1}{0.001131} \Rightarrow 0.1268x = \ln\frac{1}{0.001131} \Rightarrow x = \frac{1}{0.1268}\ln\frac{1}{0.001131} \approx 54$$

This represents the year 2024.

(c) $p'(x) = 0.001131e^{0.1268x}(0.1268) = 0.0001434108e^{0.1268x}$
$p'(40) = 0.0001434108e^{0.1268(40)} \approx 0.023$
The marginal increase in the proportion per year in 2010 is approximately 0.023.

(d) Answers will vary.

57. $I = E\left[1 - e^{-a(S-h)/E}\right]$

$$\frac{dI}{dS} = E\left(\frac{-a}{E}\right)\left(-e^{-a(S-h)/E}\right) = ae^{-a(S-h)/E}$$

59. $t(n) = 218 + 31(0.933)^n$

(a) $t(60) = 218 + 31(0.933)^{60} \approx 218.5$ sec

(b) $t'(n) = (31 \ln 0.933)(0.933)^n$
$t'(60) = (31 \ln 0.933)(0.933)^{60} \approx -0.034$
The record is decreasing by 0.034 seconds per year at the end of 2010.

(c) As $n \to \infty$, $(0.933)^n \to 0$ and $t(n) \to 218$. If the estimate is correct, then this is the least amount of time that it will ever take a human to run a mile.

61. (a) $f(t) = 0.981e^{-6.74t} + 0.0163e^{-1.019t} + 0.0031e^{-0.210t}$

$$\begin{aligned} f'(t) &= (-6.74)0.981e^{-6.74t} + (-1.019)0.0163e^{-1.019t} + (-0.210)0.0031e^{-0.210t} \\ &= -6.61194e^{-6.74t} - 0.0166097e^{-1.019t} - 0.000651e^{-0.210t} \end{aligned}$$

(i) Note that t is measured in days. Six hours is 0.25 day.

$$f(0.25) = 0.981e^{-6.74(0.25)} + 0.0163e^{-1.019(0.25)} + 0.0031e^{-0.210(0.25)} \approx 0.1975$$

$$f'(0.25) = -6.61194e^{-6.74(0.25)} - 0.0166097e^{-1.019(0.25)} - 0.000651e^{-0.210(0.25)} \approx -1.240$$

(ii)
$$f(1) = 0.981e^{-6.74(1)} + 0.0163e^{-1.019(1)} + 0.0031e^{-0.210(1)} \approx 0.009557$$

$$f'(1) = -6.61194e^{-6.74(1)} - 0.0166097e^{-1.019(1)} - 0.000651e^{-0.210(1)} \approx -0.01434$$

(iii)
$$f(5) = 0.981e^{-6.74(5)} + 0.0163e^{-1.019(5)} + 0.0031e^{-0.210(5)} \approx 0.001185$$

$$f'(5) = -6.61194e^{-6.74(5)} - 0.0166097e^{-1.019(5)} - 0.000651e^{-0.210(5)} \approx -3.296\times10^{-4}$$

(b) The fraction of radioactive iron approaches 0.

(c) The rate of change is approaching 0 as time goes on.

(d)

	$0.981e^{-6.74t}$	$0.0163e^{-1.019t}$	$0.0031e^{-0.210t}$
(i) $t = 0.25$	0.1819	0.01263	0.002941
(ii) $t = 1$	0.001160	0.005883	0.002513
(iii) $t = 5$	2.270×10^{-15}	9.988×10^{-5}	0.001085

At 6 hours, the first term, $0.981e^{-6.74t}$, has the most weight. At 1 day, the second term, $0.0163e^{-1.019t}$, has the most weight. At 5 days, the third term, 0.001085, has the most weight. Note that as t increases, the exponent on e decreases, so the value of $e^{-.210t}$ decreases, and the value of the term also decreases.

63. (a) $G_0 = 0.00369,\ m = 1,\ k = 3.5$

$$G(t) = \frac{1}{1+\left(\frac{1}{0.00369}-1\right)e^{-3.5(1)t}} = \frac{1}{1+270e^{-3.5t}}$$

(b) $G'(t) = -(1+270e^{-3.5t})^{-2}\cdot 270e^{-3.5t}(-3.5) = \dfrac{945e^{-3.5t}}{(1+270e^{-3.5t})^2}$

$$G(1) = \frac{1}{1+270e^{-3.5(1)}} \approx 0.109$$

$$G'(1) = \frac{945e^{-3.5(1)}}{\left[1+270e^{-3.5(1)}\right]^2} \approx 0.341$$

The proportion is 0.109 and the rate of growth is 0.341 per century.

(c)
$$G(2) = \frac{1}{1+270e^{-3.5(2)}} \approx 0.802$$

$$G'(2) = \frac{945e^{-3.5(2)}}{[1+270e^{-3.5(2)}]^2} \approx 0.555$$

The proportion is 0.802 and the rate of growth is 0.555 per century.

(d) $G(3) = \dfrac{1}{1+270e^{-3.5(3)}} \approx 0.993$

$G'(3) = \dfrac{945e^{-3.5(2)}}{[1+270e^{-3.5(2)}]^2} \approx 0.0256$

The proportion is 0.993 and the rate of growth is 0.0256 per century.

(e) The rate of growth increases for a while and then gradually decreases to 0. Verify by graphing $G'(t)$ in the window [0, 4] by [0, 1].

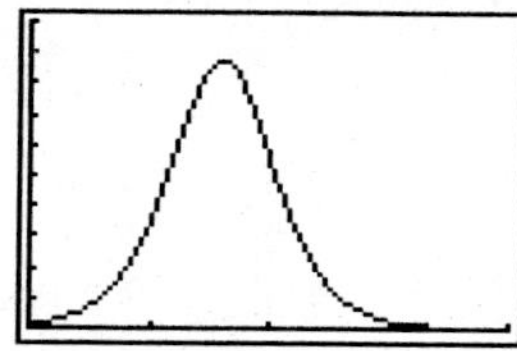

65. $H(N) = 1000(1-e^{-kN}),\ k = 0.1$

$H' = -1000e^{-0.1N}(-0.1) = 100e^{-0.1N}$

(a) $H'(10) = 100e^{-0.1(10)} \approx 36.8$

(b) $H'(100) = 100e^{-0.1(100)} \approx 0.00454$

(c) $H'(1000) = 100e^{-0.1(1000)} \approx 0$

(d) $100e^{-0.1N}$ is always positive since powers of e are never negative. This means that repetition always makes a habit stronger.

(e) $A'(t)$ will never equal zero, although as t increases, it approaches zero. Powers of e are always positive.

67. $Q(t) = CV(1-e^{-t/RC})$

(a) $I_c = \dfrac{dQ}{dt} = CV\left[0 - e^{-t/RC}\left(-\dfrac{1}{RC}\right)\right] = CV\left(\dfrac{1}{RC}\right)e^{-t/RC} = \dfrac{V}{R}e^{-t/RC}$

(b) When $C = 10^{-5}$ farads, $R = 10^7$ ohms, and $V = 10$ volts, after 200 seconds

$I_c = \dfrac{10}{10^7}e^{-200/(10^7\cdot 10^{-5})} \approx 1.35\times 10^{-7}$ amps

69. $f(t) = 5.4572\cdot 10^{-59}\cdot 10^{2.923\cdot 10^{-2}t}$

$f'(t) = 5.4572\cdot 10^{-59}\cdot \ln 10\cdot 10^{2.923\cdot 10^{-2}t}\left(\dfrac{d}{dt}\left(2.923\cdot 10^{-2}t\right)\right)$

$= 5.4572\cdot 10^{-59}\cdot 2.3026\cdot 10^{2.923\cdot 10^{-2}t}\cdot 2.923\cdot 10^{-2}$

$\approx 3.673\cdot 10^{-60}\cdot 10^{2.923\cdot 10^{-2}t}$

$f(1990) \approx 0.8029$

$f'(1990) \approx 0.0540$

In 1990, the average number of deaths was about 0.8029 per 100 students and the rate that the function was increasing was about 0.0540 deaths per 100 students per year.

4.5 Derivatives of Logarithmic Functions

1. $y = \ln(8x)$

$$\frac{dy}{dx} = \frac{d}{dx}(\ln 8x)$$
$$= \frac{d}{dx}(\ln 8 + \ln x)$$
$$= \frac{d}{dx}(\ln 8) + \frac{d}{dx}(\ln x)$$
$$= 0 + \frac{1}{x} = \frac{1}{x}$$

3. $y = \ln(8 - 3x)$

$g(x) = 8 - 3x$

$g'(x) = -3$

$$\frac{dy}{dx} = \frac{g'(x)}{g(x)} = \frac{-3}{8-3x} \text{ or } \frac{3}{3x-8}$$

5. $y = \ln|4x^2 - 9x|$

$g(x) = 4x^2 - 9x$

$g'(x) = 8x - 9$

$$\frac{dy}{dx} = \frac{g'(x)}{g(x)} = \frac{8x-9}{4x^2-9x}$$

7. $y = \ln\sqrt{x+5}$

$g(x) = \sqrt{x+5} = (x+5)^{1/2}$

$g'(x) = \frac{1}{2}(x+5)^{-1/2}$

$$\frac{dy}{dx} = \frac{\frac{1}{2}(x+5)^{-1/2}}{(x+5)^{1/2}} = \frac{1}{2(x+5)}$$

9. $y = \ln(x^4 + 5x^2)^{3/2} = \frac{3}{2}\ln(x^4 + 5x^2)$

$$\frac{dy}{dx} = \frac{3}{2}D_x[\ln(x^4 + 5x^2)]$$

$g(x) = x^4 + 5x^2$

$g'(x) = 4x^3 + 10x$

$$\frac{dy}{dx} = \frac{3}{2}\left(\frac{4x^3 + 10x}{x^4 + 5x^2}\right)$$
$$= \frac{3}{2}\left[\frac{2x(2x^2+5)}{x^2(x^2+5)}\right]$$
$$= \frac{3(2x^2+5)}{x(x^2+5)}$$

11. $y = -5x\ln(3x+2)$

Use the product rule.

$$\frac{dy}{dx} = -5x\left[\frac{d}{dx}\ln(3x+2)\right] + \ln(3x+2)\left[\frac{d}{dx}(-5x)\right]$$
$$= -5x\left(\frac{3}{3x+2}\right) + [\ln(3x+2)](-5)$$
$$= -\frac{15x}{3x+2} - 5\ln(3x+2)$$

13. $s = t^2\ln|t|$

Use the product rule.

$$\frac{ds}{dt} = t^2 \cdot \frac{1}{t} + 2t\ln|t| = t + 2t\ln|t|$$
$$= t(1 + 2\ln|t|)$$

15. $y = \frac{2\ln(x+3)}{x^2}$

Use the quotient rule.

$$\frac{dy}{dx} = \frac{x^2\left(\frac{2}{x+3}\right) - 2\ln(x+3)\cdot 2x}{(x^2)^2}$$
$$= \frac{\frac{2x^2}{x+3} - 4x\ln(x+3)}{x^4}$$
$$= \frac{2x^2 - 4x(x+3)\ln(x+3)}{x^4(x+3)}$$
$$= \frac{x[2x - 4(x+3)\ln(x+3)]}{x^4(x+3)}$$
$$= \frac{2x - 4(x+3)\ln(x+3)}{x^3(x+3)}$$

17. $y = \frac{\ln x}{4x+7}$

Use the quotient rule.

$$\frac{dy}{dx} = \frac{(4x+7)\left(\frac{1}{x}\right) - (\ln x)(4)}{(4x+7)^2}$$
$$= \frac{\frac{4x+7}{x} - 4\ln x}{(4x+7)^2}$$
$$= \frac{4x+7-4x\ln x}{x(4x+7)^2}$$

19. $y = \dfrac{3x^2}{\ln x}$

$$\frac{dy}{dx} = \frac{(\ln x)(6x) - 3x^2\left(\frac{1}{x}\right)}{(\ln x)^2} = \frac{6x \ln x - 3x}{(\ln x)^2}$$

21. $y = (\ln|x+1|)^4$

$$\frac{dy}{dx} = 4(\ln|x+1|)^3\left(\frac{1}{x+1}\right) = \frac{4(\ln|x+1|)^3}{x+1}$$

23. $y = \ln|\ln x|$

$g(x) = \ln x$

$g'(x) = \dfrac{1}{x}$

$$\frac{dy}{dx} = \frac{g'(x)}{g(x)} = \frac{\frac{1}{x}}{\ln x} = \frac{1}{x\ln x}$$

25. $y = e^{x^2} \ln x,\ x > 0$

$$\frac{dy}{dx} = e^{x^2}\left(\frac{1}{x}\right) + (\ln x)(2x)e^{x^2} = \frac{e^{x^2}}{x} + 2xe^{x^2} \ln x$$

27. $y = \dfrac{e^x}{\ln x},\ x > 0$

$$\frac{dy}{dx} = \frac{(\ln x)e^x - e^x\left(\frac{1}{x}\right)}{(\ln x)^2} \cdot \frac{x}{x} = \frac{xe^x \ln x - e^x}{x(\ln x)^2}$$

29. $g(z) = (e^{2z} + \ln z)^3$

$$g'(z) = 3(e^{2z} + \ln z)^2\left(e^{2z} \cdot 2 + \frac{1}{z}\right) = 3(e^{2z} + \ln z)^2\left(\frac{2ze^{2z} + 1}{z}\right)$$

31. $y = \log(6x)$

$g(x) = 6x$ and $g'(x) = 6.$

$$\frac{dy}{dx} = \frac{1}{\ln 10}\left(\frac{6}{6x}\right) = \frac{1}{x\ln 10}$$

33. $y = \log|1-x|$

$g(x) = 1 - x$ and $g'(x) = -1.$

$$\frac{dy}{dx} = \frac{1}{\ln 10} \cdot \frac{-1}{1-x} = -\frac{1}{(\ln 10)(1-x)} \quad \text{or} \quad \frac{1}{(\ln 10)(x-1)}$$

35. $y = \log_5 \sqrt{5x+2}$

$g(x) = \sqrt{5x+2}$ and $g'(x) = \dfrac{5}{2\sqrt{5x+2}}.$

$$\frac{dy}{dx} = \frac{1}{\ln 5} \cdot \frac{\frac{5}{2\sqrt{5x+2}}}{\sqrt{5x+2}} = \frac{5}{(2\ln 5)(5x+2)}$$

37. $y = \log_3 (x^2 + 2x)^{3/2}$

$g(x) = (x^2 + 2x)^{3/2}$ and

$$g'(x) = \frac{3}{2}(x^2 + 2x)^{1/2} \cdot (2x+2) = 3(x+1)(x^2+2x)^{1/2}$$

$$\frac{dy}{dx} = \frac{1}{\ln 3} \cdot \frac{3(x+1)(x^2+2x)^{1/2}}{(x^2+2x)^{3/2}} = \frac{3(x+1)}{(\ln 3)(x^2+2x)}$$

39. $w = \log_8(2^p - 1)$

$g(p) = 2^p - 1$

$g'(p) = (\ln 2)2^p$

$$\frac{dw}{dp} = \frac{1}{\ln 8} \cdot \frac{(\ln 2)2^p}{(2^p - 1)} = \frac{(\ln 2)2^p}{(\ln 8)(2^p - 1)}$$

41. $f(x) = e^{\sqrt{x}} \ln(\sqrt{x} + 5)$

Use the product rule.

$$f'(x) = e^{\sqrt{x}}\left(\frac{\frac{1}{2\sqrt{x}}}{\sqrt{x}+5}\right) + [\ln(\sqrt{x}+5)]e^{\sqrt{x}}\left(\frac{1}{2\sqrt{x}}\right) = \frac{e^{\sqrt{x}}}{2}\left[\frac{1}{\sqrt{x}(\sqrt{x}+5)} + \frac{\ln(\sqrt{x}+5)}{\sqrt{x}}\right]$$

43. $f(t)=\dfrac{\ln(t^2+1)+t}{\ln(t^2+1)+1}$

Use the quotient rule.

$u(t)=\ln(t^2+1)+t,\ u'(t)=\dfrac{2t}{t^2+1}+1$

$v(t)=\ln(t^2+1)+1,\ v'(t)=\dfrac{2t}{t^2+1}$

$$f'(t)=\frac{[\ln(t^2+1)+1]\left(\frac{2t}{t^2+1}+1\right)-[\ln(t^2+1)+t]\left(\frac{2t}{t^2+1}\right)}{[\ln(t^2+1)+1]^2}=\frac{\frac{2t\ln(t^2+1)}{t^2+1}+\frac{2t}{t^2+1}+\ln(t^2+1)+1-\frac{2t\ln(t^2+1)}{t^2+1}-\frac{2t^2}{t^2+1}}{[\ln(t^2+1)+1]^2}$$

$$=\frac{\frac{2t-2t^2}{t^2+1}+\ln(t^2+1)+1}{[\ln(t^2+1)+1]^2}=\frac{2t-2t^2+(t^2+1)\ln(t^2+1)+t^2+1}{(t^2+1)[\ln(t^2+1)+1]^2}=\frac{-t^2+2t+1+(t^2+1)\ln(t^2+1)}{(t^2+1)[\ln(t^2+1)+1]^2}$$

45. We use the absolute value of x or of $g(x)$ in the derivatives for the natural logarithm because logarithms are defined only for positive values.

47. Answers will vary.

49. Use the derivative of $\ln x$.

$$\frac{d\ln[u(x)v(x)]}{dx}=\frac{1}{u(x)v(x)}\cdot\frac{d[u(x)v(x)]}{dx}$$

$$\frac{d\ln u(x)}{dx}=\frac{1}{u(x)}\cdot\frac{d[u(x)]}{dx}$$

$$\frac{d\ln v(x)}{dx}=\frac{1}{v(x)}\cdot\frac{d[v(x)]}{dx}$$

Then since $\ln[u(x)v(x)]=\ln u(x)+\ln v(x)$,

$$\frac{1}{u(x)v(x)}\cdot\frac{d[u(x)v(x)]}{dx}=\frac{1}{u(x)}\cdot\frac{d[u(x)]}{dx}+\frac{1}{v(x)}\cdot\frac{d[v(x)]}{dx}.$$

Multiply both sides of this equation by $u(x)v(x)$. Then

$$\frac{d[u(x)v(x)]}{dx}=v(x)\frac{d[u(x)]}{dx}+u(x)\frac{d[v(x)]}{dx}.$$

This is the product rule.

51. Graph the function $y=\ln x$. Sketch lines tangent to the graph at $x=\frac{1}{2},1,2,3,4$.

Estimate the slopes of the tangent lines at these points.

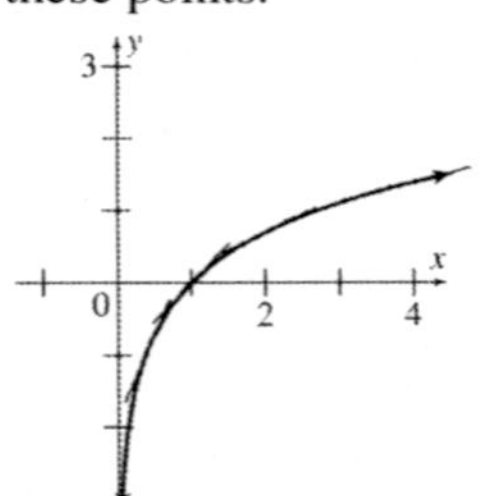

x	slope of tangent
$\frac{1}{2}$	2
1	1
2	$\frac{1}{2}$
3	$\frac{1}{3}$
4	$\frac{1}{4}$

The values of the slopes at x are $\dfrac{1}{x}$.

Thus we see that $\dfrac{d\ln x}{dx}=\dfrac{1}{x}$.

53. **(a)** $h(x)=u(x)^{v(x)}$

$$\frac{d}{dx}\ln h(x)=\frac{d}{dx}\ln[u(x)^{v(x)}]$$
$$=\frac{d}{dx}[v(x)\ln u(x)]$$
$$=v(x)\frac{d}{dx}\ln u(x)+(\ln u(x))v'(x)$$
$$=v(x)\frac{u'(x)}{u(x)}+(\ln u(x))v'(x)$$
$$=\frac{v(x)u'(x)}{u(x)}+(\ln u(x))v'(x)$$

(b) Since $\dfrac{d}{dx}\ln h(x)=\dfrac{h'(x)}{h(x)}$,

$$h'(x)=h(x)\frac{d}{dx}\ln h(x)$$
$$=h(x)\left[\frac{v(x)u'(x)}{u(x)}+(\ln u(x))v'(x)\right]$$
$$=u(x)^{v(x)}\left[\frac{v(x)u'(x)}{u(x)}+(\ln u(x))v'(x)\right]$$

55. $h(x) = (x^2+1)^{5x}$
$u(x) = x^2+1, u'(x) = 2x$
$v(x) = 5x, v'(x) = 5$

$$h'(x) = (x^2+1)^{5x}\left[\frac{5x(2x)}{x^2+1} + \ln(x^2+1)\cdot(5)\right]$$
$$= (x^2+1)^{5x}\left[\frac{10x^2}{x^2+1} + 5\ln(x^2+1)\right]$$

57. $M(t) = (0.1t+1)\ \ln\sqrt{t}$

(a) $M(15) = [0.1(15)+1]\ \ln\sqrt{15}$
≈ 3.385

When the temperature is 15°C, the number of matings is about 3.

(b) $M(25) = [0.1(25)+1]\ \ln\sqrt{25}$
≈ 5.663

When the temperature is 25°C, the number of matings is about 6.

(c)
$$M(t) = (0.1t+1)\ln\sqrt{t}$$
$$= (0.1t+1)\ln t^{1/2}$$
$$M'(t) = (0.1t+1)\left(\frac{1}{2}\cdot\frac{1}{t}\right) + (\ln t^{1/2})(0.1)$$
$$= 0.1\ln\sqrt{t} + \frac{1}{2t}(0.1t+1)$$
$$M'(15) = 0.1\ln\sqrt{15} + \frac{1}{2\cdot 15}[(0.1)(15)+1]$$
$$\approx 0.22$$

When the temperature is 15°C, the rate of change of the number of matings is about 0.22.

59. $A(w) = 4.688w^{0.8168-0.0154\log_{10} w}$

(a) $A(4000) = 4.688(4000)^{0.8168-0.0154\log_{10} 4000}$
$\approx 2590\ \text{cm}^2$

(b) $\dfrac{A(w)}{4.688} = w^{0.8166-0.0154\log_{10} w}$

$\ln A(w) - \ln 4.688$
$$= (\ln w)\left(0.8168 - 0.0154\frac{\ln w}{\ln 10}\right)$$

$$\frac{A'(w)}{A(w)} = \frac{1}{w}\left(0.8168 - 0.0154\frac{\ln w}{\ln 10}\right)$$
$$+ \ln w\left(\frac{-0.0154}{\ln 10}\right)\frac{1}{w}$$
$$= \frac{0.8168}{w} - \frac{0.0308}{\ln 10}\frac{\ln w}{w}$$
$$= \frac{1}{w}\left(0.8168 - \frac{0.0308}{\ln 10}\ln w\right)$$
$$A'(w) = \frac{1}{w}\left(0.8168 - \frac{0.0308}{\ln 10}\ln w\right)$$
$$\cdot\left(4.688w^{0.8168-0.0154\log_{10} w}\right)$$

$A'(4000) \approx 0.4571 \approx 0.46\ \text{g/cm}^2$

When the infant has a mass of 4000 g, it is gaining 0.46 square centimeters per gram of mass increase.

(c)

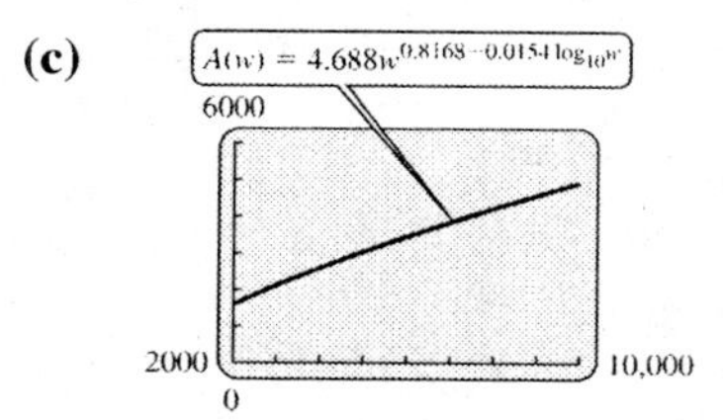

61. $F(x) = 0.774 + 0.727\log(x)$

(a) $F(25{,}000) = 0.774 + 0.727\log(25{,}000)$
$= 3.9713\ldots \approx 4$ kJ/day

(b)
$$F'(x) = 0.727\frac{1}{x\ln 10} = \frac{0.727}{\ln 10}x^{-1}$$
$$F'(25{,}000) = \frac{0.727}{\ln 10}25{,}000^{-1}$$
$$\approx 0.000012629\ldots$$
$$\approx 1.3\times 10^{-5}$$

When a fawn is 25 kg in size the rate of change of the energy expenditure of the fawn is about 1.3×10^{-5} kJ/day per gram.

(c)

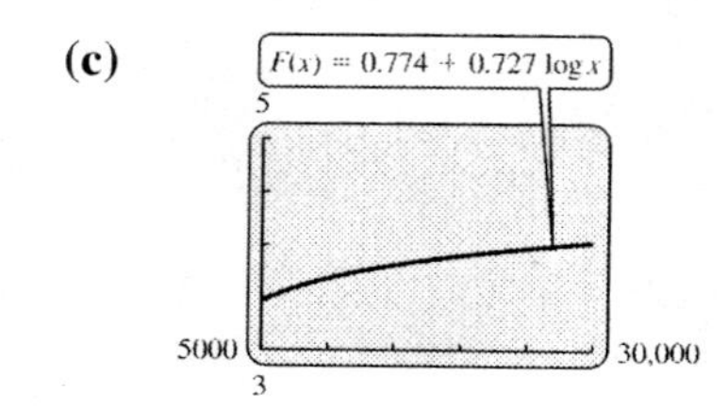

63. $J(p) = -\frac{3}{4}\ln\left(1 - \frac{4}{3}p\right)$

Use the chain rule with $u = 1 - \frac{4}{3}p$ and

$\frac{d}{du}(\log u) = \frac{1}{u}$.

$$J'(p) = -\frac{3}{4}\left(\frac{\frac{d}{dp}\left(1 - \frac{4}{3}p\right)}{1 - \frac{4}{3}p}\right) = -\frac{3}{4}\left(\frac{-\frac{4}{3}}{1 - \frac{4}{3}p}\right)$$
$$= \frac{1}{1 - \frac{4}{3}p}$$

(a) $J(0.03) = -\frac{3}{4}\ln\left(1 - \frac{4}{3}(0.03)\right) \approx 0.0306$

$J'(0.03) = \frac{1}{1 - \frac{4}{3}(0.03)} = \frac{25}{24} \approx 1.042$

The Jukes-Cantor distance is about 0.0306 and the rate that this distance is changing with respect to p is about 1.042.

(b) $J(0.10) = -\frac{3}{4}\ln\left(1 - \frac{4}{3}(0.10)\right) \approx 0.1073$

$J'(0.10) = \frac{1}{1 - \frac{4}{3}(0.10)} = \frac{15}{13} \approx 1.154$

The Jukes-Cantor distance is about 0.1073 and the rate that this distance is changing with respect to p is about 1.154.

65. $M = \frac{2}{3}\log\frac{E}{0.007}$

(a)
$$8.9 = \frac{2}{3}\log\frac{E}{0.007}$$
$$13.35 = \log\frac{E}{0.007}$$
$$10^{13.35} = \frac{E}{0.007}$$
$$E = 0.007(10^{13.35})$$
$$\approx 1.567 \times 10^{11} \text{ kWh}$$

(b) $10{,}000{,}000 \times 247$ kWh/month $= 2{,}470{,}000{,}000$ kWh/month

$$\frac{1.567 \times 10^{11} \text{ kWh}}{2{,}470{,}000{,}000 \text{ kWh/month}} \approx 63.4 \text{ months}$$

(c)
$$M = \frac{2}{3}\log E - \frac{2}{3}\log 0.007$$
$$\frac{dM}{dE} = \frac{2}{3}\left(\frac{1}{(\ln 10)E}\right) = \frac{2}{(3\ln 10)E}$$

When $E = 70{,}000$,

$$\frac{dM}{dE} = \frac{2}{(3\ln 10)70{,}000} \approx 4.14 \times 10^{-6}$$

(d) $\frac{dM}{dE}$ varies inversely with E, so as E increases, $\frac{dM}{dE}$ decreases and approaches zero.

4.6 Derivatives of Trigonometric Functions

1. $y = \frac{1}{2}\sin 8x$

$$\frac{dy}{dx} = \frac{1}{2}(\cos 8x)\cdot D_x(8x) = \frac{1}{2}(\cos 8x)\cdot 8$$
$$= 4\cos 8x$$

3. $y = 12\tan(9x+1)$

$$\frac{dy}{dx} = [12\sec^2(9x+1)\cdot D_x(9x+1)$$
$$= [12\sec(9x+1)]\cdot 9 = 108\sec^2(9x+1)$$

5. $y = \cos^4 x$

$$\frac{dy}{dx} = [4(\cos x)^3]\,D_x(\cos x)$$
$$= (4\cos^3 x)(-\sin x) = -4\sin x\cos^3 x$$

7. $y = \tan^8 x$

$$\frac{dy}{dx} = 8(\tan x)^7\cdot D_x(\tan x) = 8\tan^7 x\sec^2 x$$

9. $y = -6x\cdot\sin 2x$

$$\frac{dy}{dx} = -6x\cdot D_x(\sin 2x) + \sin 2x\cdot D_x(-6x)$$
$$= -6x(\cos 2x)\cdot D_x(2x) + (\sin 2x)(-6)$$
$$= -6x(\cos 2x)\cdot 2 - 6\sin 2x$$
$$= -12x\cos 2x - 6\sin 2x$$

11. $y = \frac{\csc x}{x}$

$$\frac{dy}{dx} = \frac{x\cdot D_x(\csc x) - (\csc x)\cdot D_x x}{x^2}$$
$$= \frac{-x\csc x\cot x - \csc x}{x^2}$$
$$= \frac{-(x\csc x\cot x + \csc x)}{x^2}$$

13. $y = \sin e^{4x}$

$$\frac{dy}{dx} = \cos e^{4x} \cdot D_x(e^{4x}) = (\cos e^{4x}) \cdot e^{4x} \cdot D_x(4x) = (\cos e^{4x}) \cdot e^{4x} \cdot 4 = 4e^{4x} \cos e^{4x}$$

15. $y = e^{\cos x}$

$$\frac{dy}{dx} = e^{\cos x} \cdot D_x(\cos x) = e^{\cos x} \cdot (-\sin x) = (-\sin x)e^{\cos x}$$

17. $y = \sin(\ln 3x^4)$

$$\frac{dy}{dx} = [\cos(\ln 3x^4)] \cdot D_x(\ln 3x^4) = \cos(\ln 3x^4) \cdot \frac{D_x(3x^4)}{3x^4}$$

$$= \cos(\ln 3x^4)\frac{12x^3}{3x^4} = \cos(\ln 3x^4) \cdot \frac{4}{x} = \frac{4}{x}\cos(\ln 3x^4)$$

19. $y = \ln|\sin x^2|$

$$\frac{dy}{dx} = \frac{D_x(\sin x^2)}{\sin x^2} = \frac{(\cos x^2) \cdot D_x(x^2)}{\sin x^2} = \frac{(\cos x^2) \cdot 2x}{\sin x^2} = \frac{2x \cos x^2}{\sin x^2} = 2x \cot x^2$$

21. $y = \dfrac{2\sin x}{3 - 2\sin x}$

$$\frac{dy}{dx} = \frac{(3-2\sin x)D_x(2\sin x) - (2\sin x) \cdot D_x(3 - 2\sin x)}{(3-2\sin x)^2} = \frac{(3-2\sin x) \cdot 2D_x(\sin x) - (2\sin x) \cdot [-2D_x(\sin x)]}{(3-2\sin x)^2}$$

$$= \frac{(3-2\sin x) \cdot 2\cos x - (2\sin x) \cdot (-2\cos x)}{(3-2\sin x)^2} = \frac{6\cos x - 4\sin x\cos x + 4\sin x\cos x}{(3-2\sin x)^2} = \frac{6\cos x}{(3-2\sin x)^2}$$

23. $y = \sqrt{\dfrac{\sin x}{\sin 3x}} = \left(\dfrac{\sin x}{\sin 3x}\right)^{1/2}$

$$\frac{dy}{dx} = \frac{1}{2}\left(\frac{\sin x}{\sin 3x}\right)^{-1/2} \cdot D_x\left(\frac{\sin x}{\sin 3x}\right) = \frac{1}{2}\left(\frac{\sin 3x}{\sin x}\right)^{1/2} \cdot \left[\frac{(\sin 3x) \cdot D_x(\sin x) - (\sin x) \cdot D_x(\sin 3x)}{(\sin 3x)^2}\right]$$

$$= \frac{1}{2}\left(\frac{\sin 3x}{\sin x}\right)^{1/2} \cdot \left[\frac{(\sin 3x)(\cos x) - (\sin x)(\cos 3x) \cdot D_x(3x)}{\sin^2 3x}\right]$$

$$= \frac{1(\sin 3x)^{1/2}}{2(\sin x)^{1/2}} \cdot \frac{\sin 3x \cos x - 3\sin x \cos 3x}{\sin^2 3x} = \frac{(\sin 3x)^{1/2}(\sin 3x \cos x - 3\sin x \cos 3x)}{2(\sin x)^{1/2}(\sin^2 3x)}$$

$$= \frac{\sqrt{\sin 3x}\,[\sin 3x \cos x - 3\sin x \cos 3x]}{2\sqrt{\sin x}(\sin^2 3x)}$$

25. $y = 3\tan\left(\dfrac{1}{4}x\right) + 4\cot 2x - 5\csc x + e^{-2x}$

$$\frac{dy}{dx} = 3\sec^2\left(\frac{1}{4}x\right) \cdot D_x\left(\frac{1}{4}x\right) + 4(-\csc^2 2x) \cdot D_x(2x) - 5(-\csc x \cot x) + e^{-2x} \cdot D_x(-2x)$$

$$= 3\sec^2\left(\frac{1}{4}x\right) \cdot \frac{1}{4} - (4\csc^2 2x) \cdot 2 + 5\csc x \cot x + e^{-2x} \cdot (-2)$$

$$= \frac{3}{4}\sec^2\left(\frac{1}{4}x\right) - 8\csc^2 2x + 5\csc x \cot x - 2e^{-2x}$$

27. $y = \sin x;\ x = 0$

Let $f(x) = \sin x$. Then $f'(x) = \cos x$, so $f'(0) = \cos 0 = 1$.

The slope of the tangent line to the graph of $y = \sin x$ at $x = 0$ is 1.

29. $y = \cos x;\ x = -\frac{5\pi}{6}$

Let $f(x) = \cos x$. Then $f'(x) = -\sin x$, so

$f'\left(-\frac{5\pi}{6}\right) = -\sin\left(-\frac{5\pi}{6}\right) = \frac{1}{2}.$

The slope of the tangent line to the graph of $y = \cos x$ at $x = -\frac{5\pi}{6}$ is $\frac{1}{2}$.

31. $y = \tan x;\ x = 0$

Let $f(x) = \tan x$. Then $f'(x) = \sec^2 x$, so

$f'(0) = \sec^2 0 = \frac{1}{\cos^2 0} = 1.$

The slope of the tangent line to the graph of $y = \tan x$ at $x = 0$ is 1.

33. $\cot x = \frac{\cos x}{\sin x}$ so, by using the quotient rule,

$$
\begin{aligned}
D_x(\cos x) &= D_x\left(\frac{\cos x}{\sin x}\right) \\
&= \frac{(\sin x)(-\sin x) - (\cos x)(\cos x)}{\sin^2 x} \\
&= \frac{-\sin^2 x - \cos^2 x}{\sin^2 x} \\
&= -\frac{\sin^2 x + \cos^2 x}{\sin^2 x} = -\frac{1}{\sin^2 x} \\
&= -\csc^2 x.
\end{aligned}
$$

35. Since $\csc x = \frac{1}{\sin x} = (\sin x)^{-1}$,

$$
\begin{aligned}
D_x(\csc x) &= D_x\left(\frac{1}{\sin x}\right) = D_x(\sin x)^{-1} \\
&= -1(\sin x)^{-2}\cos x = -\frac{\cos x}{\sin^2 x} \\
&= -\frac{1}{\sin x}\cdot\frac{\cos x}{\sin x} = -\csc x \cot x
\end{aligned}
$$

37. $L(t) = 0.022t^2 + 0.55t + 316 + 3.5\sin(2\pi t)$

(a)

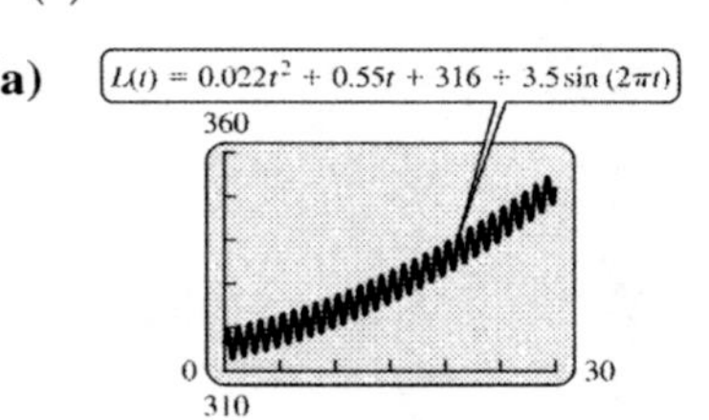

(b) $L(25) = 343.5$ parts per million
$L(35.5) \approx 363.25$ parts per million
$L(50.2) \approx 402.38$ parts per million

(c) $L'(t) = 0.044t + 0.55 + 7\pi\cos(2\pi t)$
$L'(50.2) \approx 9.55$ parts per million per year

This means that at the beginning of 2010, the level of carbon dioxide was increasing at the rate of 9.55 parts per million.

39. **(a)** $f(t) = 1000e^{2\sin(t)}$
$f(0.2) = 1000e^{2\sin(0.2)} \approx 1488$

(b) $f(t) = 1000e^{2\sin(t)}$
$f(1) = 1000e^{2\sin(1)} \approx 5381$

(c) $f'(t) = 2000\cos(t)e^{2\sin(t)}$
$f'(0) = 2000\cos(0)e^{2\sin(0)} = 2000$

(d) $f'(0.2) = 2000\cos(0.2)e^{2\sin(0.2)} \approx 2916$

(e)

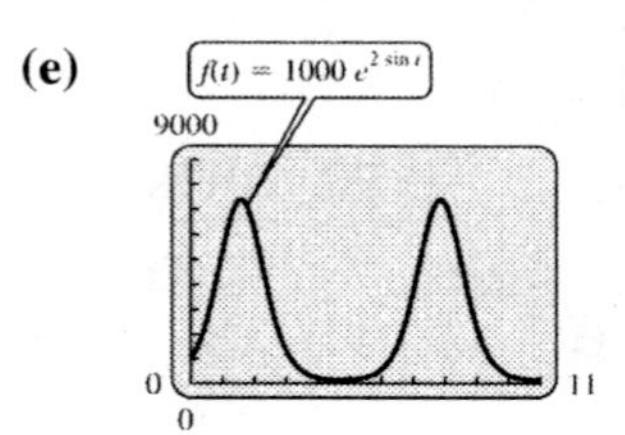

41. $s(t) = \sin t + 2\cos t$
$v(t) = s'(t) = \cos t - 2\sin t$

(a) $v(0) = 1 - 2(0) = 1$

(b) $v\left(\frac{\pi}{4}\right) = \frac{\sqrt{2}}{2} - 2\left(\frac{\sqrt{2}}{2}\right) = -\frac{\sqrt{2}}{2}$
≈ -0.7071

(c) $v\left(\frac{3\pi}{2}\right) = 0 - 2(-1) = 2$

Chapter 4 Review Exercises

1. False; the derivative of π^3 is 0 because π^3 is a constant.

2. True

3. False; the derivative of a product $u(x) \cdot v(x)$ is $u(x) \cdot v'(x) + v(x) \cdot u'(x)$.

4. False; the derivative of a quotient $\frac{u(x)}{v(x)}$ is $\frac{v(x) \cdot u'(x) - u(x) \cdot v'(x)}{[v(x)]^2}$.

5. False; the chain rule is used to take the derivative of a composition of functions.

6. False; the derivative ce^x is ce^x for any constant c.

7. False; the derivative of 10^x is $(\ln 10)10^x$.

8. True

9. True

10. False; the derivative of $\log x$ is $\frac{1}{(\ln 10)x}$, whereas the derivative of $\ln x$ is $\frac{1}{x}$.

11. False. $D_x \tan\left(x^2\right) = 2x\sec^2\left(x^2\right)$

12. True

13. $y = 5x^3 - 7x^2 - 9x + \sqrt{5}$
$$\frac{dy}{dx} = 5(3x^2) - 7(2x) - 9 + 0 = 15x^2 - 14x - 9$$

14. $y = 7x^3 - 4x^2 - 5x + \sqrt{2}$
$$\frac{dy}{dx} = 7(3x^2) - 4(2x) - 5 + 0 = 21x^2 - 8x - 5$$

15. $y = 9x^{8/3}$
$$\frac{dy}{dx} = 9\left(\frac{8}{3}x^{5/3}\right) = 24x^{5/3}$$

16. $y = -4x^{-3}$
$$\frac{dy}{dx} = -4(-3x^{-4}) = 12x^{-4} \text{ or } \frac{12}{x^4}$$

17. $f(x) = 3x^{-4} + 6\sqrt{x} = 3x^{-4} + 6x^{1/2}$
$$f'(x) = 3(-4x^{-5}) + 6\left(\frac{1}{2}x^{-1/2}\right) = -12x^{-5} + 3x^{-1/2} \text{ or } -\frac{12}{x^5} + \frac{3}{x^{1/2}}$$

18. $f(x) = 19x^{-1} - 8\sqrt{x} = 19x^{-1} - 8x^{1/2}$
$$f'(x) = 19(-x^{-2}) - 8\left(\frac{1}{2}x^{-1/2}\right) = -19x^{-2} - 4x^{-1/2} \text{ or } -\frac{19}{x^2} - \frac{4}{x^{1/2}}$$

19. $k(x) = \frac{3x}{4x+7}$
$$k'(x) = \frac{(4x+7)(3) - (3x)(4)}{(4x+7)^2} = \frac{12x + 21 - 12x}{(4x+7)^2} = \frac{21}{(4x+7)^2}$$

20. $r(x) = \frac{-8x}{2x+1}$
$$r'(x) = \frac{(2x+1)(-8) - (-8x)(2)}{(2x+1)^2} = \frac{-16x - 8 + 16x}{(2x+1)^2} = \frac{-8}{(2x+1)^2}$$

21. $y = \frac{x^2 - x + 1}{x - 1}$
$$\frac{dy}{dx} = \frac{(x-1)(2x-1) - (x^2 - x + 1)(1)}{(x-1)^2} = \frac{2x^2 - 3x + 1 - x^2 + x - 1}{(x-1)^2} = \frac{x^2 - 2x}{(x-1)^2}$$

22. $y = \frac{2x^3 - 5x^2}{x+2}$
$$\frac{dy}{dx} = \frac{(x+2)(6x^2 - 10x) - (2x^3 - 5x^2)(1)}{(x+2)^2} = \frac{6x^3 + 12x^2 - 10x^2 - 20x - 2x^3 + 5x^2}{(x+2)^2} = \frac{4x^3 + 7x^2 - 20x}{(x+2)^2}$$

23. $f(x) = (3x^2 - 2)^4$
$$f'(x) = 4(3x^2 - 2)^3[3(2x)] = 24x(3x^2 - 2)^3$$

24. $k(x) = (5x^3 - 1)^6$
$$k'(x) = 6(5x^3 - 1)^5[5(3x^2)]$$
$$= 90x^2(5x^3 - 1)^5$$

25. $y = \sqrt{2t^7 - 5} = (2t^7 - 5)^{1/2}$
$$\frac{dy}{dt} = \frac{1}{2}(2t^7 - 5)^{1/2}[2(7t^6)]$$
$$= 7t^6(2t^7 - 5)^{-1/2} \text{ or } \frac{7t^6}{(2t^7 - 5)^{1/2}}$$

26. $y = -3\sqrt{8t^4 - 1} = -3(8t^4 - 1)^{1/2}$
$$\frac{dy}{dx} = -3\left(\frac{1}{2}\right)(8t^4 - 1)^{-1/2}[8(4t^3)]$$
$$= -48t^3(8t^4 - 1)^{-1/2} \text{ or } \frac{-48t^3}{(8t^4 - 1)^{1/2}}$$

27. $y = 3x(2x + 1)^3$
$$\frac{dy}{dx} = 3x(3)(2x + 1)^2(2) + (2x + 1)^3(3)$$
$$= (18x)(2x + 1)^2 + 3(2x + 1)^3$$
$$= 3(2x + 1)^2[6x + (2x + 1)]$$
$$= 3(2x + 1)^2(8x + 1)$$

28. $y = 4x^2(3x - 2)^5$
$$\frac{dy}{dx} = (4x^2)[5(3x - 2)^4(3)] + (3x - 2)^5(8x)$$
$$= 60x^2(3x - 2)^4 + 8x(3x - 2)^5$$
$$= 4x(3x - 2)^4[15x + 2(3x - 2)]$$
$$= 4x(3x - 2)^4(15x + 6x - 4)$$
$$= 4x(3x - 2)^4(21x - 4)$$

29. $r(t) = \frac{5t^2 - 7t}{(3t + 1)^3}$
$$r'(t) = \frac{(3t + 1)^3(10t - 7) - (5t^2 - 7t)(3)(3t + 1)^2(3)}{[(3t + 1)^3]^2}$$
$$= \frac{(3t + 1)^3(10t - 7) - 9(5t^2 - 7t)(3t + 1)^2}{(3t + 1)^6}$$
$$= \frac{(3t + 1)(10t - 7) - 9(5t^2 - 7t)}{(3t + 1)^4}$$
$$= \frac{30t^2 - 11t - 7 - 45t^2 + 63t}{(3t + 1)^4}$$
$$= \frac{-15t^2 + 52t - 7}{(3t + 1)^4}$$

30. $s(t) = \frac{t^3 - 2t}{(4t - 3)^4}$
$$s'(t) = \frac{(4t - 3)^4(3t^2 - 2) - (t^3 - 2t)(4)(4t - 3)^3(4)}{[(4t - 3)^4]^2}$$
$$= \frac{(4t - 3)^4(3t^2 - 2) - 16(t^3 - 2t)(4t - 3)^3}{(4t - 3)^8}$$
$$= \frac{(4t - 3)^3[(4t - 3)(3t^2 - 2)16(t^3 - 2t)]}{(4t - 3)^8}$$
$$= \frac{(4t - 3)^3(12t^3 - 9t^2 - 8t + 6 - 16t^3 + 32t)}{(4t - 3)^8}$$
$$= \frac{-4t^3 - 9t^2 + 24t + 6}{(4t - 3)^5}$$

31. $p(t) = t^2(t^2 + 1)^{5/2}$
$$p'(t) = t^2 \cdot \frac{5}{2}(t^2 + 1)^{3/2} \cdot 2t + 2t(t^2 + 1)^{5/2}$$
$$= 5t^3(t^2 + 1)^{3/2} + 2t(t^2 + 1)^{5/2}$$
$$= t(t^2 + 1)^{3/2}[5t^2 + 2(t^2 + 1)^1]$$
$$= t(t^2 + 1)^{3/2}(7t^2 + 2)$$

32. $g(t) = t^3(t^4 + 5)^{7/2}$
$$g'(t) = t^3 \cdot \frac{7}{2}(t^4 + 5)^{5/2}(4t^3) + 3t^2 \cdot (t^4 + 5)^{7/2}$$
$$= 14t^6(t^4 + 5)^{5/2} + 3t^2(t^4 + 5)^{7/2}$$
$$= t^2(t^4 + 5)^{5/2}[14t^4 + 3(t^4 + 5)]$$
$$= t^2(t^4 + 5)^{5/2}(17t^4 + 15)$$

33. $y = -6e^{2x}$
$$\frac{dy}{dx} = -6(2e^{2x}) = -12e^{2x}$$

34. $y = 8e^{0.5x}$
$$\frac{dy}{dx} = 8(0.5e^{0.5x}) = 4e^{0.5x}$$

35. $y = e^{-2x^3}$
$$g(x) = -2x^3;\ g'(x) = -6x^2$$
$$y' = -6x^2e^{-2x^3}$$

36. $y = -4e^{x^2}$
$$g(x) = x^2;\ g'(x) = 2x$$
$$\frac{dy}{dx} = (2x)(-4e^{x^2}) = -8xe^{x^2}$$

37. $y = 5x \cdot e^{2x}$

Use the product rule.

$$\frac{dy}{dx} = 5x(2e^{2x}) + e^{2x}(5)$$
$$= 10xe^{2x} + 5e^{2x} = 5e^{2x}(2x+1)$$

38. $y = -7x^2e^{-3x}$

Use the product rule.

$$\frac{dy}{dx} = (-7x^2)(-3e^{-3x}) + e^{-3x}(-14x)$$
$$= 21x^2e^{-3x} - 14xe^{-3x} = 7xe^{-3x}(3x-2)$$

39. $y = \ln(2+x^2)$

$g(x) = 2x + x^2$

$g'(x) = 2x$

$$\frac{dy}{dx} = \frac{2x}{2+x^2}$$

40. $y = \ln(5x+3)$

$g(x) = 5x+3$

$g'(x) = 5$

$$\frac{dy}{dx} = \frac{5}{5x+3}$$

41. $y = \dfrac{\ln|3x|}{x-3}$

$$\frac{dy}{dx} = \frac{(x-3)\left(\frac{1}{3x}\right)(3) - (\ln|3x|)(1)}{(x-3)^2}$$
$$= \frac{(x-3)\left(\frac{1}{3x}\right)(3) - (\ln|3x|)(1)}{(x-3)^2} \cdot \frac{x}{x}$$
$$= \frac{x-3-x\ln|3x|}{x(x-3)^2}$$

42. $y = \dfrac{\ln|2x-1|}{x+3}$

$$\frac{dy}{dx} = \frac{(x+3)\left(\frac{2}{2x-1}\right) - (\ln|2x-1|)(1)}{(x+3)^2} \cdot \frac{2x-1}{2x-1}$$
$$= \frac{2(x+3) - (2x-1)\ln|2x-1|}{(2x-1)(x+3)^2}$$

43. $y = \dfrac{xe^x}{\ln(x^2-1)}$

$$\frac{dy}{dx} = \frac{\ln(x^2-1)[xe^x + e^x] - xe^x\left(\frac{1}{x^2-1}\right)(2x)}{[\ln(x^2-1)]^2}$$
$$= \frac{e^x(x+1)\ln(x^2-1) - \frac{2x^2e^x}{x^2-1}}{[\ln(x^2-1)]^2} \cdot \frac{x^2-1}{x^2-1}$$
$$= \frac{e^x(x+1)(x^2-1)\ln(x^2-1) - 2x^2e^x}{(x^2-1)[\ln(x^2-1)]^2}$$

44. $y = \dfrac{(x^2+1)e^{2x}}{\ln x}$

$$\frac{dy}{dx} = \frac{\left(\begin{array}{r}\ln x[(x^2+1)(2e^{2x}) + (e^{2x})(2x)] \\ -(x^2+1)e^{2x}\left(\frac{1}{x}\right)\end{array}\right)}{(\ln x)^2} \cdot \frac{x}{x}$$
$$= \frac{x\ln x[2e^{2x}(x^2+1) + 2xe^{2x}] - (x^2+1)e^{2x}}{x(\ln x)^2}$$
$$= \frac{e^{2x}[2x(\ln x)(x^2+1+x) - (x^2+1)]}{x(\ln x)^2}$$

45. $s = (t^2 + e^t)^2$

$s' = 2(t^2+e^t)(2t+e^t)$

46. $q = (e^{2p+1} - 2)^4$

$$\frac{dq}{dp} = 4(e^{2p+1}-2)^3[2e^{2p+1}]$$
$$= 8e^{2p+1}(e^{2p+1}-2)^3$$

47. $y = 3 \cdot 10^{-x^2}$

$$\frac{dy}{dx} = 3 \cdot (\ln 10)10^{-x^2}(-2x)$$
$$= -6x(\ln 10) \cdot 10^{-x^2}$$

48. $y = 10 \cdot 2^{\sqrt{x}}$

$$\frac{dy}{dx} = 10 \cdot (\ln 2) \cdot 2^{\sqrt{x}} \cdot \frac{1}{2}x^{-1/2}$$
$$= \frac{5(\ln 2)2^{\sqrt{x}}}{x^{1/2}}$$

49. $g(z) = \log_2(z^3 + z + 1)$

$$g'(z) = \frac{1}{\ln 2} \cdot \frac{3z^2+1}{z^3+z+1}$$
$$= \frac{3z^2+1}{(\ln 2)(z^3+z+1)}$$

50. $h(z) = \log(1+e^z)$

$$h'(z) = \frac{1}{\ln 10} \cdot \frac{e^z}{1+e^z} = \frac{e^z}{(\ln 10)(1+e^z)}$$

51. $f(x) = e^{2x}\ln(xe^x + 1)$

Use the product rule.

$$f'(x) = e^{2x}\left(\frac{xe^x + e^x}{xe^x+1}\right) + [\ln(xe^x+1)](2e^{2x})$$
$$= \frac{(x+1)e^{3x}}{xe^x+1} + 2e^{2x}\ln(xe^x+1)$$

52. $f(x) = \dfrac{e^{\sqrt{x}}}{\ln(\sqrt{x}+1)}$

Use the quotient rule.

$u(x) = e^{\sqrt{x}},\ u'(x) = \dfrac{e^{\sqrt{x}}}{2\sqrt{x}}$

$v(x) = \ln(\sqrt{x}+1), v'(x) = \dfrac{1}{2\sqrt{x}(\sqrt{x}+1)}$

$$f'(x) = \frac{[\ln(\sqrt{x}+1)]\left(\frac{e^{\sqrt{x}}}{2\sqrt{x}}\right) - e^{\sqrt{x}}\left(\frac{1}{2\sqrt{x}(\sqrt{x}+1)}\right)}{[\ln(\sqrt{x}+1)]^2}$$
$$= \frac{e^{\sqrt{x}}[(\sqrt{x}+1)\ln(\sqrt{x}+1)-1]}{2\sqrt{x}(\sqrt{x}+1)[\ln(\sqrt{x}+1)]^2}$$

53. $y = 2\tan 5x$

$$\frac{dy}{dx} = 2\sec^2 5x \cdot D_x(5x) = 10\sec^2 5x$$

54. $y = -4\sin 7x$

$$\frac{dy}{dx} = -4(\cos 7x)\cdot D_x(7x)$$
$$= -4(\cos 7x)7 = -28\cos 7x$$

55. $y = \cot(6-3x^2)$

$$\frac{dy}{dx} = [-\csc^2(6-3x^2)]\cdot D_x(6-3x^2)$$
$$= [-\csc^2(6-3x^2)]\cdot(-6x)$$
$$= 6x\csc^2(6-3x^2)$$

56. $y = \tan(4x^2+3)$

$$\frac{dy}{dx} = \sec^2(4x^2+3)\cdot D_x(4x^2+3)$$
$$= \sec^2(4x^2+3)\cdot(8x)$$
$$= 8x\sec^2(4x^2+3)$$

57. $y = 2\sin^4(4x^2)$

$$\frac{dy}{dx} = [8\sin^3(4x^2)]\cdot D_x[\sin(4x^2)]$$
$$= 8\sin^3(4x^2)\cdot\cos(4x^2)\cdot D_x(4x^2)$$
$$= 64x\sin^3(4x^2)\cos(4x^2)$$

58. $y = 2\cos^5 x$

$$\frac{dy}{dx} = 2D_x(\cos x)^5$$
$$= [2\cdot 5(\cos x)^4]\cdot D_x(\cos x)$$
$$= 10(\cos x)^4(-\sin x) = -10\sin x\cos^4 x$$

59. $y = \cos(1+x^2)$

$$\frac{dy}{dx} = [-\sin(1+x^2)]\cdot D_x(1+x^2)$$
$$= -2x\sin(1+x^2)$$

60. $y = \cot\left(\frac{1}{2}x^4\right)$

$$\frac{dy}{dx} = -\csc^2\left(\frac{1}{2}x^4\right)\cdot Dx\left(\frac{1}{2}x^4\right)$$
$$= -\csc^2\left(\frac{1}{2}x^4\right)\cdot 2x^3$$
$$= -2x^3\csc^2\left(\frac{1}{2}x^4\right)$$

61. $y = e^{-2x}\sin x$

$$\frac{dy}{dx} = e^{-2x}\cdot D_x(\sin x) + \sin x\cdot D_x(e^{-2x})$$
$$= e^{-2x}(\cos x) + (\sin x)(e^{-2x})\cdot D_x(-2x)$$
$$= e^{-2x}(\cos x) + (\sin x)(e^{-2x})(-2)$$
$$= e^{-2x}(\cos x - 2\sin x)$$

62. $y = x^2\csc x$

$$\frac{dy}{dx} = x^2 D_x(\csc x) + \csc x\, D_x(x^2)$$
$$= x^2(-\csc x\cot x) + \csc x(2x)$$
$$= -x^2\csc x\cot x + 2x\csc x$$

63. $y = \dfrac{\cos^2 x}{1-\cos x}$

$$\frac{dy}{dx} = \frac{(1-\cos x)(-2\cos x\sin x) - (\cos^2 x)(\sin x)}{(1-\cos x)^2}$$
$$= \frac{-2\cos x\sin x + \cos^2 x\sin x}{(1-\cos x)^2}$$

64. $y = \dfrac{\sin x - 1}{\sin x + 1}$

$$\frac{dy}{dx} = \frac{(\sin x+1)(\cos x) - (\sin x-1)(\cos x)}{(\sin x+1)^2}$$
$$= \frac{\sin x\cos x + \cos x - \sin x\cos x + \cos x}{(\sin x+1)^2}$$
$$= \frac{2\cos x}{(\sin x+1)^2}$$

65. $y = \frac{\tan x}{1+x}$

$$\frac{dy}{dx} = \frac{(1+x)(\sec^2 x) - (\tan x)(1)}{(1+x)^2} = \frac{\sec^2 x + x\sec^2 x - \tan x}{(1+x)^2}$$

66. $y = \frac{6-x}{\sec x}$

$$\frac{dy}{dx} = \frac{(\sec x)\cdot D_x(6-x) - (6-x)\cdot D_x(\sec x)}{\sec^2 x}$$
$$= \frac{(\sec x)\cdot(-1) - (6-x)\cdot(\sec x \tan x)}{\sec^2 x}$$
$$= \frac{\sec x[-1-(6-x)\tan x]}{\sec^2 x}$$
$$= \frac{1+(6-x)\tan x}{-\sec x} \text{ or } (x-6)\sin x - \cos x$$

67. $y = \ln|5\sin x|$

$$\frac{dy}{dx} = \frac{1}{5\sin x}\cdot D_x(5\sin x) = \frac{\cos x}{\sin x} = \cot x$$

68. $y = \ln|\cos x|$

$$\frac{dy}{dx} = \frac{D_x(\cos x)}{\cos x} = \frac{-\sin x}{\cos x} = -\tan x$$

69. (a) $D_x(f[g(x)])$ at $x = 2$

$$= f'[g(2)]g'(2)$$
$$= f'(1)\left(\frac{3}{10}\right) = -5\left(\frac{3}{10}\right) = -\frac{3}{2}$$

(b) $D_x(f[g(x)])$ at $x = 3$

$$= f'[g(3)]g'(3)$$
$$= f'(2)\left(\frac{4}{11}\right) = -6\left(\frac{4}{11}\right) = -\frac{24}{11}$$

70. (a) $D_x(g[f(x)])$ at $x = 2$

$$= g'[f(2)]f'(2)$$
$$= g'(4)(-6) = \frac{6}{13}(-6) = -\frac{36}{13}$$

(b) $D_x(g[f(x)])$ at $x = 3$

$$= g'[f(3)]f'(3)$$
$$= g'(2)(-7) = \frac{3}{10}(-7) = -\frac{21}{10}$$

71. $y = x^2 - 6x$; tangent at $x = 2$

$$\frac{dy}{dx} = 2x - 6$$

Slope $= y'(2) = 2(2) - 6 = -2$

Use (2, –8) and point-slope form.

$$y - (-8) = -2(x-2)$$
$$y + 8 = -2x + 4$$
$$y = -2x - 4$$

72. $y = 8 - x^2$; tangent at $x = 1$

$$y = 8 - x^2$$
$$\frac{dy}{dx} = -2x$$

slope $= y'(1) = -2(1) = -2$

Use (1, 7) and $m = -2$ and the point-slope form.

$$y - 7 = -2(x-1)$$
$$y - 7 = -2x + 2$$
$$y = -2x + 9$$

73. $y = \frac{3}{x-1}$; tangent at $x = -1$

$$y = \frac{3}{x-1} = 3(x-1)^{-1}$$
$$\frac{dy}{dx} = 3(-1)(x-1)^{-2}(1) = -3(x-1)^{-2}$$

Slope $= y'(-1) = -3(-1-1)^{-2} = -\frac{3}{4}$

Use $\left(-1, -\frac{3}{2}\right)$ and point-slope form.

$$y - \left(-\frac{3}{2}\right) = -\frac{3}{4}[x-(-1)]$$
$$y + \frac{3}{2} = -\frac{3}{4}(x+1)$$
$$y + \frac{6}{4} = -\frac{3}{4}x - \frac{3}{4}$$
$$y = -\frac{3}{4}x - \frac{9}{4}$$

74. $y = \frac{x}{x^2-1}$; tangent at $x = 2$

$$\frac{dy}{dx} = \frac{(x^2-1)\cdot 1 - x(2x)}{(x^2-1)^2} = \frac{-x^2-1}{(x^2-1)^2}$$

The value of $\frac{dy}{dx}$ when $x = 2$ is the slope.

$$m = \frac{-(2^2)-1}{(2^2-1)^2} = \frac{-5}{9} = -\frac{5}{9}$$

(*continued on next page*)

(continued)

When $x = 2$, $y = \frac{2}{4-1} = \frac{2}{3}$.

Use $m = -\frac{5}{9}$ with $P\left(2, \frac{2}{3}\right)$.

$$y - \frac{2}{3} = -\frac{5}{9}(x-2)$$
$$y - \frac{6}{9} = -\frac{5}{9}x + \frac{10}{9}$$
$$y = -\frac{5}{9}x + \frac{16}{9}$$

75. $y = \sqrt{6x-2}$; tangent at $x = 3$

$$y = \sqrt{6x-2} = (6x-2)^{1/2}$$
$$\frac{dy}{dx} = \frac{1}{2}(6x-2)^{-1/2}(6)$$
$$= 3(6x-2)^{-1/2}$$

$$\text{slope} = y'(3) = 3(6 \cdot 3 - 2)^{-1/2} = 3(16)^{-1/2}$$
$$= \frac{3}{16^{1/2}} = \frac{3}{4}$$

Use (3, 4) and point-slope form.

$$y - 4 = \frac{3}{4}(x-3)$$
$$y - \frac{16}{4} = \frac{3}{4}x - \frac{9}{4}$$
$$y = \frac{3}{4}x + \frac{7}{4}$$

76. $y = -\sqrt{8x+1}$; tangent $x = 3$

$$y = -(8x+1)^{1/2}$$
$$\frac{dy}{dx} = -\frac{1}{2}(8x+1)^{-1/2}(8) = -\frac{4}{(8x+1)^{1/2}}$$

The value of $\frac{dy}{dx}$ when $x = 3$ is the slope.

$$m = -\frac{4}{(24+1)^{1/2}} = -\frac{4}{5}$$

When $x = 3$, $y = -\sqrt{24+1} = -5$.

Use $m = -\frac{4}{5}$ with $P(3, -5)$.

$$y + 5 = -\frac{4}{5}(x-3)$$
$$y + \frac{25}{5} = -\frac{4}{5}x + \frac{12}{5}$$
$$y = -\frac{4}{5}x - \frac{13}{5}$$

77. $y = e^x$; tangent at $x = 0$

$$\frac{dy}{dx} = e^x$$

The value of $\frac{dy}{dx}$ when $x = 0$ is the slope.

$$m = e^0 = 1$$

When $x = 0$, $y = e^0 = 1$. Use $m = 1$ with $P(0, 1)$.

$$y - 1 = 1(x - 0)$$
$$y = x + 1$$

78. $y = xe^x$; tangent at $x = 1$

$$\frac{dy}{dx} = xe^x + 1 \cdot e^x \quad = e^x(x+1)$$

The value of $\frac{dy}{dx}$ when $x = 1$ is the slope.

$$m = e^1(1+1) = 2e$$

When $x = 1$, $y = 1e^1 = e$.

Use $m = 2e$ with $P(1, e)$.

$$y - e = 2e(x-1)$$
$$y = 2ex - e$$

79. $y = \ln x$; tangent at $x = 1$

$$\frac{dy}{dx} = \frac{1}{x}$$

The value of $\frac{dy}{dx}$ when $x = 1$ is the slope

$$m = \frac{1}{1} = 1.$$

When $x = 1$, $y = \ln 1 = 0$.

Use $m = 1$ with $P(1, 0)$.

$$y - 0 = 1(x-1)$$
$$y = x - 1$$

80. $y = x \ln x$; $x = e$

$$\frac{dy}{dx} = x \cdot \frac{1}{x} + 1 \cdot \ln x = 1 + \ln x$$

The value of $\frac{dy}{dx}$ when $x = e$ is the slope.

$$m = 1 + \ln e = 1 + 1 = 2$$

When $x = e$, $y = e \ln e = e \cdot 1 = e$.

Use $m = 2$ with $P(e, e)$.

$$y - e = 2(x - e)$$
$$y = 2x - e$$

81. $y = x\cos x$; tangent at $x = 0$

$$\frac{dy}{dx} = \cos x(1) + x(-\sin x)$$
$$= \cos x - x\sin x$$

Slope $= y'(0) = \cos 0 - 0\sin 0 = 1$

At $x = 0$, $y = x\cos x = 0$.

Use (0, 0) and point-slope form to find the equation of the tangent.

$$y - 0 = 1(x - 0)$$
$$y = x$$

82. $y = \tan(\pi x)$; tangent at $x = 1$

$$\frac{dy}{dx} = \pi\sec^2(\pi x)$$

Slope $= y'(1) = \pi\sec^2(\pi) = \pi(-1)^2 = \pi$

At $x = 1$, $y = \tan(\pi x) = \tan\pi = 0$

Use (1, 0) and point-slope form to find the equation of the tangent.

$$y - 0 = \pi(x - 1)$$
$$y = \pi x - \pi$$

83. The slope of the graph of $y = x + k$ is 1. First, we find the point on the graph of $f(x) = \sqrt{2x-1}$ at which the slope is also 1.

$$f(x) = (2x-1)^{1/2}$$
$$f'(x) = \frac{1}{2}(2x-1)^{-1/2}(2) = \frac{1}{\sqrt{2x-1}}$$

The slope is 1 when $\frac{1}{\sqrt{2x-1}} = 1$.

$$\frac{1}{\sqrt{2x-1}} = 1 \Rightarrow 1 = \sqrt{2x-1} \Rightarrow 1 = 2x - 1 \Rightarrow$$
$$2x = 2 \Rightarrow x = 1$$

$f(1) = 1$, so, at $P(1,1)$ on the graph of $f(x) = \sqrt{2x-1}$, the slope is 1.

An equation of the tangent line is

$$y - 1 = 1(x - 1) \Rightarrow y - 1 = x - 1 \Rightarrow y = x + 0$$

Any tangent line intersects the curve in exactly one point. From this we see that if $k = 0$, there is one point of intersection.

The graph of f is below the line $y = x + 0$. Therefore, if $k > 0$, the graph of $y = x + k$ will not intersect the graph.

Consider the point $Q\left(\frac{1}{2}, 0\right)$ on the graph. We find an equation of the line through Q with slope 1.

$$y - 0 = 1\left(x - \frac{1}{2}\right) \Rightarrow y = x - \frac{1}{2}$$

The line with a slope of 1 through $Q\left(\frac{1}{2}, 0\right)$ will intersect the graph in two points. One is Q and the other is some point on the graph to the right of Q.

The graph of $y = x + 0$ intersects the graph in one point, while the graph of $y = x - \frac{1}{2}$ intersects it in two points. If we use a value of k in $y = x + k$ with $-\frac{1}{2} < k < 0$, we will have a line with a y-intercept between $-\frac{1}{2}$ and a 0 and a slope of 1 which will intersect the graph in two points.

If k, the y-intercept, is less than $-\frac{1}{2}$, the graph of $y = x + k$ will be below point Q and will intersect the graph of f in exactly one point.

To summarize, the graph of $y = x + k$ will intersect the graph of $f(x) = \sqrt{2x-1}$ in

(1) no points if $k > 0$;

(2) exactly one point if $k = 0$ or if $k < -\frac{1}{2}$;

(3) exactly two points if $-\frac{1}{2} \le k < 0$

84. (a) Use the chain rule. Let $g(x) = \ln x$. Then $g'(x) = \frac{1}{x}$.

Let $y = g[f(x)]$. Then

$$\frac{dy}{dx} = g'[f(x)] \cdot f'(x).$$
$$\frac{d\ln f(x)}{dx} = \frac{1}{f(x)} \cdot f'(x) = \frac{f'(x)}{f(x)}.$$

(b) $\hat{f} = \frac{f'(x)}{f(x)}$, $\hat{g} = f\frac{g'(x)}{g(x)}$

$$\widehat{fg} = \frac{(fg)'(x)}{(fg)(x)}$$
$$= \frac{f(x)\cdot g'(x) + g(x)\cdot f'(x)}{f(x)g(x)}$$
$$= \frac{f(x)\cdot g'(x)}{f(x)\cdot g(x)} + \frac{g(x)\cdot f'(x)}{f(x)\cdot g(x)}$$
$$= \frac{g'(x)}{g(x)} + \frac{f'(x)}{f(x)}$$
$$= \hat{g} + \hat{f} \text{ or } \hat{f} + \hat{g}$$

(c) Using the result $\widehat{fg} = \hat{f} + \hat{g}$, the national income goes up by approximately $1\% + 2\% = 3\%$. Let P = population before the increase and I the average income per person before the increase. Then the new population is $1.01P$ and the new average income is $1.02S$, so the total national income is $(1.01P)(1.02I) = 1.0302PIS$, which is an increase of 3.02%.

85. Using the result $\widehat{fg} = \hat{f} + \hat{g}$, the total amount of tuition collected goes up by approximately $3\% + 2\% = 5\%$. Let T = tuition per person before the increase and S the number of students before the increase. Then the new tuition is $1.03T$ and the new number of students is $1.02S$, so the total amount of tuition collected is $(1.03T)(1.02S) = 1.0506TS$, which is an increase of 5.06%.

86. Answers will vary.

87. $c(x) = 0.19x$

(a) $c(20) = 0.19(20) = 3.8$ fish per hour

In two hours, an angle will catch $2(3.8) = 7.6$ fish.

(b) $c'(x) = 0.19$

The rate of change is constant. The estimated number of fish an angle can catch in an hour increases by 0.19 fish per hour as the density of fish per acre increases.

88. $W(x) = 570 + 5.6(x - 190),\ x \geq 170$

(a) $W(200) = 570 + 5.6(200 - 190) = 626$ kg

(b) Answers will vary. Sample answer: Form a tape measure that has centimeters on one side and the corresponding weight (in kilograms on the other.

(c) Answers will vary. Sample answer: The slope of the line $W(x)$ is 5.6.

$W'(x) = 5.6$, indicating that the rate of change in weight is constant.

89. $T(n) = 0.25 + \left(1.93\times10^{-2}\right)n - \left(4.78\times10^{-5}\right)n^2$

(a)

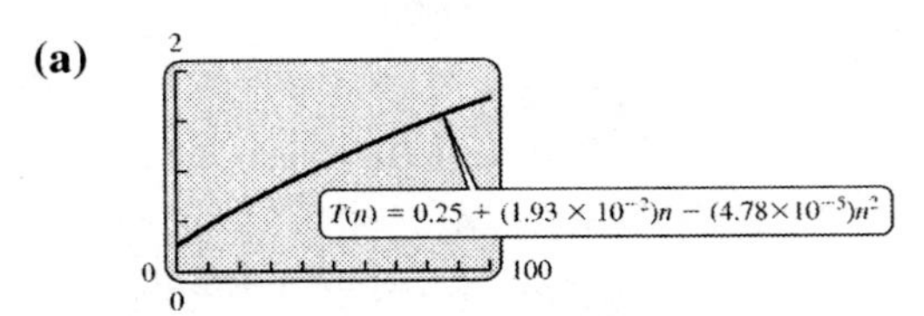

(b) Answers will vary. Sample answer: 201 fish. At this point, $T(n)$ has already begun to decrease with respect to n.

(c)
$$T'(n) = \left(1.93\times10^{-2}\right) - 2\left(4.78\times10^{-5}\right)n = \left(1.93\times10^{-2}\right) - \left(9.56\times10^{-5}\right)n$$
$$T'(40) = \left(1.93\times10^{-2}\right) - \left(9.56\times10^{-5}\right)(40) = 0.015476$$

This means that when 40 fish are caught in the net, the processing time is increasing by about 0.015 hour per additional fish caught.

90. $P(t) = ae^{0.05t}$

$P'(t) = ae^{0.05t}(0.05) = P(t)(0.05)$

If $P(t) = 1{,}000{,}000$, then

$P'(t) = 1{,}000{,}000(0.05) = 50{,}000.$

The population is growing at a rate of 50,000 per year.

91. $G(t) = \dfrac{m}{1 + \left(\dfrac{m}{G_0} - 1\right)e^{-kmt}}$, where

$m = 30{,}000,\ G_0 = 2000,$ and $k = 5\times10^{-6}.$

(a)
$$G(t) = \frac{30{,}000}{1 + \left(\frac{30{,}000}{2000} - 1\right)e^{-5\times10^{-6}(30{,}000t)}} = \frac{30{,}000}{1 + 14e^{-0.15t}}$$

(b)
$$G(t) = 30{,}000(1 + 14e^{-0.15t})^{-1}$$
$$G(6) = 30{,}000(1 + 14e^{-0.90})^{-1} \approx 4483$$
$$G'(t) = -30{,}000(1 + 14e^{-0.15t})^{-2} \cdot (-2.1e^{-0.15t}) = \frac{63{,}000e^{-0.15t}}{(1 + 14e^{-0.15t})^2}$$
$$G'(6) = \frac{63{,}000e^{-0.90}}{(1 + 14e^{-0.90})^2} \approx 572$$

The population is about 4483, and the rate of growth is about 572 people per year.

92. $L(t) = 71.5(1 - e^{-0.1t})$ and
$W(L) = 0.01289 \cdot L^{2.9}$

(a) $L(5) = 71.5(1 - e^{-0.5}) \approx 28.1$
The approximate length of a 5-year-old monkey-face is 28.1 cm.

(b) $L'(t) = 71.5(0.1e^{-0.1t})$
$L'(5) = 71.5(0.1e^{-0.5}) \approx 4.34$
The length is growing by about 4.34 cm/year.

(c) $W[L(5)] \approx 0.01289(28.1)^{2.9} \approx 205$
The approximate weight is 205 grams.

(d)
$$W'(L) = 0.01289(2.9)L^{1.9} = 0.037381L^{1.9}$$
$W'[L(5)] \approx 0.037381(28.1)^{1.9} \approx 21.2$
The rate of change of the weight with respect to length is 21.2 grams/cm.

(e) $\frac{dW}{dt} = \frac{dW}{dL} \cdot \frac{dL}{dt} \approx (21.2)(4.34) \approx 92.0$
The weight is growing at about 92.0 grams/year.

93. $M(t) = 3583e^{-e^{-0.020(t-66)}}$

(a) $M(250) = 3583e^{-e^{-0.020(250-66)}} \approx 3493.76$ grams,
or about 3.5 kilograms

(b) As $t \to \infty$, $-e^{-0.020(t-66)} \to 0$,
$e^{-e^{-0.020(t-66)}} \to 1$, and $M(t) \to 3583$
grams or about 3.6 kilograms.

(c) 50% of 3583 is 1791.5.
$$1791.5 = 3583e^{-e^{-0.020(t-66)}}$$
$$\ln\left(\frac{1791.5}{3583}\right) = -e^{-0.020(t-66)}$$
$$\ln\left(\ln\frac{3583}{1791.5}\right) = -0.020(t-66)$$
$$t = -\frac{1}{0.020}\ln\left(\ln\frac{3583}{1791.5}\right) + 66 \approx 84 \text{ days}$$

(d)
$$D_tM(t) = 3583e^{-e^{-0.020(t-66)}}D_t\left(-e^{-0.020(t-66)}\right)$$
$$= 3583e^{-e^{-0.020(t-56)}}\left(-e^{-0.020(t-66)}\right)\cdot(-0.020)$$
$$= 71.66e^{-e^{-0.020(t-66)}}\left(e^{-0.020(t-66)}\right)$$
When $t = 250$, $D_tM(t) \approx 1.76$ g/day.

(e)

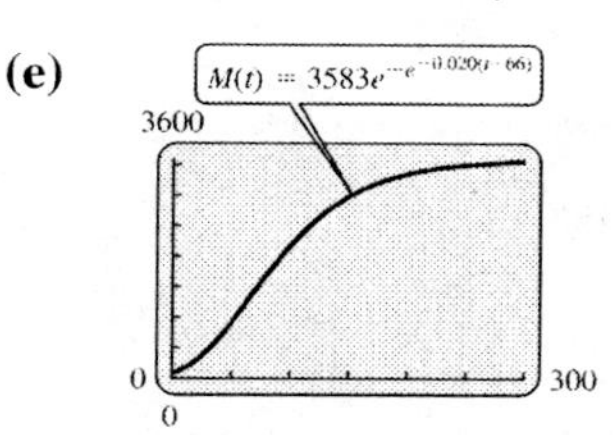

Growth is initially rapid, then tapers off.

(f)

Day	Weight	Rate
50	904	24.90
100	2159	21.87
150	2974	11.08
200	3346	4.59
250	3494	1.76
300	3550	0.66

94. $a(t) = 11.14(1.023)^t$
$a'(t) = 11.14(\ln 1.023)(1.023)^t$

(a) For 2005, $t = 5$.
$a'(5) = 11.14(\ln 1.023)(1.023)^5 \approx 0.284$
In 2005 the instantaneous rate of change is 284,000 per year.

(b) For 2025, $t = 25$.
$a'(25) = 11.14(\ln 1.023)(1.023)^{25} \approx 0.447$
In 2025 the instantaneous rate of change will be 447,000 per year.

95. $A(r) = 1000\left(1 + \frac{r}{400}\right)^{48}$
$$A'(r) = 1000 \cdot 48\left(1 + \frac{r}{400}\right)^{47} \cdot \frac{1}{400} = 120\left(1 + \frac{r}{400}\right)^{47}$$
$$A'(5) = 120\left(1 + \frac{5}{400}\right)^{47} \approx 215.15$$
The balance increases by approximately \$215.15 for every 1% increase in the interest rate when the rate is 5%.

96. $A(r) = 1000e^{12r/100}$

$A'(r) = 1000e^{12r/100} \cdot \frac{12}{100} = 120e^{12r/100}$

$A'(5) = 120e^{0.6} \approx 218.65$

The balance increases by approximately \$218.65 for every 1% increase in the interest rate when the rate is 5%.

97. $C(t) = 19{,}370(1.2557)^t$

$C'(t) = 19{,}370(\ln 1.2557)(1.2557)^t$

(a) For 2005, $t = 5$.

$C'(5) = 19{,}370(\ln 1.2557)(1.2557)^5 \approx 13{,}769$

In 2005 the rate of change in the energy capacity is about 13,769 megawatts per year.

(b) For 2010, $t = 10$.

$C'(10) = 19{,}370(\ln 1.2557)(1.2557)^{10} \approx 42{,}987$

In 2010 the rate of change in the energy capacity is 42,987 megawatts per year.

(c) For 2015, $t = 15$.

$C'(15) = 19{,}370(\ln 1.2557)(1.2557)^{15} \approx 134{,}205$

In 2015 the rate of change in the energy capacity is 134,205 megawatts per year.

98. $f(t) = \frac{8}{t+1} + \frac{20}{t^2+1}$

(a) average velocity $= \frac{f(3) - f(1)}{3-1} = \frac{\left(\frac{8}{4} + \frac{20}{10}\right) - \left(\frac{8}{2} + \frac{20}{2}\right)}{2} = \frac{4-14}{2} = -5$

Belmar's average velocity between 1 sec and 3 sec is –5 ft/sec.

(b) $f(t) = 8(t+1)^{-1} + 20(t^2+1)^{-1}$

$f'(t) = -8(t+1)^{-2} \cdot 1 - 20(t^2+1)^{-2} \cdot 2t = -\frac{8}{(t+1)^2} - \frac{40t}{(t^2+1)^2}$

$f'(3) = -\frac{8}{16} - \frac{120}{100} = -0.5 - 1.2 = -1.7$

Belmar's instantaneous velocity at 3 sec is –1.7 ft/sec.

99. $p(x) = 1.757(1.0248)^{x-1930}$

$p'(x) = 1.757(\ln 1.0248)(1.0248)^{x-1930}$

$p'(2000) = 1.757(\ln 1.0248)(1.0248)^{2010-1930} \approx 0.306$

The production of corn is increasing at a rate of 0.306 billion bushels per year in 2010.

100. (a) $N(t) = N_0 e^{-0.217t}$, where $t = 1$ and $N_0 = 210$

$N(1) = 210e^{-0.217(1)} \approx 169$

The number of words predicted to be in use in 1950 is 169, and the actual number in use was 167.

(b) $N(1.1) = 210e^{-0.217(1.1)} \approx 165$

In 2050, there will be about 165 words still being used.

(c) $N(t) = 210e^{-0.217t}$

$N'(t) = 210e^{-0.217t} \cdot (-0.217) = -45.57e^{-0.217t}$

$N'(1.1) = -45.57e^{-0.217(1.1)} \approx -36$

In the year 2050 the number of words in use will be decreasing by 36 words per millennium.

101. $f(x) = k(x-49)^6 + 0.8$

$f'(x) = k \cdot 6(x-49)^5 = (3.8 \times 10^{-9})(6)(x-49)^5 = (2.28 \times 10^{-8})(x-49)^5$

(a) $f'(20) = (2.28 \times 10^{-8})(20-49)^5 \approx -0.4677$

At the age of 20, each extra year results in a decrease of 0.4677 fatalities per 1000 licensed drivers per 100 million miles.

(b) $f'(60) = (2.28 \times 10^{-8})(60-49)^5 \approx 0.003672$

At the age of 60, each extra year results in an increase of 0.003672 fatalities per 1000 licensed drivers per 100 million miles.

102. $y = x\tan\alpha - \dfrac{16x^2}{V^2}\sec^2\alpha + h$

(a) $y = 39\tan\dfrac{\pi}{24} - \dfrac{16(39)^2}{73^2}\sec^2\dfrac{\pi}{24} + 9 \approx 9.5$

Yes, the ball will make it over the net because the height of the ball is about 9.5 feet when x, the horizontal distance, is 39 feet.

(b) Let $Y_1 = 39\tan x - 16\left(\dfrac{39}{44\cos x}\right)^2 + 9$ and

$$Y_2 = \frac{\left(\begin{array}{l}44^2\sin x\cos x \\ \quad +44^2\cos^2 x\sqrt{\tan^2 x + \dfrac{576}{(44\cos x)^2}}\end{array}\right)}{32}$$

Using the table function, we see that the ball will clear the net and travel between 39 and 60 feet for $0.10 \le \alpha \le 0.41$ radians or $5.7° \le \alpha \le 24.6°$.

X	Y_1	Y_2
.09	-.1531	38.725
.1	.21626	39.39
.11	.5838	40.06
.39	10.337	58.479
.4	10.672	59.046
.41	11.006	59.602
.42	11.339	60.146
X=.09		

(c) Using Y_2 from part (b), we have

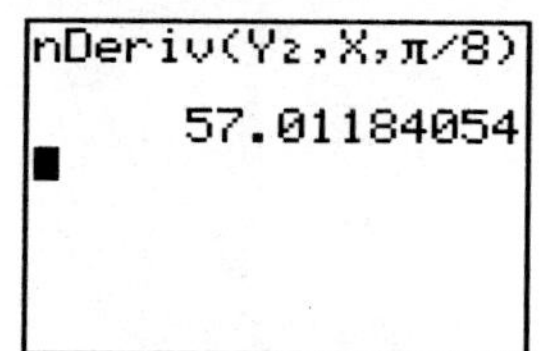

Note that 57.01 ft/radian ≈ 0.995 ft/degree. The distance the ball travels will increase by approximately 1 foot when the angle of the tennis racket is increased by one degree.

Chapter 5

GRAPHS AND THE DERIVATIVE

5.1 Increasing and Decreasing Functions

1. By reading the graph, f is

(a) increasing on $(1, \infty)$ and

(b) decreasing on $(-\infty, 1)$.

3. By reading the graph, g is

(a) increasing on $(-\infty, -2)$ and

(b) decreasing on $(-2, \infty)$.

5. By reading the graph, h is

(a) increasing on $(-\infty, -4)$ and $(-2, \infty)$ and

(b) decreasing on $(-4, -2)$.

7. By reading the graph, f is

(a) increasing on $(-7, -4)$ and $(-2, \infty)$ and

(b) decreasing on $(-\infty, -7)$ and $(-4, -2)$.

9. (a) Since the graph of $f'(x)$ is positive for $x < -1$ and $x > 3$, the intervals where $f(x)$ is increasing are $(-\infty, -1)$ and $(3, \infty)$.

(b) Since the graph of $f'(x)$ is negative for $-1 < x < 3$, the interval where $f(x)$ is decreasing is $(-1, 3)$.

11. (a) Since the graph of $f'(x)$ is positive for $x < -8$, $-6 < x < -2.5$ and $x > -1.5$, the intervals where $f(x)$ is increasing are $(-\infty, -8)$, $(-6, -2.5)$, and $(-1.5, \infty)$.

(b) Since the graph of $f'(x)$ is negative for $-8 < x < -6$ and $-2.5 < x < -1.5$, the intervals where $f(x)$ is decreasing are $(-8, -6)$ and $(-2.5, -1.5)$.

13. $y = 2.3 + 3.4x - 1.2x^2$

(a) $y' = 3.4 - 2.4x$

y' is zero when

$3.4 - 2.4x = 0 \Rightarrow x = \frac{3.4}{2.4} = \frac{17}{12}$

and there are no values of x where y' does not exist, so the only critical number is $x = \frac{17}{12}$.

Test a point in each interval.

When $x = 0$, $y' = 3.4 - 2.4(0) = 3.4 > 0$.

When $x = 2$, $y' = 3.4 - 2.4(2) = -1.4 < 0$.

(b) The function is increasing on $\left(-\infty, \frac{17}{12}\right)$.

(c) The function is decreasing on $\left(\frac{17}{12}, \infty\right)$.

15. $f(x) = \frac{2}{3}x^3 - x^2 - 24x - 4$

(a) $f'(x) = 2x^2 - 2x - 24$

$= 2(x^2 - x - 12)$

$= 2(x + 3)(x - 4)$

$f'(x)$ is zero when $x = -3$ or $x = 4$, so the critical numbers are -3 and 4.

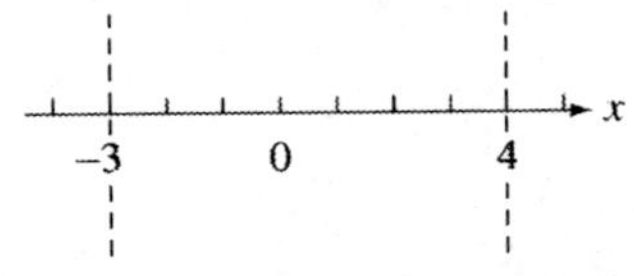

Test a point in each interval.

$f'(-4) = 16 > 0$

$f'(0) = -24 < 0$

$f'(5) = 16 > 0$

(b) f is increasing on $(-\infty, -3)$ and $(4, \infty)$.

(c) f is decreasing on $(-3, 4)$.

17. $f(x) = 4x^3 - 15x^2 - 72x + 5$

(a) $f'(x) = 12x^2 - 30x - 72$
$= 6(2x^2 - 5x - 12)$
$= 6(2x + 3)(x - 4)$

$f'(x)$ is zero when $x = -\frac{3}{2}$ or $x = 4$, so the critical numbers are $-\frac{3}{2}$ and 4.

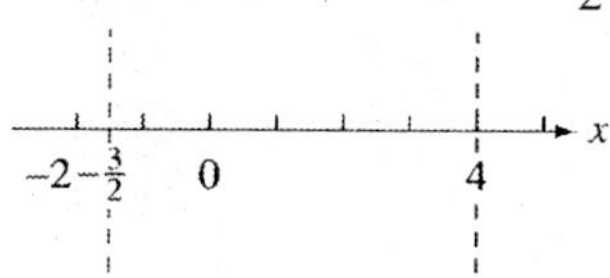

$f'(-2) = 36 > 0$
$f'(0) = -72 < 0$
$f'(5) = 78 > 0$

(b) f is increasing on $\left(-\infty, -\frac{3}{2}\right)$ and $(4, \infty)$.

(c) f is decreasing on $\left(-\frac{3}{2}, 4\right)$.

19. $f(x) = x^4 + 4x^3 + 4x^2 + 1$

(a) $f'(x) = 4x^3 + 12x^2 + 8x$
$= 4x(x^2 + 3x + 2)$
$= 4x(x + 2)(x + 1)$

$f'(x)$ is zero when $x = 0$, $x = -2$, or $x = -1$, so the critical numbers are 0, −2, and −1.

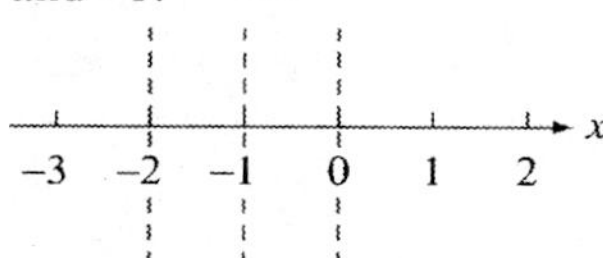

Test a point in each interval.
$f'(-3) = -12(-1)(-2) = -24 < 0$
$f'(-1.5) = -6(.5)(-.5) = 1.5 > 0$
$f'(-.5) = -2(1.5)(.5) = -1.5 < 0$
$f'(1) = 4(3)(2) = 24 > 0$

(b) f is increasing on $(-2, -1)$ and $(0, \infty)$.

(c) f is decreasing on $(-\infty, -2)$ and $(-1, 0)$.

21. $y = -3x + 6$

(a) $y' = -3 < 0$
There are no critical numbers since y' is never 0 and always exists.

(b) Since y' is always negative, the function is never increasing.

(c) y' is always negative, so the function is decreasing everywhere, or on the interval $(-\infty, \infty)$.

23. $f(x) = \frac{x+2}{x+1}$

(a) $f'(x) = \frac{(x+1)(1) - (x+2)(1)}{(x+1)^2} = \frac{-1}{(x+1)^2}$

The derivative is never 0, but it fails to exist at $x = -1$. Since −1 is not in the domain of f, −1 is not a critical number. However, the line $x = -1$ is an asymptote of the graph, so the function might change direction from one side of the asymptote to the other.

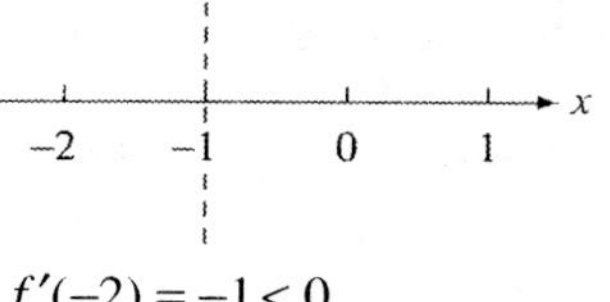

$f'(-2) = -1 < 0$
$f'(0) = -1 < 0$

(b) f is increasing on no interval.

(c) f is decreasing everywhere that it is defined, on $(-\infty, -1)$ and on $(-1, \infty)$.

25. $y = \sqrt{x^2 + 1}$
$= (x^2 + 1)^{1/2}$

(a) $y' = \frac{1}{2}(x^2 + 1)^{-1/2}(2x) = x(x^2 + 1)^{-1/2}$
$= \frac{x}{\sqrt{x^2 + 1}}$

$y' = 0$ when $x = 0$. Since y does not fail to exist for any x, and since $y' = 0$ when $x = 0$, 0 is the only critical number.

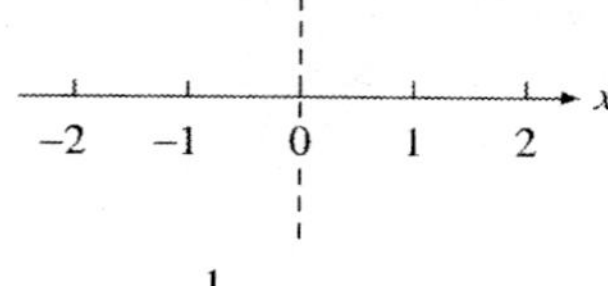

$y'(1) = \frac{1}{\sqrt{2}} > 0$
$y'(-1) = \frac{-1}{\sqrt{2}} < 0$

(b) y is increasing on $(0, \infty)$.

(c) y is decreasing on $(-\infty, 0)$.

27. $f(x) = x^{2/3}$

(a) $f'(x) = \frac{2}{3}x^{-1/3} = \frac{2}{3x^{1/3}}$

$f'(x)$ is never zero, but fails to exist when $x = 0$, so 0 is the only critical number.

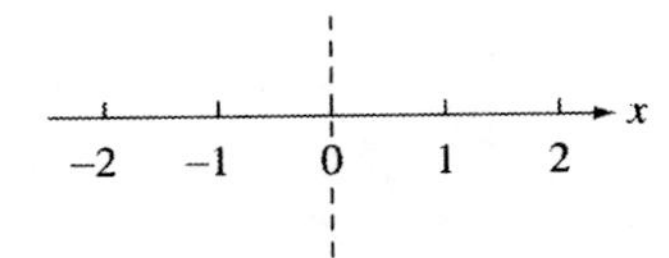

$f'(-1) = -\frac{2}{3} < 0$

$f'(1) = \frac{2}{3} > 0$

(b) f is increasing on $(0, \infty)$.

(c) f is decreasing on $(-\infty, 0)$.

29. $y = x - 4\ln(3x - 9)$

(a) $y' = 1 - \frac{12}{3x-9} = 1 - \frac{4}{x-3} = \frac{x-7}{x-3}$

y' is zero when $x = 7$. The derivative does not exist at $x = 3$, but note that the domain of f is $(3, \infty)$. Thus, the only critical number is 7.

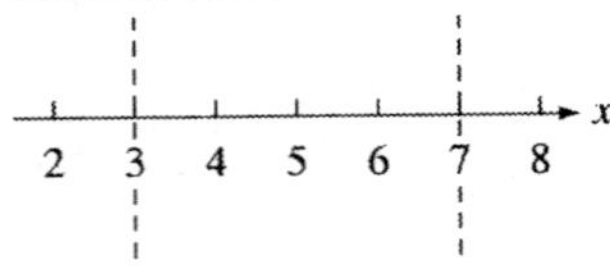

Choose a value in the intervals $(3, 7)$ and $(7, \infty)$.

$f'(4) = -3 < 0$

$f(8) = \frac{1}{5} > 0$

(b) The function is increasing on $(7, \infty)$.

(c) The function is decreasing on $(3, 7)$.

31. $f(x) = xe^{-3x}$

(a) $f'(x) = e^{-3x} + x(-3e^{-3x}) = (1 - 3x)e^{-3x}$

$= \frac{1-3x}{e^{3x}}$

$f'(x)$ is zero when $x = \frac{1}{3}$ and there are no values of x where $f'(x)$ does not exist, so the critical number is $\frac{1}{3}$.

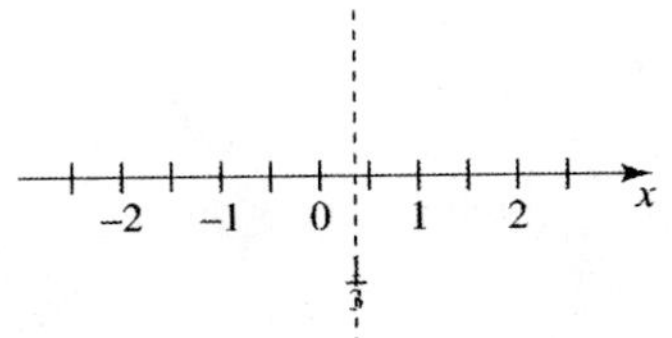

Test a point in each interval.

$f'(0) = \frac{1-3(0)}{e^{3(0)}} = 1 > 0$

$f'(1) = \frac{1-3(1)}{e^{3(1)}} = -\frac{2}{e^3} < 0$

(b) The function is increasing on $\left(-\infty, \frac{1}{3}\right)$.

(c) The function is decreasing on $\left(\frac{1}{3}, \infty\right)$.

33. $f(x) = x^2 2^{-x}$

(a) $f'(x) = x^2[\ln 2(2^{-x})(-1)] + (2^{-x})2x$

$= 2^{-x}(-x^2 \ln 2 + 2x)$

$= \frac{x(2 - x\ln 2)}{2^x}$

$f'(x)$ is zero when $x = 0$ or $x = \frac{2}{\ln 2}$ and there are no values of x where $f'(x)$ does not exist. The critical numbers are 0 and $\frac{2}{\ln 2}$.

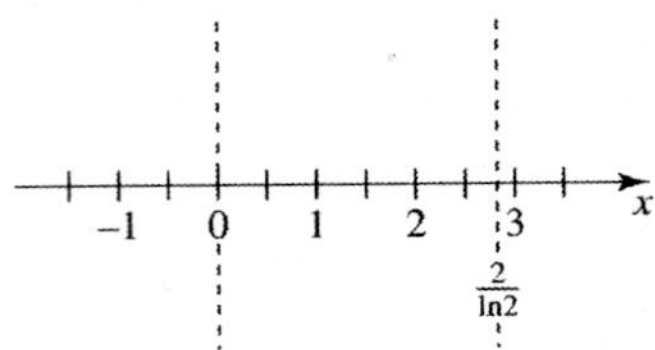

Test a point in each interval.

$f'(-1) = \frac{(-1)(2-(-1)\ln 2)}{2^{-1}}$

$= -2(2 + \ln 2) < 0$

$f'(1) = \frac{(1)(2-(1)\ln 2)}{2^1} = \frac{2 - \ln 2}{2} > 0$

$f'(3) = \frac{(3)(2-(3)\ln 2)}{2^3} = \frac{3(2 - 3\ln 2)}{8} < 0$

(b) The function is increasing on $\left(0, \frac{2}{\ln 2}\right)$.

(c) The function is decreasing on $(-\infty, 0)$ and $\left(\frac{2}{\ln 2}, \infty\right)$.

35. $y = x^{2/3} - x^{5/3}$

(a) $y' = \frac{2}{3}x^{-1/3} - \frac{5}{3}x^{2/3} = \frac{2-5x}{3x^{1/3}}$

$y' = 0$ when $x = \frac{2}{5}$. The derivative does not exist at $x = 0$. So the critical numbers are 0 and $\frac{2}{5}$.

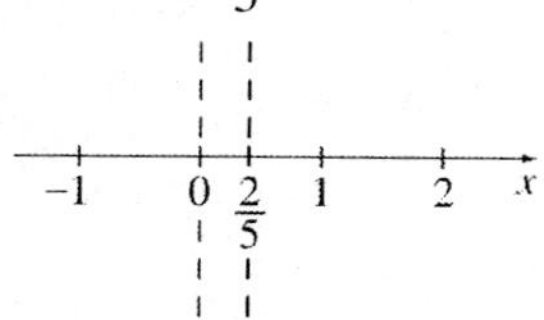

Test a point in each interval.

$y'(-1) = \frac{7}{-3} < 0$

$y'\left(\frac{1}{5}\right) = \frac{1}{3\left(\frac{1}{5}\right)^{1/3}} = \frac{5^{1/3}}{3} > 0$

$y'(1) = \frac{-3}{3} = -1 < 0$

(b) y is increasing on $\left(0, \frac{2}{5}\right)$.

(c) y is decreasing on $(-\infty, 0)$ and $\left(\frac{2}{5}, \infty\right)$.

37. $y = \sin x$

(a) $y' = \cos x$

y' is zero when

$x = \pm\frac{\pi}{2}, \pm\frac{3\pi}{2}, \pm\frac{5\pi}{2}, \cdots$. Therefore, the critical numbers are $\frac{n\pi}{2}$, where n is an odd integer.

(b) The function is increasing on

$\left(n\pi, \left(n+\frac{1}{2}\right)\pi\right) \cup \left(\left(n+\frac{3}{2}\right)\pi, (n+2)\pi\right)$,

where n is an even integer.

(c) The function is decreasing on

$\left(\left(n+\frac{1}{2}\right)\pi, \left(n+\frac{3}{2}\right)\pi\right)$, where n is an even integer.

39. $y = 3\sec x$

(a) $y' = 3\sec x \tan x$

y' is zero when $x = 0, \pm\pi, \pm 2\pi, \cdots$.

Therefore, the critical numbers are $n\pi$, where n is an integer.

(b) The function is increasing on

$\left(n\pi, \left(n+\frac{1}{2}\right)\pi\right) \cup \left(\left(n+\frac{1}{2}\right)\pi, (n+1)\pi\right)$,

where n is an even integer.

(c) The function is decreasing on

$\left((n+1)\pi, \left(n+\frac{3}{2}\right)\pi\right) \cup \left(\left(n+\frac{3}{2}\right)\pi, (n+2)\pi\right)$.

where n is an even integer.

41. Answers will vary.

43. $f(x) = ax^2 + bx + c, a < 0$

$f'(x) = 2ax + b$

Let $f'(x) = 0$ to find the critical number.

$2ax + b = 0 \Rightarrow 2ax = -b \Rightarrow x = \frac{-b}{2a}$

Choose a value in the interval $\left(-\infty, -\frac{b}{2a}\right)$.

Since $a < 0$,

$\frac{-b}{2a} - \frac{-1}{2a} = \frac{-b+1}{2a} < \frac{-b}{2a}$.

$f'\left(\frac{-b+1}{2a}\right) = 2a\left(\frac{-b+1}{2a}\right) + b$

$= 1 > 0$

Choose a value in the interval $\left(\frac{-b}{2a}, \infty\right)$.

Since $a < 0$,

$\frac{-b}{2a} - \frac{-1}{2a} = \frac{-b-1}{2a} < \frac{-b}{2a}$.

$f'\left(\frac{-b-1}{2a}\right) = 2a\left(\frac{-b-1}{2a}\right) + b$

$= -1 < 0$

f is increasing on $\left(-\infty, \frac{-b}{2a}\right)$ and decreasing on $\left(\frac{-b}{2a}, \infty\right)$. This tells us that the curve opens downward and $x = \frac{-b}{2a}$ is the x-coordinate of the vertex.

(continued on next page)

(continued)

$$f\left(\frac{-b}{2a}\right)=a\left(\frac{-b}{2a}\right)^2+b\left(\frac{-b}{2a}\right)+c$$
$$=\frac{ab^2}{4a^2}-\frac{b^2}{2a}+c=\frac{b^2}{4a}-\frac{2b^2}{4a}+\frac{4bc}{4a}$$
$$=\frac{4ac-b^2}{4a}$$

The vertex is $\left(\frac{-b}{2a}, \frac{4ac-b^2}{4a}\right)$ or

$\left(-\frac{b}{2a}, \frac{4ac-b^2}{4a}\right).$

45. $f(x)=\ln x$

$f'(x)=\frac{1}{x}$

$f'(x)$ is undefined at $x=0$. $f'(x)$ never equals zero. Note that $f(x)$ has a domain of $(0, \infty)$. Pick a value in the interval $(0, \infty)$.

$f'(2)=\frac{1}{2}>0$

$f(x)$ is increasing on $(0, \infty)$.

$f(x)$ is never decreasing.

Since $f(x)$ never equals zero, the tangent line is horizontal nowhere.

47. $f(x)=e^{0.001x}-\ln x$

$f'(x)=0.001e^{0.001x}-\frac{1}{x}$

Note that $f(x)$ is only defined for $x>0$. Use a graphing calculator to plot $f'(x)$ for $x>0$.

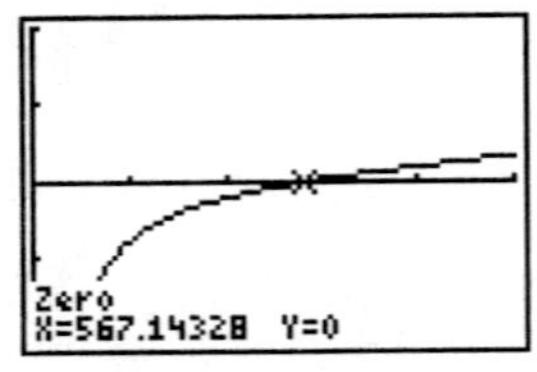

[0, 1000] by [–0.01, 0.01]

(a) $f'(x)>0$ on about $(567, \infty)$, so $f(x)$ is increasing on about $(567, \infty)$.

(b) $f'(x)<0$ on about $(0, 567)$, so $f(x)$ is decreasing on about $(0, 567)$.

49. $f(x)=x^2+3\cos x$

$f'(x)=2x-3\sin x$

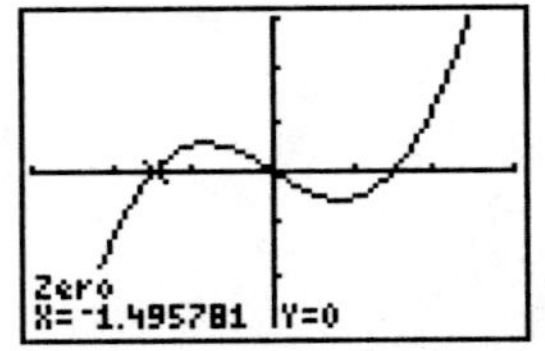

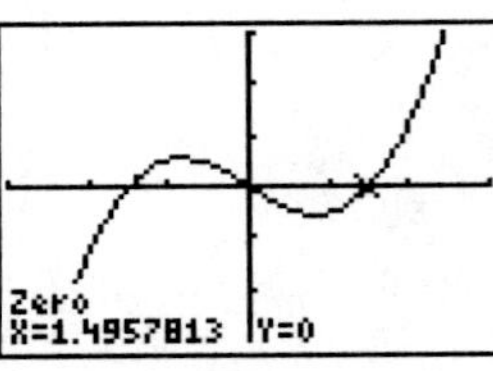

[–3, 3] by [–3, 3]

(a) $f'(x)>0$ on about $(-1.496, 0)$ and $(1.496, \infty)$, so the function is increasing on those intervals.

(b) $f'(x)<0$ on about $(-\infty, -1.496)$ and $(0, 1.496)$, so the function is decreasing on those intervals.

51. (a) These curves are graphs of functions since they all pass the vertical line test.

(b) The graph for particulates increases from April to July; it decreases from July to November; it is constant from January to April and November to December.

(c) All graphs are constant from January to April and November to December. When the temperature is low, as it is during these months, air pollution is greatly reduced.

53. $A(t)=0.003631t^3-0.03746t^2+0.1012t+0.009$

$A'(t)=0.010893t^2-0.07492t+0.1012$

Use the quadratic formula to find $A'(t)=0 \Rightarrow t\approx 1.85$ or $t\approx 5.03$

Choose $t=1$ and $t=4$ as test points.

$$A'(1)=0.010893(1)^2-0.07492(1)+0.1012$$
$$=0.037173$$
$$A'(4)=0.010893(4)^2-0.07492(4)+0.1012$$
$$=-0.024192$$

(a) The function is increasing on $(0, 1.85)$.

(b) The function is decreasing on $(1.85, 5)$.

55. $K(t)=\frac{5t}{t^2+1}$

$$K'(t)=\frac{5(t^2+1)-2t(5t)}{(t^2+1)^2}=\frac{5t^2+5-10t^2}{(t^2+1)^2}$$
$$=\frac{5-5t^2}{(t^2+1)^2}$$

(continued on next page)

(continued)

$K'(t) = 0$ when $\frac{5-5t^2}{(t^2+1)^2} = 0 \Rightarrow 5-5t^2 = 0 \Rightarrow$

$5t^2 = 5 \Rightarrow t = \pm 1.$

Since t is the time after a drug is administered, the function applies only for $[0, \infty)$, so we discard $t = -1$. Then 1 divides the domain into two intervals.

$K'(0.5) = 2.4 > 0$

$K'(2) = -0.6 < 0$

(a) K is increasing on $(0, 1)$.

(b) K is decreasing on $(1, \infty)$.

57. (a) $F(t) = -10.28 + 175.9te^{-t/1.3}$

$$F'(t) = (175.9)(e^{-t/1.3}) + (175.9.9t)\left(-\frac{1}{1.3}e^{-t/1.3}\right)$$

$$= (175.9)(e^{-t/1.3})\left(1 - \frac{t}{1.3}\right)$$

$$\approx 175.9e^{-t/1.3}(1 - 0.769t)$$

(b) $F'(t)$ is equal to 0 at $t = 1.3$ Therefore, 1.3 is a critical number. Since the domain is $(0, \infty)$, test values in the intervals from $(0, 1.3)$ and $(1.3, \infty)$.

$F'(1) \approx 18.83 > 0$ and $F'(2) \approx -20.32 < 0$

$F'(t)$ is increasing on $(0, 1.3)$ and decreasing on $(1.3, \infty)$.

59. (a) $W_2(t) = 532\left(1 - 0.911e^{-0.0021t}\right)^{1.2466}$

$W_2'(t) = 663.1912\left(1 - 0.911e^{-0.0021t}\right)^{1.2466}$

Since $W'_2(t) > 0$ for all values of $t > 0$, this function is increasing on $(0, \infty)$.

(b)

$w_1(t) = 619(1 - 0.905e^{-0.002t})^{1.2386}$

620

0 2000

0

$w_2(t) = 532(1 - 0.911e^{-0.0021t})^{1.2466}$

(c) Holstein cows grow the fastest and reach the greatest weight.

61. $C(t) = 37.29 + 0.46\cos\left(\frac{2\pi(t-16.37)}{24}\right)$

$$C'(t) = -\frac{2\pi}{24}(0.46)\sin\left(\frac{2\pi(t-16.37)}{24}\right)$$

$$= -\frac{0.46\pi}{12}\sin\left(\frac{\pi(t-16.37)}{12}\right)$$

$C'(t) = 0$ when $\frac{\pi(t-16.37)}{12} = 0 \Rightarrow t = 16.37$

or $\frac{\pi(t-16.37)}{12} = \pi \Rightarrow t = 4.37.$

Because $C'(t) > 0$ when $4.37 < t < 16.37$, the function is increasing on $(4.37, 16.37)$. $C'(t) < 0$ when $0 < t < 4.37$ or when $16.37 < t < 24$, the function is decreasing on $(0, 4.37) \cup (16.37, 24)$.

63. $f(x) = \frac{1}{\sqrt{2\pi}}e^{-x^2/2}$

$f'(x) = \frac{1}{\sqrt{2\pi}}e^{-x^2/2}(-x) = \frac{-x}{\sqrt{2\pi}}e^{-x^2/2}$

$f'(x) = 0$ when $x = 0$.

Choose a value from each of the intervals $(-\infty, 0)$ and $(0, \infty)$.

$f'(-1) = \frac{1}{\sqrt{2\pi}}e^{-1/2} > 0$

$f'(1) = \frac{-1}{\sqrt{2\pi}}e^{-1/2} < 0$

The function is increasing on $(-\infty, 0)$ and decreasing on $(0, \infty)$.

5.2 Relative Extrema

1. As shown on the graph, the relative minimum of –4 occurs when $x = 1$.

3. As shown on the graph, the relative maximum of 3 occurs when $x = -2$.

5. As shown on the graph, the relative maximum of 3 occurs when $x = -4$ and the relative minimum of 1 occurs when $x = -2$.

7. As shown on the graph, the relative maximum of 3 occurs when $x = -4$, the relative minimum of –2 occurs when $x = -7$ and $x = -2$.

9. Since the graph of the derivative is zero at $x = -1$ and $x = 3$, the critical numbers are -1 and 3. The graph of the derivative is positive on $(-\infty, -1)$ and negative on $(-1, 3)$, so there is a relative maximum at $x = -1$. Since the graph of the function is negative on $(-1, 3)$ and positive on $(3, \infty)$, there is a relative minimum at $x = 3$.

11. The graph of the derivative is zero at $x = -8$, $x = -6$, $x = -2.5$, and $x = -1.5$, so the critical numbers are -8, -6, -2.5, and -1.5. Because the graph of the derivative is positive on $(-\infty, -8)$ and negative on $(-8, -6)$, there is a relative maximum at $x = -8$. The graph of the derivative is negative on $(-8, -6)$ and positive on $(-6, -2.5)$, so there is a relative minimum at $x = -6$. The graph of the derivative is positive on $(-6, -2.5)$ and negative on $(-2.5, -1.5)$, sothere is a relative maximum at $x = -2.5$. Because the graph of the derivative is negative on $(-2.5, -1.5)$ and positive on $(-1.5, \infty)$, there is a relative minimum at $x = -1.5$.

13. $f(x) = x^2 - 10x + 33$
$f'(x) = 2x - 10$
$f'(x)$ is zero when $x = 5$.

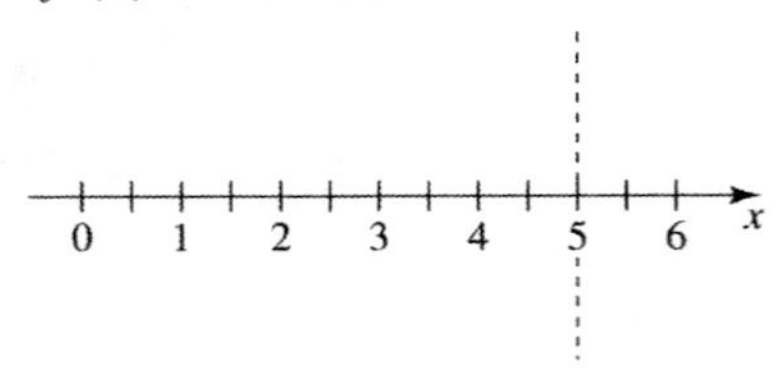

$f'(0) = -10 < 0$
$f'(6) = 2 > 0$
f is decreasing on $(-\infty, 5)$ and increasing on $(5, \infty)$. Thus, a relative minimum occurs at $x = 5$. $f(5) = 8$, so there is a relative minimum of 8 at $x = 5$.

15. $f(x) = x^3 + 6x^2 + 9x - 8$
$f'(x) = 3x^2 + 12x + 9 = 3(x^2 + 4x + 3)$
$= 3(x+3)(x+1)$
$f'(x)$ is zero when $x = -1$ or $x = -3$.

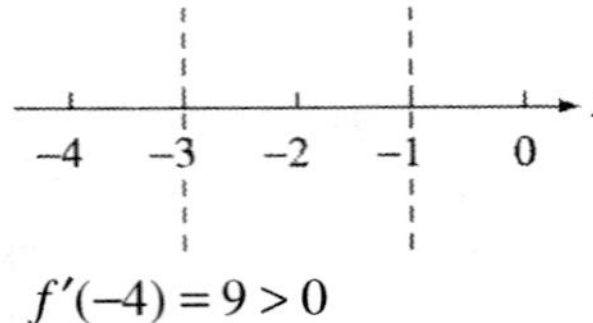

$f'(-4) = 9 > 0$
$f'(-2) = -3 < 0$
$f'(0) = 9 > 0$

Thus, f is increasing on $(-\infty, -3)$, decreasing on $(-3, -1)$, and increasing on $(-1, \infty)$.
f has a relative maximum at -3 and a relative minimum at -1.
$f(-3) = -8$; $f(-1) = -12$
Relative maximum of -8 at $x = -3$; relative minimum of -12 at $x = -1$.

17. $f(x) = -\frac{4}{3}x^3 - \frac{21}{2}x^2 - 5x + 8$
$f'(x) = -4x^2 - 21x - 5 = (-4x-1)(x+5)$
$f'(x)$ is zero when $x = -5$, or $x = -\frac{1}{4}$.

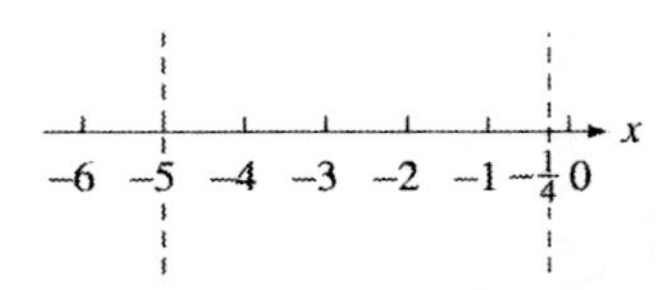

$f'(-6) = -23 < 0$
$f'(-4) = 15 > 0$
$f'(0) = -5 < 0$
f is decreasing on $(-\infty, -5)$, increasing on $\left(-5, -\frac{1}{4}\right)$, and decreasing on $\left(-\frac{1}{4}, \infty\right)$. f has a relative minimum at -5 and a relative maximum at $-\frac{1}{4}$.

$f(-5) = -\frac{377}{6}$; $f\left(-\frac{1}{4}\right) = \frac{827}{96}$

Relative maximum of $\frac{827}{96}$ at $x = -\frac{1}{4}$; relative minimum of $-\frac{377}{6}$ at $x = -5$.

19. $f(x) = x^4 - 18x^2 - 4$
$f'(x) = 4x^3 - 36x = 4x(x^2 - 9)$
$= 4x(x+3)(x-3)$
$f'(x)$ is zero when $x = 0$ or $x = -3$ or $x = 3$.

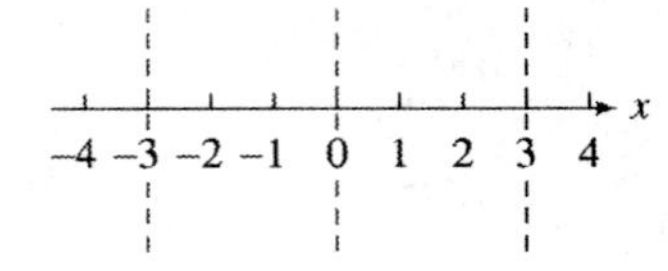

$f'(-4) = 4(-4)^3 - 36(-4) = -112 < 0$
$f'(-1) = -4 + 36 = 32 > 0$
$f'(1) = 4 - 36 = -32 < 0$
$f'(4) = 4(4)^3 - 36(4) = 112 > 0$

(continued on next page)

(*continued*)

f is decreasing on $(-\infty, -3)$ and $(0, 3)$; f is increasing on $(-3, 0)$ and $(3, \infty)$.

$f(-3) = -85; \quad f(0) = -4; \quad f(3) = -85$

Relative maximum of -4 at $x = 0$; relative minimum of -85 at $x = 3$ and $x = -3$.

21. $f(x) = 3 - (8 + 3x)^{2/3}$

$f'(x) = -\frac{2}{3}(8 + 3x)^{-1/3}(3) = -\frac{2}{(8 + 3x)^{1/3}}$

Critical number: $8 + 3x = 0 \Rightarrow x = -\frac{8}{3}$

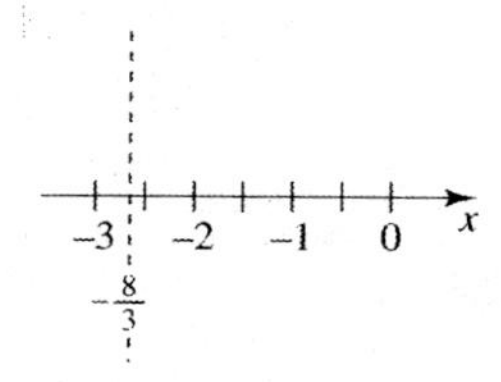

$f'(-3) = 2 > 0$

$f'(0) = -1 < 0$

f is increasing on $\left(-\infty, -\frac{8}{3}\right)$ and decreasing on $\left(-\frac{8}{3}, \infty\right)$. $f\left(-\frac{8}{3}\right) = 3$, so there is a relative maximum of 3 at $x = -\frac{8}{3}$.

23. $f(x) = 2x + 3x^{2/3}$

$f'(x) = 2 + 2x^{-1/3} = 2 + \frac{2}{\sqrt[3]{x}}$

Find the critical numbers. $f'(x) = 0$ when

$2 + \frac{2}{\sqrt[3]{x}} = 0 \Rightarrow \frac{2}{\sqrt[3]{x}} = -2 \Rightarrow x = (-1)^3 \Rightarrow$

$x = -1$.

$f'(x)$ does not exist when $\sqrt[3]{x} = 0 \Rightarrow x = 0$.

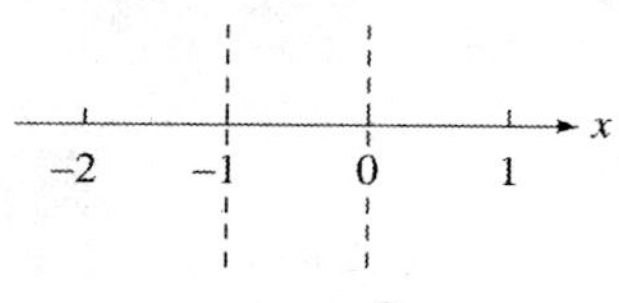

$f'(-2) = 2 + \frac{2}{\sqrt[3]{-2}} \approx 0.41 > 0$

$f'\left(-\frac{1}{2}\right) = 2 + \frac{2}{\sqrt[3]{-\frac{1}{2}}} = 2 + \frac{2\sqrt[3]{2}}{-1} \approx -0.52 < 0$

$f'(1) = 2 + \frac{2}{\sqrt[3]{1}} = 4 > 0$

f is increasing on $(-\infty, -1)$ and $(0, \infty)$. f is decreasing on $(-1, 0)$.

$f(-1) = 2(-1) + 3(-1)^{2/3} = 1$

$f(0) = 0$

Relative maximum of 1 at $x = -1$; relative minimum of 0 at $x = 0$.

25. $f(x) = x - \frac{1}{x}$

$f'(x) = 1 + \frac{1}{x^2}$ is never zero, but fails to exist at $x = 0$. Since $f(x)$ also fails to exist at $x = 0$, there are no critical numbers and no relative extrema.

27. $f(x) = \frac{x^2 - 2x + 1}{x - 3}$

$f'(x) = \frac{(x-3)(2x-2) - (x^2 - 2x + 1)(1)}{(x-3)^2}$

$= \frac{x^2 - 6x + 5}{(x-3)^2}$

Find the critical numbers:

$x^2 - 6x + 5 = 0 \Rightarrow (x-5)(x-1) = 0 \Rightarrow$

$x = 5 \quad \text{or} \quad x = 1$

Note that $f(x)$ and $f'(x)$ do not exist at $x = 3$, so the only critical numbers are 1 and 5.

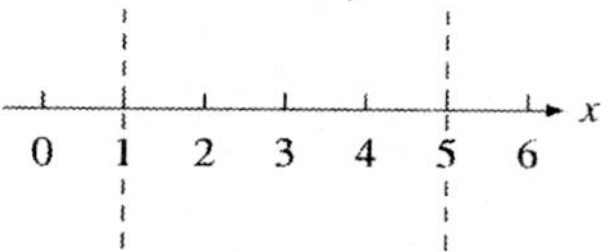

$f'(0) = \frac{5}{9} > 0$

$f'(2) = -3 < 0$

$f'(6) = \frac{5}{9} > 0$

$f(x)$ is increasing on $(-\infty, 1)$ and $(5, \infty)$.

$f(x)$ is decreasing on $(1, 5)$.

$f(1) = 0; \quad f(5) = 8$

Relative maximum of 0 at $x = 1$; relative minimum of 8 at $x = 5$.

29. $f(x) = x^2e^x - 3$

$f'(x) = x^2e^x + 2xe^x = xe^x(x + 2)$

$f'(x)$ is zero at $x = 0$ and $x = -2$.

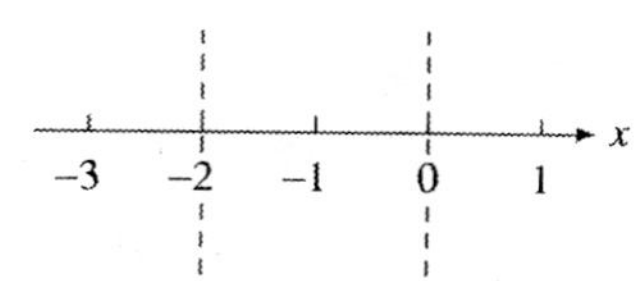

(*continued on next page*)

(continued)

$f'(-3) = 3e^{-3} = \frac{3}{e^3} > 0$

$f'(-1) = -e^{-1} = \frac{-1}{e} < 0$

$f'(1) = 3e^1 > 0$

f is increasing on $(-\infty, -2)$ and $(0, \infty)$.

f is decreasing on $(-2, 0)$.

$f(0) = 0 \cdot e^0 - 3 = -3$

$f(-2) = (-2)^2 e^{-2} - 3 = \frac{4}{e^2} - 3 \approx -2.46$

Relative minimum of –3 at $x = 0$; relative maximum of –2.46 at $x = -2$.

31. $f(x) = 2x + \ln x$

$f'(x) = 2 + \frac{1}{x} = \frac{2x+1}{x}$

$f'(x)$ is zero at $x = -\frac{1}{2}$. The domain of $f(x)$ is $(0, \infty)$. Therefore $f'(x)$ is never zero in the domain of $f(x)$. $f'(1) = 3 > 0$. Since $f(x)$ is always increasing, f has no relative extrema.

33. $f(x) = \frac{2^x}{x}$

$f'(x) = \frac{(x) \ln 2(2^x) - 2^x(1)}{x^2} = \frac{2^x(x \ln 2 - 1)}{x^2}$

Find the critical numbers:

$x \ln 2 - 1 = 0 \Rightarrow x = \frac{1}{\ln 2}$ or $x^2 = 0 \Rightarrow x = 0$

Since f is not defined for $x = 0$, 0 is not a critical number. $x = \frac{1}{\ln 2} \approx 1.44$ is the only critical number.

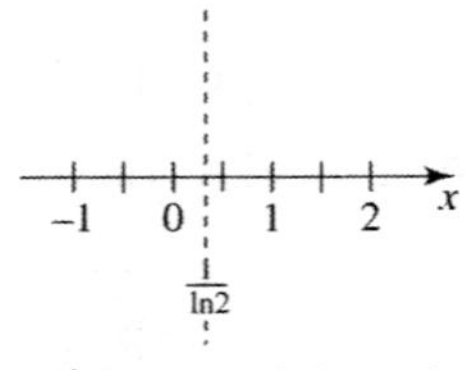

$f'(1) \approx -0.6137 < 0$

$f'(2) \approx 0.3863 > 0$

f is decreasing on $\left(0, \frac{1}{\ln 2}\right)$ and increasing on $\left(\frac{1}{\ln 2}, \infty\right)$.

$f\left(\frac{1}{\ln 2}\right) = \frac{2^{1/\ln 2}}{\frac{1}{\ln 2}} = \ln 2\left(e^{\ln 2}\right)^{1/\ln 2} = e \ln 2$

Relative minimum of $e \ln 2$ at $x = \frac{1}{\ln 2}$.

35. $f(x) = \sin(\pi x)$

$f'(x) = \pi \cos(\pi x)$

The critical numbers are

$x = \pm\frac{1}{2}, \pm\frac{3}{2}, \pm\frac{5}{2}, \ldots$

$f'(0) = \pi > 0$

$f'(1) = -\pi < 0$

$f'(2) = \pi > 0$

$\vdots$

$f(x)$ is increasing on the intervals

$\ldots, \left(-\frac{5}{2}, -\frac{3}{2}\right), \left(-\frac{1}{2}, \frac{1}{2}\right), \left(\frac{3}{2}, \frac{5}{2}\right), \ldots$

$f(x)$ is decreasing on the intervals

$\ldots, \left(-\frac{3}{2}, -\frac{1}{2}\right), \left(\frac{1}{2}, \frac{3}{2}\right), \left(\frac{5}{2}, \frac{7}{2}\right), \ldots$

There is a relative maximum of 1 at

$x = \ldots, -\frac{7}{2}, -\frac{3}{2}, \frac{1}{2}, \frac{5}{2}, \ldots$

There is a relative minimum of –1 at

$x = \ldots, -\frac{5}{2}, -\frac{1}{2}, \frac{3}{2}, \frac{7}{2}, \ldots$

37. $y = -2x^2 + 12x - 5$

$y' = -4x + 12 = -4(x - 3)$

The vertex occurs when $y' = 0$ or when $x - 3 = 0 \Rightarrow x = 3$. When $x = 3$,

$y = -2(3)^2 + 12(3) - 5 = 13$

The vertex is (3, 13).

39. $f(x) = x^5 - x^4 + 4x^3 - 30x^2 + 5x + 6$

$f'(x) = 5x^4 - 4x^3 + 12x^2 - 60x + 5$

Graph f' on a graphing calculator. A suitable choice for the viewing window is [–2, 4] by [–50, 50].

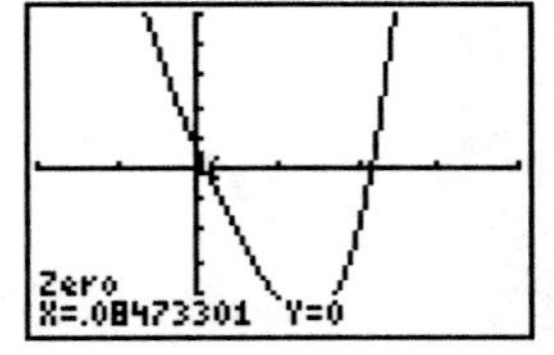

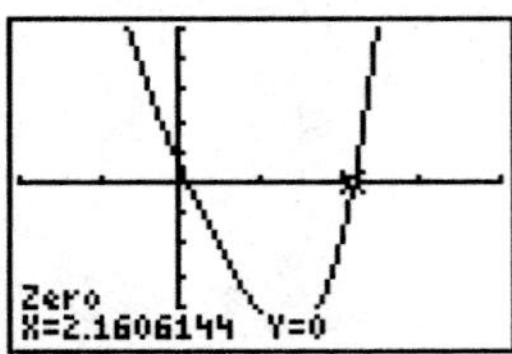

(continued on next page)

(continued)

Use the calculator to estimate the x-intercepts of this graph. These numbers are the solutions of the equation $f'(x) = 0$ and thus the critical numbers for f. Rounded to three decimal places, these x-values are 0.085 and 2.161. Examine the graph of f' near $x \approx 0.085$ and $x \approx 2.161$. Observe that $f'(x) > 0$ to the left of $x \approx 0.085$ and $f'(x) < 0$ to the right of $x \approx 0.085$. Also observe that $f'(x) < 0$ to the left of $x \approx 2.161$ and $f'(x) > 0$ to the right of $x \approx 2.161$. The first derivative test allows us to conclude that f has a relative maximum at $x \approx 0.085$ and a relative minimum at $x \approx 2.161$.

$f(0.085) \approx 6.211$
$f(2.161) \approx -57.607$

Relative maximum of 6.211 at $x \approx 0.085$; relative minimum of -57.607 at $x \approx 2.161$.

41. $f(x) = 2|x+1| + 4|x-5| - 20$

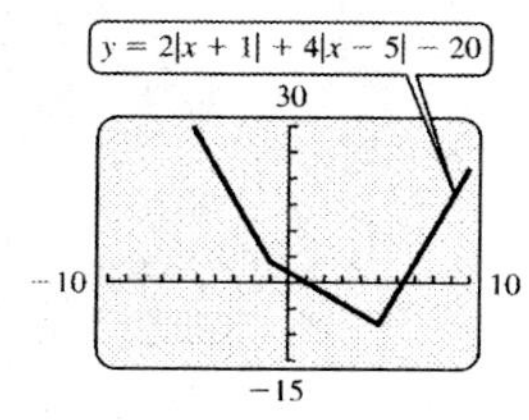

The graph shows that f has no relative maxima, but there is a relative minimum at $x = 5$. (Note that the graph has a sharp point at $(5, -8)$, indicating that $f'(-5)$ does not exist.)

43. From the graph, the zeros of the derivative are at $x = -1$ and $x = 3$. The derivative is decreasing at $x = -1$, there is a relative maximum at $x = -1$. The derivative is increasing at $x = 3$, so there is a relative minimum at $x = 3$.

45. $a(t) = 0.008t^3 - 0.288t^2 + 2.304t + 7$

$a'(t) = 0.024t^2 - 0.576t + 2.304$

Set $a' = 0$ and solve for t.

$0.024t^2 - 0.576t + 2.304 = 0 \Rightarrow 0.024(t^2 - 24t + 96) = 0 \Rightarrow t^2 - 24t + 96 = 0 \Rightarrow t \approx 5.07$ or $t \approx 18.93$

These are the critical values.

$a'(1) = 1.752 > 0$
$a'(10) = -1.056 < 0$
$a'(20) = 0.384 > 0$

Note that the problem makes sense only for $t > 0$. a is increasing on $(0, 5.07)$ and $(18.93, \infty)$. a is decreasing on $(5.07, 18.93)$. Therefore, the activity level is highest when $t = 5.07 = 5 \text{ hours} + 0.07 \cdot 60 \text{ minutes}$, which corresponds to 5:04 P.M. The activity level is lowest when $t = 18.93 = 18 \text{ hours} + 0.93 \cdot 60 \text{ minutes}$, which corresponds to 6:56 A.M.

47. $M(t) = 369(0.93)^t (t)^{0.36}$

$M'(t) = (369)(0.93)^t \ln(0.93)(t^{0.36}) + 369(0.93)^t (0.36)(t)^{-0.64}$

$$= (369t^{0.36})(0.93^t \ln 0.93) + \frac{132.84(0.93)^t}{t^{0.64}}$$

$M'(t) = 0$ when $t \approx 4.96$.

Verify that $t \approx 4.96$ gives a maximum by verifying that $M'(4) > 0$ and $M'(5) < 0$.

Now find $M(4.96)$

$M(4.96) = 369(0.93)^{4.96}(4.96)^{0.36} \approx 485.22$

The female moose reaches a maximum weight of about 458.22 kilograms at about 4.96 years.

49. $f(t)=\frac{k(n-1)n^2e^{nt}}{(n-1+e^{nt})^2}$

$$f'(t)=k(n-1)n^2\left(\frac{d}{dt}\left(\frac{e^{nt}}{(n-1+e^{nt})^2}\right)\right)=k(n-1)n^2\left(\frac{(n-1+e^{nt})^2ne^{nt}-e^{nt}(2n)(n-1+e^{nt})e^{nt}}{(n-1+e^{nt})^4}\right)$$

$$=k(n-1)n^2(ne^{nt})\left(\frac{(n-1+e^{nt})-(2e^{nt})}{(n-1+e^{nt})^3}\right)=k(n-1)n^2(ne^{nt})\left(\frac{n-1-e^{nt}}{(n-1+e^{nt})^3}\right)$$

Now solve $f'(t)=0$.

$$k(n-1)n^2(ne^{nt})\left(\frac{n-1-e^{nt}}{(n-1+e^{nt})^3}\right)=0\Rightarrow(ne^{nt})(n-1-e^{nt})=0\Rightarrow n-1-e^{nt}=0\Rightarrow e^{nt}=n-1\Rightarrow$$

$$nt=\ln(n-1)\Rightarrow t=\frac{\ln(n-1)}{n}$$

The rate of change in the number of infected individuals reaches a maximum at $t=\frac{\ln(n-1)}{n}$.

51. **(a)** Solve $C(x)=0$.

$$\frac{R}{x^n}-\frac{A}{x^m}=0\Rightarrow\frac{Rx^m}{x^nx^m}-\frac{Ax^n}{x^nx^m}=0\Rightarrow$$
$$Rx^m-Ax^n=0\Rightarrow Rx^m=Ax^n\Rightarrow$$
$$\frac{R}{A}=\frac{x^n}{x^m}=x^{n-m}\Rightarrow\left(\frac{R}{A}\right)^{1/(n-m)}=x$$

(b) First find the derivative.

$$C(x)=R(x)^{-n}-A(x)^{-m}$$
$$C'(x)=-nR(x)^{-n-1}+mA(x)^{-m-1}$$
$$=-\frac{nR}{x^{n+1}}+\frac{mA}{x^{m+1}}$$

Now solve $C'(x)=0$.

$$-\frac{nR}{x^{n+1}}+\frac{mA}{x^{m+1}}=0$$
$$\frac{-nRx^{m+1}+mAx^{n+1}}{x^{n+1}x^{m+1}}=0$$
$$-nRx^{m+1}+mAx^{n+1}=0$$
$$mAx^{n+1}=nRx^{m+1}$$
$$\frac{x^{n+1}}{x^{m+1}}=x^{n-m}=\frac{nR}{mA}$$
$$x=\left(\frac{nR}{mA}\right)^{1/(n-m)}$$

So, there is a critical point at

$$x=\left(\frac{nR}{mA}\right)^{1/(n-m)}.$$

Note that

$$C'(x)=-\frac{nR}{x^{n+1}}+\frac{mA}{x^{m+1}}$$
$$=\frac{1}{x^{m+1}}\left(mA-\frac{nR}{x^{n-m}}\right)$$

Because $n>m$, the exponent of x in the denominator is positive, as are all the constants m, A, n, and R. As x approaches infinity, the expression in parentheses approaches mA, which is positive. As x approaches 0, $C'(x)$ approaches negative infinity. So $C'(x)$ is negative to the left of the critical point and positive to the right of the critical point. By the first derivative test, the critical point is a minimum.

(c) From part (a), we have $C(x)=0$ when

$$x=\left(\frac{R}{A}\right)^{1/(n-m)}.$$

For $R=5$, $A=3$, $n=\frac{1}{2}$, and $m=\frac{1}{3}$,

$$x=\left(\frac{5}{3}\right)^{1/(1/2-1/3)}=\left(\frac{5}{3}\right)^6.$$

From part (b), we have $C'(x)=0$ when

$$x=\left(\frac{nR}{mA}\right)^{1/(n-m)}.$$

(continued on next page)

(*continued*)

For $R = 5, A = 3,\ n = \frac{1}{2},$ and $m = \frac{1}{3},$

$$x = \left(\frac{\frac{1}{2}(5)}{\frac{1}{3}(3)}\right)^{1/(1/2-1/3)} = \left(\frac{5}{2}\right)^6 \approx 244.1.$$

Use the first derivative test to verify that this is a relative minimum by evaluating $C'(200)$ and $C'(300)$.

$$C'(200) = -\frac{5}{2}(200)^{-3/2} + (200)^{-4/2} \approx -0.00003$$

$$C'(300) = -\frac{5}{2}(300)^{-3/2} + (300)^{-4/2} \approx 0.0003$$

The sign of the derivative changes from negative to positive, so there is a relative minimum at $x = \left(\frac{5}{2}\right)^6$.

53. (a) We draw θ in standard position.

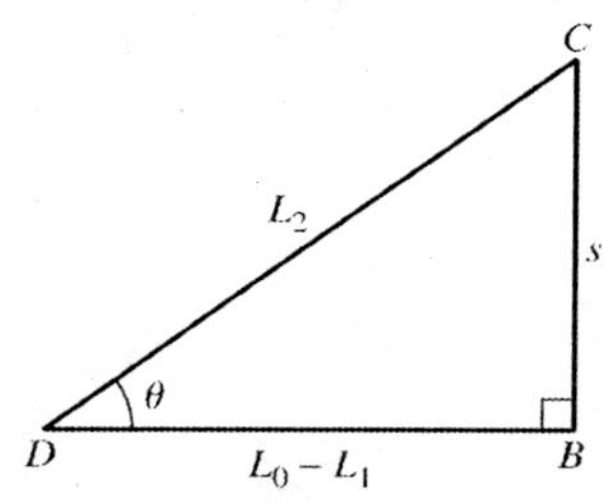

From this diagram, we see that $\sin\theta = \frac{s}{L_2}$.

(b)

$$\sin\theta = \frac{s}{L_2}$$
$$L_2 \sin\theta = s$$
$$L_2 = \frac{s}{\sin\theta}$$

(c) Using the sketch of θ in the solution to part (a) and the definition of cotangent, we see that $\cot\theta = \frac{L_0 - L_1}{s}$.

(d)

$$\cot\theta = \frac{L_0 - L_1}{s}$$
$$s\cot\theta = L_0 - L_1$$
$$L_1 + s\cot\theta = L_0$$
$$L_1 = L_0 - s\cot\theta$$

(e) The length of AD is L_1, and the radius of that section of blood vessel is r_1, so the general equation $R = K\cdot\frac{L}{r^4}$ is similar to $R_1 = k\cdot\frac{L_1}{r_1^4}$ for that particular segment of the blood vessel.

(f) $R_2 = k\cdot\frac{L_2}{r_2^4}$ where R_2 is the resistance along DC.

(g)

$$R = R_1 + R_2 = k\frac{L_1}{r_1^4} + k\frac{L_2}{r_2^4} = k\left(\frac{L_1}{r_1^4} + \frac{L_2}{r_2^4}\right)$$

(h)

$$R = k\left(\frac{L_1}{r_1^4} + \frac{L_2}{r_2^4}\right) = k\left(\frac{L_0 - s\cot\theta}{r_1^4} + \frac{s}{(\sin\theta)r_2^4}\right) = \frac{k(L_0 - s\cot\theta)}{r_1^4} + \frac{ks}{r_2^4\sin\theta}$$

(i) Since $k, L_1, L_0, s, r_1,$ and r_2 are all constants, the only letter left as a variable is θ, so the differentiation in the symbol R' must be differentiated with respect to θ.

$$R' = D_\theta R$$
$$= D_\theta\left(k\frac{L_0}{r_1^4} - \frac{sk}{r_1^4}\cot\theta + \frac{s}{r_2^4}\cdot\frac{k}{\sin\theta}\right)$$
$$= kD_\theta\left(\frac{L_0}{r_1^4} - \frac{s}{r_1^4}\cot\theta + \frac{s}{r_2^4}\cdot\frac{1}{\sin\theta}\right)$$
$$= k\left[D_\theta\left(\frac{L_0}{r_1^4}\right) - D_\theta\left(\frac{s}{r_1^4}\cot\theta\right) + D_\theta\left(\frac{s}{r_2^4}\cdot\frac{1}{\sin\theta}\right)\right]$$
$$= k\left[0 - \frac{s}{r_1^4}D_\theta(\cot\theta) + \frac{s}{r_2^4}D_\theta\left(\frac{1}{\sin\theta}\right)\right]$$
$$= k\left[-\frac{s}{r_1^4}(-\csc^2\theta) + \frac{s}{r_2^4}\left(\frac{-\cos\theta}{\sin^2\theta}\right)\right]$$
$$= k\left(\frac{s}{r_1^4}\frac{1}{\sin^2\theta} - \frac{s}{r_2^4}\cdot\frac{\cos\theta}{\sin^2\theta}\right)$$

(*continued on next page*)

(*continued*)

$$= \frac{ks}{\sin^2 \theta}\left(\frac{1}{r_1^4} - \frac{\cos \theta}{r_2^4}\right)$$
$$= \frac{ks \csc^2 \theta}{r_1^4} - \frac{ks \cos \theta}{r_2^4 \sin^2 \theta}$$

(j) Using part (i) gives

$$\frac{ks \csc^2 \theta}{r_1^4} - \frac{ks \cos \theta}{r_2^4 \sin^2 \theta} = 0.$$

(k) If the left side of the equation in the solution to part (j) is multiplied by $\frac{\sin^2 \theta}{s}$, we have

$$\frac{\sin^2 \theta}{s} \cdot \frac{ks}{\sin^2 \theta}\left(\frac{1}{r_1^4} - \frac{\cos \theta}{r_2^4}\right) = k\left(\frac{1}{r_1^4} - \frac{\cos \theta}{r_2^4}\right)$$

This gives the equation

$$\frac{dR}{d\theta} = \frac{k}{r_1^4} - \frac{k \cos \theta}{r_2^4} = 0.$$

(l) Using part (k) gives

$$k\left(\frac{1}{r_1^4} - \frac{\cos \theta}{r_2^4}\right) = 0$$
$$\frac{1}{r_1^4} - \frac{\cos \theta}{r_2^4} = 0 \quad k \neq 0$$
$$\frac{1}{r_1^4} = \frac{\cos \theta}{r_2^4} \Rightarrow \cos \theta = \frac{r_2^4}{r_1^4}$$

(m) Now use the first derivative test to verify that $\cos \theta$ is a minimum.

$r_1 > r_2$, so $\frac{r_2^3}{r_1^3} < \frac{r_2^4}{r_1^4} < \frac{r_2^5}{r_1^5}$. Evaluate

$\frac{dR}{d\theta} = \frac{k}{r_1^4} - \frac{k \cos \theta}{r_2^4}$ for $\cos \theta = \frac{r_2^3}{r_1^3}$ and $\cos \theta = \frac{r_2^5}{r_1^5}$.

$$\frac{k}{r_1^4} - \frac{k\left(\frac{r_2^3}{r_1^3}\right)}{r_2^4} = \frac{k}{r_1^4} - \frac{kr_2^3}{r_1^3 r_2^4} = \frac{k}{r_1^4} - \frac{k}{r_1^3 r_2}$$
$$= \frac{kr_2 - kr_1}{r_1^4 r_2}$$
$$= \frac{k(r_2 - r_1)}{r_1^4 r_2} < 0$$

$$\frac{k}{r_1^4} - \frac{k\left(\frac{r_2^5}{r_1^5}\right)}{r_2^4} = \frac{k}{r_1^4} - \frac{kr_2^5}{r_1^5 r_2^4} = \frac{k}{r_1^4} - \frac{kr_2}{r_1^5}$$
$$= \frac{kr_1 - kr_2}{r_1^5} = \frac{k(r_1 - r_2)}{r_1^5} > 0$$

The sign of the derivative changes from negative to positive, so there is a relative minimum at $\cos \theta = \frac{r_2^4}{r_1^4}$.

(n) If $r_1 = 1$ and $r_2 = \frac{1}{4}$, then

$$\cos \theta = \left(\frac{\frac{1}{4}}{1}\right)^4 = \left(\frac{1}{4}\right)^4 = \frac{1}{256} \approx 0.0039,$$

from which we get $\theta \approx 89.8° \approx 90°$.

(o) If $r_1 = 1.4$ and $r_2 = 0.8$, then

$$\cos \theta = \frac{(0.8)^4}{(1.4)^4} \approx 0.1066.$$

Thus, $\theta \approx 8.39° \approx 84°$.

55. $s(t) = -16t^2 + 40t + 3$
$s'(t) = -32t + 40$

(a) when $s'(t) = 0$,

$$-32t + 40 = 0 \Rightarrow 32t = 40 \Rightarrow t = \frac{40}{32} = \frac{5}{4}$$

Verify that $t = \frac{5}{4}$ gives a maximum.

$s'(1) = 8; \quad s'(2) = -24$

Now find the height when $t = \frac{5}{4}$.

$$s\left(\frac{5}{4}\right) = -16\left(\frac{5}{4}\right)^2 + 40\left(\frac{5}{4}\right) + 3 = 28$$

The maximum height of the cork is 28 feet.

(b) The cork remains in the air as long as $s(t) > 0$. Use the quadratic formula to solve $s(t) = 0$.

$$-16t^2 + 40t + 3 = 0$$
$$t = \frac{-40 \pm \sqrt{40^2 - 4(-16)(3)}}{2(-16)} = \frac{-5 \pm \sqrt{28}}{-4}$$
$$= \frac{5 \pm 2\sqrt{7}}{4} \approx -0.073, \ 2.573$$

Only the positive solution is relevant, so the cork stays in the air for about 2.57 seconds.

57. $s(\theta) = 2.625\cos\theta + 2.625(15 + \cos^2\theta)^{1/2}$

(a) $\frac{ds}{dt} = \frac{ds}{d\theta}\cdot\frac{d\theta}{dt}$

$$= \left[-2.625\sin\theta + 2.625\left(\frac{1}{2}\right)(15+\cos^2\theta)^{-1/2}\cdot D_\theta(15+\cos^2\theta)\right]\cdot\frac{d\theta}{dt}$$

$$= \left[-2.625\sin\theta + \frac{1.3125}{\sqrt{15+\cos^2\theta}}\cdot(-2\sin\theta\cos\theta)\right]\cdot\frac{d\theta}{dt} = -2.625\sin\theta\left(1+\frac{\cos\theta}{\sqrt{15+\cos^2\theta}}\right)\frac{d\theta}{dt}$$

(b) With $\theta = 4.944$ and $\frac{d\theta}{dt} = 505{,}168.1\frac{\text{radians}}{\text{hour}}$, we have

$$\frac{ds}{dt} = -2.625\sin(4.944)\times\left[1+\frac{\cos(4.944)}{\sqrt{15+\cos^2(4.944)}}\right](505{,}168.1)$$

$$\approx 1{,}367{,}018.749\frac{\text{inches}}{\text{hour}} = 1{,}367{,}018.749\frac{\text{inches}}{\text{hour}}\times\frac{1\text{ foot}}{12\text{ inches}}\times\frac{1\text{ mile}}{5280\text{ feet}}$$

$$\approx 21.6\text{ miles per hour}$$

5.3 Higher Derivatives, Concavity, and the Second Derivative Test

1. $f(x) = 5x^3 - 7x^2 + 4x + 3$

$f'(x) = 15x^2 - 14x + 4$

$f''(x) = 30x - 14$

$f''(0) = 30(0) - 14 = -14$

$f''(2) = 30(2) - 14 = 46$

3. $f(x) = 4x^4 - 3x^3 - 2x^2 + 6$

$f'(x) = 16x^3 - 9x^2 - 4x$

$f''(x) = 48x^2 - 18x - 4$

$f''(0) = 48(0)^2 - 18(0) - 4 = -4$

$f''(2) = 48(2)^2 - 18(2) - 4 = 152$

5. $f(x) = 3x^2 - 4x + 8$

$f'(x) = 6x - 4$

$f''(x) = 6$

$f''(0) = 6$

$f''(2) = 6$

7. $f(x) = \frac{x^2}{1+x}$

$$f'(x) = \frac{(1+x)(2x) - x^2(1)}{(1+x)^2} = \frac{2x+x^2}{(1+x)^2}$$

$$f''(x) = \frac{(1+x)^2(2+2x) - (2x+x^2)(2)(1+x)}{(1+x)^4}$$

$$= \frac{(1+x)(2+2x) - (2x+x^2)(2)}{(1+x)^3}$$

$$= \frac{2}{(1+x)^3}$$

$f''(0) = 2$

$f''(2) = \frac{2}{27}$

9. $f(x) = \sqrt{x^2+4} = (x^2+4)^{1/2}$

$$f'(x) = \frac{1}{2}(x^2+4)^{-1/2}\cdot 2x = \frac{x}{(x^2+4)^{1/2}}$$

$$f''(x) = \frac{(x^2+4)^{1/2}(1) - x\left[\frac{1}{2}(x^2+4)^{-1/2}\right]2x}{x^2+4}$$

$$= \frac{(x^2+4)^{1/2} - \frac{x^2}{(x^2+4)^{1/2}}}{x^2+4}$$

$$= \frac{(x^2+4) - x^2}{(x^2+4)^{3/2}} = \frac{4}{(x^2+4)^{3/2}}$$

$$f''(0) = \frac{4}{(0^2+4)^{3/2}} = \frac{4}{4^{3/2}} = \frac{4}{8} = \frac{1}{2}$$

$$f''(2) = \frac{4}{(2^2+4)^{3/2}} = \frac{4}{8^{3/2}} = \frac{4}{16\sqrt{2}} = \frac{1}{4\sqrt{2}}$$

11. $f(x)=32x^{3/4}$

$f'(x)=24x^{-1/4}$

$f''(x)=-6x^{-5/4}=-\dfrac{6}{x^{5/4}}$

$f''(0)$ does not exist.

$f''(2)=-\dfrac{6}{2^{5/4}}=-\dfrac{3}{2^{1/4}}$

13. $f(x)=5e^{-x^2}$

$f'(x)=5e^{-x^2}(-2x)=-10xe^{-x^2}$

$f''(x)=-10xe^{-x^2}(-2x)+e^{-x^2}(-10)$

$=20x^2e^{-x^2}-10e^{-x^2}$

$f''(0)=20(0^2)e^{-0^2}-10e^{-0^2}$

$=0-10=-10$

$f''(2)=20(2^2)e^{-(2^2)}-10e^{-(2^2)}$

$=80e^{-4}-10e^{-4}=70e^{-4}\approx 1.282$

15. $f(x)=\dfrac{\ln x}{4x}$

$f'(x)=\dfrac{4x\left(\frac{1}{x}\right)-(\ln x)(4)}{(4x)^2}$

$=\dfrac{4-4\ln x}{16x^2}=\dfrac{1-\ln x}{4x^2}$

$f''(x)=\dfrac{4x^2\left(-\frac{1}{x}\right)-(1-\ln x)8x}{\left(4x^2\right)^2}$

$=\dfrac{-4x-8x+8x\ln x}{16x^4}$

$=\dfrac{-12x+8x\ln x}{16x^4}=\dfrac{4x(-3+2\ln x)}{16x^4}$

$=\dfrac{-3+2\ln x}{4x^3}$

$f''(0)$ does not exist because $\ln 0$ is undefined.

$f''(2)=\dfrac{-3+2\ln 2}{4(2)^3}=\dfrac{-3+2\ln 2}{32}\approx -0.050$

17. $f(x)=\cos\left(x^3\right)$

$f'(x)=-3x^2\sin\left(x^3\right)$

$f''(x)=-6x\sin\left(x^3\right)-3x^2\left(3x^2\right)\cos\left(x^3\right)$

$=-6x\sin\left(x^3\right)-9x^4\cos\left(x^3\right)$

$f''(0)=0$

$f''(2)=-12\sin 8-144\cos 8$

19. $f(x)=7x^4+6x^3+5x^2+4x+3$

$f'(x)=28x^3+18x^2+10x+4$

$f''(x)=84x^2+36x+10$

$f'''(x)=168x+36$

$f^{(4)}(x)=168$

21. $f(x)=5x^5-3x^4+2x^3+7x^2+4$

$f'(x)=25x^4-12x^3+6x^2+14x$

$f''(x)=100x^3-36x^2+12x+14$

$f'''(x)=300x^2-72x+12$

$f^{(4)}(x)=600x-72$

23. $f(x)=\dfrac{x-1}{x+2}$

$f'(x)=\dfrac{(x+2)-(x-1)}{(x+2)^2}=\dfrac{3}{(x+2)^2}$

$f''(x)=\dfrac{-3(2)(x+2)}{(x+2)^4}=\dfrac{-6}{(x+2)^3}$

$f'''(x)=\dfrac{(-6)(-3)(x+2)^2}{(x+2)^6}$

$=18(x+2)^{-4}$ or $\dfrac{18}{(x+2)^4}$

$f^{(4)}(x)=\dfrac{-18(4)(x+2)^3}{(x+2)^8}$

$=-72(x+2)^{-5}$ or $\dfrac{-72}{(x+2)^5}$

25. $f(x)=\dfrac{3x}{x-2}$

$f'(x)=\dfrac{(x-2)(3)-3x(1)}{(x-2)^2}$

$=\dfrac{-6}{(x-2)^2}$ or $-6(x-2)^{-2}$

$f''(x)=-6(-2)(x-2)^{-3}(1)$

$=12(x-2)^{-3}$ or $\dfrac{12}{(x-2)^3}$

$f'''(x)=12(-3)(x-2)^{-4}(1)$

$=-36(x-2)^{-4}$ or $\dfrac{-36}{(x-2)^4}$

$f^{(4)}(x)=-36(-4)(x-2)^{-5}(1)$

$=144(x-2)^{-5}$ or $\dfrac{144}{(x-2)^5}$

27. **(a)** $f(x) = x^n + a_{n-1}x^{n-1} + \cdots + a_1x + a_0$
$f'(x) = nx^{n-1} + a_{n-1}(n-1)x^{n-2} + \cdots + a_1$
$f''(x) = n(n-1)x^{n-2} + a_{n-1}(n-1)(n-2)x^{n-3} + \cdots$
$f'''(x) = n(n-1)(n-2)x^{n-3} + a_{n-1}(n-1)(n-2)(n-3)x^{n-4} + \cdots$
Eventually, the coefficients of all of the terms become zero, leaving
$$f^{(n)}(x) = n(n-1)(n-2)\cdots(n-(n-1))x^{n-n}$$
$$= n(n-1)(n-2)\cdots(1)$$
$$= n!$$

(b) From part (a) $f^{(n)}(x) = n!$, a constant, so $f^{(n+1)}(x) = 0$. Therefore, if $k > n$, $f^{(k)}(x) = 0$.

29. $f(x) = e^x$
$f'(x) = e^x$
$f''(x) = e^x$
$f'''(x) = e^x$
$f^{(n)}(x) = e^x$

31. $f(x) = \sin x$
$f'(x) = \cos x$
$f''(x) = -\sin x$
$f'''(x) = -\cos x$
$f^4(x) = -(-\sin x) = \sin x$
$f^{(4n)}(x) = \sin x$

33. Concave upward on $(2, \infty)$
Concave downward on $(-\infty, 2)$
Inflection point at $(2, 3)$

35. Concave upward on $(-\infty, -1)$ and $(8, \infty)$
Concave downward on $(-1, 8)$
Inflection points at $(-1, 7)$ and $(8, 6)$

37. Concave upward on $(2, \infty)$
Concave downward on $(-\infty, 2)$
No points inflection

39. $f(x) = x^2 + 10x - 9$
$f'(x) = 2x + 10$
$f''(x) = 2 > 0$ for all x.
Always concave upward
No inflection points

41. $f(x) = -2x^3 + 9x^2 + 168x - 3$
$f'(x) = -6x^2 + 18x + 168$
$f''(x) = -12x + 18$
$f''(x) = -12x + 18 > 0$ when
$-6(2x-3) > 0 \Rightarrow 2x - 3 < 0 \Rightarrow x < \frac{3}{2}$.

Concave upward on $\left(-\infty, \frac{3}{2}\right)$

$f''(x) = -12x + 18 < 0$ when
$-6(2x-3) < 0 \Rightarrow 2x - 3 > 0 \Rightarrow x > \frac{3}{2}$.

Concave downward on $\left(\frac{3}{2}, \infty\right)$

$f''(x) = -12x + 18 = 0$ when
$-6(2x+3) = 0 \Rightarrow 2x + 3 = 0 \Rightarrow x = \frac{3}{2}$.

$f\left(\frac{3}{2}\right) = \frac{525}{2}$

Inflection point at $\left(\frac{3}{2}, \frac{525}{2}\right)$

43. $f(x) = \frac{3}{x-5}$
$f'(x) = \frac{-3}{(x-5)^2}$
$f''(x) = \frac{-3(-2)(x-5)}{(x-5)^4} = \frac{6}{(x-5)^3}$

$f''(x) = \frac{6}{(x-5)^3} > 0$ when
$(x-5)^3 > 0 \Rightarrow x - 5 > 0 \Rightarrow x > 5$.
Concave upward on $(5, \infty)$

$f''(x) = \frac{6}{(x-5)^3} < 0$ when
$(x-5)^3 < 0 \Rightarrow x - 5 < 0 \Rightarrow x < 5$.
Concave downward on $(-\infty, 5)$

(continued on next page)

(continued)

$f''(x) \neq 0$ for any value for x; it does not exist when $x = 5$. There is a change of concavity there, but no inflection point since $f(5)$ does not exist.

45. $f(x) = x(x+5)^2$

$f'(x) = x(2)(x+5) + (x+5)^2$
$= (x+5)(2x+x+5) = (x+5)(3x+5)$

$f''(x) = (x+5)(3) + (3x+5)$
$= 3x+15+3x+5 = 6x+20$

$f''(x) = 6x+20 > 0$ when

$2(3x+10) > 0 \Rightarrow 3x > -10 \Rightarrow x > -\frac{10}{3}.$

Concave upward on $\left(-\frac{10}{3}, \infty\right)$

$f''(x) = 6x+20 < 0$ when

$2(3x+10) < 0 \Rightarrow 3x < -10 \Rightarrow x < -\frac{10}{3}.$

Concave downward on $\left(-\infty, -\frac{10}{3}\right)$

$$f\left(-\frac{10}{3}\right) = -\frac{10}{3}\left(-\frac{10}{3}+5\right)^2 = \frac{-10}{3}\left(\frac{-10+15}{3}\right)^2 = -\frac{10}{3}\cdot\frac{25}{9} = -\frac{250}{27}$$

Inflection point at $\left(-\frac{10}{3}, -\frac{250}{27}\right)$

47. $f(x) = 18x - 18e^{-x}$

$f'(x) = 18 - 18e^{-x}(-1) = 18 + 18e^{-x}$

$f''(x) = 18e^{-x}(-1) = -18e^{-x}$

$f''(x) = -18e^{-x} < 0$ for all x

$f(x)$ is never concave upward and always concave downward. There are no points of inflection since $-18e^{-x}$ is never equal to 0.

49. $f(x) = x^{8/3} - 4x^{5/3}$

$f'(x) = \frac{8}{3}x^{5/3} - \frac{20}{3}x^{2/3}$

$f''(x) = \frac{40}{9}x^{2/3} - \frac{40}{9}x^{-1/3} = \frac{40(x-1)}{9x^{1/3}}$

$f''(x) = 0$ when $x = 1$

$f''(x)$ fails to exist when $x = 0$

Note that both $f(x)$ and $f'(x)$ exist at $x = 0$.

Check the sign of $f''(x)$ in the three intervals determined by $x = 0$ and $x = 1$ using test points.

$f''(-1) = \frac{40(-2)}{9(-1)} = \frac{80}{9} > 0$

$f''\left(\frac{1}{8}\right) = \frac{40\left(-\frac{7}{8}\right)}{9\left(\frac{1}{2}\right)} = -\frac{70}{9} < 0$

$f''(8) = \frac{40(7)}{9(2)} = \frac{140}{9} > 0$

Concave upward on $(-\infty, 0)$ and $(1, \infty)$; concave downward on $(0, 1)$

$f(0) = (0)^{8/3} - 4(0)^{5/3} = 0$

$f(1) = (1)^{8/3} - 4(1)^{5/3} = -3$

Inflection points at $(0, 0)$ and $(1, -3)$

51. $f(x) = \ln(x^2+1)$

$f'(x) = \frac{2x}{x^2+1}$

$f''(x) = \frac{(x^2+1)(2) - (2x)(2x)}{(x^2+1)^2} = \frac{-2x^2+2}{(x^2+1)^2}$

$f''(x) = \frac{-2x^2+2}{(x^2+1)^2} > 0$ when

$-2x^2+2 > 0 \Rightarrow -2x^2 > -2 \Rightarrow x^2 < 1 \Rightarrow$
$-1 < x < 1$

Concave upward on $(-1, 1)$

$f''(x) = \frac{-2x^2+2}{(x^2+1)^2} < 0$ when

$-2x^2+2 < 0 \Rightarrow -2x^2 < -2 \Rightarrow x^2 > 1 \Rightarrow$
$x > 1$ or $x < -1$

Concave downward on $(-\infty, -1)$ and $(1, \infty)$

$f(1) = \ln[(1)^2+1] = \ln 2$

$f(-1) = \ln[(-1)^2+1] = \ln 2$

Inflection points at $(-1, \ln 2)$ and $(1, \ln 2)$

53. $f(x) = x^2 \log|x|$

$f'(x) = 2x\log|x| + x^2\left(\frac{1}{x\ln 10}\right)$
$= 2x\log|x| + \frac{x}{\ln 10}$

$f''(x) = 2\log|x| + 2x\left(\frac{1}{x\ln 10}\right) + \frac{1}{\ln 10}$
$= 2\log|x| + \frac{3}{\ln 10}$

(continued on next page)

(continued)

$f''(x) > 0$ when $2\log|x| + \frac{3}{\ln 10} > 0 \Rightarrow$

$2\log|x| > -\frac{3}{\ln 10} \Rightarrow \log|x| > -\frac{3}{2\ln 10} \Rightarrow$

$\frac{\ln|x|}{\ln 10} > -\frac{3}{2\ln 10} \Rightarrow \ln|x| > -\frac{3}{2} \Rightarrow$

$|x| > e^{-3/2} \Rightarrow x > e^{-3/2}$ or $x < -e^{-3/2}$

Concave upward on $(-\infty, -e^{-3/2})$ and $(e^{-3/2}, \infty)$

$f''(x) < 0$ when $2\log|x| + \frac{3}{\ln 10} < 0 \Rightarrow$

$2\log|x| < -\frac{3}{\ln 10} \Rightarrow \log|x| < -\frac{3}{2\ln 10} \Rightarrow$

$\frac{\ln|x|}{\ln 10} < -\frac{3}{2\ln 10} \Rightarrow \ln|x| < -\frac{3}{2} \Rightarrow$

$|x| < e^{-3/2} \Rightarrow -e^{-3/2} < x < e^{-3/2}$

Note that $f(x)$ is not defined at $x = 0$.

Concave downward on $(-e^{-3/2}, 0)$ and $(0, e^{-3/2})$.

$$f(-e^{-3/2}) = (-e^{-3/2})^2 \log|-e^{-3/2}| = e^{-3}\log e^{-3/2} = -\frac{3e^{-3}}{2\ln 10}$$

$$f(e^{-3/2}) = (e^{-3/2})^2 \log|e^{-3/2}| = e^{-3}\log e^{-3/2} = -\frac{3e^{-3}}{2\ln 10}$$

Inflection points at $\left(-e^{-3/2}, -\frac{3e^{-3}}{2\ln 10}\right)$ and $\left(e^{-3/2}, -\frac{3e^{-3}}{2\ln 10}\right)$

55. $f(x) = \sin(2x)$
$f'(x) = 2\cos(2x)$
$f''(x) = -4\sin(2x)$
$f''(x) = 0$ when

$x = 0, \pm\frac{\pi}{2}, \pm\pi, \pm\frac{3\pi}{2}, \pm 2\pi, \ldots$

Use test points from the intervals

$\left(-\frac{\pi}{2}, 0\right), \left(0, \frac{\pi}{2}\right), \left(\frac{\pi}{2}, \pi\right), \ldots$

$f''\left(-\frac{\pi}{4}\right) = 4 > 0$

$f''\left(\frac{\pi}{4}\right) = -4 < 0$

$f''\left(\frac{3\pi}{4}\right) = 4 > 0$

Concave upward on

$\cdots \cup \left(-\frac{3\pi}{2}, -\pi\right) \cup \left(-\frac{\pi}{2}, 0\right) \cup \left(\frac{\pi}{2}, \pi\right) \cup \cdots$

Concave downward on

$\cdots \cup \left(-\pi, -\frac{\pi}{2}\right) \cup \left(0, \frac{\pi}{2}\right) \cup \left(\pi, \frac{3\pi}{2}\right) \cup \cdots$

Inflection points at $\left(\frac{n\pi}{2}, 0\right)$, where n is an integer.

57. Since the graph of $f'(x)$ is increasing on $(-\infty, 0)$ and $(4, \infty)$, the function is concave upward on $(-\infty, 0)$ and $(4, \infty)$. Since the graph of $f'(x)$ is decreasing on $(0, 4)$, the function is concave downward on $(0, 4)$. The inflection points are at $x = 0$ and $x = 4$.

59. Since the graph of $f'(x)$ is increasing on $(-7, 3)$ and $(12, \infty)$, the function is concave upward on $(-7, 3)$ and $(12, \infty)$. Since the graph of $f'(x)$ is decreasing on $(-\infty, -7)$ and $(3, 12)$, the function is concave downward on $(-\infty, -7)$ and $(3, 12)$. The inflection points are at $x = -7$, 3, and 12.

61. Choose $f(x) = x^k$, where $1 < k < 2$.

If $k = \frac{4}{3}$, then $f'(x) = \frac{4}{3}x^{1/3}$ and

$f''(x) = \frac{4}{9}x^{-2/3} = \frac{4}{9x^{2/3}}$

Critical number: 0

Since $f'(x)$ is negative when $x < 0$ and positive when $x > 0$, $f(x) = x^{4/3}$ has a relative minimum at $x = 0$.

If $k = \frac{5}{3}$, then $f'(x) = \frac{5}{3}x^{2/3}$ and

$f''(x) = \frac{10}{9}x^{-1/3} = \frac{10}{9x^{1/3}}$

$f''(x)$ is never 0, and does not exist when $x = 0$, so, the only candidate for an inflection point is at $x = 0$. Because $f''(x)$ is negative when $x < 0$ and positive when $x > 0, f(x) = x^{5/3}$ has an inflection point at $x = 0$.

63. **(a)** The slope of the tangent line to $f(x) = e^x$ as $x \to -\infty$ is close to 0 since the tangent line is almost horizontal, and a horizontal line has a slope of 0.

(b) The slope of the tangent line to $f(x) = e^x$ as $x \to 0$ is close to 1 since the first derivative represents the slope of the tangent line, $f'(x) = e^x$, and $e^0 = 1$.

65. $f(x) = -x^2 - 10x - 25$
$f'(x) = -2x - 10$
$f'(x) = 0 \Rightarrow -2(x+5) = 0 \Rightarrow x = -5$
Critical number: –5
$f''(x) = -2 < 0$ for all x.
The curve is concave downward, which means a relative maximum occurs at $x = -5$.

67. $f(x) = 3x^3 - 3x^2 + 1$
$f'(x) = 9x^2 - 6x$
$= 3x(3x-2) = 0$
Critical numbers: 0 and $\frac{2}{3}$
$f''(x) = 18x - 6$
$f''(0) = -6 < 0$, which means that a relative maximum occurs at $x = 0$.
$f''\left(\frac{2}{3}\right) = 6 > 0$, which means that a relative minimum occurs at $x = \frac{2}{3}$.

69. $f(x) = (x+3)^4$
$f'(x) = 4(x+3)^3$
$f'(x) = 0 \Rightarrow 4(x+3)^3 = 0 \Rightarrow x = -3$
Critical number: $x = -3$
$f''(x) = 12(x+3)^2$
$f''(-3) = 12(-3+3)^2 = 0$
The second derivative test fails. Use the first derivative test.

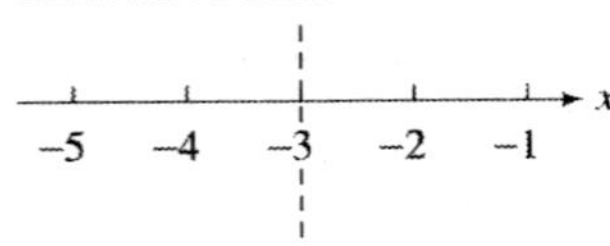

$f'(-4) = 4(-4+3)^2 = 4(-1)^3 = -4 < 0$
This indicates that f is decreasing on $(-\infty, -3)$.

$f'(0) = 4(0+3)^3 = 4(3)^3 = 108 > 0$
This indicates that f is increasing on $(-3, \infty)$.
A relative minimum occurs at –3.

71. $f(x) = x^{7/3} + x^{4/3}$
$f'(x) = \frac{7}{3}x^{4/3} + \frac{4}{3}x^{1/3}$
$f'(x) = 0$ when $\frac{7}{3}x^{4/3} + \frac{4}{3}x^{1/3} = 0 \Rightarrow$
$\frac{x^{1/3}}{3}(7x+4) = 0 \Rightarrow x = 0$ or $x = -\frac{4}{7}$.
Critical numbers: $-\frac{4}{7}, 0$
$f''(x) = \frac{28}{9}x^{1/3} + \frac{4}{9}x^{-2/3}$
$f''\left(-\frac{4}{7}\right) = \frac{28}{9}\left(-\frac{4}{7}\right)^{-1/3} + \frac{4}{9}\left(-\frac{4}{7}\right)^{-2/3}$
≈ -1.9363
Relative maximum occurs at $-\frac{4}{7}$.

$f''(0)$ does not exist, so the second derivative test fails. Use the first derivative test.

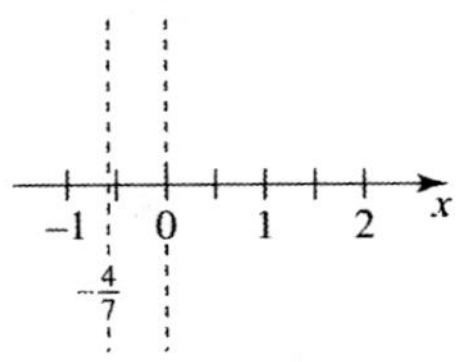

$f'\left(-\frac{1}{2}\right) = \frac{7}{3}\left(-\frac{1}{2}\right)^{4/3} + \frac{4}{3}\left(-\frac{1}{2}\right)^{1/3} \approx -0.1323$
This indicates that f is decreasing on $\left(-\frac{4}{7}, 0\right)$.
$f'(1) = \frac{7}{3}(1)^{4/3} + \frac{4}{3}(1)^{1/3} = \frac{11}{3}$
This indicates that f is increasing on $(0, \infty)$.
Relative minimum occurs at 0.

73. $f'(x) = x^3 - 6x^2 + 7x + 4$
$f''(x) = 3x^2 - 12x + 7$
Graph f' and f'' in the window $[-5, 5]$ by $[-5, 15]$.
Graph of f':

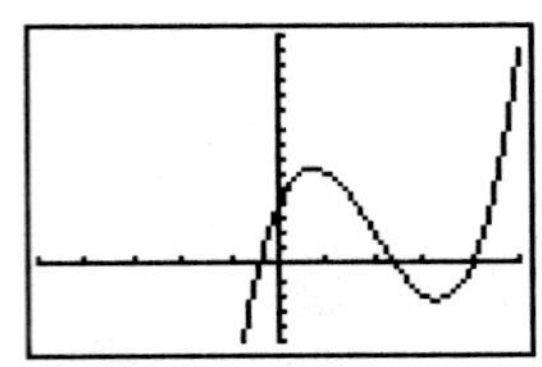

Graph of f'':

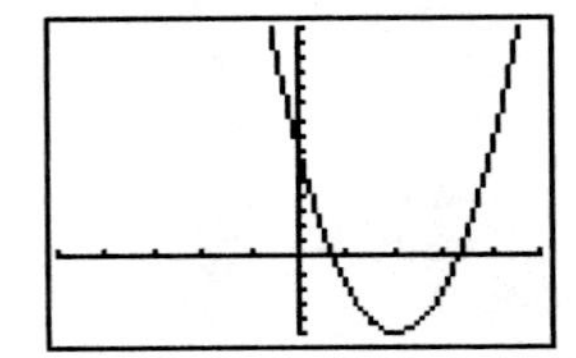

(a) f has relative extrema where $f'(x)=0$. Use the graph to approximate the x-intercepts of the graph of f'. These numbers are the solutions of the equation $f'(x)=0$.

We find that the critical numbers of f are about -0.4, 2.4, and 4.0.

By either looking at the graph of f' and applying the first derivative test or by looking at the graph of f'' and applying the second derivative test, we see that f has relative minima at about -0.4 and 4.0 and a relative maximum at about 2.4.

(b) Examine the graph of f' to determine the intervals where the graph lies above and below the x-axis. We see that $f'(x)>0$ on about $(-0.4, 2.4)$ and $(4.0, \infty)$, indicating that f is increasing on the same intervals. We also see that $f'(x)<0$ on about $(-\infty, -0.4)$ and $(2.4, 4.0)$, indicating that f is decreasing on the same intervals.

(c) Examine the graph of f''. We see that this graph has two x-intercepts, so there are two x-values where $f''(x)=0$. These x-values are about 0.7 and 3.3. Because the sign of f'' changes at these two values, we see that the x-values of the inflection points of the graph of f are about 0.7 and 3.3.

(d) We observe from the graph of f'' that $f''(x)>0$ on about $(-\infty, 0.7)$ and $(3.3, \infty)$, so f is concave upward on the same intervals. Likewise, we observe that $f''(x)<0$ on about $(0.7, 3.3)$, so f is concave downward on the same interval.

75. $$f'(x)=\frac{1-x^2}{(x^2+1)^2}$$

$$f''(x)=\frac{(x^2+1)^2(-2x)-(1-x^2)(2)(x^2+1)2x}{(x^2+1)^4}$$

$$=\frac{-2x(x^2+1)[(x^2+1)+2(1-x^2)]}{(x^2+1)^4}$$

$$=\frac{-2x(3-x^2)}{(x^2+1)^3}$$

Graph f' and f'' in the window $[-3, 3]$ by $[-1.5, 1, 5]$.

Graph of f':

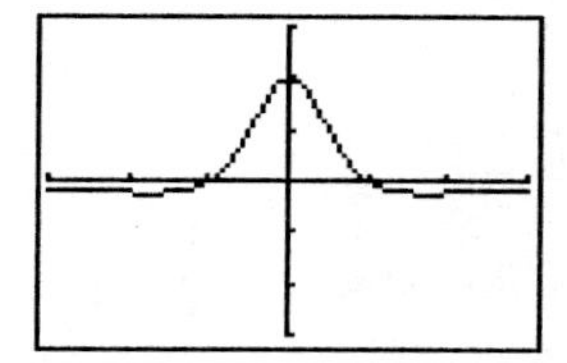

Graph of f'':

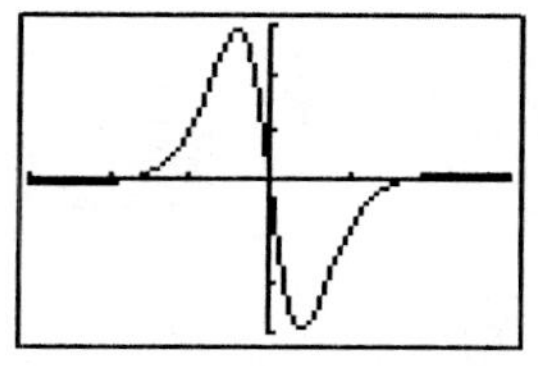

(a) The critical numbers of f are the x-intercepts of the graph of f'. (Note that there are no values where f' does not exist.) We see from the graph that these x-values are -1 and 1. By looking at the graph of f' and applying the first derivative test or by looking at the graph of f'' and applying the second derivative test, we see that f has a relative minimum at -1 and a relative maximum at 1.

(b) Examine the graph of f' to determine the intervals where the graph lies above and below the x-axis. We see that $f'(x)>0$ on $(-1, 1)$, indicating that f is increasing on the same interval. We also see that $f'(x)<0$ on $(-\infty, -1)$ and $(1, \infty)$, indicating that f is decreasing on the same intervals.

(c) Examine the graph of f''. We see that the graph has three x-intercepts, so there are three values where $f''(x)=0$. These x-values are about -1.7, 0, and about 1.7. Because the sign of f'' and thus the concavity of f changes at these three values, we see that the x-values of the inflection points of the graph of f are about -1.7, 0, and about 1.7.

(d) We observe from the graph of f'' that $f''(x) > 0$ on about $(-1.7, 0)$ and $(1.7, \infty)$, so f is concave upward on the same intervals. Likewise, we observe that $f''(x) < 0$ on about $(-\infty, -1.7)$ and $(0, 1.7)$, so f is concave downward on the same intervals.

77. There are many examples. The easiest is $f(x) = \sqrt{x}$. This graph is increasing and concave downward.

$$f'(x) = \frac{1}{2}x^{-1/2} = \frac{1}{2\sqrt{x}}$$

$f'(0)$ does not exist, while $f'(x) > 0$ for all $x > 0$. (Note that the domain of f is $[0, \infty)$.)

As x increases, the value of $f'(x)$ decreases, but remains positive. It approaches zero, but never becomes zero or negative.

79. (a) $R(t) = t^2(t-18) + 96t + 1000, \quad 0 < t < 8$

$= t^3 - 18t^2 + 96t + 1000$

$R'(t) = 3t^2 - 36t + 96$

Set $R'(t) = 0$.

$3t^2 - 36t + 96 = 0$

$t^2 - 12t + 32 = 0$

$(t-8)(t-4) = 0 \Rightarrow t = 8 \quad \text{or} \quad t = 4$

8 is not in the domain of $R(t)$.

$R''(t) = 6t - 36$

$R''(4) = -12 < 0$ implies that $R(t)$ is maximized at $t = 4$, so the population is maximized at 4 hours.

(b) $R(4) = 4^2(4-18) + 96(4) + 1000$

$= -224 + 384 + 1000 = 1160$

The maximum population is 1160 million.

81. $K(t) = \dfrac{3t}{t^2+4}$

(a) $K'(t) = \dfrac{3(t^2+4) - (2t)(3t)}{(t^2+4)^2} = \dfrac{-3t^2+12}{(t^2+4)^2}$

$K'(t) = 0 \Rightarrow -3t^2 + 12 = 0 \Rightarrow t^2 = 4 \Rightarrow t = 2 \quad \text{or} \quad t = -2$

For this application, the domain of K is $[0, \infty)$, so the only critical number is 2.

To determine whether a relative maximum or minimum occurs at $x = 2$, we find $K''(2)$ and use the second derivative test.

$$K''(t) = \frac{(t^2+4)^2(-6t) - (-3t^2+12)(2)(t^2+4)(2t)}{(t^2+4)^4} = \frac{-6t(t^2+4) - 4x(-3t^2+12)}{(t^2+4)^3} = \frac{6t^3 - 72t}{(t^2+4)^3}$$

$K''(2) = \dfrac{6(2)^3 - 72(2)}{\left(2^2+4\right)^3} = \dfrac{-96}{512} = -\dfrac{3}{16} < 0$ implies that $K(x)$ is maximized at $x = 2$. Thus, the concentration is a maximum after 2 hours.

(b) $K(2) = \dfrac{3(2)}{(2)^2+4} = \dfrac{3}{4}$

The maximum concentration is $\dfrac{3}{4}\%$.

83. $G(t)=\dfrac{10,000}{1+49e^{-0.1t}}$

$$G'(t)=\frac{(1+49e^{-0.1t})(0)-(10,000)(-4.9e^{-0.1t})}{(1+49e^{-0.1t})^2}=\frac{49,000e^{-0.1t}}{(1+49e^{-0.1t})^2}$$

To find $G''(t)$, apply the quotient rule to find the derivative of $G'(t)$.

The numerator of $G''(t)$ will be

$$\begin{aligned}
&\left(1+49e^{-0.1t}\right)^2\left(49,000e^{-0.1t}\right)(-0.1)-\left(49,000e^{-0.1t}\right)(2)\left(1+49e^{-0.1t}\right)\left(-4.9e^{-0.1t}\right)\\
&=\left(1+49e^{-0.1t}\right)^2\left(-4900e^{-0.1t}\right)-10\left(-4900e^{-0.1t}\right)(2)\left(1+49e^{-0.1t}\right)\left(-4.9e^{-0.1t}\right)\\
&=\left(1+49e^{-0.1t}\right)^2\left(-4900e^{-0.1t}\right)-20\left(-4900e^{-0.1t}\right)\left(1+49e^{-0.1t}\right)\left(-4.9e^{-0.1t}\right)\\
&=\left(1+49e^{-0.1t}\right)\left(-4900e^{-0.1t}\right)\cdot\left[\left(1+49e^{-0.1t}\right)-20\left(-4.9e^{-0.1t}\right)\right]\\
&=\left(1+49e^{-0.1t}\right)\left(-4900e^{-0.1t}\right)\left[1+49e^{-0.1t}-98e^{-0.1t}\right]\\
&=\left(1+49e^{-0.1t}\right)\left(-4900e^{-0.1t}\right)\left(1-49e^{-0.1t}\right)
\end{aligned}$$

Thus, $G''(t)=\dfrac{\left(1+49e^{-0.1t}\right)\left(-4900e^{-0.1t}\right)\left(1-49e^{-0.1t}\right)}{\left(1+49e^{-0.1t}\right)^4}=\dfrac{\left(-4900e^{-0.1t}\right)\left(1-49e^{-0.1t}\right)}{\left(1+49e^{-0.1t}\right)^3}.$

$G''(t)=0$ when $-4900e^{-0.1t}=0$ or $1-49e^{-0.1t}=0.$

$-4900e^{-0.1t}<0,$ and thus never equals zero.

$1-49e^{-0.1t}=0\Rightarrow 1=49e^{-0.1t}\Rightarrow \dfrac{1}{49}=e^{-0.1t}\Rightarrow \ln\left(\dfrac{1}{49}\right)=-0.1t\Rightarrow$

$\ln 1-\ln 49=0.1t\Rightarrow t=10\ln 49\Rightarrow t\approx 38.9182$

The point of inflection is (38.9182, 5000).

85. $L(t)=Be^{-ce^{-kt}}$

$L'(t)=Be^{-ce^{-kt}}(-ce^{-kt})'=Be^{-ce^{-kt}}[-ce^{-kt}(-kt)']=Bcke^{-ce^{-kt}-kt}$

$L''(t)=Bcke^{-ce^{-kt}-kt}(-ce^{-kt}-kt)'=Bcke^{-ce^{-kt}-kt}[-ce^{-kt}(-kt)'-k]$

$\quad=Bcke^{-ce^{-kt}-kt}(cke^{-kt}-k)=Bck^2e^{-ce^{-kt}-kt}(ce^{-kt}-1)$

$L''(t)=0$ when $ce^{-kt}-1=0\Rightarrow ce^{-kt}-1=0\Rightarrow \dfrac{c}{e^{kt}}=1\Rightarrow e^{kt}=c\Rightarrow kt=\ln c\Rightarrow t=\dfrac{\ln c}{k}$

Letting $c=7.267963$ and $k=0.670840$ gives $t=\dfrac{\ln 7.267963}{0.670840}\approx 2.96$ years

Verify that there is a point of inflection at $t=\dfrac{\ln c}{k}\approx 2.96.$ For $L''(t)=Bck^2e^{-ce^{-kt}-kt}(ce^{-kt}-1),$ we only need to test the factor $ce^{-kt}-1$ on the intervals determined by $t\approx 2.96$ since the other factors are always positive. $L''(1)$ has the same sign as $7.267963e^{-0.670840(1)}-1\approx 2.72>0.$

$L''(3)$ has the same sign as $7.267963e^{-0.670840(3)}-1\approx -0.029<0.$

Therefore L, is concave up on $\left(0,\dfrac{\ln c}{k}\approx 2.96\right)$ and concave down on $\left(\dfrac{\ln c}{k},\infty\right)$, so there is a point of inflection at $t=\dfrac{\ln c}{k}\approx 2.96$ years. This signifies the time when the rate of growth begins to slow down since L changes from concave upward to concave downward at this inflection point.

87. $v'(x) = -35.98 + 12.09x - 0.4450x^2$
$v'(x) = 12.09 - 0.89x$
$v''(x) = -0.89$
Since $-0.89 < 0$, the function is always concave downward.

89. **(a)** $l_1(v) = 0.08e^{0.33v}$
$l_1'(v) = 0.0264e^{0.33v}$
$l_1''(v) = 0.008712e^{0.33v}$
Because $l_1''(v) > 0$ for all v, $l_1(v)$ is concave upward.

$l_2(v) = -0.87v^2 + 28.17v - 211.41$
$l_2'(v) = -1.74v + 28.17$
$l_2''(v) = -1.74$
Because $l_2''(v) < 0$ for all v, $l_2(v)$ is concave downward.

(b) Answers will vary.

91. $g(x) = \frac{x}{x^2 + x + 1}$

(a) $$g'(x) = \frac{(x^2+x+1)\frac{d}{dx}x - x\frac{d}{dx}(x^2+x+1)}{(x^2+x+1)^2} = \frac{(x^2+x+1) - x(2x+1)}{(x^2+x+1)^2} = \frac{-x^2+1}{(x^2+x+1)^2}$$

Now solve $g'(x) > 0$ and $g'(x) < 0$.

$$\frac{-x^2+1}{(x^2+x+1)^2} > 0 \Rightarrow -x^2+1 > 0 \Rightarrow 1 > x^2 \Rightarrow 0 < x < 1$$

$$\frac{-x^2+1}{(x^2+x+1)^2} < 0 \Rightarrow -x^2+1 < 0 \Rightarrow 1 < x^2 \Rightarrow x > 1$$

$g'(x) > 0$ on $(0, 1)$. $g'(x) < 0$ on $(1, \infty)$.

(b) $$g''(x) = \frac{(x^2+x+1)^2\frac{d}{dx}(-x^2+1) - (-x^2+1)\frac{d}{dx}\left((x^2+x+1)^2\right)}{\left((x^2+x+1)^2\right)^2}$$

$$= \frac{(x^2+x+1)^2(-2x) - (-x^2+1)\left(2(x^2+x+1)(2x+1)\right)}{(x^2+x+1)^4} = \frac{(x^2+x+1)(-2x) - (-x^2+1)(2(2x+1))}{(x^2+x+1)^3}$$

$$= \frac{(-2x^3-2x^2-2x) - (-x^2+1)(4x+2)}{(x^2+x+1)^3} = \frac{(-2x^3-2x^2-2x) - (-4x^3-2x^2+4x+2)}{(x^2+x+1)^3}$$

$$= \frac{2x^3-6x-2}{(x^2+x+1)^3}$$

(c)

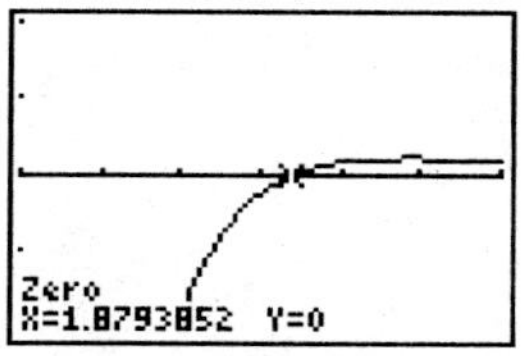

Zero
X=1.8793852 Y=0

[1.2, 2.4] by [–0.0002, 0.0002]

$g''(x) < 0$ on $(0, 1.879)$ and $g''(x) > 0$ on $(1.879, \infty)$.

93. $f(R) = \dfrac{TCR}{1 + hCR}$

The second derivative rule tells us that a curve is concave down if $f'' < 0$.

To simplify the computations, factor out the constant C from the numerator.

$$f(R) = \frac{TCR}{1+hCR} = C\left(\frac{TR}{1+hCR}\right)$$

Now, apply the quotient rule. Because T is a function of R, we must use the chain rule.

$$f'(R) = C\left(\frac{\left[\frac{d}{dR}(TR)\right](1+hCR) - TR\left[\frac{d}{dR}(1+hCR)\right]}{(1+hCR)^2}\right)$$

Next, differentiate the sum $(1 + hCR)$ and simplify.

$$f'(R) = C\left(\frac{\left[\frac{d}{dR}(TR)\right](1+hCR) - TR(hC)}{(1+hCR)^2}\right) = C\left(\frac{\left[\frac{d}{dR}(TR)\right](1+hCR) - hCTR}{(1+hCR)^2}\right)$$

Apply the product rule to find $\dfrac{d}{dR}(TR)$. The derivative of T with respect to R is $\dfrac{dT}{dR}$.

$$f'(R) = C\left(\frac{\left[\frac{d}{dR}(TR)\right](1+hCR) - hCTR}{(1+hCR)^2}\right) = C\left(\frac{\left[\frac{dT}{dR}R + T\right](1+hCR) - hCTR}{(1+hCR)^2}\right)$$

$$= C\left(\frac{\frac{dT}{dR}R + T + hCR^2\frac{dT}{dR} + hCTR - hCTR}{(1+hCR)^2}\right) = C\left(\frac{hCR^2\frac{dT}{dR} + R\frac{dT}{dR} + T}{(1+hCR)^2}\right) = C\left(\frac{T + \left(hCR^2 + R\right)\frac{dT}{dR}}{(1+hCR)^2}\right)$$

Now find $f''(R)$. Again, start with the quotient rule.

$$f''(R) = C\left(\frac{(1+hCR)^2\frac{d}{dR}\left(T + \left(hCR^2+R\right)\frac{dT}{dR}\right) - \left(T + \left(hCR^2+R\right)\frac{dT}{dR}\right)\frac{d}{dR}\left((1+hCR)^2\right)}{\left((1+hCR)^2\right)^2}\right)$$

In the numerator, we have $(1+hCR)^2\dfrac{d}{dR}\left(T + \left(hCR^2+R\right)\dfrac{dT}{dR}\right) = (1+hCR)^2\dfrac{dT}{dR} + \dfrac{d}{dR}\left(\left(hCR^2+R\right)\dfrac{dT}{dR}\right)$

Now use the product rule to differentiate the second term.

$$(1+hCR)^2\frac{d}{dR}\left(T + \left(hCR^2+R\right)\frac{dT}{dR}\right) = (1+hCR)^2\left[\frac{dT}{dR} + \frac{d}{dR}\left(\left(hCR^2+R\right)\frac{dT}{dR}\right)\right]$$

$$= (1+hCR)^2\left[\frac{dT}{dR} + \left(\frac{d}{dR}\left(hCR^2+R\right)\frac{dT}{dR} + \left(hCR^2+R\right)\frac{d^2T}{dR^2}\right)\right]$$

(continued on next page)

(continued)

$$= (1+hCR)^2\left[\frac{dT}{dR} + (1+2hCR)\frac{dT}{dR} + R(hCR+1)\frac{d^2T}{dR^2}\right]$$
$$= (1+hCR)^2\left[(2+2hCR)\frac{dT}{dR} + R(1+hCR)\frac{d^2T}{dR^2}\right]$$
$$= (1+hCR)^2\left[2(1+hCR)\frac{dT}{dR} + R(1+hCR)\frac{d^2T}{dR^2}\right]$$

$$\frac{d}{dR}\left((1+hCR)^2\right) = 2(1+hCR)\frac{d}{dR}(1+hCR) = 2hc(1+hCR)$$

Combining the terms in the numerator gives

$$(1+hCR)^2\frac{d}{dR}\left(T+\left(hCR^2+R\right)\frac{dT}{dR}\right) - \left(T+\left(hCR^2+R\right)\frac{dT}{dR}\right)\frac{d}{dR}\left((1+hCR)^2\right)$$
$$= (1+hCR)^2\left[2(1+hCR)\frac{dT}{dR} + R(1+hCR)\frac{d^2T}{dR^2}\right] - \left(T+\left(hCR^2+R\right)\frac{dT}{dR}\right)(2hC(1+hCR))$$
$$= (1+hCR)^2\left[2(1+hCR)\frac{dT}{dR} + R(1+hCR)\frac{d^2T}{dR^2}\right] - (2hC(1+hCR))\left(T+R(hCR+1)\frac{dT}{dR}\right)$$

$$f''(R) = \frac{C}{(1+hCR)^4}\left((1+hCR)^2\left[2(1+hCR)\frac{dT}{dR} + R(1+hCR)\frac{d^2T}{dR^2}\right] - (2hC(1+hCR))\left(T+R(hCR+1)\frac{dT}{dR}\right)\right)$$

Factor $(1+hCR)$ from the numerator and denominator.

$$f''(R) = C\left(\frac{(1+hCR)\left[2(1+hCR)\frac{dT}{dR} + R(1+hCR)\frac{d^2T}{dR^2}\right] - (2hC)\left(T+R(hCR+1)\frac{dT}{dR}\right)}{(1+hCR)^3}\right)$$
$$= C\left(\frac{2(1+hCR)^2\frac{dT}{dR} + R(1+hCR)^2\frac{d^2T}{dR^2} - 2hCT - \left(2hCR(hCR+1)\frac{dT}{dR}\right)}{(1+hCR)^3}\right)$$

In the numerator

$$2(1+hCR)^2\frac{dT}{dR} + R(1+hCR)^2\frac{d^2T}{dR^2} - 2hCT - \left(2hCR(hCR+1)\frac{dT}{dR}\right)$$
$$= 2(1+hCR)^2\frac{dT}{dR} - \left(2hCR(hCR+1)\frac{dT}{dR}\right) + R(1+hCR)^2\frac{d^2T}{dR^2} - 2hCT$$
$$= \left(2+4hCR+2(hCR)^2\right)\frac{dT}{dR} - \left(\left(2(hCR)^2+2hCR\right)\frac{dT}{dR}\right) + R(1+hCR)^2\frac{d^2T}{dR^2} - 2hCT$$
$$= \left(2+4hCR+2(hCR)^2 - \left(2(hCR)^2+2hCR\right)\right)\frac{dT}{dR} + R(1+hCR)^2\frac{d^2T}{dR^2} - 2hCT$$
$$= (2+2hCR)\frac{dT}{dR} + R(1+hCR)^2\frac{d^2T}{dR^2} - 2hCT = 2(1+hCR)\frac{dT}{dR} + R(1+hCR)^2\frac{d^2T}{dR^2} - 2hCT$$

Thus, $f''(R) = C\left(\frac{2(1+hCR)\frac{dT}{dR} + R(1+hCR)^2\frac{d^2T}{dR^2} - 2hCT}{(1+hCR)^3}\right)$

The denominator is always positive because h, C, and R are positive, so the curve is concave down when the numerator of $f'' < 0$, or when $2(1+hCR)\frac{dT}{dR} + R(1+hCR)^2\frac{d^2T}{dR^2} - 2hCT < 0.$

95. Since the rate of violent crimes is decreasing but at a slower rate than in previous years, we know that $f'(t) < 0$ but $f''(t) > 0$. Note that since $f'(t) < 0$, f is decreasing, and since $f''(t) > 0$, the graph of f is concave upward.

97. $s(t) = 256t - 16t^2$
$v(t) = s'(t) = 256 - 32t$
$a(t) = v'(t) = s''(t) = -32$
To find when the maximum height occurs, solve $s'(t) = 0$.
$256 - 32t = 0 \Rightarrow t = 8$
Find the maximum height.
$s(8) = 256(8) - 16(8^2) = 1024$
The maximum height of the ball is 1024 ft. The ball hits the ground when $s = 0$.
$256t - 16t^2 = 0 \Rightarrow 16t(16 - t) = 0 \Rightarrow$
$t = 0$ (initial moment)
$t = 16$ (final moment)
The ball hits the ground 16 seconds after being thrown.

99. The car was moving most rapidly when $t \approx 6$, because acceleration was positive on (0, 6) and negative after $t = 6$, so velocity was a maximum at $t = 6$.

101. $R(x) = \frac{4}{27}(-x^3 + 66x^2 + 1050x - 400)$
$0 \le x \le 25$

$R'(x) = \frac{4}{27}(-3x^2 + 132x + 1050)$

$R''(x) = \frac{4}{27}(-6x + 132)$

A point of diminishing returns occurs at a point of inflection, or where $R''(x) = 0$.

$\frac{4}{27}(-6x + 132) = 0 \Rightarrow -6x + 132 = 0 \Rightarrow$
$6x = 132 \Rightarrow x = 22$
Test $R''(x)$ to determine whether concavity changes at $x = 22$.

$R''(20) = \frac{4}{27}(-6 \cdot 20 + 132) = \frac{16}{9} > 0$

$R''(24) = \frac{4}{27}(-6 \cdot 24 + 132) = -\frac{16}{9} < 0$

$R(x)$ is concave upward on (0, 22) and concave downward on (22, 25).

$R(22) = \frac{4}{27}[-(22)^3 + 66(22)^2 + 1060(22) - 400]$
≈ 6517.9

The point of diminishing returns is (22, 6517.9).

103. $R(x) = -0.6x^3 + 3.7x^2 + 5x,\ 0 \le x \le 6$
$R'(x) = -1.8x^2 + 7.4x + 5$
$R''(x) = -3.6x + 7.4$
A point of diminishing returns occurs at a point of inflection or where $R''(x) = 0$.
$-3.6x + 7.4 = 0 \Rightarrow -3.6x = -7.4 \Rightarrow$
$x = \frac{-7.4}{-3.6} \approx 2.06$
Test $R''(x)$ to determine whether concavity changes at $x \approx 2.06$.
$R''(2) = -3.6(2) + 7.4 = 0.2 > 0$
$R''(3) = -3.6(3) + 7.4 = -3.4 < 0$
$R(x)$ is concave upward on (0, 2.06) and concave downward on (2.06, 6).

$R\left(\frac{7.4}{3.6}\right) \approx 20.7$

The point of diminishing returns is (2.06, 20.7).

5.4 Curve Sketching

1. Graph $y = x \ln |x|$ on a graphing calculator. A suitable choice for the viewing window is [−1, 1] by [−1, 1], Xscl = 0.1, Yscl = 0.1.

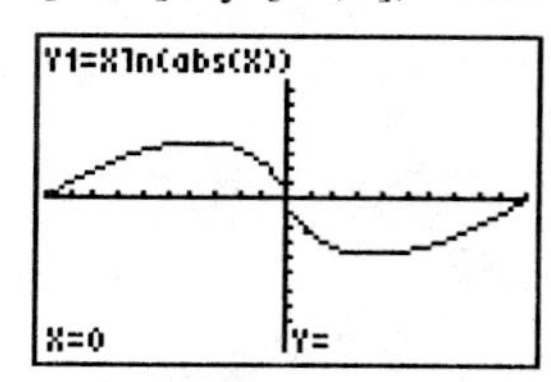

The calculator shows no y-value when $x = 0$ because 0 is not in the domain of this function. However, we see from the graph that $\lim_{x \to 0^-} x \ln |x| = 0$ and $\lim_{x \to 0^+} x \ln |x| = 0$.
Thus, $\lim_{x \to 0} x \ln |x| = 0$.

3. $f(x) = -2x^3 - 9x^2 + 108x - 10$
Domain is $(-\infty, \infty)$.

$f(-x) = -2(-x)^3 - 9(-x)^2 + 108(-x) - 10$
$= 2x^3 - 9x^2 - 108x - 10$
No symmetry
$f'(x) = -6x^2 - 18x + 108 = -6(x^2 + 3x - 18)$
$= -6(x + 6)(x - 3)$
$f'(x) = 0$ when $x = -6$ or $x = 3$.
Critical numbers: −6 and 3

(continued on next page)

(continued)

Critical points: $(-6, -550)$ and $(3, 179)$

$f''(x) = -12x - 18$
$f''(-6) = 54 > 0$
$f''(3) = -54 < 0$

Relative maximum at $x = 3$, relative minimum at $x = -6$

Increasing on $(-6, 3)$

Decreasing on $(-\infty, -6)$ and $(3, \infty)$

$f''(x) = -12x - 18 = 0 \Rightarrow -6(2x + 3) = 0 \Rightarrow$
$x = -\frac{3}{2}$

Point of inflection at $(-1.5, -185.5)$

Concave upward on $(-\infty, -1.5)$

Concave downward on $(-1.5, \infty)$

y-intercept:
$y = -2(0)^3 - 9(0)^2 + 108(0) - 10 = -10$

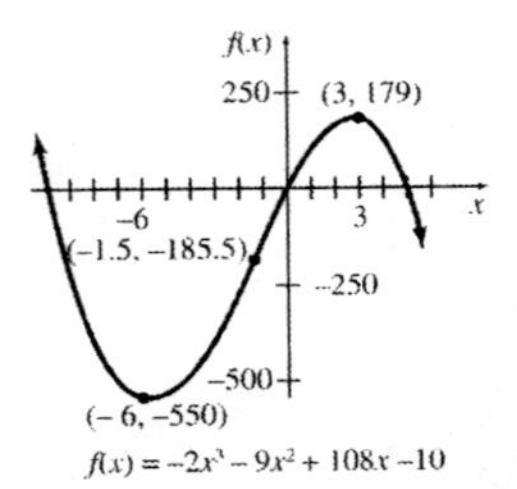

5. $f(x) = -3x^3 + 6x^2 - 4x - 1$

Domain is $(-\infty, \infty)$.

$$f(-x) = -3(-x)^3 + 6(-x)^2 - 4(-x) - 1 = 3x^3 + 6x^2 + 4x - 1$$

No symmetry

$f'(x) = 0 \Rightarrow -9x^2 + 12x - 4 = 0 \Rightarrow$
$-(3x - 2)^2 = 0 \Rightarrow x = \frac{2}{3}$

Critical number: $\frac{2}{3}$

$$f\left(\frac{2}{3}\right) = -3\left(\frac{2}{3}\right)^3 + 6\left(\frac{2}{3}\right)^2 - 4\left(\frac{2}{3}\right) - 1 = -\frac{17}{9}$$

Critical point: $\left(\frac{2}{3}, -\frac{17}{9}\right)$

$f'(0) = -9(0)^2 + 12(0) - 4 = -4 < 0$
$f'(1) = -9(1)^2 + 12(1) - 4 = -1 < 0$

No relative extremum at $\left(\frac{2}{3}, -\frac{17}{9}\right)$

Decreasing on $(-\infty, \infty)$

$f''(x) = -18x + 12 = -6(3x - 2)$
$-6(3x - 2) = 0 \Rightarrow x = \frac{2}{3}$

Point of inflection at $\left(\frac{2}{3}, -\frac{17}{9}\right)$

$f''(0) = -18(0) + 12 = 12 > 0$
$f''(1) = -18(1) + 12 = -6 < 0$

Concave upward on $\left(-\infty, \frac{2}{3}\right)$

Concave upward on $\left(\frac{2}{3}, \infty\right)$

y-intercept:
$y = -3(0)^3 + 6(0)^2 - 4(0) - 1 = -1$

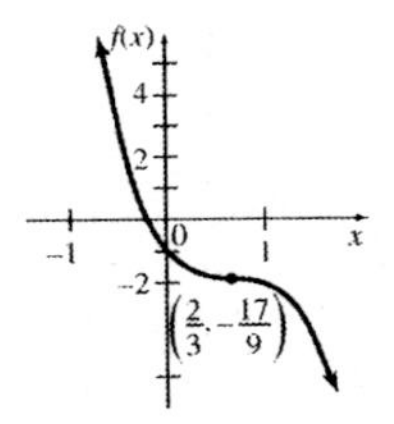

7. $f(x) = x^4 - 24x^2 + 80$

Domain is $(-\infty, \infty)$.

$$f(-x) = (-x)^4 - 24(-x)^2 + 80 = x^4 - 24x^2 + 80 = f(x)$$

The graph is symmetric about the y-axis.

$f'(x) = 4x^3 - 48x$

$f'(x) = 0 \Rightarrow 4x^3 - 48x = 0 \Rightarrow$
$4x(x^2 - 12) = 0 \Rightarrow$
$4x(x - 2\sqrt{3})(x + 2\sqrt{3}) = 0 \Rightarrow x = 0, \pm 2\sqrt{3}$

Critical numbers: $-2\sqrt{3}$, 0, and $2\sqrt{3}$

Critical points: $(-2\sqrt{3}, -64)$, $(0, 80)$, and $(2\sqrt{3}, -64)$

$f''(x) = 12x^2 - 48$
$f''(-2\sqrt{3}) = 12(-2\sqrt{3})^2 - 48 = 96 > 0$
$f''(0) = 12(0)^2 - 48 = -48 < 0$
$f''(2\sqrt{3}) = 12(2\sqrt{3})^2 - 48 = 96 > 0$

Relative maximum at 0, relative minima at $x = -2\sqrt{3}$ and $x = 2\sqrt{3}$

Increasing on $(-2\sqrt{3}, 0)$ and $(2\sqrt{3}, \infty)$

Decreasing on $(-\infty, -2\sqrt{3})$ and $(0, 2\sqrt{3})$

$12x^2 - 48 = 0 \Rightarrow 12(x^2 - 4) = 0 \Rightarrow x = \pm 2$

Points of inflection at $(-2, 0)$ and $(2, 0)$

(continued on next page)

(continued)

Concave upward on $(-\infty, -2)$ and $(2, \infty)$

Concave downward on $(-2, 2)$

x-intercepts: $0 = x^4 - 24x^2 + 80$

Let $u = x^2$.

$u^2 - 24u + 80 = 0$
$(u-4)(u-20) = 0$
$u = 4$ or $u = 20$
$x = \pm 2$ or $x = \pm 2\sqrt{5}$

y-intercept: $y = (0)^4 - 24(0)^2 + 80 = 80$

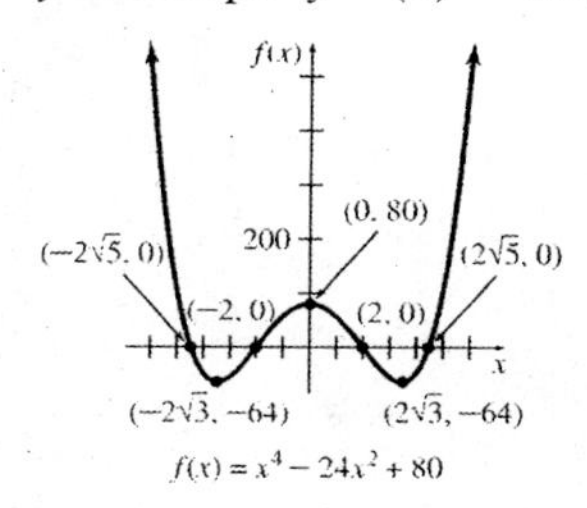

9. $f(x) = x^4 - 4x^3$

Domain is $(-\infty, \infty)$.

$f(-x) = (-x)^4 - 4(-x)^3$
$= x^4 + 4x^3 \neq f(x)$ or $-f(x)$

The graph is not symmetric about the y-axis or the origin.

$f'(x) = 4x^3 - 12x^2$

$f'(x) = 0 \Rightarrow 4x^3 - 12x^2 = 0 \Rightarrow$

$4x^2(x-3) = 0 \Rightarrow x = 0$ or $x = 3$

Critical numbers: 0 and 3

Critical points: $(0, 0)$ and $(3, -27)$

$f''(x) = 12x^2 - 24x$

$f''(0) = 12(0)^2 - 24(0) = 0$

$f''(3) = 12(3)^2 - 24(3) = 36 > 0$

Second derivative test fails for 0. Use first derivative test.

$f'(-1) = 4(-1)^3 - 12(-1)^2 = -16 < 0$

$f'(1) = 4(1)^3 - 12(1)^2 = -8 < 0$

Neither a relative minimum nor maximum at 0

Relative minimum at $x = 3$

Increasing on $(3, \infty)$

Decreasing on $(-\infty, 3)$

$f''(x) = 0 \Rightarrow 12x^2 - 24x = 0 \Rightarrow$

$12x(x-2) = 0 \Rightarrow x = 0$ or $x = 2$

Points of inflection at $(0, 0)$ and $(2, -16)$

Concave upward on $(-\infty, 0)$ and $(2, \infty)$

Concave downward on $(0, 2)$

x-intercepts:

$x^4 - 4x^3 = 0 \Rightarrow x^3(x-4) = 0 \Rightarrow x = 0$ or $x = 4$

y-intercepts: $y = (0)^4 - 4(0)^3 = 0$

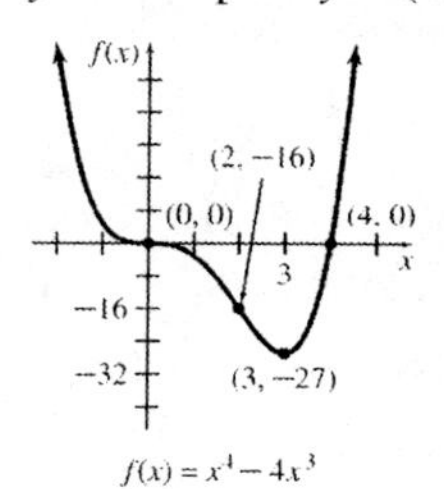

11. $f(x) = 2x + \dfrac{10}{x} = 2x + 10x^{-1}$

Since $f(x)$ does not exist when $x = 0$, the domain is $(-\infty, 0) \cup (0, \infty)$.

$f(-x) = 2(-x) + 10(-x)^{-1} = -(2x + 10x^{-1})$
$= -f(x)$

The graph is symmetric about the origin.

$f'(x) = 2 - 10x^{-2}$

$f'(x) = 0 \Rightarrow 2 - \dfrac{10}{x^2} = 0 \Rightarrow \dfrac{2(x^2-5)}{x^2} = 0 \Rightarrow$

$x = \pm\sqrt{5}$

Critical numbers: $-\sqrt{5}$ and $\sqrt{5}$

Critical points: $(-\sqrt{5}, -4\sqrt{5})$ and $(\sqrt{5}, 4\sqrt{5})$

Test a point in the intervals $(-\infty, -\sqrt{5})$, $(-\sqrt{5}, 0)$, $(0, \sqrt{5})$, and $(\sqrt{5}, \infty)$.

$f'(-3) = 2 - 10(-3)^{-2} = \dfrac{8}{9} > 0$

$f'(-1) = 2 - 10(-1)^{-2} = -8 < 0$

$f'(1) = 2 - 10(1)^{-2} = -8 < 0$

$f'(3) = 2 - 10(3)^{-2} = \dfrac{8}{9} > 0$

Relative maximum at $x = -\sqrt{5}$

Relative minimum at $x = \sqrt{5}$

Increasing on $(-\infty, -\sqrt{5})$ and $(\sqrt{5}, \infty)$

Decreasing on $(-\sqrt{5}, 0)$ and $(0, \sqrt{5})$

(Recall that $f(x)$ does not exist at $x = 0$.)

$f''(x) = 20x^{-3} = \dfrac{20}{x^3}$

$f''(x) = \dfrac{20}{x^3}$ is never equal to zero.

There are no inflection points.

Test a point in the intervals $(-\infty, 0)$ and $(0, \infty)$.

(continued on next page)

(continued)

$$f''(-1) = \frac{20}{(-1)^3} = -20 < 0$$

$$f''(1) = \frac{20}{(1)^3} = 20 > 0$$

Concave upward on $(0, \infty)$

Concave downward on $(-\infty, 0)$

$f(x)$ is never zero, so there are no x-intercepts.

$f(x)$ does not exist for $x = 0$, so there is no y-intercept.

Vertical asymptote at $x = 0$

$y = 2x$ is an oblique asymptote.

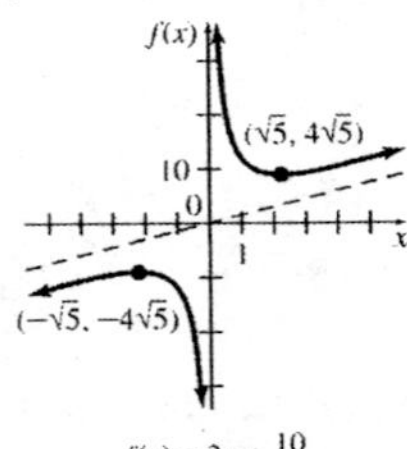

13. $f(x) = \frac{-x+4}{x+2}$

Since $f(x)$ does not exist when $x = -2$, the domain is $(-\infty, -2) \cup (-2, \infty)$.

$$f(-x) = \frac{-(-x)+4}{(-x)+2} = \frac{x+4}{-x+2}$$

The graph is not symmetric about the y-axis or the origin.

$$f'(-x) = \frac{(x+2)(-1)-(-x+4)(1)}{(x+2)^2} = \frac{-6}{(x+2)^2}$$

$f'(x) < 0$ and is never zero. $f'(x)$ fails to exist for $x = -2$.

No critical numbers; no relative extrema

Decreasing on $(-\infty, -2)$ and $(-2, \infty)$

$$f''(x) = \frac{12}{(x+2)^3}$$

$f''(x)$ fails to exist for $x = -2$.

No points of inflection

Test a point in the intervals $(-\infty, -2)$ and $(-2, \infty)$.

$f''(-3) = -12 < 0$

$f''(-1) = 12 > 0$

Concave upward on $(-2, \infty)$

Concave downward on $(-\infty, -2)$

x-intercept: $\frac{-x+4}{x+2} = 0 \Rightarrow x = 4$

y-intercept: $y = \frac{-0+4}{0+2} = 2$

Vertical asymptote at $x = -2$.

Horizontal asymptote at $y = -1$

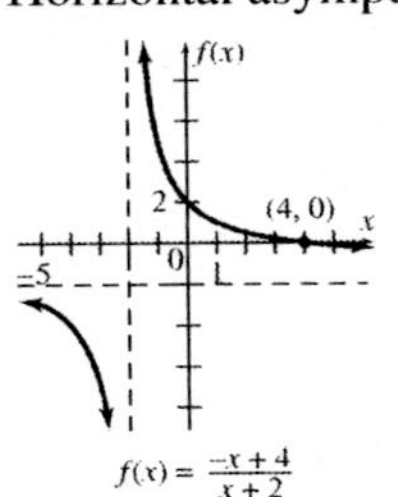

15. $f(x) = \frac{1}{x^2+4x+3} = \frac{1}{(x+3)(x+1)}$

Since $f(x)$ does not exist when $x = -3$ and $x = -1$, the domain is $(-\infty, -3) \cup (-3, -1) \cup (-1, \infty)$.

$$f(-x) = \frac{1}{(-x)^2+4(-x)+3} = \frac{1}{x^2-4x+3}$$

The graph is not symmetric about the y-axis or the origin.

$$f'(x) = \frac{0-(2x+4)}{(x^2+4x+3)^2} = \frac{-2(x+2)}{[(x+3)(x+1)]^2}$$

Critical number: -2

Test a point in the intervals $(-\infty, -3)$, $(-3, -2)$, $(-2, -1)$, and $(-1, \infty)$.

$$f'(-4) = \frac{-2(-4+2)}{[(-4+3)(-4+1)]^2} = \frac{4}{9} > 0$$

$$f'\left(-\frac{5}{2}\right) = \frac{-2\left(-\frac{5}{2}+2\right)}{\left[\left(-\frac{5}{2}+3\right)\left(-\frac{5}{2}+1\right)\right]^2} = \frac{16}{9} > 0$$

$$f'\left(-\frac{3}{2}\right) = \frac{-2\left(-\frac{3}{2}+2\right)}{\left[\left(-\frac{3}{2}+3\right)\left(-\frac{3}{2}+1\right)\right]^2} = -\frac{16}{9} < 0$$

$$f'(0) = \frac{-2(0+2)}{[(0+3)(0+1)]^2} = -\frac{4}{9} < 0$$

$$f(-2) = \frac{1}{(-2+3)(-2+1)} = -1$$

Relative maximum at $(-2, -1)$

Increasing on $(-\infty, -3)$ and $(-3, -2)$; decreasing on $(-2, -1)$ and $(-1, \infty)$

(continued on next page)

(continued)

$$f''(x)=\frac{(x^2+4x+3)^2(-2)-(-2x-4)(2)(x^2+4x+3)(2x+4)}{(x^2+4x+3)^4}$$
$$=\frac{-2(x^2+4x+3)[(x^2+4x+3)+(-2x-4)(2x+4)]}{(x^2+4x+3)^4}=\frac{-2(x^2+4x+3-4x^2-16x-16)}{(x^2+4x+3)^3}$$
$$=\frac{-2(-3x^2-12x-13)}{(x^2+4x+3)^3}=\frac{2(3x^2+12x+13)}{[(x+3)(x+1)]^3}$$

Since $3x^2+12x+13=0$ has no real solutions, there are no x-values where $f''(x)=0$. $f''(x)$ does not exist where $x=-3$ and $x=-1$. Since $f(x)$ does not exist at these x-values, there are no points of inflection.
Test a point in the intervals $(-\infty,-3)$, $(-3,-1)$, and $(-1,\infty)$.

$$f''(-4)=\frac{2[3(-4)^2+12(-4)+13]}{[(-4+3)(-4+1)]^3}=\frac{26}{27}>0$$
$$f''(-2)=\frac{2[3(-2)^2+12(-2)+13]}{[(-2+3)(-2+1)]^3}=-2<0$$
$$f''(0)=\frac{2[3(0)^2+12(0)+13]}{[(0+3)(0+1)]^3}=\frac{26}{27}>0$$

Concave upward on $(-\infty,-3)$ and $(-1,\infty)$;
concave downward on $(-3,-1)$

$f(x)$ is never zero, so there are no x-intercepts.

y-intercept: $y=\frac{1}{(0+3)(0+1)}=\frac{1}{3}$

Vertical asymptotes where $f(x)$ is undefined at $x=-3$ and $x=-1$.
Horizontal asymptote at $y=0$.

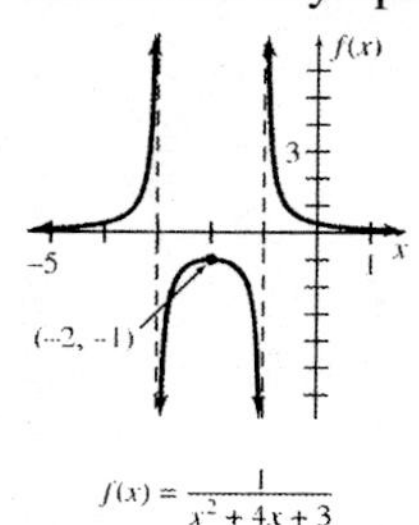

17. $f(x)=\frac{x}{x^2+1}$

Domain is $(-\infty,\infty)$

$$f(-x)=\frac{-x}{(-x)^2+1}=-\frac{x}{x^2+1}=-f(x)$$

The graph is symmetric about the origin.

$$f'(x)=\frac{(x^2+1)(1)-x(2x)}{(x^2+1)^2}=\frac{1-x^2}{(x^2+1)^2}$$

$f'(x)=0\Rightarrow 1-x^2=0\Rightarrow x=\pm1$

Critical numbers: 1 and -1

Critical points: $\left(1,\frac{1}{2}\right)$ and $\left(-1,-\frac{1}{2}\right)$

$$f''(x)=\frac{(x^2+1)^2(-2x)-(1-x^2)(2)(x^2+1)(2x)}{(x^2+1)^4}$$
$$=\frac{-2x^3-2x-4x+4x^3}{(x^2+1)^3}=\frac{2x^3-6x}{(x^2+1)^3}$$

$f''(1)=-\frac{1}{2}<0$

$f''(-1)=\frac{1}{2}>0$

Relative maximum at 1
Relative minimum at -1
Increasing on $(-1, 1)$
Decreasing on $(-\infty,-1)$ and $(1,\infty)$

$$f''(x)=\frac{2x^3-6x}{(x^2+1)^3}$$

$f''(x)=0\Rightarrow\frac{2x^3-6x}{(x^2+1)^3}=0\Rightarrow 2x^3-6x=0\Rightarrow$
$2x(x^2-3)=0\Rightarrow x=0,\ x=\pm\sqrt{3}$

Inflection points at $(0, 0)$, $\left(\sqrt{3},\frac{\sqrt{3}}{4}\right)$ and $\left(-\sqrt{3},-\frac{\sqrt{3}}{4}\right)$

Concave upward on $(-\sqrt{3},0)$ and $(\sqrt{3},\infty)$
Concave downward on $(-\infty,-\sqrt{3})$ and $(0,\sqrt{3})$

(continued on next page)

(continued)

x-intercept: $0 = \frac{x}{x^2+1} \Rightarrow 0 = x$

y-intercept: $y = \frac{0}{0^2+1} = 0$

Horizontal asymptote at $y = 0$

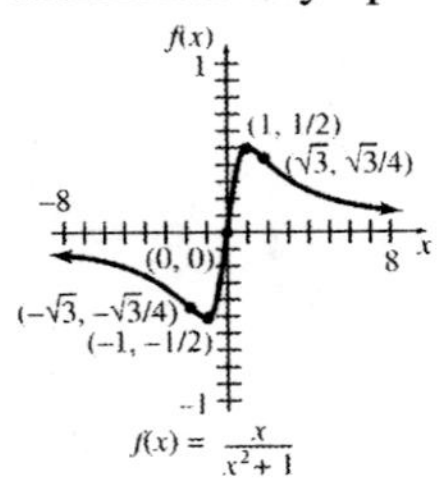

19. $f(x) = \frac{1}{x^2-9} = \frac{1}{(x+3)(x-3)}$

Since $f(x)$ does not exist when $x = -3$ and $x = 3$, the domain is $(-\infty, -3) \cup (-3, 3) \cup (3, \infty)$.

$$f(-x) = \frac{1}{(-x)^2-9} = \frac{1}{x^2-9} = f(x)$$

The graph is symmetric about the y-axis.

$$f'(x) = \frac{-2x}{(x^2-9)^2}$$

Critical number: 0

Critical point: $\left(0, -\frac{1}{9}\right)$

Test a point in the intervals $(-\infty, -3)$, $(-3, 0)$, $(0, 3)$, and $(3, \infty)$.

$$f'(-4) = \frac{-2(-4)}{[(-4)^2-9]^2} = \frac{8}{49} > 0$$

$$f'(-1) = \frac{-2(-1)}{[(-1)^2-9]^2} = \frac{1}{32} > 0$$

$$f'(1) = \frac{-2(1)}{[(1)^2-9]^2} = -\frac{1}{32} < 0$$

$$f'(4) = \frac{-2(4)}{[(4)^2-9]^2} = -\frac{8}{49} < 0$$

Relative maximum at $\left(0, -\frac{1}{9}\right)$

Increasing on $(-\infty, -3)$ and $(-3, 0)$

Decreasing on $(0, 3)$ and $(3, \infty)$

$$f''(x) = \frac{(x^2-9)^2(-2)-(-2x)(2)(x^2-9)(2x)}{(x^2-9)^4}$$

$$= \frac{-2(x^2-9)[(x^2-9)+(-2x)(2x)]}{(x^2+4)^4}$$

$$= \frac{-2(x^2-9-4x^2)}{(x^2-9)^3} = \frac{-2(-3x^2-9)}{(x^2-9)^3}$$

$$= \frac{6(x^2+3)}{[(x+3)(x-3)]^3}$$

Since $x^2+3=0$ has no solutions, there are no x-values where $f''(x)=0$. $f''(x)$ does not exist where $x = -3$ and $x = 3$. Therefore, there are no points of inflection.

Test a point in the intervals $(-\infty, -3)$, $(-3, 3)$, and $(3, \infty)$.

$$f''(-4) = \frac{6[(-4)^2+3]}{[(-4+3)(-4-3)]^3} = \frac{114}{343} > 0$$

$$f''(0) = \frac{6[(0)^2+3]}{[(0+3)(0-3)]^3} = -\frac{2}{81} < 0$$

$$f''(4) = \frac{6[(4)^2+3]}{[(4+3)(4-3)]^3} = \frac{114}{343} > 0$$

Concave upward on $(-\infty, -3)$ and $(3, \infty)$

Concave downward on $(-3, 3)$

$f(x)$ is never zero, so there are no x-intercepts.

y-intercept: $y = \frac{1}{0^2-9} = -\frac{1}{9}$

Vertical asymptotes where $f(x)$ is undefined at $x = -3$ and $x = 3$.

Horizontal asymptote at $y = 0$.

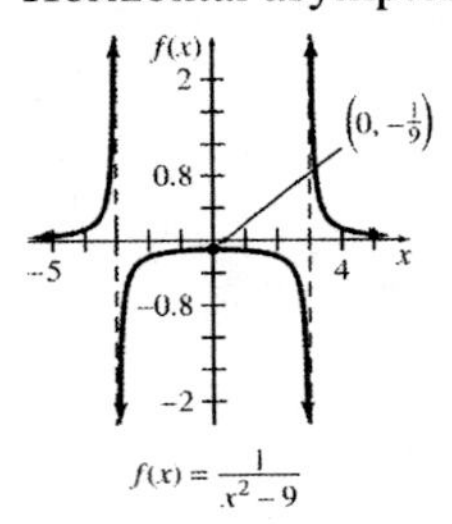

21. $f(x) = x \ln|x|$

The domain of this function is $(-\infty, 0) \cup (0, \infty)$.

$f(-x) = -x \ln|x| = -x \ln|x| = -f(x)$

The graph is symmetric about the origin.

$$f'(x) = x \cdot \frac{1}{x} + \ln|x| = 1 + \ln|x|$$

$f'(x) = 0 \Rightarrow 0 = 1 + \ln|x| \Rightarrow -1 = \ln|x| \Rightarrow$
$e^{-1} = |x| \Rightarrow x = \pm\frac{1}{e} \approx \pm 0.37.$

Critical numbers: $\pm\frac{1}{e} \approx \pm 0.37.$

(continued on next page)

(*continued*)

$$f'(-1) = 1 + \ln|-1| = 1 > 0$$
$$f'(-0.1) = 1 + \ln|-0.1| \approx -1.3 < 0$$
$$f'(0.1) = 1 + \ln|0.1| \approx -1.3 < 0$$
$$f'(1) = 1 + \ln|1| = 1 > 0$$
$$f\left(\frac{1}{e}\right) = \frac{1}{e}\ln\left|\frac{1}{e}\right| = -\frac{1}{e}$$
$$f\left(-\frac{1}{e}\right) = -\frac{1}{e}\ln\left|-\frac{1}{e}\right| = \frac{1}{e}$$

Relative maximum of $\left(-\frac{1}{e}, \frac{1}{e}\right)$; relative minimum of $\left(\frac{1}{e}, -\frac{1}{e}\right)$.

Increasing on $\left(-\infty, -\frac{1}{e}\right)$ and $\left(\frac{1}{e}, \infty\right)$ and decreasing on $\left(-\frac{1}{e}, 0\right)$ and $\left(0, \frac{1}{e}\right)$.

$$f''(x) = \frac{1}{x}$$
$$f''(-1) = \frac{1}{-1} = -1 < 0$$
$$f''(1) = \frac{1}{1} = 1 > 0$$

Concave downward on $(-\infty, 0)$;
Concave upward on $(0, \infty)$.
There is no y-intercept.
x-intercept:
$0 = x\ln|x| \Rightarrow x = 0$
or $\ln|x| = 0 \Rightarrow |x| = e^0 = 1 \Rightarrow x = \pm 1$
Since 0 is not in the domain, the only x-intercepts are -1 and 1.

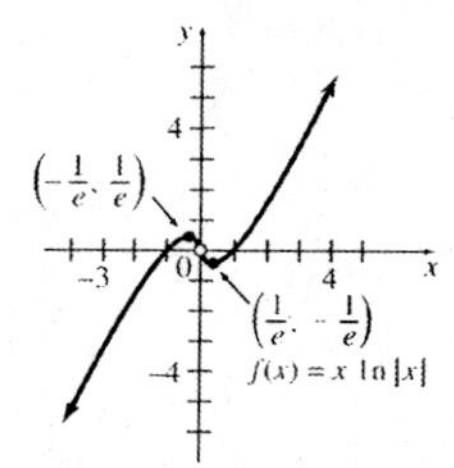

23. $f(x) = \frac{\ln x}{x}$

Note that the domain of this function is $(0, \infty)$.

$f(-x) = \frac{\ln(-x)}{-x}$ does not exist when $x \geq 0$, so there is no symmetry.

$$f'(x) = \frac{x\left(\frac{1}{x}\right) - \ln x(1)}{x^2} = \frac{1 - \ln x}{x^2}$$

Critical numbers:

$f'(x) = 0 \Rightarrow 1 - \ln x = 0 \Rightarrow 1 = \ln x \Rightarrow e = x$

$$f(e) = \frac{\ln e}{e} = \frac{1}{e}$$

Critical points: $\left(e, \frac{1}{e}\right)$

$$f'(1) = \frac{1 - \ln 1}{1^2} = \frac{1}{1} = 1 > 0$$
$$f'(3) = \frac{1 - \ln 3}{3^2} = -0.01 < 0$$

There is a relative maximum at $\left(e, \frac{1}{e}\right)$.

The function is increasing on $(0, e)$ and decreasing on (e, ∞).

$$f''(x) = \frac{x^2\left(-\frac{1}{x}\right) - (1 - \ln x)2x}{x^4}$$
$$= \frac{-x - 2x(1 - \ln x)}{x^4} = \frac{-x[1 + 2(1 - \ln x)]}{x^4}$$
$$= \frac{-(1 + 2 - 2\ln x)}{x^3} = \frac{-3 + 2\ln x}{x^3}$$

$f''(x) = 0 \Rightarrow -3 + 2\ln x = 0 \Rightarrow$
$2\ln x = 3 \Rightarrow \ln x = \frac{3}{2} = 1.5 \Rightarrow x = e^{1.5} \approx 4.48$

$$f''(1) = \frac{-3 + 2\ln 1}{1^3} = -3 < 0$$
$$f''(5) = \frac{-3 + 2\ln 5}{5^3} \approx 0.0018 > 0$$
$$f(e^{1.5}) = \frac{\ln e^{1.5}}{e^{1.5}} = \frac{1.5}{e^{1.5}} = \frac{3}{2e^{1.5}} \approx 0.33$$

Inflection point at $\left(e^{1.5}, \frac{1.5}{e^{1.5}}\right) \approx (4.48, 0.33)$

Concave downward on $(0, e^{1.5})$; concave upward on $(e^{1.5}, \infty)$
Since $x \neq 0$, there is no y-intercept.
x-intercept: $f(x) = 0 \Rightarrow \ln x = 0 \Rightarrow x = 1$
Vertical asymptote at $x = 0$
Horizontal asymptote at $y = 0$

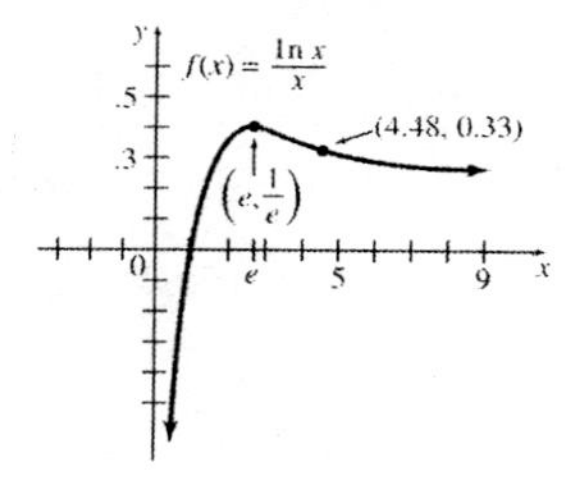

25. $f(x) = xe^{-x}$

Domain is $(-\infty, \infty)$.

$f(-x) = -xe^{x}$

The graph has no symmetry.

$f'(x) = -xe^{-x} + e^{-x} = e^{-x}(1-x)$

$f'(x) = 0 \Rightarrow e^{-x}(1-x) = 0 \Rightarrow x = 1$

Critical numbers: 1

Critical points: $\left(1, \frac{1}{e}\right)$

$f'(0) = e^{-0}(1-0) = 1 > 0$

$f'(2) = e^{-2}(1-2) = \frac{-1}{e^2} < 0$

Relative maximum at $\left(1, \frac{1}{e}\right)$

Increasing on $(-\infty, 1)$; decreasing on $(1, \infty)$

$$f''(x) = e^{-x}(-1) + (1-x)(-e^{-x})$$
$$= -e^{-x}(1+1-x) = -e^{-x}(2-x)$$

$f'' = 0 \Rightarrow -e^{-x}(2-x) = 0 \Rightarrow x = 2$

$f''(0) = -e^{-0}(2-0) = -2 < 0$

$f''(3) = -e^{-3}(2-3) = \frac{1}{e^3} > 0$

Inflection point at $\left(2, \frac{2}{e^2}\right)$

Concave downward on $(-\infty, 2)$, concave upward on $(2, \infty)$

x-intercept: $0 = xe^{-x} \Rightarrow x = 0$

y-intercept: $y = 0 \cdot e^{-0} = 0$

Horizontal asymptote at $y = 0$

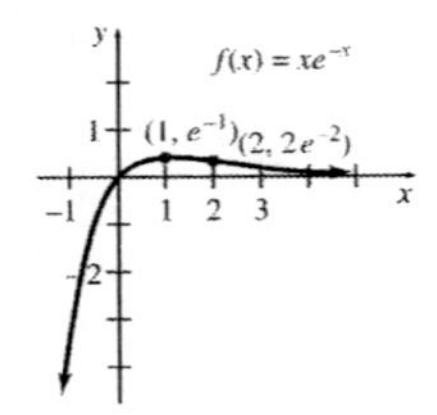

27. $f(x) = (x-1)e^{-x}$

Domain is $(-\infty, \infty)$

$f(-x) = (-x-1)e^{x}$

The graph has no symmetry.

$$f'(x) = -(x-1)e^{-x} + e^{-x}(1)$$
$$= e^{-x}[-(x-1)+1] = e^{-x}(2-x)$$

$f'(x) = 0 \Rightarrow e^{-x}(2-x) = 0 \Rightarrow x = 2$

Critical number: 2

Critical point: $\left(2, \frac{1}{e^2}\right)$

$$f''(x) = -e^{-x} + (2-x)(-e^{-x})$$
$$= -e^{-x}[1+(2-x)] = -e^{-x}(3-x)$$

$f''(2) = -e^{-2}(3-2) = \frac{-1}{e^2} < 0$

Relative maximum at $\left(2, \frac{1}{e^2}\right)$

$f'(0) = e^{-0}(2-0) = 2 > 0$

$f'(3) = e^{-3}(2-3) = \frac{-1}{e^3} < 0$

Increasing on $(-\infty, 2)$; decreasing on $(2, \infty)$.

$f''(x) = 0 \Rightarrow -e^{-x}(3-x) = 0 \Rightarrow x = 3.$

$f''(0) = -e^{-0}(3-0) = -3 < 0$

$f''(4) = -e^{-4}(3-4) = \frac{1}{e^4} > 0$

Inflection point at $\left(3, \frac{2}{e^3}\right)$

Concave downward on $(-\infty, 3)$; concave upward on $(3, \infty)$

$f(3) = (3-1)e^{-3} = \frac{2}{e^3}$

y-intercept: $y = (0-1)e^{-0} = (-1)(1) = -1$

x-intercept: $0 = (x-1)e^{-x} \Rightarrow x - 1 = 0 \Rightarrow x = 1$

Horizontal asymptote at $y = 0$

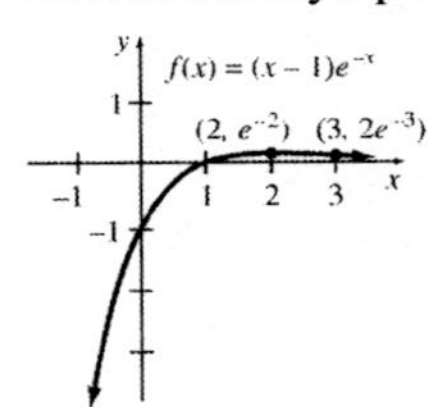

29. $f(x) = x^{2/3} - x^{5/3}$

Domain is $(-\infty, \infty)$.

$f(-x) = x^{2/3} + x^{5/3}$

The graph has no symmetry.

$f'(x) = \frac{2}{3}x^{-1/3} - \frac{5}{3}x^{2/3} = \frac{2-5x}{3x^{1/3}}$

$f'(x) = 0 \Rightarrow 2 - 5x = 0 \Rightarrow x = \frac{2}{5}$

Critical number: $x = \frac{2}{5}$

(continued on next page)

(continued)

$$f\left(\frac{2}{5}\right)=\left(\frac{2}{5}\right)^{2/3}-\left(\frac{2}{5}\right)^{5/3}=\frac{3\cdot 2^{2/3}}{5^{5/3}}\approx 0.326$$

Critical point: $(0.4, 0.326)$

$$f''(x)=\frac{3x^{1/3}(-5)-(2-5x)(3)\left(\frac{1}{3}\right)x^{-2/3}}{(3x^{1/3})^2}$$

$$=\frac{-15x^{1/3}-(2-5x)x^{-2/3}}{9x^{2/3}}$$

$$=\frac{-15x-(2-5x)}{9x^{4/3}}=\frac{-10x-2}{9x^{4/3}}$$

$$f''\left(\frac{2}{5}\right)=\frac{-10\left(\frac{2}{5}\right)-2}{9\left(\frac{2}{5}\right)^{4/3}}\approx -2.262<0$$

Relative maximum at

$$\left(\frac{2}{5},\frac{3\cdot 2^{2/3}}{5^{5/3}}\right)\approx(0.4, 0.326)$$

$f'(x)$ does not exist when $x=0$

Since $f''(0)$ is undefined, use the first derivative test.

$$f'(-1)=\frac{2-5(-1)}{3(-1)^{1/3}}=\frac{7}{-3}<0$$

$$f'\left(\frac{1}{8}\right)=\frac{2-5\left(\frac{1}{8}\right)}{3\left(\frac{1}{8}\right)^{1/3}}=\frac{11}{12}>0$$

$$f'(1)=\frac{2-5}{3\cdot 1^{1/3}}=-1<0$$

Relative minimum at $(0, 0)$

f increases on $\left(0,\frac{2}{5}\right)$.

f decreases on $(-\infty, 0)$ and $\left(\frac{2}{5},\infty\right)$.

$$f''(x)=0\Rightarrow -10x-2=0\Rightarrow x=-\frac{1}{5}$$

$f''(x)$ undefined when $9x^{4/3}=0\Rightarrow x=0$

$$f''(-1)=\frac{-10(-1)-2}{9(-1)^{4/3}}=\frac{8}{9}>0$$

$$f''\left(-\frac{1}{8}\right)=\frac{-10\left(-\frac{1}{8}\right)-2}{9\left(-\frac{1}{8}\right)^{4/3}}=-\frac{4}{3}<0$$

$$f''(1)=\frac{-10(1)-2}{9(1)^{4/3}}=-\frac{4}{3}<0$$

Concave upward on $\left(-\infty,-\frac{1}{5}\right)$

Concave downward on $\left(-\frac{1}{5},\infty\right)$

Inflection point at $\left(-\frac{1}{5},\frac{6}{5^{5/3}}\right)\approx(-0.2, 0.410)$

y-intercept: $y=0^{2/3}-0^{5/3}=0$

x-intercept:

$0=x^{2/3}-x^{5/3}\Rightarrow 0=x^{2/3}(1-x)\Rightarrow$
$x=0$ or $x=1$

31. $f(x)=x+\cos x$

Domain is $(-\infty,\infty)$.

$f(-x)=-x+\cos(-x)=-x+\cos x\neq -f(x)$

The graph has no symmetry.

$f'(x)=1-\sin x$

$f'(x)=0\Rightarrow 1-\sin x=0\Rightarrow \sin x=1\Rightarrow$

$$x=\pm\frac{\pi}{2},\ \pm\frac{3\pi}{2},\ \pm\frac{5\pi}{2},\ \ldots$$

$f'(1)=1-\sin 1\approx 0.159>0$
$f'(3)=1-\sin 3\approx 0.859>0$
$f'(6)=1-\sin 6\approx 1.279>0$

f increases on $(-\infty,\infty)$

$f''(x)=-\cos x$

$f''(x)=0\Rightarrow -\cos x=0\Rightarrow$

$$x=\pm\frac{\pi}{2},\ \pm\frac{3\pi}{2},\ \pm\frac{5\pi}{2},\ \ldots$$

$f''(0)=-\cos 0=-1<0$
$f''(2)=-\cos 2=0.416>0$
$f''(6)=-\cos 6=-0.960<0$

Concave upward on

$$\ldots,\left(-\frac{3\pi}{2},-\frac{\pi}{2}\right),\left(\frac{\pi}{2},\frac{3\pi}{2}\right),\left(\frac{5\pi}{2},\frac{7\pi}{2}\right),\ldots$$

Concave downward on

$$\ldots,\left(-\frac{\pi}{2},\frac{\pi}{2}\right),\left(\frac{3\pi}{2},\frac{5\pi}{2}\right),\left(\frac{7\pi}{2},\frac{9\pi}{2}\right),\ldots$$

Inflection points at

$$\ldots,\left(-\frac{\pi}{2},\frac{\pi}{2}\right),\left(\frac{\pi}{2},\frac{\pi}{2}\right),\left(\frac{3\pi}{2},\frac{3\pi}{2}\right),\ldots$$

y-intercept: $y=0+\cos 0=1$

(continued on next page)

(continued)

x-intercept:
$0 = x + \cos x \Rightarrow x = -\cos x \Rightarrow x \approx -0.739$

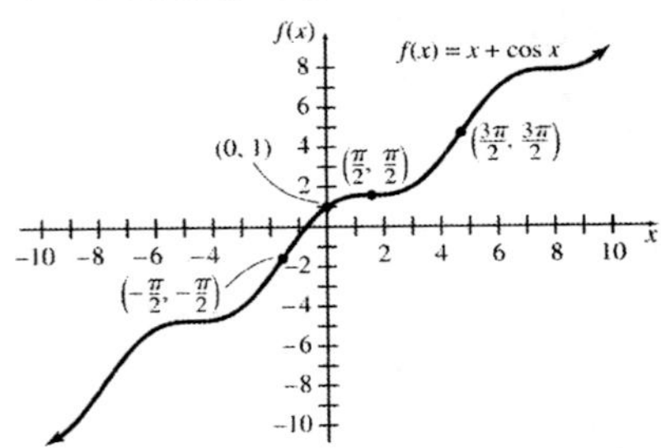

33. For Exercises 3, 7, and 9, the relative maxima or minima are outside the vertical window of $-10 \le y \le 10$.
For Exercise 11, the default window shows only a small portion of the graph.
For Exercise 15, the default window does not allow the graph to properly display the vertical asymptotes.

35. For Exercises 17, 19, 23, 25, and 27, the y-coordinate of the relative minimum, relative maximum, or inflection points is so small, it may be hard to distinguish.

For Exercises 35–39 other graphs are possible.

37. **(a)** indicates a smooth, continuous curve except where there is a vertical asymptote.

(b) indicates that the function decreases on both sides of the asymptote, so there are no relative extrema.

(c) gives the horizontal asymptote $y = 2$.

(d) and **(e)** indicate that concavity does not change left of the asymptote, but that the right portion of the graph changes concavity at $x = 2$ and $x = 4$.
There are inflection points at 2 and 4.

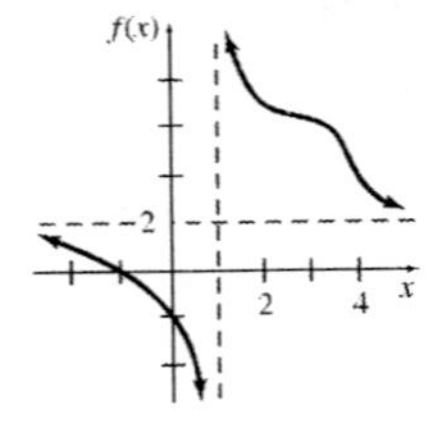

39. **(a)** indicates that there can be no asymptotes, sharp "corners", holes, or jumps. The graph must be one smooth curve.

(b) and **(c)** indicate relative maxima at –3 and 4 and a relative minimum at 1.

(d) and **(e)** are consistent with **(g)**.

(f) indicates turning points at the critical numbers –3 and 4.

(g) There are inflection points at (–1, 3) and (2, 4).

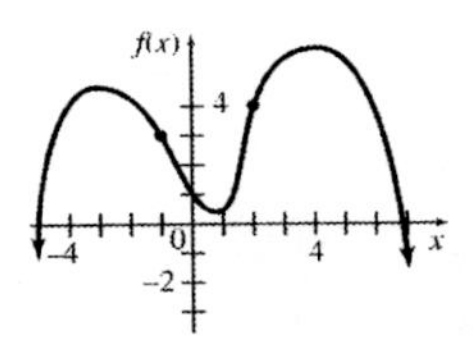

41. **(a)** indicates that the curve may not contain breaks.

(b) indicates that there is a sharp "corner" at 4.

(c) gives a point at (1, 5).

(d) shows critical numbers.

(e) and **(f)** indicate (combined with **(c)** and **(d)**) a relative maximum at (1, 5), and (combined with **(b)**) a relative minimum at 4.

(g) is consistent with **(b)**.

(h) indicates the curve is concave upward on (2, 3).

(i) indicates the curve is concave downward on $(-\infty, 2)$, (3, 4) and $(4, \infty)$.

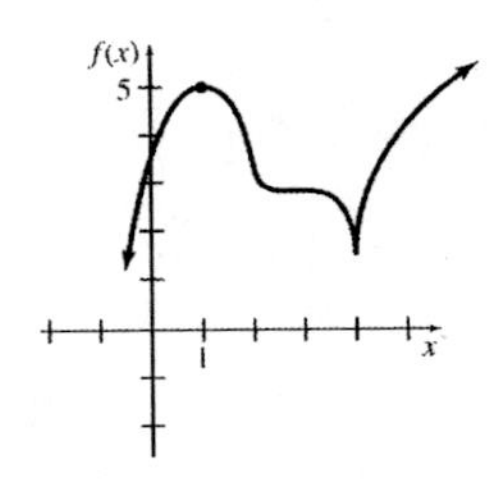

43. Sketch the curve for $l_1(v) = 0.08e^{0.33v}$

$l_1'(v) = 0.0264e^{0.33v}$
$e^{0.33v} \neq 0$
$l_1(v)$ has no critical points.

$l_1''(v) = 0.008712e^{0.33v}$
$e^{0.33v} \neq 0$
$l_1(v)$ has no inflection points.

(continued on next page)

(*continued*)

Sketch the curve for $l_2 = -0.87v^2 + 28.17v - 211.41$

$l_2'(v) = -1.74v + 28.17$

$l_2'(v) = 0 \Rightarrow -1.74v + 28.17 = 0 \Rightarrow v \approx 16.19$

Critical point: $(16.19, 16.62)$

$l_2''(v) = -1.74$

$l_2(v)$ has no inflection points.

$l_2(v)$ has a relative maximum at $(16.19, 16.62)$.

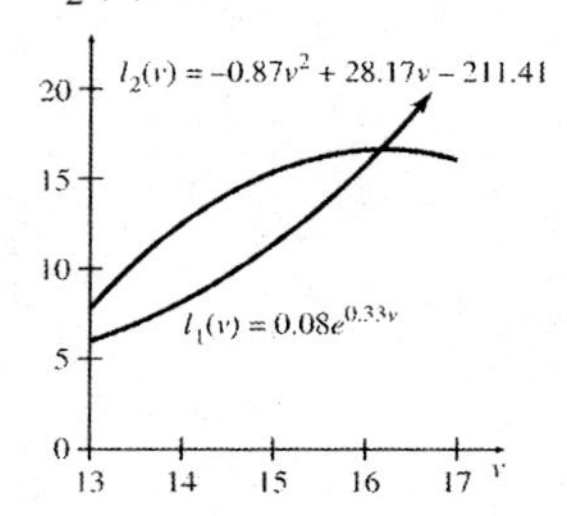

45. $I(n) = \dfrac{an^2}{b + n^2}$; $a = 0.5$, $b = 10$

$$I(n) = \frac{0.5n^2}{10 + n^2}$$

$$I'(n) = \frac{(10 + n^2)n - 0.5n^2(2n)}{(10 + n^2)^2} = \frac{10n}{(10 + n^2)^2}$$

$$I'(n) = 0 \Rightarrow \frac{10n}{(10 + n^2)^2} = 0 \Rightarrow n = 0$$

There is one critical point at (0, 0).

$$I''(n) = \frac{(10 + n^2)^2(10) - 10n(2)(10 + n^2)(2n)}{(10 + n^2)^4} = \frac{10(10 + n^2)\left[(10 + n^2) - 4n^2\right]}{(10 + n^2)^4} = \frac{100 - 30n^2}{(10 + n^2)^3}$$

$$I''(n) = 0 \Rightarrow \frac{100 - 30n^2}{(10 + n^2)^3} = 0 \Rightarrow 30n^2 = 100 \Rightarrow$$

$$n = \pm\sqrt{\frac{10}{3}} \approx \pm 1.826$$

The negative answer does not make sense in this problem so disregard it. The inflection point is about (1.83, 01.25).

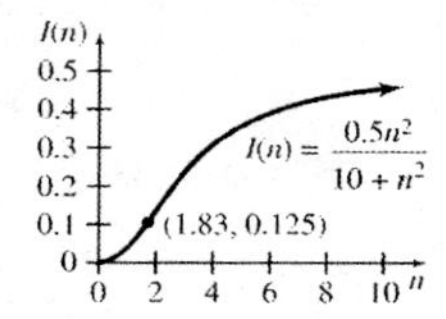

47. $L(C)=\dfrac{aC^n}{k+C^n}$, $n=4$, $k=3$, $a=100 \Rightarrow L(C)=\dfrac{100C^4}{3+C^4}$

$$L'(C)=\frac{(3+C^4)400C^3-100C^4(4C^3)}{(3+C^4)^2}=\frac{1200C^3}{(3+C^4)^2}$$

$$L'(C)=0\Rightarrow \frac{1200C^3}{(3+C^4)^2}=0\Rightarrow C=0$$

$$(3+C^4)^2\neq 0\Rightarrow C^4\neq -3$$

The only critical point is (0, 0).

$$L''(C)=\frac{(3+C^4)^2(3600C^2)-2(3+C^4)(4C^3)(1200C^3)}{(3+C^4)^4}=\frac{(3+C^4)^2(3600C^2)-9600C^6(3+C^4)}{(3+C^4)^4}$$

$$=\frac{(3+C^4)(1200C^2)\left[3(3+C^4)-8C^4\right]}{(3+C^4)^4}=\frac{(1200C^2)(9-5C^4)}{(3+C^4)^3}$$

$$L''(C)=0\Rightarrow \frac{(1200C^2)(9-5C^4)}{(3+C^4)^3}=0\Rightarrow (1200C^2)(9-5C^4)=0\Rightarrow C=0 \text{ or } C=\left(\frac{9}{5}\right)^{1/4}\approx 1.16$$

The inflection point is about (1.16, 37.5).

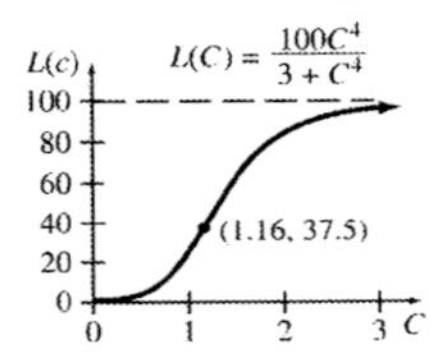

Chapter 5 Review Exercises

1. True
2. False: The function is increasing on this interval.
3. False: The function could have neither a minimum nor a maximum; consider $f(x)=x^3$ at $c=0$.
4. True
5. False: Consider $f(x)=x^{3/2}$ at $c=0$.
6. True
7. False: The function is concave upward.
8. False: Consider $f(x)=x^4$ at $c=0$.
9. False: Consider $f(x)=x^4$, which has a relative minimum at $c=0$.
10. False: Polynomials are rational functions, and nonconstant polynomials have neither vertical nor horizontal asymptotes.
11. True
12. False: Consider $f(x)=x^2$ on (–1, 1) with $c=0$.
13. The function is increasing when $f'>0$ and decreasing when $f'<0$.
14. Find the relative extrema by solving $f'(x)=0$. To test whether a relative extremum is a minimum or a maximum, use either the first derivative or second derivative test.
15. No. Answers will vary.
16. Use the second derivative to locate inflection points and to determine if the graph is concave up or concave down.

17. $f(x) = x^2 + 9x + 8$
$f'(x) = 2x + 9$

$f'(x) = 0 \Rightarrow 2x + 9 = 0 \Rightarrow x = -\frac{9}{2}$ and f' exists everywhere.

Critical number: $-\frac{9}{2}$

Test an x-value in the intervals $\left(-\infty, -\frac{9}{2}\right)$ and $\left(-\frac{9}{2}, -\infty\right)$.

$f'(-5) = -1 < 0$
$f'(-4) = 1 > 0$

f is increasing on $\left(-\frac{9}{2}, \infty\right)$ and decreasing on $\left(-\infty, -\frac{9}{2}\right)$.

18. $f(x) = -2x^2 + 7x + 14$
$f'(x) = -4x + 7$

$f'(x) = 0 \Rightarrow -4x + 7 = 0 \Rightarrow x = \frac{7}{4}$ and f' exists everywhere.

Critical number: $\frac{7}{4}$

Test an x-value in the intervals $\left(-\infty, \frac{7}{4}\right)$ and $\left(\frac{7}{4}, \infty\right)$.

$f'(0) = 7 > 0$
$f'(2) = -1 < 0$

f is increasing on $\left(-\infty, \frac{7}{4}\right)$ and decreasing on $\left(\frac{7}{4}, \infty\right)$.

19. $f(x) = -x^3 + 2x^2 + 15x + 16$
$f'(x) = -3x^2 + 4x + 15 = -(3x^2 - 4x - 15)$
$= -(3x + 5)(x - 3)$
$f'(x) = 0 \Rightarrow -(3x + 5)(x - 3) = 0 \Rightarrow$
$x = -\frac{5}{3}$ or $x = 3$
and f' exists everywhere.

Critical numbers: $-\frac{5}{3}$ and 3

Test an x-value in the intervals $\left(-\infty, -\frac{5}{3}\right)$, $\left(-\frac{5}{3}, 3\right)$, and $(3, \infty)$.

$f'(-2) = -5 < 0$
$f'(0) = 15 > 0$
$f'(4) = -17 < 0$

f is increasing on $\left(-\frac{5}{3}, 3\right)$ and decreasing on $\left(-\infty, -\frac{5}{3}\right)$ and $(3, \infty)$.

20. $f(x) = 4x^3 + 8x^2 - 16x + 11$
$f'(x) = 12x^2 + 16x - 16 = 4(3x^2 + 4x - 4)$
$= 4(3x - 2)(x + 2)$
$f'(x) = 0 \Rightarrow 4(3x - 2)(x + 2) = 0 \Rightarrow$
$x = \frac{2}{3}$ or $x = -2$
and f' exists everywhere.

Critical numbers: -2 and $\frac{2}{3}$

Test an x-value in the intervals $(-\infty, -2)$, $\left(-2, \frac{2}{3}\right)$, and $\left(\frac{2}{3}, \infty\right)$.

$f'(-3) = 44 > 0$
$f'(0) = -16 < 0$
$f'(1) = 12 > 0$

f is increasing on $(-\infty, -2)$ and $\left(\frac{2}{3}, \infty\right)$ and decreasing on $\left(-2, \frac{2}{3}\right)$.

21. $f(x) = \frac{16}{9 - 3x}$
$f'(x) = \frac{16(-1)(-3)}{(9 - 3x)^2} = \frac{48}{(9 - 3x)^2}$
$f'(x) > 0$ for all x $(x \neq 3)$, and f is not defined for $x = 3$.
f is increasing on $(-\infty, 3)$ and $(3, \infty)$ and never decreasing.

22. $f(x) = \frac{15}{2x + 7}$
$f'(x) = \frac{15(-1)(2)}{(2x + 7)^2} = \frac{-30}{(2x + 7)^2}$

$f'(x) < 0$ for all $x\left(x \neq -\frac{7}{2}\right)$, and f is not defined for $x = -\frac{7}{2}$.

(*continued on next page*)

(continued)

f is never increasing, and decreasing on $\left(-\infty, -\frac{7}{2}\right)$ and $\left(-\frac{7}{2}, \infty\right)$.

23. $f(x) = \ln|x^2 - 1|$

$f'(x) = \frac{2x}{x^2 - 1}$

f is not defined for $x = -1$ and $x = 1$.

$f'(x) = 0$ when $x = 0$.

Test an x-value in the intervals $(-\infty, -1)$, $(-1, 0)$, $(0, 1)$, and $(1, \infty)$.

$f'(-2) = -\frac{4}{3} < 0$

$f'\left(-\frac{1}{2}\right) = \frac{4}{3} > 0$

$f'\left(\frac{1}{2}\right) = -\frac{4}{3} < 0$

$f'(2) = \frac{4}{3} > 0$

f is increasing on $(-1, 0)$ and $(1, \infty)$ and decreasing on $(-\infty, -1)$ and $(0, 1)$.

24. $f(x) = 8xe^{-4x}$

$f'(x) = 8(e^{-4x}) + 8x(-4e^{-4x})$
$= 8e^{-4x} - 32xe^{-4x} = 8e^{-4x}(1 - 4x)$

$f'(x) = 0 \Rightarrow 8e^{-4x}(1 - 4x) = 0 \Rightarrow x = \frac{1}{4}$.

Test an x-value in the intervals $\left(-\infty, \frac{1}{4}\right)$ and $\left(\frac{1}{4}, \infty\right)$.

$f'(0) = 8 > 0$

$f'(1) = -24e^{-4} < 0$

f is increasing on $\left(-\infty, \frac{1}{4}\right)$ and decreasing on $\left(\frac{1}{4}, \infty\right)$.

25. $f(x) = -\tan 2x$

$f'(x) = -2\sec^2 2x$

The first derivative is always negative, so the function is never increasing.

It is decreasing on $\left(\frac{\pi}{4} + \frac{n\pi}{2}, \frac{3\pi}{4} + \frac{n\pi}{2}\right)$, where n is an integer.

26. $f(x) = -3\sin 4x$

$f'(x) = -12\cos 4x$

$f'(x) = -12\cos 4x = 0 \Rightarrow \cos 4x = 0 \Rightarrow$

$4x = \frac{\pi}{2} + 2n\pi$ | $4x = \frac{3\pi}{2} + 2n\pi$

$x = \frac{\pi}{8} + \frac{n\pi}{2}$ | $x = \frac{3\pi}{8} + \frac{n\pi}{2}$

The function is increasing on $\left(\frac{\pi}{8} + \frac{n\pi}{2}, \frac{3\pi}{8} + \frac{n\pi}{2}\right)$ and decreasing on $\left(-\frac{\pi}{8} + \frac{n\pi}{2}, \frac{\pi}{8} + \frac{n\pi}{2}\right)$, where n is an integer.

27. $f(x) = -x^2 + 4x - 8$

$f'(x) = -2x + 4 = 0$

Critical number: $x = 2$

$f''(x) = -2 < 0$ for all x, so $f(2)$ is a relative maximum.

$f(2) = -4$

Relative maximum of -4 at $x = 2$

28. $f(x) = x^2 - 6x + 4$

$f'(x) = 2x - 6$

$f'(x) = 0$ when $x = 3$.

Critical number: 3

$f''(x) = 2 > 0$ for all x, so $f(3)$ is a relative minimum.

$f(3) = -5$

Relative minimum of -5 at $x = 3$

29. $f(x) = 2x^2 - 8x + 1$

$f'(x) = 4x - 8$

$f'(x) = 0$ when $x = 2$.

Critical number: 2

$f''(x) = 4 > 0$ for all x, so $f(2)$ is a relative minimum.

$f(2) = -7$

Relative minimum of -7 at $x = 2$

30. $f(x) = -3x^2 + 2x - 5$

$f'(x) = -6x + 2$

$f'(x) = 0$ when $x = \frac{1}{3}$.

Critical number: $\frac{1}{3}$

Since $f''(x) = -6 < 0$ for all x, $f\left(\frac{1}{3}\right)$ is a relative maximum.

(continued on next page)

(*continued*)

$$f\left(\frac{1}{3}\right) = -\frac{14}{3}$$

Relative maximum of $-\frac{14}{3}$ at $x = \frac{1}{3}$

31. $f(x) = 2x^3 + 3x^2 - 36x + 20$

$f'(x) = 6x^2 + 6x - 36 = 0$

$6(x^2 + x - 6) = 0$

$(x+3)(x-2) = 0$

Critical numbers: –3 and 2

$f''(x) = 12x + 6$

$f''(-3) = -30 < 0$, so a maximum occurs at $x = -3$.

$f''(2) = 30 > 0$, so a minimum occurs at $x = 2$.

$f(-3) = 101$

$f(2) = -24$

Relative maximum of 101 at $x = -3$

Relative minimum of –24 at $x = 2$

32. $f(x) = 2x^3 + 3x^2 - 12x + 5$

$f'(x) = 6x^2 + 6x - 12 = 6(x^2 + x - 2)$

$= 6(x+2)(x-1)$

$f'(x) = 0 \Rightarrow 6(x+2)(x-1) = 0 \Rightarrow x = -2$ or $x = 1$

Critical numbers: –2, 1

$f''(x) = 12x + 6$

$f''(-2) = 18 < 0$, so a maximum occurs at $x = -2$

$f''(1) = 18 > 0$, so a minimum occurs at $x = 1$.

$f(-2) = 25$

$f(1) = -2$

Relative maximum of 25 at $x = -2$

Relative minimum of –2 at $x = 1$

33. $f(x) = \frac{xe^x}{x-1}$

$$f'(x) = \frac{(x-1)(xe^x + e^x) - xe^x(1)}{(x-1)^2}$$

$$= \frac{x^2e^x + xe^x - xe^x - e^x - xe^x}{(x-1)^2}$$

$$= \frac{x^2e^x - xe^x - e^x}{(x-1)^2} = \frac{e^x(x^2 - x - 1)}{(x-1)^2}$$

$f'(x)$ is undefined at $x = 1$, but 1 is not in the domain of $f(x)$.

$f'(x) = 0$ when $x^2 - x - 1 = 0$

$$x = \frac{1 \pm \sqrt{1 - 4(1)(-1)}}{2} = \frac{1 \pm \sqrt{5}}{2}$$

$$\frac{1+\sqrt{5}}{2} \approx 1.618 \text{ or } \frac{1-\sqrt{5}}{2} = -0.618$$

Critical numbers are –0.618 and 1.618.

$$f'(1.4) = \frac{e^{1.4}(1.4^2 - 1.4 - 1)}{(1.4-1)^2} \approx -11.15 < 0$$

$$f'(2) = \frac{e^2(2^2 - 2 - 1)}{(2-1)^2} = e^2 \approx 7.39 > 0$$

$$f'(-1) = \frac{e^{-1}[(-1)^2 - (-1) - 1]}{(-1-1)^2} \approx 0.09 > 0$$

$$f'(0) = \frac{e^0(0^2 - 0 - 1)}{(0-1)^2} = -1 < 0$$

There is a relative maximum at (–0.618, 0.206) and a relative minimum at (1.618, 13.203).

34. $y = \frac{\ln(3x)}{2x^2}$

$$y' = \frac{2x^2 \cdot \frac{1}{x} - (\ln 3x) \cdot 4x}{4x^4} = \frac{2x(1 - 2\ln 3x)}{4x^4}$$

$$= \frac{1 - 2\ln 3x}{2x^3}$$

$$y' = 0 \Rightarrow \frac{1 - 2\ln 3x}{2x^3} = 0 \Rightarrow 1 - 2\ln 3x = 0 \Rightarrow$$

$$\frac{1}{2} = \ln 3x \Rightarrow e^{1/2} = 3x \Rightarrow x = \frac{\sqrt{e}}{3} \approx 0.55$$

$$f'(1) = \frac{1 - 2\ln 3}{2} \approx -0.6 < 0$$

$$f'\left(\frac{1}{3}\right) = \frac{1 - 2\ln 1}{2\left(\frac{1}{3}\right)^3} = \frac{1}{2 \cdot \frac{1}{27}} = \frac{27}{2} > 0$$

$$f\left(\frac{\sqrt{e}}{3}\right) \approx 0.83$$

Relative maximum at $\left(\frac{\sqrt{e}}{3}, 0.83\right)$ or (0.55, 0.83)

35. $f(x) = 2\cos \pi x$

$f'(x) = -2\pi \sin \pi x$

$f''(x) = -2\pi^2 \cos \pi x$

$f'(x) = -2\pi \sin \pi x = 0 \Rightarrow \sin \pi x = 0 \Rightarrow$

$\pi x = n\pi \Rightarrow x = n$

For even values of n, we have

$f''(2n) = -2\pi^2 \cos(2n\pi) = -2\pi^2 \cos(2n\pi)$

$= -2\pi^2(1) = -2\pi^2 < 0$

(*continued on next page*)

(continued)

This indicates there is a relative maximum at $2n$.

$f(2n) = 2\cos(2n\pi) = 2\cos 2n\pi = 2$

Thus, there is a relative maximum of 2 at $x = 0, \pm 2, \pm 4, \ldots$.

For odd values of n, we have

$$\begin{aligned} f''(2n+1) &= -2\pi^2 \cos(\pi(2n+1)) \\ &= -2\pi^2 \cos(2n\pi + \pi) \\ &= -2\pi^2(-1) = 2\pi^2 > 0 \end{aligned}$$

This indicates there is a relative minimum at $2n + 1$.

$$\begin{aligned} f(2n+1) &= 2\cos(\pi(2n+1)) \\ &= 2\cos(2n\pi + \pi) = -2 \end{aligned}$$

Thus, there is a relative minimum of –2 at $x = \pm 1, \pm 3, \ldots$.

36. $f(x) = -5\sin 3x$

$f'(x) = -15\cos 3x$

$f''(x) = 45\sin 3x$

$f'(x) = -15\cos 3x = 0 \Rightarrow \cos 3x = 0 \Rightarrow$

$3x = \frac{\pi}{2} + 2n\pi \quad\Big|\quad 3x = -\frac{\pi}{2} + 2n\pi$

$x = \frac{\pi}{6} + \frac{2n\pi}{3} \quad\Big|\quad x = -\frac{\pi}{6} + \frac{2n\pi}{3}$

Let $n = 0$. Then

$f''\left(\frac{\pi}{6}\right) = 45\sin\left(\frac{\pi}{6}\right) = \frac{45}{2} > 0$

This indicates there is a relative minimum at $x = \frac{\pi}{6} + \frac{2n\pi}{3}$, where n is an integer.

$f\left(\frac{\pi}{6}\right) = -5\sin 3\left(\frac{\pi}{6}\right) = -5$

There is a relative minimum of –5 at $x = \frac{\pi}{6} + \frac{2n\pi}{3}$, where n is an integer.

$f''\left(-\frac{\pi}{6}\right) = 45\sin\left(-\frac{\pi}{6}\right) = -\frac{45}{2} < 0$

This indicates there is a relative maximum at $x = -\frac{\pi}{6} + \frac{2n\pi}{3}$, where n is an integer.

$f\left(-\frac{\pi}{6}\right) = -5\sin 3\left(-\frac{\pi}{6}\right) = 5$

There is a relative maximum of 5 at $x = -\frac{\pi}{6} + \frac{2n\pi}{3}$, where n is an integer.

37. $f(x) = 3x^4 - 5x^2 - 11x$

$f'(x) = 12x^3 - 10x - 11$

$f''(x) = 36x^2 - 10$

$f''(1) = 36(1)^2 - 10 = 26$

$f''(-3) = 36(-3)^2 - 10 = 314$

38. $f(x) = 9x^3 + \frac{1}{x} = 9x^3 + x^{-1}$

$f'(x) = 27x^2 - x^{-2}$

$f''(x) = 54x + 2x^{-3} = 54x + \frac{2}{x^3}$

$f''(1) = 54(1) + \frac{2}{(1)^3} = 56$

$$\begin{aligned} f''(-3) &= 54(-3) + \frac{2}{(-3)^3} \\ &= -162 - \frac{2}{27} = -\frac{4376}{27} \end{aligned}$$

39. $f(x) = \frac{4x+2}{3x-6}$

$$\begin{aligned} f'(x) &= \frac{(3x-6)(4) - (4x+2)(3)}{(3x-6)^2} \\ &= \frac{12x - 24 - 12x - 6}{(3x-6)^2} = \frac{-30}{(3x-6)^2} \\ &= -30(3x-6)^{-2} \end{aligned}$$

$$\begin{aligned} f''(x) &= -30(-2)(3x-6)^{-3}(3) \\ &= 180(3x-6)^{-3} \text{ or } \frac{180}{(3x-6)^3} \end{aligned}$$

$f''(1) = 180[3(1) - 6]^{-3} = -\frac{20}{3}$

$f''(-3) = 180[3(-3) - 6]^{-3} = -\frac{4}{75}$

40. $f(x) = \frac{1-2x}{4x+5}$

$$\begin{aligned} f'(x) &= \frac{(4x+5)(-2) - (1-2x)(4)}{(4x+5)^2} \\ &= \frac{-8x - 10 - 4 + 8x}{(4x+5)^2} = \frac{-14}{(4x+5)^2} \\ &= -14(4x+5)^{-2} \end{aligned}$$

$$\begin{aligned} f''(x) &= -14(-2)(4x+5)^{-3}(4) \\ &= 112(4x+5)^{-3} \text{ or } \frac{112}{(4x+5)^3} \end{aligned}$$

$f''(1) = \frac{112}{[4(1)+5]^3} = \frac{112}{729}$

$f''(-3) = \frac{112}{[4(-3)+5]^3} = -\frac{16}{49}$

41. $f(t)=\sqrt{t^2+1}=(t^2+1)^{1/2}$

$f'(t)=\frac{1}{2}(t^2+1)^{-1/2}(2t)=t(t^2+1)^{-1/2}$

$f''(t)=(t^2+1)^{-1/2}(1)+t\left[\left(-\frac{1}{2}\right)(t^2+1)^{-3/2}(2t)\right]=(t^2+1)^{-1/2}-t^2(t^2+1)^{-3/2}$

$=\frac{1}{(t^2+1)^{1/2}}-\frac{t^2}{(t^2+1)^{3/2}}=\frac{t^2+1-t^2}{(t^2+1)^{3/2}}=(t^2+1)^{-3/2}$ or $\frac{1}{(t^2+1)^{3/2}}$

$f''(1)=\frac{1}{(1+1)^{3/2}}=\frac{1}{2^{3/2}}\approx 0.354$

$f''(-3)=\frac{1}{(9+1)^{3/2}}=\frac{1}{10^{3/2}}\approx 0.032$

42. $f(t)=-\sqrt{5-t^2}=-(5-t^2)^{1/2}$

$f'(t)=-\frac{1}{2}(5-t^2)^{-1/2}(-2t)=t(5-t^2)^{-1/2}$

$f''(t)=(1)(5-t^2)^{-1/2}+t\left[-\frac{1}{2}(5-t^2)^{-3/2}(-2t)\right]=(5-t^2)^{-1/2}+t^2(5-t^2)^{-3/2}$

$=(5-t^2)^{-3/2}(5-t^2+t^2)=\frac{5}{(5-t^2)^{3/2}}$

$f''(1)=\frac{5}{(5-1)^{3/2}}=\frac{5}{8}$

$f''(-3)=\frac{5}{(5-9)^{3/2}}$

This value does not exist since $(-4)^{3/2}$ does not exist. (In fact, f is undefined at $t=-4$.)

43. $f(x)=\tan 7x$

$f'(x)=7\sec^2(7x)$

$f''(x)=7\left(2\sec(7x)\frac{d}{dx}\sec(7x)\right)$

$=14\sec(7x)(7(\sec 7x)\tan(7x))$

$=98\sec^2(7x)\tan(7x)$

$f''(1)=98\sec^2(7)\tan(7)$

$f''(-3)=98\sec^2[7(-3)]\tan[7(-3)]$

$=-98\sec^2(-21)\tan(-21)$

$=-98\sec^2(21)\tan(21)$

44. $f(x)=x\cos 3x$

$f'(x)=\cos(3x)\frac{d}{dx}x+x\frac{d}{dx}\cos(3x)$

$=\cos 3x-3x\sin 3x$

$f''(x)=\frac{d}{dx}\cos 3x-3\frac{d}{dx}(x\sin 3x)$

$=-3\sin 3x-3\left(\sin 3x\frac{d}{dx}x+x\frac{d}{dx}\sin 3x\right)$

$=-3\sin 3x-3(\sin 3x+3x\cos 3x)$

$=-3\sin 3x-3\sin 3x-9x\cos 3x$

$=-6\sin 3x-9x\cos 3x$

$f''(1)=-6\sin 3-9\cos 3$

$f''(-3)=-6\sin 3(-3)-9(-3)\cos 3(-3)$

$=-6\sin(-9)+27\cos(-9)$

$=6\sin 9+27\cos 9$

45. $f(x)=-2x^3-\frac{1}{2}x^2+x-3$

Domain is $(-\infty,\infty)$

The graph has no symmetry.

$f'(x)=-6x^2-x+1$

$f'(x)=0\Rightarrow -(3x-1)(2x+1)=0\Rightarrow$

$x=\frac{1}{3}$ or $x=-\frac{1}{2}$

Critical numbers: $\frac{1}{3}$ $\frac{1}{3}$ and $-\frac{1}{2}$

Critical points: $\left(\frac{1}{3},-2.80\right)$ and $\left(-\frac{1}{2},-3.375\right)$

$f''(x)=-12x-1$

$f''\left(\frac{1}{3}\right)=-5<0$

(*continued on next page*)

(continued)

$f''\left(-\frac{1}{2}\right) = 5 > 0$

Relative maximum at $x = \frac{1}{3}$

Relative minimum at $x = -\frac{1}{2}$

Increasing on $\left(-\frac{1}{2}, \frac{1}{3}\right)$

Decreasing on $\left(-\infty, -\frac{1}{2}\right)$ and $\left(\frac{1}{3}, \infty\right)$

$f''(x) = 0 \Rightarrow -12x - 1 = 0 \Rightarrow x = -\frac{1}{12}$

Point of inflection at $\left(-\frac{1}{12}, -3.09\right)$

Concave upward on $\left(-\infty, -\frac{1}{12}\right)$

Concave downward on $\left(-\frac{1}{12}, \infty\right)$

y-intercept:
$y = -2(0)^3 - \frac{1}{2}(0)^2 + (0) - 3 = -3$

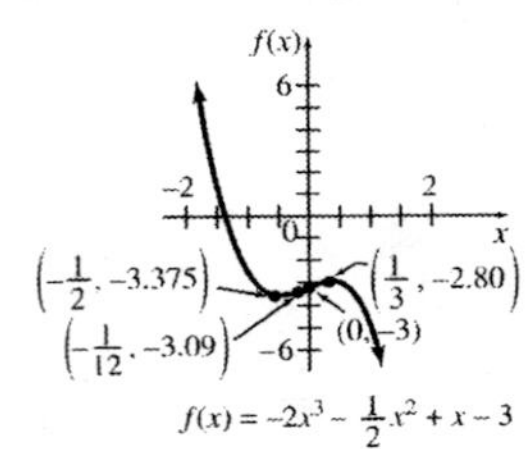

46. $f(x) = -\frac{4}{3}x^3 + x^2 + 30x - 7$

Domain is $(-\infty, \infty)$

The graph has no symmetry.

$f'(x) = -4x^2 + 2x + 30 = -2(2x^2 - x - 15)$
$= -2(2x + 5)(x - 3)$

$f'(x) = 0 \Rightarrow -2(2x + 5)(x - 3) = 0 \Rightarrow$
$x = -\frac{5}{2}$ or $x = 3$

Critical numbers: $-\frac{5}{2}$ and 3

Critical points: $\left(-\frac{5}{2}, -54.9\right)$ and $(3, 56)$

$f''(x) = -8x + 2$

$f''\left(-\frac{5}{2}\right) = 22 > 0$

$f''(3) = -22 < 0$

Relative maximum at $x = 3$

Relative minimum at $x = -\frac{5}{2}$

Increasing on $\left(-\frac{5}{2}, 3\right)$

Decreasing on $\left(-\infty, -\frac{5}{2}\right)$ and $(3, \infty)$

$f''(x) = -8x + 2$

$f''(x) = 0 \Rightarrow -8x + 2 = 0 \Rightarrow x = \frac{1}{4}$

Point of inflection at $\left(\frac{1}{4}, 0.54\right)$

Concave upward on $\left(-\infty, \frac{1}{4}\right)$

Concave downward on $\left(\frac{1}{4}, \infty\right)$

y-intercept:
$y = -\frac{4}{3}(0)^3 + (0)^3 + 30(0)^2 - 7 = -7$

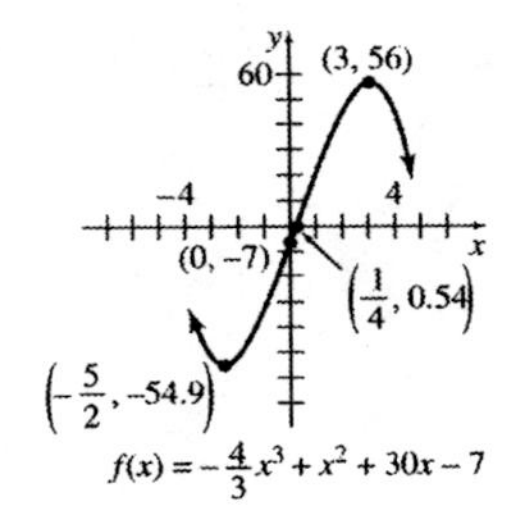

47. $f(x) = x^4 - \frac{4}{3}x^3 - 4x^2 + 1$

Domain is $(-\infty, \infty)$

The graph has no symmetry.

$f'(x) = 4x^3 - 4x^2 - 8x$

$f'(x) = 0 \Rightarrow 4x^3 - 4x^2 - 8x = 0 \Rightarrow$
$4x(x - 2)(x + 1) = 0 \Rightarrow x = 0$ or $x = 2$ or $x = -1$

Critical numbers: 0, 2, and −1

Critical points: $(0, 1)$, $\left(2, -\frac{29}{3}\right)$ and $\left(-1, -\frac{2}{3}\right)$

$f''(x) = 12x^2 - 8x - 8 = 4(3x^2 - 2x - 2)$

$f''(-1) = 12 > 0$

$f''(0) = -8 < 0$

$f''(2) = 24 > 0$

Relative maximum at $x = 0$

Relative minima at $x = -1$ and $x = 2$

Increasing on $(-1, 0)$ and $(2, \infty)$

Decreasing on $(-\infty, -1)$ and $(0, 2)$

(continued on next page)

(continued)

$f''(x) = 0 \Rightarrow 4(3x^2 - 2x - 2) = 0 \Rightarrow$

$x = \frac{2 \pm \sqrt{4-(-24)}}{6} = \frac{1 \pm \sqrt{7}}{3}$

Points of inflection at $\left(\frac{1+\sqrt{7}}{3}, -5.12\right)$ and $\left(\frac{1-\sqrt{7}}{3}, 0.11\right)$

Concave upward on $\left(-\infty, \frac{1-\sqrt{7}}{3}\right)$ and $\left(\frac{1+\sqrt{7}}{3}, \infty\right)$

Concave downward on $\left(\frac{1-\sqrt{7}}{3}, \frac{1+\sqrt{7}}{3}\right)$

y-intercept: $y = (0)^4 - \frac{4}{3}(0)^3 - 4(0)^2 + 1 = 1$

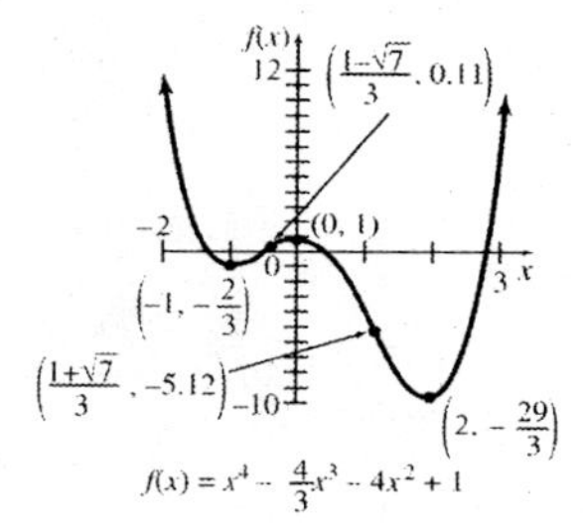

48. $f(x) = -\frac{2}{3}x^3 + \frac{9}{2}x^2 + 5x + 1$

Domain is $(-\infty, \infty)$

The graph has no symmetry.

$f'(x) = -2x^2 + 9x + 5 = -(2x+1)(x-5)$

$f'(x) = 0 \Rightarrow -(2x+1)(x-5) = 0 \Rightarrow$

$x = -\frac{1}{2}$ or $x = 5$

Critical numbers: $-\frac{1}{2}$ and 5

Critical points: $\left(-\frac{1}{2}, -0.29\right)$ and $(5, 55.17)$

$f''(x) = -4x + 9$

$f''\left(-\frac{1}{2}\right) = 11 > 0$

$f''(5) = -11 < 0$

Relative maximum at $x = 5$

Relative minimum at $x = -\frac{1}{2}$

Increasing on $\left(-\frac{1}{2}, 5\right)$

Decreasing on $\left(-\infty, -\frac{1}{2}\right)$ and $(5, \infty)$

$f''(x) = -4x + 9 = 0$

$f''(x) = 0 \Rightarrow -4x + 9 = 0 \Rightarrow x = \frac{9}{4}$

Point of inflection at $\left(\frac{9}{4}, 27.44\right)$

Concave upward on $\left(-\infty, \frac{9}{4}\right)$

Concave downward on $\left(\frac{9}{4}, \infty\right)$

y-intercept: $y = -\frac{2}{3}(0)^3 + \frac{9}{2}(0)^2 + 5(0) + 1 = 1$

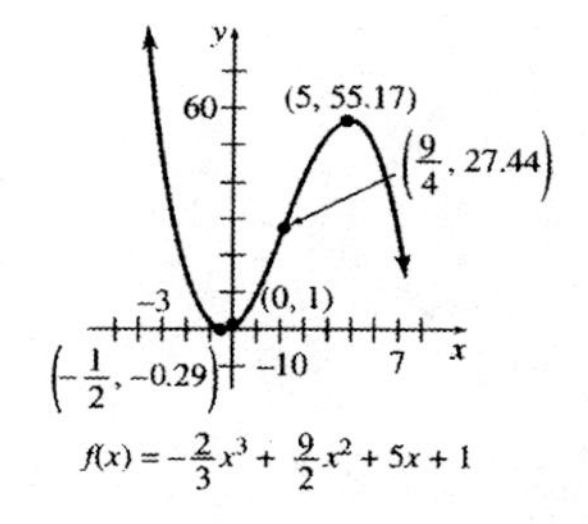

49. $f(x) = \frac{x-1}{2x+1}$

Domain is $\left(-\infty, -\frac{1}{2}\right) \cup \left(-\frac{1}{2}, \infty\right)$

The graph has no symmetry.

$f'(x) = \frac{(2x+1)(1) - (x-1)(2)}{(2x+1)^2} = \frac{3}{(2x+1)^2}$

f' is never zero.

$f'\left(-\frac{1}{2}\right)$ does not exist, but $-\frac{1}{2}$ is not a critical number because $-\frac{1}{2}$ is not in the domain of f. Thus, there are no critical numbers, so $f(x)$ has no relative extrema.

Increasing on $\left(-\infty, \frac{1}{2}\right)$ and $\left(\frac{1}{2}, \infty\right)$

$f''(x) = \frac{-12}{(2x+1)^3}$

$f''(0) = -12 < 0$

$f''(-1) = 12 > 0$

No inflection points

Concave upward on $\left(-\infty, -\frac{1}{2}\right)$

Concave downward on $\left(-\frac{1}{2}, \infty\right)$

(continued on next page)

(continued)

x-intercept: $\frac{x-1}{2x+1} = 0 \Rightarrow x = 1$

y-intercept: $y = \frac{0-1}{2(0)+1} = -1$

Vertical asymptote at $x = -\frac{1}{2}$

Horizontal asymptote at $y = \frac{1}{2}$

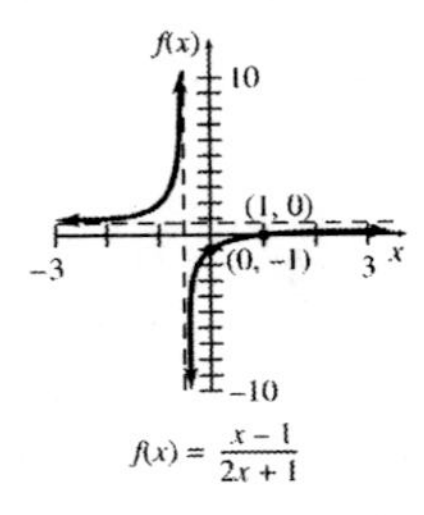

50. $f(x) = \frac{2x-5}{x+3}$

Domain is $(-\infty, -3) \cup (-3, \infty)$

$f'(x) = \frac{2(x+3)-(2x-5)}{(x+3)^2} = \frac{11}{(x+3)^2}$

f' is never zero, so $f(x)$ has no extrema.

$f''(x) = \frac{-22}{(x+3)^3}$

$f''(-4) = 22 > 0$

$f''(-2) = -22 < 0$

Increasing on $(-\infty, -3) \cup (-3, \infty)$

Concave upward on $(-\infty, -3)$

Concave downward on $(-3, \infty)$

x-intercept: $\frac{2x-5}{x+3} = 0 \Rightarrow x = \frac{5}{2}$

y-intercept: $\frac{2(0)-5}{0+3} = -\frac{5}{3}$

Vertical asymptote at $x = -3$

Horizontal asymptote at $y = 2$

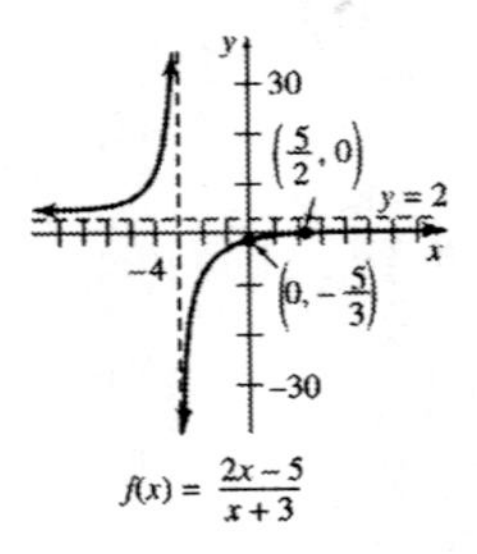

51. $f(x) = -4x^3 - x^2 + 4x + 5$

Domain is $(-\infty, \infty)$

The graph has no symmetry.

$f'(x) = -12x^2 - 2x + 4 = -2(6x^2 + x - 2)$
$= -2(3x+2)(2x-1)$

$f'(x) = 0 \Rightarrow -2(3x+2)(2x-1) = 0 \Rightarrow$
$x = -\frac{2}{3}$ or $x = \frac{1}{2}$

Critical numbers: $-\frac{2}{3}$ and $\frac{1}{2}$

Critical points: $\left(-\frac{2}{3}, 3.07\right)$ and $\left(\frac{1}{2}, 6.25\right)$

$f''(x) = -24x - 2 = -2(12x+1)$

$f''\left(-\frac{2}{3}\right) = 14 > 0$

$f''\left(\frac{1}{2}\right) = -14 < 0$

Relative maximum at $x = \frac{1}{2}$

Relative minimum at $x = -\frac{2}{3}$

Increasing on $\left(-\frac{2}{3}, \frac{1}{2}\right)$

Decreasing on $\left(-\infty, -\frac{2}{3}\right)$ and $\left(\frac{1}{2}, \infty\right)$

$f''(x) = 0 \Rightarrow -2(12x+1) = 0 \Rightarrow x = -\frac{1}{12}$

Point of inflection at $\left(-\frac{1}{12}, 4.66\right)$

Concave upward on $\left(-\infty, -\frac{1}{12}\right)$

Concave downward on $\left(-\frac{1}{12}, \infty\right)$

y-intercept: $y = -4(0)^3 - (0)^2 + 4(0) + 5 = 5$

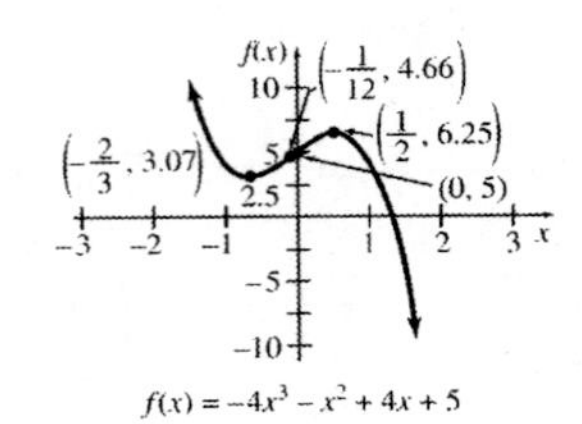

52. $f(x) = x^3 + \frac{5}{2}x^2 - 2x - 3$

Domain is $(-\infty, \infty)$

The graph has no symmetry.

$f'(x) = 3x^2 + 5x - 2 = (3x-1)(x+2)$

$f'(x) = 0 \Rightarrow (3x-1)(x+2) = 0 \Rightarrow$

$x = \frac{1}{3}$ or $x = -2$

Critical numbers: $\frac{1}{3}$ and -2

Critical points: $\left(\frac{1}{3}, -3.35\right)$ and $(-2, 3)$

$f''(x) = 6x + 5$

$f''\left(\frac{1}{3}\right) = 7 > 0$

$f''(-2) = -7 < 0$

Relative maximum at $x = -2$

Relative minimum at $x = \frac{1}{3}$

Increasing on $(-\infty, -2)$ and $\left(\frac{1}{3}, \infty\right)$

Decreasing on $\left(-2, \frac{1}{3}\right)$

$f''(x) = 0 \Rightarrow 6x + 5 = 0 \Rightarrow x = -\frac{5}{6}$

Point of inflection at $\left(-\frac{5}{6}, -0.18\right)$

Concave upward on $\left(-\frac{5}{6}, \infty\right)$

Concave downward on $\left(-\infty, -\frac{5}{6}\right)$

y-intercept: -3

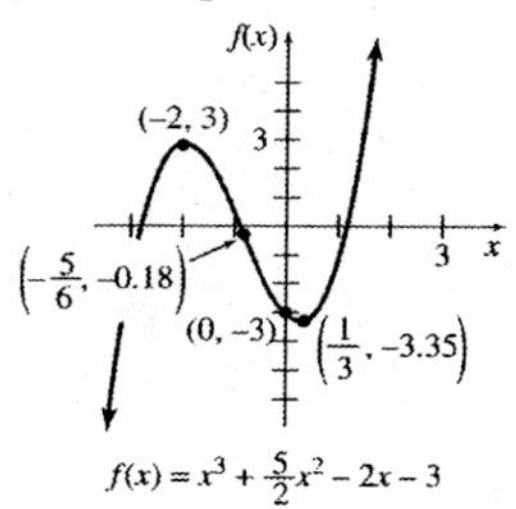

53. $f(x) = x^4 + 2x^2$

Domain is $(-\infty, \infty)$

$f(-x) = (-x)^4 - 2(-x)^2 = x^4 + 2x^2 = f(x)$

The graph is symmetric about the y-axis.

$f'(x) = 4x^3 + 4x = 4x(x^2 + 1)$

$f'(x) = 0 \Rightarrow 4x(x^2 + 1) = 0 \Rightarrow x = 0$

Critical number: 0

Critical point: $(0, 0)$

$f''(x) = 12x^2 + 4 = 4(3x^2 + 1)$

$f''(0) = 4 > 0$

Relative minimum at 0

Increasing on $(0, \infty)$

Decreasing on $(-\infty, 0)$

$f''(x) = 4(3x^2 + 1) \neq 0$ for any x

No points of inflection

$f''(1) = 16 > 0$

Concave upward on $(-\infty, \infty)$

x-intercept: 0; y-intercept: 0

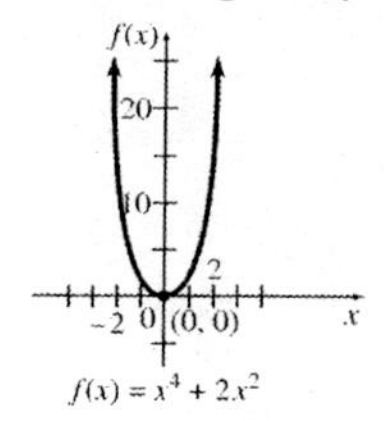

54. $f(x) = 6x^3 - x^4$

Domain is $(-\infty, \infty)$

The graph has no symmetry.

$f'(x) = 18x^2 - 4x^3 = 2x^2(9 - 2x)$

$f'(x) = 0 \Rightarrow 2x^2(9 - 2x) = 0 \Rightarrow x = 0$ or $x = \frac{9}{2}$

Critical number: 0 and $\frac{9}{2}$

Critical points: $(0, 0)$ and $\left(\frac{9}{2}, 136.7\right)$

$f''(x) = 36x - 12x^2 = 12x(3 - x)$

$f''(0) = 0$

$f''\left(\frac{9}{2}\right) = -81 < 0$

Relative maximum at $\frac{9}{2}$

No relative extrema at 0

Increasing on $\left(-\infty, \frac{9}{2}\right)$

Decreasing on $\left(\frac{9}{2}, \infty\right)$

$f''(x) = 0 \Rightarrow 12x(3 - x) = 0 \Rightarrow x = 0$ or $x = 3$

Points of inflection at $(0, 0)$ and $(3, 81)$

Concave upward on $(0, 3)$

Concave downward on $(-\infty, 0)$ and $(3, \infty)$

x-intercept:

$6x^3 - x^4 = 0 \Rightarrow x^3(6 - x) = 0 \Rightarrow x = 0, x = 6$

The x-intercepts are 0 and 6.

(*continued on next page*)

(continued)

y-intercept: 0

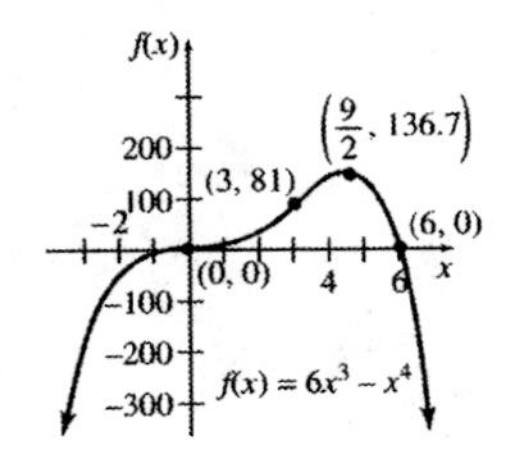

55. $f(x)=\dfrac{x^2+4}{x}$

Domain is $(-\infty, 0)\cup(0, \infty)$

$f(-x)=\dfrac{(-x)^2+4}{-x}=\dfrac{x^2+4}{-x}=-f(x)$

The graph is symmetric about the origin.

$f'(x)=\dfrac{x(2x)-(x^2+4)}{x^2}=\dfrac{x^2-4}{x^2}$

$f'(x)=0\Rightarrow\dfrac{x^2-4}{x^2}=0\Rightarrow x=\pm 2$

Critical numbers: -2 and 2

Critical points: $(-2, -4)$ and $(2, 4)$

$f''(x)=\dfrac{8}{x^3}$

$f''(-2)=-1<0$

$f''(2)=1>0$

Relative maximum at $x=-2$

Relative minimum at $x=2$

Increasing on $(-\infty, -2)$ and $(2, \infty)$

Decreasing on $(-2, 0)$ and $(0, 2)$

$f''(x)=\dfrac{8}{x^3}>0$ for all x.

No inflection points

Concave upward on $(0, \infty)$

Concave downward on $(-\infty, 0)$

No x- or y-intercepts

Vertical asymptote at $x=0$

Oblique asymptote at $y=x$

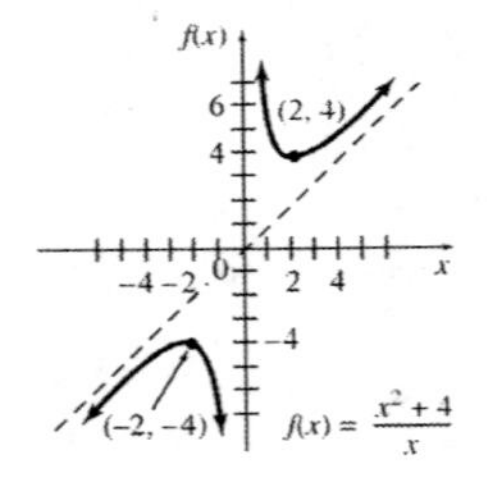

56. $f(x)=x+\dfrac{8}{x}$

Domain is $(-\infty, 0)\cup(0, \infty)$

$f(-x)=-x+\dfrac{8}{-x}=-\left(x+\dfrac{8}{x}\right)=-f(x)$

The graph is symmetric about the origin.

$f'(x)=1-\dfrac{8}{x^2}=\dfrac{x^2-8}{x^2}$

$f'(x)=0\Rightarrow\dfrac{x^2-8}{x^2}=0\Rightarrow x=\pm 2\sqrt{2}$

Critical numbers: $x=\pm 2\sqrt{2}$

Critical points: $(2\sqrt{2}, 4\sqrt{2}), (-2\sqrt{2}, -4\sqrt{2})$

$f''(x)=\dfrac{16}{x^3}$

$f''(-2\sqrt{2})=-\dfrac{\sqrt{2}}{2}<0$

$f''(2\sqrt{2})=\dfrac{\sqrt{2}}{2}>0$

Relative maximum at $x=-2\sqrt{2}$

Relative minimum at $x=2\sqrt{2}$

Increasing on $(-\infty, -2\sqrt{2})$ and $(2\sqrt{2}, \infty)$

Decreasing on $(-2\sqrt{2}, 0)$ and $(0, 2\sqrt{2})$

$f''(x)=\dfrac{16}{x^3}>0$ for all x.

No inflection points

Concave upward on $(0, \infty)$

Concave downward on $(-\infty, 0)$

Vertical asymptote at $x=0$

Oblique asymptote at $y=x$

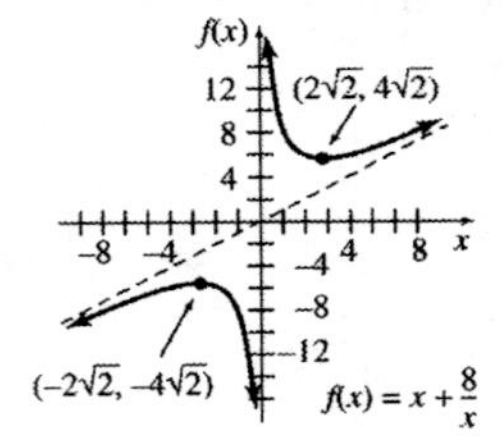

57. $f(x)=\dfrac{2x}{3-x}$

Domain is $(-\infty, 3)\cup(3, \infty)$

The graph has no symmetry.

$f'(x)=\dfrac{(3-x)(2)-(2x)(-1)}{(3-x)^2}=\dfrac{6}{(3-x)^2}$

$f'(x)$ is never zero. $f'(3)$ does not exist, but since 3 is not in the domain of f, it is not a critical number. There are no critical numbers, so there are no relative extrema.

(continued on next page)

(*continued*)

$f'(0) = \frac{2}{3} > 0$
$f'(4) = 6 > 0$
Increasing on $(-\infty, 3)$ and $(3, \infty)$

$f''(x) = \frac{12}{(3-x)^3}$

$f''(x)$ is never zero. $f''(3)$ does not exist, but since 3 is not in the domain of f, there is no inflection point at $x = 3$.

$f''(0) = \frac{12}{27} > 0$
$f''(4) = -12 < 0$
Concave upward on $(-\infty, 3)$
Concave downward on $(3, \infty)$
x-intercept: 0; y-intercept: 0
Vertical asymptote at $x = 3$
Horizontal asymptote at $y = -2$

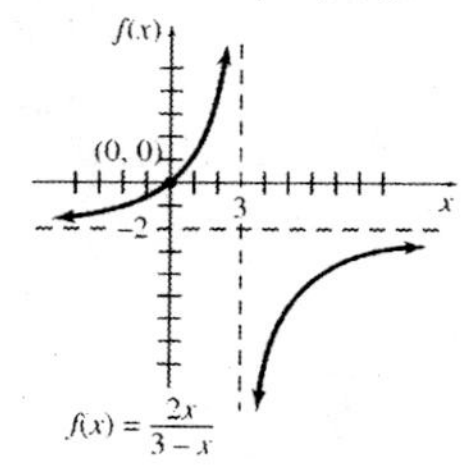

58. $f(x) = \frac{-4x}{1+2x}$

Domain is $\left(-\infty, -\frac{1}{2}\right) \cup \left(-\frac{1}{2}, \infty\right)$

The graph has no symmetry.

$f'(x) = \frac{-4(1+2x) - 2(-4x)}{(1+2x)^2} = \frac{-4-8x+8x}{(1+2x)^2}$
$= \frac{-4}{(1+2x)^2}$

$f'(x)$ is never zero.
There are no critical numbers, so there are no relative extrema.
$f'(0) = -4 < 0$
$f'(-1) = -4 < 0$

Decreasing on $\left(-\infty, -\frac{1}{2}\right)$ and $\left(-\frac{1}{2}, \infty\right)$

$f''(x) = \frac{16}{(1+2x)^3}$

$f''(x)$ is never zero; no points of inflection.
$f''(0) = 16 > 0$
$f''(-1) = -16 < 0$

Concave upward on $\left(-\frac{1}{2}, \infty\right)$

Concave downward on $\left(-\infty, -\frac{1}{2}\right)$

x- intercept: 0; y-intercept: 0

Vertical asymptote at $x = -\frac{1}{2}$

Horizontal asymptote at $y = -2$

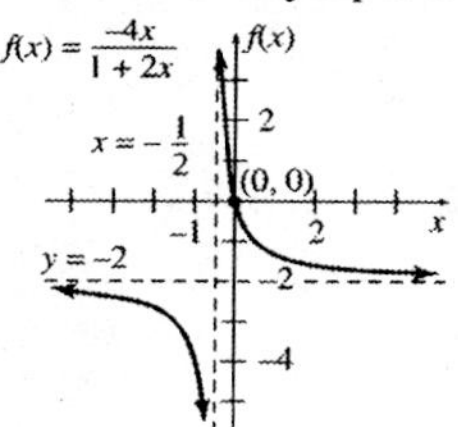

59. $f(x) = xe^{2x}$

Domain is $(-\infty, \infty)$.

$f(-x) = -xe^{-2x}$
The graph has no symmetry.
$f'(x) = (1)(e^{2x}) + (x)(2e^{2x}) = e^{2x}(2x+1)$

$f'(x) = 0$ when $x = -\frac{1}{2}$.

Critical number: $-\frac{1}{2}$

Critical point: $\left(-\frac{1}{2}, -\frac{1}{2e}\right)$

$f'(-1) = e^{2(-1)}[2(-1)+1] = -e^{-2} < 0$
$f'(0) = e^{2(0)}[2(0)+1] = 1 > 0$
No relative maximum

Relative minimum at $\left(-\frac{1}{2}, -\frac{1}{2e}\right)$

Decreasing on $\left(-\infty, -\frac{1}{2}\right)$ and increasing on $\left(-\frac{1}{2}, \infty\right)$

$f''(x) = 2e^{2x}(2x+1) + e^{2x}(2) = 4e^{2x}(x+1)$
$f''(x) = 0$ when $x = -1$.
$f''(-2) = 4e^{2(-2)}[(-2)+1] = -4e^{-4} < 0$
$f''(0) = 4e^{2(0)}[(0)+1] = 4 > 0$
Inflection point at $(-1, -e^{-2})$
Concave upward on $(-1, \infty)$
Concave downward on $(-\infty, -1)$
x-intercept: $xe^{2x} = 0 \Rightarrow x = 0$
y-intercept: $y = (0)e^{2(0)} = 0$

(*continued on next page*)

(*continued*)

Since $\lim_{x\to-\infty} xe^{2x} = 0$, there is a horizontal asymptote at $y = 0$.

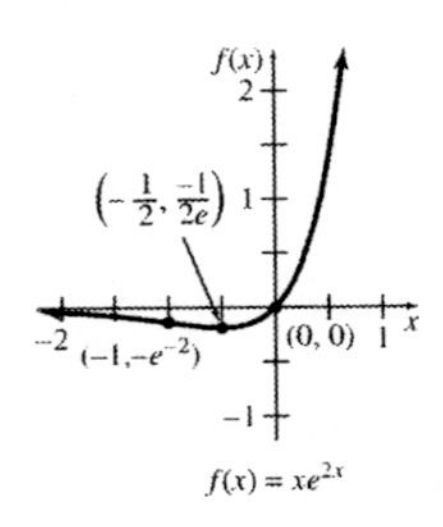

$f(x) = xe^{2x}$

60. $f(x) = x^2e^{2x}$

Domain is $(-\infty, \infty)$.

$f(-x) = (-x)^2e^{2(-x)} = x^2e^{-2x}$

The graph is not symmetric about the *y*-axis or origin.

$$f'(x) = (2x)(e^{2x}) + (x^2)(2e^{2x}) = 2(x^2 + x)e^{2x} = 2x(x+1)e^{2x}$$

$f'(x) = 0$ when $x = -1$ and $x = 0$.

Critical numbers: -1 and 0

Critical points: $(-1, e^{-2})$ and $(0, 0)$

$$f'(-2) = 2[(-2)^2 + (-2)]e^{2(-2)} = 4e^{-4} > 0$$

$$f'\left(-\frac{1}{2}\right) = 2\left[\left(-\frac{1}{2}\right)^2 + \left(-\frac{1}{2}\right)\right]e^{2(-1/2)} = -\frac{1}{2}e^{-1} < 0$$

$$f'(1) = 2[(1)^2 + (1)]e^{2(1)} = 4e^2 > 0$$

Relative maximum at $(-1, e^{-2})$

Relative minimum at $(0, 0)$

Increasing on $(-\infty, -1)$ and $(0, \infty)$

Decreasing on $(-1, 0)$

$$f''(x) = 2(2x+1)e^{2x} + 2(x^2 + x)e^{2x}(2) = 2(2x^2 + 4x + 1)e^{2x}$$

$f''(x) = 0$ when $2x^2 + 4x + 1 = 0$

$$x = \frac{-4 \pm \sqrt{16 - 4(2)(1)}}{2(2)} = \frac{-4 \pm \sqrt{8}}{4} = -1 \pm \frac{\sqrt{2}}{2}$$

$$f''(-2) = 2[2(-2)^2 + 4(-2) + 1]e^{2(-2)} = 2e^{-4} > 0$$

$$f''(-1) = 2[2(-1)^2 + 4(-1) + 1]e^{2(-1)} = -2e^{-2} < 0$$

$$f''(0) = 2[2(0)^2 + 4(0) + 1]e^{2(0)} = 2 > 0$$

Inflection points at

$$\left(-1 - \frac{\sqrt{2}}{2}, \left(\frac{3}{2} + \sqrt{2}\right)e^{-2-\sqrt{2}}\right) \approx (-1.707, 0.09588)$$

$$\left(-1 + \frac{\sqrt{2}}{2}, \left(\frac{3}{2} - \sqrt{2}\right)e^{-2+\sqrt{2}}\right) \approx (-0.293, 0.04775)$$

Concave upward on

$$\left(-\infty, -1 - \frac{\sqrt{2}}{2}\right) \text{ and } \left(-1 + \frac{\sqrt{2}}{2}, \infty\right)$$

Concave downward on

$$\left(-1 - \frac{\sqrt{2}}{2}, -1 + \frac{\sqrt{2}}{2}\right)$$

x-intercept: $x^2e^{2x} = 0 \Rightarrow x = 0$

y-intercept: $y = (0)^2e^{2(0)} = 0$

Since $\lim_{x\to-\infty} x^2e^{2x} = 0$, horizontal asymptote at $y = 0$.

$f(x) = x^2e^{2x}$

61. $f(x) = \ln(x^2 + 4)$

Domain is $(-\infty, \infty)$.

$f(-x) = \ln[(-x)^2 + 4] = \ln(x^2 + 4) = f(x)$

The graph is symmetric about the *y*-axis.

$$f'(x) = \frac{2x}{x^2 + 4}$$

$f'(x) = 0$ when $x = 0$.

Critical number: 0

Critical point: $(0, \ln 4)$

$$f'(-1) = \frac{2(-1)}{(-1)^2 + 4} = -\frac{2}{5} < 0$$

$$f'(1) = \frac{2(1)}{(1)^2 + 4} = \frac{2}{5} > 0$$

No relative maximum

Relative minimum at $(0, \ln 4)$

Increasing on $(0, \infty)$

Decreasing on $(-\infty, 0)$

$$f''(x) = \frac{(x^2 + 4)(2) - (2x)(2x)}{(x^2 + 4)^2} = \frac{-2(x^2 - 4)}{(x^2 + 4)^2}$$

(*continued on next page*)

(*continued*)

$f''(x) = 0$ when $x^2 - 4 = 0 \Rightarrow x = \pm 2$

$$f''(-3) = \frac{-2[(-3)^2 - 4]}{[(-3)^2 + 4]^2} = -\frac{10}{169} < 0$$

$$f''(0) = \frac{-2[(0)^2 - 4]}{[(0)^2 + 4]^2} = \frac{1}{2} > 0$$

$$f''(3) = \frac{-2[(3)^2 - 4]}{[(3)^2 + 4]^2} = -\frac{10}{169} < 0$$

Inflection points at $(-2, \ln 8)$ and $(2, \ln 8)$

Concave upward on $(-2, 2)$

Concave downward on $(-\infty, -2)$ and $(2, \infty)$

Since $f(x)$ never equals zero, there are no x-intercepts.

y-intercept: $y = \ln[(0)^2 + 4] = \ln 4$

No horizontal or vertical asymptotes.

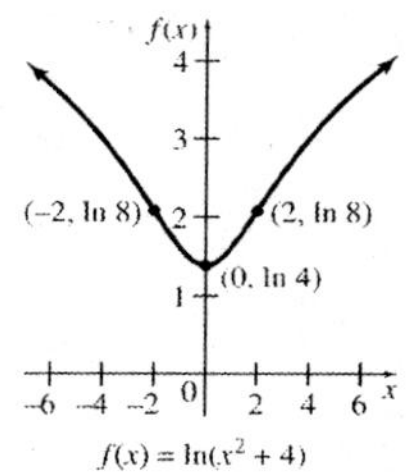

$f(x) = \ln(x^2 + 4)$

62. $f(x) = x^2 \ln x$

Domain is $(0, \infty)$.

The graph is not symmetric about the y-axis or origin since $f(x)$ is not defined for $x \le 0$.

$$f'(x) = 2x(\ln x) + x^2 \cdot \frac{1}{x} = 2x \ln x + x$$
$$= x(2 \ln x + 1)$$

$$f'(x) = 0 \Rightarrow 2 \ln x + 1 = 0 \Rightarrow \ln x = -\frac{1}{2} \Rightarrow$$
$$x = e^{-1/2}$$

Critical number: $e^{-1/2}$

Critical point: $\left(e^{-1/2}, -\frac{1}{2e}\right)$

$$f'\left(\frac{1}{2}\right) = \frac{1}{2}\left(2 \ln \frac{1}{2} + 1\right) \approx -0.1931 < 0$$

$$f'(1) = 1(2 \ln 1 + 1) = 1 > 0$$

No relative maximum

Relative minimum at $\left(e^{-1/2}, -\frac{1}{2e}\right)$

Increasing on $(e^{-1/2}, \infty)$

Decreasing on $(0, e^{-1/2})$

$$f''(x) = (1)(2 \ln x + 1) + x\left(\frac{2}{x}\right) = 2 \ln x + 3$$

$$f''(x) = 0 \Rightarrow 2 \ln x + 3 = 0 \Rightarrow \ln x = -\frac{3}{2} \Rightarrow$$
$$x = e^{-3/2}$$

$$f''(e^{-2}) = 2 \ln e^{-2} + 3 = -1 < 0$$
$$f''(1) = 2 \ln 1 + 3 = 3 > 0$$

Inflection point at $\left(e^{-3/2}, -\frac{3}{2e^3}\right)$

Concave upward on $(e^{-3/2}, \infty)$

Concave downward on $(0, e^{-3/2})$

x-intercept: $x^2 \ln x = 0 \Rightarrow \ln x = 0 \Rightarrow x = 1$

Since $f(x)$ is not defined at $x = 0$, there is no y-intercept.

No horizontal or vertical asymptotes

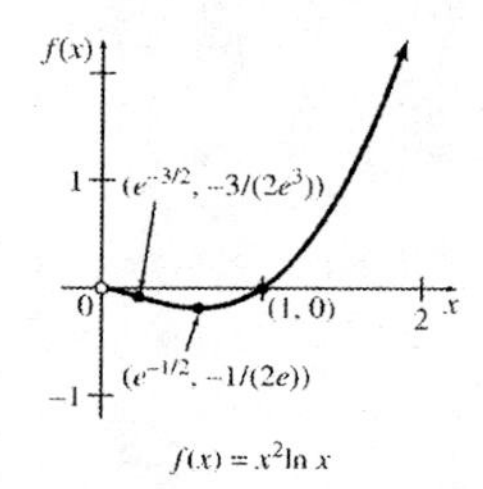

$f(x) = x^2 \ln x$

63. $f(x) = 4x^{1/3} + x^{4/3}$

Domain is $(-\infty, \infty)$.

$$f(-x) = 4(-x)^{1/3} + (-x)^{4/3} = -4x^{1/3} + x^{4/3}$$

The graph is not symmetric about the y-axis or origin.

$$f'(x) = \frac{4}{3}x^{-2/3} + \frac{4}{3}x^{1/3}$$

$$f'(x) = 0 \Rightarrow \frac{4}{3}x^{-2/3} + \frac{4}{3}x^{1/3} = 0 \Rightarrow$$

$$\frac{4}{3}x^{-2/3}(1 + x) = 0 \Rightarrow x = 0 \text{ or } x = -1$$

$f'(x)$ is not defined when $x = 0$

Critical numbers: -1 and 0

Critical points: $(-1, -3)$ and $(0, 0)$

$$f'(-8) = \frac{4}{3}(-8)^{-2/3} + \frac{4}{3}(-8)^{1/3} = -\frac{7}{3} < 0$$

$$f'\left(-\frac{1}{8}\right) = \frac{4}{3}\left(-\frac{1}{8}\right)^{-2/3} + \frac{4}{3}\left(-\frac{1}{8}\right)^{1/3} = \frac{14}{3} > 0$$

$$f'(1) = \frac{4}{3}(1)^{-2/3} + \frac{4}{3}(1)^{1/3} = \frac{8}{3} > 0$$

No relative maximum

Relative minimum at $(-1, -3)$

Increasing on $(-1, \infty)$

(*continued on next page*)

(continued)

Decreasing on $(-\infty, -1)$

$$f''(x) = -\frac{8}{9}x^{-5/3} + \frac{4}{9}x^{-2/3}$$

$$f''(x) = 0 \Rightarrow -\frac{8}{9}x^{-5/3} + \frac{4}{9}x^{-2/3} = 0 \Rightarrow$$

$$\frac{4}{9}x^{-5/3}(-2 + x) = 0 \Rightarrow x = 2$$

$f''(x)$ is not defined when $x = 0$.

$$f''(-1) = -\frac{8}{9}(-1)^{-5/3} + \frac{4}{9}(-1)^{-2/3} = \frac{4}{3} > 0$$

$$f''(1) = -\frac{8}{9}(1)^{-5/3} + \frac{4}{9}(1)^{-2/3} = -\frac{4}{9} < 0$$

$$f''(8) = -\frac{8}{9}(8)^{-5/3} + \frac{4}{9}(8)^{-2/3} = \frac{1}{12} > 0$$

Inflection points at $(0, 0)$ and $(2, 6 \cdot 2^{1/3})$

Concave upward on $(-\infty, 0)$ and $(2, \infty)$

Concave downward on $(0, 2)$

x-intercept:

$4x^{1/3} + x^{4/3} = 0 \Rightarrow x^{1/3}(4 + x) = 0 \Rightarrow$
$x = 0$ or $x = -4$

y-intercept: $y = 4(0)^{1/3} + (0)^{4/3} = 0$

No horizontal or vertical asymptotes

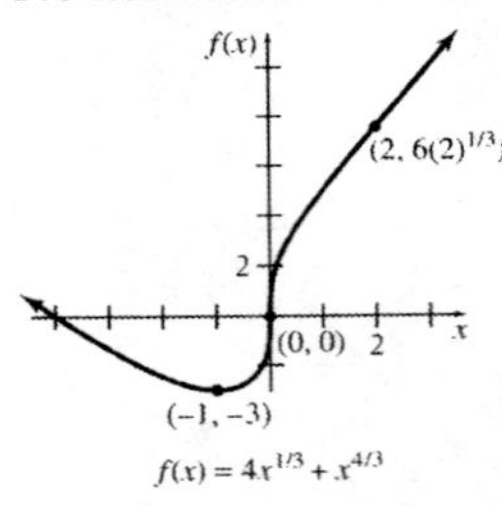

64. $f(x) = 5x^{2/3} + x^{5/3}$

Domain is $(-\infty, \infty)$.

$f(-x) = 5(-x)^{2/3} + (-x)^{5/3} = 5x^{2/3} - x^{5/3}$

The graph is not symmetric about the y-axis or origin.

$$f'(x) = \frac{10}{3}x^{-1/3} + \frac{5}{3}x^{2/3}$$

$$f'(x) = 0 \Rightarrow \frac{10}{3}x^{-1/3} + \frac{5}{3}x^{2/3} = 0 \Rightarrow$$

$$\frac{5}{3}x^{-1/3}(2 + x) = 0 \Rightarrow x = 0 \text{ or } x = -2$$

$f'(x)$ is not defined when $x = 0$.

Critical numbers: -2 and 0

Critical points: $(0, 0)$ and $(-2, 3 \cdot 2^{2/3})$

$$f'(-8) = \frac{10}{3}(-8)^{-1/3} + \frac{5}{3}(-8)^{2/3} = 5 > 0$$

$$f'(-1) = \frac{10}{3}(-1)^{-1/3} + \frac{5}{3}(-1)^{2/3} = -\frac{5}{3} < 0$$

$$f'(1) = \frac{10}{3}(1)^{-1/3} + \frac{5}{3}(1)^{2/3} = 5 > 0$$

Relative maximum at $(-2, 3 \cdot 2^{2/3})$

Relative minimum at $(0, 0)$

Increasing on $(-\infty, -2)$ and $(0, \infty)$

Decreasing on $(-2, 0)$

$$f''(x) = -\frac{10}{9}x^{-4/3} + \frac{10}{9}x^{-1/3}$$

$$f''(x) = 0 \Rightarrow -\frac{10}{9}x^{-4/3} + \frac{10}{9}x^{-1/3} = 0 \Rightarrow$$

$$\frac{10}{9}x^{-4/3}(-1 + x) = 0 \Rightarrow x = 1$$

$f''(x)$ is not defined when $x = 0$.

$$f''(-1) = -\frac{10}{9}(-1)^{-4/3} + \frac{10}{9}(-1)^{-1/3} = -\frac{20}{9} < 0$$

$$f''\left(\frac{1}{8}\right) = -\frac{10}{9}\left(\frac{1}{8}\right)^{-4/3} + \frac{10}{9}\left(\frac{1}{8}\right)^{-1/3} = -\frac{140}{9} < 0$$

$$f''(8) = -\frac{10}{9}(8)^{-4/3} + \frac{10}{9}(8)^{-1/3} = \frac{35}{72} > 0$$

Concave upward on $(1, \infty)$

Concave downward on $(-\infty, 0)$ and $(0, 1)$

Inflection point at $(1, 6)$

x-intercept:

$5x^{2/3} + x^{5/3} = 0 \Rightarrow x^{2/3}(5 + x) = 0 \Rightarrow$
$x = 0$ or $x = -5$

y-intercept: $y = 5(0)^{2/3} + (0)^{5/3} = 0$

No horizontal or vertical asymptotes.

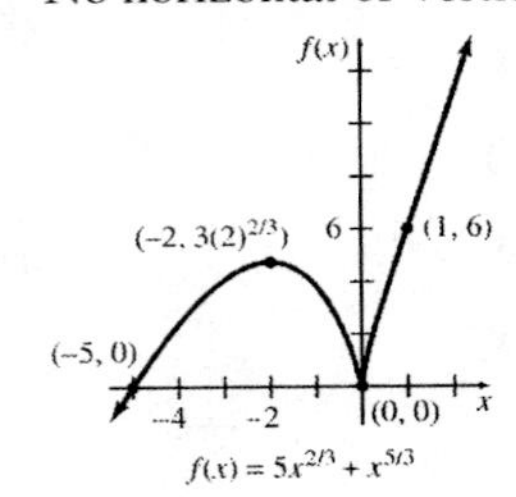

65. $f(x) = x - \sin x$

Domain is $(-\infty, \infty)$.

$f(-x) = -x - \sin(-x) = -x + \sin x = -f(x)$

The graph is symmetric with respect to the origin.

$f'(x) = 1 - \cos x$

$f'(x) = 0 \Rightarrow 1 - \cos x = 0 \Rightarrow \cos x = 1 \Rightarrow$
$x = 0, \pm 2\pi, \pm 4\pi, \ldots$

$f'(\pi) = 1 - \cos \pi = 2 > 0$

$f'(3\pi) = 1 - \cos 3\pi = 2 > 0$

$f'(5\pi) = 1 + \cos 5\pi = 2 > 0$

f increases on $(-\infty, \infty)$

$f''(x) = \sin x$

$f''(x) = 0 \Rightarrow \sin x = 0 \Rightarrow$
$x = 0, \pm\pi, \pm 2\pi, \pm 3\pi, \ldots$

$f''\left(\frac{\pi}{2}\right) = \sin\frac{\pi}{2} = 1 > 0$

$f''\left(\frac{3\pi}{2}\right) = \sin\frac{3\pi}{2} = -1 < 0$

$f''\left(\frac{5\pi}{2}\right) = \sin\frac{5\pi}{2} = 1 > 0$

Concave downward on
$\ldots, (-\pi, 0), (\pi, 2\pi), (3\pi, 4\pi), \ldots$

Concave upward on
$\ldots, (-2\pi, -\pi), (0, \pi), (2\pi, 3\pi), \ldots$

Inflection points at
$\ldots, (-\pi, -\pi), (0, 0), (\pi, \pi), \ldots$

y-intercept: $y = 0 - \sin 0 = 0$

x-intercept: $0 = x - \sin x \Rightarrow x = \sin x \Rightarrow x = 0$

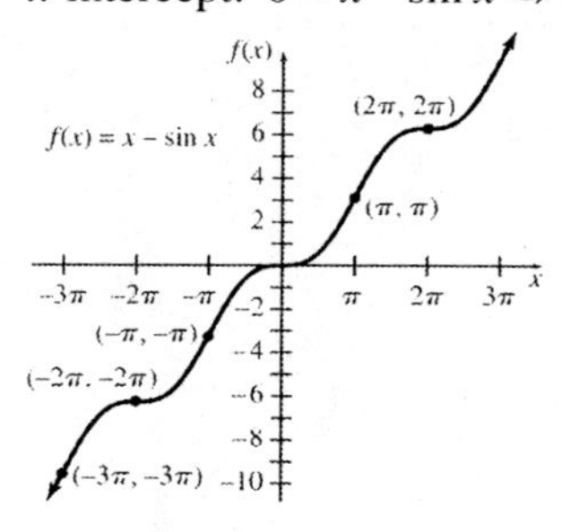

66. $f(x) = x - \cos x$

Domain is $(-\infty, \infty)$.

$f(-x) = -x - \cos(-x) = -x + \cos x \neq -f(x)$

The graph has no symmetry.

$f'(x) = 1 + \sin x$

$f'(x) = 0 \Rightarrow 1 + \sin x = 0 \Rightarrow \sin x = -1 \Rightarrow$

$x = \pm\frac{3\pi}{2}, \pm\frac{7\pi}{2}, \pm\frac{11\pi}{2}, \ldots$

$f'(0) = 1 + \sin 0 = 1 > 0$

$f'(2\pi) = 1 + \sin 2\pi = 1 > 0$

$f'(5\pi) = 1 + \sin 5\pi = 1 > 0$

f increases on $(-\infty, \infty)$

$f''(x) = \cos x$

$f''(x) = 0 \Rightarrow \cos x = 0 \Rightarrow$

$x = \pm\frac{\pi}{2}, \pm\frac{3\pi}{2}, \pm\frac{5\pi}{2}, \ldots$

$f''(0) = \cos 0 = 1 > 0$

$f''(2) = \cos 2 = -0.416 < 0$

$f''(2\pi) = \cos 2\pi = 1 > 0$

Concave downward on

$\ldots, \left(-\frac{3\pi}{2}, -\frac{\pi}{2}\right), \left(\frac{\pi}{2}, \frac{3\pi}{2}\right), \left(\frac{5\pi}{2}, \frac{7\pi}{2}\right), \ldots$

Concave upward on

$\ldots, \left(-\frac{\pi}{2}, \frac{\pi}{2}\right), \left(\frac{3\pi}{2}, \frac{5\pi}{2}\right), \left(\frac{7\pi}{2}, \frac{9\pi}{2}\right), \ldots$

Inflection points at

$\ldots, \left(-\frac{\pi}{2}, \frac{\pi}{2}\right), \left(\frac{\pi}{2}, \frac{\pi}{2}\right), \left(\frac{3\pi}{2}, \frac{3\pi}{2}\right), \ldots$

y-intercept: $y = 0 + \cos 0 = 1$

x-intercept:
$0 = x - \cos x \Rightarrow x = \cos x \Rightarrow x \approx 0.739$

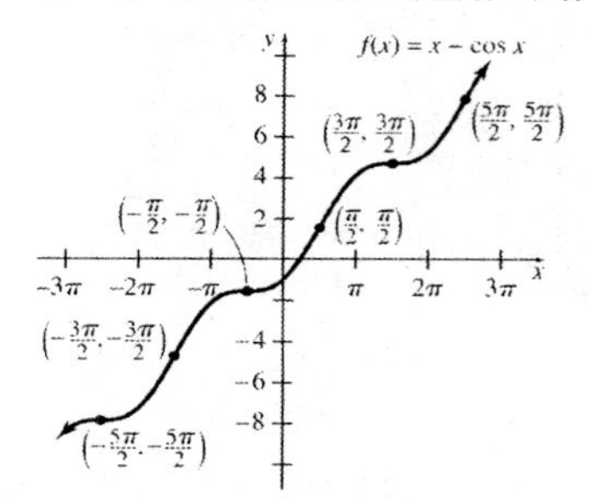

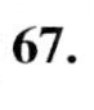

67.

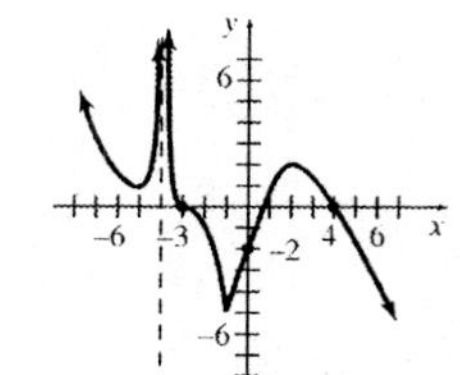

Other graphs are possible.

68.

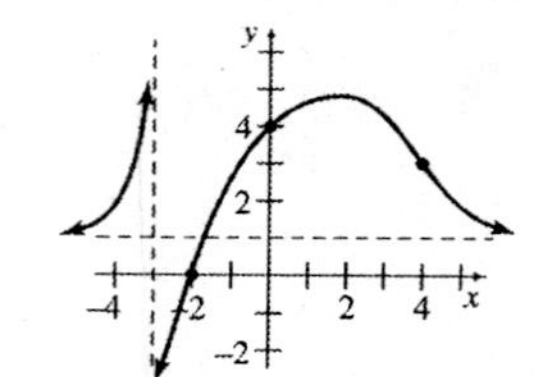

Other graphs are possible.

69. **(a)** Since the second derivative has many sign changes, the graph continually changes from concave upward to concave downward. Since there is a nonlinear decline, the graph must be one that declines, levels off, declines, levels off, etc. Therefore, the first derivative has many critical numbers where the first derivative is zero.

(b) The curve is always decreasing except at frequent points of inflection.

70. **(a)** $Y(M) = Y_0 M^b$

$Y'(M) = bY_0 M^{b-1}$

$Y''(M) = b(b-1)Y_0 M^{b-2}$

When $b > 0, Y' > 0,$ so metabolic rate and life span are increasing function of mass.

When $b < 0, Y' < 0,$ so heartbeat is a decreasing function of mass.

When $0 < b < 1, b(b-1) < 0,$ so metabolic rate and life span have graphs that are concave downward.

When $b < 0, b(b-1) > 0,$ so heartbeat has a graph that is concave upward.

(b) $\frac{dY}{dM} = bY_0 M^{b-1} = \frac{b}{M} Y_0 M^b = \frac{b}{M} Y$

71. **(a)** Set the two formulas equals to each other.

$1486S^2 - 4106S + 4514 = 1486S - 825 \Rightarrow 1486S^2 - 5592S + 5339 = 0$

Now, set the derivative equal to zero and solve for S.

$2972S - 5592 = 0 \Rightarrow S \approx 1.88$

For males with 1.88 square meters of surface area, the red cell volume increases approximately 1486 ml for each additional square meter of surface area.

(b) Set the formulas equal to each other.

$995e^{0.6085S} = 1578S \Rightarrow 995e^{0.6085S} - 1578S = 0$

Now, set the derivative equal to zero and solve for S.

$605.4575e^{0.6085S} - 1578 = 0 \Rightarrow e^{0.6085S} \approx 2.6063 \Rightarrow 0.6085S \approx \ln 2.6063 \Rightarrow S \approx 1.57 \text{ m}^2$

By plugging the exact value of S into the two formulas given for PV, we get about 2593 ml (Hurley) and 2484 for Pearson et al.

(c) For males with 1.57 square meters of surface area, the red cell volume increases approximately 1578 ml for each additional square meter of surface area.

(d) When f and g are closest together, their absolute difference is minimized.

$\left.\frac{d}{dx}\right|_{x=x_0} [f(x) - g(x)] = 0 \Rightarrow f'(x_0) - g'(x_0) = 0 \Rightarrow f'(x_0) = g'(x_0)$

72. **(a)** $y(S) = S\left[1 + C(1 + kS)^{f-1}\right]$

$y(0) = 0\left[1 + C(1 + kS)^{f-1}\right] = 0$

$y(1) = 1\left[1 + C(1 + k(1))^{f-1}\right] = 1 + C(1 + k)^{f-1}$

C and k are both positive numbers, so $y(1) > 1$.

(b) $$\begin{aligned} y'(S) &= S\frac{d}{dS}\left[1+C(1+kS)^{f-1}\right]+\left(\frac{d}{dS}S\right)\left(1+C(1+kS)^{f-1}\right) \\ &= Sk(f-1)C(1+kS)^{f-2}k+\left(1+C(1+kS)^{f-1}\right) \\ &= 1+\left[Sk(f-1)C(1+kS)^{f-2}k+C(1+kS)^{f-1}\right] \\ &= 1+C(1+kS)^{f-2}\left(Sk(f-1)+(1+kS)\right)=1+C(1+kS)^{f-2}(1+fkS) \end{aligned}$$

(c) Because y' is always greater than zero, the function is always increasing. Therefore, because S increases from $y(0)=0$ to a value greater than 1 at $y(1)$, once y reaches 1, it continues to increase. Thus, there can be no other value of S where $y = 1$.

(d) Let $k = 0$. Then $S\left(1+C(1)^{f-1}\right)=1 \Rightarrow S=\dfrac{1}{1+C}$

(e) Assume $k > 0$ and, to the contrary, assume $S > \dfrac{1}{1+C}$. Then,

$$y(S)=S\left[1+C(1+kS)^{f-1}\right] > \frac{1}{C+1}\left[1+C\left(1+k\cdot\frac{1}{C+1}\right)^{f-1}\right]=\frac{1}{C+1}+\frac{C}{C+1}\left(1+\frac{k}{C+1}\right)^{f-1}$$

Since $k > 0$, $\dfrac{k}{C+1}>0$, and $\left(1+\dfrac{k}{C+1}\right)^{f-1}>1$. (Note that in the context of this problem, $f > 1$.)

Therefore, $S=\dfrac{1}{C+1}+\dfrac{C}{C+1}\left(1+\dfrac{k}{C+1}\right)^{f-1}>\dfrac{1}{C+1}+\dfrac{C}{C+1}=1$, which is a contradiction. Thus, the value of S such that $y(S)=1$ must be between 0 and $\dfrac{1}{1+C}$.

(f) $$\begin{aligned} y'' &= C(1+kS)^{f-2}(fk)+(1+fkS)Ck(f-2)(1+kS)^{f-3} \\ &= Ck(1+kS)^{f-3}\left[f(1+kS)+(1+fkS)(f-2)\right] \\ &= Ck(1+kS)^{f-3}\left(f^2kS-fkS+2f-2\right) \\ &= Ck(1+kS)^{f-3}(fkS+2)(f-1) \end{aligned}$$

Let $f = 2$. Then $y''(2)=Ck(1+kS)^{2-3}(2kS+2)(2-1)=2Ck\left(\dfrac{1}{1+kS}\right)(1+kS)=2Ck$

Thus, $y''(S)=\begin{cases} 2Ck & \text{if } f=2 \\ Ck(1+kS)^{f-3}(fkS+2)(f-1) & \text{if } f\ge 3 \end{cases}$

(g) $y(S)=1 \Rightarrow S\left[1+C(1+kS)^{f-1}\right]=1 \Rightarrow CS(1+kS)^{f-1}=1-S \Rightarrow C=F(S)=\dfrac{1-S}{S(1+kS)^{f-1}}$

(h) $$F'(S)=\frac{S(1+kS)^{f-1}(-1)-(1-S)\frac{d}{dS}\left(S(1+kS)^{f-1}\right)}{S(1+kS)^{f-1}}$$

$$\begin{aligned} \frac{d}{dS}\left(S(1+kS)^{f-1}\right) &= S\frac{d}{dS}\left((1+kS)^{f-1}\right)+(1+kS)^{f-1}\frac{d}{dS}S \\ &= Sk(f-1)(1+kS)^{f-2}+(1+kS)^{f-1}=(1+kS)^{f-1}\left(Sk(f-1)(1+kS)^{-1}+1\right) \end{aligned}$$

(continued on next page)

(continued)

Continuing, we have

$$F'(S)=\frac{S(1+kS)^{f-1}(-1)-(1-S)(1+kS)^{f-1}\left(Sk(f-1)(1+kS)^{-1}+1\right)}{\left(S(1+kS)^{f-1}\right)^2}$$

$$=\frac{-S-(1-S)\left(Sk(f-1)(1+kS)^{-1}+1\right)}{S^2(1+kS)^{f-1}}=\frac{-S-(1-S)\left(Sk(f-1)(1+kS)^{-1}+1\right)}{S^2(1+kS)^{f-1}}\cdot\frac{1+kS}{1+kS}$$

$$=\frac{-S(1+kS)}{S^2(1+kS)^f}-\frac{(1-S)\left(Sk(f-1)(1+kS)^{-1}\right)(1+kS)}{S^2(1+kS)^f}-\frac{(1-S)(1+kS)}{S^2(1+kS)^f}$$

$$=\frac{-S(1+kS)}{S^2(1+kS)^f}-\frac{Sk(1-S)(f-1)}{S^2(1+kS)^f}-\frac{(1-S)(1+kS)}{S^2(1+kS)^f}$$

$$=\frac{-S-kS^2-Sk(f-1-Sf+S)-\left(1+kS-S-kS^2\right)}{S^2(1+kS)^f}$$

$$=\frac{-S-kS^2-Skf+Sk+S^2fk-S^2k-1-kS+kS^2+S}{S^2(1+kS)^f}$$

$$=\frac{-Skf+S^2fk-1-kS^2}{S^2(1+kS)^f}=-\frac{1+kS(f-Sf+S)}{S^2(1+kS)^f}=-\frac{1+kS(S+f(1-S))}{S^2(1+kS)^f}$$

(i) For $0<S<1$, F is decreasing because $F'(S)<0$.

73. (a) $z=(1-S)\left[1-(1+kS)^{-(f-1)}\right]$

$$\frac{dz}{dS}=(1-S)\frac{d}{dS}\left[1-(1+kS)^{-(f-1)}\right]+\left[1-(1+kS)^{-(f-1)}\right]\frac{d}{dS}(1-S)$$

$$=(1-S)(f-1)(1+kS)^{-(f-1)-1}(k)-\left[1-(1+kS)^{-(f-1)}\right]$$

$$=(1-S)(f-1)(1+kS)^{-f}(k)-\left[1-(1+kS)^{-(f-1)}\right]$$

$$=\frac{k(1-S)(f-1)}{(1+kS)^f}-1+\frac{1}{(1+kS)^{f-1}}=-1+\frac{k(1-S)(f-1)}{(1+kS)^f}+\frac{1+kS}{(1+kS)^f}$$

$$=-1+\frac{1+kS+k(1-S)(f-1)}{(1+kS)^f}=-1+(1+kS)^{-f}\left[1+kS+k(1-S)(f-1)\right]$$

(b) Let $f=2$ and $S=\frac{-1+\sqrt{1+k}}{k}$.

$$\frac{dz}{dS}=-1+\left(1+k\left(\frac{-1+\sqrt{1+k}}{k}\right)\right)^{-2}\left[1+k\left(\frac{-1+\sqrt{1+k}}{k}\right)+k\left(1-\left(\frac{-1+\sqrt{1+k}}{k}\right)\right)(2-1)\right]$$

$$=-1+\left(1-1+\sqrt{1+k}\right)^{-2}\left[1-1+\sqrt{1+k}+k+1-\sqrt{1+k}\right]$$

$$=-1+\left(\sqrt{1+k}\right)^{-2}(k+1)=-1+\left(\sqrt{1+k}\right)^{-1}(k+1)=-1+1=0$$

74. $T(n)=\dfrac{an}{1+bn^2};\ a=20,\ b=1$

$T(n)=\dfrac{20n}{1+n^2}$

$T'(n)=\dfrac{20(1+n^2)-2n(20n)}{(1+n^2)^2}=\dfrac{20-20n^2}{(1+n^2)^2}$

$T'(n)=0\Rightarrow \dfrac{20-20n^2}{(1+n^2)^2}=0\Rightarrow n=\pm 1$

In the context of this problem, n must be nonnegative, so disregard $n=-1$. The only critical point is $n=1$.
$T'(0.5)=9.6>0;\quad T'(2)=-2.4<0$

Therefore, $T(n)$ is increasing on $(0,\ 1)$ and decreasing on $(1,\ \infty)$. So, $T(n)$ has a relative maximum at $(1,\ T(1))=(1,\ 10)$.

$T''(n)=\dfrac{(1+n^2)^2(-40n)-(20-20n^2)(2)(1+n^2)(2n)}{(1+n^2)^4}=\dfrac{-40n(1+n^2)^2-4n(1+n^2)(20-20n^2)}{(1+n^2)^4}$

$=\dfrac{-40n(1+n^2)-4n(20-20n^2)}{(1+n^2)^3}=\dfrac{40n^3-120n}{(1+n^2)^3}$

$T''(n)=0\Rightarrow \dfrac{40n^3-120n}{(1+n^2)^3}=0\Rightarrow 40n(n^2-3)=0\Rightarrow n=0 \text{ or } n=\sqrt{3}$

If $n=0$, then $T(n)=0$. This is not an inflection point. There is an inflection point at $(\sqrt{3},\ T(\sqrt{3}))=(\sqrt{3},\ 5\sqrt{3})$.

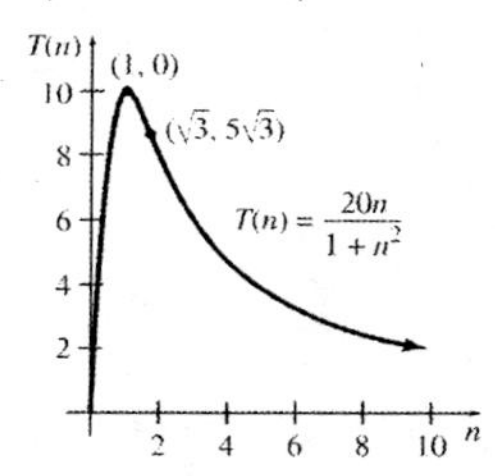

75. (a) $R(c)=\dfrac{ac^2}{1+bc^2}-kc;\ a=10,\ b=0.08,\ k=7$

$R(c)=\dfrac{10c^2}{1+0.08c^2}-7c$

$R'(c)=\dfrac{(1+0.08c^2)(20c)-10c^2(0.16c)}{(1+0.08c^2)^2}-7=\dfrac{20c}{(1+0.08c^2)^2}-7$

$=\dfrac{20c-7(1+0.08c^2)^2}{(1+0.08c^2)^2}=\dfrac{20c-7(1+0.16c^2+0.0064c^4)}{(1+0.08c^2)^2}=\dfrac{-0.0448c^4-1.12c^2+20c-7}{(1+0.08c^2)^2}$

$R'(c)=0\Rightarrow -.0448c^4-1.12c^2+20c-7=0\Rightarrow c\approx 0.36 \text{ or } c\approx 6.4$ (Use a graphing calculator to solve this.)

(*continued on next page*)

(continued)

There are two critical points, $(0.36, R(0.36)) \approx (0.36, -1.24)$ and $(6.4, R(6.4)) \approx (6.4, 50.97)$

However, in the context of this problem, $R(c)$ must be nonnegative, so disregard $c \approx 0.36$.

$R'(4) \approx 8.39 > 0; \quad R'(7) \approx -1.122 < 0$

Therefore, there is a relative maximum at $(6.4, 50.97)$.

Now find the x-intercepts:

$$R(c) = 0 \Rightarrow \frac{10c^2}{1+0.08c^2} - 7c = 0 \Rightarrow 10c^2 = 7c + 0.56c^3 \Rightarrow c\left(0.56c^2 - 10c + 7\right) = 0 \Rightarrow$$

$0.56c^2 - 10c + 7 = 0 \Rightarrow c \approx 0.73$ or $c \approx 17.13$ (using the quadratic formula)

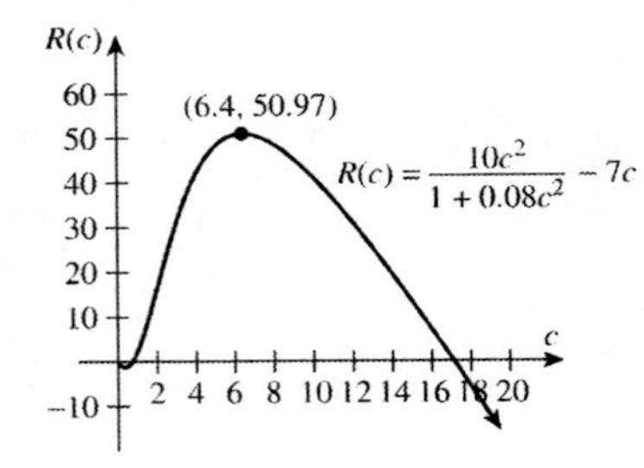

(b) The function is valid for $0.73 \le c \le 17.13$.

76. $y = \frac{\pi}{8} \cos 3\pi\left(t - \frac{1}{3}\right)$

(a) The graph should resemble the graph of $y = \cos x$ with following difference: The maximum and minimum values of y are $\frac{\pi}{8}$ and $-\frac{\pi}{8}$. The period of the graph will be $\frac{2\pi}{3\pi} = \frac{2}{3}$ units. The graph will be shifted horizontally $\frac{1}{3}$ units to the right.

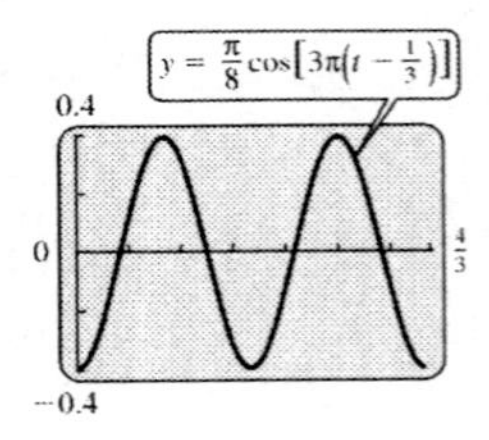

(b) velocity $= \frac{dy}{dt}$

$$\frac{dy}{dt} = D_t\left[\frac{\pi}{8}\cos 3\pi\left(t - \frac{1}{3}\right)\right] = \frac{\pi}{8} D_t\left[\cos 3\pi\left(t - \frac{1}{3}\right)\right] = \frac{\pi}{8}\left[-\sin 3\pi\left(t - \frac{1}{3}\right)\right] D_t\left[3\pi\left(t - \frac{1}{3}\right)\right]$$

$$= \frac{\pi}{8}\left[-\sin 3\pi\left(t - \frac{1}{3}\right)\right] \cdot 3\pi = -\frac{3\pi^2}{8} \sin\left[3\pi\left(t - \frac{1}{3}\right)\right]$$

acceleration $= \frac{d^2y}{dt^2}$

$$\frac{d^2y}{dt^2} = D_t\left[\frac{-3\pi^2 \sin 3\pi\left(t - \frac{1}{3}\right)}{8}\right] = \frac{-3\pi^2}{8} D_t\left[\sin 3\pi\left(t - \frac{1}{3}\right)\right]$$

$$= \frac{-3\pi^2}{8}\left[\cos 3\pi\left(t - \frac{1}{3}\right)\right] D_t\left[3\pi\left(t - \frac{1}{3}\right)\right] = \frac{-3\pi^2}{8}\left[\cos 3\pi\left(t - \frac{1}{3}\right)\right] \cdot 3\pi = -\frac{9\pi^3}{8}\cos\left[3\pi\left(t - \frac{1}{3}\right)\right]$$

(c) $$\frac{d^2y}{dt^2}+9\pi^2 y=-\frac{9\pi^3}{8}\cos 3\pi\left(t-\frac{1}{3}\right)+9\pi^2\left[\frac{\pi}{8}\cos 3\pi\left(t-\frac{1}{3}\right)\right]$$
$$=-\frac{9\pi^3}{8}\cos 3\pi\left(t-\frac{1}{3}\right)+\frac{9\pi^3}{8}\cos 3\pi\left(t-\frac{1}{3}\right)$$
$$=0$$

(d) $$a(t)=-\frac{9\pi^3}{8}\cos 3\pi\left(t-\frac{1}{3}\right)$$
$$a(1)=-\frac{9\pi^3}{8}\cos 2\pi=-\frac{9\pi^3}{8}\cdot 1=-\frac{9\pi^3}{8}<0$$
$$y(1)=\frac{\pi}{8}\cos 3\pi\left(t-\frac{1}{3}\right)=\frac{\pi}{8}\cos 2\pi=\frac{\pi}{8}\cdot 1=\frac{\pi}{8}$$

Therefore, at $t = 1$ second, the force is clockwise and the arm makes an angle of $\frac{\pi}{8}$ radians forward from the vertical. The arm is moving clockwise.

$$a\left(\frac{4}{3}\right)=-\frac{9\pi^3}{8}\cos 3\pi\left(\frac{4}{3}-\frac{1}{3}\right)=-\frac{9\pi^3}{8}\cos(3\pi)=-\frac{9\pi^3}{8}(-1)=\frac{9\pi^3}{8}>0$$
$$y\left(\frac{4}{3}\right)=\frac{\pi}{8}\cos 3\pi\left(\frac{4}{3}-\frac{1}{3}\right)=\frac{\pi}{8}\cos(3\pi)=\frac{\pi}{8}(-1)=-\frac{\pi}{8}$$

Therefore, at $t=\frac{4}{3}$ seconds, the force is counter clockwise and the arm makes an angle of $-\frac{\pi}{8}$ radians from the vertical. The arm is moving counterclockwise.

$$a\left(\frac{5}{3}\right)=-\frac{9\pi^3}{8}\cos 3\pi\left(\frac{5}{3}-\frac{1}{3}\right)=-\frac{9\pi^3}{8}\cos(4\pi)=-\frac{9\pi^3}{8}\cdot 1=-\frac{9\pi^3}{8}<0$$
$$y\left(\frac{5}{3}\right)=\frac{\pi}{8}\cos 3\pi\left(\frac{5}{3}-\frac{1}{3}\right)=\frac{\pi}{8}\cos(4\pi)=\frac{\pi}{8}\cdot 1=\frac{\pi}{8}$$

Therefore, at $t=\frac{5}{3}$ seconds, the answer corresponds to $t = 1$ second. So the arm is moving clockwise and makes an angle of $\frac{\pi}{8}$ from the vertical.

77. **(a)** The following is a table of values for $y=\frac{1}{5}\sin \pi(t-1)$.

t	0	0.5	1.0	1.5	2.0	2.5	3.0
y	0	–0.2	0	0.2	0	–0.2	0

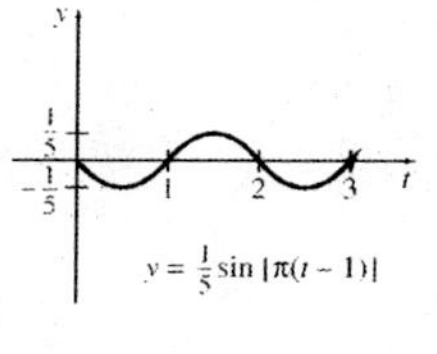

$y=\frac{1}{5}\sin[\pi(t-1)]$

(b) $$v(t)=\frac{dy}{dt}=\left[\frac{1}{5}\cos \pi(t-1)\right]\cdot D_x[\pi(t-1)]$$
$$=\frac{\pi}{5}\cos \pi(t-1)$$
$$a(t)=v'(t)=\frac{d^2y}{dt^2}$$
$$=\left[-\frac{\pi}{5}\sin \pi(t-1)\right]\cdot D_x[\pi(t-1)]$$
$$=-\frac{\pi^2}{5}\sin \pi(t-1)$$

(c) $$\frac{d^2y}{dt^2}+\pi^2 y=-\frac{\pi^2}{5}\sin[\pi(t-1)]$$
$$+(\pi^2)\frac{1}{5}\sin \pi(t-1)$$
$$=0$$

(d) Since the constant of proportionality is positive, the force and acceleration are in the same direction.
At $t = 1.5$ sec,

$$a(t) = -\frac{\pi^2}{5}\sin \pi(t-1)$$

$$a(1.5) = -\frac{\pi^2}{5}\sin \pi(0.5) = -\frac{\pi^2}{5}\sin\frac{\pi}{2}$$

$$= -\frac{\pi^2}{5}\cdot 1 = -\frac{\pi^2}{5} < 0,$$

and

$$y = \frac{1}{5}\sin \pi(t-1) = \frac{1}{5}\sin \pi(0.5)$$

$$= \frac{1}{5}\sin\frac{\pi}{2} = \frac{1}{5}\cdot 1 = \frac{1}{5}.$$

Thus, at $t = 1.5$, acceleration is negative, the arm is moving clockwise, and the arm is at an angle of $\frac{1}{5}$ radian from vertical.

At $t = 2.5$ sec,

$$a(2.5) = -\frac{\pi^2}{5}\sin \pi(1.5)$$

$$= -\frac{\pi^2}{5}\sin\frac{3\pi}{2}$$

$$= -\frac{\pi^2}{5}(-1) = \frac{\pi^2}{5} > 0,$$

and $y = \frac{1}{5}\sin\frac{3\pi}{2} = \frac{1}{5}(-1) = -\frac{1}{5}.$

Thus, at $t = 2.5$, acceleration is positive, the arm is moving counterclockwise, and the arm is at angle of $-\frac{1}{5}$ radian from vertical.

At $t = 3.5$ sec,

$$a(3.5) = -\frac{\pi^2}{5}\sin \pi(2.5)$$

$$= -\frac{\pi^2}{5}\sin\frac{5\pi}{2}$$

$$= -\frac{\pi^2}{5}\cdot 1 = -\frac{\pi^2}{5} < 0,$$

and $y = \frac{1}{5}\sin\frac{5\pi}{2} = \frac{1}{5}\cdot 1 = \frac{1}{5}.$

Thus, at $t = 3.5$, acceleration is negative, the arm is moving clockwise, and the arm is at an angle of $\frac{1}{5}$ radian from vertical.

78. $y(t) = A^{c^t}$

$$y'(t) = (\ln A)A^{c^t}\cdot\frac{d}{dt}c^t$$

$$= (\ln A)(\ln c)c^t A^{c^t}$$

$$y''(t) = (\ln A)(\ln c)\cdot[(\ln c)c^t A^{c^t} + c^t(\ln A)(\ln c)c^t A^{c^t}]$$

$$= (\ln A)(\ln c)c^t A^{c^t}[1 + (\ln A)c^t]$$

$y''(t) = 0$ when $1 + \ln(A)c^t = 0$

$$c^t = -\frac{1}{\ln A} \Rightarrow t\ln c = \ln\left(-\frac{1}{\ln A}\right) \Rightarrow$$

$$t = -\frac{\ln(-\ln A)}{\ln c} = -\frac{\ln[-\ln(0.3982\cdot 10^{-291})]}{\ln 0.4252}$$

By properties of logarithms,

$$-\ln(0.3982\cdot 10^{-291})$$

$$= -[\ln(0.3982) + \ln(10^{-291})]$$

$$= -[\ln(0.3982) - 291\ln(10)]$$

$$= -\ln(0.3982) + 291\ln(10)$$

So,

$$t = -\frac{\ln[-\ln(0.3982) + 291\ln(10)]}{\ln(0.4252)} \approx 7.6108$$

At about 7.6108 years the rate of learning to pass the test begins to slow down.

79. (a) $P(t) = 325 + 7.475(t+10)e^{-(t+10)/20}$

$$P'(t) = 7.475\left[1\cdot e^{-(t+10)/20} + (t+10)e^{-(t+10)/20}\cdot\frac{-1}{20}\right]$$

$$= 7.475\left[1 - \frac{1}{20}(t+10)\right]e^{-(t+10)/20}$$

$$= 7.475\left(\frac{1}{2} - \frac{t}{20}\right)e^{-(t+10)/20}$$

$P'(t)$ is zero when $\frac{1}{2} - \frac{t}{20} = 0 \Rightarrow t = 10$

$p'(9) \approx 0.39$
$p'(11) \approx -0.36$

$P(t)$ is increasing at 9 and decreasing at 11. So, a relative maximum occurs at $t = 10$. The population is largest in the year 2010.

$P(10) = 325 + 7.475(20)e^{-1} \approx 380.0$

The population is predicted to be 380 million in 2010.

(b) $P'(t) = 7.475\left(\frac{1}{2} - \frac{t}{20}\right)e^{-(t+10)/20}$

$$P''(t) = 7.475\left[-\frac{1}{20}e^{-(t+10)/20} + \left(\frac{1}{2} - \frac{t}{20}\right)e^{-(t+10)/20} \cdot \frac{-1}{20}\right]$$
$$= 7.475\left(-\frac{1}{20} - \frac{1}{40} + \frac{t}{400}\right)e^{-(t+10)/20}$$
$$= 7.475\left(\frac{t}{400} - \frac{3}{40}\right)e^{-(t+10)/20}$$

$P'(t)$ is zero when $\frac{t}{400} - \frac{3}{40} = 0 \Rightarrow t = 30$

$P'(29) = -0.0027$
$P'(31) = 0.0024$

P' is decreasing at 29, and increasing at 31. So, a relative minimum occurs at $t = 30$. The population is declining most rapidly in the year 2030.

(c) As time t approaches infinity, the population P approaches

$$\lim_{t\to\infty} P(t)$$
$$= \lim_{t\to\infty}\left(325 + 7.475(t+10)e^{-(t+10)/20}\right)$$
$$= 325 + 7.475 \lim_{t\to\infty} \frac{t+10}{e^{(t+10)/20}}$$
$$= 325 + 7.475(0) = 325.$$

The population is approaching 325 million.

80. (a) The U.S. total inventory was at a relative maximum in 1965, 1973, 1976, 1983, 1986, and 1988.

(b) The U.S. total inventory was at its largest relative maximum from 1965 to 1967. During this period, the Soviet total inventory was concave upward. This means that the total inventory was increasing at an increasingly rapid rate.

81. (a) $s(t) = 512t - 16t^2$
$v(t) = s'(t) = 512 - 32t$
$a(t) = v'(t) = s''(t) = -32$

(b) The maximum height is attained when $v(t) = 0 \Rightarrow 512 - 32t = 0 \Rightarrow t = 16$

$v(0) = 512 > 0$
$v(20) = 512 - 640 = -128 < 0$

The height reaches a maximum when $t = 16$.

$s(16) = 512 \cdot 16 - 16(16^2) = 4096$
The maximum height is 4096 ft.

(c) The projectile hits the ground when
$s(t) = 0 \Rightarrow 512t - 16t^2 = 0 \Rightarrow$
$16t(32 - t) = 0 \Rightarrow t = 0$ or $t = 32$
$v(32) = 512 - 32(32) = -512$
The projectile hits the ground after 32 seconds with a velocity of -512 ft/sec.

82. (a)
$$y = x\tan\alpha - \frac{16x^2}{V^2}\sec^2\alpha$$
$$= 40\tan\left(\frac{\pi}{4}\right) - \frac{16(40)^2}{44^2}\sec^2\left(\frac{\pi}{4}\right)$$
$$= 40(1) - \frac{16(40)^2}{44^2}(2) \approx 13.55 \text{ feet}$$

A piece of gravel is thrown about 13.55 ft.

(b)
$$y = x\tan\alpha - \frac{16x^2}{V^2}\sec^2\alpha$$
$$0 = x\tan\alpha - \frac{16x^2}{V^2}\sec^2\alpha$$
$$0 = x\left(\tan\alpha - \frac{16x}{V^2}\sec^2\alpha\right)$$
$$0 = \tan\alpha - \frac{16x}{V^2}\sec^2\alpha \quad (\text{for } x \neq 0)$$
$$\frac{16x}{V^2}\sec^2\alpha = \tan\alpha$$
$$x = \frac{V^2}{16} \cdot \frac{\tan\alpha}{\sec^2\alpha}$$
$$= \frac{V^2}{16} \cdot \sin\alpha\cos\alpha$$
$$= \frac{V^2}{32} \cdot 2\sin\alpha\cos\alpha$$
$$= \frac{V^2}{32}\sin(2\alpha)$$

(c)
$$x = \frac{V^2}{32}\sin(2\alpha) = \frac{44^2}{32}\sin\left[2\left(\frac{\pi}{3}\right)\right]$$
$$\approx 52.39 \text{ feet}$$

(d) $\dfrac{dx}{d\alpha} = \dfrac{V^2}{32}\cos(2\alpha) \cdot D_\alpha(2\alpha)$

$= \dfrac{V^2}{16}\cos(2\alpha)$

Find critical values.

$$\frac{V^2}{16}\cos(2\alpha) = 0$$

$$\cos(2\alpha) = 0$$

$$2\alpha = \frac{\pi}{2} + n\pi, \text{ for } n \text{ any integer}$$

$$\alpha = \frac{\pi}{4} + \frac{n\pi}{2}, \text{ for } n \text{ any integer}$$

Since $0 < \alpha < \dfrac{\pi}{2}$, $\dfrac{dx}{d\alpha} = 0$ for $\alpha = \dfrac{\pi}{4}$.

Furthermore, $\dfrac{d^2x}{d\alpha^2} = -\dfrac{V^2}{8}\sin(2\alpha)$ which

is less than zero at $\alpha = \dfrac{\pi}{4}$. Therefore, x is

maximized when $\alpha = \dfrac{\pi}{4}$.

(e) Since 60 miles per hour is 88 feet per second, evaluate

$x = \dfrac{V^2}{32}\sin(2\alpha)$ when $V = 88$ and $\alpha = \dfrac{\pi}{4}$.

$$= \frac{88^2}{32}\sin\left[2\left(\frac{\pi}{4}\right)\right] = 242 \text{ feet}$$

Chapter 6

APPLICATIONS OF THE DERIVATIVE

6.1 Absolute Extrema

1. As shown on the graph, the absolute maximum occurs at x_3; there is no absolute minimum. (There is no functional value that is less than all others.)

3. As shown on the graph, there are no absolute extrema.

5. As shown on the graph, the absolute minimum occurs at x_1; there is no absolute maximum.

7. As shown on the graph, the absolute maximum occurs at x_1; the absolute minimum occurs at x_2.

9. We will have an absolute maximum (or minimum) at $x = c$ if $f(c)$ is the largest (or smallest) value that the function will ever take over the given domain. By given domain, we mean the range of x-values that we have chosen to work with for a given problem. There may be other values of x that we can actually plug into the function but have excluded them for some reason.
We will have a relative maximum or minimum at $x = c$ if $f(c)$ is a maximum or minimum in some interval $a \le x = c \le b$. There may be larger or smaller values of the function for other values of x, but relative to $x = c$, $f(c)$ is greater than or less than all the other function values that are near it.

11. $f(x) = x^3 - 6x^2 + 9x - 8$; $[0, 5]$
Find critical numbers:

$$f'(x) = 3x^2 - 12x + 9 = 0$$
$$x^2 - 4x + 3 = 0$$
$$(x-3)(x-1) = 0$$
$$x = 1 \text{ or } x = 3$$

x	$f(x)$	
0	-8	Absolute minimum
1	-4	
3	-8	Absolute minimum
5	12	Absolute maximum

13. $f(x) = \frac{1}{3}x^3 + \frac{3}{2}x^2 - 4x + 1$; $[-5, 2]$
Find critical numbers:

$$f'(x) = x^2 + 3x - 4 = 0$$
$$(x+4)(x-1) = 0$$
$$x = -4 \text{ or } x = 1$$

x	$f(x)$	
-5	$\frac{101}{6} \approx 16.83$	
-4	$\frac{59}{3} \approx 19.67$	Absolute maximum
1	$-\frac{7}{6} \approx -1.17$	Absolute minimum
2	$\frac{5}{3} \approx 1.67$	

15. $f(x) = x^4 - 18x^2 + 1$; $[-4, 4]$

$$f'(x) = 4x^3 - 36x = 0$$
$$4x(x^2 - 9) = 0$$
$$4x(x+3)(x-3) = 0$$
$$x = 0 \text{ or } x = -3 \text{ or } x = 3$$

x	$f(x)$	
-4	-31	
-3	-80	Absolute minimum
0	1	Absolute maximum
3	-80	Absolute minimum
4	-31	

17. $f(x) = \frac{1-x}{3+x}$; $[0, 3]$

$$f'(x) = \frac{-4}{(3+x)^2}$$

No critical numbers; check the endpoints of the interval

x	$f(x)$	
0	$\frac{1}{3}$	Absolute maximum
3	$-\frac{1}{3}$	Absolute minimum

19. $f(x) = \frac{x-1}{x^2+1}$; [1, 5]

$f'(x) = \frac{-x^2+2x+1}{(x^2+1)^2}$

$f'(x) = -x^2+2x+1 = 0 \Rightarrow$
$x = 1 \pm \sqrt{2} \approx 2.414, -0.414$

Note that $1-\sqrt{2}$ is not in [1, 5].

x	$f(x)$	
1	0	Absolute minimum
$1+\sqrt{2}$	$\frac{\sqrt{2}-1}{2} \approx 0.21$	Absolute maximum
5	$\frac{2}{13} \approx 0.15$	

21. $f(x) = (x^2-4)^{1/3}$; [−2, 3]

$f'(x) = \frac{1}{3}(x^2-4)^{-2/3}(2x) = \frac{2x}{3(x^2-4)^{2/3}}$

$f'(x) = 0$ when $2x = 0 \Rightarrow x = 0$

$f'(x)$ is undefined at $x = -2$ and $x = 2$, but $f(x)$ is defined there, so −2 and 2 are also critical numbers.

x	$f(x)$	
−2	0	
0	$(-4)^{1/3} \approx -1.587$	Absolute minimum
2	0	
3	$5^{1/3} \approx 1.710$	Absolute maximum

23. $f(x) = 5x^{2/3} + 2x^{5/3}$; [−2, 1]

$f'(x) = \frac{10}{3}x^{-1/3} + \frac{10}{3}x^{2/3} = \frac{10}{3x^{1/3}} + \frac{10x^{2/3}}{3}$
$= \frac{10x+10}{3x^{1/3}} = \frac{10(x+1)}{3\sqrt[3]{x}}$

$f'(x) = 0$ when $10(x+1) = 0 \Rightarrow x+1 = 0 \Rightarrow$
$x = -1$

$f'(x)$ is undefined at $x = 0$, but $f(x)$ is defined at $x = 0$, so 0 is also a critical number.

x	$f(x)$	
−2	1.587	
−1	3	
0	0	Absolute minimum
1	7	Absolute maximum

25. $f(x) = x^2 - 8\ln x$; [1, 4]

$f'(x) = 2x - \frac{8}{x}$

$f'(x) = 0$ when $2x - \frac{8}{x} = 0 \Rightarrow 2x = \frac{8}{x} \Rightarrow$
$2x^2 = 8 \Rightarrow x^2 = 4 \Rightarrow x = -2$ or $x = 2$
but $x = -2$ is not in the given interval.
Although $f'(x)$ fails to exist at $x = 0$, 0 is not in the specified domain for $f(x)$, so 0 is not a critical number.

x	$f(x)$	
1	1	
2	−1.545	Absolute minimum
4	4.910	Absolute maximum

27. $f(x) = x + e^{-3x}$; [−1, 3]

$f'(x) = 1 - 3e^{-3x}$

$f'(x) = 0$ when $1 - 3e^{-3x} = 0 \Rightarrow$
$-3e^{-3x} = -1 \Rightarrow e^{-3x} = \frac{1}{3} \Rightarrow -3x = \ln\frac{1}{3} \Rightarrow$
$x = \frac{\ln 3}{3}$

x	$f(x)$	
−1	19.09	Absolute maximum
$\frac{\ln 3}{3}$	0.6995	Absolute minimum
3	3.000	

29. $f(x) = \frac{x}{2} - \sin x$; $[0, \pi]$

$f'(x) = \frac{1}{2} - \cos x$

Now solve $f'(x) = 0$ to find any critical points in the given interval $[0, \pi]$.

$\frac{1}{2} - \cos x = 0 \Rightarrow \cos x = \frac{1}{2} \Rightarrow x = \frac{\pi}{3}$

We must also test the endpoints $x = 0$ and $x = \pi$.

x	$f(x)$	
0	0	
$\frac{\pi}{3}$	$\frac{\pi}{6} - \frac{\sqrt{3}}{2} \approx -0.3424$	Absolute minimum
π	$\frac{\pi}{2} \approx 1.5708$	Absolute maximum

(continued on next page)

(*continued*)

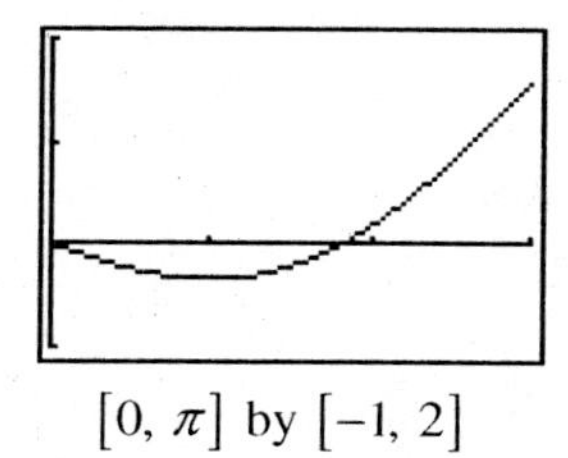

$[0, \pi]$ by $[-1, 2]$

From the graph, we see that on $[0, \pi]$ has an absolute maximum of $\frac{\pi}{2}$ at $x = \pi$ and an absolute minimum of about $\frac{\pi}{6} - \frac{\sqrt{3}}{2}$ at $x = \frac{\pi}{3}$.

31. $f(x) = \dfrac{-5x^4 + 2x^3 + 3x^2 + 9}{x^4 - x^3 + x^2 + 7}$; $[-1, 1]$

The indicated domain tells us the x-values to use for the viewing window, but we must experiment to find a suitable range for the y-values. In order to show the absolute extrema on $[-1, 1]$, we find that a suitable window is $[-1, 1]$ by $[-0.5, 1.5]$. From the graph, we see that on $[-1, 1]$, f has an absolute maximum of 1.356 at $x \approx 0.6085$ and an absolute minimum of 0.5 at $x = -1$.

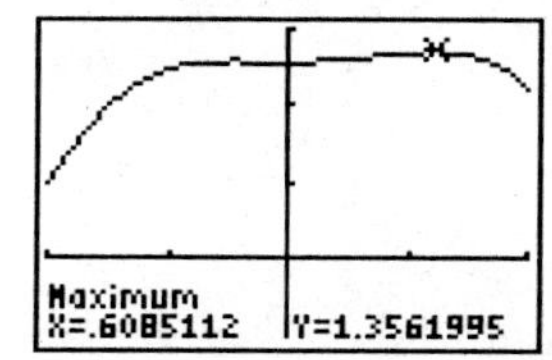

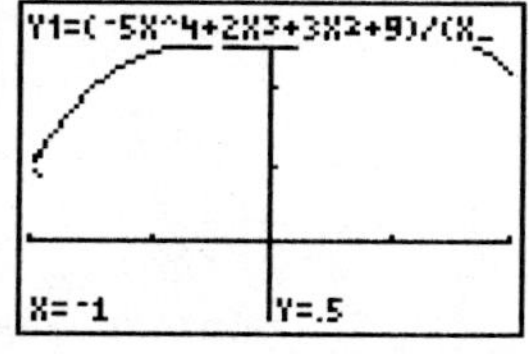

33. $f(x) = 2x + \dfrac{8}{x^2} + 1, \; x > 0$

$$f'(x) = 2 - \frac{16}{x^3} = \frac{2x^3 - 16}{x^3}$$
$$= \frac{2(x-2)(x^2 + 2x + 4)}{x^3}$$

Since the specified domain is $(0, \infty)$, a critical number is $x = 2$.

x	$f(x)$
2	7

There is an absolute minimum of 7 at $x = 2$; there is no absolute maximum, as can be seen by looking at the graph of f.

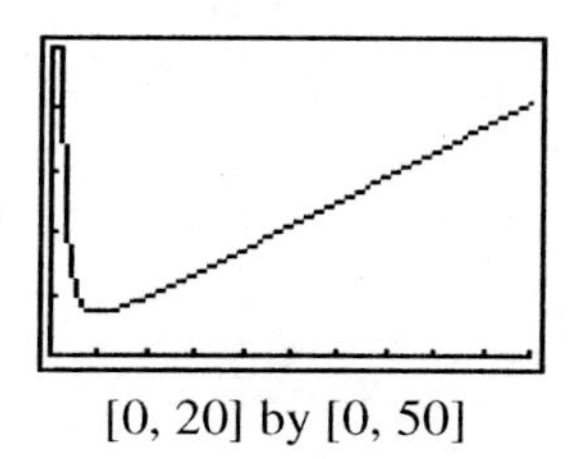

[0, 20] by [0, 50]

35. $f(x) = -3x^4 + 8x^3 + 18x^2 + 2$

$$f'(x) = -12x^3 + 24x^2 + 36x$$
$$= -12x(x^2 - 2x - 3)$$
$$= -12x(x-3)(x+1)$$

Critical numbers are 0, 3, and -1.

x	$f(x)$
-1	9
0	2
3	137

There is an absolute maximum of 137 at $x = 3$; there is no absolute minimum, as can be seen by looking at the graph of f.

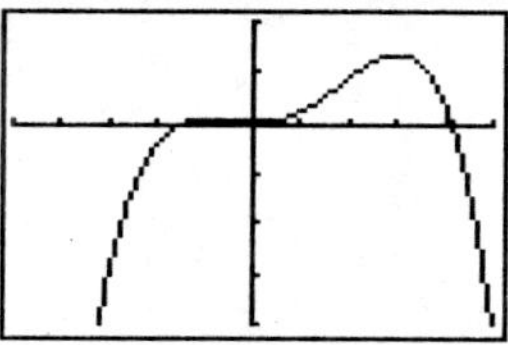

[–5, 5] by [–400, 100]

37. $f(x) = \dfrac{x-1}{x^2 + 2x + 6}$

$$f'(x) = \frac{(x^2 + 2x + 6)(1) - (x-1)(2x+2)}{(x^2 + 2x + 6)^2}$$
$$= \frac{x^2 + 2x + 6 - 2x^2 + 2}{(x^2 + 2x + 6)^2} = \frac{-x^2 + 2x + 8}{(x^2 + 2x + 6)^2}$$
$$= \frac{-(x^2 - 2x - 8)}{(x^2 + 2x + 6)^2} = \frac{-(x-4)(x+2)}{(x^2 + 2x + 6)^2}$$

Critical numbers are 4 and -2.

x	$f(x)$
-2	$-\frac{1}{2}$
4	0.1

There is an absolute maximum of 0.1 at $x = 4$ and an absolute minimum of 0.5 at $x = -2$. This can be verified by looking at the graph of f.

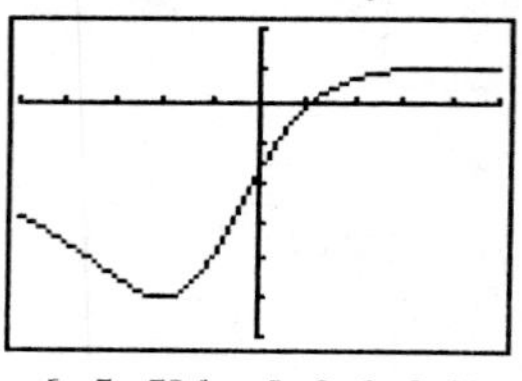

[–5, 5] by [–0.6, 0.2]

39. $f(x) = \dfrac{\ln x}{x^3}$

$$f'(x) = \frac{x^3 \cdot \frac{1}{x} - 3x^2 \ln x}{x^6} = \frac{x^2 - 3x^2 \ln x}{x^6}$$

$$= \frac{x^2(1 - 3\ln x)}{x^6} = \frac{1 - 3\ln x}{x^4}$$

$f'(x) = 0$ when $x = e^{1/3}$, and $f'(x)$ does not exist when $x \le 0$. The only critical number is $e^{1/3}$.

x	$f(x)$
$e^{1/3}$	$\frac{1}{3}e^{-1} \approx 0.1226$

There is an absolute maximum of 0.1226 at $x = e^{1/3}$. There is no absolute minimum, as can be seen by looking at the graph of f.

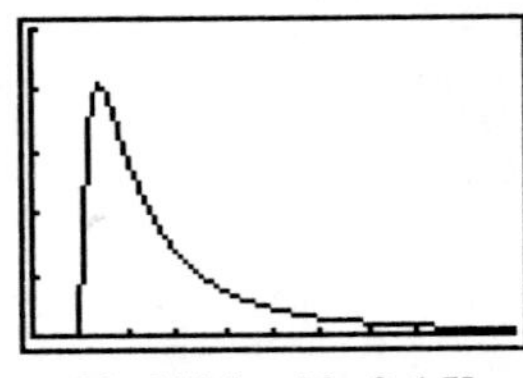

[0, 10] by [0, 0.15]

41. $f(x) = 2x - 3x^{2/3}$

$$f'(x) = 2 - 2x^{-1/3} = 2 - \frac{2}{\sqrt[3]{x}} = \frac{2\sqrt[3]{x} - 2}{\sqrt[3]{x}}$$

$f'(x) = 0$ when $2\sqrt[3]{x} - 2 = 0 \Rightarrow 2\sqrt[3]{x} = 2 \Rightarrow \sqrt[3]{x} = 1 \Rightarrow x = 1$

$f'(x)$ is undefined at $x = 0$, but $f(x)$ is defined at $x = 0$. So the critical numbers are 0 and 1.

(a) On $[-1, 0.5]$

x	$f(x)$
-1	-5
0	0
0.5	-0.88988

Absolute minimum of -5 at $x = -1$; absolute maximum of 0 at $x = 0$

(b) On $[0.5, 2]$

x	$f(x)$
0.5	-0.88988
1	-1
2	-0.7622

Absolute maximum of about -0.76 at $x = 2$; absolute minimum of -1 at $x = 1$.

43. $f(x) = \dfrac{x^2 + 36}{2x}, 1 \le x \le 12$

$$f'(x) = \frac{2x(2x) - (x^2 + 36)(2)}{(2x)^2}$$

$$= \frac{4x^2 - 2x^2 - 72}{4x^2} = \frac{2x^2 - 72}{4x^2}$$

$$= \frac{2(x^2 - 36)}{4x^2} = \frac{(x+6)(x-6)}{2x^2}$$

$f'(x) = 0$ when $x = 6$ and when $x = -6$. Only 6 is in the interval $1 \le x \le 12$.

Test for relative maximum or minimum.

$$f'(5) = \frac{(11)(-1)}{50} < 0$$

$$f'(7) = \frac{(13)(1)}{98} > 0$$

The minimum occurs at $x = 6$, or at 6 months. Since $f(6) = 6$, $f(1) = 18.5$, and $f(12) = 7.5$, the minimum percent is 6%.

45. Since we are only interested in the length during weeks 22 through 28, the domain of the function for this problem is [22, 28]. We now look for any critical numbers in this interval. We find $L'(t) = 0.788 - 0.02t$

There is a critical number at $t = \frac{0.788}{0.02} = 39.4$, which is not in the interval. Thus, the maximum value will occur at one of the endpoints.

t	$L(t)$
22	5.4
28	7.2

The maximum length is about 7.2 millimeters.

47. (a)

4.5

$M(t) = -0.5035295 + 0.0229883t + 0.0108021t^2 - 0.0003139t^3 + 0.0000025t^4$

6, 51, 0

(b) The derivative of the function is

$$M'(t) = 0.0229883 + 0.0216042t - 0.0009417t^2 + 0.00001t^3$$

Graph the derivative to find its a maximum; this gives us the point at which $M(t)$ is increasing the fastest.

(continued on next page)

(*continued*)

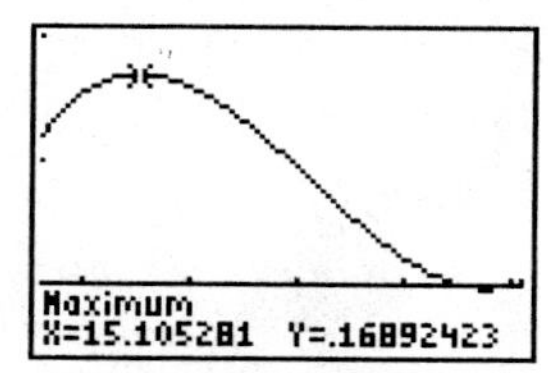

[6, 50] by [−0.05, 0.2]

The dentin is growing most rapidly $x \approx 15.11$ or about day 15.

49. **(a)**

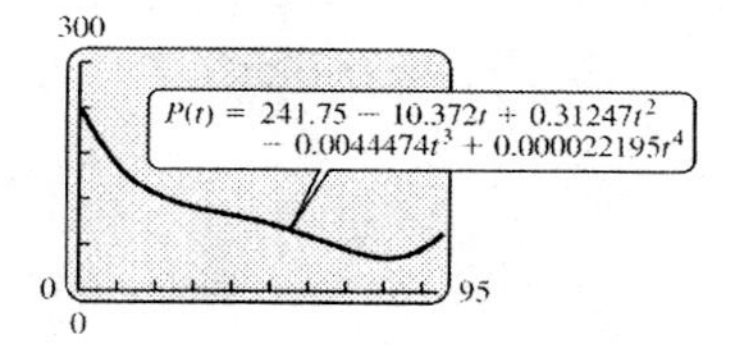

(b) Use a graphing calculator to locate any extreme points on the graph.

t	$P(t)$
0	241.75
81.51	43.60
95	71.16

The maximum number of polygons, about 242, occurs at $t = 0$. The minimum number of polygons, about 44, occurs at $t \approx 81.51$

(c) Answers will vary.

51. Total area $A(x) = \pi\left(\frac{x}{2\pi}\right)^2 + \left(\frac{12-x}{4}\right)^2$

$$= \frac{x^2}{4\pi} + \frac{(12-x)^2}{16}$$

$$A'(x) = \frac{x}{2\pi} - \frac{12-x}{8} = 0$$

$$\frac{4x - \pi(12-x)}{8\pi} = 0$$

$$x = \frac{12\pi}{4+\pi} \approx 5.28$$

x	Area
0	9
5.28	5.04
12	11.46

The total area is minimized when the piece used to form the circle is $\frac{12\pi}{4+\pi}$ feet, or about 5.28 feet long.

53. For the solution to Exercise 51, the piece of length x used to form the circle is $\frac{12\pi}{4+\pi}$ feet. The circle can be inscribed inside the square if the side of the square equals the diameter of the circle (that is, twice the radius).

side of the square = 2(radius)

$$\frac{12-x}{4} = 2\left(\frac{x}{2\pi}\right)$$

$$\frac{12-x}{4} = \frac{x}{\pi}$$

$$4x = 12\pi - \pi x$$

$$x(4+\pi) = 12\pi$$

$$x = \frac{12\pi}{4+\pi}$$

Therefore, the circle formed by piece of length $x = \frac{12\pi}{4+\pi}$ can be inscribed inside the square.

55. The value $x = 11.5$ minimizes $\frac{f(x)}{x}$ because this is the point where the line from the origin to the curve is tangent to the curve.
A production level of 11.5 units results in the minimum cost per unit.

57. The value $x = 100$ maximizes $\frac{f(x)}{x}$ because this is the point where the line from the origin to the curve is tangent to the curve.
A production level of 100 units results in the maximum profit per item produced.

6.2 Applications of Extrema

1. $x + y = 180$, $P = xy$

(a) $y = 180 - x$

(b) $P = xy = x(180 - x)$

(c) Since $y = 180 - x$ and x and y are nonnegative numbers, $x \geq 0$ and $180 - x \geq 0$ or $x \leq 180$. The domain of P is $[0, 180]$.

(d) $P'(x) = 180 - 2x$
$P'(x) = 0 \Rightarrow 180 - 2x = 0 \Rightarrow$
$2(90 - x) = 0 \Rightarrow x = 90$

(e)

x	P
0	0
90	8100
180	0

(f) From the chart, the maximum value of P is 8100; this occurs when $x = 90$ and $y = 90$.

3. $x + y = 90$; minimize $x^2 y$.

(a) $y = 90 - x$

(b) Let $P = x^2 y = x^2(90 - x)$
$= 90x^2 - x^3$.

(c) Since $y = 90 - x$ and x and y are nonnegative numbers, the domain of P is $[0, 90]$.

(d) $P' = 180x - 3x^2$
$180x - 3x^2 = 0 \Rightarrow 3x(60 - x) = 0 \Rightarrow$
$x = 0$ or $x = 60$

(e)

x	P
0	0
60	108,000
90	0

(f) The maximum value of $x^2 y$ occurs when $x = 60$ and $y = 30$. The maximum value is 108,000.

5. $p(t) = \dfrac{20t^3 - t^4}{1000}$, [0, 20]

(a) $p'(t) = \frac{3}{50}t^2 - \frac{1}{250}t^3 = \frac{1}{50}t^2\left[3 - \frac{1}{5}t\right]$

Critical numbers:

$\frac{1}{50}t^2 = 0 \Rightarrow t = 0$ or $3 - \frac{1}{5}t = 0 \Rightarrow t = 15$

t	$p(t)$
0	0
15	16.875
20	0

The number of people infected reaches a maximum in 15 days.

(b) $P(15) = 16.875\%$

7. $H(S) = f(S) - S$
$f(S) = 12S^{0.25}$
$H(S) = 12S^{0.25} - S$
$H'(S) = 3S^{-0.75} - 1$
$H'(S) = 0$ when

$3S^{-0.75} - 1 = 0 \Rightarrow S^{-0.75} = \frac{1}{3} \Rightarrow \frac{1}{S^{0.75}} = \frac{1}{3} \Rightarrow$

$S^{3/4} = 3 \Rightarrow S = 3^{4/3} \Rightarrow S = 4.327$

The number of creatures needed to sustain the population is $S_0 = 4.327$ thousand.

$H''(S) = \dfrac{-2.25}{S^{1.75}} < 0$ when $S = 4.327$, so

$H(S)$ is maximized.

$H(4.327) = 12(4.327)^{0.25} - 4.327 \approx 12.98$

The maximum sustainable harvest is 12.98 thousand.

9. $N(t) = 20\left(\frac{t}{12} - \ln\left(\frac{t}{12}\right)\right) + 30;\ 1 \le t \le 15$

$N'(t) = 20\left[\frac{1}{12} - \frac{12}{t}\left(\frac{1}{12}\right)\right] = 20\left(\frac{1}{12} - \frac{1}{t}\right)$
$= \dfrac{20(t - 12)}{12t}$

$N'(t) = 0$ when $t - 12 = 0 \Rightarrow t = 12$.

$N'(t)$ does not exist at $t = 0$, but 0 is not in the domain of N. Thus, 12 is the only critical number.

To find the absolute extrema on [1, 15], evaluate N at the critical number and at the endpoints.

t	$N(t)$
1	81.365
12	50
15	50.537

Use this table to answer the questions in (a)–(d).

(a) The number of bacteria will be a minimum at $t = 12$, which represents 12 days.

(b) The minimum number of bacteria is given by $N(12) = 50$, which represents 50 bacteria per ml.

(c) The number of bacteria will be a maximum at $t = 1$, which represents 1 day.

(d) The maximum number of bacteria is given by $N(1) = 81.365$, which represents 81.365 bacteria per ml.

11. $r = 0.1$, $P = 100$

$f(S) = Se^{r(1-S/P)}$

$f'(S) = -\frac{1}{1000} \cdot Se^{0.1(1-S/100)} + e^{0.1(1-S/100)}$

$f'(S_0) = -0.001S_0e^{0.1(1-S_0/100)} + e^{0.1(1-S_0/100)}$

(continued on next page)

(continued)

Graph $Y_1 = -0.001xe^{0.1(1-x/100)} + e^{0.1(1-x/100)}$ and $Y_2 = 1$ on the same screen. A suitable choice for the viewing window is [0, 60] by [0.5, 1.5]. By zooming or using the "intersect" option, we find the graphs intersect when $x \approx 49.37$. Thus, the maximum sustainable harvest is about 49.37.

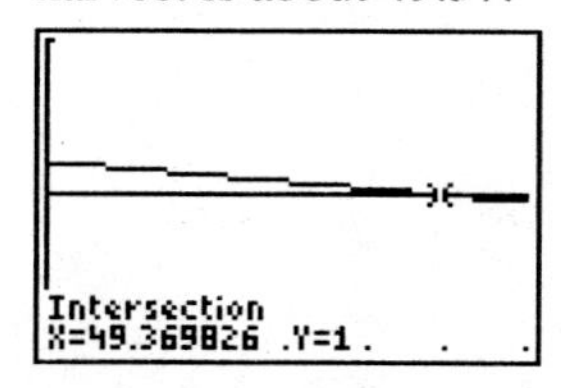

13. Let x = distance from P to A.

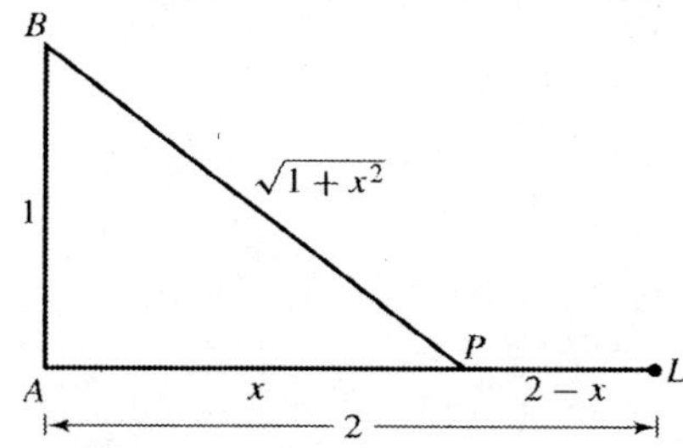

Energy used over land: 1 unit per mile

Energy used over water: $\frac{4}{3}$ units per mile

Distance over land: $(2 - x)$ mi

Distance over water: $\sqrt{1+x^2}$ mi

Find the location of P to minimize energy used.

$$E(x) = 1(2-x) + \frac{4}{3}\sqrt{1+x^2}, \text{ where } 0 \le x \le 2.$$

$$E'(x) = -1 + \frac{4}{3}\left(\frac{1}{2}\right)(1+x^2)^{-1/2}(2x)$$

If $E'(x) = 0$,

$$\frac{4}{3}x(1+x^2)^{-1/2} = 1 \Rightarrow \frac{4x}{3(1+x^2)^{1/2}} = 1 \Rightarrow$$

$$\frac{4}{3}x = (1+x^2)^{1/2} \Rightarrow \frac{16}{9}x^2 = 1 + x^2 \Rightarrow$$

$$\frac{7}{9}x^2 = 1 \Rightarrow x^2 = \frac{9}{7} \Rightarrow x = \frac{3}{\sqrt{7}} = \frac{3\sqrt{7}}{7}$$

x	$E(x)$
0	3.3333
1.134	2.8819
2	2.9814

The absolute minimum occurs at $x \approx 1.134$.

Point P is $\frac{3\sqrt{7}}{7} \approx 1.134$ mi from Point A.

15. **(a)** $f(S) = aSe^{-bS}$

$$f(S) = Se^{r(1-S/P)} = Se^{r-rS/P} = Se^re^{-rS/P} = e^rSe^{-(r/P)S}$$

Comparing the two terms, replace a with e^r and b with r/P.

(b) Shepherd:

$$f(S) = \frac{aS}{1+(S/b)^c}$$

$$f'(S) = \frac{[1+(S/b)^c](a) - (aS)[c(S/b)^{c-1}(1/b)]}{[1+(S/b)^c]^2}$$

$$= \frac{a + a(S/b)^c - (acS/b)(S/b)^{c-1}}{[1+(S/b)^c]^2}$$

$$= \frac{a + a(S/b)^c - ac(S/b)^c}{[1+(S/b)^c]^2}$$

$$= \frac{a[1+(1-c)(S/b)^c]}{[1+(S/b)^c]^2}$$

Ricker:

$$f(S) = aSe^{-bS}$$

$$f'(S) = ae^{-bS} + aSe^{-bS}(-b) = ae^{-bS}(1-bS)$$

Beverton-Holt:

$$f(S) = \frac{aS}{1+(S/b)}$$

$$f'(S) = \frac{[1+(S/b)](a) - aS(1/b)}{[1+(S/b)]^2} = \frac{a + a(S/b) - a(S/b)}{[1+(S/b)]^2} = \frac{a}{[1+(S/b)]^2}$$

(c) Shepherd:

$$f'(0) = \frac{a[1+(1-c)(0/b)^c]}{[1+(0/b)^c]^2} = a$$

Ricker:

$$f'(0) = ae^{-b(0)}[1-b(0)] = a$$

Beverton-Holt:

$$f'(0) = \frac{a}{[1+(0/b)]^2} = a$$

The constant a represents the slope of the graph of $f(S)$ at $S = 0$.

(d) First find the critical numbers by solving $f'(S) = 0$.

Shepherd:

$$f'(S) = 0$$
$$a[1 + (1-c)(S/b)^c] = 0$$
$$(1-c)(S/b)^c = -1$$
$$(c-1)(S/b)^c = 1$$

Substitute $b = 248.72$ and $c = 3.24$ and solve for S.

$$(3.24-1)(S/248.72)^{3.24} = 1$$
$$\left(\frac{S}{248.72}\right)^{3.24} = \frac{1}{2.24}$$
$$\frac{S}{248.72} = \left(\frac{1}{2.24}\right)^{1/3.24}$$
$$248.72\left(\frac{1}{2.24}\right)^{1/3.24} \approx 193.914 = S$$

Using the Shepherd model, next year's population is maximized when this year's population is about 194,000 tons. This can be verified by examing the graph of $f(S)$.

(e) First find the critical numbers by solving $f'(S) = 0$.

Ricker:

$$f'(S) = 0 \Rightarrow ae^{-bS}(1-bS) = 0 \Rightarrow$$
$$1 - bS = 0 \Rightarrow S = \frac{1}{b}$$

Substitute $b = 0.0039$ and solve for S.

$$S = \frac{1}{0.0039} \approx 256.410$$

Using the Ricker model, next year's population is maximized when this year's population is about 256,000 tons. This can be verified by examining the graph of $f(S)$.

17. $V = \pi r^2 h$ and $S = 2\pi r^2 + 2\pi rh$

We are given $V = 65$, so

$$65 = \pi r^2 h \Rightarrow h = \frac{65}{\pi r^2}.$$

Substitute this expression for h into the equation for S.

$$S = 2\pi r^2 + 2\pi r\left(\frac{65}{\pi r^2}\right) = 2\pi r^2 + \frac{130}{r}$$

There are no restriction on r other than it is a positive number, so the domain is $(0, \infty)$.

Now find the critical points for S by finding $\frac{dS}{dr}$ and then solving $\frac{dS}{dr} = 0$.

$$\frac{dS}{dr} = 4\pi r - \frac{130}{r^2}$$
$$4\pi r - \frac{130}{r^2} = 0 \Rightarrow 4\pi r = \frac{130}{r^2} \Rightarrow$$
$$r^3 = \frac{130}{4\pi} \Rightarrow r = \left(\frac{130}{4\pi}\right)^{1/3} \approx 2.179\mu\text{m}$$

Now find h.

$$h = \frac{65}{\pi(2.179)^2} \approx 4.36\mu\text{m}$$

To verify that these values give the minimum surface area, use the first derivative test for $r = 2$ and $r = 3$.

$$\left.\frac{dS}{dr}\right|_{r=2} = 4\pi(2) - \frac{130}{2^2} \approx -7.4$$
$$\left.\frac{dS}{dr}\right|_{r=3} = 4\pi(3) - \frac{130}{3^2} \approx 23.3$$

Since the sign of the first derivative changes from negative to positive, $r \approx 2.179\mu\text{m}$ and $h \approx 4.36\mu\text{m}$ give the absolute minimum surface area.

19. Graph $y = d(x)$ on a graphing calculator. Be sure that angle mode is set to degrees. A sutiable choice for the viewing window is [0, 90] by [0, 15]. The graph has a maximum at $x \approx 40$. Thus, the release angle that maximizes the distance traveled is about 40° and the maximum distance is about 12 m.

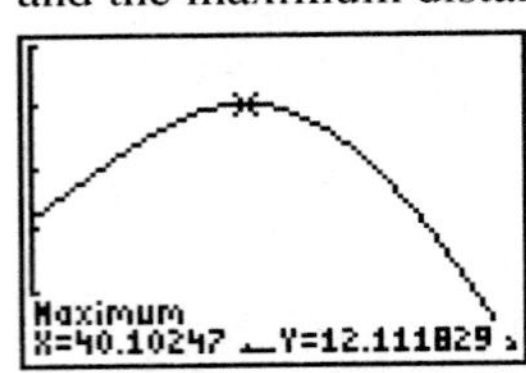

21. Let x = the width and y = the length.

Perimeter:

$$P = 2x + 2y = 300 \Rightarrow x + y = 150 \Rightarrow$$
$$y = 150 - x$$

Area: $A = xy = x(150 - x) = 150x - x^2$

Thus,

$$A(x) = 150x - x^2$$
$$A'(x) = 150 - 2x.$$

$A'(x) = 0$ when $150 - 2x = 0 \Rightarrow x = 75$.

$A''(x) = -2$, so $A''(75) = -2 < 0$, which confirms that a maximum value occurs at $x = 75$.

(*continued on next page*)

(continued)

If $x = 75$, then $y = 150 - x = 150 - 75 = 75$.
The maximum area occurs when the length is 75 m and the width is 75 m.

23. 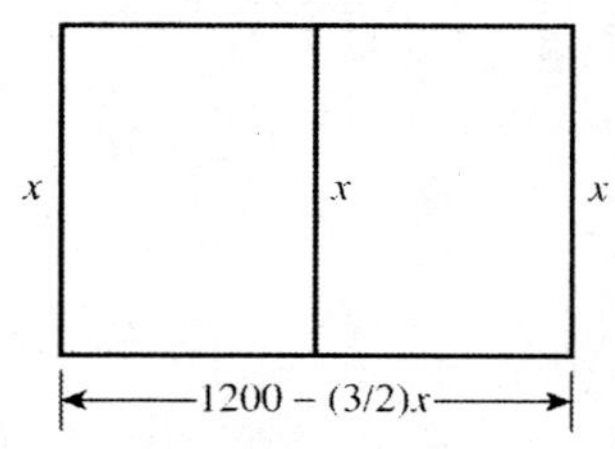

There are three fence pieces of length x, which leaves $2400 - 3x$ for the two remaining sides. Each of these remaining sides thus has length $1200 - (3/2)x$. The area enclosed is

$$A(x) = x[1200 - (3/2)x] = 1200x - \frac{3x^2}{2},$$

measured in square meters. Both x and $1200 - (3/2)x$ must be nonnegative, so the domain of A is [0, 800]. Now set the derivative equal to zero to solve for any critical numbers:

$A'(x) = 1200 - 3x$

$1200 - 3x = 0 \Rightarrow 1200 = 3x \Rightarrow x = 400$

Evaluate A at the endpoints of the domain and at the single critical value $x = 400$.

x	A
0	0
400	240,000
800	0

The maximum area enclosed by the pen is $240{,}000\,\text{m}^2$. To achieve this, the farmer will use three fence sections of length 400 m and two of length $1200 - (3/2)x = 600$ m.

25. Let x = the length at \$2.50 per foot and y = the width at \$3.20 per foot.

$xy = 20{,}000$

$$y = \frac{20{,}000}{x}$$

$$\text{Perimeter} = 2x + 2y = 2x + \frac{40{,}000}{x}$$

$$\text{Cost} = C(x) = 2x(2.5) + \frac{40{,}000}{x}(3.2) = 5x + \frac{128{,}000}{x}$$

Minimize cost:

$$C'(x) = 5 - \frac{128{,}000}{x^2}$$

$$5 - \frac{128{,}000}{x^2} = 0 \Rightarrow 5 = \frac{128{,}000}{x^2} \Rightarrow$$

$5x^2 = 128{,}000 \Rightarrow x^2 = 25{,}600 \Rightarrow x = 160$

$$y = \frac{20{,}000}{160} = 125$$

320 ft at \$2.50 per foot will cost \$800. 250 ft at \$3.20 per foot will cost \$800. The entire cost will be \$1600.

27. Let x = the width. Then $2x$ = the length and h = the height.
An equation for volume is

$V = (2x)(x)h = 2x^2h.$

So, $36 = 2x^2h \Rightarrow h = \dfrac{18}{x^2}$.

The surface area $S(x)$ is the sum of the areas of the base and the four sides.

$$S(x) = (2x)(x) + 2xh + 2(2x)h = 2x^2 + 6xh = 2x^2 + 6x\left(\frac{18}{x^2}\right) = 2x^2 + \frac{108}{x}$$

$$S'(x) = 4x - \frac{108}{x^2}$$

$$\frac{4x^3 - 108}{x^2} = 0 \Rightarrow 4(x^3 - 27) = 0 \Rightarrow x = 3$$

$$S''(x) = 4 + \frac{108(2)}{x^3} = 4 + \frac{216}{x^3} > 0 \text{ since } x > 0$$

So $x = 3$ minimizes the surface material.

If $x = 3$, $h = \dfrac{18}{x^2} = \dfrac{18}{9} = 2$.

The dimensions are 3 ft by 6 ft by 2 ft.

29. **(a)** From Example 3, the area of the base is $(12 - 2x)(12 - 2x) = 4x^2 - 48x + 144$ and the total area of all four walls is $4x(12 - 2x) = -8x^2 + 48x$. Since the box has maximum volume when $x = 2$, the area of the base is $4(2)^2 - 48(2) + 144 = 64$ square inches and the total area of all four walls is $-8(2)^2 + 48(2) = 64$ square inches. So, both are 64 square inches.

(b) From Exercise 28, the area of the base is $(3 - 2x)(8 - 2x) = 4x^2 - 22x + 24$ and the total area of all four walls is $2x(3 - 2x) + 2x(8 - 2x) = -8x^2 + 22x$.

(continued on next page)

(*continued*)

Since the box has maximum volume when $x = \frac{2}{3}$, the area of the base is

$$4\left(\frac{2}{3}\right)^2 - 22\left(\frac{2}{3}\right) + 24 = \frac{100}{9} \text{ square feet}$$

and the total area of all four walls is

$$-8\left(\frac{2}{3}\right)^2 + 22\left(\frac{2}{3}\right) = \frac{100}{9} \text{ square feet.}$$

So, both are $\frac{100}{9}$ square feet.

(c) Based on the results from parts **(a)** and **(b)**, it appears that the area of the base and the total area of the walls for the box with maximum volume are equal. (This conjecture is true.)

31. Let x = the width of printed material and y = the length of printed material. Then, the area of the printed material is $xy = 36$, so $y = \frac{36}{x}$.

Also, $x + 2$ = the width of a page and $y + 3$ = the length of a page.
The area of a page is

$$A = (x+2)(y+3) = xy + 2y + 3x + 6$$
$$= 36 + 2\left(\frac{36}{x}\right) + 3x + 6 = 42 + \frac{72}{x} + 3x.$$

$$A' = -\frac{72}{x^2} + 3 = 0 \Rightarrow x^2 = 24 \Rightarrow x = \sqrt{24} = 2\sqrt{6}$$

(We discard $x = -2\sqrt{6}$ because we must have $x > 0$.) $A'' = \frac{144}{x^3} > 0$ when $x = 2\sqrt{6}$, which implies that A is minimized when $x = 2\sqrt{6}$.

$$y = \frac{36}{x} = \frac{36}{2\sqrt{6}} = \frac{18}{\sqrt{6}} = \frac{18\sqrt{6}}{6} = 3\sqrt{6}$$

The width of a page is $x + 2 = 2\sqrt{6} + 2 \approx 6.9$ in.
The length of a page is
$y + 3 = 3\sqrt{6} + 3 \approx 10.3$ in.

33. $V = \pi r^2 h = 16 \Rightarrow h = \frac{16}{\pi r^2}$

The total cost is the sum of the cost of the top and bottom and the cost of the sides.

$$C = 2(2)(\pi r^2) + 1(2\pi r h)$$
$$= 4(\pi r^2) + 1(2\pi r)\left(\frac{16}{\pi r^2}\right) = 4\pi r^2 + \frac{32}{r}$$

Minimize cost.

$$C' = 8\pi r - \frac{32}{r^2}$$

$$8\pi r - \frac{32}{r^2} = 0 \Rightarrow 8\pi r^3 = 32 \Rightarrow \pi r^3 = 4 \Rightarrow$$

$$r = \sqrt[3]{\frac{4}{\pi}} \approx 1.08$$

$$h = \frac{16}{\pi(1.08)^2} \approx 4.34$$

The radius should be 1.08 ft and the height should be 4.34 ft. If these rounded values for the height and radius are used, the cost is

$$\$2(2)(\pi r^2) + \$1(2\pi r h)$$
$$= 4\pi(1.08)^2 + 2\pi(1.08)(4.34)$$
$$= \$44.11.$$

35. In Exercise 34, we found that the cost of the aluminum to make the can is

$$0.03\left(2\pi r^2 + \frac{2000}{r}\right) = 0.06\pi r^2 + \frac{60}{r}.$$

The cost for the vertical seam is $0.01h$. From Example 4, we see that h and r are related by the equation $h = \frac{1000}{\pi r^2}$, so the sealing cost is

$$0.01h = 0.01\left(\frac{1000}{\pi r^2}\right) = \frac{10}{\pi r^2}.$$

Thus, the total cost is given by the function

$$C(r) = 0.06\pi r^2 + \frac{60}{r} + \frac{10}{\pi r^2}$$
$$\text{or} \quad 0.06\pi r^2 + 60r^{-1} + \frac{10}{\pi} r^{-2}.$$

Then

$$C'(x) = 0.12\pi r - 60r^{-2} - \frac{20}{\pi} r^{-3}$$
$$= 0.12\pi r - \frac{60}{r^2} - \frac{20}{\pi r^3}.$$

Graph $y = 0.12\pi r - \frac{60}{r^2} - \frac{20}{\pi r^3}$ on a graphing calculator. Since r must be positive, our window should not include negative values of x. A suitable choice for the viewing window is $[0, 10]$ by $[-10, 10]$. From the graph, we find that $C'(x) = 0$ when $x \approx 5.454$.

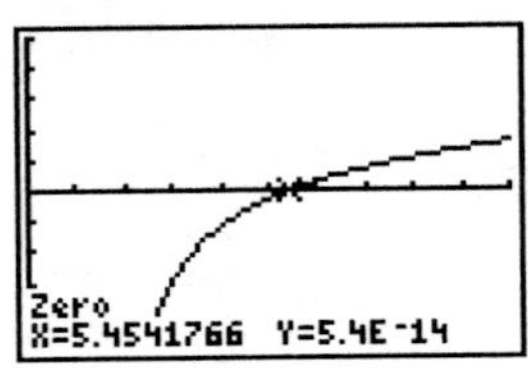

(*continued on next page*)

(continued)

Thus, the cost is minimized when the radius is about 5.454 cm.
We can find the corresponding height by using the equation $h = \frac{1000}{\pi r^2}$.

$h = \frac{1000}{\pi(5.454)^2} \approx 10.70.$

To minimize cost, the can should have radius 5.454 cm and height 10.70 cm.

37. 120 centimeters of ribbon are available; it will cover 4 heights and 8 radii.
$4h + 8r = 120 \Rightarrow h + 2r = 30 \Rightarrow h = 30 - 2r$
$V = \pi r^2 h$
$V = \pi r^2(30 - 2r) = 30\pi r^2 - 2\pi r^3$
Maximize volume.
$V' = 60\pi r - 6\pi r^2$
$60\pi r - 6\pi r^2 = 0 \Rightarrow 6\pi r(10 - r) = 0 \Rightarrow$
$r = 0 \quad \text{or} \quad r = 10$
If $r = 0$, there is no box, so we discard this value. $V'' = 60\pi - 12\pi r < 0$ for $r = 10$, which implies that $r = 10$ gives maximum volume.
When $r = 10$, $h = 30 - 2(10) = 10$.

The volume is maximum when the radius and height are both 10 cm.

39. Distance on shore: $7 - x$ miles
Cost on shore: \$400 per mile
Distance underwater: $\sqrt{x^2 + 36}$
Cost underwater: \$500 per mile
Find the distance from A, that is, $7 - x$, to minimize cost, $C(x)$.

$$C(x) = (7 - x)(400) + (\sqrt{x^2 + 36})(500)$$
$$= 2800 - 400x + 500(x^2 + 36)^{1/2}$$
$$C'(x) = -400 + 500\left(\frac{1}{2}\right)(x^2 + 36)^{1/2}(2x)$$
$$= -400 + \frac{500x}{\sqrt{x^2 + 36}}$$

If $C'(x) = 0$,

$$\frac{500x}{\sqrt{x^2 + 36}} = 400 \Rightarrow \frac{5x}{4} = \sqrt{x^2 + 36} \Rightarrow$$
$$\frac{25}{16}x^2 = x^2 + 36 \Rightarrow \frac{9}{16}x^2 = 36 \Rightarrow$$
$$x^2 = \frac{36 \cdot 16}{9} \Rightarrow x = \frac{6 \cdot 4}{3} = 8$$

(Discard the negative solution.)
$x = 8$ is impossible since Point A is only 7 miles from point C, so check the endpoints.

x	$C(x)$
0	5800
7	4610

The cost is minimized when $x = 7$.
$7 - x = 7 - 7 = 0$, so the company should angle the cable at Point A.

41. Let $8 - x$ = the distance the hunter will travel on the river.

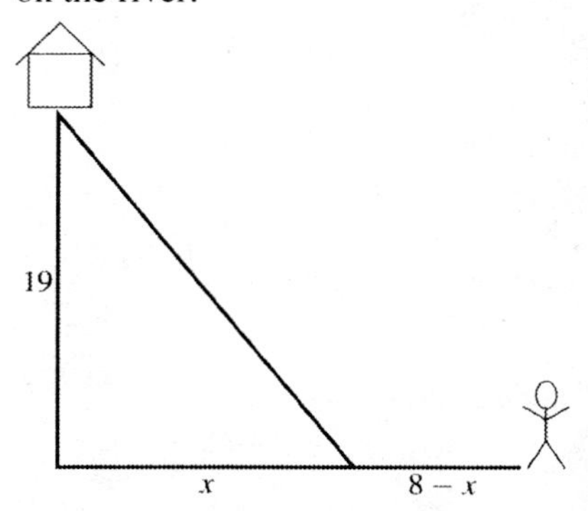

Then $\sqrt{19^2 + x^2}$ = the distance he will travel on land. Since the rate on the river is 5 mph, the rate on land is 2 mph, and $t = \frac{d}{r} \Rightarrow$

$\frac{8 - x}{5}$ = the time on the river

$\frac{\sqrt{361 + x^2}}{2}$ = the time on the land.

The total time is

$$T(x) = \frac{8 - x}{5} + \frac{\sqrt{361 + x^2}}{2}$$
$$= \frac{8}{5} - \frac{1}{5}x + \frac{1}{2}(361 + x^2)^{1/2}.$$
$$T'(x) = -\frac{1}{5} + \frac{1}{4} \cdot 2x(361 + x^2)^{-1/2}$$
$$-\frac{1}{5} + \frac{x}{2(361 + x^2)^{1/2}} = 0$$
$$\frac{1}{5} = \frac{x}{2(361 + x^2)^{1/2}} \Rightarrow 2(361 + x^2)^{1/2} = 5x \Rightarrow$$
$$4(361 + x^2) = 25x^2 \Rightarrow 1444 + 4x^2 = 25x^2 \Rightarrow$$
$$1444 = 21x^2 \Rightarrow x = \frac{38}{\sqrt{21}} \approx 8.29$$

8.29 is not possible, since the cabin is only 8 miles west. Check the endpoints.

x	$T(x)$
0	11.1
8	10.3

(continued on next page)

(continued)

$T(x)$ is minimized when $x = 8$. The distance along the river is given by $8 - x$, so the hunter should travel $8 - 8 = 0$ miles along the river. He should complete the entire trip on land.

43. Let x be the length of the ladder and let y be the distance from the fence to the bottom of the ladder. Then

$$\cos\theta = \frac{y+2}{x} \Rightarrow x\cos\theta = y+2 \text{ and}$$

$$\cot\theta = \frac{y}{9} \Rightarrow y = 9\cot\theta.$$

Thus,

$$x\cos\theta = 9\cot\theta + 2 \Rightarrow x = \frac{9\cot\theta + 2}{\cos\theta} \Rightarrow$$

$$x = 9\csc\theta + 2\sec\theta.$$

This expression gives the length of the ladder as a function of θ. Find the minimum value of this function.

$$\frac{dx}{d\theta} = -9\csc\theta\cot\theta + 2\sec\theta\tan\theta$$

$$= -9\left(\frac{1}{\sin\theta}\right)\left(\frac{\cos\theta}{\sin\theta}\right) + 2\left(\frac{1}{\cos\theta}\right)\left(\frac{\sin\theta}{\cos\theta}\right)$$

$$= \frac{-9\cos\theta}{\sin^2\theta} + \frac{2\sin\theta}{\cos^2\theta}$$

If $\frac{dx}{d\theta} = 0$, then

$$\frac{2\sin\theta}{\cos^2\theta} = \frac{9\cos\theta}{\sin^2\theta} \Rightarrow 2\sin^3\theta = 9\cos^3\theta \Rightarrow$$

$$\frac{\sin^3\theta}{\cos^3\theta} = \frac{9}{2} \Rightarrow \tan^3\theta = \frac{9}{2} \Rightarrow \tan\theta = \sqrt[3]{\frac{9}{2}} \Rightarrow$$

$\theta \approx 1.02619$ radians.

If $\theta < 1.02619$, $\frac{dx}{d\theta} < 0$. If $\theta > 1.02619$, $\frac{dx}{d\theta} > 0$. Therefore, there is a minimum when $\theta = 1.02619$. If $\theta = 1.02619$, $x \approx 14.383$.

The minimum length of the ladder is approximately 14.38 ft.

6.3 Implicit Differentiation

1. $6x^2 + 5y^2 = 36$

$$\frac{d}{dx}(6x^2 + 5y^2) = \frac{d}{dx}(36)$$

$$\frac{d}{dx}(6x^2) + \frac{d}{dx}(5y^2) = \frac{d}{dx}(36)$$

$$12x + 5\cdot 2y\frac{dy}{dx} = 0$$

$$10y\frac{dy}{dx} = -12x \Rightarrow \frac{dy}{dx} = -\frac{6x}{5y}$$

3. $8x^2 - 10xy + 3y^2 = 26$

$$\frac{d}{dx}(8x^2 - 10xy + 3y^2) = \frac{d}{dx}(26)$$

$$16x - \frac{d}{dx}(10xy) + \frac{d}{dx}(3y^2) = 0$$

$$16x - 10x\frac{dy}{dx} - y\frac{d}{dx}(10x) + 6y\frac{dy}{dx} = 0$$

$$16x - 10x\frac{dy}{dx} - 10y + 6y\frac{dy}{dx} = 0$$

$$(-10x + 6y)\frac{dy}{dx} = -16x + 10y$$

$$\frac{dy}{dx} = \frac{-16x + 10y}{-10x + 6y} = \frac{8x - 5y}{5x - 3y}$$

5. $5x^3 = 3y^2 + 4y$

$$\frac{d}{dx}(5x^3) = \frac{d}{dx}(3y^2 + 4y)$$

$$15x^2 = \frac{d}{dx}(3y^2) + \frac{d}{dx}(4y)$$

$$15x^2 = 6y\frac{dy}{dx} + 4\frac{dy}{dx} \Rightarrow \frac{15x^2}{6y+4} = \frac{dy}{dx}$$

7. $3x^2 = \frac{2-y}{2+y}$

$$\frac{d}{dx}(3x^2) = \frac{d}{dx}\left(\frac{2-y}{2+y}\right)$$

$$6x = \frac{(2+y)\frac{d}{dx}(2-y) - (2-y)\frac{d}{dx}(2+y)}{(2+y)^2}$$

$$6x = \frac{(2+y)\left(-\frac{dy}{dx}\right) - (2-y)\frac{dy}{dx}}{(2+y)^2}$$

$$6x = \frac{-4\frac{dy}{dx}}{(2+y)^2}$$

$$6x(2+y)^2 = -4\frac{dy}{dx} \Rightarrow -\frac{3x(2+y)^2}{2} = \frac{dy}{dx}$$

9. $2\sqrt{x}+4\sqrt{y}=5y$

$$\frac{d}{dx}(2x^{1/2}+4y^{1/2})=\frac{d}{dx}(5y)$$

$$x^{-1/2}+2y^{-1/2}\frac{dy}{dx}=5\frac{dy}{dx}$$

$$(2y^{-1/2}-5)\frac{dy}{dx}=-x^{-1/2}$$

$$\frac{dy}{dx}=\frac{x^{-1/2}}{5-2y^{-1/2}}\left(\frac{x^{1/2}y^{1/2}}{x^{1/2}y^{1/2}}\right)=\frac{y^{1/2}}{x^{1/2}(5y^{1/2}-2)}=\frac{\sqrt{y}}{\sqrt{x}(5\sqrt{y}-2)}$$

11. $x^4y^3+4x^{3/2}=6y^{3/2}+5$

$$\frac{d}{dx}(x^4y^3+4x^{3/2})=\frac{d}{dx}(6y^{3/2}+5)$$

$$\frac{d}{dx}(x^4y^3)+\frac{d}{dx}(4x^{3/2})=\frac{d}{dx}(6y^{3/2})+\frac{d}{dx}(5)$$

$$4x^3y^3+x^4\cdot 3y^2\frac{dy}{dx}+6x^{1/2}=9y^{1/2}\frac{dy}{dx}+0$$

$$4x^3y^3+6x^{1/2}=9y^{1/2}\frac{dy}{dx}-3x^4y^2\frac{dy}{dx}$$

$$4x^3y^3+6x^{1/2}=(9y^{1/2}-3x^4y^2)\frac{dy}{dx}$$

$$\frac{4x^3y^3+6x^{1/2}}{9y^{1/2}-3x^4y^2}=\frac{dy}{dx}$$

13. $e^{x^2y}=5x+4y+2$

$$\frac{d}{dx}(e^{x^2y})=\frac{d}{dx}(5x+4y+2)$$

$$e^{x^2y}\frac{d}{dx}(x^2y)=\frac{d}{dx}(5x)+\frac{d}{dx}(4y)+\frac{d}{dx}(2)$$

$$e^{x^2y}\left(2xy+x^2\frac{dy}{dx}\right)=5+4\frac{dy}{dx}+0$$

$$2xye^{x^2y}+x^2e^{x^2y}\frac{dy}{dx}=5+4\frac{dy}{dx}$$

$$x^2e^{x^2y}\frac{dy}{dx}-4\frac{dy}{dx}=5-2xye^{x^2y}$$

$$(x^2e^{x^2y}-4)\frac{dy}{dx}=5-2xye^{x^2y}\Rightarrow\frac{dy}{dx}=\frac{5-2xye^{x^2y}}{x^2e^{x^2y}-4}$$

15. $x+\ln y=x^2y^3$

$$\frac{d}{dx}(x+\ln y)=\frac{d}{dx}(x^2y^3)$$

$$1+\frac{1}{y}\frac{dy}{dx}=2xy^3+3x^2y^2\frac{dy}{dx}$$

$$\frac{1}{y}\frac{dy}{dx}-3x^2y^2\frac{dy}{dx}=2xy^3-1$$

$$\left(\frac{1}{y}-3x^2y^2\right)\frac{dy}{dx}=2xy^3-1$$

$$\frac{dy}{dx}=\frac{2xy^3-1}{\frac{1}{y}-3x^2y^2}=\frac{y(2xy^3-1)}{1-3x^2y^3}$$

17. $\sin(xy) = x$

$$\frac{d}{dx}[\sin(xy)] = \frac{d}{dx}(x)$$
$$\cos(xy)\left(x\frac{dy}{dx} + y\right) = 1$$
$$x\cos(xy)\frac{dy}{dx} + y\cos(xy) = 1$$
$$x\cos(xy)\frac{dy}{dx} = 1 - y\cos(xy)$$
$$\frac{dy}{dx} = \frac{1 - y\cos(xy)}{x\cos(xy)} = \frac{1}{x\cos(xy)} - \frac{y\cos(xy)}{x\cos(xy)} = \frac{\sec(xy)}{x} - \frac{y}{x}$$

19. $x^2 + y^2 = 25$; tangent at $(-3, 4)$

$$\frac{d}{dx}(x^2 + y^2) = \frac{d}{dx}(25)$$
$$2x + 2y\frac{dy}{dx} = 0 \Rightarrow 2y\frac{dy}{dx} = -2x \Rightarrow \frac{dy}{dx} = -\frac{x}{y}$$
$$m = -\frac{x}{y} = -\frac{-3}{4} = \frac{3}{4}$$
$$y - y_1 = m(x - x_1) \Rightarrow y - 4 = \frac{3}{4}[x - (-3)] \Rightarrow 4y - 16 = 3x + 9 \Rightarrow 4y = 3x + 25 \Rightarrow y = \frac{3}{4}x + \frac{25}{4}$$

21. $x^2y^2 = 1$; tangent at $(-1, 1)$

$$\frac{d}{dx}(x^2y^2) = \frac{d}{dx}(1)$$
$$x^2\frac{d}{dx}(y^2) + y^2\frac{d}{dx}(x^2) = 0$$
$$x^2(2y)\frac{dy}{dx} + y^2(2x) = 0$$
$$2x^2y\frac{dy}{dx} = -2xy^2$$
$$\frac{dy}{dx} = \frac{-2xy^2}{2x^2y} = -\frac{y}{x}$$
$$m = -\frac{y}{x} = -\frac{1}{-1} = 1$$
$$y - 1 = 1[x - (-1)] \Rightarrow y = x + 1 + 1 \Rightarrow y = x + 2$$

23. $2y^2 - \sqrt{x} = 4$; tangent at $(16, 2)$

$$\frac{d}{dx}(2y^2 - \sqrt{x}) = \frac{d}{dx}(4)$$
$$4y\frac{dy}{dx} - \frac{1}{2}x^{-1/2} = 0$$
$$4y\frac{dy}{dx} = \frac{1}{2x^{1/2}} \Rightarrow \frac{dy}{dx} = \frac{1}{8yx^{1/2}}$$
$$m = \frac{1}{8yx^{1/2}} = \frac{1}{8(2)(16)^{1/2}} = \frac{1}{8(2)(4)} = \frac{1}{64}$$
$$y - 2 = \frac{1}{64}(x - 16) \Rightarrow 64y - 128 = x - 16 \Rightarrow 64y = x + 112 \Rightarrow y = \frac{x}{64} + \frac{7}{4}$$

25. $e^{x^2+y^2} = xe^{5y} - y^2e^{5x/2}$; tangent at $(2,1)$

$$\frac{d}{dx}(e^{x^2+y^2}) = \frac{d}{dx}(xe^{5y} - y^2e^{5x/2})$$

$$e^{x^2+y^2} \cdot \frac{d}{dx}(x^2+y^2) = e^{5y} + x\frac{d}{dx}(e^{5y}) - \left[2y\frac{dy}{dx}e^{5x/2} + y^2e^{5x/2}\frac{d}{dx}\left(\frac{5x}{2}\right)\right]$$

$$e^{x^2+y^2}\left(2x+2y\frac{dy}{dx}\right) = e^{5y} + x\cdot 5e^{5y}\frac{dy}{dx} - 2ye^{5x/2}\frac{dy}{dx} - \frac{5}{2}y^2e^{5x/2}$$

$$\left(2ye^{x^2+y^2} - 5xe^{5y} + 2ye^{5x/2}\right)\frac{dy}{dx} = -2xe^{x^2+y^2} + e^{5y} - \frac{5}{2}y^2e^{5x/2}$$

$$\frac{dy}{dx} = \frac{-2xe^{x^2+y^2} + e^{5y} - \frac{5}{2}y^2e^{5x/2}}{2ye^{x^2+y^2} - 5xe^{5y} + 2ye^{5x/2}}$$

$$m = \frac{-4e^5 + e^5 - \frac{5}{2}e^5}{2e^5 - 10e^5 + 2e^5} = \frac{-\frac{11}{2}e^5}{-6e^5} = \frac{11}{12}$$

$$y - 1 = \frac{11}{12}(x-2) \Rightarrow y = \frac{11}{12}x - \frac{5}{6}$$

27. $\ln(x+y) = x^3y^2 + \ln(x^2+2) - 4$; tangent at $(1, 2)$

$$\frac{d}{dx}[\ln(x+y)] = \frac{d}{dx}[x^3y^2 + \ln(x^2+2) - 4]$$

$$\frac{1}{x+y}\cdot\frac{d}{dx}(x+y) = 3x^2y^2 + x^3\cdot 2y\frac{dy}{dx} + \frac{1}{x^2+2}\cdot\frac{d}{dx}(x^2+2) - \frac{d}{dx}(4)$$

$$\left(\frac{1}{x+y} - 2x^3y\right)\frac{dy}{dx} = 3x^2y^2 + \frac{2x}{x^2+2} - \frac{1}{x+y}$$

$$\frac{dy}{dx} = \frac{3x^2y^2 + \frac{2x}{x^2+2} - \frac{1}{x+y}}{\frac{1}{x+y} - 2x^3y}$$

$$m = \frac{3(1)^2(2)^2 + \frac{2(1)}{(1)^2+2} - \frac{1}{1+2}}{\frac{1}{1+2} - 2(1)^3(2)} = \frac{12 + \frac{1}{3}}{\frac{1}{3} - 4} = \frac{\frac{37}{3}}{-\frac{11}{3}} = -\frac{37}{11}$$

$$y - 2 = -\frac{37}{11}(x-1) \Rightarrow y = -\frac{37}{11}x + \frac{59}{11}$$

29. $x - \sin(\pi y) = 1$; tangent at $(1, 0)$

$$\frac{d}{dx}\left[x - \sin(\pi y)\right] = \frac{d}{dx}(1)$$

$$1 - \cos(\pi y)\frac{d}{dx}(\pi y) = 0$$

$$1 - \pi\cos(\pi y)\frac{dy}{dx} = 0 \Rightarrow \frac{dy}{dx} = \frac{1}{\pi\cos(\pi y)}$$

$$m = \frac{1}{\pi\cos(\pi\cdot 0)} = \frac{1}{\pi}$$

$$y - 0 = \frac{1}{\pi}(x-1) \Rightarrow y = \frac{1}{\pi}(x-1)$$

31. $y^3 + xy - y = 8x^4;\ x = 1$

First, find the y-value of the point.

$y^3 + (1)y - y = 8(1)^4 \Rightarrow y^3 = 8 \Rightarrow y = 2$

The point is (1, 2).

Find $\frac{dy}{dx}$.

$$3y^2\frac{dy}{dx} + x\frac{dy}{dx} + y - \frac{dy}{dx} = 32x^3$$
$$(3y^2 + x - 1)\frac{dy}{dx} = 32x^3 - y$$
$$\frac{dy}{dx} = \frac{32x^3 - y}{3y^2 + x - 1}$$

At (1, 2), $\frac{dy}{dx} = \frac{32(1)^3 - 2}{3(2)^2 + 1 - 1} = \frac{30}{12} = \frac{5}{2}.$

$$y - 2 = \frac{5}{2}(x - 1) \Rightarrow y - 2 = \frac{5}{2}x - \frac{5}{2} \Rightarrow$$
$$y = \frac{5}{2}x - \frac{1}{2}$$

33. $y^3 + xy^2 + 1 = x + 2y^2;\ x = 2$

Find the y-value of the point.

$y^3 + 2y^2 + 1 = 2 + 2y^2 \Rightarrow y^3 + 1 = 2 \Rightarrow$
$y^3 = 1 \Rightarrow y = 1$

The point is (2, 1).

Find $\frac{dy}{dx}$.

$$3y^2\frac{dy}{dx} + x2y\frac{dy}{dx} + y^2 = 1 + 4y\frac{dy}{dx}$$
$$3y^2\frac{dy}{dx} + 2xy\frac{dy}{dx} - 4y\frac{dy}{dx} = 1 - y^2$$
$$(3y^2 + 2xy - 4y)\frac{dy}{dx} = 1 - y^2$$
$$\frac{dy}{dx} = \frac{1 - y^2}{3y^2 + 2xy - 4y}$$

At (2, 1), $\frac{dy}{dx} = \frac{1 - 1^2}{3(1)^2 + 2(2)(1) - 4(1)} = 0.$

$y - 1 = 0(x - 2) \Rightarrow y = 1$

35. $2y^3(x - 3) + x\sqrt{y} = 3;\ x = 3$

Find the y-value of the point.

$2y^3(3 - 3) + 3\sqrt{y} = 3 \Rightarrow 3\sqrt{y} = 3 \Rightarrow y = 1$

The point is (3, 1)

Find $\frac{dy}{dx}$.

$$2y^3(1) + 6y^2(x - 3)\frac{dy}{dx}$$
$$+x\left(\frac{1}{2}\right)y^{-1/2}\frac{dy}{dx} + \sqrt{y} = 0$$
$$6y^2(x - 3)\frac{dy}{dx} + \frac{x}{2\sqrt{y}}\frac{dy}{dx} = -2y^3 - \sqrt{y}$$
$$\left[6y^2(x - 3) + \frac{x}{2\sqrt{y}}\right]\frac{dy}{dx} = -2y^3 - \sqrt{y}$$
$$\frac{dy}{dx} = \frac{-2y^3 - \sqrt{y}}{6y^2(x - 3) + \frac{x}{2\sqrt{y}}} = \frac{-4y^{7/2} - 2y}{12y^{5/2}(x - 3) + x}$$

At (3, 1),

$$\frac{dy}{dx} = \frac{-4(1) - 2}{12(1)(3 - 3) + 3} = \frac{-6}{3} = -2.$$

$y - 1 = -2(x - 3) \Rightarrow y - 1 = -2x + 6 \Rightarrow$
$y = -2x + 7$

37. $x^{2/3} + y^{2/3} = 2;\ (1, 1)$

Find $\frac{dy}{dx}$.

$$\frac{2}{3}x^{-1/3} + \frac{2}{3}y^{-1/3}\frac{dy}{dx} = 0$$
$$\frac{2}{3}y^{-1/3}\frac{dy}{dx} = -\frac{2}{3}x^{-1/3}$$
$$\frac{dy}{dx} = \frac{-\frac{2}{3}x^{-1/3}}{\frac{2}{3}y^{-1/3}} = -\frac{y^{1/3}}{x^{1/3}}$$

At (1, 1) $\frac{dy}{dx} = -\frac{1^{1/3}}{1^{1/3}} = -1.$

$y - 1 = -1(x - 1) \Rightarrow y - 1 = -x + 1 \Rightarrow y = -x + 2$

39. $y^2(x^2+y^2)=20x^2$; (1, 2)

Find $\frac{dy}{dx}$.

$$2y(x^2+y^2)\frac{dy}{dx}+y^2\left(2x+2y\frac{dy}{dx}\right)=40x$$

$$2x^2y\frac{dy}{dx}+2y^3\frac{dy}{dx}+2xy^2+2y^3\frac{dy}{dx}=40x$$

$$2x^2y\frac{dy}{dx}+4y^3\frac{dy}{dx}=-2xy^2+40x$$

$$(2x^2y+4y^3)\left(\frac{dy}{dx}\right)=-2xy^2+40x\Rightarrow\frac{dy}{dx}=\frac{-2xy^2+40x}{2x^2y+4y^3}$$

At (1, 2), $\frac{dy}{dx}=\frac{-2(1)(2)^2+40(1)}{2(1)^2(2)+4(2)^3}=\frac{32}{36}=\frac{8}{9}$

$$y-2=\frac{8}{9}(x-1)\Rightarrow y-2=\frac{8}{9}x-\frac{8}{9}\Rightarrow y=\frac{8}{9}x+\frac{10}{9}$$

41. $x^2+y^2=100$

(a) Lines are tangent at points where $x=6$. By substituting $x=6$ in the equation, we find that the points are (6, 8) and (6, –8).

$$\frac{d}{dx}(x^2+y^2)=\frac{d}{dx}(100)$$

$$2x+2y\frac{dy}{dx}=0$$

$$2y\frac{dy}{dx}=-2x\Rightarrow dy=-\frac{x}{y}$$

$$m_1=-\frac{x}{y}=-\frac{6}{8}=-\frac{3}{4}$$

$$m_2=-\frac{x}{y}=-\frac{6}{-8}=\frac{3}{4}$$

First tangent:

$$y-8=-\frac{3}{4}(x-6)\Rightarrow y=-\frac{3}{4}x+\frac{25}{2}$$

Second tangent:

$$y-(-8)=\frac{3}{4}(x-6)\Rightarrow y+8=\frac{3}{4}x-\frac{18}{4}\Rightarrow$$

$$y=\frac{3}{4}x-\frac{25}{2}$$

(b)

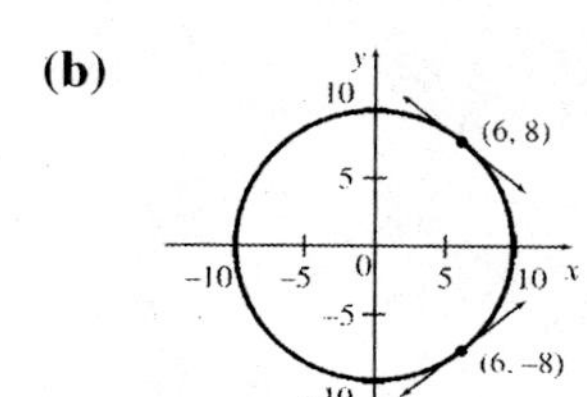

43. (a) $\sqrt{u}+\sqrt{2v+1}=5$

$$\frac{d}{dv}(\sqrt{u}+\sqrt{2v+1})=\frac{d}{dv}(5)$$

$$\frac{1}{2}u^{-1/2}\frac{du}{dv}+\frac{1}{2}(2v+1)^{-1/2}(2)=0$$

$$\frac{1}{2}u^{-1/2}\frac{du}{dv}=-\frac{1}{(2v+1)^{1/2}}$$

$$\frac{du}{dv}=-\frac{2u^{1/2}}{(2v+1)^{1/2}}$$

(b) $\sqrt{u}+\sqrt{2v+1}=5$

$$\frac{d}{du}(\sqrt{u}+\sqrt{2v+1})=\frac{d}{du}(5)$$

$$\frac{1}{2}u^{-1/2}+\frac{1}{2}(2v+1)^{-1/2}(2)\frac{dv}{du}=0$$

$$(2v+1)^{-1/2}\frac{dv}{du}=-\frac{1}{2}u^{-1/2}$$

$$\frac{dv}{du}=-\frac{(2v+1)^{1/2}}{2u^{1/2}}$$

(c) The derivatives are reciprocals.

45. $x^2+y^2+1=0$

$$\frac{d}{dx}\left(x^2+y^2\right)=\frac{d}{dx}(-1)$$

$$2x+2y\frac{dy}{dx}=0$$

$$\frac{dy}{dx}=-\frac{2x}{2y}=-\frac{x}{y}$$

If x and y are real numbers, x^2 and y^2 are nonnegative; 1 plus a nonnegative number cannot equal zero, so there is no function $y=f(x)$ that satisfies $x^2+y^2+1=0$.

47. $b-a=(b+a)^3$

$$\frac{d}{db}(b-a)=\frac{d}{db}[(b+a)^3]$$
$$1-\frac{da}{db}=3(b+a)^2\frac{d}{db}(b+a)$$
$$1-\frac{da}{db}=3(b+a)^2\left(1+\frac{da}{db}\right)$$
$$1-\frac{da}{db}=3(b+a)^2+3(b+a)^2\frac{da}{db}$$
$$-\frac{da}{db}-3(b+a)^2\frac{da}{db}=3(b+a)^2-1$$
$$[-1-3(b+a)^2]\frac{da}{db}=3(b+a)^2-1\Rightarrow\frac{da}{db}=\frac{3(b+a)^2-1}{-1-3(b+a)^2}$$
$$\frac{da}{db}=0\Rightarrow 3(b+a)^2-1=0\Rightarrow b+a=\frac{1}{\sqrt{3}}$$

Since $b-a=(b+a)^3=\left(\frac{1}{\sqrt{3}}\right)^3=\frac{1}{3\sqrt{3}}$, we have the system of equations

$$\begin{aligned} b+a&=\frac{1}{\sqrt{3}}\\ -(b-a)&=-\frac{1}{3\sqrt{3}}\\ \hline 2a&=\frac{2}{3\sqrt{3}}\Rightarrow \quad a=\frac{1}{3\sqrt{3}}\end{aligned}$$

49. **(a)** $\lambda^a-\lambda^{a-1}\phi(B)-pB=0\Rightarrow\lambda^a-\lambda^{a-1}\phi(B)=pB$

$$\frac{d}{dB}\left(\lambda^a-\lambda^{a-1}\phi(B)\right)=\frac{d}{dB}(pB)$$
$$\frac{d}{dB}\left(\lambda^a\right)-\frac{d}{dB}\left(\lambda^{a-1}\phi(B)\right)=p$$
$$a\lambda^{a-1}\frac{d\lambda}{dB}-\left(\phi(B)\frac{d}{dB}\left(\lambda^{a-1}\right)+\lambda^{a-1}\frac{d}{dB}\phi(B)\right)=p$$
$$a\lambda^{a-1}\frac{d\lambda}{dB}-\left(\phi(B)(a-1)\lambda^{a-2}\frac{d\lambda}{dB}+\lambda^{a-1}\phi'(B)\right)=p$$
$$a\lambda^{a-1}\frac{d\lambda}{dB}-(a-1)\lambda^{a-2}\phi(B)\frac{d\lambda}{dB}-\lambda^{a-1}\phi'(B)=p$$
$$\frac{d\lambda}{dB}\left(a\lambda^{a-1}-(a-1)\lambda^{a-2}\phi(B)\right)=\lambda^{a-1}\phi'(B)+p$$
$$\frac{d\lambda}{dB}=\frac{\lambda^{a-1}\phi'(B)+p}{a\lambda^{a-1}-(a-1)\lambda^{a-2}\phi(B)}$$

(b) $$\frac{\lambda^{a-1}\phi'(B)+p}{a\lambda^{a-1}-(a-1)\lambda^{a-2}\phi(B)}=0$$
$$\lambda^{a-1}\phi'(B)+p=0$$
$$\phi'(B)=-\frac{p}{\lambda^{a-1}}$$

51. $f(R)=\dfrac{tCR}{1+hCR}$

The second derivative rule tells us that a curve is concave down if $f''<0.$

To simplify the computations, first factor out C from the numerator.

$$f(R)=\frac{tCR}{1+hCR}=C\left(\frac{tR}{1+hCR}\right)$$

First, apply the quotient rule. Because t is a function of R, we must use the chain rule.

$$f'(R)=C\left(\frac{\left[\frac{d}{dR}(tR)\right](1+hCR)-tR\left[\frac{d}{dR}(1+hCR)\right]}{(1+hCR)^2}\right)$$

Next, differentiate the sum $(1+hCR)$ and simplify.

$$f'(R)=C\left(\frac{\left[\frac{d}{dR}(tR)\right](1+hCR)-tR(hC)}{(1+hCR)^2}\right)=C\left(\frac{\left[\frac{d}{dR}(tR)\right](1+hCR)-htCR}{(1+hCR)^2}\right)$$

Apply the product rule to find $\dfrac{d}{dR}(tR)$. The derivative of t with respect to R is $\dfrac{dt}{dR}$.

$$f'(R)=C\left(\frac{\left[\frac{d}{dR}(tR)\right](1+hCR)-hCtR}{(1+hCR)^2}\right)=C\left(\frac{\left[\frac{dt}{dR}R+t\right](1+hCR)-hCtR}{(1+hCR)^2}\right)$$

$$=C\left(\frac{\frac{dt}{dR}R+t+hCR^2\frac{dt}{dR}+hCtR-hCtR}{(1+hCR)^2}\right)=C\left(\frac{hCR^2\frac{dt}{dR}+R\frac{dt}{dR}+t}{(1+hCR)^2}\right)=C\left(\frac{t+\left(hCR^2+R\right)\frac{dt}{dR}}{(1+hCR)^2}\right)$$

Now find $f''(R)$. Again, start with the quotient rule.

$$f''(R)=C\left(\frac{(1+hCR)^2\frac{d}{dR}\left(t+\left(hCR^2+R\right)\frac{dt}{dR}\right)-\left(t+\left(hCR^2+R\right)\frac{dt}{dR}\right)\frac{d}{dR}\left((1+hCR)^2\right)}{\left((1+hCR)^2\right)^2}\right)$$

In the numerator, we have

$$(1+hCR)^2\frac{d}{dR}\left(t+\left(hCR^2+R\right)\frac{dt}{dR}\right)=(1+hCR)^2\frac{dt}{dR}+\frac{d}{dR}\left(\left(hCR^2+R\right)\frac{dt}{dR}\right)$$

Now use the product rule to differentiate the second term.

$$\begin{aligned}(1+hCR)^2\frac{d}{dR}\left(t+\left(hCR^2+R\right)\frac{dt}{dR}\right)&=(1+hCR)^2\left[\frac{dt}{dR}+\frac{d}{dR}\left(\left(hCR^2+R\right)\frac{dt}{dR}\right)\right]\\
&=(1+hCR)^2\left[\frac{dt}{dR}+\left(\frac{d}{dR}\left(hCR^2+R\right)\frac{dt}{dR}+\left(hCR^2+R\right)\frac{d^2t}{dR^2}\right)\right]\\
&=(1+hCR)^2\left[\frac{dt}{dR}+(1+2hCR)\frac{dt}{dR}+R(hCR+1)\frac{d^2t}{dR^2}\right]\\
&=(1+hCR)^2\left[(2+2hCR)\frac{dt}{dR}+R(1+hCR)\frac{d^2t}{dR^2}\right]\\
&=(1+hCR)^2\left[2(1+hCR)\frac{dt}{dR}+R(1+hCR)\frac{d^2t}{dR^2}\right]\end{aligned}$$

(continued on next page)

(continued)

$$\frac{d}{dR}\left((1+hCR)^2\right)=2(1+hCR)\frac{d}{dR}(1+hCR)=2hC(1+hCR)$$

Combining the terms in the numerator gives

$$(1+hCR)^2\frac{d}{dR}\left(t+\left(hCR^2+R\right)\frac{dt}{dR}\right)-\left(t+\left(hCR^2+R\right)\frac{dt}{dR}\right)\frac{d}{dR}\left((1+hCR)^2\right)$$

$$=(1+hCR)^2\left[2(1+hCR)\frac{dt}{dR}+R(1+hCR)\frac{d^2t}{dR^2}\right]-\left(t+\left(hCR^2+R\right)\frac{dt}{dR}\right)(2hC(1+hCR))$$

$$=(1+hCR)^2\left[2(1+hCR)\frac{dt}{dR}+R(1+hCR)\frac{d^2t}{dR^2}\right]-(2hC(1+hCR))\left(t+R(hCR+1)\frac{dt}{dR}\right)$$

$$f''(R)=\frac{C}{(1+hCR)^4}\left((1+hCR)^2\left[2(1+hCR)\frac{dt}{dR}+R(1+hCR)\frac{d^2T}{dR^2}\right]-(2hC(1+hCR))\left(T+R(hCR+1)\frac{dt}{dR}\right)\right)$$

Factor $(1+hCR)$ from the numerator and denominator.

$$f''(R)=C\left(\frac{(1+hCR)\left[2(1+hCR)\frac{dt}{dR}+R(1+hCR)\frac{d^2t}{dR^2}\right]-(2hC)\left(t+R(hCR+1)\frac{dt}{dR}\right)}{(1+hCR)^3}\right)$$

$$=C\left(\frac{2(1+hCR)^2\frac{dt}{dR}+R(1+hCR)^2\frac{d^2t}{dR^2}-2hCt-\left(2hCR(hCR+1)\frac{dt}{dR}\right)}{(1+hCR)^3}\right)$$

In the numerator

$$2(1+hCR)^2\frac{dt}{dR}+R(1+hCR)^2\frac{d^2t}{dR^2}-2hCt-\left(2hCR(hCR+1)\frac{dt}{dR}\right)$$

$$=2(1+hCR)^2\frac{dt}{dR}-\left(2hCR(hCR+1)\frac{dt}{dR}\right)+R(1+hCR)^2\frac{d^2t}{dR^2}-2hCt$$

$$=\left(2+4hCR+2(hCR)^2\right)\frac{dt}{dR}-\left(\left(2(hCR)^2+2hCR\right)\frac{dt}{dR}\right)+R(1+hCR)^2\frac{d^2t}{dR^2}-2hCt$$

$$=\left(2+4hCR+2(hCR)^2-\left(2(hCR)^2+2hCR\right)\right)\frac{dt}{dR}+R(1+hCR)^2\frac{d^2t}{dR^2}-2hCt$$

$$=(2+2hCR)\frac{dt}{dR}+R(1+hCR)^2\frac{d^2t}{dR^2}-2hCt$$

$$=2(1+hCR)\frac{dt}{dR}+R(1+hCR)^2\frac{d^2t}{dR^2}-2hCt$$

Thus, $f''(R)=C\left(\dfrac{2(1+hCR)\frac{dt}{dR}+R(1+hCR)^2\frac{d^2t}{dR^2}-2hCt}{(1+hCR)^3}\right)$

The denominator is always positive because h, C, and R are positive, so the curve is concave down when the numerator of $f''<0$, or when $2(1+hCR)\frac{dt}{dR}+R(1+hCR)^2\frac{d^2t}{dR^2}-2hCt<0.$

53. $2s^2 + \sqrt{st} - 4 = 3t$

$$4s\frac{ds}{dt} + \frac{1}{2}(st)^{-1/2}\left(s + t\frac{ds}{dt}\right) = 3$$

$$4s\frac{ds}{dt} + \frac{s + t\frac{ds}{dt}}{2\sqrt{st}} = 3$$

$$\frac{8s(\sqrt{st})\frac{ds}{dt} + s + t\frac{ds}{dt}}{2\sqrt{st}} = 3$$

$$\frac{(8s\sqrt{st} + t)\frac{ds}{dt} + s}{2\sqrt{st}} = 3$$

$$(8s\sqrt{st} + t)\frac{ds}{dt} = 6\sqrt{st} - s \Rightarrow \frac{ds}{dt} = \frac{-s + 6\sqrt{st}}{8s\sqrt{st} + t}$$

6.4 Related Rates

1. $y^2 - 8x^3 = -55;\ \frac{dx}{dt} = -4,\ x = 2,\ y = 3$

$$2y\frac{dy}{dt} - 24x^2\frac{dx}{dt} = 0 \Rightarrow y\frac{dy}{dt} = 12x^2\frac{dx}{dt} \Rightarrow 3\frac{dy}{dt} = 48(-4) \Rightarrow \frac{dy}{dt} = -64$$

3. $2xy - 5x + 3y^3 = -51;\ \frac{dx}{dt} = -6,\ x = 3,\ y = -2$

$$2x\frac{dy}{dt} + 2y\frac{dx}{dt} - 5\frac{dx}{dt} + 9y^2\frac{dy}{dt} = 0$$

$$(2x + 9y^2)\frac{dy}{dt} + (2y - 5)\frac{dx}{dt} = 0$$

$$(2x + 9y^2)\frac{dy}{dt} = (5 - 2y)\frac{dx}{dt} \Rightarrow \frac{dy}{dt} = \frac{5 - 2y}{2x + 9y^2}\cdot\frac{dx}{dt} = \frac{5 - 2(-2)}{2(3) + 9(-2)^2}\cdot(-6) = -\frac{9}{7}$$

5. $\frac{x^2 + y}{x - y} = 9;\ \frac{dx}{dt} = 2,\ x = 4,\ y = 2$

$$\frac{(x - y)\left(2x\frac{dx}{dt} + \frac{dy}{dt}\right) - (x^2 + y)\left(\frac{dx}{dt} - \frac{dy}{dt}\right)}{(x - y)^2} = 0$$

$$\frac{2x(x - y)\frac{dx}{dt} + (x - y)\frac{dy}{dt} - (x^2 + y)\frac{dx}{dt} + (x^2 + y)\frac{dy}{dt}}{(x - y)^2} = 0$$

$$[2x(x - y) - (x^2 + y)]\frac{dx}{dt} + [(x - y) + (x^2 + y)]\frac{dy}{dt} = 0$$

$$\frac{dy}{dt} = \frac{[(x^2 + y) - 2x(x - y)]\frac{dx}{dt}}{(x - y) + (x^2 + y)} = \frac{(-x^2 + y + 2xy)\frac{dx}{dt}}{x + x^2}$$

$$= \frac{[-(4)^2 + 2 + 2(4)(2)](2)}{4 + 4^2} = \frac{4}{20} = \frac{1}{5}$$

7. $xe^y = 2 - \ln 2 + \ln x;\ \frac{dx}{dt} = 6,\ x = 2,\ y = 0$

$$e^y\frac{dx}{dt} + xe^y\frac{dy}{dt} = 0 - 0 + \frac{1}{x}\frac{dx}{dt}$$

$$xe^y\frac{dy}{dt} = \left(\frac{1}{x} - e^y\right)\frac{dx}{dt} \Rightarrow \frac{dy}{dt} = \frac{\left(\frac{1}{x} - e^y\right)\frac{dx}{dt}}{xe^y} = \frac{(1 - xe^y)\frac{dx}{dt}}{x^2e^y} = \frac{[1 - (2)e^0](6)}{2^2e^0} = \frac{-6}{4} = -\frac{3}{2}$$

9. $\cos(\pi xy) + 2x + y^2 = 2;\ \frac{dx}{dt} = -\frac{2}{\pi},\ x = \frac{1}{2},\ y = 1$

$$\frac{d}{dt}\left(\cos(\pi xy) + 2x + y^2\right) = \frac{d}{dt}(2)$$

$$\frac{d}{dt}(\cos(\pi xy)) + \frac{d}{dt}(2x) + \frac{d}{dt}(y^2) = \frac{d}{dt}(2)$$

$$-\pi \sin(\pi xy)\frac{d}{dt}(xy) + 2\frac{dx}{dt} + 2y\frac{dy}{dt} = 0$$

$$-\pi \sin(\pi xy)\left(y\frac{dx}{dt} + x\frac{dy}{dt}\right) + 2\frac{dx}{dt} + 2y\frac{dy}{dt} = 0$$

$$-\pi y\sin(\pi xy)\frac{dx}{dt} - \pi x\sin(\pi xy)\frac{dy}{dt} + 2\frac{dx}{dt} + 2y\frac{dy}{dt} = 0$$

$$-\pi x\sin(\pi xy)\frac{dy}{dt} + 2y\frac{dy}{dt} = \pi y\sin(\pi xy)\frac{dx}{dt} - 2\frac{dx}{dt}$$

$$\frac{dy}{dt}(2y - \pi x\sin(\pi xy)) = (\pi y\sin(\pi xy) - 2)\frac{dx}{dt}$$

$$\frac{dy}{dt} = \frac{\pi y\sin(\pi xy) - 2}{2y - \pi x\sin(\pi xy)}\frac{dx}{dt}$$

Now substitute the given values.

$$\frac{dy}{dt} = \left(\frac{\pi\sin\left(\frac{\pi}{2}\right) - 2}{2 - \frac{\pi}{2}\sin\left(\frac{\pi}{2}\right)}\right)\left(-\frac{2}{\pi}\right) = \frac{\pi - 2}{2 - \frac{\pi}{2}}\left(-\frac{2}{\pi}\right) = \frac{2(\pi - 2)}{(4 - \pi)}\left(-\frac{2}{\pi}\right) = \frac{8 - 4\pi}{\pi(4 - \pi)}$$

11. $V = k(R^2 - r^2);\ k = 555.6,\ R = 0.02$ mm, $\frac{dR}{dt} = 0.003$ mm per minute; r is constant.

$$V = k(R^2 - r^2)$$

$$V = 555.6(R^2 - r^2)$$

$$\frac{dV}{dt} = 555.6\left(2R\frac{dR}{dt} - 0\right)$$

$$= 555.6(2)(0.02)(0.003)$$

$$= 0.067 \text{ mm/min}$$

13. $b = 0.22m^{0.87}$

$$\frac{db}{dt} = 0.22(0.87)m^{-0.13}\frac{dm}{dt}$$

$$= 0.1914m^{-0.13}\frac{dm}{dt}$$

$$\frac{dm}{dt} = \frac{m^{0.13}}{0.1914}\frac{db}{dt} = \frac{25^{0.13}}{0.1914}(0.25) \approx 1.9849$$

The rate of change of the total weight is about 1.9849 g/day.

15. $m = 85.65w^{0.54}$

(a) $$\frac{dm}{dt} = \frac{d}{dt}\left(85.65w^{0.54}\right)$$

$$= 85.65(0.54)w^{-0.46}\frac{dw}{dt}$$

$$= 43.251w^{-0.46}\frac{dw}{dt}$$

(b) $w = 0.25,\ \frac{dw}{dt} = 0.01$ kg/day

$$\frac{dm}{dt} = 43.251(0.25)^{-0.46}(0.01) \approx 0.875$$

The rate of change of the average daily metabolic rate is about 0.875 kcal/day^2.

17. $E = 26.5w^{-0.34}$

$$\frac{dE}{dt} = 26.5(-0.34)w^{-1.34}\frac{dw}{dt}$$

$$= -9.01w^{-1.34}\frac{dw}{dt}$$

$$= -9.01(5)^{-1.34}(0.05) \approx -0.0521$$

The rate of change of the energy expenditure is about –0.0521 kcal/kg/km/day.

19. $C = \frac{1}{10}(T-60)^2 + 100$

$\frac{dC}{dt} = \frac{1}{5}(T-60)\frac{dT}{dt}$

If $T = 76°$ and $\frac{dT}{dt} = 8$,

$\frac{dC}{dt} = \frac{1}{5}(76-60)(8) = \frac{1}{5}(16)(8) = 25.6.$

The crime rate is rising at the rate of 25.6 crimes/month.

21. Let x = the distance to the base of the ladder from the base of the building.
Let y = the distance up the side of the building to the top of the ladder.

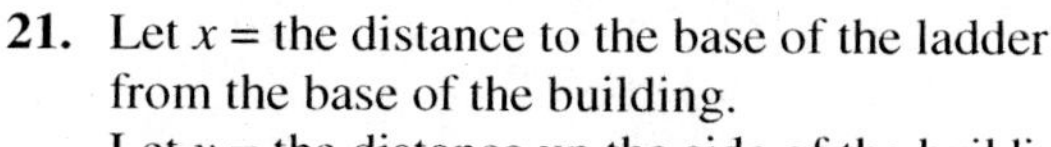

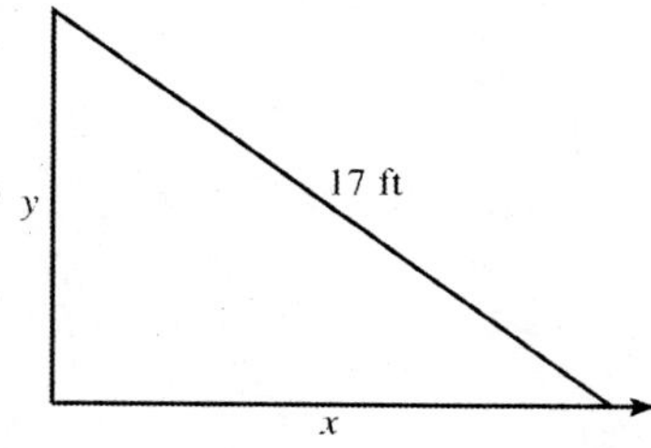

Find $\frac{dy}{dt}$ when $x = 8$ ft and $\frac{dx}{dt} = 9$ ft/min.

Since $y = \sqrt{17^2 - x^2}$, when $x = 8$, $y = 15$.

By the Pythagorean theorem, $x^2 + y^2 = 17^2$.

$$\frac{d}{dx}(x^2 + y^2) = \frac{d}{dt}(17^2)$$

$$2x\frac{dx}{dt} + 2y\frac{dy}{dt} = 0$$

$$2y\frac{dy}{dt} = -2x\frac{dx}{dt}$$

$$\frac{dy}{dt} = \frac{-2x}{2y}\cdot\frac{dx}{dt} = -\frac{x}{y}\cdot\frac{dx}{dt}$$

$$= -\frac{8}{15}(9) = -\frac{24}{5}$$

The ladder is sliding down the building at the rate of $\frac{24}{5}$ ft/min.

23. Let r = the radius of the circle formed by the ripple.

Find $\frac{dA}{dt}$ when $r = 4$ ft and $\frac{dr}{dt} = 2$ ft/min.

$A = \pi r^2$

$\frac{dA}{dt} = 2\pi r\frac{dr}{dt} = 2\pi(4)(2) = 16\pi$

The area is changing at the rate of 16π ft^2/min.

25. $V = x^3$, $x = 3$ cm, and $\frac{dV}{dt} = -2$ cm^3/min

$\frac{dV}{dt} = 3x^2\frac{dx}{dt}$

$\frac{dx}{dt} = \frac{1}{3x^2}\frac{dV}{dt} = \frac{1}{3\cdot 3^2}(-2) = -\frac{2}{27}$ cm/min

27. Let y = the length of the man's shadow; let x = the distance to the man from the lamp post; let h = the height of the lamp post.

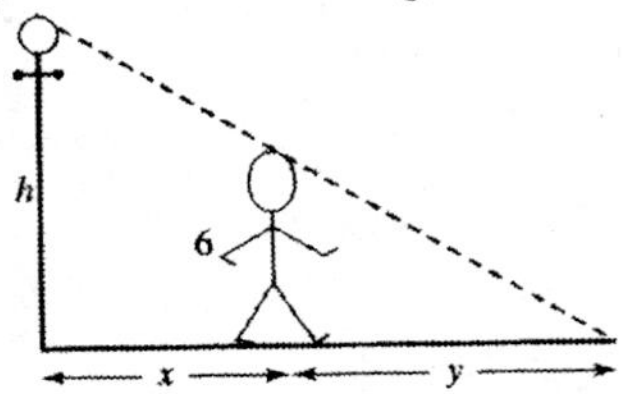

$\frac{dx}{dt} = 50$ ft/min

We must find $\frac{dy}{dt}$ when $x = 25$ ft.

We know that $\frac{h}{x+y} = \frac{6}{y}$, by similar triangles.

When $x = 8$, $y = 10$, $\frac{h}{18} = \frac{6}{10} \Rightarrow h = 10.8$

$\frac{10.8}{x+y} = \frac{6}{y} \Rightarrow 10.8y = 6x + 6y \Rightarrow 4.8y = 6x \Rightarrow$

$y = 1.25x$

$\frac{dy}{dt} = 1.25\frac{dx}{dt} = 1.25(50) \Rightarrow \frac{dy}{dt} = 62.5$

The length of the shadow is increasing at the rate of 62.5 ft/min.

29. Let x = the distance from the docks and let s = the length of the rope.

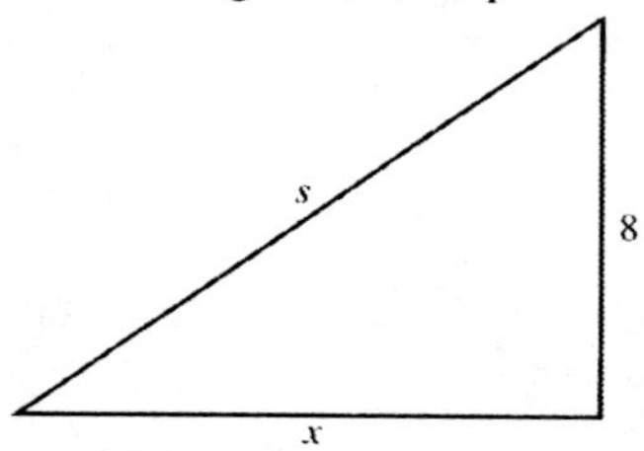

$\frac{ds}{dt} = -1$ ft/sec

$s^2 = x^2 + (8)^2$

$2s\frac{ds}{dt} = 2x\frac{dx}{dt} + 0 \Rightarrow s\frac{ds}{dt} = x\frac{dx}{dt}$

If $x = 8$, $s = \sqrt{(8)^2 + (8)^2} = \sqrt{128} = 8\sqrt{2}$.

Then, $8\sqrt{2}(-1) = 8\frac{dx}{dt} \Rightarrow \frac{dx}{dt} = -\sqrt{2} \approx -1.41$

The boat is approaching the dock at $\sqrt{2} \approx 1.41$ ft/sec.

31. $C = 0.2x^2 + 10,000;\ x = 80,\ \frac{dx}{dt} = 12$

$$\frac{dC}{dt} = 0.2(2x)\frac{dx}{dt} = 0.2(160)(12) = 384$$

The cost is changing at a rate of \$384 per month.

33. $R = 50x - 0.4x^2;\ C = 5x + 15;$

$x = 40;\ \frac{dx}{dt} = 10$

(a) $\frac{dR}{dt} = 50\frac{dx}{dt} - 0.8x\frac{dx}{dt}$
$= 50(10) - 0.8(40)(10) = 180$

Revenue is increasing at a rate of \$180 per day.

(b) $\frac{dC}{dt} = 5\frac{dx}{dt} = 5(10) = 50$

Cost is increasing at a rate of \$50 per day.

(c) Profit = Revenue − Cost

$P = R - C$

$\frac{dP}{dt} = \frac{dR}{dt} - \frac{dC}{dt} = 180 - 50 = 130$

Profit is increasing at a rate of \$130 per day.

35. (a) $\tan\theta = \frac{x}{50}$

Differentiate both sides with respect to time, t.

$$D_t(\tan\theta) = D_t\left(\frac{x}{50}\right)$$

$$\sec^2\theta \cdot \frac{d\theta}{dt} = \frac{1}{50}\cdot\frac{dx}{dt}$$

Since the light rotates twice per minute,

$\frac{d\theta}{dt} = \frac{2(2\pi \text{ radians})}{1 \text{ min}} = 4\pi$ radians per minute.

When the light beam and shoreline are at right angles, $\theta = 0$ and $\sec\theta = 1$. Thus,

$$(1)^2(4\pi) = \frac{1}{50}\cdot\frac{dx}{dt} \Rightarrow \frac{dx}{dt} = 200\pi.$$

The beam is moving along the shoreline at 200π m/min.

(b) When the beam hits the shoreline 50 m from the point on the shoreline closest to the lighthouse, $\theta = \frac{\pi}{4}$ and $\sec\theta = \sqrt{2}$.

Thus,

$$(\sqrt{2})^2(4\pi) = \frac{1}{50}\cdot\frac{dx}{dt} \Rightarrow \frac{dx}{dt} = 400\pi.$$

The beam is moving along the shoreline at 400π m/min.

6.5 Differentials: Linear Approximation

1. $y = 2x^3 - 5x;\ x = -2,\ \Delta x = 0.1$

$dy = (6x^2 - 5)dx$

$\Delta y \approx (6x^2 - 5)\Delta x \approx [6(-2)^2 - 5](0.1) \approx 1.9$

3. $y = x^3 - 2x^2 + 3,\ x = 1,\ \Delta x = -0.1$

$dy = (3x^2 - 4x)dx$

$\Delta y \approx (3x^2 - 4x)\Delta x$
$= [3(1^2) - 4(1)](-0.1) = 0.1$

5. $y = \sqrt{3x+2},\ x = 4,\ \Delta x = 0.15$

$$dy = 3\left(\frac{1}{2}(3x+2)^{-1/2}\right)dx$$

$$\Delta y \approx \frac{3}{2\sqrt{3x+2}}\Delta x \approx \frac{3}{2(3.74)}(0.15) \approx 0.060$$

7. $y = \frac{2x-5}{x+1};\ x = 2,\ \Delta x = -0.03$

$$dy = \frac{(x+1)(2) - (2x-5)(1)}{(x+1)^2}dx = \frac{7}{(x+1)^2}dx$$

$$\Delta y \approx \frac{7}{(x+1)^2}\Delta x = \frac{7}{(2+1)^2}(-0.03) = -0.023$$

9. $\sqrt{145}$

We know $\sqrt{144} = 12$, so $f(x) = \sqrt{x},\ x = 144$, $dx = 1$.

$$\frac{dy}{dx} = \frac{1}{2}x^{-1/2} \Rightarrow dy = \frac{1}{2\sqrt{x}}dx = \frac{1}{2\sqrt{144}}(1) = \frac{1}{24}$$

$$\sqrt{145} \approx f(x) + dy = 12 + \frac{1}{24} \approx 12.0417$$

By calculator, $\sqrt{145} \approx 12.0416$.

The difference is $|12.0417 - 12.0416| = 0.0001$.

11. $\sqrt{0.99}$

We know $\sqrt{1} = 1$, so $f(x) = \sqrt{x}$, $x = 1,\ dx = -0.01$.

$$\frac{dy}{dx} = \frac{1}{2}x^{-1/2} = \frac{1}{2\sqrt{x}}dx \Rightarrow$$

$$dy = \frac{1}{2\sqrt{1}}(-0.01) = -0.005$$

$$\sqrt{0.99} \approx f(x) + dy = 1 - 0.005 = 0.995$$

By calculator, $\sqrt{0.99} \approx 0.9950$.

The difference is $|0.995 - 0.9950| = 0$.

13. $e^{0.01}$

We know $e^0 = 1$, so $f(x) = e^x$, $x = 0$, $dx = 0.01$.

$$\frac{dy}{dx} = e^x \Rightarrow dy = e^x dx$$
$$dy = e^0(0.01) = 0.01$$
$$e^{0.01} \approx f(x) + dy = 1 + 0.01 = 1.01$$

By calculator, $e^{0.01} \approx 1.0101$.
The difference is $|1.01 - 1.0101| = 0.0001$.

15. ln 1.05

We know $\ln 1 = 0$, so $f(x) = \ln x$, $x = 1$, $dx = 0.05$.

$$\frac{dy}{dx} = \frac{1}{x} \Rightarrow dy = \frac{1}{x}dx = \frac{1}{1}(0.05) = 0.05$$
$$\ln 1.05 \approx f(x) + dy = 0 + 0.05 = 0.05$$

By calculator, $\ln 1.05 \approx 0.0488$.

The difference is $|0.05 - 0.0488| = 0.0012$.

17. $\sin(0.03)$

We know that $\sin(0) = 0$, so $f(x) = \sin x$, $x = 0$, and $dx = 0.03$

$$\frac{dy}{dx} = \cos x \Rightarrow dy = (\cos x)dx = 0.03\cos x$$
$$\sin(0.03) = f(x) + 0.03\cos 0 = 0 + 0.03 = 0.03$$

By calculator, $\sin(0.03) \approx 0.0300$. The difference is $|0.03 - 0.0300| = 0$.

19. **(a)** $A(t) = y = 0.003631t^3 - 0.03746t^2 + 0.1012t + 0.009$

Let $t = 1$, $dt = 0.2$.

$$\frac{dy}{dt} = 0.010893t^2 - 0.07492t + 0.1012$$
$$dy = (0.010893t^2 - 0.07492t + 0.1012)dt$$
$$\Delta y \approx (0.010893t^2 - 0.07492t + 0.1012)\Delta t \approx (0.010893 \cdot 1^2 - 0.07492 \cdot 1 + 0.1012) \cdot 0.2 \approx 0.007435$$

The alcohol concentration increases by about 0.74 percent.

(b) $\Delta y \approx (0.010893 \cdot 3^2 - 0.07492 \cdot 3 + 0.1012) \cdot 0.2 \approx -0.005105$

The alcohol concentration decreases by about 0.51 percent.

21. $P(t) = \dfrac{25t}{8 + t^2}$

$$dP = \frac{(8+t^2)(25) - 25t(2t)}{(8+t^2)^2}dt \approx \frac{(8+t^2)(25) - 25t(2t)}{(8+t^2)^2}\Delta t$$

(a) $t = 2$, $\Delta t = 0.5$

$$dP \approx \frac{[(8+4)(25) - (25)(2)(4)](0.5)}{(8+4)^2} \approx 0.347 \text{ million}$$

(b) $t = 3$, $\Delta t = 0.25$

$$dP \approx \frac{[(8+9)(25) - 25(3)(6)]0.25}{(8+9)^2} \approx -0.022 \text{ million}$$

23. r changes from 14 mm to 16 mm, so $\Delta r = 2.$

$$V = \frac{4}{3}\pi r^3$$
$$dV = \frac{4}{3}(3)\pi r^2\, dr$$
$$\Delta V \approx 4\pi r^2\, \Delta r = 4\pi(14)^2(2) = 1568\pi \text{ mm}^3$$

25. r increases from 20 mm to 22 mm, so $\Delta r = 2.$

$$A = \pi r^2$$
$$dA = 2\pi r dr$$
$$\Delta A \approx 2\pi r\, \Delta r = 2\pi(20)(2) = 80\pi \text{ mm}^2$$

27. $W(t) = -3.5 + 197.5e^{-e^{-0.01394(t-108.4)}}$

(a) $dW = 197.5e^{-e^{-0.01394(t-108.4)}}(-1)e^{-0.01394(t-108.4)}(-0.01394)dt$
$= 2.75315e^{-e^{-0.01394(t-108.4)}}e^{-0.01394(t-108.4)}dt$

We are given $t = 80$ and $dt = 90 - 80 = 10$, so $dW \approx 9.258$

The pig will gain about 9.3 kg.

(b) The actual weight gain is calculated as $W(90) - W(80) \approx 50.736 - 41.202 = 9.534$ or about 9.5 kg.

29. $r = 3$ cm, $\Delta r = -0.2$ cm

$$V = \frac{4}{3}\pi r^3$$
$$dV = 4\pi r^2 dr$$
$$\Delta V \approx 4\pi r^2 \Delta r = 4\pi(9)(-0.2) = -7.2\pi \text{ cm}^3$$

31. $V = \frac{1}{3}\pi r^2 h;\ h = 13,\ dh = 0.2$

$$V = \frac{1}{3}\pi\left(\frac{h}{15}\right)^2 h = \frac{\pi}{775}h^3$$
$$dV = \frac{\pi}{775}\cdot 3h^2 dh = \frac{\pi}{225}h^2 dh$$
$$\Delta V \approx \frac{\pi}{225}h^2\Delta h \approx \frac{\pi}{225}(13^2)(0.2) \approx 0.472 \text{ cm}^3$$

33. $A = x^2;\ x = 4,\ dA = 0.01$

$$dA = 2x\, dx$$
$$\Delta A \approx 2x\, \Delta x$$
$$\Delta x \approx \frac{\Delta A}{2x} \approx \frac{0.01}{2(4)} \approx 0.00125 \text{ cm}$$

35. $V = \frac{4}{3}\pi r^3;\ r = 5.81,\ \Delta r = \pm 0.003$

$$dV = \frac{4}{3}\pi(3r^2)dr$$
$$\Delta V \approx \frac{4}{3}\pi(3r^2)\Delta r = 4\pi(5.81)^2(\pm 0.003)$$
$$= \pm 0.405\pi \approx \pm 1.273 \text{ in.}^3$$

37. $h = 7.284$ in., $r = 1.09$, $\Delta r = \pm 0.007$ in.

$$V = \frac{1}{3}\pi r^2 h$$
$$dV = \frac{2}{3}\pi rh\, dr$$
$$\Delta V \approx \frac{2}{3}\pi rh\, \Delta r = \frac{2}{3}\pi(1.09)(7.284)(\pm 0.007)$$
$$= \pm 0.116 \text{ in.}^3$$

39. Let x = the number of beach balls, and let V = the volume of x beach balls. Then $\frac{dV}{dr} \approx$ the volume of material in the beach balls since they are hollow.

$$V = \frac{4}{3}\pi r^3 x$$
$r = 6$ in., $x = 5000$, $\Delta r = 0.03$ in.
$$dV = \frac{4}{3}\pi(3r^2 x + r^3)\,\Delta r$$
$$= \frac{4}{3}\pi(3\cdot 36\cdot 5000 + 216)(0.03)$$
$$= 21{,}608.64\pi$$

$21{,}608\pi$ in.^{3} of material would be needed.

Chapter 6 Review Exercises

1. False: The absolute maximum might occur at the endpoint of the interval of interest.

2. True

3. False: It could have either. For example $f(x) = 1/(1-x^2)$ has an absolute minimum of 1 on $(-1, 1)$.

4. True

5. True

6. True

7. True

8. True

9. $f(x) = -x^3 + 6x^2 + 1; [-1, 6]$
$f'(x) = -3x^2 + 12x = 0$ when $x = 0, 4$.
$f(-1) = 8$
$f(0) = 1$
$f(4) = 33$
$f(6) = 1$
Absolute maximum of 33 at 4; absolute minimum of 1 at $x = 0$ and $x = 6$.

10. $f(x) = 4x^3 - 9x^2 - 3; [-1, 2]$
$f'(x) = 12x^2 - 18x = 0$ when $x = 0, \frac{3}{2}$.
$f(-1) = -16$
$f(0) = -3$
$f\left(\frac{3}{2}\right) = -9.75$
$f(2) = -7$
Absolute maximum of -3 at $x = 0$; absolute minimum of -16 at $x = -1$.

11. $f(x) = x^3 + 2x^2 - 15x + 3; [-4, 2]$
$f'(x) = 3x^2 + 4x - 15 = 0 \Rightarrow$
$(3x-5)(x+3) = 0 \Rightarrow x = \frac{5}{3}$ or $x = -3$
$f(-4) = 31$
$f(-3) = 39$
$f\left(\frac{5}{3}\right) = -\frac{319}{27}$
$f(2) = -11$
Absolute maximum of 39 at $x = -3$; absolute minimum of $-\frac{319}{27}$ at $x = \frac{5}{3}$.

12. $f(x) = -2x^3 - 2x^2 + 2x - 1; [-3, 1]$
$f'(x) = -6x^2 - 4x + 2$
$f'(x) = 0 \Rightarrow 3x^2 + 2x - 1 = 0 \Rightarrow$
$(3x-1)(x+1) = 0 \Rightarrow x = \frac{1}{3}$ or $x = -1$
$f(-3) = 29$
$f(-1) = -3$
$f\left(\frac{1}{3}\right) = -\frac{17}{27}$
$f(1) = -3$
Absolute maximum of 29 at $x = -3$; absolute minimum of -3 at $x = -1$ and $x = 1$.

13. $f(x) = \cos x + x$
$f'(x) = -\sin x + 1$
Now solve $f'(x) = 0$ to find any critical points in the given interval $[0, \pi]$.
$-\sin x + 1 = 0 \Rightarrow -\sin x = -1 \Rightarrow x = \frac{\pi}{2}$
We must also test the endpoints $x = 0$ and $x = \pi$.

x	$f(x)$	
0	1	Absolute minimum
$\frac{\pi}{2}$	$\frac{\pi}{2} \approx 1.5708$	
π	$\pi - 1 \approx 2.1416$	Absolute maximum

14. $f(x) = \sin x - \cos x + 1$
$f'(x) = \cos x + \sin x$
Now solve $f'(x) = 0$ to find any critical points in the given interval $[0, \pi]$.
$\cos x + \sin x = 0 \Rightarrow \cos x = -\sin x \Rightarrow x = \frac{3\pi}{4}$
We must also test the endpoints $x = 0$ and $x = \pi$.

x	$f(x)$	
0	0	Absolute minimum
$\frac{3\pi}{4}$	$1 + \sqrt{2} \approx 2.4142$	Absolute maximum
π	2	

15. Answers will vary.

16. No.

17. (a) $f(x) = \frac{2 \ln x}{x^2}; [1, 4]$

$$f'(x) = \frac{x^2\left(\frac{2}{x}\right) - (2\ln x)(2x)}{x^4} = \frac{2x - 4x\ln x}{x^4} = \frac{2 - 4\ln x}{x^3}$$

$f'(x) = 0 \Rightarrow 2 - 4\ln x = 0 \Rightarrow 2 = 4\ln x \Rightarrow$
$0.5 = \ln x \Rightarrow e^{0.5} = x \Rightarrow x \approx 1.6487.$
(Both the derivative and the original function are undefined at $x = 0$.)

x	$f(x)$
1	0
$e^{0.5}$	0.36788
4	0.17329

Maximum is 0.37; minimum is 0.

(b) [2, 5]

Note that the critical number of f is not in the domain, so we only test the endpoints.

x	$f(x)$
2	0.34657
5	0.12876

Maximum is 0.35, minimum is 0.13.

18. $f(x) = \frac{e^{2x}}{x^2}$

First find the critical numbers.

$$f'(x) = \frac{2e^{2x}x^2 - e^{2x}(2x)}{(x^2)^2} = \frac{2e^{2x}(x^2 - x)}{x^4}$$

$f'(x) = 0$ when $x^2 - x = (x)(x-1) = 0,$ that is, at $x = 1$. (Both the derivative and the original function are undefined at $x = 0$.)

(a) The function f is continuous on the interval $[1/2, 2]$ and the critical number lies in this interval, so we evaluate the function at three points:

x	$f(x)$
1/2	10.873
1	7.389
2	13.650

The absolute minimum is 7.39 and the absolute maximum is 13.65.

(b) The function f is continuous on the interval [1, 3]. The critical number coincides with an endpoint so we need to look at only two function values:

x	$f(x)$
1	7.389
3	44.825

The absolute minimum is 7.39 and the absolute maximum is 44.83.

19. Answers will vary.

20. Answers will vary.

21. $x^2 - 4y^2 = 3x^3y^4$

$$\frac{d}{dx}(x^2 - 4y^2) = \frac{d}{dx}(3x^3y^4)$$
$$2x - 8y\frac{dy}{dx} = 9x^2y^4 + 3x^3 \cdot 4y^3\frac{dy}{dx}$$
$$(-8y - 3x^3 \cdot 4y^3)\frac{dy}{dx} = 9x^2y^4 - 2x$$
$$\frac{dy}{dx} = \frac{2x - 9x^2y^4}{8y + 12x^3y^3}$$

22. $x^2y^3 + 4xy = 2$

$$\frac{d}{dx}(x^2y^3 + 4xy) = \frac{d}{dx}(2)$$
$$2xy^3 + 3y^2\left(\frac{dy}{dx}\right)x^2 + 4y + 4x\frac{dy}{dx} = 0$$
$$(3x^2y^2 + 4x)\frac{dy}{dx} = -2xy^3 - 4y$$
$$\frac{dy}{dx} = \frac{-2xy^3 - 4y}{3x^2y^2 + 4x}$$

23. $2\sqrt{y-1} = 9x^{2/3} + y$

$$\frac{d}{dx}[2(y-1)^{1/2}] = \frac{d}{dx}(9x^{2/3} + y)$$
$$2 \cdot \frac{1}{2} \cdot (y-1)^{-1/2}\frac{dy}{dx} = 6x^{-1/3} + \frac{dy}{dx}$$
$$[(y-1)^{-1/2} - 1]\frac{dy}{dx} = 6x^{-1/3}$$
$$\frac{1 - \sqrt{y-1}}{\sqrt{y-1}} \cdot \frac{dy}{dx} = \frac{6}{x^{1/3}}$$
$$\frac{dy}{dx} = \frac{6\sqrt{y-1}}{x^{1/3}(1 - \sqrt{y-1})}$$

24. $9\sqrt{x}+4y^3=2\sqrt{y}$

$$\frac{d}{dx}(9\sqrt{x}+4y^3)=2\cdot\frac{d}{dx}(y^{1/2})$$
$$\frac{9}{2}x^{-1/2}+12y^2\frac{dy}{dx}=2\cdot\frac{1}{2}y^{-1/2}\cdot\frac{dy}{dx}$$
$$\frac{9}{2x^{1/2}}=\left(\frac{1}{y^{1/2}}-12y^2\right)\frac{dy}{dx}$$
$$\frac{9}{2x^{1/2}}=\frac{1-12y^{5/2}}{y^{1/2}}\frac{dy}{dx}$$
$$\frac{dy}{dx}=\frac{9y^{1/2}}{2x^{1/2}(1-12y^{5/2})}=\frac{9\sqrt{y}}{2\sqrt{x}(1-12y^{5/2})}$$

25. $\frac{6+5x}{2-3y}=\frac{1}{5x}$

$$5x(6+5x)=2-3y$$
$$30x+25x^2=2-3y$$
$$\frac{d}{dx}(30x+25x^2)=\frac{d}{dx}(2-3y)$$
$$30+50x=-3\frac{dy}{dx}\Rightarrow-\frac{30+50x}{3}=\frac{dy}{dx}$$

26.

$$\frac{x+2y}{x-3y}=y^{1/2}$$
$$x+2y=y^{1/2}(x-3y)$$
$$\frac{d}{dx}(x+2y)=\frac{d}{dx}[y^{1/2}(x-3y)]$$
$$1+2\frac{dy}{dx}=y^{1/2}\left(1-3\frac{dy}{dx}\right)+\frac{1}{2}(x-3y)y^{-1/2}\frac{dy}{dx}$$
$$1+2\frac{dy}{dx}=y^{1/2}-3y^{1/2}\frac{dy}{dx}+\frac{1}{2}xy^{-1/2}\frac{dy}{dx}-\frac{3}{2}y^{1/2}\frac{dy}{dx}$$
$$(2+3y^{1/2}-\frac{1}{2}xy^{-1/2}+\frac{3}{2}y^{1/2})\frac{dy}{dx}=y^{1/2}-1$$
$$\frac{2y^{1/2}\left(2+\frac{9}{2}y^{1/2}-\frac{1}{2}xy^{-1/2}\right)}{2y^{1/2}}\frac{dy}{dx}=y^{1/2}-1$$
$$\left(\frac{4y^{1/2}+9y-x}{2y^{1/2}}\right)\frac{dy}{dx}=y^{1/2}-1\Rightarrow\frac{dy}{dx}=\frac{2y-2y^{1/2}}{4y^{1/2}+9y-x}$$

27. $\ln(xy+1) = 2xy^3 + 4$

$$\frac{d}{dx}[\ln(xy+1)] = \frac{d}{dx}(2xy^3+4)$$
$$\frac{1}{xy+1}\cdot\frac{d}{dx}(xy+1) = 2y^3 + 2x\cdot 3y^2\frac{dy}{dx} + \frac{d}{dx}(4)$$
$$\frac{1}{xy+1}\left(y + x\frac{dy}{dx} + \frac{d}{dx}(1)\right) = 2y^3 + 6xy^2\frac{dy}{dx}$$
$$\frac{y}{xy+1} + \frac{x}{xy+1}\cdot\frac{dy}{dx} = 2y^3 + 6xy^2\frac{dy}{dx}$$
$$\left(\frac{x}{xy+1} - 6xy^2\right)\frac{dy}{dx} = 2y^3 - \frac{y}{xy+1}$$
$$\frac{dy}{dx} = \frac{2y^3 - \frac{y}{xy+1}}{\frac{x}{xy+1} - 6xy^2} = \frac{2y^3(xy+1) - y}{x - 6xy^2(xy+1)} = \frac{2xy^4 + 2y^3 - y}{x - 6x^2y^3 - 6xy^2}$$

28. $\ln(x+y) = 1 + x^2 + y^3$

$$\frac{d}{dx}[\ln(x+y)] = \frac{d}{dx}(1 + x^2 + y^3)$$
$$\frac{1}{x+y}\cdot\frac{d}{dx}(x+y) = 2x + 3y^2\cdot\frac{dy}{dx}$$
$$\frac{1}{x+y}\left(1 + \frac{dy}{dx}\right) = 2x + 3y^2\cdot\frac{dy}{dx}$$
$$\left(\frac{1}{x+y} - 3y^2\right)\frac{dy}{dx} = 2x - \frac{1}{x+y}$$
$$\frac{dy}{dx} = \frac{2x - \frac{1}{x+y}}{\frac{1}{x+y} - 3y^2} = \frac{2x(x+y) - 1}{1 - 3y^2(x+y)} = \frac{2x^2 + 2xy - 1}{1 - 3xy^2 - 3y^3} = \frac{1 - 2x^2 - 2xy}{3xy^2 + 3y^3 - 1}$$

29. $x + \cos(x+y) = y^2$

$$\frac{d}{dx}(x) + \frac{d}{dx}(\cos(x+y)) = \frac{d}{dx}(y^2)$$
$$1 - \sin(x+y)\frac{d}{dx}(x+y) = 2y\frac{dy}{dx}$$
$$1 - (\sin(x+y))\left(1 + \frac{dy}{dx}\right) = 2y\frac{dy}{dx}$$
$$1 - \sin(x+y) - \sin(x+y)\frac{dy}{dx} = 2y\frac{dy}{dx}$$
$$1 - \sin(x+y) = (2y + \sin(x+y))\frac{dy}{dx} \Rightarrow \frac{1 - \sin(x+y)}{2y + \sin(x+y)} = \frac{dy}{dx}$$

30.
$$\tan(xy) + x^3 = y$$
$$\frac{d}{dx}(\tan(xy)) + \frac{d}{dx}(x^3) = \frac{d}{dx}y$$
$$\sec^2(xy)\frac{d}{dx}(xy) + 3x^2 = \frac{dy}{dx}$$

Use the product rule to find $\frac{d}{dx}(xy)$.

$$\sec^2(xy)\left(y\frac{d}{dx}(x) + x\frac{d}{dx}(y)\right) + 3x^2 = \frac{dy}{dx}$$
$$\sec^2(xy)\left(y + x\frac{dy}{dx}\right) + 3x^2 = \frac{dy}{dx}$$
$$y\sec^2(xy) + x\sec^2(xy)\frac{dy}{dx} + 3x^2 = \frac{dy}{dx}$$
$$y\sec^2(xy) + 3x^2 = \left(1 - x\sec^2(xy)\right)\frac{dy}{dx}$$
$$\frac{y\sec^2(xy) + 3x^2}{1 - x\sec^2(xy)} = \frac{dy}{dx}$$

31. $\sqrt{2y} - 4xy = -22$, tangent line at $(3, 2)$.

$$\frac{d}{dx}\left(\sqrt{2y} - 4xy\right) = \frac{d}{dx}(-22)$$
$$\frac{1}{2}(2)(2y)^{-1/2}\frac{dy}{dx} - \left(4y + 4x\frac{dy}{dx}\right) = 0$$
$$((2y)^{-1/2} - 4x)\frac{dy}{dx} = 4y \Rightarrow \frac{dy}{dx} = \frac{4y}{\frac{1}{\sqrt{2y}} - 4x}$$

To find the slope m of the tangent line, substitute 3 for x and 2 for y.

$$m = \frac{4y}{\frac{1}{2\sqrt{2y}} - 4x} = \frac{4(2)}{\frac{1}{\sqrt{2(2)}} - 4(3)} = \frac{8}{\frac{1}{2} - 12} = \frac{16}{1 - 24} = -\frac{16}{23}$$

The equation of the tangent line is

$$y - y_1 = m(x - x_1)$$
$$y - 2 = -\frac{16}{23}(x - 3)$$
$$y - 2 - \frac{48}{23} = -\frac{16}{23}x \Rightarrow y = -\frac{16}{23}x + \frac{94}{23}.$$

We can also write this equation as $16x + 23y = 94$.

32. $8y^3 - 4xy^2 = 20$, tangent line at $(-3, 1)$.

$$\frac{d}{dx}\left(8y^3 - 4xy^2\right) = \frac{d}{dx}(20)$$
$$24y^2\frac{dy}{dx} - \left(8xy\frac{dy}{dx} + 4y^2\right) = 0$$
$$(24y^2 - 8xy)\frac{dy}{dx} = 4y^2$$
$$\frac{dy}{dx} = \frac{4y^2}{24y^2 - 8xy}$$
$$= \frac{y^2}{6y^2 - 2xy}$$

To find the slope m of the tangent line, substitute -3 for x and 1 for y.

$$m = \frac{y^2}{6y^2 - 2xy} = \frac{1^2}{6(1^2) - 2(-3)(1)} = \frac{1}{12}$$

The equation of the tangent line is

$$y - y_1 = m(x - x_1)$$
$$y - 1 = \frac{1}{12}(x - (-3))$$
$$y - 1 - \frac{3}{12} = \frac{1}{12}x \Rightarrow y = \frac{1}{12}x + \frac{5}{4}$$

We can also write this as $x - 12y = -15$.

33. Answers will vary.

34. Answers will vary.

35. $y = 8x^3 - 7x^2,\ \frac{dx}{dt} = 4,\ x = 2$

$$\frac{dy}{dt} = \frac{d}{dt}(8x^3 - 7x^2)$$
$$= 24x^2\frac{dx}{dt} - 14x\frac{dx}{dt}$$
$$= 24(2)^2(4) - 14(2)(4) = 272$$

36. $y = \frac{9-4x}{3+2x};\ \frac{dx}{dt} = -1,\ x = -3$

$$\frac{dy}{dt} = \frac{(-4)(3+2x) - (2)(9-4x)}{(3+2x)^2}\frac{dx}{dt}$$
$$= \frac{-30}{(3+2x)^2}\frac{dx}{dt}$$
$$= \frac{-30}{[3+2(-3)]^2}(-1) = \frac{30}{9} = \frac{10}{3}$$

37. $y = \frac{1+\sqrt{x}}{1-\sqrt{x}},\ \frac{dx}{dt} = -4,\ x = 4$

$$\frac{dy}{dt} = \frac{d}{dt}\left[\frac{1+\sqrt{x}}{1-\sqrt{x}}\right]$$
$$= \frac{\left[\left(1-\sqrt{x}\right)\left(\frac{1}{2}x^{-1/2}\frac{dx}{dt}\right) - 1\left(1+\sqrt{x}\right)\left(-\frac{1}{2}\right)\left(x^{-1/2}\frac{dx}{dt}\right)\right]}{\left(1-\sqrt{x}\right)^2}$$
$$= \frac{\left[(1-2)\left(\frac{1}{2\cdot 2}\right)(-4) - (1+2)\left(\frac{-1}{2\cdot 2}\right)(-4)\right]}{(1-2)^2}$$
$$= \frac{1-3}{1} = -2$$

38. $\frac{x^2+5y}{x-2y} = 2;$

$\frac{dx}{dt} = 1,\ x = 2,\ y = 0$

$$x^2 + 5y = 2(x - 2y)$$
$$x^2 + 5y = 2x - 4y$$
$$9y = -x^2 + 2x$$
$$y = \frac{1}{9}(-x^2 + 2x) = -\frac{1}{9}x^2 + \frac{2}{9}x$$
$$\frac{dy}{dt} = \left(-\frac{2}{9}x + \frac{2}{9}\right)\frac{dx}{dt}$$
$$= \left[\left(-\frac{2}{9}\right)(2) + \frac{2}{9}\right](1)$$
$$= -\frac{4}{9} + \frac{2}{9} = -\frac{2}{9}$$

39. $y = xe^{3x};\ \frac{dx}{dt} = -2,\ x = 1$

$$\frac{dy}{dt} = \frac{d}{dt}(xe^{3x})$$
$$= \frac{dx}{dt}\cdot e^{3x} + x\cdot\frac{d}{dt}(e^{3x})$$
$$= \frac{dx}{dt}\cdot e^{3x} + xe^{3x}\cdot 3\frac{dx}{dt}$$
$$= (1+3x)e^{3x}\frac{dx}{dt}$$
$$= (1+3\cdot 1)e^{3(1)}(-2) = -8e^3$$

40. $y = \frac{1}{e^{x^2}+1};\ \frac{dx}{dt} = 3,\ x = 1$

$$\frac{dy}{dt} = \frac{d}{dt}\left(\frac{1}{e^{x^2}+1}\right)$$
$$= \frac{-1}{(e^{x^2}+1)^2}\cdot\frac{d}{dt}(e^{x^2}+1)$$
$$= \frac{-1}{(e^{x^2}+1)^2}\left[e^{x^2}\cdot\frac{d}{dt}(x^2) + \frac{d}{dt}(1)\right]$$
$$= \frac{-1}{(e^{x^2}+1)^2}\left[e^{x^2}\cdot 2x\frac{dx}{dt}\right]$$
$$= \frac{-1}{(e+1)^2}[e\cdot 2\cdot 1\cdot 3] = \frac{-6e}{(e+1)^2}$$

41. Answers will vary.

42. Answers will vary.

43. $y = \frac{3x-7}{2x+1};\ x = 2,\ \Delta x = 0.003$

$$dy = \frac{(3)(2x+1) - (2)(3x-7)}{(2x+1)^2}dx$$
$$dy = \frac{17}{(2x+1)^2}dx$$
$$\Delta y \approx \frac{17}{(2x+1)^2}\Delta x$$
$$= \frac{17}{(2[2]+1)^2}(0.003) = 0.00204$$

44. $y = 8 - x^2 + x^3,\ x = -1,\ \Delta x = 0.02$

$$dy = (-2x + 3x^2)dx$$
$$\Delta y \approx (-2x + 3x^2)\Delta x$$
$$= [-2(-1) + 3(-1)^2](0.02) = 0.1$$

45. $y = \sin x;\ x = \frac{\pi}{3};\ \Delta x = -0.01$

$$dy = \cos x\,dx$$
$$\approx \cos\left(\frac{\pi}{3}\right)(-0.01) = \frac{1}{2}(-0.01) = -0.005$$

46. $y = \tan x;\ x = \frac{\pi}{4};\ \Delta x = -0.02$

$$dy = \sec^2 x dx$$
$$\approx \sec^2\left(\frac{\pi}{4}\right)(-0.02) = 2(-0.02) = -0.04$$

47. $-12x + x^3 + y + y^2 = 4$

$$\frac{dy}{dx}(-12x + x^3 + y + y^2) = \frac{d}{dx}(4)$$
$$-12 + 3x^2 + \frac{dy}{dx} + 2y\frac{dy}{dx} = 0$$
$$(1+2y)\frac{dy}{dx} = 12 - 3x^2$$
$$\frac{dy}{dx} = \frac{12-3x^2}{1+2y}$$

(a) $\frac{dy}{dx} = 0 \Rightarrow 12 - 3x^2 = 0 \Rightarrow 12 = 3x^2 \Rightarrow$
$\pm 2 = x$
$x = 2$:
$-24 + 8 + y + y^2 = 4 \Rightarrow y + y^2 = 20 \Rightarrow$
$y^2 + y - 20 = 0 \Rightarrow (y+5)(y-4) = 0 \Rightarrow$
$y = -5$ or $y = 4$
(2, –5) and (2, 4) are critical points.
$x = -2$:

$$24 - 8 + y + y^2 = 4$$
$$y + y^2 = -12$$
$$y^2 + y + 12 = 0$$
$$y = \frac{-1 \pm \sqrt{1^2 - 48}}{2}$$

This leads to imaginary roots. Thus, $x = -2$ does not produce critical points.

(b)

x	y_1	y_2
1.9	–4.99	3.99
2	–5	4
2.1	–4.99	3.99

The point (2, –5) is a relative minimum.
The point (2, 4) is a relative maximum.

(c) There is no absolute maximum or minimum for x or y.

48. Answers will vary.

49. $A = \pi r^2;\ \frac{dr}{dt} = 4$ ft/min, $r = 7$ ft

$$\frac{dA}{dt} = 2\pi r\frac{dr}{dt}$$
$$\frac{dA}{dt} = 2\pi(7)(4) = 56\pi$$

The rate of change of the area is 56π ft^2/min.

50. $\frac{dx}{dt} = rx(N - x) = rxN - rx^2$

$$\frac{d^2x}{dt^2} = rN\frac{dx}{dt} - 2rx\frac{dx}{dt} = r\frac{dx}{dt}(N - 2x)$$
$$= r[rx(N-x)](N-2x)$$
$$= r^2x(N-x)(N-2x)$$

$\frac{d^2x}{dt^2} = 0$ when $x = 0,\ x = N$, or $x = \frac{N}{2}$.

On $\left(0, \frac{N}{2}\right)$, $\frac{d^2x}{dt^2} > 0$; therefore, the curve is concave upward.

On $\left(\frac{N}{2}, N\right)$, $\frac{d^2x}{dt^2} < 0$; therefore, the curve is concave downward.

Hence $x = \frac{N}{2}$ is a point of inflection.

51. **(a)**

7.5

$M(t) = 1.3386309 - 0.4321173t + 0.0564512t^2 - 0.0020506t^3 + 0.0000315t^4 - 0.0000001785t^5$

3 0 51

(b) We use a graphing calculator to graph
$M'(t) = -0.4321173 + 0.1129024t - 0.0061518t^2 + 0.0001260t^3 - 0.0000008925t^4$
on $[3, 51]$ by $[0, 0.3]$. We find the maximum value of $M'(t)$ on this graph at about 15.41, or on about the 15th day.

52. **(a)**

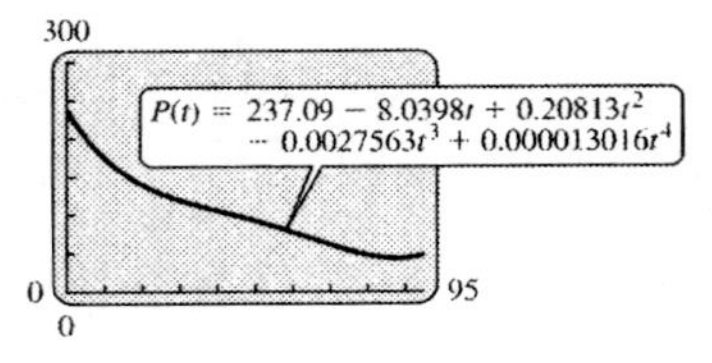

(b) To find where the maximum and minimum numbers occur, use a graphing calculator to locate any extreme points on the graph. One critical number is formed at about 87.78.

t	$P(t)$
0	237.09
87.78	43.56
95	48.66

The maximum number of polygons is about 237 at birth. The minimum number is about 44.

(c) Answers will vary.

53. $R(T) = -0.0011T^3 + 0.1005T^2 - 2.6876T + 98.171$

To find when the relative humidity is minimized, use a graphing calculator to locate any extreme points on the graph. A suitable window is [15, 46] by [70, 85].

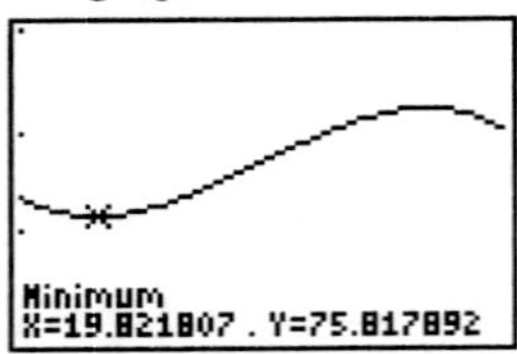

T	$R(T)$
15	76.76
19.82	75.82
46	80.13

The minimum relative humidity occurs at 19.82° C.

54. $W(t) = 0.109 + 4.47e^{-e^{(-0.01708(t-107.4))}}$

(a) $dW = 4.47e^{-e^{(-0.01708(t-107.4))}}(-1)e^{(-0.01708(t-107.4))}(-0.01708)\,dt$

$= 0.0763476e^{-e^{(-0.01708(t-107.4))}}e^{(-0.01708(t-107.4))}dt$

$t = 70$ and $dt = 2$, so $dW = 0.0763476e^{-e^{(-0.01708(70-107.4))}}e^{(-0.01708(70-107.4))}(2) \approx 0.0435$

The pig will gain about 0.04 kg.

(b) The actual weight gain is $W(72) - W(70) \approx 0.8256 - 0.7815 = 0.0441$ or about 0.04 kg.

55. Let x = the distance from the base of the ladder to the building and let y = the height on the building at the top of the ladder.

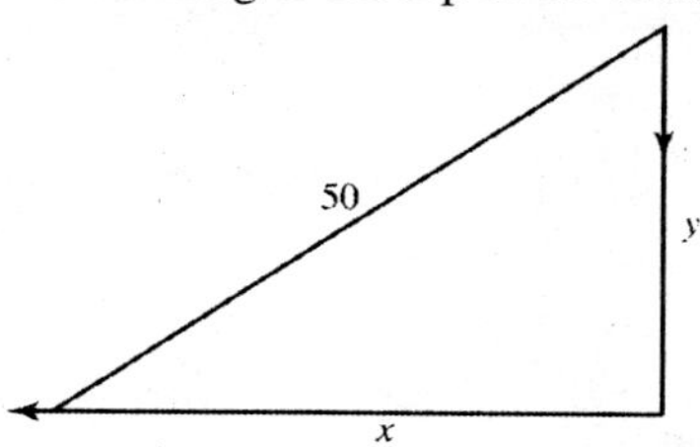

$$\frac{dy}{dt} = -2$$
$$50^2 = x^2 + y^2$$
$$0 = 2x\frac{dx}{dt} + 2y\frac{dy}{dt}$$
$$\frac{dx}{dt} = -\frac{y}{x}\frac{dy}{dt}$$

When $x = 30$, $y = \sqrt{2500 - (30)^2} = 40$.

So, $\frac{dx}{dt} = \frac{-40}{30}(-2) = \frac{80}{30} = \frac{8}{3}$

The base of the ladder is slipping away from the building at a rate of $\frac{8}{3}$ ft/min.

56. $\frac{dV}{dt} = 0.9 \text{ ft}^3/\text{min}$

Find $\frac{dr}{dt}$ when $r = 1.7$ ft

$$V = \frac{4}{3}\pi r^3$$
$$\frac{dV}{dt} = \frac{4}{3}\pi(3)r^2\frac{dr}{dt} = 4\pi r^2\frac{dr}{dt}$$
$$0.9 = 4\pi(1.7^2)\frac{dr}{dt}$$
$$\frac{dr}{dt} = \frac{0.9}{4\pi(1.7^2)} = \frac{0.9}{11.56\pi} \approx 0.0248$$

The radius is changing at the rate of $\frac{0.9}{11.56\pi} \approx 0.0248$ ft/min.

57. Let x = one-half of the width of the triangular cross section, h = the height of the water, and V = the volume of the water.

$\frac{dV}{dt} = 3.5 \text{ ft}^3/\text{min}$.

Find $\frac{dV}{dt}$ when $h = \frac{1}{3}$.

$$V = A_{\text{triangular side}} \times \text{length}$$
$$= \frac{1}{2}(\text{base})(\text{altitude}) = \frac{1}{2}(2x)(h) = xh$$

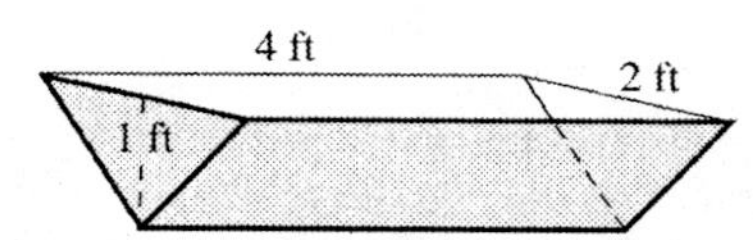

Area of triangular cross section:

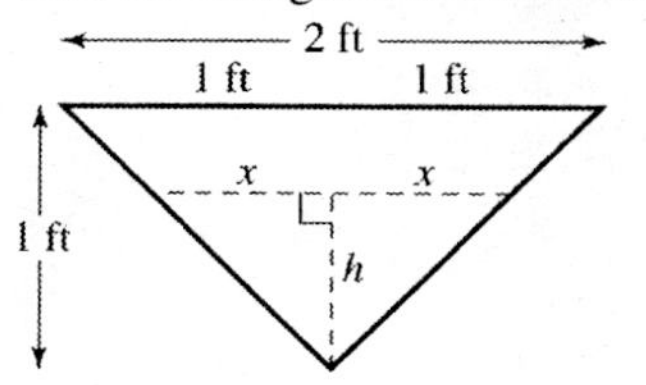

By similar triangles, $\frac{2x}{h} = \frac{2}{1}$, so $x = h$.

$$V = (xh)(4) = h^2 \cdot 4 = 4h^2$$
$$\frac{dV}{dt} = 8h\frac{dh}{dt}$$
$$\frac{1}{8h}\cdot\frac{dV}{dt} = \frac{dh}{dt}$$
$$\frac{1}{8\left(\frac{1}{3}\right)}(3.5) = \frac{dh}{dt}$$
$$\frac{dh}{dt} = \frac{21}{16} = 1.3125$$

The depth of water is changing at the rate of 1.3125 ft/min.

58. $V = \frac{4}{3}\pi r^3$, $r = 4$ in., $\Delta r = 0.02$ in.

$$dV = 4\pi r^2 dr$$
$$\Delta V \approx 4\pi r^2 \Delta r$$
$$= 4\pi(4)^2(0.02) = 1.28\pi \text{ or about } 4.021$$

The volume of the coating is 1.28π in.^{3} or about 4.021 in.^{3}.

59. $A = s^2$; $s = 9.2$, $\Delta s = \pm 0.04$

$$ds = 2s\,ds$$
$$\Delta A \approx 2s\Delta s$$
$$= 2(9.2)(\pm 0.04) = \pm 0.736 \text{ in.}^2$$

60.
$$V = l \cdot w \cdot h$$
$$w = 4 + h$$
$$l + g = 130;\ g = 2(w + h)$$
$$l + 2(w + h) = 130$$
$$l + 2w + 2h = 130$$
$$l = 130 - 2w - 2h = 130 - 2(4 + h) - 2h$$
$$= 122 - 4h$$
$$V = l \cdot w \cdot h$$
$$= (122 - 4h)(4 + h)h$$
$$= 488h + 106h^2 - 4h^3$$
$$\frac{dV}{dh} = 488 + 212h - 12h^2$$

(*continued on next page*)

(continued)

Set $\frac{dV}{dh} = 0.$

$488 + 212h - 12h^2 = 0$

$12h^2 - 212h - 488 = 0$

$3h^2 - 53h - 122 = 0$

$$h = \frac{53 \pm \sqrt{2809 + 1464}}{6}$$

≈ -2.06 or $h = 19.73$

h can't be negative, so $h \approx 19.73$.
Thus, $l \approx 122 - 4h \approx 43.1$.
The length that produces the maximum volume is about 43.1 inches.

61. We need to minimize y. Note that $x > 0$.

$\frac{dy}{dx} = \frac{x}{8} - \frac{2}{x}$

Set the derivative equal to 0 and solve for x.

$\frac{x}{8} - \frac{2}{x} = 0 \Rightarrow \frac{x}{8} = \frac{2}{x} \Rightarrow x^2 = 16 \Rightarrow x = 4$

Since $\lim_{x \to 0} y = \infty$, $\lim_{x \to \infty} y = \infty$, and $x = 4$ is the only critical value in $(0, \infty)$, $x = 4$ produces a minimum value.

$y = \frac{4^2}{16} - 2\ln 4 + \frac{1}{4} + 2\ln 6$
$= 1.25 + 2(\ln 6 - \ln 4) = 1.25 + 2\ln 1.5$

The y coordinate of the southern most point of the second boat's path is $1.25 + 2 \ln 1.5$.

62. Let x = width of play area;
y = length of play area.

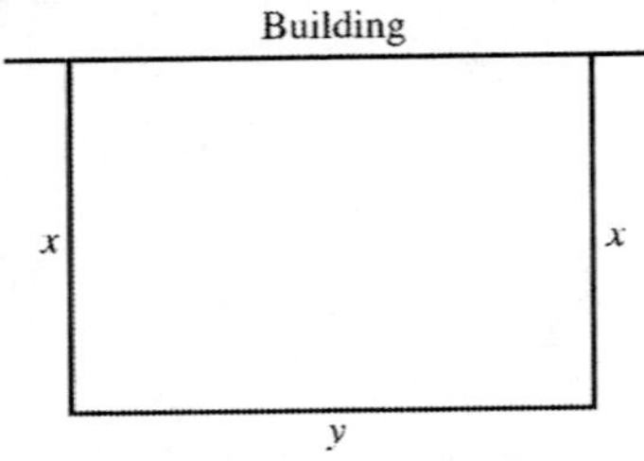

An equation describing the amount of fencing is $900 = 2x + y$ or $y = 900 - 2x$.

Then $A = xy$ so,

$A(x) = x(900 - 2x) = 900x - 2x^2.$

If $A'(x) = 900 - 4x = 0$, $x = 225$.

Then $y = 900 - 2(225) = 450$.

$A''(x) = -4 < 0$, so the area is maximized if the dimensions are 225 m by 450 m.

63. Distance on shore: $40 - x$ feet
Speed on shore: 5 feet per second
Distance in water: $\sqrt{x^2 + 40^2}$ feet
Speed in water: 3 feet for second
The total travel time t is $t = t_1 + t_2 = \frac{d_1}{v_1} + \frac{d_2}{v_2}.$

$$t(x) = \frac{40 - x}{5} + \frac{\sqrt{x^2 + 40^2}}{3}$$
$$= 8 - \frac{x}{5} + \frac{\sqrt{x^2 + 1600}}{3}$$
$$t'(x) = -\frac{1}{5} + \frac{1}{3} \cdot \frac{1}{2}(x^2 + 1600)^{-1/2}(2x)$$
$$= -\frac{1}{5} + \frac{x}{3\sqrt{x^2 + 1600}}$$

Minimize the travel time $t(x)$. If $t'(x) = 0$:

$$\frac{x}{3\sqrt{x^2 + 1600}} = \frac{1}{5}$$
$$5x = 3\sqrt{x^2 + 1600}$$
$$\frac{5x}{3} = \sqrt{x^2 + 1600}$$
$$\frac{25}{9}x^2 = x^2 + 1600$$
$$x^2 = \frac{1600 \cdot 9}{16} \Rightarrow x = \frac{40 \cdot 3}{4} = 30$$

(Discard the negative solution.)
To minimize the time, he should walk $40 - x = 40 - 30 = 10$ ft along the shore before paddling toward the desired destination. The minimum travel time is

$$\frac{40 - 30}{5} + \frac{\sqrt{30^2 + 40^2}}{3} \approx 18.67 \text{ seconds.}$$

64. Distance on shore: $25 - x$ feet
Speed on shore: 5 feet per second
Distance in water: $\sqrt{x^2 + 40^2}$ feet
Speed in water: 3 feet per second
The total travel time t is $t = t_1 + t_2 = \frac{d_1}{v_1} + \frac{d_2}{v_2}.$

$$t(x) = \frac{25 - x}{5} + \frac{\sqrt{x^2 + 40^2}}{3}$$
$$= 5 - \frac{x}{5} + \frac{\sqrt{x^2 + 1600}}{3}$$
$$t'(x) = -\frac{1}{5} + \frac{1}{3} \cdot \frac{1}{2}(x^2 + 1600)^{-1/2}(2x)$$
$$= -\frac{1}{5} + \frac{x}{3\sqrt{x^2 + 1600}}$$

(continued on next page)

(continued)

Minimize the travel time $t(x)$. If $t'(x)=0$:

$$\frac{x}{3\sqrt{x^2+1600}}=\frac{1}{5}$$
$$5x=3\sqrt{x^2+1600}$$
$$\frac{5x}{3}=\sqrt{x^2+1600}$$
$$\frac{25}{9}x^2=x^2+1600$$
$$\frac{16}{9}x^2=1600$$
$$x^2=\frac{1600\cdot 9}{16}$$
$$x=\frac{40\cdot 3}{4}=30$$

(Discard the negative solution.)
$x = 30$ is impossible since the closest point on the shore to the desired destination is only 25 ft from where he is standing.
Check the end points.

x	$t(x)$
0	18.33
25	15.72

The time is minimized when $x = 25$.
$25-x=25-25=0$ ft, so the mathematician should start paddling where he is standing.
The travel time is $t(25)=15.72$ sec.

65. Let x = the length and width of a side of the base and let h = the height.

The volume is 32 m^3; the base is square and there is no top. Find the height, length, and width for minimum surface area.

Volume $= x^2h$

$$x^2h=32\Rightarrow h=\frac{32}{x^2}$$

Surface area $= x^2+4xh$

$$A=x^2+4x\left(\frac{32}{x^2}\right)=x^2+128x^{-1}$$
$$A'=2x-128x^{-2}$$
$$A'=0\Rightarrow \frac{2x^3-128}{x^2}=0\Rightarrow x^3=64\Rightarrow x=4$$
$$A''(x)=2+2(128)x^{-3}$$
$$A''(4)=6>0$$

The minimum is at $x = 4$, where $h=\frac{32}{4^2}=2.$

The dimensions are 2 m by 4 m by 4 m.

66. Volume of cylinder $=\pi r^2h$
Surface area of cylinder open at one end $=2\pi rh+\pi r^2$.

$$V=\pi r^2h=27\pi\Rightarrow h=\frac{27\pi}{\pi r^2}=\frac{27}{r^2}$$
$$A=2\pi r\left(\frac{27}{r^2}\right)+\pi r^2=54\pi r^{-1}+\pi r^2$$
$$A'=-54\pi r^{-2}+2\pi r$$
$$A'=0,\Rightarrow 2\pi r=\frac{54\pi}{r^2}\Rightarrow r^3=27\Rightarrow r=3$$

If $r=3$, $A''=108\pi r^{-3}+2\pi>0$, so the value at $r = 3$ is a minimum.
For the minimum cost, the radius of the bottom should be 3 inches.

67. $V=\pi r^2h=40$, so $h=\frac{40}{\pi r^2}$.

$$A=2\pi r^2+2\pi rh=2\pi r^2+2\pi r\left(\frac{40}{\pi r^2}\right)$$
$$=2\pi r^2+\frac{80}{r}$$
$$\text{Cost}=C(r)=4(2\pi r^2)+3\left(\frac{80}{r}\right)$$
$$=8\pi r^2+\frac{240}{r}$$
$$C'(r)=16\pi r-\frac{240}{r^2}$$
$$16\pi r-\frac{240}{r^2}=0$$
$$16\pi r^3=240$$
$$r^3=\frac{15}{\pi}\Rightarrow r\approx 1.684$$
$$C''(r)=16\pi+\frac{480}{r^3}>0,\text{ so } r=1.684$$

minimizes cost.

$$h=\frac{40}{\pi r^2}=\frac{40}{\pi(1.684)^2}=4.490$$

The radius should be 1.684 in. and the height should be 4.490 in.

68. Refer to the figure below.

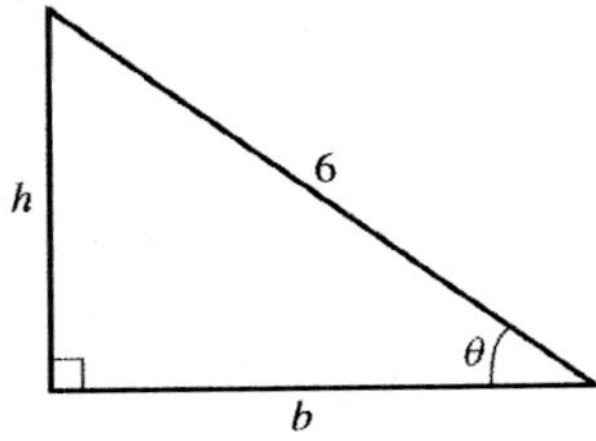

Let A be the area of the triangle.

$A = \frac{1}{2}bh$

In this triangle,

$\sin\theta = \frac{h}{6}$ and $\cos\theta = \frac{b}{6}$

$h = 6\sin\theta$ and $b = 6\cos\theta.$

Thus,

$$A = \frac{1}{2}(6\cos\theta)(6\sin\theta) = 18\sin\theta\cos\theta$$
$$= 9(2\sin\theta\cos\theta)$$
$$A = 9\sin(2\theta)$$

$$\frac{dA}{d\theta} = 9\cos(2\theta)\cdot 2 = 18\cos(2\theta)$$

$$\frac{dA}{d\theta} = 0 \Rightarrow 18\cos(2\theta) = 0 \Rightarrow \cos 2\theta = 0 \Rightarrow$$

$$2\theta = \frac{\pi}{2} \Rightarrow \theta = \frac{\pi}{4}$$

If $\theta < \frac{\pi}{4}, \frac{dA}{d\theta} > 0.$ If $\theta > \frac{\pi}{4}, \frac{dA}{d\theta} < 0.$

Thus, A is maximum when $\theta = \frac{\pi}{4}$ or 45°.

Chapter 7

INTEGRATION

7.1 Antiderivatives

1. If $F(x)$ and $G(x)$ are both antiderivatives of $f(x)$, then there is a constant C such that $F(x)-G(x)=C$. The two functions can differ only by a constant.

3. Answers will vary.

5. $$\int 6\,dk = 6\int 1\,dk = 6\int k^0 dk = 6\cdot\frac{1}{1}k^{0+1}+C = 6k+C$$

7. $$\int (2z+3)\,dz = 2\int z\,dz + 3\int z^0 dz = 2\cdot\frac{1}{1+1}z^{1+1}+3\cdot\frac{1}{0+1}z^{0+1}+C = z^2+3z+C$$

9. $$\int (6t^2-8t+7)\,dt = 6\int t^2 dt - 8\int t\,dt + 7\int t^0 dt = \frac{6t^3}{3}-\frac{8t^2}{2}+7t+C = 2t^3-4t^2+7t+C$$

11. $$\int (4z^3+3z^2+2z-6)\,dz = 4\int z^3 dz + 3\int z^2 dz + 2\int z\,dz - 6\int z^0\,dz = \frac{4z^4}{4}+\frac{3z^3}{3}+\frac{2z^2}{2}-6z+C = z^4+z^3+z^2-6z+C$$

13. $$\int (5\sqrt{z}+\sqrt{2})\,dz = 5\int z^{1/2}dz + \sqrt{2}\int dz = \frac{5z^{3/2}}{\frac{3}{2}}+\sqrt{2}z+C = 5\left(\frac{2}{3}\right)z^{3/2}+\sqrt{2}z+C = \frac{10z^{3/2}}{3}+\sqrt{2}z+C$$

15. $$\int 5x(x^2-8)\,dx = \int (5x^3-40x)\,dx = \frac{5x^4}{4}-\frac{40x^2}{2}+C = \frac{5x^4}{4}-20x^2+C$$

17. $$\int (4\sqrt{v}-3v^{3/2})\,dv = 4\int v^{1/2}dv - 3\int v^{3/2}dv = \frac{4v^{3/2}}{\frac{3}{2}}-\frac{3v^{5/2}}{\frac{5}{2}}+C = \frac{8v^{3/2}}{3}-\frac{6v^{5/2}}{5}+C$$

19. $$\int (10u^{3/2}-14u^{5/2})\,du = 10\int u^{3/2}du - 14\int u^{5/2}\,du = \frac{10u^{5/2}}{\frac{5}{2}}-\frac{14u^{7/2}}{\frac{7}{2}}+C = 10\left(\frac{2}{5}\right)u^{5/2}-14\left(\frac{2}{7}\right)u^{7/2}+C = 4u^{5/2}-4u^{7/2}+C$$

21. $$\int\left(\frac{7}{z^2}\right)dz = \int 7z^{-2}dz = 7\int z^{-2}dz = 7\left(\frac{z^{-2+1}}{-2+1}\right)+C = \frac{7z^{-1}}{-1}+C = -\frac{7}{z}+C$$

23. $$\int\left(\frac{\pi^3}{y^3}-\frac{\sqrt{\pi}}{\sqrt{y}}\right)dy = \int \pi^3 y^{-3}dy - \int \sqrt{\pi}y^{-1/2}dy = \pi^3\int y^{-3}dy - \sqrt{\pi}\int y^{-1/2}dy = \pi^3\left(\frac{y^{-2}}{-2}\right)-\sqrt{\pi}\left(\frac{y^{1/2}}{\frac{1}{2}}\right)+C = -\frac{\pi^3}{2y^2}-2\sqrt{\pi y}+C$$

25. $$\int (-9t^{-2.5}-2t^{-1})\,dt = -9\int t^{-2.5}dt - 2\int t^{-1}dt = \frac{-9t^{-1.5}}{-1.5}-2\int\frac{dt}{t} = 6t^{-1.5}-2\ln|t|+C$$

27. $$\int\frac{1}{3x^2}dx = \int\frac{1}{3}x^{-2}dx = \frac{1}{3}\int x^{-2}dx = \frac{1}{3}\left(\frac{x^{-1}}{-1}\right)+C = -\frac{1}{3}x^{-1}+C = -\frac{1}{3x}+C$$

29. $\int 3e^{-0.2x}dx = 3\int e^{-0.2x}dx$
$= 3\left(\frac{1}{-0.2}\right)e^{-0.2x} + C$
$= \frac{3(e^{-0.2x})}{-0.2} + C$
$= -15e^{-0.2x} + C$

31. $\int\left(-\frac{3}{x} + 4e^{-0.4x} + e^{0.1}\right)dx$
$= -3\int\frac{dx}{x} + 4\int e^{-0.4x}dx + e^{0.1}\int dx$
$= -3\ln|x| + \frac{4e^{-0.4x}}{-0.4} + e^{0.1}x + C$
$= -3\ln|x| - 10e^{-0.4x} + e^{0.1}x + C$

33. $\int\left(\frac{1+2t^3}{4t}\right)dt = \int\left(\frac{1}{4t} + \frac{t^2}{2}\right)dt$
$= \frac{1}{4}\int\frac{1}{t}\,dt + \frac{1}{2}\int t^2 dt$
$= \frac{1}{4}\ln|t| + \frac{1}{2}\left(\frac{t^3}{3}\right) + C$
$= \frac{1}{4}\ln|t| + \frac{t^3}{6} + C$

35. $\int(e^{2u} + 4u)\,du = \frac{e^{2u}}{2} + \frac{4u^2}{2} + C$
$= \frac{e^{2u}}{2} + 2u^2 + C$

37. $\int(x+1)^2dx = \int(x^2 + 2x + 1)\,dx$
$= \frac{x^3}{3} + \frac{2x^2}{2} + x + C$
$= \frac{x^3}{3} + x^2 + x + C$

39. $\int\frac{\sqrt{x}+1}{\sqrt[3]{x}}\,dx = \int\left(\frac{\sqrt{x}}{\sqrt[3]{x}} + \frac{1}{\sqrt[3]{x}}\right)dx$
$= \int\left(x^{(1/2-1/3)} + x^{-1/3}\right)dx$
$= \int x^{1/6}dx + \int x^{-1/3}dx$
$= \frac{x^{7/6}}{\frac{7}{6}} + \frac{x^{2/3}}{\frac{2}{3}} + C$
$= \frac{6x^{7/6}}{7} + \frac{3x^{2/3}}{2} + C$

41. $\int 10^x dx = \frac{10^x}{\ln 10} + C$

43. $\int(3\cos x - 4\sin x)dx$
$= \int 3\cos x\,dx - \int 4\sin x\,dx$
$= 3\int\cos x\,dx - 4\int\sin x\,dx$
$= 3\sin x + 4\cos x + C$

45. Find $f(x)$ such that $f'(x) = x^{2/3}$, and $\left(1, \frac{3}{5}\right)$ is on the curve.

$\int x^{2/3}dx = \frac{x^{5/3}}{\frac{5}{3}} + C \Rightarrow f(x) = \frac{3x^{5/3}}{5} + C$

Since $\left(1, \frac{3}{5}\right)$ is on the curve,

$f(1) = \frac{3}{5}.$

$f(1) = \frac{3(1)^{5/3}}{5} + C = \frac{3}{5} \Rightarrow \frac{3}{5} + C = \frac{3}{5} \Rightarrow$
$C = 0.$

Thus, $f(x) = \frac{3x^{5/3}}{5}.$

47. (a) $f'(t) = 0.01e^{-0.01t}$
$f(t) = \int 0.01e^{-0.01t}dt$
$= -\frac{0.01e^{-0.01t}}{0.01} + C = -e^{-0.01t} + C$

(b) $f(0) = -e^{-0.01(0)} + C = -e^0 + C = -1 + C$
$f(0) = 0 \Rightarrow 0 = -1 + C \Rightarrow C = 1$

Thus, $f(t) = -e^{-0.01t} + 1.$

$f(10) = -e^{-0.01(10)} + 1 = -e^{-0.1} + 1$
$= -0.905 + 1 = 0.095$

0.095 unit is excreted in 10 min.

49. (a) $c(t) = (c_0 - C)e^{-kAt/V} + C$
$c'(t) = (c_0 - C)\left(\frac{-kA}{V}\right)e^{-kAt/V}$
$= \frac{-kA}{V}(c_0 - C)e^{-kAt/V}$

(b) According to (1) and (2),
$c'(t) = \frac{kA}{V}\left[C - (c_0 - C)e^{-kAt/V} - C\right]$
$= \frac{kA}{V}(C - c_0)e^{-kAt/V}.$

This is also what we get for $c'(t)$ by differentiating Equation (2).

51. $V'(t) = -kP(t);\quad P(t) = P_0e^{-mt}$

$$V'(t) = -kP_0e^{-mt} \Rightarrow V(t) = \frac{k}{m}P_0e^{-mt} + C$$

$$V(0) = \frac{k}{m}P_0e^0 + C \Rightarrow V_0 - \frac{k}{m}P_0 = C$$

Therefore,

$$V(t) = \frac{k}{m}P_0e^{-mt} + V_0 - \frac{k}{m}P_0 = \frac{kP_0}{m}e^{-mt} + V_0 - \frac{kP_0}{m}.$$

53. $D'(t) = 29.25e^{0.03572t}$

(a) $D(t) = \int 29.25e^{0.03572t}\,dt$

$$= \frac{29.25}{0.03572}e^{0.03572t} + C$$
$$= 818.9e^{0.03572t} + C$$

In 1980, when $t = 0$, $D = 700$, so

$$818.9e^{0.03572(0)} + C = 700$$
$$818.9 + C = 700 \Rightarrow C = -118.9$$

Thus, $D(t) = 818.9e^{0.03572t} - 118.9.$

(b) To project the number of dentistry degrees conferred in 2015 we set t equal to 35 and evaluate $D(35)$.

$$D(35) = 818.9e^{0.03572(35)} - 118.9 \approx 2740$$

The formula predicts that 2740 dentistry degrees will be conferred in 2015.

55. $a(t) = 5t^2 + 4$

$$v(t) = \int (5t^2 + 4)\,dt = \frac{5t^3}{3} + 4t + C$$
$$v(0) = \frac{5(0)^3}{3} + 4(0) + C$$
$$v(0) = 6 \Rightarrow C = 6$$

Thus, $v(t) = \frac{5t^3}{3} + 4t + 6.$

57. $a(t) = -32$

$$v(t) = \int -32\,dt = -32t + C_1$$
$$v(0) = -32(0) + C_1$$

Since $v(0) = 0$, $C_1 = 0$.

$$v(t) = -32t$$
$$s(t) = \int -32t\,dt = \frac{-32t^2}{2} + C_2 = -16t^2 + C_2$$

At $t = 0$, the plane is at 6400 ft. That is, $s(0) = 6400$.

$$s(0) = -16(0)^2 + C_2 \Rightarrow 6400 = 0 + C_2 \Rightarrow$$
$$C_2 = 6400 \Rightarrow s(t) = -16t^2 + 6400$$

When the object hits the ground, $s(t) = 0$.

$$-16t^2 + 6400 = 0 \Rightarrow -16t^2 = -6400 \Rightarrow$$
$$t^2 = 400 \Rightarrow t = \pm 20$$

Discard –20 since time must be positive. The object hits the ground in 20 sec.

59. $a(t) = \frac{15}{2}\sqrt{t} + 3e^{-t}$

$$v(t) = \int\left(\frac{15}{2}\sqrt{t} + 3e^{-t}\right)dt$$
$$= \int\left(\frac{15}{2}t^{1/2} + 3e^{-t}\right)dt$$
$$= \frac{15}{2}\left(\frac{t^{3/2}}{\frac{3}{2}}\right) + 3\left(\frac{1}{-1}e^{-t}\right) + C_1$$
$$= 5t^{3/2} - 3e^{-t} + C_1$$
$$v(0) = 5(0)^{3/2} - 3e^{-0} + C_1 = -3 + C_1$$

Since $v(0) = -3$, $C_1 = 0$.

$$v(t) = 5t^{3/2} - 3e^{-t}$$
$$s(t) = \int (5t^{3/2} - 3e^{-t})\,dt$$
$$= 5\left(\frac{t^{5/2}}{\frac{5}{2}}\right) - 3\left(-\frac{1}{1}e^{-t}\right) + C_2$$
$$= 2t^{5/2} + 3e^{-t} + C_2$$
$$s(0) = 2(0)^{5/2} + 3e^{-0} + C_2 = 3 + C_2$$

Since $s(0) = 4$, $C_2 = 1$.

Thus, $s(t) = 2t^{5/2} + 3e^{-t} + 1.$

61. First find $v(t)$ by integrating $a(t)$:

$$v(t) = \int(-32)dt = -32t + k.$$

When $t = 5, v(t) = 0$:

$$0 = -32(5) + k \Rightarrow 160 = k \text{ and}$$
$$v(t) = -32t + 160.$$

Now integrate $v(t)$ to find $h(t)$.

$$h(t) = \int(-32t + 160)dt = -16t^2 + 160t + C$$

Since $h(t) = 412$ when $t = 5$, we can substitute these values into the equation for $h(t)$ to get $C = 12$ and

$$h(t) = -16t^2 + 160t + 12.$$

Therefore, from the equation given in Exercise 60, the initial velocity v_0 is 160 ft/sec and the initial height of the rocket h_0 is 12 ft.

7.2 Substitution

1. Answers will vary.

3. $\int 4(2x+3)^4\,dx$

Let $u = 2x+3$, so that $du = 2\,dx$.

$$\int 4(2x+3)^4\,dx = \frac{4}{2}\int (2x+3)^4(2)\,dx$$

$$2\int (2x+3)^4(2)\,dx = 2\int u^4 du = \frac{2\cdot u^5}{5} + C$$

$$= \frac{2(2x+3)^5}{5} + C$$

5. $\int \frac{2\,dm}{(2m+1)^3} = \int (2m+1)^{-3}\,(2)\,dm$

Let $u = 2m+1$, so that $du = 2\,dm$.

$$\int (2m+1)^{-3}\,(2)\,dm = \int u^{-3} du = \frac{u^{-2}}{-2} + C$$

$$= \frac{-(2m+1)^{-2}}{2} + C$$

7. $\int \frac{2x+2}{(x^2+2x-4)^4}dx$

$$= \int (x^2+2x-4)^{-4}(2x+2)dx$$

Let $w = x^2+2x-4$, so that $dw = (2x+2)\,dx$.

$$\int (x^2+2x-4)^{-4}(2x+2)dx$$

$$= \int w^{-4} dw = \frac{w^{-3}}{-3} + C$$

$$= -\frac{(x^2+2x-4)^{-3}}{3} + C$$

$$= -\frac{1}{3(x^2+2x-4)^3} + C$$

9. $\int z\sqrt{4z^2-5}\,dz = \int z(4z^2-5)^{1/2}dz$

Let $u = 4z^2-5$, so that $du = 8z\,dz$.

$$\frac{1}{8}\int (4z^2-5)^{1/2}(8z)\,dz = \frac{1}{8}\int u^{1/2} du$$

$$= \frac{1}{8}\cdot\frac{u^{3/2}}{\frac{3}{2}} + C$$

$$= \frac{1}{8}\cdot\left(\frac{2}{3}\right)u^{3/2} + C$$

$$= \frac{(4z^2-5)^{3/2}}{12} + C$$

11. $\int 3x^2 e^{2x^3}\,dx = \frac{1}{2}\int 2\cdot 3x^2 e^{2x^3}\,dx$

Let $u = 2x^3$, so that $du = 6x^2 dx$

$$\frac{1}{2}\int e^{2x^3}\left(6x^2\right)dx = \frac{1}{2}\int e^u du$$

$$= \frac{1}{2}e^u + C = \frac{e^{2x^3}}{2} + C$$

13. $\int (1-t)e^{2t-t^2}\,dt$

Let $u = 2t - t^2$, so that $du = (2-2t)dt$.

$$\int (1-t)e^{2t-t^2}\,dt = \frac{1}{2}\int e^{2t-t^2}\,(2(1-t))\,dt$$

$$= \frac{1}{2}\int e^u du$$

$$= \frac{e^u}{2} + C = \frac{e^{2t-t^2}}{2} + C$$

15. $\int \frac{e^{1/z}}{z^2}dz$

Let $u = \frac{1}{z}$, so that $du = \frac{-1}{z^2}dx$.

$$\int \frac{e^{1/z}}{z^2}dz = -\int e^{1/z}\cdot\frac{-1}{z^2}\,dz = -\int e^u du$$

$$= -e^u + C = -e^{1/z} + C$$

17. $\int \frac{t}{t^2+2}dt$

Let $u = t^2+2$, so that $du = 2t\,dt$.

$$\int \frac{t}{t^2+2}dt = \frac{1}{2}\int \frac{du}{u} = \frac{1}{2}\ln|u| + C$$

$$= \frac{\ln(t^2+2)}{2} + C$$

19. $\int \frac{x^3+2x}{x^4+4x^2+7}dx$

Let $u = x^4+4x^2+7$.
Then $du = (4x^3+8x)dx = 4(x^3+2x)dx$.

$$\int \frac{x^3+2x}{x^4+4x^2+7}dx = \frac{1}{4}\int \frac{(4x^3+8x)}{x^4+4x^2+7}dx$$

$$= \frac{1}{4}\int \frac{1}{u}du = \frac{1}{4}\ln|u| + C$$

$$= \frac{1}{4}\ln(x^4+4x^2+7) + C$$

Since $x^4+4x^2+7 > 0$ for all x, we can write this answer as $\frac{1}{4}\ln(x^4+4x^2+7)+C$.

(*continued*)

$$A=\sum_{i=1}^{4} f(x_i)\Delta x=\sum_{i=1}^{4} f(x_i)(1)=\sum_{i=1}^{4} f(x_i)$$
$$=(e^{-2}+1)(1)+(e^{-1}+1)(1)+2(1)+\left(e^{1}+1\right)(1)$$
$$\approx 8.2215\approx 8.22$$

(b) Using the right endpoints:

i	x_i	$f(x_i)$
1	-1	$e^{-1}+1$
2	0	2
3	1	$e+1$
4	2	e^2+1

$$\text{Area}=(e^{-1}+1)(1)+(2)(1)+(e+1)(1)+(e^2+1)(1)$$
$$\approx 15.4752\approx 15.48$$

(c) $\text{Average}=\dfrac{8.2215+15.4752}{2}\approx 11.85$

(d) Using the midpoints:

i	x_i	$f(x_i)$
1	$-\frac{3}{2}$	$e^{-3/2}+1$
2	$-\frac{1}{2}$	$e^{-1/2}+1$
3	$\frac{1}{2}$	$e^{1/2}+1$
4	$\frac{3}{2}$	$e^{3/2}+1$

$$A=\sum_{i=1}^{4} f(x_i)\Delta x$$
$$=(e^{-3/2}+1)(1)+(e^{-1/2}+1)(1)+(e^{1/2}+1)(1)+(e^{3/2}+1)(1)$$
$$\approx 10.9601\approx 10.96$$

11. $f(x)=\dfrac{2}{x}$ from $x=1$ to $x=9$

For $n=4$ rectangles: $\Delta x=\dfrac{9-1}{4}=2$

(a) Using the left endpoints:

i	x_i	$f(x_i)$
1	1	$\frac{2}{1}=2$
2	3	$\frac{2}{3}$
3	5	$\frac{2}{5}=0.4$
4	7	$\frac{2}{7}$

$$A=\sum_{i=1}^{4} f(x_i)\Delta x$$
$$=(2)(2)+\frac{2}{3}(2)+(0.4)(2)+\left(\frac{2}{7}\right)(2)$$
$$\approx 6.7048\approx 6.70$$

(b) Using the right endpoints:

i	x_i	$f(x_i)$
1	3	$\frac{2}{3}$
2	5	$\frac{2}{5}$
3	7	$\frac{2}{7}$
4	9	$\frac{2}{9}$

$$\text{Area}=\left(\frac{2}{3}\right)(2)+\left(\frac{2}{5}\right)(2)+\left(\frac{2}{7}\right)(2)+\left(\frac{2}{9}\right)(2)$$
$$=\frac{4}{3}+\frac{4}{5}+\frac{4}{7}+\frac{4}{9}\approx 3.1492\approx 3.15$$

(c) $\text{Average}=\dfrac{6.7+3.15}{2}=4.93$

(d) Using the midpoints:

i	x_i	$f(x_i)$
1	2	1
2	4	$\frac{1}{2}$
3	6	$\frac{1}{3}$
4	8	$\frac{1}{4}$

$$A = \sum_{i=1}^{4} f(x_i)\Delta x = 1(2) + \frac{1}{2}(2) + \frac{1}{3}(2) + \frac{1}{4}(2) \approx 4.17$$

13. $f(x) = \sin x$ from $x = 0$ to $x = \pi$

For $n = 4$ rectangles, $\Delta x = \frac{\pi - 0}{4} = \frac{\pi}{4}$

(a) Using the left endpoints:

i	x_i	$f(x_i)$
1	0	0
2	$\frac{\pi}{4}$	$\frac{\sqrt{2}}{2}$
3	$\frac{\pi}{2}$	1
4	$\frac{3\pi}{4}$	$\frac{\sqrt{2}}{2}$

$$A = \sum_{i=1}^{4} f(x_i)\Delta x = 0\left(\frac{\pi}{4}\right) + \frac{\sqrt{2}}{2}\left(\frac{\pi}{4}\right) + 1\left(\frac{\pi}{4}\right) + \frac{\sqrt{2}}{2}\left(\frac{\pi}{4}\right) = \frac{\pi\sqrt{2} + \pi}{4} \approx 1.896$$

(b) Using the right endpoints:

i	x_i	$f(x_i)$
1	$\frac{\pi}{4}$	$\frac{\sqrt{2}}{2}$
2	$\frac{\pi}{2}$	1
3	$\frac{3\pi}{4}$	$\frac{\sqrt{2}}{2}$
4	π	0

$$A = \sum_{i=1}^{4} f(x_i)\Delta x = \frac{\sqrt{2}}{2}\left(\frac{\pi}{4}\right) + 1\left(\frac{\pi}{4}\right) + \frac{\sqrt{2}}{2}\left(\frac{\pi}{4}\right) + 0\left(\frac{\pi}{4}\right) = \frac{\pi\sqrt{2} + \pi}{4} \approx 1.896$$

(c) $$\text{Average} = \frac{\frac{\pi\sqrt{2} + \pi}{4} + \frac{\pi\sqrt{2} + \pi}{4}}{2} \approx 1.896$$

(d) Using the midpoints:

i	x_i	$f(x_i)$
1	$\frac{\pi}{8}$	0.3827
2	$\frac{3\pi}{8}$	0.9239
3	$\frac{5\pi}{8}$	0.9239
4	$\frac{7\pi}{8}$	0.3827

$$A = \sum_{i=1}^{4} f(x_i)\Delta x = \left(\sin\frac{\pi}{8}\right)\left(\frac{\pi}{4}\right) + \left(\sin\frac{3\pi}{8}\right)\left(\frac{\pi}{4}\right) + \left(\sin\frac{5\pi}{8}\right)\left(\frac{\pi}{4}\right) + \left(\sin\frac{7\pi}{8}\right)\left(\frac{\pi}{4}\right) \approx 2.052$$

15. (a) Width $= \frac{4-0}{4} = 1$; $f(x) = \frac{x}{2}$

$$\text{Area} = 1 \cdot f\left(\frac{1}{2}\right) + 1 \cdot f\left(\frac{3}{2}\right) + 1 \cdot f\left(\frac{5}{2}\right) + 1 \cdot f\left(\frac{7}{2}\right) = \frac{1}{4} + \frac{3}{4} + \frac{5}{4} + \frac{7}{4} = \frac{16}{4} = 4$$

(b)

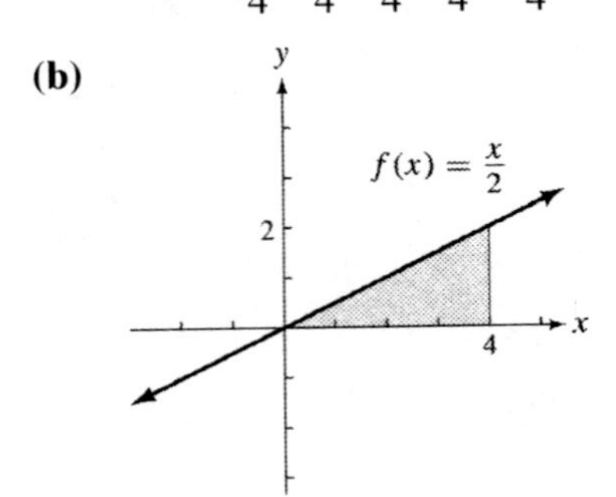

(continued on next page)

(continued)

$$\int_0^4 f(x)dx = \int_0^4 \frac{x}{2}dx = \frac{1}{2}(\text{base})(\text{height})$$
$$= \frac{1}{2}(4)(2) = 4$$

17. (a) Area of triangle is $\frac{1}{2}\cdot\text{base}\cdot\text{height}$.

The base is 4; the height is 2.

$$\int_0^4 f(x)\,dx = \frac{1}{2}\cdot 4\cdot 2 = 4$$

(b) The larger triangle has an area of $\frac{1}{2}\cdot 3\cdot 3 = \frac{9}{2}$. The smaller triangle has an area of $\frac{1}{2}\cdot 1\cdot 1 = \frac{1}{2}$. The sum is $\frac{9}{2}+\frac{1}{2}=5$.

19. $\int_{-4}^{0}\sqrt{16-x^2}\,dx$

Graph $y=\sqrt{16-x^2}$.

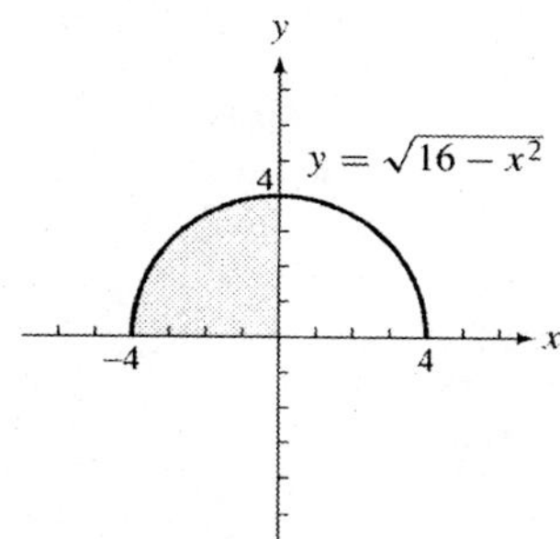

$\int_{-4}^{0}\sqrt{16-x^2}\,dx$ is the area of the portion of the circle in the second quadrant, which is one-fourth of a circle. The circle has radius 4.

$$\text{Area} = \frac{1}{4}\pi r^2 = \frac{1}{4}\pi(4)^2 = 4\pi$$

21. $\int_2^5 (1+2x)\,dx$

Graph $y=1+2x$.

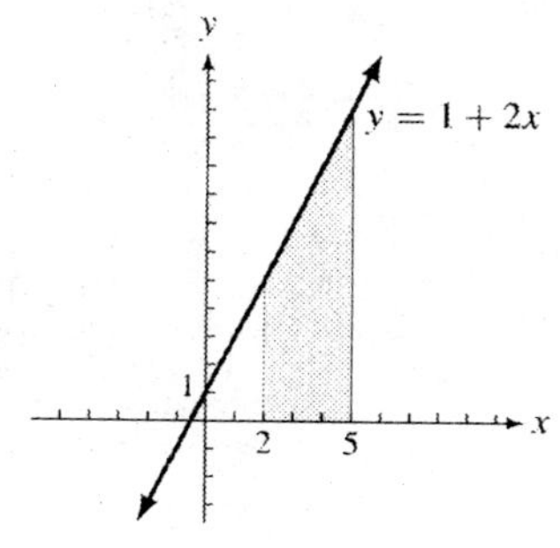

$\int_2^5 (1+2x)\,dx$ is the area of the trapezoid with $B=11, b=5$, and $h=3$. The formula for the area is $A=\frac{1}{2}(B+b)h$, so we have

$$A=\frac{1}{2}(11+5)(3)=24.$$

23. (a) With $n=10, \Delta x=\frac{1-0}{10}=0.1$, and $x_1=0+0.1=0.1$, use the command seq (X^2, X, 0.1, 1, 0.1) →L1. The resulting screen is:

```
seq(X²,X,.1,1,.1
)→L1
{.01 .04 .09 .1…
```

(b) Since $\sum_{i=1}^{n} f(x_i)\Delta x = \Delta x\left(\sum_{i=1}^{n} f(x_i)\right)$, use the command 0.1*sum (L1) to approximate $\int_0^1 x^2dx$. The resulting screen is:

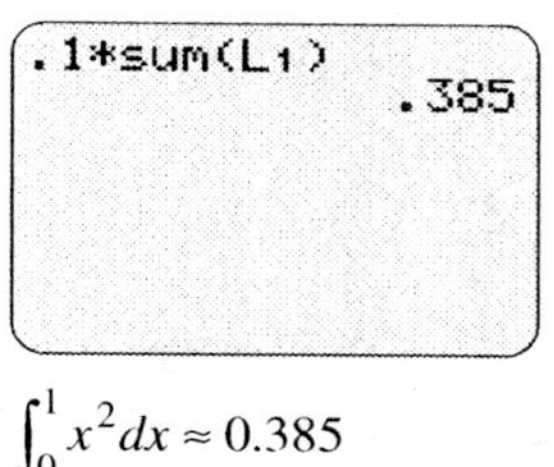

```
.1*sum(L1)
                .385
```

$$\int_0^1 x^2 dx \approx 0.385$$

(c) With $n=100, \Delta x=\frac{1-0}{100}=0.01$ and $x_1=0+0.01=0.01$, use the command seq (X^2, X, 0.01, 1, 0.01) → L1. The resulting screen is:

```
seq(X²,X,.01,1,.
01)→L1
{1E-4 4E-4 9E-4…
```

(continued on next page)

(continued)

Use the command 0.01*sum(L1) to approximate $\int_0^1 x^2 dx$. The resulting screen is:

```
.01*sum(L1)
          .33835
```

$\int_0^1 x^2 dx \approx 0.33835$

(d) With $n = 500, \Delta x = \frac{1-0}{500} = 0.002,$ and $x_1 = 0 + 0.002 = 0.002,$ use the command seq $(X^2, X, 0.002, 1, 0.002) \to L1$. The resulting screen is:

```
seq(X²,X,.002,1,
.002)→L1
{4E-6 1.6E-5 3...
```

Use the command 0.002*sum(L1) to approximate $\int_0^1 x^2 dx$. The resulting screen is:

```
.002*sum(L1)
          .334334
```

$\int_0^1 x^2 dx \approx 0.334334$

(e) As n gets larger the approximation for $\int_0^1 x^2 dx$ seems to be approaching 0.333333. We estimate $\int_0^1 x^2 dx = 0.333333.$

25. Answers will vary.

For Exercises 26–34, the readings on the graphs and answers may vary.

27. First read approximate data values from the graph. These readings are just estimates, and you may get different answers if your estimated readings differ from these. Month 1 represents mid-February.

Cows		Pigs	
Month	Cases	Month	Cases
1	3000	1	2000
2	165,000	2	62,000
3	267,000	3	68,000
4	54,000	4	3000
5	44,000	5	1000
6	21,000	6	9000
7	16,500	7	1000
8	11,500	8	0
9	1000	9	0

(a) Left endpoints: Add up the values corresponding to months 1 through 8 in the Cows table The total is 582,000 cases.
Right endpoints: Add up the values corresponding to months 2 through 9. The total is 580,000 cases.
The average of these two values is 581,000 cases.

(b) Left endpoints: Add up the values corresponding to months 1 through 8 in the Pigs table. The total is 146,000 cases.
Right endpoints: Add up the values corresponding to months 2 through 9. The total is 144,000 cases.
The average of these two values is 145,000 cases.

29. (a) Left endpoints: Read values of the function from the graph for every hour from 8:15 to 2:15 and for the half-hour beginning at 3:15. The values give the heights of 7 rectangles, each with width $\Delta x = 1,$ and one rectangle with width $\Delta x = \frac{1}{2}$. Estimate the area under the curve as

$$\begin{aligned} A &= \sum_{i=1}^{8} f(x_i)\Delta x \\ &= 0(1) + 16(1) + 23(1) + 29(1) + 30(1) \\ &\quad + 28(1) + 20(1) + 12\left(\frac{1}{2}\right) \\ &= 152 \end{aligned}$$

Right endpoints: Read values of the function from the graph for every hour from 9:15 to 3:15 and for the half-hour ending at 3:45.

(continued on next page)

(*continued*)

The values give the heights of 7 rectangles, each with width $\Delta x = 1$, and one rectangle with width $\Delta x = \frac{1}{2}$. Estimate the area under the curve as

$$A = \sum_{i=1}^{8} f(x_i)\Delta x = 16(1) + 23(1) + 29(1) + 30(1) + 28(1) + 20(1) + 12(1) + 0\left(\frac{1}{2}\right) = 158$$

Average: $\dfrac{152+158}{2} = 155$

The total is about 155 cal/cm^2.

(b) Left endpoints: Read values of the function from the graph for every hour from 7:30 to 3:30. The values give the heights of 9 rectangles, each with width $\Delta x = 1$. Estimate the area under the curve as

$$A = \sum_{i=1}^{9} f(x_i)\Delta x = 0(1) + 42(1) + 70(1) + 80(1) + 90(1) + 90(1) + 80(1) + 70(1) + 42(1) = 564$$

Right endpoints: Read values of the function from the graph for every hour from 9:30 to 4:30. The values give the heights of 7 rectangles, each with width $\Delta x = 1$. Estimate the area under the curve as

$$A = \sum_{i=1}^{9} f(x_i)\Delta x = 42(1) + 70(1) + 80(1) + 90(1) + 90(1) + 80(1) + 70(1) + 42(1) + 0(1) = 564$$

Since both estimates are the same, the total is about 564 cal/cm^2.

31. Left endpoints: Read values of the function on the graph every two years from 2002 to 2008. These values give us the heights of four rectangles. The width of each rectangle is $\Delta x = 2$. We estimate the area under the graph as follows: $A = \sum_{i=1}^{4} f(x_i)\Delta x = 3650(2) + 3701(2) + 3793(2) + 3113(2) = 28{,}514$

Right endpoints: Read values of the function on the graph every two years from 2004 to 2010. We estimate the area under the graph as follows. $A = \sum_{i=1}^{4} f(x_i)\Delta x = 3701(2) + 3793(2) + 3113(2) + 2520(2) = 26{,}254$

Average: $\dfrac{28{,}514 + 26{,}254}{2} = 27{,}384$ accidents

The area under the graph represents the number of fatal automobile accidents in California from 2002 to 2010. We estimate this number as 27,384 fatal accidents.

33. Read the value for the speed every 4 sec from $x = 0$ to $x = 12$. The midpoints of these rectangles are $x = 2$, $x = 6$, and $x = 10$. Then read the speed for $x = 13.75$, which is the midpoint of a rectangle with width $\Delta x = 3.5$.

$$\sum_{i=1}^{4} f(x_i)\Delta x \approx 40(4) + 75(4) + 102(4) + 120(3.5) = 1288$$

$$\frac{1288}{3600}(5280) \approx 1900$$

The Alfa Romeo traveled about 1900 ft. Yes, this seems correct.

35. Left endpoints: Read values of the function from the table for every number of seconds from 2.4 to 19.2. These values give the heights of 8 rectangles. The width of each rectangle varies. We estimate the area under the curve as

$$\sum_{i=1}^{8} f(x_i)\Delta x = 30(3.5-2.4)+40(5.1-3.5)+50(6.9-5.1)+60(8.9-6.9)+70(11.2-8.9)$$
$$+80(14.9-11.2)+90(19.2-14.9)+100(24.4-19.2)$$
$$=1671$$

$$\frac{5280}{3600}(1671)\approx 2451$$

Right endpoints: Read values of the function from the table for every number of seconds from 2.4 to 24.4. These values give the heights of 9 rectangles. The width of each rectangle varies. We estimate the area under the curve as

$$\sum_{i=1}^{9} f(x_i)\Delta x = 30(2.4-0)+40(3.5-2.4)+50(5.1-3.5)+60(6.9-5.1)+70(8.9-6.9)$$
$$+80(11.2-8.9)+90(14.9-11.2)+100(19.2-14.9)+110(24.4-19.2)$$
$$=1963$$

$$\frac{5280}{3600}(1963)\approx 2879$$

Average: $\dfrac{2451+2879}{2}=\dfrac{5330}{2}=2665$

The distance traveled by the Chevrolet Malibu Maxx SS is about 2665 ft.

The answers for exercises 36 and 37 may vary slightly depending on how the graphs are read.

37. (a) Read the value for a plain glass window facing south for every 2 hr from 6 to 6. These are the heights, at the midpoints, of rectangles with width $\Delta x = 2$.

$$\sum f(x_i)\Delta x \approx 6(2)+18(2)+41(2)+54(2)+41(2)+18(2)+8(2)\approx 372\approx 370$$

The heat gain is about 370 BTUs per square foot.

(b) Read the value for a window with triple-glazed glass facing south for every 2 hr from 6 to 6. These are the heights, at the midpoints, of rectangles with width $\Delta x = 2$.

$$\sum f(x_i)\Delta x \approx 3(2)+8(2)+19(2)+25(2)+19(2)+8(2)+3(2)\approx 170$$

The heat gain is about 170 BTUs per square foot.

39. Using the left endpoints: Distance $= v_0(1)+v_1(1)+v_2(1)+v_3(1)=0+8+13+17=38$ ft

Using the right endpoints: Distance $= v_1(1)+v_2(1)+v_3(1)+v_4(1)=8+13+17+18=56$ ft

41. (a) Using the left endpoints:

$$\text{Distance} = \sum_{i=1}^{n} f(x_i)\Delta x_i = 0(1.84)+12.9(1.96)+23.8(2.58)+26.3(0.85)+26.3(1.73)+26.0(0.87)$$
$$=177.162$$

Since we multiplied the units of seconds by miles per hour, we need to divide by 3600 (the number of seconds in an hour) to get a distance in miles.

$$\frac{177.162}{3600}\approx 0.0492$$

The estimate of the distance is 0.0492 mile.

(b) Using the right endpoints:

$$\text{Distance} = \sum_{i=1}^{n} f(x_i)\Delta x_i = 12.9(1.84) + 23.8(1.96) + 26.3(2.58) + 26.3(0.85) + 26.0(1.73) + 25.7(0.87)$$
$$= 227.932$$

Divide by 3600 (the number of seconds in an hour) to get a distance in miles.

$$\frac{227.932}{3600} \approx 0.0633$$

The estimate of the distance is 0.0633 mile.

(c) $\dfrac{100}{1609} \approx 0.0622$

Johnson actually ran 0.0622 mile. The answer to part **b** is closer.

7.4 The Fundamental Theorem of Calculus

1. $\int_{-2}^{4}(-3)\,dp = -3\int_{-2}^{4} dp = -3 \cdot p\Big|_{-2}^{4}$
$= -3[4-(-2)] = -18$

3. $\int_{-1}^{2}(5t-3)dt = 5\int_{-1}^{2} t\,dt - 3\int_{-1}^{2} dt$
$= \frac{5}{2}t^2\Big|_{-1}^{2} - 3t\Big|_{-1}^{2}$
$= \frac{5}{2}[2^2-(-1)^2] - 3[2-(-1)]$
$= \frac{5}{2}(4-1) - 3(2+1) = -\frac{3}{2}$

5. $\int_0^2 (5x^2-4x+2)\,dx$
$= 5\int_0^2 x^2 dx - 4\int_0^2 x\,dx + 2\int_0^2 dx$
$= \frac{5x^3}{3}\Big|_0^2 - 2x^2\Big|_0^2 + 2x\Big|_0^2$
$= \frac{5}{3}(2^3-0^3) - 2(2^2-0^2) + 2(2-0)$
$= \frac{5}{3}(8) - 2(4) + 2(2) = \frac{28}{3}$

7. $\int_0^2 3\sqrt{4u+1}\,du$

Let $4u+1 = x$, so that $4\,du = dx$.

When $u = 0$, $x = 4(0)+1 = 1$.
When $u = 2$, $x = 4(2)+1 = 9$.

$\int_0^2 3\sqrt{4u+1}\,du = \frac{3}{4}\int_0^2 \sqrt{4u+1}\,(4\,du)$
$= \frac{3}{4}\int_1^9 x^{1/2}dx = \frac{3}{4}\cdot\frac{x^{3/2}}{3/2}\Big|_1^9$
$= \frac{3}{4}\cdot\frac{2}{3}(9^{3/2} - 1^{3/2}) = 13$

9. $\int_0^4 2(t^{1/2}-t)\,dt = 2\int_0^4 t^{1/2}dt - 2\int_0^4 t\,dt$
$= 2\cdot\frac{t^{3/2}}{\frac{3}{2}}\Big|_0^4 - 2\cdot\frac{t^2}{2}\Big|_0^4$
$= \frac{4}{3}(4^{3/2}-0^{3/2}) - (4^2-0^2)$
$= \frac{32}{3} - 16 = -\frac{16}{3}$

11. $\int_1^4 (5y\sqrt{y} + 3\sqrt{y})\,dy$
$= 5\int_1^4 y^{3/2}dy + 3\int_1^4 y^{1/2}dy$
$= 5\left(\frac{y^{5/2}}{\frac{5}{2}}\right)\Bigg|_1^4 + 3\left(\frac{y^{3/2}}{\frac{3}{2}}\right)\Bigg|_1^4$
$= 2y^{5/2}\Big|_1^4 + 2y^{3/2}\Big|_1^4$
$= 2(4^{5/2}-1) + 2(4^{3/2}-1)$
$= 2(32-1) + 2(8-1) = 76$

13. $\int_4^6 \frac{2}{(2x-7)^2}\,dx$

Let $u = 2x-7$, so that $du = 2dx$.

When $x = 6$, $u = 2\cdot 6 - 7 = 5$.
When $x = 4$, $u = 2\cdot 4 - 7 = 1$.

$\int_4^6 \frac{2}{(2x-7)^2}\,dx = \int_1^5 u^{-2}du$
$= \frac{u^{-1}}{-1}\Big|_1^5 = -\frac{1}{u}\Big|_1^5$
$= -\left(\frac{1}{5}-1\right) = \frac{4}{5}$

15. $\int_1^5 (6n^{-2} - n^{-3})\,dn = 6\int_1^5 n^{-2}dn - \int_1^5 n^{-3}dn$

$$= 6 \cdot \frac{n^{-1}}{-1}\bigg|_1^5 - \frac{n^{-2}}{-2}\bigg|_1^5$$

$$= \frac{-6}{n}\bigg|_1^5 + \frac{1}{2n^2}\bigg|_1^5$$

$$= \frac{-6}{5} - \left(\frac{-6}{1}\right) + \left[\frac{1}{2(25)} - \frac{1}{2(1)}\right]$$

$$= \frac{108}{25}$$

17. $\int_{-3}^{-2} \left(2e^{-0.1y} + \frac{3}{y}\right) dy$

$$= 2\int_{-3}^{-2} e^{-0.1y}dy + \int_{-3}^{-2} \frac{3}{y}dy$$

$$= 2 \cdot \frac{e^{-0.1y}}{-0.1}\bigg|_{-3}^{-2} + 3\ln|y|\bigg|_{-3}^{-2}$$

$$= -20e^{-0.1y}\bigg|_{-3}^{-2} + 3\ln|y|\bigg|_{-3}^{-2}$$

$$= 20e^{0.3} - 20e^{0.2} + 3\ln 2 - 3\ln 3$$

$$\approx 1.353$$

19. $\int_1^2 \left(e^{4u} - \frac{1}{(u+1)^2}\right) du$

$$= \int_1^2 e^{4u}du - \int_1^2 \frac{1}{(u+1)^2} du$$

$$= \frac{e^{4u}}{4}\bigg|_1^2 - \frac{-1}{u+1}\bigg|_1^2$$

$$= \frac{e^8}{4} - \frac{e^4}{4} + \frac{1}{2+1} - \frac{1}{1+1} \approx 731.4$$

21. $\int_{-1}^{0} y(2y^2 - 3)^5\, dy$

Let $u = 2y^2 - 3$, so that $du = 4y\,dy$ and $\frac{1}{4}du = y\,dy$.

When $y = -1$, $u = 2(-1)^2 - 3 = -1$.

When $y = 0$, $u = 2(0)^2 - 3 = -3$.

$$\frac{1}{4}\int_{-1}^{-3} u^5 du = \frac{1}{4} \cdot \frac{u^6}{6}\bigg|_{-1}^{-3} = \frac{1}{24}u^6\bigg|_{-1}^{-3}$$

$$= \frac{1}{24}(-3)^6 - \frac{1}{24}(-1)^6$$

$$= \frac{729}{24} - \frac{1}{24} = \frac{91}{3}$$

23. $\int_1^{64} \frac{\sqrt{z} - 2}{\sqrt[3]{z}}\, dz = \int_1^{64} \left(\frac{z^{1/2}}{z^{1/3}} - 2z^{-1/3}\right) dz$

$$= \int_1^{64} z^{1/6}dz - 2\int_1^{64} z^{-1/3}dz$$

$$= \frac{z^{7/6}}{\frac{7}{6}}\bigg|_1^{64} - 2\frac{z^{2/3}}{\frac{2}{3}}\bigg|_1^{64}$$

$$= \frac{6z^{7/6}}{7}\bigg|_1^{64} - 3z^{2/3}\bigg|_1^{64}$$

$$= \left(\frac{6(64)^{7/6}}{7} - \frac{6(1)^{7/6}}{7}\right)$$

$$- \left(3(64)^{2/3} - 3(1)^{2/3}\right)$$

$$\approx 63.86$$

25. $\int_1^2 \frac{\ln x}{x}\, dx$

Let $u = \ln x$, so that $du = \frac{1}{x}dx$.

When $x = 1$, $u = \ln 1 = 0$.

When $x = 2$, $u = \ln 2$.

$$\int_0^{\ln 2} u\, du = \frac{u^2}{2}\bigg|_0^{\ln 2} = \frac{(\ln 2)^2}{2} - 0$$

$$= \frac{(\ln 2)^2}{2} \approx 0.2402$$

27. $\int_0^8 x^{1/3}\sqrt{x^{4/3} + 9}\, dx$

Let $u = x^{4/3} + 9$, so that $du = \frac{4}{3}x^{1/3}dx$ and $\frac{3}{4}du = x^{1/3}dx$.

When $x = 0$, $u = 0^{4/3} + 9 = 9$.

When $x = 8$, $u = 8^{4/3} + 9 = 25$.

$$\frac{3}{4}\int_9^{25} \sqrt{u}du = \frac{3}{4}\int_9^{25} u^{1/2}du$$

$$= \frac{3}{4} \cdot \frac{u^{3/2}}{\frac{3}{2}}\bigg|_9^{25} = \frac{1}{2}u^{3/2}\bigg|_9^{25}$$

$$= \frac{1}{2}(25)^{3/2} - \frac{1}{2}(9)^{3/2} = 49$$

29. $\int_0^1 \frac{e^{2t}}{(3+e^{2t})^2}\,dt$

Let $u = 3+e^{2t}$, so that $du = 2e^{2t}\,dt$.

When $x = 1,\ u = 3+e^{2\cdot 1} = 3+e^2$.

When $x = 0,\ u = 3+e^{2\cdot 0} = 4$.

$$\int_0^1 \frac{e^{2t}}{(3+e^{2t})^2}\,dt = \frac{1}{2}\int_4^{3+e^2} u^{-2}\,du$$

$$= \frac{1}{2}\cdot\frac{u^{-1}}{-1}\Bigg|_4^{3+e^2} = \frac{-1}{2u}\Bigg|_4^{3+e^2}$$

$$= \frac{1}{8} - \frac{1}{2(3+e^2)} \approx 0.07687$$

31. $\int_0^{\pi/4} \sin x\,dx = -\cos x\Big|_0^{\pi/4} = -\cos\frac{\pi}{4} - (-\cos 0)$

$$= -\frac{\sqrt{2}}{2} + 1 = 1 - \frac{\sqrt{2}}{2}$$

33. $\int_0^{\pi/6} \tan x\,dx = -\ln|\cos x|\Big|_0^{\pi/6}$

$$= -\ln\left|\cos\frac{\pi}{6}\right| - (-\ln|\cos 0|)$$

$$= -\ln\frac{\sqrt{3}}{2} + \ln 1 = -\ln\frac{\sqrt{3}}{2} + 0$$

$$= -\ln\frac{\sqrt{3}}{2}$$

35. $\int_{\pi/2}^{2\pi/3} \cos x\,dx = \sin x\Big|_{\pi/2}^{2\pi/3}$

$$= \sin\frac{2\pi}{3} - \sin\frac{\pi}{2}$$

$$= \frac{\sqrt{3}}{2} - 1$$

37. $f(x) = 2x - 14;\ [6, 10]$

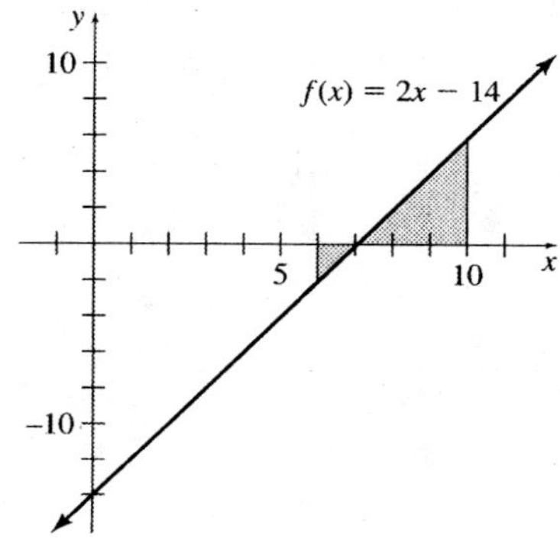

The graph crosses the x-axis at $0 = 2x - 14 \Rightarrow 2x = 14 \Rightarrow x = 7$.
This location is in the interval. The area of the region is

$$\left|\int_6^7 (2x-14)\,dx\right| + \int_7^{10}(2x-14)\,dx$$

$$= \left|x^2 - 14x\right|_6^7\Big| + (x^2 - 14x)\Big|_7^{10}$$

$$= |(7^2 - 98) - (6^2 - 84)|$$
$$+ (10^2 - 140) - (7^2 - 98)$$
$$= |-1| + (-40) - (-49) = 10$$

39. $f(x) = 2 - 2x^2;\ [0, 5]$

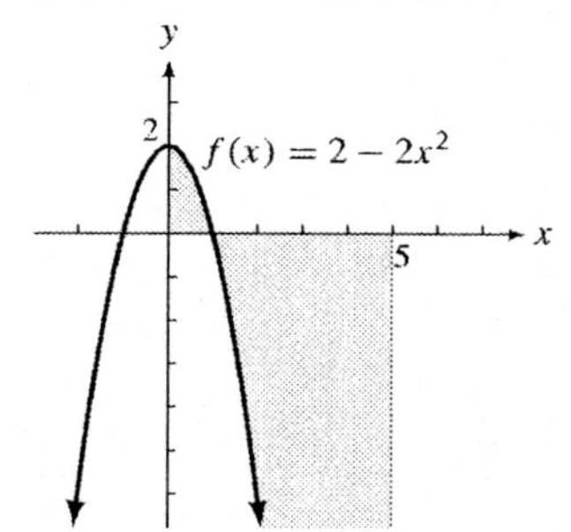

Find the points where the graph crosses the x-axis by solving $2 - 2x^2 = 0$.
$2 - 2x^2 = 0 \Rightarrow 2x^2 = 2 \Rightarrow x^2 = 1 \Rightarrow x = \pm 1.$
The only solution in the interval $[0, 5]$ is 1.
The total area is

$$\int_0^1 (2-2x^2)\,dx + \left|\int_1^5 (2-2x^2)\,dx\right|$$

$$= \left(2x - \frac{2x^3}{3}\right)\Bigg|_0^1 + \left|\left(2x - \frac{2x^3}{3}\right)\Bigg|_1^5\right|$$

$$= 2 - \frac{2}{3} + \left|10 - \frac{2(5^3)}{3} - 2 + \frac{2}{3}\right|$$

$$= \frac{4}{3} + \left|\frac{-224}{3}\right| = 76$$

41. $f(x) = x^3;\ [-1, 3]$

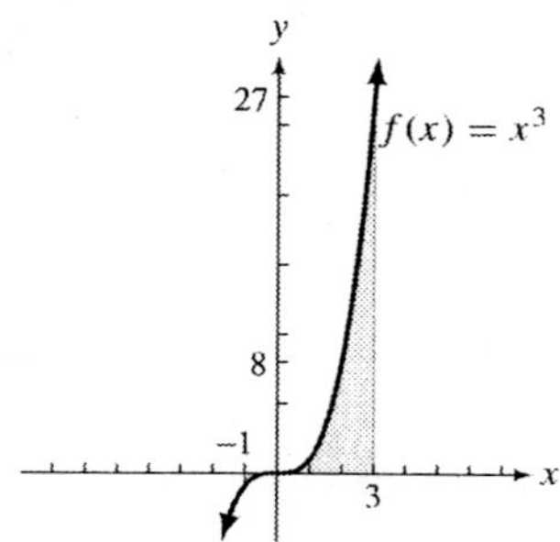

The solution $x^3 = 0 \Rightarrow x = 0$ indicates that the graph crosses the x-axis at 0 in the given interval $[-1, 3]$.

(continued on next page)

(continued)

The total area is

$$\left|\int_{-1}^{0} x^3 dx\right| + \int_0^3 x^3 dx = \left|\frac{x^4}{4}\right|_{-1}^{0} + \left(\frac{x^4}{4}\right)\Big|_0^3$$
$$= \left|\left(0 - \frac{1}{4}\right)\right| + \left(\frac{3^4}{4} - 0\right)$$
$$= \frac{41}{2}$$

43. $f(x) = e^x - 1;\ [-1, 2]$

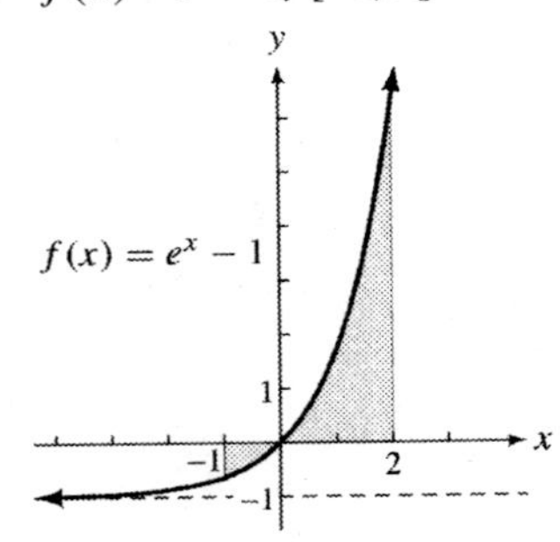

Solve $e^x - 1 = 0$.

$e^x = 1 \Rightarrow x \ln e = \ln 1 \Rightarrow x = 0$

The graph crosses the x-axis at 0 in the given interval $[-1, 2]$. The total area is

$$\left|\int_{-1}^{0} (e^x - 1)\, dx\right| + \int_0^2 (e^x - 1)\, dx$$
$$= \left|(e^x - x)\Big|_{-1}^{0}\right| + (e^x - x)\Big|_0^2$$
$$= |(1-0) - (e^{-1} + 1)| + (e^2 - 2) - (1 - 0)$$
$$= |1 - e^{-1} - 1| + e^2 - 2 - 1$$
$$= \frac{1}{e} + e^2 - 3 \approx 4.757$$

45. $f(x) = \frac{1}{x} - \frac{1}{e};\ [1, e^2]$

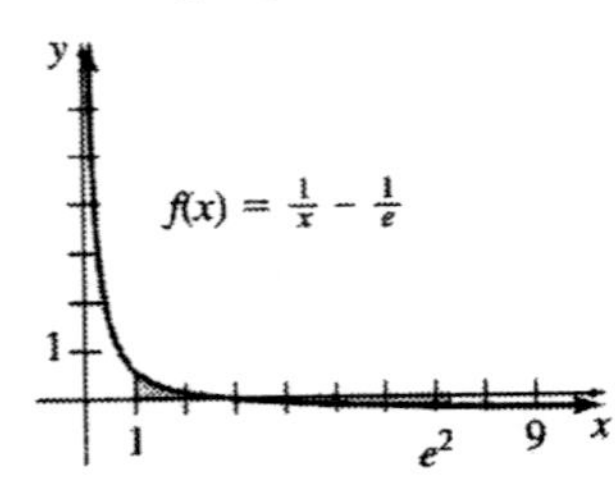

The graph crosses the x-axis at

$0 = \frac{1}{x} - \frac{1}{e} \Rightarrow \frac{1}{x} = \frac{1}{e} \Rightarrow x = e.$

This location is in the interval.

The area of the region is

$$\int_1^e \left(\frac{1}{x} - \frac{1}{e}\right) dx + \left|\int_e^{e^2} \left(\frac{1}{x} - \frac{1}{e}\right) dx\right|$$
$$= \left(\ln|x| - \frac{x}{e}\right)\Big|_1^e + \left|\left(\ln|x| - \frac{x}{e}\right)\Big|_e^{e^2}\right|$$
$$= 0 - \left(-\frac{1}{e}\right) + |(2-e) - 0|$$
$$= \frac{1}{e} + |2 - e| = e - 2 + \frac{1}{e} \approx 1.086$$

47. $f(x) = \sin x;\ \left[0, \frac{3\pi}{2}\right]$

The graph crosses the x-axis at $x = \pi$, which is in the given interval.

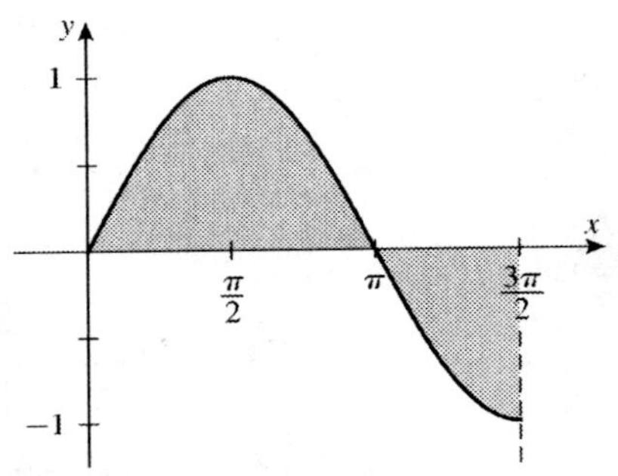

The area of the region is

$$\int_0^{\pi} \sin x\, dx + \left|\int_{\pi}^{3\pi/2} \sin x\, dx\right|$$
$$= (-\cos x)\Big|_0^{\pi} + \left|(-\cos x)\Big|_{\pi}^{3\pi/2}\right|$$
$$= \left|-\cos\pi - (-\cos 0)\right| + \left|-\cos\frac{3\pi}{2} - (-\cos\pi)\right|$$
$$= (1 + 1) + |0 - 1| = 3$$

49. $y = 4 - x^2;\ [0, 3]$

From the graph, we see that the total area is

$$\int_0^2 (4 - x^2)\, dx + \left|\int_2^3 (4 - x^2)\, dx\right|$$
$$= \left(4x - \frac{x^3}{3}\right)\Big|_0^2 + \left|\left(4x - \frac{x^3}{3}\right)\Big|_2^3\right|$$
$$= \left[\left(8 - \frac{8}{3}\right) - 0\right] + \left|\left[(12 - 9) - \left(8 - \frac{8}{3}\right)\right]\right|$$
$$= \frac{16}{3} + \left|3 - \frac{16}{3}\right| = \frac{23}{3}$$

51. $y = e^x - e;\ [0, 2]$
From the graph, we see that total area is

$$\left|\int_0^1 (e^x - e)\,dx\right| + \int_1^2 (e^x - e)\,dx$$
$$= \left|(e^x - xe)\Big|_0^1\right| + (e^x - xe)\Big|_1^2$$
$$= |(e^1 - e) - (e^0 - 0)| + (e^2 - 2e) - (e^1 - e)$$
$$= |-1| + e^2 - 2e \approx 2.952$$

53. $\int_a^c f(x)\,dx = \int_a^b f(x)\,dx + \int_b^c f(x)\,dx$

55.
$$\int_0^{16} f(x)\,dx = \int_0^2 f(x)\,dx + \int_2^5 f(x)\,dx + \int_5^8 f(x)\,dx + \int_8^{16} f(x)\,dx$$
$$= \frac{1}{2}\cdot 2(1+3) + \frac{\pi(3^2)}{4} - \frac{\pi(3^2)}{4} - \frac{1}{2}(3)(8)$$
$$= 4 + \frac{9}{4}\pi - \frac{9}{4}\pi - 12 = -8$$

57. Prove: $\int_a^b f(x)\,dx = \int_a^c f(x)\,dx + \int_c^b f(x)\,dx.$
Let $F(x)$ be an antiderivative of $f(x)$.

$$\int_a^c f(x)\,dx + \int_c^b f(x)\,dx$$
$$= F(x)\Big|_a^c + F(x)\Big|_c^b$$
$$= [F(c) - F(a)] + [F(b) - F(c)]$$
$$= F(c) - F(a) + F(b) - F(c)$$
$$= F(b) - F(a)$$
$$= \int_a^b f(x)\,dx$$

59.
$$\int_{-1}^4 f(x)\,dx = \int_{-1}^0 (2x+3)\,dx + \int_0^4\left(-\frac{x}{4} - 3\right)dx$$
$$= (x^2 + 3x)\Big|_{-1}^0 + \left(-\frac{x^2}{8} - 3x\right)\Bigg|_0^4$$
$$= -(1-3) + (-2-12) = -12$$

61. (a) $g(t) = t^4$ and $c = 1$, use substitution.

$$f(x) = \int_c^x g(t)\,dt = \int_1^x t^4\,dt = \frac{t^5}{5}\Bigg|_1^x$$
$$= \frac{x^5}{5} - \frac{(1)^5}{5} = \frac{x^5}{5} - \frac{1}{5}$$

(b)
$$f'(x) = \frac{d}{dx}(f(x)) = \frac{d}{dx}\left(\frac{x^5}{5} - \frac{1}{5}\right)$$
$$= \frac{1}{5}\cdot\frac{d}{dx}(x^5) - \frac{d}{dx}\left(\frac{1}{5}\right)$$
$$= \frac{1}{5}\cdot 5x^4 - 0 = x^4$$

Since $g(t) = t^4$, then $g(x) = x^4$ and we see $f'(x) = g(x)$.

(c) Let $g(t) = e^{t^2}$ and $c = 0$, then

$f(x) = \int_0^x e^{t^2}\,dt.$

$f(1) = \int_0^1 e^{t^2}\,dt$ and $f(1.01) = \int_0^{1.01} e^{t^2}\,dt.$

Use the fnInt command in the Math menu of your calculator to find $\int_0^1 e^{x^2}\,dx$ and $\int_0^{1.01} e^{x^2}\,dx$. The resulting screens are:

```
fnInt(e^(X²),X,0
,1)
            1.462651746
```

```
fnInt(e^(X²),X,0
,1.01)
            1.490109133
```

$f(1) \approx 1.46265$
$f(1.01) \approx 1.49011$

Use $\frac{f(1+h) - f(1)}{h}$ to approximate $f'(1)$ with $h = 0.01$

$$\frac{f(1+h) - f(1)}{h} = \frac{f(1.01) - f(1)}{0.01}$$
$$\approx \frac{1.49011 - 1.46265}{0.01}$$
$$= 2.746$$

So $f'(1) \approx 2.746$, and

$g(1) = e^{1^2} = e \approx 2.718.$

63. $P'(t) = 140t^{5/2}$

$$\int_0^4 140t^{5/2}\,dt = 140 \cdot \left.\frac{t^{7/2}}{\frac{7}{2}}\right|_0^4 = 40t^{7/2}\Big|_0^4 = 5120$$

Since 5120 is above the total level of acceptable pollution (4850), the factory cannot operate for 4 years without killing all the fish in the lake.

65. Growth rate is $0.6 + \dfrac{4}{(t+1)^3}$ ft/yr.

(a) Total growth in the second year is

$$\int_1^2 \left[0.6 + \frac{4}{(t+1)^3}\right] dt$$
$$= \left[0.6t + \frac{4}{-2(t+1)^2}\right]\Bigg|_1^2$$
$$= \left[0.6(2) - \frac{2}{(2+1)^2}\right] - \left[0.6(1) - \frac{2}{(1+1)^2}\right]$$
$$= \frac{44}{45} - \frac{1}{10} \approx 0.8778 \text{ ft}$$

(b) Total growth in the third year is

$$\int_2^3 \left[0.6 + \frac{4}{(t+1)^3}\right] dt$$
$$= \left[0.6t + \frac{4}{-2(t+1)^2}\right]\Bigg|_2^3$$
$$= \left[0.6(3) - \frac{2}{(3+1)^2}\right] - \left[0.6(2) - \frac{2}{(2+1)^2}\right]$$
$$= \frac{67}{40} - \frac{44}{45} \approx 0.6972 \text{ ft.}$$

67. $R'(t) = \dfrac{5}{t+1} + \dfrac{2}{\sqrt{t+1}}$

(a) Total reaction from $t = 1$ to $t = 12$ is

$$\int_1^{12} \left(\frac{5}{t+1} + \frac{2}{\sqrt{t+1}}\right) dt$$
$$= \left[5\ln(t+1) + 4\sqrt{t+1}\right]\Big|_1^{12}$$
$$= (5\ln 13 + 4\sqrt{13}) - (5\ln 2 + 4\sqrt{2})$$
$$\approx 18.12$$

(b) Total reaction from $t = 12$ to $t = 24$ is

$$\int_{12}^{24} \left(\frac{5}{t+1} + \frac{2}{\sqrt{t+1}}\right) dt$$
$$= \left[5\ln(t+1) + 4\sqrt{t+1}\right]\Big|_{12}^{24}$$
$$= (5\ln 25 + 4\sqrt{25}) - (5\ln 13 + 4\sqrt{13})$$
$$\approx 8.847$$

69. **(a)** Answers will vary.

(b) $\displaystyle\int_0^{60} n(x)\,dx$

(c) $\displaystyle\int_5^{10} \sqrt{5x+1}\,dx$

Let $u = 5x + 1$. Then $du = 5\,dx$.
When $x = 5$, $u = 26$; when $x = 10$, $u = 51$.

$$\frac{1}{5}\int_{26}^{51} u^{1/2}\,du = \frac{1}{5} \cdot \left.\frac{u^{3/2}}{\frac{3}{2}}\right|_{26}^{51} = \frac{2}{15}u^{3/2}\Big|_{26}^{51}$$
$$= \frac{2}{15}(51^{3/2} - 26^{3/2})$$
$$\approx 30.89 \text{ million}$$

71. **(a)** $v = k(R^2 - r^2)$

$$Q(R) = \int_0^R 2\pi v r\,dr = \int_0^R 2\pi k(R^2 - r^2)r\,dr$$
$$= 2\pi k \int_0^R (R^2 r - r^3)\,dr$$
$$= 2\pi k\left(\frac{R^2 r^2}{2} - \frac{r^4}{4}\right)\Bigg|_0^R$$
$$= 2\pi k\left(\frac{R^4}{2} - \frac{R^4}{4}\right) = 2\pi k\left(\frac{R^4}{4}\right) = \frac{\pi k R^4}{2}$$

(b) $Q(0.4) = \dfrac{\pi k(0.4)^4}{2} = 0.04k$ mm/min

73. **(a)** $E(t) = 753t^{-0.1321}$

Since t is the age of the beagle in years, to convert the formula to days, let $T = 365t$, or $t = \dfrac{T}{365}$.

$$E(T) = 753\left(\frac{T}{365}\right)^{-0.1321} \approx 1642T^{-0.1321}$$

Now, replace T with t:
$E(t) = 1642t^{-0.1321}$

(b) The beagle's age in days after one year is 365 days and after 3 years she is 1095 days old.

$$\int_{365}^{1095} 1642t^{-0.1321}dt$$
$$= 1642\frac{1}{0.8679}t^{0.8679}\Big|_{365}^{1095}$$
$$\approx 1892\,(1{,}095^{0.8679} - 365^{0.8679})$$
$$\approx 505{,}155$$

The beagle's total energy requirements are about $505{,}000 \text{ kJ/W}^{0.67}$, where W represents weight.

75. (a) $f(x) = 40.2 + 3.50x - 0.897x^2$

$$\int_0^9 (40.2 + 3.50x - 0.897x^2)\,dx$$
$$= (40.2x + 1.75x^2 - 0.299x^3)\Big|_0^9$$
$$\approx 286$$

The integral represents the population aged 0 to 90, which is about 286 million.

(b) $\int_{4.5}^{6.5} (40.2 + 3.50x - 0.897x^2)\,dx$

$$= (40.2x + 1.75x^2 - 0.299x^3)\Big|_{4.5}^{6.5}$$
$$\approx 64$$

The number of baby boomers is about 64 million.

77. $\int_2^1 \left(x - \frac{1}{x}\right)dx = -\int_0^{T/2} dt$

First evaluate the integral on the left side.

$$\int_2^1 \left(x - \frac{1}{x}\right)dx = \left(\frac{x^2}{2} - \ln|x|\right)\Bigg|_2^1$$
$$= \left(\frac{1}{2} - \ln|1|\right) - \left(\frac{4}{2} - \ln|2|\right)$$
$$= \ln 2 - \frac{3}{2}$$

Now evaluate the integral on the right side.

$$-\int_0^{T/2} dt = -t\Big|_0^{T/2} = -\frac{T}{2}$$

Substitute the values for the integrals and solve for T.

$$\int_2^1 \left(x - \frac{1}{x}\right)dx = -\int_0^{T/2} dt$$
$$\ln 2 - \frac{3}{2} = -\frac{T}{2} \Rightarrow T = 3 - 2\ln 2$$

79. $T(t) = 50 + 50\cos\left(\frac{\pi}{6}t\right)$

Since T is periodic, the number of animals passing the checkpoint is equal to the area under the curve for any 12-month period. Let t vary from 0 to 12.

$$\text{Total} = \int_0^{12}\left[50 + 50\cos\left(\frac{\pi}{6}t\right)\right]dt$$
$$= \int_0^{12} 50\,dt + \frac{6}{\pi}\int_0^{12} 50\cos\left(\frac{\pi}{6}t\right)\left(\frac{\pi}{6}dt\right)$$
$$= 50t\Big|_0^{12} + \frac{300}{\pi}\sin\left(\frac{\pi}{6}t\right)\Big|_0^{12}$$
$$= (600 - 0) + \frac{300}{\pi}(0 - 0)$$
$$= 600 \text{ (in hundreds)}$$

The total number of animals is 60,000.

81. $\int_{2.5}^{5} (0.0353x^3 - 0.541x^2 + 3.78x + 4.29)\,dx$

$$= (0.008825x^4 - 0.18033x^3 + 1.8900x^2 + 4.2900x)\Big|_{2.5}^{5}$$
$$\approx 0.316$$

About 32% of families have incomes between \$25,000 and \$50,000.

83. $C'(t) = 1.2e^{0.04t}$

$$C(T) = \int_0^T 1.2e^{0.04t} = \frac{1.2}{0.04}e^{0.04t}\Big|_0^T$$
$$= 30(e^{0.04T} - e^0) = 30(e^{0.04T} - 1)$$

In 5 yr,

$$C(5) = 30(e^{0.04(5)} - 1) = 30(e^{0.2} - 1)$$
$$\approx 6.64 \text{ billion barrels.}$$

85. The total amount of daylight is given by

$$\int_0^{365} N(t)dt = \int_0^{365} [183.549\sin(0.0172t - 1.329) + 728.124]dt$$
$$= \left[-\frac{183.549}{0.0172}\cos(0.0172t - 1.329) + 728.124t\right]\Bigg|_0^{365}$$
$$\approx 265{,}819.0192 \text{ minutes}$$
$$\approx 4430 \text{ hours.}$$

The result is relatively close to the actual value.

7.5 The Area Between Two Curves

1. $x = -2,\ x = 1,\ y = 2x^2 + 5,\ y = 0$

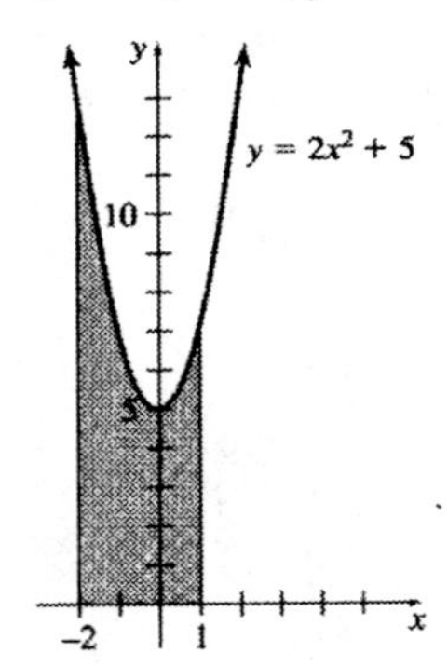

$$\int_{-2}^{1} [(2x^2 + 5) - 0] = \left(\frac{2x^3}{3} + 5x\right)\Bigg|_{-2}^{1}$$
$$= \left(\frac{2}{3} + 5\right) - \left(-\frac{16}{3} - 10\right)$$
$$= 21$$

3. $x = -3,\ x = 1,\ y = x^3 + 1,\ y = 0$

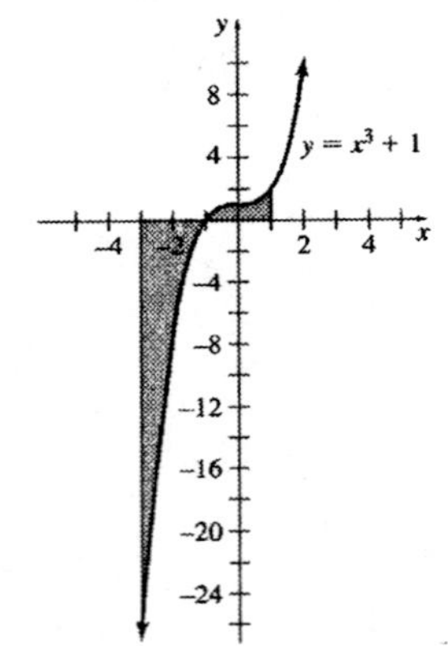

To find the points of intersection of the graphs, substitute for y.

$x^3 + 1 = 0 \Rightarrow x^3 = -1 \Rightarrow x = -1$

The region is composed of two separate regions because $y = x^3 + 1$ intersects $y = 0$ at $x = -1$. Let $f(x) = x^3 + 1$, $g(x) = 0$. In the interval $[-3, -1]$, $g(x) \ge f(x)$. In the interval $[-1, 1]$, $f(x) \ge g(x)$.

$$\int_{-3}^{-1} [0 - (x^3 + 1)\, dx] + \int_{-1}^{1} [(x^3 + 1) - 0]\, dx$$
$$= \left(\frac{-x^4}{4} - x\right)\Bigg|_{-3}^{-1} + \left(\frac{x^4}{4} + x\right)\Bigg|_{-1}^{1}$$
$$= \left(-\frac{1}{4} + 1\right) - \left(-\frac{81}{4} + 3\right) + \left(\frac{1}{4} + 1\right) - \left(\frac{1}{4} - 1\right)$$
$$= 20$$

5. $x = -2,\ x = 1,\ y = 2x,\ y = x^2 - 3$

Find the points of intersection of the graphs of $y = 2x$ and $y = x^2 - 3$ by substituting for y.

$2x = x^2 - 3 \Rightarrow 0 = x^2 - 2x - 3 \Rightarrow$
$0 = (x - 3)(x + 1)$

The only intersection in $[-2, 1]$ is at $x = -1$.

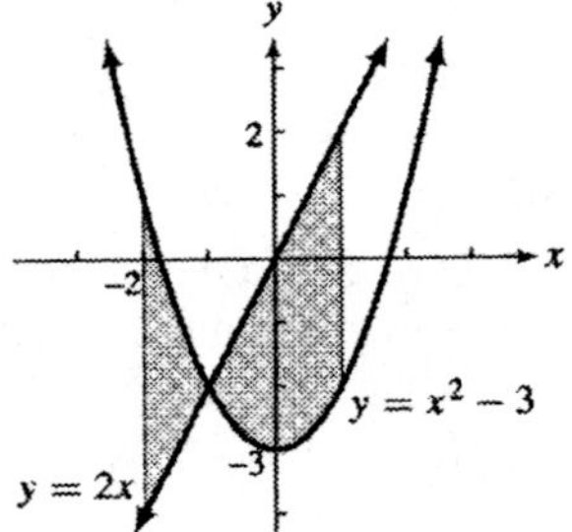

In the interval $[-2, -1]$, $(x^2 - 3) \ge 2x$. In the interval $[-1, 1]$, $2x \ge (x^2 - 3)$.

$$\int_{-2}^{-1} [(x^2 - 3) - (2x)]\, dx + \int_{-1}^{1} [(2x) - (x^2 - 3)]\, dx$$
$$= \int_{-2}^{-1} (x^2 - 3 - 2x)\, dx + \int_{-1}^{1} (2x - x^2 + 3)\, dx$$
$$= \left(\frac{x^3}{3} - 3x - x^2\right)\Bigg|_{-2}^{-1} + \left(x^2 - \frac{x^3}{3} + 3x\right)\Bigg|_{-1}^{1}$$
$$= -\frac{1}{3} + 3 - 1 - \left(-\frac{8}{3} + 6 - 4\right) + 1 - \frac{1}{3} + 3 - \left(1 + \frac{1}{3} - 3\right)$$
$$= \frac{5}{3} + 6 = \frac{23}{3}$$

7. $y = x^2 - 30$
$y = 10 - 3x$

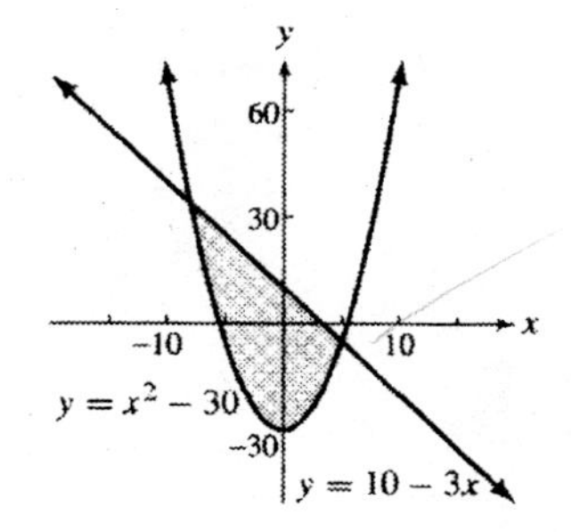

Find the points of intersection.

$$x^2 - 30 = 10 - 3x$$
$$x^2 + 3x - 40 = 0$$
$$(x+8)(x-5) = 0$$
$$x = -8 \quad \text{or} \quad x = 5$$

Let $f(x) = 10 - 3x$ and $g(x) = x^2 - 30$. The area between the curves is given by

$$\int_{-8}^{5} [f(x) - g(x)]\, dx$$
$$= \int_{-8}^{5} [(10 - 3x) - (x^2 - 30)]\, dx$$
$$= \int_{-8}^{5} (-x^2 - 3x + 40)\, dx$$
$$= \left(\frac{-x^3}{3} - \frac{3x^3}{2} + 40x\right)\Bigg|_{-8}^{5}$$
$$= \frac{-5^3}{3} - \frac{3(5)^2}{2} + 40(5)$$
$$- \left[\frac{-(-8)^3}{3} - \frac{3(-8)^2}{2} + 40(-8)\right]$$
$$= \frac{-125}{3} - \frac{75}{2} + 200 - \frac{512}{3} + \frac{192}{2} + 320$$
$$\approx 366.1667.$$

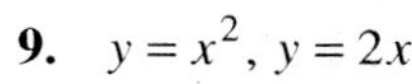

9. $y = x^2$, $y = 2x$

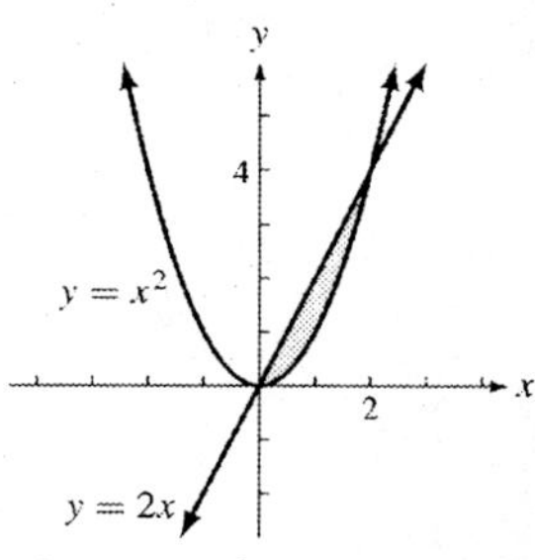

Find the points of intersection.

$x^2 = 2x \Rightarrow x^2 - 2x = 0 \Rightarrow x(x-2) = 0 \Rightarrow$
$x = 0 \quad \text{or} \quad x = 2$

Let $f(x) = 2x$ and $g(x) = x^2$. The area between the curves is given by

$$\int_0^2 [f(x) - g(x)]\, dx = \int_0^2 (2x - x^2)\, dx$$
$$= \left(\frac{2x^2}{2} - \frac{x^3}{3}\right)\Bigg|_0^2$$
$$= 4 - \frac{8}{3} = \frac{4}{3}.$$

11. $x = 1$, $x = 6$, $y = \frac{1}{x}$, $y = \frac{1}{2}$

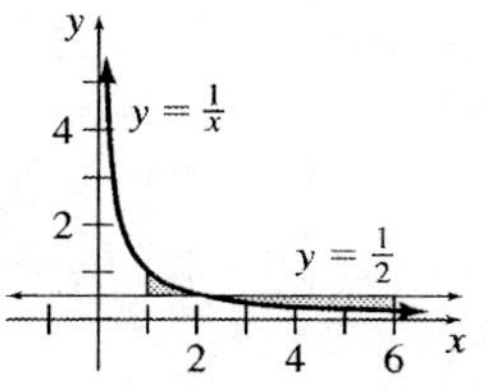

To find the points of intersection of the graphs, substitute for y.

$\frac{1}{x} = \frac{1}{2} \Rightarrow x = 2$

The region is composed of two separate regions because $y = \frac{1}{x}$ intersects $y = \frac{1}{2}$ at $x =$ 2.

Let $f(x) = \frac{1}{x}$, $g(x) = \frac{1}{2}$. In the interval $[1, 2]$, $f(x) \ge g(x)$. In the interval $[2, 6]$, $g(x) \ge f(x)$.

$$\int_1^2 \left(\frac{1}{x} - \frac{1}{2}\right) dx + \int_2^6 \left(\frac{1}{2} - \frac{1}{x}\right) dx$$
$$= \left(\ln|x| - \frac{x}{2}\right)\Bigg|_1^2 + \left(\frac{x}{2} - \ln|x|\right)\Bigg|_2^6$$
$$= (\ln 2 - 1) - \left(0 - \frac{1}{2}\right) + (3 - \ln 6) - (1 - \ln 2)$$
$$= 2\ln 2 - \ln 6 + \frac{3}{2} \approx 1.095$$

13. $x=-1,\ x=1,\ y=e^x,\ y=3-e^x$

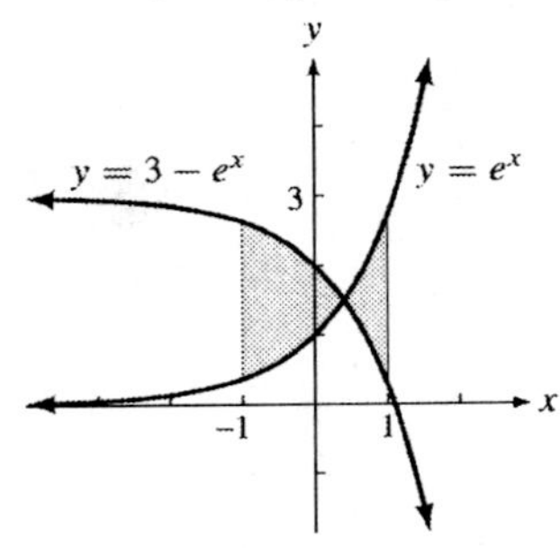

To find the point of intersection, set $e^x = 3-e^x$ and solve for x.

$e^x = 3-e^x \Rightarrow 2e^x = 3 \Rightarrow e^x = \frac{3}{2} \Rightarrow \ln e^x = \ln\frac{3}{2} \Rightarrow x = \ln\frac{3}{2}$

The area of the region between the curves from $x=-1$ to $x=1$ is

$$\int_{-1}^{\ln 3/2}[(3-e^x)-e^x]\,dx + \int_{\ln 3/2}^{1}[e^x-(3-e^x)]\,dx = \int_{-1}^{\ln 3/2}(3-2e^x)\,dx + \int_{\ln 3/2}^{1}(2e^x-3)\,dx$$

$$= (3x-2e^x)\Big|_{-1}^{\ln 3/2} + (2e^x-3x)\Big|_{\ln 3/2}^{1}$$

$$= \left[\left(3\ln\frac{3}{2} - 2e^{\ln 3/2}\right) - [3(-1)-2e^{-1}]\right] + \left[2e^1 - 3(1) - \left(2e^{\ln 3/2} - 3\ln\frac{3}{2}\right)\right]$$

$$= \left[\left(3\ln\frac{3}{2} - 3\right) - \left(-3-\frac{2}{e}\right)\right] + \left[2e-3-\left(3-3\ln\frac{3}{2}\right)\right]$$

$$= 6\ln\frac{3}{2} + \frac{2}{e} + 2e - 6 \approx 2.605.$$

15. $x=-1,\ x=2,\ y=2e^{2x},\ y=e^{2x}+1$

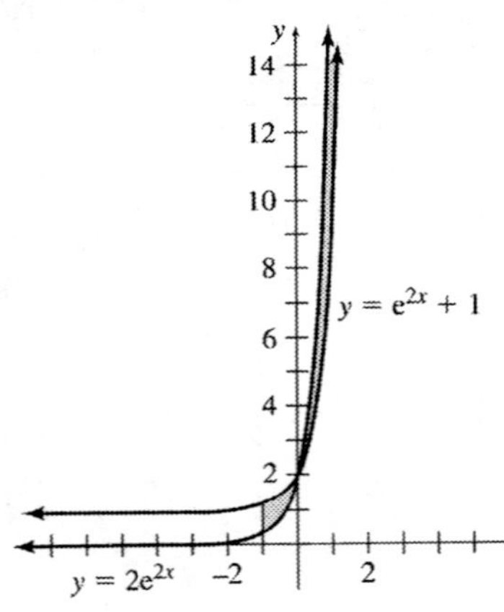

To find the points of intersection of the graphs, substitute for y.

$2e^{2x} = e^{2x}+1 \Rightarrow e^{2x} = 1 \Rightarrow 2x = 0 \Rightarrow x = 0$

The region is composed of two separate regions because $y = 2e^{2x}$ intersects $y = e^{2x}+1$ at $x=0$.

Let $f(x) = 2e^{2x}$, $g(x) = e^{2x}+1$. In the interval $[-1, 0]$, $g(x) \ge f(x)$. In the interval $[0, 2]$, $f(x) \ge g(x)$.

$$\int_{-1}^{0}(e^{2x}+1-2e^{2x})dx + \int_{0}^{2}[2e^{2x}-(e^{2x}+1)]\,dx = \left(-\frac{e^{2x}}{2}+x\right)\Bigg|_{-1}^{0} + \left(\frac{e^{2x}}{2}-x\right)\Bigg|_{0}^{2}$$

$$= \left(-\frac{1}{2}+0\right) - \left(-\frac{e^{-2}}{2}-1\right) + \left(\frac{e^4}{2}-2\right) - \left(\frac{1}{2}-0\right)$$

$$= \frac{e^{-2}+e^4}{2} - 2 \approx 25.37$$

17. $y = x^3 - x^2 + x + 1,\ y = 2x^2 - x + 1$

Find the points of intersection.

$x^3 - x^2 + x + 1 = 2x^2 - x + 1 \Rightarrow x^3 - 3x^2 + 2x = 0 \Rightarrow x(x^2 - 3x + 2) = 0 \Rightarrow$
$x(x-1)(x-2) = 0 \Rightarrow x = 0 \text{ or } x = 1 \text{ or } x = 2$

The points of intersection are at $x = 0$, $x = 1$, and $x = 2$.

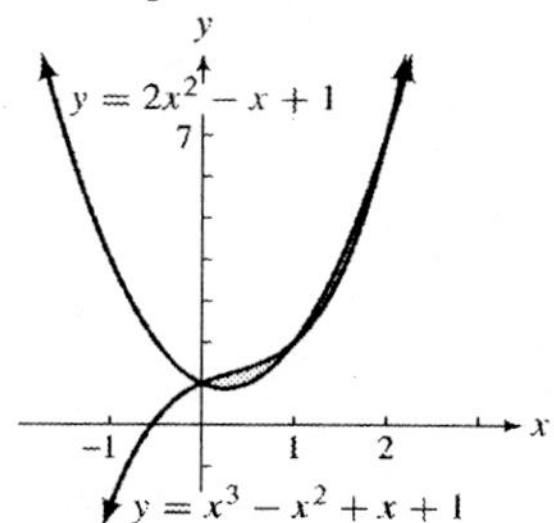

Area between the curves is

$$\int_0^1 [(x^3 - x^2 + x + 1) - (2x^2 - x + 1)]\,dx + \int_1^2 [(2x^2 - x + 1) - (x^3 - x^2 + x + 1)]\,dx$$

$$= \int_0^1 (x^3 - 3x^2 + 2x)\,dx + \int_1^2 (-x^3 + 3x^2 - 2x)]\,dx$$

$$= \left(\frac{x^4}{4} - x^3 + x^2\right)\Bigg|_0^1 + \left(\frac{-x^4}{4} + x^3 - x^2\right)\Bigg|_1^2$$

$$= \left[\left(\frac{1}{4} - 1 + 1\right) - (0)\right] + \left[(-4 + 8 - 4) - \left(-\frac{1}{4} + 1 - 1\right)\right]$$

$$= \frac{1}{4} + \frac{1}{4} = \frac{1}{2}$$

19. $y = x^4 + \ln(x+10),\ \ y = x^3 + \ln(x+10)$

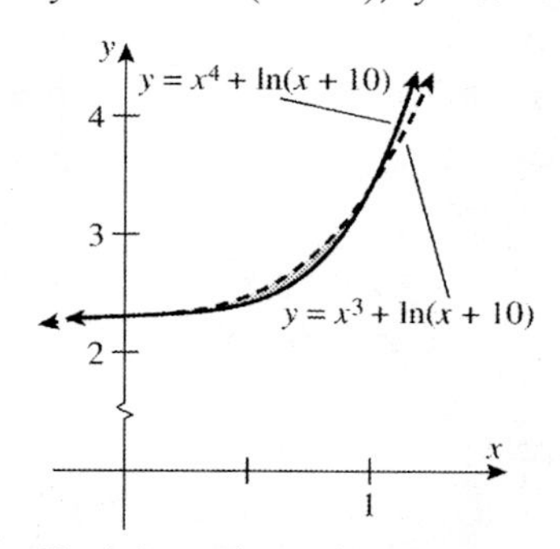

Find the points of intersection.

$x^4 + \ln(x+10) = x^3 + \ln(x+10) \Rightarrow x^4 - x^3 = 0 \Rightarrow x^3(x-1) = 0 \Rightarrow x = 0 \text{ or } x = 1$

The points of intersection are at $x = 0$ and $x = 1$. The area between the curves is

$$\int_0^1 [(x^3 + \ln(x+10)) - (x^4 + \ln(x+10))]\,dx = \int_0^1 (x^3 - x^4)\,dx = \left(\frac{x^4}{4} - \frac{x^5}{5}\right)\Bigg|_0^1 = \left(\frac{1}{4} - \frac{1}{5}\right) - (0) = \frac{1}{20}$$

21. $y = x^{4/3}$, $y = 2x^{1/3}$

Find the points of intersection.

$x^{4/3} = 2x^{1/3} \Rightarrow x^{4/3} - 2x^{1/3} = 0 \Rightarrow x^{1/3}(x-2) = 0 \Rightarrow x = 0 \quad \text{or} \quad x = 2$

The points of intersection are at $x = 0$ and $x = 2$.

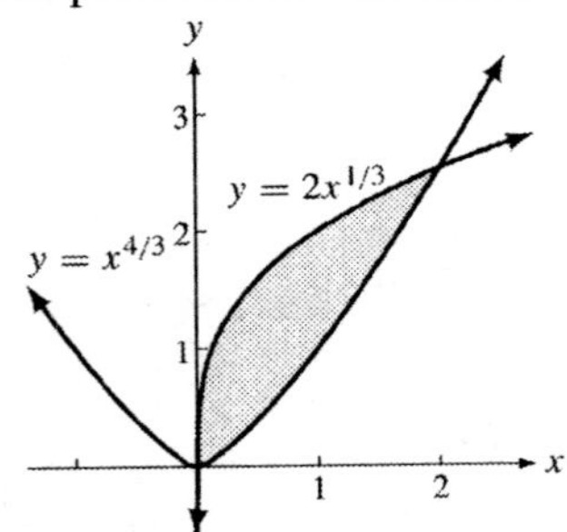

The area between the curves is

$$\int_0^2 (2x^{1/3} - x^{4/3})\,dx = 2\frac{x^{4/3}}{\frac{4}{3}} - \frac{x^{7/3}}{\frac{7}{3}}\Bigg|_0^2 = \frac{3}{2}x^{4/3} - \frac{3}{7}x^{7/3}\Bigg|_0^2 = \left[\frac{3}{2}(2)^{4/3} - \frac{3}{7}(2)^{7/3}\right] - 0$$

$$= \frac{3(2^{4/3})}{2} - \frac{3(2^{7/3})}{7} \approx 1.62$$

23. $x = 0,\ x = 3,\ y = 2e^{3x},\ y = e^{3x} + e^6$

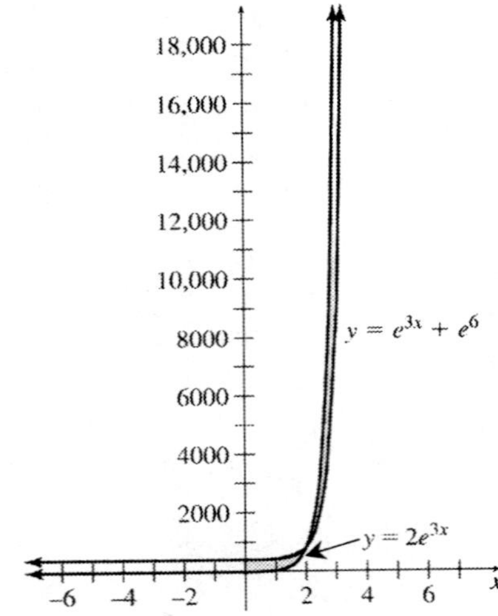

To find the points of intersection of the graphs, substitute for y and solve for x.

$2e^{3x} = e^{3x} + e^6 \Rightarrow e^{3x} = e^6 \Rightarrow 3x = 6 \Rightarrow x = 2$

The region is composed of two separate regions because $y = 2e^{3x}$ intersects $y = e^{3x} + e^6$ at $x = 2$.

Let $f(x) = 2e^{3x}$, $g(x) = e^{3x} + e^6$. In the interval $[0, 2]$, $g(x) \ge f(x)$. In the interval $[2, 3]$, $f(x) \ge g(x)$.

$$\int_0^2 (e^{3x} + e^6 - 2e^{3x})\,dx + \int_2^3 [2e^{3x} - (e^{3x} + e^6)]\,dx = \left(-\frac{e^{3x}}{3} + e^6 x\right)\Bigg|_0^2 + \left(\frac{e^{3x}}{3} - e^6 x\right)\Bigg|_2^3$$

$$= \left(-\frac{e^6}{3} + 2e^6\right) - \left(-\frac{1}{3} + 0\right) + \left(\frac{e^9}{3} - 3e^6\right) - \left(\frac{e^6}{3} - 2e^6\right)$$

$$= \frac{e^9 + e^6 + 1}{3} \approx 2836$$

25. $x = 0, x = \frac{\pi}{4}, y = \cos x, y = \sin x$

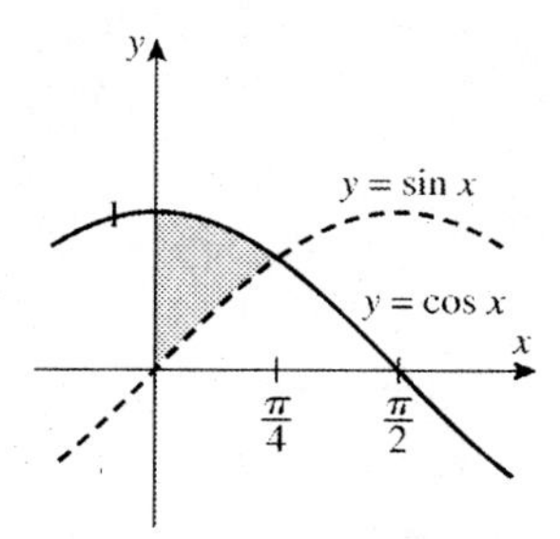

Find the points of intersection.

$\sin x = \cos x \Rightarrow \frac{\sin x}{\cos x} = 1 \Rightarrow \tan x = 1 \Rightarrow x = \frac{\pi}{4} + \pi n,$ where n is an integer. The curves intersect at $x = \frac{\pi}{4}$.

The area between the curves is

$$\int_0^{\pi/4} (\cos x - \sin x)\,dx = (\sin x + \cos x)\Big|_0^{\pi/4} = \left(\sin\frac{\pi}{4} + \cos\frac{\pi}{4}\right) - (\sin 0 + \cos 0)$$
$$= \left(\frac{\sqrt{2}}{2} + \frac{\sqrt{2}}{2}\right) - (0 + 1) = \sqrt{2} - 1$$

27. $x = \frac{\pi}{4}, y = \tan x, y = \sin x$

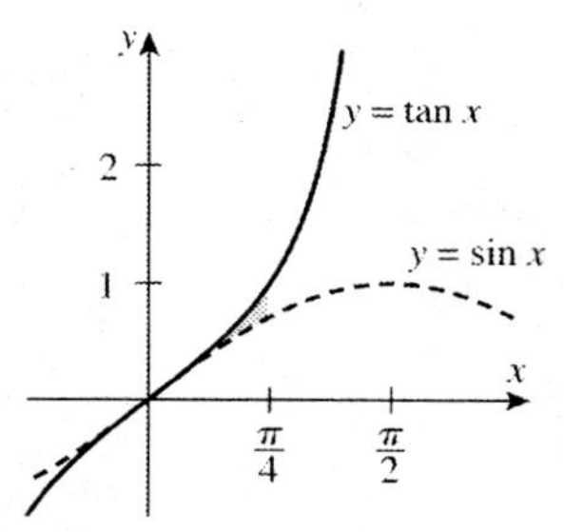

Find the points of intersection.

$\sin x = \tan x \Rightarrow \sin x - \tan x = 0 \Rightarrow \sin x - \frac{\sin x}{\cos x} = 0 \Rightarrow \sin x\left(1 - \frac{1}{\cos x}\right) = 0 \Rightarrow$

$\sin x = 0 \Rightarrow x = 0 \pm \pi n$ or $1 - \frac{1}{\cos x} = 0 \Rightarrow 1 = \frac{1}{\cos x} \Rightarrow \cos x = 1 \Rightarrow x = 0 \pm \pi n,$ where n is an integer. The curves intersect at $x = 0$. The area between the curves is

$$\int_0^{\pi/4} (\tan x - \sin x)\,dx = \left(\ln|\sec x| + \cos x\right)\Big|_0^{\pi/4} = \left(-\ln|\cos x| + \cos x\right)\Big|_0^{\pi/4}$$
$$= \left(-\ln\left(\cos\frac{\pi}{4}\right) + \cos\frac{\pi}{4}\right) - \left(-\ln(\cos 0) + \cos 0\right)$$
$$= \left[-\ln\left(\frac{\sqrt{2}}{2}\right) + \frac{\sqrt{2}}{2}\right] - (-\ln 1 + 1) = -\left(\ln\sqrt{2} - \ln 2\right) + \frac{\sqrt{2}}{2} - 1$$
$$= \frac{\sqrt{2}}{2} - 1 - \frac{1}{2}\ln 2 + \ln 2 = \frac{\sqrt{2}}{2} - 1 + \frac{\ln 2}{2}$$

29. Graph $y_1 = e^x$ and $y_2 = -x^2 - 2x$ on your graphing calculator. Use the intersect command to find the two intersection points. The resulting screens are:

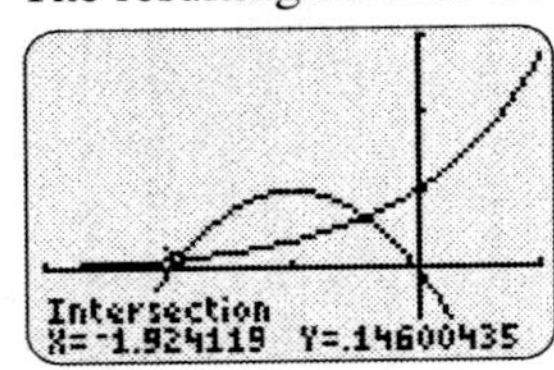

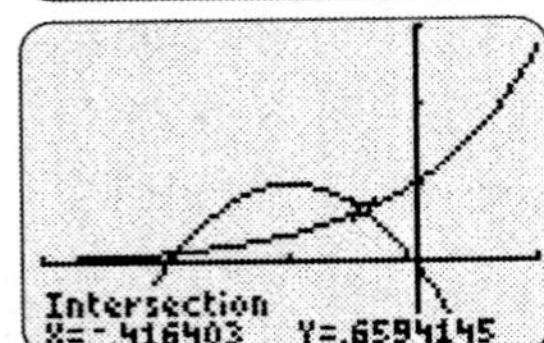

These screens show that $e^x = -x^2 - 2x$ when $x \approx -1.9241$ and $x \approx -0.4164$. In the interval $[-1.9241, -0.4164]$, $e^x < -x^2 - 2x$. The area between the curves is given by

$$\int_{-1.9241}^{-0.4164} [(-x^2 - 2x) - e^x]\,dx.$$ Use the fnInt command to approximate this definite integral. The resulting screen is:

The last screen shows that the area is approximately 0.6650.

31. (a) The pollution level in the lake is changing at the rate $f(t) - g(t)$ at any time t. We find the amount of pollution by integrating.

$$\int_0^{12} [f(t) - g(t)]\,dt$$
$$= \int_0^{12} [10(1 - e^{-0.5t}) - 0.4t]\,dt$$
$$= \left(10t - 10 \cdot \frac{1}{-0.5} e^{-0.5t} - 0.4 \cdot \frac{1}{2} t^2\right)\Bigg|_0^{12}$$
$$= (20e^{-0.5t} + 10t - 0.2t^2)\Big|_0^{12}$$
$$= [20e^{-0.5(12)} + 10(12) - 0.2(12)^2] - [20e^{-0.5(0)} + 10(0) - 0.2(0)^2]$$
$$= (20e^{-6} + 91.2) - (20)$$
$$= 20e^{-6} + 71.2 \approx 71.25$$

After 12 hours, there are about 71.25 gallons.

(b) The graphs of the functions intersect at about 25.00. So the rate that pollution enters the lake equals the rate the pollution is removed at about 25 hours.

(c) $$\int_0^{25} [f(t) - g(t)]\,dt$$
$$= (20e^{-0.5t} + 10t - 0.2t^2)\Big|_0^{25}$$
$$= [20e^{-0.5(25)} + 10(25) - 0.2(25)^2)] - 20$$
$$= 20e^{-12.5} + 105 \approx 105$$

After 25 hours, there are about 105 gallons.

(d) For $t > 25$, $g(t) > f(t)$, and pollution is being removed at the rate $g(t) - f(t)$. So, we want to solve for c, where

$$\int_0^c [f(t) - g(t)]\,dt = 0.$$

Alternatively, we could solve for c in

$$\int_{25}^c [g(t) - f(t)]\,dt = 105.$$

One way to do this with a graphing calculator is to graph the function $y = \int_0^x [f(t) - g(t)]\,dt$ and determine the values of x for which $y = 0$. The first window shows how the function can be defined.

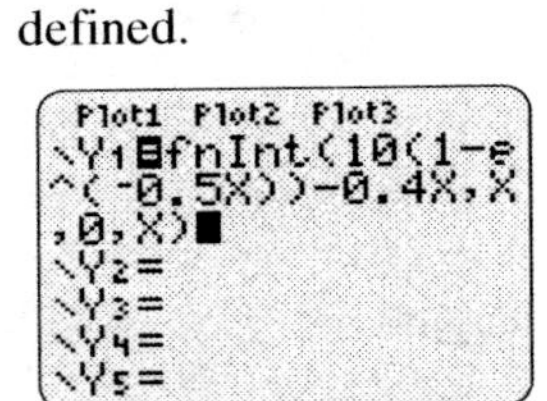

A suitable window for the graph is [0, 50] by [0, 110].

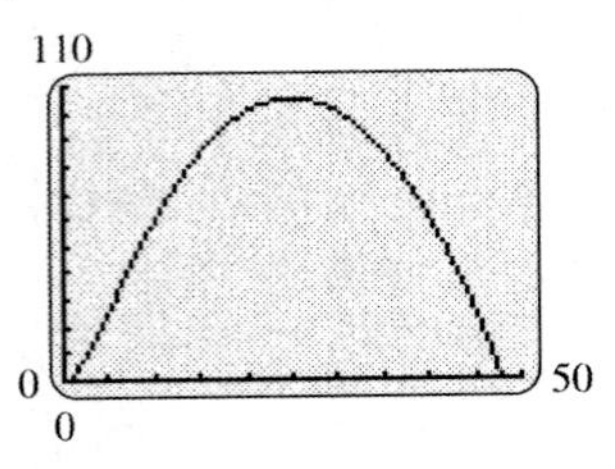

Use the calculator's features to approximate where the graph intersects the x-axis. These are at 0 and about 47.91. Therefore, the pollution will be removed from the lake after about 47.91 hours.

33. $I(x) = 0.9x^2 + 0.1x$

(a) $I(0.1) = 0.9(0.1)^2 + 0.1(0.1) = 0.019$
The lower 10% of income producers earn 1.9% of total income of the population.

(b) $I(0.4) = 0.9(0.4)^2 + 0.1(0.4) = 0.184$
The lower 40% of income producers earn 18.4% of total income of the population.

(c) The graph of $I(x) = x$ is a straight line through the points (0, 0) and (1, 1). The graph of $I(x) = 0.9x^2 + 0.1x$ is a parabola with vertex $\left(-\frac{1}{18}, -\frac{1}{360}\right)$. Restrict the domain to $0 \le x \le 1$.

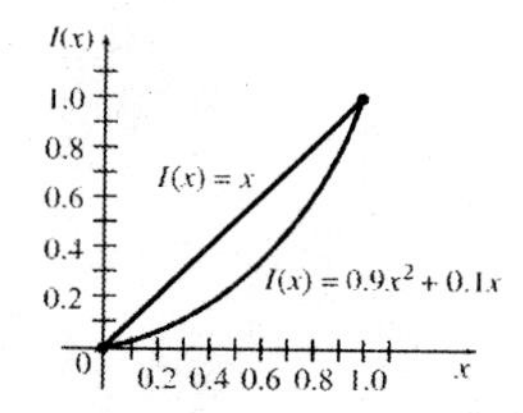

(d) To find the points of intersection, solve $x = 0.9x^2 + 0.1x$.
$0.9x^2 - 0.9x = 0 \Rightarrow 0.9x(x-1) = 0 \Rightarrow$
$x = 0$ or $x = 1$
The area between the curves is given by

$$\int_0^1 [x - (0.9x^2 + 0.1x)]\,dx$$
$$= \int_0^1 (0.9x - 0.9x^2)\,dx$$
$$= \left(\frac{0.9x^2}{2} - \frac{0.9x^3}{3}\right)\Bigg|_0^1 = \frac{0.9}{2} - \frac{0.9}{3}$$
$$= 0.15$$

(e) Income is distributed less equally in 2008 than in 1968.

35. (a) It is profitable to use the machine until $S'(t) = C'(t)$.

$$150 - t^2 = t^2 + \frac{11}{4}t$$
$$2t^2 + \frac{11}{4}t - 150 = 0$$
$$8t^2 + 11t - 600 = 0$$
$$t = \frac{-11 \pm \sqrt{121 - 4(8)(-600)}}{16} = \frac{-11 \pm 139}{16}$$
$t = 8$ or $t = -9.375$

It will be profitable to use this machine for 8 years. Reject the negative solution.

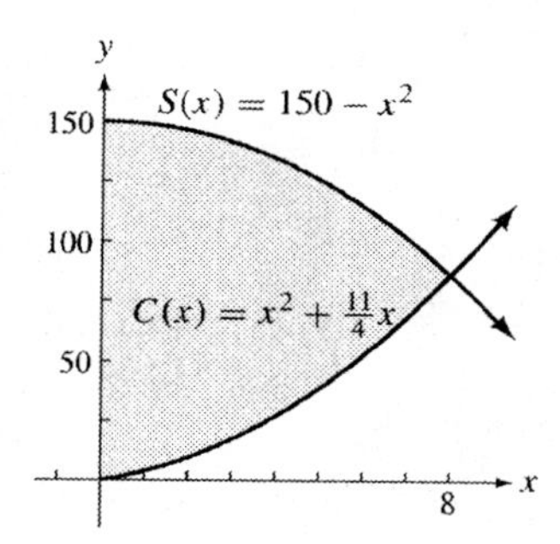

(b) Since $150 - t^2 > t^2 + \frac{11}{4}t$, in the interval [0, 8], the net total saving in the first year are

$$\int_0^1 \left[(150 - t^2) - \left(t^2 + \frac{11}{4}t\right)\right] dt$$
$$= \int_0^1 \left(-2t^2 - \frac{11}{4}t + 150\right) dt$$
$$= \left(\frac{-2t^3}{3} - \frac{11t^2}{8} + 150t\right)\Bigg|_0^1$$
$$= -\frac{2}{3} - \frac{11}{8} + 150 \approx \$148.$$

(c) The net total savings over the entire period of use are

$$\int_0^8 \left[(150 - t^2) - \left(t^2 + \frac{11}{4}t\right)\right] dt$$
$$= \left(\frac{-2t^3}{3} - \frac{11t^2}{8} + 150t\right)\Bigg|_0^8$$
$$= \frac{-2(8^3)}{3} - \frac{11(8^2)}{8} + 150(8)$$
$$= \frac{-1024}{3} - \frac{704}{8} + 1200 \approx \$771.$$

37. (a) $E'(t) = e^{0.1t}$ and $I'(t) = 98.8 - e^{0.1t}$
To find the point of intersection, where profit will be maximized, set the functions equal to each other and solve for t.
$e^{0.1t} = 98.8 - e^{0.1t} \Rightarrow 2e^{0.1t} = 98.8 \Rightarrow$
$e^{0.1t} = 49.4 \Rightarrow 0.1t = \ln 49.4 \Rightarrow$
$t = \frac{\ln 49.4}{0.1} \approx 39$
The optimum number of days for the job to last is 39.

(b) The total income for 39 days is

$$\int_0^{39} (98.8 - e^{0.1t})\, dt$$
$$= \left(98.8t - \frac{e^{0.1t}}{0.1}\right)\Bigg|_0^{39}$$
$$= \left(98.8t - 10e^{0.1t}\right)\Big|_0^{39}$$
$$= [98.8(39) - 10e^{3.9}] - (0 - 10)$$
$$= \$3369.18$$

(c) The total expenditure for 39 days is

$$\int_0^{39} e^{0.1t} dt = \frac{e^{0.1t}}{0.1}\Bigg|_0^{39} = 10e^{0.1t}\Big|_0^{39}$$
$$= 10e^{3.9} - 10 = \$484.02$$

(d) Profit $=$ Income $-$ Expense $= 3369.18 - 484.02 = \$2885.16$

Chapter 7 Review Exercises

1. True
2. False. The statement is false for $n = -1$.
3. False. For example, if $f(x) = 1$ the first expression is equal to $x^2/2 + C$ and the second is equal to $x^2 + C$.
4. True
5. True
6. False. The derivative gives the instantaneous rate of change.
7. False. If the function is positive over the interval of integration the definite integral gives the exact area.
8. True
9. True
10. False: The definite integral may be positive, negative, or zero.
11. True
12. Answers will vary. See sections 7.1 and 7.3 in the text.
13. Answers will vary. See section 7.2 in the text.
14. Answers will vary. See section 7.2 in the text.
15. $\int (2x+3)\, dx = \frac{2x^2}{2} + 3x + C = x^2 + 3x + C$
16. $\int (5x-1)\, dx = \frac{5x^2}{2} - x + C$
17. $\int (x^2 - 3x + 2)\, dx = \frac{x^3}{3} - \frac{3x^2}{2} + 2x + C$
18. $\int (6 - x^2)\, dx = 6x - \frac{x^3}{3} + C$
19. $\int 3\sqrt{x} dx = 3\int x^{1/2}\, dx = \frac{3x^{3/2}}{\frac{3}{2}} + C$
$$= 2x^{3/2} + C$$
20. $\int \frac{\sqrt{x}}{2} dx = \int \frac{1}{2} x^{1/2} dx = \frac{\frac{1}{2}x^{3/2}}{\frac{3}{2}} + C = \frac{x^{3/2}}{3} + C$
21. $\int (x^{1/2} + 3x^{-2/3})\, dx = \frac{x^{3/2}}{\frac{3}{2}} + \frac{3x^{1/3}}{\frac{1}{3}} + C$
$$= \frac{2x^{3/2}}{3} + 9x^{1/3} + C$$
22. $\int (2x^{4/3} + x^{-1/2})\, dx = \frac{2x^{7/3}}{\frac{7}{3}} + \frac{x^{1/2}}{\frac{1}{2}} + C$
$$= \frac{6x^{7/3}}{7} + 2x^{1/2} + C$$
23. $\int \frac{-4}{x^3} dx = \int -4x^{-3}\, dx = \frac{-4x^{-2}}{-2} + C = 2x^{-2} + C$
24. $\int \frac{5}{x^4} dx = \int 5x^{-4}\, dx = \frac{5x^{-3}}{-3} + C = -\frac{5}{3x^3} + C$
25. $\int -3e^{2x} dx = \frac{-3e^{2x}}{2} + C$
26. $\int 5e^{-x} dx = -5e^{-x} + C$
27. $\int xe^{3x^2} dx$

    Let $u = 3x^2$, so that $du = 6x\,dx$.
$$\int xe^{3x^2} dx = \frac{1}{6}\int e^{3x^2}(6x)\,dx = \frac{1}{6}\int e^u du$$
$$= \frac{1}{6}e^u + C = \frac{e^{3x^2}}{6} + C$$
28. $\int 2xe^{x^2} dx$

    Let $u = x^2$, so $du = 2x dx$.
$$\int e^{x^2}(2x)\,dx = \int e^u du = e^u + C = e^{x^2} + C$$

29. $\int \frac{3x}{x^2-1}dx$

Let $u = x^2 - 1$, so that $du = 2x\,dx$.

$$\int \frac{3x}{x^2-1}dx = 3\left(\frac{1}{2}\right)\int \frac{2x\,dx}{x^2-1} = \frac{3}{2}\int \frac{du}{u}$$
$$= \frac{3}{2}\ln|u| + C = \frac{3\ln|x^2-1|}{2} + C$$

30. $\int \frac{-x}{2-x^2}dx$

Let $u = 2 - x^2$, so that $du = -2x\,dx$.

$$\int \frac{-x}{2-x^2}dx = \frac{1}{2}\int \frac{-2x\,dx}{2-x^2}$$
$$= \frac{1}{2}\int \frac{du}{u} = \frac{1}{2}\ln|u| + C$$
$$= \frac{1}{2}\ln|2-x^2| + C$$

31. $\int \frac{x^2\,dx}{(x^3+5)^4}$

Let $u = x^3 + 5$, so that $du = 3x^2dx$.

$$\int \frac{x^2\,dx}{(x^3+5)^4} = \frac{1}{3}\int \frac{3x^2\,dx}{(x^3+5)^4} = \frac{1}{3}\int \frac{du}{u^4}$$
$$= \frac{1}{3}\int u^{-4}\,du = \frac{1}{3}\left(\frac{u^{-3}}{-3}\right) + C$$
$$= \frac{-(x^3+5)^{-3}}{9} + C$$

32. $\int (x^2-5x)^4(2x-5)\,dx$

Let $u = x^2 - 5x$, so that $du = (2x-5)\,dx$.

$$\int (x^2-5x)^4(2x-5)\,dx = \int u^4\,du = \frac{u^5}{5} + C$$
$$= \frac{(x^2-5x)^5}{5} + C$$

33. $\int \frac{x^3}{e^{3x^4}}dx = \int x^3e^{-3x^4}$

Let $u = -3x^4$, so that $du = -12x^3\,dx$.

$$\int \frac{x^3}{e^{3x^4}}dx = \int x^3e^{-3x^4} = -\frac{1}{12}\int e^{-3x^4}\left(-12x^3\right)dx$$
$$= -\frac{1}{12}\int e^u\,du = -\frac{1}{12}e^u + C$$
$$= \frac{-e^{-3x^4}}{12} + C$$

34. $\int e^{3x^2+4}x\,dx$

Let $u = 3x^2 + 4$ so that $du = 6x\,dx$.

$$\int e^{3x^2+4}x\,dx = \frac{1}{6}\int (e^{3x^2})(6x)dx = \frac{1}{6}\int e^u\,du$$
$$= \frac{1}{6}e^u + C = \frac{e^{3x^2+4}}{6} + C$$

35. $\int \frac{(3\ln x+2)^4}{x}dx$

Let $u = 3\ln x + 2$ so that $du = \frac{3}{x}dx$.

$$\int \frac{(3\ln x+2)^4}{x}dx = \frac{1}{3}\int (3\ln x+2)^4\left(\frac{3}{x}\right)dx$$
$$= \frac{1}{3}\int u^4\,du = \frac{1}{3}\cdot\frac{u^5}{5} + C$$
$$= \frac{(3\ln x+2)^5}{15} + C$$

36. $\int \frac{\sqrt{5\ln x+3}}{x}dx$

Let $u = 5\ln x + 3$ so that $du = \frac{5}{x}dx$.

$$\int \frac{\sqrt{(5\ln x+3)}}{x}dx = \frac{1}{5}\int \sqrt{5\ln x+3}\left(\frac{5}{x}\right)dx$$
$$= \frac{1}{5}\int u^{1/2}\,du = \frac{1}{5}\cdot\frac{2u^{3/2}}{3} + C$$
$$= \frac{2(5\ln x+3)^{3/2}}{15} + C$$

37. $\int \sin 2x\,dx$

Let $u = 2x$. Then $du = 2dx \Rightarrow \frac{1}{2}du = dx$.

$$\int \sin 2x\,dx = \int (\sin u)\left(\frac{1}{2}du\right)$$
$$= \frac{1}{2}\int \sin u\,du$$
$$= \frac{1}{2}(-\cos u) + C$$
$$= -\frac{1}{2}\cos 2x + C$$

38. $\int \cos 5x\,dx$

Let $u = 5x$, so $du = 5dx$ or $\frac{1}{5}du = dx$.

$$\int \cos 5x\,dx = \int \cos u\cdot\frac{1}{5}du = \frac{1}{5}\int \cos u\,du$$
$$= \frac{1}{5}\sin u + C = \frac{1}{5}\sin 5x + C$$

39. $\int \tan 7x\,dx$

Let $u = 7x$, so $du = 7dx$ or $\frac{1}{7}du = dx$.

$$\int \tan 7x\,dx = \int (\tan u)\cdot\frac{1}{7}du = \frac{1}{7}\int \tan u\,du = -\frac{1}{7}\ln|\cos u| + C = -\frac{1}{7}\ln|\cos 7x| + C$$

40. $\int \sec^2 5x\,dx$

Let $u = 5x$, so that $du = 5\,dx$.

$$\int \sec^2 5x\,dx = \frac{1}{5}\int (\sec^2 5x)(5\,dx) = \frac{1}{5}\int \sec^2 u\,du = \frac{1}{5}\tan u + C = \frac{1}{5}\tan 5x + C$$

41. $\int 8\sec^2 x\,dx = 8\int \sec^2 x\,dx = 8\tan x + C$

42. $\int 4\csc^2 x\,dx = -4\int -\csc^2 x\,dx = -4\cot x + C$

43. $\int x^2 \sin 4x^3\,dx$

Let $u = 4x^3$, so $du = 12x^2\,dx$ or

$\frac{1}{12}du = x^2\,dx$.

$$\int \sin 4x^3\left(x^2\right)dx = \int (\sin u)\cdot\frac{1}{12}du = \frac{1}{12}\int \sin u\,du = -\frac{1}{12}\cos u + C = -\frac{1}{12}\cos 4x^3 + C$$

44. $\int 5x\sec 2x^2\tan 2x^2\,dx$

Let $u = 2x^2$, so that $du = 4x\,dx$.

$$\int 5x\sec 2x^2\tan 2x^2\,dx = \frac{5}{4}\int \sec 2x^2\tan 2x^2(4x\,dx) = \frac{5}{4}\int \sec u\tan u\,du = \frac{5}{4}\sec u + C = \frac{5}{4}\sec 2x^2 + C$$

45. $\int \sqrt{\cos x}\sin x\,dx$

Let $u = \cos x$. Then

$du = -\sin x\,dx \Rightarrow -du = \sin x\,dx$.

$$\int \sqrt{\cos x}\sin x\,dx = \int \sqrt{u}(-du) = -\int u^{1/2}\,du = -\frac{2}{3}u^{3/2} + C = -\frac{2}{3}(\cos x)^{3/2} + C$$

46. $\int \cos^8 x\sin x\,dx$

Let $u = \cos x$, so $du = -\sin x\,dx$.

$$\int \cos^8 x\sin x\,dx = -\int u^8\,du = -\frac{1}{9}u^9 + C = -\frac{1}{9}\cos^9 x + C$$

47. $\int x\tan 11x^2\,dx$

Let $u = 11x^2$. Then $du = 22x\,dx$.

$$\int x\tan 11x^2\,dx = \frac{1}{22}\int (\tan 11x^2)\cdot(22x\,dx) = \frac{1}{22}\int \tan u\,du = \frac{1}{22}(-\ln|\cos u|) + C = -\frac{1}{22}\ln|\cos 11x^2| + C$$

48. $\int x^2\cot 8x^3\,dx$

Let $u = 8x^3$, so that $du = 24x^3\,dx$.

$$\int x^2\cot 8x^3\,dx = \frac{1}{24}\int (\cot 8x^3)(24x^2)\,dx = \frac{1}{24}\int \cot u\,du = \frac{1}{24}\ln|\sin u| + C = \frac{1}{24}\ln|\sin 8x^3| + C$$

49. $\int (\sin x)^{3/2}\cos x\,dx$

Let $u = \sin x$, so $du = \cos x\,dx$.

$$\int (\sin x)^{3/2}\cos x\,dx = \int u^{3/2}\,du = \frac{u^{3/2+1}}{\frac{3}{2}+1} + C = \frac{u^{5/2}}{\frac{5}{2}} + C = \frac{2}{5}u^{5/2} + C = \frac{2}{5}(\sin x)^{5/2} + C$$

50. $\int (\cos x)^{-4/3} \sin x\, dx$

Let, $u = \cos x$ so that $du = -\sin x\, dx$.

$$\int (\cos x)^{-4/3} \sin x\, dx = -\int (\cos x)^{-4/3}(-\sin x)\, dx$$
$$= -\int u^{-4/3}\, du$$
$$= 3u^{-1/3} + C$$
$$= 3(\cos x)^{-1/3} + C$$

51. $\int \sec^2 5x \tan 5x\, dx$

Let $u = \tan 5x$, so $du = 5\sec^2 5x\, dx$.

$$\int \sec^2 5x \tan 5x\, dx = \frac{1}{5}\int u\, du = \frac{1}{5}\cdot\frac{u^2}{2} + C$$
$$= \frac{1}{10}u^2 + C = \frac{1}{10}\tan^2 5x + C$$

52. $f(x) = 3x + 1,\ x_1 = -1,\ x_2 = 0,\ x_3 = 1,$ $x_4 = 2,\ x_5 = 3$

$f(x_1) = -2,\ f(x_2) = 1,\ f(x_3) = 4,$ $f(x_4) = 7,\ f(x_5) = 10$

$$\sum_{i=1}^{5} f(x_i) = f(1) + f(2) + f(3) + f(4) + f(5)$$
$$= -2 + 1 + 4 + 7 + 10 = 20$$

53. **(a)** $\int_0^4 f(x)\, dx = 0$, since the area above the x-axis from 0 to 2 is identical to the area below the x-axis from 2 to 4.

(b) $\int_0^4 f(x)\, dx$ can be computed by calculating the area of the rectangle and triangle that make up the region shown in graph.

Area of rectangle = (length)(width)
$= (3)(1) = 3$

Area of triangle $= \frac{1}{2}$(base)(height)
$= \frac{1}{2}(1)(3) = \frac{3}{2}$

$$\int_0^4 f(x)\, dx = 3 + \frac{3}{2} = \frac{9}{2} = 4.5$$

54. $f(x) = 2x + 3$, from $x = 0$ to $x = 4$

$$\Delta x = \frac{4-0}{4} = 1$$

i	x_i	$f(x_i)$
1	0	3
2	1	5
3	2	7
4	3	9

$$A = \sum_{i=1}^{4} f(x_i)\Delta x$$
$$= 3(1) + 5(1) + 7(1) + 9(1) = 24$$

55. $\int_0^4 (2x + 3)\, dx$

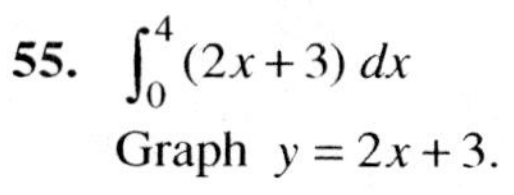

Graph $y = 2x + 3$.

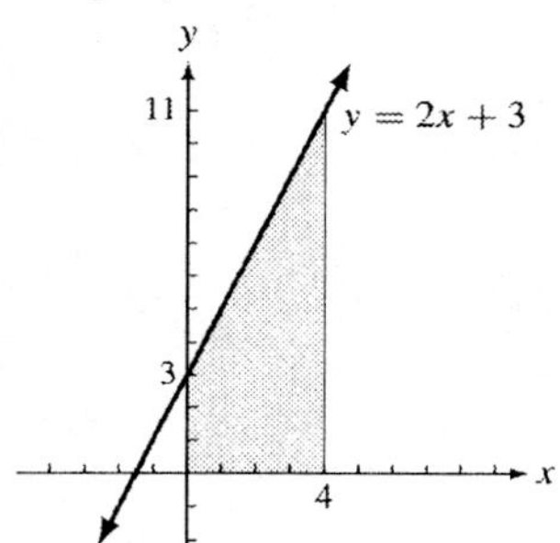

$\int_0^4 (2x + 3)\, dx$ is the area of a trapezoid with $B = 11,\ b = 3,\ h = 4$. The formula for the area is $A = \frac{1}{2}(B + b)h$.

$$A = \frac{1}{2}(11 + 3)(4) = 28, \text{ so } \int_0^4 (2x + 3)\, dx = 28.$$

56. **(a)** Since $s(t)$ represents the odometer reading, the distance traveled between $t = 0$ and $t = T$ will be $s(T) - s(0)$.

(b) $\int_0^T v(t)\, dt = s(T) - s(0)$ is equivalent to the Fundamental Theorem of Calculus with $a = 0$, and $b = T$ because $s(t)$ is an antiderivative of $v(t)$.

57. The Fundamental Theorem of Calculus states that $\int_a^b f(x)\, dx = F(x)\Big|_a^b = F(b) - F(a)$, where f is continuous on $[a, b]$ and F is any antiderivative of f.

58. $\int_1^2 (3x^2+5)\,dx = \left(\frac{3x^3}{3}+5x\right)\Big|_1^2$
$= (2^3+10)-(1+5)$
$= 18-6 = 12$

59. $\int_1^6 (2x^2+x)\,dx = \left(\frac{2x^3}{3}+\frac{x^2}{2}\right)\Big|_1^6$
$= \left[\frac{2(6)^3}{3}+\frac{(6)^2}{2}\right] - \left[\frac{2(1)^3}{3}+\frac{(1)^2}{2}\right]$
$= 144+18-\frac{2}{3}-\frac{1}{2}$
$= 162-\frac{2}{3}-\frac{1}{2} = \frac{965}{6} \approx 160.83$

60. $\int_1^5 (3x^{-1}+x^{-3})\,dx = \left(3\ln|x|+\frac{x^{-2}}{-2}\right)\Big|_1^5$
$= \left(3\ln 5-\frac{1}{50}\right)-\left(3\ln 1-\frac{1}{2}\right)$
$= 3\ln 5+\frac{12}{25} \approx 5.308$

61. $\int_1^3 (2x^{-1}+x^{-2})\,dx = \left(2\ln|x|+\frac{x^{-1}}{-1}\right)\Big|_1^3$
$= \left(2\ln 3-\frac{1}{3}\right)-(2\ln 1-1)$
$= 2\ln 3+\frac{2}{3} \approx 2.864$

62. $\int_0^1 x\sqrt{5x^2+4}\,dx$

Let $u = 5x^2+4,$ so that

$du = 10x\,dx$ and $\frac{1}{10}\,du = x\,dx.$

When $x=0,\ u = 5(0^2)+4 = 4.$

When $x=1,\ u = 5(1^2)+4 = 9.$

$\int_0^1 x\sqrt{5x^2+4}\,dx = \frac{1}{10}\int_4^9 \sqrt{u}\,du = \frac{1}{10}\int_4^9 u^{1/2}du$
$= \left(\frac{1}{10}\cdot\frac{u^{3/2}}{3/2}\right)\Big|_4^9 = \frac{1}{15}u^{3/2}\Big|_4^9$
$= \frac{1}{15}(9)^{3/2}-\frac{1}{15}(4)^{3/2}$
$= \frac{27}{15}-\frac{8}{15} = \frac{19}{15}$

63. $\int_0^2 x^2(3x^3+1)^{1/3}\,dx = \frac{(3x^3+1)^{4/3}}{12}\Big|_0^2$
$= \frac{25^{4/3}}{12}-\frac{1^{4/3}}{12} = \frac{25^{4/3}-1}{12}$
≈ 6.008

64. $\int_0^2 3e^{-2x}\,dx = \frac{-3e^{-2x}}{2}\Big|_0^2 = \frac{-3e^{-4}}{2}+\frac{3}{2}$
$= \frac{3(1-e^{-4})}{2} \approx 1.473$

65. $\int_1^5 \frac{5}{2}e^{0.4x}\,dx = \frac{5}{2}\cdot\frac{5}{2}\int_1^5 e^{0.4x}\,(0.4)\,dx$
$= \frac{5}{2}\cdot\frac{5}{2}\cdot e^{2x/5}\Big|_1^5$
$= \frac{25}{4}(e^2-e^{0.4}) = \frac{25(e^2-e^{0.4})}{4}$
≈ 36.86

66. $\int_0^{\pi/2} \cos x\,dx = \sin x\Big|_0^{\pi/2} = \sin\frac{\pi}{2}-\sin 0$
$= 1-0 = 1$

67. $\int_{-\pi}^{2\pi/3} -\sin x\,dx = \cos x\Big|_{-\pi}^{2\pi/3}$
$= \cos\frac{2\pi}{3}-\cos(-\pi)$
$= -\frac{1}{2}-(-1) = \frac{1}{2}$

68. $\int_0^{2\pi} (10+10\cos x)\,dx$
$= \int_0^{2\pi} 10\,dx+\int_0^{2\pi} 10\cos x\,dx$
$= 10(2\pi)+10\int_0^{2\pi}\cos x\,dx$
$= 20\pi+10(\sin x)\Big|_0^{2\pi}$
$= 20\pi+10(\sin 2\pi-\sin 0)$
$= 20\pi+10(0-0) = 20\pi$

69. $\int_0^{\pi/3} (3-3\sin x)\,dx$
$= \int_0^{\pi/3} 3\,dx-\int_0^{\pi/3} 3\sin x\,dx$
$= 3x\Big|_0^{\pi/3}+3\cos x\Big|_0^{\pi/3}$
$= (\pi-0)+\left(\frac{3}{2}-3\right) = \pi-\frac{3}{2}$

70. $\int_0^{1/2} x\sqrt{1-16x^4}\,dx$

Let $u = 4x^2$. Then $du = 8x\,dx$.

When $x = 0, u = 0,$ and when $x = \frac{1}{2}, u = 1.$

Thus, $\int_0^{1/2} x\sqrt{1-16x^4}\,dx = \frac{1}{8}\int_0^1 \sqrt{1-u^2}\,du.$

Note that this integral represents the area of right upper quarter of a circle centered at the origin with a radius of 1.

Area of circle $= \pi r^2 = \pi(1^2) = \pi$

$$\int_0^1 \sqrt{1-u^2}\,du = \frac{\pi}{4} \Rightarrow$$

$$\frac{1}{8}\int_0^1 \sqrt{1-u^2}\,du = \frac{1}{8}\cdot\frac{\pi}{4} = \frac{\pi}{32}$$

71. $\int_0^{\sqrt{2}} 4x\sqrt{4-x^2}\,dx$

Let $u = x^2$. Then $du = 2x\,dx$.

When $x = 0, u = 0.$ When $x = \sqrt{2}, u = 2.$

Substitute:

$$\int_0^{\sqrt{2}} 4x\sqrt{4-x^2}\,dx = 2\int_0^2 \sqrt{4-u^2}\,du$$

The integral on the right gives the area of a semicircle of radius 2 (the top half of a circle of radius 2 centered at the origin). This area is $\frac{1}{2}\pi(2)^2 = 2\pi$. Thus

$$\int_0^{\sqrt{2}} 4x\sqrt{4-x^2}\,dx = 2\pi.$$

72. $\int_1^{e^5} \frac{\sqrt{25-(\ln x)^2}}{x}\,dx$

Let $u = \ln x$. Then $du = \frac{1}{x}\,dx$.

When $x = e^5$, $u = \ln(e^5) = 5$.

When $x = 1$, $u = \ln(1) = 0$.

Thus, $\int_1^{e^5} \frac{\sqrt{25-(\ln x)^2}}{x}\,dx = \int_0^5 \sqrt{25-u^2}\,du.$

Note that this integral represents the area of a right upper quarter of a circle centered at the origin with a radius of 5.

Area of circle $= \pi r^2 = \pi(5)^2 = 25\pi$

$$\int_0^5 \sqrt{25-u^2}\,du = \frac{25\pi}{4}$$

73. $\int_1^{\sqrt{7}} 2x\sqrt{36-(x^2-1)^2}\,dx$

Let $u = x^2 - 1$. Then $du = 2x\,dx$.

When $x = \sqrt{7}, u = (\sqrt{7})^2 - 1 = 6.$

When $x = 1, u = (\sqrt{1})^2 - 1 = 0.$

Thus,

$$\int_1^{\sqrt{7}} \sqrt{36-(x^2-1)^2}\,(2x)\,dx = \int_0^6 \sqrt{36-u^2}\,du.$$

Note that this integral represents the area of a right upper quarter of a circle centered at the origin with a radius of 6.

Area of circle $= \pi r^2 = \pi(6)^2 = 36\pi$

$$\int_0^6 \sqrt{36-u^2}\,du = \frac{36\pi}{4} = 9\pi$$

74. $f(x) = \sqrt{4x-3};\ [1, 3]$

$$\begin{aligned}\text{Area} &= \int_1^3 \sqrt{4x-3}\,dx = \int_1^3 (4x-3)^{1/2}\,dx\\ &= \frac{2}{3}\cdot\frac{1}{4}\cdot(4x-3)^{3/2}\Big|_1^3\\ &= \frac{1}{6}(9)^{3/2} - \frac{1}{6}(1)^{3/2} = \frac{1}{6}(26) = \frac{13}{3}\end{aligned}$$

75. $f(x) = (3x+2)^6;\ [-2, 0]$

$$\begin{aligned}\text{Area} &= \int_{-2}^0 (3x+2)^6\,dx = \frac{(3x+2)^7}{21}\Bigg|_{-2}^0\\ &= \frac{2^7}{21} - \frac{(-4)^7}{21} = \frac{5504}{7}\end{aligned}$$

76. $f(x) = xe^{x^2};\ [0, 2]$

$$\begin{aligned}\text{Area} &= \int_0^2 xe^{x^2}\,dx = \frac{e^{x^2}}{2}\Bigg|_0^2 = \frac{e^4}{2} - \frac{1}{2}\\ &= \frac{e^4-1}{2} \approx 26.80\end{aligned}$$

77. $f(x) = 1 + e^{-x};\ [0, 4]$

$$\begin{aligned}\int_0^4 (1+e^{-x})\,dx &= (x - e^{-x})\Big|_0^4\\ &= (4 - e^{-4}) - (0 - e^0)\\ &= 5 - e^{-4} \approx 4.982\end{aligned}$$

78. $f(x) = 5 - x^2,\ g(x) = x^2 - 3$

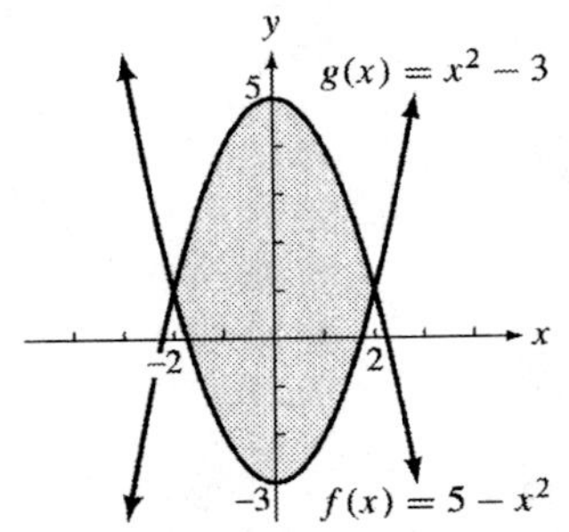

Points of intersection:

$5 - x^2 = x^2 - 3 \Rightarrow 2x^2 - 8 = 0 \Rightarrow$

$2(x^2 - 4) = 0 \Rightarrow x = \pm 2$

Since $f(x) \ge g(x)$ in $[-2, 2]$, the area between the graphs is

$$\int_{-2}^{2} [f(x) - g(x)]\,dx$$
$$= \int_{-2}^{2} [(5 - x^2) - (x^2 - 3)]\,dx$$
$$= \int_{-2}^{2} (-2x^2 + 8)\,dx$$
$$= \left(\frac{-2x^3}{3} + 8x\right)\Bigg|_{-2}^{2}$$
$$= -\frac{2}{3}(8) + 16 + \frac{2}{3}(-8) - 8(-2)$$
$$= \frac{-32}{3} + 32 = \frac{64}{3}$$

79. $f(x) = x^2 - 4x;\ g(x) = x - 6$

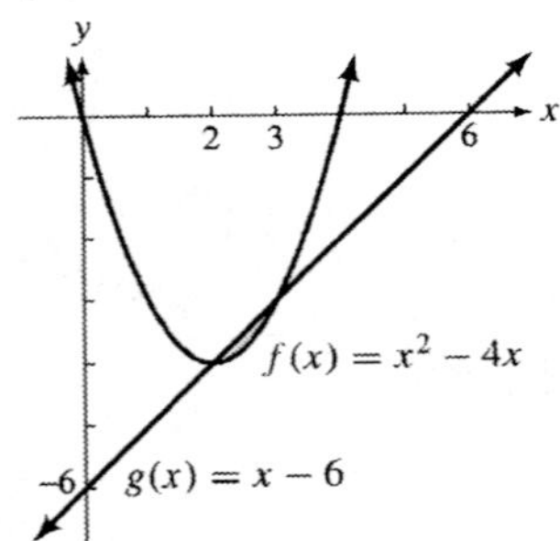

Find the points of intersection.

$x^2 - 4x = x - 6 \Rightarrow x^2 - 5x + 6 = 0 \Rightarrow$

$(x - 3)(x - 2) = 0 \Rightarrow x = 2 \quad \text{or} \quad x = 3$

Since $g(x) \ge f(x)$ in the interval $[2, 3]$, the area between the graphs is

$$\int_{2}^{3} [g(x) - f(x)]\,dx$$
$$= \int_{2}^{3} [(x - 6) - (x^2 - 4x)]\,dx$$
$$= \int_{2}^{3} (-x^2 + 5x - 6)\,dx$$
$$= \left(\frac{-x^3}{3} + \frac{5x^2}{2} - 6x\right)\Bigg|_{2}^{3}$$
$$= \frac{-27}{3} + \frac{5(9)}{2} - 6(3) - \frac{-8}{3} - \frac{5(4)}{2} + 6(2)$$
$$= -\frac{19}{3} + \frac{25}{2} - 6 = \frac{1}{6}.$$

80. $f(x) = x^2 - 4x,\ g(x) = x + 6,\ \ x = -2,\ x = 4$

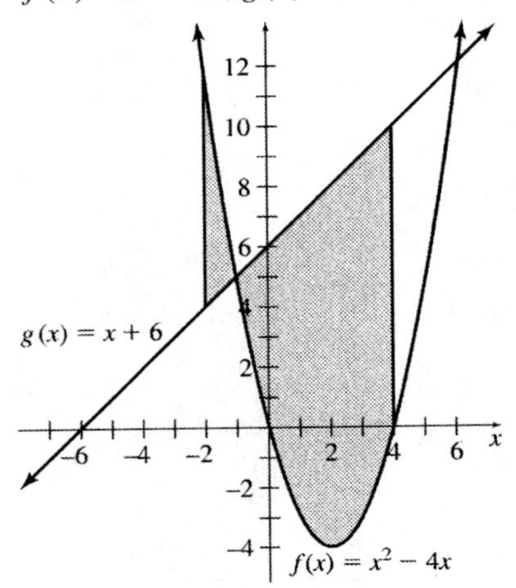

Points of intersection:

$x^2 - 4x = x + 6 \Rightarrow x^2 - 5x - 6 = 0 \Rightarrow$

$(x + 1)(x - 6) = 0 \Rightarrow x = -1 \text{ or } x = 6$

Thus, the area is

$$\int_{-2}^{-1} [x^2 - 4x - (x + 6)]\,dx + \int_{-1}^{4} [x + 6 - (x^2 - 4x)]\,dx$$
$$= \left(\frac{x^3}{3} - \frac{5x^2}{2} - 6x\right)\Bigg|_{-2}^{-1} + \left(-\frac{x^3}{3} + \frac{5x^2}{2} + 6x\right)\Bigg|_{-1}^{4}$$
$$= \left(\frac{19}{6} + \frac{2}{3}\right) + \left(\frac{128}{3} + \frac{19}{6}\right) = \frac{149}{3}$$

81. $f(x)=5-x^2,\ g(x)=x^2-3,\ x=0,\ x=4$

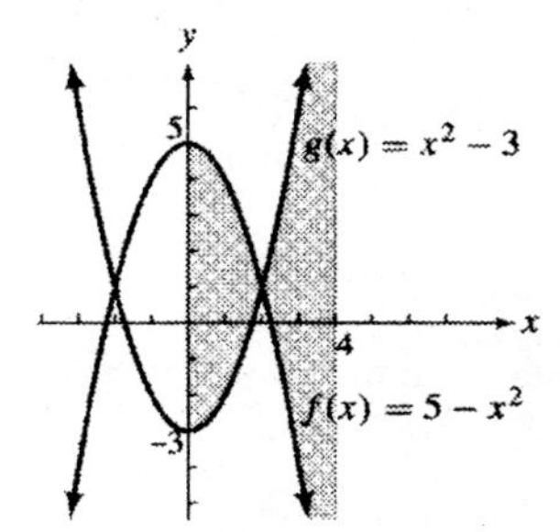

Find the points of intersection.

$5-x^2=x^2-3 \Rightarrow 8=2x^2 \Rightarrow 4=x^2 \Rightarrow \pm 2=x$

The curves intersect at $x = 2$ and $x = -2$.

Thus, area is

$$\int_0^2 [(5-x^2)-(x^2-3)]\,dx + \int_2^4 [(x^2-3)-(5-x^2)]\,dx$$

$$= \int_0^2 (-2x^2+8)\,dx + \int_2^4 (2x^2-8)\,dx$$

$$= \left(\frac{-2x^3}{3}+8x\right)\Bigg|_0^2 + \left(\frac{2x^3}{3}-8x\right)\Bigg|_2^4$$

$$= \frac{-16}{3}+16+\left(\frac{128}{3}-32\right)-\left(\frac{16}{3}-16\right)$$

$$= \frac{32}{3}+\frac{128}{3}-32-\frac{16}{3}+16=32$$

82. $f(t)=100-t\sqrt{0.4t^2+1}$

The total number of additional spiders in the first ten months is $\int_0^{10}(100-t\sqrt{0.4t^2+1})\,dt$, where t is the time in months.

$$\int_0^{10}(100-t\sqrt{0.4t^2+1})\,dt = \int_0^{10}100dt - \int_0^{10} t\sqrt{0.4t^2+1}\,dt.$$

Let $u=0.4t^2+1$, so that $du=0.8t\,dt$ and $\frac{1}{0.8}du=t\,dt$.

When $t=10,\ u=41$. When $t=0,\ u=1$.

$$\int_0^{10}(100-t\sqrt{0.4t^2+1})\,dt$$

$$= \int_0^{10}100dt - \int_0^{10} t\sqrt{0.4t^2+1}\,dt$$

$$= \int_0^{10}100\,dt - \frac{1}{0.8}\int_1^{41} u^{1/2}du$$

$$= 100t\Bigg|_0^{10} - \frac{5}{4}\cdot\frac{u^{3/2}}{\frac{3}{2}}\Bigg|_1^{41}$$

$$= 1000-\frac{5}{6}u^{3/2}\Bigg|_1^{41} \approx 782$$

The total number of additional spiders in the first 10 months is about 782.

83. The total number of infected people over the first four months is $\int_0^4 \frac{100t}{t^2+1}dt$, where t is time in months. Let $u=t^2+1$, so that $du = 2t\,dt$ and $50\,du = 100t\,dt$.

If $t=4,\ u=17$, so

$$\int_0^4 \frac{100t}{t^2+1}dt = 50\int_1^{17}\frac{1}{u}du = 50\ln|u|\Bigg|_1^{17}$$

$$= 50\ \ln 17 - 50\ \ln|1|$$

$$= 50\ln 17 \approx 141.66.$$

Approximately 142 people are infected.

84. **(a)** The total area is the area of the triangle on [0, 12] with height 0.024 plus the area of the rectangle on [12, 17.6] with height 0.024.

$$A=\frac{1}{2}(12-0)(0.024)+(17.6-12)(0.024)$$
$$=0.2784$$

(b) On [0, 12] we define the function $f(x)$ with slope $\frac{0.024-0}{12-0}=0.002$ and y-intercept 0: $f(x)=0.002x$

On [12, 17.6], define $g(x)$ as the constant value: $g(x)=0.024$.

The area is the sum of the integrals of these two functions.

$$A=\int_0^{12}0.002x\,dx+\int_{12}^{17.6}0.024\,dx$$

$$=0.001x^2\Big|_0^{12}+0.024x\Big|_{12}^{17.6}$$

$$=0.001(12^2-0^2)+0.024(17.6-12)$$
$$=0.2784$$

85. Since answers are found by estimating values on the graph, exact answers may vary slightly; however when rounded to the nearest hundred, all answers should be the same. Sample solution:

(a) Left endpoints: Read the values of the function from the graph, using the open circles for the functional values. The values of $f(x)$ are listed in the table.

x	0	2	5	15	30	45	60
$f(x)$	30	50	60	105	85	70	55

The values give the heights of 6 rectangles. The width of each rectangle is found by subtracting subsequent values of x. We estimate the area under the curve as

$$\sum_{i=1}^{6} f(x_i)\,\Delta x_i = 30(2)+50(3)+60(10)+105(15)+85(15)+70(15)=4710$$

$$\sum_{i=1}^{6} f(x_i)\,\Delta x_i = 50(2)+60(3)+105(10)+85(15)+70(15)+55(15)=4480.$$

Average: $\dfrac{4710+4480}{2}=4595 \approx 4600$ pM.

(b) Read the values of the function from the graph, using the closed circles for the functional values. The values of x and $g(x)$ are listed in the table.

x	0	2	5	15	30	45	60
$g(x)$	20	42	42	70	52	40	20

The values give the heights of 6 rectangles. The width of each rectangle is found by subtracting subsequent values of x. We estimate the area under the curve as

$$\sum_{i=1}^{6} g(x_i)\,\Delta x_i = 20(2)+42(3)+42(10)+70(15)+52(15)+40(15)=3016.$$

Right endpoints: We estimate the area under the curve as

$$\sum_{i=1}^{6} g(x_i)\,\Delta x_i = 42(2)+42(3)+70(10)+52(15)+40(15)+20(15)=2590.$$

Average: $\dfrac{3016+2590}{2}=2803 \approx 2800$ pM.

(c) $\dfrac{4600-2800}{2800} \approx 0.6428$

The area under the curve is about 64% more for the fasting sheep.

86. (a) $E=\int_{-\infty}^{\infty}\left(\dfrac{|t+x-\theta|}{\theta}+\dfrac{a(t+x)^2}{\theta^2}\right)h(t;\,\theta)\,dt$

Letting $\theta=1$ gives $E=\int_{-\infty}^{\infty}\left((|t+x-1|)+a(t+x)^2\right)h(t;\,1)\,dt.$

If $t<0.5$ or $t>1.5$, $h(t,\,\theta)=0,$ so we must find $\int_{0.5}^{1.5}\left((|t+x-1|)+a(t+x)^2\right)h(t;\,1)\,dt.$ Using the hint and the fact that for $0.5\theta \le t \le 1.5\theta,\ h(t,\,1)=\dfrac{1}{\theta}=1,$ we have

$$\int_{0.5}^{1.5}\left((|t+x-1|)+a(t+x)^2\right)h(t;\,1)\,dt$$
$$=\int_{0.5}^{1-x}\left(-(t+x-1)+a(t+x)^2\right)dt+\int_{1-x}^{1.5}\left((t+x-1)+a(t+x)^2\right)dt$$

(continued on next page)

(continued)

Let $u = t + x$. Then $t + x - 1 = u - 1$ and $du = dt$. When $t = 0.5$, $u = x + 0.5$; when $t = 1 - x$, $u = 1$; and when $t = 1.5$, $u = x + 1.5$

$$\int_{0.5}^{1.5}\left(\left(|t+x-1|\right)+a(t+x)^2\right)h(t;1)\,dt$$
$$=\int_{0.5}^{1-x}\left(-(t+x-1)+a(t+x)^2\right)dt+\int_{1-x}^{1.5}\left((t+x-1)+a(t+x)^2\right)dt$$
$$=\int_{x+0.5}^{1}\left(-(u-1)+au^2\right)du+\int_{1}^{x+1.5}\left(u-1+au^2\right)du$$
$$=\int_{x+0.5}^{1}\left(-u+1+au^2\right)du+\int_{1}^{x+1.5}\left(u-1+au^2\right)du$$
$$=\left(-\frac{u^2}{2}+u+\frac{au^3}{3}\right)\Bigg|_{x+0.5}^{1}+\left(\frac{u^2}{2}-u+\frac{au^3}{3}\right)\Bigg|_{1}^{x+1.5}$$
$$=\left(-\frac{1}{2}+1+\frac{a}{3}\right)-\left(-\frac{(x+0.5)^2}{2}+(x+0.5)+\frac{a(x+0.5)^3}{3}\right)$$
$$+\left(\frac{(x+1.5)^2}{2}-(x+1.5)+\frac{a(x+1.5)^3}{3}\right)-\left(\frac{1}{2}-1+\frac{a}{3}\right)$$
$$=\frac{(x+0.5)^2}{2}+\frac{(x+1.5)^2}{2}-\frac{a(x+0.5)^3}{3}+\frac{a(x+1.5)^3}{3}-(x+0.5)-(x+1.5)$$
$$+\left(-\frac{1}{2}+1+\frac{a}{3}\right)-\left(\frac{1}{2}-1+\frac{a}{3}\right)$$
$$=\frac{x^2+x+0.25+x^2+3x+2.25}{2}+\frac{a}{3}\left[(x+1.5)^3-(x+0.5)^3\right]-2x-1$$
$$=\frac{2x^2+4x+2.5}{2}-2x-1+\frac{a}{3}\left[(x+1.5)^3-(x+0.5)^3\right]$$
$$=\frac{2x^2+4x+2.5-4x-2}{2}+\frac{a}{3}\left[(x+1.5)^3-(x+0.5)^3\right]$$
$$=x^2+\frac{1}{4}+\frac{a}{3}\left[(x+1.5)^3-(x+0.5)^3\right]$$

(b) $\frac{dE}{dx}=2x+\frac{a}{3}\left[3(x+1.5)^2-3(x+0.5)^2\right]=2x+a\left[\left(x^2+3x+2.25\right)-\left(x^2+x+0.25\right)\right]=2x+2ax+2a$

$2x+2ax+2a=0\Rightarrow x+ax=-a\Rightarrow x(1+a)=-a\Rightarrow x=\frac{-a}{1+a}$

The first derivative test shows that this is a minimum.

(c) Now let $h(t, \theta) = \frac{1}{\theta}$. Follow the same reasoning as in part (a). If $0.5\theta \le t < \theta - x$, $|t + x - \theta| = -(t + x - \theta) = -t - x + \theta$. If $\theta - x \le t \le 1.5\theta$, $|t + x - \theta| = t + x - \theta$.

$$E = \int_{0.5\theta}^{1.5\theta}\left((|t + x - \theta|) + \frac{a(t + x)^2}{\theta^2}\right)\frac{1}{\theta}dt$$
$$= \frac{1}{\theta^2}\int_{0.5\theta}^{\theta-x}(-t - x + \theta)\,dt + \frac{1}{\theta^2}\int_{\theta-x}^{1.5\theta}(t + x - \theta)\,dt + \frac{1}{\theta^3}\int_{0.5\theta}^{1.5\theta}\left(a(t + x)^2\right)dt$$
$$= \frac{1}{\theta^2}\int_{0.5\theta}^{\theta-x}(-t + (\theta - x))\,dt + \frac{1}{\theta^2}\int_{\theta-x}^{1.5\theta}(t - (\theta - x))\,dt + \frac{1}{\theta^3}\int_{0.5\theta}^{1.5\theta}\left(a(t + x)^2\right)dt$$
$$= \frac{1}{\theta^2}\left[-\frac{t^2}{2} + (\theta - x)t\right]_{0.5\theta}^{\theta-x} + \frac{1}{\theta^2}\left[\frac{t^2}{2} - (\theta - x)t\right]_{\theta-x}^{1.5\theta} + \frac{a}{3\theta^3}(t + x)^3\Big|_{0.5\theta}^{1.5\theta}$$
$$= \frac{1}{\theta^2}\left[-\frac{(\theta - x)^2}{2} + (\theta - x)^2\right] - \frac{1}{\theta^2}\left[-\frac{(0.5\theta)^2}{2} + (\theta - x)(0.5\theta)\right]$$
$$+ \frac{1}{\theta^2}\left[\frac{(1.5\theta)^2}{2} - (\theta - x)(1.5\theta)\right] - \frac{1}{\theta^2}\left[\frac{(\theta - x)^2}{2} - (\theta - x)^2\right] + \frac{a}{3\theta^3}\left[(1.5\theta + x)^3 - (0.5\theta + x)^3\right]$$
$$= \frac{1}{\theta^2}\left(\left[-\frac{(\theta - x)^2}{2} + (\theta - x)^2\right] - \left[\frac{(\theta - x)^2}{2} - (\theta - x)^2\right]\right)$$
$$- \frac{1}{\theta^2}\left(\left[-\frac{(0.5\theta)^2}{2} + (\theta - x)(0.5\theta)\right] - \left[\frac{(1.5\theta)^2}{2} - (\theta - x)(1.5\theta)\right]\right)$$
$$+ \frac{a}{3\theta^3}\left[\left(3.375\theta^3 + 6.75\theta^2x + 4.5\theta x^2 + x^3\right) - \left(0.125\theta^3 + 0.75\theta^2x + 1.5\theta x^2 + x^3\right)\right]$$
$$= \frac{1}{\theta^2}\left((\theta - x)^2 - \left(2\theta(\theta - x) - 1.25\theta^2\right)\right) + \frac{a}{3\theta^3}\left(3.25\theta^3 + 6\theta^2x + 3\theta x^2\right)$$
$$= \frac{(\theta - x)^2}{\theta^2} - 2 + \frac{2x}{\theta} - 1.25 + \frac{a}{3\theta^3}\left(3.25\theta^3 + 6\theta^2x + 3\theta x^2\right)$$

Now find $\frac{dE}{dx}$.

$$\frac{dE}{dx} = -\frac{2(\theta - x)}{\theta^2} + \frac{2}{\theta} + \frac{2a}{\theta} + \frac{2ax}{\theta^2} = -\frac{2}{\theta^2}((\theta - x) - \theta - a\theta - ax)$$
$$= -\frac{2}{\theta^2}(-x - a\theta - ax) = \frac{2}{\theta^2}(x + ax) + \frac{2a}{\theta}$$

$$\frac{2}{\theta^2}(x + ax) + \frac{2a}{\theta} = 0 \Rightarrow 2x + 2ax + 2a\theta = 0 \Rightarrow x(1 + a) = -a\theta \Rightarrow x = -\frac{a\theta}{1 + a}$$

87. (a) Total amount

$$\approx \frac{1}{2}(2.097) + 2.060 + 1.989 + 1.893 + 1.857 + 1.853 + 1.830 + 1.954 + 2.000 + 2.063 + \frac{1}{2}(2.377)$$
$$\approx 19.736$$

The estimate is 19.736 billion barrels.

(b) The left endpoint sum is 19.596 and the right endpoint sum is 19.876. Their average is $\dfrac{19.596+19.876}{2}=19.736$

The sum in part (a) is

$$\frac{1}{2}(n_1+n_{11})+(n_2+n_3+\cdots+n_9+n_{10}).$$

The left endpoint sum is $\sum_{a=1}^{a=10} n_a$, while the right endpoint sum is $\sum_{a=2}^{a=11} n_a$.

$$\sum_{a=1}^{a=10} n_a+\sum_{a=2}^{a=11} n_a = n_1+2n_2+2n_3+\cdots+2n_{10}+n_{11}$$

Dividing by 2 gives

$$\frac{1}{2}(n_1+n_{11})+(n_2+n_3+\cdots+n_9+n_{10}).$$

(c) $y=0.001382t^3-0.0138t^2-0.0206t+2.1915$

Using the fnInt function on a TI-84 Plus calculator, we have

$$\int_2^{12}\left(0.001382t^3-0.0138t^2-0.0206t+2.1915\right)dt \approx 19.72$$

This is close to the estimate in part (a).

88. (a) Total amount $\approx \frac{1}{2}(335{,}869)+331{,}055+331{,}208+330{,}195+325{,}453+313{,}357+278{,}986+259{,}899+\frac{1}{2}(252{,}876) \approx 2{,}464{,}526$

The estimate of the total amount of property damage is \$2,464,526.

(b) The left endpoint sum is 2,506,022 and the right endpoint sum is 2,423,029. Their average is

$$\frac{2{,}506{,}022+2{,}423{,}029}{2}\approx 2{,}464{,}526$$

The sum in part (a) is

$$\frac{1}{2}(n_1+n_9)+(n_2+n_3+\cdots+n_7+n_8).$$

The left endpoint sum is $\sum_{a=1}^{a=8} n_a$, while the right endpoint sum is $\sum_{a=2}^{a=9} n_a$.

$$\sum_{a=1}^{a=8} n_a+\sum_{a=2}^{a=9} n_a = n_1+2n_2+2n_3+\cdots+2n_7+2n_8+n_9$$

Dividing by 2 gives

$$\frac{1}{2}(n_1+n_9)+(n_2+n_3+\cdots+n_7+n_8).$$

(c) $y=-11{,}112t+373{,}216$

$$\int_2^{10}(-11{,}112t+373{,}216)\,dt = \left(-\frac{11{,}112}{2}t^2+373{,}216t\right)\Bigg|_2^{10}$$
$$=\left(-\frac{11{,}112}{2}(100)+373{,}216(10)\right)-\left(-\frac{11{,}112}{2}(4)+373{,}216(2)\right)$$
$$=\$2{,}452{,}352$$

This is close to the estimate in part (a).

89. $v(t)=t^2-2t$

$$s(t)=\int_0^t(t^2-2t)\,dt=\frac{t^3}{3}-t^2+s_0$$

If $t=3$, $s=8$. So, $8=9-9+s_0 \Rightarrow 8=s_0$

Thus, $s(t)=\dfrac{t^3}{3}-t^2+8.$

90. (a) Residential usage of natural gas is probably periodic.

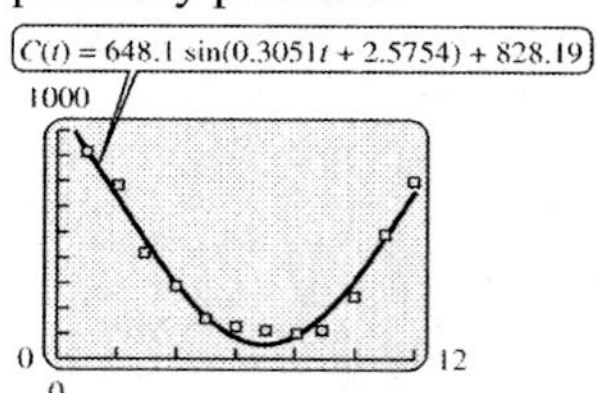

(b) $C(t)=648.1\sin(0.3051t+2.5754)+828.19$

(c) Integrating $C(t)$ from 0 to 12 using a calculator gives the estimate 6024 million cubic feet; the sum of the values in the table is about 5933 million cubic feet. (Note that in carrying out the integration we used the unrounded coefficients as given by sine regression on the calculator. When possible, all intermediate results should be kept in full precision until the final answer is obtained.)

(d) The function in the model has a period of
$T = \frac{2\pi}{b} = \frac{2\pi}{0.3051} \approx 20.6$ months.
Yes, we might expect a period of 12 months.

91. (a) Enter the data into a graphing calculator and plot.

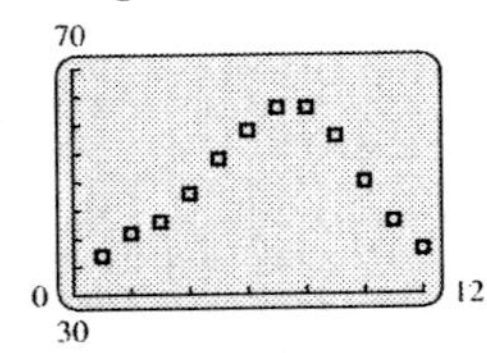

(b) Use the graphing calculator to find a trigonometric function.
$y = 12.5680 \sin(0.547655t - 2.34990) + 50.2671$

(c) Graph the function with the data.

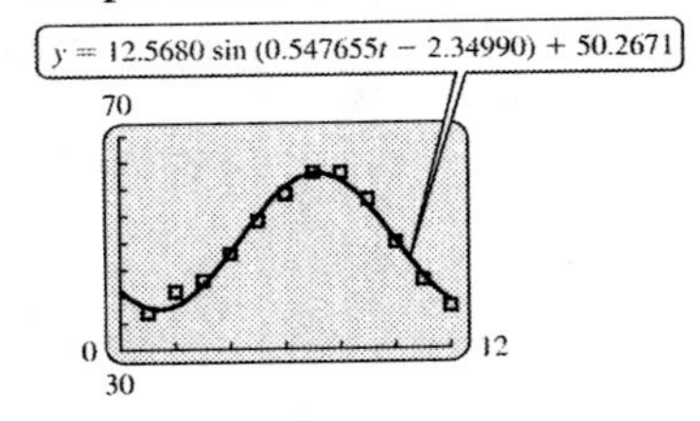

(d) $T = \frac{2\pi}{b} \approx 11.4729$
The period is about 11.5 months. Yes, this period is reasonable.

92. (a) $\frac{d}{dx} \ln|\sec x + \tan x| = \frac{\sec x \tan x + \sec^2 x}{\sec x + \tan x}$
$= \frac{\sec x(\tan x + \sec x)}{(\sec x + \tan x)}$
$= \sec x$

(b) $\frac{d}{dx}(-\ln|\sec x - \tan x|)$
$= -\frac{\sec x \tan x - \sec^2 x}{\sec x - \tan x}$
$= -\frac{\sec x(\tan x - \sec x)}{(\sec x - \tan x)}$
$= -(-\sec x) = \sec x$

(c) $\int \sec x = \ln|\sec x + \tan x| + C$ or
$\int \sec x = -\ln|\sec x - \tan x| + C$

(d) $D(\theta) = k\int_0^\theta \sec x\, dx = k \ln|\sec x + \tan x|\Big|_0^\theta$
$= k \ln|\sec\theta + \tan\theta| - k\ln|\sec 0 + \tan 0|$
$= k \ln|\sec\theta + \tan\theta| - k\ln|1+0|$
$= k \ln|\sec\theta + \tan\theta| - k\ln 1$
$= k \ln|\sec\theta + \tan\theta| - 0$
$= k \ln|\sec\theta + \tan\theta|$

Since $D(34°03') = 7$,
$7 = k\ln|\sec 34°03' + \tan 34°03'|$
$k = \frac{7}{\ln|\sec 34°03' + \tan 34°03'|}$
$\approx 11.0635.$
$D(40°45')$
$\approx 11.0635 \ln|\sec 40°45' + \tan 40°45'|$
≈ 8.63
New York City should be placed approximately 8.63 inches from the equator.

(e) $D(25°46')$
$\approx 11.0635 \ln|\sec 25°46' + \tan 25°46'|$
≈ 5.15
Miami should be placed approximately 5.15 inches from the equator.

93. $S'(t) = 225 - t^2,\ C'(t) = t^2 + 25t + 150$
$S'(t) = C'(t) \Rightarrow 225 - t^2 = t^2 + 25t + 150 \Rightarrow$
$2t^2 + 25t - 75 = 0 \Rightarrow (2t-5)(t+15) = 0 \Rightarrow$
$t = \frac{5}{2} = 2.5$
The company should use the machinery for 2.5 years.
$\int_0^{2.5}[(225 - t^2) - (t^2 + 25t + 150)]dt$
$= \int_0^{2.5}(-2t^2 - 25t + 75)\,dt$
$= \left(\frac{-2t^3}{3} - \frac{25t^2}{2} + 75t\right)\Bigg|_0^{2.5}$
$= \frac{-2(2.5^3)}{3} - \frac{25(2.5^2)}{2} + 75(2.5)$
$\approx 98.95833 \approx 99{,}000$
The net savings are about $99,000.

Chapter 8

FURTHER TECHNIQUES AND APPLICATIONS OF INTEGRATION

8.1 Numerical Integration

1. $\int_0^2 (3x^2 + 2)\,dx$

$n = 4, b = 2, a = 0, f(x) = 3x^2 + 2$

i	x_i	$f(x_i)$
0	0	2
1	$\frac{1}{2}$	2.75
2	1	5
3	$\frac{3}{2}$	8.75
4	2	14

(a) Trapezoidal rule:

$$\int_0^2 (3x^2 + 2)\,dx \approx \frac{2-0}{4}\left[\frac{1}{2}(2) + 2.75 + 5 + 8.75 + \frac{1}{2}(14)\right] = 0.5(24.5) = 12.25$$

(b) Simpson's rule:

$$\int_0^2 (3x^2 + 2)\,dx \approx \frac{2-0}{3(4)}[2 + 4(2.75) + 2(5) + 4(8.75) + 14] = \frac{2}{12}(72) = 12$$

(c) Exact value:

$$\int_0^2 (3x^2 + 2)\,dx = (x^3 + 2x)\Big|_0^2 = (8 + 4) - 0 = 12$$

3. $\int_{-1}^3 \frac{3}{5-x}\,dx$

$n = 4, b = 3, a = -1, f(x) = \frac{3}{5-x}$

i	x_i	$f(x_i)$
0	−1	0.5
1	0	0.6
2	1	0.75
3	2	1
4	3	1.5

(a) Trapezoidal rule:

$$\int_{-1}^3 \frac{3}{5-x}\,dx \approx \frac{3-(-1)}{4}\left[\frac{1}{2}(0.5) + 0.6 + 0.75 + 1 + \frac{1}{2}(1.5)\right] = 1(3.35) = 3.35$$

(b) Simpson's rule:

$$\int_{-1}^3 \frac{3}{5-x}\,dx \approx \frac{3-(-1)}{3(4)}[0.5 + 4(0.6) + 2(0.75) + 4(1) + 1.5] = \frac{1}{3}\left(\frac{99}{10}\right) = \frac{33}{10} \approx 3.3$$

(c) Exact value:

$$\int_{-1}^3 \frac{3}{5-x}\,dx = -3\ln|5-x|\Big|_{-1}^3 = -3(\ln|2| - \ln|6|) = 3\ln 3 \approx 3.296$$

5. $\int_{-1}^2 (2x^3 + 1)\,dx$

$n = 4, b = 2, a = -1, f(x) = 2x^3 + 1$

i	x_i	$f(x)$
0	−1	−1
1	$-\frac{1}{4}$	$\frac{31}{32}$
2	$\frac{1}{2}$	$\frac{5}{4}$
3	$\frac{5}{4}$	$\frac{157}{32}$
4	2	17

(a) Trapezoidal rule:

$$\int_{-1}^{2}(2x^3+1)\,dx$$
$$\approx \frac{2-(-1)}{4}\left[\frac{1}{2}(-1)+\frac{31}{32}+\frac{5}{4}+\frac{157}{32}+\frac{1}{2}(17)\right]$$
$$=0.75(15.125)\approx 11.34$$

(b) Simpson's rule:

$$\int_{-1}^{2}(2x^3+1)\,dx$$
$$\approx \frac{2-(-1)}{3(4)}\left[-1+4\left(\frac{31}{32}\right)+2\left(\frac{5}{4}\right)+4\left(\frac{157}{32}\right)+17\right]$$
$$=\frac{1}{4}(42)\ =10.5$$

(c) Exact value:

$$\int_{-1}^{2}(2x^3+1)\,dx=\left(\frac{x^4}{2}+x\right)\Bigg|_{-1}^{2}$$
$$=(8+2)-\left(\frac{1}{2}-1\right)=10.5$$

7. $\int_1^5 \frac{1}{x^2}\,dx$

$n=4,\ b=5,\ a=1,\ f(x)=\frac{1}{x^2}$

i	x_i	$f(x_i)$
0	1	1
1	2	0.25
2	3	0.1111
3	4	0.0625
4	5	0.04

(a) Trapezoidal rule:

$$\int_1^5 \frac{1}{x^2}\,dx$$
$$\approx \frac{5-1}{4}\left[\frac{1}{2}(1)+0.25+0.1111+0.0625+\frac{1}{2}(0.04)\right]$$
$$\approx 0.9436$$

(b) Simpson's rule:

$$\int_1^5 \frac{1}{x^2}\,dx$$
$$\approx \frac{5-1}{12}[1+4(0.25)+2(0.1111)+4(0.0625)+0.04)]$$
$$\approx 0.8374$$

(c) Exact value:

$$\int_1^5 x^{-2}dx=-x^{-1}\Big|_1^5=-\frac{1}{5}+1=0.8$$

9. $\int_0^1 4xe^{-x^2}\,dx$

$n=4,\ b=1,\ a=0,\ f(x)=4xe^{-x^2}$

i	x_i	$f(x_i)$
0	0	0
1	$\frac{1}{4}$	$e^{-1/16}$
2	$\frac{1}{2}$	$2e^{-1/4}$
3	$\frac{3}{4}$	$3e^{-9/16}$
4	1	$4e^{-1}$

(a) Trapezoidal rule:

$$\int_0^1 4xe^{-x^2}\,dx$$
$$\approx \frac{1-0}{4}\left[\frac{1}{2}(0)+e^{-1/16}+2e^{-1/4}+3e^{-9/16}+\frac{1}{2}(4e^{-1})\right]$$
$$=\frac{1}{4}(e^{-1/16}+2e^{-1/4}+3e^{-9/16}+2e^{-1})$$
$$\approx 1.236$$

(b) Simpson's rule:

$$\int_0^1 4xe^{-x^2}\,dx$$
$$\approx \frac{1-0}{3(4)}[0+4(e^{-1/16})+2(2e^{-1/4})+4(3e^{-9/16})+4e^{-1}]$$
$$=\frac{1}{12}(4e^{-1/16}+4e^{-1/4}+12e^{-9/16}+4e^{-1})$$
$$\approx 1.265$$

(c) Exact value:

$$\int_0^1 4xe^{-x^2}\,dx=-2e^{-x^2}\Big|_0^1=(-2e^{-1})-(-2)$$
$$=2-2e^{-1}\approx 1.264$$

11. $\int_0^{\pi} \sin x\, dx$

$n = 4, b = \pi, a = 0, f(x) = \sin x$

i	x_i	$f(x_i)$
0	0	0
1	$\frac{\pi}{4}$	0.70711
2	$\frac{\pi}{2}$	1
3	$\frac{3\pi}{4}$	0.70711
4	π	0

(a) Trapezoidal rule:

$$\int_0^{\pi} \sin x\, dx \approx \frac{\pi - 0}{4}\left[\frac{1}{2}(0) + 0.70711 + 1 + 0.70711 + \frac{1}{2}(0)\right] \approx 1.8961$$

(b) Simpson's rule:

$$\int_0^{\pi} \sin x\, dx \approx \frac{\pi - 0}{3(4)}[0 + 4(0.70711) + 2(1) + 4(0.70711) + 0] \approx 2.0046$$

(c) Exact value:

$$\int_0^{\pi} \sin x\, dx = -\cos x\Big|_0^{\pi} = -\cos \pi + \cos 0 = -(-1) + 1 = 2$$

13. $y = \sqrt{4 - x^2}$

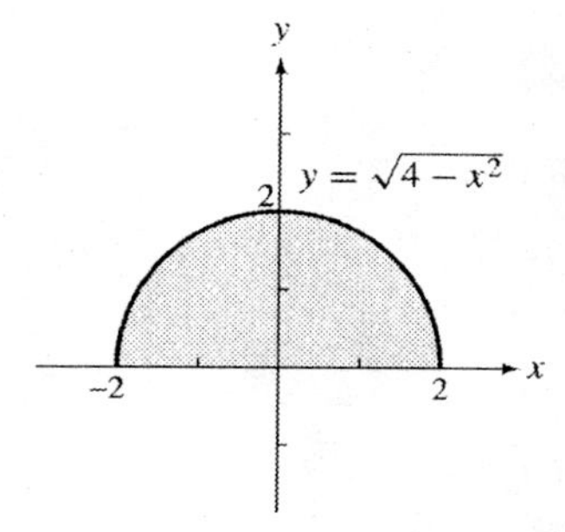

$n = 8, b = 2, a = -2, f(x) = \sqrt{4 - x^2}$

i	x_i	y
0	−2.0	0
1	−1.5	1.32288
2	−1.0	1.73205
3	−0.5	1.93649
4	0	2
5	0.5	1.93649
6	1.0	1.73205
7	1.5	1.32288
8	2.0	0

(a) Trapezoidal rule:

$$\int_{-2}^{2} \sqrt{4 - x^2}\, dx \approx \frac{2 - (-2)}{8} \cdot \left[\frac{1}{2}(0) + 1.32288 + 1.73205 + \cdots + \frac{1}{2}(0)\right] \approx 5.991$$

(b) Simpson's rule:

$$\int_{-2}^{2} \sqrt{4 - x^2}\, dx \approx \frac{2 - (-2)}{3(8)} \cdot [0 + 4(1.32288) + 2(1.73205) + 4(1.93649) + 2(2) + 4(1.93649) + 2(1.73205) + 4(1.32288) + 0] \approx 6.167$$

(c) Area of semicircle $= \frac{1}{2}\pi r^2 = \frac{1}{2}\pi(2)^2 \approx 6.283$

Simpson's rule is more accurate.

15. Since $f(x) > 0$ and $f''(x) > 0$ for all x between a and b, we know the graph of $f(x)$ on the interval from a to b is concave upward. Thus, the trapezoid that approximates the area will have an area greater than the actual area Thus, $T > \int_a^b f(x)\, dx$. The correct choice is **(b)**.

17. **(a)** $\int_0^1 x^4 dx = \left(\frac{1}{5}\right)x^5\Big|_0^1 = \frac{1}{5} = 0.2$

(b) $n = 4, b = 1, a = 0, f(x) = x^4$

$$\int_0^1 x^4 dx \approx \frac{1-0}{4}\left[\frac{1}{2}(0) + \frac{1}{256} + \frac{1}{16} + \frac{81}{256} + \frac{1}{2}(1)\right] = \frac{1}{4}\left(\frac{226}{256}\right) \approx 0.220703$$

$n = 8, b = 1, a = 0, f(x) = x^4$

$$\int_0^1 x^4 dx \approx \frac{1-0}{8}\left[\frac{1}{2}(0) + \frac{1}{4096} + \frac{1}{256} + \frac{81}{4096} + \frac{1}{16} + \frac{625}{4096} + \frac{81}{256} + \frac{2401}{4096} + \frac{1}{2}(1)\right] = \frac{1}{8}\left(\frac{6724}{4096}\right) \approx 0.205200$$

$n = 16, b = 1, a = 0, f(x) = x^4$

$$\int_0^1 x^4 dx \approx \frac{1-0}{16}\left[\frac{1}{2}(0) + \frac{1}{65,536} + \frac{1}{4096} + \frac{81}{65,536} + \frac{1}{256} + \frac{625}{65,536} + \frac{81}{4096} + \frac{2401}{65,536} + \frac{1}{16} + \frac{6561}{65,536} + \frac{625}{4096} + \frac{14,641}{65,536} + \frac{81}{256} + \frac{28,561}{65,536} + \frac{2401}{4096} + \frac{50,625}{65,536} + \frac{1}{2}(1)\right]$$

$$\approx \frac{1}{16}\left(\frac{211,080}{65,536}\right) \approx 0.201302$$

$n = 32, b = 1, a = 0, f(x) = x^4$

$$\int_0^1 x^4\, dx \approx \frac{1-0}{32}\left[\frac{1}{2}(0) + \frac{1}{1,048,576} + \frac{1}{65,536} + \frac{81}{1,048,576} + \frac{1}{4096} + \frac{625}{1,048,576} + \frac{81}{65,536} + \frac{2401}{1,048,576} + \frac{1}{256} + \frac{6561}{1,048,576} + \frac{625}{65,536} + \frac{14,641}{1,048,576} + \frac{81}{4096} + \frac{28,561}{1,048,576} + \frac{2401}{65,536} + \frac{50,625}{1,048,576} + \frac{1}{16} + \frac{83,521}{1,048,576} + \frac{6561}{65,536} + \frac{130,321}{1,048,576} + \frac{625}{4096} + \frac{194,481}{1,048,576} + \frac{14,641}{65,536} + \frac{279,841}{1,048,576} + \frac{81}{256} + \frac{390,625}{1,048,576} + \frac{28,561}{65,536} + \frac{531,441}{1,048,576} + \frac{2401}{4096} + \frac{707,281}{1,048,576} + \frac{50,625}{65,536} + \frac{923,521}{1,048,576} + \frac{1}{2}(1)\right]$$

$$\approx \frac{1}{32}\left(\frac{6,721,808}{1,048,576}\right) \approx 0.200325$$

To find error for each value of n, subtract as indicated.
$n = 4$: $(0.220703 - 0.2) = 0.020703$
$n = 8$: $(0.205200 - 0.2) = 0.005200$
$n = 16$: $(0.201302 - 0.2) = 0.001302$
$n = 32$: $(0.200325 - 0.2) = 0.000325$

(c) $p = 1$

$4^1(0.020703) = 4(0.020703) = 0.082812$
$8^1(0.005200) = 8(0.005200) = 0.0416$
Since these are not the same, try $p = 2$.
$4^2(0.020703) = 16(0.020703) = 0.331248$
$8^2(0.005200) = 64(0.005200) = 0.3328$
$16^2(0.001302) = 256(0.001302) = 0.333312$
$32^2(0.000325) = 1024(0.000325) = 0.3328$
Since these values are all approximately the same, the correct choice is $p = 2$.

19. **(a)** $\int_0^1 x^4 dx = \frac{1}{5}x^5\Big|_0^1 = \frac{1}{5} = 0.2$

(b) $n = 4, b = 1, a = 0, f(x) = x^4$

$$\int_0^1 x^4 dx \approx \frac{1-0}{3(4)}\left[0+4\left(\frac{1}{256}\right)+2\left(\frac{1}{16}\right)+4\left(\frac{81}{256}\right)+1\right]=\frac{1}{12}\left(\frac{77}{32}\right)\approx 0.2005208$$

$n = 8, b = 1, a = 0, f(x) = x^4$

$$\int_0^1 x^4\, dx \approx \frac{1-0}{3(8)}\left[0+4\left(\frac{1}{4096}\right)+2\left(\frac{1}{256}\right)+4\left(\frac{81}{4096}\right)+2\left(\frac{1}{16}\right)+4\left(\frac{625}{4096}\right)+2\left(\frac{18}{256}\right)+4\left(\frac{2401}{4096}\right)+1\right]$$

$$=\frac{1}{24}\left(\frac{4916}{1024}\right)\approx 0.2000326$$

$n = 16, b = 1, a = 0, f(x) = x^4$

$$\int_0^1 x^4 dx \approx \frac{1-0}{3(16)}\left[0+4\left(\frac{1}{65,536}\right)+2\left(\frac{1}{4096}\right)+4\left(\frac{81}{65,536}\right)+2\left(\frac{1}{256}\right)+4\left(\frac{625}{65,536}\right)+2\left(\frac{81}{4096}\right)\right.$$

$$+4\left(\frac{2401}{65,536}\right)+2\left(\frac{1}{16}\right)+4\left(\frac{6561}{65,536}\right)+2\left(\frac{625}{4096}\right)+4\left(\frac{14,641}{65,536}\right)+2\left(\frac{81}{256}\right)$$

$$\left.+4\left(\frac{28,561}{65,536}\right)+2\left(\frac{2401}{4096}\right)+4\left(\frac{50,625}{65,536}\right)+1\right]$$

$$=\frac{1}{48}\left(\frac{157,288}{16,384}\right)\approx 0.2000020$$

$n = 32, b = 1, a = 0, f(x) = x^4$

$$\int_0^1 x^4\, dx \approx \frac{1-0}{3(32)}\left[0+4\left(\frac{1}{1,048,576}\right)+2\left(\frac{1}{65,536}\right)+4\left(\frac{81}{1,048,576}\right)+2\left(\frac{1}{4096}\right)+4\left(\frac{625}{1,048,576}\right)\right.$$

$$+2\left(\frac{625}{65,536}\right)+4\left(\frac{14,641}{1,048,576}\right)+2\left(\frac{81}{4096}\right)+4\left(\frac{28,561}{1,048,576}\right)+2\left(\frac{2401}{65,536}\right)$$

$$+4\left(\frac{50,625}{1,048,576}\right)+2\left(\frac{1}{16}\right)+4\left(\frac{83,521}{1,048,576}\right)+2\left(\frac{6561}{65,536}\right)+4\left(\frac{130,321}{1,048,576}\right)$$

$$+2\left(\frac{625}{4096}\right)+4\left(\frac{194,481}{1,048,576}\right)+2\left(\frac{14,641}{65,536}\right)+4\left(\frac{279,841}{1,048,576}\right)+2\left(\frac{81}{256}\right)$$

$$+4\left(\frac{390,625}{1,048,576}\right)+2\left(\frac{28,561}{65,536}\right)+4\left(\frac{531,441}{1,048,576}\right)+2\left(\frac{2401}{4096}\right)$$

$$\left.+4\left(\frac{707,281}{1,048,576}\right)+2\left(\frac{50,625}{65,536}\right)+4\left(\frac{923,521}{1,048,576}\right)+1\right]$$

$$=\frac{1}{96}\left(\frac{50,033,168}{262,144}\right)\approx 0.2000001$$

To find error for each value of n, subtract as indicated.

$n = 4$: $(0.2005208 - 0.2) = 0.0005208$

$n = 8$: $(0.2000326 - 0.2) = 0.0000326$

$n = 16$: $(0.2000020 - 0.2) = 0.0000020$

$n = 32$: $(0.2000001 - 0.2) = 0.0000001$

(c) $p = 1$

$4^1(0.0005208) = 4(0.0005208) = 0.0020832$

$8^1(0.0000326) = 8(0.0000326) = 0.0002608$

Try $p = 2$.

$4^2(0.0005208) = 16(0.0005208) = 0.0083328$

$8^2(0.0000326) = 64(0.0000326) = 0.0020864$

(*continued on next page*)

(continued)

Try $p = 3$.

$4^3(0.0005208) = 64(0.0005208)$
$= 0.0333312$

$8^3(0.0000326) = 512(0.0000326)$
$= 0.0166912$

Try $p = 4$.

$4^4(0.0005208) = 256(0.0005208)$
$= 0.1333248$

$8^4(0.0000326) = 4096(0.0000326)$
$= 0.1335296$

$16^4(0.0000020) = 65536(0.0000020)$
$= 0.131072$

$32^4(0.0000001) = 1048576(0.0000001)$
$= 0.1048576$

These are the closest values we can get; thus, $p = 4$.

21. Midpoint rule:

$n = 4, b = 5, a = 1, f(x) = \frac{1}{x^2}, \Delta x = 1$

i	x_i	$f(x_i)$
0	$\frac{3}{2}$	$\frac{4}{9}$
1	$\frac{5}{2}$	$\frac{4}{25}$
2	$\frac{7}{2}$	$\frac{4}{49}$
3	$\frac{9}{2}$	$\frac{4}{81}$

$$\int_1^5 \frac{1}{x^2}dx \approx \sum_{i=1}^{4} f(x_i)\Delta x$$
$$= \frac{4}{9}(1) + \frac{4}{25}(1) + \frac{4}{49}(1) + \frac{4}{81}(1)$$
$$\approx 0.7355$$

Simpson's rule:

$m = 8, b = 5, a = 1, f(x) = \frac{1}{x^2}$

i	x_i	$f(x_i)$
0	1	1
1	$\frac{3}{2}$	$\frac{4}{9}$
2	2	$\frac{1}{4}$
3	$\frac{5}{2}$	$\frac{4}{25}$
4	3	$\frac{1}{9}$
5	$\frac{7}{2}$	$\frac{4}{49}$
6	4	$\frac{1}{16}$
7	$\frac{9}{2}$	$\frac{4}{81}$
8	5	$\frac{1}{25}$

$$\int_1^5 \frac{1}{x^2}dx$$
$$\approx \frac{5-1}{3(8)}\left[1 + 4\left(\frac{4}{9}\right) + 2\left(\frac{1}{4}\right) + 4\left(\frac{4}{25}\right) + 2\left(\frac{1}{9}\right) + 4\left(\frac{4}{49}\right) + 2\left(\frac{1}{16}\right) + 4\left(\frac{4}{81}\right) + \frac{1}{25}\right]$$
$$\approx \frac{1}{6}(4.82906) \approx 0.8048$$

From #7 part a, $T \approx 0.9436$, when $n = 4$.

To verify the formula evaluate $\frac{2M + T}{3}$.

$$\frac{2M + T}{3} \approx \frac{2(0.7355) + 0.9436}{3} \approx 0.8048$$

23. $y = e^{-t^2} + \frac{1}{t+1}$

The total reaction is $\int_1^9 \left(e^{-t^2} + \frac{1}{t+1}\right)dt$.

$n = 8, b = 9, a = 1, f(t) = e^{-t^2} + \frac{1}{t+1}$

i	x_i	$f(x_i)$
0	1	0.8679
1	2	0.3516
2	3	0.2501
3	4	0.2000
4	5	0.1667
5	6	0.1429
6	7	0.1250
7	8	0.1111
8	9	0.1000

(a) Trapezoidal rule:

$$\int_1^9 \left(e^{-t^2} + \frac{1}{t+1}\right) dt$$

$$\approx \frac{9-1}{8}\left[\frac{1}{2}(0.8679) + 0.3516 + 0.2501 + \cdots + \frac{1}{2}(0.1000)\right]$$

$$\approx 1.831$$

(b) Simpson's rule:

$$\int_1^9 \left(e^{-t^2} + \frac{1}{t+1}\right) dt$$

$$\approx \frac{9-1}{3(8)}[0.8679 + 4(0.3516) + 2(0.2501) + 4(0.2000) + 2(0.1667) + 4(0.1429) + 2(0.1250) + 4(0.1111) + 0.1000]$$

$$= \frac{1}{3}(5.2739) \approx 1.758$$

25. Note that heights may differ depending on the readings of the graph. Thus, answers may vary. $n = 10$, $b = 20$, $a = 0$

i	x_i	$f(x_i)$
0	0	0
1	2	5
2	4	3
3	6	2
4	8	1.5
5	10	1.2
6	12	1
7	14	0.5
8	16	0.3
9	18	0.2
10	20	0.2

Area under curve for Formulation A

$$= \frac{20-0}{10}\left[\frac{1}{2}(0) + 5 + 3 + 2 + 1.5 + 1.2 + 1 + 0.5 + 0.3 + 0.2 + \frac{1}{2}(0.2)\right]$$

$= 2(14.8) \approx 30$ mcg(h)/ml

This represents the total amount of drug available to the patient for each ml of blood.

27. As in Exercise 25, readings on the graph may vary, so answers may vary. The area both under the curve for Formulation A and above the minimum effective concentration line is on the interval $\left[\frac{1}{2}, 6\right]$.

Area under curve for Formulation A on $\left[\frac{1}{2}, 1\right]$, with $n = 1$

$$= \frac{1-\frac{1}{2}}{1}\left[\frac{1}{2}(2+6)\right] = \frac{1}{2}(4) = 2$$

Area under curve for Formulation A on [1, 6], with $n = 5$

$$= \frac{6-1}{5}\left[\frac{1}{2}(6) + 5 + 4 + 3 + 2.4 + \frac{1}{2}(2)\right] = 18.4$$

Area under minimum effective concentration line $\left[\frac{1}{2}, 6\right] = 5.5(2) = 11.0$

Area under the curve for Formulation A and above minimum effective concentration line $= 2 + 18.4 - 110 \approx 9$ mcg(h)/ml

This represents the total effective amount of drug available to the patient for each ml of blood.

29. $y = b_0 w^{b_1} e^{-b_2 w}$

(a) If $t = 7w$ then $w = \frac{t}{7}$.

$$y = b_0\left(\frac{t}{7}\right)^{b_1} e^{-b_2 t/7}$$

(b) Replacing the constants with the given values, we have

$$y = 5.955\left(\frac{t}{7}\right)^{0.233} e^{-0.027t/7} dt$$

In 25 weeks, there are 175 days.

$$\int_0^{175} 5.955\left(\frac{t}{7}\right)^{0.233} e^{-0.027t/7} dt$$

$n = 10$, $b = 175$, $a = 0$,

$$f(t) = 5.955\left(\frac{t}{7}\right)^{0.233} e^{-0.027t/7}$$

i	t_i	$f(t_i)$	i	t_i	$f(t_i)$
0	0	0	6	105	7.46
1	17.5	6.89	7	122.5	7.23
2	35	7.57	8	140	6.97
3	52.5	7.78	9	157.5	6.70
4	70	7.77	10	175	6.42
5	87.5	7.65			

(*continued on next page*)

(continued)

Trapezoidal rule:

$$\int_0^{175} 5.955\left(\frac{t}{7}\right)^{0.233} e^{-0.027t/7}\,dt$$
$$\approx \frac{175-0}{10}\left[\frac{1}{2}(0)+6.89+7.57+7.78 +7.77+7.65+7.46+7.23+6.97 +6.70+\frac{1}{2}(6.42)\right]$$
$$=1211.525$$

The total milk consumed is about 1212 kg.

Simpson's rule:

$$\int_0^{175} 5.955\left(\frac{t}{7}\right)^{0.233} e^{-0.027t/7}\,dt$$
$$\approx \frac{175-0}{3(10)}[0+4(6.89)+2(7.57) +4(7.78)+2(7.77)+4(7.65) +2(7.46)+4(7.23)+2(6.97) +4(6.70)+6.42]$$
$$=1230.6$$

The total milk consumed is about 1231 kg.

(c) Replacing the constants with the given values, we have

$$y=8.409\left(\frac{t}{7}\right)^{0.143} e^{-0.037t/7}.$$

In 25 weeks, there are 175 days.

$$\int_0^{175} 8.409\left(\frac{t}{7}\right)^{0.143} e^{-0.037t/7}\,dt$$

$n=10, b=175, a=0,$

$$f(t)=8.409\left(\frac{t}{7}\right)^{0.143} e^{-0.037t/7}$$

i	t_i	$f(t_i)$
0	0	0
1	17.5	8.74
2	35	8.80
3	52.5	8.50
4	70	8.07
5	87.5	7.60
6	105	7.11
7	122.5	6.63
8	140	6.16
9	157.5	5.71
10	175	5.28

Trapezoidal rule:

$$\int_0^{175} 8.409\left(\frac{t}{7}\right)^{0.143} e^{-0.037t/7}\,dt$$
$$\approx \frac{175-0}{10}\left[\frac{1}{2}(0)+8.74+8.80+8.50 +8.07+7.60+7.11+6.63 +6.16+5.71+\frac{1}{2}(5.28)\right]$$
$$=1224.30$$

The total milk consumed is about 1224 kg.

Simpson's rule:

$$\int_0^{175} 8.409\left(\frac{t}{7}\right)^{0.143} e^{-0.037t/7}\,dt$$
$$\approx \frac{175-0}{3(10)}[0+4(8.74)+2(8.80) +4(8.50)+2(8.07)+4(7.60) +2(7.11)+4(6.63)+2(6.16) +4(5.71)+5.28]$$
$$=\frac{35}{6}(214.28)=1249.97$$

The total milk consumed is about 1250 kg.

31. We need to evaluate

$$\int_7^{182} 3.922t^{0.242}e^{-0.00357t}\,dt.$$

Using a calculator program for Simpson's rule with $n = 20$, we obtain 1400.88 as the value of this integral. This indicates that the total amount of milk consumed by a calf from 7 to 182 days is about 1401 kg.

33. (a)

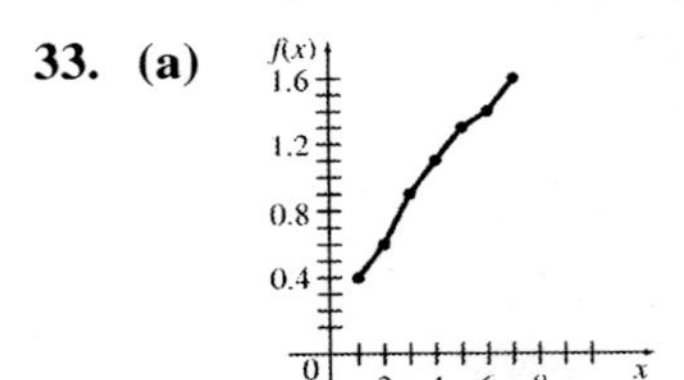

(b) $$A=\frac{7-1}{6}\left[\frac{1}{2}(0.4)+0.6+0.9+1.1 +1.3+1.4+\frac{1}{2}(1.6)\right]$$
$$=6.3$$

(c) $A = \frac{7-1}{3(6)}[0.4 + 4(0.6) + 2(0.9) + 4(1.1) + 2(1.3) + 4(1.4) + 1.6] \approx 6.27$

35. (a)

(b) $A = \frac{7-1}{6}\left[\frac{1}{2}(12) + 16 + 18 + 21 + 24 + 27 + \frac{1}{2}(32)\right]$

$A = 128$

(c) $A = \frac{7-1}{3(6)}[12 + 4(16) + 2(18) + 4(21) + 2(24) + 4(27) + 32]$

$A = 128$

37. Use a calculator program for Simpson's rule with $n = 20$ to evaluate each of the integrals in this exercise.

(a) $\int_{-1}^{1}\left(\frac{1}{\sqrt{2\pi}}e^{-x^2/2}\right)dx \approx 0.6827$

The probability that a normal random variable is within 1 standard deviation of the mean is about 0.6827.

(b) $\int_{-2}^{2}\left(\frac{1}{\sqrt{2\pi}}e^{-x^2/2}\right)dx \approx 0.9545$

The probability that a normal random variable is within 2 standard deviation of the mean is about 0.9545.

(c) $\int_{-3}^{3}\left(\frac{1}{\sqrt{2\pi}}e^{-x^2/2}\right)dx \approx 0.9973$

The probability that a normal random variable is within 3 standard deviations of the mean is about 0.9973.

8.2 Integration by Parts

1. $\int xe^x dx$

Let $dv = e^x dx$ and $u = x$.

Then $v = \int e^x dx \Rightarrow v = e^x$ and $du = dx$.

Use the formula $\int u\,dv = uv - \int v\,du$.

$\int xe^x dx = xe^x - \int e^x dx = xe^x - e^x + C$

3. $\int (5x-9)e^{-3x} dx$

Let $dv = e^{-3x}\,dx$ and $u = 5x - 9$

Then $v = \int e^{-3x}\,dx \Rightarrow v = \frac{e^{-3x}}{-3}$ and $du = 5dx$.

$$\begin{aligned}\int (5x-9)e^{-3x}dx &= (5x-9)\left(\frac{e^{-3x}}{-3}\right) - \int\left(\frac{5}{-3}\right)e^{-3x}dx \\ &= \frac{-5xe^{-3x}}{3} + 3e^{-3x} + \frac{5e^{-3x}}{-9} + C \\ &= \frac{-5xe^{-3x}}{3} - \frac{5e^{-3x}}{9} + 3e^{-3x} + C \\ &= \frac{-5xe^{-3x}}{3} + \frac{22e^{-3x}}{9} + C\end{aligned}$$

5. $\int_0^1 \frac{2x+1}{e^x}dx = \int_0^1 (2x+1)e^{-x}\,dx$

Let $dv = e^{-x}\,dx$ and $u = 2x + 1$.

Then $v = \int e^{-x}\,dx \Rightarrow v = -e^{-x}$ and $du = 2\,dx$.

$$\begin{aligned}\int \frac{2x+1}{e^x}dx &= -(2x+1)e^{-x} + \int 2e^{-x}dx \\ &= -(2x+1)e^{-x} - 2e^{-x} + C\end{aligned}$$

$$\begin{aligned}\int_0^1 \frac{2x+1}{e^x}dx &= [-(2x+1)e^{-x} - 2e^{-x}]\Big|_0^1 \\ &= [-3e^{-1} - 2e^{-1}] - (-1-2) \\ &= -5e^{-1} + 3 \approx 1.161\end{aligned}$$

7. $\int_1^4 \ln 2x\,dx$

Let $dv = dx$ and $u = \ln 2x$.

Then $v = x$ and $du = \frac{1}{x}dx$.

$$\begin{aligned}\int \ln 2x\,dx &= x\ln 2x - \int x\left(\frac{1}{x}dx\right) \\ &= x\ln 2x - x + C\end{aligned}$$

$$\begin{aligned}\int_1^4 \ln 2x\,dx &= (x\ln 2x - x)\Big|_1^4 \\ &= 4\ln 8 - 4 - \ln 2 + 1 \\ &= 4\ln 2^3 - \ln 2 - 4 + 1 \\ &= 12\ln 2 - \ln 2 - 3 = 11\ln 2 - 3 \\ &\approx 4.6246\end{aligned}$$

9. $\int x \ln x dx$

Let $dv = x\,dx$ and $u = \ln x$.

Then $v = \frac{x^2}{2}$ and $du = \frac{1}{x}\,dx$.

$$\int x \ln x dx = \frac{x^2}{2} \ln x - \int \frac{x}{2}\,dx = \frac{x^2 \ln x}{2} - \frac{x^2}{4} + C$$

11. $\int -6x \cos 5x\,dx$

Let $u = -6x$ and $dv = \cos 5x\,dx$.

Then $du = -6\,dx$ and $v = \frac{1}{5}\sin 5x$.

$$\int -6x \cos 5x\,dx$$
$$= (-6x)\left(\frac{1}{5}\sin 5x\right) - \int\left(\frac{1}{5}\sin 5x)\right)(-6\,dx)$$
$$= -\frac{6}{5}x \sin 5x + \frac{6}{5}\int \sin 5x\,dx$$
$$= -\frac{6}{5}x \sin 5x + \frac{6}{5}\cdot\frac{1}{5}(-\cos 5x) + C$$
$$= -\frac{6}{5}x \sin 5x - \frac{6}{25}\cos 5x + C$$

13. $\int 8x \sin x\,dx$

Let $u = 8x$ and $dv = \sin x\,dx$.

Then $du = 8\,dx$ and $v = -\cos x$.

$$\int 8x \sin x\,dx$$
$$= 8x(-\cos x) - \int(-\cos x)\cdot 8\,dx$$
$$= -8x \cos x + 8\int \cos x\,du$$
$$= -8x \cos x + 8\sin x + C$$

15. $\int -6x^2 \cos 8x\,dx$

Let $u = -6x^2$ and $dv = \cos 8x\,dx$.

Then $du = -12x\,dx$ and $v = \frac{1}{8}\sin 8x$.

$$\int -6x^2 \cos 8x\,dx$$
$$= (-6x^2)\left(\frac{1}{8}\sin 8x\right) - \int\left(\frac{1}{8}\sin 8x\right)(-12x\,dx)$$
$$= -\frac{3}{4}x^2 \sin 8x + \frac{3}{2}\int x \sin 8x\,dx$$

In $\int x \sin 8x\,dx$, let $u = x$ and $dv = \sin 8x\,dx$.

Then $du = dx$ and $v = -\frac{1}{8}\cos 8x$.

$$\int -6x^2 \cos 8x\,dx$$
$$= -\frac{3}{4}x^2 \sin 8x + \frac{3}{2}\left[-\frac{1}{8}x \cos 8x - \int\left(-\frac{1}{8}\cos 8x\right)dx\right]$$
$$= -\frac{3}{4}x^2 \sin 8x - \frac{3}{16}x \cos 8x + \frac{3}{16}\int \cos 8x\,dx$$
$$= -\frac{3}{4}x^2 \sin 8x - \frac{3}{16}x \cos 8x + \frac{3}{16}\cdot\frac{1}{8}\sin 8x + C$$
$$= -\frac{3}{4}x^2 \sin 8x - \frac{3}{16}x \cos 8x + \frac{3}{128}\sin 8x + C$$

17. The area is $\int_2^4 (x-2)e^x dx$.

Let $dv = e^x dx$ and $u = x - 2$.

Then $v = e^x$ and $du = dx$.

$$\int (x-2)e^x\,dx = (x-2)e^x - \int e^x dx$$
$$\int_1^4 (x-2)e^x\,dx = [(x-2)e^x - e^x]\Big|_2^4$$
$$= (2e^4 - e^4) - (0 - e^2)$$
$$= e^4 + e^2 \approx 61.99$$

19. $\int x^2 e^{2x}\,dx$

Let $u = x^2$ and $dv = e^{2x}\,dx$.

Use column integration.

D		I
x^2	+	e^{2x}
$2x$	−	$\frac{e^{2x}}{2}$
2	+	$\frac{e^{2x}}{4}$
0		$\frac{e^{2x}}{8}$

Thus,

$$\int x^2 e^{2x}\,dx = x^2\left(\frac{e^{2x}}{2}\right) - 2x\left(\frac{e^{2x}}{4}\right) + \frac{2e^{2x}}{8} + C$$
$$= \frac{x^2 e^{2x}}{2} - \frac{xe^{2x}}{2} + \frac{e^{2x}}{4} + C$$

21. $\int x^2\sqrt{x+4}\,dx$

Let $u = x^2$ and $dv = (x+4)^{1/2}dx$. Use column integration.

D		I
x^2	+	$(x+4)^{1/2}$
$2x$	−	$\frac{2}{3}(x+4)^{3/2}$
2	+	$\left(\frac{2}{3}\right)\left(\frac{2}{5}\right)(x+4)^{5/2}$
0		$\left(\frac{2}{3}\right)\left(\frac{2}{5}\right)\left(\frac{2}{7}\right)(x+4)^{7/2}$

Thus,

$$\int x^2\sqrt{x+4}\,dx$$
$$= x^2(x+4)^{3/2}\left(\frac{2}{3}\right) - 2x(x+4)^{5/2}\left(\frac{2}{3}\right)\left(\frac{2}{5}\right) + 2(x+4)^{7/2}\left(\frac{2}{3}\right)\left(\frac{2}{5}\right)\left(\frac{2}{7}\right) + C$$
$$= \frac{2}{3}x^2(x+4)^{3/2} - \frac{8}{15}x(x+4)^{5/2} + \frac{16}{105}(x+4)^{7/2} + C$$

23. $\int (8x+10)\ln(5x)\,dx$

Use integration by parts.
Let $dv = (8x+10)\,dx$ and $u = \ln(5x)$.
Then $v = 4x^2 + 10x$ and $du = \frac{1}{x}dx.$

$$\int (8x+10)\ln(5x)\,dx$$
$$= (4x^2+10x)\ln(5x) - \int (4x^2+10x)\left(\frac{1}{x}\right)dx$$
$$= (4x^2+10x)\ln(5x) - \int (4x+10)\,dx$$
$$= (4x^2+10x)\ln(5x) - 2x^2 - 10x + C$$

25. $\int_1^2 (1-x^2)e^{2x}\,dx$

Let $u = 1-x^2$ and $dv = e^{2x}\,dx.$
Use column integration.

D		I
$1-x^2$	+	e^{2x}
$-2x$	−	$\frac{e^{2x}}{2}$
-2	+	$\frac{e^{2x}}{4}$
0		$\frac{e^{2x}}{8}$

Thus,

$$\int (1-x^2)e^{2x}dx$$
$$= \frac{(1-x^2)e^{2x}}{2} - \frac{(-2x)e^{2x}}{4} + \frac{(-2)e^{2x}}{8} + C$$
$$= \frac{(1-x^2)e^{2x}}{2} + \frac{xe^{2x}}{2} - \frac{e^{2x}}{4} + C$$
$$= \frac{e^{2x}}{2}\left(1 - x^2 + x - \frac{1}{2}\right) + C$$
$$= \frac{e^{2x}}{2}\left(\frac{1}{2} - x^2 + x\right) + C$$

$$\int_1^2 (1-x^2)e^{2x}\,dx = \frac{e^{2x}}{2}\left(\frac{1}{2} - x^2 + x\right)\Bigg|_1^2$$
$$= \frac{e^4}{2}\left(-\frac{3}{2}\right) - \frac{e^2}{2}\left(\frac{1}{2}\right)$$
$$= -\frac{e^2}{4}(3e^2+1) \approx -42.80$$

27. $\int_0^1 \frac{x^3dx}{\sqrt{3+x^2}} = \int_0^1 x^3(3x+x^2)^{-1/2}\,dx$

Use integration by parts.
Let $dv = x(3+x^2)^{-1/2}\,dx$ and $u = x^2.$
Then $v = \frac{2(3+x^2)^{1/2}}{2} \Rightarrow v = (3+x^2)^{1/2}$
and $du = 2x\,dx.$

$$\int \frac{x^3dx}{\sqrt{3+x^2}} = x^2(3+x^2)^{1/2} - \int 2x(3+x^2)^{1/2}\,dx$$
$$= x^2(3+x^2)^{1/2} - \frac{2}{3}(3+x^2)^{3/2}$$

$$\int_0^1 \frac{x^3dx}{\sqrt{3+x^2}} = \left[x^2(3+x^2)^{1/2} - \frac{2}{3}(3+x^2)^{3/2}\right]\Bigg|_0^1$$
$$= 4^{1/2} - \frac{2}{3}(4^{3/2}) - 0 + \frac{2}{3}(3^{3/2})$$
$$= 2 - \frac{2}{3}(8) + \frac{2}{3}(3^{3/2}) = -\frac{10}{3} + 2\sqrt{3}$$
$$\approx 0.1308$$

29. $\int e^x \cos e^x\,dx$

Use substitution. Let $u = e^x$, so that $du = e^x dx$. Then

$$\int e^x \cos e^x\,dx = \int \cos u\,du = \sin u + C$$
$$= \sin e^x + C$$

31. $\int \frac{16}{\sqrt{x^2+16}}\,dx$

Use formula 5 from the table of integrals with $a = 4$.

$$\int \frac{16}{\sqrt{x^2+16}}\,dx = 16\int \frac{1}{\sqrt{x^2+4^2}}\,dx$$
$$= 16\ln\left|x+\sqrt{x^2+16}\right| + C$$

33. $\int \frac{3}{x\sqrt{121-x^2}}\,dx = 3\int \frac{dx}{x\sqrt{11^2-x^2}}$

If $a = 11$, this integral matches formula 9 in the table.

$$3\int \frac{dx}{x\sqrt{11^2-x^2}}$$
$$= 3\left(-\frac{1}{11}\ln\left|\frac{11+\sqrt{121-x^2}}{x}\right|\right) + C$$
$$= -\frac{3}{11}\ln\left|\frac{11+\sqrt{121-x^2}}{x}\right| + C$$

35. $\int \frac{-6}{x(4x+6)^2}\,dx$

Use formula 14 from the table of integrals with $a = 4$ and $b = 6$.

$$\int \frac{-6}{x(4x+6)^2}\,dx$$
$$= -6\int \frac{1}{x(4x+6)^2}\,dx$$
$$= -6\left[\frac{1}{6(4x+6)} + \frac{1}{6^2}\ln\left|\frac{x}{4x+6}\right|\right] + C$$
$$= -\frac{1}{(4x+6)} - \frac{1}{6}\ln\left|\frac{x}{4x+6}\right| + C$$

37. $\int e^{4x}\sin(3x)\,dx$

Use formula 37 from the Table of Integrals with $a = 4$ and $b = 3$.

$$\int e^{4x}\sin(3x)\,dx$$
$$= \frac{e^{4x}}{4^2+3^2}\left(4\sin(3x) - 3\cos(3x)\right) + C$$
$$= \frac{e^{4x}}{25}\left(4\sin(3x) - 3\cos(3x)\right) + C$$

39. $\int \sec x\,dx$

Use formula 26 from the Table of Integrals.

$$\int \sec x\,dx = \ln|\sec x + \tan x| + C$$

41. The product rule.

43. First find the indefinite integral using integration by parts.

$$\int u\,dv = uv - \int v\,du$$

Now substitute the given values.

$$\int_0^1 u\,dv = uv\Big|_0^1 - \int_0^1 v\,du$$
$$= [u(1)v(1) - u(0)v(0)] - 4$$
$$= (3)(-4) - (2)(1) - 4 = -18$$

45. $\int r\,ds = rs - \int s\,dr$

$$\int_0^2 r\,ds = rs\Big|_0^2 - \int_0^2 s\,dr$$
$$10 = r(2)s(2) - r(0)s(0) - 5$$
$$15 = r(s)s(2)$$

47. $\int x^n \cdot \ln|x|\,dx,\ n \neq -1$

Let $u = \ln|x|$ and $dv = x^n\,dx$.

Use column integration.

D		I
$\ln\|x\|$	+	x^n
$\frac{1}{x}$		$\frac{1}{n+1}x^{n+1}$

$$\int x^n \cdot \ln|x|\,dx$$
$$= \frac{1}{n+1}x^{n+1}\ln|x| - \int\left[\frac{1}{x}\cdot\frac{1}{n+1}x^{n+1}\right]dx$$
$$= \frac{1}{n+1}x^{n+1}\ln|x| - \int \frac{1}{n+1}x^n\,dx$$
$$= \frac{1}{n+1}x^{n+1}\ln|x| - \frac{1}{(n+1)^2}x^{n+1} + C$$
$$= x^{n+1}\left[\frac{\ln|x|}{n+1} - \frac{1}{(n+1)^2}\right] + C$$

49. $\int x\sqrt{x+1}\,dx$

(a) Let $u = x$ and $dv = \sqrt{x+1}\,dx$.

Use column integration.

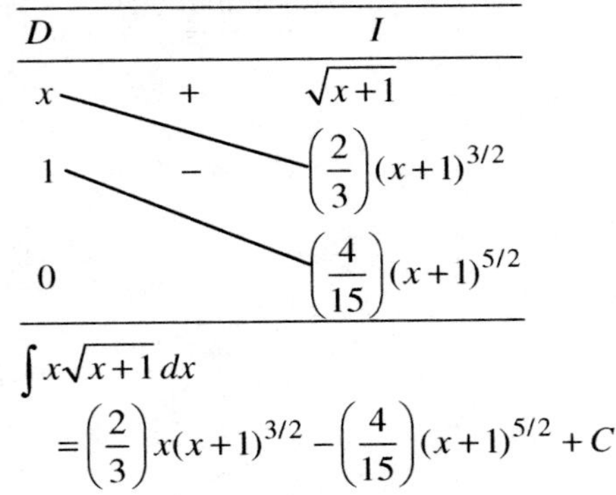

D		I
x	+	$\sqrt{x+1}$
1	−	$\left(\frac{2}{3}\right)(x+1)^{3/2}$
0		$\left(\frac{4}{15}\right)(x+1)^{5/2}$

$$\int x\sqrt{x+1}\,dx$$
$$= \left(\frac{2}{3}\right)x(x+1)^{3/2} - \left(\frac{4}{15}\right)(x+1)^{5/2} + C$$

(b) Let $u = x + 1$. Then $u - 1 = x$ and $du = dx$.

$$\int x\sqrt{x+1}\,dx = \int (u-1)u^{1/2}du = \int (u^{3/2} - u^{1/2})\,du = \frac{2}{5}u^{5/2} - \frac{2}{3}u^{3/2} + C = \frac{2}{5}(x+1)^{5/2} - \frac{2}{3}(x+1)^{3/2} + C$$

(c) Both results factor as $\frac{2}{15}(x+1)^{3/2}(3x-2)+C,$ so they are equivalent.

51. $r(t) = \int_1^6 2t^2 e^{-t}\,dt$

Let $dv = e^{-t}\,dt$ and $u = 2t^2$.

Then $v = -e^{-t}$ and $du = 4t\,dt$.

Use column integration.

D		I
$2t^2$	+	e^{-t}
$4t$	−	$-e^{-t}$
4	+	e^{-t}
0		$-e^{-t}$

$$\int 2t^2e^{-t}dt = 2t^2(-e^{-t}) - 4t(e^{-t}) + 4(-e^{-t}) + C = -2t^2e^{-t} - 4te^{-t} - 4e^{-t} + C = -2e^{-t}(t^2+2t+2) + C$$

$$\int_1^6 2t^2e^{-t}\,dt = -2e^{-t}(t^2+2t+2)\Big|_1^6 = -100e^{-6} + 10e^{-1} \approx 3.431$$

The total reaction to the drug from $t = 1$ to $t = 6$ is about 3.431.

53. $A = \int_4^9 \sqrt{t}\ln t\,dt$

Let $u = \ln t$ and $dv = \sqrt{t}\,dt = t^{1/2}\,dt$. Then $du = \frac{1}{t}dt$ and $v = \frac{2}{3}t^{3/2}$.

$$\int \sqrt{t}\ln t\,dt = \frac{2}{3}t^{3/2}\ln t - \int\left(\frac{2}{3}t^{3/2}\cdot\frac{1}{t}\right)dt = \frac{2}{3}t^{3/2}\ln t - \int \frac{2}{3}t^{1/2}\,dt = \frac{2}{3}t^{3/2}\ln t - \frac{4}{9}t^{3/2} + C$$

$$\int_4^9 \sqrt{t}\ln t\,dt = 18\ln 9 - \frac{16}{3}\ln 4 - \frac{76}{9} \approx 23.71 \text{ sq cm}$$

55. (a) $\int_0^1 ke^{-kt}(1-t)dt$

Let $u = 1 - t$ and $dv = e^{-kt}dt$. Then, $du = -dt$ and $v = -\frac{1}{k}e^{-kt}$.

$$\int_0^1 ke^{-kt}(1-t)\,dt = k\left[(1-t)\left(-\frac{1}{k}e^{-kt}\right) - \int\left(-\frac{1}{k}e^{-kt}\right)(-dt)\right]\Bigg|_0^1$$

$$= k\left[-\frac{1}{k}e^{-kt}(1-t) - \frac{1}{k}\int e^{-kt}dt\right]\Bigg|_0^1 = -e^{-kt}(1-t) + \frac{1}{k}e^{-kt}\Big|_0^1$$

$$= -e^{-kt}\left(1 - t - \frac{1}{k}\right)\Bigg|_0^1 = \left[-e^{-k}\left(-\frac{1}{k}\right)\right] - \left[-e^0\left(1-\frac{1}{k}\right)\right] = 1 - \frac{1}{k} + \frac{1}{k}e^{-k}$$

$$k = \frac{1}{12}:\ \int_0^1 \frac{1}{12}e^{-t/12}(1-t)dt = 1 - \frac{1}{1/12} + \frac{1}{1/12}e^{-1/12} = 12e^{-1/12} - 11 \approx 0.0405$$

$$k = \frac{1}{24}:\ \int_0^1 \frac{1}{24}e^{-t/24}(1-t)\,dt = 1 - \frac{1}{1/24} + \frac{1}{1/24}e^{-1/24} = 24e^{-1/24} - 23 \approx 0.0205$$

$$k = \frac{1}{48}:\ \int_0^1 \frac{1}{48}e^{-t/48}(1-t)dt = 1 - \frac{1}{1/48} + \frac{1}{1/48}e^{-1/48} = 48e^{-1/48} - 47 \approx 0.0103$$

(b) $\int_1^6 ke^{-kt}\frac{6-t}{5}\,dt$

The integral, easily found by comparing it to the integral in part (a), is

$$-e^{-kt}\left(\frac{6}{5}-\frac{t}{5}-\frac{1}{5k}\right)\Bigg|_1^6=\left[-e^{-6k}\left(\frac{6}{5}-\frac{6}{5}-\frac{1}{5k}\right)\right]-\left[-e^{-k}\left(\frac{6}{5}-\frac{1}{5}-\frac{1}{5k}\right)\right]=\frac{1}{5k}e^{-6k}+\left(1-\frac{1}{5k}\right)e^{-k}$$

$$k=\frac{1}{12}:\ \int_1^6\frac{1}{12}e^{-t/12}\frac{6-t}{5}dt=\frac{1}{5(1/12)}e^{-6(1/12)}+\left[1-\frac{1}{5(1/12)}\right]e^{-1/12}=\frac{12}{5}e^{-1/2}-\frac{7}{5}e^{-1/12}\approx 0.1676$$

$$k=\frac{1}{24}:\ \int_1^6\frac{1}{24}e^{-t/24}\frac{6-t}{5}dt=\frac{1}{5(1/24)}e^{-6(1/24)}+\left[1-\frac{1}{5(1/24)}\right]e^{-1/24}=\frac{24}{5}e^{-1/4}-\frac{19}{5}e^{-1/24}\approx 0.0933$$

$$k=\frac{1}{48}:\ \int_1^6\frac{1}{48}e^{-t/48}\frac{6-t}{5}dt=\frac{1}{5(1/48)}e^{-6(1/48)}+\left[1-\frac{1}{5(1/48)}\right]e^{-1/48}=\frac{48}{5}e^{-1/8}-\frac{43}{5}e^{-1/48}\approx 0.0493$$

8.3 Volume and Average Value

1. $f(x)=x,\ y=0,\ x=0,\ x=3$

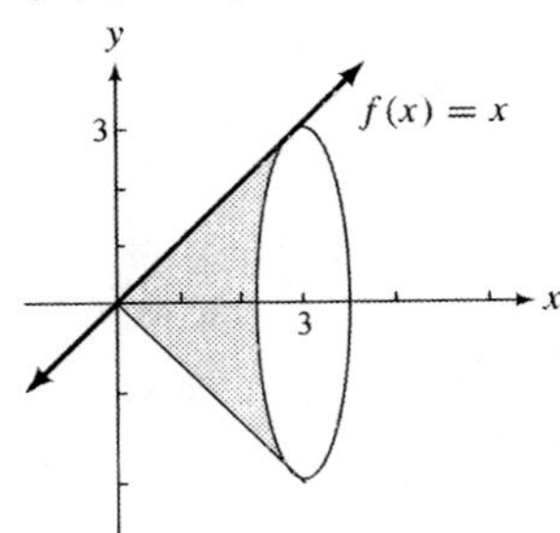

$$V=\pi\int_0^3 x^2dx=\frac{\pi x^3}{3}\Bigg|_0^3=\frac{\pi(27)}{3}-0=9\pi$$

3. $f(x)=2x+1,\ y=0,\ x=0,\ x=4$

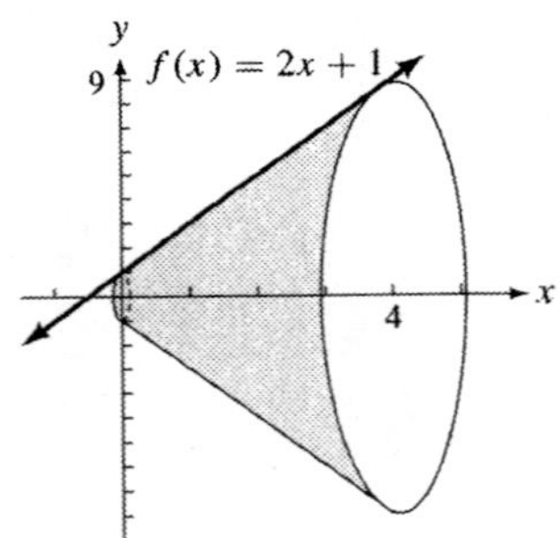

$$V=\pi\int_0^4(2x+1)^2dx$$

Let $u=2x+1$. Then $du=2\,dx$. If $x=4,\ u=9$. If $x=0,\ u=1$.

$$V=\frac{1}{2}\pi\int_0^4(2x+1)^2(2)\,dx=\frac{1}{2}\pi\int_1^9u^2du$$

$$=\frac{\pi}{2}\left(\frac{u^3}{3}\right)\Bigg|_1^9=\frac{\pi}{2}\left(\frac{729}{3}-\frac{1}{3}\right)=\frac{364\pi}{3}$$

5. $f(x)=\frac{1}{3}x+2,\ y=0,\ x=1,\ x=3$

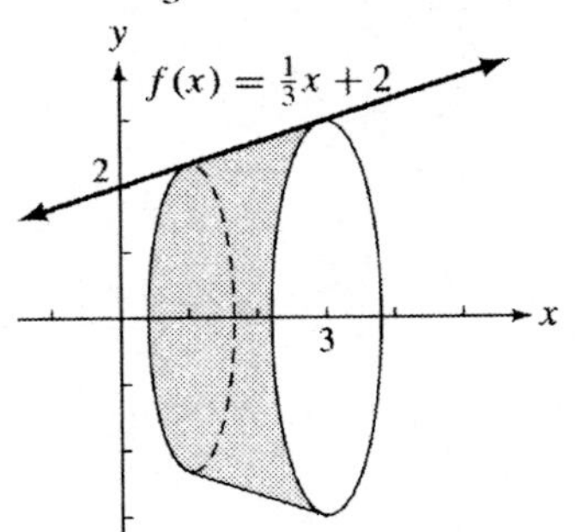

$$V=\pi\int_1^3\left(\frac{1}{3}x+2\right)^2dx=3\pi\int_1^3\frac{1}{3}\left(\frac{1}{3}x+2\right)^2dx$$

$$=3\pi\frac{\left(\frac{1}{3}x+2\right)^3}{3}\Bigg|_1^3=\pi\left(\frac{1}{3}x+2\right)^3\Bigg|_1^3$$

$$=27\pi-\frac{343\pi}{27}=\frac{386\pi}{27}$$

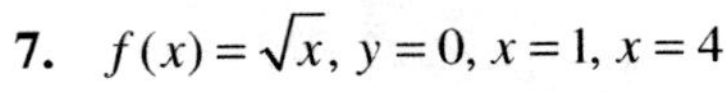

7. $f(x)=\sqrt{x},\ y=0,\ x=1,\ x=4$

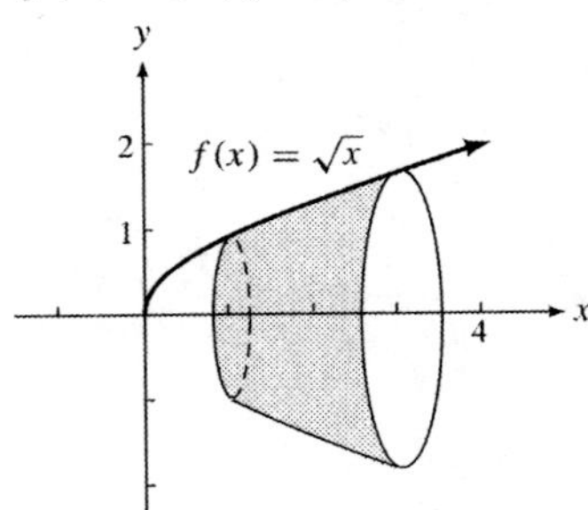

$$V=\pi\int_1^4(\sqrt{x})^2dx=\pi\int_1^4x\,dx=\frac{\pi x^2}{2}\Bigg|_1^4$$

$$=8\pi-\frac{\pi}{2}=\frac{15\pi}{2}$$

9. $f(x) = \sqrt{2x+1},\ y = 0,\ x = 1,\ x = 4$

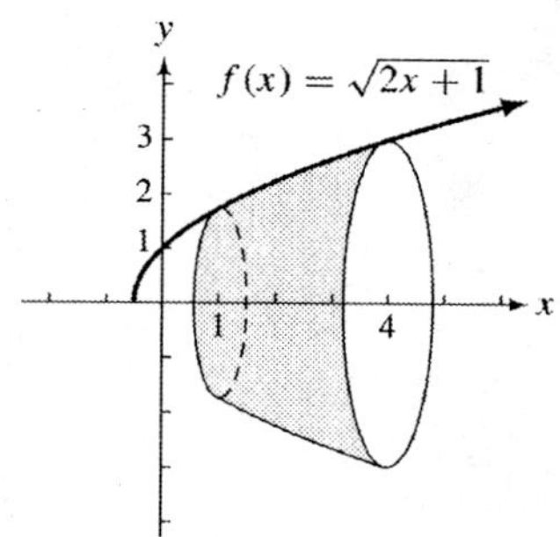

$$V = \pi\int_1^4 (\sqrt{2x+1})^2\,dx = \pi\int_1^4 (2x+1)\,dx$$
$$= \pi\left(\frac{2x^2}{2} + x\right)\Bigg|_1^4 = \pi[(16+4) - 2] = 18\pi$$

11. $f(x) = e^x;\ y = 0,\ x = 0,\ x = 2$

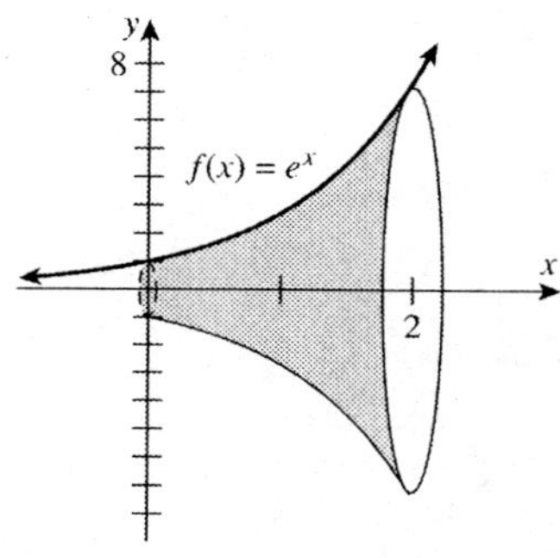

$$V = \pi\int_0^2 e^{2x}\,dx = \frac{\pi e^{2x}}{2}\Bigg|_0^2 = \frac{\pi e^4}{2} - \frac{\pi}{2}$$
$$= \frac{\pi}{2}(e^4 - 1) \approx 84.19$$

13. $f(x) = \dfrac{2}{\sqrt{x}},\ y = 0,\ x = 1,\ x = 3$

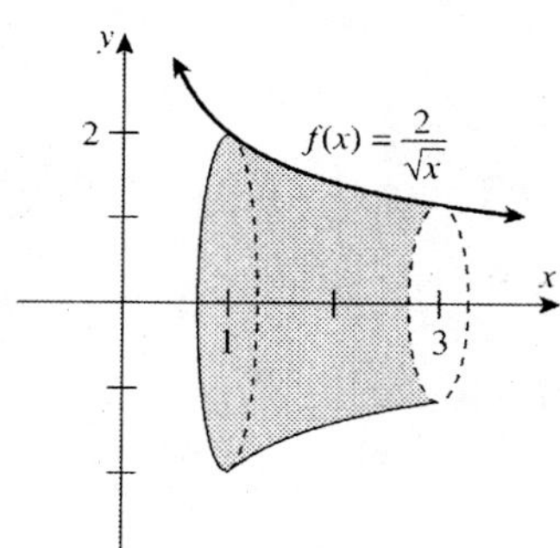

$$V = \pi\int_1^3 \left(\frac{2}{\sqrt{x}}\right)^2 dx = \pi\int_1^3 \frac{4}{x}\,dx = 4\pi \ln|x|\Big|_1^3$$
$$= 4\pi(\ln 3 - \ln 1) = 4\pi \ln 3 \approx 13.81$$

15. $f(x) = x^2,\ y = 0,\ x = 1,\ x = 5$

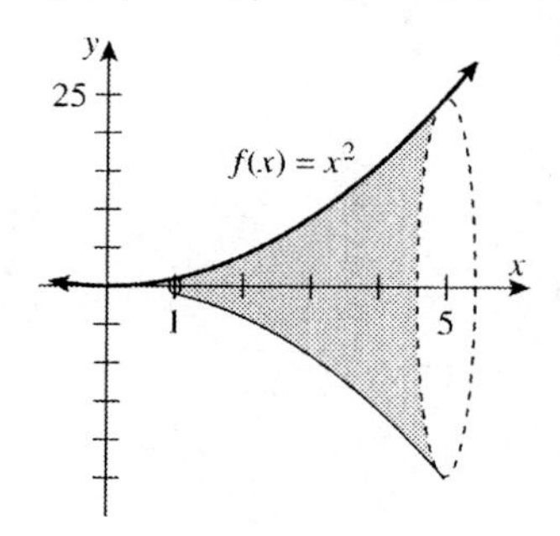

$$V = \pi\int_1^5 x^4\,dx = \frac{\pi x^5}{5}\Bigg|_1^5 = 625\pi - \frac{\pi}{5} = \frac{3124\pi}{5}$$

17. $f(x) = 1 - x^2,\ y = 0$

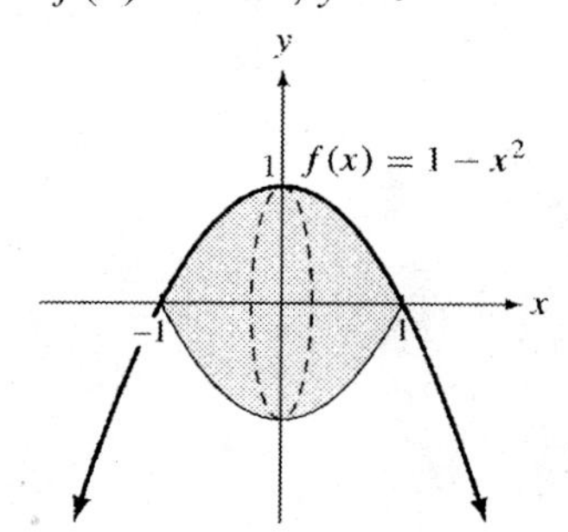

Since $f(x) = 1 - x^2$ intersects $y = 0$ where $1 - x^2 = 0 \Rightarrow x = \pm 1$, $a = -1$ and $b = 1$.

$$V = \pi\int_{-1}^1 (1 - x^2)^2\,dx = \pi\int_{-1}^1 (1 - 2x^2 + x^4)\,dx$$
$$= \pi\left(x - \frac{2x^3}{3} + \frac{x^5}{5}\right)\Bigg|_{-1}^1$$
$$= \pi\left(1 - \frac{2}{3} + \frac{1}{5}\right) - \pi\left(-1 + \frac{2}{3} - \frac{1}{5}\right)$$
$$= 2\pi - \frac{4\pi}{3} + \frac{2\pi}{5} = \frac{16\pi}{15}$$

19. $f(x) = \sec x,\ y = 0,\ x = 0,\ x = \dfrac{\pi}{4}$

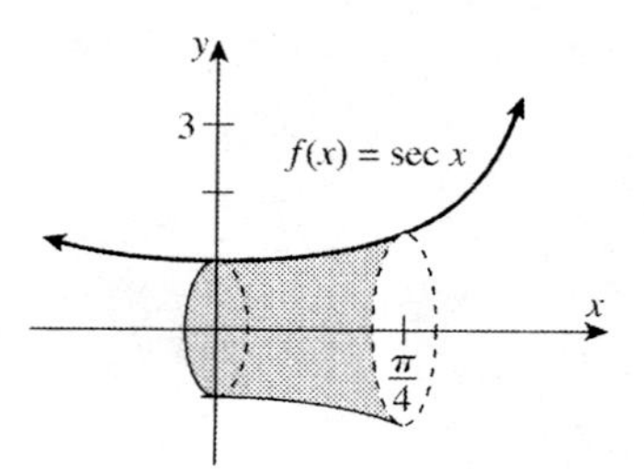

$$V = \pi\int_0^{\pi/4} (\sec x)^2\,dx = \pi\int_0^{\pi/4} \sec^2 x\,dx$$
$$= \pi \tan x\Big|_0^{\pi/4} = \pi\left(\tan\frac{\pi}{4} - \tan 0\right) = \pi$$

21. $f(x)=\sqrt{1-x^2};\ r=\sqrt{1}=1$

$$V=\pi\int_{-1}^{1}(\sqrt{1-x^2})^2\,dx=\pi\int_{-1}^{1}(1-x^2)\,dx$$
$$=\pi\left(x-\frac{x^3}{3}\right)\Bigg|_{-1}^{1}=\pi\left(1-\frac{1}{3}\right)-\pi\left(-1+\frac{1}{3}\right)$$
$$=2\pi-\frac{2}{3}\pi=\frac{4\pi}{3}$$

23. $f(x)=\sqrt{r^2-x^2}$

$$V=\pi\int_{-r}^{r}(\sqrt{r^2-x^2})^2\,dx=\pi\int_{-r}^{r}(r^2-x^2)\,dx$$
$$=\pi\left(r^2x-\frac{x^3}{3}\right)\Bigg|_{-r}^{r}$$
$$=\pi\left(r^3-\frac{r^3}{3}\right)-\pi\left(-r^3+\frac{r^3}{3}\right)$$
$$=2r^3\pi-\left(\frac{2r^3\pi}{3}\right)=\frac{4\pi r^3}{3}$$

25. $f(x)=r,\ x=0,\ x=h$

Graph $f(x)=r$; then show the solid of revolution formed by rotating about the x-axis the region bounded by $f(x)$, $x=0$, $x=h$.

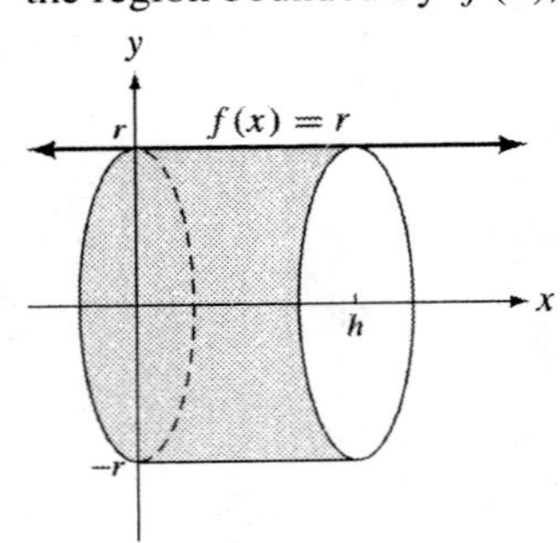

$$\int_0^h \pi r^2\,dx=\pi r^2x\Big|_0^h=\pi r^2h-0=\pi r^2h$$

27. $f(x)=x^2-4;\ [0,5]$

$$\text{Average value}=\frac{1}{5-0}\int_0^5(x^2-4)dx$$
$$=\frac{1}{5}\left(\frac{x^3}{3}-4x\right)\Bigg|_0^5$$
$$=\frac{1}{5}\left[\left(\frac{125}{3}-20\right)-0\right]$$
$$=\frac{13}{3}\approx 4.333$$

29. $f(x)=\sqrt{x+1};\ [3,8]$

$$\text{Average value}=\frac{1}{8-3}\int_3^8\sqrt{x+1}\,dx$$
$$=\frac{1}{5}\int_3^8(x+1)^{1/2}\,dx$$
$$=\frac{1}{5}\cdot\frac{2}{3}(x+1)^{3/2}\Big|_3^8$$
$$=\frac{2}{15}(9^{3/2}-4^{3/2})=\frac{38}{15}\approx 2.533$$

31. $f(x)=e^{x/7};[0,7]$

$$\text{Average value}=\frac{1}{7-0}\int_0^7 e^{x/7}\,dx$$
$$=\frac{1}{7}\cdot 7e^{x/7}\Big|_0^7$$
$$=e^{x/7}\Big|_0^7=e^1-e^0\approx 1.718$$

33. $f(x)=x^2e^{2x};\ [0,2]$

$$\text{Average value}=\frac{1}{2-0}\int_0^2 x^2e^{2x}\,dx$$

Let $u=x^2$ and $dv=e^{2x}\,dx$.

Use column integration.

D		I
x^2	+	e^{2x}
$2x$	−	$\frac{1}{2}e^{2x}$
2	+	$\frac{1}{4}e^{2x}$
0		$\frac{1}{8}e^{2x}$

$$\frac{1}{2-0}\int_0^2 x^2e^{2x}\,dx$$
$$=\frac{1}{2}\left[(x^2)\left(\frac{1}{2}\right)e^{2x}-(2x)\left(\frac{1}{4}\right)e^{2x}+2\left(\frac{1}{8}\right)e^{2x}\right]\Bigg|_0^2$$
$$=\frac{1}{2}\left(2e^4-e^4+\frac{1}{4}e^4-\frac{1}{4}\right)=\frac{5e^4-1}{8}\approx 34.00$$

35. $f(x)=\sec^2 x;\ \left[0,\frac{\pi}{4}\right]$

$$\text{Average value}=\frac{1}{\frac{\pi}{4}-0}\int_0^{\pi/4}\sec^2 x\,dx$$
$$=\frac{4}{\pi}(\tan x)\Big|_0^{\pi/4}$$
$$=\frac{4}{\pi}\left(\tan\frac{\pi}{4}-\tan 0\right)=\frac{4}{\pi}$$

37. $f(x) = e^{-x^2}$, $y = 0$, $x = -1$, $x = 1$

$$V = \pi\int_{-1}^{1} (e^{-x^2})^2 dx = \pi\int_{-1}^{2} e^{-2x^2}\, dx$$

Using an integration feature on a graphing calculator to evaluate the integral, we obtain $3.758249634 \approx 3.758$.

39. $R(t) = te^{-0.1t}$

"During the nth hour" corresponds to the interval $(n-1, n)$. The average intensity during nth hour is

$$\frac{1}{n-(n-1)}\int_{n-1}^{n} te^{-0.1t} dt = \int_{n-1}^{n} te^{-0.1t} dt$$

Let $u = t$ and $dv = e^{-0.1t} dt$.

D		I
t	$+$	$e^{-0.1t}$
1	$-$	$-10e^{-0.1t}$
0		$100e^{-0.1t}$

$$\int_{n-1}^{n} te^{-0.1t}\, dt$$
$$= (-10te^{-0.1t} - 100e^{-0.1t})\Big|_{n-1}^{n}$$

(a) Second hour, $n = 2$
Average intensity
$$= -10e^{-0.2}(12) + 10e^{-0.1}(11)$$
$$= 110e^{-0.1} - 120e^{-0.2} \approx 1.284$$

(b) Twelfth hour, $n = 12$
Average intensity
$$= -10e^{-1.2}(12+10) + 10e^{-1.1}(11+10)$$
$$= 210e^{-1.1} - 220e^{-1.2} \approx 3.640$$

(c) Twenty-fourth hour, $n = 24$
Average intensity
$$= -10e^{-2.4}(24+10) + 10e^{-2.3}(23+10)$$
$$= 330e^{-2.3} - 340e^{-2.4} \approx 2.241$$

41. For each part below, use

$$\text{Average value} = \frac{1}{b-a}\int_a^b 45\ln(t+1)dt = \frac{45}{b-a}\int_a^b \ln(t+1)dt.$$

Evaluate the integral using integration by parts.

Let $u = \ln(t+1)$ and $dv = dt$.

Then $du = \frac{1}{t+1}dt$ and $v = t$.

$$\int \ln(t+1)\, dt = t\ln(t+1) - \int \frac{t}{t+1}\, dt$$
$$= t\ln(t+1) - \int\left(1 - \frac{1}{t+1}\right)dt$$
$$= t\ln(t+1) - t + \ln(t+1) + C$$
$$= (t+1)\ln(t+1) - t + C$$

Therefore ,

$$\text{average value} = \frac{1}{b-a}\int_a^b 45\ln(t+1)\, dt = \frac{45}{b-a}[(t+1)\ln(t+1) - t]\Big|_a^b.$$

(a) The average number of items produced daily after 5 days is
$$\frac{45}{5-0}[(t+1)\ln(t+1) - t]\Big|_0^5$$
$$= 9[(6\ln 6 - 5) - (\ln 1 - 0)]$$
$$= 9(6\ln 6 - 5) \approx 51.76.$$

(b) The average number of items produced daily after 9 days is
$$\frac{45}{9-0}[(t+1)\ln(t+1) - t]\Big|_0^9$$
$$= 5(10\ln 10 - 9) \approx 70.13.$$

(c) The average number of items produced daily after 30 days is
$$\frac{45}{30-0}[(t+1)\ln(t+1) - t]\Big|_0^{30}$$
$$= \frac{3}{2}(31\ln 31 - 30) \approx 114.7.$$

43. From Exercise 24, the volume of an ellipsoid with horizontal axis of length $2a$ and vertical axis of length $2b$ is $V = \frac{4ab^2\pi}{3}$. For the Earth, $a = 6,356,752.3141$ and $b = 6,378,137$.

$$V = \frac{4(6,356,752.3141)(6,378,137)^2\pi}{3}$$
$$\approx 1.083\times 10^{21}$$

The volume of the Earth is about 1.083×10^{21} cubic meters (m^3).

45. Use the formula for average value with $a = 0$ and $b = 6$.

$$\frac{1}{6-0}\int_0^6 (37+6e^{-0.03t})dt$$
$$=\frac{1}{6}\left(37t+\frac{6}{-0.03}e^{-0.03t}\right)\Bigg|_0^6$$
$$=\frac{1}{6}(37t-200e^{-0.03t})\Bigg|_0^6$$
$$=\frac{1}{6}[(222-200e^{-0.18})-(0-200)]$$
$$=\frac{1}{6}(422-200e^{-0.18})\approx 42.49$$

The average price is about \$42.49.

47. Use the formula for average value with $a = 0$ and $b = 30$. The average daily inventory is

$$\frac{1}{30-0}\int_0^{30}(600-20\sqrt{30t})\,dt$$
$$=\frac{1}{30}\left(600t-20\sqrt{30}\cdot\frac{2}{3}t^{3/2}\right)\Bigg|_0^{30}$$
$$=\frac{1}{30}\left(600t-\frac{40\sqrt{30}}{3}t^{3/2}\right)\Bigg|_0^{30}$$
$$=\frac{1}{30}(18{,}000-12{,}000)=200\text{ cases}$$

8.4 Improper Integrals

1. $$\int_3^\infty \frac{1}{x^2}dx=\lim_{b\to\infty}\int_3^b x^{-2}dx=\lim_{b\to\infty}-x^{-1}\Big|_3^b$$
$$=\lim_{b\to\infty}\left(-\frac{1}{b}+\frac{1}{3}\right)$$
$$=\lim_{b\to\infty}\left(-\frac{1}{b}\right)+\lim_{b\to\infty}\frac{1}{3}$$

As $b\to\infty, -\frac{1}{b}\to 0$. The integral is convergent and its value is $0+\frac{1}{3}=\frac{1}{3}$.

3. $$\int_4^\infty \frac{2}{\sqrt{x}}dx=\lim_{b\to\infty}\int_4^b 2x^{-1/2}dx$$
$$=\lim_{b\to\infty}4x^{1/2}\Big|_4^b$$
$$=\lim_{b\to\infty}(4\sqrt{b}-4\sqrt{4})$$
$$=\lim_{b\to\infty}4\sqrt{b}-8$$

As $b\to\infty, 4\sqrt{b}\to\infty$. The integral diverges.

5. $$\int_{-\infty}^{-1}\frac{2}{x^3}dx=\int_{-\infty}^{-1}2x^{-3}dx=\lim_{a\to-\infty}\int_a^{-1}2x^{-3}dx$$
$$=\lim_{a\to-\infty}\left(\frac{2x^{-2}}{-2}\right)\Bigg|_a^{-1}=\lim_{a\to-\infty}\left(-1+\frac{1}{a^2}\right)$$

As $a\to-\infty, \frac{1}{a^2}\to 0$. The integral is convergent and its value is $-1+0=-1$.

7. $$\int_1^\infty \frac{1}{x^{1.0001}}dx=\int_1^\infty x^{-1.0001}\,dx$$
$$=\lim_{b\to\infty}\int_1^b x^{-1.0001}\,dx$$
$$=\lim_{b\to\infty}\left(\frac{x^{-0.0001}}{-0.0001}\right)\Bigg|_1^b$$
$$=\lim_{b\to\infty}\left(-\frac{1}{(0.0001)b^{0.0001}}+\frac{1}{0.0001}\right)$$

As $b\to\infty, -\frac{1}{0.0001b^{0.0001}}\to 0$. The integral is convergent and its value is

$$0+\frac{1}{0.0001}=10{,}000.$$

9. $$\int_{-\infty}^{-10}x^{-2}dx=\lim_{a\to-\infty}\int_a^{-10}x^{-2}dx$$
$$=\lim_{a\to-\infty}(-x^{-1})\Big|_a^{-10}$$
$$=\lim_{a\to-\infty}\left(\frac{1}{10}+\frac{1}{a}\right)=\frac{1}{10}+0=\frac{1}{10}$$

The integral is convergent and its value is $\frac{1}{10}$.

11. $$\int_{-\infty}^{-1}x^{-8/3}dx=\lim_{a\to-\infty}\int_a^{-1}x^{-8/3}dx$$
$$=\lim_{a\to-\infty}\left(-\frac{3}{5}x^{-5/3}\right)\Bigg|_a^{-1}$$
$$=\lim_{a\to-\infty}\left(\frac{3}{5}+\frac{3}{5a^{5/3}}\right)=\frac{3}{5}+0=\frac{3}{5}$$

The integral is convergent, and its value is $\frac{3}{5}$.

13. $\int_0^{\infty} 8e^{-8x}dx = \lim_{b\to\infty} \int_0^b 8e^{-8x}dx$

$= \lim_{b\to\infty} \left(\frac{8e^{-8x}}{-8}\right)\Bigg|_0^b$

$= \lim_{b\to\infty} (-e^{-8b} + 1)$

$= \lim_{b\to\infty} \left(-\frac{1}{e^{8b}} + 1\right) = 0 + 1 = 1$

The integral is convergent, and its value is 1.

15. $\int_{-\infty}^{0} 1000e^x dx = \lim_{a\to-\infty} \int_a^0 1000e^x dx$

$= \lim_{a\to-\infty} (1000e^x)\Big|_a^0$

$= \lim_{a\to-\infty} (1000 - 1000e^a)$

As $a \to \infty$, $-1000e^a \to 0$. The integral is convergent and its value is $1000 - 0 = 1000$.

17. $\int_{-\infty}^{-1} \ln |x| dx = \lim_{a\to-\infty} \int_a^{-1} \ln |x| dx$

Let $u = \ln |x|$ and $dv = dx$.

Then $du = \frac{1}{x}dx$ and $v = x$.

$\int \ln |x| dx = x \ln |x| - \int \frac{x}{x}dx$

$= x \ln |x| - x + C$

$\int_{-\infty}^{-1} \ln |x|\, dx = \lim_{a\to-\infty} (x \ln |x| - x)\Big|_a^{-1}$

$= \lim_{a\to-\infty} (-\ln 1 + 1 - a \ln |a| + a)$

$= \lim_{a\to-\infty} (1 + a - a \ln |a|)$

The integral is divergent, because as $a \to -\infty$ $(a - a \ln |a|) = -a(-1 + \ln |a|) \to \infty$.

19. $\int_0^{\infty} \frac{dx}{(x+1)^2}$

$= \lim_{b\to\infty} \int_0^b \frac{dx}{(x+1)^2}$ *Use substitution*

$= \lim_{b\to\infty} -(x+1)^{-1}\Big|_0^b$

$= \lim_{b\to\infty} \left(\frac{-1}{b+1} + 1\right)$

As $b \to \infty$, $-\frac{1}{b+1} \to 0$. The integral is convergent and its value is $0 + 1 = 1$.

21. $\int_{-\infty}^{-1} \frac{2x-1}{x^2 - x}\, dx$

$= \lim_{a\to-\infty} \int_0^{-1} \frac{2x-1}{x^2-x}dx$ *Use substitution*

$= \lim_{a\to-\infty} \ln |x^2 - x|\Big|_a^{-1}$

$= \lim_{a\to-\infty} (\ln 2 - \ln |a^2 - a|)$

As $a \to -\infty$, $\ln |a^2 - a| \to \infty$. The integral is divergent.

23. $\int_2^{\infty} \frac{1}{x \ln x}\, dx$

$= \lim_{b\to\infty} \int_2^b \frac{1}{x \ln x}dx$ *Use substitution*

$= \lim_{b\to\infty} \left[\ln (\ln x)\Big|_2^b\right]$

$= \lim_{b\to\infty} [\ln (\ln b) - \ln (\ln 2)]$

As $b \to \infty$, $\ln (\ln b) \to \infty$. The integral is divergent.

25. $\int_0^{\infty} xe^{4x}\, dx = \lim_{b\to\infty} \int_0^b xe^{4x}\, dx$

Let $dv = e^{4x}\, dx$ and $u = x$.

Then $v = \frac{1}{4}e^{4x}dx$ and $du = dx$.

$\int xe^{4x}\, dx = \frac{x}{4}e^{4x} - \int \frac{1}{4}e^{4x}dx$

$= \frac{x}{4}e^{4x} - \frac{1}{16}e^{4x} + C$

$= \frac{1}{16}(4x-1)e^{4x} + C$

$\int_0^{\infty} xe^{4x}\, dx$

$= \lim_{b\to\infty} \left[\frac{1}{16}(4x-1)e^{4x}\right]\Bigg|_0^b$

$= \lim_{b\to\infty} \left[\frac{1}{16}(4b-1)e^{4b} - \frac{1}{16}(-1)(1)\right]$

$= \lim_{b\to\infty} \left[\frac{1}{16}(4b-1)e^{4b} + \frac{1}{16}\right]$

As $b \to \infty$, $\frac{1}{16}(4b-1)e^{4b} \to \infty$. The integral is divergent.

27. $\int_{-\infty}^{\infty} x^3 e^{-x^4}\, dx$

$= \int_{-\infty}^{0} x^3 e^{-x^4}\, dx + \int_{0}^{\infty} x^3 e^{-x^4}\, dx$

We evaluate each of two improper integrals on the right.

$\int_{-\infty}^{0} x^3 e^{-x^4}\, dx$

$= \lim_{a\to-\infty} \int_{b}^{0} x^3 e^{-x^4}\, dx$ *Use substitution*

$= \lim_{a\to-\infty} \left[-\frac{1}{4} e^{-x^4} \Big|_a^0 \right]$

$= \lim_{a\to-\infty} \left[-\frac{1}{4} + \frac{1}{4e^{a^4}} \right]$

As $a \to -\infty$, $\frac{1}{4e^{a^4}} \to 0$. The integral is convergent and its value is $-\frac{1}{4} + 0 = -\frac{1}{4}$.

$\int_{0}^{\infty} x^3 e^{-x^4}\, dx$

$= \lim_{b\to\infty} \int_{0}^{b} x^3 e^{-x^4}\, dx$ *Use substitution*

$= \lim_{b\to\infty} \left[-\frac{1}{4} e^{-x^4} \Big|_0^b \right] = \lim_{b\to\infty} \left[-\frac{1}{4e^{b^4}} + \frac{1}{4} \right]$

As $b \to \infty, -\frac{1}{4e^{b^4}} \to 0$. The integral is convergent and its value is $0 + \frac{1}{4} = \frac{1}{4}$.

Since each of the improper integrals converges, the original improper integral converges and its value is $-\frac{1}{4} + \frac{1}{4} = 0$.

29. $\int_{-\infty}^{\infty} \frac{x}{x^2+1} dx$

$= \int_{-\infty}^{0} \frac{x}{x^2+1} dx + \int_{0}^{\infty} \frac{x}{x^2+1} dx$

We evaluate the first improper integrals on the right.

$\int_{-\infty}^{0} \frac{x}{x^2+1} dx$

$= \lim_{a\to-\infty} \int_{b}^{0} \frac{x}{x^2+1} dx$ *Use substitution*

$= \lim_{a\to-\infty} \left[\frac{1}{2} \ln(x^2+1) \Big|_a^0 \right]$

$= \lim_{a\to-\infty} \left[0 - \frac{1}{2} \ln(a^2+1) \right]$

As $a \to -\infty$, $\ln(a^2+1) \to \infty$. The integral is divergent. Since one of the two improper integrals on the right diverges, the original improper integral diverges.

31. $f(x) = \frac{1}{x-1}$ for $(-\infty, 0]$

$\int_{-\infty}^{0} \frac{1}{x-1} dx = \lim_{a\to-\infty} \int_{a}^{0} \frac{dx}{x-1}$

$= \lim_{a\to-\infty} \left(\ln|x-1| \Big|_a^0 \right)$

$= \lim_{a\to-\infty} (\ln|-1| - \ln|a-1|)$

But $\lim_{a\to-\infty} (\ln|a-1|) = \infty$. The integral is divergent, so the area cannot be found.

33. $f(x) = \frac{1}{(x-1)^2}$ for $(-\infty, 0]$

$\int_{-\infty}^{0} \frac{1}{(x-1)^2}$

$= \lim_{a\to-\infty} \int_{a}^{0} \frac{1}{(x-1)^2}$ *Use substitution*

$= \lim_{a\to-\infty} -(x-1)^{-1} \Big|_a^0 = \lim_{a\to-\infty} \left(-\frac{1}{-1} + \frac{1}{a-1} \right)$

As $a \to -\infty$, $\frac{1}{a-1} \to 0$. The integral is convergent and its value is $1 + 0 = 1$. Therefore, the area is 1.

35. $\int_{-\infty}^{\infty} xe^{-x^2}\, dx$

Let $u = -x^2$, so that $du = -2x\, dx$.

$\int_{-\infty}^{\infty} xe^{-x^2}\, dx = \lim_{a\to-\infty} \left(-\frac{1}{2} \int_{a}^{0} -2xe^{-x^2}\, dx \right) + \lim_{b\to\infty} \left(-\frac{1}{2} \int_{0}^{b} -2xe^{-x^2}\, dx \right)$

$= \lim_{a\to-\infty} \left(-\frac{1}{2} e^{-x^2} \right) \Big|_a^0 + \lim_{b\to\infty} \left(-\frac{1}{2} e^{-x^2} \right) \Big|_0^b$

$= \lim_{a\to-\infty} \left(-\frac{1}{2} + \frac{1}{2e^{-a^2}} \right) + \lim_{b\to\infty} \left(-\frac{1}{2e^{b^2}} + \frac{1}{2} \right)$

$= -\frac{1}{2} + \frac{1}{2} = 0$

37. $\int_1^\infty \frac{1}{x^p}dx$

Case 1a $p<1$:

$$\int_1^\infty \frac{1}{x^p}dx = \int_1^\infty x^{-p}dx = \lim_{b\to\infty}\int_1^b x^{-p}dx$$
$$= \lim_{b\to\infty}\left[\frac{x^{-p+1}}{(-p+1)}\bigg|_1^b\right]$$
$$= \lim_{b\to\infty}\left[\frac{1}{(-p+1)}(b^{-p+1}-1)\right]$$
$$= \lim_{b\to\infty}\left[\frac{1}{(-p+1)}b^{1-p} - \frac{1}{(-p+1)}\right]$$

Since $p<1$, $1-p$ is positive and, as $b\to\infty$, $b^{1-p}\to\infty$. The integral diverges.

Case 1b $p=1$:

$$\int_1^\infty \frac{1}{x^p}dx = \int_1^\infty \frac{1}{x}dx = \lim_{a\to\infty}\int_1^b \frac{1}{x}dx$$
$$= \lim_{b\to\infty}\left(\ln|x|\big|_1^b\right) = \lim_{b\to\infty}(\ln|b| - \ln 1)$$
$$= \lim_{b\to\infty}\ln|b|$$

As $b\to\infty$, $\ln|b|\to\infty$. The integral diverges.

Therefore, $\int_1^\infty \frac{1}{x^p}$ diverges when $p\le 1$.

Case 2 $p>1$:

$$\int_1^\infty \frac{1}{x^p}dx = \lim_{b\to\infty}\int_1^b x^{-p}dx = \lim_{b\to\infty}\left(\frac{x^{-p+1}}{-p+1}\bigg|_1^b\right)$$
$$= \lim_{b\to\infty}\left[\frac{b^{-p+1}}{(-p+1)} - \frac{1}{(-p+1)}\right]$$

Since $p>1$, $-p+1<0$; thus as $a\to\infty$, $\frac{a^{-p+1}}{(-p+1)}\to 0$. Therefore,

$$\lim_{a\to\infty}\left[\frac{a^{-p+1}}{(-p+1)} - \frac{1}{(-p+1)}\right] = 0 - \frac{1}{(-p+1)}$$
$$= \frac{-1}{-p+1} = \frac{1}{p-1}.$$

The integral converges.

39. **(a)** Use the *fnInt* feature on a graphing utility to obtain

$$\int_1^{20}\frac{1}{\sqrt{1+x^2}}dx \approx 2.808;$$
$$\int_1^{50}\frac{1}{\sqrt{1+x^2}}dx \approx 3.724;$$
$$\int_1^{100}\frac{1}{\sqrt{1+x^2}}dx \approx 4.417;$$
$$\int_1^{1000}\frac{1}{\sqrt{1+x^2}}dx \approx 6.720;$$
$$\int_1^{10,000}\frac{1}{\sqrt{1+x^2}}dx \approx 9.022.$$

(b) Since the values of the integrals in part a do not appear to be approaching some fixed finite number but get bigger, the integral $\int_1^\infty \frac{1}{\sqrt{1+x^2}}dx$ appears to be divergent.

(c) Use the *fnInt* feature on a graphing utility to obtain

$$\int_1^{20}\frac{1}{\sqrt{1+x^4}}dx \approx 0.8770;$$
$$\int_1^{50}\frac{1}{\sqrt{1+x^4}}dx \approx 0.9070;$$
$$\int_1^{100}\frac{1}{\sqrt{1+x^4}}dx \approx 0.9170;$$
$$\int_1^{1000}\frac{1}{\sqrt{1+x^4}}dx \approx 0.9260;$$
$$\int_1^{10,000}\frac{1}{\sqrt{1+x^4}}dx \approx 0.9269.$$

(d) Since the values of the integrals in part c appear to be approaching some fixed finite number, the integral $\int_1^\infty \frac{1}{\sqrt{1+x^4}}dx$ appears to be convergent.

(e) For large x, we may consider $1+x^2\approx x^2$ and $1+x^4\approx x^4$. Thus,

$$\frac{1}{\sqrt{1+x^2}} \approx \frac{1}{\sqrt{x^2}} = \frac{1}{x} \text{ and}$$
$$\frac{1}{\sqrt{1+x^4}} \approx \frac{1}{\sqrt{x^4}} = \frac{1}{x^2}.$$

In Example 1(a) we showed that $\int_1^\infty \frac{1}{x}dx$ diverges. Thus, we might guess that $\int_1^\infty \frac{1}{\sqrt{1+x^2}}dx$ diverges as well. In Exercise 1, we saw that $\int_3^\infty \frac{1}{x^2}dx$ converges. Thus, we might guess that $\int_1^\infty \frac{1}{\sqrt{1+x^4}}dx$ converges as well.

41. **(a)** Use the *fnInt* feature on a graphing utility to obtain

$$\int_0^{10} e^{-.00001x}dx \approx 9.9995;$$
$$\int_0^{50} e^{-.00001x}dx \approx 49.9875;$$
$$\int_0^{100} e^{-.00001x}dx \approx 99.9500;$$
$$\int_0^{1000} e^{-.00001x}dx \approx 995.0166.$$

(b) Since the values of the integrals in part a do not appear to be approaching some fixed finite number, the integral $\int_0^{\infty} e^{-0.00001x}dx$ appears to be divergent.

(c) $\int_0^{\infty} e^{-0.00001x}dx$

$$= \lim_{b\to\infty} \int_0^b e^{-0.00001x}dx$$
$$= \lim_{b\to\infty} \left[\frac{e^{-0.00001x}}{-0.00001}\Big|_0^b\right]$$
$$= \lim_{b\to\infty} \left[-\frac{1}{0.00001e^{0.00001b}} + \frac{1}{0.00001}\right]$$
$$= 0 + 100{,}000 = 100{,}000$$

43. Use the *fnInt* feature on a graphing calculoatr to obtain

$$\int_0^1 e^{-x}\cos x dx \approx 0.555396883$$
$$\int_0^{10} e^{-x}\cos x dx \approx 0.500006698$$
$$\int_0^{100} e^{-x}\cos x dx \approx 0.5000000000$$

Based on these results, it appears that the integral is convergent and its value is $\frac{1}{2}$.

45. $S = N\int_0^{\infty} \frac{a(1-e^{-kt})}{k} e^{-bt}dt$

$$= \frac{Na}{k} \lim_{c\to\infty} \int_0^c (1-e^{-kt})(e^{-bt})dt$$
$$= \frac{Na}{k} \lim_{c\to\infty} \int_0^c (e^{-bt} - e^{-(b+k)t})dt$$
$$= \frac{Na}{k} \lim_{c\to\infty} \left[-\frac{1}{b}e^{-bt} + \frac{1}{b+k}e^{-(b+k)t}\right]\Bigg|_0^c$$
$$= \frac{Na}{k} \lim_{c\to\infty} \left[\left(-\frac{1}{b}e^{-bc} + \frac{1}{b+k}e^{-(b+k)c}\right) - \left(-\frac{1}{b}e^0 + \frac{1}{b+k}e^0\right)\right]$$
$$= \frac{Na}{k}\left(0+0+\frac{1}{b}-\frac{1}{b+k}\right)$$
$$= \frac{Na}{k}\cdot\frac{(b+k)-b}{b(b+k)} = \frac{Na}{b(b+k)}$$

47. $\int_0^{\infty} 50e^{-0.06t}dt = 50 \lim_{b\to\infty} \int_0^b e^{-0.06t}dt$

$$= 50 \lim_{b\to\infty} \frac{e^{-0.06t}}{-0.06}\Bigg|_0^b$$
$$= \frac{50}{-0.06} \lim_{b\to\infty} (e^{-0.06b} - e^0)$$
$$= -\frac{50}{0.06}(0-1) \approx 833.3$$

Chapter 8 Review Exercises

1. False: The trapezoidal rule allows any number of intervals.

2. True

3. False: This integral is best evaluated by substitution. Let $u = x^3 + 1$. Then $du = 3x^2dx$.

4. True. Let $u = x$ and $dv = e^{10x}dx$. Then $du = dx$ and $v = \frac{1}{10}e^{10x}$.

5. False: Using the substitution $u = x^2$, $dv = xe^{-x^2}$ this integral requires only one integration by parts.

6. True. Let $u = \ln(4x)$ and $dv = dx$. Then $du = \frac{1}{x}dx$ and $v = x$.

7. False: The integrand should be just $2x^2 + 3$.

8. False: The integrand should be $\pi(x^2 + 1)$.

9. True

10. False: We must write the integral as $\lim_{a\to-\infty} \int_a^c xe^{-2x}dx + \lim_{b\to\infty} \int_c^b xe^{-2x}dx$. The first of these integrals diverges so $\int_{-\infty}^{\infty} xe^{-2x}dx$ diverges.

11. Use numerical integration for those integrals that cannot be evaluated by any technique.

12. Use integration by parts to reduce a complicated integral to a simpler integral.

13. Answers will vary.

14. An improper integral is a definite integral that has either or both limits infinite or an integrand that approaches infinity at one or more points in the range of integration. We use limits to evaluate these integrals.

15. $\int_1^3 \frac{\ln x}{x}\,dx$

Trapezoidal Rule:

$n = 4,\ b = 3,\ a = 1,\ f(x) = \frac{\ln x}{x}$

i	x_1	$f(x_i)$
0	1	0
1	1.5	0.27031
2	2	0.34657
3	2.5	0.36652
4	3	0.3662

$$\int_1^3 \frac{\ln x}{x}\,dx \approx \frac{3-1}{4}\left[\frac{1}{2}(0) + 0.27031 + 0.34657 + 0.36652 + \frac{1}{2}(0.3662)\right]$$
$$= 0.5833$$

Exact Value:

$$\int_1^3 \frac{\ln x}{x}\,dx$$
$$= \frac{1}{2}(\ln x)^2\Big|_1^3 = \frac{1}{2}(\ln 3)^2 - \frac{1}{2}(\ln 1)^2$$
$$\approx 0.6035$$

16. $\int_2^{10} \frac{x\,dx}{x-1}$

Trapezoidal Rule:

$n = 4,\ b = 10,\ a = 2,\ f(x) = \frac{x}{x-1}$

i	x_i	$f(x_i)$
0	2	2
1	4	$\frac{4}{3}$
2	6	$\frac{6}{5}$
3	8	$\frac{8}{7}$
4	10	$\frac{10}{9}$

$$\int_2^{10} \frac{x}{x-1}\,dx \approx \frac{10-2}{4}\left[\frac{1}{2}(2) + \frac{4}{3} + \frac{6}{5} + \frac{8}{7} + \frac{1}{2}\left(\frac{10}{9}\right)\right]$$
$$\approx 10.46$$

Exact Value:

Let $u = x - 1$, so that $du = dx$ and $x = u + 1$. Then

$$\int_2^{10} \frac{x}{x-1}\,dx = \int_1^9 \frac{u+1}{u}\,du$$
$$= \int_1^9 \left(1 + \frac{1}{u}\right) du = \int_1^9 du + \int_1^9 \frac{1}{u}\,du$$
$$= u\Big|_1^9 + \ln|u|\Big|_1^9 = (9-1) + (\ln 9 - \ln 1)$$
$$= 8 + \ln 9 \approx 10.20.$$

17. $\int_0^1 e^x\sqrt{e^x + 4}\,dx$

Trapezoidal Rule:

$n = 4,\ b = 1,\ a = 0,\ f(x) = e^x\sqrt{e^x + 4}$

i	x_i	$f(x_i)$
0	0	2.236
1	0.25	2.952
2	0.5	3.919
3	0.75	5.236
4	1	7.046

$$\int_0^1 e^x\sqrt{e^x + 4}\,dx$$
$$= \frac{1-0}{4}\left[\frac{1}{2}(2.236) + 2.952 + 3.919 + 5.236 + \frac{1}{2}(7.046)\right]$$
$$\approx 4.187$$

Exact value:

$$\int_0^1 e^x\sqrt{e^x + 4}\,dx = \int_0^1 e^x(e^x + 4)^{1/2}dx$$
$$= \frac{2}{3}(e^x + 4)^{3/2}\Big|_0^1$$
$$= \frac{2}{3}(e + 4)^{3/2} - \frac{2}{3}(5)^{3/2}$$
$$\approx 4.155$$

18. $\int_0^2 xe^{-x^2}\, dx$

Trapezoidal rule:

$n = 4, b = 2, a = 0, f(x) = xe^{-x^2}$

i	x_i	$f(x_i)$
0	0	0
1	0.5	0.3894
2	1	0.3679
3	1.5	0.1581
4	2	0.0366

$$\int_0^2 xe^{-x^2}\, dx \approx \frac{2-0}{4}\left[\frac{1}{2}(0) + 0.3894 + 0.3679 + 0.1581 + \frac{1}{2}(0.0366)\right] = 0.4668$$

Exact value:

$$\int_0^2 xe^{-x^2}\, dx = -\frac{1}{2}\int_0^2 e^{-x^2}(-2x\, dx) = -\frac{1}{2}e^{-x^2}\Big|_0^2 = -\frac{1}{2}e^{-4} + \frac{1}{2}e^0 = \frac{1}{2}(1 - e^{-4}) \approx 0.4908$$

19. $\int_1^3 \frac{\ln x}{x}\, dx$

Simpson's rule: $n = 4, b = 3, a = 1, f(x) = \frac{\ln x}{x}$

i	x_i	$f(x_i)$
0	1	0
1	1.5	0.27031
2	2	0.34657
3	2.5	0.36652
4	3	0.3662

$$\int_1^3 \frac{\ln x}{x}\, dx \approx \frac{3-1}{3(4)}[0 + 4(0.27031) + 2(0.34657) + 4(0.36652) + 0.3662] \approx 0.6011$$

This answer is close to the value of 0.6035 obtained from the exact integral in Exercise 15.

20. $\int_2^{10} \frac{x\, dx}{x-1}$

Simpson's Rule:

i	x_i	$f(x_i)$
0	2	2
1	4	$\frac{4}{3}$
2	6	$\frac{6}{5}$
3	8	$\frac{8}{7}$
4	10	$\frac{10}{9}$

$$\int_2^{10} \frac{x}{x-1}\, dx \approx \frac{10-2}{3(4)}\left[2 + 4\left(\frac{4}{3}\right) + 2\left(\frac{6}{5}\right) + 4\left(\frac{8}{7}\right) + \left(\frac{10}{9}\right)\right] \approx 10.28$$

This answer is close to the answer of 10.20 obtained from the exact integral in Exercise 16.

21. $\int_0^1 e^x\sqrt{e^x + 4}\, dx$

Simpson's rule:

$n = 4, b = 1, a = 0, f(x) = e^x\sqrt{e^x + 4}$

i	x_i	$f(x_i)$
0	0	2.236
1	0.25	2.952
2	0.5	3.919
3	0.75	5.236
4	1	7.046

$$\int_0^1 e^x\sqrt{e^x + 4}\, dx = \frac{1-0}{3(4)}[2.236 + 4(2.952) + 2(3.919) + 4(5.236) + 7.046] \approx 4.156$$

This answer is close to the answer of 4.155 obtained from the exact integral in Exercise 17.

22. $\int_0^2 xe^{-x^2}\,dx$

Simpson's rule:

$n = 4, b = 2, a = 0, f(x) = xe^{-x^2}$

i	x_i	$f(x_i)$
0	0	0
1	0.5	0.3894
2	1	0.3679
3	1.5	0.1581
4	2	0.0366

$$\int_0^2 xe^{-x^2}\,dx \approx \frac{2-0}{3(4)}[0+4(0.3894)+2(0.3679)+4(0.1581)+0.0366] \approx 0.4937$$

The answer is close to the exact value obtained in Exercise 18, which is approximately 0.4908.

23. (a) $\int_1^5 \left[\sqrt{x-1}-\left(\frac{x-1}{2}\right)\right]dx$

$$= \int_1^5 \left(\sqrt{x-1}-\frac{x}{2}+\frac{1}{2}\right)dx$$

$$= \left(\frac{2}{3}(x-1)^{3/2}-\frac{x^2}{4}+\frac{x}{2}\right)\Bigg|_1^5$$

$$= \left(\frac{16}{3}-\frac{25}{4}+\frac{5}{2}\right)-\left(0-\frac{1}{4}+\frac{1}{2}\right)$$

$$= \frac{16}{3}-6+2=\frac{4}{3}$$

(b) $n = 4, b = 5, a = 1, \; f(x)=\sqrt{x-1}-\frac{x}{2}+\frac{1}{2}$

i	x_i	$f(x_i)$
0	1	0
1	2	0.5
2	3	0.41421
3	4	0.23205
4	5	0

$$\int_1^5 \left(\sqrt{x-1}-\frac{x}{2}+\frac{1}{2}\right)dx = \left(\frac{5-1}{4}\right)\left[\frac{1}{2}(0)+0.5+0.41421+0.23205+\frac{1}{2}(0)\right] = 1.146$$

(c) $\int_1^5 \left(\sqrt{x-1}-\frac{x}{2}+\frac{1}{2}\right)dx$

$$= \left(\frac{5-1}{3(4)}\right)[0+4(0.5)+2(0.41421)+4(0.23205)+0]$$

$$= \left(\frac{1}{3}\right)(3.75662) = 1.252$$

24. (a) $\int_0^4 \left(\frac{x+2}{2}-\frac{1}{x+1}\right)dx$

$$= \left(\frac{(x+2)^2}{4}-\ln|x+1|\right)\Bigg|_0^4$$

$$= (9-\ln 5)-(1-0) \approx 6.391$$

(b) Trapezoidal rule:

$n = 4, \; b = 4, \; a = 0, \; f(x) = \frac{x+2}{2}-\frac{1}{x+1}$

i	x_i	$f(x_i)$
0	0	0
1	1	1
2	2	1.6667
3	3	2.25
4	4	2.8

$$\int_0^4 \left(\frac{x+2}{2}-\frac{1}{x+1}\right)dx \approx \frac{4-0}{4}\left[\frac{1}{2}(0)+1+1.6667+2.25+\frac{1}{2}(2.8)\right] \approx 6.317$$

(c) Simpson's rule:

$n = 4, \; b = 4, \; a = 0, \; f(x) = \frac{x+2}{2}-\frac{1}{x+1}$

i	x_i	$f(x_i)$
0	0	0
1	1	1
2	2	1.6667
3	3	2.25
4	4	2.8

$$\int_0^4 \left(\frac{x+2}{2}-\frac{1}{x+1}\right)dx \approx \frac{4-0}{4(3)}[0+4(1)+2(1.6667)+4(2.25)+2.8] \approx 6.378$$

25. $\int_{-2}^{2} [x(x-1)(x+1)(x-2)(x+2)]^2 dx$

(a) Trapezoidal Rule: $n = 4, b = -2, a = 2,$

$f(x) = [x(x-1)(x+1)(x-2)(x+2)]^2$

i	x_i	$f(x_i)$
0	−2	0
1	−1	0
2	0	0
3	1	0
4	2	0

$$\int_{-2}^{2} [x(x-1)(x+1)(x-2)(x+2)]^2 dx \approx \frac{2-(-2)}{4}\left[\frac{1}{2}(0)+0+0+0+\frac{1}{2}(0)\right] = 0$$

(b) Simpson's Rule: $n = 4, b = -2, a = 2,$

$f(x) = [x(x-1)(x+1)(x-2)(x+2)]^2$

i	x_i	$f(x_i)$
0	−2	0
1	−1	0
2	0	0
3	1	0
4	2	0

$$\int_{-2}^{2} [x(x-1)(x+1)(x-2)(x+2)]^2 dx \approx \frac{2-(-2)}{3(4)}[0+4(0)+2(0)+4(0)+0] = 0$$

26. First find the indefinite integral using integration by parts.

$\int u\,dv = uv - \int v\,du$

Now substitute the given values.

$$\int_0^1 u\,dv = uv\Big|_0^1 - \int_0^1 v\,du = [u(1)v(1) - u(0)v(0)] - 2 = (5)(-1) - (-3)(4) - 2 = 5$$

27. $\int x(8-x)^{3/2} dx$

Let $u = x$ and $dv = (8-x)^{3/2}$.

Then $du = dx$ and $v = -\frac{2}{5}(8-x)^{5/2}$.

$$\int x(8-x)^{3/2} dx = -\frac{2}{5}x(8-x)^{5/2} + \int \frac{2}{5}(8-x)^{5/2} dx = -\frac{2}{5}x(8-x)^{5/2} - \frac{2}{5}\left(\frac{2}{7}\right)(8-x)^{7/2} + C = -\frac{2x}{5}(8-x)^{5/2} - \frac{4}{35}(8-x)^{7/2} + C$$

28. $\int \frac{3x}{\sqrt{x-2}} dx = \int 3x(x-2)^{-1/2} dx$

Let $u = 3x$ and $dv = (x-2)^{-1/2} dx$.

Then $du = 3\,dx$ and $v = 2(x-2)^{1/2}$.

$$\int \frac{3x}{\sqrt{x-2}} dx = 6x(x-2)^{1/2} - 6\int (x-2)^{1/2} dx = 6x(x-2)^{1/2} - \frac{6(x-2)^{3/2}}{\frac{3}{2}} + C = 6x(x-2)^{1/2} - 4(x-2)^{3/2} + C$$

29. $\int xe^x dx$

Let $u = x$ and $dv = e^x dx$.

Then $du = dx$ and $v = e^x$.

$$\int xe^x dx = xe^x - \int e^x dx = xe^x - e^x + C$$

30. $\int (3x+6)e^{-3x} dx$

Let $u = 3x+6$ and $dv = e^{-3x} dx$.

Then $du = 3dx$ and $v = -\frac{1}{3}e^{-3x}$.

$$\int (3x+6)e^{-3x} dx = (3x+6)\left(-\frac{1}{3}e^{-3x}\right) - \int \left(-\frac{1}{3}e^{-3x}\right) 3\,dx = -(x+2)e^{-3x} + \int e^{-3x} dx = -(x+2)e^{-3x} - \frac{1}{3}e^{-3x} + C$$

31. $\int \ln|4x+5|\,dx$

First, use substitution. Let $w = 4x+5$. Then $dw = 4\,dx$.

$$\int \ln|4x+5|\,dx = \frac{1}{4}\int \ln|w|\,dw$$

Now use integration parts.

Let $u = \ln|w|$ and $dv = dw$.

Then $du = \frac{1}{w}dw$ and $v = w$.

(continued on next page)

(continued)

$$\int \ln|w|\,dw = w\ln|w| - \int w\frac{1}{w}\,dw$$
$$= w\ln|w| - \int dw$$
$$= w\ln|w| - w + C$$

Finally, resubstitute $4x+5$ for w.

$$\int \ln|4x+5|\,dx$$
$$= \frac{1}{4}[(4x+5)\ln|4x+5| - (4x+5)] + C$$
$$= \frac{1}{4}(4x+5)(\ln|4x+5| - 1) + C$$

32. $\int (x-1)\ln|x|\,dx$

Let $u = \ln|x|$ and $dv = (x-1)\,dx$.

Then $du = \frac{1}{x}dx$ and $v = \frac{x^2}{2} - x$.

$$\int (x-1)\ln|x|\,dx$$
$$= \left(\frac{x^2}{2} - x\right)\ln|x| - \int\left(\frac{x}{2} - 1\right)dx$$
$$= \left(\frac{x^2}{2} - x\right)\ln|x| - \frac{x^2}{4} + x + C$$

33. $\int \frac{x}{25-9x^2}dx$

Use substitution. Let $u = 25-9x^2$. Then $du = -18x\,dx$.

$$\int \frac{x}{25-9x^2}dx = -\frac{1}{18}\int\frac{-18x}{25-9x^2}dx$$
$$= -\frac{1}{18}\int\frac{du}{u} = -\frac{1}{18}\ln|u| + C$$
$$= -\frac{1}{18}\ln|25-9x^2| + C$$

34. $\int \frac{x}{\sqrt{16+8x^2}}dx$

Use substitution. Let $u = 16+8x^2$. Then $du = 16x\,dx$.

$$\int \frac{x}{\sqrt{16+8x^2}}dx = \frac{1}{16}\int\frac{16x}{\sqrt{16+8x^2}}dx$$
$$= \frac{1}{16}\int\frac{1}{\sqrt{u}}du = \frac{1}{16}\int u^{-1/2}du$$
$$= \frac{1}{16}(2)u^{1/2} + C$$
$$= \frac{1}{8}(16+8x^2)^{1/2} + C$$
$$= \frac{1}{8}\sqrt{16+8x^2} + C$$

35. $\int_1^e x^3 \ln x\,dx$

Let $u = \ln x$ and $dv = x^3\,dx$. Use column integration.

D		I
$\ln x$	+	x^3
$\frac{1}{x}$	−	$\frac{1}{4}x^4$

$$\int x^3 \ln x\,dx$$
$$= \frac{1}{4}x^4\ln x - \int\left(\frac{1}{4}x^4\cdot\frac{1}{x}\right)dx$$
$$= \frac{1}{4}x^4\ln x - \frac{1}{4}\int x^3\,dx$$
$$= \frac{1}{4}x^4\ln x - \frac{1}{16}x^4 + C$$

$$\int_1^e x^3\ln x\,dx = \left(\frac{1}{4}x^4\ln x - \frac{1}{16}x^4\right)\Big|_1^e$$
$$= \left(\frac{e^4}{4}\right)(1) - \frac{e^4}{16} - 0 + \frac{1}{16}$$
$$= \frac{e^4}{4} - \frac{e^4}{16} + \frac{1}{16} = \frac{3e^4+1}{16} \approx 10.30$$

36. $\int_0^1 x^2 e^{x/2}dx$

Let $u = x^2$ and $dv = e^{x/2}\,dx$. Use column integration.

D		I
x^2	+	$e^{x/2}$
$2x$	−	$2e^{x/2}$
2	+	$4e^{x/2}$
0		$8e^{x/2}$

$$\int_0^1 x^2e^{x/2}\,dx = (2x^2e^{x/2} - 8xe^{x/2} + 16e^{x/2})\Big|_0^1$$
$$= 2e^{1/2} - 8e^{1/2} + 16e^{1/2} - 16$$
$$= 10e^{1/2} - 16 \approx 0.4872$$

37. $\int (x+2)\sin x\,dx = \int (x\sin x + 2\sin x)\,dx$
$$= \int x\sin x\,dx + 2\int \sin x\,dx$$
$$= \int x\sin x\,dx - 2\cos x + C$$

For $\int x\sin x\,dx$, integrate by parts with $u = x$, $dv = \sin x\,dx$, $du = dx$, and $v = -\cos x$.

$$\int x\sin x\,dx = -x\cos x + \int \cos dx$$
$$= -x\cos x + \sin x$$

(continued on next page)

(continued)

So, we have

$$\int(x+2)\sin x\,dx = \int x\sin x\,dx - 2\cos x + C$$
$$= -x\cos x + \sin x - 2\cos x + C$$
$$= \sin x - (x+2)\cos x + C$$

38. $\int x^2\cos(2x)\,dx$

Integrate by parts with $u = x^2$, $du = 2x dx$, $dv = \cos(2x)\,dx$, and $v = \frac{1}{2}\sin(2x)$.

$$\int x^2\cos(2x)\,dx = \frac{1}{2}x^2\sin(2x) - \int x\sin(2x)\,dx$$

For $\int x\sin(2x)\,dx$, integrate again by parts, this time with $u = x$, $du = dx$, $dv = \sin(2x)\,dx$, and $v = -\frac{1}{2}\cos(2x)$.

$$\int x\sin(2x)\,dx = -\frac{1}{2}x\cos(2x) + \frac{1}{2}\int\cos(2x)\,dx$$

So, we have

$$\int x^2\cos(2x)\,dx$$
$$= \frac{1}{2}x^2\sin(2x) - \int x\sin(2x)\,dx$$
$$= \frac{1}{2}x^2\sin(2x) - \left(-\frac{1}{2}x\cos(2x) + \frac{1}{2}\int\cos(2x)\,dx\right)$$
$$= \frac{1}{2}x^2\sin(2x) + \frac{1}{2}x\cos(2x) - \frac{1}{2}\int\cos(2x)\,dx$$
$$= \frac{1}{2}x^2\sin(2x) + \frac{1}{2}x\cos(2x) - \frac{1}{2}\left(\frac{1}{2}\sin(2x)\right)$$
$$= \frac{1}{2}x^2\sin(2x) + \frac{1}{2}x\cos(2x) - \frac{1}{4}\sin(2x) + C$$

39. $A = \int_0^1 (3+x^2)e^{2x}\,dx$

$$= \int_0^1 3e^{2x}\,dx + \int_0^1 x^2e^{2x}\,dx$$

$$\int 3e^{2x}\,dx = \frac{3}{2}e^{2x} + C$$

For the second integral, $\int x^2e^{2x}\,dx$, let $u = x^2$ and $dv = e^{2x}\,dx$. Use column integration.

D		I
x^2	+	e^{2x}
$2x$	−	$\frac{e^{2x}}{2}$
2	+	$\frac{e^{2x}}{4}$
0		$\frac{e^{2x}}{8}$

$$\int x^2e^{2x}\,dx = x^2\frac{e^{2x}}{2} - 2x\frac{e^{2x}}{4} + 2\frac{e^{2x}}{8} + C$$
$$= \frac{x^2e^{2x}}{2} - \frac{xe^{2x}}{2} + \frac{e^{2x}}{4} + C$$

$$A = \left(\frac{3}{2}e^{2x} + \frac{x^2e^{2x}}{2} - \frac{xe^{2x}}{2} + \frac{e^{2x}}{4}\right)\Bigg|_0^1$$
$$= \frac{3}{2}e^2 + \frac{e^2}{2} - \frac{e^2}{2} + \frac{e^2}{4} - \left(\frac{3}{2} + \frac{1}{4}\right)$$
$$= \left(\frac{6+2-2+1}{4}\right)e^2 - \left(\frac{7}{4}\right)$$
$$= \frac{7}{4}(e^2 - 1) \approx 11.18$$

40. $A = \int_1^3 x^3(x^2-1)^{1/3}\,dx$

Let $u = x^2$ and $dv = x(x^2-1)^{1/3}\,dx$.
Then $du = 2x\,dx$ and $v = \frac{3}{8}(x^2-1)^{4/3}$.

$$\int x^3(x^2-1)^{1/3}\,dx$$
$$= \frac{3x^2}{8}(x^2-1)^{4/3} - \frac{3}{4}\int x(x^2-1)^{4/3}\,dx$$
$$= \frac{3x^2}{8}(x^2-1)^{4/3} - \frac{3}{4}\left[\frac{1}{2}\cdot\frac{3}{7}(x^2-1)^{7/3}\right]$$
$$= \frac{3x^2}{8}(x^2-1)^{4/3} - \frac{9}{56}(x^2-1)^{7/3} + C$$

$$A = \left[\frac{3x^2}{8}(x^2-1)^{4/3} - \frac{9}{56}(x^2-1)^{7/3}\right]\Bigg|_1^3$$
$$= \frac{3}{8}(144) - \frac{9}{56}(128)$$
$$= 54 - \frac{144}{7} = \frac{234}{7} \approx 33.43$$

41. $f(x) = 3x - 1,\ y = 0,\ x = 2$

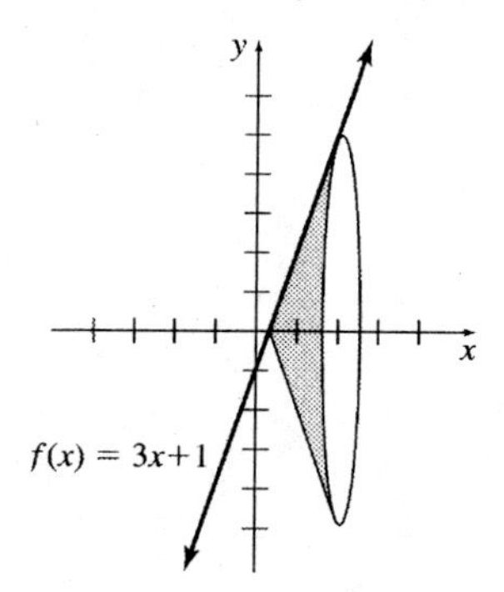

Since $f(x) = 3x - 1$ intersects $y = 0$ at $x = \frac{1}{3}$, the integral has a lower bound $a = \frac{1}{3}$.

$$\begin{aligned} V &= \pi\int_{1/3}^{2}(3x-1)^2\,dx \\ &= \pi\int_{1/3}^{2}(9x^2 - 6x + 1)dx \\ &= \pi(3x^3 - 3x^2 + x)\Big|_{1/3}^{2} \\ &= \pi\left[(24 - 12 + 2) - \left(\frac{1}{9} - \frac{1}{3} + \frac{1}{3}\right)\right] \\ &= \pi\left(14 - \frac{1}{9}\right) = \frac{125\pi}{9} \approx 43.63 \end{aligned}$$

42. $f(x) = \sqrt{x-4};\ y = 0;\ x = 13$

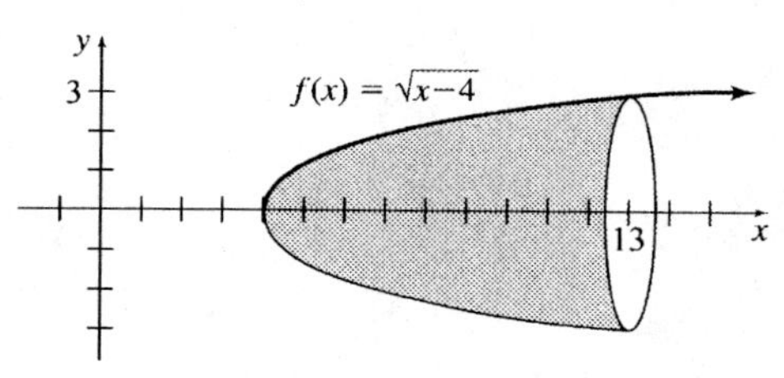

Since $f(x) = \sqrt{x-4}$ intersects $y = 0$ at $x = 4$, the integral has lower bound $a = 4$.

$$\begin{aligned} V &= \pi\int_{4}^{13}(\sqrt{x-4})^2\,dx \\ &= \pi\int_{4}^{13}(x-4)dx \\ &= \pi\left(\frac{x^2}{2} - 4x\right)\Bigg|_{4}^{13} \\ &= \pi\left[\left(\frac{169}{2} - 52\right) - (8 - 16)\right] \\ &= \pi\left(\frac{65}{2} + 8\right) = \frac{81}{2}\pi \approx 127.2 \end{aligned}$$

43. $f(x) = e^{-x},\ y = 0,\ x = -2,\ x = 1$

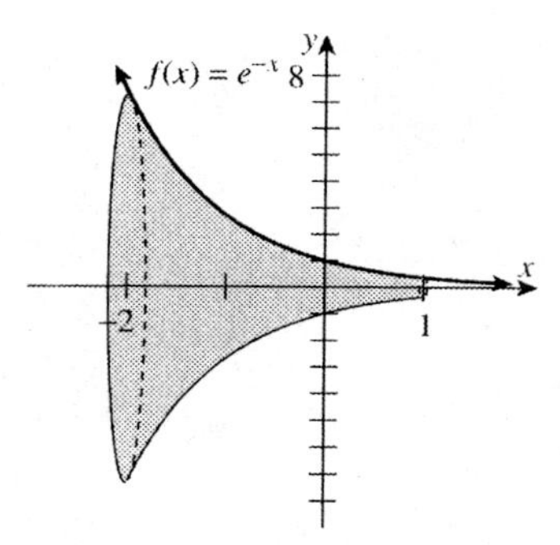

$$\begin{aligned} V &= \pi\int_{-2}^{1} e^{-2x}dx = \frac{\pi e^{-2x}}{-2}\Bigg|_{-2}^{1} \\ &= \frac{\pi e^{-2}}{-2} + \frac{\pi e^{4}}{2} = \frac{\pi(e^4 - e^{-2})}{2} \\ &\approx 85.55 \end{aligned}$$

44. $f(x) = \dfrac{1}{\sqrt{x-1}},\ y = 0,\ x = 2,\ x = 4$

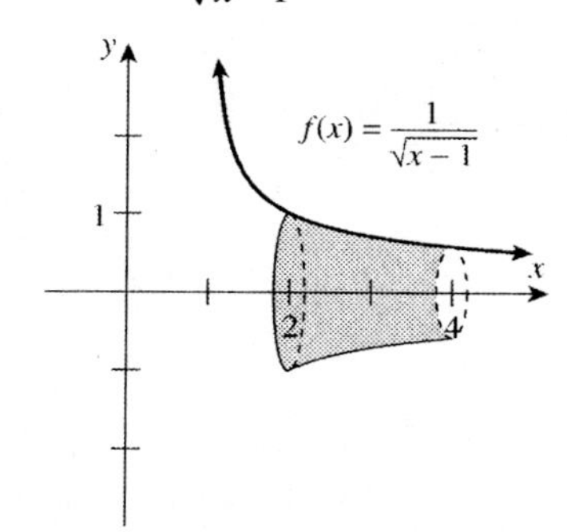

$$\begin{aligned} V &= \pi\int_{2}^{4}\left(\frac{1}{\sqrt{x-1}}\right)^2 dx \\ &= \pi\int_{2}^{4}\frac{dx}{x-1} \\ &= \pi\,(\ln|x-1|)\big|_{2}^{4} \\ &= \pi\ln 3 \approx 3.451 \end{aligned}$$

45. $f(x) = 4 - x^2,\ y = 0,\ x = -1,\ x = 1$

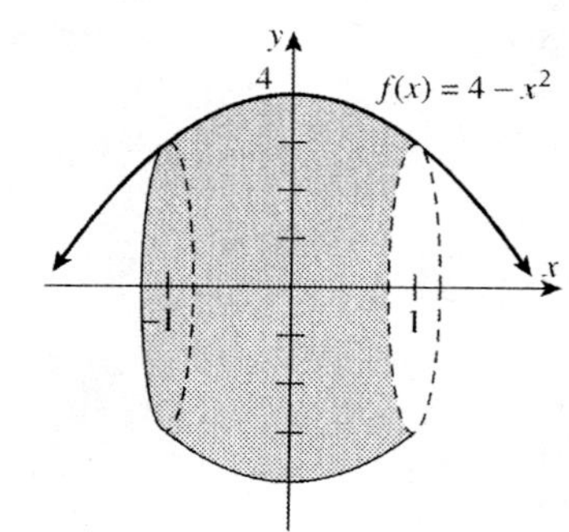

(*continued on next page*)

(continued)

$$V = \pi\int_{-1}^{1}(4-x^2)^2dx$$
$$= \pi\int_{-1}^{1}(16-8x^2+x^4)\,dx$$
$$= \pi\left(16x - \frac{8x^3}{3} + \frac{x^5}{5}\right)\Bigg|_{-1}^{1}$$
$$= \pi\left(16 - \frac{8}{3} + \frac{1}{5} + 16 - \frac{8}{3} + \frac{1}{5}\right)$$
$$= \pi\left(32 - \frac{16}{3} + \frac{2}{5}\right)$$
$$= \frac{406\pi}{15} \approx 85.03$$

46. $f(x) = \frac{x^2}{4}$, $y = 0$, $x = 4$

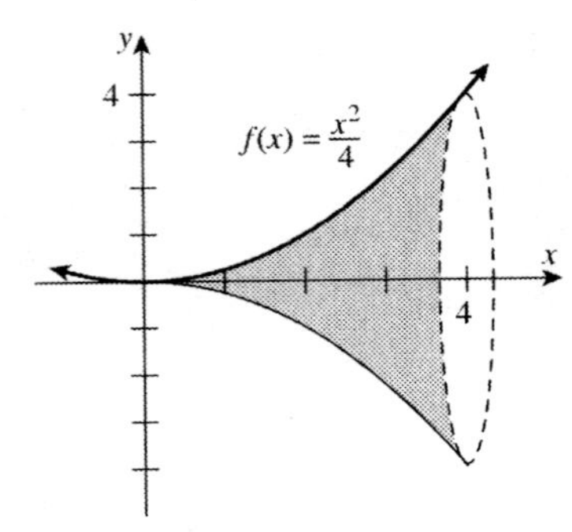

Since $f(x) = \frac{x^2}{4}$ intersects $y = 0$ at $x = 0$, the integral has a lower bound, $a = 0$.

$$V = \pi\int_0^4\left(\frac{x^2}{4}\right)^2dx = \pi\int_0^4\frac{x^4}{16} = \frac{\pi}{16}\left(\frac{x^5}{5}\right)\Bigg|_0^4$$
$$= \frac{\pi}{16}\left(\frac{1024}{5}\right) = \frac{64\pi}{5} \approx 40.21$$

47. The frustum may be shown as follows.

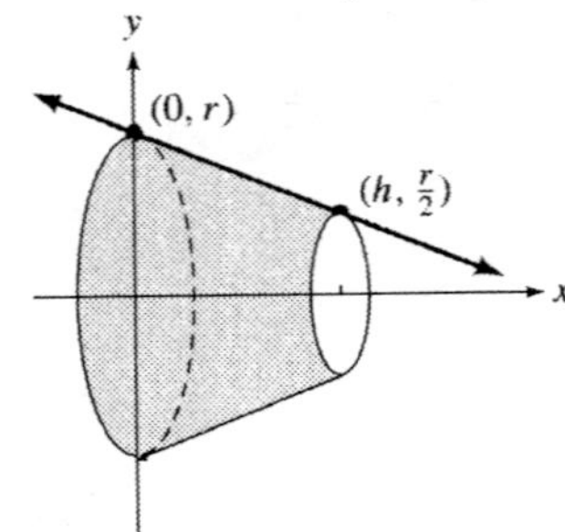

Use the two points given to find $f(x) = -\frac{r}{2h}x + r.$

$$V = \pi\int_0^h\left(-\frac{r}{2h}x + r\right)^2dx$$
$$= -\frac{2\pi h}{3r}\left(-\frac{r}{2h}x + r\right)^3\Bigg|_0^h$$
$$= -\frac{2\pi h}{3r}\left[\left(-\frac{r}{2} + r\right)^3 - (0+r)^3\right]$$
$$= -\frac{2\pi h}{3r}\left[\left(\frac{r}{2}\right)^3 - r^3\right] = -\frac{2\pi h}{3r}\left(\frac{r^3}{8} - r^3\right)$$
$$= -\frac{2\pi h}{3r}\left(-\frac{7r^3}{8}\right) = \frac{7\pi r^2h}{12}$$

48. The average value of a function f on the interval $[a, b]$ is $\frac{1}{b-a}\int_a^b f(x)\,dx,$ if the indicated definite integral exists.

49. $f(x) = \sqrt{x+1}$

$$\frac{1}{b-a}\int_a^b f(x)\,dx = \frac{1}{8-0}\int_0^8\sqrt{x+1}\,dx$$
$$= \frac{1}{8}\int_0^8(x+1)^{1/2}dx$$
$$= \frac{1}{8}\left(\frac{2}{3}\right)(x+1)^{3/2}\Bigg|_0^8$$
$$= \frac{1}{12}(9)^{3/2} - \frac{1}{12}(1)$$
$$= \frac{27}{12} - \frac{1}{12} = \frac{13}{6}$$

50. Average value $= \frac{1}{2-0}\int_0^2 7x^2(x^3+1)^6dx$
$$= \frac{7}{2}\int_0^2 x^2(x^3+1)^6dx$$

Let $u = x^3 + 1$. Then $du = 3x^2dx$.

$$\int x^2(x^3+1)^6dx = \frac{1}{3}\int 3x^2(x^3+1)^6dx$$
$$= \frac{1}{3}\int u^6du = \frac{1}{3}\cdot\frac{1}{7}u^7 + C$$
$$= \frac{1}{21}(x^3+1)^7 + C$$

$$\frac{7}{2}\int_0^2 x^2(x^3+1)^6dx = \frac{7}{2}\cdot\frac{1}{21}(x^3+1)^7\Bigg|_0^2$$
$$= \frac{1}{6}(9^7 - 1^7) = \frac{2{,}391{,}484}{3}$$

51. $\int_{10}^{\infty}x^{-1}dx = \lim\limits_{b\to\infty}\int_{10}^{b}x^{-1}dx = \lim\limits_{b\to\infty}\ln|x|\Big|_{10}^{b}$
$$= \lim_{b\to\infty}\ln b - \ln 10$$

As $b \to \infty$, $\ln|b| \to \infty$. The integral diverges.

52. $\int_{-\infty}^{-5} x^{-2}dx = \lim_{a\to-\infty}\int_a^{-5} x^{-2}dx = \lim_{a\to-\infty}\left(\frac{x^{-1}}{-1}\right)\Big|_a^{-5}$

$= \lim_{a\to-\infty}\left(-\frac{1}{x}\right)\Big|_a^{-5} = \frac{1}{5} + \lim_{a\to-\infty}\left(\frac{1}{a}\right)$

As $a \to -\infty$, $\frac{1}{a} \to 0$. The integral converges.

$\int_{-\infty}^{-5} x^{-2}dx = \frac{1}{5} + 0 = \frac{1}{5}$

53. $\int_0^{\infty} \frac{dx}{(3x+1)^2} = \lim_{b\to\infty}\int_0^b (3x+1)^{-2}dx$

$= \lim_{b\to\infty}\left[\frac{1}{3}\cdot\frac{(3x+1)^{-1}}{-1}\right]\Big|_0^b$

$= \lim_{b\to\infty}\left[\frac{-1}{3(3x+1)}\right]\Big|_0^b$

$= \lim_{b\to\infty}\left(\frac{-1}{3(3b+1)}\right) + \frac{1}{3}$

As $b \to \infty$, $\frac{-1}{3(3b+1)} \to 0$. The integral converges.

$\int_0^{\infty} \frac{dx}{(3x+1)^2} = 0 + \frac{1}{3} = \frac{1}{3}$

54. $\int_1^{\infty} 6e^{-x}dx = \lim_{b\to\infty}\int_1^b 6e^{-x}dx = \lim_{b\to\infty}\left(-6e^{-x}\right)\Big|_1^b$

$= \lim_{b\to\infty}(-6e^{-b} + 6e^{-1})$

$= \lim_{b\to\infty}\left(\frac{-6}{e^b} + \frac{6}{e}\right)$

As $b \to \infty$, $e^b \to \infty$, so $\frac{-6}{e^b} \to 0$. The integral converges.

$\int_1^{\infty} 6e^{-x}dx = 0 + \frac{6}{e} = \frac{6}{e} \approx 2.207$

55. $\int_{-\infty}^{0} \frac{x}{x^2+3}dx = \lim_{a\to-\infty}\frac{1}{2}\int_a^0 \frac{2x\,dx}{x^2+3}$

$= \lim_{a\to-\infty}\frac{1}{2}(\ln|x^2+3|)\Big|_a^0$

$= \lim_{a\to-\infty}\frac{1}{2}(\ln 3 - \ln|a^2+3|)$

As $a \to -\infty$, $\frac{1}{2}(\ln 3 - \ln|a^2+3|) \to -\infty$.

The integral diverges.

56. $\int_4^{\infty} \ln(5x)dx = \lim_{b\to\infty}\int_4^b \ln(5x)dx$

Let $u = \ln(5x)$ and $dv = dx$.

Then $du = \frac{1}{x}dx$ and $v = x$.

$\int \ln(5x)dx = x\ln(5x) - \int x\cdot\frac{1}{x}dx$

$= x\ln(5x) - \int dx = x\ln(5x) - x + C$

$\lim_{b\to\infty}\int_4^b \ln(5x)dx$

$= \lim_{b\to\infty}[x\ln(5x) - x]\Big|_4^b$

$= \lim_{b\to\infty}[b\ln(5b) - b] - (4\ln 20 - 4)$

As $b \to \infty$, $b\ln(5b) - b \to \infty$. The integral diverges.

57. $A = \int_{-\infty}^{1} \frac{5}{(x-2)^2}dx = \lim_{a\to-\infty}\int_a^1 5(x-2)^{-2}dx$

$= \lim_{a\to-\infty}\left[\frac{5(x-2)^{-1}}{-1}\right]\Big|_a^1 = \lim_{a\to-\infty}\left[\frac{-5}{x-2}\right]\Big|_a^1$

$= 5 + \lim_{a\to-\infty}\left(\frac{5}{a-2}\right)$

As $a \to -\infty$, $\frac{5}{a-2} \to 0$. The integral converges. $A = 5 + 0 = 5$

58. $f(x) = 3e^{-x}$ for $[0, \infty)$

$A = \int_0^{\infty} 3e^{-x}dx = \lim_{b\to\infty}\int_0^b 3e^{-x}dx$

$= \lim_{b\to\infty}(-3e^{-x})\Big|_0^b = \lim_{b\to\infty}\left(\frac{-3}{e^b} + 3\right)$

As $b \to \infty$, $\frac{-3}{e^b} \to 0$. The integral converges.

$A = 0 + 3 = 3$

59. (a) $\int_0^{321} 1.87t^{1.49}e^{-0.189(\ln t)^2}\,dt$

Trapezoidal rule: $n = 8, b = 321, a = 1, \quad f(t) = 1.87t^{1.49}e^{-0.189(\ln t)^2}$

i	t_i	$f(t_i)$
0	1	1.87
1	41	34.9086
2	81	33.9149
3	121	30.7147
4	161	27.5809
5	201	24.8344
6	241	22.4794
7	281	20.4622
8	321	18.7255

$$\text{Total amount} \approx \frac{321-1}{8}\left[\frac{1}{2}(1.87) + 34.9086 + 33.9149 + 30.7147 + 27.5809 + 24.8344 + 22.4794 + 20.4622 + \frac{1}{2}(18.7255)\right]$$

$$\approx 8208$$

The total milk production from $t = 1$ to $t = 321$ is approximately 8208 kg.

(b) Simpson's rule: $n = 8, b = 321, a = 1, \quad f(t) = 1.87t^{1.49}e^{-0.189(\ln t)^2}$

i	t_i	$f(t_i)$
0	1	1.87
1	41	34.9086
2	81	33.9149
3	121	30.7147
4	161	27.5809
5	201	24.8344
6	241	22.4794
7	281	20.4622
8	321	18.7255

$$\text{Total amount} \approx \frac{321-1}{8(3)}[1.87 + 4(34.9086) + 2(33.9149) + 4(30.7147) + 2(27.5809) + 4(24.8344) + 2(22.4794) + 4(20.4622) + 18.7255]$$

$$\approx 8430$$

The total milk production from $t = 1$ to $t = 321$ is approximately 8430 kg.

(c) Numerical evaluation gives $\int_0^{321} 1.87t^{1.49}e^{-0.189(\ln t)^2}\,dt \approx 8558$, or 8558 kg

60. $\operatorname{erfc}(x) = \frac{2}{\sqrt{\pi}} \int_x^{\infty} e^{-t^2}\, dt$

(a) $\operatorname{erfc}(0) = \frac{2}{\sqrt{\pi}} \int_0^{\infty} e^{-t^2}\, dt$

Use the program in *The Graphing Calculator Manual*, or use a spreadsheet as described in *The Spreadsheet Manual*, to obtain the values for each of the following.

For $k = 2$, $n = 20$, $f(t) = e^{-t^2}$:

$$\frac{2}{\sqrt{\pi}} \int_0^{2} e^{-t^2}\, dt \approx 0.9953222650$$

For $k = 3$, $n = 20$, $f(t) = e^{-t^2}$:

$$\frac{2}{\sqrt{\pi}} \int_0^{3} e^{-t^2}\, dt \approx 0.9999779095$$

For $k = 4$, $n = 20$, $f(t) = e^{-t^2}$:

$$\frac{2}{\sqrt{\pi}} \int_0^{4} e^{-t^2}\, dt \approx 0.9999999846$$

For $k = 5$, $n = 20$, $f(t) = e^{-t^2}$:

$$\frac{2}{\sqrt{\pi}} \int_0^{5} e^{-t^2}\, dt \approx 1.0000000000$$

Based on these results, $\operatorname{erfc}(0) = 1$.

(b) For any function $f(t)$ and $0 < x < \infty$,

$\int_0^{\infty} f(t)\,dt = \int_0^{x} f(t)\,dt + \int_x^{\infty} f(t)\,dt$ or

$\int_x^{\infty} f(t)\,dt = \int_0^{\infty} f(t)\,dt - \int_0^{x} f(t)\,dt$. In particular,

$$\frac{2}{\sqrt{\pi}} \int_x^{\infty} e^{-t^2}\, dt = \frac{2}{\sqrt{\pi}} \int_0^{\infty} e^{-t^2}\, dt - \frac{2}{\sqrt{\pi}} \int_0^{x} e^{-t^2}\, dt$$

or $\operatorname{erfc}(x) = 1 - \operatorname{erf}(x)$.

61. $\int_0^5 0.5te^{-t}dt = 0.5\int_0^5 te^{-t}dt$

Let $u = t$ and $dv = e^{-t}\, dt$.

Then $du = dt$ and $v = \frac{e^{-t}}{-1}$.

$$\int te^{-t}dt = \frac{te^{-t}}{-1} + \int e^{-t}dt = -te^{-t} + \frac{e^{-t}}{-1}$$

$$0.5\int_0^5 te^{-t}dt = 0.5(-te^{-t} - e^{-t})\Big|_0^5$$
$$= 0.5(-5e^{-5} - e^{-5} + e^{0}) \approx 0.4798$$

The total reaction over the first 5 hr is 0.4798.

62. $f(t) = 125e^{-0.025t}$

The total amount of oil is found by evaluating the improper integral.

$$\int_0^{\infty} 125e^{-0.025t}dt$$
$$= \lim_{b\to\infty} \int_0^{b} 125e^{-0.025t}dt$$
$$= \lim_{b\to\infty} \left(\frac{125}{-0.025}e^{-0.025t}\right)\Bigg|_0^b$$
$$= \lim_{b\to\infty} (-5000)(e^{-0.025b} - e^{0})$$
$$= -5000\left(\lim_{b\to\infty} \frac{1}{e^{0.025b}} - 1\right)$$

As $b \to \infty$, $\frac{1}{e^{0.025b}} \to 0$. The integral converges.

$$\int_0^{\infty} 125e^{-0.025t}dt = -5000(0-1) = 5000$$

The total amount of oil that will enter the bay is 5000 gallons.

63. (a) $\overline{T} = \frac{1}{10-0}\int_0^{10} (160 - 0.05x^2)\, dx$

$$= \frac{1}{10}\left(160x - \frac{0.05x^3}{3}\right)\Bigg|_0^{10}$$
$$= \frac{1}{10}\left[160(10) - \frac{0.05}{3}(10)^3\right]$$
$$\approx \frac{1}{10}(1583.3) \approx 158.3°$$

(b) $\overline{T} = \frac{1}{40-10}\int_{10}^{40} (160 - 0.05x^2)dx$

$$= \frac{1}{30}\left(160x - \frac{0.05x^3}{3}\right)\Bigg|_{10}^{40}$$
$$= \frac{1}{30}\left[\left(160(40) - \frac{0.05(40)^3}{3}\right) - \left(160(10) - \frac{0.05(10)^3}{3}\right)\right]$$
$$\approx \frac{1}{30}(5333.33 - 1583.33) = 125°$$

(c) $\overline{T} = \frac{1}{40-0}\int_0^{40}(160-0.05x^2)dx = \frac{1}{40}\left(160x - \frac{0.05x^3}{3}\right)\Bigg|_0^{40} = \frac{1}{40}\left[\left((160)(40) - \frac{(0.05)(40)^3}{3}\right)\right] \approx 133.3°$

64. $R' = x(x-50)^{1/2}$

$R = \int_{50}^{75} x(x-50)^{1/2}dx$

Let $u = x$ and $dv = (x-50)^{1/2}$. Then $du = dx$ and $v = \frac{2}{3}(x-50)^{3/2}$.

$\int x(x-50)^{1/2}dx = \frac{2}{3}x(x-50)^{3/2} - \frac{2}{3}\int (x-50)^{3/2}dx = \frac{2}{3}x(x-50)^{3/2} - \frac{2}{3}\cdot\frac{2}{5}(x-50)^{5/2}$

$R = \left[\frac{2}{3}x(x-50)^{3/2} - \frac{4}{15}(x-50)^{5/2}\right]\Bigg|_{50}^{75} = \frac{2}{3}(75)(25^{3/2}) - \frac{4}{15}(25^{5/2}) = 6250 - \frac{2500}{3} \approx \$5416.67$

Chapter 9

MULTIVARIABLE CALCULUS

9.1 Functions of Several Variables

1. $f(x, y) = 2x - 3y + 5$

(a) $f(2, -1) = 2(2) - 3(-1) + 5 = 12$

(b) $f(-4, 1) = 2(-4) - 3(1) + 5 = -6$

(c) $f(-2, -3) = 2(-2) - 3(-3) + 5 = 10$

(d) $f(0, 8) = 2(0) - 3(8) + 5 = -19$

3. $h(x, y) = \sqrt{x^2 + 2y^2}$

(a) $h(5, 3) = \sqrt{25 + 2(9)} = \sqrt{43}$

(b) $h(2, 4) = \sqrt{4 + 32} = 6$

(c) $h(-1, -3) = \sqrt{1 + 18} = \sqrt{19}$

(d) $h(-3, -1) = \sqrt{9 + 2} = \sqrt{11}$

5. $f(x, y) = e^x + \ln(x + y)$

(a) $f(1, 0) = e^1 + \ln(1 + 0) = e$

(b) $f(2, -1) = e^2 + \ln(2 - 1) = e^2$

(c) $f(0, e) = e^0 + \ln(0 + e) = 1 + 1 = 2$

(d) $f\left(0, e^2\right) = e^0 + \ln\left(0 + e^2\right) = 1 + 2\ln e$
$= 1 + 2 = 3$

7. $f(x, y) = x\sin\left(x^2 y\right)$

(a) $f\left(1, \frac{\pi}{2}\right) = 1\sin\left(1^2 \cdot \frac{\pi}{2}\right) = 1$

(b) $f\left(\frac{1}{2}, \pi\right) = \frac{1}{2}\sin\left(\left(\frac{1}{2}\right)^2 \pi\right) = \frac{1}{2}\sin\frac{\pi}{4}$
$= \frac{1}{2\sqrt{2}} = \frac{\sqrt{2}}{4}$

(c) $f\left(\sqrt{\pi}, \frac{1}{2}\right) = \sqrt{\pi}\sin\left(\sqrt{\pi}^2 \cdot \frac{1}{2}\right)$
$= \sqrt{\pi}\sin\frac{\pi}{2} = \sqrt{\pi}$

(d) $f\left(-1, -\frac{\pi}{2}\right) = -1\sin\left((-1)^2\left(-\frac{\pi}{2}\right)\right)$
$= -\sin\left(-\frac{\pi}{2}\right) = 1$

9. $x + y + z = 9$
If $x = 0$ and $y = 0$, $z = 9$. If $x = 0$ and $z = 0$, $y = 9$. If $y = 0$ and $z = 0$, $x = 9$.

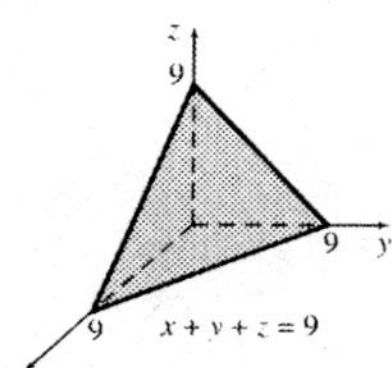

11. $2x + 3y + 4z = 12$
If $x = 0$ and $y = 0$, $z = 3$. If $x = 0$ and $z = 0$, $y = 4$. If $y = 0$ and $z = 0$, $x = 6$.

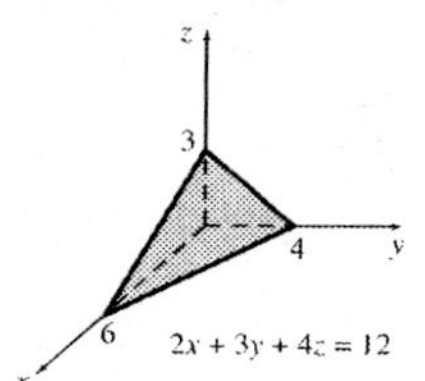

13. $x + y = 4$
If $x = 0$, $y = 4$. If $y = 0$, $x = 4$. There is no z-intercept.

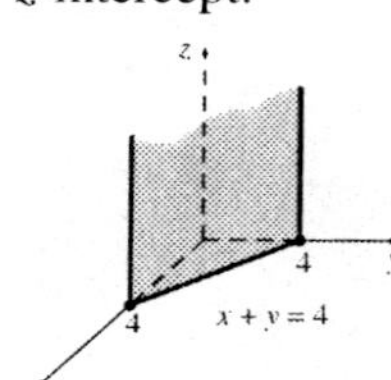

15. $x=5$

The point $(5,0,0)$ is on the graph. There are no y- or z-intercepts. The plane is parallel to the yz-plane.

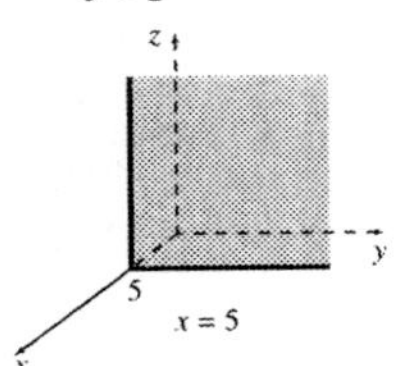

17. $3x+2y+z=24$

For $z=0$, $3x+2y=24$. Graph the line $3x+2y=24$ in the xy-plane.

For $z=2$, $3x+2y=22$. Graph the line $3x+2y=22$ in the plane $z=2$.

For $z=4$, $3x+2y=20$. Graph the line $3x+2y=20$ in the plane $z=4$.

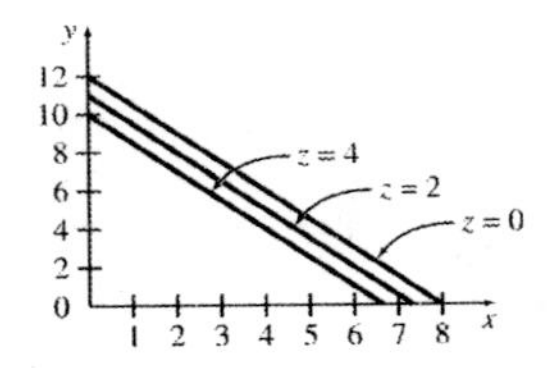

19. $y^2-x=-z$

For $z=0$, $x=y^2$. Graph $x=y^2$ in the xy-plane.

For $z=2$, $x=y^2+2$. Graph $x=y^2+2$ in the plane $z=2$.

For $z=4$, $x=y^2+4$. Graph $x=y^2+4$ in the plane $z=4$.

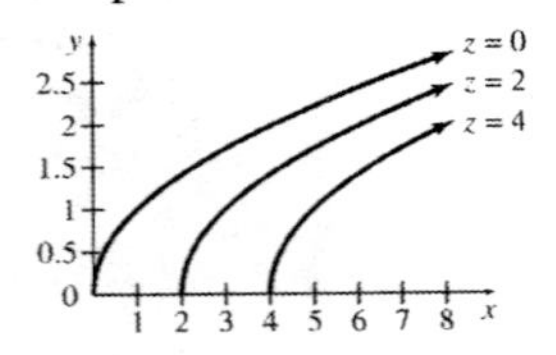

21. Answers will vary.

23. Answers will vary.

25. $z=x^2+y^2$

The xz-trace is $z=x^2+\ 0=x^2$.

The yz-trace is $z=0+y^2=y^2$.

Both are parabolas with vertices at the origin that open upward.

The xy-trace is $0=x^2+y^2$.

This is a point, the origin. The equation represents a paraboloid, as shown in **(c)**.

27. $x^2-y^2=z$

The xz-trace is $x^2=z$, which is a parabola with vertex at the origin that opens upward.

The yz-trace is $-y^2=z$, which is a parabola with vertex at the origin that opens downward.

The xy-trace is

$x^2-y^2=0 \Rightarrow x^2=y^2 \Rightarrow x=y$ or $x=-y$,

which are two lines that intersect at the origin. The equation represents a hyperbolic paraboloid, as shown in **(e)**.

29. $\dfrac{x^2}{16}+\dfrac{y^2}{25}+\dfrac{z^2}{4}=1$

xz-trace: $\dfrac{x^2}{16}+\dfrac{z^2}{4}=1$, an ellipse

yz-trace: $\dfrac{y^2}{25}+\dfrac{z^2}{4}=1$, an ellipse

xy-trace: $\dfrac{x^2}{16}+\dfrac{y^2}{25}=1$, an ellipse

The graph is an ellipsoid, as shown in **(b)**.

31. $f(x, y)=4x^2-2y^2$

(a) $$\frac{f(x+h,\ y)-f(x,\ y)}{h} = \frac{[4(x+h)^2-2y^2]-[4x^2-2y^2]}{h} = \frac{4x^2+8xh+4h^2-2y^2-4x^2+2y^2}{h} = \frac{h(8x+4h)}{h}=8x+4h$$

(b) $$\frac{f(x,\ y+h)-f(x,\ y)}{h} = \frac{[4x^2-2(y+h)^2]-[4x^2-2y^2]}{h} = \frac{4x^2-2y^2-4yh-2h^2-4x^2+2y^2}{h} = \frac{h(-4y-2h)}{h}=-4y-2h$$

(c) $$\lim_{h\to 0}\frac{f(x+h,\ y)-f(x,\ y)}{h}=\lim_{h\to 0}(8x+4h) = 8x+4(0)=8x$$

(d) $$\lim_{h\to 0}\frac{f(x,\ y+h)-f(x,\ y)}{h} = \lim_{h\to 0}(-4y-2h)=-4y-2(0)=-4y$$

33. $f(x, y) = xye^{x^2+y^2}$

(a) $\lim_{h\to 0} \frac{f(1+h, 1) - f(1, 1)}{h}$

$= \lim_{h\to 0} \frac{(1+h)(1)e^{1+2h+h^2+1} - (1)(1)e^{1+1}}{h}$

$= \lim_{h\to 0} \frac{(1+h)e^{2+2h+h^2} - e^2}{h}$

$= e^2 \lim_{h\to 0} \frac{(1+h)e^{2h+h^2} - 1}{h}$

X	Y1
.001	3.005
1E-4	3.0005
1E-5	3.0001
1E-6	3
-1E-5	3
-1E-4	2.9995
-.001	2.995

Y1=((1+X)e^(2X+...

The graphing calculator indicates that

$\lim_{h\to 0} \frac{(1+h)e^{2h+h^2} - 1}{h} = 3$; thus

$\lim_{h\to 0} \frac{f(1+h, 1) - f(1, 1)}{h} = 3e^2.$

This means that the slope of the tangent line in the direction of x at $(1, 1)$ is $3e^2$.

(b) $\lim_{h\to 0} \frac{f(1, 1+h) - f(1, 1)}{h}$

$= \lim_{h\to 0} \frac{(1)(1+h)e^{1+1+2h+h^2} - (1)(1)e^{1+1}}{h}$

$= \lim_{h\to 0} \frac{(1+h)e^{2+2h+h^2} - e^2}{h}$

$= e^2 \lim_{h\to 0} \frac{(1+h)e^{2h+h^2} - 1}{h}$

So, this limit reduces to the exact same limit as in part (a). Therefore, since

$\lim_{h\to 0} \frac{(1+h)e^{2h+h^2} - 1}{h} = 3$, then

$\lim_{h\to 0} \frac{f(1, 1+h) - f(1, 1)}{h} = 3e^2.$

This means that the slope of the tangent line in the direction of y at $(1, 1)$ is $3e^2$.

35. $L(E, P) = 23E^{0.6}P^{-0.267}$

(a) $L(7.35, 150) = 23(7.35)^{0.6}(150)^{-0.267}$
≈ 20.0 years

(b) $L(14{,}100, 68{,}700)$
$= 23(14{,}100)^{0.6}(68{,}700)^{-0.267}$
≈ 363 years

It seems reasonable that a human should live longer than a black rat, but the estimated life spans do not seem reasonable.

37. $A = 0.024265h^{0.3964}m^{0.5378}$

(a) $A = 0.024265(178)^{0.3964}(72)^{0.5378}$
$\approx 1.89 \text{ m}^2$

(b) $A = 0.024265(140)^{0.3964}(65)^{0.5378}$
$\approx 1.62 \text{ m}^2$

(c) $A = 0.024265(160)^{0.3964}(70)^{0.5378}$
$\approx 1.78 \text{ m}^2$

(d) Answers will vary.

39. $P(W, R, A) = 48 - 2.43W - 1.81R - 1.22A$

(a) $P(5, 15, 0)$
$= 48 - 2.43(5) - 1.81(15) - 1.22(0) = 8.7$

8.7% of fish will be intolerant to pollution.

(b) The maximum percentage will occur when the variable factors are a minimum, or when $W = 0$, $R = 0$, and $A = 0$.

$P(0, 0, 0) = 48 - 2.43(0) - 1.81(0) - 1.22(0)$
$= 48$

48% of fish will be intolerant to pollution.

(c) Any combination of values of W, R, and A that result in $P = 0$ is a scenario that will drive the percentage of fish intolerant to pollution to zero. Two examples are given. If $R = 0$ and $A = 0$,

$P(W, 0, 0)$
$= 48 - 2.43W - 1.81(0) - 1.22(0)$
$= 48 - 2.43W$

$48 - 2.43W = 0 \Rightarrow W = \frac{48}{2.43} \approx 19.75$

So $W = 19.75$, $R = 0$, $A = 0$ is one scenario.

If $W = 10$ and $R = 10$,

$P(10, 10, A)$
$= 48 - 2.43(10) - 1.81(10) - 1.22A$
$= 5.6 - 1.22A$

$5.6 - 1.22A = 0 \Rightarrow A = \frac{5.6}{1.22} \approx 4.59$

So $W = 10$, $R = 10$, $A = 4.59$ is another scenario.

(d) Since the coefficient of W is greater than the coefficients of R and A, a change in W will affect the value of P more than an equal change in R or A. Thus, the percentage of wetland (W) has the greatest influence on P.

41. $A(L,T,U,C) = 53.02 + 0.383L + 0.0015T + 0.0028U - 0.0003C$

(a) $A(266,\ 107{,}484,\ 31{,}697,\ 24{,}870)$
$= 53.02 + 0.383(266) + 0.0015(107{,}484) + 0.0028(31{,}697) - 0.0003(24{,}870) \approx 397$

The estimated number of accidents is 397.

(b) Since the coefficient of L is greater than the coefficients of T, U, and C, a change in L will affect the value of A more than an equal change in T, U, or C.

43. (a) $\ln(T) = 5.49 - 3.00\ln(F) + 0.18\ln(C)$

$e^{\ln(T)} = e^{5.49 - 3.00\ln(F) + 0.18\ln(C)}$

$$T = e^{5.49}e^{-3.00\ln(F)}e^{0.18\ln(C)} = \frac{e^{5.49}e^{\ln(C^{0.18})}}{e^{\ln(F^3)}} \Rightarrow T \approx \frac{242.257C^{0.18}}{F^3}$$

(b) Replace F with 2 and C with 40 in the preceding formula.

$$T \approx \frac{242.257(40)^{0.18}}{(2)^3} \approx 58.82$$

T is about 58.8%. In other words, a tethered sow spends nearly 59% of the time doing repetitive behavior when she is fed 2 kg of food per day and neighboring sows spend 40% of the time doing repetitive behavior.

45. $T(A,\ W,\ B) = 23.04 - 0.03A + 0.50W - 0.62B$
$T(55,\ 75,\ 30) = 23.04 - 0.03(55) + 0.50(75) - 0.62(30) = 40.29$ liters

47. (a) $P = \dfrac{\lambda_1 E_1}{1 + \lambda_1 h_1} = \dfrac{0.2(162)}{1 + 0.2(3.6)} \approx 18.84$

The profitability for a blue jay to eat worms only is about 18.8 kcal/min.

(b) $P = \dfrac{\lambda_1 E_1 + \lambda_2 E_2}{1 + \lambda_1 h_1 + \lambda_2 h_2} = \dfrac{0.2(162) + 3.0(24)}{1 + 0.2(3.6) + 3.0(0.6)} \approx 29.66$

The profitability for a blue jay to eat worms and moths is about 29.7 kcal/min.

(c) $P(\text{worms}) \approx 18.84$

$P(\text{moths}) = \dfrac{3.0(24)}{1 + 3.0(0.6)} \approx 25.71$

$P(\text{grubs}) = \dfrac{3.0(40)}{1 + 3.0(1.6)} \approx 20.69$

$P(\text{worms, moths}) \approx 29.66$

$P(\text{worms, grubs}) = \dfrac{0.2(162) + 3.0(40)}{1 + 0.2(3.6) + 3.0(1.6)} \approx 23.37$

$P(\text{moths, grubs}) = \dfrac{3.0(24) + 3.0(40)}{1 + 3.0(0.6) + 3.0(1.6)} \approx 25.26$

$P(\text{worms, moths, grubs}) = \dfrac{0.2(162) + 3.0(24) + 3.0(40)}{1 + 0.2(3.6) + 3.0(0.6) + 3.0(1.6)} \approx 26.97$

The combination that provides maximum profitability is worms and moths.

49. Let the area be given by $g(L,W,H)$. Then, $g(L,W,H) = 2LW + 2WH + 2LH$ ft^2.

9.2 Partial Derivatives

1. $z = f(x, y) = 6x^2 - 4xy + 9y^2$

(a)
$$\frac{\partial z}{\partial x} = \lim_{h\to 0} \frac{f(x+h, y) - f(x, y)}{h}$$
$$= \lim_{h\to 0} \frac{\left[6(x+h)^2 - 4(x+h)y + 9y^2\right] - \left(6x^2 - 4xy + 9y^2\right)}{h}$$
$$= \lim_{h\to 0} \frac{6x^2 + 12xh + 6h^2 - 4xy - 4hy + 9y^2 - 6x^2 + 4xy - 9y^2}{h}$$
$$= \lim_{h\to 0} \frac{12xh + 6h^2 - 4hy}{h} = \lim_{h\to 0} (12x + 6h - 4y) = 12 - 4y$$

(b)
$$\frac{\partial z}{\partial y} = \lim_{h\to 0} \frac{f(x, y+h) - f(x, y)}{h}$$
$$= \lim_{h\to 0} \frac{\left[6x^2 - 4x(y+h) + 9(y+h)^2\right] - \left(6x^2 - 4xy + 9y^2\right)}{h}$$
$$= \lim_{h\to 0} \frac{6x^2 - 4xy - 4xh + 9y^2 + 18yh + 9h^2 - 6x^2 + 4xy - 9y^2}{h}$$
$$= \lim_{h\to 0} \frac{-4xh + 18yh + 9h^2}{h} = \lim_{h\to 0} (-4x + 18y + 9h) = -4x + 18y$$

(c) $\frac{\partial f}{\partial x}(2, 3) = 12(2) - 4(3) = 12$

(d) $f_y(1, -2) = -4(1) + 18(-2) = -40$

3.
$$f(x, y) = -4xy + 6y^3 + 5$$
$$f_x(x, y) = -4y$$
$$f_y(x, y) = -4x + 18y^2$$
$$f_x(2, -1) = -4(-1) = 4$$
$$f_y(-4, 3) = -4(-4) + 18(3)^2 = 178$$

5.
$$f(x, y) = 5x^2y^3$$
$$f_x(x, y) = 10xy^3$$
$$f_y(x, y) = 15x^2y^2$$
$$f_x(2, -1) = 10(2)(-1)^3 = -20$$
$$f_y(-4, 3) = 15(-4)^2(3)^2 = 2160$$

7.
$$f(x, y) = e^{x+y}$$
$$f_x(x, y) = e^{x+y}$$
$$f_y(x, y) = e^{x+y}$$
$$f_x(2, -1) = e^{2-1} = e^1 = e$$
$$f_y(-4, 3) = e^{-4+3} = e^{-1} = \frac{1}{e}$$

9.
$$f(x, y) = -6e^{4x-3y}$$
$$f_x(x, y) = -24e^{4x-3y}$$
$$f_y(x, y) = 18e^{4x-3y}$$
$$f_x(2, -1) = -24e^{4(2)-3(-1)} = -24e^{11}$$
$$f_y(-4, 3) = 18e^{4(-4)-3(3)} = 18e^{-25}$$

11. $f(x, y) = \dfrac{x^2 + y^3}{x^3 - y^2}$

$$f_x(x, y) = \frac{2x(x^3 - y^2) - 3x^2(x^2 + y^3)}{(x^3 - y^2)^2}$$
$$= \frac{2x^4 - 2xy^2 - 3x^4 - 3x^2y^3}{(x^3 - y^2)^2}$$
$$= \frac{-x^4 - 2xy^2 - 3x^2y^3}{(x^3 - y^2)^2}$$

(continued on next page)

(continued)

$$f_y(x, y) = \frac{3y^2(x^3 - y^2) - (-2y)(x^2 + y^3)}{(x^3 - y^2)^2}$$
$$= \frac{3x^3y^2 - 3y^4 + 2x^2y + 2y^4}{(x^3 - y^2)^2}$$
$$= \frac{3x^3y^2 - y^4 + 2x^2y}{(x^3 - y^2)^2}$$
$$f_x(2, -1) = \frac{-2^4 - 2(2)(-1)^2 - 3(2^2)(-1)^3}{[2^3 - (-1)^2]^2}$$
$$= -\frac{8}{49}$$
$$f_y(-4, 3) = \frac{3(-4)^3(3)^2 - 3^4 + 2(-4)^2(3)}{[(-4)^3 - 3^2]^2}$$
$$= -\frac{1713}{5329}$$

13. $f(x, y) = \ln|1 + 5x^3y^2|$

$$f_x(x, y) = \frac{1}{1 + 5x^3y^2} \cdot 15x^2y^2 = \frac{15x^2y^2}{1 + 5x^3y^2}$$
$$f_y(x, y) = \frac{1}{1 + 5x^3y^2} \cdot 10x^3y = \frac{10x^3y}{1 + 5x^3y^2}$$
$$f_x(2, -1) = \frac{15(2)^2(-1)^2}{1 + 5(2)^3(-1)^2} = \frac{60}{41}$$
$$f_y(-4, 3) = \frac{10(-4)^3(3)}{1 + 5(-4)^3(3)^2} = \frac{1920}{2879}$$

15. $f(x, y) = xe^{x^2y}$

$$f_x(x, y) = e^{x^2y} \cdot 1 + x(2xy)(e^{x^2y})$$
$$= e^{x^2y}(1 + 2x^2y)$$
$$f_y(x, y) = x^3e^{x^2y}$$
$$f_x(2, -1) = e^{-4}(1 - 8) = -7e^{-4}$$
$$f_y(-4, 3) = -64e^{48}$$

17. $f(x, y) = \sqrt{x^4 + 3xy + y^4 + 10}$

$$f_x(x, y) = \frac{4x^3 + 3y}{2\sqrt{x^4 + 3xy + y^4 + 10}}$$
$$f_y(x, y) = \frac{3x + 4y^3}{2\sqrt{x^4 + 3xy + y^4 + 10}}$$
$$f_x(2, -1) = \frac{4(2)^3 + 3(-1)}{2\sqrt{2^4 + 3(2)(-1) + (-1)^4 + 10}}$$
$$= \frac{29}{2\sqrt{21}}$$
$$f_y(-4, 3) = \frac{3(-4) + 4(3)^3}{2\sqrt{(-4)^4 + 3(-4)(3) + 3^4 + 10}}$$
$$= \frac{48}{\sqrt{311}}$$

19. $f(x, y) = \dfrac{3x^2y}{e^{xy} + 2}$

$$f_x(x, y) = \frac{6xy(e^{xy} + 2) - ye^{xy}(3x^2y)}{(e^{xy} + 2)^2}$$
$$= \frac{6xy(e^{xy} + 2) - 3x^2y^2e^{xy}}{(e^{xy} + 2)^2}$$
$$f_y(x, y) = \frac{3x^2(e^{xy} + 2) - xe^{xy}(3x^2y)}{(e^{xy} + 2)^2}$$
$$= \frac{3x^2(e^{xy} + 2) - 3x^3ye^{xy}}{(e^{xy} + 2)^2}$$
$$f_x(2, -1)$$
$$= \frac{6(2)(-1)(e^{2(-1)} + 2) - 3(2)^2(-1)^2e^{2(-1)}}{(e^{2(-1)} + 2)^2}$$
$$= \frac{-12e^{-2} - 24 - 12e^{-2}}{(e^{-2} + 2)^2} = \frac{-24(e^{-2} + 1)}{(e^{-2} + 2)^2}$$
$$f_y(-4, 3)$$
$$= \frac{3(-4)^2\left(e^{(-4)(3)} + 2\right) - 3(-4)^3(3)e^{(-4)(3)}}{(e^{(-4)(3)} + 2)^2}$$
$$= \frac{48e^{-12} + 96 + 576e^{-12}}{(e^{-12} + 2)^2} = \frac{624e^{-12} + 96}{(e^{-12} + 2)^2}$$

21. $f(x, y) = x\sin(\pi y)$

$$f_x(x, y) = \sin(\pi y)$$
$$f_y(x, y) = \pi x\cos(\pi y)$$
$$f_x(2, -1) = \sin(-\pi) = 0$$
$$f_y(-4, 3) = \pi(-4)\cos(3\pi) = 4\pi$$

23. $f(x, y) = 4x^2y^2 - 16x^2 + 4y$

$$f_x(x, y) = 8xy^2 - 32x$$
$$f_y(x, y) = 8x^2y + 4$$
$$f_{xx}(x, y) = 8y^2 - 32$$
$$f_{yy}(x, y) = 8x^2$$
$$f_{xy}(x, y) = f_{yx}(x, y) = 16xy$$

25. $R(x, y) = 4x^2 - 5xy^3 + 12y^2x^2$

$R_x(x, y) = 8x - 5y^3 + 24y^2x$

$R_y(x, y) = -15xy^2 + 24yx^2$

$R_{xx}(x, y) = 8 + 24y^2$

$R_{yy}(x, y) = -30xy + 24x^2$

$R_{xy}(x, y) = -15y^2 + 48xy$
$= R_{yx}(x, y)$

27. $r(x, y) = \dfrac{6y}{x+y}$

$r_x(x, y) = \dfrac{(x+y)(0) - 6y(1)}{(x+y)^2} = -6y(x+y)^{-2}$

$r_y(x, y) = \dfrac{(x+y)(6) - 6y(1)}{(x+y)^2} = 6x(x+y)^{-2}$

$r_{xx}(x, y) = -6y(-2)(x+y)^{-3}(1) = \dfrac{12y}{(x+y)^3}$

$r_{yy}(x, y) = 6x(-2)(x+y)^{-3}(1) = -\dfrac{12x}{(x+y)^3}$

$r_{xy}(x, y) = r_{yx}(x, y)$
$= -6y(-2)(x+y)^{-3}(1) + (x+y)^{-2}(-6)$
$= \dfrac{12y - 6(x+y)}{(x+y)^3} = \dfrac{6y - 6x}{(x+y)^3}$

29. $z = 9ye^x$

$z_x = 9ye^x$

$z_y = 9e^x$

$z_{xx} = 9ye^x$

$z_{yy} = 0$

$z_{xy} = z_{yx} = 9e^x$

31. $r = \ln|x+y|$

$r_x = \dfrac{1}{x+y}$

$r_y = \dfrac{1}{x+y}$

$r_{xx} = \dfrac{-1}{(x+y)^2}$

$r_{yy} = \dfrac{-1}{(x+y)^2}$

$r_{xy} = r_{yx} = \dfrac{-1}{(x+y)^2}$

33. $z = x\ln|xy|$

$z_x = \ln|xy| + 1$

$z_y = \dfrac{x}{y}$

$z_{xx} = \dfrac{1}{x}$

$z_{yy} = -xy^{-2} = \dfrac{-x}{y^2}$

$z_{xy} = z_{yx} = \dfrac{1}{y}$

35. $z = e^x \sin y$

$z_x\left(e^x \sin y\right) = \dfrac{\partial}{\partial x}\left(e^x \sin y\right) = \sin y \dfrac{\partial}{\partial x}\left(e^x\right)$
$= e^x \sin y$

$z_y\left(e^x \sin y\right) = \dfrac{\partial}{\partial y}\left(e^x \sin y\right) = e^x \dfrac{\partial}{\partial y}(\sin y)$
$= e^x \cos y$

$z_{xx}\left(e^x \sin y\right) = \dfrac{\partial^2 z}{\partial x^2} = \dfrac{\partial}{\partial x}\left(e^x \sin y\right)$
$= e^x \sin y$

$z_{yy}\left(e^x \sin y\right) = \dfrac{\partial^2 z}{\partial y^2} = \dfrac{\partial}{\partial y}\left(e^x \cos y\right)$
$= e^x \dfrac{\partial}{\partial y}(\cos y) = -e^x \sin y$

$z_{xy}\left(e^x \sin y\right) = \dfrac{\partial^2 z}{\partial y \partial x} = \dfrac{\partial}{\partial y}\left(e^x \sin y\right)$
$= e^x \dfrac{\partial}{\partial y}(\sin y) = e^x \cos y$
$= z_{yx}\left(e^x \sin y\right)$

37. $f(x, y) = 6x^2 + 6y^2 + 6xy + 36x - 5$

First, $f_x = 12x + 6y + 36$ and $f_y = 12y + 6x$.

We must solve the system
$12x + 6y + 36 = 0$
$12y + 6x = 0.$

Multiply both sides of the first equation by –2 and add.

$$\begin{aligned} -24x - 12y - 72 &= 0 \\ 6x + 12y \quad\quad &= 0 \\ \hline -18x \quad\quad - 72 &= 0 \Rightarrow x = -4 \end{aligned}$$

Substitute into either equation to get $y = 2$.
The solution is $x = -4$, $y = 2$.

39. $f(x, y) = 9xy - x^3 - y^3 - 6$

First, $f_x = 9y - 3x^2$ and $f_y = 9x - 3y^2$.

We must solve the system

$9y - 3x^2 = 0$

$9x - 3y^2 = 0.$

From the first equation, $y = \frac{1}{3}x^2$.

Substitute into the second equation to get

$$9x - 3\left(\frac{1}{3}x^2\right)^2 = 0 \Rightarrow 9x - 3\left(\frac{1}{9}x^4\right) = 0 \Rightarrow$$

$$9x - \frac{1}{3}x^4 = 0 \Rightarrow 27x - x^4 = 0 \Rightarrow$$

$$x\left(27 - x^3\right) = 0 \Rightarrow x = 0 \text{ or } 27 - x^3 = 0 \Rightarrow$$

$$x^3 = 27 \Rightarrow x = 3$$

Substitute into $y = \frac{x^2}{3}$.

$$y = \frac{0^3}{3} = 0 \text{ or } y = \frac{3^3}{3} = 3$$

The solutions are $x = 0$, $y = 0$ and $x = 3$, $y = 3$.

41. $f(x, y, z) = x^4 + 2yz^2 + z^4$

$f_x(x, y, z) = 4x^3$

$f_y(x, y, z) = 2z^2; \quad f_z(x, y, z) = 4yz + 4z^3$

$f_{yz}(x, y, z) = 4z$

43. $f(x, y, z) = \frac{6x - 5y}{4z + 5}$

$f_x(x, y, z) = \frac{6}{4z + 5}; \quad f_y(x, y, z) = \frac{-5}{4z + 5}$

$f_z(x, y, z) = \frac{-4(6x - 5y)}{(4z + 5)^2}$

$f_{yz}(x, y, z) = \frac{20}{(4z + 5)^2}$

45. $f(x, y, z) = \ln|x^2 - 5xz^2 + y^4|$

$f_x(x, y, z) = \frac{2x - 5z^2}{x^2 - 5xz^2 + y^4}$

$f_y(x, y, z) = \frac{4y^3}{x^2 - 5xz^2 + y^4}$

$f_z(x, y, z) = \frac{-10xz}{x^2 - 5xz^2 + y^4}$

$$f_{yz}(x, y, z) = \frac{4y^3(10zx)}{(x^2 - 5xz^2 + y^4)^2} = \frac{40xy^3z}{(x^2 - 5xz^2 + y^4)^2}$$

47. $f(x, y) = \left(x + \frac{y}{2}\right)^{x + y/2}$

(a) $f_x(1, 2) = \lim_{h \to 0} \frac{f(1 + h, 2) - f(1, 2)}{h}$

We will use a small value for h. Let $h = 0.00001$.

$$f_x(1, 2) \approx \frac{f(1.00001, 2) - f(1, 2)}{0.00001} \approx \frac{\left(1.00001 + \frac{2}{2}\right)^{1.00001 + 2/2} - \left(1 + \frac{2}{2}\right)^{1 + 2/2}}{0.00001} \approx \frac{2.00001^{2.00001} - 2^2}{0.00001} \approx 6.773$$

(b) $f_y(1, 2) = \lim_{h \to 0} \frac{f(1, 2 + h) - f(1, 2)}{h}$

Again, let $h = 0.00001$.

$$f_y(1, 2) \approx \frac{f(1, 200001) - f(1, 2)}{0.00001} \approx \frac{\left(1 + \frac{2.00001}{2}\right)^{1 + 2.00001/2} - \left(1 + \frac{2}{2}\right)^{1 + 2/2}}{0.00001} \approx \frac{2.000005^{2.000005} - 2^2}{0.00001} \approx 3.386$$

49. $f(m, v) = 25.92m^{0.68} + \frac{3.62m^{0.75}}{v}$

(a) $f(300, 10) = 25.92(300)^{0.68} + \frac{3.62(300)^{0.75}}{10} \approx 1279.46$

The value is about 1279 kcal/hr.

(b)
$$f_m(m, v) = 25.92(0.68)m^{-.32} + \frac{3.62(0.75)m^{-0.25}}{v} = \frac{17.6256}{m^{0.32}} + \frac{2.715}{m^{0.25}v}$$

$$f_m(300, 10) = \frac{17.6256}{(300)^{0.32}} + \frac{2.715}{(300)^{0.25}(10)} \approx 2.906$$

The value is about 2.906 kcal/hr/g. This means the instantaneous rate of change of energy usage for a 300-kg animal traveling at 10 kilometers per hour to walk or run 1 kilometer is about 2.9 kcal/hr/g.

51. $A = 0.024265h^{0.3964}m^{0.5378}$

(a) $A_m = (0.024265)(0.5378)h^{0.3964}m^{(0.5378-1)}$
$= 0.013050h^{0.3964}m^{-0.4622}$
When the mass m increases from 72 to 73 while the height h remains at 180 cm, the approximate change in body surface area is $0.013050(180)^{0.3964}(72)^{-0.4622} \approx 0.0142$ or about 0.0142 m^2.

(b) $A_h = (0.024265)(0.3964)h^{(0.3964-1)}m^{0.5378}$
$= 0.0096186h^{-0.6036}m^{0.5378}$
When the height h increases from 160 to 161 while the mass m remains at 70, the approximate change in body surface area is $0.0096186(160)^{-0.6036}(70)^{0.5378} \approx 0.00442$ or about 0.00442 m^2.

53. $f(n,c) = \frac{1}{8}n^2 - \frac{1}{5}c + \frac{1937}{8}$

(a) $f(4,1200) = \frac{1}{8}(4)^2 - \frac{1}{5}(1200) + \frac{1937}{8}$
$= 2 - 240 + \frac{1937}{8} = 4.125$
The client could expect to lose 4.125 lb.

(b) $\frac{\partial f}{\partial n} = \frac{1}{8}(2n) - \frac{1}{5}(0) + 0 = \frac{1}{4}n$
This represents the rate of change of weight loss per unit change in number of workouts.

(c) $f_n(3,1100) = \frac{1}{4}(3) = \frac{3}{4}$ lb
This represents an additional weight loss by adding the fourth workout.

55. $ABSI = \frac{w}{b^{2/3}h^{1/2}}$

(a) $ABSI = \frac{0.864}{23.1^{2/3}\left(1.85^{1/2}\right)} \approx 0.0783$

It's a little lower.

(b) $\frac{\partial}{\partial w}\left(\frac{w}{b^{2/3}h^{1/2}}\right) = \frac{1}{b^{2/3}h^{1/2}}$

$\frac{1}{23.1^{2/3}\left(1.85^{1/2}\right)} \approx 0.0906$

The *ABSI* is going up at a rate of 0.0906 per meter with respect to his waist.

(c) $\frac{\partial}{\partial h}\left(\frac{w}{b^{2/3}h^{1/2}}\right) = \frac{w}{b^{2/3}}\frac{\partial}{\partial h}\left(\frac{1}{h^{1/2}}\right)$
$= -\frac{w}{2b^{2/3}h^{3/2}}$

$-\frac{0.864}{2\left(23.1^{2/3}\right)\left(1.85^{3/2}\right)} \approx -0.0212$

The *ABSI* is going down at a rate of –0.0212 per meter with respect to his height.

(d) An increase in waist size has a greater effect on his *ABSI* than does an increase in height.

57. $R(x,t) = x^2(a-x)t^2e^{-t} = (ax^2 - x^3)t^2e^{-t}$

(a) $\frac{\partial R}{\partial x} = (2ax - 3x^2)t^2e^{-t}$

(b) $\frac{\partial R}{\partial t} = x^2(a-x)\cdot[t^2\cdot(-e^{-t}) + e^{-t}\cdot 2t]$
$= x^2(a-x)(-t^2 + 2t)e^{-t}$

(c) $\frac{\partial^2 R}{\partial x^2} = (2a - 6x)t^2e^{-t}$

(d) $\frac{\partial^2 R}{\partial x \partial t} = (2ax - 3x^2)(-t^2 + 2t)e^{-t}$

(e) $\frac{\partial R}{\partial x}$ gives the rate of change of the reaction per unit of change in the amount of drug administered.
$\frac{\partial R}{\partial t}$ gives the rate of change of the reaction for a 1-hour change in the time after the drug is administered.

59. $W(V,T) = 91.4 - \dfrac{(10.45 + 6.69\sqrt{V} - 0.447V)(91.4 - T)}{22}$

(a) $W(20,10) = 91.4 - \dfrac{(10.45 + 6.69\sqrt{20} - 0.447(20)) \cdot (91.4 - 10)}{22} \approx -24.9$

The wind chill is –24.9°F when the wind speed is 20 mph and the temperature is 10°F.

(b) Solve $-25 = 91.4 - \dfrac{(10.45 + 6.69\sqrt{V} - 0.447V) \cdot (91.4 - 5)}{22}$ for V.

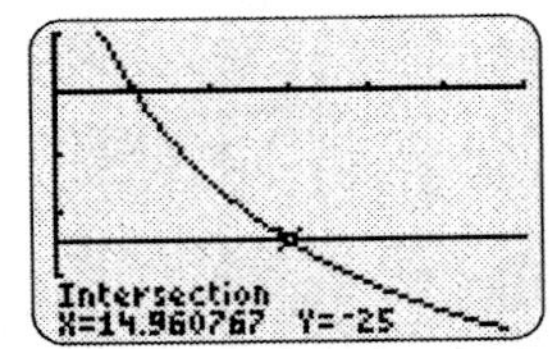

The wind speed is approximately 15 mph.

(c) $W_V = -\dfrac{1}{22}\left(\dfrac{6.69}{2\sqrt{V}} - 0.447\right)(91.4 - T)$

$W_T = -\dfrac{1}{22}(10.45 + 6.69\sqrt{V} - 0.447V)(-1) = \dfrac{1}{22}(10.45 + 6.69\sqrt{V} - 0.447V)$

$W_V(20, 10) = \dfrac{1}{22}\left(\dfrac{6.69}{2\sqrt{20}} - 0.447\right) \cdot (91.4 - 10) \approx -1.114$

When the temperature is held fixed at 10°F, the wind chill decreases approximately 1.1°F when the wind velocity increases by 1 mph.

$W_T(20,10) = \dfrac{1}{22}[10.45 + 6.69\sqrt{20} - 0.447(20)] \approx 1.429$

When the wind velocity is held fixed at 20 mph, the wind chill increases approximately 1.429°F when the temperature increases from 10°F to 11°F.

(d) A sample table is

$T \backslash V$	5	10	15	20
30	27	16	9	4
20	16	3	–5	–11
10	6	–9	–18	–25
0	–5	–21	–32	–39

61. $\ln D = 4.890179 + 0.005163R - 0.004345L - 0.000019F$

(a) From exercise 46 in section 9.1: $D(R, L, F) = 132.977375e^{0.005163R - 0.004345L - 0.000019F}$.

$$D_F(R, L, F) = 132.977375(-0.000019)e^{0.005163R - 0.004345L - 0.000019F}$$
$$= -0.002527e^{0.005163R - 0.004345L - 0.000019F}$$

(b) $D_F(150, 75, 3500) = -0.002527e^{0.005163(150) - 0.004345(75) - 0.000019(3500)} \approx -0.00370$

For a 1000 m^3/s increase in water flow, the date of passage will be about $1000D_F(150, 75, 3500) \approx 1000(-0.00370) \approx -3.7$, or about 3.7 days earlier.

63. $C(x, y) = \dfrac{k}{a + (x - by)^2}$

$$C_x(x, y) = k\frac{\partial}{\partial x}\left(\frac{1}{a + (x - by)^2}\right) = k\left(\frac{-\frac{\partial}{\partial x}\left(a + (x - by)^2\right)}{\left(a + (x - by)^2\right)^2}\right) = \frac{-2k(x - by)}{\left(a + (x - by)^2\right)^2}$$

$$C_y(x, y) = k\frac{\partial}{\partial y}\left(\frac{1}{a + (x - by)^2}\right) = k\left(\frac{-\frac{\partial}{\partial y}\left(a + (x - by)^2\right)}{\left(a + (x - by)^2\right)^2}\right) = \frac{-2k(x - by)\frac{\partial}{\partial y}(x - by)}{\left(a + (x - by)^2\right)^2} = \frac{2bk(x - by)}{\left(a + (x - by)^2\right)^2}$$

The signs are opposite. This makes sense because as the amount of zooplankton increases, the amount of phytoplankton decreases.

65. $w = -\dfrac{1}{a}\ln\left[1 - \dfrac{(N - 1)b\left(1 - e^{-kn}\right)}{N}\right]$

(a) $$\frac{\partial w}{\partial n} = -\frac{1}{a}\cdot\frac{1}{1 - \frac{(N - 1)b\left(1 - e^{-kn}\right)}{N}}\cdot\frac{-(N - 1)b\left(-e^{-kn}\right)}{N}\cdot(-k) = \frac{(N - 1)bke^{-kn}}{aN\left[1 - \frac{(N - 1)b\left(1 - e^{-kn}\right)}{N}\right]}$$

This represents the rate of change of the number of wrong attempts with respect to the number of trials.

(b) $$\frac{\partial w}{\partial N} = -\frac{1}{a}\cdot\frac{1}{1 - \frac{(N - 1)b\left(1 - e^{-kn}\right)}{N}}\cdot\frac{-\left[Nb\left(1 - e^{-kn}\right) - (N - 1)b\left(1 - e^{-kn}\right)\right]}{N^2}$$

$$= \frac{b\left(1 - e^{-kn}\right)(N - (N - 1))}{aN^2\left[1 - \frac{(N - 1)b\left(1 - e^{-kn}\right)}{N}\right]} = \frac{b\left(1 - e^{-kn}\right)}{aN^2\left[1 - \frac{(N - 1)b\left(1 - e^{-kn}\right)}{N}\right]}$$

This represents the rate of change of the number of wrong attempts with respect to the number of stimuli.

67. The rate of change in lung capacity with respect to age can be found by comparing the change in two lung capacity measurements to the difference in the respective ages when the height is held constant. So for a woman 58 inches tall, at age 20 the measured lung capacity is 1900 ml, and at age 25 the measured lung capacity is 1850 ml. So the rate of change in lung capacity with respect to age is $\dfrac{1900 - 1850}{20 - 25} = \dfrac{50}{-5} = -10$ ml per year. The rate of change in lung capacity with respect to height can be found by comparing the change in two lung capacity measurements to the difference in the respective heights when the age is held constant. So for a 20-year old woman the measured lung capacity for a woman 58 inches tall is 1900 ml and the measured lung capacity for a woman 60 inches tall is 2100 ml. So the rate of change in lung capacity with respect to height is $\dfrac{1900 - 2100}{58 - 60} = \dfrac{-200}{-2} = 100$ ml per in.
The two rates of change remain constant throughout the table.

69. $w(x, y) = \dfrac{x+y}{1+\dfrac{xy}{c^2}}$

(a) $w(50,000, 150,000)$
$$= \frac{50,000+150,000}{1+\dfrac{(50,000)(150,000)}{(186,282)^2}} \approx 164,456$$

The rocket is traveling about 164,500 mi/sec relative to the stationary observer.

(b) $$w_x = \frac{\left(1+\dfrac{xy}{c^2}\right) - \dfrac{y}{c^2}(x+y)}{\left(1+\dfrac{xy}{c^2}\right)^2}$$
$$= \frac{1+\dfrac{xy}{c^2} - \dfrac{xy}{c^2} - \dfrac{y^2}{c^2}}{\left(1+\dfrac{xy}{c^2}\right)^2} = \frac{1-\left(\dfrac{y}{c}\right)^2}{\left(1+\dfrac{xy}{c^2}\right)^2}$$

$W_x(50,000, 150,000)$
$$= \frac{1-\left(\dfrac{150,000}{186,282}\right)^2}{\left(1+\dfrac{(50,000)(150,000)}{(186,282)^2}\right)^2} \approx 0.2377$$

The instantaneous rate of change is 0.238 m/sec per m/sec.

(c) $$w(c, c) = \frac{c+c}{1+\frac{(c)(c)}{c^2}} = \frac{2c}{2} = c$$

The speed is the speed of light, c.

71. (a) $$u(x,t) = T_0 + A_0 e^{-ax} \cos\left(\frac{\pi}{6}t - ax\right)$$
$$= 16 + 11e^{-0.00706x} \cdot \cos\left(\frac{\pi}{6}t - 0.00706x\right)$$

The amplitude of $\mu(x,t)$ is given by $11e^{-0.00706x}$. We need to find where $11e^{-0.00706x} \le 1.$

$$11e^{-0.00706x} \le 1 \Rightarrow e^{-0.00706x} \le \frac{1}{11} \Rightarrow$$
$$-0.00706x \le \ln\left(\frac{1}{11}\right) \Rightarrow$$
$$x \ge \frac{\ln\left(\frac{1}{11}\right)}{-0.00706} \approx 340 \text{ cm}$$

The amplitude is at most 1°C at a minimum depth of about 340 centimeters.

(b) We wish to find x for which $14 \le u(x,t) \le 18$. Since $-1 \le \cos\left(\frac{\pi}{6}t - 0.00706x\right) \le 1$, we have $16 - 11e^{-0.00706x} \le u(x,t) \le 16 + 11e^{-0.00706x}$. For $14 \le u(x,t) \le 18$, we need to find where $16 - 11e^{-0.00706x} = 14$ and $16 + 11e^{-0.00706x} = 18$.

These two conditions are equivalent to
$$11e^{-0.00706x} = 2 \Rightarrow e^{-0.00706x} = \frac{2}{11} \Rightarrow$$
$$-0.00706x = \ln\left(\frac{2}{11}\right) \Rightarrow$$
$$x = \frac{\ln\left(\frac{2}{11}\right)}{-0.00706} \approx 242 \text{ cm}$$

A minimum depth of about 242 centimeters will keep the wine at a temperature between 14°C and 18° C.

(c) The phase shift will correspond to $\frac{1}{2}$ year or 6 months when
$$\frac{0.00706}{\frac{\pi}{6}} = 6 \Rightarrow 0.00706x = \pi \Rightarrow$$
$$x = \frac{\pi}{0.00706} \approx 445 \text{ cm}$$

A depth of about 445 centimeters gives a ground temperature prediction of winter when it is summer and vice versa.

(d) We show that the more general function $u(x,t) = T_0 + A_0 e^{-ax}\cos(wt - ax)$ satisfies the heat equation.

$$\frac{\partial u}{\partial t} = A_0 e^{-ax}[-\sin(wt-ax)](w)$$
$$= -wA_0 e^{-ax}\sin(wt-ax)$$
$$\frac{\partial u}{\partial x} = -aA_0 e^{-ax}\cos(wt-ax) + A_0 e^{-ax}[-\sin(wt-ax)](-a)$$
$$= -aA_0 e^{-ax} \cdot [\cos(wt-ax) - \sin(wt-ax)]$$

$$\frac{\partial^2 u}{\partial x^2}$$
$$= a^2 A_0 e^{-ax}[\cos(wt-ax) - \sin(wt-ax)] - aA_0 e^{-x}[a\sin(wt-ax) + a\cos(wt-ax)]$$
$$= a^2 A_0 e^{-ax}[\cos(wt-ax) - \sin(wt-ax)] - a^2 A_0 e^{-ax}[\sin(wt-ax) + \cos(wt-ax)]$$
$$= -2a^2 A_0 e^{-ax}\sin(wt-ax)$$

(continued on next page)

(continued)

$$\frac{\frac{\partial u}{\partial t}}{\frac{\partial^2 u}{\partial x^2}} = \frac{-wA_0e^{-ax}\sin(wt-ax)}{-2a^2A_0e^{-ax}\sin(wt-ax)} = \frac{w}{2a^2} = k$$

Thus, $\frac{\partial u}{\partial t} = k\frac{\partial^2 u}{\partial x^2}$.

9.3 Maxima and Minima

1. $f(x, y) = xy + y - 2x$

$f_x(x, y) = y - 2$, $f_y(x, y) = x + 1$

If $f_x(x, y) = 0$, $y = 2$.

If $f_y(x, y) = 0$, $x = -1$.

Therefore, $(-1, 2)$ is the critical point.

$f_{xx}(x, y) = 0$
$f_{yy}(x, y) = 0$
$f_{xy}(x, y) = 1$

For $(-1, 2)$, $D = 0\cdot 0 - 1^2 = -1 < 0$.
A saddle point is at $(-1, 2)$.

3. $f(x, y) = 3x^2 - 4xy + 2y^2 + 6x - 10$
$f_x(x, y) = 6x - 4y + 6$
$f_y(x, y) = -4x + 4y$

Solve the system $f_x(x, y) = 0$, $f_y(x, y) = 0$.

$$\begin{array}{l} 6x - 4y + 6 = 0 \\ \underline{-4x + 4y \quad\; = 0} \\ 2x \qquad + 6 = 0 \Rightarrow x = -3 \end{array}$$

$-4(-3) + 4y = 0 \Rightarrow y = -3$

Therefore, $(-3, -3)$ is a critical point.

$f_{xx}(x, y) = 6$
$f_{yy}(x, y) = 4$
$f_{xy}(x, y) = -4$

$D = 6\cdot 4 - (-4)^2 = 8 > 0$

Since $f_{xx}(x, y) = 6 > 0$, there is a relative minimum at $(-3, -3)$.

5. $f(x, y) = x^2 - xy + y^2 + 2x + 2y + 6$
$f_x(x, y) = 2x - y + 2$,
$f_y(x, y) = -x + 2y + 2$

Solve the system $f_x(x, y) = 0$, $f_y(x, y) = 0$.

$$\begin{array}{l} 2x - y + 2 = 0 \\ \underline{-x + 2y + 2 = 0} \end{array} \Rightarrow$$

$$\begin{array}{l} 2x - y + 2 = 0 \\ \underline{-2x + 4y + 4 = 0} \\ \qquad 3y + 6 = 0 \Rightarrow y = -2 \end{array}$$

$-x + 2(-2) + 2 = 0 \Rightarrow x = -2$

$(-2, -2)$ is the critical point.

$f_{xx}(x, y) = 2$, $f_{yy}(x, y) = 2$,
$f_{xy}(x, y) = -1$

For $(-2, -2)$, $D = (2)(2) - (-1)^2 = 3 > 0$.

Since $f_{xx}(x, y) > 0$, a relative minimum is at $(-2, -2)$.

7. $f(x, y) = x^2 + 3xy + 3y^2 - 6x + 3y$
$f_x(x, y) = 2x + 3y - 6$,
$f_y(x, y) = 3x + 6y + 3$

Solve the system $f_x(x, y) = 0$, $f_y(x, y) = 0$.

$$\begin{array}{l} 2x + 3y - 6 = 0 \\ 3x + 6y + 3 = 0 \end{array} \Rightarrow$$

$$\begin{array}{l} -4x - 6y + 12 = 0 \\ \underline{\;\;3x + 6y + \;\;3 = 0} \\ -x \qquad + 15 = 0 \Rightarrow x = 15 \end{array}$$

$3(15) + 6y + 3 = 0 \Rightarrow 6y = -48 \Rightarrow y = -8$

$(15, -8)$ is the critical point.

$f_{xx}(x, y) = 2$
$f_{yy}(x, y) = 6$
$f_{xy}(x, y) = 3$

For $(15, -8)$, $D = 2\cdot 6 - 9 = 3 > 0$.

Since $f_{xx}(x, y) > 0$, a relative minimum is at $(15, -8)$.

9. $f(x, y) = 4xy - 10x^2 - 4y^2 + 8x + 8y + 9$
$f_x(x, y) = 4y - 20x + 8$
$f_y(x, y) = 4x - 8y + 8$

Solve the system $f_x(x, y) = 0$, $f_y(x, y) = 0$.

$$\begin{array}{l} 4y - 20x + 8 = 0 \\ 4x - 8y + 8 = 0 \end{array} \Rightarrow$$

$$\begin{array}{l} 4y - 20x + 8 = 0 \\ \underline{-4y + \;2x + 4 = 0} \\ \quad -18x + 12 = 0 \Rightarrow x = \frac{2}{3} \end{array}$$

$4y - 20\left(\frac{2}{3}\right) + 8 = 0 \Rightarrow y = \frac{4}{3}$

The critical point is $\left(\frac{2}{3}, \frac{4}{3}\right)$.

$f_{xx}(x, y) = -20$
$f_{yy}(x, y) = -8$
$f_{xy}(x, y) = 4$

For $\left(\frac{2}{3}, \frac{4}{3}\right)$, $D = (-20)(-8) - 16 = 144 > 0$.

Since $f_{xx}(x, y) < 0$, a relative maximum is at $\left(\frac{2}{3}, \frac{4}{3}\right)$.

11. $f(x, y) = x^2 + xy - 2x - 2y + 2$
$f_x(x, y) = 2x + y - 2$
$f_y(x, y) = x - 2$
If $f_y(x, y) = 0,\ x = 2.$ Substitute 2 for x in
$f_x(x, y) = 2x + y - 2 = 0$ and solve for x.
$2(2) + y - 2 = 0 \Rightarrow y = -2$
The critical point is (2, –2).
$f_{xx}(x, y) = 2$
$f_{yy}(x, y) = 0$
$f_{xy}(x, y) = 1$

For (2, –2), $D = 2 \cdot 0 - 1^2 = -1 < 0.$
A saddle point is at (2, –2).

13. $f(x, y) = 3x^2 + 2y^3 - 18xy + 42$
$f_x(x, y) = 6x - 18y$
$f_y(x, y) = 6y^2 - 18x$
If $f_x(x, y) = 0,\ 6x - 18y = 0$, or $x = 3y$.
Substitute $3y$ for x in $f_y(x, y) = 0$ and solve for y.
$6y^2 - 18(3y) = 0 \Rightarrow 6y(y - 9) = 0 \Rightarrow$
$y = 0$ or $y = 9$
If $y = 0$, $x = 0$. If $y = 9$, then $x = 27$. Therefore, (0, 0) and (27, 9) are critical points.
$f_{xx}(x, y) = 6$
$f_{yy}(x, y) = 12y$
$f_{xy}(x, y) = -18$

For (0, 0), $D = 6 \cdot 12(0) - (-18)^2 = -324 < 0.$
There is a saddle point at (0, 0).
For (27, 9), $D = 6 \cdot 12(9) - (-18)^2 = 324 > 0.$
Since $f_{xx}(x, y) = 6 > 0$, there is a relative minimum at (27, 9).

15. $f(x, y) = x^2 + 4y^3 - 6xy - 1$
$f_x(x, y) = 2x - 6y,\ f_y(x, y) = 12y^2 - 6x$
Solve $f_x(x, y) = 0$ for x.
$2x - 6y = 0 \Rightarrow x = 3y$

Substitute for x in $12y^2 - 6x = 0$.
$12y^2 - 6(3y) = 0 \Rightarrow 6y(2y - 3) = 0 \Rightarrow$
$y = 0$ or $y = \frac{3}{2}$

If $y = 0$, $x = 0$. If $y = \frac{3}{2},\ x = \frac{9}{2}.$

The critical points are (0, 0) and $\left(\frac{9}{2}, \frac{3}{2}\right).$

$f_{xx}(x, y) = 2$
$f_{yy}(x, y) = 24y$
$f_{xy}(x, y) = -6$

For (0, 0), $D = 2 \cdot 24(0) - (-6)^2 = -36 < 0.$
A saddle point is at (0, 0).
For $\left(\frac{9}{2}, \frac{3}{2}\right),\ D = 2 \cdot 24\left(\frac{3}{2}\right) - (-6)^2 = 36 > 0.$
Since $f_{xx}(x, y) > 0$, a relative minimum is at $\left(\frac{9}{2}, \frac{3}{2}\right).$

17. $f(x, y) = e^{x(y+1)}$
$f_x(x, y) = (y + 1)e^{x(y+1)}$
$f_y(x, y) = xe^{x(y+1)}$

$f_x(x, y) = 0 \Rightarrow (y + 1)e^{x(y+1)} = 0 \Rightarrow$
$y + 1 = 0 \Rightarrow y = -1$

$f_y(x, y) = 0 \Rightarrow xe^{x(y+1)} = 0 \Rightarrow x = 0$
Therefore, (0, –1) is a critical point.
$f_{xx}(x, y) = (y + 1)^2 e^{x(y+1)}$
$f_{yy}(x, y) = x^2 e^{x(y+1)}$
$f_{xy}(x, y) = (y + 1)e^{x(y+1)} \cdot x + e^{x(y+1)} \cdot 1$
$= (xy + x + 1)e^{x(y+1)}$
For (0, –1),
$f_{xx}(0, -1) = (0)^2 e^0 = 0$
$f_{yy}(0, -1) = (0)^2 e^0 = 0$
$f_{xy}(0, -1) = (0 + 0 + 1)e^0 = 1$
$D = 0 \cdot 0 - 1^2 = -1 < 0$
There is a saddle point at (0, –1).

19. Given a function $f(x, y)$. To find the critical points, solve the system
$f_x(x, y) = 0,\ f_y(x, y) = 0.$

21. $z = -3xy + x^3 - y^3 + \frac{1}{8}$
$f_x(x, y) = -3y + 3x^2,\ f_y(x, y) = -3x - 3y^2$
Solve the system $f_x = 0,\ f_y = 0.$

$$\begin{matrix} -3y + 3x^2 = 0 \\ -3x - 3y^2 = 0 \end{matrix} \Rightarrow \begin{matrix} -y + x^2 = 0 \\ -x - y^2 = 0 \end{matrix}$$

Solve the first equation for y: $y = x^2$
Substitute into the second, and solve for x.
$-x - x^4 = 0 \Rightarrow x(1 + x^3) = 0 \Rightarrow x = 0$ or $x = -1$
If $x = 0$, $y = 0$. If $x = -1$, $y = 1$.
The critical points are (0, 0) and (–1, 1).
(continued on next page)

(continued)

$f_{xx}(x, y) = 6x$
$f_{yy}(x, y) = -6y$
$f_{xy}(x, y) = -3$

For (0, 0), $D = 0 \cdot 0 - (-3)^2 = -9 < 0.$
A saddle point is at (0, 0).

For (−1, 1), $D = -6(-6) - (-3)^2 = 27 > 0.$
$f_{xx}(x, y) = 6(-1) = -6 < 0.$

$f(-1, 1) = -3(-1)(1) + (-1)^3 - 1^3 + \frac{1}{8} = \frac{9}{8}$

A relative maximum of $\frac{9}{8}$ is at (−1, 1).
The equation matches graph **(a)**.

23. $z = y^4 - 2y^2 + x^2 - \frac{17}{16}$

$f_x(x, y) = 2x, \; f_y(x, y) = 4y^3 - 4y$

Solve the system $f_x = 0, f_y = 0.$

$2x = 0 \quad (1)$
$4y^3 - 4y = 0 \quad (2)$
$4y(y^2 - 1) = 0 \Rightarrow 4y(y+1)(y-1) = 0$

Equation (1) gives $x = 0$ and equation (2) gives $y = 0$, $y = -1$, or $y = 1$.
The critical points are (0, 0), (0, −1), and (0, 1).

$f_{xx}(x, y) = 2,$
$f_{yy}(x, y) = 12y^2 - 4,$
$f_{xy}(x, y) = 0$

For (0, 0), $D = 2(12 \cdot 0^2 - 4) - 0 = -8 < 0.$
A saddle point is at (0, 0).

For (0, −1), $D = 2[12(-1)^2 - 4] - 0 = 16 > 0.$
$f_{xx}(x, y) = 2 > 0$

$f(0, -1) = (-1)^4 - 2(-1)^2 + 0^2 - \frac{17}{16} = -\frac{33}{16}$

A relative minimum of $-\frac{33}{16}$ is at (0, −1).

For (0, 1), $D = 2(12 \cdot 1^2 - 4) - 0 = 16 > 0$
$f_{xx}(x, y) = 2 > 0$

$f(0, 1) = 1^4 - 2 \cdot 1^2 + 0^2 - \frac{17}{16} = -\frac{33}{16}$

A relative minimum of $-\frac{33}{16}$ is at (0, 1).
The equation matches graph **(b)**.

25. $z = -x^4 + y^4 + 2x^2 - 2y^2 + \frac{1}{16}$

$f_x(x, y) = -4x^3 + 4x, \; f_y(x, y) = 4y^3 - 4y$

Solve $f_x(x, y) = 0, \; f_y(x, y) = 0.$

$-4x^3 + 4x = 0 \Rightarrow -4x(x+1)(x-1) = 0 \Rightarrow$
$x = 0$ or $x = -1$ or $x = 1$

$4y^3 - 4y = 0 \Rightarrow$
$4y(y+1)(y-1) = 0 \Rightarrow y = 0$ or $y = 1$ or $y = -1$

Critical points are (0, 0), (0, −1), (0, 1), (−1, 0), (−1, −1), (−1, 1), (1, 0), (1, −1), (1, 1).

$f_{xx}(x, y) = -12x^2 + 4,$
$f_{yy}(x, y) = 12y^2 - 4$
$f_{xy}(x, y) = 0$

For (0, 0), $D = 4(-4) - 0 = -16 < 0.$

For (0, −1), $D = 4(8) - 0 = 32 > 0,$
and $f_{xx}(x, y) = 4 > 0.$

$f(0, -1) = -\frac{15}{16}$

For (0, 1), $D = 4(8) - 0 = 32 > 0,$
and $f_{xx}(x, y) = 4 > 0.$

$f(0, 1) = -\frac{15}{16}$

For (−1, 0), $D = -8(-4) - 0 = 32 > 0,$
and $f_{xx}(x, y) = -8 < 0.$

$f(-1, 0) = \frac{17}{16}$

For (−1, −1), $D = -8(8) - 0 = -64 < 0.$
For (−1, 1), $D = -8(8) - 0 = -64 < 0.$
For (1, 0), $D = -8(-4) = 32 > 0,$
and $f_{xx}(x, y) = -8 < 0.$

$f(1, 0) = \frac{17}{16}$

For (1, −1), $D = -8(8) - 0 = -64 < 0.$
For (1, 1), $D = -8(8) - 0 = -64 < 0.$
Saddle points are at (0, 0), (−1, −1), (−1, 1), (1, −1), and (1, 1).

Relative maximum of $\frac{17}{16}$ is at (−1, 0) and (1, 0). Relative minimum of $-\frac{15}{16}$ is at (0, −1) and (0, 1). The equation matches graph **(e)**.

27. $f(x, y) = 1 - x^4 - y^4$
$f_x(x, y) = -4x^3$, $f_y(x, y) = -4y^3$
The system
$f_x(x, y) = -4x^3 = 0$, $f_y(x, y) = -4y^3 = 0$
gives the critical point (0, 0).
$f_{xx}(x, y) = -12x^2$
$f_{yy}(x, y) = -12y^2$
$f_{xy}(x, y) = 0$

For (0, 0), $D = 0 \cdot 0 - 0^2 = 0$.
Therefore, the test gives no information.
Examine a graph of the function drawn by using level curves. (See next page.)
If $f(x, y) = 1$, then $x^4 + y^4 = 0$. The level curve is the point (0, 0, 1).
If $f(x, y) = 0$, then $x^4 + y^4 = 1$. The level curve is the circle with center (0, 0, 0) and radius 1.

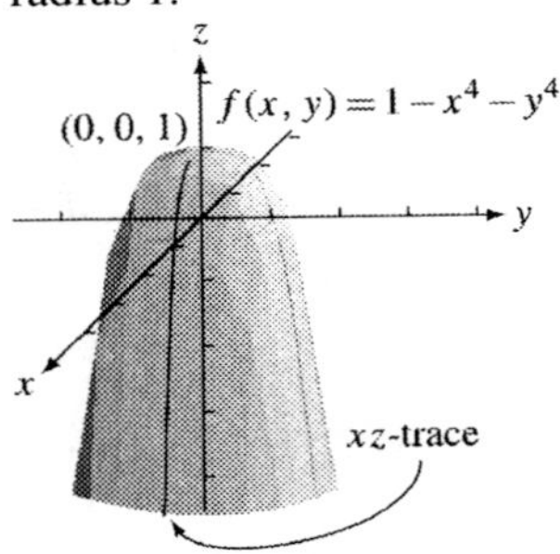

If $f(x, y) = -15$, then $x^4 + y^4 = 16$. The level curve is the curve with center (0, 0, –15) and radius 2.

The *xz*-trace is $z = 1 - x^4$.
This curve has a maximum at (0, 0, 1) and opens downward.

The *yz*-trace is $z = 1 - y^4$.
This curve also has a maximum at (0, 0, 1) and opens downward.

If $f(x, y) > 1$, then $x^4 + y^4 < 0$, which is impossible, so the function does not exist.
Thus, the function has a relative maximum of 1 at (0, 0).

29. Answers will vary.

31. $f(x, y) = x^2(y+1)^2 + k(x+1)^2 y^2$

(a) $f_x(x, y) = 2x(y+1)^2 + 2ky^2(x+1)$
$f_y(x, y) = 2x^2(y+1) + 2k(x+1)^2 y$
$f_x(0,0) = 2(0)(0+1)^2 + 2k(0)^2(0+1) = 0$
$f_y(0,0) = 2(0)^2(0+1) + 2k(0+1)^2(0) = 0$
Thus, (0, 0) is a critical point for all values of *k*.

(b) $f_{xx}(x, y) = 2(y+1)^2 + 2ky^2$
$f_{yy}(x, y) = 2x^2 + 2k(x+1)^2$
$f_{xy}(x, y) = 4x(y+1) + 4ky(x+1)$
$f_{xx}(0,0) = 2 + 2k(0)^2 = 2$
$f_{yy}(0,0) = 2(0)^2 + 2k(0+1)^2 = 2k$
$f_{xy}(0,0) = 4(0)(0+1)4k(0)(0+1) = 0$
$D = 2 \cdot 2k - 0^2 = 4k$
(0, 0) is a relative minimum when $4k > 0$, hence when $k > 0$. When $k = 0$, $D = 0$, so the test for relative extrema gives no information. But if $k = 0$,
$f(x, y) = x^2(y+1)^2$, which is always greater than or equal to $f(0,0) = 0$. So (0, 0) is a relative minimum for $k \geq 0$.

33. $f(a, b)$ is a saddle point.

35. $P(\alpha, r, s) = \alpha(3r^2(1-r) + r^3) + (1-\alpha)(3s^2(1-s) + s^3)$

(a) $P(0.9, 0.5, 0.6)$
$= 0.9[3(0.5)^2(1-0.5) + (0.5)^3] + (1-0.9)[3(0.6)^2(1-0.6) + (0.6)^3]$
$= 0.5148$
$P(0.1, 0.8, 0.4)$
$= 0.1[3(0.8)^2(1-0.8) + (0.8)^3] + (1-0.1)[3(0.4)^2(1-0.4) + (0.4)^3]$
$= 0.4064$
The jury is less likely to make the correct decision in the second situation.

(b) If $r = s = 1$, then $P(\alpha, 1, 1) = 1$, so the jury always makes a correct decision. These values do not depend on α, but in a real-life situation α is likely to influence *r* and *s*.

(c) When P reaches a maximum, P_α, P_r, and P_s equal 0.

$$P_\alpha(\alpha, r, s) = 3r^2(1-r) + r^3 - (3s^2(1-s) + s^3)$$
$$= 3r^2 - 2r^3 - (3s^2 - 2s^3)$$
$$P_r(\alpha, r, s) = \alpha(6r(1-r) - 3r^2 + 3r^2)$$
$$= 6\alpha r(1-r)$$
$$P_s(\alpha, r, s) = (1-\alpha)(6s(1-s) - 3s^2 + 3s^2)$$
$$= 6s(1-\alpha)(1-s)$$
$P_\alpha(\alpha, r, s) = 0$ when $r = s$.

Since $P_r(\alpha, r, s) = 6\alpha r(1-r)$, and $P_s(\alpha, r, s) = 6(1-\alpha)(1-s)$, then P_α, P_r, and P_s are simultaneously 0 at the points $(\alpha, 1, 1)$ and $(\alpha, 0, 0)$. So $(\alpha, 1, 1)$ and $(\alpha, 0, 0)$ are critical points.

$P(\alpha, 0, 0) = 0$ while $P(\alpha, 1, 1) = 1$

Since $P(\alpha, r, s)$ represents a probability, $0 \le P(\alpha, r, s) \le 1$. Thus, $P(\alpha, 1, 1) = 1$ is a maximum value of the function.

37. $E(t, T) = 436.16 - 10.57t - 5.46T - 0.02t^2 + 0.02T^2 + 0.08Tt$

(a) $E(0, 0) = 436.16$.

The value of E before cooking is 436.16 kJ/ mol.

(b) $E(10, 180)$
$$= 436.16 - 10.57(10) - 5.46(180) - 0.02(10)^2 + 0.02(180)^2 + 0.08(180)(10)$$
$$= 137.66$$

After cooking for 10 minutes at 180°C, the total change in color is 137.66 kJ/mol.

(c) $E_t = -10.57 - 0.04t + 0.08T$
$E_T = -5.46 + 0.04T + 0.08t$

Solve the system $E_t = 0, E_T = 0$.

$$-0.04t + 0.08T - 10.57 = 0$$
$$0.08t + 0.04T - 5.46 = 0 \Rightarrow$$
$$-0.04t + 0.08T - 10.57 = 0$$
$$\underline{-0.16t - 0.08T + 10.92 = 0}$$
$$-0.20t \qquad + 0.35 = 0 \Rightarrow t = 1.75$$
$$-0.04(1.75) + 0.08T - 10.57 = 0$$
$$0.08T - 10.64 = 0 \Rightarrow T = 133$$

(1.75, 133) is a critical point.

$E_{tt} = -0.04, \quad E_{TT} = 0.04, \quad E_{tT} = 0.08,$
$D = (-0.04)(0.04) - (0.08)^2 = -0.008 < 0$

(1.75, 133) is a saddle point.

39. $L(x, y) = \frac{3}{2}x^2 + y^2 - 2x - 2y - 2xy + 68,$

where x is the number of skilled hours and y is the number of semiskilled hours.

$L_x(x, y) = 3x - 2 - 2y,$
$L_y(x, y) = 2y - 2 - 2x$

$$3x - 2 - 2y = 0$$
$$\underline{-2x - 2 + 2y = 0}$$
$$x - 4 \qquad = 0$$
$$x = 4$$

$$-2(4) - 2 + 2y = 0$$
$$2y = 10$$
$$y = 5$$

Let $L_{xx}(x, y) = 3, \quad L_{yy}(x, y) = 2,$
$L_{xy}(x, y) = -2.$

$D = 3(2) - (-2)^2 = 2 > 0$ and $L_{xx}(x, y) > 0$.

Relative minimum at (4, 5) is

$$L(4, 5) = \frac{3}{2}(4)^2 + (5)^2 - 2(4) - 2(5) - 2(4)(5) + 68$$
$$= 59$$

So \$59 is a minimum cost, when $x = 4$ and $y = 5$.

41. $R(x, y) = 15 + 169x + 182y - 5x^2 - 7y^2 - 7xy$

$R_x(x, y) = 169 - 10x - 7y$
$R_y(x, y) = 182 - 14y - 7x$

Solve the system $R_x = 0, R_y = 0$.

$$-10x - 7y + 169 = 0$$
$$-7x - 14y + 182 = 0$$

$$20x + 14y - 338 = 0$$
$$\underline{-7x - 14y + 182 = 0}$$
$$13x \qquad - 156 = 0 \Rightarrow x = 12$$
$$-10(12) - 7y + 169 = 0 \Rightarrow -7y = -49 \Rightarrow y = 7$$

(12, 7) is a critical point.

$R_{xx} = -10$
$R_{yy} = -14$
$R_{xy} = -7$

$$D = (-10)(-14) - (-7)^2 = 91 > 0$$

Since $R_{xx} = -10 < 0$, there is a relative maximum at (12, 7).

(continued on next page)

(continued)

$$R(12,7) = 15 + 169(12) + 182(7) - 5(12)^2 - 7(7)^2 - 7(12)(7) = 1666 \text{ (hundred dollars)}$$

12 X-ray machines and 7 bone density scanners should be sold to produce a maximum revenue of \$166,600.

9.4 Total Differentials and Approximations

1. $z = f(x, y) = 2x^2 + 4xy + y^2$
$x = 5, dx = 0.03, y = -1, dy = -0.02$

$f_x(x, y) = 4x + 4y$
$f_y(x, y) = 4x + 2y$

$$\begin{aligned} dz &= (4x + 4y)dx + (4x + 2y)dy \\ &= [4(5) + 4(-1)](0.03) + [4(5) + 2(-1)](-0.02) \\ &= 0.48 - 0.36 = 0.12 \end{aligned}$$

3. $z = \dfrac{y^2 + 3x}{y^2 - x}$, $x = 4$, $y = -4$,
$dx = 0.01$, $dy = 0.03$

$$\begin{aligned} dz &= \frac{(y^2 - x)\cdot 3 - (y^2 + 3x)\cdot(-1)}{(y^2 - x)^2}dx + \frac{(y^2 - x)\cdot 2y - (y^2 + 3x)\cdot 2y}{(y^2 - x)^2}dy \\ &= \frac{4y^2}{(y^2 - x)^2}dx - \frac{8xy}{(y^2 - x)^2}dy \\ &= \frac{4(-4)^2}{[(-4)^2 - 4]^2}(0.01) - \frac{8(4)(-4)}{[(-4)^2 - 4]^2}(0.03) \\ &\approx 0.0311 \end{aligned}$$

5. $w = \dfrac{5x^2 + y^2}{z + 1}$
$x = -2,\ y = 1,\ z = 1$
$dx = 0.02,\ dy = -0.03,\ dz = 0.02$

$$f_x(x, y) = \frac{(z + 1)10x - (5x^2 + y^2)(0)}{(z + 1)^2} = \frac{10x}{z + 1}$$

$$f_y(x, y) = \frac{(z + 1)(2y) - (5x^2 + y^2)(0)}{(z + 1)^2} = \frac{2y}{z + 1}$$

$$f_z(x, y) = \frac{(z + 1)(0) - (5x^2 + y^2)(1)}{(z + 1)^2} = \frac{-5x^2 - y^2}{(z + 1)^2}$$

$$dw = \frac{10x}{z + 1}dx + \frac{2y}{z + 1}dy + \frac{-5x^2 - y^2}{(z + 1)^2}dz$$

Substitute the given values.

$$\begin{aligned} dw &= \frac{-20}{2}(0.02) + \frac{2}{2}(-0.03) + \frac{[-5(4) - 1](0.02)}{(2)^2} \\ &= -0.2 - 0.03 - \frac{21}{4}(0.02) = -0.335 \end{aligned}$$

7. Let $z = f(x, y) = \sqrt{x^2 + y^2}$.
Then

$$\begin{aligned} dz &= f_x(x, y)dx + f_y(x, y)dy \\ &= \frac{1}{2}(x^2 + y^2)^{-1/2}(2x)dx + \frac{1}{2}(x^2 + y^2)^{-1/2}(2y)dy \\ &= \frac{xdx + ydy}{\sqrt{x^2 + y^2}}. \end{aligned}$$

To approximate $\sqrt{8.05^2 + 5.97^2}$, we let $x = 8$, $dx = 0.05$, $y = 6$ and $dy = -0.03$.

$$dz = \frac{8(0.05) + 6(-0.03)}{\sqrt{8^2 + 6^2}} = 0.022$$

$$\begin{aligned} f(8.05, 5.97) &= f(8, 6) + \Delta z \approx f(8, 6) + dz \\ &= \sqrt{8^2 + 6^2} + 0.022 \\ &= 10.022 \end{aligned}$$

Thus, $\sqrt{8.05^2 + 5.97^2} \approx 10.022$.

Using a calculator, $\sqrt{8.05^2 + 5.97^2} \approx 10.0221$. The absolute value of the difference of the two results is $|10.022 - 10.0221| = 0.0001$.

9. Let $z = f(x, y) = (x^2 + y^2)^{1/3}$.
Then

$$dz = f_x(x, y)dx + f_y(x, y)dy$$

$$\begin{aligned} dz &= \frac{1}{3}(x^2 + y^2)^{-2/3}(2x)dx + \frac{1}{3}(x^2 + y^2)^{-2/3}(2y)dy \\ &= \frac{2x}{3(x^2 + y^2)^{2/3}}dx + \frac{2y}{3(x^2 + y^2)^{2/3}}dy \end{aligned}$$

(continued on next page)

(continued)

To approximate $(1.92^2 + 2.1^2)^{1/3}$, we let $x = 2$, $dx = -0.08$, $y = 2$, and $dy = 0.1$.

$$dz = \frac{2(2)}{3[(2)^2 + (2)^2]^{2/3}}(-0.08) + \frac{2(2)}{3[(2)^2 + (2)^2]^{2/3}}(0.1)$$
$$= \frac{4}{12}(-0.08) + \frac{4}{12}(0.1) = 0.00\overline{6}$$
$$f(1.92, 2.1) = f(2, 2) + \Delta z \approx f(2, 2) + dz$$
$$= 2 + 0.00\overline{6} \approx 2.0067$$

Using a calculator, $(1.92^2 + 2.1^2)^{1/3} \approx 2.0080$.
The absolute value of the difference of the two results is $|2.0067 - 2.0080| = 0.0013$.

11. Let $z = f(x, y) = xe^y$.
Then
$$dz = f_x(x, y)dx + f_y(x, y)dy$$
$$= e^y dx + xe^y dy.$$

To approximate $1.03e^{0.04}$, we let $x = 1$, $dx = 0.03$, $y = 0$, and $dy = 0.04$.

$$dz = e^0(0.03) + 1 \cdot e^0(0.04) = 0.07$$
$$f(1.03, 0.04) = f(1, 0) + \Delta z \approx f(1, 0) + dz$$
$$= 1 \cdot e^0 + 0.07 = 1.07$$

Thus, $1.03e^{0.04} \approx 1.07$.

Using a calculator, $1.03e^{0.04} \approx 1.0720$.
The absolute value of the difference of the two results is $|1.07 - 1.0720| = 0.0020$.

13. Let $z = f(x, y) = x \ln y$.
Then
$$dz = f_x(x, y)\,dx + f_y(x, y)\,dy$$
$$= \ln y\,dx + \frac{x}{y}dy$$

To approximate $0.99 \ln 0.98$, we let $x = 1$, $dx = -0.01$, $y = 1$, and $dy = -0.02$.

$$dz = \ln(1) \cdot (-0.01) + \frac{1}{1}(-0.02) = -0.02$$
$$f(0.99, 0.98) = f(1, 1) + \Delta z \approx f(1, 1) + dz$$
$$= 1 \cdot \ln(1) - 0.02 \approx -0.02$$

Thus, $0.99 \ln 0.98 \approx -0.02$.
Using a calculator, $0.99 \ln 0.98 \approx -0.0200$.
The absolute value of the difference of the two results is $|-0.02 - (-0.0200)| = 0$.

15. Let $z = f(x, y) = e^x \cos y$.

$$f_x(x, y) = e^x \cos y$$
$$f_y(x, y) = -e^x \sin y$$

Then,
$$dz = f_x(x, y)dx + f_y(x, y)\,dy$$
$$= \left(e^x \cos y\right)dx - \left(e^x \sin y\right)dy$$

To approximate $e^{0.02} \cos 0.03$, let $x = 0$, $dx = 0.02$, $y = 0$, and $dy = 0.03$.

$$dz = 0.02e^0 \cos 0 - 0.03e^0 \sin 0 = 0.02$$

Thus,
$$f(0.02, 0.03) \approx f(0, 0) + dz \approx e^0 \cos 0 + 0.02$$
$$\approx 1.02$$

Using a calculator $e^{0.02} \cos 0.03 \approx 1.0197$.
The absolute value of the difference of the two results is $|1.02 - 1.0197| = 0.0003$.

17. The volume of the bone is $V = \pi r^2 h$, with $h = 7$, $r = 1.4$, $dr = 0.09$, $dh = 2(0.09) = 0.18$

$$dV = 2\pi rh\,dr + \pi r^2 dh$$
$$= 2\pi(1.4)(7)(0.09) + \pi(1.4)^2(0.18)$$
$$= 6.65$$

6.65 cm^3 of preservative are used.

19. $C = \frac{b}{a - v} = b(a - v)^{-1}$
$a = 160$, $b = 200$, $v = 125$, $da = 145 - 160 = -15$, $db = 190 - 200 = -10$, $dv = 130 - 125 = 5$

$$dC = -b(a - v)^{-2} da + \frac{1}{a - v} db + b(a - v)^{-2} dv$$
$$= \frac{-b}{(a - v)^2} da + \frac{1}{a - v} db + \frac{b}{(a - v)^2} dv$$
$$= \frac{-200}{(160 - 125)^2}(-15) + \frac{1}{160 - 125}(-10) + \frac{200}{(160 - 125)^2}(5)$$
$$\approx 2.98 \text{ liters}$$

21. $C(t,g) = 0.6(0.96)^{(210t/1500)-1} + \dfrac{gt}{126t-900}\left[1-(0.96)^{(210t/1500)-1}\right]$

(a) $C(180,8) = 0.6(0.96)^{(210(180)/1500)-1} + \dfrac{(8)(180)}{126(180)-900}\left[1-(0.96)^{(210(180)/1500)-1}\right] \approx 0.2649$

(b) $C_t(t,g) = 0.6(\ln 0.96)\left(\dfrac{210}{1500}\right)(0.96)^{(210t/1500)-1} + \dfrac{g(126t-900)-126(gt)}{(126t-900)^2}$

$\times[1-(0.96)^{(210t/1500)-1}] - \dfrac{gt}{126t-900}(\ln 0.96)\left(\dfrac{210}{1500}\right)(0.96)^{(210t/1500)-1}$

$C_g(t,g) = \dfrac{t}{126t-900}\left[1-(0.96)^{(210t/1500)-1}\right]$

$C(180-10, 8+1) \approx C(180,8) + C_t(180,8)\cdot(-10) + C_g(180,8)\cdot(1)$

$\approx 0.2649 + (-0.00115)(-10) + 0.00519(1) \approx 0.2816$

$C(170,9) \approx 0.2817$

The approximation is very good.

23. $P(A,B,D) = \dfrac{1}{1+e^{3.68-0.016A-0.77B-0.12D}}$

(a) Since bird pecking is present, $B = 1$.

$P(150,1,20) = \dfrac{1}{1+e^{3.68-0.016(150)-0.77(1)-0.12(20)}} = \dfrac{1}{1+e^{-1.89}} \approx 0.8688$

The probability is about 87%.

(b) Since bird pecking is not present, $B = 0$.

$P(150,0,20) = \dfrac{1}{1+e^{3.68-0.016(150)-0.77(0)-0.12(20)}} = \dfrac{1}{1+e^{-1.12}} \approx 0.7540$

The probability is about 75%.

(c) Let $B = 0$. To simplify the notation, let $X = 3.68 - 0.016A - 0.12D$. Then

$P(A,0,D) = \dfrac{1}{1+e^{3.68-0.016A-0.12D}} = \dfrac{1}{1+e^X}.$

Some other values that we will need are $dA = 160-150 = 10$, $dD = 25-20 = 5$,

$X(150,20) = 3.68 - 0.016(150) - 0.12(20) = -1.12$, $X_A = \dfrac{\partial X}{\partial A} = -0.016$, $X_D = \dfrac{\partial X}{\partial D} = -0.12$.

$P_A(A,0,D) = \dfrac{X_A e^X}{(1+e^X)^2} = \dfrac{0.016e^X}{(1+e^X)^2}$

$P_D(A,0,D) = \dfrac{X_D e^X}{(1+e^X)^2} = \dfrac{0.12e^X}{(1+e^X)^2}$

$dP = P_A(A,0,D)\,dA + P_D(A,0,D)\,dD = \dfrac{0.016e^X}{(1+e^X)^2}dA + \dfrac{0.12e^X}{(1+e^X)^2}dD$

Substituting the given and calculated values,

$dP = \dfrac{0.016e^{-1.12}}{(1+e^{-1.12})^2}(10) + \dfrac{0.12e^{-1.12}}{(1+e^{-1.12})^2}(5) = (0.016\cdot 10 + 0.12\cdot 5)\dfrac{e^{-1.12}}{(1+e^{-1.12})^2}$

$\approx 0.76 \cdot 0.1855 \approx 0.14.$

Therefore, $P(160,0,25) = P(150,0,20) + \Delta P \approx P(150,0,20) + dP = 0.75 + 0.14 = 0.89.$

The probability is about 89%.

Using a calculator, $P(160,0,25) \approx 0.8676$, or about 87%.

25. (a) $P = 100\left(\frac{\frac{100a}{b}}{\frac{100a}{b}+\frac{100c}{g}}\right) = 100\left(\frac{\frac{100a}{b}}{\frac{100a}{b}+\frac{100c}{g}} \cdot \frac{\frac{bg}{100}}{\frac{bg}{100}}\right) = 100\left(\frac{ag}{ag+bc}\right) = \frac{100ag}{ag+bc}$

(b) $P = \frac{100(130)(2.0)}{(130)(2.0)+(3.6)(167)} \approx 30.19$

(c) $P_a = \frac{\partial}{\partial a}\left(\frac{100ag}{ag+bc}\right) = 100g\frac{\partial}{\partial a}\left(\frac{a}{ag+bc}\right) = 100g\left(\frac{(ag+bc) - a\frac{\partial}{\partial a}(ag+bc)}{(ag+bc)^2}\right)$

$= 100g\left(\frac{ag+bc-ag}{(ag+bc)^2}\right) = \frac{100bcg}{(ag+bc)^2}$

$P_b = \frac{\partial}{\partial b}\left(\frac{100ag}{ag+bc}\right) = 100ag\frac{\partial}{\partial b}\left(\frac{1}{ag+bc}\right) = 100ag\left(\frac{-\frac{\partial}{\partial b}(ag+bc)}{(ag+bc)^2}\right)$

$= -100ag\left(\frac{c}{(ag+bc)^2}\right) = -\frac{100acg}{(ag+bc)^2}$

$P_c = \frac{\partial}{\partial c}\left(\frac{100ag}{ag+bc}\right) = 100ag\frac{\partial}{\partial c}\left(\frac{1}{ag+bc}\right) = 100ag\left(\frac{-\frac{\partial}{\partial c}(ag+bc)}{(ag+bc)^2}\right)$

$= -100ag\left(\frac{b}{(ag+bc)^2}\right) = -\frac{100abg}{(ag+bc)^2}$

$P_g = \frac{\partial}{\partial c}\left(\frac{100ag}{ag+bc}\right) = 100a\frac{\partial}{\partial g}\left(\frac{g}{ag+bc}\right) = 100a\left(\frac{(ag+bc)\frac{\partial}{\partial g}(g) - g\frac{\partial}{\partial g}(ag+bc)}{(ag+bc)^2}\right)$

$= 100a\left(\frac{(ag+bc)-ag}{(ag+bc)^2}\right) = \frac{100abc}{(ag+bc)^2}$

At this point, evaluate P_a, P_b, P_c, and P_g for the given values of a, b, c, and g.

$P_a = \left.\frac{100bcg}{(ag+bc)^2}\right|_{a=130,\ b=3.6,\ c=167,\ g=2.0} \approx 0.1621$

$P_b = -\left.\frac{100acg}{(ag+bc)^2}\right|_{a=130,\ b=3.6,\ c=167,\ g=2.0} \approx -5.854$

$P_c = -\left.\frac{100abg}{(ag+bc)^2}\right|_{a=130,\ b=3.6,\ c=167,\ g=2.0} \approx -0.1262$

$P_g = \left.\frac{100abc}{(ag+bc)^2}\right|_{a=130,\ b=3.6,\ c=167,\ g=2.0} \approx 10.5379$

(continued on next page)

(continued)

Now find da, db, dc, and dg.

$$\frac{da}{a} = 0.07 \Rightarrow da = 0.07(130) = 9.1$$

$$\frac{db}{b} = 0.09 \Rightarrow db = 0.09(3.6) = 0.324$$

$$\frac{dc}{c} = 0.02 \Rightarrow dc = 0.02(167) = 3.34$$

$$\frac{dg}{g} = 0.02 \Rightarrow dg = 0.02(2.0) = 0.04$$

$$\frac{dP}{P} = \frac{P_a da + P_b db + P_c dc + P_g dg}{P}$$

$$\approx \frac{9.1(0.1621) + 0.324(-5.854) + 3.34(-0.1262) + 0.04(10.5379)}{30.19}$$

$$= -\frac{0.4216}{30.19} \approx -0.0140$$

(d) $$\frac{dP}{P} = \frac{\sqrt{(P_a da)^2 + (P_b db)^2 + (P_c dc)^2 + (P_g dg)^2}}{P}$$

$$= \frac{\sqrt{(9.1(0.1621))^2 + (0.324(-5.854))^2 + (3.34(-0.1262))^2 + (0.04(10.5379))^2}}{30.19}$$

$$\approx 0.0820$$

27. The area is $A = \frac{1}{2}bh$ with $b = 15.8$ cm, $h = 37.5$ cm, $db = 1.1$ cm, and $dh = 0.8$ cm.

$$dA = \frac{1}{2}b\,dh + \frac{1}{2}h\,db$$
$$= \frac{1}{2}(15.8)(0.8) + \frac{1}{2}(37.5)(1.1)$$
$$= 26.945$$

The maximum possible error is 26.945 cm^2.

29. Let $z = f(L, W, H) = LWH$

Then

$$dz = f_L(L, W, H)\,dL + f_W(L, W, H)\,dW + f_H(L, W, H)\,dH$$
$$= WHdL + LHdW + LWdH.$$

A maximum 1% error in each measurement means that the maximum values of dL, dW, and dH are given by

$dL = 0.01L$, $dW = 0.01W$, and $dH = 0.01H$.

Therefore,

$$dz = WH(0.01L) + LH(0.01W) + LW(0.01H)$$
$$= 0.01LWH + 0.01LWH + 0.01LWH$$
$$= 0.03LWH.$$

Thus, an estimate of the maximum error in calculating the volume is 3%.

31. The volume of a cone is $V = \frac{\pi}{3}r^2h$.

$$\frac{dV}{V} = \frac{\frac{2\pi}{3}rh}{\frac{\pi}{3}r^2h}dr + \frac{\frac{\pi}{3}r^2}{\frac{\pi}{3}r^2h}dh = \frac{2}{r}dr + \frac{1}{h}dh$$

When $r = 1$ and $h = 4$, a 1% change in radius changes the volume by 2%, and a 1% change in height changes the volume by $\frac{1}{4}$%. So the change produced by changing the radius is 8 times the change produced by changing the height.

33. The volume of the can is $V = \pi r^2 h$, with $r = 2.5$ cm, $h = 14$ cm, $dr = 0.08$, $dh = 0.16$.

$$dV = 2\pi rh\,dr + \pi r^2\,dh$$
$$= 2\pi(2.5)(14)(0.08) + \pi(2.5)^2(0.16)$$
$$\approx 20.73$$

Approximately 20.73 cm^3 of aluminum are needed.

35. The volume of the box is $V = LWH$ with $L = 10$, $W = 9$, and $H = 18$. Since 0.1 inch is applied to each side and each dimension has a side at each end,

$$dL = dW = dH = 2(0.1) = 0.2$$
$$dV = WH\,dL + LH\,dW + LW\,dH.$$

Substitute.

$$dV = (9)(18)(0.2) + (10)(18)(0.2) + (10)(9)(0.2) = 86.4$$

Approximately 86.4 in^3 are needed.

9.5 Double Integrals

1. $$\int_0^5 (x^4y + y)dx = \left(\frac{x^5y}{5} + xy\right)\Bigg|_0^5 = (625y + 5y) - 0 = 630y$$

3. $$\int_4^5 x\sqrt{x^2+3y}dy = \int_4^5 x(x^2+3y)^{1/2}dy = \frac{2x}{9}[(x^2+3y)^{3/2}\Big|_4^5 = \frac{2x}{9}[(x^2+15)^{3/2} - (x^2+12)^{3/2}]$$

5. $$\int_4^9 \frac{3+5y}{\sqrt{x}}dx = (3+5y)\int_4^9 x^{-1/2}dx = (3+5y)2x^{1/2}\Big|_4^9 = (3+5y)2\left[\sqrt{9} - \sqrt{4}\right] = 6+10y$$

7. $$\int_2^6 e^{2x+3y}dx = \frac{1}{2}e^{2x+3y}\Big|_2^6 = \frac{1}{2}(e^{12+3y} - e^{4+3y})$$

9. $$\int_0^3 ye^{4x+y^2}dy$$

Let $u = 4x + y^2$; then $du = 2y\,dy$.
If $y = 0$ then $u = 4x$.
If $y = 3$ then $u = 4x + 9$.

$$\int_{4x}^{4x+9} e^u \cdot \frac{1}{2}du = \frac{1}{2}e^u\Big|_{4x}^{4x+9} = \frac{1}{2}(e^{4x+9} - e^{4x})$$

11. $$\int_1^2\int_0^5 (x^4y + y)\,dx\,dy$$

From Exercise 1

$$\int_0^5 (x^4y + y)\,dx = 630y.$$

Therefore,

$$\int_1^2\left[\int_0^5 (x^4y + y)\,dx\right]dy = \int_1^2 630y\,dy = 315y^2\Big|_1^2 = 315(4-1) = 945.$$

13. $$\int_0^1\left[\int_3^6 x\sqrt{x^2+3y}dx\right]dy$$

From Exercise 4,

$$\int_3^6 x\sqrt{x^2+3y}\,dx = \frac{1}{3}[(36+3y)^{3/2} - (9+3y)^3].$$

$$\int_0^1\left[\int_3^6 x\sqrt{x^2+3y}\,dx\right]dy = \int_0^1 \frac{1}{3}[(36+3y)^{3/2} - (9+3y)^{3/2}]\,dy$$

Let $u = 36 + 3y$. Then $du = 3dy$.
When $y = 0$, $u = 36$. When $y = 1$, $u = 39$.
Let $z = 9 + 3y$. Then $dz = 3dy$.
When $y = 0$, $z = 9$. When $y = 1$, $z = 12$.

$$\frac{1}{9}\left[\int_{36}^{39} u^{3/2}du - \int_9^{12} z^{3/2}dz\right] = \frac{1}{9}\cdot\frac{2}{5}[(39)^{5/2} - (36)^{5/2} - (12)^{5/2} + (9)^{5/2}]$$
$$= \frac{2}{45}[(39)^{5/2} - (12)^{5/2} - 6^5 + 3^5] = \frac{2}{45}(39^{5/2} - 12^{5/2} - 7533)$$

15. $$\int_1^2\left[\int_4^9 \frac{3+5y}{\sqrt{x}}dx\right]dy$$

From Exercise 5,

$$\int_4^9 \frac{3+5y}{\sqrt{x}}dx = 6+10y.$$

$$\int_1^2\left[\int_4^9 \frac{3+5y}{\sqrt{x}}dx\right]dy = \int_1^2 (6+10y)dy = 6y\Big|_1^2 + 5y^2\Big|_1^2 = 6(2-1) + 5(4-1) = 21$$

17. $\int_1^3 \int_1^3 \frac{dy\,dx}{xy} = \int_1^3 \left[\int_1^3 \frac{1}{xy} dy \right] dx$

$= \int_1^3 \left(\frac{1}{x} \ln |y| \right) \Big|_1^3 dx$

$= \int_1^3 \frac{\ln 3}{x} dx = (\ln 3) \ln |x| \Big|_1^3$

$= (\ln 3)(\ln 3 - 0) = (\ln 3)^2$

19. $\int_2^4 \int_3^5 \left(\frac{x}{y} + \frac{y}{3} \right) dx\,dy$

$= \int_2^4 \left(\frac{x^2}{2y} + \frac{yx}{3} \right) \Big|_3^5 dy$

$= \int_2^4 \left[\frac{25}{2y} + \frac{5y}{3} - \left(\frac{9}{2y} + \frac{3y}{3} \right) \right] dy$

$= \int_2^4 \left(\frac{16}{2y} + \frac{2y}{3} \right) dy = \left(8 \ln |y| + \frac{y^2}{3} \right) \Big|_2^4$

$= 8(\ln 4 - \ln 2) + \frac{16}{3} - \frac{4}{3}$

$= 8 \ln \frac{4}{2} + \frac{12}{3} = 8 \ln 2 + 4$

21. $\iint_R (3x^2 + 4y) dx\,dy;$

$0 \le x \le 3, \ 1 \le y \le 4$

$\iint_R (3x^2 + 4y) dx\,dy$

$= \int_1^4 \int_0^3 (3x^2 + 4y) dx\,dy$

$= \int_1^4 (x^3 + 4xy) \Big|_0^3 dy = \int_1^4 (27 + 12y) dy$

$= (27y + 6y^2) \Big|_1^4 = (108 + 96) - (27 + 6) = 171$

23. $\iint_R \sqrt{x+y}\,dy\,dx; 1 \le x \le 3, \ 0 \le y \le 1$

$\iint_R \sqrt{x+y}\,dy\,dx$

$= \int_1^3 \int_0^1 (x+y)^{1/2} dy\,dx$

$= \int_1^3 \left[\frac{2}{3}(x+y)^{3/2} \right] \Big|_0^1 dx$

$= \int_1^3 \frac{2}{3} [(x+1)^{3/2} - x^{3/2}] dx$

$= \frac{2}{3} \cdot \frac{2}{5} \left[(x+1)^{5/2} - x^{5/2} \right] \Big|_1^3$

$= \frac{4}{15} (4^{5/2} - 3^{5/2} - 2^{5/2} + 1^{5/2})$

$= \frac{4}{15} (32 - 3^{5/2} - 2^{5/2} + 1)$

$= \frac{4}{15} (33 - 3^{5/2} - 2^{5/2})$

25. $\iint_R \frac{3}{(x+y)^2} dy\,dx; 2 \le x \le 4, 1 \le y \le 6$

$\iint_R \frac{3}{(x+y)^2} dy\,dx = 3 \int_2^4 \int_1^6 (x+y)^{-2} dy\,dx$

$= -3 \int_2^4 (x+y)^{-1} \Big|_1^6 dx$

$= -3 \int_2^4 \left(\frac{1}{x+6} - \frac{1}{x+1} \right) dx$

$= -3(\ln |x+6| - \ln |x+1|) \Big|_2^4$

$= -3 \left(\ln \left| \frac{x+6}{x+1} \right| \right) \Big|_2^4$

$= -3 \left(\ln 2 - \ln \frac{8}{3} \right)$

$= -3 \ln \frac{2}{\frac{8}{3}} = -3 \ln \frac{3}{4}$ or $3 \ln \frac{4}{3}$

27. $\iint_R ye^{x+y^2} dx\,dy; \ 2 \le x \le 3, 0 \le y \le 2$

$\iint_R ye^{x+y^2} dx\,dy = \int_0^2 \int_2^3 ye^{x+y^2} dx\,dy$

$= \int_0^2 ye^{x+y^2} \Big|_2^3 dy$

$= \int_0^2 (ye^{3+y^2} - ye^{2+y^2}) dy$

$= e^3 \int_0^2 ye^{y^2} dy - e^2 \int_0^2 ye^{y^2} dy$

$= \frac{e^3}{2} (e^{y^2}) \Big|_0^2 - \frac{e^2}{2} (e^{y^2}) \Big|_0^2$

$= \frac{e^3}{2} (e^4 - e^0) - \frac{e^2}{2} (e^4 - e^0)$

$= \frac{1}{2} (e^7 - e^6 - e^3 + e^2)$

29. $\iint_R x\cos(xy)\,dy\,dx;\quad \frac{\pi}{2} \le x \le \pi,\ 0 \le y \le 1$

$$\iint_R x\cos(xy)\,dy\,dx = \int_{\pi/2}^{\pi}\int_0^1 x\cos(xy)\,dy\,dx$$
$$= \int_{\pi/2}^{\pi} \sin(xy)\Big|_0^1\,dx$$
$$= \int_{\pi/2}^{\pi}[\sin x - \sin 0]\,dx$$
$$= \int_{\pi/2}^{\pi} \sin x\,dx = -\cos x\Big|_{\pi/2}^{\pi}$$
$$= -\cos\pi + \cos\frac{\pi}{2} = 1$$

31. $z = 8x + 4y + 10; -1 \le x \le 1,\ 0 \le y \le 3$

$$V = \int_{-1}^{1}\int_0^3 (8x + 4y + 10)\,dy\,dx$$
$$= \int_{-1}^{1} (8xy + 2y^2 + 10y)\Big|_0^3\,dx$$
$$= \int_{-1}^{1} (24x + 18 + 30 - 0)\,dx$$
$$= \int_{-1}^{1} (24x + 48)\,dx = (12x^2 + 48x)\Big|_{-1}^{1}$$
$$= (12 + 48) - (12 - 48) = 96$$

33. $z = x^2; 0 \le x \le 2,\ 0 \le y \le 5$

$$V = \int_0^2\int_0^5 x^2\,dy\,dx = \int_0^2 x^2 y\Big|_0^5\,dx$$
$$= \int_0^2 5x^2\,dx = \frac{5}{3}x^3\Big|_0^2 = \frac{40}{3}$$

35. $z = x\sqrt{x^2 + y}; 0 \le x \le 1,\ 0 \le y \le 1$

$$V = \int_0^1\int_0^1 x\sqrt{x^2 + y}\,dx\,dy$$

Let $u = x^2 + y$. Then $du = 2x\,dx$.
When $x = 0, u = y$.
When $x = 1, u = 1 + y$.

$$V = \int_0^1\int_0^1 x\sqrt{x^2 + y}\,dx\,dy$$
$$= \int_0^1\left[\int_y^{1+y} \frac{1}{2}u^{1/2}\,du\right]dy = \int_0^1 \frac{1}{2}\left(\frac{2}{3}u^{3/2}\right)\Big|_y^{1+y}\,dy$$
$$= \int_0^1 \frac{1}{3}[(1 + y)^{3/2} - y^{3/2}]\,dy$$
$$= \frac{1}{3}\cdot\frac{2}{5}[(1 + y)^{5/2} - y^{5/2}]\Big|_0^1$$
$$= \frac{2}{15}(2^{5/2} - 1 - 1) = \frac{2}{15}(2^{5/2} - 2)$$

37. $z = \dfrac{xy}{(x^2 + y^2)^2}; 1 \le x \le 2,\ 1 \le y \le 4$

$$V = \int_1^2\int_1^4 \frac{xy}{(x^2 + y^2)^2}\,dy\,dx$$
$$= \int_1^2\left[\int_1^4 xy(x^2 + y^2)^{-2}\,dy\right]dx$$
$$= \int_1^2\left[\int_1^4 \frac{1}{2}x(x^2 + y^2)^{-2}(2y)\,dy\right]dx$$
$$= \int_1^2\left[-\frac{1}{2}x(x^2 + y^2)^{-1}\right]\Bigg|_1^4\,dx$$
$$= \int_1^2\left[-\frac{1}{2}x(x^2 + 16)^{-1} + \frac{1}{2}x(x^2 + 1)^{-1}\right]dx$$
$$= -\frac{1}{2}\int_1^2 \frac{1}{2}(x^2 + 16)^{-1}(2x)\,dx$$
$$+ \frac{1}{2}\int_1^2 \frac{1}{2}(x^2 + 1)^{-1}(2x)\,dx$$
$$= -\frac{1}{2}\cdot\frac{1}{2}\ln\left|x^2 + 16\right|\Big|_1^2 + \frac{1}{2}\cdot\frac{1}{2}\ln\left|x^2 + 1\right|\Big|_1^2$$
$$= -\frac{1}{4}\cdot\ln 20 + \frac{1}{4}\ln 17 + \frac{1}{4}\ln 5 - \frac{1}{4}\ln 2$$
$$= \frac{1}{4}(-\ln 20 + \ln 17 + \ln 5 - \ln 2)$$
$$= \frac{1}{4}\ln\frac{(17)(5)}{(20)(2)} = \frac{1}{4}\ln\frac{17}{8}$$

39. $\iint_R xe^{xy}\,dx\,dy; 0 \le x \le 2,\ 0 \le y \le 1$

$$\iint_R xe^{xy}\,dx\,dy = \int_0^2\int_0^1 xe^{xy}\,dy\,dx$$
$$= \int_0^2 \frac{x}{x}e^{xy}\Big|_0^1\,dx = \int_0^2 (e^x - e^0)\,dx$$
$$= (e^x - x)\Big|_0^2 = e^2 - 2 - e^0 + 0$$
$$= e^2 - 3$$

41. $\int_2^4 \int_2^{x^2} (x^2 + y^2) dy\, dx$

$$= \int_2^4 \left(x^2 y + \frac{y^3}{3} \right)\Bigg|_2^{x^2} dx$$
$$= \int_2^4 \left(x^4 + \frac{x^6}{3} - 2x^2 - \frac{8}{3} \right) dx$$
$$= \left(\frac{x^5}{5} + \frac{x^7}{21} - \frac{2}{3}x^3 - \frac{8}{3}x \right)\Bigg|_2^4$$
$$= \frac{1024}{5} + \frac{16,384}{21} - \frac{2}{3}(64) - \frac{8}{3}(4) - \left(\frac{32}{5} + \frac{128}{21} - \frac{16}{3} - \frac{16}{3} \right)$$
$$= \frac{1024}{5} - \frac{32}{5} + \frac{16,384 - 128}{21} - \frac{128}{3} - \frac{32}{3} - \left(\frac{-32}{3} \right)$$
$$= \frac{97,632}{105}$$

43. $\int_0^4 \int_0^x \sqrt{xy}\, dy\, dx$

$$= \int_0^4 \int_0^x (xy)^{1/2}\, dy\, dx$$
$$= \int_0^4 \left[\frac{2(xy)^{3/2}}{3x} \right]\Bigg|_0^x dx$$
$$= \frac{2}{3}\int_0^4 \left[\frac{\left(\sqrt{x^2}\right)^3}{x} - \frac{0}{x} \right] dx$$
$$= \frac{2}{3}\int_0^4 x^2 dx = \frac{2}{3} \cdot \frac{x^3}{3}\Bigg|_0^4 = \frac{2}{9}(64) = \frac{128}{9}$$

45. $\int_2^6 \int_{2y}^{4y} \frac{1}{x} dx\, dy = \int_2^6 (\ln|x|)\Big|_{2y}^{4y} dy$

$$= \int_2^6 (\ln|4y| - \ln|2y|)\, dy$$
$$= \int_2^6 \ln\left|\frac{4y}{2y}\right| dy = \int_2^6 \ln 2\, dy$$
$$= (\ln 2)y\Big|_2^6 = (\ln 2)(6-2)$$
$$= 4 \ln 2$$

Note: We can write 4 ln 2 as $\ln 2^4$, or ln 16.

47. $\int_0^4 \int_1^{e^x} \frac{x}{y} dy\, dx = \int_0^4 (x \ln|y|)\Big|_1^{e^x} dx$

$$= \int_0^4 (x \ln e^x - x \ln 1)\, dx$$
$$= \int_0^4 x^2\, dx = \frac{x^3}{3}\Bigg|_0^4 = \frac{64}{3}$$

49. $\iint_R (5x+8y)\, dy\, dx;\ 1 \le x \le 3,\ 0 \le y \le x-1$

$$\iint_R (5x+8y)\, dy\, dx$$
$$= \int_1^3 \int_0^{x-1} (5x+8y)\, dy\, dx$$
$$= \int_1^3 (5xy + 4y^2)\Big|_0^{x-1} dx$$
$$= \int_1^3 [5x(x-1) + 4(x-1)^2 - 0]\, dx$$
$$= \int_1^3 (9x^2 - 13x + 4)\, dx$$
$$= \left(3x^3 - \frac{13}{2}x^2 + 4x \right)\Bigg|_1^3$$
$$= \left(81 - \frac{117}{2} + 12 \right) - \left(3 - \frac{13}{2} + 4 \right) = 34$$

51. $\iint_R (4 - 4x^2)\, dy\, dx;\ 0 \le x \le 1,\ 0 \le y \le 2 - 2x$

$$\iint_R (4-4x^2)\, dy\, dx = \int_0^1 \int_0^{2-2x} 4(1-x^2)\, dy\, dx$$
$$= \int_0^1 [4(1-x^2)y]\Big|_0^{2(1-x)} dx$$
$$= \int_0^1 4(1-x^2)(2)(1-x)\, dx$$
$$= 8\int_0^1 (1 - x - x^2 + x^3)\, dx$$
$$= 8\left(x - \frac{x^2}{2} - \frac{x^3}{3} + \frac{x^4}{4} \right)\Bigg|_0^1$$
$$= 8\left(1 - \frac{1}{2} - \frac{1}{3} + \frac{1}{4} \right) = \frac{10}{3}$$

53. $\iint_R e^{x/y^2}\, dx\, dy;\ 1 \le y \le 2,\ 0 \le x \le y^2$

$$\iint_R e^{x/y^2}\, dx\, dy = \int_1^2 \int_0^{y^2} e^{x/y^2}\, dx\, dy$$
$$= \int_1^2 [y^2 e^{x/y^2}]\Big|_0^{y^2} dy$$
$$= \int_1^2 (y^2 e^{y^2/y^2} - y^2 e^0)\, dy$$
$$= \int_1^2 (ey^2 - y^2)\, dy = (e-1)\frac{y^3}{3}\Bigg|_1^2$$
$$= (e-1)\left(\frac{8}{3} - \frac{1}{3} \right) = \frac{7(e-1)}{3}$$

55. $\iint\limits_R x^3y\,dy\,dx$; R bounded by $y = x^2$, $y = 2x$.

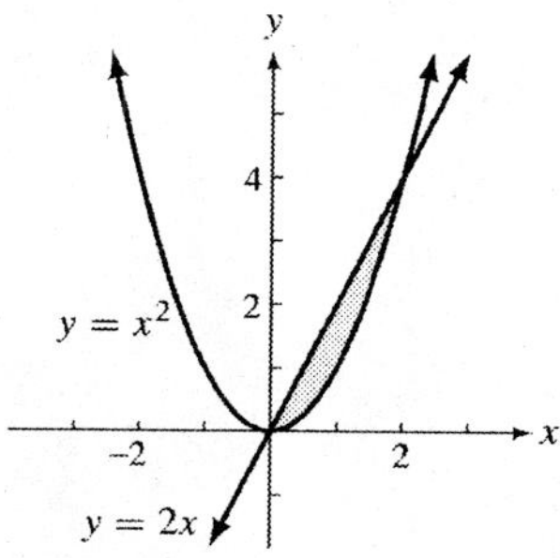

The points of intersection can be determined by solving the following system for x.

$y = x^2$
$y = 2x$

$x^2 = 2x \Rightarrow x(x-2) = 0 \Rightarrow x = 0$ or $x = 2$

Therefore,

$$\iint\limits_R x^3y\,dx\,dy$$

$$= \int_0^2\int_{x^2}^{2x} x^3y\,dy\,dx = \int_0^2\left(x^3\frac{y^2}{2}\right)\Bigg|_{x^2}^{2x} dx$$

$$= \int_0^2\left[x^3\frac{4x^2}{2} - x^3\frac{x^4}{2}\right]dx$$

$$= \int_0^2\left(2x^5 - \frac{x^7}{2}\right)dx = \left(\frac{1}{3}x^6 - \frac{1}{16}x^8\right)\Bigg|_0^2$$

$$= \frac{1}{3}\cdot 2^6 - \frac{1}{16}\cdot 2^8 = \frac{16}{3}$$

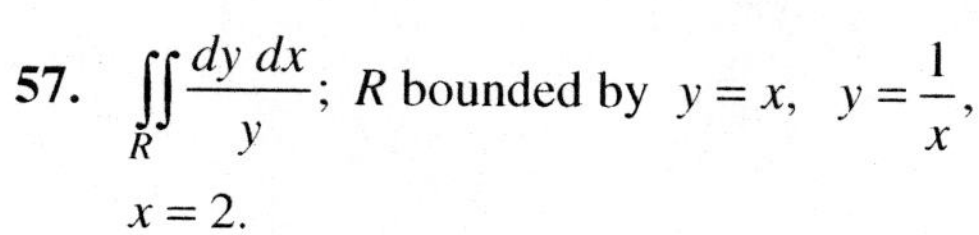

57. $\iint\limits_R \frac{dy\,dx}{y}$; R bounded by $y = x$, $y = \frac{1}{x}$, $x = 2$.

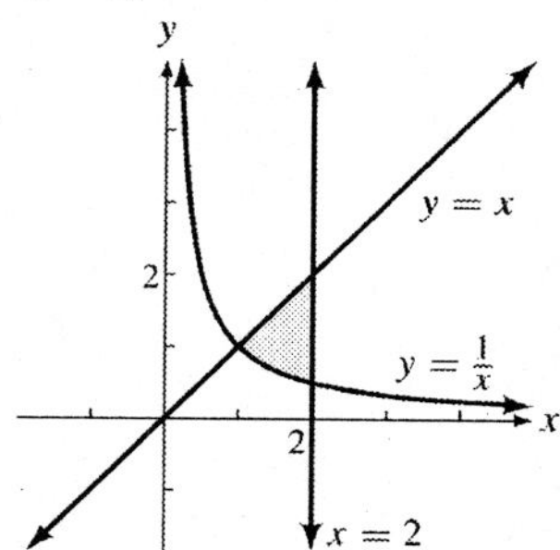

The graphs of $y = x$ and $y = \frac{1}{x}$ intersect at $(1, 1)$.

$$\int_1^2\int_{1/x}^{x}\frac{dy}{y}dx = \int_1^2 \ln y\Big|_{1/x}^{x} dx$$

$$= \int_1^2\left(\ln x - \ln\frac{1}{x}\right)dx$$

$$= \int_1^2 2\ln x\,dx = 2(x\ln x - x)\Big|_1^2$$

$$= 2[(2\ln 2 - 2) - (\ln 1 - 1)]$$

$$= 4\ln 2 - 2$$

59. $\int_0^{\ln 2}\int_{e^y}^{2}\frac{1}{\ln x}dx\,dy$

Change the order of integration.

$$\int_0^{\ln 2}\int_{e^y}^{2}\frac{1}{\ln x}dx\,dy = \int_1^2\int_0^{\ln x}\frac{1}{\ln x}dy\,dx$$

$$= \int_1^2\left[\frac{1}{\ln x}y\Big|_0^{\ln x}\right]dx$$

$$= \int_1^2(1-0)\,dx = x\Big|_1^2$$

$$= 2 - 1 = 1$$

61. Answers will vary.

63. $f(x, y) = 6xy + 2x$; $2 \le x \le 5$, $1 \le y \le 3$

The area of region R is

$A = (5-2)(3-1) = 6.$

The average value of f over R is

$$\frac{1}{A}\iint\limits_R f(x, y)\,dx\,dy$$

$$= \frac{1}{6}\int_1^3\int_2^5(6xy + 2x)\,dx\,dy$$

$$= \frac{1}{6}\int_1^3(3x^2y + x^2)\Big|_2^5 dy$$

$$= \frac{1}{6}\int_1^3[(75y + 25) - (12y + 4)]\,dx$$

$$= \frac{1}{6}\int_1^3(63y + 21)\,dy = \frac{1}{6}\left(\frac{63y^2}{2} + 21y\right)\Bigg|_1^3$$

$$= \frac{1}{6}\left(\frac{567}{2} + 63 - \frac{63}{2} - 21\right) = \frac{1}{6}(294) = 49$$

65. $f(x, y) = e^{-5y+3x}$; $0 \le x \le 2$, $0 \le y \le 2$

The area of region R is $(2-0)(2-0) = 4$.

The average value of f over R is

$$\frac{1}{4}\int_0^2\int_0^2 e^{-5y+3x}\,dx\,dy$$

$$= \frac{1}{4}\int_0^2\left(\frac{1}{3}e^{-5y+3x}\right)\Bigg|_0^2 dy$$

$$= \frac{1}{4}\int_0^2\left(\frac{1}{3}\left[e^{-5y+6} - e^{-5y}\right]\right)dy$$

$$= \frac{1}{12}\left[-\frac{1}{5}e^{-5y+6} - \frac{1}{5}e^{-5y}\right]\Bigg|_0^2$$

$$= -\frac{1}{60}[e^{-4} - e^{-10} - e^6 + 1] = \frac{e^6 + e^{-10} - e^{-4} - 1}{60}$$

67. The plane that intersects the axes has the equation $z = 6 - 2x - 2y$.

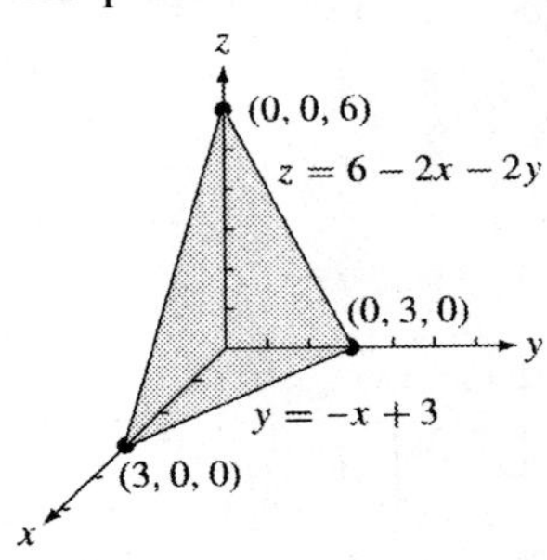

$$V = \iint_R f(x, y)\,dA = \int_0^3 \int_0^{-x+3} (6-2x-2y)\,dy\,dx = \int_0^3 (6y - 2xy - y^2)\Big|_0^{-x+3} dx$$
$$= \int_0^3 [-6x + 18 - 2x(-x+3) - (3-x)^2]\,dx = \int_0^3 (-6x + 18 + 2x^2 - 6x - 9 + 6x - x^2)\,dx$$
$$= \int_0^3 (x^2 - 6x + 9)\,dx = \left(\frac{x^3}{3} - 3x^2 + 9x\right)\Bigg|_0^3 = (9 - 27 + 27) - 0 = 9$$

The volume is 9 in^3.

69. $P(x, y) = -(x-100)^2 - (y-50)^2 + 2000$
$Area = (150-100)(80-40) = (50)(40) = 2000$
The average weekly profit is

$$\frac{1}{2000}\iint_R [-(x-100)^2 - (y-50)^2 + 2000]\,dy\,dx$$
$$= \frac{1}{2000}\int_{100}^{150}\int_{40}^{80} [-(x-100)^2 - (y-50)^2 + 2000]\,dy dx$$
$$= \frac{1}{2000}\int_{100}^{150}\left[-(x-100)^2 y - \frac{(y-50)^3}{3} + 2000y\right]\Bigg|_{40}^{80} dx$$
$$= \frac{1}{2000}\int_{100}^{150}\left[-(x-100)^2(80-40) - \frac{(80-50)^3}{3} + \frac{(40-50)^3}{3} + 2000(80-40)\right] dx$$
$$= \frac{1}{2000}\int_{100}^{150}\left[-40(x-100)^2 - \frac{30^3}{3} + \frac{(-10)^3}{3} + 2000(40)\right] dx$$
$$= \frac{1}{2000}\int_{100}^{150}\left[-40(x-100)^2 - \frac{28,000}{3} + 80,000\right] dx$$
$$= \frac{1}{2000}\cdot\left[\frac{-40(x-100)^3}{3} - \frac{28,000}{3}x + 80,000x\right]\Bigg|_{100}^{150}$$
$$= \frac{1}{2000}\left[-\frac{40}{3}(150-100)^3 + \frac{40}{3}(100-100)^3 - \frac{28,000}{3}(150-100) + 80,000(150-100)\right]$$
$$= \frac{1}{2000}\left[-\frac{40}{3}(50)^3 + \frac{40}{3}\cdot 0 - \frac{28,000}{3}(50) + 80,000(50)\right] = \$933.33.$$

Chapter 9 Review Exercises

1. True
2. True
3. True
4. True
5. False.
   $f(x+h, y) = 3(x+h)^2 + 2(x+h)y + y^2$
6. False. (a, b) could be a saddle point.
7. False: No; near a saddle point the function takes on values both larger and smaller than its value at the saddle point.

8. True

9. False: When dx and dy are interchanged, the limits on the first integral must be exchanged with the limits on the second integral.

10. True

11. False: The two integrals are over different regions, and neither region is a simple region of the sort that we deal with in this chapter.

12. The partial derivative of f with respect to x is the derivative of f obtained by treating x as a variable and y as a constant. The partial derivative of f with respect to y is the derivative of f obtained by treating y as a variable and x as a constant.

13. The partial derivative of f with respect to x, $f_x(a, b)$ gives the rate of change of the surface $z = f(x, y)$ in the x-direction at the point $(a, b, f(a, b))$. In the same way, the partial derivative with respect to y will give the slope of the line tangent to the surface in the y-direction at the point $(a, b, f(a, b))$.

14. Let $z = f(x, y)$ be a function of x and y. Let dx and dy be real numbers. Then the total differential of z is $dz = f_x(x, y)\cdot dx + f_y(x, y)\cdot dy$. It can be used to approximate the value of a function at a point.

15.
$$f(x, y) = -4x^2 + 6xy - 3$$
$$f(-1, 2) = -4(-1)^2 + 6(-1)(2) - 3 = -19$$
$$f(6, -3) = -4(6)^2 + 6(6)(-3) - 3 = -255$$

16.
$$f(x, y) = 2x^2y^2 - 7x + 4y$$
$$f(-1, 2) = 2(1)(4) - 7(-1) + 4(2) = 23$$
$$f(6, -3) = 2(36)(9) - 7(6) + 4(-3) = 594$$

17.
$$f(x, y) = \frac{x - 2y}{x + 5y}$$
$$f(-1, 2) = \frac{(-1) - 2(2)}{(-1) + 5(2)} = \frac{-5}{9} = -\frac{5}{9}$$
$$f(6, -3) = \frac{(6) - 2(-3)}{(6) + 5(-3)} = \frac{12}{-9} = -\frac{4}{3}$$

18.
$$f(x, y) = \frac{\sqrt{x^2 + y^2}}{x - y}$$
$$f(-1, 2) = \frac{\sqrt{1+4}}{-1-2} = -\frac{\sqrt{5}}{3}$$
$$f(6, -3) = \frac{\sqrt{36+9}}{6+3} = \frac{\sqrt{45}}{9} = \frac{\sqrt{5}}{3}$$

19. The plane $x + y + z = 4$ intersects the axes at (4, 0, 0), (0, 4, 0), and (0, 0, 4).

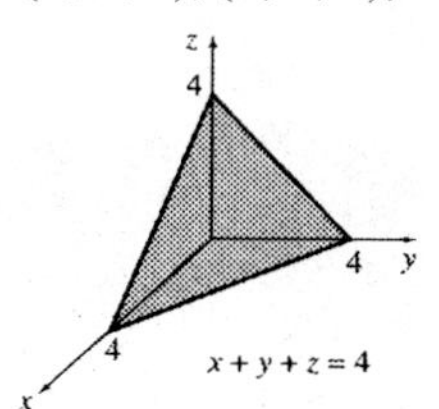

20. $x + 2y + 6z = 6$
x-intercept: $x = 6$, $y = 0$, $z = 0$
y-intercept: $x = 0$, $y = 3$, $z = 0$
z-intercept: $x = 0$, $y = 0$, $z = 1$

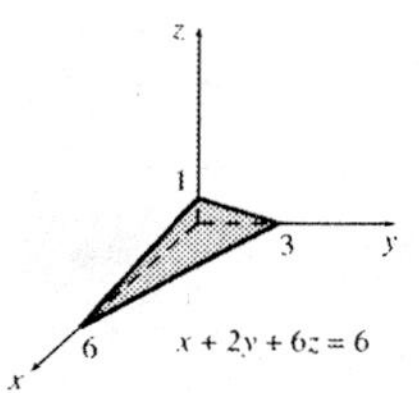

21. The plane $5x + 2y = 10$ intersects the x- and y-axes at (2, 0, 0) and (0, 5, 0). Note that there is no z-intercept since $x = y = 0$ is not a solution of the equation of the plane.

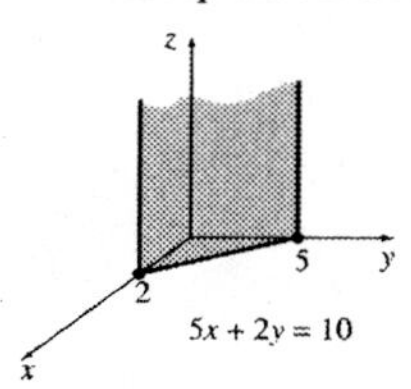

22. $4x + 3z = 12$
No y-intercept
x-intercept: $x = 3$, $y = 0$, $z = 0$
z-intercept: $x = 0$, $y = 0$, $z = 4$

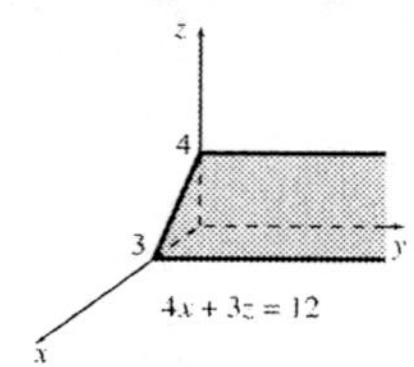

23. $x = 3$
The plane is parallel to the yz-plane. It intersects the x-axis at $(3, 0, 0)$.

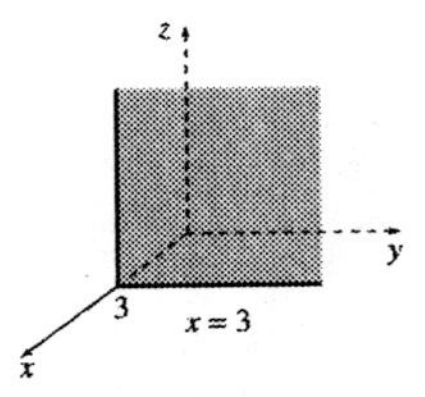

24. $y = 4$
No x-intercept, no z-intercept
The graph is a plane parallel to the xz-plane.

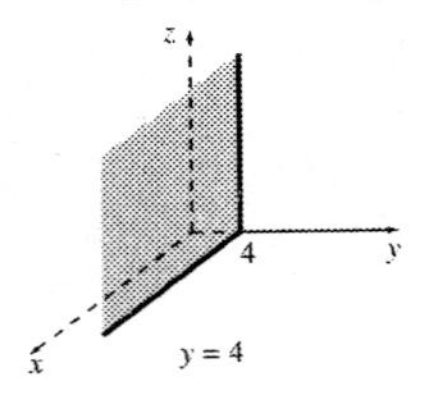

25. $z = f(x, y) = 3x^3 + 4x^2y - 2y^2$

(a) $\dfrac{\partial z}{\partial x} = 9x^2 + 8xy$

(b) $\dfrac{\partial z}{\partial y} = 4x^2 - 4y$

$$\left(\frac{\partial z}{\partial y}\right)(-1, 4) = 4(-1)^2 - 4(4) = -12$$

(c) $f_{xy}(x, y) = 8x$

$$f_{xy}(x, y)(2, -1) = 8(2) = 16$$

26. $z = f(x, y) = \dfrac{x + y^2}{x - y^2}$

(a)
$$\frac{\partial z}{\partial y} = \frac{(x - y^2) \cdot 2y - (x + y^2)(-2y)}{(x - y^2)^2} = \frac{4xy}{(x - y^2)^2}$$

(b)
$$\frac{\partial z}{\partial x} = \frac{(x - y^2) \cdot 1 - (x + y^2) \cdot 1}{(x - y^2)^2} = \frac{-2y^2}{(x - y^2)^2} = -2y^2(x - y^2)^{-2}$$

$$\left(\frac{\partial z}{\partial x}\right)(0, 2) = \frac{-8}{(-4)^2} = -\frac{1}{2}$$

(c)
$$f_{xx}(x, y) = 4y^2(x - y^2)^{-3} = \frac{4y^2}{(x - y^2)^3}$$

$$f_{xx}(-1, 0) = \frac{0}{-1} = 0$$

27. $f(x, y) = 6x^2y^3 - 4y$
$f_x(x, y) = 12xy^3$
$f_y(x, y) = 18x^2y^2 - 4$

28. $f(x, y) = 5x^4y^3 - 6x^5y$
$f_x(x, y) = 20x^3y^3 - 30x^4y$
$f_y(x, y) = 15x^4y^2 - 6x^5$

29. $f(x, y) = \sqrt{4x^2 + y^2}$

$$f_x(x, y) = \frac{1}{2}(4x^2 + y^2)^{-1/2}(8x) = \frac{4x}{(4x^2 + y^2)^{1/2}}$$

$$f_y(x, y) = \frac{1}{2}(4x^2 + y^2)^{-1/2}(2y) = \frac{y}{(4x^2 + y^2)^{1/2}}$$

30. $f(x, y) = \dfrac{2x + 5y^2}{3x^2 + y^2}$

$$f_x(x, y) = \frac{(3x^2 + y^2) \cdot 2 - (2x + 5y^2) \cdot 6x}{(3x^2 + y^2)^2} = \frac{2y^2 - 6x^2 - 30xy^2}{(3x^2 + y^2)^2}$$

$$f_y(x, y) = \frac{(3x^2 + y^2) \cdot 10y - (2x + 5y^2) \cdot 2y}{(3x^2 + y^2)^2} = \frac{30x^2y - 4xy}{(3x^2 + y^2)^2}$$

31. $f(x, y) = x^3e^{3y}$
$f_x(x, y) = 3x^2e^{3y}$
$f_y(x, y) = 3x^3e^{3y}$

32. $f(x, y) = (y - 2)^2e^{x+2y}$
$f_x(x, y) = (y - 2)^2e^{x+2y}$

$$\begin{aligned} f_y(x, y) &= e^{x+2y} \cdot 2(y - 2) + (y - 2)^2 \cdot 2e^{x+2y} \\ &= 2(y - 2)[1 + (y - 2)]e^{x+2y} \\ &= 2(y - 2)(y - 1)e^{x+2y} \end{aligned}$$

33. $f(x, y) = \ln|2x^2 + y^2|$

$$f_x(x, y) = \frac{1}{2x^2+y^2} \cdot 4x = \frac{4x}{2x^2+y^2}$$

$$f_y(x, y) = \frac{1}{2x^2+y^2} \cdot 2y = \frac{2y}{2x^2+y^2}$$

34. $f(x, y) = \ln|2 - x^2y^3|$

$$f_x(x, y) = \frac{1}{2-x^2y^3} \cdot (-2xy^3) = \frac{-2xy^3}{2-x^2y^3}$$

$$f_y(x, y) = \frac{1}{2-x^2y^3} \cdot (-3x^2y^2) = \frac{-3x^2y^2}{2-x^2y^3}$$

35. $f(x, y) = \cos\left(3x^2 + y^2\right)$

$$f_x(x, y) = -6x\sin\left(3x^2 + y^2\right)$$

$$f_y(x, y) = -2y\sin\left(3x^2 + y^2\right)$$

36. $f(x, y) = x\tan\left(7x^2 + 4y^2\right)$

$f_x(x, y)$

$$= \tan\left(7x^2 + 4y^2\right) + x\sec^2\left(7x^2 + 4y^2\right) \cdot 14x$$

$$= \tan\left(7x^2 + 4y^2\right) + 14x^2\sec^2\left(7x^2 + 4y^2\right)$$

$$f_y(x, y) = x\sec^2\left(7x^2 + 4y^2\right) \cdot 8y$$

$$= 8xy\sec^2\left(7x^2 + 4y^2\right)$$

37. $f(x, y) = 5x^3y - 6xy^2$

$$f_x(x, y) = 15x^2y - 6y^2$$

$$f_{xx}(x, y) = 30xy$$

$$f_{xy}(x, y) = 15x^2 - 12y$$

38. $f(x, y) = -3x^2y^3 + x^3y$

$$f_x(x, y) = -6xy^3 + 3x^2y$$

$$f_{xx}(x, y) = -6y^3 + 6xy$$

$$f_{xy}(x, y) = -18xy^2 + 3x^2$$

39. $f(x, y) = \frac{3x}{2x - y}$

$$f_x(x, y) = \frac{(2x-y) \cdot 3 - 3x \cdot 2}{(2x-y)^2} = \frac{-3y}{(2x-y)^2}$$

$$f_{xx}(x, y) = \frac{(2x-y)^2 \cdot 0 - (-3y) \cdot 2(2x-y) \cdot 2}{(2x-y)^4}$$

$$= \frac{12y}{(2x-y)^3}$$

$$f_{xy}(x, y) = \frac{\left[\begin{array}{l}(2x-y)^2 \cdot (-3) \\ \quad -(-3y) \cdot 2(2x-y) \cdot (-1)\end{array}\right]}{(2x-y)^4}$$

$$= \frac{-6x - 3y}{(2x-y)^3}$$

40. $f(x, y) = \frac{3x + y}{x - 1}$

$$f_x(x, y) = \frac{(x-1) \cdot 3 - (3x+y) \cdot 1}{(x-1)^2}$$

$$= \frac{-3-y}{(x-1)^2} = (-3-y)(x-1)^{-2}$$

$$f_{xx}(x, y) = -2(-3-y)(x-1)^{-3} = \frac{2(3+y)}{(x-1)^3}$$

$$f_{xy}(x, y) = \frac{-1}{(x-1)^2}$$

41. $f(x, y) = 4x^2e^{2y}$

$$f_x(x, y) = 8xe^{2y}$$

$$f_{xx}(x, y) = 8e^{2y}$$

$$f_{xy}(x, y) = 16xe^{2y}$$

42. $f(x, y) = ye^{x^2}$

$$f_x(x, y) = 2xye^{x^2}$$

$$f_{xx}(x, y) = 2xy \cdot 2xe^{x^2} + e^{x^2} \cdot 2y$$

$$= 2ye^{x^2}(2x^2 + 1)$$

$$f_{xy}(x, y) = 2xe^{x^2}$$

43. $f(x, y) = \ln|2 - x^2y|$

$$f_x(x, y) = \frac{1}{2-x^2y} \cdot (-2xy) = \frac{2xy}{x^2y-2}$$

$$f_{xx}(x, y) = \frac{(x^2y-2)2y - 2xy(2xy)}{(x^2y-2)^2}$$

$$= \frac{2y[(x^2y-2) - 2x^2y]}{(x^2y-2)^2}$$

$$= \frac{2y(-x^2y-2)}{(x^2y-2)^2} = \frac{-2x^2y^2 - 4y}{(2-x^2y)^2}$$

$$f_{xy}(x, y) = \frac{2x(x^2y-2) - x^2(2xy)}{(x^2y-2)^2}$$

$$= \frac{2x[(x^2y-2) - x^2y]}{(x^2y-2)^2}$$

$$= \frac{2x(-2)}{(x^2y-2)^2} = \frac{-4x}{(2-x^2y)^2}$$

44. $f(x, y) = \ln|1+3xy^2|$

$$f_x(x, y) = \frac{1}{1+3xy^2} \cdot 3y^2 = \frac{3y^2}{1+3xy^2}$$
$$= 3y^2(1+3xy^2)^{-1}$$
$$f_{xx}(x, y) = 3y^2 \cdot (-3y^2)(1+3xy^2)^{-2}$$
$$= \frac{-9y^4}{(1+3xy^2)^2}$$
$$f_{xy}(x, y) = \frac{(1+3xy^2) \cdot 6y - 3y^2(6xy)}{(1+3xy^2)^2}$$
$$= \frac{6y}{(1+3xy^2)^2}$$

45. $z = 2x^2 - 3y^2 + 12y$

$z_x(x, y) = 4x$
$z_y(x, y) = -6y + 12$

If $z_x(x, y) = 0$, $x = 0$. If $z_y(x, y) = 0$, $y = 2$.

Therefore, (0, 2) is a critical point.

$z_{xx}(x, y) = 4$
$z_{yy}(x, y) = -6$
$z_{xy}(x, y) = 0$
$D = 4(-6) - 0^2 = -24 < 0$

There is a saddle point at (0, 2).

46. $z = x^2 + y^2 + 9x - 8y + 1$

$z_x(x, y) = 2x + 9$, $z_y(x, y) = 2y - 8$

$2x + 9 = 0 \Rightarrow x = -\frac{9}{2}$

$2y - 8 = 0 \Rightarrow y = 4$

$z_{xx}(x, y) = 2$, $z_{yy}(x, y) = 2$, $z_{xy}(z, y) = 0$

$D = 2(2) - (0)^2 = 4 > 0$ and $z_{xx}(x, y) > 0$.

Relative minimum at $\left(-\frac{9}{2}, 4\right)$

47. $f(x, y) = x^2 + 3xy - 7x + 5y^2 - 16y$

$f_x(x, y) = 2x + 3y - 7$
$f_y(x, y) = 3x + 10y - 16$

Solve the system $f_x(x, y) = 0, f_y(x, y) = 0$.

$$2x + 3y - 7 = 0$$
$$3x + 10y - 16 = 0$$

$$-6x - 9y + 21 = 0$$
$$6x + 20y - 32 = 0$$
$$11y - 11 = 0 \Rightarrow y = 1$$

$2x + 3(1) - 7 = 0 \Rightarrow 2x = 4 \Rightarrow x = 2$

Therefore, (2, 1) is a critical point.

$f_{xx}(x, y) = 2$
$f_{yy}(x, y) = 10$
$f_{xy}(x, y) = 3$
$D = 2 \cdot 10 - 3^2 = 11 > 0$

Since $f_{xx} = 2 > 0$, there is a relative minimum at (2, 1).

48. $z = x^3 - 8y^2 + 6xy + 4$

$z_x(x, y) = 3x^2 + 6y$, $z_y(x, y) = -16y + 6x$

Setting $z_x = z_y = 0$ and solving yields

$$3x^2 + 6y = 0 \Rightarrow x^2 + 2y = 0 \Rightarrow y = -\frac{x^2}{2}$$
$$-16y + 6x = 0 \Rightarrow -8y + 3x = 0$$

Substituting, we have

$$-8\left(-\frac{x^2}{2}\right) + 3x = 0 \Rightarrow 4x^2 + 3x = 0 \Rightarrow$$
$$x(4x + 3) = 0 \Rightarrow x = 0 \text{ or } x = -\frac{3}{4}$$

If $x = 0$, $y = 0$. If $x = -\frac{3}{4}$, $y = -\frac{9}{32}$.

$z_{xx}(x, y) = 6x$, $z_{yy}(x, y) = -16$, $z_{xy}(x, y) = 6$

$D = 6x(-16) - (6)^2 = -96x - 36$

At (0, 0), $D = -36 < 0$.
Saddle point at (0, 0).

At $\left(-\frac{3}{4}, -\frac{9}{32}\right)$, $D = 36 > 0$ and $z_{xx}(x, y) = -\frac{9}{2} < 0$.

Relative maximum at $\left(-\frac{3}{4}, -\frac{9}{32}\right)$

49. $z = \frac{1}{2}x^2 + \frac{1}{2}y^2 + 2xy - 5x - 7y + 10$

$z_x(x, y) = x + 2y - 5$
$z_y(x, y) = y + 2x - 7$

Setting $z_x = z_y = 0$ and solving yields

$$x + 2y = 5$$
$$2x + y = 7$$

$$-2x - 4y = -10$$
$$2x + y = 7$$
$$-3y = -3 \Rightarrow y = 1$$

If $y = 1$, $x = 3$.

$z_{xx}(x, y) = 1$, $z_{yy}(x, y) = 1$, $z_{xy}(x, y) = 2$

For (3, 1), $D = 1 \cdot 1 - 4 = -3 < 0$.

Therefore, z has a saddle point at (3, 1).

50. $f(x, y) = 2x^2 + 4xy + 4y^2 - 3x + 5y - 15$
$f_x(x, y) = 4x + 4y - 3$
$f_y(x, y) = 4x + 8y + 5$

Solve the system $f_x(x, y) = 0,\ f_y(x, y) = 0.$

$$4x + 4y - 3 = 0$$
$$4x + 8y + 5 = 0$$

$$-4x - 4y + 3 = 0$$
$$4x + 8y + 5 = 0$$
$$4y + 8 = 0 \Rightarrow y = -2$$

$$4x + 4(-2) - 3 = 0 \Rightarrow 4x = 11 \Rightarrow x = \frac{11}{4}$$

Therefore, $\left(\frac{11}{4}, -2\right)$ is a critical point.

$f_{xx}(x, y) = 4,\quad f_{yy}(x, y) = 8,\quad f_{xy}(x, y) = 4$
$D = 4 \cdot 8 - 4^2 = 16 > 0$
Since $f_{xx} = 4 > 0$, there is a relative minimum at $\left(\frac{11}{4}, -2\right)$.

51. $z = x^3 + y^2 + 2xy - 4x - 3y - 2$

$z_x(x, y) = 3x^2 + 2y - 4$
$z_y(x, y) = 2y + 2x - 3$
Setting $z_x(x, y) = z_y(x, y) = 0$ yields

$3x^2 + 2y - 4 = 0 \quad (1)$
$2y + 2x - 3 = 0 \quad (2)$
Solving for $2y$ in equation (2) gives $2y = -2x + 3$. Substitute into equation (1).

$$3x^2 + (-2x + 3) - 4 = 0 \Rightarrow 3x^2 - 2x - 1 = 0 \Rightarrow$$
$$(3x + 1)(x - 1) = 0 \Rightarrow x = -\frac{1}{3} \text{ or } x = 1$$

If $x = -\frac{1}{3},\ y = \frac{11}{6}$. If $x = 1,\ y = \frac{1}{2}$.

$z_{xx}(x, y) = 6x,\quad z_{yy}(x, y) = 2,\quad z_{xy}(x, y) = 2$

For $\left(-\frac{1}{3}, \frac{11}{6}\right)$, $D = 6\left(-\frac{1}{3}\right)(2) - 4 = -8 < 0$,

so z has a saddle point at $\left(-\frac{1}{3}, \frac{11}{6}\right)$.

For $\left(1, \frac{1}{2}\right)$, $D = 6(1)(2) - 4 = 8 > 0$.

$z_{xx}\left(1, \frac{1}{2}\right) = 6 > 0$, so z has a relative minimum at $\left(1, \frac{1}{2}\right)$.

52. $f(x, y) = 7x^2 + y^2 - 3x + 6y - 5xy$
$f_x(x, y) = 14x - 3 - 5y$
$f_y(x, y) = 2y + 6 - 5x$
Setting $z_x(x, y) = z_y(x, y) = 0$ yields

$$\begin{aligned} 14x - 5y - 3 &= 0 \\ -5x + 2y + 6 &= 0 \end{aligned} \Rightarrow$$
$$28x - 10y - 6 = 0$$
$$-25x + 10y + 30 = 0$$
$$3x + 24 = 0 \Rightarrow x = -8$$
$$-5(-8) + 2y + 6 = 0 \Rightarrow 2y = -46 \Rightarrow y = -23$$

$f_{xx}(x, y) = 14,\ f_{yy}(x, y) = 2,\ f_{xy}(x, y) = -5$
$D = 14(2) - (-5)^2 = 3 > 0$ and $f_{xx}(x, y) > 0.$
Relative minimum at $(-8, -23)$

53. If $f_x(x, y) = f_y(x, y) = 0$, then a relative minimum, a relative maximum, or a saddle point might occur.

54. Answers will vary.

55. $z = f(x, y) = 6x^2 - 7y^2 + 4xy$
$x = 3,\ y = -1,\ dx = 0.03,\ dy = 0.01$
$f_x(x, y) = 12x + 4y$
$f_y(x, y) = -14y + 4x$

$$\begin{aligned} dz &= (12x + 4y)\,dx + (-14y + 4x)\,dy \\ &= [12(3) + 4(-1)](0.03) \\ &\quad + [-14(-1) + 4(3)](0.01) \\ &= 0.96 + 0.26 = 1.22 \end{aligned}$$

56. $z = (x, y) = \dfrac{x + 5y}{x - 2y}$
$x = 1,\ y = -2,\ dx = -0.04,\ dy = 0.02$

$$f_x(x, y) = \frac{(x - 2y)(1) - (x + 5y)(1)}{(x - 2y)^2} = \frac{-7y}{(x - 2y)^2}$$

$$f_y(x, y) = \frac{(x - 2y)(5) - (x + 5y)(-2)}{(x - 2y)^2} = \frac{7x}{(x - 2y)^2}$$

$$\begin{aligned} dz &= \frac{-7y}{(x - 2y)^2}dx + \frac{7x}{(x - 2y)^2}dy \\ &= \frac{-7(-2)}{[1 - 2(-2)]^2}(-0.04) + \frac{7(1)}{[1 - 2(-2)]^2}(0.02) \\ &= -0.0224 + 0.0056 = -0.0168 \end{aligned}$$

57. Let $z = f(x, y) = \sqrt{x^2 + y^2}$.

Then,

$$dz = f_x(x, y)dx + f_y(x, y)\,dy.$$

$$dz = \frac{1}{2}(x^2 + y^2)^{-1/2}(2x)\,dx + \frac{1}{2}(x^2 + y^2)(2y)\,dy$$

$$= \frac{x}{\sqrt{x^2 + y^2}}dx + \frac{y}{\sqrt{x^2 + y^2}}dy$$

To approximate $\sqrt{5.1^2 + 12.05^2}$, we let $x = 5$, $dx = 0.1$, $y = 12$, and $dy = 0.05$.

Then,

$$dz = \frac{5}{\sqrt{5^2 + 12^2}}(0.1) + \frac{12}{\sqrt{5^2 + 12^2}}(0.05)$$

$$= \frac{5}{13}(0.1) + \frac{12}{13}(0.05) \approx 0.0846.$$

Therefore,

$$f(5.1, 12.05) = f(5,12) + \Delta z \approx f(5,12) + dz$$

$$= \sqrt{5^2 + 12^2} + 0.0846 \approx 13.0846$$

Using a calculator, $\sqrt{5.1^2 + 12.05^2} \approx 13.0848$. The absolute value of the difference of the two results is $|13.0846 - 13.0848| = 0.0002$.

58. Let $z = f(x, y) = \sqrt{x}e^y$.

Then

$$dz = f_x(x, y)\,dx + f_y(x, y)\,dy$$

$$= \frac{1}{2}x^{-1/2}e^y dx + x^{1/2}e^y dy$$

$$= \frac{e^y}{2\sqrt{x}}dx + \sqrt{x}e^y dy.$$

To approximate $\sqrt{4.06}e^{0.04}$, let $x = 4$, $dx = 0.06$, $y = 0$, and $dy = 0.04$. Therefore,

$$dz = \frac{e^0}{2\sqrt{4}}(0.06) + \sqrt{4}e^0(0.04)$$

$$= \frac{1}{4}(0.06) + 2(0.04) = 0.095$$

$$f(4.06, 0.04) = f(4,0) + \Delta z \approx f(4,0) + dz$$

$$= \sqrt{4}e^0 + 0.095 \approx 2.095$$

Using a calculator, $\sqrt{4.06}e^{0.04} \approx 2.0972$. The absolute value of the difference of the two results is $|2.095 - 2.0972| = 0.0022$.

59. $$\int_1^4 \frac{4y - 3}{\sqrt{x}}dx = (4y - 3)(2\sqrt{x})\Big|_1^4$$

$$= (4y - 3)(2 \cdot 2 - 2 \cdot 1) = 8y - 6$$

60. $$\int_1^5 e^{3x+5y}dx = \frac{1}{3}e^{3x+5y}\Big|_1^5 = \frac{1}{3}(e^{15+5y} - e^{3+5y})$$

61. $$\int_0^5 \frac{6x}{\sqrt{4x^2 + 2y^2}}dx$$

Let $u = 4x^2 + 2y^2$; then $du = 8x\,dx$.

When $x = 0$, $u = 2y^2$.

When $x = 5$, $u = 100 + 2y^2$.

$$\int_0^5 \frac{6x}{\sqrt{4x^2 + 2y^2}}dx$$

$$= \frac{3}{4}\int_{2y^2}^{100+2y^2} u^{-1/2}du = \frac{3}{4}(2u^{1/2})\Big|_{2y^2}^{100+2y^2}$$

$$= \frac{3}{4} \cdot 2[(100 + 2y^2)^{1/2} - (2y^2)^{1/2}]$$

$$= \frac{3}{2}[(100 + 2y^2)^{1/2} - (2y^2)^{1/2}]$$

62. $$\int_1^3 y^2(7x + 11y^3)^{-1/2}dy$$

$$= \frac{2}{33}(7x + 11y^3)^{1/2}\Big|_1^3$$

$$= \frac{2}{33}[(7x + 297)^{1/2} - (7x + 11)^{1/2}]$$

63. $$\int_0^2\left[\int_0^4 (x^2y^2 + 5x)\,dx\right]dy$$

$$= \int_0^2\left(\frac{1}{3}x^3y^2 + \frac{5}{2}x^2\right)\Big|_0^4 dy$$

$$= \int_0^2\left(\frac{64}{3}y^2 + 40\right)dy = \left(\frac{64y^3}{9} + 40y\right)\Big|_0^2$$

$$= \frac{64}{9}(8) + 40(2) = \frac{1232}{9}$$

64. $$\int_0^3\left(\int_0^5 (2x + 6y + y^2)dy\right)dx$$

$$= \int_0^3 \left(2xy + 3y^2 + \frac{1}{3}y^3\right)\Big|_0^5 dx$$

$$= \int_0^3 \left[\left(10x + 75 + \frac{125}{3}\right) - 0\right]dx$$

$$= \int_0^3 \left(10x + \frac{350}{3}\right)dx = \left(5x^2 + \frac{350}{3}x\right)\Big|_0^3$$

$$= (45 + 350) - 0 = 395$$

65. $\int_3^4\left[\int_2^5 \sqrt{6x+3y}\,dx\right]dy$

$= \int_3^4 \frac{1}{9}(6x+3y)^{3/2}\Big|_2^5\, dx$

$= \int_3^4 \frac{1}{9}[(30+3y)^{3/2} - (12+3y)^{3/2}]\,dx$

$= \frac{1}{3}\cdot\frac{1}{9}\cdot\frac{2}{5}\cdot[(30+3y)^{5/2} - (12+3y)^{5/2}]\Big|_3^4$

$= \frac{2}{135}[(42)^{5/2} - (24)^{5/2} - (39)^{5/2} + (21)^{5/2}]$

66. $\int_1^2\left[\int_3^5 (e^{2x-7y})\,dx\right]dy$

$= \int_1^2 e^{2x-7y}\Big|_3^5\, dy$

$= \int_1^2 \frac{1}{2}(e^{10-7y} - e^{6-7y})\,dy$

$= -\frac{1}{14}(e^{10-7y} - e^{6-7y})\Big|_1^2$

$= -\frac{1}{14}(e^{-4} - e^{-8} - e^{3} + e^{-1})$

$= \frac{e^3 + e^{-8} - e^{-4} - e^{-1}}{14}$

67. $\int_2^4\int_2^4 \frac{dx\,dy}{y} = \int_2^4\left(\frac{1}{y}x\right)\Big|_2^4\, dy = \int_2^4\left[\frac{1}{y}(4-2)\right]dy$

$= 2\ln|y|\Big|_2^4 = 2\ln\left|\frac{4}{2}\right|$

$= 2\ln 2 \text{ or } \ln 4$

68. $\int_1^2\int_1^2 \frac{dx\,dy}{x} = \int_1^2 \ln x\Big|_1^2\, dy = \int_1^2 \ln 2\,dy$

$= y\ln 2\Big|_1^2 = 2\ln 2 - \ln 2 = \ln 2$

69. $\iint_R (x^2+2y^2)\,dx\,dy;\ 0 \le x \le 5,\ 0 \le y \le 2$

$\iint_R (x^2+2y^2)\,dx\,dy = \int_0^2\left[\int_0^5 (x^2+2y^2)\,dx\right]dy$

$= \int_0^2\left(\frac{1}{3}x^3 + 2xy^2\right)\Big|_0^5\, dy$

$= \int_0^2\left[\left(\frac{125}{3} + 10y^2\right) - 0\right]dy$

$= \int_0^2\left(\frac{125}{3} + 10y^2\right)dy$

$= \left(\frac{125}{3}y + \frac{10}{3}y^3\right)\Big|_0^2$

$= \frac{250}{3} + \frac{80}{3} = 110$

70. $\iint_R \sqrt{2x+y}\,dx\,dy;\ 1 \le x \le 3,\ 2 \le y \le 5$

$\iint_R \sqrt{2x+y}\,dx\,dy$

$= \int_1^3\left(\int_2^5 (2x+y)^{1/2}\,dy\right)dx$

$= \int_1^3 \frac{2}{3}(2x+y)^{3/2}\Big|_2^5\, dx$

$= \int_1^3 \frac{2}{3}[(2x+5)^{3/2} - (2x+2)^{3/2}]\,dx$

$= \frac{2}{15}[(2x+5)^{5/2} - (2x+2)^{5/2}]\Big|_1^3$

$= \frac{2}{15}(11^{5/2} - 8^{5/2} - 7^{5/2} + 4^{5/2})$

$= \frac{2}{15}(11^{5/2} - 8^{5/2} - 7^{5/2} + 32)$

71. $\iint_R \sqrt{y+x}\,dx\,dy;\ 0 \le x \le 7,\ 1 \le y \le 9$

$\iint_R \sqrt{y+x}\,dx\,dy$

$= \int_1^9\left(\int_0^7 \sqrt{y+x}\,dx\right)dy$

$= \int_1^9\left[\frac{2}{3}(y+x)^{3/2}\right]\Big|_0^7\, dy$

$= \frac{2}{3}\int_1^9 [(y+7)^{3/2} - y^{3/2}]\,dx$

$= \frac{2}{3}\cdot\frac{2}{5}[(7+y)^{5/2} - y^{5/2}]\Big|_1^9$

$= \frac{4}{15}[(16)^{5/2} - (9)^{5/2} - (8)^{5/2} + (1)^{5/2}]$

$= \frac{4}{15}[4^5 - 3^5 - (2\sqrt{2})^5 + 1]$

$= \frac{4}{15}(1024 - 243 - 32(4\sqrt{2}) + 1)$

$= \frac{4}{15}(782 - 128\sqrt{2}) = \frac{4}{15}(782 - 8^{5/2})$

72. $\iint_R ye^{y^2+x}dx\,dy;\ 0\le x\le 1,\ 0\le y\le 1$

$$\iint_R ye^{y^2+x}dx\,dy = \int_0^1\left(\int_0^1 ye^{y^2+x}dy\right)dx$$
$$= \int_0^1 \frac{1}{2}e^{y^2+x}\Big|_0^1 dx$$
$$= \int_0^1 \frac{1}{2}[e^{1+x} - e^x]\,dx$$
$$= \frac{1}{2}[e^{1+x} - e^x]\Big|_0^1$$
$$= \frac{1}{2}[e^2 - e - e + 1]$$
$$= \frac{e^2 - 2e + 1}{2}$$

73. $\iint_R xy\sin\left(xy^2\right)dy\,dx;\ 0\le x\le \frac{\pi}{2},\ 0\le y\le 1$

$$\iint_R xy\sin\left(xy^2\right)dy\,dx$$
$$= \int_0^{\pi/2}\left(\int_0^1 xy\sin\left(xy^2\right)dy\right)dx$$
$$= \int_0^{\pi/2}\frac{1}{2}\left(-\cos\left(xy^2\right)\right)\Big|_0^1 dx$$
$$= -\frac{1}{2}\int_0^{\pi/2}(\cos x - \cos 0)\,dx$$
$$= -\frac{1}{2}\int_0^{\pi/2}(\cos x - 1)\,dx$$
$$= -\frac{1}{2}(\sin x - x)\Big|_0^{\pi/2}$$
$$= -\frac{1}{2}\left[\left(1-\frac{\pi}{2}\right)-0\right] = -\frac{1}{2}+\frac{\pi}{4}$$

74. $\iint_R ye^x\cos\left(ye^x\right)dx\,dy;\ 0\le x\le 1,\ 0\le y\le \pi$

$$\iint_R ye^x\cos\left(ye^x\right)dxdy$$
$$= \int_0^{\pi}\left(\int_0^1 ye^x\cos\left(ye^x\right)dx\right)dy$$
$$= \int_0^{\pi}\sin\left(ye^x\right)\Big|_0^1 dy$$
$$= \int_0^{\pi}\left(\sin(ey) - \sin y\right)dy$$
$$= \left(-\frac{1}{e}\cos(ey) + \cos y\right)\Big|_0^{\pi}$$
$$= \left[-\frac{1}{e}\cos(e\pi) + \cos\pi\right] - \left[-\frac{1}{e}\cos 0 + \cos 0\right]$$
$$= \frac{1}{e} - 2 - \frac{1}{e}\cos(e\pi)$$

75. $z = x + 8y + 4;\ 0\le x\le 3,\ 1\le y\le 2$

$$V = \int_0^3\left(\int_1^2 (x+8y+4)\,dy\right)dx$$
$$= \int_0^3 (xy + 4y^2 + 4y)\Big|_1^2 dx$$
$$= \int_0^3 [(2x+16+8) - (x+4+4)]\,dx$$
$$= \int_0^3 (x+16)\,dx = \left(\frac{1}{2}x^2 + 16x\right)\Big|_0^3$$
$$= \left(\frac{9}{2}+48\right) - 0 = \frac{105}{2}$$

76. $z = x^2 + y^2;\ 3\le x\le 5,\ 2\le y\le 4$

$$V = \iint_R (x^2+y^2)\,dy\,dx = \int_3^5\left(\int_2^4 (x^2+y^2)\,dy\right)dx$$
$$= \int_3^5\left[x^2y + \frac{y^3}{3}\right]\Big|_2^4 dx$$
$$= \int_3^5\left[4x^2 + \frac{64}{3} - 2x^2 - \frac{8}{3}\right]dx$$
$$= \int_3^5\left(2x^2 + \frac{56}{3}\right)dx = \left(\frac{2x^3}{3} + \frac{56x}{3}\right)\Big|_3^5$$
$$= \frac{250}{3} + \frac{280}{3} - 18 - 56 = \frac{308}{3}$$

77. $$\int_0^1\int_0^{2x} xy\,dy\,dx = \int_0^1\left(\frac{xy^2}{2}\right)\Big|_0^{2x} dx$$
$$= \int_0^1 \frac{x}{2}(4x^2 - 0)\,dx$$
$$= \int_0^1 2x^3dx = \left(\frac{1}{2}x^4\right)\Big|_0^1 = \frac{1}{2}$$

78. $$\int_1^2\int_2^{2x^2} y\,dy\,dx = \int_1^2 \frac{1}{2}y^2\Big|_2^{2x^2} dx$$
$$= \int_1^2 (2x^4 - 2)\,dx$$
$$= \left(\frac{2}{5}x^5 - 2x\right)\Big|_1^2$$
$$= \left(\frac{64}{5} - 4\right) - \left(\frac{2}{5} - 2\right) = \frac{52}{5}$$

79. $\int_0^1 \int_{x^2}^x x^3 y\, dy\, dx = \int_0^1 \left(\frac{x^3}{2} y^2 \right) \Big|_{x^2}^{x} dx$

$= \int_0^1 \frac{x^3}{2}(x^2 - x^4)\, dx$

$= \frac{1}{2}\int_0^1 (x^5 - x^7)\, dx$

$= \frac{1}{2}\left(\frac{x^6}{6} - \frac{x^8}{8} \right)\Big|_0^1$

$= \frac{1}{2}\left(\frac{1}{6} - \frac{1}{8} \right) = \frac{1}{2} \cdot \frac{1}{24} = \frac{1}{48}$

80. $\int_0^1 \int_y^{\sqrt{y}} x\, dx\, dy = \int_0^1 \frac{x^2}{2}\Big|_y^{\sqrt{y}} dy = \int_0^1 \frac{1}{2}(y - y^2)\, dy$

$= \frac{1}{2}\left(\frac{y^2}{2} - \frac{y^3}{3} \right)\Big|_0^1$

$= \frac{1}{2}\left(\frac{1}{2} - \frac{1}{3} \right) = \frac{1}{12}$

81. $\int_0^2 \left(\int_{x/2}^1 \frac{1}{y^2+1} dy \right) dx$

Change the order of integration.

$\int_0^2 \int_{x/2}^1 \frac{1}{y^2+1} dy\, dx$

$= \int_0^1 \left(\int_0^{2y} \frac{1}{y^2+1} dx \right) dy = \int_0^1 \frac{x}{y^2+1}\Big|_0^{2y} dy$

$= \int_0^1 \left[\frac{1}{y^2+1}(2y) - \frac{1}{y^2+1}(0) \right] dy$

$= \int_0^1 \frac{2y}{y^2+1} dy = \ln\, (y^2+1)\Big|_0^1$

$= \ln 2 - \ln 1 = \ln 2 - 0 = \ln 2$

82. $\int_0^8 \left(\int_{x/2}^4 \sqrt{y^2+4}\, dy \right) dx$

Change the order of integration.

$\int_0^8 \left(\int_{x/2}^4 \sqrt{y^2+4}\, dy \right) dx$

$= \int_0^4 \left(\int_0^{2y} \sqrt{y^2+4}\, dx \right) dy$

$= \int_0^4 (y^2+4)^{1/2} x\Big|_0^{2y} dy$

$= \int_0^4 [(y^2+4)^{1/2}(2y)] - (y^2+4)(0)]\, dy$

$= \int_0^4 (y^2+4)^{1/2}(2y)\, dy$

$= \frac{(y^2+4)^{3/2}}{\frac{3}{2}}\Bigg|_0^4 = \frac{2}{3}(4^2+4)^{3/2} - \frac{2}{3}(0^2+4)^{3/2}$

$= \frac{2}{3}(20^{3/2} - 4^{3/2}) = \frac{2}{3}(20\sqrt{20} - 8)$

$= \frac{2}{3}(40\sqrt{5} - 8) = \frac{16}{3}(5\sqrt{5} - 1)$

83. $\iint_R (2x+3y)\, dx\, dy;\ 0 \le y \le 1,\ y \le x \le 2-y$

$\int_0^1 \int_y^{2-y} (2x+3y)\, dx\, dy$

$= \int_0^1 (x^2 + 3xy)\Big|_y^{2-y} dy$

$= \int_0^1 [(2-y)^2 - y^2 + 3y(2-y-y)]\, dy$

$= \int_0^1 (4 - 4y + y^2 - y^2 + 6y - 6y^2)\, dy$

$= \int_0^1 (4 + 2y - 6y^2)\, dy$

$= (4y + y^2 - 2y^3)\Big|_0^1 = 4 + 1 - 2 = 3$

84. $\iint_R (2 - x^2 - y^2)\, dy\, dx;\ 0 \le x \le 1,\ x^2 \le y \le x$

$\int_0^1 \left(\int_{x^2}^x (2 - x^2 - y^2)\, dy \right) dx$

$= \int_0^1 \left(2y - x^2 y - \frac{y^3}{3} \right)\Big|_{x^2}^{x} dx$

$= \int_0^1 \left(2x - x^3 - \frac{x^3}{3} - 2x^2 + x^4 + \frac{x^6}{3} \right) dx$

$= \int_0^1 \left(2x - 2x^2 - \frac{4x^3}{3} + x^4 + \frac{x^6}{3} \right) dx$

$= \left(x^2 - \frac{2x^3}{3} - \frac{x^4}{3} + \frac{x^5}{5} + \frac{x^7}{21} \right)\Big|_0^1$

$= 1 - \frac{2}{3} - \frac{1}{3} + \frac{1}{5} + \frac{1}{21} = \frac{26}{105}$

85. Assume that blood vessels are cylindrical.

$V = \pi r^2 h,\ r = 0.7,\ h = 2.7,\ dr = dh = \pm 0.1$

$dV = 2\pi rh\, dr + \pi r^2 dh$

$= 2\pi(0.7)(2.7)(\pm 0.1) + \pi(0.7)^2(\pm 0.1)$

$\approx \pm 1.341$

The possible error is 1.341 cm^3.

86. $T(A, M, S) = -18.37 - 0.09A + 0.34M + 0.25S$

(a) $T(65, 85, 180) = -18.37 - 0.09(65) + 0.34(85) + 0.25(180) = 49.68$

The total body water is 49.68 liters.

(b) $T_A(A, M, S) = -0.09$

The approximate change in total body water if age is increased by 1 yr and mass and height are held constant is –0.09 liter.

$T_M(A, M, S) = 0.34$

The approximate change in total body water if mass is increased by 1 kg and age and height are held constant is 0.34 liter.

$T_S(A, M, S) = 0.25$

The approximate change in total body water if height is increased by 1 cm and age and mass are held constant is 0.25 liter.

87. $L(m, t) = (0.00082t + 0.0955)e^{(\ln m + 10.49)/2.842}$

(a) $L(450, 4) = [0.00082(4) + 0.0955] \cdot e^{(\ln(450)+10.49)/2.842} \approx 33.982$

The length is about 33.98 cm.

(b) $L_m(m, t) = (0.00082t + 0.0955) \cdot e^{(\ln m+10.49)/2.842} \cdot \dfrac{1}{2.842m}$

$L_m(450, 7) \approx 0.02723$

The approximate change in the length of a trout if its mass increases form 450 to 451 g while age is held constant at 7 yr is 0.027 cm.

$L_t(m, t) = 0.00082e^{(\ln m+10.49)/2.842}$

$L_t(450, 7) \approx 0.2821$

The approximate change in the length of a trout if its age increases from 7 to 8 yr while mass is held constant at 450 g is 0.28 cm.

88. (a) $f(60, 1900) \approx 50$

In 1900, 50% of those born 60 years earlier are still alive.

(b) $f(70, 2006) \approx 80$

In 2006, about 80% of those born 70 years earlier are still alive.

(c) $f_x(60, 1900) \approx -1.25$

In 1900, the percent of those born 60 years earlier who are still alive was dropping at a rate of 1.25 percent per additional year of life.

(d) $f_x(70, 2006) = -2$

The percent of those born 70 years earlier who are still alive was dropping at a rate of about 2% per additional year of life.

89. $f(a, b) = \dfrac{1}{4}b\sqrt{4a^2 - b^2}$

(a) $f(3, 2) = \dfrac{1}{4}(2)\sqrt{4(3)^2 - 2^2} = \dfrac{1}{2}\sqrt{32} = 2\sqrt{2} \approx 2.828$

The area of the bottom of the planter is a approximately 2.828 ft^2.

(b) $A = \dfrac{1}{4}b\sqrt{4a^2 - b^2}$

$$dA = \frac{1}{4}b \cdot \frac{1}{2}(4a^2 - b^2)^{-1/2}(8a)da + \left[\frac{1}{4}b \cdot \frac{1}{2}(4a^2 - b^2)^{-1/2}(-2b) + \frac{1}{4}(4a^2 - b^2)^{1/2}\right] db$$

$$dA = \frac{ab}{\sqrt{4a^2 - b^2}}da + \frac{1}{4}\left(\frac{-b^2}{\sqrt{4a^2 - b^2}} + \sqrt{4a^2 - b^2}\right) db$$

If $a = 3$, $b = 2$, $da = 0$, and $db = 0.5$,

$$dA = \frac{1}{4}\left(\frac{-2^2}{\sqrt{4(3)^2 - 2^2}} + \sqrt{4(3)^2 - 2^2}\right)(0.5)$$

$dA \approx 0.6187$.

The approximate effect on the area is an increase of 0.6187 ft^2.

90. Let x be the length of each of the square faces of the box and y be the length of the box.

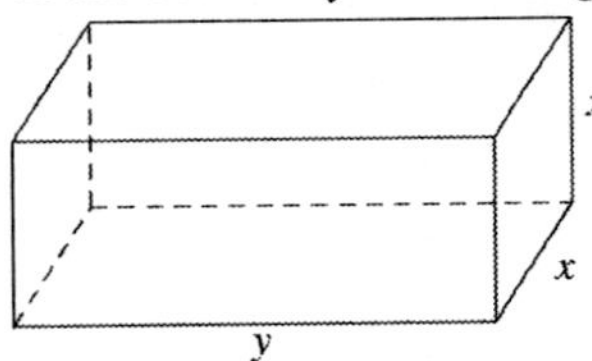

Since the volume must be 125, we have $125 = x^2y$ or $y = \dfrac{125}{x^2}$.

$f(x, y) = 2x^2 + 4xy$ is the surface area of the box.

(*continued on next page*)

(continued)

$f_x(x, y) = 4x + 4y$
$f_y(x, y) = 4x$

Solving the system $f_x = f_y = 0$ leads to the solution $x = 0$, $y = 0$, which is not a valid solution. Instead substitute $y = \frac{125}{x^2}$ into *f*.

$$f(x, y) = 2x^2 + 4xy = 2x^2 + 4x\left(\frac{125}{x^2}\right)$$
$$= 2x^2 + 500x^{-1}$$

$f_x(x, y) = 4x - 500x^{-2}$

Now solve $f_x = 0$.

$$4x - \frac{500}{x^2} = 0 \Rightarrow 4x^3 - 500 = 0 \Rightarrow$$
$$x^3 = 125 \Rightarrow x = 5$$

If $x = 5$, $y = 5$.

$$f_{xx}(x, y) = 4 + 1000x^{-3} \Rightarrow f_{xx}(5, 5) = 12$$
$$f_{yy}(x, y) = \frac{500}{y^3} \Rightarrow f_{yy}(5, 5) = 4$$
$$f_{xy}(x, y) = 0$$

$$D = f_{xx}(5, 5) \cdot f_{yy}(5, 5) - [f_{xy}(5, 5)]^2$$
$$= 12 \cdot 4 - 0 = 48 > 0$$

Therefore, (5, 5) is a minimum. The dimensions are 5 inches by 5 inches by 5 inches.

91. Let *x* be the width of the area and let *y* be the length. Since no fencing is needed along one side, we have $2x + y = 400$. We want to maximize

$$f(x, y) = xy = x(400 - 2x) = 400x - 2x^2.$$
$$f'(x) = 400 - 4x$$

Solving $f'(x) = 0 \Rightarrow x = 100$. Then $200 + y = 400 \Rightarrow y = 200$.

$f''(x) = -4 < 0$, so $f(100, 200)$ is a maximum.

Dimensions are 100 feet by 200 feet for maximum area of 20,000 ft^2.

92. $C(x, y) = 4x^2 + 5y^2 - 4xy + \sqrt{x}$

(a) $C(10, 5)$
$$= 4(10)^2 + 5(5)^2 - 4(10)(5) + \sqrt{10}$$
$$= 400 + 125 - 200 + \sqrt{10}$$
$$= 325 + \sqrt{10} \approx 328.16$$

The cost is about \$328.16.

(b) $C(15, 10)$
$$= 4(15)^2 + 5(10)^2 - 4(15)(10) + \sqrt{15}$$
$$= 900 + 500 - 600 + \sqrt{15}$$
$$= 800 + \sqrt{15} \approx 803.87$$

The cost is about \$803.87.

(c) $C(20, 20)$
$$= 4(20)^2 + 5(20)^2 - 4(20)(20) + \sqrt{20}$$
$$= 1600 + 2000 - 1600 + \sqrt{20}$$
$$= 2000 + \sqrt{20} \approx 2004.47$$

The cost is about \$2004.47.

93. $c(x, y) = 2x + y^2 + 4xy + 25$

(a) $c_x = 2 + 4y$
$c_x(640, 6) = 2 + 4(6) = 26$

For an additional 1 MB of memory, the approximate change in cost is \$26.

(b) $c_y = 2y + 4x$
$c_y(640, 6) = 2(6) + 4(640) = 2572$

For an additional hour of labor, the approximate change in cost is \$2572.

94. (a) Minimize
$c(x, y)$
$$= x^2 + 5y^2 + 4xy - 70x - 164y + 1800$$
$$c_x = 2x + 4y - 70$$
$$c_y = 10y + 4x - 164$$

$$\begin{aligned} 2x + 4y - 70 &= 0 \\ 4x + 10y - 164 &= 0 \end{aligned} \Rightarrow$$

$$\begin{aligned} -4x - 8y + 140 &= 0 \\ 4x + 10y - 164 &= 0 \\ \hline 2y - 24 &= 0 \Rightarrow y = 12 \end{aligned}$$

$$4x + 10(12) - 164 = 0 \Rightarrow 4x = 44 \Rightarrow$$
$$x = 11$$

Extremum at (11, 12)

$c_{xx} = 2$, $c_{yy} = 10$, $c_{xy} = 4$

For (11, 12),
$D = (2)(10) - 16 = 4 > 0$ and
$c_{xx}(11, 12) = 2 > 0$.

There is a relative minimum at (11, 12).

(b) $c(11, 12)$
$$= (11)^2 + 5(12)^2 + 4(11)(12) - 70(11) - 164(12) + 1800$$
$$= \$431$$

95. $C(x, y) = 100\ln(x^2 + y) + e^{xy/20}$
$x = 15,\ y = 9,\ dx = 1,\ dy = -1$

$$dC = \left(\frac{200x}{x^2 + y} + \frac{y}{20}e^{xy/20}\right)dx + \left(\frac{100}{x^2 + y} + \frac{x}{20}e^{xy/20}\right)dy$$

$$dC(15,9) = \left(\frac{200(15)}{15^2 + 9} + \frac{9}{20}e^{(15)(9)/20}\right)(1) + \left(\frac{100}{15^2 + 9} + \frac{15}{20}e^{(15)(9)/20}\right)(-1)$$

$$= \frac{1450}{117} - \frac{3}{10}e^{27/4} = -243.82$$

Costs decrease by \$243.82.

96. $V = \frac{1}{3}\pi r^2 h,\ r = 2\text{ cm},\ h = 8\text{ cm},$
$dr = 0.21\text{ cm},\ dh = 0.21\text{ cm}$

$$dV = \frac{\pi}{3}(2rh\,dr + r^2dh) = \frac{\pi}{3}[2(2)(8)(0.21) + 4(0.21)] = \frac{\pi}{3}(6.72 + 0.84) \approx 7.92\text{ cm}^3$$

97. $V = \frac{4}{3}\pi r^3,\ r = 2\text{ ft},\ dr = 1\text{ in} = \frac{1}{12}\text{ ft}$

$$dV = 4\pi r^2 dr = 4\pi(2)^2\left(\frac{1}{12}\right) \approx 4.19\text{ ft}^3$$

98. $V = \frac{1}{3}\pi r^2 h,\ r = 2.9\text{ cm},\ h = 11.4\text{ cm},$
$dr = dh = 0.2\text{ cm}$

$$dV = \frac{\pi}{3}(2rh\,dr + r^2dh)$$
$$= \frac{\pi}{3}[2(2.9)(11.4)(0.2) + (2.9)^2(0.2)] = \frac{\pi}{3}(13.224 + 1.682) \approx 15.6\text{ cm}^3$$

99. $P(x, y) = 0.01(-x^2 + 3xy + 160x - 5y^2 + 200y + 2600)$ with $x + y = 280$

(a) $y = 280 - x$

$$P(x) = 0.01[-x^2 + 3x(280 - x) + 160x - 5(280 - x)^2 + 200(280 - x) + 2600]$$
$$= 0.01(-x^2 + 840x - 3x^2 + 160x - 392{,}000 + 2800x - 5x^2 + 56{,}000 - 200x + 2600)$$
$$= 0.01(-9x^2 + 3600x - 333{,}400)$$

$P'(x) = 0.01(-18x + 3600)$
$P'(x) = 0 \Rightarrow 0.01(-18x + 3600) = 0 \Rightarrow -18x = -3600 \Rightarrow x = 200$
If $x < 200, P'(x) > 0,$ and if $x > 200,\ P'(x) < 0.$
Therefore, P is maximum when $x = 200$. If $x = 200$, $y = 80$.

$$P(x, y) = 0.01\left[-200^2 + 3(200)(80) + 160(200) - 5(80)^2 + 200(80) + 2600\right] = 0.01(26{,}600) = 266$$

Thus, \$200 spent on fertilizer and \$80 spent on seed will produce a maximum profit of \$266 per acre.

(b) $P(x, y) = 0.01(-x^2 + 3xy + 160x - 5y^2 + 200y + 2600)$

$P_x(x, y) = 0.01(-2x + 3y + 160)$

$P_y(x, y) = 0.01(3x - 10y + 200)$

Solve the system $P_x = P_y = 0.$

$$\begin{aligned} 0.01(-2x+3y+160) &= 0 \\ 0.01(3x-10y+200) &= 0 \end{aligned} \Rightarrow \begin{aligned} -2x + 3y &= -160 \\ 3x - 10y &= -200 \end{aligned}$$

$$\begin{aligned} -6x + 9y &= -480 \\ 6x - 20y &= -400 \\ \hline -11y &= -880 \Rightarrow y = 80 \end{aligned}$$

If $y = 80,\ 3x - 10(80) = -200 \Rightarrow 3x = 600 \Rightarrow x = 200.$

$P_{xx}(x, y) = 0.01(-2) = -0.02$

$P_{yy}(x, y) = 0.01(-10) = -0.1$

$P_{xy}(x, y) = 0.03$

For (200,80), $D = (-0.02)(-0.1) - 0.03^2 = 0.0011 > 0,$ and $P_{xx} < 0,$ so there is a relative maximum at (200,80). $P(200, 80) = 266,$ as in part **(a)**. Thus, \$200 spent on fertilizer and \$80 spent on seed will produce a maximum profit of \$266 per acre.

Chapter 10

MATRICES

10.1 Solution of Linear Systems

1. $3x + y = 6$
$2x + 5y = 15$

The equations are already in standard form. The augmented matrix obtained from the coefficients and the constants is $\left[\begin{array}{cc|c} 3 & 1 & 6 \\ 2 & 5 & 15 \end{array}\right]$.

3. $2x + y + z = 3$
$3x - 4y + 2z = -7$
$x + y + z = 2$

leads to the augmented matrix

$$\left[\begin{array}{ccc|c} 2 & 1 & 1 & 3 \\ 3 & -4 & 2 & -7 \\ 1 & 1 & 1 & 2 \end{array}\right].$$

5. We are given the augmented matrix

$$\left[\begin{array}{cc|c} 1 & 0 & 2 \\ 0 & 1 & 3 \end{array}\right].$$

This is equivalent to the system of equations $x = 2$, $y = 3$.

7. $\left[\begin{array}{ccc|c} 1 & 0 & 0 & 4 \\ 0 & 1 & 0 & -5 \\ 0 & 0 & 1 & 1 \end{array}\right]$

The system associated with this matrix is $x = 4$, $y = -5$, $z = 1$.

9. *Row operations* on a matrix correspond to transformations of a system of equations.

11. $\left[\begin{array}{ccc|c} 3 & 7 & 4 & 10 \\ 1 & 2 & 3 & 6 \\ 0 & 4 & 5 & 11 \end{array}\right]$

Find $R_1 + (-3)R_2$.
In row 2, column 1, 3 + (–3)1 = 0.
In row 2, column 2, 7 + (–3)2 = 1.
In row 2, column 3, 4 + (–3)3 = –5.
In row 2, column 4, 10 + (–3)6 + –8.
Replace R_2 with these values.

The new matrix is

$$\left[\begin{array}{ccc|c} 3 & 7 & 4 & 10 \\ 0 & 1 & -5 & -8 \\ 0 & 4 & 5 & 11 \end{array}\right].$$

13. $\left[\begin{array}{ccc|c} 1 & 6 & 4 & 7 \\ 0 & 3 & 2 & 5 \\ 0 & 5 & 3 & 7 \end{array}\right]$

Find $(-2)R_2 + R_1 \to R_1$

$$\left[\begin{array}{ccc|c} (-2)0+1 & (-2)3+6 & (-2)2+4 & (-2)5+7 \\ 0 & 3 & 2 & 5 \\ 0 & 5 & 3 & 7 \end{array}\right]$$

$$= \left[\begin{array}{ccc|c} 1 & 0 & 0 & -3 \\ 0 & 3 & 2 & 5 \\ 0 & 5 & 3 & 7 \end{array}\right]$$

15. $\left[\begin{array}{ccc|c} 3 & 0 & 0 & 18 \\ 0 & 5 & 0 & 9 \\ 0 & 0 & 4 & 8 \end{array}\right]$

$$\tfrac{1}{3}R_1 \to R_1 \left[\begin{array}{ccc|c} \frac{1}{3}(3) & \frac{1}{3}(0) & \frac{1}{3}(0) & \frac{1}{3}(18) \\ 0 & 5 & 0 & 9 \\ 0 & 0 & 4 & 8 \end{array}\right]$$

$$= \left[\begin{array}{ccc|c} 1 & 0 & 0 & 6 \\ 0 & 5 & 0 & 9 \\ 0 & 0 & 4 & 8 \end{array}\right]$$

In Exercises 17–40, check each solution by substituting it in the original equation of the system.

17. $x + y = 5$
$3x + 2y = 12$

Write the augmented matrix and use row operations.

$$\left[\begin{array}{cc|c} 1 & 1 & 5 \\ 3 & 2 & 12 \end{array}\right]$$

$$-3R_1 + R_2 \to R_2 \left[\begin{array}{cc|c} 1 & 1 & 5 \\ 0 & -1 & -3 \end{array}\right]$$

$$-1R_2 \to R_2 \left[\begin{array}{cc|c} 1 & 1 & 5 \\ 0 & 1 & 3 \end{array}\right]$$

(*continued on next page*)

(*continued*)

$$-1R_2 + R_1 \to R_1 \left[\begin{array}{cc|c} 1 & 0 & 2 \\ 0 & 1 & 3 \end{array}\right]$$

The solution is (2, 3).

19. $x + y = 7$
$4x + 3y = 22$

Write the augmented matrix and use row operations.

$$\left[\begin{array}{cc|c} 1 & 1 & 7 \\ 4 & 3 & 22 \end{array}\right]$$

$$-4R_1 + R_2 \to R_2 \left[\begin{array}{cc|c} 1 & 1 & 7 \\ 0 & -1 & -6 \end{array}\right]$$

$$-1R_2 \to R_2 \left[\begin{array}{cc|c} 1 & 1 & 7 \\ 0 & 1 & 6 \end{array}\right]$$

$$-1R_2 + R_1 \to R_1 \left[\begin{array}{cc|c} 1 & 0 & 1 \\ 0 & 1 & 6 \end{array}\right]$$

The solution is (1, 6).

21. $2x - 3y = 2$
$4x - 6y = 1$

Write the augmented matrix and use row operations.

$$\left[\begin{array}{cc|c} 2 & -3 & 2 \\ 4 & -6 & 1 \end{array}\right]$$

$$-2R_1 + R_2 \to R_2 \left[\begin{array}{cc|c} 2 & -3 & 2 \\ 0 & 0 & -3 \end{array}\right]$$

The system associated with the last matrix is
$2x - 3y = 2$
$0x + 0y = -3.$

Since the second equation, $0 = -3$, is false, the system is inconsistent and therefore has no solution.

23. $6x - 3y = 1$
$-12x + 6y = -2$

Write the augmented matrix of the system and use row operations.

$$\left[\begin{array}{cc|c} 6 & -3 & 1 \\ -12 & 6 & -2 \end{array}\right]$$

$$2R_1 + R_2 \to R_2 \left[\begin{array}{cc|c} 6 & -3 & 1 \\ 0 & 0 & 0 \end{array}\right]$$

$$\tfrac{1}{6}R_1 \to R_1 \left[\begin{array}{cc|c} 1 & -\frac{1}{2} & \frac{1}{6} \\ 0 & 0 & 0 \end{array}\right]$$

This is as far as we can go with the Gauss-Jordan method. To complete the solution, write the equation that corresponds to the first row of the matrix.

$$x - \frac{1}{2}y = \frac{1}{6}$$

Solve this equation for x in terms of y.

$$x = \frac{1}{2}y + \frac{1}{6} = \frac{3y+1}{6}$$

The solution is $\left(\frac{3y+1}{6}, y\right)$, where y is any real number.

25. $y = x - 3$
$y = 1 + z$
$z = 4 - x$

First write the system in standard form.
$-x + y = -3$
$y - z = 1$
$x + z = 4$

Write the augmented matrix and use row operations.

$$\left[\begin{array}{ccc|c} -1 & 1 & 0 & -3 \\ 0 & 1 & -1 & 1 \\ 1 & 0 & 1 & 4 \end{array}\right]$$

$$-1R_1 \to R_1 \left[\begin{array}{ccc|c} 1 & -1 & 0 & 3 \\ 0 & 1 & -1 & 1 \\ 1 & 0 & 1 & 4 \end{array}\right]$$

$$-1R_1 + R_3 \to R_3 \left[\begin{array}{ccc|c} 1 & -1 & 0 & 3 \\ 0 & 1 & -1 & 1 \\ 0 & 1 & 1 & 1 \end{array}\right]$$

$$\begin{array}{r} R_2 + R_1 \to R_1 \\ \\ -1R_2 + R_3 \to R_3 \end{array} \left[\begin{array}{ccc|c} 1 & 0 & -1 & 4 \\ 0 & 1 & -1 & 1 \\ 0 & 0 & 2 & 0 \end{array}\right]$$

$$\begin{array}{r} R_3 + 2R_1 \to R_1 \\ R_3 + 2R_2 \to R_2 \\ \\ \end{array} \left[\begin{array}{ccc|c} 2 & 0 & 0 & 8 \\ 0 & 2 & 0 & 2 \\ 0 & 0 & 2 & 0 \end{array}\right]$$

$$\begin{array}{r} \frac{1}{2}R_1 \to R_1 \\ \frac{1}{2}R_2 \to R_2 \\ \frac{1}{2}R_3 \to R_3 \end{array} \left[\begin{array}{ccc|c} 1 & 0 & 0 & 4 \\ 0 & 1 & 0 & 1 \\ 0 & 0 & 1 & 0 \end{array}\right]$$

The solution is (4, 1, 0).

27. $2x-2y=-5$
$2y+z=0$
$2x+z=-7$

Write the augmented matrix and use row operations.

$$\left[\begin{array}{ccc|c} 2 & -2 & 0 & -5 \\ 0 & 2 & 1 & 0 \\ 2 & 0 & 1 & -7 \end{array}\right]$$

$$\begin{array}{r} \\ \\ -1R_1+R_3\to R_3 \end{array}\left[\begin{array}{ccc|c} 2 & -2 & 0 & -5 \\ 0 & 2 & 1 & 0 \\ 0 & 2 & 1 & -2 \end{array}\right]$$

$$\begin{array}{r} R_2+R_1\to R_1 \\ \\ -1R_2+R_3\to R_3 \end{array}\left[\begin{array}{ccc|c} 2 & 0 & 1 & -5 \\ 0 & 2 & 1 & 0 \\ 0 & 0 & 0 & -2 \end{array}\right]$$

This matrix corresponds to the system of equations
$2x+z=-5$
$2y+z=0$
$0=-2.$

This false statement $0=-2$ indicates that the system is inconsistent and therefore has no solution.

29. $4x+4y-4z=24$
$2x-y+z=-9$
$x-2y+3z=1$

Write the augmented matrix and use row operations.

$$\left[\begin{array}{ccc|c} 4 & 4 & -4 & 24 \\ 2 & -1 & 1 & -9 \\ 1 & -2 & 3 & 1 \end{array}\right]$$

$$\begin{array}{r} \\ R_1+(-2)R_2\to R_2 \\ R_1+(-4)R_3\to R_3 \end{array}\left[\begin{array}{ccc|c} 4 & 4 & -4 & 24 \\ 0 & 6 & -6 & 42 \\ 0 & 12 & -16 & 20 \end{array}\right]$$

$$\begin{array}{r} 2R_2+(-3)R_1\to R_1 \\ \\ -2R_2+R_3\to R_3 \end{array}\left[\begin{array}{ccc|c} -12 & 0 & 0 & 12 \\ 0 & 6 & -6 & 42 \\ 0 & 0 & -4 & -64 \end{array}\right]$$

$$\begin{array}{r} \\ -3R_3+2R_2\to R_2 \\ \\ \end{array}\left[\begin{array}{ccc|c} -12 & 0 & 0 & 12 \\ 0 & 12 & 0 & 276 \\ 0 & 0 & -4 & -64 \end{array}\right]$$

$$\begin{array}{r} -\frac{1}{12}R_1\to R_1 \\ \frac{1}{12}R_2\to R_2 \\ -\frac{1}{4}R_3\to R_3 \end{array}\left[\begin{array}{ccc|c} 1 & 0 & 0 & -1 \\ 0 & 1 & 0 & 23 \\ 0 & 0 & 1 & 16 \end{array}\right]$$

The solution is $(-1, 23, 16)$.

31. $3x+5y-z=0$
$4x-y+2z=1$
$7x+4y+z=1$

Write the augmented matrix and use row operations.

$$\left[\begin{array}{ccc|c} 3 & 5 & -1 & 0 \\ 4 & -1 & 2 & 1 \\ 7 & 4 & 1 & 1 \end{array}\right]$$

$$\begin{array}{r} \\ 4R_1+(-3)R_2\to R_2 \\ 7R_1+(-3)R_3\to R_3 \end{array}\left[\begin{array}{ccc|c} 3 & 5 & -1 & 0 \\ 0 & 23 & -10 & -3 \\ 0 & 23 & -10 & -3 \end{array}\right]$$

$$\begin{array}{r} 23R_1+(-5)R_2\to R_1 \\ \\ R_2+(-1)R_3\to R_3 \end{array}\left[\begin{array}{ccc|c} 69 & 0 & 27 & 15 \\ 0 & 23 & -10 & -3 \\ 0 & 0 & 0 & 0 \end{array}\right]$$

$$\begin{array}{r} \frac{1}{69}R_1\to R_1 \\ \frac{1}{23}R_2\to R_2 \\ \\ \end{array}\left[\begin{array}{ccc|c} 1 & 0 & \frac{9}{23} & \frac{5}{23} \\ 0 & 1 & -\frac{10}{23} & -\frac{3}{23} \\ 0 & 0 & 0 & 0 \end{array}\right]$$

The row of zeros indicates dependent equations. Solve the first two equations respectively for x and y in terms of z to obtain

$x=-\frac{9}{23}z+\frac{5}{23}=\frac{-9z+5}{23}$ and

$y=\frac{10}{23}z-\frac{3}{23}=\frac{10z-3}{23}.$

The solution is $\left(\frac{-9z+5}{23}, \frac{10z-3}{23}, z\right)$.

33. $5x-4y+2z=6$
$5x+3y-z=11$
$15x-5y+3z=23$

Write the augmented matrix and use row operations.

$$\left[\begin{array}{ccc|c} 5 & -4 & 2 & 6 \\ 5 & 3 & -1 & 11 \\ 15 & -5 & 3 & 23 \end{array}\right]$$

$$\begin{array}{r} \\ -1R_1+R_2\to R_2 \\ -3R_1+R_3\to R_3 \end{array}\left[\begin{array}{ccc|c} 5 & -4 & 2 & 6 \\ 0 & 7 & -3 & 5 \\ 0 & 7 & -3 & 5 \end{array}\right]$$

$$\begin{array}{r} 4R_2+7R_1\to R_1 \\ \\ -1R_2+R_3\to R_3 \end{array}\left[\begin{array}{ccc|c} 35 & 0 & 2 & 62 \\ 0 & 7 & -3 & 5 \\ 0 & 0 & 0 & 0 \end{array}\right]$$

$$\begin{array}{r} \frac{1}{35}R_1\to R_1 \\ \frac{1}{7}R_2\to R_2 \\ \\ \end{array}\left[\begin{array}{ccc|c} 1 & 0 & \frac{2}{35} & \frac{62}{35} \\ 0 & 1 & -\frac{3}{7} & \frac{5}{7} \\ 0 & 0 & 0 & 0 \end{array}\right]$$

(continued on next page)

(continued)

The row of zeros indicates dependent equations. Solve the first two equations respectively for x and y in terms of z to obtain
$x = -\frac{2}{35}z + \frac{62}{35} = \frac{-2z+62}{35}$ and
$y = \frac{3}{7}z + \frac{5}{7} = \frac{3z+5}{7}$.

The solution is $\left(\frac{-2z+62}{35}, \frac{3z+5}{7}, z\right)$.

35.
$$\begin{aligned} 2x+3y+\ \ z &= 9 \\ 4x+6y+2z &= 18 \\ -\frac{1}{2}x-\frac{3}{4}y-\frac{1}{4}z &= -\frac{9}{4} \end{aligned}$$

Write the augmented matrix and use row operations.

$$\left[\begin{array}{ccc|c} 2 & 3 & 1 & 9 \\ 4 & 6 & 2 & 18 \\ -\frac{1}{2} & -\frac{3}{4} & -\frac{1}{4} & -\frac{9}{4} \end{array}\right]$$

$$\begin{array}{r} \\ -2R_1+R_2 \to R_2 \\ \frac{1}{4}R_1+R_3 \to R_3 \end{array} \left[\begin{array}{ccc|c} 2 & 3 & 1 & 9 \\ 0 & 0 & 0 & 0 \\ 0 & 0 & 0 & 0 \end{array}\right]$$

The rows of zeros indicate dependent equations. Since the equation involves x, y, and z, let y and z be parameters. Solve the equation given by the first row for x to obtain $x = \frac{9-3y-z}{2}$. The solution is

$\left(\frac{9-3y-z}{2}, y, z\right)$, where y and z are any real numbers.

37.
$$\begin{aligned} x+2y\qquad -w &= 3 \\ 2x+\qquad 4z+2w &= -6 \\ x+2y-\ z\qquad &= 6 \\ 2x-\ y+\ z+\ w &= -3 \end{aligned}$$

Write the augmented matrix and use row operations.

$$\left[\begin{array}{cccc|c} 1 & 2 & 0 & -1 & 3 \\ 2 & 0 & 4 & 2 & -6 \\ 1 & 2 & -1 & 0 & 6 \\ 2 & -1 & 1 & 1 & -3 \end{array}\right]$$

$$\begin{array}{r} \\ -2R_1+R_2 \to R_2 \\ -1R_1+R_3 \to R_3 \\ -2R_1+R_4 \to R_4 \end{array} \left[\begin{array}{cccc|c} 1 & 2 & 0 & -1 & 3 \\ 0 & -4 & 4 & 4 & -12 \\ 0 & 0 & -1 & 1 & 3 \\ 0 & -5 & 1 & 3 & -9 \end{array}\right]$$

$$\begin{array}{r} R_2+2R_1 \to R_1 \\ \\ \\ -5R_2+4R_4 \to R_4 \end{array} \left[\begin{array}{cccc|c} 2 & 0 & 4 & 2 & -6 \\ 0 & -4 & 4 & 4 & -12 \\ 0 & 0 & -1 & 1 & 3 \\ 0 & 0 & -16 & -8 & 24 \end{array}\right]$$

$$\begin{array}{r} 4R_3+R_1 \to R_1 \\ 4R_3+R_2 \to R_2 \\ \\ 16R_3+(-1)R_4 \to R_4 \end{array} \left[\begin{array}{cccc|c} 2 & 0 & 0 & 6 & 6 \\ 0 & -4 & 0 & 8 & 0 \\ 0 & 0 & -1 & 1 & 3 \\ 0 & 0 & 0 & 24 & 24 \end{array}\right]$$

$$\begin{array}{r} R_4+(-4R_1) \to R_1 \\ R_4+(-3R_2) \to R_2 \\ R_4+(-24R_3) \to R_3 \\ \\ \end{array} \left[\begin{array}{cccc|c} -8 & 0 & 0 & 0 & 0 \\ 0 & 12 & 0 & 0 & 24 \\ 0 & 0 & 24 & 0 & -48 \\ 0 & 0 & 0 & 24 & 24 \end{array}\right]$$

$$\begin{array}{r} -\frac{1}{8}R_1 \to R_1 \\ \frac{1}{12}R_2 \to R_2 \\ \frac{1}{24}R_3 \to R_3 \\ \frac{1}{24}R_4 \to R_4 \end{array} \left[\begin{array}{cccc|c} 1 & 0 & 0 & 0 & 0 \\ 0 & 1 & 0 & 0 & 2 \\ 0 & 0 & 1 & 0 & -2 \\ 0 & 0 & 0 & 1 & 1 \end{array}\right]$$

The solution is $x = 0$, $y = 2$, $z = -2$, $w = 1$, or $(0, 2, -2, 1)$.

39.
$$\begin{aligned} x+\ y-z+2w &= -20 \\ 2x-\ y+z+\ w &= 11 \\ 3x-2y+z-2w &= 27 \end{aligned}$$

Write the augmented matrix and use row operations.

$$\left[\begin{array}{cccc|c} 1 & 1 & -1 & 2 & -20 \\ 2 & -1 & 1 & 1 & 11 \\ 3 & -2 & 1 & -2 & 27 \end{array}\right]$$

$$\begin{array}{r} \\ -2R_1+R_2 \to R_2 \\ -3R_1+R_3 \to R_3 \end{array} \left[\begin{array}{cccc|c} 1 & 1 & -1 & 2 & -20 \\ 0 & -3 & 3 & -3 & 51 \\ 0 & -5 & 4 & -8 & 87 \end{array}\right]$$

$$\begin{array}{r} \\ -\frac{1}{3}R_2 \to R_2 \\ \\ \end{array} \left[\begin{array}{cccc|c} 1 & 1 & -1 & 2 & -20 \\ 0 & 1 & -1 & 1 & -17 \\ 0 & -5 & 4 & -8 & 87 \end{array}\right]$$

$$\begin{array}{r} -1R_2+R_1 \to R_1 \\ \\ 5R_2+R_3 \to R_3 \end{array} \left[\begin{array}{cccc|c} 1 & 0 & 0 & 1 & -3 \\ 0 & 1 & -1 & 1 & -17 \\ 0 & 0 & -1 & -3 & 2 \end{array}\right]$$

$$\begin{array}{r} \\ \\ -1R_3 \to R_3 \end{array} \left[\begin{array}{cccc|c} 1 & 0 & 0 & 1 & -3 \\ 0 & 1 & -1 & 1 & -17 \\ 0 & 0 & 1 & 3 & -2 \end{array}\right]$$

$$\begin{array}{r} \\ R_3+R_2 \to R_2 \\ \\ \end{array} \left[\begin{array}{cccc|c} 1 & 0 & 0 & 1 & -3 \\ 0 & 1 & 0 & 4 & -19 \\ 0 & 0 & 1 & 3 & -2 \end{array}\right]$$

(continued on next page)

(continued)

This is as far as we can go using row operations. To complete the solution, write the equations that correspond to the matrix.

$$x + w = -3$$
$$y + 4w = -19$$
$$z + 3w = -2$$

Let w be the parameter and express x, y, and z in terms of w. From the equations above, $x = -w - 3$, $y = -4w - 19$, and $z = -3w - 2$. The solution is $(-w - 3, -4w - 19, -3w - 2, w)$, where w is any real number.

41.
$$\begin{aligned} 10.47x + 3.52y + 2.58z - 6.42w &= 218.65 \\ 8.62x - 4.93y - 1.75z + 2.83w &= 157.03 \\ 4.92x + 6.83y - 2.97z + 2.65w &= 462.3 \\ 2.86x + 19.10y - 6.24z - 8.73w &= 398.4 \end{aligned}$$

Write the augmented matrix of the system.

$$\left[\begin{array}{cccc|c} 10.47 & 3.52 & 2.58 & -6.42 & 218.65 \\ 8.62 & -4.93 & -1.75 & 2.83 & 157.03 \\ 4.92 & 6.83 & -2.97 & 2.65 & 462.3 \\ 2.86 & 19.10 & -6.24 & -8.73 & 398.4 \end{array}\right]$$

This exercise should be solved by graphing calculator or computer methods. The solution, which may vary slightly, is $x \approx 28.9436$, $y \approx 36.6326$, $z \approx 9.6390$, and $w \approx 37.1036$, or $(28.9436, 36.6326, 9.6390, 37.1036)$.

43. Insert the given values, introduce variables, and the table is as follows.

$\frac{3}{8}$	a	b
c	d	$\frac{1}{4}$
e	f	g

From this, we obtain the following system of equations.

$$\begin{aligned} a + b + \tfrac{3}{8} &= 1 \\ c + d + \tfrac{1}{4} &= 1 \\ e + f + g &= 1 \\ c + e + \tfrac{3}{8} &= 1 \\ a + d + f &= 1 \\ b + g + \tfrac{1}{4} &= 1 \\ d + g + \tfrac{3}{8} &= 1 \\ b + d + e &= 1 \end{aligned}$$

The augmented matrix and the final form after row operations are as follows.

$$\left[\begin{array}{ccccccc|c} 1 & 1 & 0 & 0 & 0 & 0 & 0 & \frac{5}{8} \\ 0 & 0 & 1 & 1 & 0 & 0 & 0 & \frac{3}{4} \\ 0 & 0 & 0 & 0 & 1 & 1 & 1 & 1 \\ 0 & 0 & 1 & 0 & 1 & 0 & 0 & \frac{5}{8} \\ 1 & 0 & 0 & 1 & 0 & 1 & 0 & 1 \\ 0 & 1 & 0 & 0 & 0 & 0 & 1 & \frac{3}{4} \\ 0 & 0 & 0 & 1 & 0 & 0 & 1 & \frac{5}{8} \\ 0 & 1 & 0 & 1 & 1 & 0 & 0 & 1 \end{array}\right] \to \left[\begin{array}{ccccccc|c} 1 & 0 & 0 & 0 & 0 & 0 & 0 & \frac{1}{6} \\ 0 & 1 & 0 & 0 & 0 & 0 & 0 & \frac{11}{24} \\ 0 & 0 & 1 & 0 & 0 & 0 & 0 & \frac{5}{12} \\ 0 & 0 & 0 & 1 & 0 & 0 & 0 & \frac{1}{3} \\ 0 & 0 & 0 & 0 & 1 & 0 & 0 & \frac{5}{24} \\ 0 & 0 & 0 & 0 & 0 & 1 & 0 & \frac{1}{2} \\ 0 & 0 & 0 & 0 & 0 & 0 & 1 & \frac{7}{12} \\ 0 & 0 & 0 & 0 & 0 & 0 & 0 & 0 \end{array}\right]$$

(continued on next page)

(continued)

The solution to the system is read from the last column.
$a=\frac{1}{6}, b=\frac{11}{24}, c=\frac{5}{12}, d=\frac{1}{3}, e=\frac{5}{24}, f=\frac{1}{2},$ and $g=\frac{7}{24}$
So the magic square is:

$\frac{3}{8}$	$\frac{1}{6}$	$\frac{11}{24}$
$\frac{5}{12}$	$\frac{1}{3}$	$\frac{1}{4}$
$\frac{5}{24}$	$\frac{1}{2}$	$\frac{7}{24}$

45. Let x = the number of units of corn, let y = the number of units of soybeans, and let z = the number of units of cottonseed.
The system to be solved is

$$\begin{aligned}0.25x+0.4y+0.2z&=22\\0.4x+0.2y+0.3z&=28\\0.3x+0.2y+0.1z&=18\end{aligned}$$

The augmented matrix for this system is

$$\left[\begin{array}{ccc|c}0.25&0.4&0.2&22\\0.4&0.2&0.3&28\\0.3&0.2&0.1&18\end{array}\right].$$

Multiply each row by 100 to eliminate the decimals, then use row operations to find the solution.

$$\left[\begin{array}{ccc|c}25&40&20&2200\\40&20&30&2800\\30&20&10&1800\end{array}\right]$$

$$\begin{array}{r}\\8R_1-5R_2\to R_2\\-4R_3+3R_2\to R_3\end{array}\left[\begin{array}{ccc|c}25&40&20&2200\\0&220&10&3600\\0&-20&50&1200\end{array}\right]$$

$$\begin{array}{r}\\\\11R_3+R_2\to R_3\end{array}\left[\begin{array}{ccc|c}25&40&20&2200\\0&220&10&3600\\0&0&560&16800\end{array}\right]$$

$$\begin{array}{r}\\\\\frac{1}{560}R_3\to R_3\end{array}\left[\begin{array}{ccc|c}25&40&20&2200\\0&220&10&3600\\0&0&1&30\end{array}\right]$$

$$\begin{array}{r}-20R_3+R_1\to R_1\\-10R_3+R_2\to R_2\\\\\end{array}\left[\begin{array}{ccc|c}25&40&0&1600\\0&220&0&3300\\0&0&1&30\end{array}\right]$$

$$\begin{array}{r}\\\frac{1}{220}R_2\to R_2\\\\\end{array}\left[\begin{array}{ccc|c}25&40&0&1600\\0&1&0&15\\0&0&1&30\end{array}\right]$$

$$\frac{1}{25}(R_1-40R_2)\to R_1\left[\begin{array}{ccc|c}1&0&0&40\\0&1&0&15\\0&0&1&30\end{array}\right]$$

The solution is 40 units of corn, 15 units of soybeans,and 30 units of cottonseed.

47. Let x = the number of grams of group A, y = the number of grams of group B, and z = the number of grams of group C.

(a) The system to be solved is

$$\begin{aligned}x+y+z&=400\quad(1)\\x&=\frac{1}{3}y\quad(2)\\x+z&=2y\quad(3)\end{aligned}$$

Rewrite equations (2) and (3) in standard form and multiply both sides of equation (2) by 3 to eliminate the fraction.

$$\begin{aligned}x+\ y+z&=400\\3x-\ y\qquad&=0\\x-2y+z&=0\end{aligned}$$

Write the augmented matrix.

$$\left[\begin{array}{ccc|c}1&1&1&400\\3&-1&0&0\\1&-2&1&0\end{array}\right]$$

$$\begin{array}{r}\\-3R_1+R_2\to R_2\\-1R_1+R_3\to R_3\end{array}\left[\begin{array}{ccc|c}1&1&1&400\\0&-4&-3&-1200\\0&-3&0&-400\end{array}\right]$$

$$\begin{array}{r}\\\\-\frac{1}{3}R_3\to R_3\end{array}\left[\begin{array}{ccc|c}1&1&1&400\\0&-4&-3&-1200\\0&1&0&\frac{400}{3}\end{array}\right]$$

(continued on next page)

(continued)

Interchange rows 2 and 3.

$$\left[\begin{array}{ccc|c} 1 & 1 & 1 & 400 \\ 0 & 1 & 0 & \frac{400}{3} \\ 0 & -4 & -3 & -1200 \end{array}\right]$$

$$\begin{array}{l} -1R_2+R_1 \to R_1 \\ \\ 4R_2+R_3 \to R_3 \end{array} \left[\begin{array}{ccc|c} 1 & 0 & 1 & \frac{800}{3} \\ 0 & 1 & 0 & \frac{400}{3} \\ 0 & 0 & -3 & -\frac{2000}{3} \end{array}\right]$$

$$-\tfrac{1}{3}R_3 \to R_3 \left[\begin{array}{ccc|c} 1 & 0 & 1 & \frac{800}{3} \\ 0 & 1 & 0 & \frac{400}{3} \\ 0 & 0 & 1 & \frac{2000}{9} \end{array}\right]$$

$$-1R_3+R_1 \to R_1 \left[\begin{array}{ccc|c} 1 & 0 & 0 & \frac{400}{9} \\ 0 & 1 & 0 & \frac{400}{3} \\ 0 & 0 & 1 & \frac{2000}{9} \end{array}\right]$$

The solution is $\left(\frac{400}{9}, \frac{400}{3}, \frac{2000}{9}\right)$.

Include $\frac{400}{9}$ g of group A, $\frac{400}{3}$ g of group B, and $\frac{2000}{9}$ g of group C.

(b) If the requirement that the diet include one-third as much of A as of B is dropped, refer to the first two rows of the fifth augmented matrix in part (a).

$$\left[\begin{array}{ccc|c} 1 & 0 & 1 & \frac{800}{3} \\ 0 & 1 & 0 & \frac{400}{3} \end{array}\right]$$

This gives

$$x = \frac{800}{3} - z$$
$$y = \frac{400}{3}.$$

Therefore, for any positive number z of grams of group C, there should be z grams less than $\frac{800}{3}$ g of group A and $\frac{400}{3}$ g of group B.

(c) Since there was a unique solution for the original problem, by adding an additional condition. the only possible solution would be the one from part (a) However, by substituting those values of A, B, and C for x, y, and z in the equation for the additional condition, $0.02x+0.02y+0.03z=8.00$. the values do not work. Therefore, a solution is not possible.

49. Let x = the number of species A, y = the number of species B, and z = the number of species C.
Use a chart to organize the information.

		Species			
		A	B	C	Totals
Food	I	1.32	2.1	0.86	490
	II	2.9	0.95	1.52	897
	III	1.75	0.6	2.01	653

The system to be solved is

$$\begin{aligned} 1.32x + 2.1y + 0.86z &= 490 \\ 2.9x + 0.95y + 1.52z &= 897 \\ 1.75x + 0.6y + 2.01z &= 653. \end{aligned}$$

Use graphing calculator or computer methods to solve this system. The solution, which may vary slightly, is to stock about 244 fish of species A, 39 fish of species B, and 101 fish of species C.

51. Let x = the number of kilograms of the first chemical, y = the number of kilograms of the second chemical, and z = the number of kilograms of the third chemical.
The system to be solved is

$$\begin{aligned} x+y+z &= 750 \\ x &= 0.108(750) \\ \frac{y}{z} &= \frac{4}{3} \end{aligned} \quad \text{or} \quad \begin{aligned} x+y+z &= 750 \\ x &= 81 \\ 3y-4z &= 0 \end{aligned}$$

Use graphing calculator or computer methods to solve this system. The solution, which may vary slightly, is to use 81 kg of the first chemical, about 382.286 kg of the second chemical, and about 286.714 kg of the third chemical.

53. (a) Bulls: The number of white ones was one half plus one third the number of black greater than the brown.

$$X = \left(\frac{1}{2}+\frac{1}{3}\right)Y + T$$
$$X = \frac{5}{6}Y + T$$
$$6X = 5Y + 6T$$
$$6X - 5Y = 6T$$

The number of the black, one quarter plus one fifth the number of the spotted greater than the brown.

(continued on next page)

(*continued*)

$$Y = \left(\frac{1}{4} + \frac{1}{5}\right)Z + T$$
$$Y = \frac{9}{20}Z + T$$
$$20Y = 9Z + 20T$$
$$20Y - 9Z = 20T$$

The number of the spotted, one sixth and one seventh the number of the white greater than the brown.

$$Z = \left(\frac{1}{6} + \frac{1}{7}\right)X + T$$
$$Z = \frac{13}{42}X + T$$
$$42Z = 13X + 42T$$
$$42Z - 13X = 42T$$

So the system of equations for the bulls is

$$6X - 5Y = 6T$$
$$20Y - 9Z = 20T$$
$$42Z - 13X = 42T.$$

Cows: The number of white ones was one third plus one quarter of the total black cattle.

$$x = \left(\frac{1}{3} + \frac{1}{4}\right)(Y + y)$$
$$x = \frac{7}{12}(Y + y)$$
$$12x = 7Y + 7y$$
$$12x - 7y = 7Y$$

The number of the black, one quarter plus one fifth the total of the spotted cattle.

$$y = \left(\frac{1}{4} + \frac{1}{5}\right)(Z + z)$$
$$y = \frac{9}{20}(Z + z)$$
$$20y = 9Z + 9z$$
$$20y - 9z = 9Z$$

The number of the spotted, one fifth plus one sixth the total of the brown cattle.

$$z = \left(\frac{1}{5} + \frac{1}{6}\right)(T + t)$$
$$z = \frac{11}{30}(T + t)$$
$$30z = 11T + 11t$$
$$30z - 11t = 11T$$

The number of the brown, one sixth plus one seventh the total of the white cattle.

$$t = \left(\frac{1}{6} + \frac{1}{7}\right)(X + x)$$
$$t = \frac{13}{42}(X + x)$$
$$42t = 13X + 13x$$
$$42t - 13x = 13X$$

So the system of equations for the cows is

$$12x - 7y = 7Y$$
$$20y - 9z = 9Z$$
$$30z - 11t = 11T$$
$$-13x + 42t = 13X$$

(b) For $T = 4,149,387$, the 3×3 system to be solved is

$$\begin{aligned} 6X - 5Y \quad\quad &= 24,896,322 \\ 20Y - 9Z &= 82,987,740 \\ -13X \quad + 42Z &= 174,274,254 \end{aligned}$$

Write the augmented matrix of the system.

$$\left[\begin{array}{ccc|c} 6 & -5 & 0 & 24,896,322 \\ 0 & 20 & -9 & 82,987,740 \\ -13 & 0 & 42 & 174,274,254 \end{array}\right]$$

This exercise should be solved by graphing calculator or computer methods. The solution is $X = 10,366,482$ white bulls, $Y = 7,460,514$ black bulls, and $Z = 7,358,060$ spotted bulls.

For $X = 10,366,482$, $Y = 7,460,514$, and $Z = 7,358,060$, the 4×4 system to be solved is

$$12x - 7y = 52,223,598$$
$$20y - 9z = 66,222,540$$
$$30z - 11t = 45,643,257$$
$$-13x + 42t = 134,764,266$$

Write the augmented matrix of the system.

$$\left[\begin{array}{cccc|c} 12 & -7 & 0 & 0 & 52,223,598 \\ 0 & 20 & -9 & 0 & 66,222,540 \\ 0 & 0 & 30 & -11 & 45,643,257 \\ -13 & 0 & 0 & 42 & 134,764,266 \end{array}\right]$$

This exercise should be solved by graphing calculator or computer methods. The solution is x = 7,206,360 white cows, y = 4,893,256 black cows, z = 3,515,820 spotted cows, and t = 5,439,213 brown cows.

55. Let x = the number of hours to hire the Garcia firm, and y = the number of hours to hire the Wong firm. The system to be solved is

$10x+20y=500$ (1)
$30x+10y=750$ (2)
$5x+10y=250$ (3)

Write the augmented matrix and use row operations.

$$\left[\begin{array}{cc|c} 10 & 20 & 500 \\ 30 & 10 & 750 \\ 5 & 10 & 250 \end{array}\right]$$

$$\begin{array}{l} \frac{1}{10}R_1 \to R_1 \\ \frac{1}{10}R_2 \to R_2 \\ \frac{1}{5}R_3 \to R_3 \end{array} \left[\begin{array}{cc|c} 1 & 2 & 50 \\ 3 & 1 & 75 \\ 1 & 2 & 50 \end{array}\right]$$

$$\begin{array}{l} -3R_1 + R_2 \to R_2 \\ -1R_1 + R_3 \to R_3 \end{array} \left[\begin{array}{cc|c} 1 & 2 & 50 \\ 0 & -5 & -75 \\ 0 & 0 & 0 \end{array}\right]$$

$$-\frac{1}{5}R_2 \to R_2 \left[\begin{array}{cc|c} 1 & 2 & 50 \\ 0 & 1 & 15 \\ 0 & 0 & 0 \end{array}\right]$$

$$-2R_2 + R_1 \to R_1 \left[\begin{array}{cc|c} 1 & 0 & 20 \\ 0 & 1 & 15 \\ 0 & 0 & 0 \end{array}\right]$$

The solution is (20, 15). The Garcia firm should be hired for 20 hr and the Wong firm for 15 hr.

57. (a) Let x = the number of deluxe models
y = the number of super-deluxe models
z = the number of ultra models.
Make a table to organize the information.

	Deluxe	Super-Deluxe	Ultra	Totals
Electronic	2	1	2	54
Assembly	3	3	2	72
Finishing	5	2	6	148

We want to solve the following system.

$2x+\ \ y+2z=54$
$3x+3y+2z=72$
$5x+2y+6z=148$

Write the augmented matrix and transform the matrix.

$$\left[\begin{array}{ccc|c} 2 & 1 & 2 & 54 \\ 3 & 3 & 2 & 72 \\ 5 & 2 & 6 & 148 \end{array}\right]$$

$$\frac{1}{2}R_1 \to R_1 \left[\begin{array}{ccc|c} 1 & \frac{1}{2} & 1 & 27 \\ 3 & 3 & 2 & 72 \\ 5 & 2 & 6 & 148 \end{array}\right]$$

$$\begin{array}{l} -3R_1 + R_2 \to R_2 \\ -5R_1 + R_3 \to R_3 \end{array} \left[\begin{array}{ccc|c} 1 & \frac{1}{2} & 1 & 27 \\ 0 & \frac{3}{2} & -1 & -9 \\ 0 & -\frac{1}{2} & 1 & 13 \end{array}\right]$$

$$\frac{2}{3}R_2 \to R_2 \left[\begin{array}{ccc|c} 1 & \frac{1}{2} & 1 & 27 \\ 0 & 1 & -\frac{2}{3} & -6 \\ 0 & -\frac{1}{2} & 1 & 13 \end{array}\right]$$

$$\begin{array}{l} -\frac{1}{2}R_2 + R_1 \to R_1 \\ \frac{1}{2}R_2 + R_3 \to R_3 \end{array} \left[\begin{array}{ccc|c} 1 & 0 & \frac{4}{3} & 30 \\ 0 & 1 & -\frac{2}{3} & -6 \\ 0 & 0 & \frac{2}{3} & 10 \end{array}\right]$$

$$\frac{3}{2}R_3 \to R_3 \left[\begin{array}{ccc|c} 1 & 0 & \frac{4}{3} & 30 \\ 0 & 1 & -\frac{2}{3} & -6 \\ 0 & 0 & 1 & 15 \end{array}\right]$$

$$\begin{array}{l} -\frac{4}{3}R_3 + R_1 \to R_1 \\ \frac{2}{3}R_3 + R_2 \to R_2 \end{array} \left[\begin{array}{ccc|c} 1 & 0 & 0 & 10 \\ 0 & 1 & 0 & 4 \\ 0 & 0 & 1 & 15 \end{array}\right]$$

The solution is (10, 4, 15). Each week 10 deluxe models, 4 super-deluxe models, and 15 ultra-models should be produced.

(b) We want to solve the following system.

$2x+\ \ y+2z=54$
$3x+3y+2z=72$
$5x+\ \ y\ +6z=148$

Write the augmented matrix and transform the matrix.

$$\left[\begin{array}{ccc|c} 2 & 1 & 2 & 54 \\ 3 & 3 & 2 & 72 \\ 5 & 1 & 6 & 148 \end{array}\right]$$

$$\frac{1}{2}R_1 \to R_1 \left[\begin{array}{ccc|c} 1 & \frac{1}{2} & 1 & 27 \\ 3 & 3 & 2 & 72 \\ 5 & 1 & 6 & 148 \end{array}\right]$$

(continued on next page)

(continued)

$$\begin{array}{r} \\ -3R_1+R_2\to R_2 \\ -5R_1+R_3\to R_3 \end{array} \left[\begin{array}{ccc|c} 1 & \frac{1}{2} & 1 & 27 \\ 0 & \frac{3}{2} & -1 & -9 \\ 0 & -\frac{3}{2} & 1 & 13 \end{array}\right]$$

$$\frac{2}{3}R_2\to R_2 \left[\begin{array}{ccc|c} 1 & \frac{1}{2} & 1 & 27 \\ 0 & 1 & -\frac{2}{3} & -6 \\ 0 & -\frac{3}{2} & 1 & 13 \end{array}\right]$$

$$\begin{array}{r} -\frac{1}{2}R_2+R_1\to R_1 \\ \\ \frac{3}{2}R_2+R_3\to R_3 \end{array} \left[\begin{array}{ccc|c} 1 & 0 & \frac{4}{3} & 30 \\ 0 & 1 & -\frac{2}{3} & -6 \\ 0 & 0 & 0 & 4 \end{array}\right]$$

The last row indicates that there is no solution to the system.

(c) We want to solve the following system.

$$\begin{aligned} 2x+\ y+2z &= 54 \\ 3x+3y+2z &= 72 \\ 5x+\ y+6z &= 144 \end{aligned}$$

Write the augmented matrix and transform the matrix.

$$\left[\begin{array}{ccc|c} 2 & 1 & 2 & 54 \\ 3 & 3 & 2 & 72 \\ 5 & 1 & 6 & 144 \end{array}\right]$$

$$\frac{1}{2}R_1\to R_1 \left[\begin{array}{ccc|c} 1 & \frac{1}{2} & 1 & 27 \\ 3 & 3 & 2 & 72 \\ 5 & 1 & 6 & 144 \end{array}\right]$$

$$\begin{array}{r} \\ -3R_1+R_2\to R_2 \\ -5R_1+R_3\to R_3 \end{array} \left[\begin{array}{ccc|c} 1 & \frac{1}{2} & 1 & 27 \\ 0 & \frac{3}{2} & -1 & -9 \\ 0 & -\frac{3}{2} & 1 & 9 \end{array}\right]$$

$$\frac{2}{3}R_2\to R_2 \left[\begin{array}{ccc|c} 1 & \frac{1}{2} & 1 & 27 \\ 0 & 1 & -\frac{2}{3} & -6 \\ 0 & -\frac{3}{2} & 1 & 9 \end{array}\right]$$

$$\begin{array}{r} -\frac{1}{2}R_2+R_1\to R_1 \\ \\ \frac{3}{2}R_2+R_3\to R_3 \end{array} \left[\begin{array}{ccc|c} 1 & 0 & \frac{4}{3} & 30 \\ 0 & 1 & -\frac{2}{3} & -6 \\ 0 & 0 & 0 & 0 \end{array}\right]$$

The system is dependent. Let z be the parameter and solve the first two equations for x and y, which gives

$$x=30-\frac{4}{3}z \text{ and } y=\frac{2}{3}z-6.$$

Since x, y, and z must be nonnegative integers, we have

$$30-\frac{4}{3}z\ge 0\Rightarrow 30\ge\frac{4}{3}z\Rightarrow 22.5\le z$$

$$\frac{2}{3}z-6\ge 0\Rightarrow\frac{2}{3}z\ge 6\Rightarrow z\ge 9$$

Thus, z is an integer such that $9\le z\le 22.5$, and z must be a multiple of 3 so that x and y are integers. The permissible values of z are 9, 12, 15, 18, and 21. There are 5 solutions.

59. Let x = the amount borrowed at 8%, y = the amount borrowed at 9%, and z = the amount borrowed at 10%.

(a) The system to be solved is

$$\begin{aligned} x+y+z &= 25{,}000 \\ 0.08x+0.09y+0.10z &= 2190 \\ y &= z+1000 \end{aligned}$$

Multiply the second equation by 100 and rewrite the equations in standard form.

$$\begin{aligned} x+\ y+\ \ z &= 25{,}000 \\ 8x+9y+10z &= 219{,}000 \\ y-\ \ z &= 1000 \end{aligned}$$

Write the augmented matrix and use row operations.

$$\left[\begin{array}{ccc|c} 1 & 1 & 1 & 25{,}000 \\ 8 & 9 & 10 & 219{,}000 \\ 0 & 1 & -1 & 1000 \end{array}\right]$$

$$-8R_1+R_2\to R_2 \left[\begin{array}{ccc|c} 1 & 1 & 1 & 25{,}000 \\ 0 & 1 & 2 & 19{,}000 \\ 0 & 1 & -1 & 1000 \end{array}\right]$$

$$\begin{array}{r} -1R_2+R_1\to R_1 \\ \\ -1R_2+R_3\to R_3 \end{array} \left[\begin{array}{ccc|c} 1 & 0 & -1 & 6000 \\ 0 & 1 & 2 & 19{,}000 \\ 0 & 0 & -3 & -18{,}000 \end{array}\right]$$

$$\begin{array}{r} -1R_3+3R_1\to R_1 \\ 2R_3+3R_2\to R_2 \\ \\ \end{array} \left[\begin{array}{ccc|c} 3 & 0 & 0 & 36{,}000 \\ 0 & 3 & 0 & 21{,}000 \\ 0 & 0 & -3 & -18{,}000 \end{array}\right]$$

$$\begin{array}{r} \frac{1}{3}R_1\to R_1 \\ \frac{1}{3}R_2\to R_2 \\ -\frac{1}{3}R_3\to R_3 \end{array} \left[\begin{array}{ccc|c} 1 & 0 & 0 & 12{,}000 \\ 0 & 1 & 0 & 7000 \\ 0 & 0 & 1 & 6000 \end{array}\right]$$

The solution is (12,000, 7000, 6000). The company borrowed \$12,000 at 8%, \$7000 at 9%, and \$6000 at 10%.

(b) If the condition is dropped, the initial augmented matrix and solution is found as before.

$$\left[\begin{array}{ccc|c} 1 & 1 & 1 & 25{,}000 \\ 8 & 9 & 10 & 219{,}000 \end{array}\right]$$

$$-8R_1 + R_2 \rightarrow R_2 \left[\begin{array}{ccc|c} 1 & 1 & 1 & 25{,}000 \\ 0 & 1 & 2 & 19{,}000 \end{array}\right]$$

$$-1R_2 + R_1 \rightarrow R_1 \left[\begin{array}{ccc|c} 1 & 0 & -1 & 6000 \\ 0 & 1 & 2 & 19{,}000 \end{array}\right]$$

This gives the system of equations $\begin{aligned} x &= z + 6000 \\ y &= -2x + 19{,}000 \end{aligned}$

Since all values must be nonnegative, $z + 6000 \geq 0 \Rightarrow z \geq -6000$ and $-2z + 19{,}000 \geq 0 \Rightarrow z \leq 9500$. The second inequality produces the condition that the amount borrowed at 10% must be less than or equal to \$9500. If \$5000 is borrowed at 10%, $z = 5000$, and

$x = 5000 + 6000 = 11{,}000$

$y = -2(5000) + 19{,}000 = 9000.$

This means \$11,000 is borrowed at 8% and \$9000 is borrowed at 9%.

(c) The original conditions resulted in \$12,000 borrowed at 8%. So, if the bank sets a maximum of \$10,000 at the 8% rate, no solution is possible.

(d) The total interest would be $0.08(10{,}000) + 0.09(8000) + 0.1(7000) = 800 + 720 + 700 = 2220$ or \$2220, which is not the \$2190 interest as specified as one of the conditions of the problem.

61. (a) The system to be solved is

$43{,}500x - y = 1{,}295{,}000 \quad (1)$

$27{,}000x - y = 440{,}000 \quad (2)$

Write augmented matrix and use row operations.

$$\left[\begin{array}{cc|c} 43{,}500 & -1 & 1{,}295{,}000 \\ 27{,}000 & -1 & 440{,}000 \end{array}\right]$$

$$-\frac{27{,}000}{43{,}500}R_1 + R_2 \rightarrow R_2 \left[\begin{array}{cc|c} 43500 & -1 & 1{,}295{,}000 \\ 0 & -\frac{11}{29} & -\frac{10{,}550{,}000}{29} \end{array}\right]$$

$$-\frac{29}{11}R_1 + R_2 \rightarrow R_2 \left[\begin{array}{cc|c} 43.500 & -1 & 1{,}295{,}000 \\ 0 & 1 & \frac{10{,}550{,}000}{11} \end{array}\right]$$

$$R_1 + R_2 \rightarrow R_1 \left[\begin{array}{cc|c} 43500 & 0 & \frac{24{,}795{,}000}{11} \\ 0 & 1 & \frac{10{,}550{,}000}{11} \end{array}\right]$$

$$\frac{1}{43500}R_1 \rightarrow R_1 \left[\begin{array}{cc|c} 1 & 0 & \frac{570}{11} \\ 0 & 1 & \frac{10{,}550{,}000}{11} \end{array}\right]$$

The profit/loss will be equal after $\dfrac{570}{11}$ weeks or about 51.8 weeks. At that point, the profit will be $\dfrac{10{,}550{,}000}{11}$ or about \$959,091.

(b) If the show lasts longer than 51.8 weeks, Broadway is a more profitable venue. If it lasts less than 51.8 weeks, off Broadway is a more profitable venue.

63. Let x = number of field goals, and let y = number of foul shots.
Then

$$x+y=64$$
$$2x+y=100$$

Write the augmented matrix and use row operations to transform the matrix.

$$\left[\begin{array}{cc|c} 1 & 1 & 64 \\ 2 & 1 & 100 \end{array}\right]$$

$$\begin{array}{r} \\ -2R_1+R_2 \to R_2 \end{array} \left[\begin{array}{cc|c} 1 & 1 & 64 \\ 0 & -1 & -28 \end{array}\right]$$

$$\begin{array}{r} R_1+R_2 \to R_1 \\ \\ \end{array} \left[\begin{array}{cc|c} 1 & 0 & 36 \\ 0 & -1 & -28 \end{array}\right]$$

$$\begin{array}{r} \\ -R_2 \to R_2 \end{array} \left[\begin{array}{cc|c} 1 & 0 & 36 \\ 0 & 1 & 28 \end{array}\right]$$

Wilt Chamberlain made 36 field goals and 28 foul shots.

65. Let x = the number of singles, y = the number of doubles, z = the number of triples, and w = the number of home runs hit by Ichiro Suzuki. The system to be solved is

$$x+y+z+w=262$$
$$z=w-3$$
$$y=3w$$
$$x=45z$$

Write the equations in standard form, obtain the augmented matrix, and use row operations to solve.

$$\left[\begin{array}{cccc|c} 1 & 1 & 1 & 1 & 262 \\ 0 & 0 & 1 & -1 & -3 \\ 0 & 1 & 0 & -3 & 0 \\ 1 & 0 & -45 & 0 & 0 \end{array}\right]$$

$$\begin{array}{r} \\ \\ \\ -1R_1+R_4 \to R_4 \end{array} \left[\begin{array}{cccc|c} 1 & 1 & 1 & 1 & 262 \\ 0 & 0 & 1 & -1 & -3 \\ 0 & 1 & 0 & -3 & 0 \\ 0 & -1 & -46 & -1 & -262 \end{array}\right]$$

$$\begin{array}{r} \\ \\ R_2 \leftrightarrow R_3 \\ \\ \end{array} \left[\begin{array}{cccc|c} 1 & 1 & 1 & 1 & 262 \\ 0 & 1 & 0 & -3 & 0 \\ 0 & 0 & 1 & -1 & -3 \\ 0 & -1 & -46 & -1 & -262 \end{array}\right]$$

$$\begin{array}{r} -1R_2+R_1 \to R_1 \\ \\ \\ R_2+R_4 \to R_4 \end{array} \left[\begin{array}{cccc|c} 1 & 0 & 1 & 4 & 262 \\ 0 & 1 & 0 & -3 & 0 \\ 0 & 0 & 1 & -1 & -3 \\ 0 & 0 & -46 & -4 & -262 \end{array}\right]$$

$$\begin{array}{r} -1R_3+R_1 \to R_1 \\ \\ \\ 46R_3+R_4 \to R_4 \end{array} \left[\begin{array}{cccc|c} 1 & 0 & 0 & 5 & 265 \\ 0 & 1 & 0 & -3 & 0 \\ 0 & 0 & 1 & -1 & -3 \\ 0 & 0 & 0 & -50 & -400 \end{array}\right]$$

$$\begin{array}{r} R_4+10R_1 \to R_1 \\ -3R_4+50R_2 \to R_2 \\ -1R_4+50R_3 \to R_3 \\ \\ \end{array} \left[\begin{array}{cccc|c} 10 & 0 & 0 & 0 & 2250 \\ 0 & 50 & 0 & 0 & 1200 \\ 0 & 0 & 50 & 0 & 250 \\ 0 & 0 & 0 & -50 & -400 \end{array}\right]$$

$$\begin{array}{r} \frac{1}{10}R_1 \to R_1 \\ \frac{1}{50}R_2 \to R_2 \\ \frac{1}{50}R_3 \to R_3 \\ -\frac{1}{50}R_4 \to R_4 \end{array} \left[\begin{array}{cccc|c} 1 & 0 & 0 & 0 & 225 \\ 0 & 1 & 0 & 0 & 24 \\ 0 & 0 & 1 & 0 & 5 \\ 0 & 0 & 0 & 1 & 8 \end{array}\right]$$

Ichiro Suzuki hit 225 singles, 24 doubles, 5 triples, and 8 home runs during the 2004 season.

67. (a) $5.4 = a(8)^2 + b(8) + c$
$5.4 = 64a + 8b + c$
$6.3 = a(13)^2 + b(13) + c$
$6.3 = 169a + 13b + c$
$5.6 = a(18)^2 + b(18) + c$
$5.6 = 324a + 18b + c$

The linear system to be solved is

$$64a+8b+c=5.4$$
$$169a+13b+c=6.3$$
$$324a+18b+c=5.6.$$

Use a graphing calculator or computer methods to solve this system. The solution is $a = -0.032$, $b = 0.852$, and $c = 0.632$. Thus, the equation is

$$y = -0.032x^2 + 0.852x + 0.632.$$

(b)

QuadReg
y=ax²+bx+c
a=-.032
b=.852
c=.632
R²=1

The answer obtained using Gauss-Jordan elimination is the same as the answer obtained using the quadratic regression feature on a graphing calculator.

69. (a) The other two equations are

$$x_2 + x_3 = 700$$
$$x_3 + x_4 = 600.$$

(b) The augmented matrix is

$$\left[\begin{array}{cccc|c} 1 & 0 & 0 & 1 & 1000 \\ 1 & 1 & 0 & 0 & 1100 \\ 0 & 1 & 1 & 0 & 700 \\ 0 & 0 & 1 & 1 & 600 \end{array}\right].$$

$$-1R_1 + R_2 \to R_2 \left[\begin{array}{cccc|c} 1 & 0 & 0 & 1 & 1000 \\ 0 & 1 & 0 & -1 & 100 \\ 0 & 1 & 1 & 0 & 700 \\ 0 & 0 & 1 & 1 & 600 \end{array}\right]$$

$$-1R_2 + R_3 \to R_3 \left[\begin{array}{cccc|c} 1 & 0 & 0 & 1 & 1000 \\ 0 & 1 & 0 & -1 & 100 \\ 0 & 0 & 1 & 1 & 600 \\ 0 & 0 & 1 & 1 & 600 \end{array}\right]$$

$$-1R_3 + R_4 \to R_4 \left[\begin{array}{cccc|c} 1 & 0 & 0 & 1 & 1000 \\ 0 & 1 & 0 & -1 & 100 \\ 0 & 0 & 1 & 1 & 600 \\ 0 & 0 & 0 & 0 & 0 \end{array}\right]$$

Let x_4 be arbitrary. Solve the first three equations for x_1, x_2, and x_3.

$x_1 = 1000 - x_4$
$x_2 = 100 + x_4$
$x_3 = 600 - x_4$

The solution is
$(1000 - x_4,\ 100 + x_4,\ 600 - x_4,\ x_4)$.

(c) For x_4, we see that $x_4 \geq 0$ and $x_4 \leq 600$ since $600 - x_4$ must be nonnegative. Therefore, $0 \leq x_4 \leq 600$.

(d) x_1: If $x_4 = 0$, then $x_1 = 1000$.
If $x_4 = 600$, then
$x_1 = 1000 - 600 = 400$.
Therefore, $400 \leq x_1 \leq 1000$.

x_2: If $x_4 = 0$, then $x_2 = 100$.
If $x_4 = 600$, then
$x_2 = 100 + 600 = 700$.
Therefore, $100 \leq x_2 \leq 700$.

x_3: If $x_4 = 0$, then $x_3 = 600$.
If $x_4 = 600$, then
$x_3 = 600 - 600 = 0$.
Therefore, $0 \leq x_3 \leq 600$.

(e) If you know the number of cars entering or leaving three of the intersections, then the number entering or leaving the fourth is automatically determined because the number leaving must equal the number entering.

10.2 Addition and Subtraction of Matrices

1. $\begin{bmatrix} 1 & 3 \\ 5 & 7 \end{bmatrix} = \begin{bmatrix} 1 & 5 \\ 3 & 7 \end{bmatrix}$

This statement is false, since not all corresponding elements are equal.

3. $\begin{bmatrix} x \\ y \end{bmatrix} = \begin{bmatrix} -2 \\ 8 \end{bmatrix}$ if $x = -2$ and $y = 8$.

This statement is true. The matrices are the same size and corresponding elements are equal.

5. $\begin{bmatrix} 1 & 9 & -4 \\ 3 & 7 & 2 \\ -1 & 1 & 0 \end{bmatrix}$ is a square matrix.

This statement is true. The matrix has 3 rows and 3 columns.

7. $\begin{bmatrix} -4 & 8 \\ 2 & 3 \end{bmatrix}$ is a 2×2 square matrix.

9. $\begin{bmatrix} -6 & 8 & 0 & 0 \\ 4 & 1 & 9 & 2 \\ 3 & -5 & 7 & 1 \end{bmatrix}$ is a 3×4 matrix.

11. $\begin{bmatrix} -7 \\ 5 \end{bmatrix}$ is a 2×1 column matrix.

13. The sum of an $n \times m$ matrix and an $m \times n$ matrix, where $m \neq n$, is undefined.

15. $\begin{bmatrix} 3 & 4 \\ -8 & 1 \end{bmatrix} = \begin{bmatrix} 3 & x \\ y & z \end{bmatrix}$

Corresponding elements must be equal for the matrices to be equal. Therefore, $x = 4$, $y = -8$, and $z = 1$.

17. $\begin{bmatrix} s-4 & t+2 \\ -5 & 7 \end{bmatrix} = \begin{bmatrix} 6 & 2 \\ -5 & r \end{bmatrix}$

Corresponding elements must be equal.
$s - 4 = 6 \Rightarrow s = 10;\quad t + 2 = 2 \Rightarrow t = 0;\quad r = 7$

19. $\begin{bmatrix} a+2 & 3b & 4c \\ d & 7f & 8 \end{bmatrix} + \begin{bmatrix} -7 & 2b & 6 \\ -3d & -6 & -2 \end{bmatrix} = \begin{bmatrix} 15 & 25 & 6 \\ -8 & 1 & 6 \end{bmatrix}$

Add the two matrices on the left side to obtain

$$\begin{bmatrix} a+2 & 3b & 4c \\ d & 7f & 8 \end{bmatrix} + \begin{bmatrix} -7 & 2b & 6 \\ -3d & -6 & -2 \end{bmatrix} = \begin{bmatrix} (a+2)+(-7) & 3b+2b & 4c+6 \\ d+(-3d) & 7f+(-6) & 8+(-2) \end{bmatrix} = \begin{bmatrix} a-5 & 5b & 4c+6 \\ -2d & 7f-6 & 6 \end{bmatrix}$$

So, $\begin{bmatrix} a-5 & 5b & 4c+6 \\ -2d & 7f-6 & 6 \end{bmatrix} = \begin{bmatrix} 15 & 25 & 6 \\ -8 & 1 & 6 \end{bmatrix}$.

Corresponding elements of this matrix and the matrix on the right side of the original equation must be equal.

$a-5=15 \Rightarrow a=20;\ 5b=25 \Rightarrow b=5$

$4c+6=6 \Rightarrow c=0$

$-2d=-8 \Rightarrow d=4;\ 7f-6=1 \Rightarrow f=1$

Thus, $a = 20$, $b = 5$, $c = 0$, $d = 4$, $f = 1$.

21. $\begin{bmatrix} 2 & 4 & 5 & -7 \\ 6 & -3 & 12 & 0 \end{bmatrix} + \begin{bmatrix} 8 & 0 & -10 & 1 \\ -2 & 8 & -9 & 11 \end{bmatrix} = \begin{bmatrix} 2+8 & 4+0 & 5+(-10) & -7+1 \\ 6+(-2) & -3+8 & 12+(-9) & 0+11 \end{bmatrix} = \begin{bmatrix} 10 & 4 & -5 & -6 \\ 4 & 5 & 3 & 11 \end{bmatrix}$

23. $\begin{bmatrix} 1 & 3 & -2 \\ 4 & 7 & 1 \end{bmatrix} + \begin{bmatrix} 3 & 0 \\ 6 & 4 \\ -5 & 2 \end{bmatrix}$

These matrices cannot be added because the first matrix has size 2×3, while the second has size 3×2. Only matrices that are the same size can be added.

25. The matrices have the same size, so the subtraction can be done. Let A and B represent the given matrices. Using the definition of subtraction, we have

$$A-B=A+(-B)=\begin{bmatrix} 2 & 8 & 12 & 0 \\ 7 & 4 & -1 & 5 \\ 1 & 2 & 0 & 10 \end{bmatrix} + \begin{bmatrix} -1 & -3 & -6 & -9 \\ -2 & 3 & 3 & -4 \\ -8 & 0 & 2 & -17 \end{bmatrix} = \begin{bmatrix} 1 & 5 & 6 & -9 \\ 5 & 7 & 2 & 1 \\ -7 & 2 & 2 & -7 \end{bmatrix}$$

27. $\begin{bmatrix} 2 & 3 \\ -2 & 4 \end{bmatrix} + \begin{bmatrix} 4 & 3 \\ 7 & 8 \end{bmatrix} - \begin{bmatrix} 3 & 2 \\ 1 & 4 \end{bmatrix} = \begin{bmatrix} 2+4-3 & 3+3-2 \\ -2+7-1 & 4+8-4 \end{bmatrix} = \begin{bmatrix} 3 & 4 \\ 4 & 8 \end{bmatrix}$

29. $\begin{bmatrix} 2 & -1 \\ 0 & 13 \end{bmatrix} - \begin{bmatrix} 4 & 8 \\ -5 & 7 \end{bmatrix} + \begin{bmatrix} 12 & 7 \\ 5 & 3 \end{bmatrix} = \begin{bmatrix} 2 & -1 \\ 0 & 13 \end{bmatrix} + \begin{bmatrix} -4 & -8 \\ 5 & -7 \end{bmatrix} + \begin{bmatrix} 12 & 7 \\ 5 & 3 \end{bmatrix}$

$$= \begin{bmatrix} 2+(-4)+12 & -1+(-8)+7 \\ 0+5+5 & 13+(-7)+3 \end{bmatrix} = \begin{bmatrix} 10 & -2 \\ 10 & 9 \end{bmatrix}$$

31. $\begin{bmatrix} -4x+2y & -3x+y \\ 6x-3y & 2x-5y \end{bmatrix} + \begin{bmatrix} -8x+6y & 2x \\ 3y-5x & 6x+4y \end{bmatrix} = \begin{bmatrix} (-4x+2y)+(-8x+6y) & (-3x+y)+2x \\ (6x-3y)+(3y-5x) & (2x-5y)+(6x+4y) \end{bmatrix}$

$$= \begin{bmatrix} -12x+8y & -x+y \\ x & 8x-y \end{bmatrix}$$

33. The additive inverse of $X = \begin{bmatrix} x & y \\ z & w \end{bmatrix}$ is

$$-X = \begin{bmatrix} -x & -y \\ -z & -w \end{bmatrix}.$$

35. Show that $X + (T + P) = (X + T) + P$.

On the left side, the sum $T + P$ is obtained first, and then $X + (T + P)$.

This gives the matrix

$$\begin{bmatrix} x + (r + m) & y + (s + n) \\ z + (t + p) & w + (u + q) \end{bmatrix}.$$

For the right side, first the sum $X + T$ is obtained, and then $(X + T) + P$.

This gives the matrix

$$\begin{bmatrix} (x + r) + m & (y + s) + n \\ (z + t) + p & (w + u) + q \end{bmatrix}.$$

Comparing corresponding elements, we see that they are equal by the associative property of addition of real numbers. Thus, $X + (T + P) = (X + T) + P$.

37. Show that $P + O = P$.

$$P + O = \begin{bmatrix} m & n \\ p & q \end{bmatrix} + \begin{bmatrix} 0 & 0 \\ 0 & 0 \end{bmatrix} = \begin{bmatrix} m + 0 & n + 0 \\ p + 0 & q + 0 \end{bmatrix}$$

$$= \begin{bmatrix} m & n \\ p & q \end{bmatrix} = P$$

Thus, $P + O = P$.

39. **(a)** There are four food groups and three meals. To represent the data by a 3×4 matrix, we must use the rows to correspond to the meals, breakfast, lunch, and dinner, and the columns to correspond to the four food groups. Thus, we obtain the matrix

$$\begin{bmatrix} 2 & 1 & 2 & 1 \\ 3 & 2 & 2 & 1 \\ 4 & 3 & 2 & 1 \end{bmatrix}.$$

(b) There are four food groups. These will correspond to the four rows. There are three components in each food group: fat, carbohydrates, and protein. These will correspond to the three columns. The matrix is

$$\begin{bmatrix} 5 & 0 & 7 \\ 0 & 10 & 1 \\ 0 & 15 & 2 \\ 10 & 12 & 8 \end{bmatrix}.$$

(c) The matrix is

$$\begin{bmatrix} 8 \\ 4 \\ 5 \end{bmatrix}.$$

41.

	Obtained Pain Relief: Yes	No
Painfree	22	3
Placebo	8	17

(a) Of the 25 patients who took the placebo, 8 got relief.

(b) Of the 25 patients who took Painfree, 3 got no relief.

(c)
$$\begin{bmatrix} 22 & 3 \\ 8 & 17 \end{bmatrix} + \begin{bmatrix} 21 & 4 \\ 6 & 19 \end{bmatrix} + \begin{bmatrix} 19 & 6 \\ 10 & 15 \end{bmatrix} + \begin{bmatrix} 23 & 2 \\ 3 & 22 \end{bmatrix} = \begin{bmatrix} 85 & 15 \\ 27 & 73 \end{bmatrix}$$

(d) Yes, it appears that Painfree is effective. Of the 100 patients who took the medication, 85% got relief.

43. **(a)** The matrix for the life expectancy of African Americans is

	M	F
1970	60.0	68.3
1980	63.8	72.5
1990	64.5	73.6
2000	68.2	75.1
2010	71.8	78.0

(b) The matrix for the life expectancy of White Americans is

	M	F
1970	68.0	75.6
1980	70.7	78.1
1990	72.7	79.4
2000	74.7	79.9
2010	76.5	81.3

(c) The matrix showing the difference in the life expectancy between the two groups is

$$\begin{bmatrix} 60.0 & 68.3 \\ 63.8 & 72.5 \\ 64.5 & 73.6 \\ 68.2 & 75.1 \\ 71.8 & 78.0 \end{bmatrix} - \begin{bmatrix} 68.0 & 75.6 \\ 70.7 & 78.1 \\ 72.7 & 79.4 \\ 74.7 & 79.9 \\ 76.5 & 81.3 \end{bmatrix} = \begin{bmatrix} -8.0 & -7.3 \\ -6.9 & -5.6 \\ -8.2 & -5.8 \\ -6.5 & -4.8 \\ -4.7 & -3.3 \end{bmatrix}$$

45. (a) The matrix for the educational attainment of African Americans is

$$\begin{array}{c} \\ 1980 \\ 1985 \\ 1990 \\ 1995 \\ 2000 \\ 2010 \end{array} \begin{array}{c} \begin{array}{cc} \text{HS} & \text{College} \end{array} \\ \begin{bmatrix} 51.2 & 7.9 \\ 59.8 & 11.1 \\ 66.2 & 11.3 \\ 73.8 & 13.2 \\ 78.5 & 16.5 \\ 84.2 & 19.8 \end{bmatrix} \end{array}$$

(b) The matrix for the educational attainment of Hispanic Americans is

$$\begin{array}{c} \\ 1980 \\ 1985 \\ 1990 \\ 1995 \\ 2000 \\ 2010 \end{array} \begin{array}{c} \begin{array}{cc} \text{HS} & \text{College} \end{array} \\ \begin{bmatrix} 45.3 & 7.9 \\ 47.9 & 8.5 \\ 50.8 & 9.2 \\ 53.4 & 9.3 \\ 57.0 & 10.6 \\ 62.9 & 13.9 \end{bmatrix} \end{array}$$

(c) The matrix showing the difference in the educational attainment between African and Hispanic Americans is

$$\begin{bmatrix} 51.2 & 7.9 \\ 59.8 & 11.1 \\ 66.2 & 11.3 \\ 73.8 & 13.2 \\ 78.5 & 16.5 \\ 84.2 & 19.8 \end{bmatrix} - \begin{bmatrix} 45.3 & 7.9 \\ 47.9 & 8.5 \\ 50.8 & 9.2 \\ 53.4 & 9.3 \\ 57.0 & 10.6 \\ 62.9 & 13.9 \end{bmatrix} = \begin{bmatrix} 5.9 & 0 \\ 11.9 & 2.6 \\ 15.4 & 2.1 \\ 20.4 & 3.9 \\ 21.5 & 5.9 \\ 21.3 & 5.9 \end{bmatrix}$$

47. (a) The matrix for the death rate of male drivers is

$$\begin{array}{l} \\ \\ \text{Age 16} \\ \text{Age 17} \\ \text{Ages 30}-59 \end{array} \begin{array}{c} \text{Number of Passengers} \\ \begin{array}{cccc} 0 & 1 & 2 & \geq 3 \end{array} \\ \begin{bmatrix} 2.61 & 4.39 & 6.29 & 9.08 \\ 1.63 & 2.77 & 4.61 & 6.92 \\ 0.92 & 0.75 & 0.62 & 0.54 \end{bmatrix} \end{array}$$

(b) The matrix for the death rate of female drivers is

$$\begin{array}{l} \\ \\ \text{Age 16} \\ \text{Age 17} \\ \text{Ages 30}-59 \end{array} \begin{array}{c} \text{Number of Passengers} \\ \begin{array}{cccc} 0 & 1 & 2 & \geq 3 \end{array} \\ \begin{bmatrix} 1.38 & 1.72 & 1.94 & 3.31 \\ 1.26 & 1.48 & 2.82 & 2.28 \\ 0.41 & 0.33 & 0.27 & 0.40 \end{bmatrix} \end{array}$$

(c) The matrix showing the difference between the death rates of males and females is

$$\begin{bmatrix} 2.61 & 4.39 & 6.29 & 9.08 \\ 1.63 & 2.77 & 4.61 & 6.92 \\ 0.92 & 0.75 & 0.62 & 0.54 \end{bmatrix} - \begin{bmatrix} 1.38 & 1.72 & 1.94 & 3.31 \\ 1.26 & 1.48 & 2.82 & 2.28 \\ 0.41 & 0.33 & 0.27 & 0.40 \end{bmatrix} = \begin{bmatrix} 1.23 & 2.67 & 4.35 & 5.77 \\ 0.37 & 1.29 & 1.79 & 4.64 \\ 0.51 & 0.42 & 0.35 & 0.14 \end{bmatrix}$$

(d) Answers will vary.

49. (a)

	I	II	III
Bread	88	105	60
Milk	48	72	40
PB	16	21	0
Cold cuts	112	147	50

(b)
$$\begin{bmatrix} 88+0.25(88) & 105+\frac{1}{3}(105) & 60+0.1(60) \\ 48+0.25(48) & 72+\frac{1}{3}(72) & 40+0.1(40) \\ 16+0.25(16) & 21+\frac{1}{3}(21) & 0+0.1(0) \\ 112+0.25(112) & 147+\frac{1}{3}(147) & 50+0.1(50) \end{bmatrix} = \begin{bmatrix} 110 & 140 & 66 \\ 60 & 96 & 44 \\ 20 & 28 & 0 \\ 140 & 196 & 55 \end{bmatrix}$$

(c) Add the final matrices from parts (a) and (b).

$$\begin{bmatrix} 88 & 105 & 60 \\ 48 & 72 & 40 \\ 16 & 21 & 0 \\ 112 & 147 & 50 \end{bmatrix} + \begin{bmatrix} 110 & 140 & 66 \\ 60 & 96 & 44 \\ 20 & 28 & 0 \\ 140 & 196 & 55 \end{bmatrix} = \begin{bmatrix} 198 & 245 & 126 \\ 108 & 168 & 84 \\ 36 & 49 & 0 \\ 252 & 343 & 105 \end{bmatrix}$$

10.3 Multiplication of Matrices

In Exercises 1-6, let

$$A = \begin{bmatrix} -2 & 4 \\ 0 & 3 \end{bmatrix} \text{ and } B = \begin{bmatrix} -6 & 2 \\ 4 & 0 \end{bmatrix}.$$

1. $2A = 2\begin{bmatrix} -2 & 4 \\ 0 & 3 \end{bmatrix} = \begin{bmatrix} -4 & 8 \\ 0 & 6 \end{bmatrix}$

3. $-6A = -6\begin{bmatrix} -2 & 4 \\ 0 & 3 \end{bmatrix} = \begin{bmatrix} 12 & -24 \\ 0 & -18 \end{bmatrix}$

5.
$$\begin{aligned} -4A + 5B &= -4\begin{bmatrix} -2 & 4 \\ 0 & 3 \end{bmatrix} + 5\begin{bmatrix} -6 & 2 \\ 4 & 0 \end{bmatrix} \\ &= \begin{bmatrix} 8 & -16 \\ 0 & -12 \end{bmatrix} + \begin{bmatrix} -30 & 10 \\ 20 & 0 \end{bmatrix} \\ &= \begin{bmatrix} -22 & -6 \\ 20 & -12 \end{bmatrix} \end{aligned}$$

7. Matrix A size Matrix B size

$2\times\underline{2}$ $\underline{2}\times 2$

The number of columns of A is the same as the number of rows of B, so the product AB exists. The size of the matrix AB is 2×2.

Matrix B size Matrix A size

$2\times\underline{2}$ $\underline{2}\times 2$

Since the number of columns of B is the same as the number of rows of A, the product BA also exists and has size 2×2.

9. Matrix A size Matrix B size

$3\times\underline{4}$ $\underline{4}\times 4$

Since matrix A has 4 columns and matrix B has 4 rows, the product AB exists and has size 3×4.

Matrix B size Matrix A size

$4\times\underline{4}$ $\underline{3}\times 4$

Since B has 4 columns and A has 3 rows, the product BA does not exist.

11. Matrix A size Matrix B size

$4\times\underline{2}$ $\underline{3}\times 4$

The number of columns of A is not the same as the number of rows of B, so the product AB does not exist.

Matrix B size Matrix A size

$3\times\underline{4}$ $\underline{4}\times 2$

The number of columns of B is the same as the number of rows of A, so the product BA exists and has size 3×2.

13. To find the product matrix AB, the number of columns of A must be the same as the number of rows of B.

15. Call the first matrix A and the second matrix B. The product matrix AB will have size 2×1.
Step 1: Multiply the elements of the first row of A by the corresponding elements of the column of B and add.

$$\begin{bmatrix} \mathbf{2} & \mathbf{-1} \\ 5 & 8 \end{bmatrix}\begin{bmatrix} \mathbf{3} \\ \mathbf{-2} \end{bmatrix} \quad \mathbf{2}(\mathbf{3}) + (\mathbf{-1})(\mathbf{-2}) = 8$$

Therefore, 8 is the first row entry of the product matrix AB.
Step 2: Multiply the elements of the second row of A by the corresponding elements of the column of B and add.

$$\begin{bmatrix} 2 & -1 \\ \mathbf{5} & \mathbf{8} \end{bmatrix}\begin{bmatrix} \mathbf{3} \\ \mathbf{-2} \end{bmatrix} \quad \mathbf{5}(\mathbf{3}) + \mathbf{8}(\mathbf{-2}) = -1$$

The second row entry of the product is -1.

Step 3: Write the product using the two entries found above: $AB = \begin{bmatrix} 2 & -1 \\ 5 & 8 \end{bmatrix}\begin{bmatrix} 3 \\ -2 \end{bmatrix} = \begin{bmatrix} 8 \\ -1 \end{bmatrix}$

17. $$\begin{bmatrix} 2 & -1 & 7 \\ -3 & 0 & -4 \end{bmatrix}\begin{bmatrix} 5 \\ 10 \\ 2 \end{bmatrix} = \begin{bmatrix} 2\cdot 5 + (-1)\cdot 10 + 7\cdot 2 \\ (-3)\cdot 5 + 0\cdot 10 + (-4)\cdot 2 \end{bmatrix} = \begin{bmatrix} 14 \\ -23 \end{bmatrix}$$

19. $$\begin{bmatrix} 2 & -1 \\ 3 & 6 \end{bmatrix}\begin{bmatrix} -1 & 0 & 4 \\ 5 & -2 & 0 \end{bmatrix} = \begin{bmatrix} 2\cdot(-1) + (-1)\cdot 5 & 2\cdot 0 + (-1)\cdot(-2) & 2\cdot 4 + (-1)\cdot 0 \\ 3\cdot(-1) + 6\cdot 5 & 3\cdot 0 + 6\cdot(-2) & 3\cdot 4 + 6\cdot 0 \end{bmatrix} = \begin{bmatrix} -7 & 2 & 8 \\ 27 & -12 & 12 \end{bmatrix}$$

21. $$\begin{bmatrix} 2 & 2 & -1 \\ 3 & 0 & 1 \end{bmatrix}\begin{bmatrix} 0 & 2 \\ -1 & 4 \\ 0 & 2 \end{bmatrix} = \begin{bmatrix} 2\cdot 0 + 2(-1) + (-1)0 & 2\cdot 2 + 2\cdot 4 + (-1)2 \\ 3\cdot 0 + 0(-1) + 1(0) & 3\cdot 2 + 0\cdot 4 + 1\cdot 2 \end{bmatrix} = \begin{bmatrix} -2 & 10 \\ 0 & 8 \end{bmatrix}$$

23. $$\begin{bmatrix} 1 & 2 \\ 3 & 4 \end{bmatrix}\begin{bmatrix} -1 & 5 \\ 7 & 0 \end{bmatrix} = \begin{bmatrix} 1(-1) + 2\cdot 7 & 1\cdot 5 + 2\cdot 0 \\ 3(-1) + 4\cdot 7 & 3\cdot 5 + 4\cdot 0 \end{bmatrix} = \begin{bmatrix} 13 & 5 \\ 25 & 15 \end{bmatrix}$$

25. $$\begin{bmatrix} -2 & -3 & 7 \\ 1 & 5 & 6 \end{bmatrix}\begin{bmatrix} 1 \\ 2 \\ 3 \end{bmatrix} = \begin{bmatrix} -2(1) + (-3)2 + 7\cdot 3 \\ 1\cdot 1 + 5\cdot 2 + 6\cdot 3 \end{bmatrix} = \begin{bmatrix} 13 \\ 29 \end{bmatrix}$$

27. $$\left(\begin{bmatrix} 2 & 1 \\ -3 & -6 \\ 4 & 0 \end{bmatrix}\begin{bmatrix} 1 & -2 \\ 2 & -1 \end{bmatrix}\right)\begin{bmatrix} 3 \\ 1 \end{bmatrix} = \begin{bmatrix} 4 & -5 \\ -15 & 12 \\ 4 & -8 \end{bmatrix}\begin{bmatrix} 3 \\ 1 \end{bmatrix} = \begin{bmatrix} 7 \\ -33 \\ 4 \end{bmatrix}$$

29. $$\begin{bmatrix} 2 & -2 \\ 1 & -1 \end{bmatrix}\left(\begin{bmatrix} 4 & 3 \\ 1 & 2 \end{bmatrix} + \begin{bmatrix} 7 & 0 \\ -1 & 5 \end{bmatrix}\right) = \begin{bmatrix} 2 & -2 \\ 1 & -1 \end{bmatrix}\begin{bmatrix} 11 & 3 \\ 0 & 7 \end{bmatrix} = \begin{bmatrix} 22 & -8 \\ 11 & -4 \end{bmatrix}$$

31. **(a)** $AB = \begin{bmatrix} -2 & 4 \\ 1 & 3 \end{bmatrix}\begin{bmatrix} -2 & 1 \\ 3 & 6 \end{bmatrix} = \begin{bmatrix} 16 & 22 \\ 7 & 19 \end{bmatrix}$

(b) $BA = \begin{bmatrix} -2 & 1 \\ 3 & 6 \end{bmatrix}\begin{bmatrix} -2 & 4 \\ 1 & 3 \end{bmatrix} = \begin{bmatrix} 5 & -5 \\ 0 & 30 \end{bmatrix}$

(c) No, AB and BA are not equal here.

(d) No, AB does not always equal BA.

33. Verify that $P(X+T)=PX+PT$.

Find $P(X+T)$ and $PX+PT$ separately and compare their values to see if they are the same.

$$P(X+T)=\begin{bmatrix} m & n \\ p & q \end{bmatrix}\left(\begin{bmatrix} x & y \\ z & w \end{bmatrix}+\begin{bmatrix} r & s \\ t & u \end{bmatrix}\right)=\begin{bmatrix} m & n \\ p & q \end{bmatrix}\left(\begin{bmatrix} x+r & y+s \\ z+t & w+u \end{bmatrix}\right)$$

$$=\begin{bmatrix} m(x+r)+n(z+t) & m(y+s)+n(w+u) \\ p(x+r)+q(z+t) & p(y+s)+q(w+u) \end{bmatrix}=\begin{bmatrix} mx+mr+nz+nt & my+ms+nw+nu \\ px+pr+qz+qt & py+ps+qw+qu \end{bmatrix}$$

$$PX+PT=\begin{bmatrix} m & n \\ p & q \end{bmatrix}\begin{bmatrix} x & y \\ z & w \end{bmatrix}+\begin{bmatrix} m & n \\ p & q \end{bmatrix}\begin{bmatrix} r & s \\ t & u \end{bmatrix}=\begin{bmatrix} mx+nz & my+nw \\ px+qz & py+qw \end{bmatrix}+\begin{bmatrix} mr+nt & ms+nu \\ pr+qt & ps+qu \end{bmatrix}$$

$$=\begin{bmatrix} (mx+nz)+(mr+nt) & (my+nw)+(ms+nu) \\ (px+qz)+(pr+qt) & (py+qw)+(ps+qu) \end{bmatrix}=\begin{bmatrix} mx+nz+mr+nt & my+nw+ms+nu \\ px+qz+pr+qt & py+qw+ps+qu \end{bmatrix}$$

$$=\begin{bmatrix} mx+mr+nz+nt & my+ms+nw+nu \\ px+pr+qz+qt & py+ps+qw+qu \end{bmatrix}$$

Observe that the two results are identical. Thus, $P(X+T)=PX+PT$.

35. Verify that $(k+h)P=kP+hP$ for any real numbers k and h.

$$(k+h)P=(k+h)\begin{bmatrix} m & n \\ p & q \end{bmatrix}=\begin{bmatrix} (k+h)m & (k+h)n \\ (k+h)p & (k+h)q \end{bmatrix}=\begin{bmatrix} km+hm & kn+hn \\ kp+hp & kq+hq \end{bmatrix}$$

$$=\begin{bmatrix} km & kn \\ kp & kq \end{bmatrix}+\begin{bmatrix} hm & hn \\ hp & hq \end{bmatrix}=k\begin{bmatrix} m & n \\ p & q \end{bmatrix}+h\begin{bmatrix} m & n \\ p & q \end{bmatrix}=kP+hP$$

Thus, $(k+h)P=kP+hP$ for any real numbers k and h.

37. $$\begin{bmatrix} 2 & 3 & 1 \\ 1 & -4 & 5 \end{bmatrix}\begin{bmatrix} x_1 \\ x_2 \\ x_3 \end{bmatrix}=\begin{bmatrix} 2x_1+3x_2+x_3 \\ x_1-4x_2+5x_3 \end{bmatrix},$$

and $$\begin{bmatrix} 2x_1+3x_2+x_3 \\ x_1-4x_2+5x_3 \end{bmatrix}=\begin{bmatrix} 5 \\ 8 \end{bmatrix}.$$

This is equivalent to

$2x_1+3x_2+x_3=5$

$x_1-4x_2+5x_3=8$

since corresponding elements of equal matrices must be equal. Reversing this, observe that the given system of linear equations can be written as the matrix equation

$$\begin{bmatrix} 2 & 3 & 1 \\ 1 & -4 & 5 \end{bmatrix}\begin{bmatrix} x_1 \\ x_2 \\ x_3 \end{bmatrix}=\begin{bmatrix} 5 \\ 8 \end{bmatrix}.$$

39. **(a)** Use a graphing calculator or a computer to find the product matrix. The answer is

$$AC=\begin{bmatrix} 6 & 106 & 158 & 222 & 28 \\ 120 & 139 & 64 & 75 & 115 \\ -146 & -2 & 184 & 144 & -129 \\ 106 & 94 & 24 & 116 & 110 \end{bmatrix}.$$

(b) CA does not exist.

(c) AC and CA are clearly not equal, since CA does not even exist.

41. Use a graphing calculator or computer to find the matrix products and sums. The answers are as follows.

(a) $$C+D=\begin{bmatrix} -1 & 5 & 9 & 13 & -1 \\ 7 & 17 & 2 & -10 & 6 \\ 18 & 9 & -12 & 12 & 22 \\ 9 & 4 & 18 & 10 & -3 \\ 1 & 6 & 10 & 28 & 5 \end{bmatrix}$$

(b) $$(C+D)B=\begin{bmatrix} -2 & -9 & 90 & 77 \\ -42 & -63 & 127 & 62 \\ 413 & 76 & 180 & -56 \\ -29 & -44 & 198 & 85 \\ 137 & 20 & 162 & 103 \end{bmatrix}$$

(c) $$CB=\begin{bmatrix} -56 & -1 & 1 & 45 \\ -156 & -119 & 76 & 122 \\ 315 & 86 & 118 & -91 \\ -17 & -17 & 116 & 51 \\ 118 & 19 & 125 & 77 \end{bmatrix}$$

(d) $DB = \begin{bmatrix} 54 & -8 & 89 & 32 \\ 114 & 56 & 51 & -60 \\ 98 & -10 & 62 & 35 \\ -12 & -27 & 82 & 34 \\ 19 & 1 & 37 & 26 \end{bmatrix}$

(e) $CB + DB = \begin{bmatrix} -2 & -9 & 90 & 77 \\ -42 & -63 & 127 & 62 \\ 413 & 76 & 180 & -56 \\ -29 & -44 & 198 & 85 \\ 137 & 20 & 162 & 103 \end{bmatrix}$

(f) Yes, $(C + D)B$ and $CB + DB$ are equal, as can be seen by observing that the answers to parts (b) and (e) are identical.

43. (a) $XY = \begin{bmatrix} 2 & 1 & 2 & 1 \\ 3 & 2 & 2 & 1 \\ 4 & 3 & 2 & 1 \end{bmatrix} \begin{bmatrix} 5 & 0 & 7 \\ 0 & 10 & 1 \\ 0 & 15 & 2 \\ 10 & 12 & 8 \end{bmatrix} = \begin{bmatrix} 20 & 52 & 27 \\ 25 & 62 & 35 \\ 30 & 72 & 43 \end{bmatrix}$

The rows give the amounts of fat, carbohydrates, and protein, respectively, in each of the daily meals.

(b) $YZ = \begin{bmatrix} 5 & 0 & 7 \\ 0 & 10 & 1 \\ 0 & 15 & 2 \\ 10 & 12 & 8 \end{bmatrix} \begin{bmatrix} 8 \\ 4 \\ 5 \end{bmatrix} = \begin{bmatrix} 75 \\ 45 \\ 70 \\ 168 \end{bmatrix}$

The rows give the number of calories in one exchange of each of the food groups.

(c) Use the matrices found for XY and YZ from parts (a) and (b).

$(XY)Z = \begin{bmatrix} 20 & 52 & 27 \\ 25 & 62 & 35 \\ 30 & 72 & 43 \end{bmatrix} \begin{bmatrix} 8 \\ 4 \\ 5 \end{bmatrix} = \begin{bmatrix} 503 \\ 623 \\ 743 \end{bmatrix}$

$X(YZ) = \begin{bmatrix} 2 & 1 & 2 & 1 \\ 3 & 2 & 2 & 1 \\ 4 & 3 & 2 & 1 \end{bmatrix} \begin{bmatrix} 75 \\ 45 \\ 70 \\ 168 \end{bmatrix} = \begin{bmatrix} 503 \\ 623 \\ 743 \end{bmatrix}$

The rows give the number of calories in each meal.

45. $$\frac{1}{6}\left(\begin{bmatrix} 60.0 & 68.3 \\ 63.8 & 72.5 \\ 64.5 & 73.6 \\ 68.2 & 75.1 \\ 71.8 & 78.0 \end{bmatrix} + 5 \begin{bmatrix} 68.0 & 75.6 \\ 70.7 & 78.1 \\ 72.7 & 79.4 \\ 74.7 & 79.9 \\ 76.5 & 81.3 \end{bmatrix} \right) = \frac{1}{6}\left(\begin{bmatrix} 60.0 & 68.3 \\ 63.8 & 72.5 \\ 64.5 & 73.6 \\ 68.2 & 75.1 \\ 71.8 & 78.0 \end{bmatrix} + \begin{bmatrix} 340 & 378 \\ 353.5 & 390.5 \\ 363.5 & 397 \\ 373.5 & 399.5 \\ 382.5 & 406.5 \end{bmatrix} \right)$$

$$= \frac{1}{6}\left(\begin{bmatrix} 400 & 446.3 \\ 417.3 & 463 \\ 428 & 470.6 \\ 441.7 & 474.6 \\ 454.3 & 484.5 \end{bmatrix} \right) = \begin{bmatrix} 66.7 & 74.4 \\ 69.6 & 77.2 \\ 71.3 & 78.4 \\ 73.6 & 79.1 \\ 75.7 & 80.8 \end{bmatrix}$$

47. **(a)** Let n represent the present year. Then $j_n = 900,\ s_n = 500,\ a_n = 2600.$

For the year, $n+1$, we have $\begin{bmatrix} j_{n+1} \\ s_{n+1} \\ a_{n+1} \end{bmatrix} = \begin{bmatrix} 0 & 0 & 0.33 \\ 0.18 & 0 & 0 \\ 0 & 0.71 & 0.94 \end{bmatrix} \begin{bmatrix} j_n \\ s_n \\ a_n \end{bmatrix} = \begin{bmatrix} 0 & 0 & 0.33 \\ 0.18 & 0 & 0 \\ 0 & 0.71 & 0.94 \end{bmatrix} \begin{bmatrix} 900 \\ 500 \\ 2600 \end{bmatrix} = \begin{bmatrix} 858 \\ 162 \\ 2799 \end{bmatrix}.$

For the year $n + 1$, there is a total of 858 + 162 + 2799 = 3819 female owls.
For the next year, $n + 2$, we have

$$\begin{bmatrix} j_{n+2} \\ s_{n+2} \\ a_{n+2} \end{bmatrix} = \begin{bmatrix} 0 & 0 & 0.33 \\ 0.18 & 0 & 0 \\ 0 & 0.71 & 0.94 \end{bmatrix} \begin{bmatrix} j_{n+1} \\ s_{n+1} \\ a_{n+1} \end{bmatrix} = \begin{bmatrix} 0 & 0 & 0.33 \\ 0.18 & 0 & 0 \\ 0 & 0.71 & 0.94 \end{bmatrix} \begin{bmatrix} 858 \\ 162 \\ 2799 \end{bmatrix} \approx \begin{bmatrix} 924 \\ 154 \\ 2746 \end{bmatrix}.$$

For the year $n + 2$, there is a total of approximately 924 + 154 + 2746 = 3824 female owls.
For the next year, $n + 3$, we have

$$\begin{bmatrix} j_{n+3} \\ s_{n+3} \\ a_{n+3} \end{bmatrix} = \begin{bmatrix} 0 & 0 & 0.33 \\ 0.18 & 0 & 0 \\ 0 & 0.71 & 0.94 \end{bmatrix} \begin{bmatrix} j_{n+2} \\ s_{n+2} \\ a_{n+2} \end{bmatrix} = \begin{bmatrix} 0 & 0 & 0.33 \\ 0.18 & 0 & 0 \\ 0 & 0.71 & 0.94 \end{bmatrix} \begin{bmatrix} 924 \\ 154 \\ 2746 \end{bmatrix} \approx \begin{bmatrix} 906 \\ 166 \\ 2691 \end{bmatrix}.$$

For the year $n + 3$, there is a total of approximately 906 + 166 + 2691 = 3763 female owls.
For the next year, $n + 4$, we have

$$\begin{bmatrix} j_{n+4} \\ s_{n+4} \\ a_{n+4} \end{bmatrix} = \begin{bmatrix} 0 & 0 & 0.33 \\ 0.18 & 0 & 0 \\ 0 & 0.71 & 0.94 \end{bmatrix} \begin{bmatrix} j_{n+3} \\ s_{n+3} \\ a_{n+3} \end{bmatrix} = \begin{bmatrix} 0 & 0 & 0.33 \\ 0.18 & 0 & 0 \\ 0 & 0.71 & 0.94 \end{bmatrix} \begin{bmatrix} 906 \\ 166 \\ 2691 \end{bmatrix} \approx \begin{bmatrix} 888 \\ 163 \\ 2647 \end{bmatrix}.$$

For the near $n + 4$, there is a total of approximately 888 + 163 + 2647 = 3698 female owls.
For the year $n + 5$, we have

$$\begin{bmatrix} j_{n+5} \\ s_{n+5} \\ a_{n+5} \end{bmatrix} = \begin{bmatrix} 0 & 0 & 0.33 \\ 0.18 & 0 & 0 \\ 0 & 0.71 & 0.94 \end{bmatrix} \begin{bmatrix} j_{n+4} \\ s_{n+4} \\ a_{n+4} \end{bmatrix} = \begin{bmatrix} 0 & 0 & 0.33 \\ 0.18 & 0 & 0 \\ 0 & 0.71 & 0.94 \end{bmatrix} \begin{bmatrix} 888 \\ 163 \\ 2647 \end{bmatrix} \approx \begin{bmatrix} 874 \\ 160 \\ 2604 \end{bmatrix}.$$

For the year $n + 5$, there is a total of approximately 874 + 160 + 2604 = 3638 female owls.

(b) Each year, the population is about 98 percent of the population of the previous year. In the long run, the northern spotted owl will become extinct.

(c) Change 0.18 in the original matrix equation to 0.40.
For the next year, $n + 1$, we have

$$\begin{bmatrix} j_{n+1} \\ s_{n+1} \\ a_{n+1} \end{bmatrix} = \begin{bmatrix} 0 & 0 & 0.33 \\ 0.40 & 0 & 0 \\ 0 & 0.71 & 0.94 \end{bmatrix} \begin{bmatrix} j_n \\ s_n \\ a_n \end{bmatrix} = \begin{bmatrix} 0 & 0 & 0.33 \\ 0.40 & 0 & 0 \\ 0 & 0.71 & 0.94 \end{bmatrix} \begin{bmatrix} 900 \\ 500 \\ 2600 \end{bmatrix} = \begin{bmatrix} 858 \\ 360 \\ 2799 \end{bmatrix}.$$

For the year $n + 1$, there would be a total of 858 + 360 +2799 = 4017 female owls.
For the next year, $n + 2$, we have

$$\begin{bmatrix} j_{n+2} \\ s_{n+2} \\ a_{n+2} \end{bmatrix} = \begin{bmatrix} 0 & 0 & 0.33 \\ 0.40 & 0 & 0 \\ 0 & 0.71 & 0.94 \end{bmatrix} \begin{bmatrix} j_{n+1} \\ s_{n+1} \\ a_{n+1} \end{bmatrix} = \begin{bmatrix} 0 & 0 & 0.33 \\ 0.40 & 0 & 0 \\ 0 & 0.71 & 0.94 \end{bmatrix} \begin{bmatrix} 858 \\ 360 \\ 2799 \end{bmatrix} \approx \begin{bmatrix} 924 \\ 343 \\ 2887 \end{bmatrix}.$$

For the year $n + 2$, there would be a total of approximately 924 + 343 + 2887 = 4154 female owls.
For the next year, $n + 3$, we have

$$\begin{bmatrix} j_{n+3} \\ s_{n+3} \\ a_{n+3} \end{bmatrix} = \begin{bmatrix} 0 & 0 & 0.33 \\ 0.40 & 0 & 0 \\ 0 & 0.71 & 0.94 \end{bmatrix} \begin{bmatrix} j_{n+2} \\ s_{n+2} \\ a_{n+2} \end{bmatrix} = \begin{bmatrix} 0 & 0 & 0.33 \\ 0.40 & 0 & 0 \\ 0 & 0.71 & 0.94 \end{bmatrix} \begin{bmatrix} 924 \\ 343 \\ 2887 \end{bmatrix} = \begin{bmatrix} 953 \\ 370 \\ 2957 \end{bmatrix}.$$

For the year $n + 3$, there would be a total of approximately 953 + 370 + 2957 = 4280 female owls.

(continued on next page)

(*continued*)

For the next year, $n + 4$, we have

$$\begin{bmatrix} j_{n+4} \\ s_{n+4} \\ a_{n+4} \end{bmatrix} = \begin{bmatrix} 0 & 0 & 0.33 \\ 0.40 & 0 & 0 \\ 0 & 0.71 & 0.94 \end{bmatrix} \begin{bmatrix} j_{n+3} \\ s_{n+3} \\ a_{n+3} \end{bmatrix} = \begin{bmatrix} 0 & 0 & 0.33 \\ 0.40 & 0 & 0 \\ 0 & 0.71 & 0.94 \end{bmatrix} \begin{bmatrix} 953 \\ 370 \\ 2957 \end{bmatrix} \approx \begin{bmatrix} 976 \\ 381 \\ 3042 \end{bmatrix}.$$

For the year $n + 4$, there would be a total of approximately $976 + 381 + 3042 = 4399$ female owls.
For the next year, $n + 5$, we have

$$\begin{bmatrix} j_{n+5} \\ s_{n+5} \\ a_{n+5} \end{bmatrix} = \begin{bmatrix} 0 & 0 & 0.33 \\ 0.40 & 0 & 0 \\ 0 & 0.71 & 0.94 \end{bmatrix} \begin{bmatrix} j_{n+4} \\ s_{n+4} \\ a_{n+4} \end{bmatrix} = \begin{bmatrix} 0 & 0 & 0.33 \\ 0.40 & 0 & 0 \\ 0 & 0.71 & 0.94 \end{bmatrix} \begin{bmatrix} 976 \\ 381 \\ 3042 \end{bmatrix} \approx \begin{bmatrix} 1004 \\ 390 \\ 3130 \end{bmatrix}.$$

For the year $n + 5$, there would be a total of approximately $1004 + 390 + 3130 = 4524$ female owls.
Assuming that better habitat management could increase the survival rate of juvenile female spotted owls from 18 percent to 40 percent, the overall population would increase each year and would, therefore, not become extinct.

49. Cardinals Finches

$$P = \begin{bmatrix} 25 & 31 \\ 20 & 35 \\ 22 & 29 \\ 36 & 20 \end{bmatrix} \begin{matrix} \text{Spring} \\ \text{Summer} \\ \text{Fall} \\ \text{Winter} \end{matrix}, \quad \begin{matrix} & \text{Sunflower} & \text{Corn} & \text{Millet} \end{matrix} \quad Q = \begin{bmatrix} 32 & 10 & 30 \\ 10 & 18 & 27 \end{bmatrix} \begin{matrix} \text{Cardinals} \\ \text{Finches} \end{matrix}, \quad R = \begin{bmatrix} 3 \\ 2 \\ 1 \end{bmatrix} \begin{matrix} \text{Sunflower} \\ \text{Corn} \\ \text{Millet} \end{matrix} \text{ (Cost)}$$

(a) $$QR = \begin{bmatrix} 32 & 10 & 30 \\ 10 & 18 & 27 \end{bmatrix} \begin{bmatrix} 3 \\ 2 \\ 1 \end{bmatrix} = \begin{bmatrix} 32 \cdot 3 + 10 \cdot 2 + 30 \cdot 1 \\ 10 \cdot 3 + 18 \cdot 2 + 27 \cdot 1 \end{bmatrix} = \begin{bmatrix} 146 \\ 93 \end{bmatrix} \begin{matrix} \text{Cardinals} \\ \text{Finches} \end{matrix} \text{ (Cost)}$$

QR represents the cost (in cents) to feed each type of bird.

(b)

$$P(QR) = \begin{bmatrix} 25 & 31 \\ 20 & 35 \\ 22 & 29 \\ 36 & 20 \end{bmatrix} \begin{bmatrix} 146 \\ 93 \end{bmatrix} = \begin{bmatrix} 25 \cdot 146 + 31 \cdot 93 \\ 20 \cdot 146 + 35 \cdot 93 \\ 22 \cdot 146 + 29 \cdot 93 \\ 36 \cdot 146 + 20 \cdot 93 \end{bmatrix} = \begin{bmatrix} 6533 \\ 6175 \\ 5909 \\ 7116 \end{bmatrix} \begin{matrix} \text{Spring} \\ \text{Summer} \\ \text{Fall} \\ \text{Winter} \end{matrix} \text{ (Cost)}$$

Thus, $P(QR) = (PQ)R$.

51. (a) A B

$$\begin{matrix} & \text{CC} & \text{MM} & \text{AD} \end{matrix}$$
$$\begin{matrix} \text{S} \\ \text{C} \end{matrix} \begin{bmatrix} 0.5 & 0.4 & 0.3 \\ 0.2 & 0.3 & 0.3 \end{bmatrix}$$

(b)

$$\begin{matrix} & \text{S} & \text{C} \end{matrix}$$
$$\begin{matrix} \text{SD} \\ \text{MC} \\ \text{M} \end{matrix} \begin{bmatrix} 4 & 3 \\ 2 & 5 \\ 1 & 7 \end{bmatrix}$$

(c) $$\begin{bmatrix} 4 & 3 \\ 2 & 5 \\ 1 & 7 \end{bmatrix} \begin{bmatrix} 0.5 & 0.4 & 0.3 \\ 0.2 & 0.3 & 0.3 \end{bmatrix} = \begin{matrix} \text{SD} \\ \text{MC} \\ \text{M} \end{matrix} \begin{bmatrix} 2.6 & 2.5 & 2.1 \\ 2 & 2.3 & 2.1 \\ 1.9 & 2.5 & 2.4 \end{bmatrix} \quad (\text{columns: CC, MM, AD})$$

(d) Look at the entry in row 3, column 2 of the last matrix. The cost is \$2.50.

(e) $\begin{bmatrix} 2.6 & 2.5 & 2.1 \\ 2 & 2.3 & 2.1 \\ 1.9 & 2.5 & 2.4 \end{bmatrix}\begin{bmatrix} 100 \\ 200 \\ 500 \end{bmatrix} = \begin{bmatrix} 1810 \\ 1710 \\ 1890 \end{bmatrix}$

The total sugar and chocolate cost is \$1810 in San Diego, \$1710 in Mexico City, and \$1890 in Managua, so the order can be produced for the lowest cost in Mexico City.

10.4 Matrix Inverses

1. $\begin{bmatrix} 2 & 1 \\ 5 & 3 \end{bmatrix}\begin{bmatrix} 3 & -1 \\ -5 & 2 \end{bmatrix} = \begin{bmatrix} 6-5 & -2+2 \\ 15-15 & -5+6 \end{bmatrix} = \begin{bmatrix} 1 & 0 \\ 0 & 1 \end{bmatrix} = I$

$\begin{bmatrix} 3 & -1 \\ -5 & 2 \end{bmatrix}\begin{bmatrix} 2 & 1 \\ 5 & 3 \end{bmatrix} = \begin{bmatrix} 6-5 & 3-3 \\ -10+10 & -5+6 \end{bmatrix} = \begin{bmatrix} 1 & 0 \\ 0 & 1 \end{bmatrix} = I$

Yes, the given matrices are inverses of each other.

3. $\begin{bmatrix} 2 & 6 \\ 2 & 4 \end{bmatrix}\begin{bmatrix} -1 & 2 \\ 2 & -4 \end{bmatrix} = \begin{bmatrix} 10 & -20 \\ 6 & -12 \end{bmatrix} \neq I$

No, the matrices are not inverses of each other since their product matrix is not *I*.

5. $\begin{bmatrix} 2 & 0 & 1 \\ 1 & 1 & 2 \\ 0 & 1 & 0 \end{bmatrix}\begin{bmatrix} 1 & 1 & -1 \\ 0 & 1 & 0 \\ -1 & -2 & 2 \end{bmatrix}$

$= \begin{bmatrix} 2+0-1 & 2+0-2 & -2+0+2 \\ 1+0-2 & 1+1-4 & -1+0+4 \\ 0+0+0 & 0+1+0 & 0+0+0 \end{bmatrix}$

$= \begin{bmatrix} 1 & 0 & 0 \\ -1 & -2 & 3 \\ 0 & 1 & 0 \end{bmatrix} \neq I$

No, the matrices are not inverses of each other since their product matrix is not *I*.

7. $\begin{bmatrix} 1 & 3 & 3 \\ 1 & 4 & 3 \\ 1 & 3 & 4 \end{bmatrix}\begin{bmatrix} 7 & -3 & -3 \\ -1 & 1 & 0 \\ -1 & 0 & 1 \end{bmatrix} = \begin{bmatrix} 1 & 0 & 0 \\ 0 & 1 & 0 \\ 0 & 0 & 1 \end{bmatrix} = I$

$\begin{bmatrix} 7 & -3 & -3 \\ -1 & 1 & 0 \\ -1 & 0 & 1 \end{bmatrix}\begin{bmatrix} 1 & 3 & 3 \\ 1 & 4 & 3 \\ 1 & 3 & 4 \end{bmatrix} = \begin{bmatrix} 1 & 0 & 0 \\ 0 & 1 & 0 \\ 0 & 0 & 1 \end{bmatrix} = I$

Yes, these matrices are inverses of each other.

9. No, a matrix with a row of all zeros does not have an inverse; the row of all zeros makes it impossible to get all the 1's in the main diagonal of the identity matrix.

11. Let $A = \begin{bmatrix} 1 & -1 \\ 2 & 0 \end{bmatrix}$.

Form the augmented matrix $[A|I]$.

$[A|I] = \left[\begin{array}{cc|cc} 1 & -1 & 1 & 0 \\ 2 & 0 & 0 & 1 \end{array}\right]$

Perform row operations on $[A|I]$ to get a matrix of the form $[I|B]$.

$\left[\begin{array}{cc|cc} 1 & -1 & 1 & 0 \\ 2 & 0 & 0 & 1 \end{array}\right]$

$-2R_1 + R_2 \to R_2 \quad \left[\begin{array}{cc|cc} 1 & -1 & 1 & 0 \\ 0 & 2 & -2 & 1 \end{array}\right]$

$2R_1 + R_2 \to R_1 \quad \left[\begin{array}{cc|cc} 2 & 0 & 0 & 1 \\ 0 & 2 & -2 & 1 \end{array}\right]$

$\frac{1}{2}R_1 \to R_1$, $\frac{1}{2}R_2 \to R_2 \quad \left[\begin{array}{cc|cc} 1 & 0 & 0 & \frac{1}{2} \\ 0 & 1 & -1 & \frac{1}{2} \end{array}\right] = [I|B]$

The matrix *B* in the last transformation is the desired multiplicative inverse.

$A^{-1} = \begin{bmatrix} 0 & \frac{1}{2} \\ -1 & \frac{1}{2} \end{bmatrix}$

This answer may be checked by showing that $AA^{-1} = I$ and $A^{-1}A = I$.

13. Let $A = \begin{bmatrix} 3 & -1 \\ -5 & 2 \end{bmatrix}$.

$[A|I] = \left[\begin{array}{cc|cc} 3 & -1 & 1 & 0 \\ -5 & 2 & 0 & 1 \end{array}\right]$

$5R_1 + 3R_2 \to R_2 \quad \left[\begin{array}{cc|cc} 3 & -1 & 1 & 0 \\ 0 & 1 & 5 & 3 \end{array}\right]$

$R_1 + R_2 \to R_1 \quad \left[\begin{array}{cc|cc} 3 & 0 & 6 & 3 \\ 0 & 1 & 5 & 3 \end{array}\right]$

$\frac{1}{3}R_1 \to R_1 \quad \left[\begin{array}{cc|cc} 1 & 0 & 2 & 1 \\ 0 & 1 & 5 & 3 \end{array}\right] = [I|B]$

The desired inverse is $A^{-1} = \begin{bmatrix} 2 & 1 \\ 5 & 3 \end{bmatrix}$.

15. Let $A=\begin{bmatrix}1 & -3\\ -2 & 6\end{bmatrix}$.

$[A|I]=\left[\begin{array}{cc|cc}1 & -3 & 1 & 0\\ -2 & 6 & 0 & 1\end{array}\right]$

$2R_1+R_2\to R_2\quad \left[\begin{array}{cc|cc}1 & -3 & 1 & 0\\ 0 & 0 & 2 & 1\end{array}\right]$

Because the last row has all zeros to the left of the vertical bar, there is no way to complete the desired transformation. A has no inverse.

17. Let $A=\begin{bmatrix}1 & 0 & 0\\ 0 & -1 & 0\\ 1 & 0 & 1\end{bmatrix}$.

$[A|I]=\left[\begin{array}{ccc|ccc}1 & 0 & 0 & 1 & 0 & 0\\ 0 & -1 & 0 & 0 & 1 & 0\\ 1 & 0 & 1 & 0 & 0 & 1\end{array}\right]$

$-1R_1+R_3\to R_3\quad \left[\begin{array}{ccc|ccc}1 & 0 & 0 & 1 & 0 & 0\\ 0 & -1 & 0 & 0 & 1 & 0\\ 0 & 0 & 1 & -1 & 0 & 1\end{array}\right]$

$-1R_2\to R_2\quad \left[\begin{array}{ccc|ccc}1 & 0 & 0 & 1 & 0 & 0\\ 0 & 1 & 0 & 0 & -1 & 0\\ 0 & 0 & 1 & -1 & 0 & 1\end{array}\right]$

$A^{-1}=\begin{bmatrix}1 & 0 & 0\\ 0 & -1 & 0\\ -1 & 0 & 1\end{bmatrix}$

19. Let $A=\begin{bmatrix}-1 & -1 & -1\\ 4 & 5 & 0\\ 0 & 1 & -3\end{bmatrix}$.

$[A|I]=\left[\begin{array}{ccc|ccc}-1 & -1 & -1 & 1 & 0 & 0\\ 4 & 5 & 0 & 0 & 1 & 0\\ 0 & 1 & -3 & 0 & 0 & 1\end{array}\right]$

$4R_1+R_2\to R_2\quad \left[\begin{array}{ccc|ccc}-1 & -1 & -1 & 1 & 0 & 0\\ 0 & 1 & -4 & 4 & 1 & 0\\ 0 & 1 & -3 & 0 & 0 & 1\end{array}\right]$

$\begin{array}{r}R_2+R_1\to R_1\\ \\ -1R_2+R_3\to R_3\end{array}\quad \left[\begin{array}{ccc|ccc}-1 & 0 & -5 & 5 & 1 & 0\\ 0 & 1 & -4 & 4 & 1 & 0\\ 0 & 0 & 1 & -4 & -1 & 1\end{array}\right]$

$\begin{array}{r}5R_3+R_1\to R_1\\ 4R_3+R_2\to R_2\\ \end{array}\quad \left[\begin{array}{ccc|ccc}-1 & 0 & 0 & -15 & -4 & 5\\ 0 & 1 & 0 & -12 & -3 & 4\\ 0 & 0 & 1 & -4 & -1 & 1\end{array}\right]$

$-1R_1\to R_1\quad \left[\begin{array}{ccc|ccc}1 & 0 & 0 & 15 & 4 & -5\\ 0 & 1 & 0 & -12 & -3 & 4\\ 0 & 0 & 1 & -4 & -1 & 1\end{array}\right]$

$A^{-1}=\begin{bmatrix}15 & 4 & -5\\ -12 & -3 & 4\\ -4 & -1 & 1\end{bmatrix}$

21. Let $A=\begin{bmatrix}1 & 2 & 3\\ -3 & -2 & -1\\ -1 & 0 & 1\end{bmatrix}$.

$[A|I]=\left[\begin{array}{ccc|ccc}1 & 2 & 3 & 1 & 0 & 0\\ -3 & -2 & -1 & 0 & 1 & 0\\ -1 & 0 & 1 & 0 & 0 & 1\end{array}\right]$

$\begin{array}{r}\\ 3R_1+R_2\to R_2\\ R_1+R_3\to R_3\end{array}\quad \left[\begin{array}{ccc|ccc}1 & 2 & 3 & 1 & 0 & 0\\ 0 & 4 & 8 & 3 & 1 & 0\\ 0 & 2 & 4 & 1 & 0 & 1\end{array}\right]$

$\begin{array}{r}R_2+(-2R_1)\to R_1\\ \\ R_2+(-2R_3)\to R_3\end{array}\quad \left[\begin{array}{ccc|ccc}-2 & 0 & 2 & 1 & 1 & 0\\ 0 & 4 & 8 & 3 & 1 & 0\\ 0 & 0 & 0 & 1 & 1 & -2\end{array}\right]$

Because the last row has all zeros to the left of the vertical bar, there is no way to complete the desired transformation. A has no inverse.

23. Find the inverse of $A=\begin{bmatrix}1 & 3 & -2\\ 2 & 7 & -3\\ 3 & 8 & -5\end{bmatrix}$, if it exists.

$[A|I]=\left[\begin{array}{ccc|ccc}1 & 3 & -2 & 1 & 0 & 0\\ 2 & 7 & -3 & 0 & 1 & 0\\ 3 & 8 & -5 & 0 & 0 & 1\end{array}\right]$

$\begin{array}{r}\\ -2R_1+R_2\to R_2\\ -3R_1+R_3\to R_3\end{array}\quad \left[\begin{array}{ccc|ccc}1 & 3 & -2 & 1 & 0 & 0\\ 0 & 1 & 1 & -2 & 1 & 0\\ 0 & -1 & 1 & -3 & 0 & 1\end{array}\right]$

$\begin{array}{r}-3R_2+R_1\to R_1\\ \\ R_2+R_3\to R_3\end{array}\quad \left[\begin{array}{ccc|ccc}1 & 0 & -5 & 7 & -3 & 0\\ 0 & 1 & 1 & -2 & 1 & 0\\ 0 & 0 & 2 & -5 & 1 & 1\end{array}\right]$

$\begin{array}{r}5R_3+2R_1\to R_1\\ -1R_3+2R_2\to R_2\\ \end{array}\quad \left[\begin{array}{ccc|ccc}2 & 0 & 0 & -11 & -1 & 5\\ 0 & 2 & 0 & 1 & 1 & -1\\ 0 & 0 & 2 & -5 & 1 & 1\end{array}\right]$

$\begin{array}{r}\frac{1}{2}R_1\to R_1\\ \frac{1}{2}R_2\to R_2\\ \frac{1}{2}R_3\to R_3\end{array}\quad \left[\begin{array}{ccc|ccc}1 & 0 & 0 & -\frac{11}{2} & -\frac{1}{2} & \frac{5}{2}\\ 0 & 1 & 0 & \frac{1}{2} & \frac{1}{2} & -\frac{1}{2}\\ 0 & 0 & 1 & -\frac{5}{2} & \frac{1}{2} & \frac{1}{2}\end{array}\right]$

$A^{-1}=\begin{bmatrix}-\frac{11}{2} & -\frac{1}{2} & \frac{5}{2}\\ \frac{1}{2} & \frac{1}{2} & -\frac{1}{2}\\ -\frac{5}{2} & \frac{1}{2} & \frac{1}{2}\end{bmatrix}$

25. Let $A = \begin{bmatrix} 1 & -2 & 3 & 0 \\ 0 & 1 & -1 & 1 \\ -2 & 2 & -2 & 4 \\ 0 & 2 & -3 & 1 \end{bmatrix}$.

$$[A|I] = \left[\begin{array}{rrrr|rrrr} 1 & -2 & 3 & 0 & 1 & 0 & 0 & 0 \\ 0 & 1 & -1 & 1 & 0 & 1 & 0 & 0 \\ -2 & 2 & -2 & 4 & 0 & 0 & 1 & 0 \\ 0 & 2 & -3 & 1 & 0 & 0 & 0 & 1 \end{array}\right]$$

$$2R_1 + R_3 \to R_3 \quad \left[\begin{array}{rrrr|rrrr} 1 & -2 & 3 & 0 & 1 & 0 & 0 & 0 \\ 0 & 1 & -1 & 1 & 0 & 1 & 0 & 0 \\ 0 & -2 & 4 & 4 & 2 & 0 & 1 & 0 \\ 0 & 2 & -3 & 1 & 0 & 0 & 0 & 1 \end{array}\right]$$

$$\begin{array}{r} 2R_2 + R_1 \to R_1 \\ \\ 2R_2 + R_3 \to R_3 \\ -2R_2 + R_4 \to R_4 \end{array} \left[\begin{array}{rrrr|rrrr} 1 & 0 & 1 & 2 & 1 & 2 & 0 & 0 \\ 0 & 1 & -1 & 1 & 0 & 1 & 0 & 0 \\ 0 & 0 & 2 & 6 & 2 & 2 & 1 & 0 \\ 0 & 0 & -1 & -1 & 0 & -2 & 0 & 1 \end{array}\right]$$

$$\begin{array}{r} R_3 + (-2)R_1 \to R_1 \\ R_3 + 2R_2 \to R_2 \\ \\ R_3 + 2R_4 \to R_4 \end{array} \left[\begin{array}{rrrr|rrrr} -2 & 0 & 0 & 2 & 0 & -2 & 1 & 0 \\ 0 & 2 & 0 & 8 & 2 & 4 & 1 & 0 \\ 0 & 0 & 2 & 6 & 2 & 2 & 1 & 0 \\ 0 & 0 & 0 & 4 & 2 & -2 & 1 & 2 \end{array}\right]$$

$$\begin{array}{r} -2R_1 + R_4 \to R_1 \\ R_2 + (-2)R_4 \to R_2 \\ 2R_3 + (-3)R_4 \to R_3 \\ \\ \end{array} \left[\begin{array}{rrrr|rrrr} 4 & 0 & 0 & 0 & 2 & 2 & -1 & 2 \\ 0 & 2 & 0 & 0 & -2 & 8 & -1 & -4 \\ 0 & 0 & 4 & 0 & -2 & 10 & -1 & -6 \\ 0 & 0 & 0 & 4 & 2 & -2 & 1 & 2 \end{array}\right]$$

$$\begin{array}{r} \frac{1}{4}R_1 \to R_1 \\ \frac{1}{2}R_2 \to R_2 \\ \frac{1}{4}R_3 \to R_3 \\ \frac{1}{4}R_4 \to R_4 \end{array} \left[\begin{array}{rrrr|rrrr} 1 & 0 & 0 & 0 & \frac{1}{2} & \frac{1}{2} & -\frac{1}{4} & \frac{1}{2} \\ 0 & 1 & 0 & 0 & -1 & 4 & -\frac{1}{2} & -2 \\ 0 & 0 & 1 & 0 & -\frac{1}{2} & \frac{5}{2} & -\frac{1}{4} & -\frac{3}{2} \\ 0 & 0 & 0 & 1 & \frac{1}{2} & -\frac{1}{2} & \frac{1}{4} & \frac{1}{2} \end{array}\right]$$

$$A^{-1} = \begin{bmatrix} \frac{1}{2} & \frac{1}{2} & -\frac{1}{4} & \frac{1}{2} \\ -1 & 4 & -\frac{1}{2} & -2 \\ -\frac{1}{2} & \frac{5}{2} & -\frac{1}{4} & -\frac{3}{2} \\ \frac{1}{2} & -\frac{1}{2} & \frac{1}{4} & \frac{1}{2} \end{bmatrix}$$

27. $2x + 5y = 15$
$x + 4y = 9$

First, write the system in matrix form.

$$\begin{bmatrix} 2 & 5 \\ 1 & 4 \end{bmatrix}\begin{bmatrix} x \\ y \end{bmatrix} = \begin{bmatrix} 15 \\ 9 \end{bmatrix}$$

Let $A = \begin{bmatrix} 2 & 5 \\ 1 & 4 \end{bmatrix}$, $X = \begin{bmatrix} x \\ y \end{bmatrix}$, and $B = \begin{bmatrix} 15 \\ 9 \end{bmatrix}$.

The system in matrix form is $AX = B$. We wish to find $X = A^{-1}AX = A^{-1}B$. Use row operations to find A^{-1}.

$$[A|I] = \left[\begin{array}{rr|rr} 2 & 5 & 1 & 0 \\ 1 & 4 & 0 & 1 \end{array}\right]$$

$$-1R_1 + 2R_2 \to R_2 \quad \left[\begin{array}{rr|rr} 2 & 5 & 1 & 0 \\ 0 & 3 & -1 & 2 \end{array}\right]$$

(continued on next page)

(*continued*)

$$-5R_2+3R_1 \to R_1 \quad \left[\begin{array}{cc|cc} 6 & 0 & 8 & -10 \\ 0 & 3 & -1 & 2 \end{array}\right]$$

$$\begin{array}{l} \frac{1}{6}R_1 \to R_1 \\ \frac{1}{3}R_2 \to R_2 \end{array} \left[\begin{array}{cc|cc} 1 & 0 & \frac{4}{3} & -\frac{5}{3} \\ 0 & 1 & -\frac{1}{3} & \frac{2}{3} \end{array}\right]$$

$$A^{-1} = \begin{bmatrix} \frac{4}{3} & -\frac{5}{3} \\ -\frac{1}{3} & \frac{2}{3} \end{bmatrix} = \frac{1}{3}\begin{bmatrix} 4 & -5 \\ -1 & 2 \end{bmatrix}$$

Next find the product $A^{-1}B$.

$$X = A^{-1}B = \begin{bmatrix} \frac{4}{3} & -\frac{5}{3} \\ -\frac{1}{3} & \frac{2}{3} \end{bmatrix}\begin{bmatrix} 15 \\ 9 \end{bmatrix}$$

$$= \frac{1}{3}\begin{bmatrix} 4 & -5 \\ -1 & 2 \end{bmatrix}\begin{bmatrix} 15 \\ 9 \end{bmatrix}$$

$$= \frac{1}{3}\begin{bmatrix} 15 \\ 3 \end{bmatrix} = \begin{bmatrix} 5 \\ 1 \end{bmatrix}$$

Thus, the solution is (5, 1).

29. $2x + y = 5$
$5x + 3y = 13$

Let $A = \begin{bmatrix} 2 & 1 \\ 5 & 3 \end{bmatrix}$, $X = \begin{bmatrix} x \\ y \end{bmatrix}$, $B = \begin{bmatrix} 5 \\ 13 \end{bmatrix}$.

Use row operations to find A^{-1}.

$$[A|I] = \left[\begin{array}{cc|cc} 2 & 1 & 1 & 0 \\ 5 & 3 & 0 & 1 \end{array}\right]$$

$$-3R_1+R_2 \to R_2 \left[\begin{array}{cc|cc} 2 & 1 & 1 & 0 \\ -1 & 0 & -3 & 1 \end{array}\right]$$

Interchange R_1 and R_2.

$$\left[\begin{array}{cc|cc} -1 & 0 & -3 & 1 \\ 2 & 1 & 1 & 0 \end{array}\right]$$

$$2R_1+R_2 \to R_2 \left[\begin{array}{cc|cc} -1 & 0 & -3 & 1 \\ 0 & 1 & -5 & 2 \end{array}\right]$$

$$-R_1 \to R_1 \left[\begin{array}{cc|cc} 1 & 0 & 3 & -1 \\ 0 & 1 & -5 & 2 \end{array}\right]$$

$$A^{-1} = \begin{bmatrix} 3 & -1 \\ -5 & 2 \end{bmatrix}$$

Next find the product $A^{-1}B$.

$$X = A^{-1}B = \begin{bmatrix} 3 & -1 \\ -5 & 2 \end{bmatrix}\begin{bmatrix} 5 \\ 13 \end{bmatrix} = \begin{bmatrix} 2 \\ 1 \end{bmatrix}$$

The solution is (2, 1).

31. $3x - 2y = 3$
$7x - 5y = 0$

Let $A = \begin{bmatrix} 3 & -2 \\ 7 & -5 \end{bmatrix}$, $X = \begin{bmatrix} x \\ y \end{bmatrix}$, and $B = \begin{bmatrix} 3 \\ 0 \end{bmatrix}$.

Use row operations to find A^{-1}.

$$[A|I] = \left[\begin{array}{cc|cc} 3 & -2 & 1 & 0 \\ 7 & -5 & 0 & 1 \end{array}\right]$$

$$-7R_1+3R_2 \to R_2 \left[\begin{array}{cc|cc} 3 & -2 & 1 & 0 \\ 0 & -1 & -7 & 3 \end{array}\right]$$

$$-2R_2+R_1 \to R_1 \left[\begin{array}{cc|cc} 3 & 0 & 15 & -6 \\ 0 & -1 & -7 & 3 \end{array}\right]$$

$$\begin{array}{l} \frac{1}{3}R_1 \to R_1 \\ -1R_2 \to R_2 \end{array} \left[\begin{array}{cc|cc} 1 & 0 & 5 & -2 \\ 0 & 1 & 7 & -3 \end{array}\right]$$

$$A^{-1} = \begin{bmatrix} 5 & -2 \\ 7 & -3 \end{bmatrix}$$

Next find the product $A^{-1}B$.

$$X = A^{-1}B = \begin{bmatrix} 5 & -2 \\ 7 & -3 \end{bmatrix}\begin{bmatrix} 3 \\ 0 \end{bmatrix} = \begin{bmatrix} 15 \\ 21 \end{bmatrix}$$

Thus, the solution is (15, 21).

33. $-x - 8y = 12$
$3x + 24y = -36$

Let $A = \begin{bmatrix} -1 & -8 \\ 3 & 24 \end{bmatrix}$, $X = \begin{bmatrix} x \\ y \end{bmatrix}$, $B = \begin{bmatrix} 12 \\ -36 \end{bmatrix}$.

Use row operations to find A^{-1}.

$$[A|I] = \left[\begin{array}{cc|cc} -1 & -8 & 1 & 0 \\ 3 & 24 & 0 & 1 \end{array}\right]$$

$$3R_1+R_2 \to R_2 \left[\begin{array}{cc|cc} -1 & -8 & 1 & 0 \\ 0 & 0 & 3 & 1 \end{array}\right]$$

The zeros in the second row indicate that matrix A does not have an inverse. We cannot complete the solution by this method.
Since the second equation is a multiple of the first, the equations are dependent. Solve the first equation of the system for x.

$$-x - 8y = 12 \Rightarrow -x = 8y + 12 \Rightarrow x = -8y - 12$$

The solution is $(-8y - 12, y)$, where y is any real number.

35. $$\begin{aligned} -x - y - z &= 1 \\ 4x + 5y \quad &= -2 \\ y - 3z &= 3 \end{aligned}$$

has coefficient matrix

$$A = \begin{bmatrix} -1 & -1 & -1 \\ 4 & 5 & 0 \\ 0 & 1 & -3 \end{bmatrix}.$$

In Exercise 19, we found

$$A^{-1} = \begin{bmatrix} 15 & 4 & -5 \\ -12 & -3 & 4 \\ -4 & -1 & 1 \end{bmatrix}.$$

Since $X = A^{-1}B$,

$$\begin{bmatrix} x \\ y \\ z \end{bmatrix} = \begin{bmatrix} 15 & 4 & -5 \\ -12 & -3 & 4 \\ -4 & -1 & 1 \end{bmatrix}\begin{bmatrix} 1 \\ -2 \\ 3 \end{bmatrix} = \begin{bmatrix} -8 \\ 6 \\ 1 \end{bmatrix}.$$

The solution is (–8, 6, 1).

37. $$\begin{aligned} x + 3y - 2z &= 4 \\ 2x + 7y - 3z &= 8 \\ 3x + 8y - 5z &= -4 \end{aligned}$$

has coefficient matrix

$$A = \begin{bmatrix} 1 & 3 & -2 \\ 2 & 7 & -3 \\ 3 & 8 & -5 \end{bmatrix}.$$

In Exercise 23, we found

$$A^{-1} = \begin{bmatrix} -\frac{11}{2} & -\frac{1}{2} & \frac{5}{2} \\ \frac{1}{2} & \frac{1}{2} & -\frac{1}{2} \\ -\frac{5}{2} & \frac{1}{2} & \frac{1}{2} \end{bmatrix} = \frac{1}{2}\begin{bmatrix} -11 & -1 & 5 \\ 1 & 1 & -1 \\ -5 & 1 & 1 \end{bmatrix}.$$

Since $X = A^{-1}B$,

$$\begin{bmatrix} x \\ y \\ z \end{bmatrix} = \frac{1}{2}\begin{bmatrix} -11 & -1 & 5 \\ 1 & 1 & -1 \\ -5 & 1 & 1 \end{bmatrix}\begin{bmatrix} 4 \\ 8 \\ -4 \end{bmatrix} = \frac{1}{2}\begin{bmatrix} -72 \\ 16 \\ -16 \end{bmatrix} = \begin{bmatrix} -36 \\ 8 \\ -8 \end{bmatrix}$$

Thus, the solution is (–36, 8, –8).

39. $$\begin{aligned} 2x - 2y \quad &= 5 \\ 4y + 8z &= 7 \\ x \quad + 2z &= 1 \end{aligned}$$

has coefficient matrix

$$A = \begin{bmatrix} 2 & -2 & 0 \\ 0 & 4 & 8 \\ 1 & 0 & 2 \end{bmatrix}.$$

First calculate A^{-1} using row operations.

$$[A|I] = \left[\begin{array}{ccc|ccc} 2 & -2 & 0 & 1 & 0 & 0 \\ 0 & 4 & 8 & 0 & 1 & 0 \\ 1 & 0 & 2 & 0 & 0 & 1 \end{array}\right]$$

Interchange R_1 and R_3.

$$\left[\begin{array}{ccc|ccc} 1 & 0 & 2 & 0 & 0 & 1 \\ 0 & 4 & 8 & 0 & 1 & 0 \\ 2 & -2 & 0 & 1 & 0 & 0 \end{array}\right]$$

$$-2R_1 + R_3 \to R_3 \left[\begin{array}{ccc|ccc} 1 & 0 & 2 & 0 & 0 & 1 \\ 0 & 4 & 8 & 0 & 1 & 0 \\ 0 & -2 & -4 & 1 & 0 & -2 \end{array}\right]$$

$$2R_2 + R_3 \to R_3 \left[\begin{array}{ccc|ccc} 1 & 0 & 2 & 0 & 0 & 1 \\ 0 & 4 & 8 & 0 & 1 & 0 \\ 0 & 0 & 0 & 1 & 2 & -2 \end{array}\right]$$

The zeros in the third row indicate that matrix *A* does not have an inverse. Since the original equations are not multiples of each other, the system has no solution.

41. $$\begin{aligned} x - 2y + 3z \quad &= 4 \\ y - z + w &= -8 \\ -2x + 2y - 2z + 4w &= 12 \\ 2y - 3z + w &= -4 \end{aligned}$$

has coefficient matrix

$$A = \begin{bmatrix} 1 & -2 & 3 & 0 \\ 0 & 1 & -1 & 1 \\ -2 & 2 & -2 & 4 \\ 0 & 2 & -3 & 1 \end{bmatrix}.$$

In Exercise 25,we found

$$A^{-1} = \begin{bmatrix} \frac{1}{2} & \frac{1}{2} & -\frac{1}{4} & \frac{1}{2} \\ -1 & 4 & -\frac{1}{2} & -2 \\ -\frac{1}{2} & \frac{5}{2} & -\frac{1}{4} & -\frac{3}{2} \\ \frac{1}{2} & -\frac{1}{2} & \frac{1}{4} & \frac{1}{2} \end{bmatrix}.$$

Since $X = A^{-1}B$,

$$\begin{bmatrix} x \\ y \\ z \\ w \end{bmatrix} = \begin{bmatrix} \frac{1}{2} & \frac{1}{2} & -\frac{1}{4} & \frac{1}{2} \\ -1 & 4 & -\frac{1}{2} & -2 \\ -\frac{1}{2} & \frac{5}{2} & -\frac{1}{4} & -\frac{3}{2} \\ \frac{1}{2} & -\frac{1}{2} & \frac{1}{4} & \frac{1}{2} \end{bmatrix}\begin{bmatrix} 4 \\ -8 \\ 12 \\ -4 \end{bmatrix} = \begin{bmatrix} -7 \\ -34 \\ -19 \\ 7 \end{bmatrix}.$$

The solution is (−7, −34, −19, 7).

In Exercises 43–48, let $A = \begin{bmatrix} a & b \\ c & d \end{bmatrix}$.

43. $IA = \begin{bmatrix} 1 & 0 \\ 0 & 1 \end{bmatrix}\begin{bmatrix} a & b \\ c & d \end{bmatrix} = \begin{bmatrix} a & b \\ c & d \end{bmatrix} = A$

Thus, $IA = A$.

45. $A \cdot 0 = \begin{bmatrix} a & b \\ c & d \end{bmatrix}\begin{bmatrix} 0 & 0 \\ 0 & 0 \end{bmatrix} = \begin{bmatrix} 0 & 0 \\ 0 & 0 \end{bmatrix} = 0$

Thus, $A \cdot 0 = 0$.

47. In Exercise 46, it was found that

$$A^{-1} = \frac{1}{ad - bc}\begin{bmatrix} d & -b \\ -c & a \end{bmatrix}.$$

$$A^{-1}A = \left(\frac{1}{ad - bc}\begin{bmatrix} d & -b \\ -c & a \end{bmatrix}\right)\begin{bmatrix} a & b \\ c & d \end{bmatrix} = \frac{1}{ad - bc}\left(\begin{bmatrix} d & -b \\ -c & a \end{bmatrix}\begin{bmatrix} a & b \\ c & d \end{bmatrix}\right)$$

$$= \frac{1}{ad - bc}\begin{bmatrix} ad - bc & 0 \\ 0 & ad - bc \end{bmatrix} = \begin{bmatrix} 1 & 0 \\ 0 & 1 \end{bmatrix} = I$$

Thus, $A^{-1}A = I$.

49. $AB = O \Rightarrow A^{-1}(AB) = A^{-1} \cdot O \Rightarrow (A^{-1}A)B = O \Rightarrow I \cdot B = O \Rightarrow B = O$

Thus, if $AB = O$ and A^{-1} exists, then $B = O$.

51. This exercise should be solved by graphing calculator or computer methods. The solution, which may vary slightly, is

$$C^{-1} = \begin{bmatrix} -0.0477 & -0.0230 & 0.0292 & 0.0895 & -0.0402 \\ 0.0921 & 0.0150 & 0.0321 & 0.0209 & -0.0276 \\ -0.0678 & 0.0315 & -0.0404 & 0.0326 & 0.0373 \\ 0.0171 & -0.0248 & 0.0069 & -0.0003 & 0.0246 \\ -0.0208 & 0.0740 & 0.0096 & -0.1018 & 0.0646 \end{bmatrix}.$$

(Entries are rounded to 4 places.)

53. This exercise should be solved by graphing calculator or computer methods. The solution, which may vary slightly, is

$$D^{-1} = \begin{bmatrix} 0.0394 & 0.0880 & 0.0033 & 0.0530 & -0.1499 \\ -0.1492 & 0.0289 & 0.0187 & 0.1033 & 0.1668 \\ -0.1330 & -0.0543 & 0.0356 & 0.1768 & 0.1055 \\ 0.1407 & 0.0175 & -0.0453 & -0.1344 & 0.0655 \\ 0.0102 & -0.0653 & 0.0993 & 0.0085 & -0.0388 \end{bmatrix}.$$

(Entries are rounded to 4 places.)

55. This exercise should be solved by graphing calculator or computer methods. The solution may vary slightly. The answer is, yes, $D^{-1}C^{-1} = (CD)^{-1}$.

57. This exercise should be solved by graphing calculator or computer methods. The solution, which may vary slightly, is

$$\begin{bmatrix} 1.51482 \\ 0.053479 \\ -0.637242 \\ 0.462629 \end{bmatrix}.$$

59. Let x = the number of Super Vim tablets, y = the number of Multitab tablets, and z = the number of Mighty Mix tablets.
The total number of vitamins is $x+y+z$.
The total amount of niacin is $15x+20y+25z$.
The total amount of Vitamin E is $12x+15y+35z$.

(a) The system to be solved is

$$\begin{aligned} x+\quad y+\quad z &= 225 \\ 15x+20y+25z &= 4750 \\ 12x+15y+35z &= 5225. \end{aligned}$$

Let

$$A=\begin{bmatrix} 1 & 1 & 1 \\ 15 & 20 & 25 \\ 20 & 15 & 35 \end{bmatrix},\ X=\begin{bmatrix} x \\ y \\ z \end{bmatrix},\ B=\begin{bmatrix} 225 \\ 4750 \\ 5225 \end{bmatrix}.$$

Thus, $AX=B$ and

$$\begin{bmatrix} 1 & 1 & 1 \\ 15 & 20 & 25 \\ 12 & 15 & 35 \end{bmatrix}\begin{bmatrix} x \\ y \\ z \end{bmatrix}=\begin{bmatrix} 225 \\ 4750 \\ 5225 \end{bmatrix}.$$

Use row operations to obtain the inverse of the coefficient matrix.

$$A^{-1}=\begin{bmatrix} \frac{65}{17} & -\frac{4}{17} & \frac{1}{17} \\ -\frac{45}{17} & \frac{23}{85} & -\frac{2}{17} \\ -\frac{3}{17} & -\frac{3}{85} & \frac{1}{17} \end{bmatrix}$$

Since $X=A^{-1}B$,

$$\begin{bmatrix} x \\ y \\ z \end{bmatrix}=\begin{bmatrix} \frac{65}{17} & -\frac{4}{17} & \frac{1}{17} \\ -\frac{45}{17} & \frac{23}{85} & -\frac{2}{17} \\ -\frac{3}{17} & -\frac{3}{85} & \frac{1}{17} \end{bmatrix}\begin{bmatrix} 225 \\ 4750 \\ 5225 \end{bmatrix}=\begin{bmatrix} 50 \\ 75 \\ 100 \end{bmatrix}.$$

There are 50 Super Vim tablets, 75 Multitab tablets, and 100 Mighty Mix tablets.

(b) The matrix of constants is changed to

$$B=\begin{bmatrix} 185 \\ 3625 \\ 3750 \end{bmatrix}.$$

$$\begin{bmatrix} x \\ y \\ z \end{bmatrix}=\begin{bmatrix} \frac{65}{17} & -\frac{4}{17} & \frac{1}{17} \\ -\frac{45}{17} & \frac{23}{85} & -\frac{2}{17} \\ -\frac{3}{17} & -\frac{3}{85} & \frac{1}{17} \end{bmatrix}\begin{bmatrix} 185 \\ 3625 \\ 3750 \end{bmatrix}=\begin{bmatrix} 75 \\ 50 \\ 60 \end{bmatrix}$$

There are 75 Super Vim tablets, 50 Multitab tablets, and 60 Mighty Mix tablets.

(c) The matrix of constants is changed to

$$B=\begin{bmatrix} 230 \\ 4450 \\ 4210 \end{bmatrix}.$$

$$\begin{bmatrix} x \\ y \\ z \end{bmatrix}=\begin{bmatrix} \frac{65}{17} & -\frac{4}{17} & \frac{1}{17} \\ -\frac{45}{17} & \frac{23}{85} & -\frac{2}{17} \\ -\frac{3}{17} & -\frac{3}{85} & \frac{1}{17} \end{bmatrix}\begin{bmatrix} 230 \\ 4450 \\ 4210 \end{bmatrix}=\begin{bmatrix} 80 \\ 100 \\ 50 \end{bmatrix}$$

There are 80 Super Vim tablets, 100 Multitab tablets, and 50 Mighty Mix tablets.

61. (a) The matrix is $B=\begin{bmatrix} 72 \\ 48 \\ 60 \end{bmatrix}$.

(b) The matrix equation is

$$\begin{bmatrix} 2 & 4 & 2 \\ 2 & 1 & 2 \\ 2 & 1 & 3 \end{bmatrix}\begin{bmatrix} x_1 \\ x_2 \\ x_3 \end{bmatrix}=\begin{bmatrix} 72 \\ 48 \\ 60 \end{bmatrix}.$$

(c) To solve the system, begin by using row operations to find A^{-1}.

$$[A|I]=\left[\begin{array}{ccc|ccc} 2 & 4 & 2 & 1 & 0 & 0 \\ 2 & 1 & 2 & 0 & 1 & 0 \\ 2 & 1 & 3 & 0 & 0 & 1 \end{array}\right]$$

$$\begin{array}{r} \\ R_1-1R_2\to R_2 \\ R_1-1R_3\to R_3 \end{array}\left[\begin{array}{ccc|ccc} 2 & 4 & 2 & 1 & 0 & 0 \\ 0 & 3 & 0 & 1 & -1 & 0 \\ 0 & 3 & -1 & 1 & 0 & -1 \end{array}\right]$$

$$\begin{array}{r} -4R_2+3R_1\to R_1 \\ \\ R_2-1R_3\to R_3 \end{array}\left[\begin{array}{ccc|ccc} 6 & 0 & 6 & -1 & 4 & 0 \\ 0 & 3 & 0 & 1 & -1 & 0 \\ 0 & 0 & 1 & 0 & -1 & 1 \end{array}\right]$$

$$\begin{array}{r} -6R_3+R_1\to R_1 \\ \\ \\ \end{array}\left[\begin{array}{ccc|ccc} 6 & 0 & 0 & -1 & 10 & -6 \\ 0 & 3 & 0 & 1 & -1 & 0 \\ 0 & 0 & 1 & 0 & -1 & 1 \end{array}\right]$$

$$\begin{array}{r} \frac{1}{6}R_1\to R_1 \\ \frac{1}{3}R_2\to R_2 \\ \\ \end{array}\left[\begin{array}{ccc|ccc} 1 & 0 & 0 & -\frac{1}{6} & \frac{5}{3} & -1 \\ 0 & 1 & 0 & \frac{1}{3} & -\frac{1}{3} & 0 \\ 0 & 0 & 1 & 0 & -1 & 1 \end{array}\right]$$

(*continued on next page*)

The inverse matrix is $A^{-1} = \begin{bmatrix} -\frac{1}{6} & \frac{5}{3} & -1 \\ \frac{1}{3} & -\frac{1}{3} & 0 \\ 0 & -1 & 1 \end{bmatrix}$.

Since $X = A^{-1}B$, $\begin{bmatrix} x_1 \\ x_2 \\ x_3 \end{bmatrix} = \begin{bmatrix} -\frac{1}{6} & \frac{5}{3} & -1 \\ \frac{1}{3} & -\frac{1}{3} & 0 \\ 0 & -1 & 1 \end{bmatrix} \begin{bmatrix} 72 \\ 48 \\ 60 \end{bmatrix} = \begin{bmatrix} 8 \\ 8 \\ 12 \end{bmatrix}$.

There are 8 daily orders for type I, 8 for type II, and 12 for type III.

63. Let x = the amount invested in AAA bonds, y = the amount invested in A bonds, and z = amount invested in B bonds

(a) The total investment is $x + y + z = 25{,}000$. The annual return is $0.06x + 0.065y + 0.08z = 1650$. Since twice as much is invested in AAA bonds as in B bonds, $x = 2z$. The system to be solved is

$$\begin{aligned} x + y + z &= 25{,}000 \\ 0.06x + 0.065y + 0.08z &= 1650 \\ x - 2z &= 0 \end{aligned}$$

Let $A = \begin{bmatrix} 1 & 1 & 1 \\ 0.06 & 0.065 & 0.08 \\ 1 & 0 & -2 \end{bmatrix}$, $B = \begin{bmatrix} 25{,}000 \\ 1650 \\ 0 \end{bmatrix}$, and $X = \begin{bmatrix} x \\ y \\ z \end{bmatrix}$.

First, find the inverse of the coefficient matrix.

$$\left[\begin{array}{ccc|ccc} 1 & 1 & 1 & 1 & 0 & 0 \\ 0.06 & 0.065 & 0.08 & 0 & 1 & 0 \\ 1 & 0 & -2 & 0 & 0 & 1 \end{array}\right]$$

$$\begin{array}{r} \\ 1000R_2 \to R_2 \\ -R_1 + R_3 \to R_3 \end{array} \left[\begin{array}{ccc|ccc} 1 & 1 & 1 & 1 & 0 & 0 \\ 60 & 65 & 80 & 0 & 1000 & 0 \\ 0 & -1 & -3 & -1 & 0 & 1 \end{array}\right]$$

$$-60R_1 + R_2 \to R_2 \left[\begin{array}{ccc|ccc} 1 & 1 & 1 & 1 & 0 & 0 \\ 0 & 5 & 20 & -60 & 1000 & 0 \\ 0 & -1 & -3 & -1 & 0 & 1 \end{array}\right]$$

$$\frac{1}{5}(R_2 + 5R_3) \to R_3 \left[\begin{array}{ccc|ccc} 1 & 1 & 1 & 1 & 0 & 0 \\ 0 & 5 & 20 & -60 & 1000 & 0 \\ 0 & 0 & 1 & -13 & 200 & 1 \end{array}\right]$$

$$\begin{array}{r} -R_3 + R_1 \to R_1 \\ \frac{1}{5}(-20R_3 + R_2) \to R_2 \\ \end{array} \left[\begin{array}{ccc|ccc} 1 & 1 & 0 & 14 & -200 & -1 \\ 0 & 1 & 0 & 40 & -600 & -4 \\ 0 & 0 & 1 & -13 & 200 & 1 \end{array}\right]$$

$$-R_2 + R_1 \to R_1 \left[\begin{array}{ccc|ccc} 1 & 0 & 0 & -26 & 400 & 3 \\ 0 & 1 & 0 & 40 & -600 & -4 \\ 0 & 0 & 1 & -13 & 200 & 1 \end{array}\right] \Rightarrow A^{-1} = \begin{bmatrix} -26 & 400 & 3 \\ 40 & -600 & -4 \\ -13 & 200 & 1 \end{bmatrix}.$$

(continued on next page)

(continued)

Solve the matrix equation $X = A^{-1}B$.

$$X = \begin{bmatrix} x \\ y \\ z \end{bmatrix} = \begin{bmatrix} -26 & 400 & 3 \\ 40 & -600 & -4 \\ -13 & 200 & 1 \end{bmatrix} \begin{bmatrix} 25,000 \\ 1650 \\ 0 \end{bmatrix} = \begin{bmatrix} 10,000 \\ 10,000 \\ 5000 \end{bmatrix}$$

\$10,000 should be invested at 6% in AAA bonds, \$10,000 at 6.5% in A bonds, and \$5000 at 8% in B bonds.

(b) The matrix of constants is changed to

$$B = \begin{bmatrix} 30,000 \\ 1985 \\ 0 \end{bmatrix}.$$

$$X = \begin{bmatrix} x \\ y \\ z \end{bmatrix} = \begin{bmatrix} -26 & 400 & 3 \\ 40 & -600 & -4 \\ -13 & 200 & 1 \end{bmatrix} \begin{bmatrix} 30,000 \\ 1985 \\ 0 \end{bmatrix} = \begin{bmatrix} 14,000 \\ 9000 \\ 7000 \end{bmatrix}$$

\$14,000 should be invested at 6% in AAA bonds, \$9000 at 6.5% in A bonds, and \$7000 at 8% in B bonds.

(c) The matrix of constants is changed to

$$B = \begin{bmatrix} 40,000 \\ 2660 \\ 0 \end{bmatrix}.$$

$$X = \begin{bmatrix} x \\ y \\ z \end{bmatrix} = \begin{bmatrix} -26 & 400 & 3 \\ 40 & -600 & -4 \\ -13 & 200 & 1 \end{bmatrix} \begin{bmatrix} 40,000 \\ 2660 \\ 0 \end{bmatrix} = \begin{bmatrix} 24,000 \\ 4000 \\ 12,000 \end{bmatrix}$$

\$24,000 should be invested at 6% in AAA bonds, \$4000 at 6.5% in A bonds, and \$12,000 at 8% in B bonds.

65. (a) First, divide the letters and spaces of the sentence into groups of 3, writing each group as a column vector.

$$\begin{bmatrix} A \\ l \\ l \end{bmatrix}, \begin{bmatrix} \text{(space)} \\ i \\ s \end{bmatrix}, \begin{bmatrix} \text{(space)} \\ f \\ a \end{bmatrix}, \begin{bmatrix} i \\ r \\ \text{(space)} \end{bmatrix}, \begin{bmatrix} i \\ n \\ \text{(space)} \end{bmatrix}, \begin{bmatrix} l \\ o \\ v \end{bmatrix}, \begin{bmatrix} e \\ \text{(space)} \\ a \end{bmatrix}, \begin{bmatrix} n \\ d \\ \text{(space)} \end{bmatrix}, \begin{bmatrix} w \\ a \\ r \end{bmatrix}$$

Next, convert each letter into a number, assigning 1 to A, 2 to B, and so on, with the number 27 used to represent each space between words.

$$\begin{bmatrix} 1 \\ 12 \\ 12 \end{bmatrix}, \begin{bmatrix} 27 \\ 9 \\ 19 \end{bmatrix}, \begin{bmatrix} 27 \\ 6 \\ 1 \end{bmatrix}, \begin{bmatrix} 9 \\ 18 \\ 27 \end{bmatrix}, \begin{bmatrix} 9 \\ 14 \\ 27 \end{bmatrix}, \begin{bmatrix} 12 \\ 15 \\ 22 \end{bmatrix}, \begin{bmatrix} 5 \\ 27 \\ 1 \end{bmatrix}, \begin{bmatrix} 14 \\ 4 \\ 27 \end{bmatrix}, \begin{bmatrix} 23 \\ 1 \\ 18 \end{bmatrix}$$

Now find the product of the coding matrix presented in Example 7,

$$A = \begin{bmatrix} 1 & 3 & 4 \\ 2 & 1 & 3 \\ 4 & 2 & 1 \end{bmatrix},$$ and each column vector above. This produces a new set of vectors, which represents the coded message.

$$\begin{bmatrix} 85 \\ 50 \\ 40 \end{bmatrix}, \begin{bmatrix} 130 \\ 120 \\ 145 \end{bmatrix}, \begin{bmatrix} 49 \\ 63 \\ 121 \end{bmatrix}, \begin{bmatrix} 171 \\ 117 \\ 99 \end{bmatrix}, \begin{bmatrix} 159 \\ 113 \\ 91 \end{bmatrix}, \begin{bmatrix} 145 \\ 105 \\ 100 \end{bmatrix}, \begin{bmatrix} 90 \\ 40 \\ 75 \end{bmatrix}, \begin{bmatrix} 134 \\ 113 \\ 91 \end{bmatrix}, \begin{bmatrix} 98 \\ 101 \\ 112 \end{bmatrix}$$

The message will be transmitted as 85, 50, 40, 130, 120, 145, 49, 63, 121, 171, 117, 99, 159, 113, 91, 145, 105, 100, 90, 40, 75, 134, 113, 91, 98, 101, 112.

(b) First, divide the coded message into groups of three numbers and form each group into a column vector.

$$\begin{bmatrix} 138 \\ 81 \\ 102 \end{bmatrix}, \begin{bmatrix} 101 \\ 67 \\ 109 \end{bmatrix}, \begin{bmatrix} 162 \\ 124 \\ 173 \end{bmatrix}, \begin{bmatrix} 210 \\ 150 \\ 165 \end{bmatrix}$$

(continued on next page)

(*continued*)

Next, find the product of the decoding matrix presented in Example 6, the inverse of matrix A in part (a) above,

$A^{-1} = \begin{bmatrix} -0.2 & 0.2 & 0.2 \\ 0.4 & -0.6 & 0.2 \\ 0 & 0.4 & -0.2 \end{bmatrix}$, and each of the column vectors above. This produces a new set of vectors,

which represents the decoded message.

$$\begin{bmatrix} 9 \\ 27 \\ 12 \end{bmatrix}, \begin{bmatrix} 15 \\ 22 \\ 5 \end{bmatrix}, \begin{bmatrix} 27 \\ 25 \\ 15 \end{bmatrix}, \begin{bmatrix} 21 \\ 27 \\ 27 \end{bmatrix}$$

Lastly, convert each number into a letter, assigning A to 1, B to 2, and so on, with the number 27 used to represent each space between words. The decoded message is I LOVE YOU.

67. **(a)** $[B|I] = \left[\begin{array}{ccc|ccc} 50 & 50 & 45 & 1 & 0 & 0 \\ 0 & 15 & 20 & 0 & 1 & 0 \\ 1 & 1 & 1 & 0 & 0 & 1 \end{array}\right]$

Interchange rows 1 and 3.

$$\left[\begin{array}{ccc|ccc} 1 & 1 & 1 & 0 & 0 & 1 \\ 0 & 15 & 20 & 0 & 1 & 0 \\ 50 & 50 & 45 & 1 & 0 & 0 \end{array}\right]$$

$$\begin{array}{r} \\ \\ -50R_1 + R_3 \to R_3 \end{array} \left[\begin{array}{ccc|ccc} 1 & 1 & 1 & 0 & 0 & 1 \\ 0 & 15 & 20 & 0 & 1 & 0 \\ 0 & 0 & -5 & 1 & 0 & -50 \end{array}\right]$$

$$\begin{array}{r} R_2 + (-15)R_1 \to R_1 \\ \\ \\ \end{array} \left[\begin{array}{ccc|ccc} -15 & 0 & 5 & 0 & 1 & -15 \\ 0 & 15 & 20 & 0 & 1 & 0 \\ 0 & 0 & -5 & 1 & 0 & -50 \end{array}\right]$$

$$\begin{array}{r} R_3 + R_1 \to R_1 \\ 4R_3 + R_2 \to R_2 \\ \\ \end{array} \left[\begin{array}{ccc|ccc} -15 & 0 & 0 & 1 & 1 & -65 \\ 0 & 15 & 0 & 4 & 1 & -200 \\ 0 & 0 & -5 & 1 & 0 & -50 \end{array}\right]$$

$$\begin{array}{r} -\frac{1}{15}R_1 \to R_1 \\ \frac{1}{15}R_2 \to R_2 \\ -\frac{1}{5}R_3 \to R_3 \end{array} \left[\begin{array}{ccc|ccc} 1 & 0 & 0 & -\frac{1}{15} & -\frac{1}{15} & \frac{13}{3} \\ 0 & 1 & 0 & \frac{4}{15} & \frac{1}{15} & -\frac{40}{3} \\ 0 & 0 & 1 & -\frac{1}{5} & 0 & 10 \end{array}\right] \Rightarrow B^{-1} = \begin{bmatrix} -\frac{1}{15} & -\frac{1}{15} & \frac{13}{3} \\ \frac{4}{15} & \frac{1}{15} & -\frac{40}{3} \\ -\frac{1}{5} & 0 & 10 \end{bmatrix}$$

(b) $A = \begin{bmatrix} 40 & 55 & 60 \\ 10 & 10 & 15 \\ 1 & 1 & 1 \end{bmatrix} B^{-1} = \begin{bmatrix} 40 & 55 & 60 \\ 10 & 10 & 15 \\ 1 & 1 & 1 \end{bmatrix} \begin{bmatrix} -\frac{1}{15} & -\frac{1}{15} & \frac{13}{3} \\ \frac{4}{15} & \frac{1}{15} & -\frac{40}{3} \\ -\frac{1}{5} & 0 & 10 \end{bmatrix} = \begin{bmatrix} 0 & 1 & 40 \\ -1 & 0 & 60 \\ 0 & 0 & 1 \end{bmatrix}$

(c) Denoting the original positions of the band members as x_1, x_2, and so on, the original shape was

	30	35	40	45	50	55	60	65
25								
20			x_1	x_2	x_3	x_4	x_5	
15					x_6			
10					x_7			
5					x_8			
0					x_9			

We are given that band member x_9, originally positioned at (50, 0), moved to (40, 10); band member x_6, originally at (50, 15), moved to (55, 10); band member x_2, originally at (45, 20), moved to (60, 15).

To find the new position of x_1, originally at (40, 20), multiply A by the vector $\begin{bmatrix} 40 \\ 20 \\ 1 \end{bmatrix}$.

$$A\begin{bmatrix} 40 \\ 20 \\ 1 \end{bmatrix} = \begin{bmatrix} 0 & 1 & 40 \\ -1 & 0 & 60 \\ 0 & 0 & 1 \end{bmatrix}\begin{bmatrix} 40 \\ 20 \\ 1 \end{bmatrix} = \begin{bmatrix} 60 \\ 20 \\ 1 \end{bmatrix}$$

The new position of band member x_1 is (60, 20).

To find the new position of x_3, originally at (50, 20), multiply A by the vector $\begin{bmatrix} 50 \\ 20 \\ 1 \end{bmatrix}$.

$$A\begin{bmatrix} 50 \\ 20 \\ 1 \end{bmatrix} = \begin{bmatrix} 0 & 1 & 40 \\ -1 & 0 & 60 \\ 0 & 0 & 1 \end{bmatrix}\begin{bmatrix} 50 \\ 20 \\ 1 \end{bmatrix} = \begin{bmatrix} 60 \\ 10 \\ 1 \end{bmatrix}$$

The new position of band member x_3 is (60, 10).

To find the new position of x_4, originally at (55, 20), multiply A by the vector $\begin{bmatrix} 55 \\ 20 \\ 1 \end{bmatrix}$.

$$A\begin{bmatrix} 55 \\ 20 \\ 1 \end{bmatrix} = \begin{bmatrix} 0 & 1 & 40 \\ -1 & 0 & 60 \\ 0 & 0 & 1 \end{bmatrix}\begin{bmatrix} 55 \\ 20 \\ 1 \end{bmatrix} = \begin{bmatrix} 60 \\ 5 \\ 1 \end{bmatrix}$$

The new position of band member x_4 is (60, 5).

To find the new position of x_5, originally at (60, 20), multiply A by the vector $\begin{bmatrix} 60 \\ 20 \\ 1 \end{bmatrix}$.

$$A\begin{bmatrix} 60 \\ 20 \\ 1 \end{bmatrix} = \begin{bmatrix} 0 & 1 & 40 \\ -1 & 0 & 60 \\ 0 & 0 & 1 \end{bmatrix}\begin{bmatrix} 60 \\ 20 \\ 1 \end{bmatrix} = \begin{bmatrix} 60 \\ 0 \\ 1 \end{bmatrix}$$

The new position of band member x_5 is (60, 0).

To find the new position of x_7, originally at (50, 10), multiply A by the vector $\begin{bmatrix} 50 \\ 10 \\ 1 \end{bmatrix}$.

(continued on next page)

(*continued*)

$$A\begin{bmatrix}50\\10\\1\end{bmatrix}=\begin{bmatrix}0&1&40\\-1&0&60\\0&0&1\end{bmatrix}\begin{bmatrix}50\\10\\1\end{bmatrix}=\begin{bmatrix}50\\10\\1\end{bmatrix}$$

The new position of band member x_7 is (50, 10).

To find the new position of x_8, originally at (50, 5), multiply A by the vector $\begin{bmatrix}50\\5\\1\end{bmatrix}$.

$$A\begin{bmatrix}50\\5\\1\end{bmatrix}=\begin{bmatrix}0&1&40\\-1&0&60\\0&0&1\end{bmatrix}\begin{bmatrix}50\\5\\1\end{bmatrix}=\begin{bmatrix}45\\10\\1\end{bmatrix}$$

The new position of band member x_8 is (45, 10). The new position of the band is

25								
20							x_1	
15							x_2	
10			x_9	x_8	x_7	x_6	x_3	
5							x_4	
0							x_5	
	30	35	40	45	50	55	60	65

Thus, the new shape is a sideways T whose vertical and horizontal intersection is at mark (60, 10).

10.5 Eigenvalues and Eigenvectors

1. $\det\begin{bmatrix}3&-1\\2&5\end{bmatrix}=3(5)-(-1)(2)=17$

3. To find $\det\begin{bmatrix}4&1&0\\-1&7&-2\\2&3&5\end{bmatrix}$, copy the first two columns of the matrix on the right, then add the products of the numbers along each of the three diagonals going from upper left to lower right, and subtract the products of the numbers along each of the three diagonals from upper right to lower left.

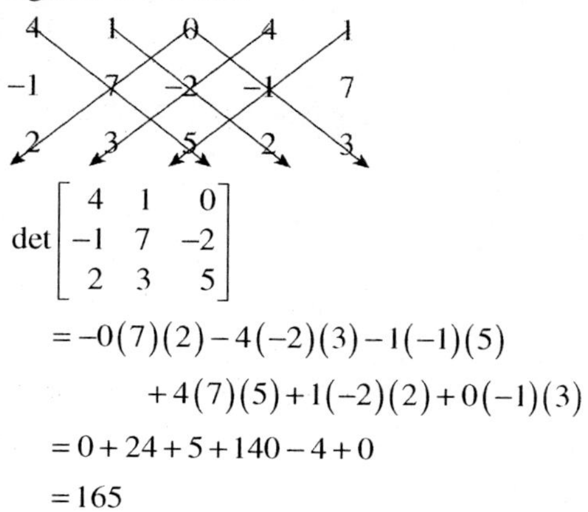

$$\det\begin{bmatrix}4&1&0\\-1&7&-2\\2&3&5\end{bmatrix}$$

$$=-0(7)(2)-4(-2)(3)-1(-1)(5)$$
$$+4(7)(5)+1(-2)(2)+0(-1)(3)$$
$$=0+24+5+140-4+0$$
$$=165$$

5. $M=\begin{bmatrix}5&0\\2&1\end{bmatrix}$

$$\det(M-\lambda I)=0$$
$$\det\begin{bmatrix}5-\lambda&0\\2&1-\lambda\end{bmatrix}=0$$
$$(5-\lambda)(1-\lambda)-0(2)=0$$
$$(5-\lambda)(1-\lambda)=0\Rightarrow$$

$5-\lambda=0\Rightarrow\lambda=5$ or $1-\lambda=0\Rightarrow\lambda=1$

The eigenvalues are 1 and 5.

Now find the eigenvector corresponding to $\lambda=1$.

$$\begin{bmatrix}0\\0\end{bmatrix}=\begin{bmatrix}5-1&0\\2&1-1\end{bmatrix}\begin{bmatrix}x_1\\x_2\end{bmatrix}=\begin{bmatrix}4&0\\2&0\end{bmatrix}\begin{bmatrix}x_1\\x_2\end{bmatrix}$$

The augmented matrix for this system is $\left[\begin{array}{cc|c}4&0&0\\2&0&0\end{array}\right]$. Note that row 1 is a multiple of row 2, so this is a dependent system. The first row indicates that $x_1+0x_2=0\Rightarrow x_1=0$. A solution is $x_1=0,\ x_2=1$. Thus, an eigenvector corresponding to $\lambda=1$ is $\begin{bmatrix}0\\1\end{bmatrix}$.

(*continued on next page*)

(*continued*)

Now find the eigenvector corresponding to $\lambda = 5$.

$$\begin{bmatrix}0\\0\end{bmatrix} = \begin{bmatrix}5-5 & 0\\ 2 & 1-5\end{bmatrix}\begin{bmatrix}x_1\\x_2\end{bmatrix} = \begin{bmatrix}0 & 0\\2 & -4\end{bmatrix}\begin{bmatrix}x_1\\x_2\end{bmatrix}$$

The augmented matrix for this system is $\left[\begin{array}{cc|c}0 & 0 & 0\\ 2 & -4 & 0\end{array}\right]$. The zeros in row 1 indicate that this is a dependent system. Since the second row indicates that $2x_1 - 4x_2 = 0 \Rightarrow x_1 = 2x_2$, a simple solution is $x_1 = 2,\ x_2 = 1$. Thus, an eigenvector corresponding to $\lambda = 5$ is $\begin{bmatrix}2\\1\end{bmatrix}$.

7. $M = \begin{bmatrix}3 & 2\\3 & 8\end{bmatrix}$

$$\det(M - \lambda I) = 0$$
$$\det\begin{bmatrix}3-\lambda & 2\\ 3 & 8-\lambda\end{bmatrix} = 0$$
$$(3-\lambda)(8-\lambda) - 2(3) = 0$$
$$\lambda^2 - 11\lambda + 18 = 0$$
$$(\lambda - 9)(\lambda - 2) = 0 \Rightarrow$$
$$\lambda - 9 = 0 \Rightarrow \lambda = 9 \text{ or } \lambda - 2 = 0 \Rightarrow \lambda = 2$$

The eigenvalues are 2 and 9.

Find the eigenvector corresponding to $\lambda = 2$.

$$\begin{bmatrix}0\\0\end{bmatrix} = \begin{bmatrix}3-2 & 2\\ 3 & 8-2\end{bmatrix}\begin{bmatrix}x_1\\x_2\end{bmatrix} = \begin{bmatrix}1 & 2\\3 & 6\end{bmatrix}\begin{bmatrix}x_1\\x_2\end{bmatrix}$$

The augmented matrix for this system is $\left[\begin{array}{cc|c}1 & 2 & 0\\ 3 & 6 & 0\end{array}\right]$. Note that row 2 is a multiple of row 1, so this is a dependent system. The first row indicates that $x_1 + 2x_2 = 0 \Rightarrow x_1 = -2x_2$. A solution is $x_1 = 2,\ x_2 = -1$. Thus, an eigenvector corresponding to $\lambda = 2$ is $\begin{bmatrix}2\\-1\end{bmatrix}$.

Now find the eigenvector corresponding to $\lambda = 9$.

$$\begin{bmatrix}0\\0\end{bmatrix} = \begin{bmatrix}3-9 & 2\\ 3 & 8-9\end{bmatrix}\begin{bmatrix}x_1\\x_2\end{bmatrix} = \begin{bmatrix}-6 & 2\\3 & -1\end{bmatrix}\begin{bmatrix}x_1\\x_2\end{bmatrix}$$

The augmented matrix for this system is $\left[\begin{array}{cc|c}-6 & 2 & 0\\ 3 & -1 & 0\end{array}\right]$. Note that row 1 is a multiple of row 2, so this is a dependent system. The second row indicates that

$$3x_1 - x_2 = 0 \Rightarrow x_1 = \frac{1}{3}x_2.$$

A solution is $x_1 = 1,\ x_2 = 3$. Thus, an eigenvector corresponding to $\lambda = 9$ is $\begin{bmatrix}1\\3\end{bmatrix}$.

9. $\begin{bmatrix}4 & -3\\2 & -1\end{bmatrix}$

$$\det(M - \lambda I) = 0$$
$$\det\begin{bmatrix}4-\lambda & -3\\ 2 & -1-\lambda\end{bmatrix} = 0$$
$$(4-\lambda)(-1-\lambda) - (-3)(2) = 0$$
$$\lambda^2 - 3\lambda + 2 = 0$$
$$(\lambda - 1)(\lambda - 2) = 0 \Rightarrow \lambda = 1 \text{ or } \lambda = 2$$

The eigenvalues are 1 and 2.

Find the eigenvector corresponding to $\lambda = 1$.

$$\begin{bmatrix}0\\0\end{bmatrix} = \begin{bmatrix}4-1 & -3\\ 2 & -1-1\end{bmatrix}\begin{bmatrix}x_1\\x_2\end{bmatrix} = \begin{bmatrix}3 & -3\\2 & -2\end{bmatrix}\begin{bmatrix}x_1\\x_2\end{bmatrix}$$

The augmented matrix for this system is $\left[\begin{array}{cc|c}3 & -3 & 0\\ 2 & -2 & 0\end{array}\right]$. Note that row 1 is a multiple of row 2, so this is a dependent system. The first row indicates that $3x_1 - 3x_2 = 0 \Rightarrow x_1 = x_2$. A solution is $x_1 = 1,\ x_2 = 1$. Thus, an eigenvector corresponding to $\lambda = 1$ is $\begin{bmatrix}1\\1\end{bmatrix}$.

Now find the eigenvector corresponding to $\lambda = 2$.

$$\begin{bmatrix}0\\0\end{bmatrix} = \begin{bmatrix}4-2 & -3\\ 2 & -1-2\end{bmatrix}\begin{bmatrix}x_1\\x_2\end{bmatrix} = \begin{bmatrix}2 & -3\\2 & -3\end{bmatrix}\begin{bmatrix}x_1\\x_2\end{bmatrix}$$

The augmented matrix for this system is $\left[\begin{array}{cc|c}2 & -3 & 0\\ 2 & -3 & 0\end{array}\right]$. Note that row 1 is the same as row 2, so this is a dependent system. The first row indicates that $2x_1 - 3x_2 = 0 \Rightarrow x_1 = \frac{3}{2}x_2$.

A solution is $x_1 = 3,\ x_2 = 2$. Thus, an eigenvector corresponding to $\lambda = 2$ is $\begin{bmatrix}3\\2\end{bmatrix}$.

11. $M = \begin{bmatrix} 4 & 0 & 0 \\ 3 & -1 & 0 \\ 2 & 5 & -3 \end{bmatrix}$

$$\det(M - \lambda I) = 0$$

$$\det \begin{bmatrix} 4-\lambda & 0 & 0 \\ 3 & -1-\lambda & 0 \\ 2 & 5 & -3-\lambda \end{bmatrix} = 0$$

$$(4-\lambda)\left[(-1-\lambda)(-3-\lambda)-0(5)\right]-0+0=0$$
$$(4-\lambda)(-1-\lambda)(-3-\lambda)=0 \Rightarrow$$

$\lambda = 4$ or $\lambda = -1$ or $\lambda = -3$

The eigenvalues are –3, –1, and 4.

Find the eigenvector corresponding to $\lambda = -3$.

$$\begin{bmatrix} 0 \\ 0 \\ 0 \end{bmatrix} = \begin{bmatrix} 4-(-3) & 0 & 0 \\ 3 & -1-(-3) & 0 \\ 2 & 5 & -3-(-3) \end{bmatrix} \begin{bmatrix} x_1 \\ x_2 \\ x_3 \end{bmatrix}$$
$$= \begin{bmatrix} 7 & 0 & 0 \\ 3 & 2 & 0 \\ 2 & 5 & 0 \end{bmatrix} \begin{bmatrix} x_1 \\ x_2 \\ x_3 \end{bmatrix}$$

The augmented matrix for this system is

$$\left[\begin{array}{ccc|c} 7 & 0 & 0 & 0 \\ 3 & 2 & 0 & 0 \\ 2 & 5 & 0 & 0 \end{array}\right].$$

Use row operations to reduce the matrix.

$$\tfrac{1}{7}R_1 \to R_1 \left[\begin{array}{ccc|c} 1 & 0 & 0 & 0 \\ 3 & 2 & 0 & 0 \\ 2 & 5 & 0 & 0 \end{array}\right]$$

$$\tfrac{1}{2}(-3R_1 + R_2) \to R_2 \left[\begin{array}{ccc|c} 1 & 0 & 0 & 0 \\ 0 & 1 & 0 & 0 \\ 2 & 5 & 0 & 0 \end{array}\right]$$

Row 1 gives $x_1 = 0$ and row 2 gives $x_2 = 0$. Since x_3 is arbitrary, let $x_3 = 1$. Thus an

eigenvector corresponding to $\lambda = -3$ is $\begin{bmatrix} 0 \\ 0 \\ 1 \end{bmatrix}$.

Find the eigenvector corresponding to $\lambda = -1$.

$$\begin{bmatrix} 0 \\ 0 \\ 0 \end{bmatrix} = \begin{bmatrix} 4-(-1) & 0 & 0 \\ 3 & -1-(-1) & 0 \\ 2 & 5 & -3-(-1) \end{bmatrix} \begin{bmatrix} x_1 \\ x_2 \\ x_3 \end{bmatrix}$$
$$= \begin{bmatrix} 5 & 0 & 0 \\ 3 & 0 & 0 \\ 2 & 5 & -2 \end{bmatrix} \begin{bmatrix} x_1 \\ x_2 \\ x_3 \end{bmatrix}$$

The augmented matrix for this system is

$$\left[\begin{array}{ccc|c} 5 & 0 & 0 & 0 \\ 3 & 0 & 0 & 0 \\ 2 & 5 & -2 & 0 \end{array}\right].$$

Use row operations to reduce the matrix.

$$\tfrac{1}{5}R_1 \to R_5 \left[\begin{array}{ccc|c} 1 & 0 & 0 & 0 \\ 3 & 0 & 0 & 0 \\ 2 & 5 & -2 & 0 \end{array}\right]$$

$$-2R_1 + R_3 \to R_3 \left[\begin{array}{ccc|c} 1 & 0 & 0 & 0 \\ 3 & 0 & 0 & 0 \\ 0 & 5 & -2 & 0 \end{array}\right]$$

Row 1 gives $x_1 = 0$ and row 3 gives

$5x_2 - 2x_3 = 0 \Rightarrow x_2 = \frac{2}{5}x_3$. A solution is

$x_2 = 2,\ x_3 = 5$. Thus an eigenvector

corresponding to $\lambda = -1$ is $\begin{bmatrix} 0 \\ 2 \\ 5 \end{bmatrix}$.

Finally, find the eigenvector corresponding to $\lambda = 4$.

$$\begin{bmatrix} 0 \\ 0 \\ 0 \end{bmatrix} = \begin{bmatrix} 4-4 & 0 & 0 \\ 3 & -1-4 & 0 \\ 2 & 5 & -3-4 \end{bmatrix} \begin{bmatrix} x_1 \\ x_2 \\ x_3 \end{bmatrix}$$
$$= \begin{bmatrix} 0 & 0 & 0 \\ 3 & -5 & 0 \\ 2 & 5 & -7 \end{bmatrix} \begin{bmatrix} x_1 \\ x_2 \\ x_3 \end{bmatrix}$$

The augmented matrix for this system is

$$\left[\begin{array}{ccc|c} 0 & 0 & 0 & 0 \\ 3 & -5 & 0 & 0 \\ 2 & 5 & -7 & 0 \end{array}\right].$$

Row 2 gives $3x_1 - 5x_2 = 0 \Rightarrow x_1 = \frac{5}{3}x_2$. A

solution is $x_1 = 5,\ x_2 = 3$. Substituting in row

3 gives $2(5) + 5(3) - 7x_3 = 0 \Rightarrow x_3 = \frac{25}{7}$.

In order to obtain integer solutions, multiply each proposed solution by 7, giving $x_1 = 35,\ x_2 = 21,\ x_3 = 25$. Thus an

eigenvector corresponding to $\lambda = 4$ is $\begin{bmatrix} 35 \\ 21 \\ 25 \end{bmatrix}$.

13. $ax + by = 0$
$cx + dy = 0$

The augmented matrix for this system is
$\left[\begin{array}{cc|c} a & b & 0 \\ c & d & 0 \end{array}\right]$.

We are given that $ad - bc = 0$, so $a = \dfrac{bc}{d}$.

Substituting into the augmented matrix gives
$\left[\begin{array}{cc|c} \frac{bc}{d} & b & 0 \\ c & d & 0 \end{array}\right]$. Now use row operations.

$\frac{1}{b}R_1 \to R_1$ $\left[\begin{array}{cc|c} \frac{c}{d} & 1 & 0 \\ c & d & 0 \end{array}\right]$

$dR_1 \to R_1$ $\left[\begin{array}{cc|c} c & d & 0 \\ c & d & 0 \end{array}\right]$

Since row 1 is the same as row 2, the system of equations represented by the matrix is dependent.

15. $M = \begin{bmatrix} 10 & -9 \\ 4 & -2 \end{bmatrix}$

$$\det(M - \lambda I) = 0$$
$$\det\begin{bmatrix} 10-\lambda & -9 \\ 4 & -2-\lambda \end{bmatrix} = 0$$
$$(10-\lambda)(-2-\lambda) - (-9)(4) = 0$$
$$\lambda^2 - 8\lambda + 16 = 0$$
$$(\lambda - 4)^2 = 0 \Rightarrow \lambda = 4$$

The eigenvalue is 4.
Find the eigenvector corresponding to $\lambda = 4$.

$$\begin{bmatrix} 0 \\ 0 \end{bmatrix} = \begin{bmatrix} 10-4 & -9 \\ 4 & -2-4 \end{bmatrix}\begin{bmatrix} x_1 \\ x_2 \end{bmatrix} = \begin{bmatrix} 6 & -9 \\ 4 & -6 \end{bmatrix}\begin{bmatrix} x_1 \\ x_2 \end{bmatrix}$$

The augmented matrix for this system is
$\left[\begin{array}{cc|c} 6 & -9 & 0 \\ 4 & -6 & 0 \end{array}\right]$.

Note that row 1 is a multiple of row 2, so this is a dependent system. The first row indicates that $6x_1 - 9x_2 = 0 \Rightarrow x_1 = \dfrac{3}{2}x_2$. A solution is $x_1 = 3,\ x_2 = 2$. Thus, an eigenvector corresponding to $\lambda = 4$ is $\begin{bmatrix} 3 \\ 2 \end{bmatrix}$.

17. $M = \begin{bmatrix} 1 & 5 \\ -2 & 3 \end{bmatrix}$

$$\det(M - \lambda I) = 0$$
$$\det\begin{bmatrix} 1-\lambda & 5 \\ -2 & 3-\lambda \end{bmatrix} = 0$$
$$(1-\lambda)(3-\lambda) - 5(-2) = 0$$
$$\lambda^2 - 4\lambda + 13 = 0$$

Solve using the quadratic formula.

$$x = \frac{-(-4) \pm \sqrt{(-4)^2 - 4(13)}}{2} = \frac{4 \pm \sqrt{-36}}{2}$$
$$= \frac{4 \pm 6i}{2} = 2 \pm 3i$$

The eigenvalues are $2 - 3i$ and $2 + 3i$.
Find the eigenvector corresponding to $\lambda = 2 - 3i$.

$$\begin{bmatrix} 0 \\ 0 \end{bmatrix} = \begin{bmatrix} 1-(2-3i) & 5 \\ -2 & 3-(2-3i) \end{bmatrix}\begin{bmatrix} x_1 \\ x_2 \end{bmatrix}$$
$$= \begin{bmatrix} -1+3i & 5 \\ -2 & 1+3i \end{bmatrix}\begin{bmatrix} x_1 \\ x_2 \end{bmatrix}$$

The augmented matrix for this system is
$\left[\begin{array}{cc|c} -1+3i & 5 & 0 \\ -2 & 1+3i & 0 \end{array}\right]$.

Using row operations, we have

$(1+3i)R_1 \to R_1$ $\left[\begin{array}{cc|c} -10 & 5(1+3i) & 0 \\ -2 & 1+3i & 0 \end{array}\right]$

$\frac{1}{5}R_1 \to R_1$ $\left[\begin{array}{cc|c} -2 & 1+3i & 0 \\ -2 & 1+3i & 0 \end{array}\right]$

Note that row 2 is the same as row 1, so this is a dependent system. The first row indicates that $-2x_1 + (1+3i)x_2 = 0 \Rightarrow x_1 = \dfrac{1+3i}{2}x_2$. A solution is $x_1 = 1+3i,\ x_2 = 2$. Thus, an eigenvector corresponding to $\lambda = 2 - 3i$ is $\begin{bmatrix} 1+3i \\ 2 \end{bmatrix}$.

Now find the eigenvector corresponding to $\lambda = 2 + 3i$.

$$\begin{bmatrix} 0 \\ 0 \end{bmatrix} = \begin{bmatrix} 1-(2+3i) & 5 \\ -2 & 3-(2+3i) \end{bmatrix}\begin{bmatrix} x_1 \\ x_2 \end{bmatrix}$$
$$= \begin{bmatrix} -1-3i & 5 \\ -2 & 1-3i \end{bmatrix}\begin{bmatrix} x_1 \\ x_2 \end{bmatrix}$$

The augmented matrix for this system is
$\left[\begin{array}{cc|c} -1-3i & 5 & 0 \\ -2 & 1-3i & 0 \end{array}\right]$.

(continued on next page)

(*continued*)

Using row operations, we have

$$(-1+3i)R_1 \to R_1 \left[\begin{array}{cc|c} 10 & 5(-1+3i) & 0 \\ -2 & 1-3i & 0 \end{array}\right]$$

$$-\tfrac{1}{5}R_1 \to R_1 \left[\begin{array}{cc|c} -2 & 1-3i & 0 \\ -2 & 1-3i & 0 \end{array}\right]$$

Note that row 2 is the same as row 1, so this is a dependent system. The first row indicates that

$-2x_1 + (1-3i)x_2 = 0 \Rightarrow x_1 = \dfrac{1-3i}{2}x_2$. A solution is $x_1 = 1-3i,\ x_2 = 2$. Thus, an eigenvector corresponding to $\lambda = 2+3i$ is $\begin{bmatrix} 1-3i \\ 2 \end{bmatrix}$.

19. First, find the eigenvalues and eigenvectors.

$$M = \begin{bmatrix} 0.5 & 0.9 \\ 1.4 & 0 \end{bmatrix}$$

$$\det(M - \lambda I) = 0$$

$$\det\begin{bmatrix} 0.5-\lambda & 0.9 \\ 1.4 & 0-\lambda \end{bmatrix} = 0$$

$$(0.5-\lambda)(-\lambda) - 0.9(1.4) = 0$$

$$\lambda^2 - 0.5\lambda - 1.26 = 0$$

Solve using the quadratic formula.

$$\lambda = \frac{-(-0.5) \pm \sqrt{(-0.5)^2 - 4(-1.26)}}{2}$$

$$= \frac{0.5 \pm \sqrt{5.29}}{2} = \frac{0.5 \pm 2.3}{2} = 1.4 \text{ or } -0.9$$

The eigenvalues are 1.4 and –0.9.

Find the eigenvector corresponding to $\lambda = 1.4$.

$$\begin{bmatrix} 0 \\ 0 \end{bmatrix} = \begin{bmatrix} 0.5-1.4 & 0.9 \\ 1.4 & 0-1.4 \end{bmatrix}\begin{bmatrix} x_1 \\ x_2 \end{bmatrix}$$

$$= \begin{bmatrix} -0.9 & 0.9 \\ 1.4 & -1.4 \end{bmatrix}\begin{bmatrix} x_1 \\ x_2 \end{bmatrix}$$

The augmented matrix for this system is $\left[\begin{array}{cc|c} -0.9 & 0.9 & 0 \\ 1.4 & -1.4 & 0 \end{array}\right]$. Using row operations gives $\left[\begin{array}{cc|c} 1 & -1 & 0 \\ 1 & -1 & 0 \end{array}\right]$. Note that row 2 is the same as row 1, so this is a dependent system. The first row indicates that $x_1 - x_2 = 0 \Rightarrow x_1 = x_2$. A solution is $x_1 = 1,\ x_2 = 1$. Thus, an eigenvector corresponding to $\lambda = 1.4$ is $\begin{bmatrix} 1 \\ 1 \end{bmatrix}$.

Now find the eigenvector corresponding to $\lambda = -0.9$.

$$\begin{bmatrix} 0 \\ 0 \end{bmatrix} = \begin{bmatrix} 0.5-(-0.9) & 0.9 \\ 1.4 & 0-(-0.9) \end{bmatrix}\begin{bmatrix} x_1 \\ x_2 \end{bmatrix}$$

$$= \begin{bmatrix} 1.4 & 0.9 \\ 1.4 & 0.9 \end{bmatrix}\begin{bmatrix} x_1 \\ x_2 \end{bmatrix}$$

The augmented matrix for this system is $\left[\begin{array}{cc|c} 1.4 & 0.9 & 0 \\ 1.4 & 0.9 & 0 \end{array}\right]$. Note that row 2 is the same as row 1, so this is a dependent system. The first row indicates that

$1.4x_1 + 0.9x_2 = 0 \Rightarrow x_1 = -\dfrac{9}{14}x_2$. Since this leads to a negative value, it cannot be considered as a population.

(a) The initial population is 10,000, so a multiple of the eigenvector is used.

$$\begin{bmatrix} x_1 \\ x_2 \end{bmatrix} = k\begin{bmatrix} 1 \\ 1 \end{bmatrix} = \begin{bmatrix} k \\ k \end{bmatrix}$$

$k + k = 10{,}000 \Rightarrow 2k = 10{,}000 \Rightarrow$
$k = 5000$

The initial population is $\begin{bmatrix} 5000 \\ 5000 \end{bmatrix}$.

(b) The growth factor is 1.4.

21. First, find the eigenvalues and eigenvectors.

$$M = \begin{bmatrix} 0.2 & 0.7 \\ 0.7 & 0.2 \end{bmatrix}$$

$$\det(M - \lambda I) = 0$$

$$\det\begin{bmatrix} 0.2-\lambda & 0.7 \\ 0.7 & 0.2-\lambda \end{bmatrix} = 0$$

$$(0.2-\lambda)^2 - 0.7^2 = 0$$

$$\lambda^2 - 0.4\lambda - 0.45 = 0$$

Solve using the quadratic formula.

$$\lambda = \frac{-(-0.4) \pm \sqrt{(-0.4)^2 - 4(-0.45)}}{2}$$

$$= \frac{0.4 \pm \sqrt{1.96}}{2} = \frac{0.4 \pm 1.4}{2} = 0.9 \text{ or } -0.5$$

The eigenvalues are 0.9 and –0.5.

Find the eigenvector corresponding to $\lambda = 0.9$.

$$\begin{bmatrix} 0 \\ 0 \end{bmatrix} = \begin{bmatrix} 0.2-0.9 & 0.7 \\ 0.7 & 0.2-0.9 \end{bmatrix}\begin{bmatrix} x_1 \\ x_2 \end{bmatrix}$$

$$= \begin{bmatrix} -0.7 & 0.7 \\ 0.7 & -0.7 \end{bmatrix}\begin{bmatrix} x_1 \\ x_2 \end{bmatrix}$$

(*continued on next page*)

(continued)

The augmented matrix for this system is $\left[\begin{array}{cc|c} -0.7 & 0.7 & 0 \\ 0.7 & -0.7 & 0 \end{array}\right]$. Note that row 2 is a multiple of row 1, so this is a dependent system. The first row indicates that $-0.7x_1 + 0.7x_2 = 0 \Rightarrow x_1 = x_2$. A solution is $x_1 = 1,\ x_2 = 1$. Thus, an eigenvector corresponding to $\lambda = 0.9$ is $\begin{bmatrix} 1 \\ 1 \end{bmatrix}$.

Now find the eigenvector corresponding to $\lambda = -0.5$.

$$\begin{bmatrix} 0 \\ 0 \end{bmatrix} = \begin{bmatrix} 0.2-(-0.5) & 0.7 \\ 0.7 & 0.2-(-0.5) \end{bmatrix}\begin{bmatrix} x_1 \\ x_2 \end{bmatrix} = \begin{bmatrix} 0.7 & 0.7 \\ 0.7 & 0.7 \end{bmatrix}\begin{bmatrix} x_1 \\ x_2 \end{bmatrix}$$

The augmented matrix for this system is $\left[\begin{array}{cc|c} 0.7 & 0.7 & 0 \\ 0.7 & 0.7 & 0 \end{array}\right]$. Note that row 1 is the same as row 2, so this is a dependent system. The first row indicates that $0.7x_1 + 0.7x_2 = 0 \Rightarrow x_1 = -x_2$. Since this leads to a negative value, it cannot be considered as a population.

(a) The initial population is 10,000, so a multiple of the eigenvector is used.

$$\begin{bmatrix} x_1 \\ x_2 \end{bmatrix} = k\begin{bmatrix} 1 \\ 1 \end{bmatrix} = \begin{bmatrix} k \\ k \end{bmatrix}$$

$k + k = 10,000 \Rightarrow 2k = 10,000 \Rightarrow$
$k = 5000$

The initial population is $\begin{bmatrix} 5000 \\ 5000 \end{bmatrix}$.

(b) The growth factor is 0.9.

23. First, find the eigenvalues and eigenvectors.

$$M = \begin{bmatrix} 0.1 & 1.2 \\ 0.4 & 0.3 \end{bmatrix}$$

$$\det(M - \lambda I) = 0$$

$$\det\begin{bmatrix} 0.1-\lambda & 1.2 \\ 0.4 & 0.3-\lambda \end{bmatrix} = 0$$

$$(0.1-\lambda)(0.3-\lambda) - 1.2(0.4) = 0$$

$$\lambda^2 - 0.4\lambda - 0.45 = 0$$

Solve using the quadratic formula.

$$\lambda = \frac{-(-0.4) \pm \sqrt{(-0.4)^2 - 4(-0.45)}}{2} = \frac{0.4 \pm \sqrt{1.96}}{2} = \frac{0.4 \pm 1.4}{2} = 0.9 \text{ or } -0.5$$

The eigenvalues are 0.9 and –0.5.

Find the eigenvector corresponding to $\lambda = 0.9$.

$$\begin{bmatrix} 0 \\ 0 \end{bmatrix} = \begin{bmatrix} 0.1-0.9 & 1.2 \\ 0.4 & 0.3-0.9 \end{bmatrix}\begin{bmatrix} x_1 \\ x_2 \end{bmatrix} = \begin{bmatrix} -0.8 & 1.2 \\ 0.4 & -0.6 \end{bmatrix}\begin{bmatrix} x_1 \\ x_2 \end{bmatrix}$$

The augmented matrix for this system is $\left[\begin{array}{cc|c} -0.8 & 1.2 & 0 \\ 0.4 & -0.6 & 0 \end{array}\right]$. Note that row 1 is a multiple of row 2, so this is a dependent system. The first row indicates that $-0.8x_1 + 1.2x_2 = 0 \Rightarrow x_1 = \frac{3}{2}x_2$. A solution is $x_1 = 3,\ x_2 = 2$. Thus, an eigenvector corresponding to $\lambda = 0.9$ is $\begin{bmatrix} 3 \\ 2 \end{bmatrix}$.

Now find the eigenvector corresponding to $\lambda = -0.5$.

$$\begin{bmatrix} 0 \\ 0 \end{bmatrix} = \begin{bmatrix} 0.1-(-0.5) & 1.2 \\ 0.4 & 0.3-(-0.5) \end{bmatrix}\begin{bmatrix} x_1 \\ x_2 \end{bmatrix} = \begin{bmatrix} 0.6 & 1.2 \\ 0.4 & 0.8 \end{bmatrix}\begin{bmatrix} x_1 \\ x_2 \end{bmatrix}$$

The augmented matrix for this system is $\left[\begin{array}{cc|c} 0.6 & 1.2 & 0 \\ 0.4 & 0.8 & 0 \end{array}\right]$. Note that row 1 is a multiple of row 2, so this is a dependent system. The first row indicates that $0.6x_1 + 1.2x_2 = 0 \Rightarrow x_1 = -2x_2$. Since this leads to a negative value, it cannot be considered as a population.

(a) The initial population is 10,000, so a multiple of the eigenvector is used.

$$\begin{bmatrix} x_1 \\ x_2 \end{bmatrix} = k\begin{bmatrix} 3 \\ 2 \end{bmatrix} = \begin{bmatrix} 3k \\ 2k \end{bmatrix}$$

$3k + 2k = 10,000 \Rightarrow 5k = 10,000 \Rightarrow$
$k = 2000$

The initial population is

$$\begin{bmatrix} 3(2000) \\ 2(2000) \end{bmatrix} = \begin{bmatrix} 6000 \\ 4000 \end{bmatrix}.$$

(b) The growth factor is 0.9.

25. The Leslie matrix, $M = \begin{bmatrix} 0.5 & 0.9 \\ 1.4 & 0 \end{bmatrix}$ has eigenvalues $\lambda_1 = -0.9$ and $\lambda_2 = 1.4$ with corresponding eigenvectors $V_1 = \begin{bmatrix} -9 \\ 14 \end{bmatrix}$ and $V_2 = \begin{bmatrix} 1 \\ 1 \end{bmatrix}$, respectively. The initial population is $X(0) = \begin{bmatrix} 3000 \\ 1000 \end{bmatrix}$. We must find values a and b such that $aV_1 + bV_2 = X(0)$.

The equation $a\begin{bmatrix} -9 \\ 14 \end{bmatrix} + b\begin{bmatrix} 1 \\ 1 \end{bmatrix} = \begin{bmatrix} 3000 \\ 1000 \end{bmatrix}$ corresponds to the augmented matrix $\left[\begin{array}{cc|c} -9 & 1 & 3000 \\ 14 & 1 & 1000 \end{array}\right]$. Using row operations, we have

$$-\tfrac{1}{23}(-R_2 + R_1) \to R_1 \quad \left[\begin{array}{cc|c} 1 & 0 & -\frac{2000}{23} \\ 14 & 1 & 1000 \end{array}\right]$$

$$-14R_1 + R_2 \to R_2 \quad \left[\begin{array}{cc|c} 1 & 0 & -\frac{2000}{23} \\ 0 & 1 & \frac{51{,}000}{23} \end{array}\right]$$

$$\begin{aligned} X(n) &= M^n X(0) \\ &= M^n\left(-\frac{2000}{23}V_1 + \frac{51{,}000}{23}V_2\right) \\ &= -\frac{2000}{23}M^n V_1 + \frac{51{,}000}{23}M^n V_2 \\ &= -\frac{2000}{23}\lambda_1^n V_1 + \frac{51{,}000}{23}\lambda_2^n V_2 \\ &= -\frac{2000}{23}(-0.9)^n\begin{bmatrix} -9 \\ 14 \end{bmatrix} \\ &\quad + \frac{51{,}000}{23}(1.4)^n\begin{bmatrix} 1 \\ 1 \end{bmatrix} \end{aligned}$$

Since $\lim_{n\to\infty}(-0.9)^n = 0$, the first term approaches zero as time passes. Therefore, when n is large, $X(n) \approx \frac{51{,}000}{23}(1.4)^n\begin{bmatrix} 1 \\ 1 \end{bmatrix}$.

Chapter 10 Review Exercises

1. True
2. False; a system with three equations and four unknowns has an infinite number of solutions.
3. True
4. False; only row operations can be used.
5. False; matrix A is a 2×2 matrix and matrix B is a 3×2 matrix. Only matrices having the same dimension can be added.
6. False; only matrices having the same dimension can be added.
7. True
8. False; in general, matrix multiplication is not commutative.
9. False; any $n \times n$ zero matrix does not have an inverse; the matrix $\begin{bmatrix} 2 & -4 \\ 1 & -2 \end{bmatrix}$ is an example of a matrix that doesn't have an inverse.
10. False; only square matrices can have inverses.
11. False; if $AB = C$ and A has an inverse, then $B = A^{-1}C$.
12. True
13. False; $AB = CB$ implies $A = C$ only if B is the identity matrix or B has an inverse.
14. False. The determinant of the matrix $\begin{bmatrix} 1 & 4 \\ 2 & 5 \end{bmatrix}$ is $1(5) - 4(2) = -3$.
15. True
16. True
17. For a system of m linear equations in n unknowns and $m = n$, there could be one, none, or an infinite number of solutions. If $m < n$, there are an infinite number of solutions. If $m > n$, there could be one, none, or an infinite number of solutions.
18. Answers will vary.
19. $$\begin{aligned} 2x + 4y &= -6 \\ -3x - 5y &= 12 \end{aligned}$$

    Write the augmented matrix and use row operations.

    $$\left[\begin{array}{cc|c} 2 & 4 & -6 \\ -3 & -5 & 12 \end{array}\right]$$

    $$3R_1 + 2R_2 \to R_2 \quad \left[\begin{array}{cc|c} 2 & 4 & -6 \\ 0 & 2 & 6 \end{array}\right]$$

    $$-2R_2 + R_1 \to R_1 \quad \left[\begin{array}{cc|c} 2 & 0 & -18 \\ 0 & 2 & 6 \end{array}\right]$$

    $$\begin{array}{l} \tfrac{1}{2}R_1 \to R_1 \\ \tfrac{1}{2}R_2 \to R_2 \end{array} \quad \left[\begin{array}{cc|c} 1 & 0 & -9 \\ 0 & 1 & 3 \end{array}\right]$$

    The solution is $(-9, 3)$.

20. $x-4y=10$
$5x+3y=119$

Write the augmented matrix and use row operations.

$$\left[\begin{array}{cc|c} 1 & -4 & 10 \\ 5 & 3 & 119 \end{array}\right]$$

$$-5R_1+R_2 \to R_2 \quad \left[\begin{array}{cc|c} 1 & -4 & 10 \\ 0 & 23 & 69 \end{array}\right]$$

$$4R_2+23R_1 \to R_1 \quad \left[\begin{array}{cc|c} 23 & 0 & 506 \\ 0 & 23 & 69 \end{array}\right]$$

$$\begin{array}{r} \frac{1}{23}R_1 \to R_1 \\ \frac{1}{23}R_2 \to R_2 \end{array} \quad \left[\begin{array}{cc|c} 1 & 0 & 22 \\ 0 & 1 & 3 \end{array}\right]$$

The solution is (22, 3).

21. $x-y+3z=13$
$4x+y+2z=17$
$3x+2y+2z=1$

Write the augmented matrix and use row operations.

$$\left[\begin{array}{ccc|c} 1 & -1 & 3 & 13 \\ 4 & 1 & 2 & 17 \\ 3 & 2 & 2 & 1 \end{array}\right]$$

$$\begin{array}{r} \\ -4R_1+R_2 \to R_2 \\ -3R_1+R_3 \to R_3 \end{array} \quad \left[\begin{array}{ccc|c} 1 & -1 & 3 & 13 \\ 0 & 5 & -10 & -35 \\ 0 & 5 & -7 & -38 \end{array}\right]$$

$$\begin{array}{r} R_2+5R_1 \to R_1 \\ \\ -1R_2+R_3 \to R_3 \end{array} \quad \left[\begin{array}{ccc|c} 5 & 0 & 5 & 30 \\ 0 & 5 & -10 & -35 \\ 0 & 0 & 3 & -3 \end{array}\right]$$

$$\begin{array}{r} 5R_3+(-3R_1) \to R_1 \\ 10R_3+3R_2 \to R_2 \\ \\ \end{array} \quad \left[\begin{array}{ccc|c} -15 & 0 & 0 & -105 \\ 0 & 15 & 0 & -135 \\ 0 & 0 & 3 & -3 \end{array}\right]$$

$$\begin{array}{r} -\frac{1}{15}R_1 \to R_1 \\ \frac{1}{15}R_2 \to R_2 \\ \frac{1}{3}R_3 \to R_3 \end{array} \quad \left[\begin{array}{ccc|c} 1 & 0 & 0 & 7 \\ 0 & 1 & 0 & -9 \\ 0 & 0 & 1 & -1 \end{array}\right]$$

The solution is $(7, -9, -1)$.

22. $x+2y+3z=9$
$x-2y=4$
$3x+2z=12$

Write the augmented matrix and use row operations.

$$\left[\begin{array}{ccc|c} 1 & 2 & 3 & 9 \\ 1 & -2 & 0 & 4 \\ 3 & 0 & 2 & 12 \end{array}\right]$$

$$\begin{array}{r} \\ -1R_1+R_2 \to R_2 \\ -3R_1+R_3 \to R_3 \end{array} \quad \left[\begin{array}{ccc|c} 1 & 2 & 3 & 9 \\ 0 & -4 & -3 & -5 \\ 0 & -6 & -7 & -15 \end{array}\right]$$

$$\begin{array}{r} R_1+R_2 \to R_1 \\ \\ \frac{1}{5}(3R_2+(-2)R_3) \to R_3 \end{array} \quad \left[\begin{array}{ccc|c} 1 & -2 & 0 & 4 \\ 0 & -4 & -3 & -5 \\ 0 & 0 & 1 & 3 \end{array}\right]$$

$$-\frac{1}{4}(3R_3+R_2) \to R_2 \quad \left[\begin{array}{ccc|c} 1 & -2 & 0 & 4 \\ 0 & 1 & 0 & -1 \\ 0 & 0 & 1 & 3 \end{array}\right]$$

$$-2R_2+R_1 \to R_1 \quad \left[\begin{array}{ccc|c} 1 & 0 & 0 & 2 \\ 0 & 1 & 0 & -1 \\ 0 & 0 & 1 & 3 \end{array}\right]$$

The solution is $(2, -1, 3)$.

23. $3x-6y+9z=12$
$-x+2y-3z=-4$
$x+y+2z=7$

Write the augmented matrix and use row operations.

$$\left[\begin{array}{ccc|c} 3 & -6 & 9 & 12 \\ -1 & 2 & -3 & -4 \\ 1 & 1 & 2 & 7 \end{array}\right]$$

$$\begin{array}{r} \\ R_1+3R_2 \to R_2 \\ -1R_1+3R_3 \to R_3 \end{array} \quad \left[\begin{array}{ccc|c} 3 & -6 & 9 & 12 \\ 0 & 0 & 0 & 0 \\ 0 & 9 & -3 & 9 \end{array}\right]$$

The zero in row 2, column 2 is an obstacle. To proceed, interchange the second and third rows.

$$\left[\begin{array}{ccc|c} 3 & -6 & 9 & 12 \\ 0 & 9 & -3 & 9 \\ 0 & 0 & 0 & 0 \end{array}\right]$$

$$3R_1+2R_2 \to R_1 \quad \left[\begin{array}{ccc|c} 9 & 0 & 21 & 54 \\ 0 & 9 & -3 & 9 \\ 0 & 0 & 0 & 0 \end{array}\right]$$

$$\begin{array}{r} \frac{1}{9}R_1 \to R_1 \\ \frac{1}{9}R_2 \to R_2 \\ \\ \end{array} \quad \left[\begin{array}{ccc|c} 1 & 0 & \frac{7}{3} & 6 \\ 0 & 1 & -\frac{1}{3} & 1 \\ 0 & 0 & 0 & 0 \end{array}\right]$$

The row of zeros indicates dependent equations.

(*continued on next page*)

(*continued*)

Solve the first two equations respectively for x and y in terms of z to obtain

$x = 6 - \frac{7}{3}z$ and $y = 1 + \frac{1}{3}z$

The solution of the system is

$\left(6 - \frac{7}{3}z,\ 1 + \frac{1}{3}z,\ z\right)$, where z is any real number.

24. $x - 2z = 5$
$3x + 2y = 8$
$-x + 2z = 10$

Write the system in augmented matrix form and apply row operations.

$$\left[\begin{array}{ccc|c} 1 & 0 & -2 & 5 \\ 3 & 2 & 0 & 8 \\ -1 & 0 & 2 & 10 \end{array}\right]$$

$$\begin{array}{r} -3R_1 + R_2 \to R_2 \\ R_1 + R_3 \to R_3 \end{array} \left[\begin{array}{ccc|c} 1 & 0 & -2 & 5 \\ 0 & 2 & 6 & -7 \\ 0 & 0 & 0 & 15 \end{array}\right]$$

The last row says that 0 = 15, which is false, so the system is inconsistent and there is no solution.

In Exercises 25–28, corresponding elements must be equal.

25. $\begin{bmatrix} 2 & 3 \\ 5 & q \end{bmatrix} = \begin{bmatrix} a & b \\ c & 9 \end{bmatrix}$

Size: 2×2; $a = 2, b = 3, c = 5, q = 9$; square matrices

26. $\begin{bmatrix} 2 & x \\ y & 6 \\ 5 & z \end{bmatrix} = \begin{bmatrix} a & -1 \\ 4 & 6 \\ p & 7 \end{bmatrix}$

Each matricx has size 3×2. For matrices to be equal, corresponding elements must be equal, so $a = 2$, $x = -1$, $y = 4$, $p = 5$, and $z = 7$.

27. $\begin{bmatrix} 2m & 4 & 3z & -12 \end{bmatrix} = \begin{bmatrix} 12 & k+1 & -9 & r-3 \end{bmatrix}$

Size: 1×4; $m = 6, k = 3, z = -3, r = -9$; row matrices

28. $\begin{bmatrix} a+5 & 3b & 6 \\ 4c & 2+d & -3 \\ -1 & 4p & q-1 \end{bmatrix} = \begin{bmatrix} -7 & b+2 & 2k-3 \\ 3 & 2d-1 & 4\ell \\ m & 12 & 8 \end{bmatrix}$

These are 3×3 square matrices. Since corresponding elements must be equal, $a + 5 = -7$, so $a = -12$; $3b = b + 2$, so $b = 1$; $6 = 2k - 3$, so $k = \frac{9}{2}$; $4c = 3$, so $c = \frac{3}{4}$; $2 + d = 2d - 1$, so $d = 3$; $-3 = 4\ell$, so $\ell = -\frac{3}{4}$; $m = -1$; $4p = 12$, so $p = 3$; and $q - 1 = 8$, so $q = 9$

29. $A + C = \begin{bmatrix} 4 & 10 \\ -2 & -3 \\ 6 & 9 \end{bmatrix} + \begin{bmatrix} 5 & 0 \\ -1 & 3 \\ 4 & 7 \end{bmatrix} = \begin{bmatrix} 9 & 10 \\ -3 & 0 \\ 10 & 16 \end{bmatrix}$

30. $2G - 4F = 2\begin{bmatrix} -2 & 0 \\ 1 & 5 \end{bmatrix} - 4\begin{bmatrix} -1 & 4 \\ 3 & 7 \end{bmatrix}$

$= \begin{bmatrix} -4 & 0 \\ 2 & 10 \end{bmatrix} + \begin{bmatrix} 4 & -16 \\ -12 & -28 \end{bmatrix}$

$= \begin{bmatrix} 0 & -16 \\ -10 & -18 \end{bmatrix}$

31. $3C + 2A = 3\begin{bmatrix} 5 & 0 \\ -1 & 3 \\ 4 & 7 \end{bmatrix} + 2\begin{bmatrix} 4 & 10 \\ -2 & -3 \\ 6 & 9 \end{bmatrix}$

$= \begin{bmatrix} 15 & 0 \\ -3 & 9 \\ 12 & 21 \end{bmatrix} + \begin{bmatrix} 8 & 20 \\ -4 & -6 \\ 12 & 18 \end{bmatrix}$

$= \begin{bmatrix} 23 & 20 \\ -7 & 3 \\ 24 & 39 \end{bmatrix}$

32. Since B is a 3×3 matrix, and C is a 3×2 matrix, the calculation of $B - C$ is not possible.

33. $2A - 5C = 2\begin{bmatrix} 4 & 10 \\ -2 & -3 \\ 6 & 9 \end{bmatrix} - 5\begin{bmatrix} 5 & 0 \\ -1 & 3 \\ 4 & 7 \end{bmatrix}$

$= \begin{bmatrix} 8 & 20 \\ -4 & -6 \\ 12 & 18 \end{bmatrix} - \begin{bmatrix} 25 & 0 \\ -5 & 15 \\ 20 & 35 \end{bmatrix}$

$= \begin{bmatrix} -17 & 20 \\ 1 & -21 \\ -8 & -17 \end{bmatrix}$

34. A has size 3×2 and G has 2×2, so AG will have size 3×2.

$AG = \begin{bmatrix} 4 & 10 \\ -2 & -3 \\ 6 & 9 \end{bmatrix}\begin{bmatrix} -2 & 0 \\ 1 & 5 \end{bmatrix} = \begin{bmatrix} 2 & 50 \\ 1 & -15 \\ -3 & 45 \end{bmatrix}$

35. A is 3×2 and C is 3×2, so finding the product AC is not possible.

$$\begin{matrix} A & C \\ 3\times \mathbf{2} & \mathbf{3}\times 2 \end{matrix}$$

(The inner 2 numbers must match.)

36. D has size 3×1 and E has size 1×3, so DE will have size 3×3.

$$DE = \begin{bmatrix} 6 \\ 1 \\ 0 \end{bmatrix}\begin{bmatrix} 1 & 3 & -4 \end{bmatrix} = \begin{bmatrix} 6 & 18 & -24 \\ 1 & 3 & -4 \\ 0 & 0 & 0 \end{bmatrix}$$

37. $ED = [1 \quad 3 \quad -4]\begin{bmatrix} 6 \\ 1 \\ 0 \end{bmatrix}$

$= [1\cdot 6 + 3\cdot 1 + (-4)0] = [9]$

38. B has size 3×3 and D has size 3×1, so BD will have size 3×1.

$$BD = \begin{bmatrix} 2 & 3 & -2 \\ 2 & 4 & 0 \\ 0 & 1 & 2 \end{bmatrix}\begin{bmatrix} 6 \\ 1 \\ 0 \end{bmatrix} = \begin{bmatrix} 15 \\ 16 \\ 1 \end{bmatrix}$$

39. $EC = \begin{bmatrix} 1 & 3 & -4 \end{bmatrix}\begin{bmatrix} 5 & 0 \\ -1 & 3 \\ 4 & 7 \end{bmatrix}$

$= [1\cdot 5 + 3(-1) + (-4)\cdot 4 \quad 1\cdot 0 + 3\cdot 3 + (-4)\cdot 7]$

$= \begin{bmatrix} -14 & -19 \end{bmatrix}$

40. $F = \begin{bmatrix} -1 & 4 \\ 3 & 7 \end{bmatrix}$

$$[F|I] = \left[\begin{array}{cc|cc} -1 & 4 & 1 & 0 \\ 3 & 7 & 0 & 1 \end{array}\right]$$

$$3R_1 + R_2 \to R_2 \quad \left[\begin{array}{cc|cc} -1 & 4 & 1 & 0 \\ 0 & 19 & 3 & 1 \end{array}\right]$$

$$4R_2 + (-19R_1) \to R_1 \quad \left[\begin{array}{cc|cc} 19 & 0 & -7 & 4 \\ 0 & 19 & 3 & 1 \end{array}\right]$$

$$\begin{matrix} \frac{1}{19}R_1 \to R_1 \\ \frac{1}{19}R_2 \to R_2 \end{matrix} \quad \left[\begin{array}{cc|cc} 1 & 0 & -\frac{7}{19} & \frac{4}{19} \\ 0 & 1 & \frac{3}{19} & \frac{1}{19} \end{array}\right]$$

$$F^{-1} = \begin{bmatrix} -\frac{7}{19} & \frac{4}{19} \\ \frac{3}{19} & \frac{1}{19} \end{bmatrix}$$

41. Find the inverse of $B = \begin{bmatrix} 2 & 3 & -2 \\ 2 & 4 & 0 \\ 0 & 1 & 2 \end{bmatrix}$, if it exists. Write the augmented matrix to obtain

$$[B|I] = \left[\begin{array}{ccc|ccc} 2 & 3 & -2 & 1 & 0 & 0 \\ 2 & 4 & 0 & 0 & 1 & 0 \\ 0 & 1 & 2 & 0 & 0 & 1 \end{array}\right]$$

$$-1R_1 + R_2 \to R_2 \quad \left[\begin{array}{ccc|ccc} 2 & 3 & -2 & 1 & 0 & 0 \\ 0 & 1 & 2 & -1 & 1 & 0 \\ 0 & 1 & 2 & 0 & 0 & 1 \end{array}\right]$$

$$\begin{matrix} -3R_2 + R_1 \to R_1 \\ \\ -1R_2 + R_3 \to R_3 \end{matrix} \quad \left[\begin{array}{ccc|ccc} 2 & 0 & -8 & 4 & -3 & 0 \\ 0 & 1 & 2 & -1 & 1 & 0 \\ 0 & 0 & 0 & 1 & -1 & 1 \end{array}\right]$$

No inverse exists, since the third row is all zeros to the left of the vertical bar.

42. A and C are 3×2 matrices, so their sum $A + C$ is a 3×2 matrix. Only square matrices have inverses. Therefore, $(A+C)^{-1}$ does not exist.

43. Find the inverse of $A = \begin{bmatrix} 1 & 3 \\ 2 & 7 \end{bmatrix}$, if it exists.

Write the augmented matrix $[A|I]$.

$$[A \mid I] = \left[\begin{array}{cc|cc} 1 & 3 & 1 & 0 \\ 2 & 7 & 0 & 1 \end{array}\right]$$

Perform row operations on $[A|I]$ to obtain a matrix of the form $[I|B]$.

$$-2R_1 + R_2 \to R_2 \quad \left[\begin{array}{cc|cc} 1 & 3 & 1 & 0 \\ 0 & 1 & -2 & 1 \end{array}\right]$$

$$-3R_2 + R_1 \to R_1 \quad \left[\begin{array}{cc|cc} 1 & 0 & 7 & -3 \\ 0 & 1 & -2 & 1 \end{array}\right]$$

The last augmented matrix is of the form $[I|B]$, so the desired inverse is

$$A^{-1} = \begin{bmatrix} 7 & -3 \\ -2 & 1 \end{bmatrix}.$$

44. $A = \begin{bmatrix} -4 & 2 \\ 0 & 3 \end{bmatrix}$.

$$[A|I] = \left[\begin{array}{cc|cc} -4 & 2 & 1 & 0 \\ 0 & 3 & 0 & 1 \end{array}\right]$$

$$2R_2 + (-3R_1) \to R_1 \left[\begin{array}{cc|cc} 12 & 0 & -3 & 2 \\ 0 & 3 & 0 & 1 \end{array}\right]$$

$$\begin{array}{l} \frac{1}{12}R_1 \to R_1 \\ \frac{1}{3}R_2 \to R_2 \end{array} \left[\begin{array}{cc|cc} 1 & 0 & -\frac{1}{4} & \frac{1}{6} \\ 0 & 1 & 0 & \frac{1}{3} \end{array}\right]$$

$$A^{-1} = \begin{bmatrix} -\frac{1}{4} & \frac{1}{6} \\ 0 & \frac{1}{3} \end{bmatrix}$$

45. Find the inverse of $A = \begin{bmatrix} 3 & -6 \\ -4 & 8 \end{bmatrix}$, if it exists.

Write the augmented matrix $[A|I]$.

$$[A|I] = \left[\begin{array}{cc|cc} 3 & -6 & 1 & 0 \\ -4 & 8 & 0 & 1 \end{array}\right]$$

Perform row operations on $[A|I]$ to obtain a matrix of the form $[I|B]$.

$$4R_1 + 3R_2 \to R_2 \left[\begin{array}{cc|cc} 3 & -6 & 1 & 0 \\ 0 & 0 & 4 & 3 \end{array}\right]$$

Since the entries to the left of the vertical bar in the second row are zeros, no inverse exists.

46. $A = \begin{bmatrix} 6 & 4 \\ 3 & 2 \end{bmatrix}$

$$[A|I] = \left[\begin{array}{cc|cc} 6 & 4 & 1 & 0 \\ 3 & 2 & 0 & 1 \end{array}\right]$$

$$R_1 + (-2)R_2 \to R_2 \left[\begin{array}{cc|cc} 6 & 4 & 1 & 0 \\ 0 & 0 & 1 & -2 \end{array}\right]$$

The zeros in the second row indicate that the original matrix has no inverse.

47. Find the inverse of $A = \begin{bmatrix} 2 & -1 & 0 \\ 1 & 0 & 1 \\ 1 & -2 & 0 \end{bmatrix}$, if it exists. The augmented matrix is

$$[A|I] = \left[\begin{array}{ccc|ccc} 2 & -1 & 0 & 1 & 0 & 0 \\ 1 & 0 & 1 & 0 & 1 & 0 \\ 1 & -2 & 0 & 0 & 0 & 1 \end{array}\right].$$

$$\begin{array}{l} \\ R_1 + (-2)R_2 \to R_2 \\ R_1 + (-2)R_3 \to R_3 \end{array} \left[\begin{array}{ccc|ccc} 2 & -1 & 0 & 1 & 0 & 0 \\ 0 & -1 & -2 & 1 & -2 & 0 \\ 0 & 3 & 0 & 1 & 0 & -2 \end{array}\right]$$

$$\begin{array}{l} -1R_2 + R_1 \to R_1 \\ \\ 3R_2 + R_3 \to R_3 \end{array} \left[\begin{array}{ccc|ccc} 2 & 0 & 2 & 0 & 2 & 0 \\ 0 & -1 & -2 & 1 & -2 & 0 \\ 0 & 0 & -6 & 4 & -6 & -2 \end{array}\right]$$

$$\begin{array}{l} R_3 + 3R_1 \to R_1 \\ R_3 + (-3)R_2 \to R_2 \\ \end{array} \left[\begin{array}{ccc|ccc} 6 & 0 & 0 & 4 & 0 & -2 \\ 0 & 3 & 0 & 1 & 0 & -2 \\ 0 & 0 & -6 & 4 & -6 & -2 \end{array}\right]$$

$$\begin{array}{l} \frac{1}{6}R_1 \to R_1 \\ \frac{1}{3}R_2 \to R_2 \\ -\frac{1}{6}R_3 \to R_3 \end{array} \left[\begin{array}{ccc|ccc} 1 & 0 & 0 & \frac{2}{3} & 0 & -\frac{1}{3} \\ 0 & 1 & 0 & \frac{1}{3} & 0 & -\frac{2}{3} \\ 0 & 0 & 1 & -\frac{2}{3} & 1 & \frac{1}{3} \end{array}\right]$$

$$A^{-1} = \begin{bmatrix} \frac{2}{3} & 0 & -\frac{1}{3} \\ \frac{1}{3} & 0 & -\frac{2}{3} \\ -\frac{2}{3} & 1 & \frac{1}{3} \end{bmatrix}$$

48. $A = \begin{bmatrix} 2 & 0 & 4 \\ 1 & -1 & 0 \\ 0 & 1 & -2 \end{bmatrix}$.

$$[A|I] = \left[\begin{array}{ccc|ccc} 2 & 0 & 4 & 1 & 0 & 0 \\ 1 & -1 & 0 & 0 & 1 & 0 \\ 0 & 1 & -2 & 0 & 0 & 1 \end{array}\right]$$

$$-2R_2 + R_1 \to R_2 \left[\begin{array}{ccc|ccc} 2 & 0 & 4 & 1 & 0 & 0 \\ 0 & 2 & 4 & 1 & -2 & 0 \\ 0 & 1 & -2 & 0 & 0 & 1 \end{array}\right]$$

$$-2R_3 + R_2 \to R_3 \left[\begin{array}{ccc|ccc} 2 & 0 & 4 & 1 & 0 & 0 \\ 0 & 2 & 4 & 1 & -2 & 0 \\ 0 & 0 & 8 & 1 & -2 & -2 \end{array}\right]$$

$$\begin{array}{l} -1R_3 + 2R_1 \to R_1 \\ -1R_3 + 2R_2 \to R_2 \\ \end{array} \left[\begin{array}{ccc|ccc} 4 & 0 & 0 & 1 & 2 & 2 \\ 0 & 4 & 0 & 1 & -2 & 2 \\ 0 & 0 & 8 & 1 & -2 & -2 \end{array}\right]$$

$$\begin{array}{l} \frac{1}{4}R_1 \to R_1 \\ \frac{1}{4}R_2 \to R_2 \\ \frac{1}{8}R_3 \to R_3 \end{array} \left[\begin{array}{ccc|ccc} 1 & 0 & 0 & \frac{1}{4} & \frac{1}{2} & \frac{1}{2} \\ 0 & 1 & 0 & \frac{1}{4} & -\frac{1}{2} & \frac{1}{2} \\ 0 & 0 & 1 & \frac{1}{8} & -\frac{1}{4} & -\frac{1}{4} \end{array}\right]$$

$$A^{-1} = \begin{bmatrix} \frac{1}{4} & \frac{1}{2} & \frac{1}{2} \\ \frac{1}{4} & -\frac{1}{2} & \frac{1}{2} \\ \frac{1}{8} & -\frac{1}{4} & -\frac{1}{4} \end{bmatrix}.$$

49. Find the inverse of $A = \begin{bmatrix} 1 & 3 & 6 \\ 4 & 0 & 9 \\ 5 & 15 & 30 \end{bmatrix}$, if it exists.

$$[A|I] = \left[\begin{array}{ccc|ccc} 1 & 3 & 6 & 1 & 0 & 0 \\ 4 & 0 & 9 & 0 & 1 & 0 \\ 5 & 15 & 30 & 0 & 0 & 1 \end{array}\right]$$

$$\begin{array}{l} \\ -4R_1 + R_2 \to R_2 \\ -5R_1 + R_3 \to R_3 \end{array} \left[\begin{array}{ccc|ccc} 1 & 3 & 6 & 1 & 0 & 0 \\ 0 & -12 & -15 & -4 & 1 & 0 \\ 0 & 0 & 0 & -5 & 0 & 1 \end{array}\right]$$

The last row is all zeros to the left of the bar line, so no inverse exists.

50. Find the inverse of $A = \begin{bmatrix} 2 & -3 & 4 \\ 1 & 5 & 7 \\ -4 & 6 & -8 \end{bmatrix}$, if it exists.

$$[A|I] = \left[\begin{array}{ccc|ccc} 2 & -3 & 4 & 1 & 0 & 0 \\ 1 & 5 & 7 & 0 & 1 & 0 \\ -4 & 6 & -8 & 0 & 0 & 1 \end{array}\right]$$

$$\begin{array}{r} \\ -1R_1 + 2R_2 \to R_2 \\ 2R_1 + R_3 \to R_3 \end{array} \left[\begin{array}{ccc|ccc} 2 & -3 & 4 & 1 & 0 & 0 \\ 0 & 13 & 10 & -1 & 2 & 0 \\ 0 & 0 & 0 & 2 & 0 & 1 \end{array}\right]$$

The zeros in the third row to the left of the vertical bar indicate that the original matrix has no inverse.

51. $A = \begin{bmatrix} 5 & 1 \\ -2 & -2 \end{bmatrix}, B = \begin{bmatrix} -8 \\ 24 \end{bmatrix}$

The matrix equation to be solved is $AX = B$, or $\begin{bmatrix} 5 & 1 \\ -2 & -2 \end{bmatrix}\begin{bmatrix} x \\ y \end{bmatrix} = \begin{bmatrix} -8 \\ 24 \end{bmatrix}$.

Calculate the inverse of the coefficient matrix A to obtain $\begin{bmatrix} 5 & 1 \\ -2 & -2 \end{bmatrix}^{-1} = \begin{bmatrix} \frac{1}{4} & \frac{1}{8} \\ -\frac{1}{4} & -\frac{5}{8} \end{bmatrix}$.

Now $X = A^{-1}B$, so

$$\begin{bmatrix} x \\ y \end{bmatrix} = \begin{bmatrix} \frac{1}{4} & \frac{1}{8} \\ -\frac{1}{4} & -\frac{5}{8} \end{bmatrix}\begin{bmatrix} -8 \\ 24 \end{bmatrix} = \begin{bmatrix} 1 \\ -13 \end{bmatrix}.$$

52. $A = \begin{bmatrix} 1 & 2 \\ 2 & 4 \end{bmatrix}, B = \begin{bmatrix} 5 \\ 10 \end{bmatrix}$

Row operations may be used to see that matrix A has no inverse. The matrix equation $AX = B$ may be written as the system of equations

$$\begin{aligned} x + 2y &= 5 \quad (1) \\ 2x + 4y &= 10 \quad (2) \end{aligned}$$

Use the elimination method to solve this system. Begin by eliminating x in equation (2).

$$\begin{array}{lll} & x + 2y = 5 & (1) \\ -2R_1 + R_2 \to R_2 & 0 = 0 & (3) \end{array}$$

The true statement in equation (3) indicates that the equations are dependent. Solve equation (1) for x in terms of y: $x = -2y + 5$

The solution is $(-2y + 5, y)$, where y is any real number.

53. $A = \begin{bmatrix} 1 & 0 & 2 \\ -1 & 1 & 0 \\ 3 & 0 & 4 \end{bmatrix}, B = \begin{bmatrix} 8 \\ 4 \\ -6 \end{bmatrix}$

The inverse of the coefficient matrix is

$$A^{-1} = \begin{bmatrix} -2 & 0 & 1 \\ -2 & 1 & 1 \\ \frac{3}{2} & 0 & -\frac{1}{2} \end{bmatrix}.$$

Since $X = A^{-1}B$,

$$X = \begin{bmatrix} -2 & 0 & 1 \\ -2 & 1 & 1 \\ \frac{3}{2} & 0 & -\frac{1}{2} \end{bmatrix}\begin{bmatrix} 8 \\ 4 \\ -6 \end{bmatrix} = \begin{bmatrix} -22 \\ -18 \\ 15 \end{bmatrix}.$$

54. $A = \begin{bmatrix} 2 & 4 & 0 \\ 1 & -2 & 0 \\ 0 & 0 & 3 \end{bmatrix}, B = \begin{bmatrix} 72 \\ -24 \\ 48 \end{bmatrix}$

Use row operations to find the inverse of A, which is

$$A^{-1} = \begin{bmatrix} \frac{1}{4} & \frac{1}{2} & 0 \\ \frac{1}{8} & -\frac{1}{4} & 0 \\ 0 & 0 & \frac{1}{3} \end{bmatrix}.$$

Since $X = A^{-1}B$,

$$X = \begin{bmatrix} \frac{1}{4} & \frac{1}{2} & 0 \\ \frac{1}{8} & -\frac{1}{4} & 0 \\ 0 & 0 & \frac{1}{3} \end{bmatrix}\begin{bmatrix} 72 \\ -24 \\ 48 \end{bmatrix} = \begin{bmatrix} 6 \\ 15 \\ 16 \end{bmatrix}.$$

55. $x+2y=4$
$2x-3y=1$

The coefficient matrix is $A=\begin{bmatrix}1 & 2\\ 2 & -3\end{bmatrix}$.

Calculate the inverse of A.

$$A^{-1}=\begin{bmatrix}\frac{3}{7} & \frac{2}{7}\\ \frac{2}{7} & -\frac{1}{7}\end{bmatrix}$$

Use $X=A^{-1}B$ to solve.

$$\begin{bmatrix}x\\ y\end{bmatrix}=\begin{bmatrix}\frac{3}{7} & \frac{2}{7}\\ \frac{2}{7} & -\frac{1}{7}\end{bmatrix}\begin{bmatrix}4\\ 1\end{bmatrix}=\begin{bmatrix}2\\ 1\end{bmatrix}$$

The solution is (2, 1).

56. $5x+10y=80$
$3x-2y=120$

Let $A=\begin{bmatrix}5 & 10\\ 3 & -2\end{bmatrix}$, $X=\begin{bmatrix}x\\ y\end{bmatrix}$, $B=\begin{bmatrix}80\\ 120\end{bmatrix}$.

Use row operations to find the inverse of A, which is

$$A^{-1}=\begin{bmatrix}\frac{1}{20} & \frac{1}{4}\\ \frac{3}{40} & -\frac{1}{8}\end{bmatrix}.$$

Since $X=A^{-1}B$,

$$\begin{bmatrix}x\\ y\end{bmatrix}=\begin{bmatrix}\frac{1}{20} & \frac{1}{4}\\ \frac{3}{40} & -\frac{1}{8}\end{bmatrix}\begin{bmatrix}80\\ 120\end{bmatrix}=\begin{bmatrix}34\\ -9\end{bmatrix}$$

The solution is $(34, -9)$.

57. $x+y+z=1$
$2x+y=-2$
$3y+z=2$

The coefficient matrix is

$$A=\begin{bmatrix}1 & 1 & 1\\ 2 & 1 & 0\\ 0 & 3 & 1\end{bmatrix}.$$

Find that the inverse of A is

$$A^{-1}=\begin{bmatrix}\frac{1}{5} & \frac{2}{5} & -\frac{1}{5}\\ -\frac{2}{5} & \frac{1}{5} & \frac{2}{5}\\ \frac{6}{5} & -\frac{3}{5} & -\frac{1}{5}\end{bmatrix}.$$

Now $X=A^{-1}B$, so

$$\begin{bmatrix}x\\ y\\ z\end{bmatrix}=\begin{bmatrix}\frac{1}{5} & \frac{2}{5} & -\frac{1}{5}\\ -\frac{2}{5} & \frac{1}{5} & \frac{2}{5}\\ \frac{6}{5} & -\frac{3}{5} & -\frac{1}{5}\end{bmatrix}\begin{bmatrix}1\\ -2\\ 2\end{bmatrix}=\begin{bmatrix}-1\\ 0\\ 2\end{bmatrix}.$$

The solution is $(-1, 0, 2)$.

58. $x-4y+2z=-1$
$-2x+y-3z=-9$
$3x+5y-2z=7$

Let $A=\begin{bmatrix}1 & -4 & 2\\ -2 & 1 & -3\\ 3 & 5 & -2\end{bmatrix}$, $X=\begin{bmatrix}x\\ y\\ z\end{bmatrix}$, $B=\begin{bmatrix}-1\\ -9\\ 7\end{bmatrix}$.

Use row operations to find the inverse of A, which is

$$A^{-1}=\begin{bmatrix}\frac{1}{3} & \frac{2}{39} & \frac{10}{39}\\ -\frac{1}{3} & -\frac{8}{39} & -\frac{1}{39}\\ -\frac{1}{3} & -\frac{17}{39} & -\frac{7}{39}\end{bmatrix}=\frac{1}{39}\begin{bmatrix}13 & 2 & 10\\ -13 & -8 & -1\\ -13 & -17 & -7\end{bmatrix}.$$

Since $X=A^{-1}B$,

$$\begin{bmatrix}x\\ y\\ z\end{bmatrix}=\frac{1}{39}\begin{bmatrix}13 & 2 & 10\\ -13 & -8 & -1\\ -13 & -17 & -7\end{bmatrix}\begin{bmatrix}-1\\ -9\\ 7\end{bmatrix}=\frac{1}{39}\begin{bmatrix}39\\ 78\\ 117\end{bmatrix}=\begin{bmatrix}1\\ 2\\ 3\end{bmatrix}.$$

The solution is (1, 2, 3).

59. $M=\begin{bmatrix}-3 & 12\\ -2 & 7\end{bmatrix}$

$$\det(M-\lambda I)=0$$
$$\det\begin{bmatrix}-3-\lambda & 12\\ -2 & 7-\lambda\end{bmatrix}=0$$
$$(-3-\lambda)(7-\lambda)-12(-2)=0$$
$$\lambda^2-4\lambda+3=0$$
$$(\lambda-3)(\lambda-1)=0\Rightarrow\lambda=3\text{ or }\lambda=1$$

The eigenvalues are 1 and 3.

(continued on next page)

(continued)

Find the eigenvector corresponding to $\lambda = 1$.

$$\begin{bmatrix}0\\0\end{bmatrix}=\begin{bmatrix}-3-1 & 12\\ -2 & 7-1\end{bmatrix}\begin{bmatrix}x_1\\x_2\end{bmatrix}=\begin{bmatrix}-4 & 12\\ -2 & 6\end{bmatrix}\begin{bmatrix}x_1\\x_2\end{bmatrix}$$

The augmented matrix for this system is $\left[\begin{array}{cc|c}-4 & 12 & 0\\ -2 & 6 & 0\end{array}\right]$. Note that row 1 is a multiple of row 2, so this is a dependent system. The first row indicates that $-4x_1+12x_2=0 \Rightarrow x_1=3x_2$. A solution is $x_1=3,\ x_2=1$. Thus, an eigenvector corresponding to $\lambda=1$ is $\begin{bmatrix}3\\1\end{bmatrix}$.

Find the eigenvector corresponding to $\lambda = 3$.

$$\begin{bmatrix}0\\0\end{bmatrix}=\begin{bmatrix}-3-3 & 12\\ -2 & 7-3\end{bmatrix}\begin{bmatrix}x_1\\x_2\end{bmatrix}=\begin{bmatrix}-6 & 12\\ -2 & 4\end{bmatrix}\begin{bmatrix}x_1\\x_2\end{bmatrix}$$

The augmented matrix for this system is $\left[\begin{array}{cc|c}-6 & 12 & 0\\ -2 & 4 & 0\end{array}\right]$. Note that row 1 is a multiple of row 3, so this is a dependent system. The first row indicates that $-6x_1+12x_2=0 \Rightarrow x_1=2x_2$. A solution is $x_1=2,\ x_2=1$. Thus, an eigenvector corresponding to $\lambda=3$ is $\begin{bmatrix}2\\1\end{bmatrix}$.

60. $M=\begin{bmatrix}5 & 3\\ 3 & 5\end{bmatrix}$

$$\det(M-\lambda I)=0$$
$$\det\begin{bmatrix}5-\lambda & 3\\ 3 & 5-\lambda\end{bmatrix}=0$$
$$(5-\lambda)^2-3^2=0$$
$$\lambda^2-10\lambda+16=0$$
$$(\lambda-2)(\lambda-8)=0 \Rightarrow \lambda=2 \text{ or } \lambda=8$$

The eigenvalues are 2 and 8.

Find the eigenvector corresponding to $\lambda = 2$.

$$\begin{bmatrix}0\\0\end{bmatrix}=\begin{bmatrix}5-2 & 3\\ 3 & 5-2\end{bmatrix}\begin{bmatrix}x_1\\x_2\end{bmatrix}=\begin{bmatrix}3 & 3\\ 3 & 3\end{bmatrix}\begin{bmatrix}x_1\\x_2\end{bmatrix}$$

The augmented matrix for this system is $\left[\begin{array}{cc|c}3 & 3 & 0\\ 3 & 3 & 0\end{array}\right]$. Note that row 1 is the same as row 2, so this is a dependent system. The first row indicates that $3x_1+3x_2=0 \Rightarrow x_1=-x_2$. A solution is $x_1=1,\ x_2=-1$. Thus, an eigenvector corresponding to $\lambda=2$ is $\begin{bmatrix}1\\-1\end{bmatrix}$.

Find the eigenvector corresponding to $\lambda = 8$.

$$\begin{bmatrix}0\\0\end{bmatrix}=\begin{bmatrix}5-8 & 3\\ 3 & 5-8\end{bmatrix}\begin{bmatrix}x_1\\x_2\end{bmatrix}=\begin{bmatrix}-3 & 3\\ 3 & -3\end{bmatrix}\begin{bmatrix}x_1\\x_2\end{bmatrix}$$

The augmented matrix for this system is $\left[\begin{array}{cc|c}-3 & 3 & 0\\ 3 & -3 & 0\end{array}\right]$. Note that row 2 is a multiple of row 1, so this is a dependent system. The first row indicates that $-3x_1+3x_2=0 \Rightarrow x_1=x_2$. A solution is $x_1=1,\ x_2=1$. Thus, an eigenvector corresponding to $\lambda=8$ is $\begin{bmatrix}1\\1\end{bmatrix}$.

61. $M=\begin{bmatrix}1 & 0 & 0\\ 2 & 2 & 0\\ 2 & 1 & -3\end{bmatrix}$

$$\det(M-\lambda I)=0$$
$$\det\begin{bmatrix}1-\lambda & 0 & 0\\ 2 & 2-\lambda & 0\\ 2 & 1 & -3-\lambda\end{bmatrix}=0$$
$$(1-\lambda)\left[(2-\lambda)(-3-\lambda)-0(1)\right]-0+0=0$$
$$(1-\lambda)(2-\lambda)(-3-\lambda)=0 \Rightarrow$$
$$\lambda=1 \text{ or } \lambda=2 \text{ or } \lambda=-3$$

The eigenvalues are 1, 2, and –3.

Find the eigenvector corresponding to $\lambda=-3$.

$$\begin{bmatrix}0\\0\\0\end{bmatrix}=\begin{bmatrix}1-(-3) & 0 & 0\\ 2 & 2-(-3) & 0\\ 2 & 1 & -3-(-3)\end{bmatrix}\begin{bmatrix}x_1\\x_2\\x_3\end{bmatrix}=\begin{bmatrix}4 & 0 & 0\\ 2 & 5 & 0\\ 2 & 1 & 0\end{bmatrix}\begin{bmatrix}x_1\\x_2\\x_3\end{bmatrix}$$

The augmented matrix for this system is $\left[\begin{array}{ccc|c}4 & 0 & 0 & 0\\ 2 & 5 & 0 & 0\\ 2 & 1 & 0 & 0\end{array}\right]$.

Use row operations to reduce the matrix.

$$\tfrac{1}{4}R_1 \to R_1 \quad \left[\begin{array}{ccc|c}1 & 0 & 0 & 0\\ 2 & 5 & 0 & 0\\ 2 & 1 & 0 & 0\end{array}\right]$$

$$\begin{array}{r}\\ \tfrac{1}{4}(-R_3+R_2)\to R_2\\ -2R_1+R_3\to R_3\end{array} \left[\begin{array}{ccc|c}1 & 0 & 0 & 0\\ 0 & 1 & 0 & 0\\ 0 & 1 & 0 & 0\end{array}\right]$$

(continued on next page)

(continued)

Row 1 gives $x_1 = 0$ and row 2 gives $x_2 = 0$. Since x_3 is arbitrary, let $x_3 = 1$. Thus an eigenvector corresponding to $\lambda = -3$ is $\begin{bmatrix} 0 \\ 0 \\ 1 \end{bmatrix}$.

Find the eigenvector corresponding to $\lambda = 1$.

$$\begin{bmatrix} 0 \\ 0 \\ 0 \end{bmatrix} = \begin{bmatrix} 1-1 & 0 & 0 \\ 2 & 2-1 & 0 \\ 2 & 1 & -3-1 \end{bmatrix} \begin{bmatrix} x_1 \\ x_2 \\ x_3 \end{bmatrix} = \begin{bmatrix} 0 & 0 & 0 \\ 2 & 1 & 0 \\ 2 & 1 & -4 \end{bmatrix} \begin{bmatrix} x_1 \\ x_2 \\ x_3 \end{bmatrix}$$

The augmented matrix for this system is

$$\left[\begin{array}{ccc|c} 0 & 0 & 0 & 0 \\ 2 & 1 & 0 & 0 \\ 2 & 1 & -4 & 0 \end{array}\right].$$

Use row operations to reduce the matrix.

$$\begin{array}{r} \\ \\ \frac{1}{4}(-R_3 + R_2) \to R_3 \end{array} \left[\begin{array}{ccc|c} 0 & 0 & 0 & 0 \\ 2 & 1 & 0 & 0 \\ 0 & 0 & 1 & 0 \end{array}\right]$$

Row 3 gives $x_3 = 0$ and row 2 gives

$2x_1 + x_2 = 0 \Rightarrow x_1 = -\frac{1}{2}x_2.$

A solution is $x_1 = 1,\ x_2 = -2$. Thus an eigenvector corresponding to $\lambda = 1$ is $\begin{bmatrix} 1 \\ -2 \\ 0 \end{bmatrix}$.

Finally, find the eigenvector corresponding to $\lambda = 2$.

$$\begin{bmatrix} 0 \\ 0 \\ 0 \end{bmatrix} = \begin{bmatrix} 1-2 & 0 & 0 \\ 2 & 2-2 & 0 \\ 2 & 1 & -3-2 \end{bmatrix} \begin{bmatrix} x_1 \\ x_2 \\ x_3 \end{bmatrix} = \begin{bmatrix} -1 & 0 & 0 \\ 2 & 0 & 0 \\ 2 & 1 & -5 \end{bmatrix} \begin{bmatrix} x_1 \\ x_2 \\ x_3 \end{bmatrix}$$

The augmented matrix for this system is

$$\left[\begin{array}{ccc|c} -1 & 0 & 0 & 0 \\ 2 & 0 & 0 & 0 \\ 2 & 1 & -5 & 0 \end{array}\right].$$

Row 1 gives $x_1 = 0$. Substituting in row 3 gives $2(0) + x_2 - 5x_3 = 0 \Rightarrow x_2 = 5x_3$.

A solution is $x_2 = 5,\ x_3 = 1$. Thus an eigenvector corresponding to $\lambda = 2$ is $\begin{bmatrix} 0 \\ 5 \\ 1 \end{bmatrix}$.

62. $M = \begin{bmatrix} -2 & 0 & 0 \\ 1 & 3 & 0 \\ -1 & 1 & 2 \end{bmatrix}$

$$\det(M - \lambda I) = 0$$

$$\det \begin{bmatrix} -2-\lambda & 0 & 0 \\ 1 & 3-\lambda & 0 \\ -1 & 1 & 2-\lambda \end{bmatrix} = 0$$

$$(-2-\lambda)\left[(3-\lambda)(2-\lambda) - 0(1)\right] - 0 + 0 = 0$$

$$(-2-\lambda)(3-\lambda)(2-\lambda) = 0 \Rightarrow$$

$\lambda = -2$ or $\lambda = 3$ or $\lambda = 2$

The eigenvalues are –2, 2, and 3.

Find the eigenvector corresponding to $\lambda = -2$.

$$\begin{bmatrix} 0 \\ 0 \\ 0 \end{bmatrix} = \begin{bmatrix} -2-(-2) & 0 & 0 \\ 1 & 3-(-2) & 0 \\ -1 & 1 & 2-(-2) \end{bmatrix} \begin{bmatrix} x_1 \\ x_2 \\ x_3 \end{bmatrix} = \begin{bmatrix} 0 & 0 & 0 \\ 1 & 5 & 0 \\ -1 & 1 & 4 \end{bmatrix} \begin{bmatrix} x_1 \\ x_2 \\ x_3 \end{bmatrix}$$

The augmented matrix for this system is

$$\left[\begin{array}{ccc|c} 0 & 0 & 0 & 0 \\ 1 & 5 & 0 & 0 \\ -1 & 1 & 4 & 0 \end{array}\right].$$

Row 2 gives $x_1 + 5x_2 = 0 \Rightarrow x_1 = -5x_2$ A solution is $x_1 = 5,\ x_2 = -1$. Substitute these values in row 3 and solve for x_3.

$-5 + (-1) + 4x_3 = 0 \Rightarrow x_3 = \frac{3}{2}$. In order to obtain integer solutions, multiply each proposed solution by 2, giving $x_1 = 10,\ x_2 = -2,\ x_3 = 3$. Thus an eigenvector corresponding to $\lambda = -2$ is $\begin{bmatrix} 10 \\ -2 \\ 3 \end{bmatrix}$.

Find the eigenvector corresponding to $\lambda = 2$.

(continued on next page)

(continued)

$$\begin{bmatrix}0\\0\\0\end{bmatrix}=\begin{bmatrix}-2-2 & 0 & 0\\ 1 & 3-2 & 0\\ -1 & 1 & 2-2\end{bmatrix}\begin{bmatrix}x_1\\x_2\\x_3\end{bmatrix}=\begin{bmatrix}-4 & 0 & 0\\ 1 & 1 & 0\\ -1 & 1 & 0\end{bmatrix}\begin{bmatrix}x_1\\x_2\\x_3\end{bmatrix}$$

The augmented matrix for this system is

$$\left[\begin{array}{ccc|c}-4 & 0 & 0 & 0\\ 1 & 1 & 0 & 0\\ -1 & 1 & 0 & 0\end{array}\right].$$

Row 1 gives $x_1 = 0$ and row 2 gives $x_1 + x_2 = 0 \Rightarrow 0 + x_2 = 0 \Rightarrow x_2 = 0$. Since x_3 is arbitrary, let $x_3 = 1$. Thus an eigenvector corresponding to $\lambda = 2$ is $\begin{bmatrix}0\\0\\1\end{bmatrix}$.

Finally, find the eigenvector corresponding to $\lambda = 3$.

$$\begin{bmatrix}0\\0\\0\end{bmatrix}=\begin{bmatrix}-2-3 & 0 & 0\\ 1 & 3-3 & 0\\ -1 & 1 & 2-3\end{bmatrix}\begin{bmatrix}x_1\\x_2\\x_3\end{bmatrix}=\begin{bmatrix}-5 & 0 & 0\\ 1 & 0 & 0\\ -1 & 1 & -1\end{bmatrix}\begin{bmatrix}x_1\\x_2\\x_3\end{bmatrix}$$

The augmented matrix for this system is

$$\left[\begin{array}{ccc|c}-5 & 0 & 0 & 0\\ 1 & 0 & 0 & 0\\ -1 & 1 & -1 & 0\end{array}\right].$$

Row 2 gives $x_1 = 0$. Substituting in row 3 gives $0 + x_2 - x_3 = 0 \Rightarrow x_2 = x_3$. A solution is $x_2 = 1,\ x_3 = 1$. Thus an eigenvector corresponding to $\lambda = 3$ is $\begin{bmatrix}0\\1\\1\end{bmatrix}$.

63. The given information can be written as the following 4×3 matrix.

$$\begin{bmatrix}8 & 8 & 8\\ 10 & 5 & 9\\ 7 & 10 & 7\\ 8 & 9 & 7\end{bmatrix}$$

64. **(a)** The X-ray passes through cells *B* and *C*, so the attenuation value for beam 3 is $b + c$.

(b) Beam 1: $a + b = 0.8$
Beam 2: $a + c = 0.55$
Beam 3: $b + c = 0.65$

$$\begin{bmatrix}1 & 1 & 0\\ 1 & 0 & 1\\ 0 & 1 & 1\end{bmatrix}\begin{bmatrix}a\\b\\c\end{bmatrix}=\begin{bmatrix}0.8\\0.55\\0.65\end{bmatrix}$$

$$\begin{bmatrix}a\\b\\c\end{bmatrix}=\begin{bmatrix}1 & 1 & 0\\ 1 & 0 & 1\\ 0 & 1 & 1\end{bmatrix}^{-1}\begin{bmatrix}0.8\\0.55\\0.65\end{bmatrix}=\begin{bmatrix}\frac{1}{2} & \frac{1}{2} & -\frac{1}{2}\\ \frac{1}{2} & -\frac{1}{2} & \frac{1}{2}\\ -\frac{1}{2} & \frac{1}{2} & \frac{1}{2}\end{bmatrix}\begin{bmatrix}0.8\\0.55\\0.65\end{bmatrix}=\begin{bmatrix}0.35\\0.45\\0.2\end{bmatrix}$$

The solution is (0.35, 0.45, 0.2), so *A* is tumorous, *B* is bone, and *C* is healthy.

(c) For patient X,

$$\begin{bmatrix}a\\b\\c\end{bmatrix}=\begin{bmatrix}\frac{1}{2} & \frac{1}{2} & -\frac{1}{2}\\ \frac{1}{2} & -\frac{1}{2} & \frac{1}{2}\\ -\frac{1}{2} & \frac{1}{2} & \frac{1}{2}\end{bmatrix}\begin{bmatrix}0.54\\0.40\\0.52\end{bmatrix}=\begin{bmatrix}0.21\\0.33\\0.19\end{bmatrix}.$$

A and *C* are healthy; *B* is tumorous.

For patient Y,

$$\begin{bmatrix}a\\b\\c\end{bmatrix}=\begin{bmatrix}\frac{1}{2} & \frac{1}{2} & -\frac{1}{2}\\ \frac{1}{2} & -\frac{1}{2} & \frac{1}{2}\\ -\frac{1}{2} & \frac{1}{2} & \frac{1}{2}\end{bmatrix}\begin{bmatrix}0.65\\0.80\\0.75\end{bmatrix}=\begin{bmatrix}0.35\\0.3\\0.45\end{bmatrix}.$$

A and *B* are tumorous; *C* is bone.

For patient Z,

$$\begin{bmatrix}a\\b\\c\end{bmatrix}=\begin{bmatrix}\frac{1}{2} & \frac{1}{2} & -\frac{1}{2}\\ \frac{1}{2} & -\frac{1}{2} & \frac{1}{2}\\ -\frac{1}{2} & \frac{1}{2} & \frac{1}{2}\end{bmatrix}\begin{bmatrix}0.51\\0.49\\0.44\end{bmatrix}=\begin{bmatrix}0.28\\0.23\\0.21\end{bmatrix}.$$

A could be healthy or tumorous; *B* and *C* are healthy.

65. **(a)** $a+b=0.60$ (1)
$c+d=0.75$ (2)
$a+c=0.65$ (3)
$b+d=0.70$ (4)

The augmented matrix of the system is

$$\left[\begin{array}{cccc|c} 1 & 1 & 0 & 0 & 0.60 \\ 0 & 0 & 1 & 1 & 0.75 \\ 1 & 0 & 1 & 0 & 0.65 \\ 0 & 1 & 0 & 1 & 0.70 \end{array}\right].$$

$$-1R_1+R_3 \to R_3 \left[\begin{array}{cccc|c} 1 & 1 & 0 & 0 & 0.60 \\ 0 & 0 & 1 & 1 & 0.75 \\ 0 & -1 & 1 & 0 & 0.05 \\ 0 & 1 & 0 & 1 & 0.70 \end{array}\right]$$

Interchange rows 2 and 4.

$$\left[\begin{array}{cccc|c} 1 & 1 & 0 & 0 & 0.60 \\ 0 & 1 & 0 & 1 & 0.70 \\ 0 & -1 & 1 & 0 & 0.05 \\ 0 & 0 & 1 & 1 & 0.75 \end{array}\right]$$

$$\begin{array}{l} -1R_2+R_1 \to R_1 \\ \\ R_2+R_3 \to R_3 \\ \\ \end{array} \left[\begin{array}{cccc|c} 1 & 0 & 0 & -1 & -0.10 \\ 0 & 1 & 0 & 1 & 0.70 \\ 0 & 0 & 1 & 1 & 0.75 \\ 0 & 0 & 1 & 1 & 0.75 \end{array}\right]$$

Since R_3 and R_4 are identical, there will be infinitely many solutions. We do not have enough information to determine the values of a, b, c, and d.

(b) **i.** If $d=0.33$, the system of equations in part (a) becomes

$a+b=0.60$ (1)
$c+0.33=0.75$ (2)
$a+c=0.65$ (3)
$b+0.33=0.70$ (4)

Equation (2) gives $c=0.42$, and equation (4) gives $b=0.37$. Substituting $c=0.42$ into equation (3) gives $a=0.23$. Therefore, $a=0.23$, $b=0.37$, $c=0.42$, and $d=0.33$. Thus, A is healthy, B and D are tumorous, and C is bone.

ii. If $d=0.43$, the system of equations in part (a) becomes

$a+b=0.60$ (1)
$c+0.43=0.75$ (2)
$a+c=0.65$ (3)
$b+0.43=0.70.$ (4)

Equation (2) gives $c=0.32$, and equation (4) gives $b=0.27$. Substituting $c=0.32$ into equation (3) gives $a=0.33$. Therefore, $a=0.33$, $b=0.27$, $c=0.32$, and $d=0.43$. Thus, A and C are tumorous, B could be healthy or tumorous, and D is bone.

(c) The original system now has two additional equations.

$a+b=0.60$ (1)
$c+d=0.75$ (2)
$a+c=0.65$ (3)
$b+d=0.70$ (4)
$b+c=0.85$ (5)
$a+d=0.50$ (6)

The augmented matrix of this system is

$$\left[\begin{array}{cccc|c} 1 & 1 & 0 & 0 & 0.60 \\ 0 & 0 & 1 & 1 & 0.75 \\ 1 & 0 & 1 & 0 & 0.65 \\ 0 & 1 & 0 & 1 & 0.70 \\ 0 & 1 & 1 & 0 & 0.85 \\ 1 & 0 & 0 & 1 & 0.50 \end{array}\right].$$

Using the Gauss-Jordan method we obtain

$$\left[\begin{array}{cccc|c} 1 & 0 & 0 & 0 & 0.20 \\ 0 & 1 & 0 & 0 & 0.40 \\ 0 & 0 & 1 & 0 & 0.45 \\ 0 & 0 & 0 & 1 & 0.30 \\ 0 & 0 & 0 & 0 & 0 \\ 0 & 0 & 0 & 0 & 0 \end{array}\right].$$

Therefore, $a=0.20$, $b=0.40$, $c=0.45$, and $d=0.30$. Thus, A is healthy, B and C are bone, and D is tumorous.

(d) As we saw in part (c), the six equations reduced to four independent equations. We need only four beams, correctly chosen, to obtain a solution. The four beams must pass through all four cells and must lead to independent equations. One such choice would be beams 1, 2, 3, and 6. Another choice would be beams 1, 2, 4, and 5.

(e) Answers will vary.

66. The matrix representing the rates per 1000 athlete-exposures for specific injuries that caused a player wearing either shield to miss one or more events is

$$\begin{bmatrix} 3.54 & 1.41 \\ 1.53 & 1.57 \\ 0.34 & 0.29 \\ 7.53 & 6.21 \end{bmatrix}.$$

(*continued on next page*)

(*continued*)

Since an equal number of players wear each type of shield and the total number of athlete-exposures for the league in a season is 8000, each type of shield is worn by 4000 players. Since the rates are given per 1000 athletic-exposures, the matrix representing the number of 1000 athlete-exposures for each type of shield is $\begin{bmatrix} 4 \\ 4 \end{bmatrix}$. The product of these matrices is

$\begin{bmatrix} 20 \\ 12 \\ 3 \\ 55 \end{bmatrix}$. (Values have been rounded.)

There would be about 20 head and face injuries, 12 concussions, 3 neck injuries, and 55 other injuries.

67. First, find the eigenvalues and eigenvectors.

$$M = \begin{bmatrix} 0.3 & 0.2 \\ 0.3 & 0.8 \end{bmatrix}$$

$$\det(M - \lambda I) = 0$$

$$\det \begin{bmatrix} 0.3-\lambda & 0.2 \\ 0.3 & 0.8-\lambda \end{bmatrix} = 0$$

$$(0.3-\lambda)(0.8-\lambda) - 0.2(0.3) = 0$$

$$\lambda^2 - 1.1\lambda + 0.18 = 0$$

Solve using the quadratic formula.

$$\lambda = \frac{-(-1.1) \pm \sqrt{(-1.1)^2 - 4(0.18)}}{2}$$

$$= \frac{1.1 \pm \sqrt{0.49}}{2} = \frac{1.1 \pm 0.7}{2} = 0.9 \text{ or } 0.2$$

The eigenvalues are 0.9 and 0.2.
Find the eigenvector corresponding to $\lambda = 0.9$.

$$\begin{bmatrix} 0 \\ 0 \end{bmatrix} = \begin{bmatrix} 0.3-0.9 & 0.2 \\ 0.3 & 0.8-0.9 \end{bmatrix} \begin{bmatrix} x_1 \\ x_2 \end{bmatrix}$$

$$= \begin{bmatrix} -0.6 & 0.2 \\ 0.3 & -0.1 \end{bmatrix} \begin{bmatrix} x_1 \\ x_2 \end{bmatrix}$$

The augmented matrix for this system is $\left[\begin{array}{cc|c} -0.6 & 0.2 & 0 \\ 0.3 & -0.1 & 0 \end{array}\right]$. Note that row 2 is a multiple of row 1, so this is a dependent system. The first row indicates that

$-0.6x_1 + 0.2x_2 = 0 \Rightarrow x_1 = \frac{1}{3}x_2$. A solution is $x_1 = 1,\ x_2 = 3$. Thus, an eigenvector corresponding to $\lambda = 0.9$ is $\begin{bmatrix} 1 \\ 3 \end{bmatrix}$.

Now find the eigenvector corresponding to $\lambda = 0.2$.

$$\begin{bmatrix} 0 \\ 0 \end{bmatrix} = \begin{bmatrix} 0.3-0.2 & 0.2 \\ 0.3 & 0.8-0.2 \end{bmatrix} \begin{bmatrix} x_1 \\ x_2 \end{bmatrix}$$

$$= \begin{bmatrix} 0.1 & 0.2 \\ 0.3 & 0.6 \end{bmatrix} \begin{bmatrix} x_1 \\ x_2 \end{bmatrix}$$

The augmented matrix for this system is $\left[\begin{array}{cc|c} 0.1 & 0.2 & 0 \\ 0.3 & 0.6 & 0 \end{array}\right]$. Note that row 2 is a multiple of row 1, so this is a dependent system. The first row indicates that

$0.1x_1 + 0.2x_2 = 0 \Rightarrow x_1 = -2x_2$. Since this leads to a negative value, it cannot be considered as a population.

(a) The initial population is 10,000, so a multiple of the eigenvector is used.

$$\begin{bmatrix} x_1 \\ x_2 \end{bmatrix} = k \begin{bmatrix} 1 \\ 3 \end{bmatrix} = \begin{bmatrix} k \\ 3k \end{bmatrix}$$

$k + 3k = 10{,}000 \Rightarrow 4k = 10{,}000 \Rightarrow$
$k = 2500$

The initial population is

$$\begin{bmatrix} 2500 \\ 3(2500) \end{bmatrix} = \begin{bmatrix} 2500 \\ 7500 \end{bmatrix}.$$

(b) The growth factor is 0.9.

68. First, find the eigenvalues and eigenvectors.

$$M = \begin{bmatrix} 0.2 & 0.3 \\ 0.4 & 0.1 \end{bmatrix}$$

$$\det(M - \lambda I) = 0$$

$$\det \begin{bmatrix} 0.2-\lambda & 0.3 \\ 0.4 & 0.1-\lambda \end{bmatrix} = 0$$

$$(0.2-\lambda)(0.1-\lambda) - 0.3(0.4) = 0$$

$$\lambda^2 - 0.3\lambda - 0.1 = 0$$

Solve using the quadratic formula.

$$\lambda = \frac{-(-0.3) \pm \sqrt{(-0.3)^2 - 4(-0.1)}}{2}$$

$$= \frac{0.3 \pm \sqrt{0.49}}{2} = \frac{0.3 \pm 0.7}{2} = 0.5 \text{ or } -0.2$$

The eigenvalues are 0.5 and –0.2.
Find the eigenvector corresponding to $\lambda = 0.5$.

(*continued on next page*)

(continued)

$$\begin{bmatrix}0\\0\end{bmatrix}=\begin{bmatrix}0.2-0.5 & 0.3\\0.4 & 0.1-0.5\end{bmatrix}\begin{bmatrix}x_1\\x_2\end{bmatrix}=\begin{bmatrix}-0.3 & 0.3\\0.4 & -0.4\end{bmatrix}\begin{bmatrix}x_1\\x_2\end{bmatrix}$$

The augmented matrix for this system is $\left[\begin{array}{cc|c}-0.3 & 0.3 & 0\\0.4 & -0.4 & 0\end{array}\right]$. Note that row 2 is a multiple of row 1, so this is a dependent system. The first row indicates that $-0.3x_1+0.3x_2=0 \Rightarrow x_1=x_2$. A solution is $x_1=1,\ x_2=1$.

Thus, an eigenvector corresponding to $\lambda=0.5$ is $\begin{bmatrix}1\\1\end{bmatrix}$.

Now find the eigenvector corresponding to $\lambda=-0.2$.

$$\begin{bmatrix}0\\0\end{bmatrix}=\begin{bmatrix}0.2-(-0.2) & 0.3\\0.4 & 0.1-(-0.2)\end{bmatrix}\begin{bmatrix}x_1\\x_2\end{bmatrix}=\begin{bmatrix}0.4 & 0.3\\0.4 & 0.3\end{bmatrix}\begin{bmatrix}x_1\\x_2\end{bmatrix}$$

The augmented matrix for this system is $\left[\begin{array}{cc|c}0.4 & 0.3 & 0\\0.4 & 0.3 & 0\end{array}\right]$. Note that row 2 is the same as row 1, so this is a dependent system. The first row indicates that $0.4x_1+0.3x_2=0 \Rightarrow x_1=-\frac{3}{4}x_2$. Since this leads to a negative value, it cannot be considered as a population.

(a) The initial population is 10,000, so a multiple of the eigenvector is used.

$$\begin{bmatrix}x_1\\x_2\end{bmatrix}=k\begin{bmatrix}1\\1\end{bmatrix}=\begin{bmatrix}k\\k\end{bmatrix}$$

$k+k=10{,}000 \Rightarrow 2k=10{,}000 \Rightarrow k=5000$

The initial population is $\begin{bmatrix}5000\\5000\end{bmatrix}$.

(b) The growth factor is 0.5.

69. (a) Each row gives the fraction of each class of the herd that will become memebers of the class represented by that row in the next year.

(b) Let $V=\begin{bmatrix}k\\0\\0\\0\\0\\0\end{bmatrix}$.

$$MV=\begin{bmatrix}0.95 & 0 & 0.75 & 0 & 0 & 0\\0 & 0.95 & 0 & 0.75 & 0 & 0\\0 & 0 & 0 & 0 & 0.6 & 0\\0 & 0 & 0 & 0 & 0 & 0.6\\0 & 0.48 & 0 & 0 & 0 & 0\\0 & 0.42 & 0 & 0 & 0 & 0\end{bmatrix}\begin{bmatrix}k\\0\\0\\0\\0\\0\end{bmatrix}=\begin{bmatrix}0.95k\\0\\0\\0\\0\\0\end{bmatrix}=0.95V$$

Therefore, $\lambda=0.5$. This tells us that if the herd consists of only adult males, then next year it will consist of only adulat males with population decreasing by a factor of 0.95.

(c) Let X = the population and Q = the harvest. Then next year's herd – the harvest = next year's herd without harvest.

$MX-Q=X \Rightarrow MX-X=Q \Rightarrow (M-I)X=Q \Rightarrow X=(M-I)^{-1}Q$

(d) $X = (M - I)^{-1} Q$

$$= \begin{bmatrix} 0.95-1 & 0 & 0.75 & 0 & 0 & 0 \\ 0 & 0.95-1 & 0 & 0.75 & 0 & 0 \\ 0 & 0 & 0-1 & 0 & 0.6 & 0 \\ 0 & 0 & 0 & 0-1 & 0 & 0.6 \\ 0 & 0.48 & 0 & 0 & 0-1 & 0 \\ 0 & 0.42 & 0 & 0 & 0 & 0-1 \end{bmatrix}^{-1} \begin{bmatrix} 100 \\ 100 \\ 10 \\ 10 \\ 0 \\ 0 \end{bmatrix}$$

$$= \begin{bmatrix} -0.05 & 0 & 0.75 & 0 & 0 & 0 \\ 0 & -0.05 & 0 & 0.75 & 0 & 0 \\ 0 & 0 & -1 & 0 & 0.6 & 0 \\ 0 & 0 & 0 & -1 & 0 & 0.6 \\ 0 & 0.48 & 0 & 0 & -1 & 0 \\ 0 & 0.42 & 0 & 0 & 0 & -1 \end{bmatrix}^{-1} \begin{bmatrix} 100 \\ 100 \\ 10 \\ 10 \\ 0 \\ 0 \end{bmatrix}$$

$$X = \begin{bmatrix} -20.0 & 31.0791 & -15.0 & 23.3094 & -9.0 & 13.9856 \\ 0 & 7.1942 & 0 & 5.3957 & 0 & 3.2374 \\ 0 & 2.0719 & -1.0 & 1.5540 & -0.6 & 0.9324 \\ 0 & 1.8126 & 0 & 0.3597 & 0 & 0.2158 \\ 0 & 3.4532 & 0 & 2.5899 & -1.0 & 1.5540 \\ 0 & 3.0216 & 0 & 2.2662 & 0 & 0.3597 \end{bmatrix} \begin{bmatrix} 100 \\ 100 \\ 10 \\ 10 \\ 0 \\ 0 \end{bmatrix} \approx \begin{bmatrix} 1191 \\ 773 \\ 213 \\ 185 \\ 371 \\ 325 \end{bmatrix}$$

(e) Let X = this year's herd and Q = the harvest. Then next year's herd – the harvest = next year's herd with 10% increase.

$MX - Q = 1.1X \Rightarrow MX - 1.1X = Q \Rightarrow (M - 1.1I)X = Q \Rightarrow X = (M - 1.1I)^{-1} Q$

(f) $X = (M - I)^{-1} Q$

$$= \begin{bmatrix} 0.95-1.1 & 0 & 0.75 & 0 & 0 & 0 \\ 0 & 0.95-1.1 & 0 & 0.75 & 0 & 0 \\ 0 & 0 & 0-1.1 & 0 & 0.6 & 0 \\ 0 & 0 & 0 & 0-1.1 & 0 & 0.6 \\ 0 & 0.48 & 0 & 0 & 0-1.1 & 0 \\ 0 & 0.42 & 0 & 0 & 0 & 0-1.1 \end{bmatrix}^{-1} \begin{bmatrix} 100 \\ 100 \\ 10 \\ 10 \\ 0 \\ 0 \end{bmatrix}$$

$$= \begin{bmatrix} -0.15 & 0 & 0.75 & 0 & 0 & 0 \\ 0 & -0.15 & 0 & 0.75 & 0 & 0 \\ 0 & 0 & -1 & 0 & 0.6 & 0 \\ 0 & 0 & 0 & -1 & 0 & 0.6 \\ 0 & 0.48 & 0 & 0 & -1 & 0 \\ 0 & 0.42 & 0 & 0 & 0 & -1 \end{bmatrix}^{-1} \begin{bmatrix} 100 \\ 100 \\ 10 \\ 10 \\ 0 \\ 0 \end{bmatrix}$$

$$= \begin{bmatrix} -6.6667 & 192.00 & -4.544 & 130.9091 & -2.4793 & 71.4050 \\ 0 & 161.33 & 0 & 110 & 0 & 60 \\ 0 & 38.40 & -0.9091 & 26.1818 & -0.4959 & 14.2810 \\ 0 & 33.60 & 0 & 22 & 0 & 12 \\ 0 & 70.40 & 0 & 48 & -0.9091 & 26.1818 \\ 0 & 61.60 & 0 & 42 & 0 & 22 \end{bmatrix}^{-1} \begin{bmatrix} 100 \\ 100 \\ 10 \\ 10 \\ 0 \\ 0 \end{bmatrix} \approx \begin{bmatrix} 19,797 \\ 17,233 \\ 4093 \\ 3580 \\ 7520 \\ 6580 \end{bmatrix}$$

70. $C = at^2 + bt + c$

Use the values for C from the table.

(a) For 1960, $t = 0$ and $C = 317$.

$317 = a(0)^2 + b(0) + c \Rightarrow 317 = c$

For 1980, $t = 20$ and $C = 339$.

$339 = a(20)^2 + b(20) + 317 \Rightarrow$
$22 = 400a + 20b$

For 2010, $t = 50$ and $C = 390$.

$390 = a(50)^2 + b(50) + 317 \Rightarrow$
$73 = 2500a + 50b$

Thus, we need to solve the system

$400a + 20b = 22$
$2500a + 50b = 73$

$$\begin{bmatrix} 400 & 20 \\ 2500 & 50 \end{bmatrix}\begin{bmatrix} a \\ b \end{bmatrix} = \begin{bmatrix} 22 \\ 73 \end{bmatrix}$$

$$\begin{bmatrix} a \\ b \end{bmatrix} = \begin{bmatrix} 400 & 20 \\ 2500 & 50 \end{bmatrix}^{-1}\begin{bmatrix} 22 \\ 73 \end{bmatrix} = \begin{bmatrix} -\frac{1}{600} & \frac{1}{1500} \\ \frac{1}{12} & -\frac{1}{75} \end{bmatrix}\begin{bmatrix} 22 \\ 73 \end{bmatrix} = \begin{bmatrix} \frac{3}{250} \\ \frac{43}{50} \end{bmatrix} = \begin{bmatrix} 0.012 \\ 0.86 \end{bmatrix}$$

Therefore, $C = 0.012t^2 + 0.86t + 317$.

(b) In 1960, $C = 317$. So, double that level would be $C = 634$.

$0.012t^2 + 0.86t + 317 = 634$
$0.012t^2 + 0.86t - 317 = 0$

Use the quadratic formula with $a = 0.012$, $b = 0.86$, and $c = -317$.

$$t = \frac{-0.86 \pm \sqrt{0.86^2 - 4(0.012)(-317)}}{2(0.012)} = \frac{-0.86 \pm \sqrt{15.9556}}{2(0.012)} \approx \frac{-0.86 \pm 3.9944}{0.024} \approx 130.6 \text{ or } -202.3$$

Ignore the negative value. If $t = 130.6$, then $1960 + 130.6 = 2090.6$ The 1960 CO_2 level will double in the year 2091.

71. (a)
$$\begin{bmatrix} 1 \\ 1 \end{bmatrix}x + \begin{bmatrix} 0 \\ 2 \end{bmatrix}y = \begin{bmatrix} 1 \\ 2 \end{bmatrix}$$

$$\begin{bmatrix} x \\ x \end{bmatrix} + \begin{bmatrix} 0 \\ 2y \end{bmatrix} = \begin{bmatrix} 1 \\ 2 \end{bmatrix} \Rightarrow \begin{bmatrix} x \\ x+2y \end{bmatrix} = \begin{bmatrix} 1 \\ 2 \end{bmatrix}$$

Since corresponding elements must be equal, $x = 1$ and $x + 2y = 2$. Substituting $x = 1$ in the second equation gives $y = \frac{1}{2}$.

Note that $x = 1$ and $y = \frac{1}{2}$ are the values that balance the equation.

(b) $xCO_2 + yH_2 + zCO = H_2O$

$$\begin{bmatrix} 1 \\ 0 \\ 2 \end{bmatrix}x + \begin{bmatrix} 0 \\ 2 \\ 0 \end{bmatrix}y + \begin{bmatrix} 1 \\ 0 \\ 1 \end{bmatrix}z = \begin{bmatrix} 0 \\ 2 \\ 1 \end{bmatrix}$$

$$\begin{bmatrix} x \\ 0 \\ 2x \end{bmatrix} + \begin{bmatrix} 0 \\ 2y \\ 0 \end{bmatrix} + \begin{bmatrix} z \\ 0 \\ z \end{bmatrix} = \begin{bmatrix} 0 \\ 2 \\ 1 \end{bmatrix}$$

$$\begin{bmatrix} x+z \\ 2y \\ 2x+z \end{bmatrix} = \begin{bmatrix} 0 \\ 2 \\ 1 \end{bmatrix}$$

Since corresponding elements must be equal, $x + z = 0$, $2y = 2$, and $2x + z = 1$. Solving $2y = 2$ gives $y = 1$. Solving the system $\begin{cases} x + z = 0 \\ 2x + z = 1 \end{cases}$ gives $x = 1$ and $z = -1$. Thus, the values that balance the equation are $x = 1$, $y = 1$, $z = -1$.

72.
$$\frac{\sqrt{3}}{2}(W_1 + W_2) = 100 \quad (1)$$
$$W_1 - W_2 = 0 \quad (2)$$

Equation (2) gives $W_1 = W_2$. Substitute W_1 for W_2 in equation (1).

$$\frac{\sqrt{3}}{2}(W_1 + W_1) = 100 \Rightarrow \frac{\sqrt{3}}{2}(2W_1) = 100 \Rightarrow$$
$$\sqrt{3}W_1 = 100 \Rightarrow W_1 = \frac{100}{\sqrt{3}} = \frac{100\sqrt{3}}{3} \approx 58$$

Therefore, $W_1 = W_2 \approx 58$ lb.

73.
$$\frac{1}{2}W_1 + \frac{\sqrt{2}}{2}W_2 = 150 \quad (1)$$
$$\frac{\sqrt{3}}{2}W_1 - \frac{\sqrt{2}}{2}W_2 = 0 \quad (2)$$

Adding equations (1) and (2) gives

$$\left(\frac{1}{2} + \frac{\sqrt{3}}{2}\right)W_1 = 150.$$

Multiply by 2.

$$(1 + \sqrt{3})W_1 = 300 \Rightarrow W_1 = \frac{300}{1 + \sqrt{3}} \approx 110$$

From equation (2),

$$\frac{\sqrt{3}}{2}W_1 = \frac{\sqrt{2}}{2}W_2 \Rightarrow W_2 = \frac{\sqrt{3}}{\sqrt{2}}W_1.$$

(continued on next page)

(continued)

Substitute $\frac{300}{1+\sqrt{3}}$ from above for W_1.

$$W_2 = \frac{\sqrt{3}}{\sqrt{2}} \cdot \frac{300}{1+\sqrt{3}} = \frac{300\sqrt{3}}{(1+\sqrt{3})\sqrt{2}} \approx 134$$

Therefore, $W_1 \approx 110$ lb and $W_2 \approx 134$ lb.

74. Use a table to organize the information.

	Standard	Extra Large	Time Available
Hours Cutting	$\frac{1}{4}$	$\frac{1}{3}$	4
Hours Shaping	$\frac{1}{2}$	$\frac{1}{3}$	6

Let x = the number of standard paper clips (in thousands), and let y = the number of extra large paper clips (in thousands).
The given information leads to the system

$$\frac{1}{4}x + \frac{1}{3}y = 4$$
$$\frac{1}{2}x + \frac{1}{3}y = 6.$$

Solve this system by any method to get $x = 8$, $y = 6$. The manufacturer can make 8 thousand (8000) standard and 6 thousand (6000) extra large paper clips.

75. Let x_1 = the number of blankets, x_2 = the number of rugs, and x_3 = the number of skirts. The given information leads to the system

$$24x_1 + 30x_2 + 12x_3 = 306 \quad (1)$$
$$4x_1 + 5x_2 + 3x_3 = 59 \quad (2)$$
$$15x_1 + 18x_2 + 9x_3 = 201 \quad (3)$$

Simplify equations (1) and (3).

$$\tfrac{1}{6}R_1 \to R_1 \quad 4x_1 + 5x_2 + 2x_3 = 51 \quad (4)$$
$$4x_1 + 5x_2 + 3x_3 = 59 \quad (2)$$
$$\tfrac{1}{3}R_3 \to R_3 \quad 5x_1 + 6x_2 + 3x_3 = 67 \quad (5)$$

Solve this system by the Gauss-Jordan method. Write the augmented matrix and use row operations.

$$\left[\begin{array}{ccc|c} 4 & 5 & 2 & 51 \\ 4 & 5 & 3 & 59 \\ 5 & 6 & 3 & 67 \end{array}\right]$$

$$\begin{array}{r} \\ -1R_1 + R_2 \to R_2 \\ -4R_3 + 5R_1 \to R_3 \end{array} \left[\begin{array}{ccc|c} 4 & 5 & 2 & 51 \\ 0 & 0 & 1 & 8 \\ 0 & 1 & -2 & -13 \end{array}\right]$$

Interchange the second and third rows.

$$\left[\begin{array}{ccc|c} 4 & 5 & 2 & 51 \\ 0 & 1 & -2 & -13 \\ 0 & 0 & 1 & 8 \end{array}\right]$$

$$-5R_2 + R_1 \to R_1 \left[\begin{array}{ccc|c} 4 & 0 & 12 & 116 \\ 0 & 1 & -2 & -13 \\ 0 & 0 & 1 & 8 \end{array}\right]$$

$$\begin{array}{r} -12R_3 + R_1 \to R_1 \\ 2R_3 + R_2 \to R_2 \\ \\ \end{array} \left[\begin{array}{ccc|c} 4 & 0 & 0 & 20 \\ 0 & 1 & 0 & 3 \\ 0 & 0 & 1 & 8 \end{array}\right]$$

$$\tfrac{1}{4}R_1 \to R_1 \left[\begin{array}{ccc|c} 1 & 0 & 0 & 5 \\ 0 & 1 & 0 & 3 \\ 0 & 0 & 1 & 8 \end{array}\right]$$

The solution of the system is $x = 5$, $y = 3$, $z = 8$. 5 blankets, 3 rugs, and 8 skirts can be made.

76. Let x = Tulsa's number of gallons, y = New Orleans' number of gallons, and z = Ardmore's number of gallons.

The given information leads to the system

$$\begin{aligned} 0.5x+0.4y+0.3z &= 219{,}000 \quad \textit{Chicago} \\ 0.2x+0.4y+0.4z &= 192{,}000 \quad \textit{Dallas} \\ 0.3x+0.2y+0.3z &= 144{,}000. \quad \textit{Atlanta} \end{aligned}$$

Write the augmented matrix and use row operations.

$$\left[\begin{array}{ccc|c} 0.5 & 0.4 & 0.3 & 219{,}000 \\ 0.2 & 0.4 & 0.4 & 192{,}000 \\ 0.3 & 0.2 & 0.3 & 144{,}000 \end{array}\right]$$

$$\begin{array}{r} \\ 2R_1+(-5)R_2 \to R_2 \\ 3R_1+(-5)R_3 \to R_3 \end{array} \left[\begin{array}{ccc|c} 0.5 & 0.4 & 0.3 & 219{,}000 \\ 0 & -1.2 & -1.4 & -522{,}000 \\ 0 & 0.2 & -0.6 & -63{,}000 \end{array}\right]$$

$$\begin{array}{r} -2R_3+R_1 \to R_1 \\ \\ R_2+6R_3 \to R_3 \end{array} \left[\begin{array}{ccc|c} 0.5 & 0 & 1.5 & 345{,}000 \\ 0 & -1.2 & -1.4 & -522{,}000 \\ 0 & 0 & -5 & -900{,}000 \end{array}\right]$$

$$\begin{array}{r} 0.3R_3+R_1 \to R_1 \\ -14R_3+50R_2 \to R_2 \\ \\ \end{array} \left[\begin{array}{ccc|c} 0.5 & 0 & 0 & 75{,}000 \\ 0 & -60 & 0 & -13{,}500{,}000 \\ 0 & 0 & -5 & -900{,}000 \end{array}\right]$$

$$\begin{array}{r} 2R_1 \to R_1 \\ -\frac{1}{60}R_2 \to R_2 \\ -\frac{1}{5}R_3 \to R_3 \end{array} \left[\begin{array}{ccc|c} 1 & 0 & 0 & 150{,}000 \\ 0 & 1 & 0 & 225{,}000 \\ 0 & 0 & 1 & 180{,}000 \end{array}\right]$$

Thus, 150,000 gal were produced at Tulsa, 225,000 gal at New Orleans, and 180,000 gal at Ardmore.

77. The 4×5 matrix of stock reports is

	div	*ratio*	*sales*	*price*	*change*
AT & T	1.33	17.6	152,000	26.75	1.88
GE	1.00	20.0	238,200	32.36	−1.50
SaraLee	0.79	25.4	39,110	16.51	−0.89
Disney	0.27	21.2	122,500	28.60	0.75

78. **(a)** $\begin{array}{l} \text{High} \\ \text{Medium} \\ \text{Coated} \end{array} \begin{bmatrix} 3170 \\ 2360 \\ 1800 \end{bmatrix}$ **(b)** $\begin{bmatrix} x \\ y \\ z \end{bmatrix}$

(c) $\begin{bmatrix} 10 & 5 & 8 \\ 12 & 0 & 4 \\ 0 & 10 & 5 \end{bmatrix} \begin{bmatrix} x \\ y \\ z \end{bmatrix} = \begin{bmatrix} 3170 \\ 2360 \\ 1800 \end{bmatrix}$

(d) $\begin{bmatrix} x \\ y \\ z \end{bmatrix} = \begin{bmatrix} 10 & 5 & 8 \\ 12 & 0 & 4 \\ 0 & 10 & 5 \end{bmatrix}^{-1} \begin{bmatrix} 3170 \\ 2360 \\ 1800 \end{bmatrix} = \begin{bmatrix} -0.154 & 0.212 & 0.0769 \\ -0.231 & 0.192 & 0.2154 \\ 0.462 & -0.385 & -0.231 \end{bmatrix} \begin{bmatrix} 3170 \\ 2360 \\ 1800 \end{bmatrix} = \begin{bmatrix} 150 \\ 110 \\ 140 \end{bmatrix}$

79. Let x = the number of boys and let y = the number of girls. The given information leads to the system

$0.2x + 0.3y = 500$ (1)
$0.6x + 0.9y = 1500$ (2)

Write the augmented matrix and use row operations to solve the system.

$$\left[\begin{array}{cc|c} 0.2 & 0.3 & 500 \\ 0.6 & 0.9 & 1500 \end{array}\right]$$

$$5R_1 \to R_1 \left[\begin{array}{cc|c} 1 & 1.5 & 2500 \\ 0.6 & 0.9 & 1500 \end{array}\right]$$

$$-0.6R_1 + R_2 \to R_2 \left[\begin{array}{cc|c} 1 & 1.5 & 2500 \\ 0 & 0 & 0 \end{array}\right]$$

Thus, $x + 1.5y = 2500 \Rightarrow x = 2500 - 1.5y$

There are y girls and $2500 - 1.5y$ boys, where y is any even integer between 0 and 1666 because $y \geq 0$ and $2500 - 1.5y \geq 0 \Rightarrow$ $-1.5y \geq -2500 \Rightarrow y \leq 1666.\overline{6}$.

80. Let x = the number of singles, y = the number of doubles, z = the number of triples, and w = the number of home runs hit by Ichiro Suzuki. The number of singles he hit was 11 more than four times the total of doubles and home runs, so $x = 4(y + w) + 11$, or $x - 4y - 4w = 11$. The number of doubles he hit was 1 more than twice this total of triples and home runs, so $y = 2(z + w) + 1$, or $-2z + y - 2w = 1$. The total of singles and home runs he hit was 15 more than five times the total of doubles and triples, so $x + w = 15 + 5(y + z)$, or $x - 5y - 5z + w = 15$.

Write the augmented matrix and use row operations to solve the system.

$$\left[\begin{array}{cccc|c} 1 & 1 & 1 & 1 & 225 \\ 1 & -4 & 0 & -4 & 11 \\ 0 & 1 & -2 & -2 & 1 \\ 1 & -5 & -5 & 1 & 15 \end{array}\right]$$

$$R_2 \leftrightarrow R_3 \left[\begin{array}{cccc|c} 1 & 1 & 1 & 1 & 225 \\ 0 & 1 & -2 & -2 & 1 \\ 1 & -4 & 0 & -4 & 11 \\ 1 & -5 & -5 & 1 & 15 \end{array}\right]$$

$$\begin{array}{r} \\ \\ -R_1 + R_3 \to R_3 \\ -R_1 + R_4 \to R_4 \end{array} \left[\begin{array}{cccc|c} 1 & 1 & 1 & 1 & 225 \\ 0 & 1 & -2 & -2 & 1 \\ 0 & -5 & -1 & -5 & -214 \\ 0 & -6 & -6 & 0 & -210 \end{array}\right]$$

$$-R_3 \to R_3 \left[\begin{array}{cccc|c} 1 & 1 & 1 & 1 & 225 \\ 0 & 1 & -2 & -2 & 1 \\ 0 & 5 & 1 & 5 & 214 \\ 0 & -6 & -6 & 0 & -210 \end{array}\right]$$

$$\begin{array}{r} -R_2 + R_1 \to R_1 \\ \\ -5R_2 + R_3 \to R_3 \\ 6R_2 + R_4 \to R_4 \end{array} \left[\begin{array}{cccc|c} 1 & 0 & 3 & 3 & 224 \\ 0 & 1 & -2 & -2 & 1 \\ 0 & 0 & 11 & 15 & 209 \\ 0 & 0 & -18 & -12 & -204 \end{array}\right]$$

$$R_3 \leftrightarrow R_4 \left[\begin{array}{cccc|c} 1 & 0 & 3 & 3 & 224 \\ 0 & 1 & -2 & -2 & 1 \\ 0 & 0 & -18 & -12 & -204 \\ 0 & 0 & 11 & 15 & 209 \end{array}\right]$$

$$-\tfrac{1}{18}R_3 \to R_3 \left[\begin{array}{cccc|c} 1 & 0 & 3 & 3 & 224 \\ 0 & 1 & -2 & -2 & 1 \\ 0 & 0 & 1 & \frac{2}{3} & \frac{34}{3} \\ 0 & 0 & 11 & 15 & 209 \end{array}\right]$$

$$\begin{array}{r} -3R_3 + R_1 \to R_1 \\ 2R_3 + R_2 \to R_2 \\ \\ -11R_3 + R_4 \to R_4 \end{array} \left[\begin{array}{cccc|c} 1 & 0 & 0 & 1 & 190 \\ 0 & 1 & 0 & -\frac{2}{3} & \frac{71}{3} \\ 0 & 0 & 1 & \frac{2}{3} & \frac{34}{3} \\ 0 & 0 & 0 & \frac{23}{3} & \frac{253}{3} \end{array}\right]$$

$$\tfrac{3}{23}R_4 \to R_4 \left[\begin{array}{cccc|c} 1 & 0 & 0 & 1 & 190 \\ 0 & 1 & 0 & -\frac{2}{3} & \frac{71}{3} \\ 0 & 0 & 1 & \frac{2}{3} & \frac{34}{3} \\ 0 & 0 & 0 & 1 & 11 \end{array}\right]$$

$$\begin{array}{r} -1R_4 + R_1 \to R_1 \\ \frac{2}{3}R_4 + R_2 \to R_2 \\ -\frac{2}{3}R_4 + R_3 \to R_3 \\ \\ \end{array} \left[\begin{array}{cccc|c} 1 & 0 & 0 & 0 & 179 \\ 0 & 1 & 0 & 0 & 31 \\ 0 & 0 & 1 & 0 & 4 \\ 0 & 0 & 0 & 1 & 11 \end{array}\right]$$

Ichiro Suzuki hit 179 singles, 31 doubles, 4 triples, and 11 home runs.

81. Let x = the weight of a single chocolate wafer and let y = the weight of a single layer of vanilla crème. A serving of regular Oreo cookies is three cookies so $3(2x + y) = 34$. A serving of Double Stuf is two cookies so $2(2x + 2y) = 29$. Write the equations in standard form, obtain the augmented matrix, and use row operations to solve.

$$\left[\begin{array}{cc|c} 6 & 3 & 34 \\ 4 & 4 & 29 \end{array}\right]$$

$$-2R_1 + 3R_2 \to R_2 \quad \left[\begin{array}{cc|c} 6 & 3 & 34 \\ 0 & 6 & 19 \end{array}\right]$$

$$-1R_2 + 2R_1 \to R_1 \quad \left[\begin{array}{cc|c} 12 & 0 & 49 \\ 0 & 6 & 19 \end{array}\right]$$

$$\begin{array}{r} \frac{1}{12}R_1 \to R_1 \\ \frac{1}{6}R_2 \to R_2 \end{array} \quad \left[\begin{array}{cc|c} 1 & 0 & \frac{49}{12} \\ 0 & 1 & \frac{19}{6} \end{array}\right]$$

The solution is $\left(\frac{49}{12}, \frac{19}{6}\right)$, or about (4.08, 3.17). A chocolate wafer weighs 4.08 g and a single layer of vanilla creme weighs 3.17g.

Chapter 11

DIFFERENTIAL EQUATIONS

11.1 Solutions of Elementary and Separable Differential Equations

1. $\frac{dy}{dx} = -4x + 6x^2$

$$y = \int(-4x + 6x^2)\,dx$$
$$= -2x^2 + 2x^3 + C$$

3. $4x^3 - 2\frac{dy}{dx} = 0$

Solve for $\frac{dy}{dx}$.

$$\frac{dy}{dx} = 2x^3$$
$$y = 2\int x^3 dx = 2\left(\frac{x^4}{4}\right) + C$$
$$= \frac{x^4}{2} + C$$

5. $y\frac{dy}{dx} = x^2$

Separate the variables and take antiderivatives.

$$\int y\,dy = \int x^2\,dx$$
$$\frac{y^2}{2} = \frac{x^3}{3} + K$$
$$y^2 = \frac{2}{3}x^3 + 2K$$
$$y^2 = \frac{2}{3}x^3 + C$$

7. $\frac{dy}{dx} = 2xy$

$$\int\frac{dy}{y} = \int 2x\,dx$$
$$\ln|y| = \frac{2x^2}{2} + C = x^2 + C$$
$$e^{\ln|y|} = e^{x^2 + C}$$
$$y = \pm e^{x^2 + C} = \pm e^{x^2}\cdot e^C = ke^{x^2}$$

9. $\frac{dy}{dx} = 3x^2y - 2xy = y(3x^2 - 2x)$

$$\int\frac{dy}{y} = \int(3x^2 - 2x)\,dx$$
$$\ln|y| = \frac{3x^3}{3} - \frac{2x^2}{2} + C$$
$$e^{\ln|y|} = e^{x^3 - x^2 + C}$$
$$y = \pm\left(e^{x^3 - x^2}\right)e^C = ke^{x^3 - x^2}$$

11. $\frac{dy}{dx} = \frac{y}{x},\ x > 0$

$$\int\frac{dy}{dx} = \int\frac{dx}{x}$$
$$\ln|y| = \ln x + C_1$$
$$e^{\ln|y|} = e^{\ln x + C_1}$$
$$y = \pm e^{\ln x}\cdot e^{C_1} = Ce^{\ln x} = Cx$$

13. $\frac{dy}{dx} = \frac{y^2 + 6}{2y}$

$$\frac{2y}{y^2 + 6}dy = dx$$
$$\int\frac{2y}{y^2 + 6}dy = \int dx$$
$$\ln|(y^2 + 6)| = x + C$$

Since $y^2 + 6$ is always greater than 0 we can write this as $\ln(y^2 + 6) = x + C$.

15. $\frac{dy}{dx} = y^2e^{2x}$

$$\int y^{-2}dy = \int e^{2x}dx$$
$$-y^{-1} = \frac{1}{2}e^{2x} + C$$
$$-\frac{1}{y} = \frac{1}{2}e^{2x} + C \Rightarrow y = \frac{-1}{\frac{1}{2}e^{2x} + C}$$

17. $\frac{dy}{dx} = \frac{\cos x}{\sin y}$

$$\sin y\,dy = \cos x\,dx$$
$$\int\sin y\,dy = \int\cos x\,dx$$
$$-\cos y = \sin x + C \Rightarrow \cos y = -\sin x + C$$

19. $\frac{dy}{dx}+3x^2=2x$

$$\frac{dy}{dx}=2x-3x^2$$

$$y=\frac{2x^2}{2}-\frac{3x^3}{3}+C=x^2-x^3+C$$

Since $y=5$ when $x=0$,
$5=0-0+C \Rightarrow C=5.$

Thus, $y=x^2-x^3+5.$

21. $2\frac{dy}{dx}=4xe^{-x}$

$$\frac{dy}{dx}=2xe^{-x}$$

Use the table of integrals or integrate by parts.

$y=2(-x-1)e^{-x}+C$

Since $y=42$ when $x=0$,
$42=2(0-1)(1)+C \Rightarrow 42=-2+C \Rightarrow C=44$

Thus, $y=-2xe^{-x}-2e^{-x}+44.$

23. $\frac{dy}{dx}=\frac{x^3}{y};\ y=5$ when $x=0.$

$$\int y\,dy=\int x^3\,dx$$

$$\frac{y^2}{2}=\frac{x^4}{4}+C$$

$$y^2=\frac{1}{2}x^4+2C=\frac{1}{2}x^4+k$$

Since $y=5$ when $x=0$, $25=0+k \Rightarrow k=25.$

So $y^2=\frac{1}{2}x^4+25.$

25. $(2x+3)y=\frac{dy}{dx};\ y=1$ when $x=0.$

$$\int(2x+3)dx=\int\frac{dy}{y}$$

$$\frac{2x^2}{2}+3x+C=\ln|y|$$

$$e^{x^2+3x+C}=e^{\ln|y|}$$

$$y=(e^{x^2+3x})(\pm e^C)=ke^{x^2+3x}$$

Since $y=1$ when $x=0$, $1=ke^{0+0} \Rightarrow k=1.$ So $y=e^{x^2+3x}.$

27. $\frac{dy}{dx}=\frac{2x+1}{y-3};\ y=4$ when $x=0.$

$$\int(y-3)\,dy=\int(2x+1)dx$$

$$\frac{y^2}{2}-3y=\frac{2x^2}{2}+x+C$$

Since $y=4$ when $x=0$,

$$\frac{16}{2}-12=0+0+C \Rightarrow C=-4.$$

So, $\frac{y^2}{2}-3y=x^2+x-4.$

29. $\frac{dy}{dx}=\frac{y^2}{x};\ y=3$ when $x=e.$

$$\int y^{-2}dy=\int\frac{dx}{x}$$

$$-y^{-1}=\ln|x|+C$$

$$-\frac{1}{y}=\ln|x|+C$$

$$y=\frac{-1}{\ln|x|+C}$$

Since $y=3$ when $x=e$,

$$3=\frac{-1}{\ln e+C} \Rightarrow 3=\frac{-1}{1+C} \Rightarrow 3+3C=-1 \Rightarrow$$

$$3C=-4 \Rightarrow C=-\frac{4}{3}.$$

So $y=\frac{-1}{\ln|x|-\frac{4}{3}}=\frac{-3}{3\ln|x|-4}.$

31. $\frac{dy}{dx}=(y-1)^2e^{x-1};\ y=2$ when $x=1.$

$$\frac{dy}{(y-1)^2}=e^{x-1}\,dx$$

$$\int(y-1)^{-2}dy=\int e^{x-1}dx$$

$$\frac{(y-1)^{-1}}{-1}=-\frac{1}{y-1}=e^{x-1}+C$$

$$-(y-1)=-y+1=\frac{1}{e^{x-1}+C}$$

$$1-\frac{1}{e^{x-1}+C}=y \Rightarrow$$

$$y=\frac{e^{x-1}+C}{e^{x-1}+C}-\frac{1}{e^{x-1}+C}=\frac{e^{x-1}+C-1}{e^{x-1}+C}$$

$$2=\frac{e^0+C-1}{e^0+C}=\frac{C}{C+1} \Rightarrow 2C+2=C \Rightarrow$$

$C=-2$

$$y=\frac{e^{x-1}+(-2)-1}{e^{x-1}+(-2)}=\frac{e^{x-1}-3}{e^{x-1}-2}$$

33. $\frac{dy}{dx} = y\cos x; \quad y(0) = 3$

$\frac{1}{y}dy = \cos x\, dx$

$\int \frac{1}{y}dy = \int \cos x\, dx$

$\ln|y| = \sin x + C$

$y = e^{\sin x + C} = e^C e^{\sin x}$

$3 = e^C e^{\sin 0} \Rightarrow 3 = e^C$

Thus, $y = 3e^{\sin x}$.

35. Find the equilibrium points by setting the right side of the equation equal to 0 and solving for y.

$y(y^2 - 1) = 0 \Rightarrow y = -1, 0, 1$

Draw a number line with arrows pointing to the right if $dy/dx > 0$ in the interval and pointing to the left if $dy/dx < 0$ in the interval.

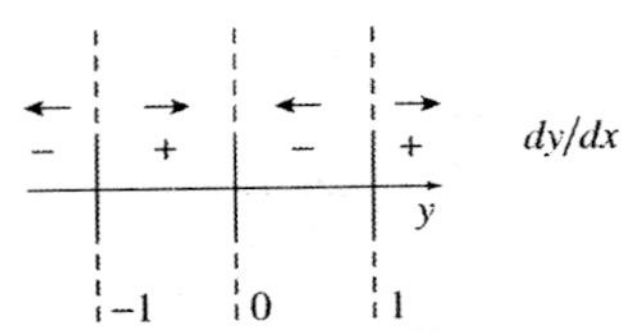

Since the arrows point away from $y = -1$ and $y = 1$, these two equilibrium points are unstable. The arrows point towards $y = 0$, so this equilibrium point is stable.

37. Find the equilibrium points by setting the right side of the equation equal to 0 and solving for y.

$(e^y - 1)(y - 3) = 0 \Rightarrow y = 0, 3$

Draw a number line with arrows pointing to the right if $dy/dx > 0$ in the interval and pointing to the left if $dy/dx < 0$ in the interval.

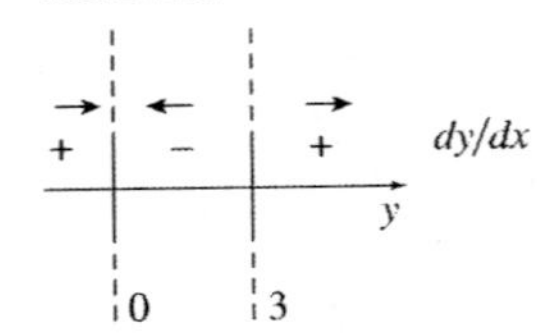

The arrows point towards $y = 0$, so this equilibrium point is stable. The arrows point away from $y = 3$, so this equilibrium point is unstable.

39. $\frac{dy}{dt} = \frac{k}{N}(N - y)y$

(a) $\frac{N\,dy}{(N-y)y} = k\,dt$

Since $\frac{1}{y} + \frac{1}{N-y} = \frac{N}{(N-y)y}$,

$\int \frac{dy}{y} + \int \frac{dy}{N-y} = k\,dt$

$\ln\left|\frac{y}{N-y}\right| = kt + C \Rightarrow \frac{y}{N-y} = Ce^{kt}$

For $0 < y < N$, $Ce^{kt} > 0$.
For $0 < N < y$, $Ce^{kt} < 0$.

Solve for y: $y = \frac{Ce^{kt}N}{1 + Ce^{kt}} = \frac{N}{1 + C^{-1}e^{-kt}}$

Let $b = C^{-1} > 0$ for $0 < y < N$.

$y = \frac{N}{1 + be^{-kt}}$

Let $-b = C^{-1} < 0$ for $0 < N < y$.

$y = \frac{N}{1 - be^{-kt}}$

(b) For $0 < y < N; t = 0, y = y_0$.

$y_0 = \frac{N}{1 + be^0} = \frac{N}{1 + b}$

Solve for b: $b = \frac{N - y_0}{y_0}$

(c) For $0 < N < y; t = 0, y = y_0$.

$y_0 = \frac{N}{1 - be^0} = \frac{N}{1 - b}$

Solve for b: $b = \frac{y_0 - N}{y_0}$

41. (a) $0 < y_0 < N$ implies that $y_0 > 0$, $N > 0$, and $N - y_0 > 0$. Therefore, $b = \frac{N - y_0}{y_0} > 0$. Also, $e^{-kx} > 0$ for all x, which implies that $1 + be^{-kx} > 1$.

(1) $y(x) = \frac{N}{1 + be^{-kx}} < N$ since $1 + be^{-kx} > 1$.

(2) $y(x) = \frac{N}{1 + be^{-kx}} > 0$ since $N > 0$ and $1 + be^{-kx} > 0$.

(continued on next page)

(*continued*)

Combining statements (1) and (2), we have

$$0<\frac{N}{1+be^{-kx}}=y(x)=\frac{N}{1+be^{-kx}}<N$$

or $0<y(x)<N$ for all x.

(b) $\lim_{x\to\infty}\frac{N}{1+be^{-kx}}=\frac{N}{1+b(0)}=N$

$\lim_{x\to-\infty}\frac{N}{1+be^{-kx}}=0$

Note that as $x\to-\infty$, $1+be^{-kx}$ becomes infinitely large. Therefore, the horizontal asymptotes are $y = N$ and $y = 0$.

(c) $y'(x)=\frac{(1+be^{-kx})(0)-N(-kbe^{-kx})}{(1+be^{-kx})^2}$

$=\frac{Nkbe^{-kx}}{(1+be^{-kx})^2}>0$ for all x.

Therefore, $y(x)$ is an increasing function.

(d) To find $y''(x)$, apply the quotient rule to find the derivation of $y'(x)$. The numerator of $y''(x)$, is

$$\begin{aligned}y''(x)&=(1+be^{-kx})^2\,(-Nk^2be^{-kx})\\&\quad-Nkbe^{-kx}[-2kbe^{-kx}(1+be^{-kx})]\\&=-Nk^2be^{-kx}(1-be^{-kx})(1+be^{-kx}),\end{aligned}$$

and the denominator is

$[(1+be^{-kt})^2]^2=(1+be^{-kx})^4$. Thus,

$$y''(x)=\frac{-Nk^2be^{-kx}(1-be^{-kx})}{(1+be^{-kx})^3}.$$

$y''(x)=0$ when

$k-kbe^{-kx}=0\Rightarrow be^{-kx}=1\Rightarrow$

$e^{-kx}=\frac{1}{b}\Rightarrow -kx=\ln\left(\frac{1}{b}\right)\Rightarrow$

$$x=\frac{\ln\left(\frac{1}{b}\right)}{k}=\frac{\ln\left(\frac{1}{b}\right)^{-1}}{k}=\frac{\ln b}{k}.$$

When $x=\frac{\ln b}{k}$,

$$y=\frac{N}{1+be^{-k\left(\frac{\ln b}{k}\right)}}=\frac{N}{1+be^{(-\ln b)}}$$

$$=\frac{N}{1+be^{\ln(1/b)}}=\frac{N}{1+b\left(\frac{1}{b}\right)}=\frac{N}{2}.$$

Therefore, $\left(\frac{\ln b}{k},\frac{N}{2}\right)$ is a point of inflection.

(e) To locate the maximum of $\frac{dy}{dx}$, we must consider, from part (d),

$$\frac{d}{dx}\left(\frac{dy}{dx}\right)=\frac{-Nkbe^{-kx}(k-kbe^{-kx})}{(1+be^{-kx})^3}.$$

Since $y''(x)>0$ for $x<\frac{\ln b}{k}$ and $y''(x)<0$ for $x>\frac{\ln b}{k}$, we know that $x=\frac{\ln b}{k}$ locates a relative maximum of $\frac{dy}{dx}$.

43. $\frac{dy}{dt}=-0.03y$

$\int\frac{dy}{y}=-0.03\int dt$

$\ln|y|=-0.03t+C$

$e^{\ln|y|}=e^{-0.03t+C}\Rightarrow y=Me^{-0.03t}$

Since $y = 6$ when $t = 0$, $6=Me^0\Rightarrow M=6$.

If $t = 10$, $y=6e^{-0.03(10)}\approx 4.4$

After 10 minutes, about 4.4 cc of dye will be present.

45. $\int\frac{dy}{4000-y}=\int 0.02\,dx$

$-\ln|4000-y|=0.02x+C$

$\ln|4000-y|=-0.02x+C$

$|4000-y|=e^{-0.02x+C}=e^{-0.02x}\cdot e^C$

$4000-y=Me^{-0.02x}$

$y=4000-Me^{-0.02x}$

Since $y=320$ when $x=0$,

$320=4000-Me^0\Rightarrow M=3680.$

So $y=4000-3680e^{-0.02x}$.

If $x=10$, $y=4000-3680e^{-0.02(10)}\approx 987$.

At the end of 10 years, there will be about 987 fish.

47. (a) $\frac{dw}{dt} = \frac{1}{3500}(2500 - 17.5w)$; $w = 180$ when $t = 0$.

$$\frac{3500}{2500 - 17.5w} dw = dt$$

$$\int \frac{3500}{2500 - 17.5w} dw = \int dt$$

$$\frac{3500}{-17.5} \int \frac{-17.5}{2500 - 17.5w} dw = \int dt$$

$$-200 \ln|2500 - 17.5w| = t + C_1$$

$$\ln|2500 - 17.5w| = -0.005t + C_2$$

$$|2500 - 17.5w| = e^{-0.005t + C_2}$$

$$|2500 - 17.5w| = e^{C_2} e^{-0.005t}$$

$$2500 - 17.5w = \pm e^{C_2} e^{-0.005t}$$

$$-17.5w = -2500 + C_3 e^{-0.005t}$$

$$w = 143 + C_4 e^{-0.005t}$$

Since $w = 180$ when $t = 0$,

$180 = 143 + C_4(1) \Rightarrow C_4 = 37$

The weight function is

$w = 143 + 37e^{-0.005t}$.

(b) $\lim_{t \to \infty} = \lim_{t \to \infty} \left(143 + 37e^{-0.005t}\right)$

$= 143 + 37(0) = 143$

The asymptote is $w = 143$.

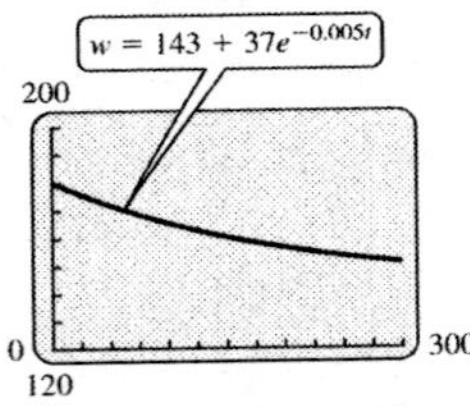

According to the model, the value 143 will never be attained.

(c)

$$145 = 143 + 37e^{-0.005t}$$

$$2 = 37e^{-0.005t}$$

$$e^{-0.005t} = \frac{2}{37}$$

$$-0.005t = \ln\left(\frac{2}{37}\right)$$

$$t = \frac{\ln\left(\frac{2}{37}\right)}{-0.005} \approx 583.55$$

About 584 days (about 19.5 months) would be required to reach a weight of 145.

49. (a)

(b) $y = \frac{1383}{1 + 1.867e^{-0.05615t}}$

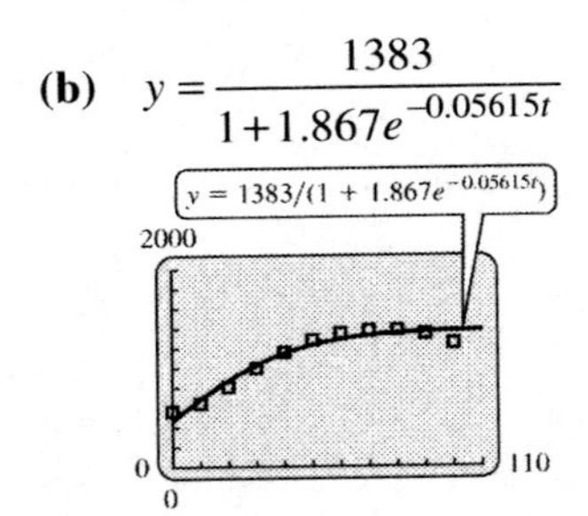

(c) According to the model, the limiting size of the Chinese population is 1383 million.

(d)

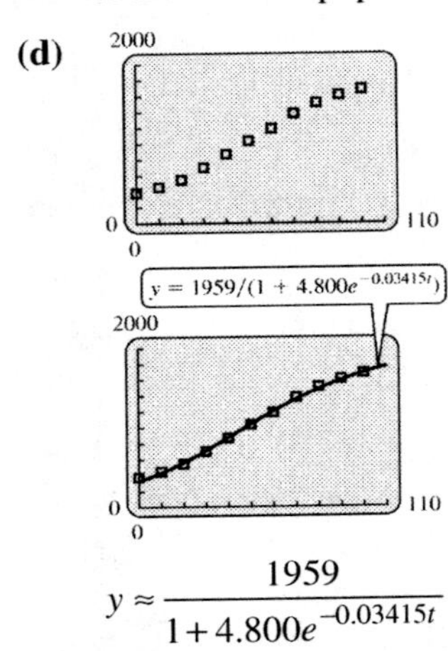

$$y \approx \frac{1959}{1 + 4.800e^{-0.03415t}}$$

According to the model, the limiting size of the Indian population is 1959 million.

51. $\frac{dy}{dt} = ky$

First separate the variables and integrate.

$$\frac{dy}{y} = k\,dt$$

$$\int \frac{dy}{y} = \int k\,dt$$

$\ln|y| = kt + C_1$.

Solve for y.

$|y| = e^{kt + C_1} = e^{C_1} e^{kt}$

$y = Ce^{kt}$, where $C = \pm e^{C_1}$.

$y(0) = 10.7$ so $10.7 = Ce^0 = C$, and

$y = 10.7\,e^{kt}$.

(continued on next page)

(continued)

Solve for k. Since $y(50) = 33.4$, then

$$33.4 = 10.7\,e^{50k}$$
$$e^{50k} = \frac{33.4}{10.7}$$
$$50k = \ln\left(\frac{33.4}{10.7}\right)$$
$$k = \frac{\ln\left(\frac{33.4}{10.7}\right)}{50} \approx 0.02277,$$

so $y = 10.7e^{0.02277t}$

53. (a)

$$\frac{dN}{dt} = mN + i = m\left(N + \frac{i}{m}\right)$$
$$\frac{1}{N + \frac{i}{m}}dN = m\,dt$$
$$\int \frac{dN}{N + \frac{i}{m}} = \int m\,dt$$
$$\ln\left(N + \frac{i}{m}\right) = mt + C$$
$$N + \frac{i}{m} = e^{mt+C} = e^{mt}e^{c}$$
$$N = e^{mt}e^{C} - \frac{i}{m}$$

For $N(0) = N_0$,

$$N_0 = e^{m(0)}e^{C} - \frac{i}{m} \Rightarrow e^{C} = N_0 + \frac{i}{m}$$

Substituting gives

$$N = \left(N_0 + \frac{i}{m}\right)e^{mt} - \frac{i}{m}.$$

(b) $N_0 = 0$, so

$$N = \left(0 + \frac{i}{m}\right)e^{mt} - \frac{i}{m} = \frac{i}{m}\left(e^{mt} - 1\right)$$
$$N(8) = \frac{i}{m}\left(e^{8m} - 1\right)$$
$$mN(8) = i\left(e^{8m} - 1\right) \Rightarrow i = \frac{mN(8)}{e^{8m} - 1}$$

(c) $m = \ln F_{sd} = \ln 0.709 \approx -0.344$

$$i = \frac{mN(8)}{e^{8m} - 1} = \frac{(\ln 0.709)(4.5)}{e^{8\ln 0.709} - 1} \approx 1.65$$

55. (a) A calculator with a logistic regression function determined that

$$y = \frac{11.74}{1 + (1.423 \times 10^{22})e^{-0.02554t}}$$

best fits the data.

(b) From the graph, the function from part (a) seems to fit the data from 1927 on very well. For the year 1804, the function does not fit the data very well.

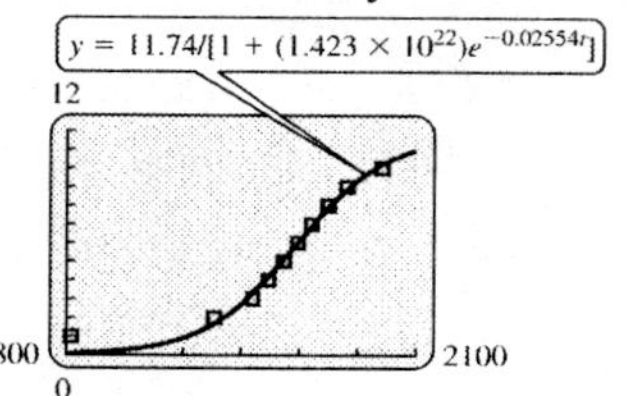

(c) After subtracting 0.99 from the y-values in the list, a calculator with a logistic regression function determines that

$$y = \frac{9.803}{1 + (2.612 \times 10^{29})e^{-0.03391t}}$$

best fits the data.

(d)

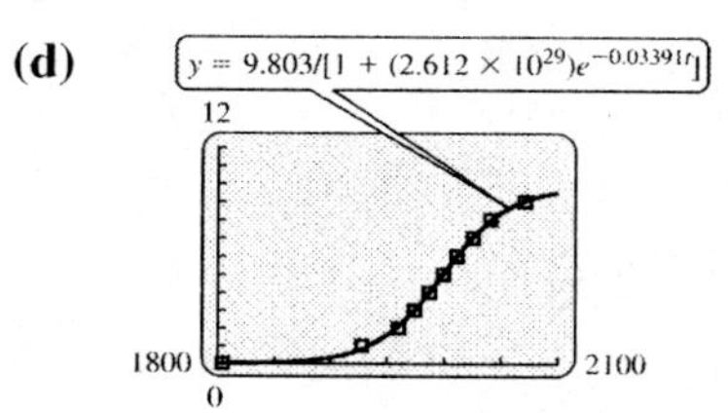

From the graph, the function in part (c) does seem to fit the data better than the graph found in part (b).

(e) As x gets larger and larger y_i approaches 0 so that y approaches $\frac{9.803}{1+0} = 9.803$. If you add back the 0.99 that was subtracted from the y-values, the result is approximately 10.79 billion.

(f) For the function found in part (c), as x gets smaller and approaches negative infinity, the denominator of this logistic function approaches infinity so that y approaches 0. After adding back the 0.99 that was subtracted earlier, this would imply that the limiting value for the world population as you go further and further back in time is 0.99 billion. This does not seem reasonable though because the world population was not always more than 990 million.

57. (a) $\frac{dy}{dt} = -0.03y$

(b) $\int \frac{dy}{y} = -\int 0.03\,dt$

$\ln|y| = -0.03t + C$

$e^{\ln|y|} = e^{-0.03+C}$

$y = \pm e^{-0.03t} \cdot e^{C} = Me^{-0.03t}$

(c) Since $y = 75$ when $t = 0$,

$75 = Me^{0} \Rightarrow M = 75.$

So $y = 75e^{-0.03t}$.

(d) At $t = 10$, $y = 75e^{-0.03(10)} \approx 56.$

After 10 months, about 56 g are left.

59. If $T = Ce^{-kt} + T_M$,

$$\lim_{t\to\infty} T = \lim_{t\to\infty}(Ce^{-kt} + T_M)$$
$$= \lim_{t\to\infty}(Ce^{-kt}) + T_M = T_M$$

(The exponential term has limit 0 since $k > 0$.) Therefore, the temperature of the object approaches the temperature of the surrounding medium.

61. Use the formula from Exercise 58:

$T = Ce^{-kt} + T_M$.

(a) In this problem, $T_M = 10, C = 88.6,$ and $k = 0.24$. Therefore,

$T = 88.6e^{-0.24t} + 10.$

(b)

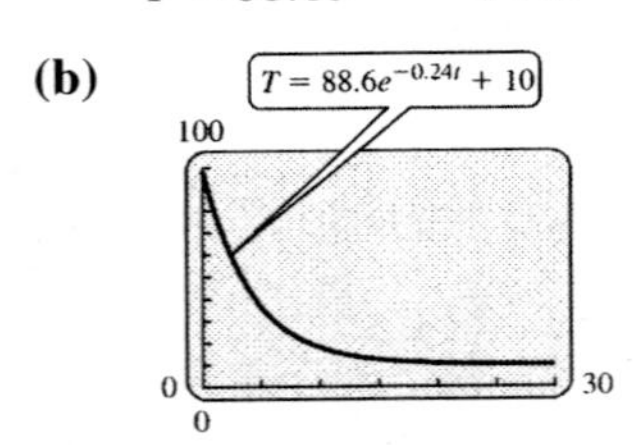

(c) The graph shows the most rapid decrease in the first few hours which is just after death.

(d) If $t = 4$, $T = 88.6e^{-0.24(4)} + 10 \approx 43.9.$ The temperature of the body will be 43.9°F after 4 hours.

(e) $40 = 88.6e^{-0.24t} + 10$

$88.6e^{-0.24t} = 30$

$e^{-0.24t} = \frac{30}{88.6}$

$-0.24t = \ln\left(\frac{30}{88.6}\right)$

$t = \frac{\ln\left(\frac{30}{88.6}\right)}{-0.24} \approx 4.5$

The body will reach a temperature of 40°F in 4.5 hours.

11.2 Linear First-Order Differential Equations

1. $\frac{dy}{dx} + 3y = 6$

$I(x) = e^{3\int dx} = e^{3x}$

Multiply each term by e^{3x}.

$e^{3x}\frac{dy}{dx} + 3e^{3x}y = 6e^{3x}$

$D_x(e^{3x}y) = 6e^{3x}$

Integrate both sides.

$e^{3x}y = \int 6e^{3x}dx = 2e^{3x} + C$

$y = 2 + Ce^{-3x}$

3. $\frac{dy}{dx} + 2xy = 4x$

$I(x) = e^{\int 2x\,dx} = e^{x^2}$

$e^{x^2}\frac{dy}{dx} + 2xe^{x^2}y = 4xe^{x^2}$

$D_x\left(e^{x^2}y\right) = 4xe^{x^2}$

$e^{x^2}y = \int 4xe^{x^2}dx = 2e^{x^2} + C$

$y = 2 + Ce^{-x^2}$

5. $x\frac{dy}{dx} - y - x = 0, \ x > 0$

$\frac{dy}{dx} - \frac{1}{x}y = 1$

$I(x) = e^{-\int 1/x\,dx} = e^{-\ln x} = \frac{1}{x}$

$\frac{1}{x}\frac{dy}{dx} - \frac{1}{x^2}y = \frac{1}{x}$

$D_x\left(\frac{1}{x}y\right) = \frac{1}{x}$

$\frac{y}{x} = \int \frac{1}{x}dx = \ln x + C$

$y = x\ln x + Cx$

7. $2\dfrac{dy}{dx}-2xy-x=0$

$$\frac{dy}{dx}-xy=\frac{x}{2}$$

$$I(x)=e^{-\int x\,dx}=e^{-x^2/2}$$

$$e^{-x^2/2}\frac{dy}{dx}-xe^{-x^2/2}y=\frac{x}{2}e^{-x^2/2}$$

$$D_x(e^{-x^2/2}y)=\frac{x}{2}e^{-x^2/2}$$

$$e^{-x^2/2}y=\int\frac{x}{2}e^{-x^2/2}dx=\frac{-1}{2}e^{-x^2/2}+C$$

$$y=-\frac{1}{2}+Ce^{x^2/2}$$

9. $x\dfrac{dy}{dx}+2y=x^2+6x;\ x>0$

$$\frac{dy}{dx}+\frac{2}{x}y=x+6$$

$$I(x)=e^{\int 2/x\,dx}=e^{2\ln x}=x^2$$

$$x^2\frac{dy}{dx}+2xy=x^3+6x^2$$

$$D_x(x^2y)=x^3+6x^2$$

$$x^2y=\int(x^3+6x^2)\,dx=\frac{x^4}{4}+2x^3+C$$

$$y=\frac{x^2}{4}+2x+\frac{C}{x^2}$$

11. $y-x\dfrac{dy}{dx}=x^3;\ x>0$

$$\frac{dy}{dx}-\frac{y}{x}=-x^2$$

$$I(x)=e^{-\int 1/x\,dx}=e^{-\ln x}=x^{-1}$$

$$\frac{1}{x}\frac{dy}{dx}-\frac{y}{x^2}=-x$$

$$D_x\left(\frac{1}{x}y\right)=-x$$

$$\frac{y}{x}=\int -x\,dx=\frac{-x^2}{2}+C$$

$$y=\frac{-x^3}{2}+Cx$$

13. $\dfrac{dy}{dx}+y\cot x=x$

$$I(x)=e^{\int\cot x dx}=e^{\ln\sin x}=\sin x$$

$$\sin x\frac{dy}{dx}+(\sin x)\,y\cot x=x\sin x$$

$$\sin x\frac{dy}{dx}+y\cos x=x\sin x$$

$$D_x(y\sin x)=x\sin x$$

$$y\sin x=\int x\sin x dx=\sin x-x\cos x+C$$

$$y=1-x\cot x+\frac{C}{\sin x}$$

15. $\dfrac{dy}{dx}+y=4e^x;\ y=50$ when $x=0.$

$$I(x)=e^{\int dx}=e^x$$

$$e^x\frac{dy}{dx}+ye^x=4e^{2x}$$

$$D_x(e^xy)=4e^{2x}$$

$$e^xy=\int 4e^{2x}dx=2e^{2x}+C$$

$$y=2e^x+Ce^{-x}$$

Since $y=50$ when $x=0$,

$50=2e^0+Ce^0\Rightarrow 50=2+C\Rightarrow C=48.$

Therefore, $y=2e^x+48e^{-x}.$

17. $\dfrac{dy}{dx}-2xy-4x=0;\ y=20$ when $x=1.$

$$\frac{dy}{dx}-2xy=4x$$

$$I(x)=e^{-\int 2x\,dx}=e^{-x^2}$$

$$e^{-x^2}\frac{dy}{dx}-2xe^{-x^2}y=4xe^{-x^2}$$

$$D_x(e^{-x^2}y)=4xe^{-x^2}$$

$$e^{-x^2}y=\int 4xe^{-x^2}dx$$

$$e^{-x^2}y=-2e^{-x^2}+C$$

$$y=-2+Ce^{x^2}$$

Since $y=20$ when $x=1$,

$20=-2+Ce^1\Rightarrow 22=Ce\Rightarrow C=\dfrac{22}{e}.$

Therefore, $y=-2+\dfrac{22}{e}\left(e^{x^2}\right)=-2+22e^{(x^2-1)}$

19. $x\frac{dy}{dx}+5y=x^2;\ y=12$ when $x=2$.

$$\frac{dy}{dx}+\frac{5}{x}y=x$$
$$I(x)=e^{\int 5/x\,dx}=e^{5\ln x}=x^5$$
$$x^5\frac{dy}{dx}+5x^4y=x^6$$
$$D_x(x^5y)=x^6$$
$$x^5y=\int x^6dx$$
$$x^5y=\frac{x^7}{7}+C$$
$$y=\frac{x^2}{7}+\frac{C}{x^5}$$

Since $y = 12$ when $x = 2$,

$$12=\frac{4}{7}+\frac{C}{32}\Rightarrow\frac{80}{7}=\frac{C}{32}\Rightarrow C=\frac{2560}{7}$$

Therefore, $y=\frac{x^2}{7}+\frac{2560}{7x^5}$.

21. $x\frac{dy}{dx}+(1+x)y=3;\ y=50$ when $x=4$.

$$\frac{dy}{dx}+\left(\frac{1+x}{x}\right)y=\frac{3}{x}$$
$$I(x)=e^{\int(1+x)\,dx/x}=e^{\int(1/x)\,dx+dx}=e^{(\ln x)+x}$$
$$=e^{\ln x}\cdot e^x=xe^x$$
$$xe^x\frac{dy}{dx}+(1+x)e^xy=3e^x$$
$$D_x(xe^xy)=3e^x$$
$$xe^xy=\int 3e^xdx$$
$$xe^xy=3e^x+C$$
$$y=\frac{3}{x}+\frac{C}{xe^x}$$

Since $y = 50$ when $x = 4$,

$$50=\frac{3}{4}+\frac{C}{4e^4}\Rightarrow\frac{197}{4}=\frac{C}{4e^4}\Rightarrow C=197e^4.$$

Therefore,

$$y=\frac{3}{x}+\frac{197e^4}{xe^x}=\frac{3}{x}+\frac{197}{x}e^{4-x}$$
$$=\frac{3+197e^{4-x}}{x}.$$

23. $\frac{dy}{dx}=cy-py^2$

(a) Let $y=\frac{1}{z}$ and $\frac{dy}{dx}=-\frac{z'}{z^2}$.

$$-\frac{z'}{z^2}=c\left(\frac{1}{z}\right)-p\left(\frac{1}{z^2}\right)$$
$$z'=-cz+p\Rightarrow z'+cz=p$$
$$I(x)=e^{\int c\,dx}=e^{cx}$$
$$D_x(e^{cx}\cdot z)=\int pe^{cx}dx$$
$$e^{cx}\cdot z=\frac{p}{c}e^{cx}+K$$
$$z=\frac{p}{c}+Ke^{-cx}=\frac{p+Kce^{-cx}}{c}$$

Therefore, $y=\frac{c}{p+Kce^{-cx}}$.

(b) Let $z(0)=\frac{1}{y_0}$.

$$\frac{1}{y_0}=\frac{p+Kce^0}{c}=\frac{p+Kc}{c}$$
$$\frac{c}{y_0}=p+Kc$$
$$Kc=\frac{c}{y_0}-p=\frac{c-py_0}{y_0}\Rightarrow K=\frac{c-py_0}{cy_0}$$

From part (a),

$$y=\frac{c}{p+\left(\frac{c-py_0}{cy_0}\right)ce^{-cx}}$$
$$=\frac{cy_0}{py_0+(c-py_0)e^{-cx}}$$

(c) $$\lim_{x\to\infty}y=\lim_{x\to\infty}\left(\frac{cy_0}{py_0+(c-py_0)e^{-cx}}\right)$$
$$=\frac{cy_0}{py_0-0}=\frac{c}{p}$$

25. $\frac{dy}{dt}=\alpha(1-y)-\beta y;\ \ y(0)=y_0$

(a) We must write the equation in the linear form $\frac{dy}{dt}+P(t)\,y=Q(t)$.

$$\frac{dy}{dt}=\alpha(1-y)-\beta y$$
$$=\alpha-\alpha y-\beta y$$
$$=\alpha-(\alpha+\beta)\,y$$

(continued on next page)

(*continued*)

Now, rewrite the equation as $\frac{dy}{dt}+(\alpha+\beta)\,y=\alpha$

$P(t)=\alpha+\beta,$ so the integrating factor is $e^{\int(\alpha+\beta)dt}=e^{(\alpha+\beta)t}.$

Multiply each side of the equation by the integrating factor.

$$e^{(\alpha+\beta)t}\frac{dy}{dt}+(\alpha+\beta)e^{(\alpha+\beta)t}y=\alpha e^{(\alpha+\beta)t}$$

The sum of the terms on the left can be replaced by $D_t\left(e^{(\alpha+\beta)t}y\right).$

$$D_t\left(e^{(\alpha+\beta)t}y\right)=\alpha e^{(\alpha+\beta)t}$$

Integrate both sides and solve for y.

$$e^{(\alpha+\beta)t}y=\alpha\int e^{(\alpha+\beta)t}\,dt\Rightarrow e^{(\alpha+\beta)t}y=\alpha\left(\frac{1}{\alpha+\beta}\right)e^{(\alpha+\beta)t}+C\Rightarrow y=\frac{\alpha}{\alpha+\beta}+Ce^{-(\alpha+\beta)t}$$

Now substitute $y(0)=y_0$ to find C.

$$y_0=\frac{\alpha}{\alpha+\beta}+Ce^{-(\alpha+\beta)(0)}=\frac{\alpha}{\alpha+\beta}+C\Rightarrow C=y_0-\frac{\alpha}{\alpha+\beta}$$

The particular solution is

$$y=\frac{\alpha}{\alpha+\beta}+\left(y_0-\frac{\alpha}{\alpha+\beta}\right)e^{-(\alpha+\beta)t}$$

(b) As $t\to\infty,\ e^{-(\alpha+\beta)t}\to 0,$ so the limit is $\frac{\alpha}{\alpha+\beta}.$

27. $$\frac{dN}{dt}=rN-\frac{\alpha r(\alpha+b+v)}{\beta\left[\alpha-r\left(1+\frac{v}{b+\gamma}\right)\right]}$$

$$\frac{dN}{dt}-rN=-\frac{\alpha r(\alpha+b+v)}{\beta\left[\alpha-r\left(1+\frac{v}{b+\gamma}\right)\right]}$$

Integrating factor: $I=e^{\int -r\,dt}=e^{-rt}.$

$$e^{-rt}\frac{dN}{dt}-e^{-rt}rN=e^{-rt}\left(-\frac{\alpha r(\alpha+b+v)}{\beta\left[\alpha-r\left(1+\frac{v}{b+\gamma}\right)\right]}\right)$$

$$D_t(Ne^{-rt})=-\frac{\alpha r(\alpha+b+v)}{\beta\left[\alpha-r\left(1+\frac{v}{b+\gamma}\right)\right]}e^{-rt}$$

$$Ne^{-rt}=-\frac{\alpha r(\alpha+b+v)}{\beta\left[\alpha-r\left(1+\frac{v}{b+\gamma}\right)\right]}\int e^{-rt}dt=-\frac{\alpha r(\alpha+b+v)}{\beta\left[\alpha-r\left(1+\frac{v}{b+\gamma}\right)\right]}\left(-\frac{1}{r}\right)e^{-rt}+C$$

$$=\frac{\alpha(\alpha+b+v)}{\beta\left[\alpha-r\left(1+\frac{v}{b+\gamma}\right)\right]}e^{-rt}+C$$

$$N(t)=\frac{\alpha+b+v}{\beta}\cdot\frac{\alpha}{\left[\alpha-r\left(1+\frac{v}{b+\gamma}\right)\right]}+Ce^{rt}$$

(*continued on next page*)

(continued)

Use the initial condition $N(0)=\frac{\alpha+b+v}{\beta}$.

$$N(0)=\frac{\alpha+b+v}{\beta}\cdot\frac{\alpha}{\alpha-r\left(1+\frac{v}{b+\gamma}\right)}+Ce^{r(0)}\Rightarrow\frac{\alpha+b+v}{\beta}=\frac{\alpha+b+v}{\beta}\cdot\frac{\alpha}{\alpha-r\left(1+\frac{v}{b+\gamma}\right)}+C$$

$$C=\frac{\alpha+b+v}{\beta}-\frac{\alpha+b+v}{\beta}\cdot\frac{\alpha}{\alpha-r\left(1+\frac{v}{b+\gamma}\right)}=\frac{\alpha+b+v}{\beta}\left(1-\frac{\alpha}{\alpha-r\left(1+\frac{v}{b+\gamma}\right)}\right)$$

Replace this in the anti-derivative previously found.

$$N(t)=\frac{\alpha+b+v}{\beta}\cdot\frac{\alpha}{\alpha-r\left(1+\frac{v}{b+\gamma}\right)}+\frac{\alpha+b+v}{\beta}\left(1-\frac{\alpha}{\alpha-r\left(1+\frac{v}{b+\gamma}\right)}\right)e^{rt}$$

Now use the substitution $R=\alpha-r\left(1+\frac{v}{b+\gamma}\right)$.

$$N(t)=\frac{\alpha+b+v}{\beta}\cdot\frac{\alpha}{\alpha-r\left(1+\frac{v}{b+\gamma}\right)}+\frac{\alpha+b+v}{\beta}\cdot\left(1-\frac{\alpha}{\alpha-r\left(1+\frac{v}{b+\gamma}\right)}\right)e^{rt}$$

$$=\frac{\alpha+b+v}{\beta}\cdot\frac{\alpha}{R}+\frac{\alpha+b+v}{\beta}\left(1-\frac{\alpha}{R}\right)e^{rt}=\frac{\alpha+b+v}{\beta R}[\alpha+(R-\alpha)e^{rt}]=\frac{\alpha+b+v}{\beta R}[(R-\alpha)e^{rt}+\alpha]$$

29. $\frac{dy}{dt}=0.02y+e^{t}$; $y=10{,}000$ when $t=0$.

$$\frac{dy}{dt}-0.02y=e^{t}$$

$$I(t)=e^{\int -0.02dt}=e^{-0.02t}$$

$$e^{-0.02t}\frac{dy}{dt}-0.02e^{-0.02t}y=e^{-0.02t}\cdot e^{t}$$

$$D_t(e^{-0.02t}y)=e^{0.98t}$$

$$e^{-0.02t}y=\int e^{0.98t}dt=\frac{e^{0.98t}}{0.98}+C$$

$$y=\frac{e^{t}}{0.98}+Ce^{0.02t}$$

$$10{,}000=\frac{1}{0.98}+C\Rightarrow C\approx 9999$$

$$y\approx\frac{e^{t}}{0.98}+9999e^{0.02t}=1.02e^{t}+9999e^{0.02t}$$

31. $\frac{dy}{dt}=0.02y-t$; $y=10{,}000$ when $t=0$.

$$\frac{dy}{dt}-0.02y=-t$$

$$I(t)=e^{\int -0.02}=e^{-0.02t}$$

$$e^{-0.02t}\frac{dy}{dt}-0.02e^{-0.02t}y=-te^{-0.02t}$$

$$D_t(e^{-0.02t}y)=-te^{-0.02t}$$

$$e^{-0.02t}y=\int -te^{-0.02t}dt$$

Use integration by parts:

$u=-t\Rightarrow du=-dt$

$dv=e^{-0.02t}dt\Rightarrow v=\frac{e^{-0.02t}}{-0.02}$

$$e^{-0.02t}y=\frac{te^{-0.02t}}{0.02}-\int\frac{e^{-0.02t}}{-0.02}(-dt)$$

$$e^{-0.02t}y=\frac{te^{-0.02t}}{0.02}+\frac{e^{-0.02t}}{0.0004}+C$$

$$y=50t+2500+Ce^{0.02t}$$

$$10{,}000=2500+C\Rightarrow C=7500$$

$$y=50t+2500+7500e^{0.02t}$$

33. $$\frac{dT}{dt} = -k(T - T_M)$$
$$\frac{dT}{dt} = -kT + kT_M$$
$$\frac{dT}{dt} + kT = kT_M$$
$$I(t) = e^{\int k\,dt} = e^{kt}$$
Multiply both sides by e^{kt}.
$$e^{kt}\frac{dT}{dt} + ke^{kt}T = kT_M e^{kt}$$
$$D_t(Te^{kt}) = kT_M e^{kt}$$
$$Te^{kt} = \int kT_M e^{kt}\,dt$$
$$Te^{kt} = T_M e^{kt} + C$$
$$T = T_M + Ce^{-kt} = Ce^{-kt} + T_M$$

11.3 Euler's Method

Note: In each step of the calculation shown in this section, all digits should be kept in your calculator as you proceed through Euler's method. Do not round intermediate results.

1. $\frac{dy}{dx} = x^2 + y^2$; $y(0) = 2$, $h = 0.1$. Find $y(0.5)$.

$g(x, y) = x^2 + y^2$
$x_0 = 0$; $y_0 = 2$
$g(x_0, y_0) = 0 + 4 = 4$
$x_1 = 0.1$
$y_1 = 2 + 4(0.1) = 2.4$
$g(x_1, y_1) = (0.1)^2 + (2.4)^2 = 5.77$
$x_2 = 0.2$
$y_2 = 2.4 + 5.77(0.1) = 2.977$
$g(x_2, y_2) = (0.2)^2 + (2.977)^2 \approx 8.903$
$x_3 = 0.3$
$y_3 = 2.977 + 8.903(0.1) \approx 3.867$
$g(x_3, y_3) = (0.3)^2 + (3.867)^2 \approx 15.046$
$x_4 = 0.4$
$y_4 = 3.867 + 15.046(0.1) \approx 5.372$
$g(x_4, y_4) = (0.4)^2 + (5.372)^2 \approx 29.016$
$x_5 = 0.5$
$y_5 = 5.372 + 29.016(0.1) \approx 8.273$

These results are tabulated as follows.

x_i	y_i
0	2
0.1	2.4
0.2	2.977
0.3	3.867
0.4	5.372
0.5	8.273

$y(0.5) \approx 8.273$.

Use Euler's method as outlined in the solution for Exercise 1 in the following exercises. The results are tabulated.

3. $\frac{dy}{dx} = 1 + y$; $y(0) = 2, h = 0.1$; find $y(0.6)$.

x_i	y_i
0	2
0.1	2.3
0.2	2.63
0.3	2.993
0.4	3.3923
0.5	3.8315
0.6	4.31468

$y(0.6) \approx 4.315$

5. $\frac{dy}{dx} = x + \sqrt{y}$; $y(0) = 1$, $h = 0.1$; find $y(0.4)$.

x_i	y_i
0	1
0.1	1.1
0.2	1.215
0.3	1.345
0.4	1.491

$y(0.4) \approx 1.491$

7. $\frac{dy}{dx} = 2x\sqrt{1 + y^2}$; $y(1) = 2$, $h = 0.1$; find $y(1.5)$.

x_i	y_i
1	2
1.1	2.447
1.2	3.029
1.3	3.794
1.4	4.815
1.5	6.191

$y(1.5) \approx 6.191$.

9. $\frac{dy}{dx} = y + \cos x;\ y(0) = 0;$ find $y(0.5)$

$x_0 = 0;\ y_0 = 0$
$g(x_0,\ y_0) = 0 + 1.0 = 1.0$
$x_1 = 0.1;\ y_0 = 0 + 1.0(0.1) = 0.1$
$g(x_1,\ y_1) = 0.1 + \cos(0.1) = 1.095$
$x_2 = 0.2;\ y_2 = 0.1 + 1.095(0.1) = 0.210$
$g(x_2,\ y_2) = 0.210 + \cos(0.2) = 1.190$
$x_3 = 0.3;\ y_3 = 0.210 + 1.190(0.1) = 0.329$
$g(x_3,\ y_3) = 0.329 + \cos(0.3) = 1.284$
$x_4 = 0.4;\ y_4 = 0.329 + 1.284(0.1) = 0.457$
$g(x_4,\ y_4) = 0.457 + \cos(0.4) = 1.378$
$x_5 = 0.5;\ y_5 = 0.457 + 1.378(0.1) = 0.595$

These results are tabulated as follows.

x_i	y_i
0	0.000
0.1	0.100
0.2	0.210
0.3	0.328
0.4	0.457
0.5	0.595

$y(0.5) \approx 0.595.$

11. $\frac{dy}{dx} = -4 + x;\ y(0) = 1,\ h = 0.1,$ find $y(0.4)$.

x_i	y_i
0	1
0.1	0.6
0.2	0.21
0.3	–0.17
0.4	–0.540

$y(0.4) \approx -0.540$

Exact solution:

$$\frac{dy}{dx} = -4 + x \Rightarrow y = -4x + \frac{x^2}{2} + C$$

At $y(0) = 1,\ 1 = -4(0) + \frac{0}{2} + C \Rightarrow C = 1$

Therefore,

$$y = -4x + \frac{x^2}{2} + 1$$

$$y(0.4) = -4(0.4) + \frac{(0.4)^2}{2} + 1 = -0.520$$

13. $\frac{dy}{dx} = x^3;\ y(0) = 4,\ h = 0.1,$ find $y(0.5)$.

x_i	y_i
0	4
0.1	4
0.2	4.0001
0.3	4.0009
0.4	4.0036
0.5	4.010

$y(0.5) \approx 4.010$

Exact solution: $\frac{dy}{dx} = x^3 \Rightarrow y = \frac{x^4}{4} + C$

At $y(0) = 4,\ 4 = \frac{0}{4} + C \Rightarrow C = 4.$

Therefore,

$$y = \frac{x^4}{4} + 4$$

$$y(0.5) = \frac{(0.5)^4}{4} + 4 \approx 4.016.$$

15. $\frac{dy}{dx} = 2xy;\ y(1) = 1,\ h = 0.1,$ find $y(1.6)$.

x_i	y_i
1	1
1.1	1.2
1.2	1.464
1.3	1.815
1.4	2.287
1.5	2.928
1.6	3.806

$y(1.6) \approx 3.806$

Exact solution:

$$\frac{dy}{y} = 2x\,dx$$

$$\int \frac{dy}{y} = \int 2x\,dx$$

$$\ln|y| = x^2 + C \Rightarrow |y| = e^{x^2 + C} \Rightarrow y = ke^{x^2}$$

At $y(1) = 1,\ 1 = ke^1 = ke \Rightarrow k = \frac{1}{e}.$

Therefore,

$$y = \frac{1}{e}\left(e^{x^2}\right) = e^{x^2 - 1}$$

$$y(1.6) = e^{(1.6)^2 - 1} = 4.759.$$

17. $\frac{dy}{dx} = ye^x$; $y(0) = 2$, $h = 0.1$, find $y(0.4)$.

x_i	y_i
0	2
0.1	2.2
0.2	2.443
0.3	2.742
0.4	3.112

So, $y(0.4) \approx 3.112$.

Exact solution:

$$\frac{dy}{y} = e^x dx$$

$$\int \frac{dy}{y} = \int e^x + c$$

$$\ln|y| = e^x + c \Rightarrow |y| = e^{e^x + c} = e^c e^{e^x} \Rightarrow$$

$$y = ke^{e^x}, \text{ where } k = \pm e^c.$$

At $y(0) = 2$, $2 = ke^{e^0} = ke$, so $k = \frac{2}{e}$.

Therefore, $y = \frac{2}{e} e^{e^x} = 2e^{e^x - 1}$, so

$$y(0.4) = 2e^{e^{0.4} - 1} \approx 3.271.$$

19. $\frac{dy}{dx} + y = 2e^x$; $y(0) = 100$, $h = 0.1$.

Find $y(0.3)$.

x_i	y_i
0	100
0.1	90.2
0.2	81.401
0.3	73.505

$y(0.3) \approx 73.505$.

Exact solution:

$$I(x) = e^{\int dx} = e^x$$

$$e^x \frac{dy}{dx} + e^x y = 2e^x e^x$$

$$D_x(e^x y) = 2e^{2x}$$

$$e^x y = \int 2e^{2x} dx + C = e^{2x} + C$$

$$y = e^x + Ce^{-x}$$

$$100 = 1 + C \Rightarrow C = 99$$

$$y = e^x + 99e^{-x}$$

$$y(0.3) = e^{0.3} + 99e^{-0.3} \approx 74.691$$

21. $\frac{dy}{dx} = ye^x$; $y(0) = 2$, $h = 0.05$, find $y(0.4)$.

Using the program for Euler's method in the Graphing Calculator Manual, the following values are obtained:

x_i	y_i	$y(x_i)$	$y_i - y(x_i)$
0	2	2	2.1
0.05	2.1	2.207669302	2.210383465
0.10	2.210383465	2.442851523	2.332526041
0.15	2.332526041	2.710008627	2.468026473
0.20	2.468026473	3.014454341	2.61874919
0.25	2.61874919	3.362540519	2.786876216
0.30	2.786876216	3.761889405	2.974970686
0.35	2.974970686	4.221684358	3.186054904
0.40	3.186054904		

So, $y(0.4) \approx 3.186$.

From exercise 17, the estimated value is $y(0.4) \approx 3.112$ and the actual value is $y(0.4) \approx 3.271$. This gives an error of $3.271 - 3.112 = 0.159$. Using the smaller value of h leads to an error of $3.271 - 3.186 = 0.085$.

23. $\frac{dy}{dx} = \sqrt[3]{x}$, $y(0) = 0$

Using the program for Euler's method in the Graphing Calculator Manual, the following values are obtained:

x_i	y_i	$y(x_i)$	$y_i - y(x_i)$
0	0	0	0
0.2	0	0.08772053	−0.08772053
0.4	0.11696071	0.22104189	−0.10408118
0.6	0.26432197	0.37954470	−0.11522273
0.8	0.43300850	0.55699066	−0.12398216
1.0	0.61867206	0.75000000	−0.13132794

25. $\frac{dy}{dx} = 4 - y$, $y(0) = 0$

Using the program for Euler's method in the Graphing Calculator Manual, the following values are obtained:

x_i	y_i	$y(x_i)$	$y_i - y(x_i)$
0	0	0	0
0.2	0.8	0.725077	0.07492
0.4	1.44	1.3187198	0.12128
0.6	1.952	1.8047535	0.14725
0.8	2.3616	2.2026841	0.15892
1.0	2.68928	2.5284822	0.16080

27. $\frac{dy}{dx} = \sqrt[3]{x};\ y(0) = 0$

See Exercise 23.

$dy = x^{1/3}dx$

$y = \int x^{1/3}dx = \frac{3}{4}x^{4/3} + C$

$0 = \frac{3}{4}(0)^{4/3} + C \Rightarrow C = 0$, so $y = \frac{3}{4}x^{4/3}$.

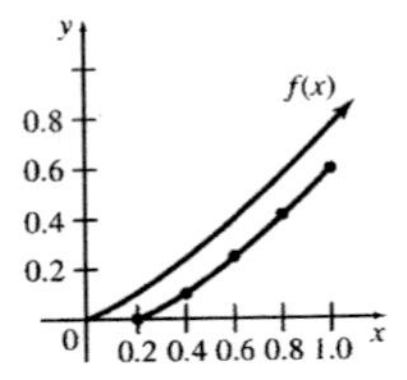

29. $\frac{dy}{dx} = 4 - y;\ y(0) = 0$

See Exercise 25.

$$\frac{dy}{4-y} = dx$$

$$\int \frac{dy}{4-y} = \int dx$$

$$-\ln|4-y| = x + C$$

$$e^{-\ln|4-y|} = e^{x+C} = e^x e^C$$

$$\frac{1}{4-y} = e^x e^C$$

$$4 - y = e^{-x}e^C \Rightarrow y = 4 - e^{-x}e^C$$

$$0 = 4 - e^{-0}e^C \Rightarrow 4 = e^C \Rightarrow \ln 4 = C$$

$$y = 4 - e^{-x}e^{\ln 4} = 4 - 4e^{-x} = 4\left(1 - e^{-x}\right)$$

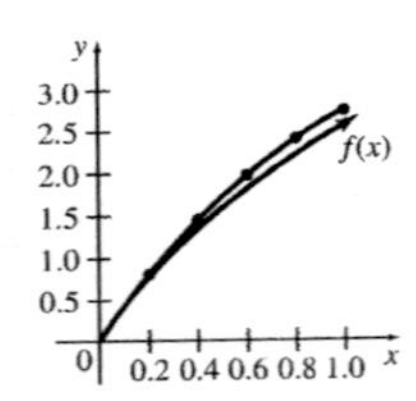

31. $\frac{dy}{dx} = y^2;\ y(0) = 1$

(a)

x_i	y_i
0	1
0.2	1.2
0.4	1.488
0.6	1.9308288
0.8	2.676448771
1.0	4.109124376

Thus, $y(1.0) \approx 4.109$.

(b) $\frac{dy}{dx} = y^2;\ y = 1$ when $x = 0$

$$\frac{dy}{dx} = y^2 \Rightarrow \frac{1}{y^2}dy = dx$$

$$\int \frac{1}{y^2}dy = \int dx$$

$$-\frac{1}{y} = x + C$$

When $x = 0$, $y = 1$.

$$-\frac{1}{1} = 0 + C \Rightarrow C = -1$$

$$-\frac{1}{y} = x - 1 \Rightarrow -1 = (x-1)y \Rightarrow$$

$$y = \frac{-1}{x-1} = \frac{1}{1-x}$$

As x approaches 1 from the left, y approaches ∞.

33. $\frac{dy}{dt} = 0.02(100 - y^{1/2}) = 2 - 0.02\sqrt{y}$

$t_0 = 0;\ y_0 = 10;\ h = 0.5$

Using the program in the Graphing Calculator Manual, the following values are obtained:

x_i	y_i	$y(x_i)$	$y_i - y(x_i)$
0	10	1.936754447	10.96837722
0.50	10.96837722	1.933762919	11.93525868
1.00	11.93525868	1.930905113	12.90071124
1.50	12.90071124	1.92816488	13.86479368
2.00	13.86479368	1.925529083	14.82755822
2.50	14.82755822	1.922986863	15.78905165
3.00	15.78905165	1.920529121	16.74931621
3.50	16.74931621	1.918148143	17.70839028
4.00	17.70839028	1.915837324	18.66630895
4.50	18.66630895	1.913590952	19.62310442
5.00	19.62310442		

There will be about 20 species.

35. $\frac{dy}{dt} = -y + 0.02y^2 + 0.003y^3$; for $[0, 4]$

$h = 1,\ t_0 = 0,\ y_0 = 15$

$$g(t, y) = -y + 0.02y^2 + 0.003y^3$$

$$g(t_0, y_0) = -15 + 0.02(15)^2 + 0.003(15)^3 = -0.375$$

$t_1 = 1;\ y_1 = 15 + (-0.375)(1) = 14.625$

$$g(t_1, y_1) = -14.625 + 0.02(14.625)^2 + 0.003(14.625)^3 = -0.963$$

(continued on next page)

(*continued*)

$t_2 = 2;\ y_2 = 14.625 + (-0.963)(1) = 13.662$

$g(t_2, y_2) = -13.662 + 0.02(13.662)^2 + 0.003(13.662)^2 = -2.279$

$t_3 = 3;\ y_3 = 13.662 + (-2.279)(1) = 11.383$

$g(t_3, y_3) = -11.383 + 0.02(11.383)^2 + 0.003(11.383)^3 = -4.367$

$t_4 = 4$

$y_4 = 11.383 + (-4.367)(1) = 7.016$ thousand

There will be about 7000 whales.

37. (a) Using Method 2 described after Example 1 of the text, store $\frac{dy}{dt}$ for the function variable Y_1 with $k = 0.5$ and $m = 4$. That is, Y_1 should equal $0.5\left(P - P^2\right)^{1.5}$. Store 5 to H (use keystrokes $5 \to$ H), -5 to T ($-5 \to$ T), and $p_0 = 0.1$ to P ($0.1 \to$ P). Next enter the keystrokes $\text{T} + \text{H} \to \text{T:P} + \text{Y}_1\text{H} \to \text{P}$. Each time the ENTER key is pressed, the subsequent values for t_i will be stored into T and the corresponding values for p_i will appear on the screen. This summarized in the table below.

t_i	p_i
0	0.1
5	0.1675
10	0.297678
15	0.536660
20	0.846644
25	0.963605
30	0.980024

(b) By continuing to press the ENTER key, it appears that the values for p_i are approaching 1.

11.4 Linear Systems of Differential Equations

1. $\frac{dx_1}{dt} = 3x_1 - 2x_2$

$\frac{dx_2}{dt} = x_1$

The system can be represented in matrix form as $\begin{bmatrix} \frac{dx_1}{dt} \\ \frac{dx_2}{dt} \end{bmatrix} = \begin{bmatrix} 3 & -2 \\ 1 & 0 \end{bmatrix}\begin{bmatrix} x_1 \\ x_2 \end{bmatrix}$.

Find the eigenvalues and eigenvectors.

$$\begin{aligned} 0 &= \det(M - \lambda I) \\ &= \det\begin{bmatrix} 3-\lambda & -2 \\ 1 & 0-\lambda \end{bmatrix} \\ &= (3-\lambda)(-\lambda) - (-2)(1) \\ &= \lambda^2 - 3\lambda + 2 \\ &= (\lambda - 1)(\lambda - 2) \Rightarrow \lambda = 1 \text{ or } \lambda = 2 \end{aligned}$$

For $\lambda = 1$ we have

$$\begin{bmatrix} 0 \\ 0 \end{bmatrix} = \begin{bmatrix} 3-1 & -2 \\ 1 & 0-1 \end{bmatrix}\begin{bmatrix} x_1 \\ x_2 \end{bmatrix} = \begin{bmatrix} 2 & -2 \\ 1 & -1 \end{bmatrix}\begin{bmatrix} x_1 \\ x_2 \end{bmatrix}$$

One solution to this system is $\begin{bmatrix} 1 \\ 1 \end{bmatrix}$.

For $\lambda = 2$ we have

$$\begin{bmatrix} 0 \\ 0 \end{bmatrix} = \begin{bmatrix} 3-2 & -2 \\ 1 & 0-2 \end{bmatrix}\begin{bmatrix} x_1 \\ x_2 \end{bmatrix} = \begin{bmatrix} 1 & -2 \\ 1 & -2 \end{bmatrix}\begin{bmatrix} x_1 \\ x_2 \end{bmatrix}$$

One solution to this system is $\begin{bmatrix} 2 \\ 1 \end{bmatrix}$.

The matrix of eigenvectors is $P = \begin{bmatrix} 1 & 2 \\ 1 & 1 \end{bmatrix}$.

$P^{-1} = \begin{bmatrix} -1 & 2 \\ 1 & -1 \end{bmatrix}$ and $P^{-1}MP = \begin{bmatrix} 1 & 0 \\ 0 & 2 \end{bmatrix}$.

Now let $X = PY$ and multiply both sides of the differential equation by P^{-1}, yielding $\frac{dY}{dt} = P^{-1}MPY$, or

$$\begin{bmatrix} \frac{dy_1}{dt} \\ \frac{dy_2}{dt} \end{bmatrix} = \begin{bmatrix} 1 & 0 \\ 0 & 2 \end{bmatrix}\begin{bmatrix} y_1 \\ y_2 \end{bmatrix} \Rightarrow \frac{dy_1}{dt} = y_1,\ \frac{dy_2}{dt} = 2y_2.$$

Integrate these to obtain $y_1 = C_1e^t$ and $y_2 = C_2e^{2t}$. Finally, calculate $X = PY$.

$$\begin{bmatrix} x_1 \\ x_2 \end{bmatrix} = \begin{bmatrix} 1 & 2 \\ 1 & 1 \end{bmatrix}\begin{bmatrix} C_1e^t \\ C_2e^{2t} \end{bmatrix} = \begin{bmatrix} C_1e^t + 2C_2e^{2t} \\ C_1e^t + C_2e^{2t} \end{bmatrix}, \text{ or}$$

$x_1 = C_1e^t + 2C_2e^{2t},\ x_2 = C_1e^t + C_2e^{2t}$.

3. $\frac{dx_1}{dt} = 3x_1 + 2x_2$

$\frac{dx_2}{dt} = 3x_1 + 8x_2$

The system can be represented in matrix form as

$$\begin{bmatrix} \frac{dx_1}{dt} \\ \frac{dx_2}{dt} \end{bmatrix} = \begin{bmatrix} 3 & 2 \\ 3 & 8 \end{bmatrix} \begin{bmatrix} x_1 \\ x_2 \end{bmatrix}.$$

Find the eigenvalues and eigenvectors.

$$0 = \det(M - \lambda I) = \det \begin{bmatrix} 3-\lambda & 2 \\ 3 & 8-\lambda \end{bmatrix}$$
$$= (3-\lambda)(8-\lambda) - 2(3)$$
$$= \lambda^2 - 11\lambda + 18$$
$$= (\lambda - 2)(\lambda - 9) \Rightarrow \lambda = 2 \text{ or } \lambda = 9$$

For $\lambda = 2$ we have

$$\begin{bmatrix} 0 \\ 0 \end{bmatrix} = \begin{bmatrix} 3-2 & 2 \\ 3 & 8-2 \end{bmatrix} \begin{bmatrix} x_1 \\ x_2 \end{bmatrix} = \begin{bmatrix} 1 & 2 \\ 3 & 6 \end{bmatrix} \begin{bmatrix} x_1 \\ x_2 \end{bmatrix}$$

One solution to this system is $\begin{bmatrix} 2 \\ -1 \end{bmatrix}$.

For $\lambda = 9$ we have

$$\begin{bmatrix} 0 \\ 0 \end{bmatrix} = \begin{bmatrix} 3-9 & 2 \\ 3 & 8-9 \end{bmatrix} \begin{bmatrix} x_1 \\ x_2 \end{bmatrix} = \begin{bmatrix} -6 & 2 \\ 3 & -1 \end{bmatrix} \begin{bmatrix} x_1 \\ x_2 \end{bmatrix}$$

One solution to this system is $\begin{bmatrix} 1 \\ 3 \end{bmatrix}$.

The matrix of eigenvectors is $P = \begin{bmatrix} 2 & 1 \\ -1 & 3 \end{bmatrix}$.

$P^{-1} = \frac{1}{7}\begin{bmatrix} 3 & -1 \\ 1 & 2 \end{bmatrix}$ and $P^{-1}MP = \begin{bmatrix} 2 & 0 \\ 0 & 9 \end{bmatrix}$.

Now let $X = PY$ and multiply both sides of the differential equation by P^{-1}, yielding

$$\frac{dY}{dt} = P^{-1}MPY, \text{ or } \begin{bmatrix} \frac{dy_1}{dt} \\ \frac{dy_2}{dt} \end{bmatrix} = \begin{bmatrix} 2 & 0 \\ 0 & 9 \end{bmatrix} \begin{bmatrix} y_1 \\ y_2 \end{bmatrix} \Rightarrow$$

$\frac{dy_1}{dt} = 2y_1, \frac{dy_2}{dt} = 9y_2$. Integrate these to obtain $y_1 = C_1e^{2t}$ and $y_2 = C_2e^{9t}$. Finally, calculate $X = PY$.

$$\begin{bmatrix} x_1 \\ x_2 \end{bmatrix} = \begin{bmatrix} 2 & 1 \\ -1 & 3 \end{bmatrix} \begin{bmatrix} C_1e^{2t} \\ C_2e^{9t} \end{bmatrix}$$
$$= \begin{bmatrix} 2C_1e^{2t} + C_2e^{9t} \\ -C_1e^{2t} + 3C_2e^{9t} \end{bmatrix}, \text{ or}$$

$x_1 = 2C_1e^{2t} + C_2e^{9t}$, $x_2 = -C_1e^{2t} + 3C_2e^{9t}$.

5. $\frac{dx_1}{dt} = 3x_1$

$\frac{dx_2}{dt} = 5x_1 + 2x_2$

$\frac{dx_3}{dt} = 2x_1 + 4x_2 - x_3$

The system can be represented in matrix form as

$$\begin{bmatrix} \frac{dx_1}{dt} \\ \frac{dx_2}{dt} \\ \frac{dx_3}{dt} \end{bmatrix} = \begin{bmatrix} 3 & 0 & 0 \\ 5 & 2 & 0 \\ 2 & 4 & -1 \end{bmatrix} \begin{bmatrix} x_1 \\ x_2 \\ x_3 \end{bmatrix}.$$

Find the eigenvalues and eigenvectors.

$$0 = \det(M - \lambda I)$$
$$= \det \begin{bmatrix} 3-\lambda & 0 & 0 \\ 5 & 2-\lambda & 0 \\ 2 & 4 & -1-\lambda \end{bmatrix}$$
$$= (3-\lambda)[(2-\lambda)(-1-\lambda) - 0(4)] - 0 + 0$$
$$= (3-\lambda)(2-\lambda)(-1-\lambda) \Rightarrow$$

$\lambda = 3$ or $\lambda = 2$ or $\lambda = -1$

For $\lambda = 3$ we have

$$\begin{bmatrix} 0 \\ 0 \\ 0 \end{bmatrix} = \begin{bmatrix} 3-3 & 0 & 0 \\ 5 & 2-3 & 0 \\ 2 & 4 & -1-3 \end{bmatrix} \begin{bmatrix} x_1 \\ x_2 \\ x_3 \end{bmatrix}$$
$$= \begin{bmatrix} 0 & 0 & 0 \\ 5 & -1 & 0 \\ 2 & 4 & -4 \end{bmatrix} \begin{bmatrix} x_1 \\ x_2 \\ x_3 \end{bmatrix}$$

One solution to this system is $\begin{bmatrix} 2 \\ 10 \\ 11 \end{bmatrix}$.

For $\lambda = 2$ we have

$$\begin{bmatrix} 0 \\ 0 \\ 0 \end{bmatrix} = \begin{bmatrix} 3-2 & 0 & 0 \\ 5 & 2-2 & 0 \\ 2 & 4 & -1-2 \end{bmatrix} \begin{bmatrix} x_1 \\ x_2 \\ x_3 \end{bmatrix}$$
$$= \begin{bmatrix} 1 & 0 & 0 \\ 5 & 0 & 0 \\ 2 & 4 & -3 \end{bmatrix} \begin{bmatrix} x_1 \\ x_2 \\ x_3 \end{bmatrix}$$

One solution to this system is $\begin{bmatrix} 0 \\ 3 \\ 4 \end{bmatrix}$.

(continued on next page)

(continued)

For $\lambda = -1$ we have

$$\begin{bmatrix} 0 \\ 0 \\ 0 \end{bmatrix} = \begin{bmatrix} 3-(-1) & 0 & 0 \\ 5 & 2-(-1) & 0 \\ 2 & 4 & -1-(-1) \end{bmatrix} \begin{bmatrix} x_1 \\ x_2 \\ x_3 \end{bmatrix} = \begin{bmatrix} 4 & 0 & 0 \\ 5 & 3 & 0 \\ 2 & 4 & 0 \end{bmatrix} \begin{bmatrix} x_1 \\ x_2 \\ x_3 \end{bmatrix}$$

One solution to this system is $\begin{bmatrix} 0 \\ 0 \\ 1 \end{bmatrix}$.

The matrix of eigenvectors is $P = \begin{bmatrix} 2 & 0 & 0 \\ 10 & 3 & 0 \\ 11 & 4 & 1 \end{bmatrix}$.

$P^{-1} = \frac{1}{6} \begin{bmatrix} 3 & 0 & 0 \\ -10 & 2 & 0 \\ 7 & -8 & 6 \end{bmatrix}$ and

$P^{-1}MP = \begin{bmatrix} 3 & 0 & 0 \\ 0 & 2 & 0 \\ 0 & 0 & -1 \end{bmatrix}$.

Now let $X = PY$ and multiply both sides of the differential equation by P^{-1}, yielding $\frac{dY}{dt} = P^{-1}MPY$, or

$$\begin{bmatrix} \frac{dy_1}{dt} \\ \frac{dy_2}{dt} \\ \frac{dy_3}{dt} \end{bmatrix} = \begin{bmatrix} 3 & 0 & 0 \\ 0 & 2 & 0 \\ 0 & 0 & -1 \end{bmatrix} \begin{bmatrix} y_1 \\ y_2 \\ y_3 \end{bmatrix} \Rightarrow$$

$\frac{dy_1}{dt} = 3y_1, \frac{dy_2}{dt} = 2y_2, \frac{dy_3}{dt} = -y_3$

Integrate these to obtain

$y_1 = C_1e^{3t}$, $y_2 = C_2e^{2t}$, and $y_3 = C_3e^{-t}$.

Finally, calculate $X = PY$.

$$\begin{bmatrix} x_1 \\ x_2 \\ x_3 \end{bmatrix} = \begin{bmatrix} 2 & 0 & 0 \\ 10 & 3 & 0 \\ 11 & 4 & 1 \end{bmatrix} \begin{bmatrix} C_1e^{3t} \\ C_2e^{2t} \\ C_3e^{-t} \end{bmatrix} = \begin{bmatrix} 2C_1e^{3t} \\ 10C_1e^{3t} + 3C_2e^{2t} \\ 11C_1e^{3t} + 4C_2e^{2t} + C_3e^{-t} \end{bmatrix}, \text{ or}$$

$x_1 = 2C_1e^{3t}$, $x_2 = 10C_1e^{3t} + 3C_2e^{2t}$,
$x_3 = 11C_1e^{3t} + 4C_2e^{2t} + C_3e^{-t}$

$\frac{dy_1}{dt} = 2y_1, \frac{dy_2}{dt} = 5y_2, \frac{dy_3}{dt} = -4y_3$

Integrate these to obtain

$y_1 = C_1e^{2t}$, $y_2 = C_2e^{5t}$, and $y_3 = C_3e^{-4t}$.

Finally, calculate $X = PY$.

$$\begin{bmatrix} x_1 \\ x_2 \\ x_3 \end{bmatrix} = \begin{bmatrix} 6 & 0 & 0 \\ -6 & 9 & 0 \\ 1 & 5 & 1 \end{bmatrix} \begin{bmatrix} C_1e^{2t} \\ C_2e^{5t} \\ C_3e^{-4t} \end{bmatrix} = \begin{bmatrix} 6C_1e^{2t} \\ -6C_1e^{2t} + 9C_2e^{5t} \\ C_1e^{2t} + 5C_2e^{5t} + C_3e^{-4t} \end{bmatrix}, \text{ or}$$

$x_1 = 6C_1e^{2t}$, $x_2 = -6C_1e^{2t} + 9C_2e^{5t}$,
$x_3 = C_1e^{2t} + 5C_2e^{5t} + C_3e^{-4t}$

7. $\frac{dx_1}{dt} = 4x_1 - 2x_2 + e^t$

$\frac{dx_2}{dt} = -3x_1 + 9x_2 + e^{2t}$

The system can be represented in matrix form as $\begin{bmatrix} \frac{dx_1}{dt} \\ \frac{dx_2}{dt} \end{bmatrix} = \begin{bmatrix} 4 & -2 \\ -3 & 9 \end{bmatrix} \begin{bmatrix} x_1 \\ x_2 \end{bmatrix} + \begin{bmatrix} e^t \\ e^{2t} \end{bmatrix}$. Find the eigenvalues and eigenvectors.

$$0 = \det(M - \lambda I) = \det \begin{bmatrix} 4-\lambda & -2 \\ -3 & 9-\lambda \end{bmatrix}$$
$$= (4-\lambda)(9-\lambda) - (-2)(-3)$$
$$= \lambda^2 - 13\lambda + 30 = (\lambda - 3)(\lambda - 10) \Rightarrow$$

$\lambda = 3$ or $\lambda = 10$

For $\lambda = 3$ we have

$$\begin{bmatrix} 0 \\ 0 \end{bmatrix} = \begin{bmatrix} 4-3 & -2 \\ -3 & 9-3 \end{bmatrix} \begin{bmatrix} x_1 \\ x_2 \end{bmatrix} = \begin{bmatrix} 1 & -2 \\ -3 & 6 \end{bmatrix} \begin{bmatrix} x_1 \\ x_2 \end{bmatrix}$$

One solution to this system is $\begin{bmatrix} 2 \\ 1 \end{bmatrix}$.

For $\lambda = 10$ we have

$$\begin{bmatrix} 0 \\ 0 \end{bmatrix} = \begin{bmatrix} 4-10 & -2 \\ -3 & 9-10 \end{bmatrix} \begin{bmatrix} x_1 \\ x_2 \end{bmatrix} = \begin{bmatrix} -6 & -2 \\ -3 & -1 \end{bmatrix} \begin{bmatrix} x_1 \\ x_2 \end{bmatrix}$$

One solution to this system is $\begin{bmatrix} 1 \\ -3 \end{bmatrix}$.

The matrix of eigenvectors is $P = \begin{bmatrix} 2 & 1 \\ 1 & -3 \end{bmatrix}$.

(continued on next page)

(continued)

$$P^{-1}=\frac{1}{7}\begin{bmatrix}3 & 1\\ 1 & -2\end{bmatrix} \text{ and } P^{-1}MP=\begin{bmatrix}3 & 0\\ 0 & 10\end{bmatrix}.$$

Now let $X = PY$ and multiply both sides of the differential equation by P^{-1}, yielding

$$\frac{dY}{dt}=P^{-1}MPY+P^{-1}\begin{bmatrix}e^t\\ e^{2t}\end{bmatrix}, \text{ or } \begin{bmatrix}\frac{dy_1}{dt}\\ \frac{dy_2}{dt}\end{bmatrix}=\begin{bmatrix}3 & 0\\ 0 & 10\end{bmatrix}\begin{bmatrix}y_1\\ y_2\end{bmatrix}+\frac{1}{7}\begin{bmatrix}3 & 1\\ 1 & -2\end{bmatrix}\begin{bmatrix}e^t\\ e^{2t}\end{bmatrix}\Rightarrow$$

$$\frac{dy_1}{dt}=3y_1+\frac{3e^t}{7}+\frac{e^{2t}}{7},\ \frac{dy_2}{dt}=10y_2+\frac{e^t}{7}-\frac{2e^{2t}}{7}.$$

Multiply the first differential equation by an integrating factor $e^{\int -3dt}=e^{-3t}$.

$$\frac{dy_1}{dt}-3y_1=\frac{3e^t}{7}+\frac{e^{2t}}{7}$$

$$e^{-3t}\frac{dy_1}{dt}-3e^{-3t}y_1=\frac{3e^{-2t}}{7}+\frac{e^{-t}}{7}$$

$$D_t\left(e^{-3t}y_1\right)=\frac{3e^{-2t}}{7}+\frac{e^{-t}}{7}$$

$$e^{-3t}y_1=\int\left(\frac{3e^{-2t}}{7}+\frac{e^{-t}}{7}\right)dt=-\frac{3e^{-2t}}{14}-\frac{e^{-t}}{7}+C_1\Rightarrow y_1=-\frac{3e^t}{14}-\frac{e^{2t}}{7}+C_1e^{3t}$$

Multiply the second differential equation by an integrating factor $e^{\int -10dt}=e^{-10t}$.

$$\frac{dy_2}{dt}-10y_2=\frac{e^t}{7}-\frac{2e^{2t}}{7}$$

$$e^{-10t}\frac{dy_2}{dt}-10e^{-10t}y_2=\frac{e^{-9t}}{7}-\frac{2e^{-8t}}{7}$$

$$D_t\left(e^{-10t}y_2\right)=\frac{e^{-9t}}{7}-\frac{2e^{-8t}}{7}$$

$$e^{-10t}y_2=\int\left(\frac{e^{-9t}}{7}-\frac{2e^{-8t}}{7}\right)dt=-\frac{e^{-9t}}{63}+\frac{e^{-8t}}{28}+C_2\Rightarrow y_2=-\frac{e^t}{63}+\frac{e^{2t}}{28}+C_2e^{10t}$$

Finally, calculate $X = PY$.

$$\begin{bmatrix}x_1\\ x_2\end{bmatrix}=\begin{bmatrix}2 & 1\\ 1 & -3\end{bmatrix}\begin{bmatrix}-\frac{3e^t}{14}-\frac{e^{2t}}{7}+C_1e^{3t}\\ -\frac{e^t}{63}+\frac{e^{2t}}{28}+C_2e^{10t}\end{bmatrix}=\begin{bmatrix}2\left(-\frac{3e^t}{14}-\frac{e^{2t}}{7}+C_1e^{3t}\right)+\left(-\frac{e^t}{63}+\frac{e^{2t}}{28}+C_2e^{10t}\right)\\ \left(-\frac{3e^t}{14}-\frac{e^{2t}}{7}+C_1e^{3t}\right)-3\left(-\frac{e^t}{63}+\frac{e^{2t}}{28}+C_2e^{10t}\right)\end{bmatrix}$$

$$=\begin{bmatrix}-\frac{4}{9}e^t-\frac{1}{4}e^{2t}+2C_1e^{3t}+C_2e^{10t}\\ -\frac{1}{6}e^t-\frac{1}{4}e^{2t}+C_1e^{3t}-3C_2e^{10t}\end{bmatrix} \text{ or}$$

$$x_1=-\frac{4}{9}e^t-\frac{1}{4}e^{2t}+2C_1e^{3t}+C_2e^{10t}$$

$$x_2=-\frac{1}{6}e^t-\frac{1}{4}e^{2t}+C_1e^{3t}-3C_2e^{10t}$$

9. From exercise 1, the general solution is
$x_1 = C_1e^t + 2C_2e^{2t}$, $x_2 = C_1e^t + C_2e^{2t}$.
Setting $x_1(0) = 7$ and $x_2(0) = 5$ gives

$$7 = C_1e^0 + 2C_2e^{2(0)} = C_1 + 2C_2$$
$$5 = C_1e^0 + C_2e^{2(0)} = C_1 + C_2$$

Solving this system gives $C_1 = 3$, $C_2 = 2$.
Therefore, the particular solution is
$x_1 = 3e^t + 4e^{2t}$, $x_2 = 3e^t + 2e^{2t}$.

11. From exercise 3, the general solution is
$x_1 = 2C_1e^{2t} + C_2e^{9t}$, $x_2 = -C_1e^{2t} + 3C_2e^{9t}$.
Setting $x_1(0) = 3$ and $x_2(0) = 23$ gives

$$3 = 2C_1e^{2(0)} + C_2e^{9(0)} = 2C_1 + C_2$$
$$23 = -C_1e^{2(0)} + 3C_2e^{9(0)} = -C_1 + 3C_2$$

Solving this system gives $C_1 = -2$, $C_2 = 7$.
Therefore, the particular solution is
$x_1 = -4e^{2t} + 7e^{9t}$, $x_2 = 2e^{2t} + 21e^{9t}$.

13. From exercise 5, the general solution is
$x_1 = 2C_1e^{3t}$, $x_2 = 10C_1e^{3t} + 3C_2e^{2t}$,
$x_3 = 11C_1e^{3t} + 4C_2e^{2t} + C_3e^{-t}$.
Setting $x_1(0) = 6$, $x_2(0) = 15$, and $x_3(0) = -1$ gives the system

$$6 = 2C_1e^{3(0)}$$
$$15 = 10C_1e^{3(0)} + 3C_2e^{2(0)}$$
$$-1 = 11C_1e^{3(0)} + 4C_2e^{2(0)} + C_3e^{-(0)}$$

or

$$6 = 2C_1$$
$$15 = 10C_1 + 3C_2$$
$$-1 = 11C_1 + 4C_2 + C_3$$

Solving this system gives
$C_1 = 3$, $C_2 = -5$, $C_3 = -14$.
Therefore, the particular solution is
$x_1 = 6e^{3t}$, $x_2 = 30e^{3t} - 15e^{2t}$,
$x_3 = 33e^{3t} - 20e^{2t} - 14e^{-t}$.

15. (a)
$$\frac{dx_1}{dt} = x_2 - 2$$
$$\frac{dx_2}{dt} = -x_1 + 2$$

The system can be represented in matrix form as
$$\begin{bmatrix} \frac{dx_1}{dt} \\ \frac{dx_2}{dt} \end{bmatrix} = \begin{bmatrix} 0 & 1 \\ -1 & 0 \end{bmatrix}\begin{bmatrix} x_1 \\ x_2 \end{bmatrix} + \begin{bmatrix} -2 \\ 2 \end{bmatrix}.$$

Find the eigenvalues.
$$0 = \det(M - \lambda I)$$
$$= \det\begin{bmatrix} 0-\lambda & 1 \\ -1 & 0-\lambda \end{bmatrix}$$
$$= (0-\lambda)(0-\lambda) - 1(-1)$$
$$= \lambda^2 + 1$$

(b) $\lambda^2 + 1 = 0 \Rightarrow \lambda^2 = -1 \Rightarrow \lambda = \pm\sqrt{-1} = \pm i$

(c) For $\lambda = -i$ we have
$$\begin{bmatrix} 0 \\ 0 \end{bmatrix} = \begin{bmatrix} i & 1 \\ -1 & i \end{bmatrix}\begin{bmatrix} x_1 \\ x_2 \end{bmatrix}$$
The augmented matrix for this system is
$$\left[\begin{array}{cc|c} i & 1 & 0 \\ -1 & i & 0 \end{array}\right].$$
$$-iR_2 \to R_2 \left[\begin{array}{cc|c} i & 1 & 0 \\ i & 1 & 0 \end{array}\right]$$
A solution to this system is $\begin{bmatrix} x \\ y \end{bmatrix} = \begin{bmatrix} 1 \\ -i \end{bmatrix}$.

For $\lambda = i$ we have
$$\begin{bmatrix} 0 \\ 0 \end{bmatrix} = \begin{bmatrix} -i & 1 \\ -1 & -i \end{bmatrix}\begin{bmatrix} x_1 \\ x_2 \end{bmatrix}$$
The augmented matrix for this system is
$$\left[\begin{array}{cc|c} -i & 1 & 0 \\ -1 & -i & 0 \end{array}\right].$$
$$iR_2 \to R_2 \left[\begin{array}{cc|c} -i & 1 & 0 \\ -i & 1 & 0 \end{array}\right]$$
A solution to this system is $\begin{bmatrix} x \\ y \end{bmatrix} = \begin{bmatrix} 1 \\ i \end{bmatrix}$.

The matrix of eigenvectors is
$$P = \begin{bmatrix} 1 & 1 \\ i & -i \end{bmatrix}.$$

(d)
$$PP^{-1} = \begin{bmatrix} 1 & 1 \\ i & -i \end{bmatrix}\begin{bmatrix} \frac{1}{2} & -\frac{i}{2} \\ \frac{1}{2} & \frac{i}{2} \end{bmatrix}$$
$$= \begin{bmatrix} 1\left(\frac{1}{2}\right) + 1\left(\frac{1}{2}\right) & 1\left(-\frac{i}{2}\right) + 1\left(\frac{i}{2}\right) \\ i\left(\frac{1}{2}\right) - i\left(\frac{1}{2}\right) & i\left(-\frac{i}{2}\right) - i\left(\frac{i}{2}\right) \end{bmatrix}$$
$$= \begin{bmatrix} 1 & 0 \\ 0 & 1 \end{bmatrix} = I$$

(e) $P^{-1}MP = \begin{bmatrix} i & 0 \\ 0 & -i \end{bmatrix}$.

Now let $X = PY$ and multiply both sides of the differential equation by P^{-1}, yielding $\frac{dY}{dt} = P^{-1}MPY + P^{-1}\begin{bmatrix} -2 \\ 2 \end{bmatrix}$, or

$$\begin{bmatrix} \frac{dy_1}{dt} \\ \frac{dy_2}{dt} \end{bmatrix} = \begin{bmatrix} i & 0 \\ 0 & -i \end{bmatrix}\begin{bmatrix} y_1 \\ y_2 \end{bmatrix} + \begin{bmatrix} \frac{1}{2} & -\frac{i}{2} \\ \frac{1}{2} & \frac{i}{2} \end{bmatrix}\begin{bmatrix} -2 \\ 2 \end{bmatrix}$$
$$= \begin{bmatrix} iy_1 \\ -iy_2 \end{bmatrix} + \begin{bmatrix} -1-i \\ -1+i \end{bmatrix}, \text{ or}$$

$$\frac{dy_1}{dt} = iy_1 + (-1-i),$$
$$\frac{dy_2}{dt} = -iy_2 + (-1+i).$$

(f) Multiply the first differential equation by an integrating factor $e^{\int -idt} = e^{-it}$.

$$\frac{dy_1}{dt} - iy_1 = (-1-i)$$
$$e^{-it}\frac{dy_1}{dt} - ie^{-it}y_1 = e^{-it}(-1-i)$$
$$D_t\left(e^{-it}y_1\right) = e^{-it}(-1-i)$$
$$e^{-it}y_1 = (-1-i)\int e^{-it}dt$$
$$= (-1-i)\left(\frac{1}{-i}e^{-it} + C\right)$$
$$= (-1-i)\left(ie^{-it} + C\right)$$
$$y_1 = (-1-i)\left(i + Ce^{it}\right)$$
$$= (-1-i)Ce^{it} + (1-i)$$

Let $C_1 = (-1-i)C$. Then

$y_1 = C_1e^{it} + (1-i)$.

Multiply the second differential equation by an integrating factor $e^{\int idt} = e^{it}$.

$$\frac{dy_2}{dt} + iy_2 = -1+i$$
$$e^{it}\frac{dy_2}{dt} + ie^{it}y_2 = e^{it}(-1+i)$$
$$D_t\left(e^{it}y_2\right) = e^{it}(-1+i)$$
$$e^{it}y_2 = (-1+i)\int e^{it}dt$$
$$= (-1+i)\left(\frac{1}{i}e^{-it} + C\right)$$
$$= (-1+i)\left(-ie^{it} + C\right)$$
$$y_2 = (-1+i)\left(-i + Ce^{-it}\right)$$
$$= (-1+i)Ce^{-it} + (1+i)$$

Let $C_2 = (-1+i)C$. Then

$y_2 = C_2e^{-it} + (1+i)$.

(g) Calculate $X = PY$.

$$\begin{bmatrix} x_1 \\ x_2 \end{bmatrix}\begin{bmatrix} 1 & 1 \\ i & -i \end{bmatrix} = \begin{bmatrix} C_1e^{it} + (1-i) \\ C_2e^{-it} + (1+i) \end{bmatrix}$$
$$= \begin{bmatrix} C_1e^{it} + C_2e^{-it} + 2 \\ iC_1e^{it} - iC_2e^{-it} + 2 \end{bmatrix}$$

(h) Setting $x_1(0) = 2$ and $x_2(0) = 3$ gives

$$C_1e^{i(0)} + C_2e^{-i(0)} + 2 = 2 \Rightarrow C_1 + C_2 = 0$$
$$iC_1e^{i(0)} - iC_2e^{-i(0)} + 2 = 3 \Rightarrow iC_1 - iC_2 = 1$$

The solution to this system is

$C_1 = -\frac{i}{2}$, $C_2 = \frac{i}{2}$.

(i) From parts (g) and (h),

$$x_1 = -\frac{i}{2}e^{it} + \frac{i}{2}e^{-it} + 2$$
$$x_2 = i\left(-\frac{i}{2}\right)e^{it} - i\left(\frac{i}{2}\right)e^{-it} + 2$$
$$= \frac{1}{2}e^{it} + \frac{1}{2}e^{-it} + 2$$

Using the identity of Euler, we have

$$x_1 = -\frac{i}{2}(\cos t + i\sin t) + \frac{i}{2}(\cos t - i\sin t) + 2$$
$$= -\frac{i}{2}\cos t + \frac{1}{2}\sin t + \frac{i}{2}\cos t + \frac{1}{2}\sin t + 2$$
$$= \sin t + 2$$
$$x_2 = \frac{1}{2}(\cos t + i\sin t) + \frac{1}{2}(\cos t - i\sin t) + 2$$
$$= \frac{1}{2}\cos t + \frac{i}{2}\sin t + \frac{1}{2}\cos t - \frac{i}{2}\sin t + 2$$
$$= \cos t + 2$$

(j) The answer is biologically reasonable since x_1 and x_2 are real-valued.

11.5 Nonlinear Systems of Differential Equations

1. $\frac{dx_1}{dt} = 4x_1 - 2x_1x_2$

$\frac{dx_2}{dt} = 3x_2 - x_1x_2$

Find the equilibrium point.

$\frac{dx_1}{dt} = 4x_1 - 2x_1x_2 = 0 \Rightarrow 4x_1 = 2x_1x_2 \Rightarrow$
$2 = x_2$

$\frac{dx_2}{dt} = 3x_2 - x_1x_2 = 0 \Rightarrow 3x_2 = x_1x_2 \Rightarrow 3 = x_1$

The equilibrium point is $x_1 = 3,\ x_2 = 2$.

Region	1	2	3	4
dx_1/dt	–	–	+	+
dx_2/dt	–	+	+	–

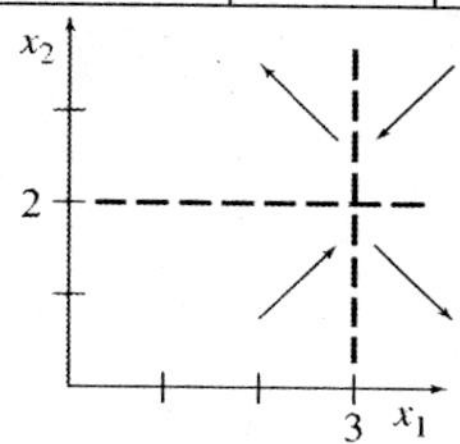

3. $\frac{dx_1}{dt} = 3x_1x_2 - x_1x_2^2$

$\frac{dx_2}{dt} = 2x_1^2x_2 - 2x_1x_2$

Find the equilibrium point.

$\frac{dx_1}{dt} = 3x_1x_2 - x_1x_2^2 = 0 \Rightarrow 3x_1x_2 = x_1x_2^2 \Rightarrow$
$x_2 = 3$

$\frac{dx_2}{dt} = 2x_1^2x_2 - 2x_1x_2 = 0 \Rightarrow 2x_1^2x_2 = 2x_1x_2 \Rightarrow$
$x_1 = 1$

The equilibrium point is $x_1 = 1,\ x_2 = 3$.

Region	1	2	3	4
dx_1/dt	–	–	+	+
dx_2/dt	+	–	–	+

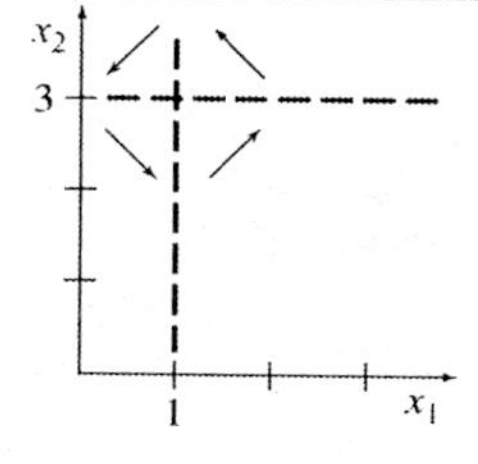

5. (a) $\frac{dx_1}{dt} = x_1^2 + x_1x_2 - 2x_1$

$\frac{dx_2}{dt} = 2x_1x_2 - x_2^2 - x_2$

Find the equilibrium point.

$\frac{dx_1}{dt} = x_1^2 + x_1x_2 - 2x_1 = 0 \Rightarrow$
$x_1(x_1 + x_2 - 2) = 0 \Rightarrow x_2 = -x_1 + 2 \quad (1)$

$\frac{dx_2}{dt} = 2x_1x_2 - x_2^2 - x_2 = 0 \Rightarrow$
$x_2(2x_1 - x_2 - 1) = 0 \Rightarrow x_2 = 2x_1 - 1 \quad (2)$

Solving equations (1) and (2) gives
$-x_1 + 2 = 2x_1 - 1 \Rightarrow x_1 = 1, x_2 = 1.$

(b) Plot equations (1) and (2) and test each region.

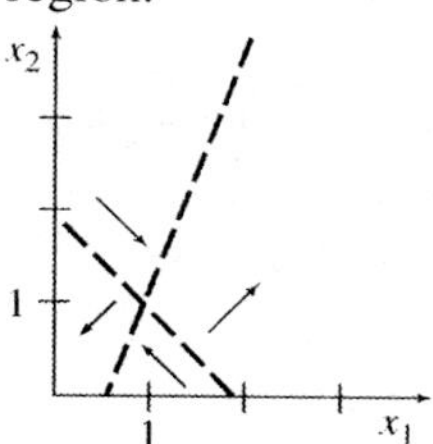

7. (a) $\frac{dx_1}{dt} = x_1^2x_2 + x_1x_2^2 - 3x_1x_2$

$\frac{dx_2}{dt} = 2x_1x_2^2 - x_1^2x_2 - 3x_1x_2$

Find the equilibrium point.

$\frac{dx_1}{dt} = x_1^2x_2 + x_1x_2^2 - 3x_1x_2 = 0 \Rightarrow$
$x_1x_2(x_1 + x_2 - 3) = 0 \Rightarrow x_2 = -x_1 + 3 \quad (1)$

$\frac{dx_2}{dt} = 2x_1x_2^2 - x_1^2x_2 - 3x_1x_2 = 0 \Rightarrow$
$x_1x_2(2x_2 - x_1 - 3) = 0 \Rightarrow$

$x_2 = \frac{x_1}{2} + \frac{3}{2} \quad (2)$

Solving equations (1) and (2) gives

$-x_1 + 3 = \frac{x_1}{2} + \frac{3}{2} \Rightarrow x_1 = 1,\ x_2 = 2.$

(b) Plot equations (1) and (2) and test each region.

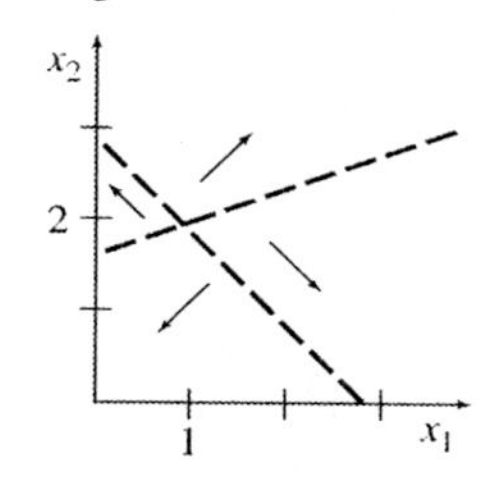

9. (a) $\frac{dx_1}{dt} = -2x_1 + 3x_1x_2$

$\frac{dx_2}{dt} = 3x_2 - 2x_1x_2$

$$\frac{dx_2}{dx_1} = \frac{\frac{dx_2}{dt}}{\frac{dx_1}{dt}} = \frac{3x_2 - 2x_1x_2}{-2x_1 + 3x_1x_2} = \frac{x_2(3-2x_1)}{x_1(-2+3x_2)}$$

Separating variables yields

$$\frac{-2+3x_2}{x_2}dx_2 = \frac{3-2x_1}{x_1}dx_1$$

$$\int\left(-\frac{2}{x_2}+3\right)dx_2 = \int\left(\frac{3}{x_1}-2\right)dx_1$$

$$-2\ln x_2 + 3x_2 = 3\ln x_1 - 2x_1 + C$$

Use the initial conditions $x_2 = 2$ when $x_1 = 1$ to find C.

$$-2\ln 2 + 3(2) = 3\ln 1 - 2(1) + C$$
$$-\ln 4 + 6 = 0 - 2 + C$$
$$C = -\ln 4 + 8$$

Thus, the desired equation is

$3\ln x_1 - 2x_1 + 2\ln x_2 - 3x_2 = \ln 4 - 8.$

(b) $\frac{dx_1}{dt} = -2x_1 + 3x_1x_2 = 0 \Rightarrow$

$x_1(-2+3x_2) = 0 \Rightarrow x_2 = \frac{2}{3}$

$\frac{dx_2}{dt} = 3x_2 - 2x_1x_2 = 0 \Rightarrow$

$x_2(3-2x_1) = 0 \Rightarrow x_1 = \frac{3}{2}$

The equilibrium point is $x_1 = \frac{3}{2},\ x_2 = \frac{2}{3}$.

(c)

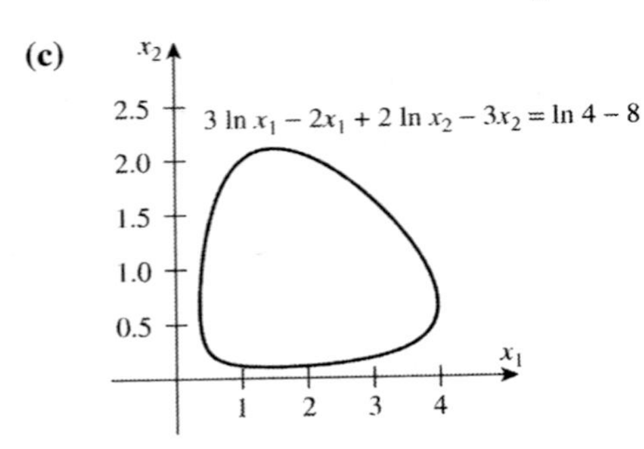

A graph of the equation shows that the variations in the populations are cyclic. We know from Example 1 that the solution moves about this cycle in a clockwise direction.

11. $\frac{dy_1}{dt} = k_1y_1(1 - F_1 - y_1 - by_2)$

$\frac{dy_2}{dt} = k_2y_2\left(1 - F_2 - \frac{y_2}{y_1}\right)$

$\frac{dy_1}{dt} = 0 = k_1y_1(1 - F_1 - y_1 - by_2) \Rightarrow$

$y_1 = 1 - F_1 - by_2$

$\frac{dy_2}{dt} = 0 = k_2y_2\left(1 - F_2 - \frac{y_2}{y_1}\right) \Rightarrow$

$$1 = F_2 + \frac{y_2}{y_1} \Rightarrow y_1 = F_2y_1 + y_2 \Rightarrow$$

$y_2 = y_1 - F_2y_1$

Substitute the expression for y_2 into the expression for y_1 and solve for y_1.

$$y_1 = 1 - F_1 - b(y_1 - F_2y_1)$$
$$= 1 - F_1 - by_1 + bF_2y_1$$

$$y_1(1 + b - bF_2) = 1 - F_1 \Rightarrow y_1 = \frac{1-F_1}{1+b(1-F_2)}$$

Now find y_2.

$$y_2 = y_1 - F_2y_1 = y_1(1-F_2) = \frac{(1-F_1)(1-F_2)}{1+b(1-F_2)}$$

13. $\frac{dx_1}{dt} = a - bx_1 - \beta x_1x_2 + \gamma x_3$

$\frac{dx_2}{dt} = \beta x_1x_2 - (b + \alpha + \nu)x_2$

$\frac{dx_3}{dt} = \nu x_2 - (\beta + \gamma)x_3$

Set the derivatives to zero.

$a - bx_1 - \beta x_1x_2 + \gamma x_3 = 0 \quad (1)$

$\beta x_1x_2 - (b + \alpha + \nu)x_2 = 0 \quad (2)$

$\nu x_2 - (\beta + \gamma)x_3 = 0 \quad (3)$

Begin by solving equation (2) for x_1.

$\beta x_1x_2 - (b + \alpha + \nu)x_2 = 0 \Rightarrow$

$x_2[\beta x_1 - (b + \alpha + \nu)] = 0 \Rightarrow x_2 = 0$

or $\beta x_1 - (b + \alpha + \nu) = 0 \Rightarrow x_1 = \frac{b + \alpha + \nu}{\beta}$

Notice that, since all constants are positive, x_1 must be positive.

Now solve equation (3) for x_3 in terms of x_2.

$$\nu x_2 - (\beta + \gamma)x_3 = 0 \Rightarrow x_3 = \frac{\nu}{\beta + \gamma}x_2$$

(continued on next page)

(continued)

Replace x_1 and x_3 in equation (1) with the expressions just found.

$$a - bx_1 - \beta x_1 x_2 + \gamma x_3 = 0$$

$$a - b\left(\frac{b+\alpha+\nu}{\beta}\right) - \beta\left(\frac{b+\alpha+\nu}{\beta}\right)x_2 + \gamma\left(\frac{\nu}{\beta+\gamma}x_2\right) = 0$$

$$a - \frac{b(b+\alpha+\nu)}{\beta} - (b+\alpha+\nu)x_2 + \frac{\gamma\nu}{\beta+\gamma}x_2 = 0$$

$$x_2\left[-(b+\alpha+\nu) + \frac{\gamma\nu}{\beta+\gamma}\right] = -\left[a - \frac{b(b+\alpha+\nu)}{\beta}\right]$$

$$x_2\left[-(b+\alpha+\nu) + \frac{\gamma\nu}{\beta+\gamma}\right]\beta(\beta+\gamma) = -\left[a - \frac{b(b+\alpha+\nu)}{\beta}\right]\beta(\beta+\gamma)$$

$$x_2\left[-\beta(\beta+\gamma)(b+\alpha+\nu) + \beta\gamma\nu\right] = -\left[a\beta(\beta+\gamma) - b(b+\alpha+\nu)(\beta+\gamma)\right]$$

$$x_2 = \frac{a\beta(\beta+\gamma) - b(\beta+\gamma)(b+\alpha+\nu)}{\beta(\beta+\gamma)(b+\alpha+\nu) - \beta\gamma\nu}$$

$$= \frac{(\beta+\gamma)\left[a\beta - b(b+\alpha+\nu)\right]}{\beta(\beta+\gamma)(b+\alpha+\nu) - \beta\gamma\nu}$$

$$x_3 = \frac{\nu}{\beta+\gamma}x_2 = \frac{\nu(\beta+\gamma)\left[a\beta - b(b+\alpha+\nu)\right]}{(\beta+\gamma)\left[\beta(\beta+\gamma)(b+\alpha+\nu) - \beta\gamma\nu\right]} = \frac{\nu\left[a\beta - b(b+\alpha+\nu)\right]}{\left[\beta(\beta+\gamma)(b+\alpha+\nu) - \beta\gamma\nu\right]}$$

Sice all constants are positive, for x_2 and x_3 to be positive, the numerator of the last rational expression must be positive.

$$(\beta+\gamma)\left[a\beta - b(b+\alpha+\nu)\right] > 0 \Rightarrow a\beta - b(b+\alpha+\nu) > 0 \Rightarrow a\beta > b(b+\alpha+\nu) \Rightarrow \frac{a}{b} > \frac{b+\alpha+\nu}{\beta}$$

15. (a) $\delta = 0$, so the system becomes

$$\frac{dx}{dt} = \mu - \mu x - N\beta xy$$

$$\frac{dy}{dt} = N\beta xy - (\mu+\nu)y$$

We are finding the equilibrium point, so $\frac{dx}{dt} = \frac{dy}{dt} = 0.$

$$\mu - \mu x - N\beta xy = 0 \qquad (1)$$

$$N\beta xy - (\mu+\nu)y = 0 \qquad (2)$$

Starting with equation (2), we have

$$N\beta xy - (\mu+\nu)y = 0$$

$$y\left(N\beta x - (\mu+\nu)\right) = 0 \Rightarrow y = 0 \text{ or } N\beta x - (\mu+\nu) = 0 \Rightarrow N\beta x = \mu+\nu \Rightarrow x = \frac{\mu+\nu}{N\beta}$$

If $y = 0$, then, using equation (1), we have $\mu - \mu x = 0 \Rightarrow x = 1$. This means that the entire population is susceptible and none of the population is infected, definitely not an equilibrium point.

If $x = \frac{\mu+\nu}{N\beta}$, then, using equation (1), we have

$$\mu - x(\mu + N\beta y) = 0 \Rightarrow x(\mu + N\beta y) = \mu \Rightarrow \left(\frac{\mu+\nu}{N\beta}\right)(\mu + N\beta y) = \mu \Rightarrow \mu + N\beta y = \frac{\mu N\beta}{\mu+\nu} \Rightarrow$$

$$N\beta y = \frac{\mu N\beta - \mu(\mu+\nu)}{\mu+\nu} \Rightarrow y = \frac{\mu(N\beta - \mu - \nu)}{N\beta(\mu+\nu)}$$

Thus, the equilibrium point is $x_0 = \frac{\mu+\nu}{N\beta}$, $y_0 = \frac{\mu(N\beta-\mu-\nu)}{N\beta(\mu+\nu)}$.

(b) Substitute $x_0 + x_1$ for x and $y_0 + y_1$ for y. Note that $\frac{dx}{dt}$ becomes $\frac{dx_1}{dt}$ because x_0 is a constant and $d(x_0 + x_1) = dx_1$. Similarly, $\frac{dy}{dt}$ becomes $\frac{dy_1}{dt}$.

$$\begin{aligned}\frac{dx_1}{dt} &= \mu - \mu(x_0 + x_1) - N\beta(x_0 + x_1)(y_0 + y_1)(1 + \delta\beta \sin \omega t)\\ &= \mu - \mu x_0 - \mu x_1 - N\beta(x_0 y_0 + x_0 y_1 + x_1 y_0 + x_1 y_1)(1 + \delta\beta \sin \omega t)\end{aligned}$$

We are told that $x_1 y_1$ is very small, so that this term can be ignored.

$$\begin{aligned}\frac{dx_1}{dt} &= \mu - \mu x_0 - \mu x_1 - N\beta(x_0 y_0 + x_0 y_1 + x_1 y_0)(1 + \delta\beta \sin \omega t)\\ &= \mu - \mu x_0 - \mu x_1 - N\beta x_0 y_0 - N\beta x_0 y_1 - N\beta x_1 y_0 - N\beta^2 \delta x_0 y_0 \sin \omega t\\ &\qquad - N\beta^2 \delta x_0 y_1 \sin \omega t - N\beta^2 \delta x_1 y_0 \sin \omega t\end{aligned}$$

We are also told that δx_1 and δy_1 are very small, so that these terms can be ignored.

$$\frac{dx_1}{dt} = \mu - \mu x_0 - \mu x_1 - N\beta x_0 y_0 - N\beta x_0 y_1 - N\beta x_1 y_0 - N\beta^2 \delta x_0 y_0 \sin \omega t$$

Now substitute the expressions found in part (a) for x_0 and y_0, then simplify.

$$\begin{aligned}\frac{dx_1}{dt} &= \mu - \mu\left(\frac{\mu+\nu}{N\beta}\right) - \mu x_1 - N\beta\left(\frac{\mu+\nu}{N\beta}\right)\left(\frac{\mu(N\beta-\mu-\nu)}{N\beta(\mu+\nu)}\right) - N\beta\left(\frac{\mu+\nu}{N\beta}\right)y_1\\ &\qquad - N\beta x_1\left(\frac{\mu(N\beta-\mu-\nu)}{N\beta(\mu+\nu)}\right) - N\beta^2\delta\left(\frac{\mu+\nu}{N\beta}\right)y_0 \sin \omega t\\ &= \mu - \mu\left(\frac{\mu+\nu}{N\beta}\right) - \mu x_1 - \frac{\mu(N\beta-\mu-\nu)}{N\beta} - (\mu+\nu)y_1 - x_1\left(\frac{\mu(N\beta-\mu-\nu)}{\mu+\nu}\right) - \beta\delta(\mu+\nu)y_0 \sin \omega t\\ &= \mu\left(1 - \frac{\mu+\nu}{N\beta} - \frac{(N\beta-\mu-\nu)}{N\beta}\right) - \mu x_1 - (\mu+\nu)y_1 - x_1\left(\frac{\mu(N\beta-\mu-\nu)}{\mu+\nu}\right) - \beta\delta(\mu+\nu)y_0 \sin \omega t\\ &= -\mu x_1 - (\mu+\nu)y_1 - N\beta x_1\left(\frac{\mu(N\beta-\mu-\nu)}{N\beta(\mu+\nu)}\right) - (\mu+\nu)\beta\delta y_0 \sin \omega t\\ &= -\mu x_1 - (\mu+\nu)y_1 - N\beta x_1 y_0 - (\mu+\nu)\beta\delta y_0 \sin \omega t\\ &= -(N\beta y_0 + \mu)x_1 - (\mu+\nu)y_1 - (\mu+\nu)\beta\delta y_0 \sin \omega t\end{aligned}$$

$$\begin{aligned}\frac{dy_1}{dt} &= N\beta(x_0 + x_1)(y_0 + y_1)(1 + \delta\beta \sin \omega t) - (\mu+\nu)(y_0 + y_1)\\ &= N\beta x_0 y_0 + N\beta x_0 y_1 + N\beta x_1 y_0 + N\beta x_1 y_1 + N\beta^2 \delta x_0 y_0 \sin \omega t\\ &\qquad + N\beta^2 \delta x_0 y_1 \sin \omega t + N\beta^2 \delta x_1 y_0 \sin \omega t + N\beta^2 \delta x_1 y_1 \sin \omega t - (\mu+\nu)(y_0 + y_1)\end{aligned}$$

Ignore all terms containing $x_1 y_1$, δx_1, and δy_1.

$$\begin{aligned}\frac{dy_1}{dt} &= N\beta x_0 y_0 + N\beta x_0 y_1 + N\beta x_1 y_0 + N\beta^2 \delta x_0 y_0 \sin \omega t - (\mu+\nu)(y_0 + y_1)\\ &= N\beta\left(\frac{\mu+\nu}{N\beta}\right)y_0 + N\beta\left(\frac{\mu+\nu}{N\beta}\right)y_1 + N\beta x_1 y_0 + N\beta^2 \delta x_0 y_0 \sin \omega t - y_0(\mu+\nu) - y_1(\mu+\nu)\\ &= (\mu+\nu)y_0 + (\mu+\nu)y_1 + N\beta x_1 y_0 + N\beta^2 \delta x_0 y_0 \sin \omega t - y_0(\mu+\nu) - y_1(\mu+\nu)\\ &= N\beta x_1 y_0 + N\beta^2\delta\left(\frac{\mu+\nu}{N\beta}\right)y_0 \sin \omega t\\ &= N\beta x_1 y_0 + (\mu+\nu)\beta\delta y_0 \sin \omega t\end{aligned}$$

17. (a) Find the derivative of $f(T) = \frac{dN}{dt} = b_1C_0RNT - C_1PNT^2 - D_1N$ and then use either the first or second derivative test.

$f'(T) = b_1C_0RN - 2C_1PNT$

$$0 = b_1C_0RN - 2C_1PNT \Rightarrow T = \frac{b_1C_0R}{2C_1P}$$

$f''(T) = -2C_1PN < 0 \Rightarrow f$ is concave down and thus, T is a maximum point.

(b) $\frac{dR}{dt} = F - C_0NRT$

$\frac{dN}{dt} = b_1C_0RNT - C_1PNT^2 - D_1N$

$\frac{dP}{dt} = b_2C_1PNT^2 - D_2P$

We are finding the equilibrium point, so begin by setting the derivatives equal to zero.

$$F - C_0NRT = 0 \quad (1)$$
$$b_1C_0RNT - C_1PNT^2 - D_1N = 0 \quad (2)$$
$$b_2C_1PNT^2 - D_2P = 0 \quad (3)$$

Substitute $T = \frac{b_1C_0R}{2C_1P}$ into each equation and simplify.

$$F - C_0NR\left(\frac{b_1C_0R}{2C_1P}\right) = 0 \quad (1)$$
$$b_1C_0RN\left(\frac{b_1C_0R}{2C_1P}\right) - C_1PN\left(\frac{b_1C_0R}{2C_1P}\right)^2 - D_1N = 0 \quad (2)$$
$$b_2C_1PN\left(\frac{b_1C_0R}{2C_1P}\right)^2 - D_2P = 0 \quad (3)$$

$$F - \frac{b_1C_0^2NR^2}{2C_1P} = 0 \quad (1)$$
$$\frac{b_1^2C_0^2R^2N}{2C_1P} - \frac{b_1^2C_0^2R^2N}{4C_1P} - D_1N = 0 \quad (2)$$
$$\frac{b_1^2C_0^2R^2b_2N}{4C_1P} - D_2P = 0 \quad (3)$$

Use the hint and solve equations (1) and (2) for $\frac{R^2}{P}$.

$$F - \frac{b_1C_0^2NR^2}{2C_1P} = 0 \Rightarrow \frac{R^2}{P} = \frac{2C_1F}{b_1C_0^2N}$$

$$\frac{b_1^2C_0^2R^2N}{2C_1P} - \frac{b_1^2C_0^2R^2N}{4C_1P} - D_1N = 0$$
$$\frac{R^2}{P}\left(\frac{2b_1^2C_0^2N - b_1^2C_0^2N}{4C_1}\right) = D_1N$$
$$\frac{R^2}{P}\left(\frac{b_1^2C_0^2N}{4C_1}\right) = D_1N \Rightarrow \frac{R^2}{P} = \frac{4C_1D_1}{b_1^2C_0^2}$$

(*continued on next page*)

(continued)

Now set the two expressions for $\frac{R^2}{P}$ equal and solve for N.

$$\frac{2C_1F}{b_1C_0^2N} = \frac{4C_1D_1}{b_1^2C_0^2} \Rightarrow 4b_1C_1D_1C_0^2N = 2C_1Fb_1^2C_0^2 \Rightarrow N = \frac{Fb_1}{2D_1}$$

Next, substitute the expressions for $\frac{R^2}{P}$ and N into equation (3) and solve for P.

$$\frac{b_1^2C_0^2R^2b_2N}{4C_1P} - D_2P = 0 \Rightarrow \frac{R^2}{P}(N)\left(\frac{b_1^2C_0^2b_2}{4C_1}\right) - D_2P = 0 \Rightarrow \frac{4C_1D_1}{b_1^2C_0^2}\left(\frac{Fb_1}{2D_1}\right)\left(\frac{b_1^2C_0^2b_2}{4C_1}\right) = D_2P \Rightarrow$$

$$\frac{Fb_1b_2}{2D_2} = P$$

Finally, substitute the expression for P into the expression for $\frac{R^2}{P}$ and solve for R.

$$\frac{R^2}{P} = \frac{4C_1D_1}{b_1^2C_0^2} \Rightarrow R^2 = \frac{4C_1D_1}{b_1^2C_0^2}(P) = \frac{4C_1D_1}{b_1^2C_0^2}\left(\frac{Fb_1b_2}{2D_2}\right) = \frac{2C_1D_1Fb_2}{C_0^2D_2} \Rightarrow R = \sqrt{\frac{2C_1D_1Fb_2}{b_1C_0^2D_2}}$$

11.6 Applications of Differential Equations

1. (a) Let y = the number of individuals infected. The differential equation is

$$\frac{dy}{dt} = a(N-y)y.$$

The solution is Equation 3 in Example 1, which is $y = \frac{N}{1+(N-1)e^{-aNt}}$, where $a = \frac{k}{N}$.

The number of individuals uninfected at time t is

$$y = N - \frac{N}{1+(N-1)e^{-aNt}} = \frac{N+N(N-1)e^{-aNt} - N}{1+(N-1)e^{-aNt}} = \frac{N(N-1)}{N-1+e^{aNt}}.$$

Now substitute $N = 5000$ and $a = 0.00005$.

$$y = \frac{5000(5000-1)}{5000-1+e^{(0.00005)(5000)t}} = \frac{24,995,000}{4999+e^{0.25t}}$$

(b) $t = 30$

$$y = \frac{24,995,000}{4999+e^{0.25(30)}} = 3672$$

(c) $t = 50$

$$\frac{24,995,000}{4999+e^{0.25(50)}} = 91$$

(d) From Example 1,

$$t_m = \frac{\ln(N-1)}{aN} = \frac{\ln(5000-1)}{(0.00005)(5000)} = 34.$$

The maximum infection rate will occur on the 34th day.

3. (a) The differential equation is

$$\frac{dy}{dt} = a(N-y)y.$$

The solution is Equation 2 in Example 1, which is $y = \frac{N}{1+be^{-kt}}$, where $b = \frac{N-y_0}{y_0}$ and $k = aN$. Since $y_0 = 100$ and $N = 20,000$,

$$b = \frac{20,000-100}{100} = 199;\ k = 20,000a.$$

Therefore, $y = \frac{20,000}{1+199e^{-20,000at}}$.

Since $y = 400$ when $t = 10$, we have

$$\frac{20,000}{1+199e^{-20,000(10)a}} = 400$$

$$400 + 400(199)e^{-200,000a} = 20,000$$

$$e^{-200,000a} = \frac{19,600}{400(199)} = 0.2462312$$

(continued on next page)

(*continued*)

$$a = \frac{\ln(0.2462312)}{-200{,}000} = 7\times10^{-6}$$

$$k = 20{,}000a = 20{,}000(7\times10^{-6}) = 0.14$$

Therefore, $y = \dfrac{20{,}000}{1+199e^{-0.14t}}$ or $\dfrac{20{,}000e^{0.14t}}{e^{0.14t}+199}$.

(b) Half the community is $y = 10{,}000$. Find t for $y = 10{,}000$.

$$\frac{20{,}000}{1+19e^{-0.14t}} = 10{,}000 \Rightarrow 10{,}000 + 10{,}000(199)e^{-0.14t} = 20{,}000$$

$$e^{-0.14t} = \frac{10{,}000}{10{,}000(199)} \approx 0.005 \Rightarrow t = \frac{\ln(0.005)}{-0.14} \approx 37.77$$

Half the community will be infected in about 38 days.

5. (a) $\dfrac{dy}{dt} = -ay + b(f-y)Y$

$a = 1$, $b = 1$, $f = 0.5$, $Y = 0.01$; $y = 0.02$ when $t = 0$.

$$\frac{dy}{dt} = -y + 1(0.5-y)(0.01) = -1.010y + 0.005$$

$$\int \frac{dy}{-1.010y+0.005} = \int dt$$

$$\frac{1}{-1.010}\ln\left|-1.010y+0.005\right| = t + C_2$$

$$\ln\left|-1.010y+0.005\right| = -1.010t + C_1$$

$$\left|-1.010y+0.005\right| = e^{-1.010t+C_1} = e^{C_1}e^{-1.010t}$$

$$-1.010y + 0.005 = Ce^{-1.010t} \Rightarrow y = 0.005 - 0.990Ce^{-1.010t}$$

Since $y = 0.02$ when $t = 0$, $0.02 = 0.005 - 0.990Ce^{0} \Rightarrow -0.990C = 0.015$.

Therefore, $y = 0.005 + 0.015e^{-1.010t}$.

(b) $\dfrac{dY}{dt} = -AY + B(F-Y)y$

$A = 1$, $B = 1$, $y = 0.1$, $F = 0.03$; $Y = 0.01$ when $t = 0$.

$$\frac{dY}{dt} = -Y + 1(0.03-Y)(0.1) = -1.1Y + 0.003$$

$$\frac{dY}{-1.1Y+0.003} = dt$$

$$\int \frac{dY}{-1.1Y+0.003} = \int dt$$

$$-\frac{1}{1.1}\ln\left|-1.1Y+0.003\right| = t + C_2$$

$$\ln\left|-1.1Y+0.003\right| = -1.1t + C_1$$

$$\left|-1.1Y+0.003\right| = e^{-1.1t+C_1} = e^{C_1}e^{-1.1t}$$

$$-1.1Y + 0.003 = Ce^{-1.1t}$$

$$Y = \frac{C}{-1.1}e^{-1.1t} - \frac{0.003}{-1.1} = -0.909Ce^{-1.1t} + 0.00273$$

Since $Y = 0.01$ when $t = 0$, $0.01 = -0.909Ce^{0} + 0.00273 \Rightarrow -0.909C = 0.00727$.

Therefore, $Y = 0.00727e^{-1.1t} + 0.00273$.

7. (a) $\frac{dy}{dt} = k\left(1 - \frac{y}{N}\right)y$

$y_0 = 3$; $y = 12$ when $t = 3$; $N = 45$

From equations (1) and (2) in Example 1 of this section, the solution is

$y = \frac{N}{1 + be^{-kt}}$, where y = number of people who have heard the information,

$b = \frac{N - y_0}{y_0}$, and $k = aN$.

Since $y_0 = 3$ and $N = 45$,

$b = \frac{45-3}{3} = 14$; $k = 45a$.

$y = \frac{45}{1 + 14e^{-45at}}$.

$y = 12$ when $t = 3$ so

$$12 = \frac{45}{1 + 14e^{-135a}}$$
$$12 + 168e^{-135a} = 45$$
$$e^{-135a} = \frac{33}{168} = \frac{11}{56}$$
$$-135a = \ln \frac{11}{56} = -1.627$$
$$-45a = -0.542$$

Therefore, $y \approx \frac{45}{1 + 14e^{-0.54t}}$.

(b) When $y = 30$,

$$30 = \frac{45}{1 + 14e^{-0.54t}}$$
$$30 + 420e^{-0.54t} = 45$$
$$e^{-0.54t} = \frac{15}{420} = \frac{1}{28}$$
$$t = -\frac{1}{0.54} \ln \frac{1}{28} = 6.17.$$

In about 6 days, 30 employees have heard the rumor.

9. (a) $\frac{dy}{dt} = kye^{-at}$; $a = 0.1$; $y = 3$ when $t = 0$; $y = 12$ when $t = 3$.

$$\int \frac{dy}{y} = k\int e^{-0.1t}\,dt$$
$$\ln|y| = -10ke^{-0.1t} + C_1$$
$$|y| = e^{-10ke^{-0.1t} + C_1} = e^{C_1}e^{-10ke^{-0.1t}}$$
$$y = Ce^{-10ke^{-0.1t}}$$

Since $y = 3$ when $t = 0$,

$3 = Ce^{-10k} \Rightarrow C = 3e^{10k}$.

Since $y = 12$ when $t = 3$,

$12 = Ce^{-10ke^{-0.3}} = Ce^{-7.41k} \Rightarrow$

$C = 12e^{7.41k}$.

Solve the system

$C = 3e^{10k}$

$C = 12e^{7.41k}$.

$$3e^{10k} = 12e^{7.41k}$$
$$e^{10k} = 4e^{7.41k}$$
$$10k \ln e = \ln 4 + 7.41k$$
$$2.59k = \ln 4$$
$$k = \frac{1}{2.59} \ln 4 = 0.535$$

Thus, $C = 3e^{10(0.535)} \approx 631$ and

$y \approx 631e^{-10(0.535)e^{-0.1t}} = 631e^{-5.35e^{-0.1t}}$.

(b) If $y = 30$,

$$30 = 631e^{-5.35e^{-0.1t}}$$
$$e^{-5.35e^{-0.1t}} = \frac{30}{631}$$
$$-5.35e^{-0.1t} \ln e = \ln \frac{30}{631}$$
$$e^{-0.1t} = -\frac{1}{5.35} \ln \frac{30}{631}$$
$$-0.1t \ln e = \ln\left(-\frac{1}{5.35} \ln \frac{30}{631}\right)$$
$$t = -10 \ln\left(-\frac{1}{5.35} \ln \frac{30}{631}\right)$$
$$\approx 5.632.$$

30 employees have heard the rumor in about 6 days.

11. Let y = the amount of salt present at time t.

(a)
$$\frac{dy}{dt} = (\text{rate of salt in}) - (\text{rate of salt out})$$

rate of salt in $= (3 \text{ gal/min})(2 \text{ lb/gal}) = 6$ lb/min

rate of salt out $= \left(\frac{y}{V} \text{ lb/gal}\right)(2 \text{ gal/min}) = \left(\frac{2y}{V} \text{ lb/min}\right)$

$$\frac{dy}{dt} = 6 - \frac{2y}{V};\ y(0) = 20 \text{ lb}$$

$$\frac{dV}{dt} = (\text{rate of liquid in}) - (\text{rate of liquid out}) = 3 \text{ gal/min} - 2 \text{ gal/min} = 1 \text{ gal/min}$$

$$\frac{dV}{dt} = 1 \Rightarrow V = t + C_1$$

(continued on next page)

(continued)

When $t = 0$, $V = 100$. Thus, $C_1 = 100$ and $V = t + 100$. Therefore,

$$\frac{dy}{dt} = 6 - \frac{2y}{t+100}$$

$$\frac{dy}{dt} + \frac{2}{t+100}y = 6.$$

$$I(t) = e^{\int 2\,dt/(t+100)} = e^{2\ln|t+100|} = (t+100)^2$$

$$\frac{dy}{dt}(t+100)^2 + 2y(t+100) = 6(t+100)^2$$

$$D_t[y(t+100)^2] = 6(t+100)^2$$

$$y(t+100)^2 = 6\int (t+100)^2\,dt$$

$$y(t+100)^2 = 2(t+100)^3 + C$$

$$y = 2(t+100) + \frac{C}{(t+100)^2}$$

Since $t = 0$ when $y = 20$,

$$20 = 2(100) + \frac{C}{100^2}$$

$$C = -1{,}800{,}000.$$

$$y = 2(t+100) - \frac{1{,}800{,}000}{(t+100)^2} = \frac{2(t+100)^3 - 1{,}800{,}000}{(t+100)^2}.$$

(b) $t = 1$ hr $= 60$ min

$$y = \frac{2(160)^3 - 1{,}800{,}000}{(160)^2} = 249.69$$

After 1 hr, about 250 lb of salt are present.

(c) As time increases, salt concentration continues to increase.

13. Let y = the amount of salt present at time t minutes.

(a) $\frac{dy}{dt}$ = (rate of salt in) – (rate of salt out)

rate of salt in = 0

rate of salt out $= \left(\frac{y}{V}\text{ lb/gal}\right)(2\text{ gal/min}) = \frac{2y}{V}\text{ lb/min}$

$$\frac{dV}{dt} = -\frac{2y}{V};\ y(0) = 20$$

$\frac{dV}{dt}$ = (rate of liquid in) – (rate of liquid out) = 2 gal/min – 2 gal/ min = 0

$$\frac{dV}{dt} = 0 \Rightarrow V = C_1$$

When $t = 0$, $V = 100$, so $C_1 = 100$. Therefore,

$$\frac{dy}{dt} = -\frac{2y}{100} = -0.02y$$

$$\frac{dy}{y} = -0.02\,dt$$

$$\int \frac{dy}{y} = -0.02\int dt$$

$$\ln|y| = -0.02t + C_1$$

$$|y| = e^{-0.02t + C_1} = e^{C_1}e^{-0.02t} = Ce^{-0.02t}$$

Since $t = 0$ when $y = 20$,

$20 = Ce^0 \Rightarrow C = 20$, and $y = 20e^{-0.02t}$.

(b) $t = 1$ hr $= 60$ min.

$$y = 20e^{-0.02(60)} = 6.024$$

After 1 hr, about 6 lb of salt are present.

(c) As time increases, salt concentration continues to decrease.

15. Let y = amount of the chemical at time t.

(a) $\frac{dy}{dt}$ = (rate of chemical in) – (rate of chemical out)

rate of chemical in = (2 liters/min) (0.25 g/liter) = 0.5 g/min

rate of chemical out $= \left(\frac{y}{V}\text{ g/liter}\right)(1\text{ liter/min}) = \frac{y}{V}\text{ g/liter}$

$$\frac{dy}{dt} = 0.5 - \frac{y}{V};\ y(0) = 5$$

$\frac{dV}{dt}$ = (rate of liquid in) – (rate of liquid out) = 2 liter/min – 1 liter/min = 1 liter/min

$$\frac{dV}{dt} = 1 \Rightarrow V = t + C_1$$

(continued on next page)

(continued)

When $t = 0, V = 100$, so $C_1 = 100$ and $V = t + 100$.
Therefore,

$$\frac{dy}{dt} = 0.5 - \frac{y}{t+100}$$

$$\frac{dy}{dt} + \frac{1}{t+100} \cdot y = 0.5$$

$$I(t) = e^{\int dt/(t+100)} = e^{\ln|t+100|}$$
$$= t + 100$$

$$\frac{dy}{dt}(t+100) + y = 0.5(t+100)$$
$$D_x(t+100)y = 0.5(t+100)$$
$$(t+100)y = \int 0.5(t+100)\, dt$$
$$(t+100)y = 0.25(t+100)^2 + C$$
$$y = 0.25(t+100) + \frac{C}{t+100}$$

$t = 0, \; y = 5$

$$5 = 0.25(100) + \frac{C}{100} \Rightarrow 500 = 2500 + C \Rightarrow$$

$C = -2000$
Therefore,

$$y = 0.25(t+100) + \frac{-2000}{t+100}$$
$$= \frac{0.25(t+100)^2 - 2000}{t+100}.$$

(b) When $t = 30$ min,

$$y = \frac{0.25(130)^2 - 2000}{130} = 17.115.$$

After 30 min, about 17.1 g of chemical are present.

Chapter 11 Review Exercises

1. True
2. False: No; $y = e^{2x}$ satisfies the given differential equation.
3. True
4. False: No; the term $y(dy/dx)$ makes the equation nonlinear.
5. False: Many (most) differential equations are neither separable nor linear.
6. True
7. False: There is no way to separate the variables in this equation.
8. True
9. False: The integrating factor is x^5.
10. False: Euler's method finds a numerical approximation to a particular solution over some interval.
11. True
12. False
13. Answers will vary. A differential equation is an equation that involves an unknown function $y = f(x)$ and a finite number of its derivatives. Solving the differential equation for y gives the unknown function.
14. Answers will vary. The **general solution** to a differential equation is the most general form that the solution can take and doesn't take any initial conditions into account. The **particular solution** to a differential equation is the specific solution that not only satisfies the differential equation, but also satisfies the given initial condition(s).
15. Answers will vary. A separable differential equation is any differential equation that can be written in the form $q(y)dy = p(x)dx$. A linear differential equation can be written in the form $\frac{dy}{dx} + P(x)y = Q(x)$.
16. Yes, a diffential equation can be both separable and linear. An example is $\frac{dy}{dx} + x = xy$.
17. $y\frac{dy}{dx} = 2x + y$
    The equation cannot be rewritten in the form $g(y)dy = f(x)dx$ where g is a function of y alone and f is a function of x alone, so the equation is not separable. The equation cannot be rewritten in the form $\frac{dy}{dx} + P(x)y = Q(x)$, so the equation is not a linear first-order differential equation. Therefore, the equation is neither linear nor separable.

18. $\frac{dy}{dx} + y^2 = xy^2$

The equation can be rewritten in the form $\frac{dy}{y^2} = (x-1)dx,$ so the equation is separable, but since it cannot be rewritten in the form $\frac{dy}{dx} + P(x)y = Q(x),$ the equation is not linear.

19. $\sqrt{x}\frac{dy}{dx} = \frac{1+\ln x}{y}$

The equation can be rewritten in the form $y\,dy = \frac{1+\ln x^x}{\sqrt{x}}dx,$ so the equation is separable, but since it cannot be rewritten in the form $\frac{dy}{dx} + P(x)y = Q(x),$ the equation is not linear.

20. $\frac{dy}{dx} = xy + e^x$

The equation can be rewritten in the form $\frac{dy}{dx} + (-x)y = e^x,$ so the equation is linear. Since the equation cannot be rewritten in the form $g(y)dy = f(x)dx,$ it is not separable.

21. $\frac{dy}{dx} + x = xy$

The equation can be rewritten in the form $\frac{dy}{dx} + (-x)y = -x,$ so the equation is linear. The equation can be rewritten in the form $\frac{dy}{y-1} = x\,dx,$ so it is also separable. Therefore, it is both linear and separable.

22. $\frac{x}{y}\frac{dy}{dx} = 4 + x^{3/2}$

The equation can be rewritten in the form $\frac{dy}{dx} + \left(-\frac{4+x^{3/2}}{x}\right)y = 0,$ so the equation is linear. Since it can be rewritten in the form $\frac{dy}{y} = \frac{4+x^{3/2}}{x}dx,$ it is also separable. Therefore, it is both linear and separable.

23. $x\frac{dy}{dx} + y = e^x(1+y)$

The equation can be rewritten in the form $\frac{dy}{dx} + y\left(\frac{1-e^x}{x}\right) = \frac{e^x}{x},$ so the equation is linear. Since the equation cannot be rewritten in the form $g(y)dy = f(x)dx,$ the equation is not separable.

24. $\frac{dy}{dx} = x^2 + y^2$

Since the equation cannot be rewritten in either form $\frac{dy}{dx} + P(x)y = Q(x)$ or the form $g(y)dy = f(x)dx,$ then the equation is neither nor separable.

25. $\frac{dy}{dx} = 3x^2 + 6x$

$$dy = (3x^2 + 6x)dx$$
$$y = x^3 + 3x^2 + C$$

26. $\frac{dy}{dx} = 4x^3 + 6x^5$

$$y = x^4 + x^6 + C$$

27. $\frac{dy}{dx} = 4e^{2x}$

$$dy = 4e^{2x}dx$$
$$y = 2e^{2x} + C$$

28. $\frac{dy}{dx} = \frac{1}{3x+2}$

$$y = \frac{1}{3}\ln|3x+2| + C$$

29. $\frac{dy}{dx} = \frac{3x+1}{y}$

$$y\,dy = (3x+1)\,dx$$
$$\frac{y^2}{2} = \frac{3x^2}{2} + x + C_1$$
$$y^2 = 3x^2 + 2x + C$$

30. $\frac{dy}{dx} = \frac{e^x + x}{y-1}$

$$(y-1)\,dy = (e^x + x)\,dx$$
$$\frac{y^2}{2} - y = e^x + \frac{x^2}{2} + C$$

31. $\frac{dy}{dx}=\frac{2y+1}{x}$

$\frac{dy}{2y+1}=\frac{dx}{x}$

$\frac{1}{2}\left(\frac{2\,dy}{2y+1}\right)=\frac{dx}{x}$

$\frac{1}{2}\int\frac{2\,dy}{2y+1}=\int\frac{dx}{x}$

$\frac{1}{2}\ln|2y+1|=\ln|x|+C_1$

$\ln|2y+1|^{1/2}=\ln|x|+\ln k$

Let $\ln k=C_1$

$\ln|2y+1|^{1/2}=\ln k|x|$

$|2y+1|^{1/2}=k|x|$

$2y+1=k^2x^2$

$2y+1=Cx^2 \Rightarrow y=\frac{Cx^2-1}{2}$

32. $\frac{dy}{dx}=\frac{3-y}{e^x}$

$\frac{1}{3-y}\,dy=e^{-x}dx$

$\int\frac{1}{3-y}\,dy=\int e^{-x}dx$

$-\ln|3-y|=-e^{-x}+C$

$\ln|3-y|=e^{-x}-C$

$3-y=ke^{e^{-x}}$

$y=3-ke^{e^{-x}}$ or $y=3+Me^{e^{-x}}$

33. $\frac{dy}{dx}+y=x$

$I(x)=e^{\int dx}=e^x$

$e^x\frac{dy}{dx}+e^xy=xe^x$

$D_x(e^xy)=xe^x$

Integrate both sides, integrating the right side by parts.

$e^xy=xe^x-e^x+C$

$y=x-1+Ce^{-x}$

34. $x^4\frac{dy}{dx}+3x^3y=1$

$\frac{dy}{dx}+\frac{3}{x}y=\frac{1}{x^4}$

$I(x)=e^{\int\frac{3}{x}dx}=x^3$

$x^3\frac{dy}{dx}+3x^2y=\frac{1}{x}$

$D_x(x^3y)=\frac{1}{x}$

Integrate both sides.

$x^3y=\ln|x|+C \Rightarrow y=\frac{\ln|x|+C}{x^3}$

35. $x\ln x\frac{dy}{dx}+y=2x^2$

$\frac{dy}{dx}+\frac{1}{x\ln x}y=\frac{2x}{\ln x}$

$I(x)=e^{\int\frac{1}{x\ln x}dx}=e^{\int\frac{1}{\ln x}\left(\frac{1}{x}dx\right)}=e^{\ln(\ln x)}=\ln x$

Multiply each term by the integrating factor, express the left side as the derivative of a product and integrate on both sides.

$\ln x\frac{dy}{dx}+\frac{1}{x}y=2x$

$D_x[(\ln x)y]=2x$

$(\ln x)y=x^2+C \Rightarrow y=\frac{x^2+C}{\ln x}$

36. $x\frac{dy}{dx}+2y-e^{2x}=0$

$\frac{dy}{dx}+\frac{2}{x}y=\frac{1}{x}e^{2x}$

$I(x)=e^{\int 2dx/x}=e^{2\ln x}=x^2$

$x^2\frac{dy}{dx}+2xy=xe^{2x}$

$D_x(x^2y)=xe^{2x}$

$x^2y=\int xe^{2x}dx$

$=\int xe^{2x}dx=\frac{x}{2}e^{2x}-\frac{1}{4}e^{2x}+C$

$y=\frac{e^{2x}}{2x}-\frac{e^{2x}}{4x^2}+\frac{C}{x^2}$

37. $\frac{dy}{dx}=x^2-6x;\ y(0)=3$

$dy=(x^2-6x)dx$

$y=\frac{x^3}{3}-3x^2+C$

When $x=0,\ y=3$.

$3=0-0+C \Rightarrow C=3$

$y=\frac{x^3}{3}-3x^2+3$

38. $\frac{dy}{dx} = 5(e^{-x} - 1);\; y(0) = 17$

$dy = 5(e^{-x} - 1)\,dx$

$y = 5(-e^{-x} - x) + C = -5e^{-x} - 5x + C$

$17 = -5e^{0} - 0 + C \Rightarrow C = 22$

$y = -5e^{-x} - 5x + 22$

39. $\frac{dy}{dx} = (x+2)^3 e^{y};\; y(0) = 0$

$e^{-y}dy = (x+2)^3 dx$

$-e^{-y} = \frac{1}{4}(x+2)^4 + C$

$-1 = \frac{1}{4}(2)^4 + C \Rightarrow C = -5$

$e^{-y} = 5 - \frac{1}{4}(x+2)^4 \Rightarrow y = -\ln\left[5 - \frac{1}{4}(x+2)^4\right]$

Notice that $x < \sqrt[4]{20} - 2$.

40. $(3-2x)y = \frac{dy}{dx};\; y(0) = 5$

$(3-2x)dx = \frac{dy}{y}$

$3x - x^2 + C = \ln|y|$

$e^{3x-x^2+C} = |y|$

$e^{3x-x^2} \cdot e^{C} = |y| \Rightarrow Me^{3x-x^2} = y$

$Me^{0} = 5 \Rightarrow M = 5$

$y = 5e^{3x-x^2}$

41. $\frac{dy}{dx} = \frac{1-2x}{y+3};\; y(0) = 16$

$(y+3)\,dy = (1-2x)\,dx$

$\frac{y^2}{2} + 3y = x - x^2 + C$

$\frac{16^2}{2} + 3(16) = 0 + C \Rightarrow 176 = C$

$\frac{y^2}{2} + 3y = x - x^2 + 176 \Rightarrow$

$y^2 + 6y = 2x - 2x^2 + 352$

42. $\sqrt{x}\frac{dy}{dx} = xy;\; y(1) = 4$

$\frac{1}{y}dy = \frac{x}{\sqrt{x}}dx$

$\int \frac{1}{y}dy = \int x^{1/2}\,dx$

$\ln|y| = \frac{x^{3/2}}{\frac{3}{2}} + C_1$

$\ln|y| = \frac{2}{3}x^{3/2} + C_1$

$|y| = e^{(2/3)x^{3/2} + C_1}$

$|y| = e^{C_1}e^{(2/3)x^{3/2}}$

$y = \pm e^{C_1}e^{(2/3)x^{3/2}} \Rightarrow y = Ce^{(2/3)x^{3/2}}$

Since $y = 4$ when $x = 1$,

$4 = Ce^{(2/3)(1)^{3/2}} \Rightarrow C = \frac{4}{e^{2/3}} \approx 2.054$

$y = 2.054e^{(2/3)x^{3/2}}$.

43. $e^{x}\frac{dy}{dx} - e^{x}y = x^2 - 1;\; y(0) = 42$

$\frac{dy}{dx} - y = (x^2 - 1)e^{-x}$

$I(x) = e^{\int -1\,dx} = e^{-x}$

$e^{-x}y = \int (x^2 - 1)e^{-x} \cdot e^{-x}dx$

$= \int (x^2 - 1)e^{-2x}\,dx$

Integration by parts:

Let $u = x^2 - 1 \Rightarrow du = 2x\,dx$

$dv = e^{-2x}dx \Rightarrow v = -\frac{1}{2}e^{-2x}$.

$e^{-x}y = \frac{-(x^2-1)}{2}e^{-2x} + \int xe^{-2x}dx$

Let $u = x \Rightarrow du = dx$

$dv = e^{-2x}dx \Rightarrow v = -\frac{1}{2}e^{-2x}$.

$e^{-x}y = -\frac{(x^2-1)}{2}e^{-2x} - \frac{x}{2}e^{-2x} + \int \frac{1}{2}e^{-2x}\,dx$

$= -\frac{(x^2-1)}{2}e^{-2x} - \frac{x}{2}e^{-2x} - \frac{1}{4}e^{-2x} + C$

$y = -\frac{(x^2-1)}{2}e^{-x} - \frac{x}{2}e^{-x} - \frac{1}{4}e^{-x} + Ce^{x}$

$42 = \frac{1}{2} - 0 - \frac{1}{4} + C \Rightarrow C = 41.75$

$y = -\frac{(x^2-1)}{2}e^{-x} - \frac{x}{2}e^{-x} - \frac{1}{4}e^{-x} + 41.75e^{x}$

$= e^{-x}\left[-\frac{x^2}{2} - \frac{x}{2} + \frac{1}{4}\right] + 41.75e^{x}$

$= \frac{-x^2e^{-x}}{2} - \frac{xe^{-x}}{2} + \frac{e^{-x}}{4} + 41.75e^{x}$

44. $\frac{dy}{dx}+3x^2y=x^2;\ y(0)=2$

$$I(x)=e^{\int 3x^2dx}=e^{x^3}$$

$$e^{x^3}\frac{dy}{dx}+3x^2e^{x^3}y=x^2e^{x^3}$$

$$D_x(e^{x^3}y)=x^2e^{x^3}$$

$$e^{x^3}y=\int x^2e^{x^3}dx=\frac{1}{3}e^{x^3}+C$$

$$y=\frac{1}{3}+Ce^{-x^3}$$

Since $x = 0$ when $y = 2$, $2=\frac{1}{3}+Ce^0\Rightarrow C=\frac{5}{3}$.

Therefore, $y=\frac{1}{3}+\frac{5}{3}e^{-x^3}$.

45. $x\frac{dy}{dx}-2x^2y+3x^2=0;\ y(0)=15$

$$x\frac{dy}{dx}-2x^2y+3x^2=0\Rightarrow\frac{dy}{dx}-2xy=-3x$$

$$I(x)=e^{\int -2x\,dx}=e^{-x^2}$$

$$e^{-x^2}\frac{dy}{dx}-2xe^{-x^2}y=-3xe^{-x^2}$$

$$D_x(e^{-x^2}y)=-3xe^{-x^2}$$

$$e^{-x^2}y=\int -3xe^{-x^2}dx=\frac{3}{2}e^{-x^2}+C$$

$$y=\frac{3}{2}+Ce^{x^2}$$

Since $x = 0$ when $y = 15$,

$15=\frac{3}{2}+Ce^0\Rightarrow C=\frac{27}{2}$.

Therefore, $y=\frac{3}{2}+\frac{27e^{x^2}}{2}$.

46. $x^2\frac{dy}{dx}+4xy-e^{2x^3}=0;\ y(1)=e^2$

$$x^2\frac{dy}{dx}+4xy-e^{2x^3}=0\Rightarrow\frac{dy}{dx}+\frac{4}{x}y=\frac{1}{x^2}e^{2x^3}$$

$$I(x)=e^{\int 4/x\,dx}=e^{4\ln x}=x^4$$

$$x^4y=\int x^4\cdot\frac{1}{x^2}e^{2x^3}dx$$

$$=\int x^2e^{2x^3}dx=\frac{1}{6}e^{2x^3}+C$$

$$1\cdot e^2=\frac{1}{6}e^2+C\Rightarrow C=\frac{5}{6}e^2$$

$$y=\frac{e^{2x^3}+5e^2}{6x^4}$$

47. Answers will vary.

48. $\frac{dy}{dx}=x+y^{-1};\ y(0)=1;\ h=0.2;$

$g(x,y)=x+\frac{1}{y}$

$x_0=0;\ y_0=1$

$g(x_0,y_0)=0+\frac{1}{1}=1$

$x_1=0.2;\ y_1=1+1(0.2)=1.2$

$g(x_1,y_1)=0.2+\frac{1}{1.2}=1.0333$

$x_2=0.4;\ y_2=1.2+1.0333(0.2)=1.4067$

$g(x_2,y_2)=0.4+\frac{1}{1.4067}=1.1109$

$x_3=0.6;\ y_3=1.4067+1.1109(0.2)=1.6289$

$g(x_3,y_3)=0.6+\frac{1}{1.6289}=1.2139$

$x^4=0.8;\ y_4=1.6289+1.2139(0.2)=1.8717$

$g(x_4,y_4)=0.8+\frac{1}{1.8717}=1.3343$

$x_5=1.0;\ y_5=1.8719+1.3343(0.2)=2.13855$

$y(1)\approx 2.138$

49. $\frac{dy}{dx}=e^x+y;\ y(0)=1,\ h=0.2;$ find $y(0.6)$.

$g(x,y)=e^x+y$

$x_0=0;\ y_0=1$

$g(x_0,y_0)=e^0+1=2$

$x_1=0.2;\ y_1=y_0+g(x_0,y_0)h=1+2(0.2)=1.4$

$g(x_1,y_1)=e^{.2}+1.4\approx 2.6214$

$x_2=0.4;$

$y_2=y_1+g(x_1,y_1)h=1.4+2.6214(0.2)\approx 1.9243$

$g(x_2,y_2)=e^{0.4}+1.9243\approx 3.4161$

$x_3=0.6;$

$y_3=y_2+g(x_2,y_2)h=1.9243+3.4161(0.2)\approx 2.6075$

x_i	y_i
0	1
0.2	1.4
0.4	1.9243
0.6	2.608

So, $y(0.6)\approx 2.608$.

50. $\frac{dy}{dx} = \frac{x}{2} + 4$; $y(0) = 0$; $h = 0.1$,

$g(x, y) = \frac{x}{2} + 4$

$x_0 = 0$; $y_0 = 0$

$g(x_0, y_0) = \frac{0}{2} + 4 = 4$

$x_1 = 0.1$; $y_1 = 0 + 4(0.1) = 0.4$

$g(x_1, y_1) = \frac{0.1}{2} + 4 = 4.05$

$x_2 = 0.2$; $y_2 = 0.4 + 4.05(0.1) = 0.805$

$g(x_2, y_2) = \frac{0.2}{2} + 4 = 4.1$

$x_3 = 0.3$; $y_3 = 0.805 + 4.1\,(0.1) = 1.215$

Solving the differential equation gives

$$\frac{dy}{dx} = \frac{x}{2} + 4$$

$$\int dy = \int\left(\frac{x}{2} + 4\right)dx$$

$$y = \frac{x^2}{4} + 4x + C.$$

Since $x = 0$ when $y = 0$, $C = 0$.

$$y = \frac{x^2}{4} + 4x$$

$$y(x_3) = y(0.3) = \frac{0.3^2}{4} + 4(0.3) = 1.223$$

$$y_3 - y(x_3) = 1.215 - 1.223 = -0.008$$

51. $\frac{dy}{dx} = 3 + \sqrt{y}$, $y(0) = 0$, $h = 0.2$, find $y(1)$.

x_i	y_i
0	0
0.2	0.6
0.4	1.354919
0.6	2.187722
0.8	3.083541
1	4.034741

Therefore, $y(1) \approx 4.035$.

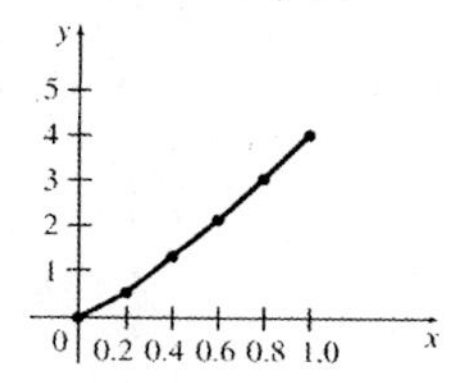

52. The logistic equation is $\frac{dy}{dt} = k\left(1 - \frac{y}{N}\right)y$. The general solution is $y = \frac{N}{1 + be^{-kt}}$.

53. $\frac{dx_1}{dt} = 2x_1 + 7x_2 + t$

$\frac{dx_2}{dt} = 7x_1 + 2x_2 + 2t$

The system can be represented in matrix form as $\begin{bmatrix} \frac{dx_1}{dt} \\ \frac{dx_2}{dt} \end{bmatrix} = \begin{bmatrix} 2 & 7 \\ 7 & 2 \end{bmatrix}\begin{bmatrix} x_1 \\ x_2 \end{bmatrix} + \begin{bmatrix} t \\ 2t \end{bmatrix}$.

Find the eigenvalues.

$$0 = \det(M - \lambda I)$$
$$= \det\begin{bmatrix} 2-\lambda & 7 \\ 7 & 2-\lambda \end{bmatrix}$$
$$= (2-\lambda)(2-\lambda) - 7(7)$$
$$= \lambda^2 - 4\lambda - 45 \Rightarrow (\lambda - 9)(\lambda + 5) \Rightarrow \lambda = 9, -5$$

For $\lambda = -5$ we have

$$(M - \lambda I) = \begin{bmatrix} 7 & 7 \\ 7 & 7 \end{bmatrix}\begin{bmatrix} x_1 \\ x_2 \end{bmatrix} = \begin{bmatrix} 0 \\ 0 \end{bmatrix}.$$

A solution to this system is $\begin{bmatrix} x \\ y \end{bmatrix} = \begin{bmatrix} 1 \\ -1 \end{bmatrix}$.

For $\lambda = 9$ we have

$$(M - \lambda I) = \begin{bmatrix} -7 & 7 \\ 7 & -7 \end{bmatrix}\begin{bmatrix} x_1 \\ x_2 \end{bmatrix} = \begin{bmatrix} 0 \\ 0 \end{bmatrix}.$$

A solution to this system is $\begin{bmatrix} x \\ y \end{bmatrix} = \begin{bmatrix} 1 \\ 1 \end{bmatrix}$.

The matrix of eigenvectors is $P = \begin{bmatrix} 1 & 1 \\ -1 & 1 \end{bmatrix}$.

$P^{-1} = \frac{1}{2}\begin{bmatrix} 1 & -1 \\ 1 & 1 \end{bmatrix}$ and $P^{-1}MP = \begin{bmatrix} -5 & 0 \\ 0 & 9 \end{bmatrix}$.

Now let $X = PY$ and multiply both sides of the differential equation by P^{-1}, yielding

$$\frac{dY}{dt} = P^{-1}MPY + P^{-1}\begin{bmatrix} t \\ 2t \end{bmatrix}, \text{ or}$$

$$\begin{bmatrix} \frac{dy_1}{dt} \\ \frac{dy_2}{dt} \end{bmatrix} = \begin{bmatrix} -5 & 0 \\ 0 & 9 \end{bmatrix}\begin{bmatrix} y_1 \\ y_2 \end{bmatrix} + \frac{1}{2}\begin{bmatrix} 1 & -1 \\ 1 & 1 \end{bmatrix}\begin{bmatrix} t \\ 2t \end{bmatrix} \Rightarrow$$

$$\frac{dy_1}{dt} = -5y_1 + \frac{t}{2} - t = -5y_1 - \frac{t}{2}$$

$$\frac{dy_2}{dt} = 9y_2 + \frac{t}{2} + t = 9y_2 + \frac{3t}{2}$$

(continued on next page)

(continued)

Multiply the first differential equation by an integrating factor, $e^{\int 5dt} = e^{5t}$.

$$\frac{dy_1}{dt} + 5y_1 = -\frac{t}{2}$$
$$e^{5t}\frac{dy_1}{dt} + 5e^{5t}y_1 = -\frac{te^{5t}}{2}$$
$$D_t\left(e^{5t}y_1\right) = -\frac{te^{5t}}{2}$$

Using integration by parts with

$u = t \Rightarrow du = dt$ and $dv = e^{5t}dt \Rightarrow v = \frac{1}{5}e^{5t}$,

we have

$$e^{5t}y_1 = -\frac{e^{5t}}{10}\left(t - \frac{1}{5}\right) + C_1$$
$$y_1 = -\frac{1}{10}\left(t - \frac{1}{5}\right) + C_1e^{-5t}$$

Multiply the second differential equation by an integrating factor, $e^{\int -9dt} = e^{-9t}$.

$$\frac{dy_2}{dt} = 9y_2 + \frac{3t}{2}$$
$$e^{-9t}\frac{dy_2}{dt} - 9e^{-9t}y_2 = \frac{3t}{2}e^{-9t}$$
$$D_t\left(e^{-9t}y_2\right) = \frac{3t}{2}e^{-9t}$$

Using integration by parts with
$u = t \Rightarrow du = dt$ and
$dv = e^{-9t}dt \Rightarrow v = -\frac{1}{9}e^{-9t}$, we have

$$e^{-9t}y_2 = \frac{3e^{-9t}}{-18}\left(t + \frac{1}{9}\right) + C_2$$
$$y_2 = -\frac{1}{6}\left(t + \frac{1}{9}\right) + C_2e^{9t}$$

Finally, calculate $X = PY$.

$$\begin{bmatrix} x_1 \\ x_2 \end{bmatrix} = \begin{bmatrix} 1 & 1 \\ -1 & 1 \end{bmatrix}\begin{bmatrix} -\frac{1}{10}\left(t - \frac{1}{5}\right) + C_1e^{-5t} \\ -\frac{1}{6}\left(t + \frac{1}{9}\right) + C_2e^{9t} \end{bmatrix}$$
$$= \begin{bmatrix} C_1e^{-5t} + C_2e^{9t} - \frac{4t}{15} + \frac{1}{675} \\ -C_1e^{-5t} + C_2e^{9t} - \frac{t}{15} - \frac{26}{675} \end{bmatrix}, \text{ or}$$

$$x_1 = C_1e^{-5t} + C_2e^{9t} - \frac{4t}{15} + \frac{1}{675},$$
$$x_2 = -C_1e^{-5t} + C_2e^{9t} - \frac{t}{15} - \frac{26}{675}$$

54. $\frac{dx_1}{dt} = 7x_1 + 2x_2 + 3t$
$\frac{dx_2}{dt} = -4x_1 + x_2 - t$

The system can be represented in matrix form as $\begin{bmatrix} \frac{dx_1}{dt} \\ \frac{dx_2}{dt} \end{bmatrix} = \begin{bmatrix} 7 & 2 \\ -4 & 1 \end{bmatrix}\begin{bmatrix} x_1 \\ x_2 \end{bmatrix} + \begin{bmatrix} 3t \\ -t \end{bmatrix}$.

Find the eigenvalues.

$$0 = \det(M - \lambda I)$$
$$= \det\begin{bmatrix} 7-\lambda & 2 \\ -4 & 1-\lambda \end{bmatrix}$$
$$= (7-\lambda)(1-\lambda) - 2(-4)$$
$$= \lambda^2 - 8\lambda + 15 \Rightarrow (\lambda - 3)(\lambda - 5) \Rightarrow \lambda = 3, 5$$

For $\lambda = 3$ we have

$$(M - \lambda I) = \begin{bmatrix} 4 & 2 \\ -4 & -2 \end{bmatrix}\begin{bmatrix} x_1 \\ x_2 \end{bmatrix} = \begin{bmatrix} 0 \\ 0 \end{bmatrix}.$$

A solution to this system is $\begin{bmatrix} x \\ y \end{bmatrix} = \begin{bmatrix} 1 \\ -2 \end{bmatrix}$.

For $\lambda = 5$ we have

$$(M - \lambda I) = \begin{bmatrix} 2 & 2 \\ -4 & -4 \end{bmatrix}\begin{bmatrix} x_1 \\ x_2 \end{bmatrix} = \begin{bmatrix} 0 \\ 0 \end{bmatrix}.$$

A solution to this system is $\begin{bmatrix} x \\ y \end{bmatrix} = \begin{bmatrix} 1 \\ -1 \end{bmatrix}$.

The matrix of eigenvectors is $P = \begin{bmatrix} 1 & 1 \\ -2 & -1 \end{bmatrix}$.

$P^{-1} = \begin{bmatrix} -1 & -1 \\ 2 & 1 \end{bmatrix}$ and $P^{-1}MP = \begin{bmatrix} 3 & 0 \\ 0 & 5 \end{bmatrix}$.

Now let $X = PY$ and multiply both sides of the differential equation by P^{-1}, yielding

$$\frac{dY}{dt} = P^{-1}MPY + P^{-1}\begin{bmatrix} 3t \\ -t \end{bmatrix}, \text{ or}$$

$$\begin{bmatrix} \frac{dy_1}{dt} \\ \frac{dy_2}{dt} \end{bmatrix} = \begin{bmatrix} 3 & 0 \\ 0 & 5 \end{bmatrix}\begin{bmatrix} y_1 \\ y_2 \end{bmatrix} + \begin{bmatrix} -1 & -1 \\ 2 & 1 \end{bmatrix}\begin{bmatrix} 3t \\ -t \end{bmatrix} \Rightarrow$$

$$\frac{dy_1}{dt} = 3y_1 - 2t$$
$$\frac{dy_2}{dt} = 5y_2 + 5t$$

Multiply the first differential equation by an integrating factor, $e^{\int -3dt} = e^{-3t}$.

(continued on next page)

(*continued*)

$$\frac{dy_1}{dt} - 3y_1 = -2t$$
$$e^{-3t}\frac{dy_1}{dt} - 3e^{-3t}y_1 = -2te^{-3t}$$
$$D_t\left(e^{-3t}y_1\right) = -2te^{-3t}$$

Using integration by parts with
$u = t \Rightarrow du = dt$ and
$dv = e^{-3t}dt \Rightarrow v = -\frac{1}{3}e^{-3t}$, we have

$$e^{-3t}y_1 = \frac{2e^{-3t}}{3}\left(t + \frac{1}{3}\right) + C_1$$
$$y_1 = \frac{2}{3}\left(t + \frac{1}{3}\right) + C_1e^{3t}$$

Multiply the second differential equation by an integrating factor, $e^{\int -5dt} = e^{-5t}$.

$$\frac{dy_2}{dt} - 5y_2 = 5t$$
$$e^{-5t}\frac{dy_2}{dt} - 5e^{-5t}y_2 = 5te^{-5t}$$
$$D_t\left(e^{-5t}y_2\right) = 5te^{-5t}$$

Using integration by parts with
$u = t \Rightarrow du = dt$ and
$dv = e^{-5t}dt \Rightarrow v = -\frac{1}{5}e^{-5t}$, we have

$$e^{-5t}y_2 = -e^{-5t}\left(t + \frac{1}{5}\right) + C_2$$
$$y_2 = -t - \frac{1}{5} + C_2e^{5t}$$

Finally, calculate $X = PY$.

$$\begin{bmatrix} x_1 \\ x_2 \end{bmatrix} = \begin{bmatrix} 1 & 1 \\ -2 & -1 \end{bmatrix}\begin{bmatrix} \frac{2}{3}\left(t + \frac{1}{3}\right) + C_1e^{3t} \\ -t - \frac{1}{5} + C_2e^{5t} \end{bmatrix}$$
$$= \begin{bmatrix} C_1e^{3t} + C_2e^{5t} - \frac{t}{3} + \frac{1}{45} \\ -2C_1e^{3t} - C_2e^{5t} - \frac{t}{3} - \frac{11}{45} \end{bmatrix}, \text{ or}$$

$$x_1 = C_1e^{3t} + C_2e^{5t} - \frac{t}{3} + \frac{1}{45},$$
$$x_2 = -2C_1e^{3t} - C_2e^{5t} - \frac{t}{3} - \frac{11}{45}$$

55. (a) Set the derivatives equal to zero.

$4x_1 - x_1^2 - x_1x_2 = 0 \Rightarrow x_2 = 4 - x_1$ (1)
$2x_1^2 - x_1x_2 - 2x_1 = 0 \Rightarrow x_2 = 2x_1 - 2$ (2)

Solving the system gives
$4 - x_1 = 2x_1 - 2 \Rightarrow x_1 = 2,\ x_2 = 2$

(b) Graph equations (1) and (2) and test each region.

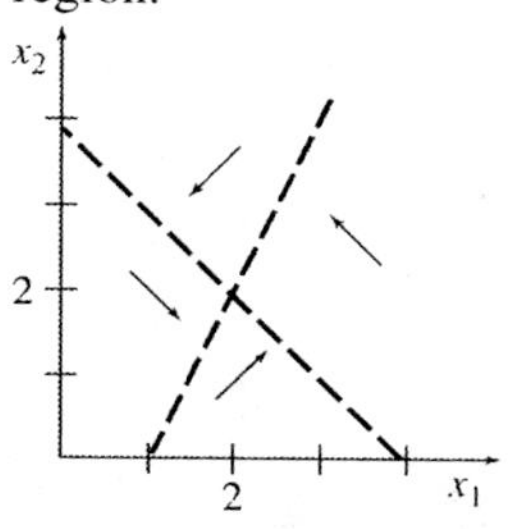

56. (a) Set the derivatives equal to zero.

$x_1x_2 - x_2^2 - x_2 = 0 \Rightarrow x_2 = x_1 - 1$ (1)
$4x_1x_2 - x_1^2x_2 - 2x_1x_2^2 = 0 \Rightarrow$
$x_2 = -\frac{x_1}{2} + 2$ (2)

Solving the system gives
$x_1 - 1 = -\frac{x_1}{2} + 2 \Rightarrow x_1 = 2,\ x_2 = 1$

(b) Graph equations (1) and (2) and test each region.

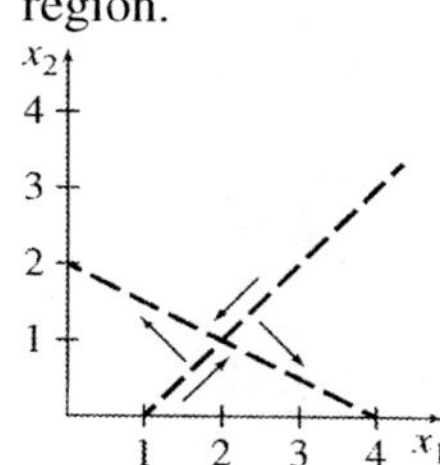

57. $\frac{dy}{dt} = \frac{-10}{1+5t}$; $y = 50$ when $t = 0$.

$y = -2\ln(1+5t) + C$
$50 = -2\ln 1 + C = C$
$y = 50 - 2\ln(1+5t)$

(a) If $t = 24$,
$y = 50 - 2\ln[1 + 5(24)]$
≈ 40 insects.

(b) If $y = 0$,
$50 = 2\ln(1+5t)$
$1 + 5t = e^{25}$
$t = \frac{e^{25} - 1}{5} \approx 1.44 \times 10^{10}$ hours
$\approx 6 \times 10^8$ days
≈ 1.6 million years.

58. $\frac{dy}{dt} = ky; k = 0.1, t = 0, y = 120$

$\frac{dy}{y} = k\,dt$

$\ln|y| = kt + C_1$

$|y| = e^{kt+C_1} = e^{C_1}e^{kt} \Rightarrow y = Me^{kt}$

$y = Me^{0.1t}$

$120 = Me^0 \Rightarrow M = 120$

$y = 120e^{0.1t}$

Let $t = 6$ and find y.

$y = 120e^{0.6} \approx 219$

After 6 weeks, about 219 are present.

59. $\frac{dx}{dt} = 0.2x - 0.5xy$

$\frac{dy}{dt} = -0.3y + 0.4xy$

$\frac{dy}{dt} = \frac{\frac{dy}{dt}}{\frac{dx}{dt}} = \frac{-0.3y + 0.4xy}{0.2x - 0.5xy}$

$= \frac{y(-0.3 + 0.4x)}{x(0.2 - 0.5y)}$

$\frac{0.2 - 0.5y}{y}dy = \frac{-0.3 + 0.4x}{x}dx$

$\left(\frac{0.2}{y} - 0.5\right)dy = \left(\frac{-0.3}{x} + 0.4\right)dx$

$0.2 \ln y - 0.5y = -0.3 \ln x + 0.4x + C \Rightarrow$

$0.3 \ln x + 0.2 \ln y - 0.4x - 0.5y = C$

Both growth rates are 0 if $0.2x - 0.5xy = 0$ and $-0.3y + 0.4xy = 0$. If $x \neq 0$ and $y \neq 0$, we have $0.2 - 0.5y = 0$ and $-0.3 + 0.4x = 0$, so $x = \frac{3}{4}$ unit and $y = \frac{2}{5}$ unit.

60. $\frac{dy}{dt}$ = rate of smoke in – rate of smoke out

rate of smoke in = 0

rate of smoke out $= \left(\frac{y}{V}\right)(1200) = \frac{1200y}{V}$

$\frac{dy}{dt} = \frac{-1200y}{V}$

$\frac{dV}{dt} = 1200 - 1200 = 0$

$V(t) = C_1$; at $t = 0, V = 15,000 \Rightarrow C_1 = 15,000$

$V(t) = 15,000$

$\frac{dy}{dt} = \frac{-1200y}{15,000} = -0.08y \Rightarrow \frac{dy}{y} = -0.08\,dt \Rightarrow$

$\ln y = -0.08t + C \Rightarrow y = ke^{-0.08t}$

At $t = 0$, $y = 20$.

$20 = ke^0 \Rightarrow k = 20$

Therefore, $y = 20e^{-0.08t}$.

At $y = 5$,

$5 = 20e^{-0.08t} \Rightarrow 0.25 = e^{-0.08t} \Rightarrow$

$t = \frac{-\ln(0.25)}{0.08} = 17.3$ min.

61. Let y = the amount in parts per million (ppm) of smoke at time t.

When $t = 0$, $y = 20$ ppm, $V = 15,000$ ft^3.

rate of smoke in = 5 ppm,

rate of smoke out

$= (1200 \text{ ft}^3/\text{min})\left(\frac{y}{V} \text{ ppm/ft}^3\right)$

Rate of air in = rate of air out,

so $V = 15,000\text{ft}^3$ for all t.

$\frac{dy}{dt} = 5 - \frac{1200y}{15,000} = 5 - \frac{2y}{25} = \frac{125 - 2y}{25}$

$\frac{1}{125 - 2y}dy = \frac{dt}{25}$

$-\frac{1}{2}\ln(125 - 2y) = \frac{t}{25} + C$

$-\frac{1}{2}\ln(125 - 2(20)) = C$

$C = -\frac{1}{2}\ln 85$

If $y = 10$,

$-\frac{1}{2}\ln[125 - 2(10)] = \frac{t}{25} - \frac{1}{2}\ln 85$

$\ln 105 = \ln 85 - \frac{2t}{25}$

$t = \frac{25}{2}[\ln 85 - \ln 105]$,

which is negative. Thus, it is impossible to reduce y to 10 ppm.

62. (a) The differential equation for y, the number of individuals infected, is

$\frac{dy}{dt} = a(N - y)y.$

From Example 1 in Section 6 of this chapter, the solution is

$y = \frac{N}{1 + (N-1)e^{-aNt}}$; t is in weeks. The number of individuals uninfected at time t is

$y = N - \frac{N}{1 + (N-1)e^{-aNt}} = \frac{N(N-1)}{N - 1 + e^{aNt}}.$

(continued on next page)

(*continued*)

Substitute $N = 700$.

$$y = \frac{700(699)}{699 + e^{700at}} = \frac{489,300}{699 + e^{700at}}$$

Substitute $t = 6, y = 300$.

$$300 = \frac{489,300}{699 + e^{700(6)a}} = \frac{489,300}{699 + e^{4200a}}$$

$$209,700 + 300e^{4200a} = 489,300$$

$$e^{4200a} = 932$$

$$a = \frac{\ln 932}{4200} \approx 0.00163$$

$$700a \approx 1.140$$

Therefore, $y = \dfrac{489,300}{699 + e^{1.140t}}$.

(b) At $t = 7$ weeks,

$$y = \frac{489,300}{699 + e^{1.140(7)}} \approx 135 \text{ people.}$$

(c) From Example 1,

$$t_m = \frac{\ln (N-1)}{aN} \approx \frac{\ln (700-1)}{(0.00163)(700)} \approx 5.7 \text{ wk.}$$

63. $y = \dfrac{N}{1 + be^{-kt}}$; $y = y_i$ when $x = x_i$, $i = 1, 2, 3$.

t_1, t_2, t_3 are equally spaced:

$t_3 = 2t_2 - t_1$, so $t_1 + t_3 = 2t_2$, or $t_2 = \dfrac{t_1 + t_3}{2}$.

Show $N = \dfrac{\frac{1}{y_1} + \frac{1}{y_3} - \frac{2}{y_2}}{\frac{1}{y_1 y_3} - \frac{1}{y_2^2}}$.

Let

$$A = \frac{1}{y_1} + \frac{1}{y_3} - \frac{2}{y_2}$$

$$= \frac{1 + be^{-kt_1}}{N} + \frac{1 + be^{-kt_3}}{N} - \frac{2\left(1 + be^{-kt_2}\right)}{N}$$

$$= \frac{1}{N}\left(1 + be^{-kt_1} + 1 + be^{-kt_3} - 2 - 2be^{-kt_2}\right)$$

$$= \frac{b}{N}\left[e^{-kt_1} + e^{-kt_3} - 2e^{-kt_2}\right]$$

Let

$$B = \frac{1}{y_1 y_3} - \frac{1}{y_2^2}$$

$$= \frac{\left(1 + be^{-kt_1}\right)\left(1 + be^{-kt_3}\right)}{N^2} - \frac{\left(1 + be^{-kt_2}\right)^2}{N^2}$$

$$= \frac{1}{N^2}\Big[1 + be^{-kt_1} + be^{-kt_3} + b^2 e^{-k(t_1 + t_3)} - 1 - 2be^{-kt_2} - b^2 e^{-2kt_2}\Big]$$

$$= \frac{b}{N^2}\Big[e^{-kt_1} + e^{-kt_3} + be^{-k(2t_2)} - 2e^{-kt_2} - be^{-2kt_2}\Big]$$

$$= \frac{b}{N^2}\left[e^{-kt_1} + e^{-kt_3} - 2e^{-kt_2}\right]$$

Clearly, $\dfrac{A}{B} = N$.

Hence, $N = \dfrac{\frac{2}{y_1} + \frac{2}{y_3} - \frac{3}{y_2}}{\frac{2}{y_1 y_3} - \frac{2}{y_2^2}}$.

64. $N = \dfrac{\frac{1}{y_1} + \frac{1}{y_3} - \frac{2}{y_2}}{\frac{1}{y_1 y_3} - \frac{1}{y_2^2}}$.

(a) Using the formula for N with $y_1 = 5.3$, $y_2 = 23.2$, and $y_3 = 76.0$, $N \approx 185$. So, the upper bound that the U.S. population was approaching during these years was approximately 185 million.

(b) Using the formula for N with $y_1 = 23.2$, $y_2 = 76.0$, and $y_3 = 150.7$, $N \approx 207$. So, the upper bound that the U.S population was approaching during these years was approximately 207 million.

(c) Using the formula for N with $y_1 = 39.8$, $y_2 = 105.7$, and $y_3 = 203.3$, $N \approx 326$. So, the upper bound that the U.S. population was approaching during these years was approximately 326 million.

(d) Answers will vary.

65. (a) From Exercise 64,

$$N = \frac{\frac{1}{39.8} + \frac{1}{203.3} - \frac{2}{105.7}}{\frac{1}{39.8(203.3)} - \frac{1}{105.7^2}} \approx 326.$$

$y_0 = 39.8$, so

$$b = \frac{N - y_0}{y_0} = \frac{326 - 39.8}{39.8} \approx 7.20.$$

Since 1920 corresponds to $t = 5$ decades,

$$105.7 = \frac{326}{1 + 7.20e^{-5k}}$$

$$1 + 7.20e^{-5k} = \frac{326}{105.7}$$

$$e^{-5k} = \frac{1}{7.20}\left(\frac{326}{105.7} - 1\right)$$

$$-5k = \ln\left[\frac{1}{7.20}\left(\frac{326}{105.7} - 1\right)\right]$$

$$k = -\frac{1}{5}\ln\left[\frac{1}{7.20}\left(\frac{326}{105.7} - 1\right)\right]$$

$$\approx 0.248$$

(b) $y = \dfrac{326}{1 + 7.20e^{-0.248t}}$

In 2010, $t = 14$, so

$$y = \frac{326}{1 + 7.20e^{-0.248t}} \approx 266 \text{ million}$$

The predicated population is 266 million which is less than the table value of 308.7 million.

(c) In 2030, $t = 16$, so

$$y = \frac{326}{1 + 7.20e^{-0.248(16)}} \approx 287 \text{ million.}$$

In 2050, $t = 18$, so

$$y = \frac{326}{1 + 7.20e^{-0.248(18)}} \approx 301 \text{ million.}$$

66. (a)

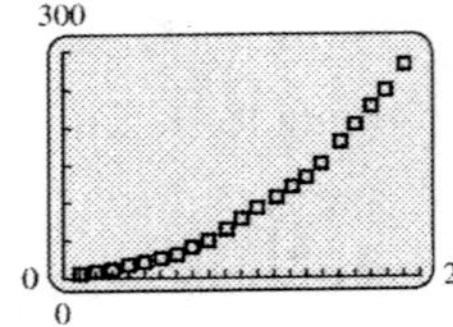

The points suggest the lower portion of a logistic growth curve, so yes, a logistic function seems appropriate.

(b) A calculator with a logistic regression function determines that

$$y = \frac{487}{1 + 58.1e^{-0.208t}}$$

best fits the data.

(c)

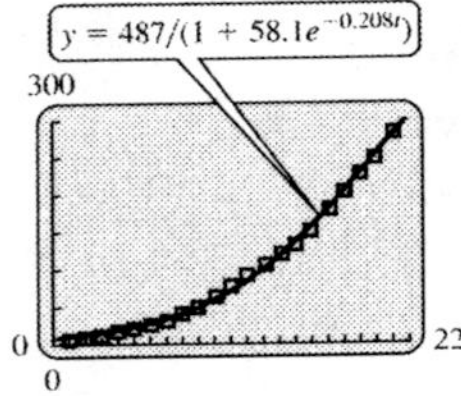

The model does produce appropriate y-values for the given t-values.

(d) As t gets larger and larger, $e^{-0.208t}$ approaches 0, so y approaches

$$\frac{487}{1 + 58.1 \cdot 0} = 487.$$

Therefore, according to this model, the U.S. population has a limiting size of 487 million.

67. (a) Exponential growth model:

$\dfrac{dA}{dt} = kA \Rightarrow A = Me^{kt}$, where $M = e^C$ and k are nonnegative constants.

Limited growth model:

$\dfrac{dy}{dt} = k(N - y) \Rightarrow y = N - Me^{-kt}$, where $M = e^{-C}$ and k are nonnegative constants. Also, $N - y > 0$

Logistic growth model:

$$\frac{dy}{dt} = k\left(1 - \frac{y}{N}\right)y \Rightarrow y = \frac{N}{1 + be^{-kt}}$$

Another way of writing this model is

$G(t) = \dfrac{m}{1 + \left(\frac{m}{G_0}\right)e^{-kmt}}$, where m is the limiting value of the population, G_0 is the initial number present, and k is a positive constant.

All three functions are increasing for all values of t.

(b) The second derivative test can be used for all three models. However, you should know the general shape of the curve for each model.

Exponential growth model: concave up for all t. (See Example 7 in section 2.1 of your text.)

Limited growth model: concave down for all t (See Example 4 in section 2.3 of your text.)

Logistic growth model: concave up for $t < \dfrac{\ln b}{k}$; concave down for $t > \dfrac{\ln b}{k}$. See the discussion in section 11.1 for an explanation.

(c) See section 11.1 along with sections 2.1 and 2.3 in your text for descriptions of each model.

68. (a) $\frac{dx}{dt} = 1 - kx$

Separate the variables and integrate.

$$\int \frac{dx}{1-kx} = \int dt$$

$-\frac{1}{k}\ln|1-kx| = t + C_1$

Solve for x.

$\ln|1-kx| = -kt - kC_1$

$|1-kx| = e^{-kC_1}e^{-kt}$

$1 - kx = Me^{-kt}$, where $M = \pm e^{-kC_1}$.

$$x = -\frac{1}{k}(Me^{-kt} - 1) = \frac{1}{k} + Ce^{-kt},$$

where $C = -\frac{M}{k}$.

(b) Write the linear first-order differential equation in the linear form

$\frac{dx}{dt} + kx = 1.$

The integrating factor is

$I(t) = e^{\int k\,dt} = e^{kt}.$

Multiply both sides of the differential equation by $I(t)$.

$\frac{dx}{dt}e^{kt} + kxe^{kt} = e^{kt}$

Replace the left side of this equation by

$D_t(xe^{kt}) = \frac{dx}{dt}e^{kt} + xke^{kt}.$

$D_t(xe^{kt}) = e^{kt}$

Integrate both sides with respect to t.

$xe^{kt} = \int e^{kt}dt = \frac{1}{k}e^{kt} + C$

Solve for x.

$x = \frac{1}{k} + Ce^{-kt}.$

(c) Since $k > 0$, then as t gets larger and larger, Ce^{-kt} approaches 0, so $\lim_{x\to\infty}\left(\frac{1}{k} + Ce^{-kt}\right) = \frac{1}{k}.$

69. (a) $\frac{dy}{dt} = a(N-y)y;\ N = 200;$

$y_0 = 10;\ y = 35$ when $t = 3.$

The solution to the differential equation is equation 2 in Example1 of Section 6 in this chapter, which is

$y = \frac{N}{1+be^{-kt}}$, where

$b = \frac{N - y_0}{y_0}$ and $k = aN.$

$b = \frac{200-10}{10} = 19;\ k = 200a$

$y = \frac{200}{1+19e^{-200at}}$

Since $y = 35$ when $t = 3$,

$$35 = \frac{200}{1+19e^{-200a(3)}}$$

$$35 + 665e^{-600a} = 200$$

$$e^{-600a} = \frac{165}{665} = \frac{33}{133}$$

$$-600a = \ln\left(\frac{33}{133}\right)$$

$$-200a = \frac{1}{3}\ln\left(\frac{33}{133}\right) \approx -0.4646.$$

Therefore, $y = \frac{200}{1+19e^{-0.4646t}}.$

(b) For $t = 5$ days,

$y = \frac{200}{1+19e^{-0.465(5)}} \approx 69.98$

In 5 days, about 70 people have heard the rumor.

70. $\frac{dT}{dt} = k(T - T_M)$

$\frac{dT}{dt} = k(T - 300°)$

$T(0) = 40°$

$T(1) = 150°$

$\frac{dT}{T-300} = k\,dt$

$\ln|T-300| = kt + C_1$

$|T-300| = e^{kt+C_1} = e^{C_1}e^{kt}$

$T - 300 = Ce^{kt}$

$T = Ce^{kt} + 300$

Since $T(0) = 40°$,

$40 = Ce^0 + 300 \Rightarrow C = -260.$

Therefore, $T = -260e^{kt} + 300.$

(continued on next page)

(continued)

At $T(1) = 150°$,

$$150 = -260e^k + 300$$
$$-150 = -260e^k$$
$$\frac{15}{26} = e^k$$
$$\ln\left(\frac{15}{26}\right) = k \ln e$$
$$k = \ln\left(\frac{15}{26}\right) = -0.55.$$

Therefore, $T = -260e^{-0.55t} + 300.$
At $t = 2$,
$T = -260e^{-0.55(2)} + 300 = 213°.$

71. $\frac{dT}{dt} = k(T - T_M);\ T_M = 300;\ T = 40$
when $t = 0$.
$T = 150$ when $t = 1$.
From Exercise 58 in Section 1 of this chapter, the solution to the differential equation is $T = Ce^{kt} + T_M$, where C is a constant ($-k$ has been replaced by k in this exercise.) Here,
$T = Ce^{kt} + 300.$

$$40 = C + 300 \Rightarrow C = -260$$
$$T = 300 - 260e^{kt}$$
$$150 = 300 - 260e^k \Rightarrow e^k = \frac{15}{26}$$
$$k = \ln\left(\frac{15}{26}\right) \approx -0.55$$
$$T = 300 - 260e^{-0.55t}$$
$$250 = 300 - 260e^{-0.55t}$$
$$e^{-0.55t} = \frac{5}{26} \Rightarrow t = -\frac{1}{0.55}\ln\left(\frac{5}{26}\right) \approx 3\text{ hr}$$

72. **(a)** $\frac{dv}{dt} = G^2 - K^2v^2$

$$dv = (G^2 - K^2v^2)\,dt$$
$$\frac{1}{G^2 - K^2v^2}dv = dt$$
$$\int \frac{1}{G^2 - K^2v^2}dv = \int dt$$
$$\frac{1}{K^2}\int \frac{1}{\left(\frac{G}{K}\right)^2 - v^2} = \int dt$$

Use formula 7 from the table of integrals.

$$\frac{1}{K^2}\cdot\frac{1}{2\frac{G}{K}}\ln\left|\frac{\frac{G}{K} + v}{\frac{G}{K} - v}\right| = t + C_1$$
$$\frac{1}{2GK}\ln\left|\frac{G + Kv}{G - Kv}\right| = t + C_1$$
$$\ln\left|\frac{G + Kv}{G - Kv}\right| = 2GKt + C_2$$

Since $v < \frac{G}{K}$, $Kv < G$ and $\frac{G + Kv}{G - Kv}$ is positive.

$$\ln\left(\frac{G + Kv}{G - Kv}\right) = 2GK(t) + C_2$$

When $t = 0$, $v = 0$, so

$$\ln\left(\frac{G + 0}{G - 0}\right) = 2GK(0) + C_2 \Rightarrow$$
$$\ln 1 = C_2 \Rightarrow C_2 = 0.$$

Thus,

$$\ln\left(\frac{G + Kv}{G - Kv}\right) = 2GKt$$
$$\frac{G + Kv}{G - Kv} = e^{2GKt}$$
$$G + Kv = Ge^{2GKt} - Kve^{2GKt}$$
$$Kv + Kve^{2GKt} = Ge^{2GKt} - G$$
$$vK(e^{2GKt} + 1) = G(e^{2GKt} - 1)$$
$$v = \frac{G(e^{2GKt} - 1)}{K(e^{2GKt} + 1)}$$
$$v = \frac{G}{K}\cdot\frac{e^{2GKt} - 1}{e^{2GKt} + 1}$$

(b) $\lim_{t\to\infty} v = \lim_{t\to\infty}\left(\frac{G}{K}\cdot\frac{e^{2GKt} - 1}{e^{2GKt} + 1}\right)$

$$\lim_{t\to\infty} v = \frac{G}{K}\lim_{t\to\infty}\frac{e^{2GKt} - 1}{e^{2GKt} + 1}$$
$$\lim_{t\to\infty} v = \frac{G}{K}\cdot 1 = \frac{G}{K}$$

A falling object in the presence of air resistance has a limiting velocity, $\frac{G}{K}$.

(c) $\lim_{t\to\infty} v = \frac{G}{k} \Rightarrow 88 = \frac{G}{k} \Rightarrow k = \frac{G}{88} \Rightarrow$

$$Gk = \frac{G^2}{88} \Rightarrow Gk = \frac{32}{88} \Rightarrow$$
$$2Gk = \frac{64}{88} \approx 0.727$$
$$v = \frac{G}{k}\frac{e^{2Gkt} - 1}{e^{2Gkt} + 1} = 88\frac{e^{0.727t} - 1}{e^{0.727t} + 1}$$

Chapter 12

PROBABILITY

12.1 Sets

1. $3 \in \{2,5,7,9,10\}$

The number 3 is not an element of the set, so the statement is false.

3. $\{2,5,8,9\} = \{2,5,9,8\}$

The sets contain exactly the same elements, so they are equal. The statement is true.

5. $0 \in \emptyset$

The empty set has no elements. The statement is false.

In Exercises 7–12,

$$A = \{2,4,6,10,12\},$$
$$B = \{2,4,8,10\},$$
$$C = \{4,8,12\},$$
$$D = \{2,10\},$$
$$E = \{6\},$$
$$U = \{2,4,6,8,10,12,14\}.$$

7. Since every element of A is also an element of U, A is a subset of U, written $A \subseteq U$.

9. A contains elements that do not belong to E, namely 2, 4, 10, and 12, so A is not a subset of E, written $A \not\subseteq E$.

11. The empty set is a subset of every set, so $\emptyset \subseteq A$.

13. Since every element of A is also an element of U, and $A \neq U$, $A \boxed{\subset} U$.

Since every element of E is also an element of A, and $E \neq A$, $E \boxed{\subset} A$.

Since every element of A is not also an element of E, $A \boxed{\not\subset} E$.

Since every element of B is not also an element of C, $B \boxed{\not\subset} C$.

Since $\emptyset$ is a subset of every set, and $\neq A$, $\emptyset \boxed{\subset} A$.

Since every element of $\{0,2\}$ is not also an element of D, $\{0,2\} \boxed{\not\subset} D$.

15. A set with n distinct elements has 2^n subsets. A has $n = 5$ elements, so there are exactly $2^5 = 32$ subsets of A.

17. Since $\{7,9\}$ is the set of elements belonging to both sets, which is the intersection of the two sets, we write $\{5,7,9,19\} \cap \{7,9,11,15\} = \{7,9\}$.

19. Since $\{1,2,5,7,9\}$ is the set of elements belonging to one or the other (or both) of the listed sets, it is their union.
$\{2,1,7\} \cup \{1,5,9\} = \{1,2,5,7,9\}$

21. Since $\emptyset$ contains no elements, there are no elements belonging to both sets. Thus, the intersection is the empty set, and we write $\{3,5,9,10\} \cap \emptyset = \emptyset$.

23. $\{1,2,4\}$ is the set of elements belonging to both sets, and $\{1,2,4\}$ is also the set of elements in the first set or in the second set or possibly both. Thus,
$\{1,2,4\} \cap \{1,2,4\} = \{1,2,4\}$ and
$\{1,2,4\} \cup \{1,2,4\} = \{1,2,4\}$ are both true statements.

25. The intersection of two sets is the set consisting of the elements that the original sets have in common. The union of two sets is the set consisting of all the elements of the original sets.

In Exercises 27–34,

$$U = \{1,2,3,4,5,6,7,8,9\},$$
$$X = \{2,4,6,8\},$$
$$Y = \{2,3,4,5,6\},$$
$$Z = \{1,2,3,8,9\}.$$

27. $X \cap Y$, the intersection of X and Y, is the set of elements belonging to both X and Y. Thus, $X \cap Y = \{2,4,6,8\} \cap \{2,3,4,5,6\} = \{2,4,6\}$.

29. X', the complement of X, consists of those elements of U that are not in X. Thus, $X' = \{1,3,5,7,9\}$.

31. First find $X \cup Z$.

$X \cup Z = \{2,4,6,8\} \cup \{1,2,3,8,9\}$
$= \{1,2,3,4,6,8,9\}$

Now find $Y \cap (X \cup Z)$.

$Y \cap (X \cup Z) = \{2,3,4,5,6\} \cap \{1,2,3,4,6,8,9\}$
$= \{2,3,4,6\}$

33. $Z' = \{4,5,6,7\}$, $Y' = \{1,7,8,9\}$.

$(X \cap Y') \cup (Z' \cap Y')$
$= (\{2,4,6,8\} \cap \{1,7,8,9\})$
$\cup (\{4,5,6,7\} \cap \{1,7,8,9\})$
$= \{8\} \cup \{7\} = \{7,8\}$

35. $(A \cap B) \cup (A \cap B')$
$= (\{3,6,9\} \cap \{2,4,6,8\})$
$\cup (\{3,6,9\} \cap \{0,1,3,5,7,9,10\})$
$= \{6\} \cup \{3,9\} = \{3,6,9\} = A$

37. $A = \{1,2,3,\{3\},\{1,4,7\}\}$

(a) $1 \in A$ is true.

(b) $\{3\} \in A$ is true.

(c) $\{2\} \in A$ is false. $(\{2\} \subseteq A)$

(d) $4 \in A$ is false. $(4 \in \{1,4,7\})$

(e) $\{\{3\}\} \subset A$ is true.

(f) $\{1,4,7\} \in A$ is true.

(g) $\{1,4,7\} \subseteq A$ is false. $(\{1,4,7\} \in A)$

39. $B \cap A'$ is the set of all elements in B *and* not in A.

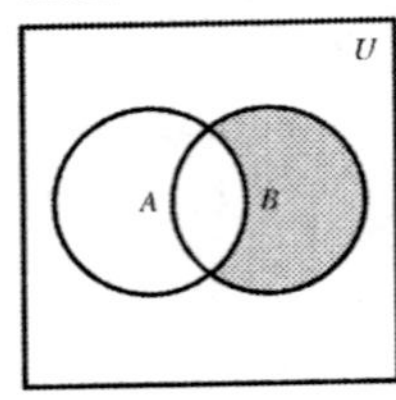

$B \cap A'$

41. $A' \cup B$ is the set of all elements that do not belong to A *or* that do belong to B, or both.

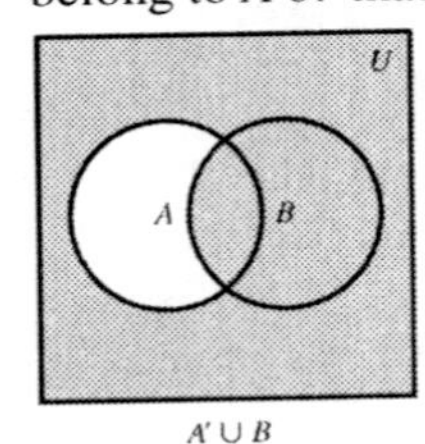

$A' \cup B$

43. First find $A' \cap B'$, the set of elements not in A *and* not in B. Then, for the union, we want those elements in B' *or* $(A' \cap B')$, or both.

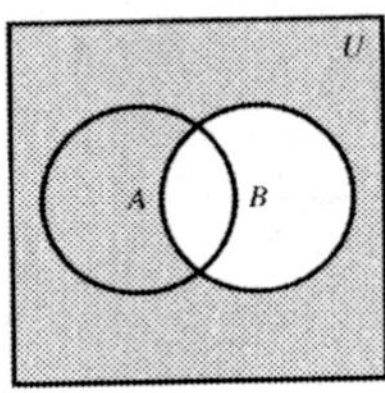

$B' \cup (A' \cap B')$

45. U' is the empty set $\emptyset$.

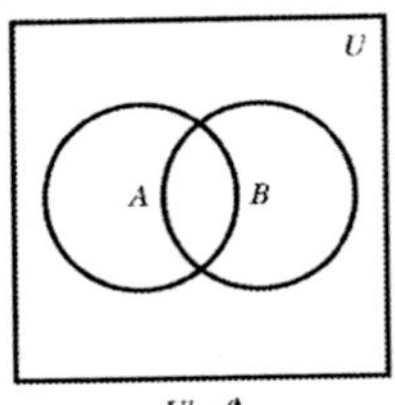

$U' = \emptyset$

47. Three sets divide the universal set into at most 8 regions. (Examples of this situation will be seen in Exercises 49–56.)

49. $(A \cap B) \cap C$

First form the intersection of A with B. Then form the intersection of $A \cap B$ with C. The result will be the set of all elements that belong to all three sets.

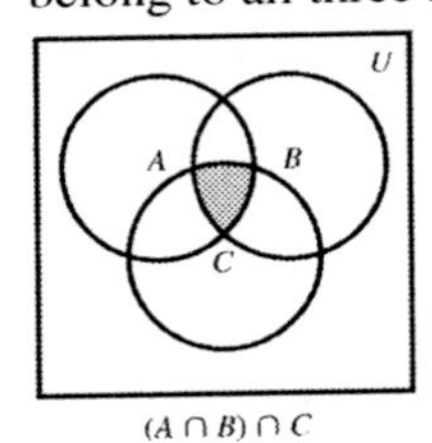

$(A \cap B) \cap C$

51. $A \cap (B \cup C')$

C' is the set of all elements in U that are not elements of C. Next, form the union of C' with B. Finally, find the intersection of this region with A.

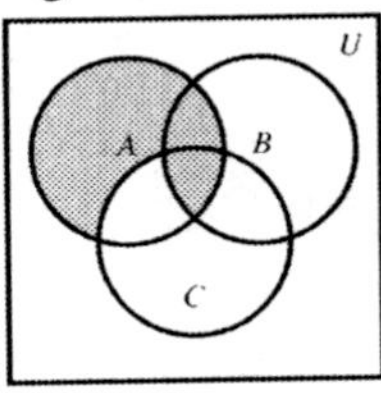

$A \cap (B \cup C')$

53. $(A' \cap B') \cap C'$

$A' \cap B'$ is the part of the universal set not in A *and* not in B. C' is the part of the universal set not in C. $(A' \cap B') \cap C'$ is the intersection of these two regions.

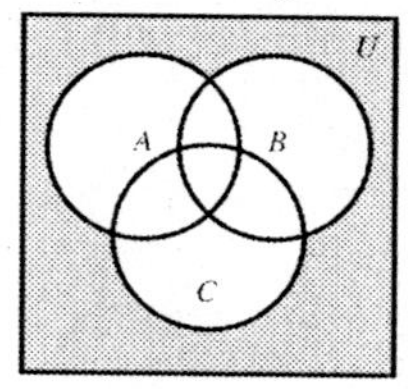

$(A' \cap B') \cap C'$

55. $A' \cap (B' \cup C)$

First find A'. Then find $B' \cup C$, the region not in B *or* in C, or both. Finally, intersect these regions.

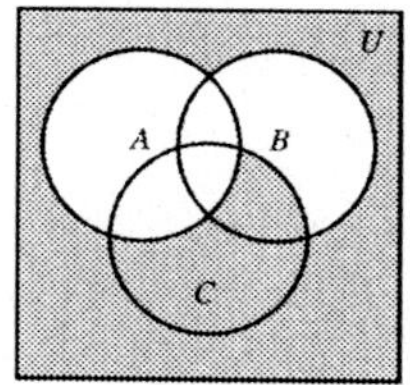

$A' \cap (B' \cup C)$

57. $n(A \cup B) = n(A) + n(B) - n(A \cap B)$
$= 5 + 12 - 4 = 13$

59. $n(A \cup B) = n(A) + n(B) - n(A \cap B)$
$22 = n(A) + 9 - 5$
$22 = n(A) + 4$
$18 = n(A)$

61. $n(U) = 41$, $n(A) = 16$, $n(A \cap B) = 12$, $n(B') = 20$

First put 12 in $A \cap B$. Since $n(A) = 16$, and 12 are in $A \cap B$, there must be 4 elements in A that are not in $A \cap B$. $n(B') = 20$, so there are 20 not in B. We already have 4 not in B (but in A), so there must be another 16 outside B *and* outside A. So far we have accounted for 32, and $n(U) = 41$, so 9 must be in B but not in any region yet identified. Thus $n(A' \cap B) = 9$.

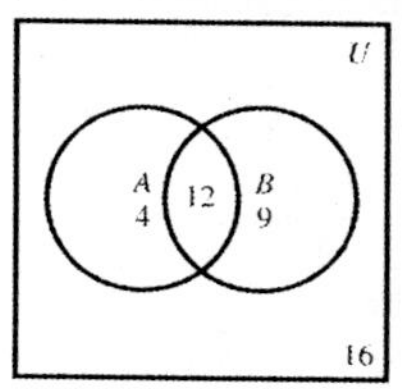

63. $n(A \cup B) = 24$, $n(A \cap B) = 6$, $n(A) = 11$, $n(A' \cup B') = 25$

Start with $n(A \cap B) = 6$. Since $n(A) = 11$, there must be 5 more in A not in B. $n(A \cup B) = 24$; we already have 11, so 13 more must be in B not yet counted. $A' \cup B'$ consists of all the region not in $A \cap B$, where we have 6. So far 5 + 13 = 18 are in this region, so another 25 – 18 = 7 must be outside both A and B.

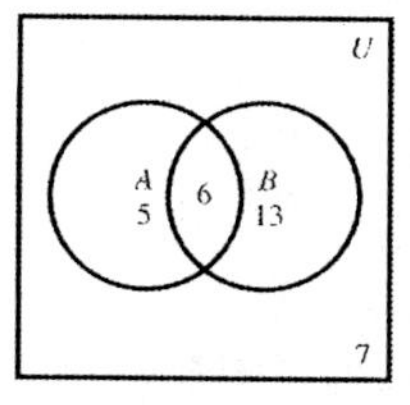

65. $n(A) = 28$, $n(B) = 34$, $n(C) = 25$
$n(A \cap B) = 14$, $n(B \cap C) = 15$, $n(A \cap C) = 11$
$n(A \cap B \cap C) = 9$, $n(U) = 59$

We start with $n(A \cap B \cap C) = 9$. If $n(A \cap B) = 14$, an additional 5 are in $A \cap B$ but not in $A \cap B \cap C$. Similarly, $n(B \cap C) = 15$, so $15 - 9 = 6$ are in $B \cap C$ but not in $A \cap B \cap C$. Also, $n(A \cap C) = 11$, so $11 - 9 = 2$are in $A \cap C$ but not in $A \cap B \cap C$. Now we turn our attention to $n(A) = 28$. So far we have 2 + 9 + 5 = 16 in A; there must be another 28 – 16 = 12 in A not yet counted. Similarly, $n(B) = 34$; we have 5 + 9 + 6 = 20 so far, and 34 – 20 = 14 more must be put in B. For C, $n(C) = 25$; we have 2 + 9 + 6 = 17 counted so far. Then there must be 8 more in C not yet counted. The count now stands at 56, and $n(U) = 59$, so 3 must be outside the three sets.

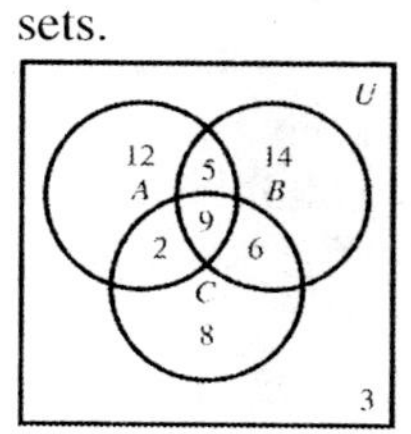

67. $n(A \cap B) = 6,\ n(A \cap B \cap C) = 4$
$n(A \cap C) = 7,\ n(B \cap C) = 4$
$n(B \cap C') = 11,\ n(B \cap C') = 8$
$n(C) = 15,\ n(A' \cap B' \cap C') = 5$

Start with $n(A \cap B) = 6$ and $n(A \cap B \cap C) = 4$ to get $6 - 4 = 2$ in that portion of $A \cap B$ outside of C. From $n(B \cap C) = 4,$ there are $4 - 4 = 0$ elements in that portion of $B \cap C$ outside of A. Use $n(A \cap C) = 7$ to get $7 - 4 = 3$ elements in that portion of $A \cap C$ outside of B. Since $n(A \cap C') = 11,$ there are $11 - 2 = 9$ elements in that part of A outside of B and C. Use $n(B \cap C') = 8$ to get $8 - 2 = 6$ elements in that part of B outside of A and C. Since $n(C) = 15,$ there are $15 - 3 - 4 - 0 = 8$ elements in C outside of A and B. Finally, 5 must be outside all three sets, since $n(A' \cap B' \cap C') = 5.$

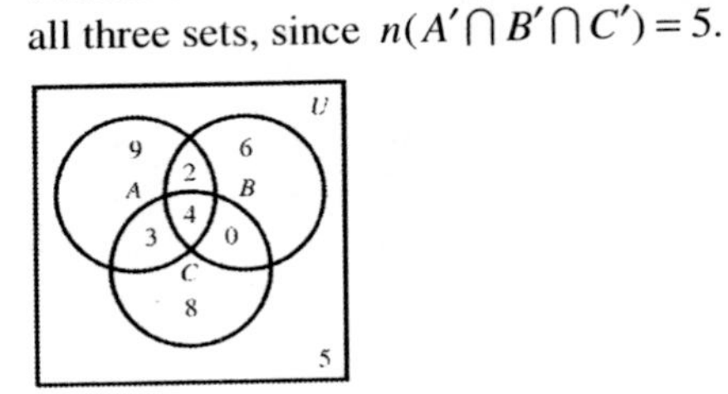

69. $(A \cup B)' = A' \cap B'$

For $(A \cup B)',$ first find $A \cup B.$ Then find $(A \cup B)',$ the region outside $A \cup B.$

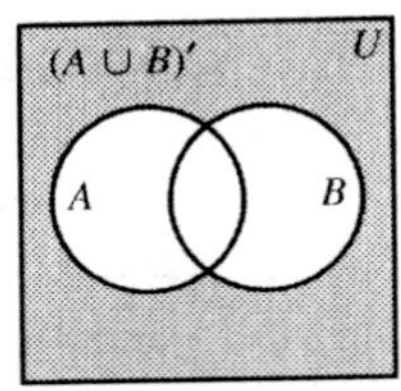

For $A' \cap B',$ first find A' and B' individually. Then $A' \cap B'$ is the region where A' and B' overlap, which is the entire region outside $A \cup B$ (the same result as before).

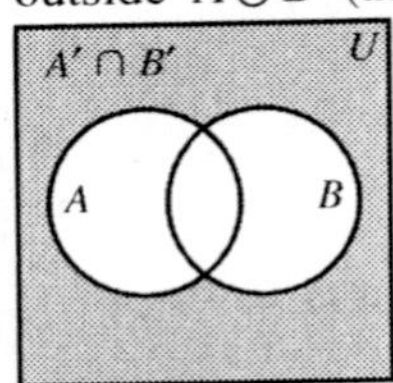

Therefore, $(A \cup B)' = A' \cap B'.$

71. $A \cap (B \cup C) = (A \cap B) \cup (A \cap C)$

First find A and $B \cup C$ individually.

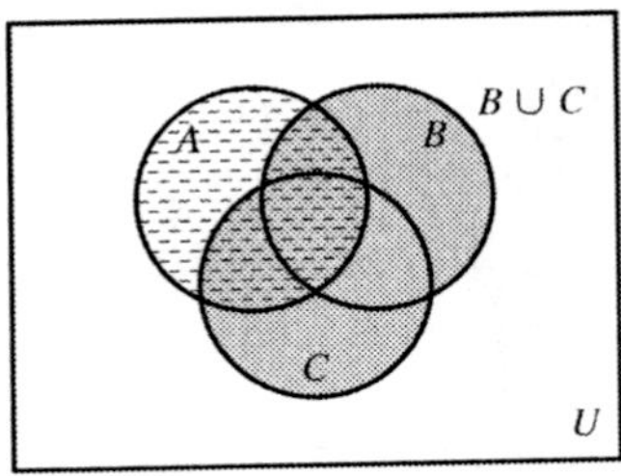

Then $A \cap (B \cup C)$ is the region where the above two diagram overlap.

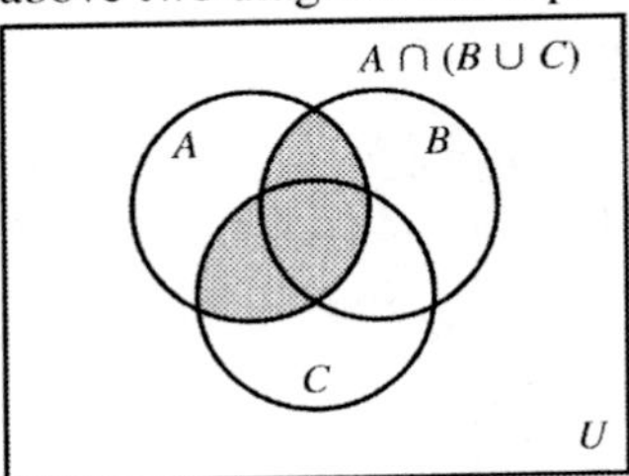

Next find $A \cap B$ and $A \cap C$ indivisually.

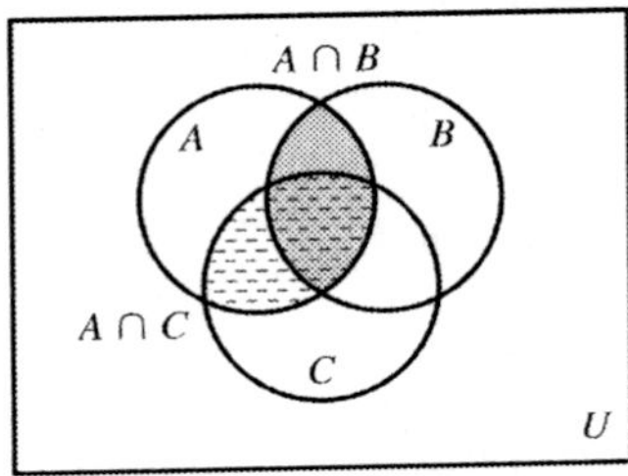

Then $(A \cap B) \cup (A \cap C)$ is the union of the above two diagrams.

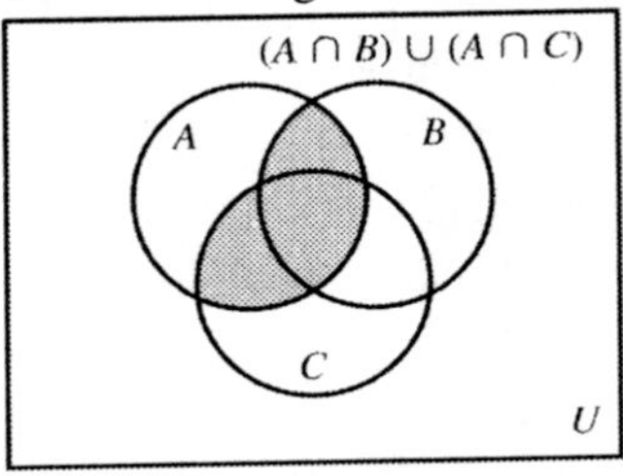

The Venn diagram for $A \cap (B \cup C)$ is identical to the Venn diagram for $(A \cap B) \cup (A \cap C)$ so conclude that $A \cap (B \cup C) = (A \cap B) \cup (A \cap C).$

73. Prove

$n(A \cup B \cup C)$
$= n(A) + n(B) + n(C) - n(A \cap B) - n(A \cap C)$
$- n(B \cap C) + n(A \cap B \cap C)$

$n(A \cup B \cup C)$
$= n[A \cup (B \cup C)]$
$= n(A) + n(B \cup C) - n[A \cap (B \cup C)]$
$= n(A) + n(B) + n(C) - n(B \cap C)$
$- n[(A \cap B) \cup (A \cap C)]$
$= n(A) + n(B) + n(C) - n(B \cap C)$
$- \{n(A \cap B) + n(A \cap C)$
$- n[(A \cap B) \cap (A \cap C)]\}$
$= n(A) + n(B) + n(C) - n(B \cap C) - n(A \cap B)$
$- n(A \cap C) + n(A \cap B \cap C)$

75. (a) The blood has the A antigen but is Rh negative and has no B antigen. This blood type is A-negative.

(b) Both A and B antigens are present and the blood is Rh negative. This blood type is AB-negative.

(c) Only the B antigen is present. This blood type is B-negative.

(d) Both A and Rh antigens are present. This is blood type is A-positive.

(e) All antigens are present. This blood type is AB-positive.

(f) Both B and Rh antigens are present. This blood type is B-positive.

(g) Only the Rh antigen is present. This blood type is O-positive.

(h) No antigens at all are present. This blood type is O-negative.

77. Extend the table to include totals for each row.

	W	B	I	A	Total
F	1,063,235	141,157	7049	24,562	1,236,003
M	1,051,514	145,802	8516	26,600	1,232,432

(a) $n(F)$ is the total for the first row in the table. $n(F) = 1,236,003$. Thus, there are 1,236,003 people in the set F.

(b) $n(F \cap (I \cup A))$ is the sum of the entries in the first row in the I and A columns. Therefore, $n(F \cap (I \cup A)) = n(F \cap I) + n(F \cap A) =$ 31,611. Therefore, there are 31,611 people in the set $F \cap (I \cup A)$.

(c) $n(M \cup B) = n(M) + n(B) - n(M \cap B)$
$= 1,232,432 + (141,157 + 145,802)$
$- 145,802$
$= 1,373,589$
There are 1,373,589 people in the set $M \cup B$.

(d) $W' \cup I' \cup A'$ is the universe, since each person is either *not* white, or *not* American Indian, or *not* Asian or Pacific Islander. Thus, there are 2,468,435 people in the set $W' \cup I' \cup A'$

(e) $F \cap (I \cup A)$ represents females who are American Indian or Asian/Pacific Islander.

79. (a) Reading directly from the table, $n(A \cap F) = 110.6$. Thus, there are 110.6 million people in the set $A \cap F$.

(b) $n(G \cup B) = n(G) + n(B) - n(G \cap B)$
$= 80.4 + 52.6 - 10.3$
$= 122.7$
Thus, there are 122.7 million people in the set $G \cup B$.

(c) $n(G \cup (C \cap H)) = n(G) + n(C \cap H)$
$= 80.4 + 5.0 = 85.4$
There are 85.4 million people in the set $G \cup (C \cap H)$.

(d) The only intersection of the set F and the set $B \cup H$ is the set $F \cap B$. Therefore, $n(F \cap (B \cup H)) = n(F \cap B) = 37.6$. There are 37.6 million people in the set $F \cap (B \cup H)$.

(e) $n(H \cup D) = n(H) + n(D) - n(H \cap D)$
$= 53.6 + 19.6 - 2.2 = 71.0$
There are 71.0 million people in the set $H \cup D$.

(f) First of all, $A' \cap C' = B \cup D \cup E$ since the only people *not* in set A and *not* in the set C are the people in set B or set D or set E. Second $G' = F \cup H$, since the only people *not* in the set G are either in the set F or the set H. Thus, the set $G' \cap (A' \cap C')$ consists of people in either F or H and also in B, D, or E.

(continued on next page)

(continued)

Therefore,
$nG' \cap (A' \cap C'))$
$= n(F \cap B) + n(F \cap D) + n(F \cap E)$
$+ n(H \cap B) + n(H \cap D) + n(H \cap E)$
$= 37.6 + 13.1 + 2.2 + 4.7 + 2.2 + 0.3$
$= 60.1$ million people

81. **(a)** $n(A \cup B)$ is the total number of female personnel serving in the Army or the Air Force.
$n(A \cup B) = n(A) + n(B)$
$= 74{,}182 + 63{,}131$
$= 137{,}313$

(b) $E \cup (C \cup D)$ includes all enlisted personnel, and all personnel in the Navy and Marines. In order not to count any person twice, we take $n(E)$ and add $n(C \cap E')$ and $n(D \cap E')$. This gives us 164,046 + (8634 + 147) + (1348 + 0) = 174,175
There are 174,175 people in the set $E \cup (C \cup D)$.

(c) $O' \cap M' = E,$ since the enlisted are the only group who are both not officers and not cadets or midshipmen. Thus,
$n(O' \cap M') = n(E) = 164{,}046$
There are 164,046 people in the set $O' \cap M'$.

83. Let *W* be the set of women, *C* be the set of those who speak Cantonese, and *F* be the set of those who set off firecrackers. We are given the following information.
$n(W) = 120,\ n(C) = 150,\ n(F) = 170$
$n(W' \cap C) = 108,\ n(W' \cap F') = 100$
$n(W \cap C' \cap F) = 18,\ n(W' \cap C' \cap F') = 78$
$n(W \cap C \cap F) = 30$
Note that
$n(W' \cap C \cap F') = n(W' \cap F') - n(W' \cap C' \cap F')$
$= 100 - 78 = 22.$
Furthermore,
$n(W' \cap C \cap F) = n(W' \cap C) - n(W' \cap C \cap F')$
$= 108 - 22 = 86.$
We now have
$n(W \cap C \cap F')$
$= n(C) - n(W' \cap C \cap F) - n(W \cap C \cap F)$
$- n(W' \cap C \cap F')$
$= 150 - 86 - 30 - 22 = 12.$
With all of the overlaps of *W*, *C*, and *F* determined, we can now compute
$n(W \cap C' \cap F') = 60$ and $n(W' \cap C' \cap F) = 36.$

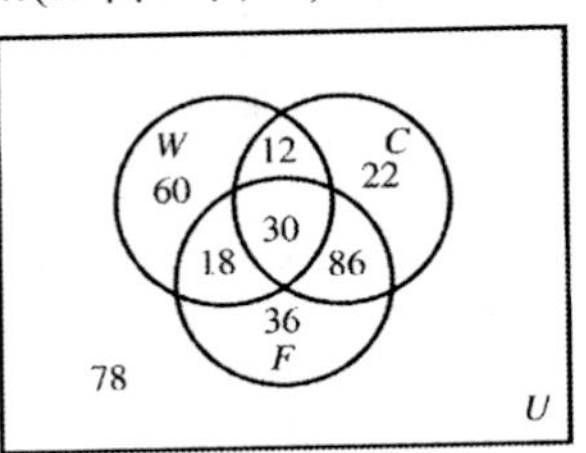

(a) Adding up the disjoint components, we find the total attendance to be
$60 + 12 + 18 + 30 + 22 + 86 + 36 + 78 = 342.$

(b) $n(C') = 342 - n(C) = 342 - 150 = 192$

(c) $n(W \cap F') = 60 + 12 = 72$

(d) $n(W' \cap C \cap F) = 86$

85. Let *A* be the set of trucks that carried early peaches, *B* be the set of trucks that carried late peaches, and *C* be the set of trucks that carried extra late peaches. We are given the following information.
$n(A) = 34,\ n(B) = 61,\ n(C) = 50$
$n(A \cap B) = 25,\ n(B \cap C) = 30$
$n(A \cap C) = 8,\ n(A \cap B \cap C) = 6$
$n(A' \cap B' \cap C') = 9$
Start with $A \cap B \cap C.$ We know that $n(A \cap B \cap C) = 6.$
Since $n(A \cap B) = 25,$ the number in $A \cap B$ but not in *C* is 25 – 6 = 19.
Since $n(B \cap C) = 30,$ the number in $B \cap C$ but not in *A* is 30 – 6 = 24.
Since $n(A \cap C) = 8,$ the number in $A \cap C$ but not in *B* is 8 – 6 = 2.
Since $n(A) = 34,$ the number in *A* but not in *B* or *C* is 34 – (19 + 6 + 2) = 7.
Since $n(B) = 61,$ the number in *B* but not in *A* or *C* is 61 – (19 + 6 + 24) = 12.
Since $n(C) = 50,$ the number in *C* but not in *A* or *B* is 50 – (24 + 6 + 2) = 18.
Since $n(A' \cap B' \cap C') = 9,$ the number outside $A \cup B \cup C$ is 9.

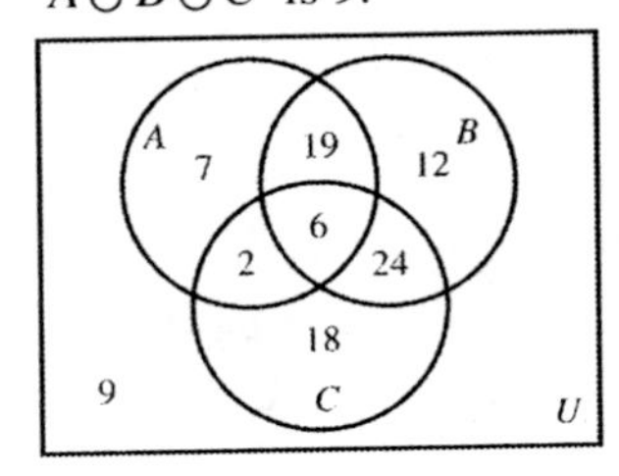

(continued on next page)

(*continued*)

(a) From the Venn diagram, 12 trucks carried only late peaches.

(b) From the Venn diagram, 18 trucks carried only extra late peaches.

(c) From the Venn diagram, 7 + 12 + 18 = 37 trucks carried only one type of peach.

(d) From the Venn diagram, $6+2+19+24+7+12+18+9=97$ trucks went out during the week.

87. Let C represent the total number of children,
B represent the number of children living with both parents,
N represent the number of children living with neither parent,
F represent the number of children living with their father only, and
M represent the number of children living with their mother only. Then,
$C = B + N + F + M$ or $C - B - N - F = M$
$M = 73{,}817 - 50{,}267 - 2634 - 2924 = 17{,}992$
In 2012, 17,992 thousand children lived with their mother only.

89. The number of subsets of a set with 51 elements (50 states plus the District of Columbia) is $2^{51} \approx 2.522 \times 10^{15}$.

91. Joe should always first choose the complement of what Dorothy chose. This will leave only two sets to choose from, and Joe will get the last choice.

93. **(a)** $(A \cup B)' \cap C$

$A \cup B$ is the set of states whose name contains the letter e or has a population over 4,000,000. Therefore, $(A \cup B)'$ is the set of states that are not among those whose name contains the letter e or has a population over 4,000,000. As a result, $(A \cup B)' \cap C$ is the set of states that are not among those whose name contains the letter e or has a population over 4,000,000, and that also have an area over 40,000 square miles.

(b) $(A \cup B)'$
= {(Kentucky, Maine, Nebraska, New Jersey} $\cup$ {Alabama, Colorado, Florida, Indiana, Kentucky, New Jersey})′
= {Alabama, Colorado, Florida, Indiana, Kentucky, Maine, Nebraska, New Jersey}′
= {Alaska, Hawaii}

$(A \cup B)' \cap C$
= {Alaska, Hawai} $\cap$ {Alabama, Alaska, Colorado, Florida, Kentucky, Nebraska}
= {Alaska}

12.2 Introduction to Probability

1. The set of all possible outcomes for an experiment is the sample space for that experiment.

3. The sample space is the set of the twelve months, {January, February, March, . . ., December}.

5. Let "surgery" be the outcome "have surgery," "medicine" be "treat with medicine," and "wait" be "wait six months and see." The sample space is the set {surgery, medicine, wait}.

7. Let h = heads and t = tails for the coin; the die can display 6 different numbers. There are 12 possible outcomes in the sample space, which is the set
$\{(h,1),(t,1),(h,2),(t,2),(h,3),(t,3),(h,4),(t,4),(h,5),(t,5),(h,6),(t,6)\}$.

9. Use the first letter of each name. The sample space is the set
$S = \{\text{AB, AC, AD, AE, BC, BD, BE, CD, CE, DE}\}$.
$n(S) = 10$. Assuming the committee is selected at random, the outcomes are equally likely.

(a) One of the committee members must be Chinn. This event is {AC, BC, CD, CE}.

(b) Alam, Bartolini, and Chinn may be on any committee; Dickson and Ellsberg may not be on the same committee. This event is {AB, AC, AD, AE, BC, BD, BE, CD, CE}.

(c) Both Alam and Chinn are on the committee. This event is {AC}.

11. Each outcome consists of two of the numbers 1, 2, 3, 4, and 5, without regard for order. For example, let (2, 5) represent the outcome that the slips of paper marked with 2 and 5 are drawn.

$S = \{(1,2), (1,3), (1,4), (1,5), (2,3), (2,4), (2,5), (3,4), (3,5), (4,5)\}.$

$n(S) = 10$. The outcomes are equally likely.

(a) Both numbers in the outcome pair are even. This event is $\{(2, 4)\}$, which is called a simple event since it consists of only one outcome.

(b) One number in the pair is even and the other number is odd. This event is $\{(1,2),(1,4),(2,3),(2,5),(3,4),(4,5)\}$.

(c) Each slip of paper has a different number written on it, so it is not possible to draw two slips marked with the same number. This event is $\emptyset$, which is called an impossible event since it contains no outcomes.

13. $S = \{HH, THH, HTH, TTHH, THTH, HTTH, TTTH, TTHT, THTT, HTTT, TTTT\}$

$n(S) = 11$. The outcomes are not equally likely.

(a) The coin is tossed four times. This event is written $\{TTHH, THTH, HTTH, TTTH, TTHT, THTT, HTTT, TTTT\}$.

(b) Exactly two heads are tossed. This event is written $\{HH, THH, HTH, TTHH, THTH, HTTH\}$.

(c) No heads are tossed. This event is written $\{TTTT\}$.

For Exercises 15–20, use the sample space

$$S = \{1, 2, 3, 4, 5, 6\}.$$

15. "Getting a 2" is the event $E = \{2\}$, so $n(E) = 1$ and $n(S) = 6$. If all the outcomes in a sample space S are equally likely, then the probability of an event E is $P(E) = \frac{n(E)}{n(S)}$.

In this problem, $P(E) = \frac{n(E)}{n(S)} = \frac{1}{6}$.

17. "Getting a number less than 5" is the event $E = \{1, 2, 3, 4\}$, so $n(E) = 4$.

$$P(E) = \frac{4}{6} = \frac{2}{3}.$$

19. "Getting a 3 or a 4" is the event $E = \{3, 4\}$, so $n(E) = 2$.

$$P(E) = \frac{2}{6} = \frac{1}{3}.$$

For Exercises 21–28, the sample space contains all 52 cards in the deck, so $n(S) = 52$.

21. Let E be the event "a 9 is drawn." There are four 9's in the deck, so $n(E) = 4$.

$$P(9) = P(E) = \frac{n(E)}{n(S)} = \frac{4}{52} = \frac{1}{13}$$

23. Let F be the event "a black 9 is drawn." There are two black 9's in the deck, so $n(F) = 2$.

$$P(\text{black } 9) = P(F) = \frac{n(F)}{n(S)} = \frac{2}{52} = \frac{1}{26}$$

25. Let H be the event "a 2 or a queen is drawn." There are four 2's and four queens in the deck, so $n(H) = 8$.

$$P(2 \text{ or queen}) = P(H) = \frac{n(H)}{n(S)} = \frac{8}{52} = \frac{2}{13}$$

27. Let E be the event "a red card or a ten is drawn." There are 26 red cards and 4 tens in the deck. But 2 tens are red cards and are counted twice.

$$n(E) = n(\text{red cards}) + n(\text{tens}) - n(\text{red tens}) = 26 + 4 - 2 = 28$$

Now calculate the probability of E.

$$P(\text{red cards or ten}) = \frac{n(E)}{n(S)} = \frac{28}{52} = \frac{7}{13}$$

For Exercises 29–32, the sample space consists of all the marbles in the jar. There are 3 + 4 + 5 + 8 = 20 marbles, so $n(S) = 20$.

29. 3 of the marbles are white, so $P(\text{white}) = \frac{3}{20}$.

31. 3 + 4 + 5 = 12 of the marbles are not black, so

$$P(\text{not black}) = \frac{12}{20} = \frac{3}{5}.$$

33. Events E and F are mutually exclusive events if $E \cap F = \emptyset$.

35. A person can own a dog and own a smartphone at the same time. No, these events are not mutually exclusive.

37. A person cannot be a teenager and be 70 years old at the same time. Yes, these events are mutually exclusive.

For exercises 39–44, refer to Example 8 in the text.

39. When two dice are rolled, there are 36 equally likely outcomes.

(a) Of the 36 ordered pairs, there is only one for which the sum is 2, namely $\{(1,1)\}$.

Thus, $P(\text{sum is } 2) = \frac{1}{36}.$

(b) $\{(1,3),(2,2),(3,1)\}$ comprise the ways of getting a sum of 4. Thus,

$P(\text{sum is } 4) = \frac{3}{36} = \frac{1}{12}.$

(c) $\{(1,4),(2,3),(3,2),(4,1)\}$ comprise the ways of getting a sum of 5. Thus,

$P(\text{sum is } 5) = \frac{4}{36} = \frac{1}{9}.$

(d) $\{(1,5),(2,4),(3,3),(4,2),(5,1)\}$ comprise the ways of getting a sum of 6. Thus,

$P(\text{sum is } 6) = \frac{5}{36}.$

41. When two dice are rolled there are 36 equally likely outcomes.

(a) Here, the event is the union of four mutually exclusive events, namely, the sum is 9, the sum is 10, the sum is 11, and the sum is 12. Hence,

$$\begin{aligned} &P(\text{sum is 9 or more}) \\ &\quad = P(\text{sum is } 9) + P(\text{sum is } 10) \\ &\qquad + P(\text{sum is } 11) + (\text{sum is } 12) \\ &\quad = \frac{4}{36} + \frac{3}{36} + \frac{2}{36} + \frac{1}{36} = \frac{10}{36} = \frac{5}{18}. \end{aligned}$$

(b)
$$\begin{aligned} &P(\text{sum is less than } 7) \\ &\quad = P(\text{sum is } 1) + P(\text{sum is } 2) \\ &\qquad + P(\text{sum is } 3) + P(\text{sum is } 4) \\ &\qquad + P(\text{sum is } 5) + P(\text{sum is } 6) \\ &\quad = \frac{0}{36} + \frac{1}{36} + \frac{2}{36} + \frac{3}{36} + \frac{4}{36} + \frac{5}{36} \\ &\quad = \frac{15}{36} = \frac{5}{12} \end{aligned}$$

(c)
$$\begin{aligned} &P(\text{sum is between 5 and 8}) \\ &\quad = P(\text{sum is } 6) + P(\text{sum is } 7) \\ &\quad = \frac{5}{36} + \frac{6}{36} \\ &\quad = \frac{11}{36} \end{aligned}$$

43.
$$\begin{aligned} &P(\text{first die is 3 or sum is 8}) \\ &\quad = P(\text{first die is 3}) + P(\text{sum is 8}) \\ &\qquad - P(\text{first die is 3 and sum is 8}) \\ &\quad = \frac{6}{36} + \frac{5}{36} - \frac{1}{36} = \frac{10}{36} = \frac{5}{18} \end{aligned}$$

45. (a) The events E, "9 is drawn," and F, "10 is drawn," are mutually exclusive, so $P(E \cap F) = 0$. Using the union rule,

$$\begin{aligned} P(9 \text{ or } 10) &= P(9) + P(10) \\ &= \frac{4}{52} + \frac{4}{52} = \frac{8}{52} = \frac{2}{13}. \end{aligned}$$

(b)
$$\begin{aligned} &P(\text{red or } 3) \\ &\quad = P(\text{red}) + P(3) - P(\text{red and } 3) \\ &\quad = \frac{26}{52} + \frac{4}{52} - \frac{2}{52} = \frac{28}{52} = \frac{7}{13} \end{aligned}$$

(c) These events are mutually exclusive.

$$\begin{aligned} P(9 \text{ or black } 10) &= P(9) + P(\text{black } 10) \\ &= \frac{4}{52} + \frac{2}{52} = \frac{6}{52} = \frac{3}{26}. \end{aligned}$$

(d)
$$\begin{aligned} &P(\text{heart or black}) \\ &\quad = P(\text{heart}) + P(\text{black}) \\ &\qquad - P(\text{heart and black}) \\ &\quad = \frac{13}{52} + \frac{26}{52} - \frac{0}{52} = \frac{39}{52} = \frac{3}{4}. \end{aligned}$$

(e)
$$\begin{aligned} &P(\text{face card or diamond}) \\ &\quad = P(\text{face card}) + P(\text{diamond}) \\ &\qquad - P(\text{face card and diamond}) \\ &\quad = \frac{12}{52} + \frac{13}{52} - \frac{3}{52} = \frac{22}{52} = \frac{11}{26} \end{aligned}$$

47. (a) Since these events are mutually exclusive,

$$\begin{aligned} P(\text{brother or uncle}) &= P(\text{brother}) + P(\text{uncle}) \\ &= \frac{2}{13} + \frac{3}{13} = \frac{5}{13}. \end{aligned}$$

(b) Since these events are mutually exclusive,

$$\begin{aligned} &P(\text{brother or cousin}) \\ &\quad = P(\text{brother}) + P(\text{cousin}) \\ &\quad = \frac{2}{13} + \frac{5}{13} = \frac{7}{13}. \end{aligned}$$

(c) Since these events are mutually exclusive,

$$\begin{aligned} &P(\text{brother or mother}) \\ &\quad = P(\text{brother}) + P(\text{mother}) \\ &\quad = \frac{2}{13} + \frac{1}{13} = \frac{3}{13}. \end{aligned}$$

49. **(a)** There are 5 possible numbers on the first slip drawn, and for each of these, 4 possible numbers on the second, so the sample space contains $5 \cdot 4 = 20$ ordered pairs. Two of these ordered pairs have a sum of 9: (4, 5) and (5, 4). Thus,

$$P(\text{sum is } 9) = \frac{2}{20} = \frac{1}{10}.$$

(b) The outcomes for which the sum is 5 or less are (1, 2), (1, 3), (1, 4), (2, 1), (2, 3), (3, 1), (3, 2), and (4,1). Thus,

$$P(\text{sum is 5 or less}) = \frac{8}{20} = \frac{2}{5}.$$

(c) Let A be the event "the first number is 2" and B the event "the sum is 6." Use the union rule.

$$P(A \cup B) = P(A) + P(B) - P(A \cap B)$$
$$= \frac{4}{20} + \frac{4}{20} - \frac{1}{20} = \frac{7}{20}$$

51. The odds in favor of an event E are defined as the ratio of $P(E)$ to $P(E')$, or $\frac{P(E)}{P(E')}$, where $P(E') \neq 0$.

53. Let E be the event "a 3 is rolled."

$P(E) = \frac{1}{6}$ and $P(E') = \frac{5}{6}$.

The odds in favor of rolling a 3 are

$\frac{P(E)}{P(E')} = \frac{\frac{1}{6}}{\frac{5}{6}} = \frac{1}{5}$, which is written "1 to 5."

55. Let E be the event "a 2, 3, 4, or 5 is rolled." Here $P(E) = \frac{4}{6} = \frac{2}{3}$ and $P(E') = \frac{1}{3}$. The odds in favor of E are $\frac{P(E)}{P(E')} = \frac{\frac{2}{3}}{\frac{1}{3}} = \frac{2}{1}$, which is written "2 to 1."

57. **(a)** Yellow: There are 3 ways to win and 15 ways to lose. The odds in favor of drawing yellow are 3 to 15, or 1 to 5.

(b) Blue: There are 11 ways to win and 7 ways to lose; the odds in favor of drawing blue are 11 to 7.

(c) White: There are 4 ways to win and 14 ways to lose; the odds in favor of drawing white are 4 to 14, or 2 to 7.

(d) From part (c), the odds of not drawing a white are 7 to 2.

59. It is possible to establish an exact probability for this event, so it is not an empirical probability.

61. This is not empirical; a formula can compute the probability exactly.

63. This is empirical, based on experience rather than probability theory.

65. A probability distribution is a table listing each possible outcome of an experiment and its corresponding probability.

67. Each of the probabilities is between 0 and 1 and the sum of all the probabilities is $0.09 + 0.32 + 0.21 + 0.25 + 0.13 = 1$, so this assignment is possible.

69. The sum of the probabilities $\frac{1}{3} + \frac{1}{4} + \frac{1}{6} + \frac{1}{8} + \frac{1}{10} = \frac{117}{120} < 1$, so this assignment is not possible.

71. This assignment is not possible because one of the probabilities is –0.08 which is not between 0 and 1. A probability cannot be negative.

73. The answers that are given are theoretical. Using the Monte Carlo method with at least 50 repetitions on a graphing calculator should give values close to these.

(a) 0.2778

(b) 0.4167

75. The answers that are given are theoretical. Using the Monte Carlo method with at least 50 repetitions on a graphing calculator should give values close to these.

(a) 0.0463

(b) 0.2963

77. Since $P(E \cap F) = 0.16$, the overlapping region $E \cap F$ is assigned the probability 0.16 in the diagram. Since $P(E) = 0.26$ and $P(E \cap F) = 0.16$, the region in E but not F is given the label 0.10. Similarly, the remaining regions are labeled.

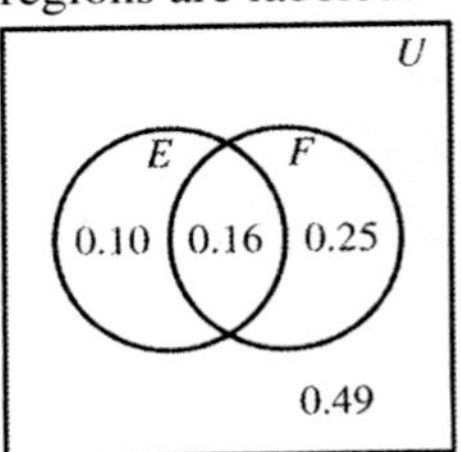

(a) $P(E \cup F) = 0.10 + 0.16 + 0.25 = 0.51$
Consequently, the part of U outside $E \cup F$ receives the label $1 - 0.51 = 0.49$.

(b) $P(E' \cap F) = P(\text{in } F \text{ but not in } E) = 0.25$

(c) The region $E \cap F'$ is that part of E which is not in F. Thus, $P(E \cap F') = 0.10$.

(d) $P(E' \cup F') = P(E') + P(F') - P(E' \cap F')$
$= 0.74 + 0.59 - 0.49 = 0.84$

79. The outcomes are not equally likely.

81. Answers will vary.

83. E: person smokes
F: person has a family history of heart disease
G: person is overweight

(a) G': "person is not overweight."

(b) $F \cap G$: "person has a family history of heart disease and is overweight."

(c) $E \cup G'$: "person smokes or is not overweight."

85. Let S be the event "the person is short," and let O be the event "the person is overweight."

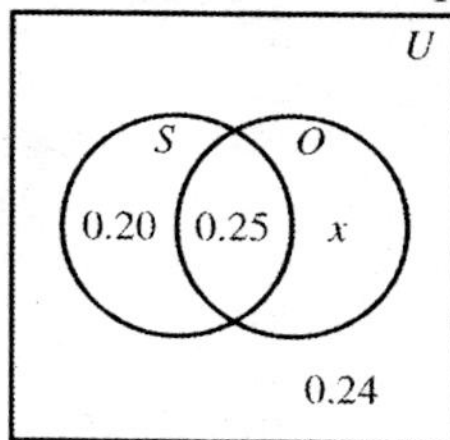

From the Venn diagram,
$0.20 + 0.25 + x + 0.24 = 1 \Rightarrow 0.69 + x = 1 \Rightarrow$
$x = 0.31$.
The probability that a person is

(a) overweight is 0.25 + 0.31 = 0.56;

(b) short, but not overweight is 0.20;

(c) tall (not short) and overweight is 0.31.

87. (a) Since red is dominant, the event "plant has red flowers" $= \{RR, RW, WR\}$;

$$P(\text{red}) = \frac{3}{4}.$$

(b) $P(\text{white}) = 1 - P(\text{red}) = \frac{1}{4}$

89. (a) $P(\text{no more than 4 good toes})$
$= 0.77 + 0.13 = 0.90$

(b) $P(5 \text{ toes}) = 0.13 + 0.10 = 0.23$

91. Let T be the event "patient visits a physical therapist" and C be the event "patient visits a chiropractor." If 22% of patients visit both a physical therapist and a chiropractor, then $P(T \cap C) = 0.22$. If 12% of patients visit neither, then $P((T \cup C)') = 0.12$. This means $P(T \cup C) = 1 - P((T \cup C)') = 1 - 0.12 = 0.88$.
Let $x = P(T)$. If the probability that a patient visits a chiropractor exceeds by 0.14 the probability that a patient visits a physical therapist, then $P(C) = P(T) + 0.14 = x + 0.14$.
Using the union rule for probability, we have

$$P(T \cup C) = P(T) + P(C) - P(T \cap C)$$
$$0.88 = x + (x + 0.14) - 0.22$$
$$0.88 = 2x - 0.08$$
$$0.96 = 2x \Rightarrow x = 0.48$$

The correct answer choice is **d**.

93. The total population for 2020 is 322,742, and the total for 2050 is 393,931.

(a) $P(\text{Hispanic in 2020}) = \dfrac{52{,}652}{322{,}742} \approx 0.1631$

(b) $P(\text{Hispanic in 2050}) = \dfrac{96{,}508}{393{,}931} \approx 0.2450$

(c) $P(\text{Black in 2020}) = \dfrac{41{,}538}{322{,}742} \approx 0.1287$

(d) $P(\text{Black in 2050}) = \dfrac{53{,}555}{393{,}931} \approx 0.1360$

95. (a) $P\begin{pmatrix}\text{civilian laborer in} \\ \text{2008 is 55 or older}\end{pmatrix} = \dfrac{27.9}{154.3} = 0.1808$

(b) $P\begin{pmatrix}\text{civilian laborer in} \\ \text{2018 is 55 or older}\end{pmatrix} = \dfrac{39.8}{166.8} = 0.2386$

(c) Answers will vary.

97. The odds are as follows:

$\dfrac{0.03}{1-0.03} = \dfrac{0.03}{0.97}$ or 3 to 97

$\dfrac{0.65}{1-0.65} = \dfrac{0.65}{0.35} = \dfrac{13}{7}$ or 13 to 7

$\dfrac{0.61}{1-0.61} = \dfrac{0.61}{0.39}$ or 61 to 39

$\dfrac{0.21}{1-0.21} = \dfrac{0.21}{0.79}$ or 21 to 79

$\dfrac{0.02}{1-0.02} = \dfrac{0.02}{0.98} = \dfrac{1}{49}$ or 1 to 49

99. **(a)** $P(\text{XI Corps}) = \dfrac{9188}{91{,}950} \approx 0.0999$

(b) $P(\text{lost in battle}) = \dfrac{22{,}803}{91{,}950} \approx 0.2480$

(c) $P(\text{I Corps lost in battle}) = \dfrac{6059}{12{,}222} \approx 0.4957$

(d) $P(\text{I Corps not lost in battle})$
$= \dfrac{12{,}222 - 6059}{12{,}222} \approx 0.5043$
$P(\text{II Corps not lost in battle})$
$= \dfrac{11{,}347 - 4369}{11{,}347} \approx 0.6150$
$P(\text{III Corps not lost in battle})$
$= \dfrac{10{,}675 - 4211}{10{,}675} \approx 0.6055$
$P(\text{V Corps not lost in battle})$
$= \dfrac{10{,}907 - 2187}{10{,}907} \approx 0.7995$
$P(\text{VI Corps not lost in battle})$
$= \dfrac{13{,}596 - 242}{13{,}596} \approx 0.9822$
$P(\text{XI Corps not lost in battle})$
$= \dfrac{9188 - 3801}{9188} \approx 0.5863$
$P(\text{XII Corps not lost in battle})$
$= \dfrac{9788 - 1082}{9788} \approx 0.8895$
$P(\text{ Calvary not lost in battle})$
$= \dfrac{11{,}851 - 610}{11{,}851} \approx 0.9485$
$P(\text{Artillery not lost in battle})$
$= \dfrac{2376 - 242}{2376} \approx 0.8981$
VI Corps had the highest probability of not being lost in battle.

(e) $P(\text{I Corps loss}) = \dfrac{6059}{12{,}222} \approx 0.4957$

$P(\text{II Corps loss }) = \dfrac{4369}{11{,}347} \approx 0.3850$

$P(\text{III Corps loss }) = \dfrac{4211}{10{,}675} \approx 0.3945$

$P(\text{V Corps loss }) = \dfrac{2187}{10{,}907} \approx 0.2005$

$P(\text{VI Corps loss }) = \dfrac{242}{13{,}596} \approx 0.0178$

$P(\text{XI Corps loss }) = \dfrac{3801}{9188} \approx 0.4137$

$P(\text{XII Corps loss }) = \dfrac{1082}{9788} \approx 0.1105$

$P(\text{Calvary loss }) = \dfrac{610}{11{,}851} \approx 0.0515$

$P(\text{Artillery loss }) = \dfrac{242}{2376} \approx 0.1019$

I Corps had the highest probability of loss.

101. **(a)** $P(\text{somewhat or extremely intolerant of Facists})$
$= P(\text{somewhat intolerant of Facists})$
$+ P(\text{extremely intolerant of Facists})$
$= \dfrac{27.1}{100} + \dfrac{59.5}{100} = \dfrac{86.6}{100} = 0.866$

(b) $P(\text{completely tolerant of Communists})$
$= P(\text{no intolerance at all of Communists})$
$= \dfrac{47.8}{100} = 0.478$

(c) Answers will vary.

103. Since 55 of the workers were women, $130 - 55 = 75$ were men. Since 3 of the women earned more than \$40,000, $55 - 3 = 52$ of them earned \$40,000 or less. Since 62 of the men earned \$40,000 or less, $75 - 62 = 13$ earned more than \$40,000. These data for the 130 workers can be summarized in the following table.

	Men	Women
\$40,000 or less	62	52
Over \$40,000	13	3

(a) $P(\text{a woman earning \$40,000 or less})$
$= \dfrac{52}{130} = 0.4$

(b) $P(\text{a man earning more than \$40,000})$
$= \dfrac{13}{130} = 0.1$

(c) $P(\text{a man or is earning more than \$40,000})$
$= \dfrac{62 + 13 + 3}{130} = \dfrac{78}{130} = 0.6$

(d) $P(\text{a woman or is earning \$40,000 or less})$
$= \dfrac{52 + 3 + 62}{130} = \dfrac{117}{130} = 0.9$

105. Let A be the set of refugees who came to escape abject poverty and B be the set of refugees who came to escape political oppression. Then $P(A) = 0.80$, $P(B) = 0.90$, and $P(A \cap B) = 0.70$.

$$\begin{aligned} P(A \cup B) &= P(A) + P(B) - P(A \cap B) \\ &= 0.80 + 0.90 - 0.70 = 1 \end{aligned}$$
$$\begin{aligned} P(A' \cap B') &= 1 - P(A \cap B) \\ &= 1 - 1 = 0 \end{aligned}$$

The probability that a refugee in the camp was neither poor nor seeking political asylum is 0.

107. There were 342 in attendance.

(a) $P(\text{speaks Cantonese}) = \frac{150}{342} = \frac{25}{57}$

(b) $P(\text{does not speaks Cantonese}) = \frac{192}{342} = \frac{32}{57}$

(c) $P(\text{woman who did not light firecracker})$.
$$= \frac{72}{342} = \frac{4}{19}$$

12.3 Conditional Probability; Independent Events; Bayes' Theorem

1. Let A be the event "the number is 2" and B be the event "the number is odd." The problem seeks the conditional probability $P(A|B)$. Use the definition $P(A|B) = \frac{P(A \cap B)}{P(B)}$. Here, $P(A \cap B) = 0$ and $P(B) = \frac{1}{2}$. Thus,

$$P(A|B) = \frac{0}{\frac{1}{2}} = 0.$$

3. Let A be the event "the number is even" and B be the event "the number is 6." Then

$$P(A|B) = \frac{P(A \cap B)}{P(B)} = \frac{\frac{1}{6}}{\frac{1}{6}} = 1.$$

5. $P(\text{sum of } 8 | \text{greater than } 7)$
$$\begin{aligned} &= \frac{P(8 \cap \text{greater than } 7)}{P(\text{greater than } 7)} \\ &= \frac{n(8 \cap \text{greater than } 7)}{n(\text{greater than } 7)} = \frac{5}{15} = \frac{1}{3} \end{aligned}$$

7. The event of getting a double given that 9 was rolled is impossible; hence, $P(\text{double}|\text{sum of } 9) = 0$.

9. Use a reduced sample space. After the first card drawn is a heart, there remain 51 cards, of which 12 are hearts. Thus,

$$P(\text{heart on 2nd}|\text{heart on 1st}) = \frac{12}{51} = \frac{4}{17}.$$

11. Use a reduced sample space. After the first card drawn is a jack, there remain 51 cards, of which 11 are face cards. Thus,

$$P(\text{face card on 2nd}|\text{ jack on 1st}) = \frac{11}{51}.$$

13. $P(\text{a jack and a } 10)$
$$\begin{aligned} &= P(\text{jack followed by } 10) \\ &\quad + P(10 \text{ followed by jack}) \\ &= \frac{4}{52} \cdot \frac{4}{51} + \frac{4}{52} \cdot \frac{4}{51} \\ &= \frac{16}{2652} + \frac{16}{2652} = \frac{32}{2652} = \frac{8}{663} \end{aligned}$$

15. $P(\text{two black cards})$
$$\begin{aligned} &= P(\text{black on 1st}) \\ &\quad \cdot P(\text{black on 2nd}|\text{black on 1st}) \\ &= \frac{26}{52} \cdot \frac{25}{51} = \frac{650}{2652} = \frac{25}{102} \end{aligned}$$

17. The conditional probability of event E given event F is $P(E|F) = \frac{P(E \cap F)}{P(F)}$, where $P(F) \neq 0$.

19. Examine a table of all possible outcomes of rolling a red die and rolling a green die (such as Figure 17 in Section 12.2). There are 9 outcomes of the 36 total outcomes that correspond to rolling "red die comes up even and green die comes up even"—in other words, corresponding to $A \cap B$. Therefore, $P(A \cap B) = \frac{9}{36} = \frac{1}{4}$. We also know that $P(A) = 1/2$ and $P(B) = 1/2$. Since $P(A \cap B) = \frac{1}{4} = \frac{1}{2} \cdot \frac{1}{2} = P(A) \cdot P(B)$, the events A and B are independent.

21. Notice that $P(F|E) \neq P(F)$: the knowledge that a person lives in Dallas affects the probability that the person lives in Dallas or Houston. Therefore, the events are dependent.

23. (a) The events that correspond to "sum is 7" are (2, 5), (3, 4), (4, 3), and (5, 2), where the first number is the number on the first slip of paper and the second number is the number on the second. Of these, only (3, 4) corresponds to "first is 3," so $P(\text{first is } 3|\text{sum is } 7) = \frac{1}{4}$.

(b) The events that correspond to "sum is 8" are (3, 5) and (5, 3). Of these, only (3, 5) corresponds to "first is 3," so $P(\text{first is } 3|\text{sum is } 8) = \frac{1}{2}$.

25. (a) Many answers are possible; for example, let B be the event that the first die is a 5. Then

$P(A \cap B) = P(\text{sum is 7 and first is 5}) = \frac{1}{36}$.

$P(A) \cdot P(B) = P(\text{sum is 7}) \cdot P(\text{first is 5}) = \frac{6}{36} \cdot \frac{1}{6} = \frac{1}{36}$, so $P(A \cap B) = P(A) \cdot P(B)$.

(b) Many answers are possible; for example, let B be the event that at least one die is a 5.

$P(A \cap B) = P(\text{sum is 7 and at least one is a 5}) = \frac{2}{36}$

$P(A) \cdot P(B) = P(\text{sum is 7}) \cdot P(\text{at least one is a 5}) = \frac{6}{36} \cdot \frac{11}{36} = \frac{11}{216}$, so $P(A \cap B) \neq P(A) \cdot P(B)$.

27. Answers will vary.

29. Since A and B are independent events, $P(A \cap B) = P(A) \cdot P(B) = \frac{1}{4} \cdot \frac{1}{5} = \frac{1}{20}$.

Thus, $P(A \cup B) = P(A) + P(B) - P(A \cap B) = \frac{1}{4} + \frac{1}{5} - \frac{1}{20} = \frac{2}{5}$.

31. Use Bayes' theorem with two possibilities M and M'.

$$P(M|N) = \frac{P(M) \cdot P(N|M)}{P(M) \cdot P(N|M) + P(M') \cdot P(N|M')}$$
$$= \frac{0.4(0.3)}{0.4(0.3) + 0.6(0.4)} = \frac{0.12}{0.12 + 0.24} = \frac{0.12}{0.36} = \frac{12}{36} = \frac{1}{3}$$

33. Using Bayes' theorem,

$$P(R_1|Q) = \frac{P(R_1) \cdot P(Q|R_1)}{P(R_1) \cdot P(Q|R_1) + P(R_2) \cdot P(Q|R_2) + P(R_3) \cdot P(Q|R_3)}$$
$$= \frac{0.15(0.40)}{(0.15)(0.40) + 0.55(0.20) + 0.30(0.70)} = \frac{0.06}{0.38} = \frac{6}{38} = \frac{3}{19}.$$

35. Using Bayes' theorem,

$$P(R_3|Q) = \frac{P(R_3) \cdot P(Q|R_3)}{P(R_1) \cdot P(Q|R_1) + P(R_2) \cdot P(Q|R_2) + P(R_3) \cdot P(Q|R_3)}$$
$$= \frac{0.30(0.70)}{(0.15)(0.40) + 0.55(0.20) + 0.30(0.70)} = \frac{0.21}{0.38} = \frac{21}{38}$$

37. We first draw the tree diagram and determine the probabilities as indicated below.

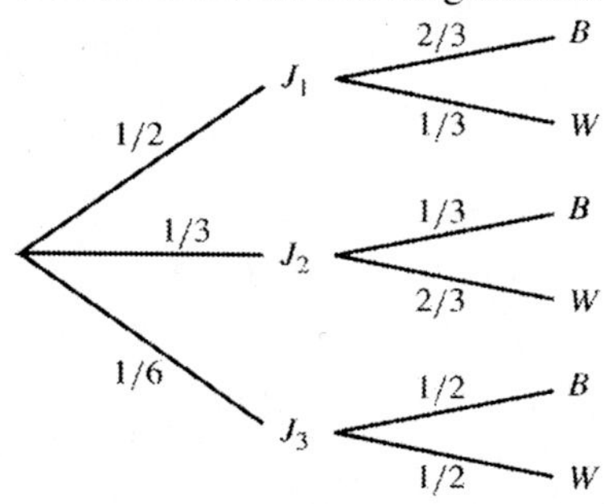

We want to determine the probability that if a white ball is drawn, it came from the second jar. This is $P(J_2 \mid W)$. Use Bayes' theorem.

$$P(J_2|W) = \frac{P(J_2)\cdot P(W\,|\,J_2)}{P(J_2)\cdot P(W\,|\,J_2) + P(J_1)\cdot P(W\,|\,J_1) + P(J_3)\cdot P(W\,|\,J_3)} = \frac{\frac{1}{3}\cdot\frac{2}{3}}{\frac{1}{3}\cdot\frac{2}{3}+\frac{1}{2}\cdot\frac{1}{3}+\frac{1}{6}\cdot\frac{1}{2}}$$

$$= \frac{\frac{2}{9}}{\frac{2}{9}+\frac{1}{6}+\frac{1}{12}} = \frac{\frac{2}{9}}{\frac{17}{36}} = \frac{8}{17}$$

39. At the first booth, there are three possibilities: shaker 1 has heads and shaker 2 has heads; shaker 1 has tails and shaker 2 has heads; shaker 1 has heads and shaker 2 has tails. We restrict ourselves to the condition that at least one head has appeared. These three possibilities are equally likely so the probability of two heads is $\frac{1}{3}$.

At the second booth we are given the condition of one head in one shaker. The probability that the second shaker has one head is $\frac{1}{2}$.

Therefore, you stand the best chance at the second booth.

41. No, these events are not independent.

43. Assume that each box is equally likely to be drawn from and that within each box each marble is equally likely to be drawn. If Laura does not redistribute the marbles, then the probability of winning the Porsche is $\frac{1}{2}$, since the event of a pink marble being drawn is equivalent to the event of choosing the first of the two boxes. If however, Laura puts 49 of the pink marbles into the second box with the 50 blue marbles, the probability of a pink marble being drawn increases to $\frac{74}{99}$. The probability of the first box being chosen is $\frac{1}{2}$, and the probability of drawing a pink marble from this box is 1.

The probability of the second box being chosen is $\frac{1}{2}$, and the probability of drawing a pink marble from this box is $\frac{49}{99}$. Thus, the probability of drawing a pink marble is $\frac{1}{2}\cdot 1 + \frac{1}{2}\cdot\frac{49}{99} = \frac{74}{99}$. Therefore Laura increases her chances of winning.

45. The sample space is $\{RW, WR, RR, WW\}$. The event "red" is $\{RW, WR, RR\}$, and the event "mixed" is $\{RW, WR\}$.

$$P(\text{mixed}|\text{red}) = \frac{n(\text{mixed and red})}{n(\text{red})} = \frac{2}{3}.$$

47. Use the following tree diagram for parts (a) through (e).

1st child	2nd child	3rd child	Branch	Probability
1/2 B	1/2 B	1/2 B	1	1/8
		1/2 G	2	1/8
	1/2 G	1/2 B	3	1/8
		1/2 G	4	1/8
1/2 G	1/2 B	1/2 B	5	1/8
		1/2 G	6	1/8
	1/2 G	1/2 B	7	1/8
		1/2 G	8	1/8

(a) $P(\text{all girls}|\text{first is a girl})$

$$= \frac{P(\text{all girls and first is a girl})}{P(\text{first is a girl})}$$

$$= \frac{n(\text{all girls and first is a girl})}{n(\text{first is a girl})} = \frac{1}{4}$$

(b) $P(\text{3 girls}|\text{3rd is a girl})$

$$= \frac{P(\text{3 girls and 3rd is a girl})}{P(\text{3rd is a girl})} = \frac{\frac{1}{8}}{\frac{1}{2}} = \frac{1}{4}$$

(c) $P(\text{all girls}|\text{second is a girl})$

$$= \frac{P(\text{all girls and second is a girl})}{P(\text{second is a girl})}$$
$$= \frac{n(\text{all girls and second is a girl})}{n(\text{second is a girl})} = \frac{1}{4}$$

(d) $P(\text{3 girls} \mid \text{at least 2 girls})$

$$= \frac{P(\text{3 girls and at least 2 girls})}{P(\text{at least 2 girls})}$$
$$= \frac{P(\text{3 girls})}{P(\text{at least two girls})}$$
$$= \frac{P(\text{3 girls})}{P(\text{2 girls}) + P(\text{3 girls})}$$
$$= \frac{\frac{1}{8}}{\frac{3}{8}+\frac{1}{8}} = \frac{\frac{1}{8}}{\frac{4}{8}} = \frac{1}{4}$$

Note that

$P(\text{3 girls}) = P(GGG) = \frac{1}{2}\cdot\frac{1}{2}\cdot\frac{1}{2} = \frac{1}{8}$ and

$$P(\text{2 girls}) = P(GGB) + P(BGG) + P(GBG)$$
$$= \frac{1}{8}+\frac{1}{8}+\frac{1}{8} = \frac{3}{8}.$$

(e) $P(\text{all girls}|\text{at least 1 girl})$

$$= \frac{P(\text{all girls and at least 1 girl})}{P(\text{at least 1 girl})}$$
$$= \frac{n(\text{all girls and at least 1 girl})}{n(\text{at least 1 girl})}$$
$$= \frac{1}{7}$$

49. (a), (b) From the table, $P(C \cap D) = 0.0008$ and

$$P(C)\cdot P(D) = 0.0400(0.0200) = 0.0008.$$

Since $P(C \cap D) = P(C)\cdot P(D)$, C and D are independent events; color blindness and deafness are independent events.

51. Let H be the event "patient has high blood pressure,"
N be the event "patient has normal blood pressure,"
L be the event "patient has low blood pressure,"
R be the event "patient has a regular heartbeat,"
and I be the event "patient has an irregular heartbeat.
We wish to determine $P(R \cap L)$.

Statement (i) tells us $P(H) = 0.14$ and statement (ii) tells us $P(L) = 0.22$. Therefore,

$$P(H) + P(N) + P(L) = 1$$
$$0.14 + P(N) + 0.22 = 1$$
$$P(N) = 0.64.$$

Statement (iii) tells us $P(I) = 0.15$. This and statement (iv) lead to

$$P(I \cap H) = \frac{1}{3}P(I) = \frac{1}{3}(0.15) = 0.05.$$

Statement (v) tells us

$$P(N \cap I) = \frac{1}{8}P(N) = \frac{1}{3}(0.64) = 0.08.$$

Make a table and fill in the data just found.

	H	*N*	*L*	**Totals**
R	–	–	–	–
I	0.05	0.08	–	0.15
Totals	0.14	0.64	0.22	1.00

To determine $P(R \cap L)$, we need to find only $P(I \cap L)$.

$$P(I) = P(I \cap H) + P(I \cap N) + P(I \cap L)$$
$$0.15 = 0.05 + 0.08 + P(I \cap L)$$
$$0.15 = 0.13 + P(I \cap L)$$
$$P(I \cap L) = 0.02$$

Now calculate $P(R \cap L)$.

$$P(L) = P(R \cap L) + P(I \cap L)$$
$$0.22 = P(R \cap L) + 0.02$$
$$P(I \cap L) = 0.20$$

The correct answer choice is **e**.

53. (a) The total number of males is $2(5844) + 6342 = 18{,}030$. The total number of infants is $2(17{,}798) = 35{,}596$. So among infants who are part of a twin pair, the proportion of males is $\frac{18{,}030}{35{,}596} = 0.5065$. To answer parts (b) through (g) we make the assumption that the event "twin comes from an identical pair" is independent of the event "twin is male." We can then construct the following tree diagram.

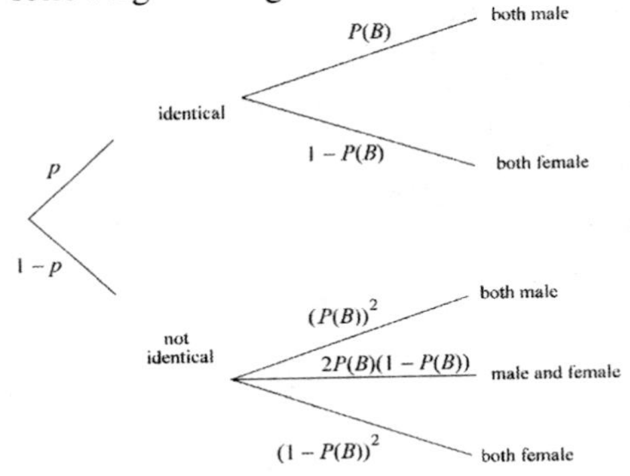

(b) The event that the pair of twins is male happens along two compound branches. Multiplying the probabilities along these branches and adding the products we get $pP(B)+(1-p)(P(B))^2$.

(c) Using the values from (a) and (b) together with the fraction of mixed twins, we solve this equation:

$$\frac{5844}{17,798}=p(0.506518)+(1-p)(0.506518)^2$$
$$0.328352=0.506518p+0.256560-0.256560p$$
$$0.071792=0.249958p$$
$$p=\frac{0.071792}{0.249958}=0.2872$$

So our estimate for p is 0.2872. Note that if you use fewer places in the value for $P(B)$ you may get a slightly different answer.

(d) Multiplying along the two branches that result in two female twins and adding the products gives $p(1-P(B))+(1-p)(1-P(B))^2$.

(e) The equation to solve will now be the following: $\frac{5612}{17,198}=p(1-0.506518)+(1-p)(1-0.506518)^2$

The answer will be the same as in part (c)

(f) Now only the "not identical" branch is involved, and the expression is $2(1-p)P(B)(1-P(B))$.

(g) The equation to solve will now be the following: $\frac{6342}{17,198}=2(1-p)(0.506518)(1-0.506518)$

Again the answer will be the same as in part (c) Note that because of our independence assumption these three estimates for p must agree.

55. Let E represent the event "hemoccult test is positive," and let F represent the event "has colorectal cancer." We are given $P(F)=0.003$, $P(E|F)=0.5$, and $P(E|F')=0.03$, and we want to find $P(F|E)$. Since $P(F)=0.003$, $P(F')=0.997$. Therefore,

$$P(F|E)=\frac{P(F)\cdot P(E|F)}{P(F)\cdot P(E|F)+P(F')\cdot P(E|F')}=\frac{0.003\cdot 0.5}{0.003\cdot 0.5+0.997\cdot 0.03}\approx 0.0478.$$

57. Let D be the event "has breast cancer" and T be the event "positive test."

$$P(T^+\mid D^+)=0.796$$
$$P(T^-\mid D^-)=0.902$$
$$P(D^+)=0.005 \text{ so } P(D^-)=0.995$$

We can now fill in the complete tree.

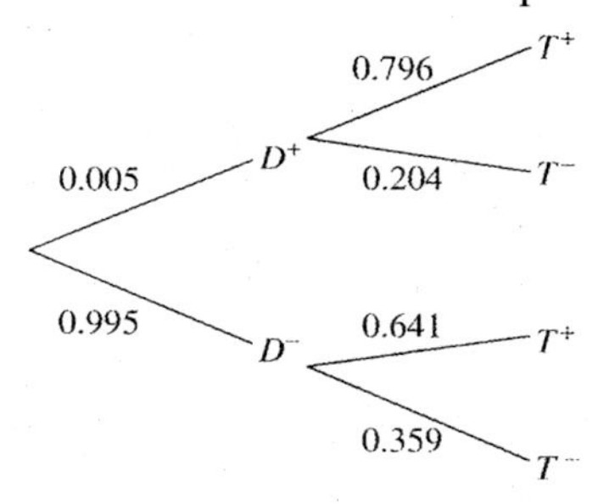

(a)
$$P(D^+|T^+)=\frac{P(D^+)P(T^+\mid D^+)}{P(D^+)P(T^+\mid D+)+P(D^-)P(T^+\mid D^-)}$$
$$=\frac{(0.005)(0.796)}{(0.005)(0.796)+(0.995)(0.098)}=0.039$$

(b) $$P(D^-|T^-) = \frac{P(D^-)P(T^-|D^-)}{P(D^-)P(T^-|D^-) + P(D^+)P(T^-|D^+)} = \frac{(0.995)(0.902)}{(0.995)(0.902) + (0.005)(0.204)} = 0.999$$

(c) $$P(D^+|T^-) = \frac{P(D^+)P(T^-|D^+)}{P(D^+)P(T^-|D^+) + P(D^-)P(T^-|D^-)} = \frac{(0.005)(0.204)}{(0.005)(0.204) + (0.995)(0.902)} = 0.001$$

Alternatively, since $P(D^-|T^-) + P(D^+|T^-) = 1$, we can subtract the answer to (b) from 1.

(d) The tree stays the same, but $P(D^+)$ is now 0.015 and $P(D^-)$ is 0.985.

$$P(D^+|T^+) = \frac{P(D^+)P(T^+|D^+)}{P(D^+)P(T^+|D^+) + P(D^-)P(T^+|D^-)} = \frac{(0.015)(0.796)}{(0.015)(0.796) + (0.985)(0.098)} = 0.110$$

59. Let H represent "heavy smoker," L be "light smoker," N be "nonsmoker," and D be "person died." Let $x = P(D|N)$, that is, let x be the probability that a nonsmoker died. Then $P(D|L) = 2x$ and $P(D|H) = 4x$. Create a table.

Level of Smoking	Probability of Level	Probability of Death for Level
H	0.2	$4x$
L	0.3	$2x$
N	0.5	x

We wish to find $P(H|D)$.

$$P(H|D) = \frac{P(H)\cdot P(D|H)}{P(H)\cdot P(D|H) + P(L)\cdot P(D|L) + P(N)\cdot P(D|N)} = \frac{0.2(4x)}{0.2(4x) + 0.3(2x) + 0.5(x)} = \frac{0.8x}{1.9x} \approx 0.42$$

The correct answer choice is **d**.

61. Let H represent "person has the disease" and R be "test indicates presence of the disease." We wish to determine $P(H|R)$. Construct a table as before.

Category of Person	Probability of Population	Probability of Presence of Disease
H	0.01	0.950
H'	0.99	0.005

$$P(H|R) = \frac{P(H)\cdot P(R|H)}{P(H)\cdot P(R|H) + P(H')\cdot P(R|H')} = \frac{0.01(0.950)}{0.01(0.950) + 0.99(0.005)} = \frac{0.00950}{0.01445} \approx 0.657$$

The correct answer choice is **b**.

63. Let S^+ and S^- stand for "wore seat belt" and "did not wear seat belt" and let E^+ and E^- stand for "was ejected" and "was not ejected." We start by constructing the tree corresponding to the given data.

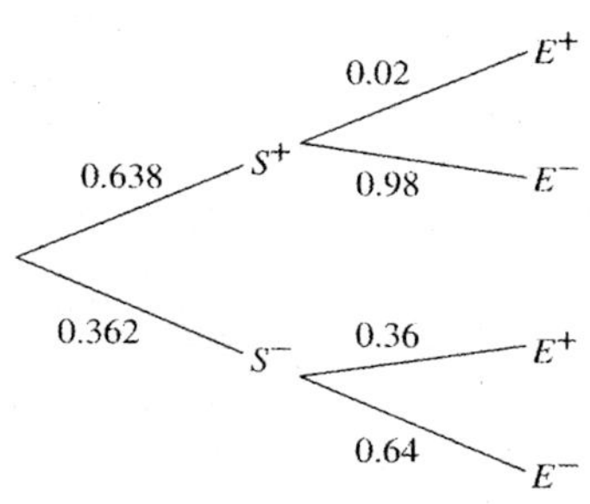

(a) $$P(S^+ \mid E^+) = \frac{P(S^+)P(E^+ \mid S^+)}{P(S^+)P(E^+ \mid S^+) + P(S^-)P(E^+ \mid S^-)} = \frac{(0.638)(0.02)}{(0.638)(0.02) + (0.362)(0.36)} = 0.0892$$

(b) $$P(S^- \mid E^-) = \frac{P(S^-)P(E^- \mid S^-)}{P(S^-)P(E^- \mid S^-) + P(S^+)P(E^- \mid S^+)} = \frac{(0.362)(0.64)}{(0.362)(0.64) + (0.638)(0.98)} = 0.2704$$

65. Let S stand for "smokes" and S^- for "does not smoke."

$P(45-64|S^-)$

$$= \frac{P(45-64)P(S^- \mid 45-64)}{P(45-64)P(S^- \mid 45-64) + P(18-44)P(S^- \mid 18-44) + P(65-74)P(S^- \mid 65-74) + P(>75)P(S^- \mid >75)}$$

$$= \frac{(0.34)(1-0.22)}{(0.34)(1-0.22) + (0.49)(1-0.23) + (0.09)(1-0.12) + (0.08)(1-0.06)}$$

$= 0.3328$

67. (a) $P(\text{forecast of rain} \mid \text{rain}) = \frac{66}{80} = 0.825 \approx 83\%$

$P(\text{forecast of no rain} \mid \text{no rain}) = \frac{764}{920} \approx 0.83 \approx 83\%$

(b) $P(\text{rain} \mid \text{forecast of rain}) = \frac{66}{222} \approx 0.2973$

(c) $P(\text{no rain} \mid \text{forecast of no rain}) = \frac{764}{778} \approx 0.9820$

(d) Answers will vary.

69. (a) $P(\text{second class}) = \frac{357}{1316} \approx 0.2713$

(b) $P(\text{surviving}) = \frac{499}{1316} \approx 0.3792$

(c) $P(\text{surviving} \mid \text{first class}) = \frac{203}{325} \approx 0.6246$

(d) $P(\text{surviving} \mid \text{child and third class}) = \frac{27}{79} \approx 0.3418$

(e) $P(\text{woman} \mid \text{first class and survived}) = \frac{140}{203} \approx 0.6897$

(f) $P(\text{third class} \mid \text{man and survived}) = \frac{75}{146} \approx 0.5137$

(g) $P(\text{survived} \mid \text{man}) = \frac{146}{805} \approx 0.1814$

$P(\text{survived} \mid \text{man and third class}) = \frac{75}{462} \approx 0.1623$

No, the events are not independent.

71. First draw the tree diagram.

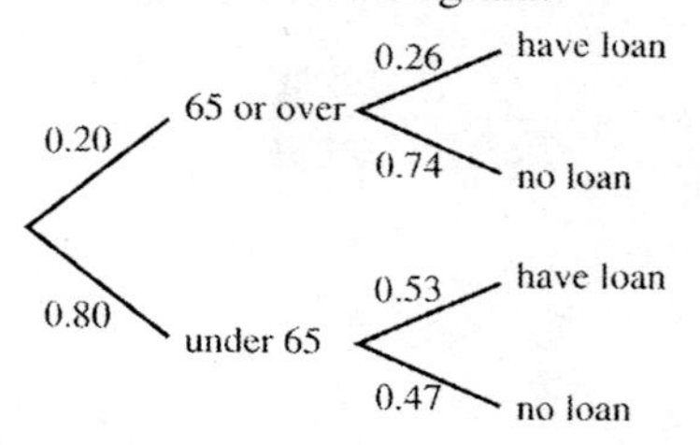

(a) $P(\text{person is 65 or over and has a loan})$
$= P(\text{65 or over}) \cdot P(\text{has loan} \mid \text{65 or over})$
$= 0.20(0.26) = 0.052$

(b) $P(\text{person has a loan}) = P(\text{65 or over and has loan}) + P(\text{under 65 and has loan})$
$= 0.20(0.26) + 0.80(0.53) = 0.476$

73. First draw the tree diagram.

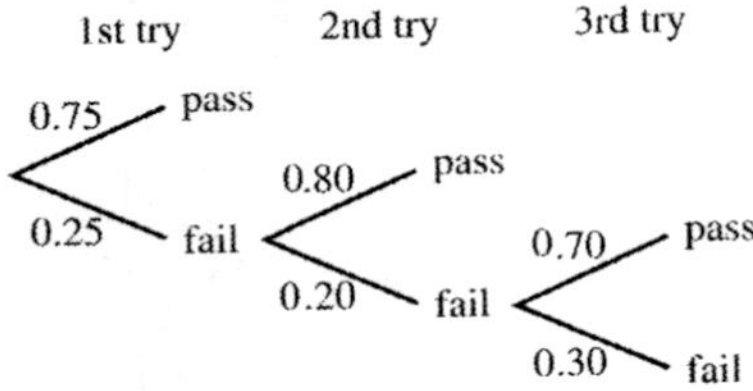

(a) $P(\text{fails both 1st and 2nd tests}) = P(\text{fails 1st}) \cdot P(\text{fails 2nd} | \text{fails 1st}) = 0.25(0.20) = 0.05$

(b) $P(\text{fails three times in a row}) = 0.25(0.20)(0.30) = 0.015$

(c) $P(\text{requires at least 2 tries}) = P(\text{does not pass on 1st try}) = 0.25$

75. Let A be the event "student studies" and B be the event "student gets a good grade." We are told that $P(A) = 0.6, P(B) = 0.7,$ and $P(A \cap B) = 0.52.$
$P(A) \cdot P(B) = 0.6(0.7) = 0.42$

(a) Since $P(A) \cdot P(B)$ is not equal to $P(A \cap B)$, A and B are not independent. Rather, they are dependent events.

(b) Let A be the event "a student studies" and B be the event "the student gets a good grade." Then
$$P(B | A) = \frac{P(B \cap A)}{P(A)} = \frac{0.52}{0.6} \approx 0.87.$$

(c) Let B be the event "a student gets a good grade" and A be the event "the student studied." Then
$$P(A | B) = \frac{P(A \cap B)}{P(B)} = \frac{0.52}{0.7} \approx 0.74.$$

77. $P(0) = (0.4)(0.4) = 0.16$
$P(1) = 2(0.4)(0.6) = 0.48$
$P(2) = (0.6)(0.6) = 0.36$

79. Let M be the event "wife was murdered" and G be the event "husband is guilty." Set up a tree diagram.

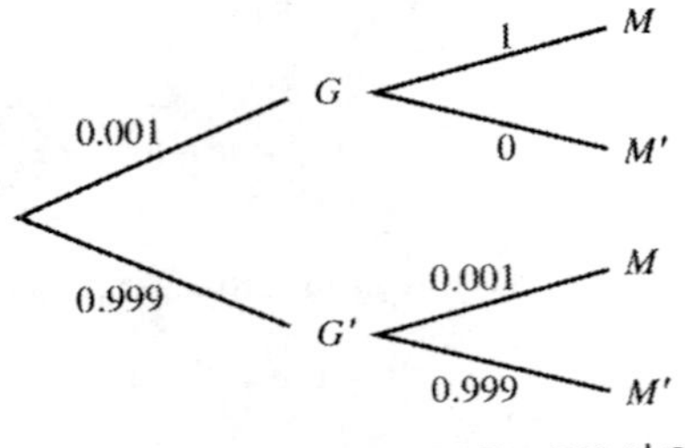

$$P(G|M) = \frac{P(G) \cdot P(M | G)}{P(G) \cdot P(M | G) + P(G') \cdot P(M | G')} = \frac{0.001(1)}{0.001(1) + 0.999(0.001)} \approx 0.500$$

81. To find the proportion of women who have been married in each age category, we subtract the numbers in the second column of the table of data for women from 1, giving the values 0.175, 0.634, 0.853, 0.908, and 0.960.
$P(\text{between 18 and 24} | \text{has been married})$ (for a randomly selected woman)
$$= \frac{(0.121)(0.175)}{(0.121)(0.175) + (0.172)(0.634) + (0.178)(0.853) + (0.345)(0.908) + (0.184)(0.960)} = 0.0274$$

83. Let Q be the event "person is qualified" and A be the event "person was approved." Set up a tree diagram.

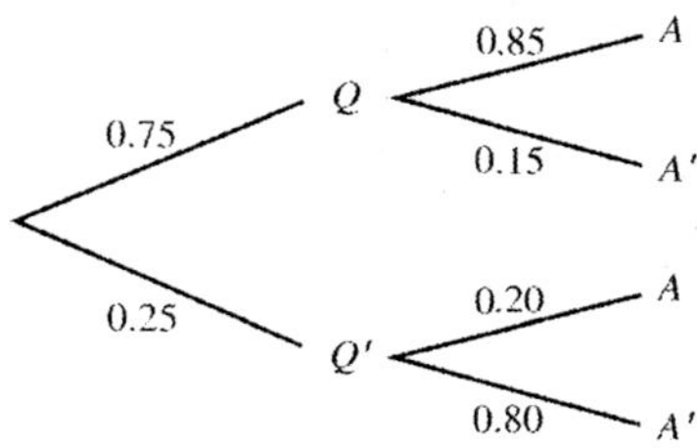

Using Bayes' theorem,

$$P(Q|A) = \frac{P(Q)\cdot P(A|Q)}{P(Q)\cdot P(A|Q) + P(Q')\cdot P(A|Q')} = \frac{0.75(0.85)}{0.75(0.85)+0.25(0.20)} = \frac{0.6375}{0.6875} = \frac{51}{55} \approx 0.9273$$

85. Let A represent "driver is 16–20," B represent "driver is 21–30," C represent "driver is 31–65," D represent "driver is 66–99," and E represent "driver has an accident."
We wish to find $P(A|E)$.

$$P(A|E) = \frac{P(A)\cdot P(E|A)}{P(A)\cdot P(E|A) + P(B)\cdot P(E|B) + P(C)\cdot P(E|C) + P(D)\cdot P(E|D)}$$
$$= \frac{0.08(0.06)}{0.08(0.06)+0.15(0.03)+0.49(0.02)+0.28(0.04)} = \frac{0.0048}{0.0303} \approx 0.16$$

The correct choice is **b**.

87. (a) Prisoner A, before hearing from the jailer, estimates his chances of being pardoned as 1/3, the same as both B and C. Then, the probability that either B or C will be executed is 1/2 for either. Thus, the probability that B will be executed is $\frac{1}{2}\cdot\frac{1}{3}=\frac{1}{6}$. The jailer says B will be executed, so either C will be pardoned (1/3 chance) or A will be pardoned (1/3 chance). After hearing that B will be executed, the estimate of A's chance of being pardoned is half that of C. This means his chances of being pardoned, now knowing B isn't, again are 1/3, but C has a 2/3 chance of being pardoned. We can verify this using Bayes' Theorem. Let A, B, and C represent the events that the corresponding prisoner will be pardoned, and let b represent the event that the jailer mentions prisoner B as not being pardoned. Then

$$P(A|b) = \frac{P(b|A)P(A)}{P(b|A)P(A)+P(b|B)P(B)+P(b|C)P(C)} = \frac{\frac{1}{2}\times\frac{1}{3}}{\frac{1}{2}\times\frac{1}{3}+0\times\frac{1}{3}+1\times\frac{1}{3}} = \frac{1}{3}$$

(b) Now we assume that the probabilities of A, B, and C being freed are 1/4, 1/4, and 1/2 respectively. Then, A's probability of being free, given that B will be executed is

$$P(A|b) = \frac{P(b|A)P(A)}{P(b|A)P(A)+P(b|B)P(B)+P(b|C)P(C)} = \frac{\frac{1}{2}\times\frac{1}{4}}{\frac{1}{2}\times\frac{1}{4}+0\times\frac{1}{4}+1\times\frac{1}{2}} = \frac{1}{5}$$

12.4 Discrete Random Variables; Applications to Decision Making

1. Let x denote the number of heads observed. Then x can take on 0, 1, 2, 3, or 4 as values. The probabilities are as follows.

$$P(x=0) = C(4,0)\left(\frac{1}{2}\right)^0\left(\frac{1}{2}\right)^4 = \frac{1}{16}$$

$$P(x=1) = C(4,1)\left(\frac{1}{2}\right)^1\left(\frac{1}{2}\right)^3 = \frac{4}{16} = \frac{1}{4}$$

$$P(x=2) = C(4,2)\left(\frac{1}{2}\right)^2\left(\frac{1}{2}\right)^2 = \frac{6}{16} = \frac{3}{8}$$

$$P(x=3) = C(4,3)\left(\frac{1}{2}\right)^3\left(\frac{1}{2}\right)^1 = \frac{4}{16} = \frac{1}{4}$$

$$P(x=4) = C(4,4)\left(\frac{1}{2}\right)^4\left(\frac{1}{2}\right)^0 = \frac{1}{16}$$

Therefore, the probability distribution is as follows.

Number of Heads	0	1	2	3	4
Probability	$\frac{1}{16}$	$\frac{1}{4}$	$\frac{3}{8}$	$\frac{1}{4}$	$\frac{1}{16}$

3. Let x denote the number of aces drawn. Then x can take on values 0, 1, 2, or 3. The probabilities are as follows.

$$P(x=0)=C(3,0)\left(\frac{48}{52}\right)\left(\frac{47}{51}\right)\left(\frac{46}{50}\right)\approx 0.7826$$

$$P(x=1)=C(3,1)\left(\frac{4}{52}\right)\left(\frac{48}{51}\right)\left(\frac{47}{50}\right)\approx 0.2042$$

$$P(x=2)=C(3,2)\left(\frac{4}{52}\right)\left(\frac{3}{51}\right)\left(\frac{48}{50}\right)\approx 0.0130$$

$$P(x=3)=C(3,3)\left(\frac{4}{52}\right)\left(\frac{3}{51}\right)\left(\frac{2}{50}\right)\approx 0.0002$$

Therefore, the probability distribution is as follows.

Number of Aces	0	1	2	3
Probability	0.7826	0.2042	0.0130	0.0002

5. Use the probabilities that were calculated in Exercise 1. Draw a histogram with 5 rectangles, corresponding to $x = 0$, $x = 1$, $x = 2$, $x = 3$, and $x = 4$. $P(x \le 2)$ corresponds to $P(x=0)+P(x=1)+P(x=2)$, so shade the first 3 rectangles in the histogram.

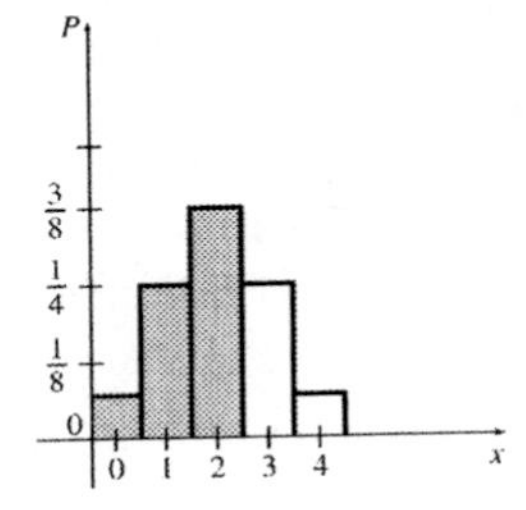

7. Use the probabilities that were calculated in Exercise 3. Draw a histogram with 4 rectangles, corresponding to $x = 0$, $x = 1$, $x = 2$, and $x = 3$. $P(\text{at least one ace}) = P(x \ge 1)$ corresponds to $P(x=1)+P(x=2)+P(x=3)$, so shade the last 3 rectangles.

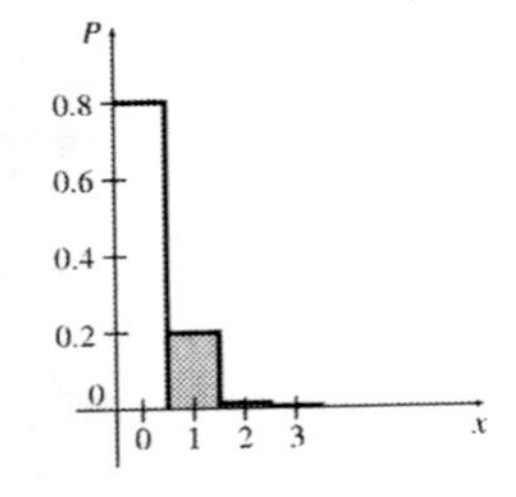

9. $E(x) = 2(0.1)+3(0.4)+4(0.3)+5(0.2) = 3.6$

$$\sigma^2=\sum_x (x-\mu)^2 P(x)$$
$$=(2-3.6)^2(0.1)+(3-3.6)^2(0.4)+(4-3.6)^2(0.3)+(5-3.6)^2(0.2)$$
$$=0.84$$
$$\sigma\approx 0.917$$

11. $E(z)=9(0.14)+12(0.22)+15(0.38)+18(0.19)+21(0.07)$
$=14.49$

$$\sigma^2=\sum_x (x-\mu)^2 P(x)$$
$$=(9-14.49)^2(0.14)+(12-14.49)^2(0.22)+(15-14.49)^2(0.38)+(18-14.49)^2(0.19)+(21-14.49)^2(0.07)$$
$$=10.9899$$
$$\sigma\approx 3.315$$

13. It is possible (but not necessary) to begin by writing the histogram's data as a probability distribution, which would look as follows.

x	1	2	3	4
$P(x)$	0.2	0.3	0.1	0.4

The expected value of x is
$E(x)=1(0.2)+2(0.3)+3(0.1)+4(0.4)=2.7.$

15. The expected value of x is
$$E(x)=6(0.1)+12(0.2)+18(0.4)+24(0.2)+30(0.1)$$
$$=18$$

17. Using the data from Example 5, the expected winnings for Mary are
$$E(x)=-1.2\left(\frac{1}{4}\right)+1.2\left(\frac{1}{4}\right)+1.2\left(\frac{1}{4}\right)+(-1.2)\left(\frac{1}{4}\right)$$
$$=0.$$
Yes, it is still a fair game if Mary tosses and Donna calls.

19. (a) Let x be the number of times 1 is rolled. Since the probability of getting a 1 on any single roll is $\frac{1}{6}$, the probability of any other outcome is $\frac{5}{6}$. Use combinations since the order of outcomes is not important.

$$P(x=0)=C(4,0)\left(\frac{1}{6}\right)^0\left(\frac{5}{6}\right)^4=\frac{625}{1296}$$

$$P(x=1)=C(4,1)\left(\frac{1}{6}\right)^1\left(\frac{5}{6}\right)^3=\frac{125}{324}$$

$$P(x=2)=C(4,2)\left(\frac{1}{6}\right)^2\left(\frac{5}{6}\right)^2=\frac{25}{216}$$

$$P(x=3)=C(4,3)\left(\frac{1}{6}\right)^3\left(\frac{5}{6}\right)^1=\frac{5}{324}$$

$$P(x=4)=C(4,4)\left(\frac{1}{6}\right)^4\left(\frac{5}{6}\right)^0=\frac{1}{1296}$$

x	0	1	2	3	4
$P(x)$	$\frac{625}{1296}$	$\frac{125}{324}$	$\frac{25}{216}$	$\frac{5}{324}$	$\frac{1}{1296}$

(b)
$$E(x)=0\left(\frac{625}{1296}\right)+1\left(\frac{125}{324}\right)+2\left(\frac{25}{216}\right)+3\left(\frac{5}{324}\right)+4\left(\frac{1}{1296}\right)$$
$$=\frac{2}{3}$$

21. Answers will vary. A probability distribution is a function or rule that assigns probabilities to each value of a random variable. The distribution may in some cases be listed. In other cases it is presented as a graph.

23. (a) First list the possible sums, 5, 6, 7, 8, and 9, and find the probabilities for each. The total possible number of results are $4\cdot 3=12$. There are two ways to draw a sum of 5 (2 then 3, and 3 then 2). The probability of 5 is $\frac{2}{12}=\frac{1}{6}$. There are two ways to draw a sum of 6 (2 then 4, and 4 then 2). The probability of 6 is $\frac{2}{12}=\frac{1}{6}$. There are four ways to draw a sum of 7 (2 then 5, 3 then 4, 4 then 3, and 5 then 2). The probability of 7 is $\frac{4}{12}=\frac{1}{3}$. There are two ways to draw a sum of 8 (3 then 5, and 5 then 3). The probability of 8 is $\frac{2}{12}=\frac{1}{6}$. There are two ways to draw a sum of 9 (4 then 5, and 5 then 4).The probability of 9 is $\frac{2}{12}=\frac{1}{6}$. The distribution is as follows.

Sum	5	6	7	8	9
Probability	$\frac{1}{6}$	$\frac{1}{6}$	$\frac{1}{3}$	$\frac{1}{6}$	$\frac{1}{6}$

(b)

P

$\frac{1}{3}$

$\frac{1}{6}$

5 6 7 8 9 x

(c) The probability that the sum is even is $\frac{1}{6}+\frac{1}{6}=\frac{1}{3}$. Thus the odds are 1 to 2.

(d)
$$E(x)=\frac{1}{6}(5)+\frac{1}{6}(6)+\frac{1}{3}(7)+\frac{1}{6}(8)+\frac{1}{6}(9)$$
$$=7$$

(e)
$$\sigma^2=\sum_x(x-\mu)^2P(x)$$
$$=(5-7)^2\left(\frac{1}{6}\right)+(6-7)^2\left(\frac{1}{6}\right)+(7-7)^2\left(\frac{1}{3}\right)+(8-7)^2\left(\frac{1}{6}\right)+(9-7)^2\left(\frac{1}{6}\right)$$
$$=\frac{5}{3}$$
$$\sigma\approx 1.291$$

25. (a) Expected cost of Amoxicillin:
$$E(x)=0.75(\$59.30)+0.25(\$96.15)$$
$$=\$68.51$$
Expected cost of Cefaclor:
$$E(x)=0.90(\$69.15)+0.10(\$106.00)$$
$$=\$72.84$$

(b) Amoxicillin should be used to minimize total expected cost.

27. With the new contamination probability, the probabilities for the branches become

Branch	1	2	3	4	5	6	7
Outcome (x)	0	0.1	1	1	0	1	0
$P(x)$	0.024	0.0066	0.0017	0.0009	0.005	0.9131	0.0531

The expected outcome for transfusion is
$E_1 = 0(0.024) + 0.1(0.0066) + 1(0.0017) + 1(0.0009) + 0(0.005) + 1(0.9131) + 0(0.0531) \approx 0.9164$
The expected outcome for no transfusion remains unchanged at 0.865.
Since the value of the outcome increases, blood transfusions should be given after the blood has been screened.

29. $E(x) = 250(0.74) = 185$
We would expect 185 low-birth-weight babies to graduate from high school.

31. **(a)** Let x represent the amount of damage in millions of dollars. For seeding, the expected value is
$E(x) = 0.038(335.8) + 0.143(191.1) + 0.392(100) + 0.255(46.7) + 0.172(16.3) \approx \$94.0$ million.
For not seeding, the expected value is
$E(x) = 0.054(335.8) + 0.206(191.1) + 0.480(100) + 0.206(46.7) + 0.054(16.3) \approx \$116.0$ million.

(b) Seed, because the total expected damage is less with that option.

33. Below is the probability distribution of x, which stands for the person's payback.

x	\$398	\$78	–\$2
$P(x)$	$\frac{1}{500} = 0.002$	$\frac{3}{500} = 0.006$	$\frac{497}{500} = 0.994$

The expected value of the person's winnings is
$E(x) = 398(0.002) + 78(0.006) + (-2)(0.994) \approx -\$0.72$ or $-72¢$.
Since the expected value of the payback is not 0, this is not a fair game.

35. There are 18 + 20 = 38 possible outcomes. In 18 cases you win a dollar and in 20 you lose a dollar; hence,
$E(x) = 1\left(\frac{18}{38}\right) + (-1)\left(\frac{20}{38}\right) = -\frac{1}{19}$, or about $-5.3¢$.

37. You have one chance in a thousand of winning \$500 on a \$1 bet for a net return of \$499. In the 999 other outcomes, you lose your dollar.
$E(x) = 499\left(\frac{1}{1000}\right) + (-1)\left(\frac{999}{1000}\right) = -\frac{500}{1000} = -50¢$

39. At any one restaurant, your expected winnings are

$$E(x) = 100{,}000\left(\frac{1}{176{,}402{,}500}\right) + 25{,}000\left(\frac{1}{39{,}200{,}556}\right) + 5000\left(\frac{1}{17{,}640{,}250}\right) + 1000\left(\frac{1}{1{,}568{,}022}\right) + 100\left(\frac{1}{288{,}244}\right) + 5\left(\frac{1}{7056}\right) + 1\left(\frac{1}{588}\right) = 0.00488.$$

Going to 25 restaurants gives you expected earnings of $25(0.00488) = 0.122$. Since you spent \$1, you lose 87.8¢ on the average, so your expected value is $-87.8¢$.

41. **(a)** Expected value of a two-point conversion: $E(x) = 2(0.45) = 0.90$
Expected value of an extra-point kick: $E(x) = 1(0.96) = 0.96$

(b) Since the expected value of an extra-point kick is greater than the expected value of a two-point conversion, the extra-point kick will maximize the number of points scored over the long run.

(c) Answers will vary.

43. We first compute the amount of money the company can expect to pay out for each kind of policy. The sum of these amounts will be the total amount the company can expect to pay out. For a single \$100,000 policy, we have the following probability distribution.

	Pay	Don't Pay
Outcome	\$100,000	\$100,000
Probability	0.0012	0.9998

$$E(\text{payoff}) = 100,000(0.0012) + 0(0.9998) = \$120$$

For all 100 such policies, the company can expect to pay out $100(120) = \$12,000$.

For a single \$50,000 policy,

$$E(\text{payoff}) = 50,000(0.0012) + 0(0.9998) = \$60.$$

For all 500 such policies, the company can expect to pay out $500(60) = \$30,000$.

Similarly, for all 1000 policies of \$10,000, the company can expect to pay out $1000(12) = \$12,000$.

Thus, the total amount the company can expect to pay out is $\$12,000 + \$30,000 + \$12,000 = \$54,000$.

Chapter 12 Review Exercises

1. True

2. True

3. False. The union of a set with itself has the same number of elements as the set.

4. False. The intersection of a set with itself has the same number of elements as the set.

5. False. If the sets share elements, this procedure gives the wrong answer.

6. True

7. False: This procedure is correct only if the two events are mutually exclusive.

8. False: We can calculate this probability by assuming a sample space in which each card in the 52-card deck is equally likely to be drawn.

9. False: If two events A and B are mutually exclusive, then $P(A \cap B) = 0$ and this will not be equal to $P(A)P(B)$ if $P(A)$ and $P(B)$ are greater than 0.

10. True

11. False: In general these two probabilities are different. For example, for a draw from a 52-card deck, $P(\text{heart}|\text{queen}) = 1/4$ and $P(\text{queen}|\text{heart}) = 1/13$.

12. True

13. True

14. False: For example, the random variable that assigns 0 to a head and 1 to a tail has expected value 1/2 for a fair coin.

15. True

16. False: The expected value of a fair game is 0.

17. $9 \in \{8, 4, -3, -9, 6\}$

Since 9 is not an element of the set, this statement is false.

18. $4 \notin \{3, 9, 7\}$

Since 4 is not an element of the given set, the statement is true.

19. $2 \notin \{0, 1, 2, 3, 4\}$

Since 2 is an element of the set, this statement is false.

20. $0 \in \{0, 1, 2, 3, 4\}$

Since 0 is an element of the given set, the statement is true.

21. $\{3, 4, 5\} \subseteq \{2, 3, 4, 5, 6\}$

Every element of {3, 4, 5} is an element of {2, 3, 4, 5, 6}, so this statement is true.

22. $\{1, 2, 5, 8\} \subseteq \{1, 2, 5, 10, 11\}$

Since 8 is an element of the first set but not of the second set, the first set cannot be a subset of the second. The statement is false.

23. $\{3, 6, 9, 10\} \subseteq \{3, 9, 11, 13\}$

10 is an element of {3, 6, 9, 10}, but 10 is not an element of {3, 9, 11, 13}. Therefore, {3, 6, 9, 10} is not a subset of {3, 9, 11, 13}. The statement is false.

24. $\emptyset \subseteq \{1\}$

The empty set is a subset of every set, so the statement is true.

25. $\{2, 8\} \not\subseteq \{2, 4, 6, 8\}$

Since both 2 and 8 are elements of {2, 4, 6, 8}, {2, 8} is a subset of {2, 4, 6, 8}. This statement is false.

26. $0 \subseteq \emptyset$

The empty set contains no elements and has no subsets except itself. Therefore, the statement is false.

In Exercises 27–36, $U = \{a, b, c, d, e, f, g, h\}$, $K = \{c, d, e, f, h\}$, and $R = \{a, c, d, g\}$.

27. K has 5 elements, so it has $2^5 = 32$ subsets.

28. $n(R) = 4$, so R has $2^4 = 16$ subsets.

29. K' (the complement of K) is the set of all elements of U that do *not* belong to K.
$K' = \{a, b, g\}$

30. $R' = \{b, e, f, h\}$ since these elements are in U but not in R.

31. $K \cap R$ (the intersection of K and R) is the set of all elements belonging to both set K and set R. $K \cap R = \{c, d\}$

32. $K \cup R = \{a, c, d, e, f, g, h\}$ since these elements are in K or R, or both.

33. $(K \cap R)' = \{a, b, e, f, g, h\}$ since these elements are in U but not in $K \cap R$. (See Exercise 31.)

34. $(K \cup R)' = \{b\}$ since this element is in U but not in $K \cup R$. (See Exercise 32.)

35. $\emptyset' = U$

36. $U' = \emptyset$, which is always true.

37. $A \cap C$ is the set of all female employees who are in the pediatrics department.

38. $B \cap D$ is the set of employees in the maternity department who have a nursing degree.

39. $A \cup D$ is the set of all employees who are in the pediatrics department *or* who have a nursing degree.

40. $A' \cap D$ is the set of all employees with nursing degrees who are not in the pediatrics department.

41. $B' \cap C'$ is the set of all male employees who are not in the maternity department.

42. $(B \cup C)'$ is the set of all employees who are neither in the maternity department nor female, that is, all male employees not in the maternity department.

43. $A \cup B'$ is the set of all elements which belong to A or do not belong to B, or both.

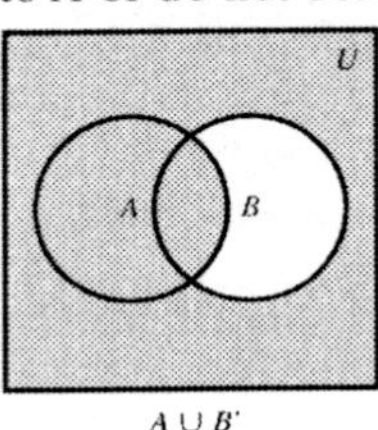

$A \cup B'$

44. $A' \cap B$ contains all elements in B and not in A.

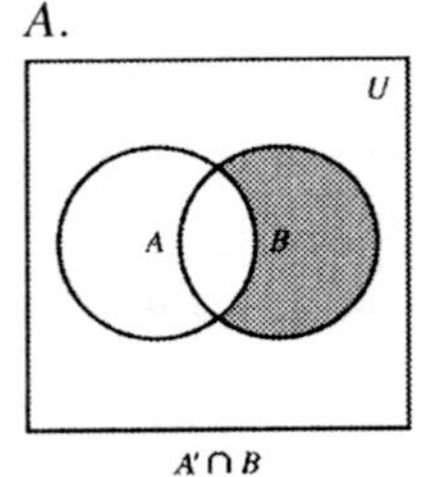

$A' \cap B$

45. $(A \cap B) \cup C$

First find $A \cap B$.

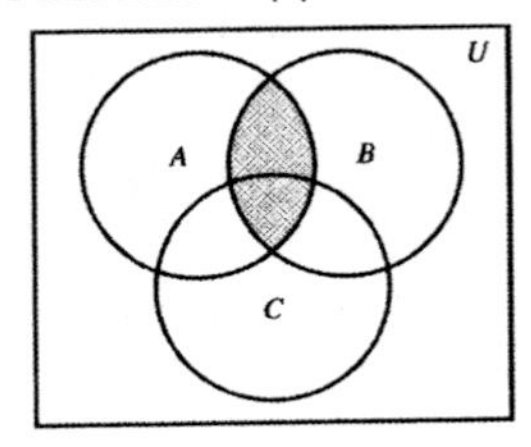

$A \cap B$

Now find the union of this region with C.

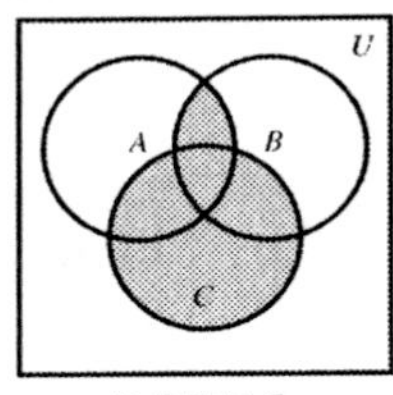

$(A \cap B) \cup C$

46. $(A \cup B)' \cap C$ includes those elements in C and not in either A or B.

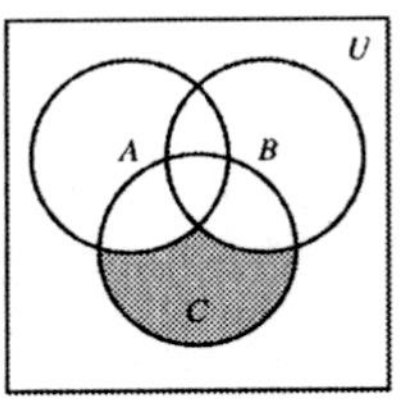

$(A \cup B)' \cap C$

47. The sample space for rolling a die is $S = \{1, 2, 3, 4, 5, 6\}$.

48. $S = \{\text{ace}, 2, 3, 4, 5, 6, 7, 8, 9, 10, J, Q, K\}$

49. The sample space of the possible weights is $S = \{0, 0.5, 1, 1.5, 2, \ldots, 299.5, 300\}$.

50. There are 16 possibilities.
$S = \{HHHH, HHHT, HHTH, HTHH, THHH, HHTT, HTHT, HTTH, THHT, TTHH, THTH, HTTT, THTT, TTHT, TTTH, TTTT\}$

51. The sample space consists of all ordered pairs (a, b) where a can be 3, 5, 7, 9, or 11, and b is either R (red) or G (green). Thus,
$S = \{(3, R), (3, G), (5, R), (5, G), (7, R), (7, G), (9, R), (9, G), (11, R), (11, G)\}$.

52. Let R = red and G = green. Then,
$E = \{(7, R), (7, G), (9, R), (9, G), (11, R), (11, G)\}$.

53. The event F that the second ball is green is $F = \{(3, G), (5, G), (7, G), (9, G), (11, G)\}$.

54. The outcomes are not equally likely since there are more red than green balls. For example, $(7, R)$ is twice as likely as $(7, G)$.

55. There are 13 hearts out of 52 cards in a deck. Thus, $P(\text{heart}) = \frac{13}{52} = \frac{1}{4}$.

56. There are 2 red queens out of 52 cards, so $P(\text{red queen}) = \frac{2}{52} = \frac{1}{26}$.

57. There are 3 face cards in each of the four suits.

$$P(\text{face card}) = \frac{12}{52},\ P(\text{heart}) = \frac{13}{52},$$

$$P(\text{face card and heart}) = \frac{3}{52}$$

$$\begin{aligned} P(\text{face card or heart}) &= P(\text{face card}) + P(\text{heart}) \\ &\quad - P(\text{face card and heart}) \\ &= \frac{12}{52} + \frac{13}{52} - \frac{3}{52} \\ &= \frac{22}{52} = \frac{11}{26} \end{aligned}$$

58. There are 26 black cards plus 6 red face cards, so $P(\text{black or a face card}) = \frac{32}{52} = \frac{8}{13}$.

59. There are 4 queens of which 2 are red, so

$$P(\text{red} \mid \text{queen}) = \frac{n(\text{red and queen})}{n(\text{queen})} = \frac{2}{4} = \frac{1}{2}.$$

60. There are 12 face cards, of which 4 are jacks, so $P(\text{jack} \mid \text{face card}) = \frac{4}{12} = \frac{1}{3}$.

61. There are 4 kings of which all 4 are face cards. Thus,

$$P(\text{face card} \mid \text{king}) = \frac{n(\text{face card and king})}{n(\text{king})} = \frac{4}{4} = 1.$$

62. Since the king is a face card, $P(\text{king} \mid \text{not face card}) = 0$.

63. Sets that have no elements in common are disjoint sets.

64. Two events are mutually exclusive if they cannot occur at the same time. An example is tossing a coin once, which can result in either heads or tails, but not both.

65. If two events are disjoint, then the probability of them both occurring at the same time is 0. If two events are mutually exclusive, then the probability of either occurring is the sum of the probabilities of each occurring.

66. Two events, A and B, are independent if the fact that A occurs does not affect the probability of B occurring.

67. If A and B are nonempty and independent, then $P(A \cap B) = P(A) \cdot P(B)$. For mutually exclusive events, $P(A \cap B) = 0$, which would mean $P(A) = 0$ or $P(B) = 0$. So independent events with nonzero probabilities are not mutually exclusive. But independent events one of which has zero probability are mutually exclusive.

68. Marilyn vos Savant's answer is that the contestant should switch doors. To understand why, recall that the puzzle begins with the contestant choosing door 1 and then the host opening door 3 to reveal a goat. When the host opens door 3 and shows the goat, that does not affect the probability of the car being behind door 1; the contestant had a 1/3 probability of being correct to begin with, and he still has a 1/3 probability after the host opens door 3. The contestant knew that the host would open another door regardless of what was behind door 1, so opening either other door gives no new information about door 1.

(*continued on next page*)

(continued)

The probability of the car being behind door 1 is still 1/3; with the goat behind door 3, the only other place the car could be is behind door 2. So, the probability that the car is behind door 2 is now 2/3. By switching to door 2, the contestant can double his chances of winning the car.

69. Let C be the event "a club is drawn." There are 13 clubs in the deck, so $n(C) = 13$,

$P(C') = \frac{13}{52} = \frac{1}{4}$, and $P(C') = 1 - P(C) = \frac{3}{4}$.

The odds in favor of drawing a club are

$\frac{P(C)}{P(C')} = \frac{\frac{1}{4}}{\frac{3}{4}} = \frac{1}{3}$, which is written "1 to 3."

70. Let E represent the event "draw a black jack." $P(E) = \frac{2}{52} = \frac{1}{26}$ and then $P(E') = \frac{25}{26}$. The odds in favor of drawing a black jack are

$\frac{P(E)}{P(E')} = \frac{\frac{1}{26}}{\frac{25}{26}} = \frac{1}{25}$, or 1 to 25.

71. Let R be the event "a red face card is drawn" and Q be the event "a queen is drawn." Use the union rule for probability to find $P(R \cup Q)$.

$$P(R \cup Q) = P(R) + P(Q) - P(R \cap Q) = \frac{6}{52} + \frac{4}{52} - \frac{2}{52} = \frac{8}{52} = \frac{2}{13}$$

$$P(R \cup Q)' = 1 - P(R \cup Q) = 1 - \frac{2}{13} = \frac{11}{13}$$

The odds in favor of drawing a red face card or a queen are $\frac{P(R \cup Q)}{P(R \cup Q)'} = \frac{\frac{2}{13}}{\frac{11}{13}} = \frac{2}{11}$, which is written "2 to 11."

72. Let E be the event "ace or club."

$$P(E) = P(\text{ace}) + P(\text{club}) - P(\text{ace and club}) = \frac{4}{52} + \frac{13}{52} - \frac{1}{52} = \frac{16}{52} = \frac{4}{13}$$

Therefore $P(E') = 1 - \frac{4}{13} = \frac{9}{13}$. The odds in favor of E are $\frac{4/13}{9/13} = \frac{4}{9}$ or 4 to 9.

73. The sum is 8 for each of the 5 outcomes (2, 6), (3, 5), (4, 4), (5, 3), and (6, 2). There are 36 outcomes in all in the sample space.

$P(\text{sum is } 8) = \frac{5}{36}$

74. A sum of 0 is impossible, so $P(\text{sum is } 0) = 0$.

75.
$$P(\text{sum is at least } 10) = P(\text{sum is } 10) + P(\text{sum is } 11) + P(\text{sum is } 12) = \frac{3}{36} + \frac{2}{36} + \frac{1}{36} = \frac{6}{36} = \frac{1}{6}$$

76.
$$P(\text{sum is no more than } 5) = P(2) + P(3) + P(4) + P(5) = \frac{1}{36} + \frac{2}{36} + \frac{3}{36} + \frac{4}{36} = \frac{10}{36} = \frac{5}{18}$$

77. The sum can be 9 or 11. $P(\text{sum is } 9) = \frac{4}{36}$ and

$P(\text{sum is } 11) = \frac{2}{36}$.

$$P(\text{sum is odd number greater than } 8) = \frac{4}{36} + \frac{2}{36} = \frac{6}{36} = \frac{1}{6}$$

78. A roll greater than 10 means 11 or 12. There are 3 ways to get 11 or 12, 2 for 11 and 1 for 12. Hence, $P(12 | \text{sum greater than } 10) = \frac{1}{3}$.

79. Consider the reduced sample space of the 11 outcomes in which at least one die is a four. Of these, 2 have a sum of 7, (3, 4) and (4, 3). Therefore,

$P(\text{sum is } 7 | \text{at least one die is a } 4) = \frac{2}{11}$.

80. For $P(\text{at least } 9 | \text{one die is a } 5)$, the sample space is reduced to

$\{(5,1), (5,2), (5,3), (5,4), (5,5), (5,6), (1,5), (2,5), (3,5), (4,5), (6,5)\}$.

Of these 11 outcomes, 5 give a sum of 9 or more, so

$P(\text{at least } 9 | \text{at least one die is } 5) = \frac{5}{11}$.

81. $P(E) = 0.51$, $P(F) = 0.37$, $P(E \cap F) = 0.22$

(a) $P(E \cup F) = P(E) + P(F) - P(E \cap F) = 0.51 + 0.37 - 0.22 = 0.66$

(b) Draw a Venn diagram.

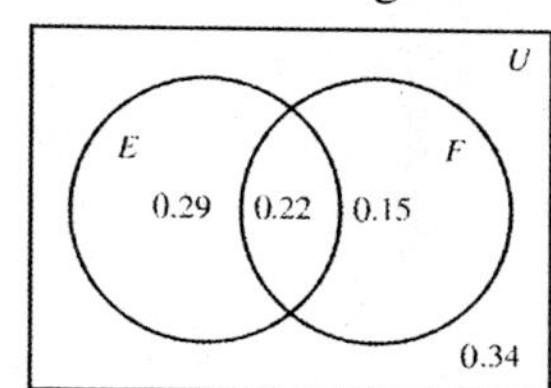

$E \cap F'$ is the portion of the diagram that is inside E and outside F. $P(E \cap F') = 0.29$

(c) $E' \cup F$ is outside E or inside F, or both. $P(E' \cup F) = 0.22 + 0.15 + 0.34 = 0.71$.

(d) $E' \cap F'$ is outside E and outside F. $P(E' \cap F') = 0.34$

82. Let M represent "first urn," N represent "second urn," R represent "red ball," and B represent "blue ball." Let x be the number of blue balls in the second urn. The probability that both balls drawn are the same color is

$$P(R|M) \cdot P(R|N) + P(B|M) \cdot P(B|N) = \frac{4}{10} \cdot \frac{16}{x+16} + \frac{6}{10} \cdot \frac{x}{x+16} = \frac{6x+64}{10(x+16)} = \frac{3x+32}{5(x+16)}$$

Now set this expression equal to 0.44 and solve for x.

$$\frac{3x+32}{5(x+16)} = 0.44 \Rightarrow 3x + 32 = 2.2x + 35.2 \Rightarrow 0.8x = 3.2 \Rightarrow x = 4$$

The correct answer choice is **a**.

83. First make a tree diagram. Let A represent "box A" and K represent "black ball."

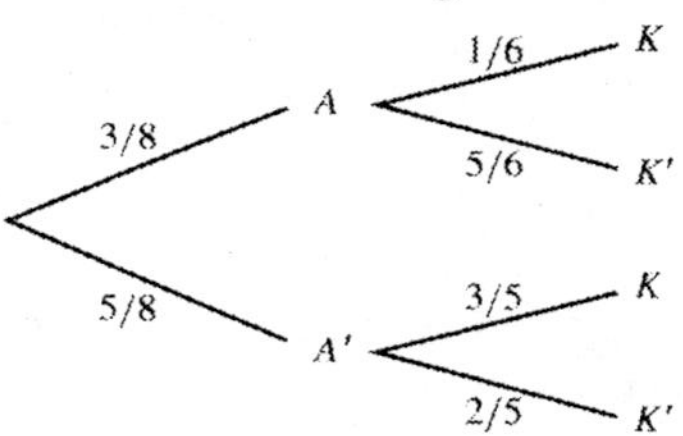

Use Bayes' theorem:. $P(A|K) = \dfrac{P(A) \cdot P(K|A)}{P(A) \cdot P(K|A) + P(A') \cdot P(K|A')} = \dfrac{\frac{3}{8} \cdot \frac{1}{6}}{\frac{3}{8} \cdot \frac{1}{6} + \frac{5}{8} \cdot \frac{3}{5}} = \dfrac{\frac{1}{16}}{\frac{7}{16}} = \dfrac{1}{7}$

84. The probability that the ball came from box B, given that it is red, is

$$P(B|\text{red}) = \frac{P(B) \cdot P(\text{red}|B)}{P(B) \cdot P(\text{red}|B) + P(A) \cdot P(\text{red}|A)} = \frac{\frac{5}{8}\left(\frac{2}{5}\right)}{\frac{5}{8}\left(\frac{2}{5}\right) + \frac{3}{8}\left(\frac{5}{6}\right)} = \frac{4}{9}.$$

85. (a) $P(\text{success}) = P(\text{head}) = \frac{1}{2}$. Hence, $n = 3$ and $p = \frac{1}{2}$.

Number of Heads	Probability
0	$C(3,0)\left(\frac{1}{2}\right)^0\left(\frac{1}{2}\right)^3 = 0.125$
1	$C(3,1)\left(\frac{1}{2}\right)^1\left(\frac{1}{2}\right)^2 = 0.375$
2	$C(3,2)\left(\frac{1}{2}\right)^2\left(\frac{1}{2}\right)^1 = 0.375$
3	$C(3,3)\left(\frac{1}{2}\right)^3\left(\frac{1}{2}\right)^0 = 0.125$

(b)

(c) $E(x) = 0(0.125) + 1(0.375) + 2(0.375) + 3(0.125) = 1.5$

(d) $\sigma^2 = \sum_x (x-\mu)^2 P(x) = (0-1.5)^2(0.125) + (1-1.5)^2(0.375) + (2-1.5)^2(0.375) + (3-1.5)^2(0.125)$

$= 0.75$

$\sigma \approx 0.866$

86. (a) There are $n = 36$ possible outcomes. Let x represent the sum of the dice, and note that the possible values of x are the whole numbers from 2 to 12. The probability distribution is as follows.

x	2	3	4	5	6	
$P(x)$	$\frac{1}{36}$	$\frac{2}{36}=\frac{1}{18}$	$\frac{3}{36}=\frac{1}{12}$	$\frac{4}{36}=\frac{1}{9}$	$\frac{5}{36}$	
x	7	8	9	10	11	12
$P(x)$	$\frac{6}{36}=\frac{1}{6}$	$\frac{5}{36}$	$\frac{4}{36}=\frac{1}{9}$	$\frac{3}{36}=\frac{1}{12}$	$\frac{2}{36}=\frac{1}{18}$	$\frac{1}{36}$

(b) The histogram consists of 11 rectangles.

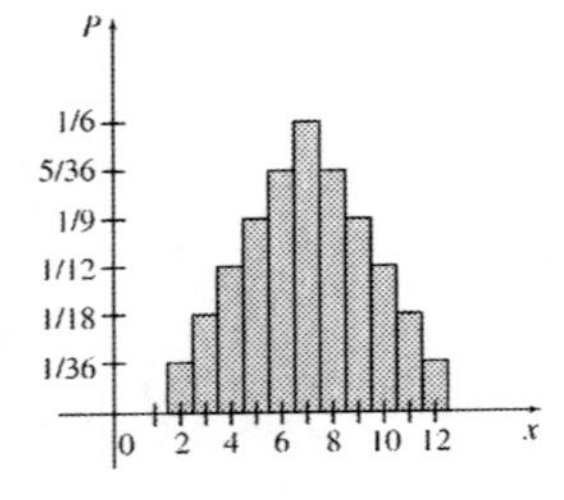

(c) The expected value is

$$E(x)=2\left(\frac{1}{36}\right)+3\left(\frac{2}{36}\right)+4\left(\frac{3}{36}\right)+5\left(\frac{4}{36}\right)+6\left(\frac{5}{36}\right)+7\left(\frac{6}{36}\right)+8\left(\frac{5}{36}\right)+9\left(\frac{4}{36}\right)+10\left(\frac{3}{36}\right)+11\left(\frac{2}{36}\right)+12\left(\frac{1}{36}\right)$$

$$=\frac{252}{36}=7.$$

$$\sigma^2=\sum_x (x-\mu)^2 P(x)$$

$$=(2-7)^2\left(\frac{1}{36}\right)+(3-7)^2\left(\frac{1}{18}\right)+(4-7)^2\left(\frac{1}{12}\right)+(5-7)^2\left(\frac{1}{9}\right)+(6-7)^2\left(\frac{5}{36}\right)+(7-7)^2\left(\frac{1}{6}\right)+(8-7)^2\left(\frac{5}{36}\right)+(9-7)^2\left(\frac{1}{9}\right)+(10-7)^2\left(\frac{1}{12}\right)+(11-7)^2\left(\frac{1}{18}\right)+(12-7)^2\left(\frac{1}{36}\right)$$

$$=\frac{35}{6}$$

$$\sigma\approx 2.415$$

87. The probability that corresponds to the shaded region of the histogram is the total of the shaded areas, that is, $1(0.3)+1(0.2)+1(0.1)=0.6.$

88. The probability that corresponds to the shaded region of the histogram is the total of the shaded areas, that is, $1(0.1)+1(0.3)+1(0.2)=0.6.$

89. The probability of rolling a 6 is $\frac{1}{6}$, and your net winnings would be \$2. The probability of rolling a 5 is $\frac{1}{6}$, and your net winnings would be \$1. The probability of rolling something else is $\frac{4}{6}$, and your net winnings would be $-\$2$. Let x represent your winnings. The expected value is

$$E(x)=2\left(\frac{1}{6}\right)+1\left(\frac{1}{6}\right)+(-2)\left(\frac{4}{6}\right)=-\frac{5}{6}\approx -\$0.833 \text{ or } -83.3¢.$$

This is not a fair game since the expected value is not 0.

90. Let x represent the number of girls. The probability distribution is as follows.

x	0	1	2	3
$P(x)$	$\frac{1}{8}$	$\frac{3}{8}$	$\frac{3}{8}$	$\frac{1}{8}$

The expected value is $E(x)=0\left(\frac{1}{8}\right)+1\left(\frac{3}{8}\right)+2\left(\frac{3}{8}\right)+3\left(\frac{1}{8}\right)=\frac{12}{8}=1.5$ girls.

91. (a)

Number of Aces	Probability
0	$\frac{C(4,0)C(48,3)}{C(52,3)} = \frac{17,296}{22,100}$
1	$\frac{C(4,1)C(48,2)}{C(52,3)} = \frac{4512}{22,100}$
2	$\frac{C(4,2)C(48,1)}{C(52,3)} = \frac{288}{22,100}$
3	$\frac{C(4,3)C(48,0)}{C(52,3)} = \frac{4}{22,100}$

$$E(x) = 0\left(\frac{17,296}{22,100}\right) + 1\left(\frac{4512}{22,100}\right) + 2\left(\frac{288}{22,100}\right) + 3\left(\frac{4}{22,100}\right) = \frac{5100}{22,100} = \frac{51}{221} = \frac{3}{13} \approx 0.231$$

(b)

Number of Clubs	Probability
0	$\frac{C(13,0)C(39,3)}{C(52,3)} = \frac{9139}{22,100}$
1	$\frac{C(13,1)C(39,2)}{C(52,3)} = \frac{9633}{22,100}$
2	$\frac{C(13,2)C(39,1)}{C(52,3)} = \frac{3042}{22,100}$
3	$\frac{C(13,3)C(39,0)}{C(52,3)} = \frac{286}{22,100}$

$$E(x) = 0\left(\frac{9139}{22,100}\right) + 1\left(\frac{9633}{22,100}\right) + 2\left(\frac{3042}{22,100}\right) + 3\left(\frac{286}{22,100}\right) = \frac{16,575}{22,100} = \frac{3}{4} = 0.75$$

92. $P(3\text{clubs}) = \frac{C(13,3)}{C(52,3)} = \frac{286}{22,100} \approx 0.0129$

Thus, $P(\text{win}) = 0.0129$ and
$P(\text{lose}) = 1 - 0.0129 = 0.9871$.
Let x represent the amount you should pay. Your net winnings are $100 - x$ if you win and $-x$ if you lose. If it is a fair game, your expected winnings will be 0. Thus, $E(x) = 0$ becomes

$$0.0129(100 - x) + 0.9871(-x) = 0$$
$$1.29 - 0.0129x - 0.9871x = 0$$
$$1.29 - x = 0 \Rightarrow x = 1.29.$$

You should pay \$1.29.

93. (a)

	N_2	T_2
N_1	N_1N_2	N_1T_2
T_1	T_1N_2	T_1T_2

Since the four combinations are equally likely, each has probability $\frac{1}{4}$.

(b) $P(\text{two trait cells}) = P(T_1T_2) = \frac{1}{4}$

(c) $P(\text{one normal cell and one trait cell})$
$= P(N_1T_2) + P(T_1N_2) = \frac{1}{4} + \frac{1}{4} = \frac{1}{2}$

(d) $P(\text{not a carrier and does not have disease})$
$= P(N_1N_2) = \frac{1}{4}$

94. Let $P(E)$ be the probability the random donor has blood type E.

(a) $P(O^+) + P(O^-) = 0.38 + 0.08$
$= 0.46$, or 46%

(b) $P(A^+) + P(A^-) + P(O^+) + P(O^-)$
$= 0.32 + 0.07 + 0.38 + 0.08$
$= 0.85$. or 85%

(c) $P(B^+) + P(B^-) + P(O^+) + P(O^-)$
$= 0.09 + 0.02 + 0.38 + 0.08$
$= 0.57$, or 57%

(d) $P(A^-) + P(O^-) = 0.07 + 0.08$
$= 0.15$, or 15%

(e) $P(B^-) + P(O^-) = 0.02 + 0.08$
$= 0.10$, or 10%

(f) $P(AB^-) + P(A^-) + P(B^-) + P(O^-)$
$= 0.01 + 0.07 + 0.02 + 0.08$
$= 0.18$, or 18%

95. We want to find $P(A' \cap B' \cap C' | A')$. Use a Venn diagram, fill in the information given, and use the diagram to help determine the missing values.

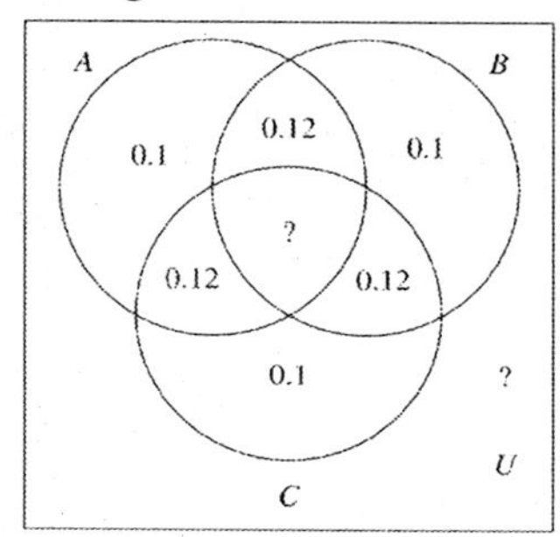

To determine $P(A \cap B \cap C)$, we are told that "The probability that a woman has all three risk factors, given that she has A and B, is 1/3." Therefore, $P(A \cap B \cap C \mid A \cap B) = 1/3$.

Let $x = P(A \cap B)$; then, using the diagram as a guide,

$$P(A \cap B \cap C) + P(A \cap B \cap C') = P(A \cap B)$$
$$\frac{1}{3}x + 0.12 = x$$
$$0.12 = \frac{2}{3}x$$
$$x = 0.18$$

So, $P(A \cap B \cap C) = (1/3)(0.16) = 0.06$.

It can be shown that

$A' \cap B' \cap C' = (A \cup B \cup C)'$, so that

$$\begin{aligned} P(A' \cap B' \cap C') &= P[(A \cup B \cup C)'] \\ &= 1 - P(A \cup B \cup C) \\ &= 1 - [3(0.10) + 3(0.12) + 0.06] \\ &= 0.28. \end{aligned}$$

Therefore,

$$\begin{aligned} P(A' \cap B' \cap C' | A') &= \frac{P(A' \cap B' \cap C' \cap A')}{P(A')} \\ &= \frac{P(A' \cap B' \cap C')}{P(A')} \\ &= \frac{0.28}{0.6} \approx 0.467. \end{aligned}$$

The correct answer choice is **c**.

96. In calculating the probability that two babies in a family would die of SIDS is $(1/8543)^2$, he assumed that the events that either infant died of SIDS are independent. There may be a genetic factor, in which case the events are dependent.

97. (a) $M \cap C$ is the set of overweight males.
$n(M \cap C) = 44{,}970$ thousand

(b) $F \cup B$ is the set of females or those who have a healthy weight. Add the number of females and the total number of those who have a healthy weight, then subtract the number of females who have a healthy weight (so they are not counted twice).

$$\begin{aligned} n(F \cup B) &= 112{,}446 + 78{,}847 - 46{,}640 \\ &= 144{,}653 \text{ thousand} \end{aligned}$$

(c) $M' \cap (C \cup D)$ is the set of overweight or obese females.

$$\begin{aligned} n(M' \cap (C \cup D)) &= 31{,}800 + 31{,}357 \\ &= 62{,}957 \text{ thousand} \end{aligned}$$

(d) $F' \cap (C' \cap D')$ represents the number of underweight or healthy weight males.

$$\begin{aligned} n(F' \cap (C' \cap D')) &= 1071 + 32{,}207 \\ &= 33{,}278 \text{ thousand} \end{aligned}$$

98. (a) The probability that a person is overweight is given by the total number of overweight people divided by the total number of people.

$$\frac{76{,}770}{221{,}562} \approx 0.3465$$

(b) The probability that a person is a female and at a healthy weight is given by the number of females at a healthy weight divided by the total number of people.

$$\frac{46{,}640}{221{,}562} \approx 0.2105$$

(c) The probability that a person is overweight or obese is given by the sum of the total number of overweight people and the total number of obese people, divided by the total number of people.

$$\frac{76,770+62,025}{221,562} \approx 0.6264$$

(d) The probability that a person is overweight given that the person is male is given by the number of overweight males divided by the total number of males.

$$\frac{44,970}{109,116} \approx 0.4121$$

(e) The probability that a person is female given that the person is overweight is given by the number of overweight females divided by the total number of overweight people.

$$\frac{31,800}{76,770} \approx 0.4142$$

(f) No, the events "male" and "overweight" are not independent, because a person can be both male and overweight, and $P(C) \neq P(C \mid M)$.

99. We will use Bayes' theorem. First draw a tree diagram to help visualize the problem. Since 0.2% of men have prostate cancer,

$P(D^+) = 0.002$ and

$P(D^-) = 1 - 0.002 = 0.998$

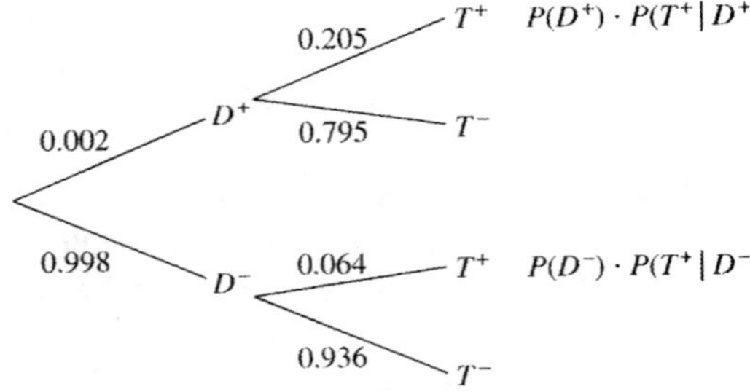

The sensitivity of the PSA test is 0.205, so the probability that a man will test positive given that he has the disease is $P(T^+ \mid D^+) = 0.205$.

By the complement rule, the probability that he will test negative, given that he has the disease, is $P(T^- \mid D^+) = 1 - 0.205 = 0.795$.

The specificity of the test is 0.936, so the probability that a man will test negative when he does not have the disease is

$P(T^- \mid D^-) = 0.936$. The probability that he will test positive when he does not have the disease is $P(T^+ \mid D^-) = 1 - 0.936 = 0.064$.

Using Bayes' theorem, we have

$$\begin{aligned} P(D^+ \mid T^+) &= \frac{P(D^+) \cdot P(T^+ \mid D^+)}{P(D^+) \cdot P(T^+ \mid D^+) + P(D^-) \cdot P(T^+ \mid D^-)} \\ &= \frac{0.002(0.205)}{0.002(0.205) + 0.998(0.064)} \\ &= 0.006378 \end{aligned}$$

100. (a) $E = 0.5(1 \cdot 0.8 + 0.98 \cdot 0.2) + 0.5(0.7 \cdot 0.1 + .06 \cdot 0.8 + 0) = 0.773$

(b) $E = 0.7$

Using antibiotics gives the higher expected quality of life.

101. (a) The percentage of voters who voted for Obama is given by the percentage of male voters who voted for him added to the percentage of female voters who voted for him.

$O = 0.47 \cdot 0.45 + 0.53 \cdot 0.55 = 0.503$

Thus, 50.3% of voters voted for Obama.

(b) The probability that a randomly selected voter for Obama was male is given by the percent of all voters who were male and voted for Obama divided by the percent of all voters who voted for Obama.

$$M = \frac{0.47 \cdot 0.45}{0.503} \approx 0.4205$$

(c) The probability that a randomly selected voter for Obama was female is given by 1 – the probability that a randomly selected voter for Obama was male.

$F = 1 - 0.4205 = 0.5795$

102. Let C be the set of viewers who watch situation comedies, G be the set of viewers who watch game shows, and M be the set of viewers who watch movies.
We are given the following information.
$n(C) = 20$ $n(G) = 19$ $n(M) = 27$
$n(M \cap G') = 19$
$n(C \cap G') = 15$
$n(C \cap M) = 10$
$n(C \cap G \cap M) = 3$
$n(C' \cap G' \cap M') = 7$
Start with $C \cap G \cap M : n(C \cap G \cap M) = 3$. Since $n(C \cap M) = 10$, the number of people who watched comedies and movies but not game shows, or $n(C \cap G' \cap M)$, is $10 - 3 = 7$. Since $n(M \cap G') = 19$, $n(C' \cap G' \cap M) = 19 - 7 = 12$. Since $n(M) = 27$, $n(C' \cap G \cap M) = 27 - 3 - 7 - 12 = 5$. Since $n(C \cap G') = 15$, $n(C \cap G' \cap M') = 15 - 7 = 8$. Since $n(C) = 20$, $n(C \cap G \cap M') = 20 - 8 - 3 - 7 = 2$. Finally, since $n(G) = 19$, $n(C' \cap G \cap M') = 19 - 2 - 3 - 5 = 9$.

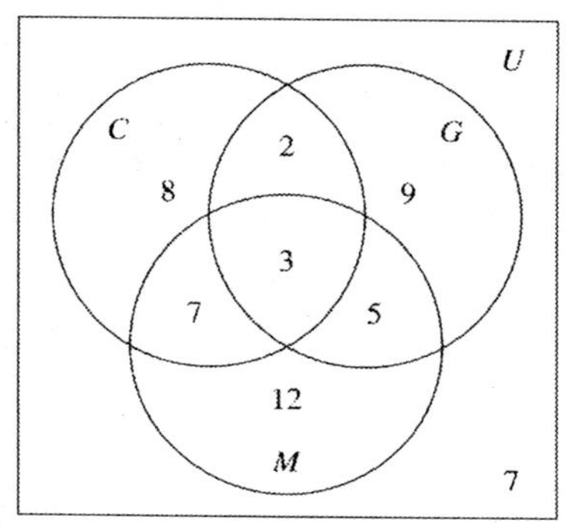

(a) $n(U) = 8 + 2 + 9 + 7 + 3 + 5 + 12 + 7 = 53$

(b) $n(C \cap G' \cap M) = 7$

(c) $n(C' \cap G' \cap M) = 12$

(d) $n(M') = n(U) - n(M) = 53 - 27 = 26$

103. **(a)** $P(\text{answer yes}) = P(\text{question } B) \cdot P(\text{answer yes} | \text{answer } B) + P(\text{answer } A) \cdot P(\text{answer yes} | \text{question } A)$
Divide by $P(\text{question } B)$.
$$\frac{P(\text{answer yes})}{P(\text{question } B)} = P(\text{answer yes} | \text{question } B) + \frac{P(\text{question } A) \cdot P(\text{answer yes} | \text{question } A)}{P(\text{question } B)}$$
Solve for $P(\text{answer yes} | \text{question } B)$.
$$P(\text{answer yes} \mid \text{question } B) = \frac{P(\text{answer yes}) - P(\text{question } A) \cdot P(\text{answer yes} | \text{question } A)}{P(\text{question } B)}$$

(b) Using the formula from part (a), $\dfrac{0.6 - \frac{1}{2}\left(\frac{1}{2}\right)}{\frac{1}{2}} = \dfrac{7}{10}$.

104. Let C be the event "the culprit penny is chosen." Then $P(C|HHH) = \dfrac{P(C \cap HHH)}{P(HHH)}$.

These heads will result two different ways. The culprit coin is chosen $\frac{1}{3}$ of the time and the probability of a head on any one flip is $\frac{3}{4}$: $P(C \cap HHH) = \frac{1}{3}\left(\frac{3}{4}\right)^3 \approx 0.1406$. If a fair (innocent) coin is chosen, the probability of a head on any one flip is $\frac{1}{2}$: $P(C'|HHH) = \frac{2}{3}\left(\frac{1}{2}\right)^3 \approx 0.0833$. Therefore,

$$P(C|HHH) = \frac{P(C \cap HHH)}{P(HHH)} = \frac{P(C \cap HHH)}{P(C \cap HHH) + P(C' \cap HHH)} \approx \frac{0.1406}{0.1406 + 0.0833} \approx 0.6279$$

105. $P(\text{earthquake}) = \dfrac{9}{9+1} = \dfrac{9}{10} = 0.90$

106. **(a)** $P(\text{making a 1st down with } n \text{ yards to go}) = \dfrac{\text{number of successes}}{\text{number of trials}}$

n	**Trials**	**Successes**	**Probability of Making First Down with n Yards to Go**
1	543	388	$\frac{388}{543} \approx 0.7145$
2	327	186	$\frac{186}{327} \approx 0.5688$
3	356	146	$\frac{146}{356} \approx 0.4101$
4	302	97	$\frac{97}{302} \approx 0.3212$
5	336	91	$\frac{91}{336} \approx 0.2708$

(b) Answers will vary.

107. Let W be the set of western states, S be the set of small states, and E be the set of early states. We are given the following information.

$n(W) = 24$ $\quad n(S) = 22$ $\quad n(E) = 26$

$n(W' \cap S' \cap E') = 9$ $\quad n(W \cap S) = 14$ $\quad n(S \cap E) = 11$ $\quad n(W \cap S \cap E) = 7$

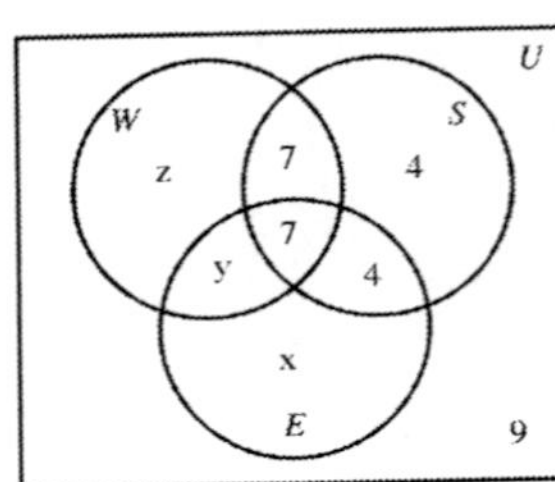

First, put 7 in $W \cap S \cap E$ and 9 in $W' \cap S' \cap E'$. Complete $S \cap E$ with 4 for a total of 11. Complete $W \cap S$ with 7 for a total of 14. Complete S with 4 for a total of 22.

To complete the rest of the diagram requires solving some equations. Let the incomplete region of E be x, the incomplete region of $W \cap E$ be y, and the incomplete region of W be z. Then, using the given values and the fact that $n\ (U) = 50$,

$$\begin{aligned} x + y \quad &= 15 \\ y + z &= 10 \\ x + y + z &= 50 - 22 - 9 = 19. \end{aligned}$$

(continued on next page)

(continued)

The solution to the system is $x = 9$, $y = 6$, $z = 4$.
Complete $W \cap E$ with 6 for a total of 13. Complete E with 9 for a total of 26. Complete W with 4 for a total of 24. The completed diagram is as follows.

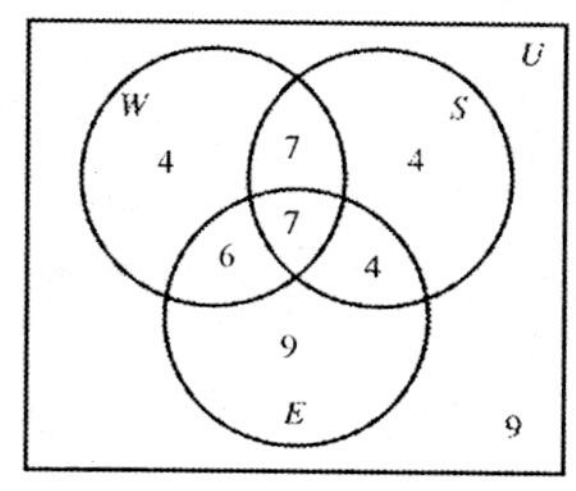

(a) $n(W \cap S' \cap E') = 4$

(b) $n(W' \cap S') = n((W \cup S)') = 18$

108. Let L be the set of songs about love, P be the set of songs about prison, and T be the set of songs about trucks. We are given the following information.

$n(L \cap P \cap T) = 12$ $\quad n(L \cap P) = 13$ $\quad n(L) = 28$ $\quad n(L \cap T) = 18$
$n(P') = 33$ $\quad n(P) = 18$ $\quad n(P \cap T) = 15$ $\quad n(P' \cap T) = 16$

Start with $L \cap P \cap T : n(L \cap P \cap T) = 12$. Since $n(L \cap P) = 13$, $n(L \cap P \cap T') = 1$.
Since $n(L \cap T) = 18$, $n(L \cap P' \cap T) = 6$. Since $n(L) = 28$, $n(L \cap P' \cap T') = 28 - 12 - 1 - 6 = 9$.
Since $n(P \cap T) = 15$, $n(L' \cap P \cap T) = 3$. Since $n(P) = 18$, $n(L' \cap P \cap T') = 18 - 1 - 12 - 3 = 2$.
Since $n(P' \cap T) = 16$, $n(L' \cap P' \cap T) = 16 - 6 = 10$. Finally, $n(L' \cap P' \cap T') = 8$.

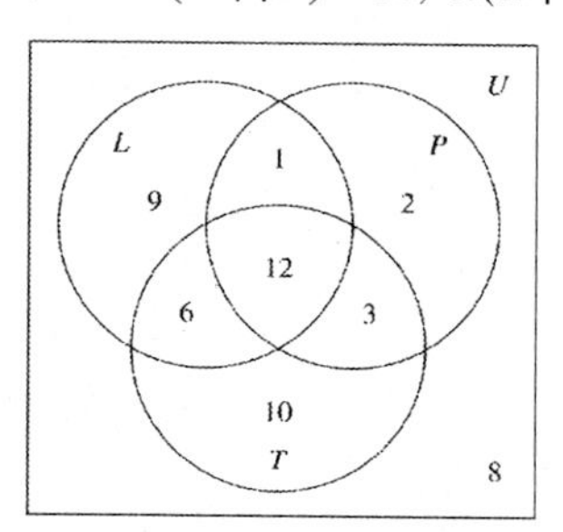

(a) $n(U) = 9 + 1 + 2 + 6 + 12 + 3 + 10 + 8 = 51$

(b) $n(T) = 6 + 12 + 3 + 10 = 31$

(c) $n(P \cap T' \cap L') = 2$

(d) $n(P \cap T' \cap L) = 1$

(e) $n(P \cap T') = 1 + 2 = 3$

(f) $n(L') = n(U) - n(L) = 51 - 28 = 23$

109. Let R be "a red side is facing up" and RR be "the 2-sided red card is chosen."
If a red side is facing up, we want to find $P(RR|R)$ since the other possibility would be a green side is facing down.

$$P(RR|R) = \frac{P(RR)}{P(R)} = \frac{\frac{1}{3}}{\frac{1}{2}} = \frac{2}{3}$$

No, the bet is not a good bet.

110. **(a)** $P(\text{double miss}) = 0.05(0.05) = 0.0025$

(b) $P(\text{specific silo destroyed}) = 1 - P(\text{double miss}) = 1 - 0.0025 = 0.9975$

(c) $P(\text{all ten destroyed}) = (0.9975)^{10} \approx 0.9753$

(d) $P(\text{at least one survived}) = 1 - P(\text{none survived}) = 1 - P(\text{all ten destroyed})$
$= 1 - 0.9753 = 0.0247$ or 2.47%

This does not agree with the quote of a 5% chance that at least one would survive.

(e) The events that each of the two bombs hit their targets are assumed to be independent. The events that each silo is destroyed are assumed to be independent.

111. Let G be the set of people who watched gymnastics, B be the set of people who watched baseball, and S be the set of people who watched soccer. We want to find $P(G' \cap B' \cap S')$, which is equivalent to $P[(G \cup B \cup S)']$.

We are given the following information.

$P(G) = 0.28$ $\quad P(B) = 0.29$ $\quad P(S) = 0.19$ $\quad P(G \cap B) = 0.14$
$P(B \cap S) = 0.12$ $\quad P(G \cap S) = 0.10$ $\quad P(G \cap B \cap S) = 0.08$

Start with $P(G \cap B \cap S) = 0.08$ and work from the inside out.

Since $P(G \cap S) = 0.10$, $P(G \cap B' \cap S) = 0.02$. Since $P(B \cap S) = 0.12$, $P(G' \cap B \cap S) = 0.04$.

Since $P(G \cap B) = 0.14$, $P(G \cap B \cap S') = 0.06$. Since $P(S) = 0.19, P(G' \cap B' \cap S) = 0.19 - 0.14 = 0.05$.

Since $P(B) = 0.29, P(G' \cap B \cap S') = 0.29 - 0.18 = 0.11$.

Since $P(G) = 0.28, P(G \cap B' \cap S') = 0.28 - 0.16 = 0.12$.

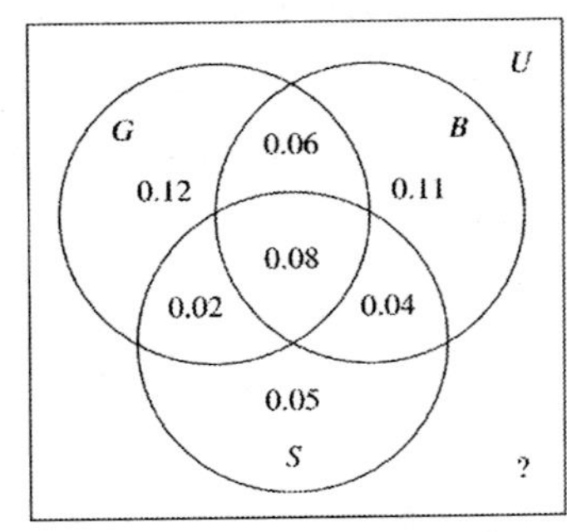

Therefore,

$P(G' \cap B' \cap S') = P[(G \cup B \cup S)'] = 1 - P(G \cup B \cup S)$
$= 1 - (0.12 + 0.06 + 0.11 + 0.02 + 0.08 + 0.04 + 0.05) = 0.52.$

The correct answer choice is **d**.

112. (a) $P(5 \text{ or more}) = P(5) + P(6) + P(7) + P(8) + P(9)$
$= 0.0040 + 0.0018 + 0.0007 + 0.0003 + 0.0001 = 0.0069$

(b) $P(\text{less than } 2) = P(0) + P(1) = 0.7345 + 0.1489 = 0.8834$

(c) The expected number of runs is

$E(x) = 0(0.7345) + 1(0.1489) + 2(0.0653) + 3(0.0306) + 4(0.0137) + 5(0.0040) + 6(0.0018)$
$+ 7(0.0007) + 8(0.0003) + 9(0.0001)$
$= 0.4651$

Chapter 13

PROBABILITY AND CALCULUS

13.1 Continuous Probability Models

1. $f(x) = \frac{1}{9}x - \frac{1}{18}$; [2, 5]

Show that condition 1 holds.

Since $2 \le x \le 5$,

$$\frac{2}{9} \le \frac{1}{9}x \le \frac{5}{9}$$

$$\frac{1}{6} \le \frac{1}{9}x - \frac{1}{18} \le \frac{1}{2}.$$

Hence, $f(x) \ge 0$ on [2, 5].

Show that condition 2 holds.

$$\int_2^5 \left(\frac{1}{9}x - \frac{1}{18}\right) dx = \frac{1}{9}\int_2^5 \left(x - \frac{1}{2}\right) dx$$

$$= \frac{1}{9}\left(\frac{x^2}{2} - \frac{1}{2}x\right)\Bigg|_2^5$$

$$= \frac{1}{9}\left(\frac{25}{2} - \frac{5}{2} - \frac{4}{2} + 1\right)$$

$$= \frac{1}{9}(8+1) = 1$$

Yes, $f(x)$ is a probability density function.

3. $f(x) = \frac{1}{21}x^2$; [1, 4]

Since $x^2 \ge 0$, $f(x) \ge 0$ on [1, 4].

$$\frac{1}{21}\int_1^4 x^2\,dx = \frac{1}{21}\left(\frac{x^3}{3}\right)\Bigg|_1^4 = \frac{1}{21}\left(\frac{64}{3} - \frac{1}{3}\right) = 1$$

Yes, $f(x)$ is a probability density function.

5. $f(x) = 4x^3$; [0, 3]

$$4\int_0^3 x^3\,dx = 4\left(\frac{x^4}{4}\right)\Bigg|_0^3 = 4\left(\frac{81}{4} - 0\right) = 81 \ne 1$$

No, $f(x)$ is not a probability density function.

7. $f(x) = \frac{x^2}{16}$; [−2, 2]

$$\frac{1}{16}\int_{-2}^2 x^2\,dx = \frac{1}{16}\left(\frac{x^3}{3}\right)\Bigg|_{-2}^2 = \frac{1}{16}\left(\frac{8}{3} + \frac{8}{3}\right)$$

$$= \frac{1}{3} \ne 1$$

No, $f(x)$ is not a probability density function.

9. $f(x) = \frac{5}{3}x^2 - \frac{5}{90}$; [−1, 1]

Let $x = 0$. Then $f(x) = f(0) = -\frac{5}{90} < 0$.

So, $f(x) < 0$ for at least one x-value in [−1, 1]. Therefore, $f(x)$ is not a probability density function.

11. $f(x) = kx^{1/2}$; [1, 4]

$$\int_1^4 kx^{1/2}dx = \frac{2}{3}kx^{3/2}\Bigg|_1^4 = \frac{2}{3}k(8-1) = \frac{14}{3}k$$

$$\frac{14}{3}k = 1 \Rightarrow k = \frac{3}{14}.$$

Notice that $f(x) = \frac{3}{14}x^{1/2} \ge 0$ for all x in [1, 4].

13. $f(x) = kx^2$; [0, 5]

$$\int_0^5 kx^2\,dx = k\frac{x^3}{3}\Bigg|_0^5 = k\left(\frac{125}{3} - 0\right) = k\left(\frac{125}{3}\right)$$

$$k\left(\frac{125}{3}\right) = 1 \Rightarrow k = \frac{3}{125}.$$

Notice that $f(x) = \frac{3}{125}x^2 \ge 0$ for all x in [0, 5].

15. $f(x) = kx$; [0, 3]

$$\int_0^3 kx\,dx = k\frac{x^2}{2}\Bigg|_0^3 = k\left(\frac{9}{2} - 0\right) = \frac{9}{2}k$$

$$\frac{9}{2}k = 1 \Rightarrow k = \frac{2}{9}.$$

Notice that $f(x) = \frac{2}{9}x \ge 0$ for all x in [0, 3].

17. $f(x) = kx;\ [1, 5]$

$$\int_1^5 kx\,dx = k\frac{x^2}{2}\bigg|_1^5 = k\left(\frac{25}{2} - \frac{1}{2}\right) = 12k$$

$$12k = 1 \Rightarrow k = \frac{1}{12}.$$

Notice that $f(x) = \frac{1}{12}x \geq 0$ for all x in $[1, 5]$.

19. For the probability density function $f(x) = \frac{1}{9}x - \frac{1}{18}$ on $[2, 5]$, the cumulative distribution function is

$$F(x) = \int_a^x f(t)\,dt = \int_a^x \left(\frac{1}{9}t - \frac{1}{18}\right) dt$$

$$= \left(\frac{1}{18}t^2 - \frac{1}{18}t\right)\bigg|_2^x = \frac{1}{18}[(x^2 - x) - (4 - 2)]$$

$$= \frac{1}{18}(x^2 - x - 2),\ 2 \leq x \leq 5.$$

21. For the probability density function $f(x) = \frac{x^2}{21}$ on $[1, 4]$, the cumulative distribution function is

$$F(x) = \int_1^x \frac{t^2}{21}dt = \frac{t^3}{63}\bigg|_1^x = \frac{1}{63}(x^3 - 1),\ 1 \leq x \leq 4.$$

23. The value of k was found to be $\frac{3}{14}$. For the probability density function $f(x) = \frac{3}{14}x^{1/2}$ on $[1, 4]$, the cumulative distribution function is

$$F(x) = \int_1^x \frac{3}{14}t^{1/2}\,dt = \frac{3}{14}\cdot\frac{2}{3}t^{3/2}\bigg|_1^x$$

$$= \frac{1}{7}(x^{3/2} - 1),\ 1 \leq x \leq 4.$$

25. The total area under the graph of a probability density function always equals 1.

27. A discrete probability function has a finite domain or an infinite domain that can be listed. For a probability density function, only probabilities of intervals can be found.

29. $f(x) = \frac{1}{2}(1+x)^{-3/2};\ [0, \infty)$

$$\frac{1}{2}\int_0^\infty (1+x)^{-3/2}\,dx$$

$$= \lim_{b\to\infty} \frac{1}{2}\int_0^b (1+x)^{-3/2}\,dx$$

$$= \lim_{b\to\infty} \frac{1}{2}(1+x)^{-1/2}\left(\frac{-2}{1}\right)\bigg|_0^b$$

$$= \lim_{b\to\infty} [-(1+b)^{-1/2} + 1]$$

$$= \lim_{b\to\infty} \left(\frac{-1}{\sqrt{1+b}} + 1\right) = 0 + 1 = 1$$

Since $x \geq 0,\ f(x) \geq 0.$

$f(x)$ is a probability density function.

(a) $P(0 \leq X \leq 2) = \frac{1}{2}\int_0^2 (1+x)^{-3/2}dx$

$$= -(1+x)^{-1/2}\Big|_0^2$$

$$= -3^{-1/2} + 1 \approx 0.4226$$

(b) $P(1 \leq X \leq 3) = \frac{1}{2}\int_1^3 (1+x)^{-3/2}dx$

$$= -(1+x)^{-1/2}\Big|_1^3$$

$$= -4^{-1/2} + 2^{-1/2} \approx 0.2071$$

(c) $P(X \geq 5) = \frac{1}{2}\int_5^\infty (1+x)^{-3/2}\,dx$

$$= \lim_{a\to\infty} \frac{1}{2}\int_5^a (1+x)^{-3/2}\,dx$$

$$= \lim_{a\to\infty} [-(1+x)^{-1/2]}\Big|_5^a$$

$$= \lim_{a\to\infty} [-(1+a)^{-1/2} + 6^{-1/2}]$$

$$= \lim_{a\to\infty} \left(\frac{-1}{\sqrt{1+a}} + 6^{-1/2}\right)$$

$$\approx 0 + 0.4082 = 0.4082$$

31. $f(x) = \frac{1}{2}e^{-x/2};\ [0, \infty)$

$$\frac{1}{2}\int_0^\infty e^{-x/2}\,dx = \lim_{a\to\infty} \frac{1}{2}\int_0^a e^{-x/2}\,dx$$

$$= \lim_{a\to\infty} \frac{1}{2}\left(\frac{-2}{1}e^{-x/2}\right)\bigg|_0^a$$

$$= \lim_{a\to\infty} -e^{-x/2}\Big|_0^a$$

$$= \lim_{a\to\infty} \left(\frac{-1}{e^{a/2}} + 1\right) = 0 + 1 = 1$$

$f(x) > 0$ for all x. Thus, $f(x)$ is a probability density function.

(a) $P(0 \le X \le 1) = \frac{1}{2}\int_0^1 e^{-x/2}\,dx = -e^{-x/2}\Big|_0^1$
$= \frac{-1}{e^{x/2}} + 1 \approx 0.3935$

(b) $P(1 \le X \le 3) = \frac{1}{2}\int_1^3 e^{-x/2}\,dx = -e^{-x/2}\Big|_1^3$
$= \frac{-1}{e^{3/2}} + \frac{1}{e^{1/2}} \approx 0.3834$

(c) $P(X \ge 2) = \frac{1}{2}\int_2^\infty e^{-x/2}\,dx$
$= \lim_{a\to\infty} \frac{1}{2}\int_2^a e^{-x/2}\,dx$
$= \lim_{a\to\infty} (-e^{-x/2})\Big|_2^a$
$= \lim_{a\to\infty} \left(\frac{-1}{e^{a/2}} + \frac{1}{e}\right) \approx 0.3679$

33. $f(x) = \begin{cases} \frac{x^3}{12} & \text{if } 0 \le x \le 2 \\ \frac{16}{3x^3} & \text{if } x > 2 \end{cases}$

First, note that $f(x) > 0$ for $x > 0$. Next,

$\int_0^\infty f(x)dx$
$= \int_0^2 \frac{x^3}{12}dx + \lim_{a\to\infty}\int_2^a \frac{16}{3x^3}dx$
$= \left(\frac{x^4}{48}\right)\Big|_0^2 + \lim_{a\to\infty}\left(-\frac{8}{3x^2}\right)\Big|_2^a$
$= \left(\frac{1}{3} - 0\right) + \left[\lim_{a\to\infty}\left(-\frac{8}{3a^2}\right) - \left(-\frac{8}{12}\right)\right]$
$= \frac{1}{3} + \frac{2}{3} = 1.$

Therefore, $f(x)$ is a probability density function.

(a) $P(0 \le X \le 2) = \int_0^2 f(x)dx = \left(\frac{x^4}{48}\right)\Big|_0^2 = \frac{1}{3}$

(b) $P(X \ge 2) = P(X > 2) = \int_2^\infty \frac{16}{3x^3}dx$
$= \lim_{a\to\infty}\int_2^a \frac{16}{3x^3}dx$
$= \lim_{a\to\infty}\left(-\frac{8}{3x^2}\right)\Big|_2^a$
$= \lim_{a\to\infty}\left(-\frac{8}{3a^2}\right) - \left(-\frac{8}{3\cdot 2^2}\right)$
$= 0 - \left(-\frac{2}{3}\right) = \frac{2}{3}$

(c) $P(1 \le X \le 3) = \int_1^2 \frac{x^3}{12}dx + \int_2^3 \frac{16}{3x^3}dx$
$= \left(\frac{x^4}{48}\right)\Big|_1^2 + \left(-\frac{8}{3x^2}\right)\Big|_2^3$
$= \left(\frac{1}{3} - \frac{1}{48}\right) + \left(-\frac{8}{27} + \frac{2}{3}\right)$
$= \frac{295}{432}$

35. $f(x) = \frac{1}{2\sqrt{x}}$; [1, 4]

(a) $P(3 \le X \le 4) = \int_3^4 \left(\frac{1}{2\sqrt{x}}\right)dx$
$= \frac{1}{2}\int_3^4 x^{-1/2}\,dx$
$= \frac{1}{2}(2)x^{1/2}\Big|_3^4 = 2 - 3^{1/2}$
≈ 0.2679

(b) $P(1 \le X \le 2) = \int_1^2 \left(\frac{1}{2\sqrt{x}}\right)dx = \frac{1}{2}(2)x^{1/2}\Big|_1^2$
$= 2^{1/2} - 1 = 0.4142$

(c) $P(2 \le X \le 3) = \int_2^3 \left(\frac{1}{2\sqrt{x}}\right)dx = \frac{1}{2}(2)x^{1/2}\Big|_2^3$
$= 3^{1/2} - 2^{1/2} = 0.3178$

37. $f(x) = 1.185 \cdot 10^{-9} x^{4.5222} e^{-0.049846x}$

(a) $P(0 \le X \le 150)$
$= \int_0^{150} 1.185 \cdot 10^{-9} x^{4.5222} e^{-0.049846x}\,dx$
≈ 0.8131

(b) $P(100 \le x \le 200)$
$= \int_{100}^{200} 1.185 \cdot 10^{-9} x^{4.5222} e^{-0.049846x}\,dx$
≈ 0.4901

39. $f(t) = \frac{8}{7(t-2)^2}$; [3, 10]

(a) $P(3 \le T < 4) = \frac{8}{7}\int_3^4 (t-2)^{-2}\,dt$
$= -\frac{8}{7}(t-2)^{-1}\Big|_3^4$
$= -\frac{8}{7}\left(\frac{1}{2} - 1\right) = \frac{4}{7} \approx 0.5714$

(b) $P(5 < T \le 10) = \frac{8}{7}\int_5^{10} (t-2)^{-2}\,dt$
$= -\frac{8}{7}(t-2)^{-1}\Big|_5^{10}$
$= -\frac{8}{7}\left(\frac{1}{8} - \frac{1}{3}\right) = \frac{5}{21} \approx 0.2381$

41. $f(t) = 0.06049e^{-0.03211t}$; $[16, 84]$

(a) $P(16 \le T \le 25) = \int_{16}^{25} f(t)dt$
$= \int_{16}^{25} 0.06049e^{-0.03211t}dt$
$= \frac{0.06049}{-0.03211}(e^{-0.03211t})\Big|_{16}^{25}$
$\approx -1.88384(e^{-0.03211t})\Big|_{16}^{25}$
≈ 0.2829

(b) $P(35 \le T \le 84) = \int_{35}^{84} 0.06049e^{-0.03211t}dt$
$\approx -1.88384(e^{-0.03211t})\Big|_{35}^{84}$
≈ 0.4853

(c) $P(21 \le T \le 30) = \int_{21}^{30} 0.06049e^{-0.03211t}\,dt$
$\approx -1.88384(e^{-0.03211t})\Big|_{21}^{30}$
≈ 0.2409

(d) $F(t) = \int_{16}^{t} 0.06049e^{-0.03211s}\,ds$
$= 0.06049 \cdot \frac{1}{-0.03211} e^{-0.03211s}\Big|_{16}^{t}$
$= -1.8838(e^{-0.03211t} - e^{-0.03211\cdot 16})$
$= 1.8838(0.5982 - e^{-0.03211t}),$
$16 \le t \le 84$

(e) $F(21) = 1.8838(0.5982 - e^{-0.03211\cdot 21})$
$= 1.8838(0.0887) = 0.1671$
The probability is 0.1671.

43. $f(t) = 3t^{-4}$; $[1, \infty)$

(a) $P(1 \le T \le 2) = \int_1^2 3t^{-4}\,dt = -t^{-3}\Big|_1^2$
$= 1 - \frac{1}{8} = 0.875$

(b) $P(3 \le T \le 5) = \int_3^5 3t^{-4}\,dt = -t^{-3}\Big|_3^5$
$= \frac{1}{27} - \frac{1}{125} \approx 0.0290$

(c) $P(T > 3) = \int_3^{\infty} 3t^{-4}\,dt = \lim_{b\to\infty} \int_3^b 3t^{-4}\,dt$
$= \lim_{b\to\infty} (-t^{-3})\Big|_3^b = \lim_{b\to\infty}\left(\frac{1}{27} - \frac{1}{b^3}\right)$
$= \frac{1}{27} \approx 0.0370$

45. We are told that the payment is the loss minus the deductible. Therefore,
$P(\text{payment} < 0.5) = P(X - C < 0.5)$
$= P(X < C + 0.5)$
$= \int_0^{C+0.5} 2x\,dx = x^2\Big|_0^{C+0.5}$
$= (C+0.5)^2 - 0$
$= C^2 + C + 0.25.$
We are given that the probability is 0.64.
$C^2 + C + 0.25 = 0.64$
$C^2 + C - 0.39 = 0$
$(C+1.3)(C-0.3) = 0 \Rightarrow C = -1.3$ or $C = 0.3$
The solution $C = -1.3$ is extraneous since $0 < C$. Thus, $C = 0.3$. The correct answer choice is **b**.

47. $f(x) = \frac{5.5 - x}{15}$; $[0, 5]$

(a) $P(3 < X \le 5) = \int_3^5 \frac{5.5 - x}{15}dx$
$= \left(\frac{5.5}{15}x - \frac{1}{15}\cdot\frac{x^2}{2}\right)\Bigg|_3^5$
$= \left(\frac{5.5}{15}\cdot 5 - \frac{1}{15}\cdot\frac{5^2}{2}\right)$
$- \left(\frac{5.5}{15}\cdot 3 - \frac{1}{15}\cdot\frac{3^2}{2}\right)$
$= 0.2$

(b) $P(0 \le X < 2) = \int_0^2 \frac{5.5 - x}{15}dx$
$= \left(\frac{5.5}{15}x - \frac{1}{15}\cdot\frac{x^2}{2}\right)\Bigg|_0^2$
$= \left(\frac{5.5}{15}\cdot 2 - \frac{1}{15}\cdot\frac{2^2}{2}\right)$
$- \left(\frac{5.5}{15}\cdot 0 - \frac{1}{15}\cdot\frac{0^2}{2}\right)$
$= 0.6$

(c) $P(1 \le X \le 4) = \int_1^4 \frac{5.5 - x}{15} dx$

$$= \left(\frac{5.5}{15}x - \frac{1}{15} \cdot \frac{x^2}{2} \right) \Big|_1^4$$

$$= \left(\frac{5.5}{15} \cdot 4 - \frac{1}{15} \cdot \frac{4^2}{2} \right) - \left(\frac{5.5}{15} \cdot 1 - \frac{1}{15} \cdot \frac{1^2}{2} \right)$$

$$= 0.6$$

49. $f(t) = \frac{1}{3650.1} e^{-t/3650.1}$

(a) $P(365 < T < 1095)$

$$= \int_{365}^{1095} \frac{1}{3650.1} e^{-t/3650.1} dt$$

$$= (-e^{-t/3650.1}) \Big|_{365}^{1095}$$

$$= -e^{-1095/3650.1} + e^{-365/3650.1}$$

$$\approx 0.1640$$

(b) $P(T > 7300)$

$$= \int_{7300}^{\infty} \frac{1}{3650.1} e^{-t/3650.1} dt$$

$$= \lim_{b \to \infty} \int_{7300}^{b} \frac{1}{3650.1} e^{-t/3650.1} dt$$

$$= \lim_{b \to \infty} (-e^{-t/3650.1}) \Big|_{7300}^{b}$$

$$= \lim_{b \to \infty} (-e^{-b/3650.1} + e^{-7300/3650.1})$$

$$= 0 + e^{-7300/3650.1} \approx 0.1353$$

13.2 Expected Value and Variance of Continuous Random Variables

1. $f(x) = \frac{1}{4}$; $[3, 7]$

$$E(X) = \mu = \int_3^7 \frac{1}{4} x \, dx = \frac{1}{4} \left(\frac{x^2}{2} \right) \Big|_3^7 = \frac{49}{8} - \frac{9}{8} = 5$$

$$\text{Var}(X) = \int_3^7 (x-5)^2 \left(\frac{1}{4} \right) dx$$

$$= \frac{1}{4} \cdot \frac{(x-5)^3}{3} \Big|_3^7 = \frac{8}{12} + \frac{8}{12} \approx 1.33$$

$$\sigma \approx \sqrt{\text{Var}(X)} = \sqrt{\frac{4}{3}} \approx 1.15$$

3. $f(x) = \frac{x}{8} - \frac{1}{4}$; $[2, 6]$

$$\mu = \int_2^6 x \left(\frac{x}{8} - \frac{1}{4} \right) dx = \int_2^6 \left(\frac{x^2}{8} - \frac{x}{4} \right) dx$$

$$= \left(\frac{x^3}{24} - \frac{x^2}{8} \right) \Big|_2^6 = \left(\frac{216}{24} - \frac{36}{8} \right) - \left(\frac{8}{24} - \frac{4}{8} \right)$$

$$= \frac{208}{24} - 4 = \frac{14}{3} \approx 4.67$$

Use the alternative formula to find Var (X).

$$\text{Var}(X) = \int_2^6 x^2 \left(\frac{x}{8} - \frac{1}{4} \right) dx - \left(\frac{14}{3} \right)^2$$

$$= \int_2^6 \left(\frac{x^3}{8} - \frac{x^2}{4} \right) dx - \frac{196}{9}$$

$$= \left(\frac{x^4}{32} - \frac{x^3}{12} \right) \Big|_2^6 - \frac{196}{9}$$

$$= \left(\frac{1296}{32} - \frac{216}{12} \right) - \left(\frac{16}{32} - \frac{8}{12} \right) - \frac{196}{9}$$

$$\approx 0.89.$$

$$\sigma = \sqrt{\text{Var}(X)} \approx \sqrt{0.89} \approx 0.94$$

5. $f(x) = 1 - \frac{1}{\sqrt{x}}$; $[1, 4]$

$$\mu = \int_1^4 x(1 - x^{-1/2}) dx = \int_1^4 (x - x^{1/2}) dx$$

$$= \left(\frac{x^2}{2} - \frac{2x^{3/2}}{3} \right) \Big|_1^4 = \frac{16}{2} - \frac{16}{3} - \frac{1}{2} + \frac{2}{3} = \frac{17}{6}$$

$$\approx 2.83$$

$$\text{Var}(X) = \int_1^4 x^2 (1 - x^{-1/2}) dx - \left(\frac{17}{6} \right)^2$$

$$= \int_1^4 (x^2 - x^{3/2}) dx - \frac{289}{36}$$

$$= \left(\frac{x^3}{3} - \frac{2x^{5/2}}{5} \right) \Big|_1^4 - \frac{289}{36}$$

$$= \frac{64}{3} - \frac{64}{5} - \frac{1}{3} + \frac{2}{5} - \frac{289}{36} \approx 0.57$$

$$\sigma \approx \sqrt{\text{Var}(X)} \approx 0.76$$

7. $f(x) = 4x^{-5}$; $[1, \infty)$

$$\mu = \int_1^{\infty} x(4x^{-5})dx = \lim_{a\to\infty} \int_1^a 4x^{-4}\,dx$$

$$= \lim_{a\to\infty} \left(\frac{4x^{-3}}{-3}\right)\Bigg|_1^a = \lim_{a\to\infty}\left(\frac{-4}{3a^3} + \frac{4}{3}\right) = \frac{4}{3}$$

$$\approx 1.33$$

$$\text{Var}(X) = \int_1^{\infty} x^2(4x^{-5})dx - \left(\frac{4}{3}\right)^2$$

$$= \lim_{a\to\infty} \int_1^a 4x^{-3}\,dx - \frac{16}{9}$$

$$= \lim_{a\to\infty}\left(\frac{4x^{-2}}{-2}\right)\Bigg|_1^a - \frac{16}{9}$$

$$= \lim_{a\to\infty}\left(\frac{-2}{a^2} + 2\right) - \frac{16}{9}$$

$$= 2 - \frac{16}{9} = \frac{2}{9} \approx 0.22$$

$$\sigma = \sqrt{\text{Var}(X)} = \sqrt{\frac{2}{9}} \approx 0.47$$

9. Answers will vary. Sample answer: The mean of a random variable (expected value) is the value you expect to obtain if you carry out some experiment whose outcomes are represented by the random variable. Geometrically, the expected value (or mean) of a probability distribution represents the balancing point of the distribution.

11. $f(x) = \dfrac{\sqrt{x}}{18}$; $[0, 9]$

(a) $$E(X) = \mu = \int_0^9 \frac{x\sqrt{x}}{18}dx = \int_0^9 \frac{x^{3/2}}{18}dx$$

$$= \frac{2x^{5/2}}{90}\Bigg|_0^9 = \frac{x^{5/2}}{45}\Bigg|_0^9 = \frac{243}{45} = 5.40$$

(b) $$\text{Var}(X) = \int_0^9 \frac{x^2\sqrt{x}}{18}\,dx - \left(\frac{27}{5}\right)^2$$

$$= \int_0^9 \frac{x^{5/2}}{18}\,dx - \left(\frac{27}{5}\right)^2$$

$$= \frac{x^{7/2}}{63}\Bigg|_0^9 - \left(\frac{27}{5}\right)^2 = \frac{2187}{63} - \left(\frac{27}{5}\right)^2$$

$$\approx 5.55$$

(c) $\sigma = \sqrt{\text{Var}(X)} \approx 2.36$

(d) $$P(5.40 < X \le 9) = \int_{5.4}^9 \frac{x^{1/2}}{18}\,dx = \frac{x^{3/2}}{27}\Bigg|_{5.4}^9$$

$$= \frac{27}{27} - \frac{(5.4)^{1.5}}{27} \approx 0.5352$$

(e) $P(5.40 - 2.36 \le X \le 5.40 + 2.36)$

$$= \int_{3.04}^{7.76} \frac{x^{1/2}}{18}\,dx = \frac{x^{3/2}}{27}\Bigg|_{3.04}^{7.76}$$

$$= \frac{7.76^{3/2}}{27} - \frac{3.04^{3/2}}{27} \approx 0.6043$$

13. $f(x) = \frac{1}{4}x^3$; $[0, 2]$

(a) $$E(X) = \mu = \int_0^2 \frac{1}{4}x^4\,dx = \frac{x^5}{20}\Bigg|_0^2$$

$$= \frac{32}{20} = \frac{8}{5} = 1.6$$

(b) $$\text{Var}(X) = \int_0^2 \frac{1}{4}x^5 dx - \frac{64}{25} = \frac{x^6}{24}\Bigg|_0^2 - \frac{64}{25}$$

$$= \frac{8}{3} - \frac{64}{25} = \frac{8}{75} \approx 0.11$$

(c) $\sigma = \sqrt{\text{Var}(X)} = \sqrt{\frac{8}{75}} \approx 0.3266 \approx 0.33$

(d) $$P(8/5 < X \le 2) = \int_{8/5}^2 \frac{x^3}{4}\,dx = \frac{x^4}{16}\Bigg|_{8/5}^2$$

$$= 1 - \frac{256}{625} = \frac{369}{625} \approx 0.5904$$

(e) Use a four-place value for the standard deviation.

$P(1.6 - 0.3266 \le X \le 1.6 + 0.3266)$

$$= \int_{1.2734}^{1.9266} \frac{x^3}{4}\,dx = \frac{x^4}{16}\Bigg|_{1.2734}^{1.9266}$$

$$= \frac{1.9266^4}{16} - \frac{0.1.2734^4}{16} \approx 0.6967$$

15. $f(x) = \frac{1}{4}$; $[3, 7]$

(a) $$\int_3^m \frac{1}{4}dx = \frac{1}{2} \Rightarrow \frac{1}{4}x\Bigg|_3^m = \frac{1}{2} \Rightarrow$$

$$\frac{m}{4} - \frac{3}{4} = \frac{1}{2} \Rightarrow m - 3 = 2 \Rightarrow m = 5$$

(b) $E(X)=\mu=5$ (from Exercise 1)

$$P(X=5)=\int_5^5 \frac{1}{4}\,dx=0$$

17. $f(x)=\frac{x}{8}-\frac{1}{4};\ [2,6]$

(a) $$\int_2^m\left(\frac{x}{8}-\frac{1}{4}\right)dx=\frac{1}{2}\Rightarrow\left(\frac{x^2}{16}-\frac{x}{4}\right)\Bigg|_2^m=\frac{1}{2}\Rightarrow\frac{m^2}{16}-\frac{m}{4}-\frac{1}{4}+\frac{1}{2}=\frac{1}{2}\Rightarrow m^2-4m-4=0\Rightarrow$$

$$m=\frac{4\pm\sqrt{16+16(1)}}{2}$$

Reject $\frac{4-\sqrt{32}}{2}$ since it is not in $[2, 6]$.

$$m=\frac{4+\sqrt{32}}{2}=2+2\sqrt{2}\approx 4.8284$$

(b) $E(X)=\mu=\frac{14}{3}$ (from Exercise 3)

$$P\left(\frac{14}{3}\le X\le 2+2\sqrt{2}\right)=\int_{14/3}^{2+2\sqrt{2}}\left(\frac{x}{8}-\frac{1}{4}\right)dx=\left(\frac{x^2}{16}-\frac{x}{4}\right)\Bigg|_{14/3}^{2+2\sqrt{2}}$$

$$=\frac{\left(2+2\sqrt{2}\right)^2}{16}-\frac{2+2\sqrt{2}}{4}-\frac{(14/3)^2}{16}+\frac{14/3}{4}=\frac{1}{18}\approx 0.0556$$

If you do the integration on a calculator using rounded values for the limits you may get a slightly different answer, such as 0.0553.

19. $f(x)=4x^{-5};\ [1,\infty)$

(a) $$\int_1^m 4x^{-5}\,dx=\frac{1}{2}\Rightarrow\frac{4x^{-4}}{-4}\Bigg|_1^m=\frac{1}{2}\Rightarrow -m^{-4}+1=\frac{1}{2}\Rightarrow 1-\frac{1}{m^4}=\frac{1}{2}\Rightarrow 2m^4-2=m^4\Rightarrow m^4=2\Rightarrow$$

$$m=\sqrt[4]{2}\approx 1.189$$

(b) $E(X)=\mu=\frac{4}{3}$ (from Exercise 7)

$$P\left(1.19\le X\le\tfrac{4}{3}\right)\approx\int_{1.189}^{1.333}4x^{-5}\,dx\approx -x^{-4}\Big|_{1.189}^{1.333}\approx-\frac{1}{1.333^4}+\frac{1}{1.189^4}\approx 0.1836$$

21. $$f(x)=\begin{cases}\frac{x^3}{12} & \text{if } 0\le x\le 2\\ \frac{16}{3x^3} & \text{if } x>2\end{cases}$$

Expected value:

$$E(X)=\mu=\int_0^\infty x\,f(x)dx=\int_0^2 x\left(\frac{x^3}{12}\right)dx+\lim_{a\to\infty}\int_2^a x\left(\frac{16}{3x^3}\right)dx=\int_0^2\frac{x^4}{12}\,dx+\lim_{a\to\infty}\int_2^a\frac{16}{3x^2}\,dx$$

$$=\left(\frac{x^5}{60}\right)\Bigg|_0^2+\lim_{a\to\infty}\left(-\frac{16}{3x}\right)\Bigg|_2^a=\left(\frac{8}{15}-0\right)+\left[\lim_{a\to\infty}\left(-\frac{16}{3a}\right)-\left(-\frac{16}{6}\right)\right]=\frac{16}{5}$$

(continued on next page)

(*continued*)

Variance:

$$\text{Var}(X) = \int_0^\infty x^2 f(x)dx - \mu^2 = \int_0^2 x^2\left(\frac{x^3}{12}\right)dx + \int_2^\infty x^2\left(\frac{16}{3x^3}\right)dx - \left(\frac{16}{5}\right)^2$$

Examine the second integral.

$$\int_2^\infty x^2\left(\frac{16}{3x^3}\right)dx = \lim_{a\to\infty}\int_2^a x^2\left(\frac{16}{3x^3}\right)dx = \lim_{a\to\infty}\int_2^a \frac{16}{3x}dx = \lim_{a\to\infty}\frac{16}{3}\ln|a| - \frac{16}{3}\ln|2|$$

Since the limit diverges, neither the variance nor the standard deviation exists.

23. $f(x) = \begin{cases} \frac{|x|}{10} & \text{for } -2 \le x \le 4 \\ 0 & \text{otherwise} \end{cases}$

First, note that

$$|x| = \begin{cases} -x \text{ for } -2 \le x \le 0 \\ x \text{ for } 0 \le x \le 4 \end{cases}$$

The expected value is

$$E(X) = \mu = \int_{-2}^4 x \cdot \frac{|x|}{10}dx$$
$$= \int_{-2}^0 x \cdot \frac{-x}{10}dx + \int_0^4 x \cdot \frac{x}{10}dx$$
$$= \int_{-2}^0 -\frac{x^2}{10}dx + \int_0^4 \frac{x^2}{10}dx = -\frac{x^3}{30}\bigg|_{-2}^0 + \frac{x^3}{30}\bigg|_0^4$$
$$= -\left(0 - \frac{-8}{30}\right) + \left(\frac{64}{30} - 0\right) = \frac{56}{30} = \frac{28}{15}$$

The correct answer choice is **d.**

25. $f(x) = \frac{3}{32}(4x - x^2);\ [0, 4]$

(a) $E(X) = \int_0^4 \frac{3x}{32}(4x - x^2)dx$

$$= \int_0^4 \left(\frac{3}{8}x^2 - \frac{3x^3}{32}\right)dx$$
$$= \left(\frac{x^3}{8} - \frac{3x^4}{128}\right)\bigg|_0^4$$
$$= 8 - \frac{768}{128} = 2$$

(b) $\text{Var}(X) = \int_0^4 \frac{3x^2}{32}(4x - x^2)dx - 4$

$$= \int_0^4 \left(\frac{3x^3}{8} - \frac{3x^4}{32}\right)dx - 4$$
$$= \left(\frac{3x^4}{32} - \frac{3x^5}{160}\right)\bigg|_0^4 - 4$$
$$= \frac{3(4^4)}{32} - \frac{3(4^5)}{160} - 4 = 0.8$$

$\sigma = \sqrt{\text{Var}(X)} \approx 0.8944$

(c) $P(2 - 0.89443 \le X \le 2 + 0.89443)$

$$\approx \int_{1.10557}^{2.89443} \frac{3}{32}(4x - x^2)dx$$
$$= \left(\frac{3}{16}x^2 - \frac{x^3}{32}\right)\bigg|_{1.10557}^{2.89443}$$
$$= \frac{3(2.89443)^2}{16} - \frac{2.89443^3}{32} - \frac{3(1.10557)^2}{16} + \frac{1.10557^3}{32}$$
$$\approx 0.6261$$

27. $p(x, t) = \dfrac{e^{-x^2/(4Dt)}}{\int_0^L e^{-u^2/(4Dt)}du}$

Letting $t = 12$, $L = 6$, and $D = 38.3$,

$$p(x, 12) = \frac{e^{-x^2(4\cdot 38.3\cdot 12)}}{\int_0^6 e^{-u^2/(4\cdot 38.3\cdot 12)}du}$$
$$= \frac{e^{-x^2/1838.4}}{\int_0^6 e^{-u^2/(1838.4)}du}$$
$$\approx \frac{1}{5.9611}e^{-x^2/1838.4}$$

The integral in the denominator was evaluated using the integration feature on the calculator.

$$E(X) = \int_0^L x \cdot p(x, 12)dx$$
$$= \frac{1}{5.9611}\int_0^6 xe^{-x^2/1838.4}dx$$
$$= \frac{1}{5.9611}\cdot\left(-\frac{1}{2}\right)\cdot 1838.4e^{-x^2}\bigg|_0^6 \approx 2.990$$

The expected recapture distance is 2.990 m.

29. $f(t) = 0.07599t^{1.43}e^{-t/2.62}$

(a) $E(t) = \mu = \int_0^{40} tf(t)dt$

Using the fnInt function on the TI-84 Plus graphing calculator gives $\mu \approx 6.37$ days.

(b) $\sigma = \sqrt{\text{Var}(t)}$

$\text{Var}(t) = \int_0^{40} (t - 6.37)^2 f(t)\,dt$

Using the fnInt function on the TI-84 Plus graphing calculator gives

$\sigma = \sqrt{\text{Var}(t)} \approx 4.08$ days.

(c) $P(5 < t < 10) = \int_5^{10} f(t)\,dt$

Using the fnInt function on the TI-84 Plus graphing calculator gives

$P(5 < t < 10) \approx 0.39.$

31. $f(t) = \dfrac{4.045}{t^{1.532}}$ for t in [16, 80].

(a) $\mu = \int_{16}^{80} t\,f(t)\,dt = \int_{16}^{80} t\left(\dfrac{4.045}{t^{1.532}}\right)dt$

$$= \int_{16}^{80} 4.045t^{-0.532}\,dt = \frac{4.045}{0.468}t^{0.468}\Big|_{16}^{80}$$

$$= \frac{4.045}{0.468}(80^{0.468} - 16^{0.468}) \approx 35.55 \text{ yr}$$

(b) $\text{Var}(T) = \int_{16}^{80} t^2 f(t)\,dt - (35.55)^2$

$$= \int_{16}^{80} t^2\left(\frac{4.045}{t^{1.532}}\right)dt - (35.55)^2$$

$$= \int_{16}^{80} 4.045t^{0.468}\,dt - (35.55)^2$$

$$= \frac{4.045}{1.468}(t^{1.468})\Big|_{16}^{80} - (35.55)^2$$

$$= \frac{4.045}{1.468}(80^{1.468} - 16^{1.468}) - (35.55)^2$$

$$\approx 288.156$$

$$\sigma \approx \sqrt{288.156} \approx 16.98 \text{ years}$$

(c) $\mu - \sigma = 35.555 - 16.975 = 18.58$

$P(T < 18.58)$

$$= \int_{16}^{18.58} f(t)\,dt = \int_{16}^{18.58} \frac{4.045}{t^{1.532}}\,dt$$

$$= -\frac{4.045}{0.532}(t^{-0.532})\Big|_{16}^{18.58}$$

$$= -\frac{4.045}{0.532}(18.58^{-0.532} - 16^{-0.532})$$

$$\approx 0.1330$$

(d) The median age m satisfies

$$\int_{16}^{m} \frac{4.045}{t^{1.532}}\,dt = \frac{1}{2}.$$

Use the integral found in (c).

$$\int_{16}^{m} \frac{4.045}{t^{1.532}}\,dt = \frac{1}{2}$$

$$-\frac{4.045}{0.532}(t^{-0.532})\Big|_{16}^{m} = \frac{1}{2}$$

$$-\frac{4.045}{0.532}(m^{-0.532} - 16^{-0.532}) = \frac{1}{2}$$

$$m^{-0.532} = -\frac{0.532}{4.045}\left(\frac{1}{2}\right) + 16^{-0.532}$$

$$m^{-0.532} = 0.16301$$

$$e^{-0.532 \ln m} = 0.16301$$

$$-0.532 \ln m = \ln 0.16301$$

$$\ln m = \frac{-1.8139}{-0.532} = 3.4096$$

$$m = e^{3.4096} \approx 30.25 \text{ years}$$

33. $S(t) = \dfrac{1}{101{,}370}(-2.564t^3 + 99.11t^2 - 964.6t + 5631)$ for t in [0, 24]

$$\mu = \int_0^{24} tS(t)\,dt$$

$$= \frac{1}{101{,}370}\int_0^{24} (-2.564t^4 + 99.11t^3 - 964.6t^2 + 5631t)\,dt$$

Use a calculator to evaluate the integral: $\mu \approx 12.964$

The expected time of day at which a fatal accident will occur is about 1 pm.

35. Using the hint, we have

$$\text{loss not paid} = \begin{cases} x \text{ for } 0.6 < x < 2 \\ 2 \text{ for } x > 2 \end{cases}$$

Therefore, the mean of the manufacturer's annual losses not paid will be

$$\mu = \int_{0.6}^{2} x \cdot f(x)dx + \int_2^{\infty} 2 \cdot f(x)dx$$

$$= \int_{0.6}^{2} x\frac{2.5(0.6)^{2.5}}{x^{3.5}}\,dx + \int_2^{\infty} 2\frac{2.5(0.6)^{2.5}}{x^{3.5}}\,dx$$

$$= 2.5(0.6)^{2.5}\int_{0.6}^{2} \frac{1}{x^{2.5}}\,dx + 5(0.6)^{2.5}\int_2^{\infty} \frac{1}{x^{3.5}}\,dx$$

$$= 2.5(0.6)^{2.5}\left(\frac{1}{-1.5}\right)\frac{1}{x^{1.5}}\Big|_{0.6}^{2} + 5(0.6)^{2.5}\left(\frac{1}{-2.5}\right)\frac{1}{x^{2.5}}\Big|_{2}^{\infty}$$

(continued on next page)

(continued)

$$= -\frac{5}{3}(0.6)^{2.5}\left(\frac{1}{2^{1.5}} - \frac{1}{0.6^{1.5}}\right) - 2(0.6)^{2.5}\left(0 - \frac{1}{2^{2.5}}\right)$$

$$\approx 0.8357 + 0.0986 \approx 0.93$$

The correct answer choice is **c**.

37. Since the probability density function is proportional to $(1+x)^{-4}$, we have $f(x) = k(1+x)^{-4}, 0 < x < \infty$. To determine k, solve the equation $\int_0^\infty k\,f(x)\,dx = 1$.

$$\int_0^\infty k(1+x)^{-4}\,dx = 1$$

$$k\left(-\frac{1}{3}\right)(1+x)^{-3}\Big|_0^\infty = 1$$

$$-\frac{k}{3}(0-1) = 1 \Rightarrow \frac{k}{3} = 1 \Rightarrow k = 3$$

Thus, $f(x) = 3(1+x)^{-4}, 0 < x < \infty$.
The expected monthly claims are

$$\int_0^\infty x \cdot 3(1+x)^{-4}\,dx = 3\int_0^\infty \frac{x}{(1+x)^4}\,dx$$

The antiderivative can be found using the substitution $u = 1 + x$.

$$\int \frac{x}{(1+x)^4}\,dx = \int \frac{u-1}{u^4}\,du = \int\left(\frac{1}{u^3} - \frac{1}{u^4}\right)du = -\frac{1}{2u^2} + \frac{1}{3u^3}$$

Resubstitute $u = 1 + x$.

$$3\int_0^\infty \frac{x}{(1+x)^4}\,dx = 3\left(-\frac{1}{2(1+x)^2} + \frac{1}{3(1+x)^3}\right)\Bigg|_0^\infty = 3\left[0 - \left(-\frac{1}{2} + \frac{1}{3}\right)\right] = 3\left(\frac{1}{6}\right) = \frac{1}{2}$$

The correct answer choice is **c**.

39. $f(t) = \frac{1}{960}e^{-t/960}$; $[0, \infty)$

$$E(T) = \int_0^\infty \frac{t}{960}e^{-t/960}\,dt = \lim_{b\to\infty}\int_0^b \frac{t}{960}e^{-t/960}\,dt$$

Integrating by parts, choose $dv = e^{-t/960}dt$ and $u = \frac{t}{960}$. Then $v = -960e^{-t/960}$ and $du = \frac{1}{960}dt$. So,

$$\int u\,dv = uv - \int v\,du$$

$$\int_0^b \frac{t}{960}e^{-t/960}dt = -te^{-t/960}\Big|_0^b - \int_0^b -e^{-t/960}dt$$
$$= -te^{-t/960}\Big|_0^b - 960e^{-t/960}\Big|_0^b$$
$$= -(t+960)e^{-t/960}\Big|_0^b$$
$$= -(b+960)e^{-b/960} + 960$$

Thus,

$$E(T) = \lim_{b\to\infty}\int_0^b \frac{t}{960}e^{-t/960}dt = \lim_{b\to\infty}[-(b+960)e^{-b/960} + 960] = 960 \text{ days}$$

$$\text{Var}(T) = \int_0^\infty \frac{t^2}{960}e^{-t/960}dt - 960^2 = \frac{1}{960}\left[\lim_{b\to\infty}\int_0^b t^2e^{-t/960}dt\right] - 960^2$$

Use the column method for integration by parts. Let $u = t^2$ and $dv = e^{-t/960}dt$.

D		I
t^2	+	$e^{-t/960}$
$2t$	−	$-960e^{-t/960}$
2	+	$960^2e^{-t/960}$
0		$-960^3e^{-t/960}$

$$\int_0^b t^2e^{-t/960}dt$$
$$= -960t^2e^{-t/960} - 2t(960^2e^{-t/960}) + 2(-960^3e^{-t/960})\Big|_0^b$$
$$= -960e^{-t/960}(t^2 + 2(960)t + 2(960^2))\Big|_0^b$$
$$= -960e^{-b/960}(b^2 + 2(960)b + 2(960^2)) + 2(960^3)$$

$$\text{Var}(T) = \frac{1}{960}\left[\lim_{b\to\infty}\int_0^b t^2e^{-t/960}dt\right] - 960^2$$
$$= \frac{1}{960}[\lim_{b\to\infty}[-960e^{-b/960}(b^2 + 2(960)b + 2(960^2)) + 2(960^3)]] - 960^2$$
$$= \frac{1}{960}\cdot 2(960^3) - 960^2 = 960^2$$

$$\sigma = \sqrt{\text{Var}(T)} = \sqrt{960^2} = 960 \text{ days}$$

13.3 Special Probability Density Functions

1. $f(x)=\frac{5}{7}$ for x in $[3,4.4]$

This is a uniform distribution: $a = 3$, $b = 4.4$

(a) $\mu=\frac{1}{2}(4.4+3)=\frac{1}{2}(7.4)=3.7$ cm

(b) $\sigma=\frac{1}{\sqrt{12}}(4.4-3)=\frac{1}{\sqrt{12}}(1.4)$
≈ 0.4041 cm

(c) $P(3.7<X<3.7+0.4041)$
$=P(3.7<X<4.1041)$
$=\int_{3.7}^{4.1041}\frac{5}{7}\,dx=\frac{5}{7}x\Big|_{3.7}^{4.1041}\approx 0.2886$

3. $f(t)=4e^{-4t}$ for t in $[0,\infty)$

This is an exponential distribution: $a = 4$.

(a) $\mu=\frac{1}{4}=0.25$ year

(b) $\sigma=\frac{1}{4}=0.25$ year

(c) $P(0.25<T<0.25+0.25)$
$=P(0.25<T<0.5)$
$=\int_{0.25}^{0.5}4e^{-4t}\,dt=-e^{-4t}\Big|_{0.25}^{0.5}$
$=-\frac{1}{e^{-2}}+\frac{1}{e^{-1}}\approx 0.2325$

5. $f(t)=\frac{e^{-t/3}}{3}$ for t in $[0,\infty)$

This is an exponential distribution: $a=\frac{1}{3}$.

(a) $\mu=\frac{1}{\frac{1}{3}}=3$ days

(b) $\sigma=\frac{1}{\frac{1}{3}}=3$ days

(c) $P(3<T<3+3)=P(3<T<6)$
$=\int_{3}^{6}\frac{e^{-t/3}}{3}\,dt=e^{-t/3}\Big|_{3}^{6}$
$=-\frac{1}{e^{-2}}+\frac{1}{e^{-1}}\approx 0.2325$

In Exercises 7–14, use Table 2 in Appendix D for areas under the normal curve.

7. $z = 3.50$
The area to the left of $z = 3.50$ is 0.9998. Given mean $\mu = z = 0$, so area to left of μ is 0.5. Area between μ and z is $0.9998 - 0.5 = 0.4998$. Therefore, this area represents 49.98% of the total area under a normal curve.

9. Between $z = 1.28$ and $z = 2.05$
Area to left of $z = 2.05$ is 0.9798 and area to left of $z = 1.28$ is 0.8997.
$0.9798 - 0.8997 = 0.0801$
Percent of total area = 8.01%

11. Since 10% = 0.10, the z-score that corresponds to the area of 0.10 to the left of z is -1.28.

13. 18% of the total area to the right of z means $1 - 0.18 = 0.82 = 82\%$ of the total area is to the left of z. The closest z-score that corresponds to the area of 0.82 is 0.92

15. Answers will vary.

17. **(a)** See figure 9 in the text.

(b) See figure 10 in the text.

(c) See figure 11 in the text.

19. $P(x=5)=\frac{5^2e^{-3.5}}{5!}\approx 0.1322$

21. $P(x>3)=1-P(x\le 3)=1-0.5366=0.4634$

23. $\mu=40,\ \sigma=12,\ n=36$

$P(38\le \bar{X}\le 43)$
$=P\left(\frac{38-40}{12/\sqrt{36}}\le\frac{\bar{X}-\mu}{\sigma/\sqrt{n}}\le\frac{43-40}{12/\sqrt{36}}\right)$
$=P\left(-1.00\le\frac{\bar{X}-\mu}{\sigma/\sqrt{n}}\le 1.50\right)$

Because the sample size is greater than 30, it is approximately distributed according to the standard normal distribution, so

$P(-1.00\le Z\le 1.50)$
$=P(Z\le 1.50)-P(Z\le -1.00)$
$=0.9332-0.1587=0.7745$

25. $\mu = 40,\ \sigma = 12,\ n = 61$

$$P(38.2 \le \bar{X} \le 39.4)$$
$$= P\left(\frac{38.2-40}{12/\sqrt{61}} \le \frac{\bar{X}-\mu}{\sigma/\sqrt{n}} \le \frac{39.4-40}{12/\sqrt{61}}\right)$$
$$= P\left(-1.17 \le \frac{\bar{X}-\mu}{\sigma/\sqrt{n}} \le -0.39\right)$$

Because the sample size is greater than 30, it is approximately distributed according to the standard normal distribution, so

$$P(-1.17 \le Z \le -0.39)$$
$$= P(Z \le -0.39) - P(Z \le -1.17)$$
$$= 0.3483 - 0.1210$$
$$= 0.2273$$

27. Let m be the median of the exponential distribution $f(x) = ae^{-ax}$ for $[0, \infty)$.

$$\int_0^m ae^{-ax}dx = 0.5$$
$$-e^{-ax}\Big|_0^m = 0.5$$
$$-e^{-am} + 1 = 0.5$$
$$0.5 = e^{-am} \Rightarrow -am = \ln 0.5 \Rightarrow$$
$$m = -\frac{\ln 0.5}{a}$$

or $-am = \ln\frac{1}{2} \Rightarrow -am = -\ln 2 \Rightarrow m = \frac{\ln 2}{a}$.

29. The area that is to the left of x is

$A = \int_{-\infty}^{x} \frac{1}{\sigma\sqrt{2\pi}} e^{-\frac{(t-\mu)^2}{2\sigma^2}}\, dt.$ Let $u = \frac{t-\mu}{\sigma}$.

Then $du = \frac{1}{\sigma}dt$ and $dt = \sigma du$.

If $t = x$, $u = \frac{x-\mu}{\sigma} = z$. As $t \to -\infty$, $u \to -\infty$.

Therefore,

$$A = \int_{-\infty}^{z} \frac{1}{\sigma\sqrt{2\pi}} e^{(-1/2)u^2} \sigma\, du$$
$$= \frac{\sigma}{\sigma}\int_{-\infty}^{z} \frac{1}{\sqrt{2\pi}} e^{-u^2/2} du = \int_{-\infty}^{z} \frac{1}{\sqrt{2\pi}} e^{-u^2/2} du.$$

This is the area to the left of z for the standard normal curve.

31. Use Simpson's rule with n = 140 or use the integration feature on a graphing calculator to approximate the integrals. Answers may very slightly from those given here depending on the method that is used.

(a) $\int_0^{35} 0.5e^{-0.5x}\, dx \approx 1.00000$

(b) $\int_0^{35} 0.5xe^{-0.5x}\, dx \approx 1.99999$

(c) $\int_0^{35} 0.5x^2e^{-0.5x}\, dx = 8.00000$

33. Use Simpson's rule with n = 40 and limits of –6 and 6 to approximate the mean and standard deviation of a normal probability distribution.

(a) $\int_{-\infty}^{\infty} \frac{x}{\sqrt{2\pi}} e^{-x^2/2}\, dx$

$$\approx \int_{-6}^{6} \frac{x}{\sqrt{2\pi}} e^{-x^2/2}\, dx \Rightarrow \mu = 0$$

(For the integral of an odd function over an interval symmetric to 0, Simpson's rule will give 0; there is no need to do any calculation.)

(b) $\int_{-\infty}^{\infty} \frac{x^2}{\sqrt{2\pi}} e^{-x^2/2}\, dx$

$$\approx \int_{-6}^{6} \frac{x^2}{\sqrt{2\pi}} e^{-x^2/2}\, dx$$

$\sigma \approx 0.9999999224$ (Simpson's rule)
$\sigma \approx 0.9999999251$ (calculator)

35. The probability density function for the uniform distribution is $f(x) = \frac{1}{b-a}$ for x in $[a, b]$.

The cumulative distribution function for f is

$$F(x) = P(X \le x) = \int_a^x f(t)\, dt$$
$$= \int_a^x \frac{1}{b-a} dt = \frac{1}{b-a} t\Big|_a^x$$
$$= \frac{1}{b-a}(x-a) = \frac{x-a}{b-a},\ a \le x \le b.$$

37. For a uniform distribution, $f(x) = \frac{1}{b-a}$ for x in $[a, b]$.

$f(x) = \frac{1}{36-20} = \frac{1}{16}$ for x in $[20, 36]$

(a) $\mu = \frac{1}{2}(20+36) = \frac{1}{2}(56) = 28$ days

(b) $P(30 < X \le 36)$

$$= \int_{30}^{36} \frac{1}{16} dx = \frac{1}{16}x\Big|_{30}^{36}$$
$$= \frac{1}{16}(36-30) = 0.375$$

39. We have an exponential distribution, with $a = 1$. $f(t) = e^{-t}$, $[0, \infty)$

(a) $\mu = \frac{1}{1} = 1$ hr

(b) $P(T < 30 \text{ min}) = \int_0^{0.5} e^{-t}\,dt = -e^{-t}\Big|_0^{0.5} \Rightarrow$
$1 - e^{-0.5} \approx 0.3935$

41. $f(x) = ae^{-ax}$ for $[0, \infty]$

Since $\mu = 25$ and $\mu = \frac{1}{a}$, $a = \frac{1}{25} = 0.04$.
Thus, $f(x) = 0.04e^{-0.04x}$.

(a) We must find t such that $P(X \le t) = 0.90$.

$$\int_0^t 0.04e^{-0.04x}\,dx = 0.90$$
$$-e^{-0.04x}\Big|_0^t = 0.90$$
$$-e^{-0.04t} + 1 = 0.90$$
$$0.10 = e^{-0.04t}$$
$$-0.04t = \ln 0.10$$
$$t = \frac{\ln 0.10}{-0.04} \approx 57.56$$

The longest time within which the predator will be 90% certain of finding a prey is approximately 58 min.

(b) $P(X \ge 60) = \int_{60}^{\infty} 0.04e^{-0.04x}\,dx$
$$= \lim_{b\to\infty} \int_{60}^{b} 0.04e^{-0.04x}\,dx$$
$$= \lim_{b\to\infty} (-e^{-0.04x})\Big|_{60}^{b}$$
$$= \lim_{b\to\infty} \left[-e^{-0.04b} + e^{-0.04(60)}\right]$$
$$= 0 + e^{-2.4} \approx 0.0907$$

The probability that the predator will have to spend more than one hour looking for a prey is approximately 0.0907.

43. For an exponential distribution, $f(x) = ae^{-ax}$ for x in $[0, \infty)$. Since $\mu = \frac{1}{a} = 12.3$, $a = \frac{1}{12.3}$.

(a) $P(X \ge 20) = \int_{20}^{\infty} \frac{1}{12.3}e^{-x/12.3}\,dx$
$$= \lim_{b\to\infty} \int_{20}^{b} \frac{1}{12.3}e^{-x/12.3}\,dx$$
$$= \lim_{b\to\infty} \left(e^{-x/12.3}\Big|_{20}^{b}\right)$$
$$= \lim_{b\to\infty} (-e^{-b/12.3} + e^{-20/12.3})$$
$$= e^{-20/12.3} \approx 0.1967$$

(b) $P(10 \le X \le 20) = \int_{10}^{20} \frac{1}{12.3}e^{-x/12.3}\,dx$
$$= (-e^{-x/12.3})\Big|_{10}^{20}$$
$$= -e^{-20/12.3} + e^{-10/12.3}$$
$$\approx 0.2468$$

45. (a) $P(x = 0) = \dfrac{\left(\frac{102}{240}\right)^0 e^{-102/240}}{0!} \approx 0.6538$

(b) $P(\text{at least } 1) = 1 - P(x = 0)$
$$= 1 - 0.6538 = 0.3462$$

(c) $P(x = 3) = \dfrac{\left(\frac{102}{240}\right)^3 e^{-102/240}}{3!} \approx 0.0084$

47. We have an exponential distribution, with $a = 0.229$. So $f(t) = 0.229e^{-0.229t}$, for $[0, \infty)$.

(a) The life expectancy is
$\mu = \frac{1}{a} = \frac{1}{0.229} \approx 4.37$ millennia.
The standard deviation is
$\sigma = \frac{1}{a} = \frac{1}{0.229} \approx 4.37$ millennia.

(b) $P(T \ge 2) = \int_2^{\infty} 0.229e^{-0.229t}\,dt$
$$= 1 - \int_0^2 0.229e^{-0.229t}\,dt$$
$$= 1 + \left(e^{-0.229t}\Big|_0^2\right)$$
$$= 1 + \left[e^{-0.229(2)} - 1\right]$$
$$= e^{-0.458} \approx 0.6325$$

49. For an exponential distribution, $f(x) = ae^{-ax}$ for $[0, \infty)$. Since $\mu = \frac{1}{a} = 8$, $a = \frac{1}{8}$.

(a) $P(X \ge 10) = \int_{10}^{\infty} \frac{1}{8}e^{-x/8}dx$
$$= 1 - \int_0^{10} \frac{1}{8}e^{-x/8}dx$$
$$= 1 + \left(e^{-x/8}\Big|_0^{10}\right)$$
$$= 1 + [e^{-10/8} - 1]$$
$$= e^{-10/8} \approx 0.2865$$

(b) $P(X<2)=\int_0^2 \frac{1}{8}e^{-x/8}dx=-e^{-x/8}\Big|_0^2$
$=-e^{-2/8}+1\approx 0.2212$

51. We have an exponential distribution $f(x)=ae^{-ax}$ for $x\geq 0$. Since $a=\frac{1}{90}$,
$f(x)=\frac{1}{90}e^{-x/90}$ for $x\geq 0$.

(a) The probability that the time for a goal is no more than 71 minutes is

$$P(0<X\leq 71)=\int_0^{71}\frac{1}{90}e^{-x/90}dx$$
$$=-e^{-x/90}\Big|_0^{71}$$
$$=-e^{-71/90}+1\approx 0.5457$$

(b) The probability that the time for a goal is 499 minutes or more is

$$P(X\geq 499)=\int_{499}^{\infty}\frac{1}{90}e^{-x/90}dx$$
$$=e^{-x/90}\Big|_{499}^{\infty}$$
$$=0+e^{-499/90}\approx 0.0039$$

53. We have a normal distribution, with $\mu=54.40, \sigma=13.50.$

$P(-a<z<a)=\frac{1}{2}\Rightarrow P(z<-a)=0.25$

Since the closest value to 0.25 is 0.2514, we use $z=-0.67$.

$$-0.67<z<0.67$$
$$-0.67<\frac{x-54.40}{13.50}<0.67$$
$$45.36<x<63.45$$

Therefore, $P(45.36<X<63.45)=\frac{1}{2}$, and the middle 50% of the customers spend between \$45.36 and \$63.45.

55. Let the random variable X be the number of days that elapse between the beginning of a calendar year and the moment a high-risk driver is involved in an accident. Then it has exponential distribution $f(x)=ae^{-ax}$ for $x\geq 0$. We are given $P(0\leq X\leq 50)=0.3$ so that

$$\int_0^{50}ae^{-ax}dx=0.3\Rightarrow -e^{-ax}\Big|_0^{50}=0.3\Rightarrow$$
$$-e^{-50a}+1=0.3\Rightarrow e^{-50a}=0.7\Rightarrow$$
$$-50a=\ln 0.7\Rightarrow a=-\frac{1}{50}\ln 0.7\approx 0.007133$$

Thus, $f(x)=0.007133e^{-0.007133x}$ for $x\geq 0$. The portion expected to be involved in an accident during the first 80 days is

$$\int_0^{80}0.007133e^{-0.007133x}dx=-e^{-0.007133x}\Big|_0^{80}$$
$$=-e^{-0.007133(80)}+1$$
$$\approx 0.43$$

The correct answer choice is **c**.

57. Let the random variable X be the life time to failure. Then it has exponential distribution $f(x)=ae^{-ax}$ for $x\geq 0$. If the mean is 4 hours, then

$$\int_0^4 ae^{-ax}dx=\frac{1}{2}\Rightarrow -e^{-ax}\Big|_0^4=\frac{1}{2}\Rightarrow$$
$$-e^{-4a}+1=\frac{1}{2}\Rightarrow -e^{-4a}=-\frac{1}{2}\Rightarrow -4a=\ln\frac{1}{2}\Rightarrow$$
$$a=-\frac{1}{4}\ln\frac{1}{2}\approx 0.1733$$

So the exponential function for X is $f(x)=0.1733e^{-0.1733x}$ for $x\geq 0$. The probability that the component will work without failing for at least 5 hours is,

$$P(5\leq X\leq\infty)=\int_5^{\infty}0.1733e^{-0.1733x}dx$$
$$=-e^{-0.1733x}\Big|_5^{\infty}=0+e^{-0.1733(5)}$$
$$\approx 0.4204$$

The correct answer choice is **d**.

Chapter 13 Review Exercises

1. True
2. True
3. True
4. False: A density function is always nonnegative.
5. False: If the random variable takes on negative values the expectation may also be negative.
6. True
7. True
8. True
9. False: The normal distribution is symmetrical; the exponential distribution has a long tail to the right.
10. False: The expected value is 0 and the standard deviation is 1.

11. In a probability function, the y-values (or function values) represent probabilities.

12. A continuous random variable is a random variable where the data can take infinitely many values. For example, a random variable measuring the time taken for something to be done is continuous since there are an infinite number of possible times that can be taken.

13. A probability density function f for $[a, b]$ must satisfy the following two conditions:

(1) $f(x) \geq 0$ for all x in the interval $[a, b]$;

(2) $\int_a^b f(x)dx = 1.$

14. In a probability density function, the probability that X equals a specific value, $P(X = c)$, is zero.

15. $f(x) = \sqrt{x};\ [4, 9]$

$$\int_4^9 x^{1/2}dx = \frac{2}{3}x^{3/2}\Big|_4^9 = \frac{2}{3}(27-8) = \frac{38}{3} \neq 1$$

$f(x)$ is not a probability density function.

16. $f(x) = \frac{1}{27}(2x+4);\ [1, 4]$

$$\int_1^4 \frac{1}{27}(2x+4)dx = \frac{1}{27}(x^2+4x)\Big|_1^4 = \frac{1}{27}(32-5) = 1$$

Since $1 \leq x \leq 4$, $f(x) \geq 0$. Therefore, $f(x)$ is a probability density function.

17. $f(x) = 0.7e^{-0.7x};\ [0, \infty)$

$$\int_0^\infty 0.7e^{-0.7x}dx = -e^{-0.7x}\Big|_0^\infty = \lim_{b\to\infty}(-e^{-0.7b}) + e^0 = \lim_{b\to\infty}\left(-\frac{1}{e^{0.7b}}\right) + 1 = 0 + 1 = 1$$

$f(x) \geq 0$ for all x in $[0, \infty)$. Therefore, $f(x)$ is a probability density function.

18. $f(x) = 0.4;\ [4, 6.5]$

$$\int_4^{6.5} 0.4\,dx = 0.4x\Big|_4^{6.5} = 0.4(6.5-4) = 1$$

$f(x) \geq 0$ for all x in $[4, 6.5]$. Therefore $f(x)$ is a probability density function.

19. $f(x) = kx^2;\ [1, 4]$

$$\int_1^4 kx^2dx = \frac{kx^3}{3}\Big|_1^4 = 21k$$

Since $f(x)$ is a probability density function, $21k = 1 \Rightarrow k = \frac{1}{21}.$

20. $f(x) = k\sqrt{x};\ [4, 9]$

$$\int_4^9 k\sqrt{x}\,dx = \int_4^9 kx^{1/2}dx = \frac{2}{3}kx^{3/2}\Big|_4^9 = \frac{2}{3}k(27-8) = \frac{38}{3}k$$

Since $f(x)$ is a probability density function, $\frac{38}{3}k = 1 \Rightarrow k = \frac{3}{38}.$

21. $f(x) = \frac{1}{10}$ for $[10, 20]$

(a) $P(10 \leq X \leq 12) = \int_{10}^{12}\frac{1}{10}dx = \frac{x}{10}\Big|_{10}^{12} = \frac{1}{5} = 0.2$

(b) $P\left(\frac{31}{2} \leq X \leq 20\right) = \int_{31/2}^{20}\frac{1}{10}dx = \frac{x}{10}\Big|_{31/2}^{20} = 2 - \frac{31}{20} = \frac{9}{20} = 0.45$

(c) $P(10.8 \leq X \leq 16.2) = \int_{10.8}^{16.2}\frac{1}{10}dx = \frac{x}{10}\Big|_{10.8}^{16.2} = 0.54$

22. $f(x) = 1 - \frac{1}{\sqrt{x-1}};\ [2, 5]$

(a) $P(3 \leq X \leq 5) = \int_3^5 [1-(x-1)^{-1/2}]dx = [x - 2(x-1)^{1/2}]\Big|_3^5 = 5 - 2(2) - 3 + 2\sqrt{2} \approx 0.8284$

(b) $P(2 \leq X \leq 4) = \int_2^4 [1-(x-1)^{-1/2}]dx = [x - 2(x-1)^{1/2}]\Big|_2^4 = 4 - 2\sqrt{3} - 2 + 2 \approx 0.5359$

(c) $P(3 \le X \le 4) = \int_3^4 [1-(x-1)^{-1/2}]dx$
$= [x - 2(x-1)^{1/2}]\Big|_3^4$
$= 4 - 2\sqrt{3} - 3 + 2\sqrt{2}$
≈ 0.3643

23. If we consider the probabilities as weights, the expected value or mean of a probability distribution represents the point at which the distribution balances.

24. The distribution that is tallest or most peaked has the smallest standard deviation. This is the distribution pictured in graph **(b)**.

25. $f(x) = \frac{2}{9}(x-2)$; $[2, 5]$

(a) $\mu = \int_2^5 \frac{2x}{9}(x-2)dx = \int_2^5 \frac{2}{9}(x^2-2x)dx$
$= \frac{2}{9}\left(\frac{x^3}{3} - x^2\right)\Big|_2^5$
$= \frac{2}{9}\left(\frac{125}{3} - 25 - \frac{8}{3} + 4\right) = 4$

(b) $\text{Var}(X) = \int_2^5 \frac{2x^2}{9}(x-2)dx - (4)^2$
$= \int_2^5 \frac{2}{9}(x^3 - 2x^2)dx - 16$
$= \frac{2}{9}\left(\frac{x^4}{4} - \frac{2x^3}{3}\right)\Big|_2^5 - 16$
$= \frac{2}{9}\left(\frac{625}{4} - \frac{250}{3} - 4 + \frac{16}{3}\right) - 16$
$= 0.5$

(c) $\sigma = \sqrt{0.5} \approx 0.7071$

(d) $\int_2^m \frac{2}{9}(x-2)dx = \frac{1}{2}$
$\frac{1}{9}(m-2)^2\Big|_2^m = \frac{1}{2}$
$\frac{1}{9}[(m-2)^2 - 0] = \frac{1}{2}$
$m^2 - 4m + 4 = \frac{9}{2}$
$m^2 - 4m - \frac{1}{2} = 0$
$m = \frac{4 \pm 3\sqrt{2}}{2} \approx -0.121,\ 4.121$

We reject -0.121 since it is not in $[2, 5]$.
So, $m = 4.121$.

(e) $F(x) = \int_2^x \frac{2}{9}(t-2)dt = \frac{1}{9}(t-2)^2\Big|_2^x$
$= \frac{1}{9}[(x-2)^2 - 0]$
$= \frac{(x-2)^2}{9},\ 2 \le x \le 5$

26. $f(x) = \frac{1}{5}$; $[4, 9]$

(a) $E(X) = \mu = \int_4^9 x\left(\frac{1}{5}\right)dx = \frac{x^2}{10}\Big|_4^9$
$= \frac{81}{10} - \frac{16}{10} = \frac{65}{10} = 6.5$

(b) $\text{Var}(X) = \int_4^9 \frac{1}{5}(x-6.5)^2 dx$
$= \frac{1}{15}(x-6.5)^3\Big|_4^9$
$= \frac{1}{15}(2.5^3 + 2.5^3) \approx 2.083$

(c) $\sigma = \sqrt{\text{Var}(X)} \approx 1.443$

(d) $\int_4^m \frac{1}{5}dx = \frac{1}{2}$
$\frac{1}{5}x\Big|_4^m = \frac{1}{2}$
$\frac{1}{5}(m-4) = \frac{1}{2}$
$m - 4 = \frac{5}{2} \Rightarrow m = \frac{13}{2} = 6.5$

(e) $F(x) = \int_4^x \frac{1}{5}dt = \frac{1}{5}t\Big|_4^x$
$= \frac{x-4}{5},\ 4 \le x \le 9.$

27. $f(x) = 5x^{-6}$; $[1, \infty)$

(a) $\mu = \int_1^\infty x \cdot 5x^{-6}\ dx = \int_1^\infty 5x^{-5}\ dx$
$= \lim_{b\to\infty} \int_1^b 5x^{-5}\ dx = \lim_{b\to\infty} \frac{5x^{-4}}{-4}\Big|_1^b$
$= \lim_{b\to\infty} \frac{5}{4}\left(1 - \frac{1}{b^4}\right) = \frac{5}{4}$

(b) $\text{Var}(X) = \int_1^{\infty} x^2 \cdot 5x^{-6}\,dx - \left(\frac{5}{4}\right)^2$

$= \lim_{b\to\infty} \int_1^b 5x^{-4}\,dx - \frac{25}{16}$

$= \lim_{b\to\infty} \left.\frac{5x^{-3}}{-3}\right|_1^b - \frac{25}{16}$

$= \lim_{b\to\infty} \frac{5}{3}\left(1 - \frac{1}{b^3}\right) - \frac{25}{16}$

$= \frac{5}{3} - \frac{25}{16} = \frac{5}{48} \approx 0.1042$

(c) $\sigma \approx \sqrt{\text{Var}(X)} \approx 0.3227$

(d) $\int_1^m 5x^{-6}dx = \frac{1}{2}$

$\left.-x^{-5}\right|_1^m = \frac{1}{2}$

$-m^{-5} + 1 = \frac{1}{2}$

$m^{-5} = \frac{1}{2} \Rightarrow m^5 = 2 \Rightarrow$

$m = \sqrt[5]{2} \approx 1.149$

(e) $F(x) = \int_1^x 5t^{-6}dt = \left.-t^{-5}\right|_1^x = -x^{-5} + 1$

$= 1 - \frac{1}{x^5},\ x \ge 1$

28. $f(x) = \frac{1}{20}\left(1 + \frac{3}{\sqrt{x}}\right)$; [1, 9]

(a) $E(X) = \mu = \int_1^9 x\left[\frac{1}{20}(1 + 3x^{-1/2})\right]dx$

$= \frac{1}{20}\int_1^9 (x + 3x^{1/2})\,dx$

$= \frac{1}{20}\left.\left(\frac{x^2}{2} + 2x^{3/2}\right)\right|_1^9$

$= \frac{1}{20}\left(\frac{81}{2} + 54 - \frac{1}{2} - 2\right) = \frac{92}{20} = 4.6$

(b) $\text{Var}(X) = \int_1^9 \frac{x^2}{20}(1 + 3x^{-1/2})dx - (4.6)^2$

$= \int_1^9 \left(\frac{x^2}{20} + \frac{3}{20}x^{3/2}\right)dx - (4.6)^2$

$= \left.\left(\frac{x^3}{60} + \frac{3}{50}x^{5/2}\right)\right|_1^9 - (4.6)^2$

$= \left(\frac{729}{60} + \frac{729}{50} - \frac{1}{60} - \frac{3}{50}\right) - (4.6)^2$

≈ 5.493

(c) $\sigma = \sqrt{\text{Var}(X)} \approx 2.3438$

$\sigma \approx 2.344$

(d) $\int_1^m \frac{1}{20}\left(1 + \frac{3}{\sqrt{x}}\right)dx = \frac{1}{2}$

$\left.\frac{1}{20}(x + 6x^{1/2})\right|_1^m = \frac{1}{2}$

$\frac{1}{20}m + \frac{3}{10}m^{1/2} - \frac{7}{20} = \frac{1}{2}$

$m + 6\sqrt{m} - 17 = 0$

To solve this equation, let $u = \sqrt{m}$ so $u^2 = m$.

$u^2 + 6u - 17 = 0$

$u = \frac{-6 \pm \sqrt{36 + 68}}{2}$

$= -3 \pm \sqrt{26}$

$m = u^2 \approx 4.4059,\ 65.5941$

We reject 65.5941 since it is not in [1, 9]. So, $m \approx 4.406$.

(e) $F(x) = \int_1^x \frac{1}{20}\left(1 + \frac{3}{\sqrt{t}}\right)dt$

$= \left.\frac{1}{20}(t + 6t^{1/2})\right|_1^x$

$= \frac{1}{20}[(x + 6\sqrt{x}) - (1 + 6)]$

$= \frac{1}{20}(x + 6\sqrt{x} - 7),\ 1 \le x \le 9$

29. $f(x) = 4x - 3x^2$; [0, 1]

(a) $\mu = \int_0^1 x(4x - 3x^2)dx = \int_0^1 (4x^2 - 3x^3)dx$

$= \left.\left(\frac{4x^3}{3} - \frac{3x^4}{4}\right)\right|_0^1 = \frac{4}{3} - \frac{3}{4} = \frac{7}{12}$

≈ 0.5833

(b) $\text{Var}(X) = \int_0^1 x^2(4x - 3x^2)dx - \left(\frac{7}{12}\right)^2$

$= \int_0^1 (4x^3 - 3x^4)dx - \left(\frac{7}{12}\right)^2$

$= \left.\left(x^4 - \frac{3x^5}{5}\right)\right|_0^1 - \left(\frac{7}{12}\right)^2$

$= 1 - \frac{3}{5} - \left(\frac{7}{12}\right)^2 \approx 0.0597$

$\sigma \approx \sqrt{\text{Var}(X)} \approx 0.2444$

(c) $P\left(0 \le X < \frac{7}{12}\right) = \int_0^{7/12} (4x - 3x^2)dx$

$= (2x^2 - x^3)\Big|_0^{7/12}$

$= 2\left(\frac{7}{12}\right)^2 - \left(\frac{7}{12}\right)^3$

≈ 0.4821

(d) $P(\mu - \sigma \le X \le \mu + \sigma)$

$\approx P(0.339 \le x \le 0.828)$

$= \int_{0.339}^{0.828} (4x - 3x^2)dx$

$= (2x^2 - x^3)\Big|_{0.339}^{0.828}$

$= 2(0.828)^2 - (0.828)^3 - 2(0.339)^2 + (0.339)^3$

≈ 0.6123

30. $f(x) = 4x - 3x^2$; $[0, 1]$

$\mu = \int_0^1 x(4x - 3x^2)dx = \int_0^1 (4x^2 - 3x^2)dx$

$= \left(\frac{4x^3}{3} - \frac{3x^4}{4}\right)\Big|_0^1 = \frac{4}{3} - \frac{3}{4} = \frac{7}{12} \approx 0.5833$

Find m such that $\int_0^m (4x - 3x^2)dx = \frac{1}{2}$.

$\int_0^m (4x - 3x^2)dx = (2x^2 - x^3)\Big|_0^m$

$= 2m^2 - m^3 = \frac{1}{2}$

Therefore, $2m^3 - 4m^2 + 1 = 0$.

This equation has no rational roots, but trial and error used with synthetic division reveals that $m \approx -0.4516, 0.5970,$ and 1.855. The only one of these in $[0, 1]$ is 0.5970.

$P\left(\frac{7}{12} < X < 0.5970\right)$

$= \int_{7/12}^{0.5970} (4x - 3x^2)dx = 2x^2 - x^3\Big|_{7/12}^{0.5970}$

$= 2(0.5970)^2 - (0.5970)^3 - 2\left(\frac{7}{12}\right)^2 + \left(\frac{7}{12}\right)^3$

≈ 0.0180

31. $f(x) = 0.01e^{-0.01x}$ for $[0, \infty)$ is an exponential distribution.

(a) $\mu = \frac{1}{0.01} = 100$

(b) $\sigma = \frac{1}{0.01} = 100$

(c) $P(100 - 100 < X < 100 + 100)$

$= P(0 < X < 200)$

$= \int_0^{200} 0.01e^{-0.01x}\, dx$

$= -e^{-0.01x}\Big|_0^{200}$

$= 1 - e^{-2} \approx 0.8647$

32. $f(x) = \frac{5}{112}(1 - x^{-3/2})$; $[1, 25]$

(a) $\mu = \int_1^{25} \frac{5x}{112}(1 - x^{-3/2})dx$

$= \frac{5}{112}\int_1^{25} (x - x^{-1/2})dx$

$= \frac{5}{112}\left(\frac{x^2}{2} - 2x^{1/2}\right)\Big|_1^{25}$

$= \frac{5}{112}\left(\frac{625}{2} - 10 - \frac{1}{2} + 2\right)$

$= \frac{95}{7} \approx 13.5714 \approx 13.6$

(b) $\text{Var}(X) = \int_1^{25} \frac{5x^2}{112}(1 - x^{-3/2})\,dx - \left(\frac{95}{7}\right)^2$

$= \frac{5}{112}\int_1^{25} (x^2 - x^{1/2})\,dx - \left(\frac{95}{7}\right)^2$

$= \frac{5}{112}\left(\frac{x^3}{3} - \frac{2}{3}x^{3/2}\right)\Big|_1^{25} - \left(\frac{95}{7}\right)^2$

$= \frac{5}{112}\left(\frac{25^3}{3} - \frac{250}{3} - \frac{1}{3} + \frac{2}{3}\right) - \left(\frac{95}{7}\right)^2$

$= \frac{6560}{147}$

$\sigma = \sqrt{\text{Var}(X)} \approx 6.6803 \approx 6.7$

(c) $P(\mu - \sigma \le X \le \mu + \sigma)$

$= P(6.8911 \le X \le 20.2517)$

$= \int_{6.8911}^{20.2517} \frac{5}{112}(1 - x^{-3/2})dx$

$= \frac{5}{112}(x + 2x^{-1/2})\Big|_{6.8911}^{20.2517}$

$= \frac{5}{112}\left(20.2517 + \frac{2}{\sqrt{20.2517}} - 6.8911 - \frac{2}{\sqrt{6.8911}}\right)$

≈ 0.5823

(continued on next page)

(continued)

If you do the integration on a calculator using differently rounded values for μ, you may get a slightly different answer, such as 0.5840.

For Exercises 33–40, use Table 2 in the Appendix D for the areas under the normal curve.

33. Area to the left of $z = -0.43$ is 0.3336.
Percent of area is 33.36%

34. Area to right of $z = 1.62$ is $1 - 0.9474 = 0.526$ or 5.26%.

35. Area between $z = -1.17$ and $z = -0.09$ is $0.4641 - 0.1210 = 0.3431$.
Percent of area is 34.31%.

36. Area to left of $z = 1.28$ is 0.8997.
Area to left of $z = -1.39$ is equivalent to area to right of $z = 1.39 = 1 - 0.9177 = 0.0823$.
Area between is $0.8977 - 0.0823 = 0.8174$ or 81.74%

37. The region up to 1.2 standard deviations below the mean is the region to the left of $z = -1.2$. The area is 0.1151, so the percent of area is 11.51%.

38. $\sigma = 2.5,\ \mu = 0,\ z = 0 + 2.5$
Area to left of $z = 2.5$ is 0.9938 or 99.38%.

39. 52% of area is to the right implies that 48% is to the left. $P(z < a) = 0.48$ for $a = -0.05$
Thus, 52% of the area lies to the right of $z = -0.05$.

40. We want to find the z-score for 21% of the area under the normal curve to the left of z. We note that 21% < 50%, so z must be negative. The z-score for the value in the table nearest 0.21 is $z \approx -0.81$.

41. $$P(3 \le x \le 5) = \frac{4.2^3 e^{-4.2}}{3!} + \frac{4.2^4 e^{-4.2}}{4!} + \frac{4.2^4 e^{-4.2}}{5!} \approx 0.5429$$

42. $$P(x = 7) = \frac{4.2^7 e^{-4.2}}{7!} \approx 0.0686$$

43. $$P(x \le 5) = P(0) + P(1) + P(2) + P(3) + P(4) + P(5)$$
$$= \frac{4.2^0 e^{-4.2}}{0!} + \frac{4.2^1 e^{-4.2}}{1!} + \frac{4.2^2 e^{-4.2}}{2!} + \frac{4.2^3 e^{-4.2}}{3!} + \frac{4.2^4 e^{-4.2}}{4!} + \frac{4.2^4 e^{-4.2}}{5!}$$
$$\approx 0.7531$$

44. $$P(x > 5) = 1 - P(x \le 5) = 1 - 0.7531 = 0.2469$$

45. $\mu = 60,\ \sigma = 16,\ n = 45$
$$P(58.2 \le \overline{X} \le 61.1)$$
$$= P\left(\frac{58.2 - 60}{16/\sqrt{45}} \le \frac{\overline{X} - \mu}{\sigma/\sqrt{n}} \le \frac{61.1 - 60}{16/\sqrt{45}}\right)$$
$$= P\left(-0.75 \le \frac{\overline{X} - \mu}{\sigma/\sqrt{n}} \le 0.46\right)$$
Because the sample size is greater than 30, it is approximately distributed according to the standard normal distribution, so
$$P(-0.75 \le Z \le 0.46)$$
$$= P(Z \le 0.46) - P(Z \le -0.75)$$
$$= 0.6772 - 0.2266 = 0.4506$$

46. $\mu = 60,\ \sigma = 16,\ n = 52$
$$P(57.4 \le \overline{X} \le 60.6)$$
$$= P\left(\frac{57.4 - 60}{16/\sqrt{52}} \le \frac{\overline{X} - \mu}{\sigma/\sqrt{n}} \le \frac{60.6 - 60}{16/\sqrt{52}}\right)$$
$$= P\left(-1.17 \le \frac{\overline{X} - \mu}{\sigma/\sqrt{n}} \le 0.27\right)$$
Because the sample size is greater than 30, it is approximately distributed according to the standard normal distribution, so
$$P(-1.17 \le Z \le 0.27)$$
$$= P(Z \le 0.27) - P(Z \le -1.17)$$
$$= 0.6064 - 0.1210 = 0.4854$$

47. $\mu = 60,\ \sigma = 16,\ n = 58$
$$P(59.3 \le \overline{X} \le 61.9)$$
$$= P\left(\frac{59.3 - 60}{16/\sqrt{58}} \le \frac{\overline{X} - \mu}{\sigma/\sqrt{n}} \le \frac{61.9 - 60}{16/\sqrt{58}}\right)$$
$$= P\left(-0.33 \le \frac{\overline{X} - \mu}{\sigma/\sqrt{n}} \le 0.90\right)$$

(continued on next page)

(continued)

Because the sample size is greater than 30, it is approximately distributed according to the standard normal distribution, so

$$P(-0.33 \le Z \le 0.90) = P(Z \le 0.90) - P(Z \le -0.33) = 0.8159 - 0.3707 = 0.4452$$

48. $\mu = 60,\ \sigma = 16,\ n = 38$

$$P(56.8 \le \bar{X} \le 64.6) = P\left(\frac{56.8-60}{16/\sqrt{38}} \le \frac{\bar{X}-\mu}{\sigma/\sqrt{n}} \le \frac{64.6-60}{16/\sqrt{38}}\right) = P\left(-1.23 \le \frac{\bar{X}-\mu}{\sigma/\sqrt{n}} \le 1.77\right)$$

Because the sample size is greater than 30, it is approximately distributed according to the standard normal distribution, so

$$P(-1.23 \le Z \le 1.77) = P(Z \le 1.77) - P(Z \le -1.23) = 0.9616 - 0.1093 = 0.8523$$

49. $f(x) = 0.05$ for $[10, 30]$

(a) This is a uniform distribution.

(b) The domain of f is $[10, 30]$.
The range of f is $[0.05]$.

(c)

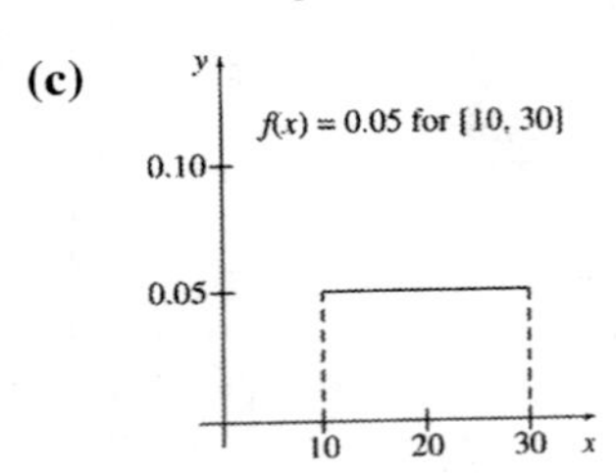

(d) For a uniform distribution, $\mu = \frac{1}{2}(b+a)$ and $\text{Var}(X) = \frac{b^2 - 2ab + a^2}{12}$. Thus,

$$\mu = \frac{1}{2}(30+10) = \frac{1}{2}(40) = 20$$

$$\text{Var}(X) = \frac{30^2 - 2(10)(30) + 10^2}{12} = \frac{400}{12}.$$

$$\sigma = \sqrt{\frac{400}{12}} \approx 5.77$$

(e)
$$P(\mu - \sigma \le X \le \mu + \sigma) = P(20 - 5.77 \le X \le 20 + 5.77) = P(14.23 \le X \le 25.77) = \int_{14.23}^{25.77} 0.05\,dx = 0.05x\Big|_{14.23}^{25.77} = 0.05(25.77 - 14.23) \approx 0.577$$

50. $f(x) = e^{-x}$ for $[0, \infty)$

(a) This is an exponential distribution with $a = 1$.

(b) The domain of f is $[0, \infty)$.
The range of f is $(0, 1]$.

(c)

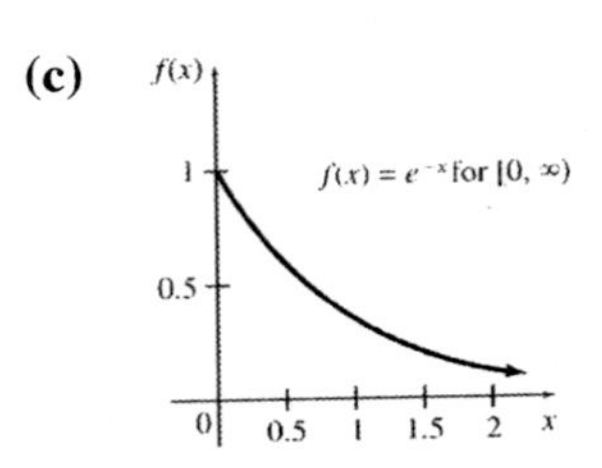

(d) For an exponential distribution, $\mu = \frac{1}{a}$ and $\sigma = \frac{1}{a}$. Thus $\mu = \frac{1}{1} = 1$ and $\sigma = \frac{1}{1} = 1$.

(e)
$$P(\mu - \sigma \le X \le \mu + \sigma) = P(1 - 1 \le X \le 1 + 1) = P(0 \le X \le 2) = \int_0^2 e^{-x}dx = -e^{-x}\Big|_0^2 = -e^{-2} + 1 \approx 0.8647$$

51. $f(x) = \frac{e^{-x^2}}{\sqrt{\pi}}$ for $(-\infty, \infty)$

(a) Since the exponent of e in $f(x)$ may be written $-x^2 = \frac{-(x-0)^2}{2\left(\frac{1}{\sqrt{2}}\right)^2}$, and $\frac{1}{\sqrt{\pi}} = \frac{1}{\frac{1}{\sqrt{2}}\sqrt{2\pi}}$, $f(x)$ is a normal distribution with $\mu = 0$ and $\sigma = \frac{1}{\sqrt{2}}$.

(b) The domain of f is $(-\infty, \infty)$.
The range of f is $\left(0, \frac{1}{\sqrt{\pi}}\right]$.

(c)

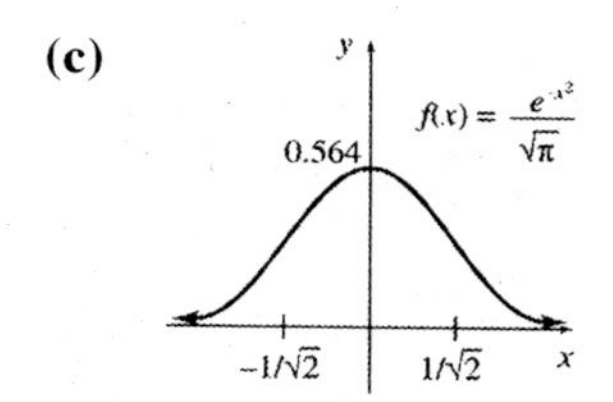

(d) For this normal distribution, $\mu = 0$ and $\sigma = \frac{1}{\sqrt{2}}$.

(e) $P(\mu - \sigma \le X \le \mu + \sigma)$
$= 2P(0 \le X \le \mu + \sigma)$
$= 2P\left(0 \le X \le \frac{1}{\sqrt{2}}\right)$

If $x = \frac{1}{\sqrt{2}}$, $z = \frac{\frac{1}{\sqrt{2}} - 0}{\frac{1}{\sqrt{2}}} = 1.00$. Thus,

$P(\mu - \sigma \le X \le \mu + \sigma)$
$= 2P(0 \le z \le 1.00)$
$= 2(0.3413)$
≈ 0.6827

52. $f(x) = \frac{xe^{-x/2}}{4}$ for x in $[0, \infty)$

(a) $P(0 \le X \le \infty) = \int_0^\infty \frac{xe^{-x/2}}{4} dx$

For all $x \ge 0$, $e^{-x/2} > 0$ so that $f(x) \ge 0$ for x in $[0, \infty)$. Evaluate $\int xe^{-x/2} dx$ using integration by parts.

$u = x \Rightarrow du = dx$ and
$dv = e^{-x/2} dx \Rightarrow v = -2e^{-x/2}$

$$\frac{1}{4}\int xe^{-x/2} dx = \frac{1}{4}\left(-2xe^{-x/2} + \int 2e^{-x/2} dx\right)$$
$$= \frac{1}{4}\left(-2xe^{-x/2} - 4e^{-x/2}\right)$$
$$= -\frac{1}{2}xe^{-x/2} - e^{-x/2}$$

Therefore,

$$\int_0^\infty \frac{xe^{-x/2}}{4} dx$$
$$= \left(-\frac{1}{2}xe^{-x/2} - e^{-x/2}\right)\Bigg|_0^\infty$$
$$= \lim_{b\to\infty}\left(-\frac{1}{2}xe^{-x/2} - e^{-x/2}\right)\Bigg|_0^b$$
$$= \lim_{b\to\infty}\left(-\frac{1}{2}be^{-b/2} - e^{-b/2}\right) - (0 - 1)$$

$\lim_{b\to\infty} be^{-b/2} = 0$. Therefore,

$$\int_0^\infty \frac{xe^{-x/2}}{4} dx = 0 - (-1) = 1.$$

(b) $P(0 \le X \le 3) = \int_0^3 \frac{xe^{-x/2}}{4} dx$

$$= \left(-\frac{1}{2}xe^{-x/2} - e^{-x/2}\right)\Bigg|_0^3$$
$$= \left(-\frac{3}{2}e^{-3/2} - e^{-3/2}\right) - (0 - 1)$$
$$= 1 - \frac{5}{2}e^{-3/2} \approx 0.4422$$

53. $f(x) = \frac{x^{-1/2}e^{-x/2}}{\sqrt{2\pi}}$ for x in $(0, \infty)$

(a) Using integration by parts:
$u = e^{-x/2} \Rightarrow du = -\frac{1}{2}e^{-x/2}$ and
$dv = x^{-1/2} dx \Rightarrow v = 2x^{1/2}$

$$\frac{1}{\sqrt{2\pi}}\int x^{-1/2}e^{-x/2} dx$$
$$= \frac{1}{\sqrt{2\pi}}\left[2x^{1/2}e^{-x/2} - \int 2x^{1/2}\left(-\frac{1}{2}\right)e^{-x/2} dx\right]$$
$$= \frac{1}{\sqrt{2\pi}}\left[2x^{1/2}e^{-x/2} + \int x^{1/2}e^{-x/2} dx\right]$$

Thus,

$P(0 < X \le b)$
$$= \frac{1}{\sqrt{2\pi}}\int_0^b x^{-1/2}e^{-x/2} dx$$
$$= \frac{1}{\sqrt{2\pi}}\left[2x^{1/2}e^{-x/2}\Big|_0^b + \int_0^b x^{1/2}e^{-x/2} dx\right].$$

(b) $P(0 < X \le 1)$
$$= \frac{1}{\sqrt{2\pi}}\left[2x^{1/2}e^{-x/2}\Big|_0^1 + \int_0^1 x^{1/2}e^{-x/2} dx\right]$$

Notice that

$$2x^{1/2}e^{-x/2}\Big|_0^1 = 2e^{-1/2} - 0 \approx 1.2131.$$

Using Simpson's rule with $n = 12$ to evaluate the improper integral, we have $\int_0^1 x^{1/2}e^{-x/2} dx \approx 0.4962$. Therefore,

$P(0 \le X \le 1)$
$$= \frac{1}{\sqrt{2\pi}}\int_0^1 x^{-1/2}e^{-x/2} dx$$
$$\approx \frac{1}{\sqrt{2\pi}}(1.2131 + 0.4962) \approx 0.6819.$$

(c) $P(0 < X \le 10)$

$$= \frac{1}{\sqrt{2\pi}}\left[2x^{1/2}e^{-x/2}\Big|_0^{10} + \int_0^{10} x^{1/2}e^{-x/2}dx\right]$$

First,

$$2x^{1/2}e^{-x/2}\Big|_0^{10} = 2\sqrt{10}e^{-5} - 0 \approx 0.0426.$$

Using Simpson's rule with $n = 12$ to evaluate the improper integral, we have $\int_0^{10} x^{1/2}e^{-x/2}dx \approx 2.3928$. Therefore,

$$P(0 \le X \le 10) = \frac{1}{\sqrt{2\pi}}\int_0^{10} x^{-1/2}e^{-x/2}dx$$
$$\approx \frac{1}{\sqrt{2\pi}}(0.0426 + 2.3928)$$
$$\approx 0.9716.$$

(d) Since $f(x)$ is a probability density function, the limit as $b \to \infty$ should be 1. The previous results support this conclusion.

54. $f(x) = \frac{8}{7}x^{-2}$ for $[1, 8]$

(a) $\mu = \int_1^8 \frac{8}{7}x^{-2}(x)\,dx = \int_1^8 \frac{8}{7}x^{-1}\,dx$

$$= \frac{8}{7}\ln x\Big|_1^8 = \frac{8}{7}(\ln 8 - \ln 1) \approx 2.3765 \text{ g}$$

(b) $\text{Var}(X) = \int_1^8 \frac{8}{7}x^{-2}(x^2)\,dx - 2.377^2$

$$= \frac{8}{7}x\Big|_1^8 - 2.377^2$$
$$= \frac{64}{7} - \frac{8}{7} - 2.377^2 \approx 2.352$$
$$\sigma = \sqrt{\text{Var}(X)} \approx 1.534 \text{ g}$$

(c) $P(\mu - \sigma \le X \le \mu + \sigma)$

$= P(0.844 \le X \le 3.91)$

$= P(1 \le X \le 3.91)$ since $[1, 8]$

$$= \int_1^{3.91} \frac{8}{7}x^{-2}\,dx = \frac{-8}{7}x^{-1}\Big|_1^{3.91}\,dx$$
$$= \frac{-8}{7}\left(\frac{1}{3.91}\right) + \frac{8}{7} \approx 0.8506$$

55. $f(x) = 0.01e^{-0.01x}$ for $[0, \infty)$ is an exponential distribution.

$$P(0 \le X \le 100) = \int_0^{100} 0.01e^{-0.01x}\,dx$$
$$= -e^{-0.01x}\Big|_0^{100} = 1 - \frac{1}{e}$$
$$\approx 0.6321$$

56. $f(x) = \frac{1}{b-a}$ for $[a, b]$

$$f(x) = \frac{1}{30-2} = \frac{1}{28}$$

This is uniform distribution for $a = 2$, $b = 30$.

(a) $E(X) = \mu = \frac{1}{2}(b + a)$

$$= \frac{1}{2}(30 + 2) = 16 \text{ in.}$$

(b) $P(20 < X \le 30) = \int_{20}^{30} \frac{1}{28}\,dx = \frac{1}{28}x\Big|_{20}^{30}$

$$= \frac{30}{28} - \frac{20}{28} \approx 0.3571$$

57. $f(x) = \frac{3}{19{,}696}(x^2 + x)$ for x in $[38, 42]$

(a) $\mu = \frac{3}{19{,}696}\int_{38}^{42} x(x^2 + x)\,dx$

$$= \frac{3}{19{,}696}\int_{38}^{42} (x^3 + x^2)\,dx$$
$$= \frac{3}{19{,}696}\left(\frac{x^4}{4} + \frac{x^3}{3}\right)\Bigg|_{38}^{42}$$
$$= \frac{3}{19{,}696}\left(\frac{(42)^4}{4} + \frac{(42)^3}{3} - \frac{(38)^4}{4} - \frac{(38)^3}{3}\right)$$
$$\approx 40.07$$

The expected body temperature of the species is 40.07°C.

(b) $P(X \le \mu)$

$$= \frac{3}{19{,}696}\int_{38}^{40.07} (x^2 + x)\,dx$$
$$= \frac{3}{19{,}696}\left(\frac{x^3}{3} + \frac{x^2}{2}\right)\Bigg|_{38}^{40.07}$$
$$= \frac{3}{19{,}696}\left(\frac{(40.07)^3}{3} + \frac{(40.07)^2}{2} - \frac{(38)^3}{3} - \frac{(38)^2}{2}\right)$$
$$\approx 0.4928$$

The probability of a body temperature below the mean is 0.4928.

58. $\mu = 7.8$ lb, $\sigma = 1.1$ lb, $x = 9$ lb

$$z = \frac{x - \mu}{\sigma} = \frac{9 - 7.8}{1.1} \approx 1.09$$

$P(X > 9) \approx P(z > 1.09) = 1 - 0.8621 \approx 0.1379$

59. Normal distribution, $\mu = 2.2$ g, $\sigma = 0.4$ g, X = tension

$$P(X < 1.9) = P\left(\frac{x-2.2}{0.4} < \frac{1.9-2.2}{0.4}\right) = P(z < -0.75) \approx 0.2266.$$

60. For an exponential distribution, $f(x) = ae^{-ax}$ for x in $[0, \infty)$. Since $\mu = \frac{1}{a} = 17.0$, $a = \frac{1}{17.0}$.

(a)
$$\begin{aligned} P(X \ge 15) &= \int_{15}^{\infty} \frac{1}{17.0} e^{-x/17.0}\,dx \\ &= \lim_{b\to\infty} \int_{15}^{b} \frac{1}{17.0} e^{-x/17.0}\,dx \\ &= \lim_{b\to\infty} \left(e^{-x/17.0} \Big|_{15}^{b} \right) \\ &= \lim_{b\to\infty} \left(-e^{-b/17.0} + e^{-15/17.0} \right) \\ &= e^{-15/17.0} \approx 0.4138 \end{aligned}$$

(b)
$$\begin{aligned} P(X < 10) &= \int_{0}^{10} \frac{1}{17.0} e^{-x/17.0}\,dx \\ &= (-e^{-x/17.0}) \Big|_{0}^{10} \\ &= -e^{-10/17.0} + e^{0} \approx 0.4447 \end{aligned}$$

61. For an exponential distribution, $f(x) = ae^{-ax}$ for x in $[0, \infty)$. Since $\mu = \frac{1}{a} = 32.5$, $a = \frac{1}{32.5}$.

(a)
$$\begin{aligned} P(X \ge 40) &= \int_{40}^{\infty} \frac{1}{32.5} e^{-x/32.5}\,dx \\ &= \lim_{b\to\infty} \int_{40}^{b} \frac{1}{32.5} e^{-x/32.5}\,dx \\ &= \lim_{b\to\infty} \left(e^{-x/32.5} \Big|_{40}^{b} \right) \\ &= \lim_{b\to\infty} \left(-e^{-b/32.5} + e^{-40/32.5} \right) \\ &= e^{-40/32.5} \approx 0.2921 \end{aligned}$$

(b)
$$\begin{aligned} P(30 \le X \le 50) &= \int_{30}^{50} \frac{1}{32.5} e^{-x/32.5}\,dx \\ &= (-e^{-x/32.5}) \Big|_{30}^{50} \\ &= -e^{-50/32.5} + e^{-30/32.5} \\ &\approx 0.1826 \end{aligned}$$

62. (a)

A polynomial function could fit the data.

Your answers to (b) through (f) may differ slightly from the answers given here, depending on how many places you keep in each of the coefficients of N and S.

(b) $N(t) = -0.001703t^4 + 0.39286t^3 - 30.9606t^2 + 887.496t - 3785.25$

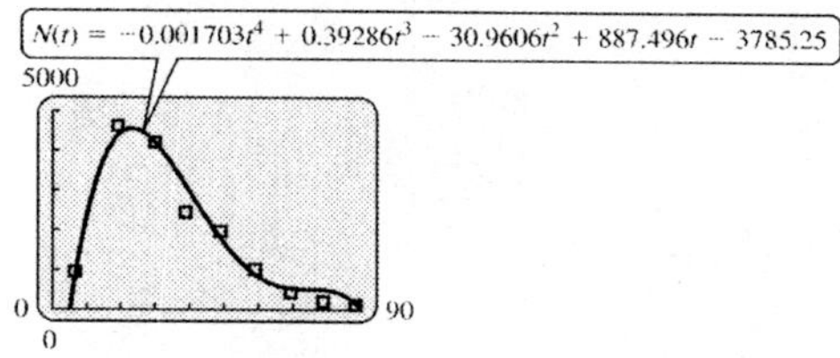

The function fits the data well.

(c) $N(t) = -0.001703t^4 + 0.39286t^3 - 30.9606t^2 + 887.496t - 3785.25$

Using the fnInt function on the TI-84 Plus graphing calculator, we obtain $\int_{5.2}^{88.9} N(t)\,dt \approx 172{,}009$. Thus the density function corresponding to the quartic fit is

$$S(t) = \frac{1}{172{,}009}\left(-0.001703t^4 + 0.39286t^3 - 30.9606t^2 + 887.496t - 3785.25\right)$$

Use a calculator to compute the integrals needed in (d), (e), and (f). Answers may vary slightly due to rounding.

(d) $P(\text{age} < 25) = \int_{5.2}^{25} S(t)\,dt \approx 0.3812$

$$\text{actual relative frequency} = \frac{957 + 4678}{16242} \approx 0.3469$$

$P(45 \le \text{age} < 65) = \int_{45}^{65} S(t)\,dt \approx 0.1447$

$$\text{actual relative frequency} = \frac{1997 + 1065}{16242} \approx 0.1885$$

$P(75 \le \text{age}) = \int_{75}^{88.9} S(t)\,dt \approx 0.0306$

$$\text{actual relative frequency} = \frac{250 + 112}{16242} \approx 0.0223$$

(e) $\mu = \int_{5.1}^{89.7} tS(t)\,dt \approx 32.76$

The expected age at which a person will die by assault is about 32.76 years.

(f) $\text{Var}(t) = \int_{5.1}^{89.7} t^2\, S(t)\,dt - (32.76)^2$

≈ 288.7254

$\sigma = \sqrt{288.7254} \approx 17.00$

The standard deviation of the distribution of age from death by assault is about 17.00 years.

63. (a) $P(\text{fewer than } 3)$

$= P(0) + P(1) + P(2)$

$= \dfrac{4.68^0 e^{-4.68}}{0!} + \dfrac{4.68^1 e^{-4.68}}{1!} + \dfrac{4.68^2 e^{-4.68}}{2!}$

≈ 0.1543

(b) $P(x = 3) = \dfrac{4.68^3 e^{-4.68}}{3!} \approx 0.1585$

(c) $P(\text{at least } 3) = 1 - P(\text{fewer than } 3)$

$= 1 - 0.1543$

$= 0.8457$

64. $f(t) = \dfrac{1}{3650.1} e^{-t/3650.1}$

This is an exponential distribution with $a = \dfrac{1}{3650.1}$. So the expected value is $\mu = \dfrac{1}{a} = 3650.1$ days. The standard deviation is $\sigma = \dfrac{1}{a} = 3650.1$ days.

65. Normal distribution, $\mu = 40$, $\sigma = 13$, $X =$ "take"

$P(X > 50) = P\left(\dfrac{X-40}{13} > \dfrac{50-40}{13}\right)$

$= P(Z > 0.77) = 1 - P(Z \le 0.77)$

$= 1 - 0.7794 = 0.2206$

66. Let the random variable X be the number of pieces of equipment. We have an exponential distribution $f(t) = ae^{-at}$ for t in $[0, \infty)$. Since $\mu = 10$, $a = \dfrac{1}{\mu} = 0.1$ so that

$f(t) = 0.1e^{-0.1t}$, $t \ge 0$.

We need to determine the portion of the pieces of equipment sold in the first year and in the second and third year. In other words, we need to calculate $P(0 \le X \le 1)$ and $P(1 \le X \le 3)$.

$P(0 \le X \le 1) = \int_0^1 0.1e^{-0.1t}\,dt = -e^{-0.1t}\Big|_0^1$

$= -e^{-0.1} + e^0 = 1 - e^{-0.1}$

$P(1 \le X \le 3) = \int_1^3 0.1e^{-0.1t}\,dt = -e^{-0.1t}\Big|_1^3$

$= -e^{-0.3} + e^{-0.1} = e^{-0.1} - e^{-0.3}$

Since

$$\text{payment} = \begin{cases} x & \text{for } 0 \le x \le 1 \\ 0.5x & \text{for } 1 \le x \le 3 \\ 0 & \text{for } x > 3 \end{cases}$$

the expected payment will be

$E(X) = (1 - e^{-0.1})(x) + (e^{-0.1} - e^{-0.3})(0.5x) + 0$

$= x - 1.5xe^{-0.1} - 0.5xe^{-0.3}$.

To determine the level x must be set for this to be 1000, solve

$E(X) = x - 1.5xe^{-0.1} - 0.5xe^{-0.3} = 1000.$

Using a calculator, we find $x \approx 5644$. The correct answer choice is **d**.

Chapter 14

DISCRETE DYNAMICAL SYSTEMS

14.1 Sequences

1. $a_n = 5n + 2$
$a_1 = 5(1) + 2 = 7$
$a_2 = 5(2) + 2 = 12$
$a_3 = 5(3) + 2 = 17$
$a_4 = 5(4) + 2 = 22$

3. $a_n = 2(3^n)$
$a_1 = 2(3^1) = 6$
$a_2 = 2(3^2) = 18$
$a_3 = 2(3^3) = 54$
$a_4 = 2(3^4) = 162$

5. $a_n = \sin n$
$a_1 = \sin 1 \approx 0.8415$
$a_2 = \sin 2 \approx 0.9093$
$a_3 = \sin 3 \approx 0.1411$
$a_4 = \sin 4 \approx -0.7568$

7. $a_n = 2a_{n-1} - 3,\ a_1 = 4$
$a_2 = 2a_{2-1} - 3 = 2(4) - 3 = 5$
$a_3 = 2a_{3-1} - 3 = 2(5) - 3 = 7$
$a_4 = 2a_{4-1} - 3 = 2(7) - 3 = 11$
$a_5 = 2a_{5-1} - 3 = 2(11) - 3 = 19$

9. $a_n = -5a_{n-1} + 2,\ a_1 = -1$
$a_2 = -5a_{2-1} + 2 = -5(-1) + 2 = 7$
$a_3 = -5a_{3-1} + 2 = -5(7) + 2 = -33$
$a_4 = -5a_{4-1} + 2 = -5(-33) + 2 = 167$
$a_5 = -5a_{5-1} + 2 = -5(167) + 2 = -833$

11. $a_n = a_{n-1}^2 - 4,\ a_1 = 3$
$a_2 = a_{2-1}^2 - 4 = 3^2 - 4 = 5$
$a_3 = a_{3-1}^2 - 4 = 5^2 - 4 = 21$
$a_4 = a_{4-1}^2 - 4 = 21^2 - 4 = 437$
$a_5 = a_{5-1}^2 - 4 = 437^2 - 4 = 190{,}965$

13. $f(x) = x + 2$
$f^2(x) = f(f(x)) = f(x+2)$
$= (x+2) + 2 = x + 4$
$f^3(x) = f(f^2(x)) = f(x+4)$
$= (x+4) + 2 = x + 6$
$f^n(x) = f(f^{n-1}(x)) = x + 2n$

15. $f(x) = 2x$
$f^2(x) = f(f(x)) = f(2x)$
$= 2(2x) = 4x$
$f^3(x) = f(f^2(x)) = f(4x)$
$= 2(4x) = 8x$
$f^n(x) = f(f^{n-1}(x)) = 2^n x$

17. $f(x) = 1$
$f^2(x) = f(f(x)) = f(1) = 1$
$f^3(x) = f(f^2(x)) = f(1) = 1$
$f^n(x) = f(f^{n-1}(x)) = 1$

19. $a_{n+1} = f(a_n) = 1.5\left(1 - \frac{a_n}{3000}\right)a_n,\ a_1 = 1500$
$a_2 = f(a_1) = 1.5\left(1 - \frac{1500}{3000}\right)1500 = 1125$
$a_3 = f(a_2) = 1.5\left(1 - \frac{1125}{3000}\right)1125 \approx 1055$
$a_4 = f(a_3) = 1.5\left(1 - \frac{1055}{3000}\right)1055 \approx 1026$
$a_5 = f(a_4) = 1.5\left(1 - \frac{1026}{3000}\right)1026 \approx 1013$

21. $a_{n+1} = f(a_n) = 2.5\left(1 - \frac{a_n}{4000}\right)a_n,\ a_1 = 1800$

$$a_2 = f(a_1) = 2.5\left(1 - \frac{1800}{4000}\right)1800 = 2475$$

$$a_3 = f(a_2) = 2.5\left(1 - \frac{2475}{4000}\right)2475 \approx 2359$$

$$a_4 = f(a_3) = 2.5\left(1 - \frac{2359}{4000}\right)2359 \approx 2419$$

$$a_5 = f(a_4) = 2.5\left(1 - \frac{2419}{4000}\right)2419 \approx 2390$$

23. $a_{n+1} = f(a_n) = 3.4\left(1 - \frac{a_n}{6000}\right)a_n,\ a_1 = 5000$

$$a_2 = f(a_1) = 3.4\left(1 - \frac{5000}{6000}\right)5000 \approx 2833$$

$$a_3 = f(a_2) = 3.4\left(1 - \frac{2833}{6000}\right)2833 \approx 5084$$

$$a_4 = f(a_3) = 3.4\left(1 - \frac{5084}{6000}\right)5084 \approx 2639$$

$$a_5 = f(a_4) = 3.4\left(1 - \frac{2639}{6000}\right)2639 \approx 5026$$

25. (a) $f(x) = xe^{r(1-x/N)}$

Using the hint, we have $\lim_{x\to\infty} xe^{-x} = 0.$

$$f(x) = xe^{r-rx/N} = xe^{r}\left(e^{-rx/N}\right)$$

$$\lim_{x\to\infty} f(x) = \lim_{x\to\infty} e^{r}\left(xe^{-rx/N}\right) = \lim_{x\to\infty} e^{r} \cdot \lim_{x\to\infty}\left(xe^{-rx/N}\right) = e^{r} \cdot 0 = 0$$

(b) $f'(x) = x\frac{d}{dx}e^{r(1-x/N)} + e^{r(1-x/N)}\frac{d}{dx}x = xe^{r(1-x/N)}\frac{d}{dx}\left(r - \frac{rx}{N}\right) + e^{r(1-x/N)}$

$$= xe^{r(1-x/N)}\left(-\frac{r}{N}\right) + e^{r(1-x/N)} = e^{r(1-x/N)}\left(1 - \frac{rx}{N}\right)$$

(c) First, find any critical points.

$f(x) = xe^{r(1-x/N)} \ge 0 \Rightarrow x \ge 0$

(Note that the model is measuring the number of salmon, which cannot be negative.)

$$f'(x) = e^{r(1-x/N)}\left(1 - \frac{rx}{N}\right) = 0$$

$$1 - \frac{rx}{N} = 0 \Rightarrow 1 = \frac{rx}{N} \Rightarrow x = \frac{N}{r}$$

Now solve using the first derivative test. Choose a value in the intervals $\left[0, \frac{N}{r}\right)$ and $\left(\frac{N}{r}, \infty\right)$.

$$f'\left(\frac{N}{2r}\right) = e^{r(1-(N/2r)/N)}\left(1 - \frac{\frac{N}{2r}r}{N}\right) = e^{r(1-(N/2r)/N)}\left(\frac{1}{2}\right) > 0$$

$$f'\left(\frac{2N}{r}\right) = e^{r(1-(2N/r)/N)}\left(1 - \frac{\frac{2N}{r}r}{N}\right) = e^{r(1-(N/2r)/N)}(-1) < 0$$

Thus, f is increasing for $0 \le x < \frac{N}{r}$ and decreasing for $x > \frac{N}{r}$. The function has a maximum value at $x = \frac{N}{r}$.

(d) $f''(x) = e^{r(1-x/N)}\left(1-\frac{rx}{N}\right)\frac{d}{dx}$

$$= \left(1-\frac{rx}{N}\right)\frac{d}{dx}e^{r(1-x/N)} + e^{r(1-x/N)}\frac{d}{dx}\left(1-\frac{rx}{N}\right)$$

$$= \left(1-\frac{rx}{N}\right)e^{r(1-x/N)}\frac{d}{dx}\left(1-\frac{rx}{N}\right) + e^{r(1-x/N)}\left(-\frac{r}{N}\right)$$

$$= \left(1-\frac{rx}{N}\right)e^{r(1-x/N)}\left(-\frac{r}{N}\right) + e^{r(1-x/N)}\left(-\frac{r}{N}\right)$$

$$= e^{r(1-x/N)}\left(-\frac{r}{N}\right)\left(1-\frac{rx}{N}+1\right)$$

$$= e^{r(1-x/N)}\left(\frac{rx}{N}-2\right)\left(\frac{r}{N}\right)$$

(e) $f''(x) = e^{r(1-x/N)}\left(\frac{r}{N}\right)\left(\frac{rx}{N}-2\right) = 0 \Rightarrow \frac{xr}{N} - 2 = 0 \Rightarrow x = \frac{2N}{r}$

Now use the second derivative test. Choose a value in each interval $\left[0, \frac{2N}{r}\right)$ and $\left(\frac{2N}{r}, \infty\right)$ to test.

$$f''\left(\frac{N}{r}\right) = e^{r(1-(N/r)/N)}\left(\frac{r}{N}\right)\left(\frac{\left(\frac{N}{r}\right)r}{N}-2\right) = e^{r(1-(N/r)/N)}\left(\frac{r}{N}\right)(-1) < 0$$

$$f''\left(\frac{4N}{r}\right) = e^{r(1-(4N/r)/N)}\left(\frac{r}{N}\right)\left(\frac{\left(\frac{4N}{r}\right)r}{N}-2\right) = e^{r(1-(4N/r)/N)}\left(\frac{r}{N}\right)(2) > 0$$

Thus, f is concave down for $0 \le x < \frac{2N}{r}$ and concave up for $x > \frac{2N}{r}$. The function has an inflection point at $x = \frac{2N}{r}$.

(f)

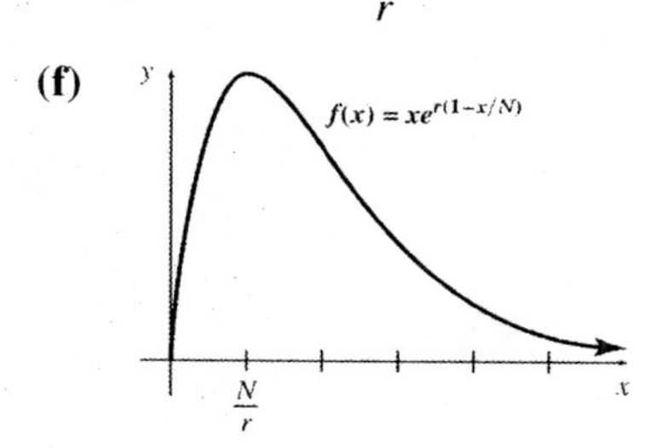

(g) The logistic growth model increases and then levels off near a positive value while the Ricker model increases and then decreases and approaches zero. The graphs of both models have parts that are concave up and concave down.

27. $a_{n+1} = f(a_n) = xe^{2.5(1-x/2000)}$, $a_1 = 500$

$a_2 = f(a_1) = 500e^{2.5(1-500/2000)} \approx 3260$

$a_3 = f(a_2) = 3260e^{2.5(1-3260/2000)} \approx 675$

$a_4 = f(a_3) = 675e^{2.5(1-675/2000)} \approx 3537$

$a_5 = f(a_4) = 3537e^{2.5(1-993/2000)} \approx 518$

29. $a_{n+1} = f(a_n) = \dfrac{3a_n}{1+\dfrac{a_n}{1000}}$; $a_1 = 500$

$$a_2 = f(a_1) = \frac{3(500)}{1+\frac{500}{1000}} = 1000$$

$$a_3 = f(a_2) = \frac{3(1000)}{1+\frac{1000}{1000}} = 1500$$

$$a_4 = f(a_3) = \frac{3(1500)}{1+\frac{1500}{1000}} = 1800$$

$$a_5 = f(a_4) = \frac{3(1800)}{1+\frac{1800}{1000}} \approx 1929$$

31. (a) $$\lim_{x\to\infty} f(x) = \lim_{x\to\infty} \frac{rx}{1+\left(\frac{x}{b}\right)^2} = \lim_{x\to\infty} \frac{\frac{rx}{x}}{\frac{1}{x}+\frac{x}{b^2}} = \lim_{x\to\infty} \frac{r}{\frac{1}{x}+\frac{x}{b^2}} = \lim_{x\to\infty} \frac{r}{0+\frac{x^2}{b^2}} = \lim_{x\to\infty} \frac{b^2 r}{x^2} = 0$$

(b) $$f'(x) = \frac{\left(1+\left(\frac{x}{b}\right)^2\right)\frac{d}{dx}(rx) - rx\frac{d}{dx}\left(1+\left(\frac{x}{b}\right)^2\right)}{\left(1+\left(\frac{x}{b}\right)^2\right)^2} = \frac{r\left(1+\left(\frac{x}{b}\right)^2\right) - rx\left(\frac{2x}{b^2}\right)}{\left(1+\left(\frac{x}{b}\right)^2\right)^2} = \frac{r\left(1+\left(\frac{x}{b}\right)^2 - \frac{2x^2}{b^2}\right)}{\left(1+\left(\frac{x}{b}\right)^2\right)^2}$$

$$= \frac{r\left(1-\left(\frac{x}{b}\right)^2\right)}{\left(1+\left(\frac{x}{b}\right)^2\right)^2}$$

(c) First, find any critical points.

$$f(x) = \frac{rx}{1+\left(\frac{x}{b}\right)^2} \ge 0 \Rightarrow x \ge 0$$

$$f'(x) = \frac{r\left(1-\left(\frac{x}{b}\right)^2\right)}{\left(1+\left(\frac{x}{b}\right)^2\right)^2} = 0 \Rightarrow 1-\left(\frac{x}{b}\right)^2 = 0 \Rightarrow 1 = \frac{x}{b} \Rightarrow x = b$$

Now solve using the first derivative test. Choose a value in each interval $[0, b)$ and (b, ∞) to determine if the function is increasing or decreasing.

$$f'\left(\frac{b}{2}\right) = \frac{r\left(1-\left(\frac{\frac{b}{2}}{b}\right)^2\right)}{\left(1+\left(\frac{\frac{b}{2}}{b}\right)^2\right)^2} = \frac{\frac{3r}{4}}{\frac{25}{16}} = \frac{12r}{25} > 0$$

(*continued on next page*)

(*continued*)

$$f'(2b)=\frac{r\left(1-\left(\frac{2b}{b}\right)^2\right)}{\left(1+\left(\frac{2b}{b}\right)^2\right)^2}=\frac{-9r}{25}<0$$

Thus, f is increasing for $0 \le x < b$ and decreasing for $x > b$. The function has a maximum value at $x = b$.

(d) $$f''(x)=r\frac{d}{dx}\frac{\left(1-\left(\frac{x}{b}\right)^2\right)}{\left(1+\left(\frac{x}{b}\right)^2\right)^2}=r\cdot\frac{\left(1+\left(\frac{x}{b}\right)^2\right)^2\frac{d}{dx}\left(1-\left(\frac{x}{b}\right)^2\right)-\left(1-\left(\frac{x}{b}\right)^2\right)\frac{d}{dx}\left(1+\left(\frac{x}{b}\right)^2\right)^2}{\left(1+\left(\frac{x}{b}\right)^2\right)^4}$$

$$=r\cdot\frac{\left(1+\left(\frac{x}{b}\right)^2\right)^2\left(-\frac{2x}{b^2}\right)-2\left(1-\left(\frac{x}{b}\right)^2\right)\left(1+\left(\frac{x}{b}\right)^2\right)\frac{d}{dx}\left(1+\left(\frac{x}{b}\right)^2\right)}{\left(1+\left(\frac{x}{b}\right)^2\right)^4}$$

$$=r\cdot\frac{\left(1+\frac{x^2}{b^2}\right)\left(-\frac{2x}{b^2}\right)-2\left(1-\frac{x^2}{b^2}\right)\left(\frac{2x}{b^2}\right)}{\left(1+\left(\frac{x}{b}\right)^2\right)^3}=r\left(-\frac{2x}{b^2}\right)\frac{\left(1+\frac{x^2}{b^2}\right)+2\left(1-\frac{x^2}{b^2}\right)}{\left(1+\left(\frac{x}{b}\right)^2\right)^3}=-\frac{2rx}{b^2}\left(\frac{3-\frac{x^2}{b^2}}{\left(1+\left(\frac{x}{b}\right)^2\right)^3}\right)$$

(e) $$f''(x)=\frac{-2rx\left(3-\left(\frac{x}{b}\right)^2\right)}{b^2\left(1+\frac{x}{b}\right)^3}=0$$

$$3-\left(\frac{x}{b}\right)^2=0\Rightarrow\left(\frac{x}{b}\right)^2=3\Rightarrow x=\sqrt{3}b$$

Now use the second derivative test. Choose a value in each interval $\left[0, \sqrt{3}b\right)$ and $\left(\sqrt{3}b, \infty\right)$ to determine if the function is concave up or concave down.

$$f''(b)=\frac{-2rx\left(3-\left(\frac{b}{b}\right)^2\right)}{b^2\left(1+\frac{b}{b}\right)^3}=\frac{-4rx}{8b^2}<0$$

$$f''(3b)=\frac{-2rx\left(3-\left(\frac{3b}{b}\right)^2\right)}{b^2\left(1+\frac{3b}{b}\right)^3}=\frac{12rx}{64b^2}>0$$

So, the function is concave down for all values of $0 \le x < \sqrt{3}b$ and concave up for $x > \sqrt{3}b$. The inflection point is at $x = \sqrt{3}b$.

(f)

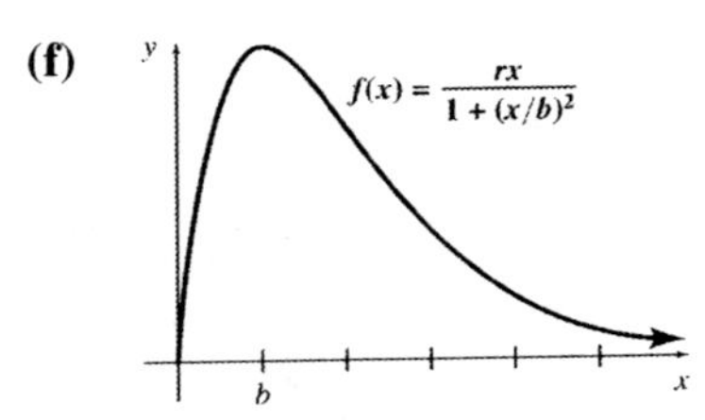

(g) This graph of this function has a similar shape to that of the graph of the Ricker model.

33. $a_{n+1} = f(a_n) = \dfrac{4.5a_n}{1+\left(\dfrac{a_n}{2000}\right)^2}$; $a_0 = 1500$

$a_2 = f(a_1) = \dfrac{4.5(1500)}{1+\left(\dfrac{1500}{2000}\right)^2} = 4320$

$a_3 = f(a_2) = \dfrac{4.5(4320)}{1+\left(\dfrac{4320}{2000}\right)^2} \approx 3431$

$a_4 = f(a_3) = \dfrac{4.5(3431)}{1+\left(\dfrac{3431}{2000}\right)^2} \approx 3916$

$a_5 = f(a_4) = \dfrac{4.5(3916)}{1+\left(\dfrac{3916}{2000}\right)^2} \approx 3646$

14.2 Equilibrium Points

1.
$$f(x) = x$$
$$1-|2x-1| = x$$
$$|2x-1| = 1-x$$

$2x-1 = 1-x$, $3x = 2$, $x = \dfrac{2}{3}$ | $2x-1 = x-1$, $x = 0$

The equilibrium points are $x = 0$ and $x = \dfrac{2}{3}$.

3.
$$f(x) = x$$
$$6x^2(1-x) = x$$
$$-x+6x^2-6x^3 = 0$$
$$-x\left(1-6x+6x^2\right) = 0$$

$-x = 0$, $x = 0$ | $1-6x+6x^2 = 0$

$$x = \frac{-(-6)\pm\sqrt{(-6)^2-4(6)(1)}}{2(6)} = \frac{6\pm\sqrt{12}}{12} = \frac{3\pm\sqrt{3}}{6} \approx 0.2113,\ 0.7887$$

The equilibrium points are $x = 0$ and $x = \dfrac{3\pm\sqrt{3}}{6}$.

5.
$$f(x) = x$$
$$2x\left(1-x^2\right) = x$$
$$2x-2x^3 = x$$
$$x-2x^3 = 0$$
$$x\left(1-2x^2\right) = 0$$

$x = 0$ | $1-2x^2 = 0$, $2x^2 = 1$, $x^2 = \dfrac{1}{2} \Rightarrow x = \pm\dfrac{\sqrt{2}}{2}$

However, $x = -\dfrac{\sqrt{2}}{2}$ is not in the given domain, so it is not an equilibrium point. The equilibrium points are $x = 0$ and $x = \dfrac{\sqrt{2}}{2} \approx 0.7071$.

7. $f(x) = 1-|2x-1|$

(a) $x_1 = 0.15$
$$x_2 = f(x_1) = 1-|2(0.15)-1| = 0.3$$
$$x_3 = f(x_2) = 1-|2(0.3)-1| = 0.6$$
$$x_4 = f(x_3) = 1-|2(0.6)-1| = 0.8$$
$$x_5 = f(x_4) = 1-|2(0.8)-1| = 0.4$$
$$x_6 = f(x_5) = 1-|2(0.4)-1| = 0.8$$
$$x_7 = f(x_6) = 1-|2(0.8)-1| = 0.4$$

(b) $x_1 = 0.4$
$$x_2 = f(x_1) = 1-|2(0.4)-1| = 0.8$$
$$x_3 = f(x_2) = 1-|2(0.8)-1| = 0.4$$
$$x_4 = f(x_3) = 1-|2(0.4)-1| = 0.8$$
$$x_5 = f(x_4) = 1-|2(0.8)-1| = 0.4$$
$$x_6 = f(x_5) = 1-|2(0.4)-1| = 0.8$$
$$x_7 = f(x_6) = 1-|2(0.8)-1| = 0.4$$

(c) $x_1 = 0.65$
$x_2 = f(x_1) = 1 - |2(0.65) - 1| = 0.7$
$x_3 = f(x_2) = 1 - |2(0.7) - 1| = 0.6$
$x_4 = f(x_3) = 1 - |2(0.6) - 1| = 0.8$
$x_5 = f(x_4) = 1 - |2(0.8) - 1| = 0.4$
$x_6 = f(x_5) = 1 - |2(0.4) - 1| = 0.8$
$x_7 = f(x_6) = 1 - |2(0.8) - 1| = 0.4$

(d) $x_1 = 0.85$
$x_2 = f(x_1) = 1 - |2(0.85) - 1| = 0.3$
$x_3 = f(x_2) = 1 - |2(0.3) - 1| = 0.6$
$x_4 = f(x_3) = 1 - |2(0.6) - 1| = 0.8$
$x_5 = f(x_4) = 1 - |2(0.8) - 1| = 0.4$
$x_6 = f(x_5) = 1 - |2(0.4) - 1| = 0.8$
$x_7 = f(x_6) = 1 - |2(0.8) - 1| = 0.4$

The equilibrium points found in exercise 1 are $x = 0$ and $x = \frac{2}{3}$. Since none of the oscillations found in parts (a)–(d) approach these equilibrium points, both are unstable.

For exercises 8–12, we use a TI-84 graphing calculator set in Sequence mode to find the list of values. The Y = screen is shown for the first value of x_1. Use the Table menu to find the list of values.

9. $f(x) = 6x^2(1 - x)$

```
Plot1 Plot2 Plot3
 nMin=1
:.u(n)=6u(n-1)²(1
-u(n-1)
 u(nMin)={.15}
\v(n)=
 v(nMin)=
\w(n)=
```

(a) $x_1 = 0.15$
$x_2 = f(x_1) = 0.1148$
$x_3 = f(x_2) = 0.6994$
$x_4 = f(x_3) = 0.0273$
$x_5 = f(x_4) = 0.00435$
$x_6 = f(x_5) = 0.0001$
$x_7 = f(x_6) = 0.0000$

(b) $x_1 = 0.4$
$x_2 = f(x_1) = 0.5760$
$x_3 = f(x_2) = 0.8440$
$x_4 = f(x_3) = 0.6666$
$x_5 = f(x_4) = 0.8889$
$x_6 = f(x_5) = 0.5267$
$x_7 = f(x_6) = 0.7879$

(c) $x_1 = 0.65$
$x_2 = f(x_1) = 0.8873$
$x_3 = f(x_2) = 0.5325$
$x_4 = f(x_3) = 0.7954$
$x_5 = f(x_4) = 0.7766$
$x_6 = f(x_5) = 0.8084$
$x_7 = f(x_6) = 0.7512$

(d) $x_1 = 0.85$
$x_2 = f(x_1) = 0.6503$
$x_3 = f(x_2) = 0.8873$
$x_4 = f(x_3) = 0.5324$
$x_5 = f(x_4) = 0.7952$
$x_6 = f(x_5) = 0.7770$
$x_7 = f(x_6) = 0.8078$

The equilibrium points found in exercise 3 are $x = 0$ and

$$x = \frac{3 \pm \sqrt{3}}{6} \approx 0.7887,\ 0.2113.$$

The sequence of values in part (a) approach 0, so that equilibrium point is stable. The other two equilibrium points are unstable.

11. $f(x) = 2x(1 - x^2)$

```
Plot1 Plot2 Plot3
 nMin=1
:.u(n)=2u(n-1)(1-
u(n-1)²)
 u(nMin)={.15}
\v(n)=■
 v(nMin)=
\w(n)=
```

(a) $x_1 = 0.15$
$x_2 = f(x_1) = 0.2933$
$x_3 = f(x_2) = 0.5361$
$x_4 = f(x_3) = 0.7640$
$x_5 = f(x_4) = 0.6361$
$x_6 = f(x_5) = 0.7575$
$x_7 = f(x_6) = 0.6457$

(b) $x_1 = 0.4$
$x_2 = f(x_1) = 0.672$
$x_3 = f(x_2) = 0.7371$
$x_4 = f(x_3) = 0.6733$
$x_5 = f(x_4) = 0.7362$
$x_6 = f(x_5) = 0.6744$
$x_7 = f(x_6) = 0.7353$

(c) $x_1 = 0.65$
$x_2 = f(x_1) = 0.7508$
$x_3 = f(x_2) = 0.6552$
$x_4 = f(x_3) = 0.7479$
$x_5 = f(x_4) = 0.6592$
$x_6 = f(x_5) = 0.7455$
$x_7 = f(x_6) = 0.6623$

(d) $x_1 = 0.85$
$x_2 = f(x_1) = 0.4718$
$x_3 = f(x_2) = 0.7335$
$x_4 = f(x_3) = 0.6777$
$x_5 = f(x_4) = 0.7329$
$x_6 = f(x_5) = 0.6785$
$x_7 = f(x_6) = 0.7323$

The equilibrium points found in exercise 5 are $x = 0$ and $x = \frac{\sqrt{2}}{2} \approx 0.7071$.

Since none of the oscillations found in parts (a)–(d) approach 0, this equilibrium point is unstable. The oscillations found in parts (a)–(d) do approach $x = \frac{\sqrt{2}}{2}$, so this equilibrium point is stable.

13.

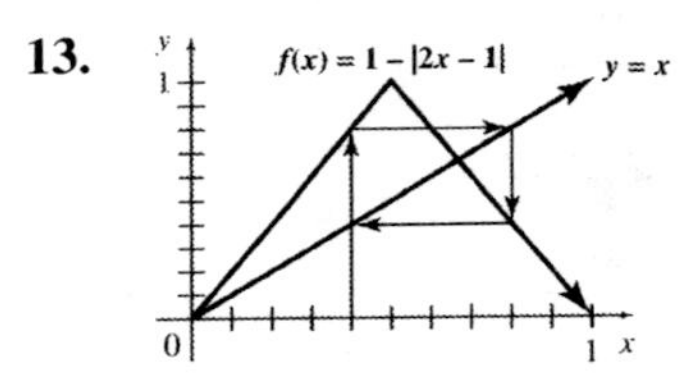

15.

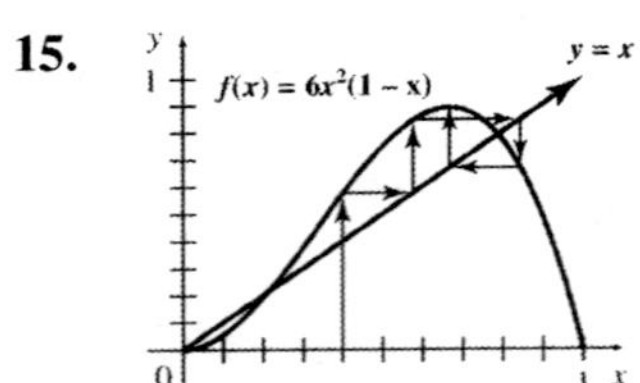

17.

f(x) = 2x(1 − x²)
y = x

19. (a) $f(x) = xe^{r(1-x)}$

To find any equilibrium points, solve $f(x) = x$.

$$xe^{r(1-x)} = x$$
$$xe^{r(1-x)} - x = 0$$
$$x\left(e^{r(1-x)} - 1\right) = 0$$

$$x = 0 \quad \Big| \quad e^{r(1-x)} - 1 = 0$$
$$e^{r(1-x)} = 1$$
$$r(1-x) = 0 \Rightarrow x = 1,\ r = 0$$

The equilibrium points are $x = 0$ and $x = 1$.

(b) i. $a_1 = 0.6,\ r = 2$
$a_2 = f(a_1) = 1.3353$
$a_3 = f(a_2) = 0.6829$
$a_4 = f(a_3) = 1.2876$
$a_5 = f(a_4) = 0.7244$
$a_6 = f(a_5) = 1.2571$

Since the oscillations approach $x = 1$, this equilibrium point is stable.

ii. $a_1 = 0.6,\ r = 3$
$a_2 = f(a_1) = 1.9921$
$a_3 = f(a_2) = 0.1016$
$a_4 = f(a_3) = 1.5042$
$a_5 = f(a_4) = 0.3314$
$a_6 = f(a_5) = 2.4630$

Since the oscillations do not approach $x = 0$ or $x = 1$, both are unstable.

(c)

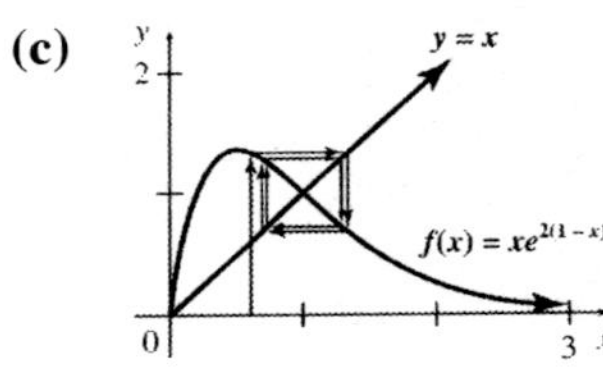

(d)

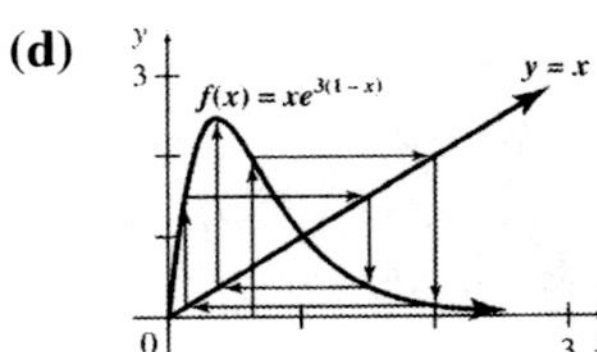

21. (a) $f(x)=\dfrac{rx}{1+x^2}$

To find any equilibrium points, solve $f(x)=x$.

$$\frac{rx}{1+x^2}=x$$
$$rx=x+x^3$$
$$x^3-rx+x=0$$
$$x^3+x(1-r)=0$$
$$x\left(x^2+1-r\right)=0$$

$x=0$ | $x^2+1-r=0$
$x^2=r-1$
$x=\sqrt{r-1}$

The equilibrium points are $x=0$ and $x=\sqrt{r-1}$.

(b) i. $a_1=4,\ r=2$
$a_2=f(a_1)=0.4706$
$a_3=f(a_2)=0.7705$
$a_4=f(a_3)=0.9670$
$a_5=f(a_4)=0.9994$
$a_6=f(a_5)=1.0000$

Since the oscillations approach $x=\sqrt{r-1}=1,$ this equilibrium point is stable.

ii. $a_1=4,\ r=10$
$a_2=f(a_1)=2.3529$
$a_3=f(a_2)=3.5998$
$a_4=f(a_3)=2.5789$
$a_5=f(a_4)=3.3708$
$a_6=f(a_5)=2.7267$

Since the oscillations approach $x=\sqrt{r-1}=3,$ this equilibrium point is stable.

(c)

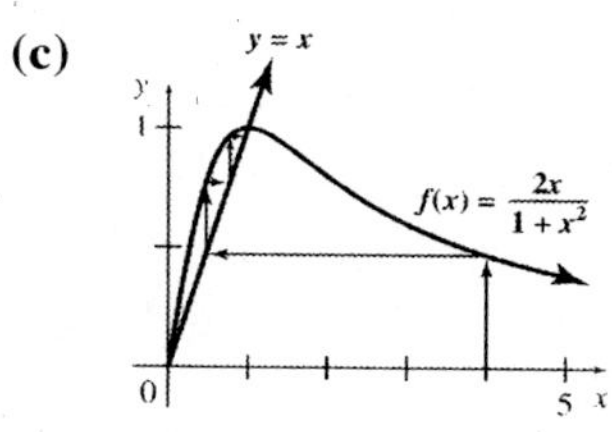

(d)

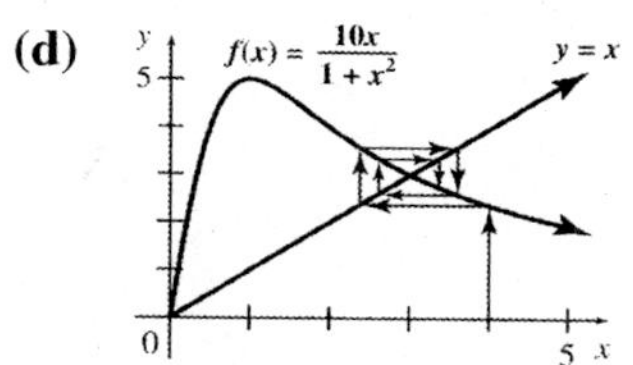

14.3 Determining Stability

1. $f(x)=1-|2x-1|$

From section 14.2, we know that the equilibrium points are $x=0$ and $x=\dfrac{2}{3}$.

$0\le x\le 1\Rightarrow 0\le 2x\le 2\Rightarrow -1\le 2x-1\le 1,$ so $|2x-1|=1-2x.$ So $f(x)=1-(1-2x)=2x$ and $f'(x)=2\Rightarrow|f'(x)|>1.$ Therefore, both equilibrium points are unstable.

3. $f(x)=6x^2(1-x)$

From section 14.2, we know that the equilibrium points are $x=0$ and $x=\dfrac{3\pm\sqrt{3}}{6}$.

$f'(x)=12x(1-x)-6x^2=12x-18x^2$

$|f'(0)|=\left|12(0)-18(0)^2\right|=0<1,$ so $x=0$ is stable.

$$\left|f'\left(\frac{3-\sqrt{3}}{6}\right)\right|=\left|12\left(\frac{3-\sqrt{3}}{6}\right)-18\left(\frac{3-\sqrt{3}}{6}\right)^2\right|\approx 1.7>1,$$

so $x=\dfrac{3-\sqrt{3}}{6}$ is unstable.

$$\left|f'\left(\frac{3+\sqrt{3}}{6}\right)\right|=\left|12\left(\frac{3+\sqrt{3}}{6}\right)-18\left(\frac{3+\sqrt{3}}{6}\right)^2\right|\approx|-1.7|>1,$$

so $x=\dfrac{3+\sqrt{3}}{6}$ is unstable.

5. $f(x)=2x\left(1-x^2\right)$

From section 14.2, we know that the equilibrium points are $x=0$ and $x=\dfrac{\sqrt{2}}{2}$.

$$f'(x)=2\left(1-x^2\right)-2x(2x)=2-2x^2-4x^2$$
$$=2-6x^2$$

$|f'(0)|=\left|2-6(0)^2\right|=2>1,$ so $x=0$ is unstable.

$$|f'(0)|=\left|2-6\left(\frac{\sqrt{2}}{2}\right)^2\right|=1,$$ so it is not possible to tell if $x=\dfrac{\sqrt{2}}{2}$ is stable using this test.

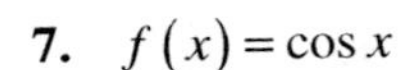

7. $f(x) = \cos x$

Plot $Y_1 = \cos x$ and $Y_2 = x$.

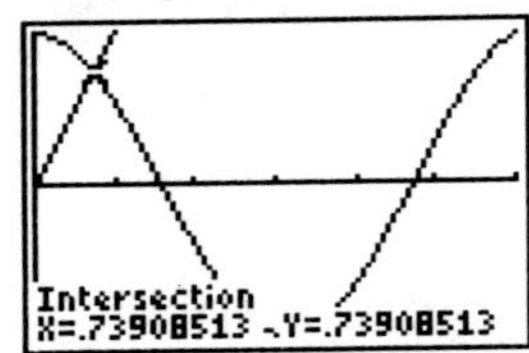

The equilibrium point is $x \approx 0.7391$. Using the nDeriv function, we have $|f'(0.7391)| \approx 0.67 < 1$, so $x = 0.7391$ is stable.

9. $f(x) = 1 - 16\left(x - \frac{1}{2}\right)^4$

Plot $Y_1 = 1 - 16\left(x - \frac{1}{2}\right)^4$ and $Y_2 = x$.

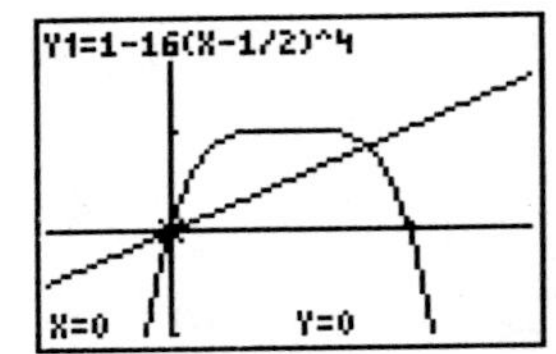

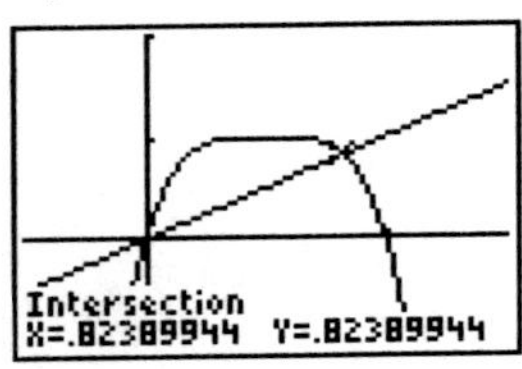

The equilibrium points are $x = 0$ and $x \approx 0.8239$. Using the nDeriv function, we have $|f'(0)| \approx 8.0 > 1$, so $x = 0$ is unstable. $|f'(0.8239)| \approx 2.17 > 1$, so $x = 0.8239$ is unstable.

11. (a) $f(x) = 4x^2(1 - x)$

Find the equilibrium points by solving $f(x) = x$.

$$4x^2(1-x) = x$$
$$-x + 4x^2 - 4x^3 = 0$$
$$-x\left(1 - 4x + 4x^2\right) = 0$$
$$-x(2x-1)^2 = 0$$

$-x = 0 \;\Rightarrow\; x = 0$ | $(2x-1)^2 = 0 \;\Rightarrow\; x = \frac{1}{2}$

The equilibrium points are $x = 0$ and $x = \frac{1}{2}$.

(b) $f'(x) = 8x(1-x) - 4x^2$
$$= 8x - 8x^2 - 4x^2$$
$$= 8x - 12x^2$$

$f'(0) = 8(0) - 12(0)^2 = 0$

$f'\left(\frac{1}{2}\right) = 8\left(\frac{1}{2}\right) - 12\left(\frac{1}{2}\right)^2 = 1$

(c) $|f'(0)| = 0 < 1$, so $x = 0$ is stable.

$\left|f'\left(\frac{1}{2}\right)\right| = 1$, so it is not possible to determine if $x = \frac{1}{2}$ is stable using the Stability of Equilibrium Points theorem.

(d) i. $a_1 = 0.4$
$a_2 = f(a_1) = 0.384$
$a_3 = f(a_2) = 0.3633$
$a_4 = f(a_3) = 0.3362$
$a_5 = f(a_4) = 0.3001$

ii. $a_1 = 0.7$
$a_2 = f(a_1) = 0.588$
$a_3 = f(a_2) = 0.5698$
$a_4 = f(a_3) = 0.5587$
$a_5 = f(a_4) = 0.5510$

(e) Values of x less than $\frac{1}{2}$ appear to move toward 0, while values of x greater than $\frac{1}{2}$ appear to move toward $\frac{1}{2}$.

13. (a) $f(x) = x^3 + x$

Find the equilibrium points by solving $f(x) = x$.

$$x^3 + x = x$$
$$x^3 = 0 \Rightarrow x = 0$$

The equilibrium point is $x = 0$.

(b) $f'(x) = 3x^2 + 1$

$f'(0) = 3(0)^2 + 1 = 1$

(c) $|f'(0)| = 1$, so it is not possible to determine if $x = 0$ is stable using the Stability of Equilibrium Points theorem.

(d) i. $a_1 = 0.1$
$a_2 = f(a_1) = 0.101$
$a_3 = f(a_2) = 0.1020$
$a_4 = f(a_3) = 0.1031$
$a_5 = f(a_4) = 0.1042$

ii. $a_1 = -0.1$
$a_2 = f(a_1) = -0.101$
$a_3 = f(a_2) = -0.102$
$a_4 = f(a_3) = -0.1031$
$a_5 = f(a_4) = -0.1042$

(e) All values of x appear to move away from 0.

15. (a) $f(x) = \frac{3}{4} - x^2$

Find the equilibrium points by solving $f(x) = x$.

$$\frac{3}{4} - x^2 = x$$
$$x^2 + x - \frac{3}{4} = 0$$
$$\frac{1}{4}(2x-1)(2x+3) = 0$$
$$x = \frac{1}{2},\ -\frac{3}{2}$$

The equilibrium points are $x = -\frac{3}{2}$ and $x = \frac{1}{2}$.

(b) $f'(x) = -2x$

$f'\left(-\frac{3}{2}\right) = 3$

$f'\left(\frac{1}{2}\right) = -1$

(c) $\left|f'\left(-\frac{3}{2}\right)\right| = 3 > 1$, so $x = -\frac{3}{2}$ is unstable.

$\left|f'\left(\frac{1}{2}\right)\right| = 1$, so it is not possible to determine if $x = \frac{1}{2}$ is stable using the Stability of Equilibrium Points theorem.

(d) i. $a_1 = 0.8$
$a_2 = f(a_1) = 0.11$
$a_3 = f(a_2) = 0.7379$
$a_4 = f(a_3) = 0.2055$
$a_5 = f(a_4) = 0.7078$

ii. $a_1 = 0.6$
$a_2 = f(a_1) = 0.39$
$a_3 = f(a_2) = 0.5979$
$a_4 = f(a_3) = 0.3925$
$a_5 = f(a_4) = 0.5959$

(e) The values of x move toward $\frac{1}{2}$, oscillating from one side to the other.

17. $f(x) = 3.1x(1-x)$

Some of the iterations are shown below.

n	u(n)
1	.5
2	.775
3	.54056
4	.7699
5	.54918
6	.7675
7	.55317

u(n)=.5

n	u(n)
8	[illegible]
9	.55527
10	.76553
11	.55643
12	.76513
13	.55709
14	.7649

u(n)=.7662357552

n	u(n)
100	[illegible]
101	.55801
102	.76457
103	.55801
104	.76457
105	.55801
106	.76457

u(n)=.76456652

It appears that the two x-values are $x \approx 0.765$ and $x \approx 0.558$.

19. $f(x) = 3.48x(1-x)$

Some of the iterations are shown below.

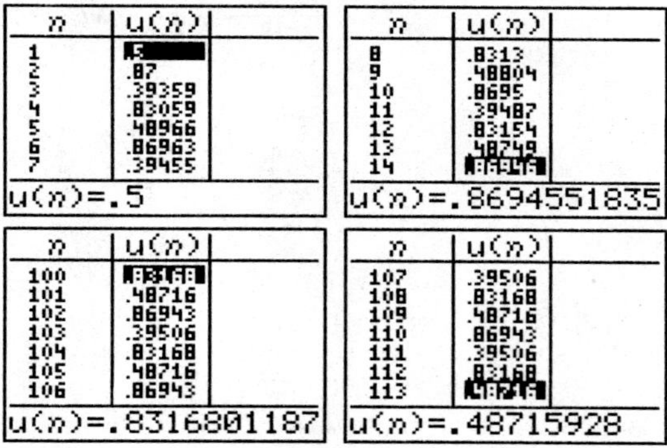

n	u(n)
1	.5
2	.87
3	.39359
4	.83059
5	.48966
6	.86963
7	.39455

u(n)=.5

n	u(n)
8	.8313
9	.48804
10	.8695
11	.39487
12	.83154
13	.48749
14	[illegible]

u(n)=.8694551835

n	u(n)
100	[illegible]
101	.48716
102	.86943
103	.39506
104	.83168
105	.48716
106	.86943

u(n)=.8316801187

n	u(n)
107	.39506
108	.83168
109	.48716
110	.86943
111	.39506
112	.83168
113	[illegible]

u(n)=.48715928

It appears that the four x-values are $x \approx 0.487$, $x \approx 0.869$, $x \approx 0.395$, and $x \approx 0.832$.

21. $f(x) = xe^{r(1-x)}$

From section 14.2, the equilibrium points are $x = 0$ and $x = 1$.

$$f'(x) = e^{r(1-x)}\frac{d}{dx}x + x\frac{d}{dx}e^{r(1-x)}$$
$$= e^{r(1-x)} + xe^{r(1-x)}\frac{d}{dx}(r - rx)$$
$$= e^{r(1-x)} - rxe^{r(1-x)}$$
$$|f'(0)| = \left|e^{r(1-0)} - r(0)e^{r(1-0)}\right|$$
$$= \left|e^{r}\right| > 1$$

Thus, $x = 0$ is unstable.

$$|f'(1)| = \left|e^{r(1-1)} - r(1)e^{r(1-1)}\right|$$
$$= |1 - r|$$

$|1 - r| > 1 \Rightarrow 1 - r > 1$ or $r - 1 > 1 \Rightarrow$ $r < 0$ or $r > 2$

$|1 - r| < 1 \Rightarrow -1 < 1 - r < 1 \Rightarrow 0 < r < 2$

Since we are given that r is a positive constant, $x = 1$ is stable if $0 < r < 2$ and unstable if $r > 2$.

23. $f(x) = \dfrac{rx}{1 + x^2}$

From section 14.2, the equilibrium points are $x = 0$ and $x = \sqrt{r - 1}$.

$$f'(x) = r \cdot \frac{(1 + x^2)\frac{d}{dx}(x) - x\frac{d}{dx}(1 + x^2)}{(1 + x^2)^2}$$
$$= r \cdot \frac{1 + x^2 - 2x^2}{(1 + x^2)^2} = \frac{r(1 - x^2)}{(1 + x^2)^2}$$
$$|f'(0)| = \left|\frac{r(1 - (0)^2)}{(1 + (0)^2)^2}\right| = r$$

Thus $x = 0$ is stable if $0 < r < 1$ and unstable if $r > 1$.

$$\left|f'\sqrt{r - 1}\right| = \left|\frac{r\left(1 - (\sqrt{r - 1})^2\right)}{\left(1 + (\sqrt{r - 1})^2\right)^2}\right| = \left|\frac{r(1 - (r - 1))}{(1 + r - 1)^2}\right|$$
$$= \left|\frac{r(2 - r)}{r^2}\right| = \left|\frac{2 - r}{r}\right|$$

Note that $\left|f'\sqrt{r - 1}\right|$ is undefined if $r < 1$.

Thus, $r > 1$ and $\left|f'\sqrt{r - 1}\right| = \dfrac{2 - r}{r} < 1$, so $x = r - 1$ is stable.

25. (a) $x_{n+1} = kx_n(1 - x_n) - Hx_n$

(b) $f(x) = 2.9x(1 - x) - 0.1x$

Find the equilibrium points by solving $f(x) = x$.

$$2.9x(1 - x) - 0.1x = x$$
$$-2.9x^2 + 1.8x = 0$$
$$x(-2.9x + 1.8) = 0$$

$x = 0$ | $-2.9x + 1.8 = 0$, $x \approx 0.6207$

$$f(x) = 2.9x(1 - x) - 0.1x$$
$$= 2.8x - 2.9x^2$$
$$f'(x) = -5.8x + 2.8$$
$$|f'(0)| = |-5.8(0) + 2.8|$$
$$= 2.8 > 1$$

Thus, $x = 0$ is unstable.

$$|f'(0.6207)| = |-5.8(0.6207) + 2.8|$$
$$\approx 0.80 < 1$$

Thus, $x = 0.6207$ is stable.

Chapter 14 Review Exercises

1. False. A sequence is a function whose *domain* is the set of natural numbers.
2. True
3. True
4. True
5. False. See Example 3b in section 14.2 in your text.
6. False. See exercise 1 in section 14.3 for an example.
7. False. If the value of the derivative equals 1, then it is not possible to determine if an equilibrium point is stable using the Stability of Equilibrium Points theorem.
8. False. If $f'(x) = -1$, then $|f'(x)| = 1$ and it is not possible to determine if an equilibrium point is stable using the Stability of Equilibrium Points theorem.
9. $a_n = 6n - 3$

$$a_1 = 6(1) - 3 = 3$$
$$a_2 = 6(2) - 3 = 9$$
$$a_3 = 6(3) - 3 = 15$$
$$a_4 = 6(4) - 3 = 21$$

10. $a_n = 8n - 7$
$a_1 = 8(1) - 7 = 1$
$a_2 = 8(2) - 7 = 9$
$a_3 = 8(3) - 7 = 17$
$a_4 = 8(4) - 7 = 25$

11. $a_n = 3 + (-2)^n$
$a_1 = 3 + (-2)^1 = 1$
$a_2 = 3 + (-2)^2 = 7$
$a_3 = 3 + (-2)^3 = -5$
$a_4 = 3 + (-2)^4 = 19$

12. $a_n = \frac{4}{2^n}$
$a_1 = \frac{4}{2^1} = 2$
$a_2 = \frac{4}{2^2} = 1$
$a_3 = \frac{4}{2^3} = \frac{1}{2}$
$a_4 = \frac{4}{2^4} = \frac{1}{4}$

13. $a_n = -2a_{n-1} - 3,\ a_1 = 0$
$a_2 = -2a_{2-1} - 3 = -2(0) - 3 = -3$
$a_3 = -2a_{3-1} - 3 = -2(-3) - 3 = 3$
$a_4 = -2a_{4-1} - 3 = -2(3) - 3 = -9$
$a_5 = -2a_{5-1} - 3 = -2(-9) - 3 = 15$

14. $a_n = 3a_{n-1} - 4,\ a_1 = 1$
$a_2 = 3a_{2-1} - 4 = 3(1) - 4 = -1$
$a_3 = 3a_{3-1} - 4 = 3(-1) - 4 = -7$
$a_4 = 3a_{4-1} - 4 = 3(-7) - 4 = -25$
$a_5 = 3a_{5-1} - 4 = 3(-25) - 4 = -79$

15. $a_n = -4a_{n-1} + 2,\ a_1 = -2$
$a_2 = -4a_{2-1} + 2 = -4(-2) + 2 = 10$
$a_3 = -4a_{3-1} + 2 = -4(10) + 2 = -38$
$a_4 = -4a_{4-1} + 2 = -4(-38) + 2 = 154$
$a_5 = -4a_{5-1} + 2 = -4(154) + 2 = -614$

16. $a_n = 5a_{n-1} - 12,\ a_1 = 2$
$a_2 = 5a_{2-1} - 12 = 5(2) - 12 = -2$
$a_3 = 5a_{3-1} - 12 = 5(-2) - 12 = -22$
$a_4 = 5a_{4-1} - 12 = 5(-22) - 12 = -122$
$a_5 = 5a_{5-1} - 12 = 5(-122) - 12 = -622$

17. $f(x) = \frac{1}{x}$
$f^2(x) = f(f(x)) = f\left(\frac{1}{x}\right) = \frac{1}{\frac{1}{x}} = x$
$f^3(x) = f(f^2(x)) = f(x) = \frac{1}{x}$
$f^n(x) = f(f^{n-1}(x)) = \begin{cases} x & \text{if } x \text{ is even} \\ \frac{1}{x} & \text{if } x \text{ is odd} \end{cases}$

18. $f(x) = x$
$f^2(x) = f(f(x)) = f(x) = x$
$f^3(x) = f(f^2(x)) = f(x) = x$
$f^n(x) = f(f^{n-1}(x)) = x$

19. $a_{n+1} = f(a_n) = 0.5\left(1 - \frac{a_n}{6000}\right)a_n,\ a_1 = 3500$
$a_2 = f(a_1) = 0.5\left(1 - \frac{3500}{6000}\right)3500 \approx 729$
$a_3 = f(a_2) = 0.5\left(1 - \frac{729}{6000}\right)729 \approx 320$
$a_4 = f(a_3) = 0.5\left(1 - \frac{320}{6000}\right)320 \approx 151$
$a_5 = f(a_4) = 0.5\left(1 - \frac{151}{6000}\right)151 \approx 74$

20. $a_{n+1} = f(a_n) = 1.2\left(1 - \frac{a_n}{7000}\right)a_n,\ a_1 = 2600$
$a_2 = f(a_1) = 1.2\left(1 - \frac{2600}{7000}\right)2600 \approx 1961$
$a_3 = f(a_2) = 1.2\left(1 - \frac{1961}{7000}\right)1961 \approx 1694$
$a_4 = f(a_3) = 1.2\left(1 - \frac{1694}{7000}\right)1694 \approx 1541$
$a_5 = f(a_4) = 1.2\left(1 - \frac{1541}{7000}\right)1541 \approx 1442$

21. $a_{n+1} = f(a_n) = 2.3\left(1 - \frac{a_n}{8000}\right)a_n,\ a_1 = 3200$
$a_2 = f(a_1) = 2.3\left(1 - \frac{3200}{8000}\right)3200 = 4416$
$a_3 = f(a_2) = 2.3\left(1 - \frac{4416}{8000}\right)4416 \approx 4550$
$a_4 = f(a_3) = 2.3\left(1 - \frac{4550}{8000}\right)4550 \approx 4513$
$a_5 = f(a_4) = 2.3\left(1 - \frac{4513}{8000}\right)4513 \approx 4524$

22. $a_{n+1} = f(a_n) = 3.6\left(1 - \frac{a_n}{9000}\right)a_n,\ a_1 = 2200$

$a_2 = f(a_1) = 3.6\left(1 - \frac{2200}{9000}\right)2200 = 5984$

$a_3 = f(a_2) = 3.6\left(1 - \frac{5984}{9000}\right)5984 \approx 7219$

$a_4 = f(a_3) = 3.6\left(1 - \frac{7219}{9000}\right)7219 \approx 5143$

$a_5 = f(a_4) = 3.6\left(1 - \frac{5143}{9000}\right)5143 \approx 7935$

23. **(a)**

$$f(x) = x$$
$$\frac{1}{4} - \left|\frac{x}{2} - \frac{1}{4}\right| = x$$
$$\left|\frac{x}{2} - \frac{1}{4}\right| = \frac{1}{4} - x$$

$$\frac{x}{2} - \frac{1}{4} = \frac{1}{4} - x \qquad \Bigg| \qquad \frac{x}{2} - \frac{1}{4} = x - \frac{1}{4}$$
$$\frac{3}{2}x = \frac{1}{2} \qquad \Bigg| \qquad -\frac{x}{2} = 0$$
$$x = \frac{1}{3} \qquad \Bigg| \qquad x = 0$$

However, $x = \frac{1}{3}$ is an extraneous solution. Thus, the equilibrium point is $x = 0$.

(b) Rewrite $f(x) = \frac{1}{4} - \left|\frac{x}{2} - \frac{1}{4}\right|$ as

$$f(x) = \begin{cases} \frac{1}{4} - \left(\frac{x}{2} - \frac{1}{4}\right) = -\frac{x}{2} + \frac{1}{2} & \text{if } x > \frac{1}{2} \\ \frac{1}{4} - \left(\frac{1}{4} - \frac{x}{2}\right) = \frac{x}{2} & \text{if } x \le \frac{1}{2} \end{cases}$$

Thus, $f'(0) = \frac{1}{2}$ using the slope of $f(x)$.

(c) $|f'(x)| = \frac{1}{2} < 1,$ so $x = 0$ is stable.

(d) $a_1 = 0.15$
$a_2 = f(a_1) = 0.075$
$a_3 = f(a_2) = 0.0375$
$a_4 = f(a_3) = 0.0188$
$a_5 = f(a_4) = 0.0934$

(e) $a_1 = 0.4$
$a_2 = f(a_1) = 0.2$
$a_3 = f(a_2) = 0.1$
$a_4 = f(a_3) = 0.05$
$a_5 = f(a_4) = 0.025$

(f) $a_1 = 0.65$
$a_2 = f(a_1) = 0.175$
$a_3 = f(a_2) = 0.0875$
$a_4 = f(a_3) = 0.0438$
$a_5 = f(a_4) = 0.0219$

(g) $a_1 = 0.85$
$a_2 = f(a_1) = 0.075$
$a_3 = f(a_2) = 0.0375$
$a_4 = f(a_3) = 0.0188$
$a_5 = f(a_4) = 0.0094$

(h)

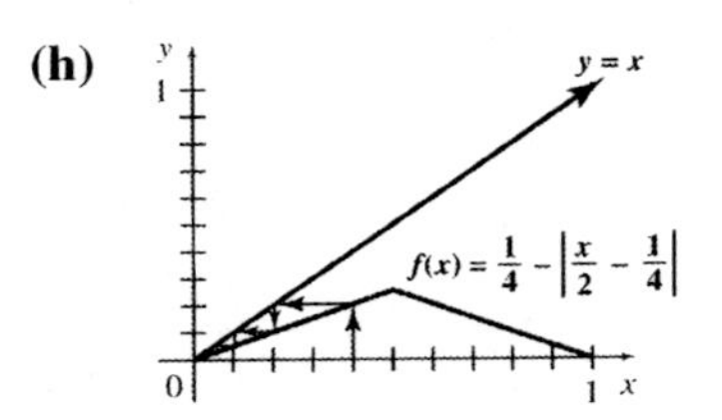

(i) The values in parts (d)– (g) all approach 0, showing that $x = 0$ is a stable equilibrium point.

24. **(a)**

$$f(x) = x$$
$$\frac{3}{5} - \left|\frac{6}{5}x - \frac{3}{5}\right| = x$$
$$\left|\frac{6}{5}x - \frac{3}{5}\right| = \frac{3}{5} - x$$

$$\frac{6}{5}x - \frac{3}{5} = \frac{3}{5} - x \qquad \Bigg| \qquad \frac{6}{5}x - \frac{3}{5} = x - \frac{3}{5}$$
$$\frac{11}{5}x = \frac{6}{5} \qquad \Bigg| \qquad \frac{1}{5}x = 0$$
$$x = \frac{6}{11} \qquad \Bigg| \qquad x = 0$$

Thus, the equilibrium points are $x = 0$ and $x = \frac{6}{11}$.

(b) Rewrite $f(x) = \frac{3}{5} - \left|\frac{6}{5}x - \frac{3}{5}\right|$ as

$$f(x) = \begin{cases} \frac{3}{5} - \left(\frac{6}{5}x - \frac{3}{5}\right) = \frac{6}{5}(1-x) & \text{if } x > \frac{1}{2} \\ \frac{3}{5} - \left(\frac{3}{5} - \frac{6}{5}x\right) = \frac{6}{5}x & \text{if } x \le \frac{1}{2} \end{cases}$$

Thus, $f'(0) = \frac{6}{5}$ and $f'\left(\frac{6}{11}\right) = -\frac{6}{5}$ using the slope of $f(x)$.

(c) $|f'(0)| = \frac{6}{5} > 1$ and $\left|f'\left(\frac{6}{11}\right)\right| = \frac{6}{5} > 1,$ so both equilibrium points are unstable.

(d) $a_1 = 0.15$
$a_2 = f(a_1) = 0.18$
$a_3 = f(a_2) = 0.216$
$a_4 = f(a_3) = 0.2592$
$a_5 = f(a_4) = 0.3110$

(e) $a_1 = 0.4$
$a_2 = f(a_1) = 0.48$
$a_3 = f(a_2) = 0.576$
$a_4 = f(a_3) = 0.5088$
$a_5 = f(a_4) = 0.5894$

(f) $a_1 = 0.65$
$a_2 = f(a_1) = 0.42$
$a_3 = f(a_2) = 0.504$
$a_4 = f(a_3) = 0.5952$
$a_5 = f(a_4) = 0.4858$

(g) $a_1 = 0.85$
$a_2 = f(a_1) = 0.18$
$a_3 = f(a_2) = 0.216$
$a_4 = f(a_3) = 0.2592$
$a_5 = f(a_4) = 0.3110$

(h)

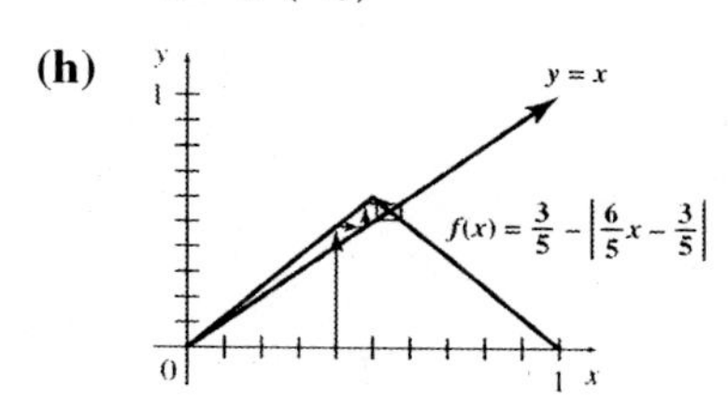

(i) The values in parts (d)–(g) all move away from 0, showing that $x = 0$ is an unstable equilibrium point. If you find additional values, you will find that they also move away from $x = \frac{6}{11} \approx 0.5455$.

25. (a)

$$f(x) = x$$
$$4.2x^2(1-x) = x$$
$$-4.2x^3 + 4.2x^2 - x = 0$$
$$-x\left(4.2x^2 - 4.2x + 1\right) = 0$$

Using the quadratic formula, we have

$x = 0,\ x = \frac{21-\sqrt{21}}{42} \approx 0.3909,$ and

$x = \frac{21+\sqrt{21}}{42} \approx 0.6091.$

Thus, the equilibrium points are $x = 0$, $x = 0.3909$, and $x = 0.6091$.

(b) $f(x) = 4.2x^2(1-x) = 4.2x^2 - 4.2x^3$
$f'(x) = 8.4x - 12.6x^2$
$f'(0) = 0$
$f'(0.3909) = 1.3583$

Note: This is the answer obtained by using the actual value $x = \frac{21-\sqrt{21}}{42}$. If the value $x = 0.3909$ is used, then we obtained $f'(0.3909) = 1.3582$.

$f'(0.6091) = 0.4417$

(c) $|f'(0)| = 0 < 1$, so $x = 0$ is stable.

$|f'(0.3909)| = 1.3583 > 1$, so $x = 0.3909$ is unstable.

$|f'(0.6091)| = 0.4417 < 1$, so $x = 0.6091$ is stable.

(d) $a_1 = 0.15$
$a_2 = f(a_1) = 0.0803$
$a_3 = f(a_2) = 0.0249$
$a_4 = f(a_3) = 0.0025$
$a_5 = f(a_4) = 0.0000$

(e) $a_1 = 0.4$
$a_2 = f(a_1) = 0.4032$
$a_3 = f(a_2) = 0.4075$
$a_4 = f(a_3) = 0.4132$
$a_5 = f(a_4) = 0.4208$

(f) $a_1 = 0.65$
$a_2 = f(a_1) = 0.6211$
$a_3 = f(a_2) = 0.6139$
$a_4 = f(a_3) = 0.6111$
$a_5 = f(a_4) = 0.6100$

(g) $a_1 = 0.85$
$a_2 = f(a_1) = 0.4552$
$a_3 = f(a_2) = 0.4741$
$a_4 = f(a_3) = 0.4965$
$a_5 = f(a_4) = 0.5213$

(h)

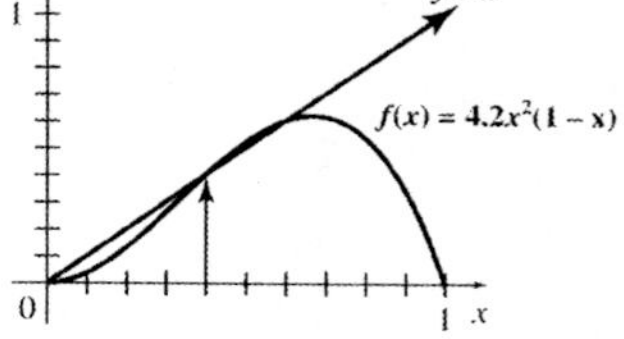

y = x
f(x) = 4.2x²(1 − x)

(i) The values in part (d) move towards 0, showing that $x = 0$ is a stable equilibrium point. The values in part (f) move towards 0.6091, showing that $x = 0.6091$ is a stable equilibrium point. The values in parts (e) and (f) move away from 0.3909, showing that $x = 0.3909$ is unstable.

26. (a)
$$f(x) = x$$
$$5.2x^2(1-x) = x$$
$$-5.2x^3 + 5.2x^2 - x = 0$$
$$-x\left(5.2x^2 - 5.2x + 1\right) = 0$$

Using the quadratic formula, we have
$x = 0,\ x = \dfrac{13-\sqrt{39}}{26} \approx 0.2598,$ and
$x = \dfrac{13+\sqrt{39}}{26} \approx 0.7402.$
Thus, the equilibrium points are $x = 0$, $x = 0.2598$, and $x = 0.7402$.

(b) $f(x) = 5.2x^2(1-x) = 5.2x^2 - 5.2x^3$
$f'(x) = 10.4x - 15.6x^2$
$f'(0) = 0$
$f'(0.2598) = 1.6490$
$f'(0.7402) = -0.8490$
Note: This is the answer obtained by using the actual value $\dfrac{13+\sqrt{39}}{26}$. If the value $x = 0.7402$ is used, then you will obtain $f'(0.7402) = -0.8491$

(c) $|f'(0)| = 0 < 1,$ so $x = 0$ is stable.
$|f'(0.2598)| = 1.6490 > 1$ so $x = 0.2598$ is unstable.
$|f'(0.7402)| = 0.8490$ so $x = 0.7402$ is stable.

(d) $a_1 = 0.15$
$a_2 = f(a_1) = 0.0995$
$a_3 = f(a_2) = 0.0463$
$a_4 = f(a_3) = 0.0106$
$a_5 = f(a_4) = 0.0006$

(e) $a_1 = 0.4$
$a_2 = f(a_1) = 0.4992$
$a_3 = f(a_2) = 0.6490$
$a_4 = f(a_3) = 0.7688$
$a_5 = f(a_4) = 0.7106$

(f) $a_1 = 0.65$
$a_2 = f(a_1) = 0.7690$
$a_3 = f(a_2) = 0.7104$
$a_4 = f(a_3) = 0.7600$
$a_5 = f(a_4) = 0.7209$

(g) $a_1 = 0.85$
$a_2 = f(a_1) = 0.5636$
$a_3 = f(a_2) = 0.7208$
$a_4 = f(a_3) = 0.7543$
$a_5 = f(a_4) = 0.7269$

(h)
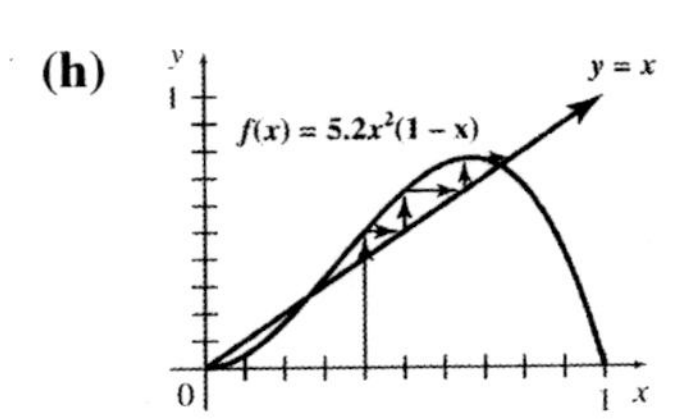

(i) The values in part (d) move towards 0, showing that $x = 0$ is a stable equilibrium point. The values in parts (e)–(g) oscillate about 0.7402, moving from one side to the other, showing that $x = 0.7402$ is a stable equilibrium point. The values in parts (d)–(f) move away from 0.2598, showing that $x = 0.2598$ is unstable.

27. (a) $a_n = 1.10a_{n-1}$

(b) $a_1 = 100{,}000$
$a_2 = f(a_1) = 110{,}000$
$a_3 = f(a_2) = 121{,}000$
$a_4 = f(a_3) = 133{,}100$
$a_5 = f(a_4) = 146{,}410$

(c) $a_n = 100{,}000(1.10)^{n-1}$

(d) $f(x) = x$
$1.10x = x \Rightarrow x = 0$
The equilibrium point is $x = 0$.

(e) $f'(x) = 1.10$ for all values of x.

(f) $|f'(0)| = 1.10 > 1,$ so $x = 0$ is unstable. In fact, any equilibrium point would be unstable.

28. (a) $a_n = ra_{n-1} + b$

(b) $f(x) = rx + b$

$rx + b = x$

$rx - x = -b$

$x(1-r) = b \Rightarrow x = \dfrac{b}{1-r}$

The equilibrium point is $x = \dfrac{b}{1-r}$.

(c) $x = \dfrac{b}{1-r}$ must be a positive value to be a valid population size. If $b \geq 0$, then $1 - r$ must be positive, so $r < 1$. If $b < 0$, then $1 - r$ must be negative, so $r > 1$.

(d) $f'(x) = r$

$|f'(x)| = |r| < 1 \Rightarrow -1 < r < 1$

The equilibrium point will be stable if $-1 < r < 1$.

(e) $a_n = a_1 + (n-1)b$

(f) Using the formula, we have

$a_{n-1} = r^{n-1} + \left(1 + r + r^2 + \cdots + r^{n-2}\right)b.$

Substitute the expression for a_{n-1} into the equation in part (a) and simplify.

$$\begin{aligned} a_n &= ra_{n-1} + b \\ &= r\left(r^{n-1} + \left(1 + r + r^2 + \cdots + r^{n-2}\right)b\right) + b \\ &= r^n + \left(r + r^2 + r^3 + \cdots + r^{n-1}\right)b + b \\ &= r^n + \left(1 + r + r^2 + \cdots + r^{n-1}\right)b \end{aligned}$$

Thus, $a_n = r^n + \left(1 + r + r^2 + \cdots + r^{n-1}\right)b$ is a solution of the equation.

29. (a) $m_n p$ represents the actual number of people with malaria in year n. $m_{n-1}p$ represents the actual number of people who had malaria in the previous year. $rm_{n-1}p$ represents the actual number of people who recovered from malaria in the previous year, so $m_{n-1}p - rm_{n-1}p$ represents the number of people who had malaria in the previous year and still have malaria. Therefore, $m_{n-1}p - rm_{n-1}p +$ number of new cases represents the actual number of people with malaria in year n, or $m_n p$.

(b) From part (a), we know that $m_{n-1}p$ represents the actual number of people with malaria in year $n - 1$. Then, $im_{n-1}p$ represents the number of people with gametids in their blood and $aim_{n-1}p$ represents the number of mosquitoes for each of these people. The number of mosquitoes that bite a person with gametids in their blood is represented by $baim_{n-1}p$, and $bsaim_{n-1}p$ represents the number of mosquitoes in which the malaria matures. If we multiply this number by b, the fraction of malaria-bearing mosquitoes that bite a person (the next person to contract malaria), we have the number of mosquitoes that succeed in infecting people represented by $b^2saim_{n-1}p$. The number of new cases will be the number of mosquitoes that succeed in infecting people times the fraction of the population that was not previously infected, or $1 - m_{n-1}$. So, the number of new cases is

$b^2saim_{n-1}p(1 - m_{n-1})$.

(c) From part (a), we have

$$m_n p = m_{n-1}p - rm_{n-1}p + \text{number of new cases}$$

Substituting from part (b) gives

$$m_n p = m_{n-1}p - rm_{n-1}p + b^2saim_{n-1}p(1 - m_{n-1})$$

Dividing by p gives

$$\begin{aligned} m_n &= m_{n-1} - rm_{n-1} + b^2saim_{n-1}(1 - m_{n-1}) \\ &= m_{n-1} - rm_{n-1} + b^2saim_{n-1} - b^2saim_{n-1}^2 \\ &= m_{n-1}\left(1 - r + b^2sai\right) - b^2saim_{n-1}^2 \end{aligned}$$

(d) $m_1 = 0.5$
$m_2 = 0.525$
$m_3 = 0.5447$
$m_4 = 0.5598$
$m_5 = 0.5710$

(e) $m_1 = 0.5$
$m_2 = 0.475$
$m_3 = 0.4548$
$m_4 = 0.4382$
$m_5 = 0.4244$

(f) $m_1 = 0.5$
$m_2 = 0.45$
$m_3 = 0.4095$
$m_4 = 0.3760$
$m_5 = 0.3477$

(g) We have

$f(x)=(1-0.2+0.005a)x-0.005ax^2.$

$$f(x)=x$$
$$(1-0.2+0.005a)x-0.005ax^2=x$$
$$-0.2x+0.005ax-0.005ax^2=0$$
$$0.005x(-40+a-ax)=0$$

$$0.005x=0 \;\Rightarrow\; x=0 \quad\Big|\quad -40+a-ax=0,\; a(1-x)=40,\; 1-x=\frac{40}{a}\Rightarrow x=1-\frac{40}{a}$$

Thus, the equilibrium points are $x=0$ and $x=1-\dfrac{40}{a}$. Since x must be nonnegative, $a\ge 40$.

(h) $f(x)=(0.8+0.005a)x-0.005ax^2$

$f'(x)=0.8+0.005a-0.010ax$

$$f'\left(1-\frac{40}{a}\right)$$
$$=0.8+0.005a-0.010a\left(1-\frac{40}{a}\right)$$
$$=0.8+0.005a-0.010a+0.4$$
$$=1.2-0.005a$$

$$\left|f'\left(1-\frac{40}{a}\right)\right|<1\Rightarrow -1<1.2-0.005a<1$$
$$-1<1.2-0.005a<1$$
$$-2.2<-0.005a<-0.2$$
$$440>a>40 \text{ or } 40<a<440$$

In part (d), $x=1-\dfrac{40}{100}=0.6$. If you examine more iterations, it is clear that $x=0.6$ is stable because the values move towards 0.6.

In part (e), $x=1-\dfrac{40}{60}=0.3333$. If you examine more iterations, it is clear that $x=0.3333$ is stable because the values move towards 0.3333.

In part (f), $x=1-\dfrac{40}{40}=0$, which is the other equilibrium point. If you examine more iterations, it is clear that $x=0$ is stable because the values move towards 0.

30. (a) Vc_n is the volume of the lungs times the concentration of the chemical in the lungs which gives the amount of the chemical in the lungs. V_ia is the volume of a breath (the amount of air inhaled or exhaled) times the concentration of chemical in the air. This gives the amount of chemical inhaled. V_rc_{n-1} is the volume remaining in the lungs after exhaling times the concentration of chemical in the lungs after the previous breath, so this gives the amount of chemical remaining in the lungs after exhaling. So, the amount of chemical in the lungs, $Vc_n=V_ia+V_rc_{n-1}$, the amount of chemical inhaled added to the amount of chemical already in the lungs.

(b) c_{n-1} $V=V_i+V_r\Rightarrow V_r=V-V_i$

$$Vc_n=V_ia+V_rc_{n-1}$$
$$Vc_n=V_ia+(V-V_i)c_{n-1}$$
$$c_n=\frac{V_i}{V}a+\left(\frac{V-V_i}{V}\right)c_{n-1}$$
$$c_n=pa+(1-p)c_{n-1}$$

(c) $V=V_i+V_r=1.5+0.5=2.0$

$$p=\frac{V_i}{V}=\frac{1.5}{2.0}=0.75$$
$$f(x)=0.75(0.2095)+(1-0.75)x$$
$$=0.75(0.2095)+0.25x$$

$c_1=0.5$
$c_2=0.2821$
$c_3=0.2277$
$c_4=0.2140$
$c_5=0.2106$

(d) $c_2=f(c_1)=pa+(1-p)c_1$

$$c_3=f(c_2)=pa+(1-p)c_2$$
$$=pa+(1-p)(pa+(1-p)c_1)$$
$$=pa+(1-p)pa+(1-p)^2c_1$$
$$=[1+(1-p)]pa+(1-p)^2c_1$$
$$c_4=f(c_3)=pa+(1-p)c_3$$
$$=pa+(1-p)\cdot\left([1+(1-p)]pa+(1-p)^2c_1\right)$$
$$=pa+(1-p)pa+(1-p)^2pa+(1-p)^3c_1$$
$$=pa\left[1+(1-p)+(1-p)^2\right]+(1-p)^3c_1$$

Continuing in the same manner, we deduce

$$c_n=pa\left[1+(1-p)+\cdots+(1-p)^{n-2}\right]+(1-p)^{n-1}c_1$$

(e) $f(x) = pa + (1-p)x$

$f(x) = x$

$pa + (1-p)x = x \Rightarrow pa + x - px = x \Rightarrow pa - px = 0 \Rightarrow a - x = 0 \Rightarrow x = a$

(f) $f(x) = pa + (1-p)x$

$f'(x) = 1 - p \Rightarrow f'(a) = 1 - p$

Since $p = \dfrac{V_i}{V} \Rightarrow 0 < p < 1$, $|f'(a)| = 1 - p < 1$. Therefore, the equilibrium point is stable.

(g) Using the formula for the sum of a geometric series,

$$1 + (1-p) + \cdots + (1-p)^{n-2} = \frac{1-(1-p)^{n-1}}{1-(1-p)} = \frac{1-(1-p)^{n-1}}{p}$$

So,

$$\begin{aligned} c_n &= pa\left[1 + (1-p) + \cdots + (1-p)^{n-2}\right] + (1-p)^{n-1} c_1 \\ &= pa\left(\frac{1-(1-p)^{n-1}}{p}\right) + (1-p)^{n-1} c_1 \\ &= a\left[1-(1-p)^{n-1}\right] + (1-p)^{n-1} c_1 \end{aligned}$$

(h) As $n \to \infty$, $(1-p)^{n-1} \to 0$. So $\lim\limits_{n\to\infty} c_n = a[1-0] + (0)^{n-1} c_1 = a$.

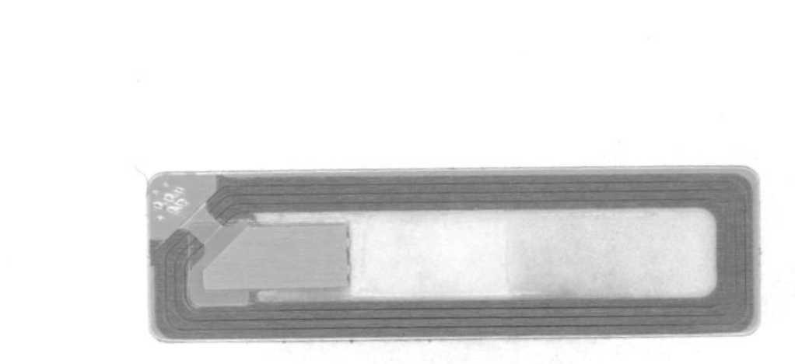